Aspects of Representation Theory and Noncommutative Harmonic Analysis

Mathematical Foundations for Computer Vision, Machine Learning, and Robotics

Series Editors: Jean Gallier (*University of Pennsylvania, USA*)
Jocelyn Quaintance (*University of Pennsylvania, USA*)

Published

Vol. 2 *Aspects of Representation Theory and Noncommutative Harmonic Analysis*
by Jean Gallier and Jocelyn Quaintance

Vol. 1 *Aspects of Harmonic Analysis on Locally Compact Abelian Groups*
by Jean Gallier and Jocelyn Quaintance

Mathematical Foundations for Computer Vision, Machine Learning, and Robotics: Vol. 2

Aspects of Representation Theory and Noncommutative Harmonic Analysis

Jean Gallier
University of Pennsylvania, USA

Jocelyn Quaintance
University of Pennsylvania, USA

World Scientific

NEW JERSEY • LONDON • SINGAPORE • GENEVA • BEIJING • SHANGHAI • TAIPEI • CHENNAI

Published by

World Scientific Publishing Co. Pte. Ltd.
5 Toh Tuck Link, Singapore 596224
USA office: 27 Warren Street, Suite 401-402, Hackensack, NJ 07601
UK office: 57 Shelton Street, Covent Garden, London WC2H 9HE

Library of Congress Control Number: 2024060684

British Library Cataloguing-in-Publication Data
A catalogue record for this book is available from the British Library.

Mathematical Foundations for Computer Vision, Machine Learning, and Robotics — Vol. 2
ASPECTS OF REPRESENTATION THEORY AND NONCOMMUTATIVE HARMONIC ANALYSIS

ISBN 978-981-98-1081-9 (hardcover)
ISBN 978-981-98-1082-6 (ebook for institutions)
ISBN 978-981-98-1083-3 (ebook for individuals)

For any available supplementary material, please visit
https://www.worldscientific.com/worldscibooks/10.1142/14244#t=suppl

Preface

The question that motivated writing this book is:

What is the Fourier transform?

We were quite surprised by how involved the answer is, and how much mathematics is needed to answer it, from measure theory, integration theory, some functional analysis, to some representation theory.

If G is a commutative locally compact group, there is a beautiful and well-understood theory of the Fourier transform based on results of Gelfand, Pontrjagin, and André Weil. Aspects of this theory are discussed in Volume I of this series.

If G is *not* commutative, things are a lot tougher. Characters no longer provide a good input domain, and instead one has to turn to *unitary representations*. A unitary representation is a homomorphism $U\colon G \to \mathbf{U}(H)$ satisfying a certain continuity property, where $\mathbf{U}(H)$ is the group of unitary operators on the Hilbert space H. Then $\widehat{G}$ is the set of equivalence classes of irreducible unitary representations of G, but it is no longer a group. Some aspects of noncommutative harmonic analysis and representation theory are presented in this book.

In the special case of a compact group, there is a deep interplay between analysis and representation theory which was first discovered by Hermann Weyl and refined by André Weil. If the group G is compact, an important theorem due to Peter and Weyl gives a nice decomposition of $\mathrm{L}^2(G)$ as a Hilbert sum of finite-dimensional matrix algebras corresponding to the irreducible unitary representations of G (see Theorem 4.2 and Theorem 4.6). As a consequence, there is a good notion of Fourier transform, such that the Fourier transform $\mathcal{F}(f)$ is a function with domain $\widehat{G}$, but its output domain is no longer $\mathbb{C}$. Instead, it is a space of matrices depending on the irreducible representation given as input (see Section 4.6). In general, it is very difficult to find the irreducible representations of a compact group, so in general this Fourier transform is not very useful. However, for certain groups, such as $\mathbf{SO}(2)$, $\mathbf{SO}(3)$ and $\mathbf{SU}(2)$, the irreducible representations can be determined explicitly, so for these groups it is practical.

We now outline the contents of this book and along the way point out what we believe is not found in other books on the subject.

Chapter 2 discusses the notion of representation of algebras with involution. Complete separable Hilbert algebras are defined and their structure is described. This result plays a crucial role in the proof of the Peter–Weyl theorem (Theorem 4.2). The representations of several algebras of continuous functions are defined. This leads to the important technical tool of projection-valued measures.

Chapter 3 is devoted to the theory of unitary representations of a locally compact group. One of the main results is the equivalence of unitary representations of a locally compact group G and the nondegenerate (algebra) representations of the algebra $\mathrm{L}^1(G)$. This equivalence plays a major role in the proof of the second version of the Peter–Weyl theorem (Theorem 4.16). It is also shown that every unitary representation of a locally compact abelian group is determined by a projection-valued measure. This result is used in Mackey's theory of induced representations (see Chapter 7). Functions of positive type are also introduced.

Chapter 4 contains the most important theorems about the structure of the function space $\mathrm{L}^2(G)$ when G is a metrizable compact group. We follow Dieudonné [Dieudonné (2003a,b)] and Folland [Folland (1995)]. It turns out that the irreducible representations of a metrizable compact group G are finite-dimensional and form a finite or countable family $(M_\rho)_{\rho\in R}$. The space $\mathrm{L}^2(G)$ is a complete Hilbert algebra that can be expressed as a Hilbert sum of algebras each isomorphic to the matrix algebra $\mathrm{M}_{n_\rho}(\mathbb{C})$. These results constitute Theorem 4.2, a deep and beautiful theorem due to Peter and Weyl that we refer to as Peter–Weyl I.

The characters of the representations M_ρ are defined. The second part of the Peter–Weyl theorem (Theorem 4.16), referred to as Peter–Weyl II, deals with unitary representations and is discussed in Section 4.3. This result allows the decomposition of an arbitrary unitary representation of a compact group G as a Hilbert sum of some of the irreducible representations M_ρ. It plays an unexpected role in Chapter 8 where it is used to construct steerable families.

In Section 4.4, we discuss tensor products of finite-dimensional representations.

In Section 4.5, we define the notion of *contragredient representation* $\overline{U}\colon G \to \mathbf{GL}(H^*)$ of a representation $U\colon G \to \mathbf{GL}(H)$ and the notion of Hom *representation* $\mathrm{Hom}(U_1, U_2)\colon G \to \mathbf{GL}(\mathrm{Hom}(H_1, H_2))$ of two representations $U_1\colon G \to \mathbf{GL}(H_1)$ and $U_2\colon G \to \mathbf{GL}(H_2)$. These notions will be needed in Chapter 8.

In Section 4.6, we define a notion of Fourier transform and Fourier cotransform for a (metrizable) compact group G. We discuss versions of Fourier inversion; see Section 4.8.

Chapter 5 deals with explicit matrix descriptions of the irreducible representations of the groups $\mathbf{SL}(2, \mathbb{C})$, $\mathbf{SU}(2)$ and $\mathbf{SO}(3)$ (unitary representation in the last two cases). Our presentation (except for Section 5.7) relies heavily on Vilenkin's exposition [Vilenkin (1968)], especially Chapter III. Among other things we derive the famous *Wigner d-matrices and $\mathcal{D}$-matrices* and we discuss the *Clebsch–Gordan Coefficients*, a standard topic in quantum mechanics.

Chapter 6 is devoted to induced representations of locally compact groups. If G is a locally compact group and if H is a closed subgroup of G, under certain conditions, it is possible to construct a Hilbert space $\mathcal{H}$ and a unitary representation $\Pi\colon G \to \mathbf{U}(\mathcal{H})$ of G in $\mathcal{H}$ from a unitary representation $U\colon H \to \mathbf{U}(E)$ of H in a (separable) Hilbert space E. The representation Π is called an *induced representation*. Interestingly, induced representations play a central role in group equivariant deep learning in convolutional networks; see Chapter 8.

There are two approaches for the construction of the Hilbert space $\mathcal{H}$:

(1) The Hilbert space $\mathcal{H}$ is a set of functions from $X = G/H$ to E.
(2) The Hilbert space $\mathcal{H}$ is a set of functions from G to E.

We give a detailed description of these two methods and explain how to pass from one to the other. This involves picking a suitable section $r\colon G/H \to G$. When the space E is a (separable) Hilbert space, there are technical difficulties. The description of induced representations in terms of certain G-bundles is also discussed. This chapter does not contain any new material but it pulls together aspects of this theory which appear in different sources.

One of the most important contributions to the theory of unitary representations is a method due to Mackey for constructing all irreducible representations of a locally compact group as induced irreducible representations from "small" subgroups H of G. This method is often referred to as the "Mackey machine." Chapter 7 presents a simple version of Mackey's method.

In its most general form the method is very complicated but in the case where G has an abelian normal subgroup N it is tractable. Mackey introduced the novel concept of *(transitive) system of imprimitivity* (see Definition 7.3). The relevance of systems of imprimitivity is Mackey's *imprimitivity theorem* (Theorem 7.3), which implies that a unitary representation U is equivalent to a representation obtained by the induction method from some irreducible representation of some "small" subgroup G_ν of G.

Unfortunately, the subgroups G_ν may still not be small enough. However, if some condition on G_ν holds (an extension condition), then for every irreducible representation $\rho\colon G_\nu/N \to \mathbf{U}(\mathcal{H}_\rho)$ of G_ν/N, there is an irreducible representation of G_ν in $\mathcal{H}_\rho$ (see Proposition 7.6). In this case, we can use irreducible representations of the "little groups" $H_\nu = G_\nu/N$ in the inducing process of Theorem 7.4.

The above extension condition is satisfied by semi-direct products $G = N \rtimes H$, where N is a normal abelian subgroup of G. Then every irreducible representation of G is obtained in terms of the characters of N and the irreducible representations of the little groups H_ν associated with the characters $\nu \in \widehat{N}$; see Theorem 7.7. Using this method, we describe all irreducible representations of $\mathbf{SE}(n)$; see Example 7.1. We also determine all irreducible representations of $\mathbf{O}(2)$ (see Example 7.2) and indicate how all irreducible representations of $\mathbf{E}(2)$ and $\mathbf{E}(3)$ can be obtained. This chapter contains material that is rarely presented in simplified form. Folland [Folland (1995)] presents Mackey's theory but at a rather advanced level.

Chapter 8 is the most original chapter of this book. The general theme is to develop a theory of equivariant convolutional neural networks (CNNs). The purpose and the need for such neural networks is very clearly articulated in the preface of the recent book by Weiler, Forré, Verlinde, and

Welling [Weiler *et al.* (2023)] that we highly recommend. Our goal in this chapter is to show how many of the fairly abstract concepts discussed earlier (representations, analysis on compact groups, Peter–Weyl theorems, Fourier transform, induced representations) are used to tackle very practical problems. Most of the material in Chapter 8 is heavily inspired by the works of Bekkers, Boomsma, Cesa, Cohen, Forré, Geiger, Lang, Verlinder, Weiler, and Welling.

However, Sections 8.7, 8.9 and 8.10 contain original results.

In Section 8.7, we provide novel decompositions of various function spaces in terms of steerable functions. In Example 8.9, we consider the space $\mathrm{L}^2(\mathbf{SE}(n))$ for any $n \geq 2$. Example 8.10 deals with a homogeneous space $X = H/H_0$, where H is a compact (metrizable and separable) group and H_0 is any closed subgroup of H. In Example 8.11, we describe a general method to decompose the space $\mathrm{L}^2(X)$, given an action of a compact group H on X, where X is a locally compact, metrizable, separable space equipped with a σ-regular, locally finite, Borel measure μ. The decomposition involves functions defined on the orbits of the action of H on X and functions defined on the orbit space $\Omega = X/H$. It can be viewed as a generalization of the decomposition of functions on $\mathrm{L}^2(\mathbf{SE}(2))$ in terms of harmonic functions and radial function.

In Section 8.9, if H is a compact group, the semi-direct product $G = \mathbb{R}^d \rtimes H$ is not compact but we can still define the Fourier transform of a function $f \in \mathrm{L}^2(\mathbb{R}^d \rtimes H)$ as an indexed family of *functions* $\widehat{f}_\rho \colon \mathbb{R}^d \to \mathrm{M}_{n_\rho}(\mathbb{C})$. These families of functions can be made into a Hilbert space $\mathrm{L}^2(\mathbb{R}^d, \widehat{H})$ and we prove a Plancherel-type theorem that shows that the Fourier transform is an isometry between the Hilbert spaces $\mathrm{L}^2(\mathbb{R}^d \rtimes H)$ and $\mathrm{L}^2(\mathbb{R}^d, \widehat{H})$; see Theorem 8.8.

In Section 8.10, we explain how to define directly a notion of correlation on the space of "Fourier coefficients with variables" $\mathrm{L}^2(\mathbb{R}^d, \widehat{H})$; see Formulae ($*_{\Phi_{\rho_2,\rho_1}}$) and ($*_{\widehat{\Phi}_{\rho_2,\rho_1}}$). A special case of this construction is the case $d = 2$ and $H = \mathbf{SO}(2)$, which is known as the *harmonic nets*, and is covered extensively in Section 8.11.

A way to deal with noncommutativity, due to Gelfand, is to work with pairs (G, K), where K is a compact subgroup of G. Instead of working with functions on G, which is "too big," we work with functions on the homogeneous space G/K, the space of left cosets. Then, under certain assumptions on G and K, which makes (G, K) a *Gelfand pair*, it is possible to consider a commutative algebra of functions on the set of double cosets KsK ($s \in G$), so that some results from the commutative theory can be used (see Chapter 9). The domain of the Fourier transform is a set of functions called *spherical functions*, and this set happens to be homeomorphic to the set of characters on the commutative algebra mentioned above. There is a very nice theory of the Fourier transform and its inverse (see Section 9.8), but how useful it is in practice remains to be seen.

Since this book is already quite long, we do not present the machinery of Lie algebras and *semisimple Lie groups* developed by Élie Cartan and Hermann Weyl involving weights and roots. If G is a connected semisimple Lie group, the finite-dimensional irreducible representations are determined by highest weights. There is a beautiful and extensive theory of representations of semisimple Lie groups, and many books have been written on the subject; see the end of Section 3.7 for some classical references.

Acknowledgments

Many thanks to the participants of the "underground" Tuesday meetings, Christine Allen-Blanchette, Carlos Esteves, Stephen Phillips, and João Sedoc, for catching mistakes and for many helpful comments. We also thank Kostas Daniilidis for being a source of inspiration. Our debt to J. Dieudonné, G. Folland, E. Hewitt and K.A. Ross, A. Knapp, S. Lang, A.A. Kirillov, Laurent Schwartz, E. Stein and R. Shakarchi, and W. Rudin, is enormous. Every result in this manuscript is found in one form or another in their seminal books.

Contents

Chapter 1

Introduction

The main two topics of this book are harmonic analysis and representation theory.

In Volume I, we discussed aspects of harmonic analysis on locally compact *abelian* groups. There is a beautiful and well-understood theory of the Fourier transform based on results of Gelfand, Pontrjagin, and André Weil presented in Volume I of this book.

If G is *not* commutative, things are a lot tougher. Characters no longer provide a good input domain for the Fourier transform, and instead one has to turn to *unitary representations*. A unitary representation is a homomorphism $U\colon G \to \mathbf{U}(H)$ satisfying a certain continuity property, where $\mathbf{U}(H)$ is the group of unitary operators on the Hilbert space H. Then $\widehat{G}$ is the set of equivalence classes of irreducible unitary representations of G, but it is no longer a group. Some aspects of representation theory and of noncommutative harmonic analysis are discussed in this book (Volume II).

Chapters 2 and 3 provide the background material needed in Chapter 4. Chapter 2 discusses representations of algebras and gives an introduction to Hilbert algebras. For our purposes, the most important examples of complete Hilbert algebras are the Hilbert–Schmidt operators on a separable Hilbert space and $\mathrm{L}^2(G)$, where G is a *compact* (metrizable, separable) group. One of the main theorems of this chapter is a structure theorem for complete separable algebras (Theorem 2.33). This theorem is the key result for proving a major part of the Peter–Weyl theorem in Chapter 4. We follow closely Dieudonné [Dieudonné (2005)].

Chapter 3 gives an introduction to the theory of unitary representations of locally compact groups. We prove that there is a bijection between unitary representations of a locally compact group G and nondegenerate representations of the algebra $L^1(G)$. We define functions and measures of positive type and prove that there is a bijection between the set of functions of positive type and cyclic unitary representations (Gelfand–Raikov, Godement). We follow Dieudonné [Dieudonné (2003a,b)] and Folland [Folland (1995)].

Chapter 4 contains the most important theorems about the structure of the function space $L^2(G)$ when G is a metrizable compact group. It turns out that the irreducible representations of a metrizable compact group G are finite-dimensional and form a finite or countable family $(M_\rho)_{\rho\in R}$. The space $L^2(G)$ is a complete Hilbert algebra that can be expressed as a Hilbert sum of topologically simple algebras $\mathfrak{a}_\rho$, which are minimal two-sided ideals, and each ideal $\mathfrak{a}_\rho$ is isomorphic to the matrix algebra $M_{n_\rho}(\mathbb{C})$. More precisely, there is a family of functions

$$\left(\frac{1}{\sqrt{n_\rho}} m_{ij}^{(\rho)}\right)_{1\leq i,j\leq n_\rho,\ \rho\in R},$$

which is a Hilbert basis of $L^2(G)$, where the subfamily corresponding to a fixed $\rho \in R$ is an orthonormal basis of the ideal $\mathfrak{a}_\rho$.

Furthermore, for every $s \in G$, if we define the $n_\rho \times n_\rho$ matrix $M_\rho(s)$ by

$$M_\rho(s) = \left(\frac{1}{n_\rho} m_{ij}(s)\right),$$

then these matrices are invertible and satisfy the equations

$$M_\rho(st) = M_\rho(s)M_\rho(t) \qquad \text{and} \qquad M_\rho(s^{-1}) = (M_\rho(s))^*.$$

Thus, the map $s \mapsto M_\rho(s)$ is a continuous unitary representation in matrix form $M_\rho\colon G \to \mathbf{U}(n_\rho)$ of G in $\mathbb{C}^{n_\rho}$. The representations $M_\rho\colon G \to \mathbf{U}(n_\rho)$ are irreducible, and every irreducible unitary representation of G is equivalent to some M_ρ.

These results constitute Theorem 4.2, a deep and beautiful theorem due to Peter and Weyl that we refer to as Peter–Weyl I.

Besides characters of groups and characters of algebras, there is one more kind of characters, namely, characters of finite-dimensional representations. For every $\rho \in R$, define the *character* χ_ρ *of* G *associated with the ideal* $\mathfrak{a}_\rho$ as the function given by

$$\chi_\rho(s) = \mathrm{tr}(M_\rho(s)), \qquad \text{for all } s \in G.$$

One of the main properties of the characters is that the family of characters $(\chi_\rho)_{\rho\in R}$ forms a Hilbert basis of the center of $L^2(G)$; see Proposition 4.10.

The second part of the Peter–Weyl theorem (Theorem 4.16), referred to as Peter–Weyl II, deals with unitary representations and is discussed in Section 4.3. This theorem asserts the following facts. Let $V\colon G \to \mathbf{U}(H)$ be a unitary representation of G in a separable Hilbert space H. Then H is a Hilbert sum of subspaces E_ρ invariant under V, and each nontrivial E_ρ is the Hilbert sum of invariant subspaces corresponding to irreducible representations of G. More precisely:

(1) For every $\rho \in R$, there is an orthogonal projection of H onto a closed subspace E_ρ (which may be reduced to (0)), and H is the Hilbert sum of the $E_\rho \neq (0)$.
(2) Every subspace $E_\rho \neq (0)$ is invariant under V, and the restriction V_ρ of V to E_ρ is a finite or countably infinite Hilbert sum of irreducible representations, all equivalent to M_ρ.

In particular, all the representations $M_\rho\colon G \to \mathbf{U}(n_\rho)$ occurring in Peter–Weyl I *are irreducible*, and since every unitary irreducible representation is equivalent to some representation of the form M_ρ, the index set R corresponds to a complete set of unitary representations of G.

In Section 4.4, we discuss tensor products of finite-dimensional representations. We begin with the definition of the tensor product representation $U_1 \otimes U_2\colon G \to \mathbf{U}(H_1 \otimes H_2)$ of two finite-dimensional unitary representations $U_1\colon G \to \mathbf{U}(H_1)$ and $U_2\colon G \to \mathbf{U}(H_2)$ of the same locally compact (metrizable, separable) group G. In general, if U_1 and U_2 are irreducible, then the tensor product representation $U_1 \otimes U_2$ is *not* irreducible. If G is compact, the representation $U_1 \otimes U_2$ splits as a sum of irreducible representations of G, but finding this decomposition is generally very difficult. In the special case $G = \mathbf{SU}(2)$, this can be done. This is an important result of quantum physics; see Section 5.17 on the Clebsch–Gordan coefficients.

Next, we define the tensor product representation $U_1 \otimes U_2\colon G_1 \times G_2 \to \mathbf{U}(H_1 \otimes H_2)$ of the finite-dimensional unitary representations $U_1\colon G_1 \to \mathbf{U}(H_1)$ and $U_2\colon G_2 \to \mathbf{U}(H_2)$ of two locally compact groups G_1 and G_2. This time it turns out that $U_1 \otimes U_2$ is irreducible iff U_1 and U_2 are irreducible. We prove this result when G is compact.

In Section 4.5, we define the notion of *contragredient representation* $\overline{U}\colon G \to \mathbf{GL}(H^*)$ of a representation $U\colon G \to \mathbf{GL}(H)$ and the notion of Hom *representation* $\mathrm{Hom}(U_1, U_2)\colon G \to \mathbf{GL}(\mathrm{Hom}(H_1, H_2))$ of two representations $U_1\colon G \to \mathbf{GL}(H_1)$ and $U_2\colon G \to \mathbf{GL}(H_2)$. These notions will be needed in Chapter 8. The main result is that if H_1 and H_2 are finite-dimensional vector spaces then the representations $\overline{U_1} \otimes U_2$ and $\mathrm{Hom}(U_1, U_2)$ are equivalent; see Proposition 4.23.

In Section 4.6, we define a notion of Fourier transform and Fourier cotransform for a (metrizable) compact group G. Since for a nonabelian compact group the set of characters is not a group, the definition of the spaces $\mathrm{L}^p(\widehat{G})$ is more complicated. The *Fourier transform* $\mathcal{F}f$ of a function $f \in \mathrm{L}^1(G)$ is now a function with domain R, a complete set of irreducible unitary representations of G, such that for every $\rho \in R$,

$$\mathcal{F}(f)(\rho) = \int f(t)(M_\rho(t))^* \, d\lambda_g(t).$$

The Fourier transform defined above is the natural generalization of the definition of the Fourier transform, when G is an abelian compact group (Vol. I, Definition 10.3),

$$\mathcal{F}(f)(\chi) = \int f(s)\overline{\chi(s)} \, d\lambda(s) = \int f(s)\chi(s^{-1}) \, d\lambda(s);$$

the character χ is replaced by the irreducible representation M_ρ.

The definition of $\mathcal{F}(f)(\rho)$ implies that $\mathcal{F}(f)(\rho)$ is a linear map from $\mathbb{C}^{n_\rho}$ to itself (since $(M_\rho(t))^*$ is a matrix). Thus, $\mathcal{F}(f) \in \prod_{\rho \in R} \mathrm{M}_{n_\rho}(\mathbb{C})$. Every element $F \in \prod_{\rho \in R} \mathrm{M}_{n_\rho}(\mathbb{C})$ is an R-indexed sequence $F = (F(\rho))_{\rho \in R}$ of $n_\rho \times n_\rho$ matrices $F(\rho)$. These sequences can be added and rescaled componentwise, so we obtain a vector space.

It is natural to define $\widehat{G}$ as R, but the vector space $\prod_{\rho \in R} \mathrm{M}_{n_\rho}(\mathbb{C})$ is too big. Thus, we define some normed vector spaces $\mathrm{L}^p(\widehat{G})$, which are subspaces of $\prod_{\rho \in R} \mathrm{M}_{n_\rho}(\mathbb{C})$. For this, we define some norms due to von Neumann; see Section 4.7.

We can also define a notion of Fourier cotransform and there are versions of Fourier inversion; see Section 4.8. For any $F \in \prod_{\rho \in R} \mathrm{M}_{n_\rho}(\mathbb{C})$, the *Fourier cotransform* $\overline{\mathcal{F}}(F)$ of F is the function on G given by

$$\overline{\mathcal{F}}(F)(s) = \sum_{\rho \in R} n_\rho \operatorname{tr}(F(\rho)M_\rho(s)), \qquad s \in G.$$

Of course, there are convergence issues. It can be shown (Theorem 4.33) that if $F \in \mathrm{L}^1(\widehat{G})$, then the map

$$s \mapsto (\overline{\mathcal{F}}(F))(s) = \sum_{\rho \in R} n_\rho \operatorname{tr}(F(\rho)M_\rho(s))$$

converges uniformly to a continuous function f. Furthermore, we have the Fourier inversion formula

$$(\overline{\mathcal{F}}(\mathcal{F}(f)))(s) = \sum_{\rho \in R} n_\rho \operatorname{tr}(\mathcal{F}(f)(\rho) M_\rho(s)), \qquad s \in G.$$

Also, Fourier inversion holds for $\mathrm{L}^2(G)$ (see Theorem 4.35).

Chapter 5 deals with explicit matrix descriptions of the irreducible representations of the groups $\mathbf{SL}(2, \mathbb{C})$, $\mathbf{SU}(2)$ and $\mathbf{SO}(3)$ (unitary representation in the last two cases). Our presentation (except for Section 5.7) relies heavily on Vilenkin's exposition [Vilenkin (1968)], especially Chapter III. To the best of our knowledge, Vilenkin contains the most detailed presentation of this material.

We begin by proving (Section 5.1) that the representations $U_m \colon \mathbf{SU}(2) \to \mathbf{GL}(\mathcal{P}_m^{\mathbb{C}}(2))$ and $W_\ell \colon \mathbf{SO}(3) \to \mathbf{GL}(\mathcal{P}_{2\ell}^{\mathbb{C}}(2))$, which were shown to be irreducible in Example 3.8 and Example 3.9, form complete sets of set of irreducible (unitary) representations. Here, $\mathcal{P}_m^{\mathbb{C}}(2)$ is the vector space of complex homogeneous polynomials of degree m in two variables (z_1 and z_2).

In Section 5.2, we give a more pleasant description of the irreducible unitary representations of $\mathbf{SO}(3)$ in terms of the spaces $\mathcal{H}_k^{\mathbb{C}}(3)$ of complex homogeneous harmonic polynomials in three variables.

It turns out that to obtain the most explicit matrix descriptions of the representations of $\mathbf{SU}(2)$ and $\mathbf{SO}(3)$, it is crucial to factor a unit quaternion q as the product of three types of unit quaternions $r_x(\varphi/2)$, $r_y(\psi/2)$, $r_z(\theta/2)$, which happen to induce the well-known rotations of $\mathbb{R}^3$ associated with the Euler angles. For example, we have the factorizations $q = r_x(\varphi/2) r_z(\theta/2) r_x(\psi/2)$ and $q = r_x(-\varphi/2) r_y(\theta/2) r_x(-\psi/2)$. This matter is treated in great detail in Section 5.3. This is a standard topic in quantum mechanics but it is also a source of confusion because different formulae are obtained depending on the method chosen for defining the rotation in $\mathbf{SO}(3)$ induced by a unit quaternion q in $\mathbf{SU}(2)$. We thoroughly discuss this issue.

Until now, the representations $U_m \colon \mathbf{SU}(2) \to \mathbf{GL}(\mathcal{P}_m^{\mathbb{C}}(2))$, which are also representations of $\mathbf{SL}(2)$, act on the vector space $\mathcal{P}_m^{\mathbb{C}}(2)$ of complex homogenous polynomials of degree m in two variables. In quantum mechanics it is preferable to use the integer or half-integer index $\ell = m/2$. The space $\mathcal{P}_m^{\mathbb{C}}(2) = \mathcal{P}_{2\ell}^{\mathbb{C}}(2)$ then has dimension $2\ell + 1$ and the monomials $c_k z_1^{\ell - k} z_2^{\ell + k}$ of a polynomial $P(z_1, z_2)$ are indexed by the index k

which ranges from $-\ell$ to ℓ. It is actually preferable to use the "dehomogenized" polynomial $Q(z) = P(z,1)$ in the single variable z. The vector space of such polynomials (of degree $2\ell+1$) is denoted $\mathcal{P}_\ell^{\mathbb{C}}$, and we define the representation $T_\ell\colon \mathbf{SL}(2,\mathbb{C}) \to \mathbf{GL}(\mathcal{P}_\ell^{\mathbb{C}})$, which yields a representation $T_\ell\colon \mathbf{SU}(2) \to \mathbf{GL}(\mathcal{P}_\ell^{\mathbb{C}})$ when restricted to the subgroup $\mathbf{SU}(2)$ of $\mathbf{SL}(2,\mathbb{C})$; see Section 5.5, Definition 5.3.

We will need to define an $\mathbf{SU}(2)$-invariant hermitian inner product on each space $\mathcal{P}_\ell^{\mathbb{C}}$, and for this it is useful to figure out what is the derivative of the representation $T_\ell\colon \mathbf{SL}(2,\mathbb{C}) \to \mathbf{GL}(\mathcal{P}_\ell^{\mathbb{C}})$ at the identity. This yields a representation $\mathfrak{t}_\ell\colon \mathfrak{sl}(2,\mathbb{C}) \to \mathrm{Hom}(\mathcal{P}_\ell^{\mathbb{C}}, \mathcal{P}_\ell^{\mathbb{C}})$, which is a representation of Lie algebras. This topic is discussed in Section 5.6.

In Section 5.7, we determine all the irreducible Lie algebra representations of $\mathfrak{sl}(2,\mathbb{C})$ (and again, of $\mathfrak{su}(2)$). This section presents key results of representation theory which occur in every book in representation theory. We follow Serre's exposition [Serre (2000)].

We return to our goal of finding explicit formulae for the matrix representations of $\mathbf{SL}(2,\mathbb{C})$, $\mathbf{SU}(2)$, and $\mathbf{SO}(3)$. In Section 5.8, we prove that if we consider the polynomials $\psi_k(z)$ given by

$$\psi_k(z) = \frac{z^{\ell-k}}{\sqrt{(\ell-k)!(\ell+k)!}}, \quad -\ell \le k \le \ell,$$

then the hermitian inner product on $\mathcal{P}_\ell^{\mathbb{C}}$ making the basis (ψ_k) orthonormal is $\mathbf{SU}(2)$-invariant (see Proposition 5.20).

In Section 5.9, we give $\mathcal{P}_\ell^{\mathbb{C}}$ the hermitian inner product making (ψ_k) an orthonormal basis and we give various expressions for the matrix entries of the matrix $t^{(\ell)}(A)$ representing $T_\ell(A)$ in this basis.

In Section 5.10, we restrict our attention to matrices in the group $\mathbf{SU}(2)$, in which case the hermitian inner product on $\mathcal{P}_\ell^{\mathbb{C}}$ making the basis (ψ_k) orthonormal is $\mathbf{SU}(2)$-invariant (see Proposition 5.20). Using the Euler angles representation of Section 5.3, we prove the important fact (see Proposition 5.23) that for any matrix $q \in \mathbf{SU}(2)$ expressed in terms of the Euler angles as $q = u(\varphi,\theta,\psi) = r_x(\varphi/2)r_z(\theta/2)r_x(\psi/2)$, with respect to the orthonormal basis (ψ_k) of $\mathcal{P}_\ell^{\mathbb{C}}$, we have

$$t_{jk}^{(\ell)}(q) = e^{-i(j\varphi+k\psi)}\, t_{jk}^{(\ell)}(r_z(\theta/2)), \quad -\ell \le j,k \le \ell.$$

Thus, we are left with finding an explicit expression for the matrix $t^{(\ell)}(r_z(\theta/2))$, which we denote as $t^{(\ell)}(\theta)$ (see Definition 5.11). Such a formula is given in Proposition 5.24.

Since $\mathbf{SU}(2)$ is the universal cover of $\mathbf{SO}(3)$, we obtain a formula for the matrix $w^{(\ell)}(R)$ of the unitary map $W_\ell(R)$ associated with the irreducible representation $W_\ell\colon \mathbf{SO}(3) \to \mathbf{U}(\mathcal{P}_\ell^{\mathbb{C}})$, where $R \in \mathbf{SO}(3)$ is expressed in terms of the Euler angles as $R = R_x(\varphi)R_z(\theta)R_x(\psi)$. With respect to the orthonormal basis (ψ_k) of $\mathcal{P}_\ell^{\mathbb{C}}$, the matrix $w^{(\ell)}(R)$ is given by

$$w_{jk}^{(\ell)}(R) = e^{-i(j\varphi+k\psi)}\, t_{jk}^{(\ell)}(\theta), \quad \ell \in \mathbb{N}.$$

We also discuss the famous Wigner d-matrices and $\mathcal{D}$-matrices.

There is one more method for computing the matrix elements $t_{jk}^{(\ell)}(A)$ (with $A \in \mathbf{SL}(2,\mathbb{C})$) based on integration. The idea is to use another representing space for the representation T_ℓ, namely the vector space $\mathfrak{F}_\ell$ (of dimension $2\ell+1$) of finite Fourier series

$$\Phi(e^{i\varphi}) = \sum_{k=-\ell}^{\ell} c_k e^{-ik\varphi},$$

with $c_k \in \mathbb{C}$. In Section 5.11, we define the (irreducible) representations $\mathcal{T}_\ell\colon \mathbf{SL}(2,\mathbb{C}) \to \mathbf{GL}(\mathfrak{F}_\ell)$. In Proposition 5.26, we obtain an integral formula for the matrix elements $t_{jk}^{(\ell)}(A)$. By specializing to the matrices $A = r_z(\theta/2)$, we obtain an integral formula for computing the matrix elements $t_{jk}^{(\ell)}(\theta)$ (see Proposition 5.27). For small values of ℓ, this equation is quite practical.

In Section 5.12, we show that the matrix elements $t_{jk}^{(\ell)}(\theta)$ can be expressed in terms of certain polynomials known as the *Jacobi polynomials*. In the special case where $k = 0$, in which case the function

$$t_{j0}^{(\ell)}(q) = e^{-ij\varphi}\, P_{j0}^{\ell}(\cos\theta)$$

is independent of the angle ψ, the function $P_{j0}^{\ell}(z)$ is a rescaling of the *associated Legendre function* $P_\ell^j(z)$. The function $t_{j0}^{(\ell)}(q)$ (with $q = r_x(\varphi/2)r_z(\theta/2)$) can be viewed as a function on the sphere S^2 and is denoted $Y_{\ell j}(\varphi,\theta)$, with $0 \le \varphi < 2\pi$ and $0 \le \theta < \pi$. The function $Y_{\ell j}(\varphi,\theta)$ is called a *spherical function*. Up to a constant, $Y_{\ell j}(\varphi,\theta)$ is the classical spherical harmonic (unfortunately) denoted $Y_\ell^j(\theta,\varphi)$ and called the *Laplace spherical harmonic* by Dieudonné.

In Section 5.14, we derive explicit formulae for the normalized Haar measures on $\mathbf{SU}(2)$ and $\mathbf{SO}(3)$ when these groups are parametrized by the Euler angles. Technically, these parametrizations are injective only on open subsets of $\mathbf{SU}(2)$ and $\mathbf{SO}(3)$, but the complements of these open sets have measure zero so from the point of view integration we obtain formulae for integrating all functions in $\mathrm{L}^2(\mathbf{SU}(2))$ and all functions in $\mathrm{L}^2(\mathbf{SO}(3))$ (respectively equipped with these left and right invariant Haar measures).

Combining results from Section 5.14 and the previous sections, in Section 5.15 we obtain explicit Fourier series expansions for the functions in $\mathrm{L}^2(\mathbf{SU}(2))$ and $\mathrm{L}^2(\mathbf{SO}(3))$ in terms of the matrix elements $t^{(\ell)}_{jk}$. The reason is that by Peter–Weyl the family of functions

$$\left(\sqrt{2\ell+1}\, t^{(\ell)}_{ij}\right)_{-\ell\leq i,j\leq \ell,\, \ell\in R}$$

with $R = \{0, 1/2, 1, 3/2, 2, \ldots\}$, is a Hilbert basis of $\mathrm{L}^2(\mathbf{SU}(2))$. Actually, we obtain explicit formulae for the Fourier transform and the Fourier cotransform (discussed in Section 4.6) on $\mathrm{L}^2(\mathbf{SU}(2))$. Similarly, the family of functions

$$\left(\sqrt{2\ell+1}\, w^{(\ell)}_{ij}\right)_{-\ell\leq i,j\leq \ell,\, \ell\in \mathbb{N}}$$

is a Hilbert basis of $\mathrm{L}^2(\mathbf{SO}(3))$. This yields another explicit example of the Fourier transform and the Fourier cotransform on $\mathrm{L}^2(\mathbf{SO}(3))$. If the functions are expressed in terms of the Euler angles, then we obtain formulae that are practically computable.

In Section 5.16, following Vilenkin, we show how to decompose not only scalar-valued but also vector-valued functions on the sphere S^2 into Fourier series that behave nicely under rotations of the sphere.

The last section of this chapter (Section 5.16) deals with the *Clebsch–Gordan coefficients*, a standard topic in quantum mechanics. In general, the tensor product $T_{\ell_1} \otimes T_{\ell_2}$ of two irreducible representations T_{ℓ_1} and T_{ℓ_2} of $\mathbf{SU}(2)$ is not irreducible, so according to the Peter–Weyl theorem (Theorem 4.16) it splits as a direct sum of irreducible representations. Since the character associated with the representation $T_{\ell_1} \otimes T_{\ell_2}$ is equal to the product $\chi_{T_{\ell_1}}\chi_{T_{\ell_2}}$ of the characters $\chi_{T_{\ell_1}}$ and $\chi_{T_{\ell_2}}$ associated with T_{ℓ_1} and T_{ℓ_2}, it turns out that the following famous result (known to H. Weyl and E. Wigner) can be obtained (see Proposition 5.46). For any two irreducible representations T_{ℓ_1} and T_{ℓ_2} of $\mathbf{SU}(2)$, we have

$$\chi_{T_{\ell_1}}(q)\chi_{T_{\ell_2}}(q) = \sum_{\ell=|\ell_1-\ell_2|}^{\ell_1+\ell_2} \chi_{T_\ell}(q), \quad q \in \mathbf{SU}(2).$$

As a consequence, we also have an isomorphism

$$\mathcal{P}_{\ell_1} \otimes \mathcal{P}_{\ell_2} \simeq \bigoplus_{\ell=|\ell_1-\ell_2|}^{\ell_1+\ell_2} \mathcal{P}_\ell.$$

The space $\mathcal{P}_{\ell_1} \otimes \mathcal{P}_{\ell_2}$ has dimension $(2\ell_1 + 1)(2\ell_2 + 1)$ and each summand $\mathcal{P}_\ell$ has dimension $2\ell + 1$.

By Proposition 5.16, each vector space $\mathcal{P}_\ell$ has an orthonormal basis (ψ_k) $(-\ell \leq k \leq \ell)$ invariant under the action of $\mathbf{SU}(2)$. Following Vilenkin [Vilenkin (1968)] (Chapter III, Section 8.2), we denote the basis of $\mathcal{P}_{\ell_1}$ as $(\mathbf{f}_j)$ $(-\ell_1 \leq j \leq \ell_1)$ and the basis of $\mathcal{P}_{\ell_2}$ as $(\mathbf{h}_k)$ $(-\ell_2 \leq k \leq \ell_2)$. Then the family of tensor products

$$\mathbf{f}_j \otimes \mathbf{h}_k, \quad -\ell_1 \leq j \leq \ell_1, \ -\ell_2 \leq k \leq \ell_2,$$

is a basis of $\mathcal{P}_{\ell_1} \otimes \mathcal{P}_{\ell_2}$. If we give $\mathcal{P}_{\ell_1} \otimes \mathcal{P}_{\ell_2}$ the inner product defined in Definition 4.10 induced by the inner products associated with the bases $(\mathbf{f}_j)$ and $(\mathbf{h}_k)$, then the vectors $(\mathbf{f}_j \otimes \mathbf{h}_k)$ form an orthonormal basis of $\mathcal{P}_{\ell_1} \otimes \mathcal{P}_{\ell_2}$.

Since we have the direct sum

$$\mathcal{P}_{\ell_1} \otimes \mathcal{P}_{\ell_2} \simeq \bigoplus_{\ell=|\ell_1-\ell_2|}^{\ell_1+\ell_2} \mathcal{P}_\ell,$$

we also have a basis of $\mathcal{P}_{\ell_1} \otimes \mathcal{P}_{\ell_2}$ consisting of the union of the bases associated with each of the summand $\mathcal{P}_l$, which Vilenkin denotes by

$$\mathbf{a}^\ell_m, \quad |\ell_1 - \ell_2| \leq \ell \leq \ell_1 + \ell_2, \ -\ell \leq m \leq \ell,$$

where for ℓ fixed, $(\mathbf{a}^\ell_m)$ $(-\ell \leq m \leq \ell)$ is the basis of $\mathcal{P}_\ell$. Since both bases are orthonormal bases of $\mathcal{P}_{\ell_1} \otimes \mathcal{P}_{\ell_2}$, there is a unitary matrix C expressing the basis $(\mathbf{f}_j \otimes \mathbf{h}_k)$ in terms of the basis $(\mathbf{a}^\ell_m)$, and the entries of the matrix C are called the *Clebsch–Gordan coefficients*. More precisely, the change of basis matrix $C = (C_{(\ell m),(jk)})$ is the unitary matrix defined such that the (jk)th column of C consists of the coefficients of $\mathbf{f}_j \otimes \mathbf{h}_k$ over the basis $(\mathbf{a}^\ell_m)$, namely

$$\mathbf{f}_j \otimes \mathbf{h}_k = \sum_{\ell=|\ell_1-\ell_2|}^{\ell_1+\ell_2} \sum_{m=-\ell}^{\ell} C_{(\ell m),(jk)} \mathbf{a}^\ell_m,$$

with $-\ell_1 \leq j \leq \ell_1$, $-\ell_2 \leq k \leq \ell_2$.

Amazingly, the coefficients $C_{(\ell m),(jk)}$ can be computed explicitly, but the formulae are very complicated and the technical details of the computations are quite involved. Complete details can be found in Vilenkin [Vilenkin (1968)] (Chapter III, Section 8). In this section, we will content ourselves by providing an outline of these computations.

Chapter 6 is devoted to induced representations. If G is a locally compact group and if H is a closed subgroup of G, under certain conditions, it is possible to construct a Hilbert space $\mathcal{H}$ and a unitary representation $\Pi\colon G \to \mathbf{U}(\mathcal{H})$ of G in $\mathcal{H}$ from a unitary representation $U\colon H \to \mathbf{U}(E)$ of H in a (separable) Hilbert space E. The representation Π is called an *induced representation*. Induced representations are very useful in cases where it is difficult to construct directly representations of a group. This is the case for $\mathbf{SL}(2,\mathbb{R})$. Interestingly, induced representations play a central role in group equivariant deep learning in convolutional networks; see Chapter 8.

There are two approaches for the construction of the Hilbert space $\mathcal{H}$:

(1) The Hilbert space $\mathcal{H}$ is a set of functions from $X = G/H$ to E.
(2) The Hilbert space $\mathcal{H}$ is a set of functions from G to E.

In the first approach, we will construct unitary representations of G in $\mathcal{H}$ using certain functions $\alpha\colon G \times (G/H) \to \mathbf{GL}(E)$ called *cocycles*. In the second approach, the construction of the Hilbert space $\mathcal{H}$ is more complicated, but the definition of the operator Π_s is simpler.

The general construction (in the first approach) consists of seven steps, where the first four are purely algebraic and do not deal with continuous unitary representations, but instead linear representations (group homomorphisms $U\colon G \to \mathbf{GL}(E)$, where G is a group not equipped with any topology and E is just a vector space with no additional structure):

(1) Let G be a group acting (on the left) on a set X, and let E be a vector space. In Section 6.1, we define the notion of *equilinear action* of G on $X \times E$ and *cocycle*. A *cocycle of G with values in* $\mathbf{GL}(E)$ is a map $\alpha\colon G \times X \to \mathbf{GL}(E)$ satisfying the conditions:

  (a) For all $x \in X$
  $$\alpha(e,x) = \mathrm{id}_E.$$
  (b) For all $x \in X$ and all $s,t \in G$,
  $$\alpha(st,x) = \alpha(s,t\cdot x)\circ\alpha(t,x).$$

  A equilinear action of G on $X \times E$ defines a cocycle and an action of G on X, and conversely. An equilinear action of G on $X \times E$ induces a homomorphism $\Pi\colon G \to \mathbf{GL}(E^X)$, where E^X is the vector space of all functions from X to E. More precisely, for every function $f\colon X \to E$, for every $s \in G$, $\Pi_s(f)\colon X \to E$ is function given by
  $$(\Pi_s(f))(x) = \alpha(s,s^{-1}\cdot x)(f(s^{-1}\cdot x)), \qquad \text{for every } x \in X.$$

(2) In Section 6.2, we specialize the construction to the homogeneous space $X = G/H$ of left cosets. Then G acts on G/H on the left by

$$s \cdot (gH) = sgH.$$

By choosing a set of representatives $(r_x)_{x \in G/H}$ in the cosets of $X = G/H$ (with $x_0 = H$ and $r_{x_0} = e$), a cocycle $\alpha\colon G \times X \to \mathbf{GL}(E)$ determines a homomorphism $\sigma\colon H \to \mathbf{GL}(E)$ given by $\sigma(h) = \alpha(h, x_0)$ and a map $\beta\colon X \to \mathbf{GL}(E)$ given by $\beta(x) = \alpha(r_x, x_0)$. Conversely, a homomorphism $\sigma\colon H \to \mathbf{GL}(E)$ and a map $\beta\colon X \to \mathbf{GL}(E)$ determine a cocycle $\alpha\colon G \times X \to \mathbf{GL}(E)$. In fact, we may restrict ourselves to the map β given by $\beta(x) = \mathrm{id}_E$, and if we define $u\colon G \times X \to H$ by $u(s, x) = r_{s \cdot x}^{-1} s r_x$, the map $\alpha\colon G \times X \to \mathbf{GL}(E)$ given by $\alpha(s, x) = \sigma(u(s, x))$ is a cocycle. The induced representation is given by

$$(\Pi_s(f))(x) = \sigma(u(s, s^{-1} \cdot x))(f(s^{-1} \cdot x)), \ \ f \in E^X, \ x \in X.$$

This step is the most important application of Step 1, and E is an arbitrary vector space.

(3) For a given homomorphism $\sigma\colon H \to \mathbf{GL}(E)$, the homomorphisms $\Pi\colon G \to \mathbf{GL}(E^X)$ corresponding to cocycles associated with different maps β are equivalent.

(4) In Section 6.3, we show that a cocycle $\alpha\colon G \times X \to \mathbf{GL}(E)$ determines a bijection between E^X and a subspace L^α of the set E^G of maps from G to E defined by

$$L^\alpha = \{f \in E^G \mid f(sh) = \sigma(h^{-1})(f(s)), \ \ s \in G, h \in H\}.$$

As a consequence, the representation $\Pi\colon G \to \mathbf{GL}(E^X)$ corresponding to a cocycle α is equivalent to the representation $\Pi_{L^\alpha}\colon G \to \mathbf{GL}(L^\alpha)$ given by

$$((\Pi_{L^\alpha})_s(g))(t) = g(s^{-1}t) \qquad \text{for all } g \in L^\alpha \text{ and all } s, t \in G.$$

Observe that this is simply the left regular representation of L^α. The issue of choosing between representations in the space E^X or representations in the space L^α comes up in Chapter 8.

This completes the purely algebraic construction. The next steps use topology and analysis to construct *unitary* representations.

(5) In Section 6.4, we assume that G is a locally compact group and H is a closed subgroup of G, in which case G/H is also locally compact. Let μ be a positive measure on $X = G/H$, and assume that E is a separable Hilbert space. We then define a Hilbert space $\mathcal{L}^2_\mu(X; E)$ consisting of measurable functions from X to E.

(6) In Section 6.5, given a unitary representation U of H in E, we assume that the measure μ on $X = G/H$ is G-invariant and if the cocycle α satisfies certain conditions, then the homomorphism $s \mapsto \Pi_s([f]) = [\Pi_s(f)]$ is a unitary representation of G in $\mathrm{L}^2_\mu(X; E) = \mathcal{H}$.
(7) In Sections 6.6 and 6.7, we generalize the previous construction to certain measure called *quasi-invariant*. If the measure μ on G/H is quasi-invariant and another technical condition is satisfied, then the homomorphism $s \mapsto \Pi_s([f]) = [\Pi_s(f)]$ is a unitary representation of G in $\mathrm{L}^2_\mu(X; E)$. Quasi-invariant measures on G/H always exist and can be constructed using rho-functions.

In Section 6.8, we illustrate the method of Section 6.7 by showing how to construct unitary representations of $\mathbf{SL}(2, \mathbb{R})$ using induced representations.

In Section 6.9, we consider a compact (metrizable) group G and a closed subgroup H of G, and our goal is to determine the canonical (unitary) representation of G in $\mathrm{L}^2_\mu(G/H; \mathbb{C})$ induced by the trivial representation of H in $E = \mathbb{C}$ (see Definition 6.13), where μ is the G-invariant measure on G/H induced by a Haar measure λ on G. For simplicity of notation, we write $\mathrm{L}^2_\mu(G/H)$ instead of $\mathrm{L}^2_\mu(G/H; \mathbb{C})$. To do this, it is necessary to understand what is the restriction of the representation $M_\rho\colon G \to \mathbf{U}(\mathbb{C}^{n_\rho})$ to H, with $\rho \in R(G)$.

In Proposition 6.18, we show that the space $\mathrm{L}^2_\mu(G/H)$ is the Hilbert sum of subspaces $L_\rho \subseteq \mathfrak{a}_\rho$. If the trivial representation σ_0 of H is contained $d = (\rho : \sigma_0) \geq 1$ times in the restriction of M_ρ to H, then L_ρ is the direct sum of the first d columns of the matrix $M_\rho^{(H)} = P^* M_\rho P$, where P is a suitable change of basis matrix.

Then, we consider the space $H\backslash G$ of right cosets Hs of G ($s \in G$). We show in Proposition 6.19 that the space $\mathrm{L}^2_{\mu'}(H\backslash G)$ is the Hilbert sum of subspaces $\check{L}_\rho \subseteq \mathfrak{a}_\rho$. If the trivial representation σ_0 of H is contained $d = (\rho : \sigma_0) \geq 1$ times in the restriction of M_ρ to H, then $\check{L}_\rho$ is the direct sum of the first d rows of $M_\rho^{(H)}$.

In preparation for Chapter 9, we consider the intersection $\mathrm{L}^2_\mu(G/H) \cap \mathrm{L}^2_{\mu'}(H\backslash G)$. This is a closed involutive subalgebra of $\mathrm{L}^2(G)$, thus a complete Hilbert algebra. We can view a function $g \in \mathrm{L}^2_\mu(G/H) \cap \mathrm{L}^2_{\mu'}(H\backslash G)$ as a function $g \in \mathrm{L}^2(G)$, such that

$$g(tst') = g(s) \qquad \text{for all } t, t' \in H \text{ and all } s \in G. \tag{$*_{H\backslash G/H}$}$$

We denote the algebra of functions in $\mathrm{L}^2(G)$ satisfying $(*_{H\backslash G/H})$ as $\mathrm{L}^2(H\backslash G/H)$. Then we show in Proposition 6.20 that the algebra $\mathrm{L}^2(H\backslash G/H)$ is the Hilbert sum of minimal two-sided ideals.

Again, in preparation for Chapter 9 on Gelfand pairs, we show in Proposition 6.21 that the algebra $\mathrm{L}^2(H\backslash G/H)$ is commutative if and only if $(\rho : \sigma_0) \leq 1$ for all $\rho \in R(G)$.

In Section 6.10, we present a nice example of the above situation for $G = \mathbf{SO}(n+1)$ and $H = \mathbf{SO}(n)$. In this case, $G/H = \mathbf{SO}(n+1)/\mathbf{SO}(n) \simeq S^n$, the sphere in $\mathbb{R}^{n+1}$. As a consequence, we obtain a decomposition of $\mathrm{L}^2(S^n)$ as a Hilbert sum of the classical spaces $\mathcal{H}_k^{\mathbb{C}}(S^n)$ of spherical harmonics on S^n.

In Section 6.11, we present a method due to Blattner to deal with the situation where G/H has no G-invariant measure. This is a modification of the construction of the Hilbert space $\mathcal{H}$ and the inner product described at the end of Section 6.5. This can be done in two ways. These constructions yield induced unitary representations of G from a unitary representation $U\colon H \to \mathbf{U}(E)$ of H and do not involve cocycles.

In Section 6.12, we explain how the spaces of functions L^α (from Definition 6.8), and the spaces $\mathcal{H}_0$ and $\mathcal{H}^0$ from Section 6.11 can be viewed as sections of spaces that are similar to vector bundles but have less structure. More precisely, such structures have no trivialization maps.

We begin with the simplest situation where we have a group G without any topology on it, a subgroup H of G, a vector space $\mathcal{H}_\sigma$, and a linear representation $\sigma\colon H \to \mathbf{GL}(\mathcal{H}_\sigma)$. As usual, write $X = G/H$ and $\pi\colon G \to G/H$ for the quotient map. Let L^σ be the subspace of $(\mathcal{H}_\sigma)^G$ consisting of all functions $f\colon G \to \mathcal{H}_\sigma$, such that

$$f(gh) = \sigma(h^{-1})(f(g)), \qquad \text{for all } g \in G \text{ and all } h \in H.$$

The key point is to construct a space $E = G \times_H \mathcal{H}_\sigma$, together with a surjective map $p\colon E \to X$, such that for every $x \in X = G/H$, the fibre $E_x = p^{-1}(x)$ is isomorphic to the vector space $\mathcal{H}_\sigma$, and the space of sections from X to E is in bijection with L^σ. This is a special case of the so-called Borel construction used to construct a vector bundle from a principal bundle; see Gallier and Quaintance [Gallier and Quaintance (2020b)] (Chapter 9, Section 9.9). Then the main point of this section is to define two maps $\mathcal{S}\colon L^\sigma \to \Gamma(E)$ and $\mathcal{L}\colon \Gamma(E) \to L^\sigma$, which are mutual inverses, where $\Gamma(E)$

is the space of sections of E, namely the set of functions $s\colon X \to E$, such that $p \circ s = \mathrm{id}_X$, where p is the projection $p\colon E \to X$.

The last important ingredient is that G acts (on the left) on $E = G \times_H \mathcal{H}_\sigma$ in an *equilinear* fashion; this is explained in Proposition 6.23.

In Section 6.13, we show how induced representations can be recovered from certain kinds of vector bundles E over the base space $X = G/H$ (actually a more basic notion of vector bundle) equipped with an equilinear action of a group G on E. Such bundles, called *G-bundles*, are equipped with an equilinear action of the group G and generalize the notion of bundle introduced in the previous section. If x_0 denotes the coset $H = eH$ in G, then the action of G on the fibre E_0 above x_0 defines a representation $\sigma\colon H \to \mathbf{GL}(E_0)$. Again, the main point is to define a space of functions L^σ and two maps $\mathcal{S}\colon L^\sigma \to \Gamma(E)$ and $\mathcal{L}\colon \Gamma(E) \to L^\sigma$, which are mutual inverses, where $\Gamma(E)$ is the space of sections of E. The induced representation of G induced by the representation σ of H can then be recovered from the action of G on sections of E in terms of $\mathcal{L}$ and $\mathcal{S}$.

The sections in $\Gamma(E)$, called *feature fields* in group equivariant deep learning in computer vision, are functions whose domain transforms under the action of G and whose codomain transforms by representations of H equivalent to $\sigma\colon H \to \mathbf{GL}(E_0)$.

The above definitions and constructions are adapted to deal with unitary representations in Section 6.14. In this case, G is a locally compact group, H is a closed subgroup of G, and $\sigma\colon H \to \mathbf{U}(\mathcal{H}_\sigma)$ is a unitary representation, where $\mathcal{H}_\sigma$ is a separable Hilbert space. These bundles are called *hermitian G-bundles*. We treat the special case where $\mathcal{H}_\sigma$ is finite-dimensional in detail.

Unfortunately, in general the maps $\mathcal{L}$ and $\mathcal{S}$ are no longer well-defined. To remedy this problem, we assume that our G-bundles are locally trivializable, that is, they are (smooth) vector bundles.

Consequently, in Section 6.15, we review principal H-bundles and hermitian vector bundles. We then define *hermitian G-vector bundles*, which are simultaneously hermitian vector bundles and hermitian G-bundles. We discuss the construction of a hermitian vector bundle from a principal H-bundle obtained by replacing the fibre H by a vector space $\mathcal{H}_\sigma$, which is the space of a unitary representation $\sigma\colon H \to \mathbf{U}(\mathcal{H}_\sigma)$; see Theorem 6.27.

The generalization to hermitian G-vector bundles of infinite rank is sketched in Section 6.16.

One of the most important contributions to the theory of unitary representations is a method due to Mackey for constructing all irreducible representations of a locally compact group as induced irreducible representations from "small" subgroups H_ν of G. This method is often referred to as the "Mackey machine." Chapter 7 presents a simple version of Mackey's method.

In its most general form the method is very complicated but in the case where G has an abelian normal subgroup N, it is tractable. The basic reason is that because N is abelian, its irreducible representations are given by the characters of N. There is also a natural action $\cdot\colon G \times \widehat{N} \to \widehat{N}$ of G on the dual group $\widehat{N}$ (the group of characters of N). The key to the construction is that because N is an abelian locally compact group, by Theorem 3.20, for any unitary representation $U\colon G \to \mathbf{U}(\mathcal{H}_U)$ of G, since the restriction of U to N is a unitary representation, there is a unique regular projection-valued measure P on the dual group $\widehat{N}$, such that

$$U(n) = \int_{\widehat{N}} \chi(n)\, dP(\chi), \quad n \in N.$$

Moreover, the projection-valued measure P on $\widehat{N}$ satisfies two properties (see Proposition 7.1):

(1) For all $s \in G$, we have

$$U(s)P(E)U(s)^{-1} = P(s \cdot E), \quad \text{for all Borel subsets } E \subseteq \widehat{N}. \quad \text{(imp)}$$

(2) If U is irreducible, then for every G-invariant Borel set $E \subseteq \widehat{N}$ (which means that $\{s\cdot\chi \mid \chi \in E\} = s\cdot E = E$ for every $s \in G$), either $P(E) = I$ or $P(E) = 0$.

If the action of G on $\widehat{N}$ is nice enough (the space of orbits of this action is countably separated, see Definition 7.2), then P is identically zero except on a single orbit $\mathcal{O}_\nu$, so we can consider P as living on G/G_ν (where G_ν is the stabilizer of ν), and G acts transitively on this space. Then the data (G, U, X, P) consisting of the unitary representation $U\colon G \to \mathbf{U}(\mathcal{H}_U)$, of a transitive action of G on the homogeneous space $X = G/G_\nu$ (for some fixed $\nu \in \widehat{N}$), and of a regular projection-valued measure P on G/G_ν, such that

$$U(s)P(E)U(s)^{-1} = P(s \cdot E), \quad \text{for all Borel sets } E \subseteq G/G_\nu \text{ and all } s \in G,$$

constitute a *transitive system of imprimitivity* (see Definition 7.3).

The relevance of systems of imprimitivity is Mackey's *imprimitivity theorem* (Theorem 7.3), which implies that U is equivalent to a representation obtained by the induction method from some irreducible representation of G_ν. Technically, Mackey's *imprimitivity theorem* says more, namely that any transitive system of imprimitivity is equivalent to a system of imprimitivity arising by induction from the subgroup defining the homogeneous space X. If the action of G on $\widehat{N}$ is regular (see Definition 7.6), then Mackey's imprimitivity theorem implies Theorem 7.4, which shows that for every irreducible representation $U\colon G \to \mathbf{U}(\mathcal{H}_U)$ of G, there is a unique orbit $\mathcal{O}$, such that for any $\nu \in \mathcal{O}$ (so that $\mathcal{O} = \mathcal{O}_\nu$), there is an irreducible unitary representation $\sigma\colon G_\nu \to \mathbf{U}(\mathcal{H}_\sigma)$, such that U is equivalent to $\mathrm{Ind}_{G_\nu}^{G}\sigma$, the induced representation obtained from σ.

Unfortunately, the subgroups G_ν may still not be small enough. However, if some condition on G_ν holds (an extension condition), then for every irreducible representation $\rho\colon G_\nu/N \to \mathbf{U}(\mathcal{H}_\rho)$ of G_ν/N, there is an irreducible representation of G_ν in $\mathcal{H}_\rho$ (see Proposition 7.6). In this case, we can use irreducible representations of the "little groups" $H_\nu = G_\nu/N$ in the inducing process of Theorem 7.4.

The above extension condition is satisfied by semi-direct products $G = N \rtimes H$, where N is a normal abelian subgroup of G. Then every irreducible representation of G is obtained in terms of the characters of N and of the irreducible representations of the little groups H_ν associated with the characters $\nu \in \widehat{N}$; see Theorem 7.7. Using this method, we describe all irreducible representations of $\mathbf{SE}(n)$; see Example 7.1. We also determine all irreducible representations of $\mathbf{O}(2)$ (see Example 7.2) and indicate how all irreducible representations of $\mathbf{E}(2)$ and $\mathbf{E}(3)$ can be obtained.

Most of the material in Chapter 8 is heavily inspired by the works of Bekkers, Boomsma, Cesa, Cohen, Forré, Geiger, Lang, Verlinder, Weiler, and Welling. The general theme is to develop a theory of equivariant convolutional neural networks (CNNs). The purpose and the need for such neural networks is very clearly articulated in the preface of the recent book by Weiler, Forré, Verlinde, and Welling [Weiler *et al.* (2023)] that we highly recommend. Our goal in this chapter is to show how many of the fairly abstract concepts discussed earlier (representations, analysis on compact groups, Peter–Weyl theorems, Fourier transform, induced representations) are used to tackle very practical problems.

In Section 8.2, motivated by the problem of matching a pattern k (also called a correlation kernel or template kernel) in an image f, we define the notion of *cross-correlation*, for short *correlation*, given by

$$(k \star f)(x) = \int_{\mathbb{R}^2} f(t)k(t - x)\,dt.$$

Since images and correlation kernels are viewed as functions from $\mathbb{R}^2$ to $\mathbb{R}$, it is natural to view the action of a group G on images as given by the regular representation $\mathbf{R}$ of G on $\mathrm{L}^2(\mathbb{R}^2)$ induced by an action of G in $\mathbb{R}^2$, namely

$$[\mathbf{R}_g(f)](x) = \lambda_g(f)(x) = f(g^{-1} \cdot x), \quad g \in G,\ x \in \mathbb{R}^2,\ f \in \mathrm{L}^2(\mathbb{R}^2).$$

If we want to be more precise, we denote this representation by $\mathbf{R}^{G \to \mathrm{L}^{\mathbb{R}^2}}$.

In the special case where $G = \mathbb{R}^2$, the group of translations of $\mathbb{R}^2$ itself, because the Lebesgue measure on $\mathbb{R}^2$ is translation-invariant, we have the following commutative diagram expressing that the linear map $\Phi\colon \mathrm{L}^2(\mathbb{R}^2) \to \mathrm{L}^2(\mathbb{R}^2)$ given by $\Phi(f) = k \star f$ is *translation-invariant*:

$$\begin{array}{ccc} \mathrm{L}^2(\mathbb{R}^2) & \xrightarrow{\ \Phi\ } & \mathrm{L}^2(\mathbb{R}^2) \\ {\scriptstyle \mathbf{R}_x^{\mathbb{R}^2 \to \mathrm{L}^2(\mathbb{R}^2)}} \big\downarrow & & \big\downarrow {\scriptstyle \mathbf{R}_x^{\mathbb{R}^2 \to \mathrm{L}^2(\mathbb{R}^2)}} \\ \mathrm{L}^2(\mathbb{R}^2) & \xrightarrow[\ \Phi\ ]{} & \mathrm{L}^2(\mathbb{R}^2) \end{array}$$

commutes for all $x \in \mathbb{R}^2$.

However, if G is the group $\mathbf{SO}(2)$, the rotations in the plane $\mathbb{R}^2$, if the image f is rotated by an angle θ, we have the new image given by

$$(\mathbf{R}_{R_\theta}^{\mathbf{SO}(2) \to \mathrm{L}^2(\mathbb{R}^2)} f)(t) = f(R_{-\theta}(t)), \quad t \in \mathbb{R}^2,\ R_\theta \in \mathbf{SO}(2),$$

but the diagram

$$\begin{array}{ccc} \mathrm{L}^2(\mathbb{R}^2) & \xrightarrow{\ \Phi\ } & \mathrm{L}^2(\mathbb{R}^2) \\ {\scriptstyle \mathbf{R}_{R_\theta}^{\mathbf{SO}(2) \to \mathrm{L}^2(\mathbb{R}^2)}} \big\downarrow & & \big\downarrow {\scriptstyle \mathbf{R}_{R_\theta}^{\mathbf{SO}(2) \to \mathrm{L}^2(\mathbb{R}^2)}} \\ \mathrm{L}^2(\mathbb{R}^2) & \xrightarrow[\ \Phi\ ]{} & \mathrm{L}^2(\mathbb{R}^2) \end{array}$$

does *not* commute. The linear map Φ is *not* rotation-equivariant.

This is unfortunate, because in general, we would like to know whether the pattern k occurs in f, translated or rotated. More generally, if G is a group of transformations of $\mathbb{R}^2$, we would like our transform Φ to be *G-equivariant*, which means that the diagram

$$\begin{CD} \mathrm{L}^2(\mathbb{R}^2) @>{\Phi}>> \mathrm{L}^2(\mathbb{R}^2) \\ @V{\mathbf{R}_g^{G\to \mathrm{L}^2(\mathbb{R}^2)}}VV @VV{\mathbf{R}_g^{G\to \mathrm{L}^2(\mathbb{R}^2)}}V \\ \mathrm{L}^2(\mathbb{R}^2) @>>{\Phi}> \mathrm{L}^2(\mathbb{R}^2) \end{CD}$$

commutes for all $g \in G$.

As we just explained, equivariance fails beyond translation-equivariance, so what can we do to remedy this problem?

A solution is to define a *lifted correlation*. The basic idea presented in Section 8.3 is that instead of rotating the input image we apply a *rotated kernel* to the image. We first illustrate this process in the case of $\mathbf{SE}(2) = \mathbb{R}^2 \rtimes \mathbf{SO}(2)$, but this method works for semi-direct products of the form $G = \mathbb{R}^d \rtimes H$. We will denote an element of $\mathbf{SE}(2)$ as $g = (x, \theta)$, where $x \in \mathbb{R}^2$ and $\theta \in \mathbb{R} \pmod{2\pi}$. Then we define the lifted correlation $k \widetilde{\star} f$ by

$$(k \widetilde{\star} f)(x, \theta) = \int_{\mathbb{R}^2} f(t)(\lambda_{(x,\theta)} k)(t)\, dt = \int_{\mathbb{R}^2} f(t) k(R_{-\theta}(t - x))\, dt.$$

We are now using the lifted (rotated) kernel $\lambda_{R_\theta} k$, but observe that our transform now takes an input function f (image, signal) in $\mathrm{L}^2(\mathbb{R}^2)$, but yields an output function $\Phi(f) = k \widetilde{\star} f$ in the larger function space $\mathrm{L}^2(\mathbf{SE}(2))$ of functions defined on the *group* $\mathbf{SE}(2)$. Such functions are called *feature maps*.

The major benefit of lifted kernels is that we recover equivariance under the group $\mathbf{SO}(2)$.

All this is generalized to semi-direct products of the form $G = \mathbb{R}^d \rtimes H$, where H is a compact group. Correlation on feature maps (functions in $\mathrm{L}^2(G)$), called group correlation, is discussed in Section 8.4. However, for $d > 2$, it is usually not practically possible to discretize the group H, so a different approach is needed. A solution is to use *steerable families*, which are discussed in Section 8.6. The notion of steerability occurred first in the seminal paper of Freeman and Adelson [Freeman and Adelson (1991)].

The idea behind steerability is that if a function f is defined on some measure space X and if there is an action of a group H on X, then it would be nice if $f(h^{-1} \cdot x)$ could be expressed in a simple way in terms of $f(x)$. In general this is asking for too much, but if we consider a *family* of linearly independent functions $(Y_1, \ldots, Y_L)$ in $\mathrm{L}^2(X)$, then we say that they form an *H-steerable family* if there is representation $\Sigma\colon H \to \mathbf{U}(L)$, such that

$$Y(h^{-1} \cdot x) = \Sigma(h)^\top Y(x), \quad h \in H,\, x \in X,$$

where $Y(x)$ denotes the column vector

$$Y(x) = \begin{pmatrix} Y_1(x) \\ \vdots \\ Y_L(x) \end{pmatrix} \in \mathbb{C}^L;$$

see Definition 8.5. If a correlation kernel k can be expressed as a linear combination of a steerable family Y, then the lifted convolution $k \mathbin{\widetilde{\star}} f$ can be computed in a cheap way in terms of the vector-valued function

$$f^Y(x) = \int_{\mathbb{R}^d} f(t) Y(t-x)\, dt.$$

We can think of $f^Y(x)$ as some kinds of Fourier coefficients.

In Section 8.7, we present a method for finding steerable families on a suitable space X equipped with a continuous action of a compact group H. The trick is to consider the unitary representation $V\colon H \to \mathbf{U}(\mathrm{L}^2(X))$ given by

$$(V(h)f)(x) = f(h^{-1} \cdot x), \quad h \in H,\, f \in \mathrm{L}^2(X),\, x \in X, \tag{V}$$

and then use the Peter–Weyl theorem, Version II to express the space $\mathrm{L}^2(X)$ as the Hilbert sum of closed subspaces E_ρ with $\rho \in R(H)$.

In Section 8.8, we introduce the notion of *feature field*, which as Cesa, Lang and Weiler [Cesa *et al.* (2022)] say, "is the fundamental design choice underlying steerable CNNs." Such functions already arise when steerable kernels are used. Feature fields are vector-valued functions $f\colon \mathbb{R}^d \to \mathcal{H}$, whose domain transforms under the action of a group $G = \mathbb{R}^d \rtimes H$ and whose codomain transforms under a representation $\sigma\colon H \to \mathbf{GL}(\mathcal{H})$, in most cases, actually a unitary representation. Thus, the space of feature fields transforms under the induced representation $\mathrm{Ind}_H^G \sigma$, namely for any feature field f,

$$[(\mathrm{Ind}_H^G \sigma)_{(x,h)} f](t) = \sigma(h)(f(h^{-1} \cdot (t-x))), \quad (x,h) \in \mathbb{R}^d \rtimes H,\, t \in \mathbb{R}^d.$$

We know how to transform G-feature maps using group correlation defined in Definition 8.4. This defines a transform Φ on $\mathrm{L}^2(G)$ (where $G = \mathbb{R}^d \rtimes H$) given by $f_{\mathrm{out}} = \Phi(f_{\mathrm{in}}) = k \star f_{\mathrm{in}}$. Since it is too expensive to compute $\Phi(f_{\text{in}}) = k \star f_{\text{in}}$, it would be nice if we could define a vector space of Fourier coefficients $\mathrm{L}^2(\mathbb{R}^d, \widehat{H})$ consisting of matrix-valued functions on $\mathbb{R}^d$, and a new Fourier transform $\mathcal{F}^{\tau} : \mathrm{L}^2(G) \to \mathrm{L}^2(\mathbb{R}^d, \widehat{H})$ and cotransform $\overline{\mathcal{F}^{\tau}} : \mathrm{L}^2(\mathbb{R}^d, \widehat{H}) \to \mathrm{L}^2(G)$ that promote the Fourier transform $\mathcal{F}$ on H (and are cheap to compute), so that we have the following diagram

$$\begin{array}{ccc} \mathrm{L}^2(G) & \xrightarrow{\;\;\Phi\;\;} & \mathrm{L}^2(G) \\ {\scriptstyle \mathcal{F}^{\tau}}\Big\downarrow \Big\uparrow{\scriptstyle \overline{\mathcal{F}^{\tau}}} & & {\scriptstyle \mathcal{F}^{\tau}}\Big\downarrow \Big\uparrow{\scriptstyle \overline{\mathcal{F}^{\tau}}} \\ \mathrm{L}^2(\mathbb{R}^d, \widehat{H}) & \xrightarrow[\;\;?\;\;]{} & \mathrm{L}^2(\mathbb{R}^d, \widehat{H}). \end{array}$$

The missing map $\widehat{\Phi}$, a notion of correlation on feature fields, would allow us to recover $k \star f_{\text{in}}$ by Fourier inversion. We simply define $\widehat{\Phi}$ as

$$\widehat{\Phi} = \mathcal{F}^{\tau} \circ \Phi \circ \overline{\mathcal{F}^{\tau}},$$

using $\mathcal{F}^{\tau}$ and $\overline{\mathcal{F}^{\tau}}$. The problem is then to define the space $\mathrm{L}^2(\mathbb{R}^d, \widehat{H})$ and the Fourier transform and Fourier cotransform on it. To do this rigorously is nontrivial.

A function $f \in \mathrm{L}^2(\mathbb{R}^d \rtimes H)$ can be viewed as a function $f^H : \mathbb{R}^d \to \mathrm{L}^2(H)$, and when H is a compact group, f^H corresponds to a family $(\widehat{f}_{\rho})$ of functions defined by the Fourier transforms of the functions $f^H(x)$. Furthermore, the functions $\widehat{f}_{\rho}$ are feature fields $\widehat{f}_{\rho} : \mathbb{R}^d \to \mathrm{M}_{n_{\rho}}(\mathbb{C})$. The original function $f \in \mathrm{L}^2(\mathbb{R}^d \rtimes H)$ can be recovered pointwise by Fourier inversion from the family of functions $\widehat{f}_{\rho}$. However, the new twist is that the Fourier coefficients of f are now tuples $(\widehat{f}_{\rho})_{\rho \in R(H)}$ of *functions* $\widehat{f}_{\rho} : \mathbb{R}^d \to \mathrm{M}_{n_{\rho}}(\mathbb{C})$. This causes new problems to reconstruct a function from its Fourier coefficients because even if the functions $\widehat{f}_{\rho}$ belong to $\mathrm{L}^2(\mathbb{R}^d, \mathrm{M}_{n_{\rho}}(\mathbb{C}))$, there is no guarantee that the function obtained from the inverse Fourier transform belongs to $\mathrm{L}^2(G)$. Some additional condition is required on the functions $\widehat{f}_{\rho}$.

We provide a solution to this problem in Section 8.9 by constructing a Hilbert space $\mathrm{L}^2(\mathbb{R}^d, \widehat{H})$, such that the new Fourier transform $\mathcal{F}^{\tau} : \mathrm{L}^2(G) \to \mathrm{L}^2(\mathbb{R}^d, \widehat{H})$ and the Fourier cotransform $\overline{\mathcal{F}^{\tau}} : \mathrm{L}^2(\mathbb{R}^d, \widehat{H}) \to \mathrm{L}^2(G)$ are mutual inverses; see Theorem 8.8.

In Section 8.10, we show how to construct explicitly the lifted correlation $\widehat{\Phi}$ in terms of a correlation Φ on $\mathrm{L}^2(G)$ given by a kernel k. We also show how to compute $\widehat{\Phi}$ when $H = \mathbf{SO}(d)$, in terms of a steerable family on $\mathrm{L}^2(\mathbf{SE}(d))$.

The above construction is performed entirely in Section 8.11 for the group $\mathbf{SE}(2) = \mathbb{R}^2 \rtimes \mathbf{SO}(2)$. The corresponding CNNs are known as *harmonic nets*.

Because the group correlation Φ is equivariant with respect to the left regular representation $\mathbf{R}$ (on $\mathrm{L}^2(G)$), the components $\widehat{\Phi}_{\rho_2,\rho_1}$ of $\widehat{\Phi}$ are equivariant with respect to the representations $\mathrm{Ind}_H^G \sigma_{\rho_1}$ and $\mathrm{Ind}_H^G \sigma_{\rho_2}$, namely the following diagram commutes

$$\begin{array}{ccc} \mathrm{L}^2(\mathbb{R}^d, \widehat{H})_{\rho_1} & \xrightarrow{\widehat{\Phi}_{\rho_2,\rho_1}} & \mathrm{L}^2(\mathbb{R}^d, \widehat{H})_{\rho_2} \\ {\scriptstyle (\mathrm{Ind}_H^G \sigma_{\rho_1})_{(x,h)}} \downarrow & & \downarrow {\scriptstyle (\mathrm{Ind}_H^G \sigma_{\rho_2})_{(x,h)}} \\ \mathrm{L}^2(\mathbb{R}^d, \widehat{H})_{\rho_1} & \xrightarrow[\widehat{\Phi}_{\rho_2,\rho_1}]{} & \mathrm{L}^2(\mathbb{R}^d, \widehat{H})_{\rho_2}. \end{array}$$

Practice shows that it is desirable to design more general group correlations that are equivariant with respect to other representations besides the left regular representation and to consider feature fields that transform under representations other than the representations $\mathrm{Hom}(M_\rho, \mathrm{id})$.

A first generalization is to have two feature fields spaces $\mathbf{FF}(\mathbb{R}^d, H, \sigma_{\mathrm{in}} \colon H \to \mathbf{U}(\mathcal{H}_{\mathrm{in}}))$ and $\mathbf{FF}(\mathbb{R}^d, H, \sigma_{\mathrm{out}} \colon H \to \mathbf{U}(\mathcal{H}_{\mathrm{out}}))$ associated with an input representation σ_{in} and an output representation σ_{out}, where $\mathcal{H}_{\mathrm{in}}$ and $\mathcal{H}_{\mathrm{out}}$ are two finite-dimensional vector spaces equipped with a hermitian inner product, and what we are seeking is a linear G-equivariant map $\widehat{\Phi}$ between these spaces. We assume that feature fields $f \colon \mathbb{R}^d \to \mathcal{H}_{\mathrm{in}}$ are functions in $\mathrm{L}^2(\mathbb{R}^d, \mathcal{H}_{\mathrm{in}})$, and similarly for feature fields $f \colon \mathbb{R}^d \to \mathcal{H}_{\mathrm{out}}$ (see Definition 6.25). To say that $\widehat{\Phi}$ is G-equivariant means that the following diagram commutes

$$\begin{array}{ccc} \mathbf{FF}(\mathbb{R}^d, H, \sigma_{\mathrm{in}}) & \xrightarrow{\widehat{\Phi}} & \mathbf{FF}(\mathbb{R}^d, H, \sigma_{\mathrm{out}}) \\ {\scriptstyle (\mathrm{Ind}_H^G \sigma_{\mathrm{in}})_{(x,h)}} \downarrow & & \downarrow {\scriptstyle (\mathrm{Ind}_H^G \sigma_{\mathrm{out}})_{(x,h)}} \\ \mathbf{FF}(\mathbb{R}^d, H, \sigma_{\mathrm{in}}) & \xrightarrow[\widehat{\Phi}]{} & \mathbf{FF}(\mathbb{R}^d, H, \sigma_{\mathrm{out}}) \end{array}$$

for all $(x, h) \in G = \mathbb{R}^d \rtimes H$.

A complete solution to this problem was given in a sequence of remarkable papers by Weiler, Geiger, Weilling, Boomsma and Cohen [Weiler *et al.* (2018)] (for $\mathbf{SE}(3)$), Weiler and Cesa [Weiler and Cesa (2019)] (for

$\mathbf{E}(2)$), Lang and Weiler [Lang and Weiler (2019)] (for a homogeneous space X induced by a transitive action of a compact group H), Cesa, Lang and Weiler [Cesa *et al.* (2022)] (for $\mathbf{E}(3)$), and Cohen, Geiger and Weiler [Cohen *et al.* (2019)] (feature fields on homogeneous spaces).

In the case where $H = \mathbf{SO}(d)$, it is shown in Section 8.12 that such a map is given by a kernel $K \colon \mathbb{R}^d \to \mathrm{Hom}(\mathcal{H}_{\mathrm{in}}, \mathcal{H}_{\mathrm{out}})$ *via*

$$\widehat{\Phi}(f)(t) = \int_{\mathbb{R}^d} K(y-t)(f(y))\,dy, \quad f \colon \mathbb{R}^d \to \mathcal{H}_{\mathrm{in}}, \quad t \in \mathbb{R}^d, \tag{K1}$$

and the kernel K satisfies the *equivariance constraint*

$$K(h \cdot t) = \sigma_{\mathrm{out}}(h) \circ K(t) \circ \sigma_{\mathrm{in}}(h)^{-1}, \quad h \in \mathbf{SO}(d),\ t \in \mathbb{R}^d. \tag{EC$_1$}$$

Functions $K \colon \mathbb{R}^d \to \mathrm{Hom}(\mathcal{H}_{\mathrm{in}}, \mathcal{H}_{\mathrm{out}})$ satisfying the equivariance constraint (EC$_1$) are called *equivariant convolution kernels* or *G-steerable kernels*. The above result is often referred to by the slogan "correlation is all you need."

Until now we have assumed that we are dealing with feature fields defined on $X = \mathbb{R}^d$ and that the group G is a semi-direct product $G = \mathbb{R}^d \rtimes H$ with $H = \mathbf{SO}(d)$, and more generally a compact group. It is possible to deal with the more general situation where X is a homogeneous space of the form $X = G/H$, with G locally compact and unimodular and H compact equipped with a unitary representation $\sigma \colon H \to \mathbf{U}(\mathcal{H}_\sigma)$. The main problem is to define the "right" notion of feature field. This issue is addressed in Section 8.13.

Cohen, Geiger and Weiler [Cohen *et al.* (2019)] propose to use the G-bundle $E = G \times_H \mathcal{H}_\sigma$ introduced in Section 6.13; see Definition 6.17. But then we might as well use the hermitian G-bundles of finite rank of Definition 6.23 (see Section 6.13) and *the natural choice for the space of feature fields is the subspace* $\mathrm{L}^2(X; E)$ *of the space of sections of the hermitian G-bundle* $p \colon E \to X$, with $X = G/H$ (see Definition 6.25).

Inspired by Cohen, Geiger and Weiler [Cohen *et al.* (2019)] we consider the more general situation in which we have two hermitian G-bundles of finite rank $p_{\mathrm{in}} \colon E_{\mathrm{in}} \to X_{\mathrm{in}}$ and $p_{\mathrm{out}} \colon E_{\mathrm{out}} \to X_{\mathrm{out}}$, where $X_{\mathrm{in}} = G/H_{\mathrm{in}}$ and $X_{\mathrm{out}} = G/H_{\mathrm{out}}$ for the *same* group G, input and output representations σ_{in} and σ_{out}, and determine what are the linear maps $\Phi \colon L^{\sigma_{\mathrm{in}}} \to L^{\sigma_{\mathrm{out}}}$ that are equivariant with respect to the representations $\mathrm{Ind}_{H_{\mathrm{in}}}^G \sigma_{\mathrm{in}}$ and $\mathrm{Ind}_{H_{\mathrm{out}}}^G \sigma_{\mathrm{out}}$, which means that the following diagram commutes

$$\begin{array}{ccc} L^{\sigma_{\rm in}} & \xrightarrow{\Phi} & L^{\sigma_{\rm out}} \\ {\scriptstyle (\mathrm{Ind}_{H_{\rm in}}^G \sigma_{\rm in})(g)} \downarrow & & \downarrow {\scriptstyle (\mathrm{Ind}_{H_{\rm out}}^G \sigma_{\rm out})(g)} \\ L^{\sigma_{\rm in}} & \xrightarrow[\Phi]{} & L^{\sigma_{\rm out}} \end{array}$$

for all $g \in G$ (for simplicity of notation, we use Φ instead of $\widehat{\Phi}$).

Proposition 8.12 generalizes results proven in Cohen, Geiger and Weiler [Cohen *et al.* (2019)] (see Theorem 3.1 and Theorem 3.2) and shows that the equivariant maps Φ as above are determined by the space of equivariant G-kernels given by

$$\begin{aligned} \mathrm{Hom}_{H_{\rm in},H_{\rm out}}(G, \mathrm{Hom}(E_0^{\rm in}, E_0^{\rm out})) = \{K \colon G \to \mathrm{Hom}(E_0^{\rm in}, E_0^{\rm out}) \mid \\ K(h_2 g h_1) = \sigma_{\rm out}(h_2) \circ K(g) \circ \sigma_{\rm in}(h_1), \\ g \in G,\ h_1 \in H_{\rm in},\ h_2 \in H_{\rm out}\}. \qquad (\mathrm{EC}_2) \end{aligned}$$

The above condition is more complicated than (EC_1), and these kernels are defined on G, which makes them rather impractical.

In Section 8.14, we give another characterizations originally due to Cohen, Geiger and Weiler [Cohen *et al.* (2019)] of the space $\mathrm{Hom}_{H_{\rm in},H_{\rm out}}(\mathrm{Ind}_{H_{\rm in}}^G \sigma_{\rm in}, \mathrm{Ind}_{H_{\rm out}}^G \sigma_{\rm out})$ in terms of kernels defined on $X_{\rm in} = G/H_{\rm in}$. More precisely, we prove that there is a bijection between the space of equivariant G-kernels $\mathrm{Hom}_{H_{\rm in},H_{\rm out}}(G, \mathrm{Hom}(E_0^{\rm in}, E_0^{\rm out}))$ and the space $\mathrm{Hom}_{H_{\rm out}}(X_{\rm in}, \mathrm{Hom}(E_0^{\rm in}, E_0^{\rm out}))$ of equivariant $X_{\rm in}$-kernels, which are maps $\kappa \colon X_{\rm in} \to \mathrm{Hom}(E_0^{\rm in}, E_0^{\rm out})$ satisfying a certain condition; see Proposition 8.13.

The G-equivariant maps in $\mathrm{Hom}_{H_{\rm in},H_{\rm out}}(\mathrm{Ind}_{H_{\rm in}}^G \sigma_{\rm in},\ \mathrm{Ind}_{H_{\rm out}}^G \sigma_{\rm out})$ are functions from $L^{\sigma_{\rm in}}$ to $L^{\sigma_{\rm out}}$ and still require integration over G to be computed using equivariant kernels in the space $\mathrm{Hom}_{H_{\rm in},H_{\rm out}}(G, \mathrm{Hom}(E_0^{\rm in}, E_0^{\rm out}))$. It would be nice if we could transform the integration over G to a more practically computable integration over $X_{\rm in}$. This can be achieved by using the maps $\mathcal{S}_{\rm out} \colon L^{\sigma_{\rm out}} \to \mathrm{L}^2(X_{\rm out}, E_{\rm out})$ and $\mathcal{L}_{\rm in} \colon \mathrm{L}^2(X_{\rm in}, E_{\rm in}) \to L^{\sigma_{\rm in}}$ given by $(\mathcal{S}_3'')$ and $(\mathcal{L}_3')$ of Section 6.13. When these maps are well-defined, which is our assumption, they can be used to define maps from $\mathrm{L}^2(X, E_{\rm in})$ to $\mathrm{L}^2(X, E_{\rm out})$, and from functions from $L^{\sigma_{\rm in}}$ to $L^{\sigma_{\rm out}}$. This process is explained in Section 8.15.

The issue of finding G-equivariant kernels still remains and is addressed in Section 8.16.

As in Lang and Weiler [Lang and Weiler (2019)] and Cesa, Lang and Weiler [Cesa *et al.* (2022)] we now assume that $H_{\mathrm{in}} = H_{\mathrm{out}} = H$, so $X_{\mathrm{in}} = X_{\mathrm{out}} = X = G/H$, and we have two Hermitian G-bundles E_{in} and E_{out}. The Hermitian G-bundles define two representations $\sigma_{\mathrm{in}}\colon H \to \mathbf{U}(\mathcal{H}_{\mathrm{in}})$ and $\sigma_{\mathrm{out}}\colon H \to \mathbf{U}(\mathcal{H}_{\mathrm{out}})$. We consider the space of *equivariant X-kernels* defined as

$$\begin{aligned}\mathrm{Hom}_H(X, \mathrm{Hom}(\mathcal{H}_{\mathrm{in}}, \mathcal{H}_{\mathrm{out}})) = \{\kappa\colon X \to \mathrm{Hom}(\mathcal{H}_{\mathrm{in}}, \mathcal{H}_{\mathrm{out}}) \mid\\ \kappa(h \cdot x) = \sigma_{\mathrm{out}}(h) \circ \kappa(x) \circ \sigma_{\mathrm{in}}(h)^{-1},\\ x \in X,\ h \in H\}.\end{aligned} \tag{EC$_6$}$$

Remarkably, Lang and Weiler [Lang and Weiler (2019)] and Cesa, Lang and Weiler [Cesa *et al.* (2022)] completely characterized the kernels in $\kappa \in \mathrm{Hom}_H(X, \mathrm{Hom}(\mathcal{H}_{\mathrm{in}}, \mathcal{H}_{\mathrm{out}}))$, when H is a compact group acting on a topological Hausdorff space X equipped with the σ-algebra of Borel sets and an H-invariant measure μ.

A key ingredient is the analog of the left regular representation $V\colon H \to \mathbf{U}(\mathrm{L}^2(X))$ of $\mathrm{L}^2(X)$ induced by the action of H on X already introduced in Section 8.7, and given by

$$(V(h)f)(x) = f(h^{-1} \cdot x), \quad h \in H,\ f \in \mathrm{L}^2(X),\ x \in X.$$

For the sake of consistency of notation, we will also denote the representation V as $\mathbf{R}^{H \to \mathrm{L}^2(X)}$.

The other key ingredient is the set of H-maps

$$\mathrm{Hom}_H(\mathbf{R}^{H \to \mathrm{L}^2(X)}, \mathrm{Hom}(\sigma_{\mathrm{in}}, \sigma_{\mathrm{out}})),$$

which is the space of linear maps $\mathcal{K}\colon \mathrm{L}^2(X) \to \mathrm{Hom}(\mathcal{H}_{\mathrm{in}}, \mathcal{H}_{\mathrm{out}})$, such that the following diagram commutes

$$\begin{array}{ccc} \mathrm{L}^2(X) & \xrightarrow{\ \mathcal{K}\ } & \mathrm{Hom}(\mathcal{H}_{\mathrm{in}}, \mathcal{H}_{\mathrm{out}}) \\ {\scriptstyle \mathbf{R}^{H \to \mathrm{L}^2(X)}(h)}\big\downarrow & & \big\downarrow{\scriptstyle \mathrm{Hom}(\sigma_{\mathrm{in}}, \sigma_{\mathrm{out}})(h)} \\ \mathrm{L}^2(X) & \xrightarrow[\ \mathcal{K}\ ]{} & \mathrm{Hom}(\mathcal{H}_{\mathrm{in}}, \mathcal{H}_{\mathrm{out}}) \end{array}$$

for every $h \in H$; see Definition 3.9. Cesa, Lang and Weiler [Cesa *et al.* (2022)] call the maps $\mathcal{K}$ *kernel operators*.

The main result is that there is a bijection between the space $\mathrm{Hom}_H(X, \mathrm{Hom}(\mathcal{H}_{\mathrm{in}}, \mathcal{H}_{\mathrm{out}}))$ of equivariant X-kernels and the space $\mathrm{Hom}_H(\mathbf{R}^{H\to \mathrm{L}^2(X)}, \mathrm{Hom}(\sigma_{\mathrm{in}}, \sigma_{\mathrm{out}}))$ of kernel operators; see Theorem 8.14. This isomorphism is a kind of linearization of the first space.

But now Proposition 4.23 tells us that the representations $\mathrm{Hom}(\sigma_{\mathrm{in}}, \sigma_{\mathrm{out}})$ and $\overline{\sigma_{\mathrm{in}}} \otimes \sigma_{\mathrm{out}}$ are equivalent, so we obtain an isomorphism

$$\mathrm{Hom}_H(X, \mathrm{Hom}(\mathcal{H}_{\mathrm{in}}, \mathcal{H}_{\mathrm{out}})) \approx \mathrm{Hom}_H(\mathbf{R}^{H\to \mathrm{L}^2(X)}, \overline{\sigma_{\mathrm{in}}} \otimes \sigma_{\mathrm{out}}). \qquad (\dagger_{15})$$

Since H is a compact group, we can now use Theorem 8.7 (a direct consequence of Peter–Weyl II) to express $\mathrm{L}^2(X)$ as a Hilbert sum of spaces corresponding to irreducible representations of H and the decomposition of the tensor product representation $\overline{\sigma_{\mathrm{in}}} \otimes \sigma_{\mathrm{out}}$ as a Hilbert sum of irreducible representations of H (see Proposition 4.18 and Equation ($\otimes$) in Section 4.4). Such a decomposition is achieved in a theorem referred to as *Wigner–Eckart theorem for steerable kernels* by Cesa, Lang and Weiler [Cesa *et al.* (2022)] (Theorem B.5); see Theorem 8.15. The more general theorem that also applies to real representations is proven in Cesa, Lang and Weiler [Cesa *et al.* (2022)].

Cesa, Lang and Weiler [Cesa *et al.* (2022)] also prove a version of the above result in which a basis of $\mathrm{Hom}_H(X, \mathrm{Hom}(\mathcal{H}_{\mathrm{in}}, \mathcal{H}_{\mathrm{out}}))$ is exhibited. The formulae are a bit messy so we will not give details here; see Theorem B.6 and Theorem B.7 in Cesa, Lang and Weiler [Cesa *et al.* (2022)]; see Section 8.16.

A way to deal with noncommutativity, due to Gelfand, is to work with pairs (G, K), where K is a compact subgroup of G. This theory is presented in Chapter 9. Then instead of working with functions on G, which is "too big," we work with functions on the homogeneous space G/K, the space of left cosets. Under certain assumptions on G and K, which makes (G, K) a *Gelfand pair*, it is possible to consider a commutative algebra of functions on the set of double cosets KsK ($s \in G$), so that some results from the commutative theory can be used. The domain of the Fourier transform is a set of functions called *spherical functions*, and this set happens to be homeomorphic to the set of characters on the commutative algebra mentioned above. There is a very nice theory of the Fourier transform and its inverse, but how useful it is in practice remains to be seen.

Even though the present document is already quite long, it is by no means complete. If a locally compact group is a *Lie group*, then the whole machinery of Lie algebras and Lie groups developed by Élie Cartan and Hermann Weyl involving weights and roots becomes available. In particular, if G is a connected *semisimple Lie group*, there is a beautiful and extensive theory of harmonic analysis due to Harish–Chandra. We lack the expertise to discuss this difficult theory and refer the ambitious reader to Warner's monographs [Warner (1972a,b)], and Helgason's treatises [Helgason (2001)], [Helgason (2000)] (especially Chapter IV), and [Helgason (1994)] (especially Chapter III, Section 12).

To keep the length of this book under control, we resigned ourselves to omit many proofs. This is unfortunate because some beautiful proofs had to be omitted. However, whenever a proof is omitted, we provide precise pointers to sources where such a proof is given.

Chapter 2

Representations of Algebras and Hilbert Algebras

In order to generalize harmonic analysis to nonabelian locally compact groups, we need to introduce group representations. However, it turns out that in order to prove the main theorem of the subject, the Peter–Weyl theorem, one needs the notion of representation of algebras, because there is a bijection between the set of unitary representations of a locally compact group G and the set of nondegenerate representations of the involutive algebra $\mathrm{L}^1(G)$. When G is compact, $\mathrm{L}^2(G)$ is actually a Hilbert algebra, and there is a beautiful structure theorem for Hilbert algebras which says that such an algebra splits as a Hilbert sum of minimal left ideals, and this result can be used to prove the Peter–Weyl theorem.

The purpose of this chapter is to define the notion of Hilbert algebra and to develop the machinery needed to prove three fundamental theorems (Theorem 2.33, Theorem 2.34, and Theorem 2.36) about complete Hilbert algebras. We also state two important theorems about commutative Hilbert algebras; the Plancherel–Godement theorem and the Bochner-Godement theorem. These theorems will be needed later when we discuss Gelfand pairs. We mostly follow Dieudonné [Dieudonné (2005)] (Chapter XV, Sections 15.5–15.9), occasionally borrowing from Folland [Folland (1995)] (Chapters 1 and 3). This is a rather technical chapter and we do not give all proofs, relying on the above references for details.

We begin with the definition of the notion of representation of an involutive algebra in a Hilbert space. We define the notion of Hilbert sum of a finite or an infinite family of Hilbert spaces. Then we introduce the crucial concepts of invariant subspace, of a topologically irreducible representation, of a nondegenerate representation, of a totalizing (or cyclic) vector, and of a topologically cyclic representation.

In Section 2.3, we define positive linear forms and positive Hilbert forms on an involutive algebra A. Positive Hilbert forms are positive hermitian forms which may fail to be positive definite but satisfy a kind of adjunction property, namely a positive hermitian form g satisfying the condition

$$g(xy, z) = g(y, x^*z), \qquad \text{for all } x, y, z \in A.$$

They can be used to define topologically cyclic representations. A good method for producing positive Hilbert forms is to use positive linear forms which satisfy the property $f(s^*s) \geq 0$, for all $s \in A$, a kind of positive semidefinite property. Then the map g given by $g(x, y) = f(y^*x)$ is a positive Hilbert form.

In Section 2.4, we introduce bitraces and Hilbert algebras. A bitrace is a positive Hilbert form $g: A \times A \to \mathbb{C}$, such that

$$g(t^*, s^*) = g(s, t), \qquad \text{for all } s, t \in A.$$

The most important concept of this section is the notion of Hilbert algebra. An involutive algebra A is a *Hilbert algebra* if its underlying vector space is a hermitian space whose hermitian inner product $\langle -, - \rangle$ is a bitrace satisfying two extra conditions (U) and (N) (see Definition 2.14). Specifically, the conditions for being a bitrace hold

$$\langle y^*, x^* \rangle = \langle x, y \rangle \tag{1}$$
$$\langle xy, z \rangle = \langle y, x^*z \rangle, \tag{2}$$

and the following two conditions hold: for every $x \in A$, there is some $M_x \geq 0$, such that

$$\langle xy, xy \rangle \leq M_x \langle y, y \rangle, \quad \text{for all } y \in A, \tag{U}$$

and

$$\text{the subspace spanned by the set } \{xy \mid x, y \in A\} \text{ is dense in } A. \tag{N}$$

In general, the map $(x, y) \mapsto xy$ is not continuous, so a Hilbert algebra is not a normed algebra in the sense of Vol. I, Definition 9.4. However, if the Hilbert algebra A is complete, then it can be shown that the map $(x, y) \mapsto xy$ is continuous, and thus that A is a normable algebra, which is a Banach space (see Proposition 2.16).

An important example of a complete Hilbert algebra is the algebra $\mathcal{L}_2(H)$ of Hilbert–Schmidt operators on a separable Hilbert space H. Hilbert–Schmidt operators are discussed in Section 2.5. In the special case where the separable Hilbert space H is a space $\mathrm{L}^2_\mu(X, \mathcal{H})$ of L^2-integrable functions with values in a finite-dimensional vector space $\mathcal{H}$, Hilbert–Schmidt operators are represented by integral operators defined by kernel functions. This a classical ("folklore") result of functional analysis asserting that a linear map $T\colon \mathrm{L}^2_\mu(X) \to \mathrm{L}^2_\mu(X)$ is a Hilbert–Schmidt operator iff there is a kernel $\mathcal{K}\colon X \times X \to \mathbb{C}$, which is a function in $\mathrm{L}^2_{\mu\otimes\mu}(X \times X)$, such that

$$(T(f))(x) = \int_X \mathcal{K}(x, y) f(y)\, d\mu(y), \quad f \in \mathrm{L}^2_\mu(X),\ x \in X. \tag{HS1}$$

Our search for a complete proof of this result in standard texts was unsuccessful. This fact is stated in Dunford and Schwartz [Dunford and Schwartz (1988)], Section 6, Chapter XI, but as Exercise 44. With some help from Lang [Lang (1985)] (Chapter I, Section 3, Theorem 2) and Lang [Lang (1996)] (Chapter XVII, Problem 9), we were able to present a fairly complete proof. Such kernels play a key role in Chapter 8.

Another very important example of a complete Hilbert algebra is $\mathrm{L}^2(G)$, where G is a compact, metrizable group.

Section 2.6, which is about complete separable Hilbert algebras, contains the most important results of this chapter regarding the structure of such algebras. To understand the structure of complete separable Hilbert algebras, we need to study minimal left ideals and the irreducible self-adjoint idempotents which generate them. Recall that an element e of an algebra A is *idempotent* if $e^2 = e$, and *self-adjoint* if $e = e^*$.

Roughly speaking, the master decomposition theorem (Theorem 2.33) states that given a complete separable Hilbert algebra A, there is an irredundant list $(\mathfrak{l}_k)_{k\in J}$ of the minimal left ideals of A, and A is the Hilbert sum of two-sided ideals $\mathfrak{a}_k$,

$$A = \bigoplus_{k\in J} \mathfrak{a}_k,$$

where each $\mathfrak{a}_k$ is the Hilbert sum obtained by picking a certain number of copies of the minimal left ideal $\mathfrak{l}_k$ of A,

$$\mathfrak{a}_k = \bigoplus_{j\in I_k} \mathfrak{l}'_j,$$

with $\mathfrak{l}'_j$ isomorphic to $\mathfrak{l}_k$.

Each two-sided ideal $\mathfrak{a}_k$ contains no closed two-sided ideal other than (0) and $\mathfrak{a}_k$. They are said to be *topologically simple*.

Theorem 2.34 gives the structure of a topologically simple Hilbert algebra. Theorem 2.34 implies that in the Hilbert sum

$$A = \bigoplus_{k\in J} \mathfrak{a}_k$$

given by Theorem 2.33, the Hilbert algebra $\mathfrak{a}_k$, which is a building block of the decomposition, is either isomorphic to the algebra $\mathcal{L}_2(\mathfrak{l}_k)$ of Hilbert–Schmidt operators on $\mathfrak{l}_k$, or to the finite-dimensional algebra $\mathrm{End}_{\mathbb{C}}(\mathfrak{l}_k)$ of all endomorphisms of the vector space $\mathfrak{l}_k$. If G is a metrizable compact group and $A = \mathrm{L}^2(G)$, then every $\mathfrak{a}_k$ in the Hilbert sum for A is isomorphic to the finite-dimensional algebra $\mathrm{End}_{\mathbb{C}}(\mathfrak{l}_k)$.

The master decomposition for a nondegenerate continuous representation $V\colon A \to \mathcal{L}(H)$ (Theorem 2.36) states that the Hilbert space H is a Hilbert sum $H = \bigoplus_{k\in J} H_k$ of subspaces invariant under V, and that the restriction V_k of V to $\mathfrak{a}_k$ can be considered as a representation of $\mathfrak{a}_k$ in H_k. Furthermore, each representation V_k is the Hilbert sum of irreducible representations, each equivalent to a representation $U_{\mathfrak{l}_k}$ canonically associated with a minimal ideal $\mathfrak{l}_k$ of $\mathfrak{a}_k$.

Section 2.9 discusses the Plancherel–Godement theorem and the Bochner–Godement theorem without proofs. These theorems apply to a *commutative* Hilbert algebra (not necessarily complete) arising from the quotient of a commutative Hilbert algebra by a left ideal induced by a bitrace satisfying two additional conditions. Therefore, we go back to positive Hilbert forms to describe the construction of a certain representation.

The idea is that if g is a positive Hilbert form on an involutive (not necessarily commutative) algebra A, it almost defines an inner product, but in general it fails to be positive definite because there may be nonzero elements $s \in A$, such that $g(s,s) = 0$. However, if we take the quotient of A by the set $\mathfrak{n} = \{s \in A \mid g(s,s) = 0\}$, which is a left ideal because g is a positive Hilbert form, then we can define an inner product on the quotient vector space $A/\mathfrak{n}$. If g is a bitrace, then $A/\mathfrak{n}$ is an involutive algebra (see Proposition 2.37).

If a positive Hilbert form g satisfies the analog of condition (U) of Definition 2.14, namely, for every $s \in A$, there is some $M_s \geq 0$, such that

$$g(st, st) \leq M_s g(t,t), \quad \text{for all } t \in A, \tag{U}$$

and if the hermitian space $A/\mathfrak{n}_g$ is separable, where $\mathfrak{n}_g = \{s \in A \mid g(s,s) = 0\}$, then g defines a unitary representation $U_g\colon A \to \mathcal{L}(H_g)$, where H_g is the Hilbert space which is the completion of $A/\mathfrak{n}_g$ (see Proposition 2.39).

In general, the representation $U_g\colon A \to \mathcal{L}(H_g)$ given by Proposition 2.39 may be degenerate. It is nondegenerate if and only if the following condition holds:

the subspace spanned by the set $\{\pi_g(st) \mid s, t \in A\}$ is dense in $A/\mathfrak{n}_g$. (N)

If A is a commutative Hilbert algebra and if property (U) holds, then the representation $U_g\colon A \to \mathcal{L}(H_g)$ is nondegenerate and the image of A under U_g is a commutative subalgebra of the C^*-algebra $\mathcal{L}(H_g)$. Let $\mathcal{A}_g$ be the closure of $U_g(A)$ in $\mathcal{L}(H_g)$, so that $\mathcal{A}_g$ is a commutative C^*-algebra.

Roughly speaking, the Plancherel–Godement theorem (Theorem 2.43) states that if A is a commutative involutive algebra, if g is a bitrace on A satisfying conditions (U) and (N), and if the hermitian space $A/\mathfrak{n}_g$ and the C*-algebra $\mathcal{A}_g \subseteq \mathcal{L}(H_g)$ are separable, then g is obtained from a positive measure by a process of integration from hermitian characters.

Sections 2.10, 2.11 and 2.12 present results generally called spectral theorems. The important notion of projection-valued measure is introduced.

Section 2.10 contains a technically crucial characterization of a representation of the algebra $\mathcal{C}_{\mathbb{C}}(K)$ of continuous functions on a compact metrizable space K, a result proven using the theorems from Section 2.9, the Plancherel–Godement theorem and the Bochner–Godement theorem. This theorem shows that every topologically cyclic representation $U\colon \mathcal{C}_{\mathbb{C}}(K) \to \mathcal{L}(H)$ of the commutative unital C^*-algebra $\mathcal{C}_{\mathbb{C}}(K)$ (for K compact) in a separable Hilbert space H is equivalent to a representation $M_\mu\colon \mathcal{C}_{\mathbb{C}}(K) \to \mathcal{L}(\mathrm{L}^2_\mu(K;\mathbb{C}))$, such that for every $u \in \mathcal{C}_{\mathbb{C}}(K)$, $M_\mu(u)\colon \mathrm{L}^2_\mu(K;\mathbb{C}) \to \mathrm{L}^2_\mu(K;\mathbb{C})$ is the continuous linear map multiplication by u (μ is some positive Radon measure on K); see Theorem 2.45.

A particularly interesting case for the space K arises if we consider a commutative unital C^*-subalgebra $\mathcal{A}$ of $\mathcal{L}(H)$. In this case, by the Gelfand–Naimark theorem (Vol. I, Theorem 9.37), the Gelfand transform $\mathcal{G}\colon \mathcal{A} \to \mathcal{C}_{\mathbb{C}}(\mathsf{X}(\mathcal{A}))$ is an isometric isomorphism between $\mathcal{A}$ and $\mathcal{C}_{\mathbb{C}}(\mathsf{X}(\mathcal{A}))$. Furthermore, $K = \mathsf{X}(\mathcal{A})$ is compact (see Vol. I, Theorem 9.19). Thus, the inverse Gelfand transform $\mathcal{G}^{-1}\colon \mathcal{C}_{\mathbb{C}}(\mathsf{X}(\mathcal{A})) \to \mathcal{A}$ is a representation of $\mathcal{C}_{\mathbb{C}}(\mathsf{X}(\mathcal{A}))$ and Theorem 2.45 can be used to prove Theorem 2.47 (Spectral Theorem I), which can be viewed as a generalization of the spectral theorem for normal linear maps on a finite-dimensional hermitian space. This result applies to any commutative unital C^*-subalgebra $\mathcal{A}$ of $\mathcal{L}(H)$. An interesting special case is the subalgebra $\mathcal{A}_T$ of $\mathcal{L}(H)$ generated by T, T^*

and I, where T is a normal continuous linear map T on a Hilbert space H. We have Theorem 2.48, a first version of a spectral theorem for normal continuous linear maps. The end of this section presents a condition for a scalar in the spectrum $\sigma(T)$ of T to be an eigenvalue.

The next step taken in Section 2.11 is to realize that a representation $U\colon \mathcal{C}_{\mathbb{C}}(K) \to \mathcal{L}(H)$ as above determines certain complex Radon measures $\mu_{u,v}$ on K, and that conversely these measures determine U. Indeed, for any two vectors $u, v \in H$, there is a unique complex Radon measure $\mu_{u,v}$ on K, such that

$$\langle U(f)(u), v\rangle = \int_K f\, d\mu_{u,v}, \quad f \in \mathcal{C}_{\mathbb{C}}(K).$$

The measure $\mu_{u,v}$ is often called a *spectral measure*.

Then it is possible to extend the representation U of $\mathcal{C}_{\mathbb{C}}(K)$ to the larger commutative unital C^*-algebra $B(K)$ of bounded Borel measurable functions on K. We obtain the representation $\widetilde{U}\colon B(K) \to \mathcal{L}(H)$, which is completely determined by the equation

$$\langle \widetilde{U}(f)(u), v\rangle = \int_K f\, d\mu_{u,v} \quad \text{for all } u, v \in H \text{ and for all } f \in B(K). \qquad (*_3)$$

The above equation defines a "weak integral" with respect to the family of measures $\mu_{u,v}$ denoted

$$\widetilde{U}(f) = \int f\, d\mu.$$

For simplicity of notation, we denote $\widetilde{U}$ as U. Since for any commutative unital C^*-subalgebra $\mathcal{A}$ of $\mathcal{L}(H)$ the inverse Gelfand transform $\mathcal{G}^{-1}\colon \mathcal{C}_{\mathbb{C}}(\mathsf{X}(\mathcal{A})) \to \mathcal{A}$ is a representation of $\mathcal{C}_{\mathbb{C}}(\mathsf{X}(\mathcal{A}))$, we obtain a representation $U\colon B(\mathsf{X}(\mathcal{A})) \to \mathcal{L}(H)$ of $\mathcal{G}^{-1}\colon \mathcal{C}_{\mathbb{C}}(\mathsf{X}(\mathcal{A})) \to \mathcal{A}$. Consequently, we obtain Theorem 2.53 which states that there is a family of complex Radon measures $(\mu_{u,v})_{(u,v)\in H\times H}$ on $\mathsf{X}(\mathcal{A})$ and we have

$$T = \int \mathcal{G}_T\, d\mu, \quad U(f) = \int f\, d\mu$$

for all $T \in \mathcal{A}$ and all $f \in B(\mathsf{X}(\mathcal{A}))$. This is another spectral theorem for a commutative unital C^*-subalgebra $\mathcal{A}$ of $\mathcal{L}(H)$.

What we gain in doing all this, is the fact that we can apply the extended representation U to the characteristic functions χ_E of Borel sets E (on K) (the functions χ_E are not continuous), and such operators $P(E) = U(\chi_E)$ turn out to be orthogonal projections in $\mathcal{L}(H)$. These families of projections have properties that make them *projection-valued measures* (also called *spectral measures*), and such measures can be used to define representations of $B(K)$ that generalize the notion of integral.

Projection-valued measures are defined and used to prove more spectral theorems in Section 2.12. The connection between a family P of projection-valued measures and families of complex Radon measures as above is that if for all $u, v \in E$ we define $P_{u,v}$ by

$$P_{u,v}(E) = \langle P(E)(u), v\rangle,$$

then the $P_{u,v}$ are complex Radon measures (with $P_{u,u}$ a positive measure) with the same properties as the $\mu_{u,v}$. This allows us to define a notion of weak integral with respect to a projection-valued measure. We obtain the important Theorem 2.56 which states that for any function $f \in B(K)$, the integral

$$U(f) = \int f\, dP$$

is defined by the equation

$$\langle U(f)(u), v\rangle = \int f\, dP_{u,v}, \quad \text{for all } u, v \in H \text{ and all } f \in B(K).$$

Furthermore, the map $U\colon B(K) \to \mathcal{L}(H)$ is a representation.

Theorem 2.56 yields more spectral theorems in terms of projection-valued measures, in particular another spectral theorem (Spectral Theorem II) for any commutative unital C^*-subalgebra $\mathcal{A}$ of $\mathcal{L}(H)$ (see Theorem 2.57).

Remarkably, Theorem 2.57 (Spectral Theorem III) can be generalized to unital commutative Banach algebras. Theorem 2.59 states that for any commutative unital involutive Banach algebra $\mathcal{A}$, for any representation $U\colon \mathcal{A} \to \mathcal{L}(H)$ of $\mathcal{A}$ in a Hilbert space H, there is a regular projection-valued measure P on $\mathsf{X}(\mathcal{A})$, such that

$$U(a) = \int \mathcal{G}_a\, dP, \quad a \in \mathcal{A},$$

where $\mathcal{G}_a$ is the Gelfand transform. In fact, the projection-valued measure P is unique.

There is one more generalization (Spectral Theorem IV) where the involutive Banach algebra $\mathcal{A}$ is not necessarily unital, but the representation $U\colon \mathcal{A} \to \mathcal{L}(H)$ is nondegenerate; see Theorem 2.60. This theorem is crucial to the proof of Theorem 3.20 characterizing the unitary representations of an *abelian* locally compact group. Intuitively, the characters of G are glued by a suitable projection-valued measure. In turn, Theorem 3.20 is a key result used in Mackey's theory for constructing induced representations; see Chapter 7, Proposition 7.1.

As corollary of Theorem 2.60, we also obtain a spectral theorem for nondegenerate representations $U\colon \mathcal{C}_0(X;\mathbb{C}) \to \mathcal{L}(H)$ of $\mathcal{C}_0(X;\mathbb{C})$; see Theorem 2.61. This theorem is used in Section 7.2 to give an alternate definition of a system of imprimitivity; see Definition 7.4.

2.1 Representations of Algebras with Involution

Let A be an algebra with an involution (not necessarily a normed algebra, nor a commutative or a unital algebra). Since representations of algebras involve Hilbert spaces, the reader may want to review Vol. I, Chapter D, especially Sections D.1 and D.2. For the reader's convenience, we quickly review some basic notions, including Hilbert bases.

If H is a complex vector space, recall that a map $\langle -,-\rangle\colon H \times H \to \mathbb{C}$ is a *hermitian form* if it satisfies the following properties for all $x, y, x_1, x_2, y_1, y_2 \in H$ and all $\lambda \in \mathbb{C}$: it is *sesquilinear*, which means that

$$\begin{aligned}
\langle x_1 + x_2, y\rangle &= \langle x_1, y\rangle + \langle x_2, y\rangle \\
\langle x, y_1 + y_2\rangle &= \langle x, y_1\rangle + \langle x, y_2\rangle \\
\langle \lambda x, y\rangle &= \lambda\langle x, y\rangle \\
\langle x, \lambda y\rangle &= \overline{\lambda}\langle x, y\rangle,
\end{aligned}$$

and satisfies the *hermitian property*,

$$\langle y, x\rangle = \overline{\langle x, y\rangle}.$$

The hermitian property implies that $\langle x, x\rangle \in \mathbb{R}$ for all $x \in H$.

A hermitian form $\langle -,-\rangle\colon H \times H \to \mathbb{C}$ is *positive* if

$$\langle x, x\rangle \geq 0, \qquad \text{for all } x \in H.$$

A positive hermitian form satisfies the *Cauchy–Schwarz inequality*:

$$|\langle x, y\rangle|^2 \leq \langle x, x\rangle\langle y, y\rangle, \qquad \text{for all } x, y \in H.$$

A positive hermitian form is *positive definite* if for all $x \in H$,

$$\langle x, x\rangle = 0 \quad \text{implies that } x = 0,$$

or equivalently,

$$\langle x, x\rangle > 0, \qquad \text{for all } x \neq 0.$$

A positive definite hermitian form on H is often called a *hermitian inner product* on H, and H is called a *hermitian space* (sometimes a *pre-Hilbert space*).

If H is a hermitian space with a hermitian inner product $\langle -, -\rangle$, then the map $x \mapsto \|x\| = \sqrt{\langle x, x\rangle}$ is a norm on H. We say that H is a *Hilbert space* if H is complete for the norm $\| \, \|$ (every Cauchy sequence converges).

Let H be a Hilbert space. An orthonormal family $(a_\alpha)_{\alpha\in\Lambda}$ of vectors $a_\alpha \in H$ (which means that $\langle a_\alpha, a_\beta\rangle = 0$ for all $\alpha \neq \beta$, and $\langle a_\alpha, a_\alpha\rangle = 1$, for all $\alpha, \beta \in \Lambda$) is a *Hilbert basis* of H if the subspace spanned by $(a_\alpha)_{\alpha\in\Lambda}$ (the set of all *finite* linear combinations of vectors in $(a_\alpha)_{\alpha\in\Lambda}$) is dense in H. Every Hilbert space admits a Hilbert basis, and the cardinality of the index set Λ is the same for any two Hilbert bases; see Vol. I, Section D.2, Rudin [Rudin (1987)] (Chapter 4) and Schwartz [Schwartz (1980)] (Chapter XXIII).

A Hilbert space is *separable* if it has a countable Hilbert basis.

Definition 2.1. Given an algebra A with involution and a Hilbert space H, a *representation of A in H*[1] is an algebra homomorphism $U\colon A \to \mathcal{L}(H)$ from A to the involutive algebra $\mathcal{L}(H)$ of continuous linear maps from H to itself, which means that U satisfies the conditions

$$\begin{aligned} U(s+t) &= U(s) + U(t)\\ U(\lambda s) &= \lambda U(s)\\ U(st) &= U(s)\circ U(t)\\ U(s^*) &= (U(s))^*, \end{aligned}$$

for all $s, t \in A$ and all $\lambda \in \mathbb{C}$. If A is unital with identity element e, we require that

$$U(e) = \mathrm{id}_H.$$

The Hilbert space H is called the *representation space*. The representation U is *faithful* if the homomorphism $s \mapsto U(s)$ is injective, which means that $U(s)(x) = 0$ for all $x \in H$ implies that $U(s) = 0$.

[1]Technically, we are defining *unitary representations*, presumably because the definition of equivalence of representations (Definition 2.2) uses isometries between Hilbert spaces, but since we shall not discuss other types of representations, we shall suppress the word "unitary."

Following common practice, the composition $U(s) \circ U(t)$ is abbreviated as $U(s)U(t)$. To simplify notation, we often write U_s instead of $U(s)$.

Remark: Folland [Folland (1995)] uses the terminology *$*$-representation* for a representation of an involutive algebra. When different representations $U\colon A \to \mathcal{L}(H)$ of the same algebra A arise, it is sometimes convenient to denote the representation space by H_U. Although a representation of an algebra A consists of a homomorphism U and of a Hilbert space H, by abuse of language, we often refer to a representation as U.

Example 2.1. Let $A = \mathrm{M}_n(\mathbb{C})$ be the algebra of $n \times n$ complex matrices with involution $X \mapsto X^*$ (the conjugate transpose). The map

$$\langle X, Y\rangle \mapsto \langle X, Y\rangle = \mathrm{tr}(Y^* X)$$

is a Hermitian inner product on A, which makes A into a Hilbert space of finite dimension denoted H. The linear maps in $\mathrm{Hom}(H, H)$ are automatically continuous, so $\mathrm{Hom}(H, H) = \mathcal{L}(H)$. The map $U\colon A \to \mathcal{L}(H)$ given by

$$U(X)(Y) = XY, \quad X, Y \in \mathrm{M}_n(\mathbb{C})$$

is a representation of A. The only property that is not obvious is the property $U(X^*) = U(X)^*$. But by definition, the adjoint $V(X) = U(X)^*$ of the linear map $U(X)$ is the unique linear map $V(X)$, such that

$$\langle U(X)(Y), Z\rangle = \langle Y, V(X)(Z)\rangle, \quad \text{for all } Y, Z \in \mathrm{M}_n(\mathbb{C}),$$

that is,

$$\mathrm{tr}(Z^* XY) = \mathrm{tr}((V(X)(Z))^* Y), \quad \text{for all } Y, Z \in \mathrm{M}_n(\mathbb{C}),$$

which implies $V(X)(Z) = (Z^* X)^* = X^* Z$. Thus $U(X)^*(Z) = V(X)(Z) = X^* Z = U(X^*)(Z)$, that is,

$$U(X)^* = U(X^*),$$

as desired.

Example 2.2. Let $A = \mathrm{M}_n(\mathbb{C})$ be the algebra of $n \times n$ complex matrices with involution $X \mapsto X^*$ (the conjugate transpose), and let $H = \mathbb{C}^n$, with the standard hermitian inner product given by $\langle x, y\rangle = y^* x$, where $x, y \in \mathbb{C}^n$. Since H is finite-dimensional, it is a Hilbert space, and the linear maps in $\mathrm{Hom}(H, H)$ are automatically continuous, so $\mathrm{Hom}(H, H) = \mathcal{L}(H)$. The map $U_1\colon A \to \mathcal{L}(H)$ given by

$$U_1(X)(y) = Xy, \quad X \in \mathrm{M}_n(\mathbb{C}), y \in \mathbb{C}^n$$

is a representation of A. The only property that is not obvious is the property $U_1(X^*) = U_1(X)^*$. But by definition, the adjoint $V(X) = U_1(X)^*$ of the linear map $U_1(X)$ is the unique linear map $V(X)$, such that

$$\langle U_1(X)(y), z\rangle = \langle y, V(X)(z)\rangle, \quad \text{for all } y, z \in \mathbb{C}^n,$$

that is,

$$z^* X y = (V(X)(z))^* y, \quad \text{for all } y, z \in \mathbb{C}^n,$$

thus $U_1(X)^*(z) = V(X)(z) = X^* z = U_1(X^*)(z)$, namely

$$U_1(X)^* = U_1(X^*).$$

Observe that $H = \mathbb{C}^n$ is isomorphic to the subspace $\mathfrak{b}$ of A consisting of all $n \times n$ complex matrices whose last $n-1$ columns are zero. The subspace $\mathfrak{b}$ is a left ideal in A, and in fact a minimal left ideal. The map $U_2 \colon A \to \mathcal{L}(\mathfrak{b})$ given by

$$U_2(X)(Y) = XY, \quad X \in \mathrm{M}_n(\mathbb{C}), Y \in \mathfrak{b}$$

is also a representation of A. Since $\mathbb{C}^n$ and $\mathfrak{b}$ are isomorphic Hilbert spaces, we say that U_1 and U_2 are equivalent representations; see Definition 2.2.

A generalization of this example occurs in Proposition 2.20.

Example 2.3. If G is a metrizable locally compact group, the space $A = \mathrm{L}^1(G)$ is an algebra under convolution (denoted $*$) and $H = \mathrm{L}^2(G)$ is a Hilbert space. It is shown in Section 3.3 that the map $\mathbf{R}_{\mathrm{ext}} \colon \mathrm{L}^1(G) \to \mathcal{L}(\mathrm{L}^2(G))$ given by

$$(\mathbf{R}_{\mathrm{ext}}(f))(g) = f * g, \qquad f \in \mathrm{L}^1(G),\ g \in \mathrm{L}^2(G)$$

is a representation of $\mathrm{L}^1(G)$ in $\mathrm{L}^2(G)$ (called *left regular representation*).

Definition 2.1 implies that if s is self-adjoint ($s^* = s$), then $U(s)$ is self-adjoint. Observe that $U(s)$ is *not* necessarily invertible. Also, if A is a normed algebra, then the map $U \colon A \to \mathcal{L}(H)$ is *not* necessarily continuous. However, if A is a unital Banach algebra with involution, then by the next proposition the map $U \colon A \to \mathcal{L}(H)$ is continuous.

Proposition 2.1. *If A is a unital Banach algebra with involution, then every representation $U \colon A \to \mathcal{L}(H)$ satisfies the condition $\|U(s)\| \leq \|s\|$, which implies that U is a continuous mapping from A to $\mathcal{L}(H)$.*

Proof. Recall that $\mathcal{L}(H)$ is a C^*-algebra, so by Vol. I, Proposition 9.31 $\rho(T) = \|T\|$ for all normal linear maps $T \in \mathcal{L}(H)$, so if we let $T = U(s)^*U(s)$, which is obviously self-adjoint, we have

$$\|U(s)\|^2 = \|U(s)^*U(s)\| = \rho(U(s)^*U(s)).$$

By Property (5) just after Vol. I, Definition 9.6,

$$\sigma(U(s^*s)) \subseteq \sigma(s^*s),$$

by Vol. I, Proposition 9.17,

$$\rho(U(s^*s)) \leq \rho(s^*s),$$

and $U(s)^*U(s) = U(s^*s)$, so we have

$$\rho(U(s)^*U(s)) = \rho(U(s^*s)) \leq \rho(s^*s) \leq \|s^*s\| \leq \|s^*\|\,\|s\| = \|s\|^2,$$

which proves our result. □

Recall that if $(H_1, \langle -,-\rangle_1)$ and $(H_2, \langle -,-\rangle_2)$ are two Hilbert spaces, a Hilbert space isomorphism is a continuous linear map $T\colon H_1 \to H_2$, whose inverse is also continuous, and T is an isometry, which means that

$$\langle T(x), T(y)\rangle_2 = \langle x, y\rangle_1, \qquad \text{for all } x, y \in H.$$

If $H_1 = H_2$, then a Hilbert space automorphism $T\colon H \to H$ is an invertible element in $\mathcal{L}(H)$, such that

$$TT^* = T^*T = \mathrm{id},$$

where T^* is the adjoint of T (defined by the property that

$$\langle Tx, y\rangle = \langle x, T^*y\rangle, \qquad \text{for all } x, y \in H).$$

Such maps are called *unitary*.

Definition 2.2. If A is an algebra (as above) and if H_1 and H_2 are two Hilbert spaces, two representations $U_1\colon A \to \mathcal{L}(H_1)$ and $U_2\colon A \to \mathcal{L}(H_2)$ are *equivalent* if there a Hilbert space isomorphism $T\colon H_1 \to H_2$, such that

$$U_2(s) = TU_1(s)T^{-1} \qquad \text{for all } s \in A,$$

as illustrated by the following diagram:

$$\begin{array}{ccc} H_1 & \xrightarrow{U_1(s)} & H_1 \\ {\scriptstyle T^{-1}}\uparrow & & \downarrow{\scriptstyle T} \\ H_2 & \xrightarrow[U_2(s)]{} & H_2. \end{array}$$

Example 2.4. The representations $U_1\colon A \to \mathcal{L}(\mathbb{C}^n)$ and $U_2\colon A \to \mathcal{L}(\mathfrak{b})$ of Example 2.2 are equivalent under the obvious isomorphism from $\mathbb{C}^n$ to $\mathfrak{b}$.

It is often useful to make a new representation from old ones using the process of constructing a Hilbert sum. We begin with the simplest case involving two Hilbert spaces. Later, we generalize this construction to an arbitrary family of Hilbert spaces.

Let H_1 and H_2 be two Hilbert spaces, and let $U_1\colon A \to \mathcal{L}(H_1)$ and $U_2\colon A \to \mathcal{L}(H_2)$ be two representations. The *Hilbert sum* H of H_1 and H_2 is the direct sum $H_1 \oplus H_2$ of H_1 and H_2, with the hermitian product given by

$$\langle x_1 + x_2, y_1 + y_2\rangle = \langle x_1, y_1\rangle_1 + \langle x_2, y_2\rangle_2,$$

for all $x_1, y_1 \in H_1$ and all $x_2, y_2 \in H_2$. We define the representation $U\colon A \to H$ by

$$U(s)(x_1 + x_2) = U_1(s)(x_1) + U_2(s)(x_2),$$

for all $x_1 \in H_1$ and all $x_2 \in H_2$. It is immediately verified that $U(s) \in \mathcal{L}(H)$ for all $s \in A$, and that U is a representation of A.

Definition 2.3. The representation U constructed from two representations $U_1\colon A \to \mathcal{L}(H_1)$ and $U_2\colon A \to \mathcal{L}(H_2)$ as above is called the *Hilbert sum* of U_1 and U_2.

We now generalize the construction of Hilbert sum to any arbitrary family of Hilbert spaces. The generalization to representations will be made in the next section.

Let $(H_\alpha, \langle -, -\rangle_\alpha)_{\alpha\in\Lambda}$ be a family of Hilbert spaces indexed by some index set Λ. In most applications, $\Lambda = \mathbb{N}$, so for simplicity the reader may assume this. We define the set H as the set of all sequences $(x_\alpha)_{\alpha\in\Lambda}$ with $x_\alpha \in H_\alpha$, such that $\sum_{\alpha\in\Lambda} \|x_\alpha\|_{H_\alpha}^2 < \infty$. Since the index set Λ may not be countable, what we are asserting is that the family $(\|x_\alpha\|_{H_\alpha}^2)_{\alpha\in\Lambda}$ is summable; see Vol. I, Definition D.6 (in particular, this implies that only countably many elements x_α are nonzero). We define a vector space structure on H by defining

$$\begin{aligned}(x_\alpha) + (y_\alpha) &= (x_\alpha + y_\alpha)\\ \lambda(x_\alpha) &= (\lambda x_\alpha),\end{aligned}$$

with $x_\alpha, y_\alpha \in H_\alpha$. It is easy to check that these operations make H into a vector space. We define the inner product $\langle -, - \rangle$ on H by

$$\langle (x_\alpha), (y_\alpha) \rangle = \sum_{\alpha \in \Lambda} \langle x_\alpha, y_\alpha \rangle_\alpha.$$

It can be verified that $\langle -, - \rangle$ is a Hermitian inner product on H. It can also be shown that H is complete, so it is a Hilbert space. For details, see Dieudonné [Dieudonné (2010)], (Chapter VI, Section 4) and Schwartz [Schwartz (1980)] (Chapter XXIII, Theorem 1 and Theorem 2).

Definition 2.4. Let $(H_\alpha, \langle -, - \rangle_\alpha)_{\alpha \in \Lambda}$ be a family of Hilbert spaces indexed by some index set Λ. The space H constructed as above is called the *Hilbert sum* of the sequence of Hilbert spaces (H_α) and is denoted by

$$H = \bigoplus_{\alpha \in \Lambda} H_\alpha.$$

We define continuous injections $j_\alpha \colon H_\alpha \to H$, such that $j_\alpha(x_\alpha) = (0, \ldots, 0, x_\alpha, 0, \ldots)$, with the αth term being x_α. Each j_α is an isomorphism of H_α onto a closed subspace of H denoted H'_α. By definition of the inner product on H, $\langle j_\alpha(x_\alpha), j_\beta(x_\beta) \rangle = 0$ for all $\alpha \neq \beta$, all $x_\alpha \in H_\alpha$, and all $x_\beta \in H_\beta$. A very useful fact is that the direct sum $\bigoplus_{\alpha \in \Lambda} H'_\alpha$ is dense in H (recall that $\bigoplus_{\alpha \in \Lambda} H'_\alpha$ consists of all sequences $(x_\alpha)_{\alpha \in \Lambda}$, such that $x_\alpha = 0$ for all but finitely many indices α).

Proposition 2.2. *Let H be the Hilbert sum of a family $(H_\alpha, \langle -, - \rangle_\alpha)_{\alpha \in \Lambda}$ of Hilbert spaces indexed by some index set Λ. For every $x = (x_\alpha)_{\alpha \in \Lambda} \in H$, the family $(j_\alpha(x_\alpha))_{\alpha \in \Lambda}$ of vectors in the direct sum $\bigoplus_{\alpha \in \Lambda} H'_\alpha$ is summable in H and we have*

$$x = \sum_{\alpha \in \Lambda} j_\alpha(x_\alpha)$$

(the convergence is not necessarily uniform). Consequently, $\bigoplus_{\alpha \in \Lambda} H'_\alpha$ is dense in H.

Proof. Following Schwartz, first we claim that if $\sum_{\alpha \in \Lambda} u_\alpha$ a summable series in a Banach space and if its sum is S, then for every subset $\Omega \subseteq \Lambda$ finite or not, the series $\sum_{\alpha \in \Omega} u_\alpha$ is also summable. This follows from the Cauchy criterion for summable families; see Vol. I, Proposition D.12. Indeed, since $\sum_{\alpha \in \Lambda} u_\alpha$ is summable, for every $\epsilon > 0$ there is some finite subset $I \subseteq \Lambda$, such that for every finite subset $J \subseteq \Lambda$ disjoint from I we have $\left\| \sum_{j \in J} u_j \right\| < \epsilon$, and this fact also applies if $J \subseteq \Omega$. By Proposition D.12, the series $\sum_{\alpha \in \Omega} u_\alpha$ is summable, and let S_Ω denote its sum.

Let I be a finite subset of Λ, such that $\|S_J\| \leq \epsilon$ for every finite subset J, such that I and J are disjoint. We claim that $\|S_K\| \leq \epsilon$ for any subset K finite or not, such that K and I are disjoint. This is because since $\sum_{\alpha \in K} u_\alpha$ is summable by the previous fact, its sum S_K belongs to the closure of the set $\{S_J \mid J \subseteq K, J \text{ finite}, \ I \cap J = \emptyset\}$. Indeed, for every $\eta > 0$, there is some finite subset $I_0 \subseteq K$, such that for every finite subset $J \subseteq K$ with $I_0 \subseteq J$ we have $\|S_J - S_K\| \leq \eta$, and since $K \cap I = \emptyset$ and $J \subseteq K$, we automatically have $J \cap I = \emptyset$.

Since $\sum_{\alpha \in \Lambda} \|x_\alpha\|_{H_\alpha}^2 < \infty$, by the Cauchy criterion, for every $\epsilon > 0$, there is finite subset I of Λ, such that $\sum_{j \in J} \|x_j\|_{H_j}^2 \leq \epsilon$ for any finite subset J disjoint from I. Since $\Lambda - I$ is disjoint from I, we have

$$\sum_{\alpha \in \Lambda - I} \|x_\alpha\|_{H_\alpha}^2 = \sum_{\alpha \in \Lambda - I} \|j_\alpha(x_\alpha)\|_{H'_\alpha}^2 \leq \epsilon,$$

where the sums exist by the first fact and their norms are bounded by ϵ by the second fact. Then for every subset K (finite or not), such that $I \subseteq K \subseteq \Lambda$, the family $(j_\alpha(x_\alpha))_{\alpha \in K}$ is summable in H, and if we let

$$x_K = \sum_{\alpha \in K} j_\alpha(x_\alpha),$$

we have

$$\begin{aligned} \|x - x_K\|_H &= \left(\sum_{\alpha \in \Lambda - K} \|x_\alpha\|^2 \right)^{1/2} = \left(\sum_{\alpha \in \Lambda - K} \|j_\alpha(x_\alpha)\|^2 \right)^{1/2} \\ &\leq \left(\sum_{\alpha \in \Lambda - I} \|j_\alpha(x_\alpha)\|^2 \right)^{1/2} \leq \epsilon, \end{aligned}$$

which, by Vol. I, Definition D.6, proves that $(j_\alpha(x_\alpha))_{\alpha \in \Lambda}$ is summable in H and that its sum is x. $\square$

Remark: By picking $\epsilon = 1/(n+1)$, we can define a sequence of *finite* subsets $K_n \subseteq K_{n+1}$, such that $\sum_{\alpha \in K_n} j_\alpha(x_\alpha)$ converges to x in H.

We often identify H_α and H'_α. The above construction defines what we might call an external Hilbert sum.

Unfortunately, the notation

$$H = \bigoplus_{\alpha \in \Lambda} H_\alpha$$

for the Hilbert sum of a family $(H_\alpha, \langle -, - \rangle_\alpha)_{\alpha \in \Lambda}$ of Hilbert spaces clashes with the notion of algebraic *direct sum*

$$\bigoplus_{\alpha \in \Lambda} H_\alpha$$

of vector spaces. The second definition refers to the subspace of sequences $(x_\alpha)_{\alpha \in \Lambda}$, such that $x_\alpha = 0$ *for all but finitely many indices* α. If we temporarily denote the Hilbert sum by

$$H = \bigoplus_{\alpha \in \Lambda}^{\mathrm{H}} H_\alpha,$$

then we see that if the index set Λ is finite, then the two notions agree. But if Λ is infinite, then the algebraic direct sum is a proper subspace of the Hilbert sum, because the Hilbert sum consists of sequences $(x_\alpha)_{\alpha \in \Lambda}$, such that $\sum_{\alpha \in \Lambda} \|x_\alpha\|^2 < \infty$, which may contain a countably infinite number of nonzero x_α. The direct sum is a dense subspace of the Hilbert sum. It is usually clear from the context which sum of spaces is intended (typically, when we refer to Hilbert spaces, we mean a Hilbert sum), so we will not use the heavier notation $\bigoplus^{\mathrm{H}}$ for Hilbert sums.

We also have the following proposition proven in Dieudonné [Dieudonné (2010)], (Chapter VI, Section 4) and Schwartz [Schwartz (1980)] (Chapter XXIII, Theorem 4 and its corollaries), which gives the definition of an internal Hilbert sum (in the sense that the H_α are subspaces of an already given space H).

Proposition 2.3. *Let H be a Hilbert space, and let $(H_\alpha)_{\alpha \in \Lambda}$ be a family of closed subspaces of H satisfying the following conditions:*

(1) For all $\alpha \neq \beta$, the subspaces H_α and H_β are orthogonal.
(2) The direct sum $\bigoplus_{\alpha \in \Lambda} H_\alpha$ is dense in H.

If E is the Hilbert sum of the sequence (H_α), then there is a unique isomorphism of H onto E which, on each H_α, coincides with the injection j_α of H_α into E.

If Λ is finite, (2) is equivalent to the fact that H is the direct sum of H_α (because, if E is any closed subspace of a Hilbert space, then $H = E \oplus E^\perp$ as a direct sum, so by picking $E = H_{\alpha_1}$, we see that the direct sum $\bigoplus_{j=2}^n H_{\alpha_j}$ is dense in $E^\perp$, and we finish by induction).

Another convenient characterization of Hilbert sums can be given in terms of orthogonal projections. Recall from Vol. I, Chapter D, Proposition D.7, that if H is a Hilbert space and if W is a *closed* subspace of

E, then for any $v \in H$, the *orthogonal projection* $p_W(v)$ *of* v *onto* W is well-defined. It is the unique vector $w \in W$, such that $v - w$ is orthogonal to W. The following proposition is proven in Schwartz [Schwartz (1980)] (Chapter XXIII, Theorem 2, Corollary 2).

Proposition 2.4. *Let H be a (separable) Hilbert space, and let $(H_\ell)_{\ell\in\Lambda}$ be a family of closed pairwise orthogonal subspaces of H. The following properties are equivalent:*

(1) *The family $(H_\ell)_{\ell\in\Lambda}$ is a Hilbert sum (that is, the algebraic direct sum $\bigoplus_{\ell\in\Lambda} H_\ell$ is dense in H).*
(2) *For every vector $v \in H$, if $(v_\ell)_{\ell\in\Lambda}$ is the family of orthogonal projections of v onto the H_ℓ, then the family $(v_\ell)_{\ell\in\Lambda}$ is summable and $v = \sum_{\ell\in\Lambda} v_\ell$.*
(3) *For every vector $v \in H$, we have the Parseval identity:*

$$\|v\|^2 = \sum_{\ell\in\Lambda} \|v_\ell\|^2 .$$

Two key notions of representation theory are invariant subspaces and (topologically) irreducible representations.

2.2 Invariant Subspaces and Irreducible Representations

Definition 2.5. Let $U\colon A \to \mathcal{L}(H)$ be a representation of an algebra A. A subspace E of H is *invariant* (or *stable*) under the representation U if $U(s)(x) \in E$ for all $s \in A$ and all $x \in E$. The representation $U_E\colon A \to \mathcal{L}(E)$ given by $U_E(s)(x) = U(s)(x)$ for all $x \in E$, is called a *subrepresentation* of A in E.

Example 2.5. In Example 2.2, the subspace $\mathfrak{b}$ of $A = \mathrm{M}_n(\mathbb{C})$ is invariant under the representation $U\colon A \to \mathcal{L}(H)$ of Example 2.1 and $U_1\colon A \to \mathcal{L}(\mathfrak{b})$ is a subrepresentation of A in $\mathfrak{b}$.

Observe a small abuse of language: if E is not a closed subspace of H, then E is not a Hilbert space, and so $U_E\colon A \to \mathcal{L}(E)$ is not a representation. Thus, the notion of subrepresentation should be defined for *closed* invariant subspaces of H. However, Proposition 2.5 shows that the closure $\overline{E}$ of an invariant subspace E is invariant, so we can define the subrepresentation $U_E\colon A \to \mathcal{L}(\overline{E})$.

The following facts hold.

Proposition 2.5. *Let $U: A \to \mathcal{L}(H)$ be a representation of an algebra A.*

(1) If the subspace E of H is invariant under U, then its closure $\overline{E}$ is also invariant under U.

(2) Let E be a closed subspace of H invariant under U. If $E^{\perp}$ is the orthogonal complement of E in H, then $E^{\perp}$ is invariant under U. If $U_1(s)$ and $U_2(s)$ are the restrictions of $U(s)$ to E and $E^{\perp}$, then the representation U is the Hilbert sum of the representations U_1 and U_2.

Proof. Part (1) is easy to prove and follows from the continuity of $U(s)$; see Dieudonné [Dieudonné (2010)], (Chapter III, Section 11). For part (2), let $x \in E$ and $y \in E^{\perp}$. For any $s \in A$, we have

$$\langle x, U(s)(y)\rangle = \langle (U(s))^*(x), y\rangle = \langle U(s^*)(x), y\rangle = 0,$$

since E is invariant under U, so $U(s^*)(x) \in E$, and since $E^{\perp}$ is the orthogonal complement of E and $y \in E^{\perp}$. Then $U(s)(y)$ is orthogonal to all $x \in E$, which means that $U(s)(y) \in E^{\perp}$, so $E^{\perp}$ is invariant under U. The last property is obvious because as E is closed, H is the (algebraic) direct sum $H = E \oplus E^{\perp}$. □

The notion of Hilbert sum of representations is generalized to arbitrary Hilbert sums as follows.

Definition 2.6. Assume a Hilbert space H is the Hilbert sum of a sequence $(H_\alpha)_{\alpha\in\Lambda}$ of subspaces invariant under a representation U of A. For every $s \in A$, let $U_\alpha(s)$ be the restriction of U to H_α, so that the map $s \mapsto U_\alpha(s)$ is a representation of A in H_α. By abuse of language, we say that U is the *Hilbert sum* of the representations U_α. For each $s \in A$ and each $x = \sum_\alpha x_\alpha \in H$, where $x_\alpha \in H_\alpha$, we have

$$U(s)(x) = \sum_\alpha U_\alpha(s)(x_\alpha),$$

and

$$\sum_\alpha \|U_\alpha(s)(x_\alpha)\|^2 = \|U(s)(x)\|^2.$$

Recall Vol. I, Definition D.2 of the orthogonal projection p_V of a Hilbert space E onto a closed subspace V. Such a map is linear and continuous. It is also called an *orthogonal projector*. In this chapter, we denote p_V by P_V to conform to Dieudonné.

The following result is not hard to prove; see Dieudonné [Dieudonné (2005)], (Chapter XV, Section 5).

Proposition 2.6. *Let H be a Hilbert space. A continuous linear P map on H is an orthogonal projector iff it is idempotent ($P^2 = P \circ P = P$) and hermitian ($P^* = P$).*

Here is a convenient way to characterize when a closed subspace is invariant.

Proposition 2.7. *Let $U : A \to \mathcal{L}(H)$ be a representation of an algebra A. A closed subspace E of H is invariant under U iff $P_E U(s) = U(s) P_E$ for all $s \in A$, as illustrated in the diagram below*

$$\begin{array}{ccc} H & \xrightarrow{U(s)} & H \\ {\scriptstyle P_E}\downarrow & & \downarrow{\scriptstyle P_E} \\ E & \xrightarrow[U(s)]{} & E, \end{array}$$

where $P_E : H \to E$ is the orthogonal projection of H onto E.

The proof of Proposition 2.7 is identical to the proof for group representations; see Proposition 3.7 and its proof. Proposition 2.7 is also proven in Dieudonné [Dieudonné (2005)], (Chapter XV, Section 5, Proposition 15.5.3).

The notion of a topologically irreducible representation is similar in spirit to the notion of prime number. Namely, a topologically irreducible representation cannot be decomposed into simpler representations. It is one of the most important concepts in representation theory.

Definition 2.7. A representation $U : A \to \mathcal{L}(H)$ of A in H is *topologically irreducible* if $H \neq (0)$ and if there is no *closed* subspace E of H other than $\{0\}$ and H, which is invariant under U.

Example 2.6. The representation $U : A \to \mathcal{L}(H)$ of Example 2.1 is reducible because the proper nonzero subspace $\mathfrak{b}$ of H is invariant under U. On the other hand, by Theorem 2.36, the representation $U_2 : A \to \mathcal{L}(\mathfrak{b})$ of Example 2.2 is topologically irreducible (in finite dimension, a subspace is automatically closed).

Proposition 2.8. *Let $U : A \to \mathcal{L}(H)$ be a representation of A in H, let E be the closure of the subspace spanned by the set*

$$\{U(s)(x) \mid s \in A,\ x \in H\},$$

and let E' be the set

$$E' = \{x \in H \mid U(s)(x) = 0,\ \textit{for all } s \in A\}.$$

Then E and E' are invariant under U, and E' is the orthogonal complement of E in H (that is, $E' = E^{\perp}$ and $H = E \oplus E'$).

Proof. Since $U(rs) = U(r)U(s)$ for all $r, s \in A$, it is clear that E and E' are invariant under U. Let us prove that $E^{\perp} \subseteq E'$, where $E^{\perp}$ is the orthogonal complement of E in H. We already know from Proposition 2.5 that $E^{\perp}$ is invariant under U, so for any $x \in E^{\perp}$ we have $U(s)(x) \in E^{\perp}$ for all $s \in A$. But by definition of E, we have $U(s)(x) \in E$, so $U(s)(x) \in E \cap E^{\perp} = (0)$, which means that $U(s)(x) = 0$ for all $s \in A$, that is, $x \in E'$. Therefore, $E^{\perp} \subseteq E'$.

Next we prove that $E' \subseteq E^{\perp}$. If $x \in E'$, for any $s \in A$ and any $y \in H$ we have

$$\langle x, U(s)(y) \rangle = \langle U(s^*)(x), y \rangle = 0,$$

since $x \in E'$ means that $U(s)(x) = 0$ for all $s \in A$, and since s and y are arbitrary, by definition of E, this means that x is orthogonal to E, that is, $x \in E^{\perp}$, and thus $E' \subseteq E^{\perp}$. In conclusion, $E' = E^{\perp}$. □

Definition 2.8. Let $U\colon A \to \mathcal{L}(H)$ be a representation of A in H. The subspace E defined in Proposition 2.8 is called the *essential subspace* for U. If $E' = (0)$, then we say that the representation is *nondegenerate*.

Although very easy to prove, the following result is important and often used.

Proposition 2.9. *A representation U of A in H is nondegenerate iff the subspace spanned by set $\{U(s)(x) \mid s \in A,\ x \in H\}$ is dense in H. If A is unital, since $U(e) = \mathrm{id}$, this is always the case.*

Definition 2.9. Let $U\colon A \to \mathcal{L}(H)$ be a representation of A in H. A vector $x_0 \in H$ is called a *totalizer* or *totalizing vector* (or *cyclic vector*) for the representation U if the subspace of H spanned by the set $\{U(s)(x_0) \mid s \in A\}$ is dense in H. Equivalently, if $\mathcal{M}_{x_0}$ denotes the closure of the set $\{U(s)(x_0) \mid s \in A\}$, called the *cyclic subspace* generated by x_0, which is invariant under U, then x_0 is a totalizer (a cyclic vector) if $\mathcal{M}_{x_0} = H$. A representation which admits a totalizer is said to be *topologically cyclic*.

The following fact follows immediately from the definitions: a representation U is topologically irreducible iff every nonzero vector $x_0 \in H$ is a totalizer. The importance of totalizers stems from the following result.

Proposition 2.10. *Assume that the algebra A is unital, and let $U: A \to \mathcal{L}(H)$ be a representation of A in $H \neq (0)$ (which must be nondegenerate). Then H is the Hilbert sum of a sequence $(H_\alpha)_{\alpha \in \Lambda}$ of closed subspaces $H_\alpha \neq (0)$ of H invariant under U, and such that the restriction of U to each H_α is topologically cyclic. If H is separable, the family Λ is countable (possibly finite).*

Proof. We prove the proposition in the separable case, following Dieudonné [Dieudonné (2005)] (Chapter XV, Proposition 15.5.6). The general case uses Zorn's lemma; see Folland [Folland (1995)] (Chapter 3, Proposition 3.3). Let (x_n) be a dense sequence (finite or countably infinite) in H. We define the sequence (H_n) by induction as follows. Let H_1 be the subspace spanned by the set of vectors $\{H(s)(x_1) \mid s \in A\}$. Since A is unital, $x_1 \in H_1$. Assuming that $H_1, \ldots, H_n$ have been defined, either H is the direct sum of the H_i, in which case we are done, or else we proceed as follows. Let $L \neq (0)$ be the orthogonal complement of the direct sum $H_1 \oplus \cdots \oplus H_n$ in H. Then let $p(n+1)$ be the smallest index, such that if y_{n+1} is the orthogonal projection of $x_{p(n+1)}$ on L, then the subspace H'_{n+1} of L generated by the subset $\{U(s)(y_{n+1}) \mid s \in A\}$ is not the trivial subspace (0). Since A is unital, $y_{n+1} \in \{U(s)(y_{n+1}) \mid s \in A\}$ because $U(\mathbf{1}) = \mathrm{id}$, so such an index must exist. By definition of $p(n+1)$, we have $x_1, \ldots, x_{p(n+1)-1} \in H_1 \oplus \cdots \oplus H_n$. Let H_{n+1} be the closure of H'_{n+1} in H. Since $H_{n+1} \subseteq L$ and L is the orthogonal complement of the direct sum $H_1 \oplus \cdots \oplus H_n$ in H, the fact that $y_{n+1} \in H_{n+1}$ implies that $x_{p(n+1)} \in H_1 \oplus \cdots \oplus H_{n+1}$, so the direct sum of the H_k contains the dense sequence (x_n), and by Proposition 2.3, the space H is indeed the Hilbert sum of the H_k. ☐

2.3 Positive Linear Forms and Positive Hilbert Forms

The Peter–Weyl theorem can be obtained from a structure theorem about certain kinds of algebras with a hermitian inner product satisfying special conditions. Such inner products are bitraces, which are special kinds of positive Hilbert forms. A good method for producing positive Hilbert forms is to use positive linear forms.

Definition 2.10. Let A be an involutive algebra (not necessarily commutative, unital, and not necessarily normed). A linear form $f: A \to \mathbb{C}$ is *positive* if

$$f(s^*s) \geq 0, \qquad \text{for all } s \in A.$$

Positive linear forms arise from representations as follows.

Definition 2.11. Let $U\colon A \to \mathcal{L}(H)$ be a representation of A in H. For any $x_0 \in H$, define the map $f_{x_0}\colon A \to \mathbb{C}$ by

$$f_{x_0}(s) = \langle U(s)(x_0), x_0\rangle, \quad s \in A.$$

Proposition 2.11. *The map $f_{x_0}\colon A \to \mathbb{C}$ is a positive linear form.*

Proof. It is clear that f_{x_0} is a linear form, and since $U(s^*s) = U(s^*)U(s) = U(s)^*U(s)$, we have

$$\begin{aligned} f_{x_0}(s^*s) &= \langle U(s^*s)(x_0), x_0\rangle = \langle (U(s)^*U(s))(x_0), x_0\rangle \\ &= \langle U(s)(x_0), U(s)(x_0)\rangle = \|U(s)(x_0)\|^2 \geq 0. \end{aligned}$$ □

A positive linear form also defines a positive hermitian form as follows.

Proposition 2.12. *Given any positive linear form f on an involutive algebra A, let*
$g\colon A \times A \to \mathbb{C}$ be the map given by

$$g(x, y) = f(y^*x), \qquad \text{for all } x, y \in A.$$

Then g is a positive hermitian form, and the following properties hold:

(1) For all $x, y \in A$, we have

$$f(x^*y) = \overline{f(y^*x)}.$$

(2) For all $x, y \in A$, we have

$$|f(y^*x)|^2 \leq f(x^*x)f(y^*y).$$

(3) If A is unital, then

$$f(x^*) = \overline{f(x)}, \qquad |f(x)|^2 \leq f(e)f(x^*x).$$

Proof. To prove (1), since f is linear, we have

$$\begin{aligned} g(x+y, x+y) &= f((x+y)^*(x+y)) \\ &= f(x^*x + x^*y + y^*x + y^*y) \\ &= f(x^*x) + f(x^*y) + f(y^*x) + f(y^*y) \\ &= g(x,x) + g(y,x) + g(x,y) + g(y,y). \end{aligned}$$

Since $g(x+y, x+y), g(x,x)$, and $g(y,y)$ are real (and nonnegative), we must have

$$\Im(g(y,x)) = -\Im(g(x,y)).$$

If we replace x by ix, this becomes

$$\Re(g(y,x)) = \Re(g(x,y)).$$

Therefore,

$$g(y,x) = f(x^*y) = \overline{f(y^*x)} = \overline{g(x,y)},$$

as claimed. Part (2) is the Cauchy–Schwartz inequality, and (3) follows from (1) and (2) by replacing y by e and the fact that $e^* = e$. ☐

The hermitian form g obtained from the positive linear form f is not arbitrary since it satisfies the condition

$$g(xy,z) = f(z^*xy) = f((x^*z)^*y) = g(y,x^*z)$$

for all $x, y, z \in A$.

This motivates the following definition.

Definition 2.12. A *positive Hilbert form* on an involutive algebra A is a positive hermitian form g satisfying the condition

$$g(xy,z) = g(y,x^*z), \qquad \text{for all } x,y,z \in A.$$

Proposition 2.13. *If A is unital with unit e, then every positive Hilbert form g comes from the positive linear form f given by $f(s) = g(s,e)$ for all $s \in A$.*

Proof. Indeed, the positive Hilbert form g' induced by f is given by

$$g'(s,t) = f(t^*s) = g(t^*s,e) = g(s,t),$$

by setting $x = t^*, y = s$, and $z = e$ in the equation of Definition 2.12. ☐

Given a representation $U\colon A \to \mathcal{L}(H)$, observe that the positive Hilbert form g_{x_0} associated with the positive linear form f_{x_0} is given by

$$g_{x_0}(s,t) = f_{x_0}(t^*s) = \langle U(s)(x_0), U(t)(x_0)\rangle, \qquad s,t \in A.$$

Remarkably, every topologically cyclic representation arises from a positive Hilbert form, but we won't need this fact until Section 2.9, so we postpone discussing this matter.

2.4 Traces, Bitraces, Hilbert Algebras

If A is an involutive algebra and if f is a positive linear form on A, in general $f(st) \neq f(ts)$.

Definition 2.13. Let A be an involutive algebra. A *trace* on A is a positive linear form $f\colon A \to \mathbb{C}$, such that

$$f(st) = f(ts), \qquad \text{for all } s, t \in A.$$

A *bitrace* is a positive Hilbert form $g\colon A \times A \to \mathbb{C}$, such that

$$g(t^*, s^*) = g(s, t), \qquad \text{for all } s, t \in A.$$

Example 2.7. If H is a finite-dimensional vector space of dimension n with a hermitian inner product, and if $A = \mathcal{L}(H)$, the algebra of linear maps from H to itself, for any orthonormal basis $(e_1, \ldots, e_n)$ of H, then for any linear map $T \in \mathcal{L}(H)$, the linear form

$$\mathrm{Tr}(T) = \sum_{i=1}^{n} \langle T(e_i), e_i \rangle$$

is a trace. In fact, $\mathrm{Tr}(T) = \sum_{i=1}^{n} a_{ii}$, the trace of the matrix (a_{ij}) representing T over the basis $(e_1, \ldots, e_n)$. If H is an infinite-dimensional Hilbert space, it can be shown that there exists no trace on $\mathcal{L}(H)$.

Proposition 2.14. *Let $f\colon A \to \mathbb{C}$ be a positive linear form on an involutive algebra A. If*

$$f(ss^*) = f(s^*s), \qquad \textit{for all } s \in A,$$

then f is a trace.

Proof. By replacing s by $s + t$, using Proposition 2.12(1), we have

$$\begin{aligned} f((s+t)(s+t)^*) &= f(ss^* + st^* + ts^* + tt^*) \\ &= f(ss^*) + f(st^*) + f(ts^*) + f(tt^*) \\ &= f(s^*s) + f(st^*) + \overline{f(st^*)} + f(t^*t), \end{aligned}$$

and similarly,

$$f((s+t)^*(s+t)) = f(s^*s) + \overline{f(t^*s)} + f(t^*s) + f(t^*t),$$

and since $f((s+t)(s+t)^*) = f((s+t)^*(s+t))$, we get

$$f(st^*) + \overline{f(st^*)} = f(t^*s) + \overline{f(t^*s)},$$

that is

$$\Re(f(t^*s)) = \Re(f(st^*)).$$

If we replace s by is, we get

$$\Im(f(t^*s)) = \Im(f(st^*)),$$

so that

$$f(t^*s) = f(st^*),$$

for all $s, t \in A$, as claimed. $\square$

Proposition 2.15. *If the positive Hilbert form g on an involutive algebra A arising from a positive linear form f as $g(s,t) = f(t^*s)$ is a bitrace, then f is a trace. Conversely, if f is a trace, then g is a bitrace.*

Proof. Expressing that g is a bitrace says that

$$f(t^*s) = g(s,t) = g(t^*, s^*) = f((s^*)^*t^*) = f(st^*),$$

namely that f is a trace. The same computation shows that if f is a trace, then g is a bitrace. $\square$

One of the most important examples of a bitrace arises when G is a compact group. In this case, the inner product on the involutive Banach algebra $L^2(G)$ is a bitrace. In fact, this bitrace satisfies two more properties that makes $L^2(G)$ into a Hilbert algebra, defined next.

Definition 2.14. An involutive algebra A is a *Hilbert algebra* if its underlying vector space is a hermitian space whose hermitian inner product $\langle -, - \rangle$ is a bitrace satisfying two extra conditions (U) and (N). Specifically, the conditions for being a bitrace hold

$$\langle y^*, x^* \rangle = \langle x, y \rangle \tag{1}$$

$$\langle xy, z \rangle = \langle y, x^*z \rangle, \tag{2}$$

and the following two conditions hold: for every $x \in A$, there is some $M_x \geq 0$, such that

$$\langle xy, xy \rangle \leq M_x \langle y, y \rangle, \quad \text{for all } y \in A, \tag{U}$$

and

$$\text{the subspace spanned by the set } \{xy \mid x, y \in A\} \text{ is dense in } A. \tag{N}$$

From conditions (1) and (2) and the hermitian property, we get

$$\langle yx, z\rangle = \langle z^*, x^*y^*\rangle = \overline{\langle x^*y^*, z^*\rangle} = \overline{\langle y^*, xz^*\rangle} = \langle xz^*, y^*\rangle = \langle y, zx^*\rangle,$$

so

$$\langle yx, z\rangle = \langle y, zx^*\rangle. \tag{2'}$$

By (1) and (U), for every $x \in A$, we also have

$$\langle yx, yx\rangle = \langle x^*y^*, x^*y^*\rangle \leq M_{x^*}\langle y^*, y^*\rangle = M_{x*}\langle y, y\rangle,$$

namely,

$$\langle yx, yx\rangle \leq M_{x^*}\langle y, y\rangle, \quad \text{for all } y \in A. \tag{U'}$$

The inequality (U) says that for x fixed, the linear map $y \mapsto xy$ is continuous, and the inequality (U') says that for y fixed, the linear map $x \mapsto xy$ is continuous.

Beware that in general, this *does not imply* that the map $(x, y) \mapsto xy$ is continuous. Thus, in general, a Hilbert algebra is *not* a normed algebra in the sense of Vol. I, Definition 9.4, because in a normed algebra, the map $(x, y) \mapsto xy$ is continuous. However, if the Hilbert algebra A is *complete*, then using Baire's theorem, it can be shown that the map $(x, y) \mapsto xy$ is continuous; see Dieudonné [Dieudonné (2005)] (Chapter XII, Section 16, Problem 8(c)). As a consequence, we can show that a complete Hilbert algebra is normable, and thus a Banach algebra.

Proposition 2.16. *If A is a complete Hilbert algebra, then there is a norm $\|\ \|_b$ equivalent to the norm $x \mapsto \|x\| = \sqrt{\langle x, x\rangle}$ induced by the inner product $\langle -, -\rangle$ on A, such that A is a Banach algebra with the norm $\|\ \|_b$.*

Proof. First, assume that A is not unital. Since A is complete, as we said earlier, the bilinear map $(x, y) \mapsto xy$ is continuous, so there is some constant $c > 0$, such that $\|xy\| \leq c\,\|x\|\,\|y\|$ for all $x, y \in A$. Since $c\,\|xy\| \leq c\,\|x\|\,c\,\|y\|$, if we let $\|x\|_b = c\,\|x\|$, we obtain a norm equivalent to $\|\ \|$, such that

$$\|xy\|_b \leq \|x\|_b\,\|y\|_b\,,$$

and with this norm, A is a Banach algebra.

If A has a multiplicative unit e, then the norm $\|\ \|_b$ also needs to satisfy the condition $\|e\|_b = 1$, so we need a different construction. For every $x \in A$, let $L_x\colon A \to A$ be the linear map given by

$$L_x(y) = xy, \qquad y \in A.$$

Since the bilinear map $(x, y) \mapsto xy$ is continuous, the linear map L_x is continuous. We check immediately that

$$\begin{aligned} L_{x+y} &= L_x + L_y \\ L_{\alpha x} &= \alpha L_x \\ L_{xy} &= L_x \circ L_y. \end{aligned}$$

Therefore, the map $L\colon A \to \mathcal{L}(A)$ given by $x \mapsto L_x$ is an algebra homomorphism. The homomorphism L is injective, because if $L_x = 0$, then $L_x(e) = xe = x = 0$. We claim that there is a constant $c > 0$, such that $\|L_x\| \le c\,\|x\|$. Recall that for the operator norm $\|L_x\|$, we have

$$\|L_x\| = \sup\{\|L_x(y)\| \mid \|y\| = 1\} = \sup\{\|xy\| \mid \|y\| = 1\},$$

and since the bilinear map $(x, y) \mapsto xy$ is continuous

$$\|L_x\| = \sup\{\|xy\| \mid \|y\| = 1\} \le \sup\{c\,\|x\|\,\|y\| \mid \|y\| = 1\} = c\,\|x\|\,.$$

On the other hand, since $\|L_x\|$ is the operator norm,

$$\|x\| = \|xe\| = \|L_x(e)\| \le \|L_x\|\,\|e\|\,.$$

Therefore, if we let $\|x\|_b = \|L_x\|$, we obtain a norm on A equivalent to the norm $\|\ \|$, and A is a normed algebra with the norm $\|\ \|_b$, since $\|L_{xy}\| = \|L_x \circ L_y\| \le \|L_x\|\,\|L_y\|$. ☐

Remark: The proof of Proposition 2.16 shows that if A is an algebra, whose topology is defined by a norm $\|\ \|$, and if the bilinear map $(x, y) \mapsto xy$ is continuous, then there is a norm $\|\ \|_b$ equivalent to the norm $\|\ \|$, such that A is a normed algebra with the norm $\|\ \|_b$.

The following result will be needed to prove that if A is a Hilbert algebra and if $x \in A$ with $x \neq 0$, then $Ax \neq (0)$.

Proposition 2.17. *Let A be a Hilbert algebra. For every $x \in A$, if $x^*x = 0$, then $x = 0$.*

Proof. Since $x^*x = 0$, from property (2) we have

$$\langle xy, xy\rangle = \langle x^*xy, y\rangle = 0,$$

hence $xy = 0$ for all $y \in A$ (since the hermitian product is positive definite). In particular, $xz^* = 0$. By (2'), we get

$$\langle x, yz\rangle = \langle xz^*, y\rangle = 0$$

for all $y, z \in A$, and since by (N) the subspace spanned by the set $\{yz \mid y, z \in A\}$ is dense in A, we conclude that $x = 0$ (since the hermitian product is positive definite). ☐

Recall from Vol. I, Proposition A.47 that if a topological space is metrizable and compact, then it is separable (which means that it contains a countable dense subset). Here is our first important example of a Hilbert algebra.

Proposition 2.18. *If G is a compact metrizable group, then the involutive (complex) Banach algebra* $\mathrm{L}^2(G)$ *is a separable Hilbert algebra.*

Proof sketch. Proposition 2.18 is proven in Dieudonné [Dieudonné (2005)] (Chapter XXI, Section 2). We may assume that G is equipped with a Haar measure λ, such that $\lambda(G) = 1$. Recall that the involution $f \mapsto f^*$ is defined by $f^* = \overline{\check{f}}$. Condition (1) follows from the definition of the inner product on $\mathrm{L}^2(G)$. Since G is compact, $\mathcal{C}_0(G;\mathbb{C}) = \mathcal{K}_{\mathbb{C}}(G) \subseteq \mathrm{L}^2(G)$. By Vol. I, Proposition 8.49, $f * g \in \mathcal{C}_0(G;\mathbb{C}) \subseteq \mathrm{L}^2(G)$, and $\|f * g\|_\infty \leq \|f\|_2 \|g\|_2$ (where $\| \, \|_2$ is the L^2 semi-norm on $\mathrm{L}^2(G)$). By Vol. I, Proposition 5.24(2) and since $\lambda(G) = 1$, we also have $\|f * g\|_2 \leq \|f * g\|_\infty$. Consequently, $\|f * g\|_2 \leq \|f\|_2 \|g\|_2$, so condition (U) follows. Condition (N) is a corollary of regularization (see Vol. I, Section 8.14). Finally, condition (2), namely

$$\langle f * g, h\rangle = \langle g, f^* * h\rangle$$

is easily shown if f is real and continuous. In this case, $f^* = \check{f}$. We obtain the formula in general by continuity and using the fact that $\mathcal{K}_{\mathbb{R}}(G)$ is dense in $\mathrm{L}^2(G)$. □

The family of Hilbert–Schmidt operator is the second important example of Hilbert algebras.

2.5 Hilbert–Schmidt Operators

Let H be a separable Hilbert space. Recall that this means H has a countable Hilbert basis, that is, a countable orthonormal basis $(a_i)_{i\geq 1}$, such that the subspace spanned by $(a_i)_{i\geq 1}$ is dense in H (for every vector $x \in H$, there is some sequence (x_n), with x_n a linear combination $\sum_{k\in I_n} \lambda_k a_k$, where I_n a finite set, and x_n converges to x).

Definition 2.15. Let H be a separable Hilbert space. A continuous linear map $u \in \mathcal{L}(H)$ is a *Hilbert–Schmidt operator* if the series $\sum_{n=1}^\infty \|u(a_n)\|^2$ converges, that is, $\sum_{n=1}^\infty \|u(a_n)\|^2 < \infty$.

We now state various properties about Hilbert–Schmidt operators whose proofs can be found in Dieudonné [Dieudonné (2005)] (Chapter XV, Section 4), Dunford and Schwartz [Dunford and Schwartz (1988)] (Chapter XI,

Section 6) and Lang [Lang (1985)]. See also Lang [Lang (1996)] (Chapter XVII, Problem 9).

The quantity $\sum_{n=1}^{\infty} \|u(a_n)\|^2$ is independent of the Hilbert basis (a_n). The set of Hilbert–Schmidt operators is denoted by $\mathcal{L}_2(H)$. Then we define the map $u \mapsto \|u\|_{\mathrm{HS}}$ on the set $\mathcal{L}_2(H)$ by

$$\|u\|_{\mathrm{HS}}^2 = \sum_{n=1}^{\infty} \|u(a_n)\|^2 .$$

It can be shown (using the Hilbert basis (a_n) and Parseval) that $\|u\|_{\mathrm{HS}} = \|u^*\|_{\mathrm{HS}}$. For any Hilbert–Schmidt operator $u \in \mathcal{L}_2(H)$ and any continuous linear map $v \in \mathcal{L}(H)$, it can also be shown that $u \circ v$ and $v \circ u$ are also in $\mathcal{L}_2(H)$, that

$$\|v \circ u\|_{\mathrm{HS}} \leq \|v\| \, \|u\|_{\mathrm{HS}} , \qquad \|u \circ v\|_{\mathrm{HS}} \leq \|v\| \, \|u\|_{\mathrm{HS}} ,$$

and if $u \in \mathcal{L}_2(H)$ and $x \in H$, then

$$\|u(x)\|^2 \leq \|x\|^2 \, \|u\|_{\mathrm{HS}}^2 ,$$

and so

$$\|u\| \leq \|u\|_{\mathrm{HS}} ,$$

where $\|u\|$ denotes the operator norm of u in $\mathcal{L}(H)$.

Then it can be shown that with the norm $\| \, \|_{\mathrm{HS}}$, the space $\mathcal{L}_2(H)$ of Hilbert–Schmidt operators is an involutive Banach algebra under composition, with the involution given by $u \mapsto u^*$ (where u^* is the adjoint of u). The space $\mathcal{L}_2(H)$ is a self-adjoint two-sided ideal in the involutive Banach algebra $\mathcal{L}(H)$, but in general, it is not closed in $\mathcal{L}(H)$.

The algebra $\mathcal{L}_2(H)$ is not unital because the identity map is not a Hilbert–Schmidt operator, and it is not a C^*-algebra.

If E and F are two normed vector spaces, a linear map $u \colon E \to F$ is a *compact operator* if the closure of $f(B)$ is compact for every bounded subset B of E. A compact operator is continuous. Every Hilbert–Schmidt operator is a compact operator, but the converse is false.

The space $\mathcal{L}_2(H)$ contains the continuous linear maps of finite rank. In fact, every Hilbert–Schmidt operator $u \in \mathcal{L}_2(H)$ is the limit of a sequence of continuous linear maps u_n of finite rank, in the sense that $\lim_{n \mapsto \infty} \|u - u_n\|_{\mathrm{HS}} = 0$; see Theorem 6 in Dunford and Schwartz [Dunford and Schwartz (1988)] (Chapter XI, Section 6).

For any two Hilbert–Schmidt operators $u, v \in \mathcal{L}_2(H)$, it can be shown that the quantity

$$g(u,v) = \sum_{n \geq 1} \langle u(a_n), v(a_n) \rangle$$

is defined and independent of the Hilbert basis (a_n). Then it can be shown that g is a hermitian inner product, which is a bitrace, such that $g(u,u) = \|u\|_{\mathrm{HS}}$; see Dieudonné [Dieudonné (2005)] (Chapter XV, Section 7).

If H is a finite-dimensional Hilbert space of dimension n, then for every linear map $u\colon H \to H$, for every orthonormal basis $(a_1, \ldots, a_n)$, the quantity $\|u\|_{\mathrm{HS}}^2 = \sum_{i=1}^n \|u(a_n)\|^2$ is defined. Since $\|u(a_i)\|^2 = \langle u(a_i), u(a_i) \rangle = \langle (u^* \circ u)(a_i), a_i \rangle$, we have

$$\|u\|_{\mathrm{HS}}^2 = \sum_{i=1}^n \langle (u^* \circ u)(a_i), a_i \rangle = \mathrm{tr}(u^* \circ u) = \mathrm{tr}(u \circ u^*),$$

which is just the *Frobenius norm* (also called *Hilbert–Schmidt norm*). Then

$$g(u,v) = \sum_{i=1}^n \langle u(a_n), v(a_n) \rangle = \sum_{i=1}^n \langle (v^* \circ u)(a_n), a_n \rangle = \mathrm{tr}(v^* \circ u)$$

is the corresponding inner product, denoted by $\langle u, v \rangle_{\mathrm{HS}}$.

If $\mathcal{L}_2(H)$ is the involutive algebra of Hilbert–Schmidt operators, then it can be shown that the bitrace g satisfies the properties (U) and (N). Consequently, *$\mathcal{L}_2(H)$ is a Hilbert algebra*; see Dieudonné [Dieudonné (2005)] (Chapter XV, Section 7).

In Chapter 8, we need a generalization of the notion of Hilbert–Schmidt operator involving two separable Hilbert spaces H_1 and H_2.

Definition 2.16. Let H_1 and H_2 be two separable Hilbert spaces. A continuous linear map $u \in \mathcal{L}(H_1, H_2)$ is a *Hilbert–Schmidt operator* if for some Hilbert basis (a_n) in H_1 the series $\sum_{n=1}^\infty \|u(a_n)\|_{H_2}^2$ converges, that is, $\sum_{n=1}^\infty \|u(a_n)\|_{H_2}^2 < \infty$.

Proposition 2.19. *Let H_1 and H_2 be two separable Hilbert spaces. For any two orthonormal Hilbert bases (a_n^1) and (a_n^2) in H_1 and any Hilbert basis (b_n) in H_2, for any $u \in \mathcal{L}(H_1, H_2)$, if $\sum_n \left\|u(a_n^1)\right\|_{H_2}^2$ converges, then we have*

$$\sum_n \left\|u(a_n^1)\right\|_{H_2}^2 = \sum_n \left\|u(a_n^2)\right\|_{H_2}^2 = \sum_n \|u^*(b_n)\|_{H_1}^2 .$$

Consequently, $\sum_n \left\|u(a_n^1)\right\|_{H_2}^2$ is independent of the Hilbert basis chosen in H_1 and $\sum_n \|u^*(b_n)\|_{H_1}^2$ is independent of the Hilbert basis chosen in H_2.

Proof. We can express $u(a_j^1)$ over the orthonormal Hilbert basis (b_i) of H_2 in terms of its Fourier coefficients $\langle u(a_j^1), b_i\rangle_{H_2}$ as

$$u(a_j^1) = \sum_i \langle u(a_j^1), b_i\rangle_{H_2}\, b_i,$$

and similarly we can express $u^*(b_i)$ over the orthonormal Hilbert basis (a_i^1) of H_1 in terms of its Fourier coefficients $\langle u^*(b_i), a_j^1\rangle_{H_1}$ as

$$u^*(b_i) = \sum_j \langle u^*(b_i), a_j^1\rangle_{H_1}\, a_j.$$

By Parseval and because the double sum is commutatively convergent, we have

$$\begin{aligned}\sum_j \left\|u(a_j^1)\right\|_{H_2}^2 &= \sum_i\sum_j |\langle u(a_j^1), b_i\rangle_{H_2}|^2 = \sum_i\sum_j |\langle a_j^1, u^*(b_i)\rangle_{H_1}|^2 \\ &= \sum_j\sum_i |\langle u^*(b_i), a_j^1\rangle_{H_1}|^2 = \sum_i \left\|u^*(b_i)\right\|_{H_1}^2,\end{aligned}$$

and the same computation with the basis (a_j^2) instead of the basis (a_j^1) shows that

$$\sum_j \left\|u(a_j^1)\right\|_{H_2}^2 = \sum_j \left\|u(a_j^2)\right\|_{H_2}^2 = \sum_i \left\|u^*(b_i)\right\|_{H_1}^2,$$

as claimed. Since by Parseval we also have

$$\begin{aligned}\sum_j \left\|u(a_j^1)\right\|_{H_2}^2 &= \sum_i\sum_j |\langle u(a_j^1), b_i^1\rangle_{H_2}|^2 = \sum_j\sum_i |\langle u^*(b_i^1), a_j^1\rangle_{H_1}|^2 \\ &= \sum_i \left\|u^*(b_i^1)\right\|_{H_1}^2 \\ \sum_j \left\|u(a_j^1)\right\|_{H_2}^2 &= \sum_i\sum_j |\langle u(a_j^1), b_i^2\rangle_{H_2}|^2 = \sum_j\sum_i |\langle u^*(b_i^2), a_j^1\rangle_{H_1}|^2 \\ &= \sum_i \left\|u^*(b_i^2)\right\|_{H_1}^2,\end{aligned}$$

for any two Hilbert bases (b_i^1) and (b_i^2) of H_2, we conclude that $\sum_n \left\|u^*(b_n)\right\|_{H_1}^2$ is independent of the Hilbert basis chosen in H_2. $\square$

Proposition 2.19 shows that the set of Hilbert–Schmidt operators $u \in \mathcal{L}(H_1, H_2)$, denoted by $\mathcal{L}_2(H_1, H_2)$, is a normed vector space with the norm $u \mapsto \|u\|_{\mathrm{HS}}$ given by

$$\|u\|_{\mathrm{HS}}^2 = \sum_{n=1}^{\infty} \left\|u(a_n)\right\|_{H_2}^2.$$

For any $x \in H_1$, if we write $x = \sum_{i=1}^{\infty} \xi_i a_i$ for some $\xi_i \in \mathbb{C}$ (where (a_i) is an orthonormal basis of H_1), then

$$\|x\|_{H_1}^2 = \sum_{i=1}^{\infty} |\xi_i|^2,$$

and if $\|u\|_{\mathrm{HS}}^2 = \sum_{i=1}^{\infty} \|u(a_i)\|_{H_2}^2 < \infty$, then because the vectors $(|\xi_i|)$ and $(\|u(a_i)\|_{H_2})$ belong to $\ell^2(\mathbb{R})$, by Proposition D.14 in Vol. I, the series $\sum_i |\xi_i| \, \|u(a_i)\|_{H_2} = \sum_i \|u(\xi_i a_i)\|_{H_2}$ is summable, so it belongs to $\ell^1(\mathbb{R})$, and since $\ell^1(\mathbb{R}) \subseteq \ell^2(\mathbb{R})$, the sum $\sum_i \|u(\xi_i a_i)\|_{H_2}^2$ is finite, so we have

$$\begin{aligned}\|u(x)\|_{H_2}^2 &= \|\sum_i u(\xi_i a_i)\|_{H_2}^2 = \sum_i |\xi_i|^2 \, \|u(a_i)\|_{H_2}^2 \\ &\leq \|x\|_{H_1}^2 \sum_i \|u(a_i)\|_{H_2}^2 = \|x\|_{H_1}^2 \, \|u\|_{\mathrm{HS}}^2 .\end{aligned}$$

In summary,

$$\|u(x)\|_{H_2} \leq \|x\|_{H_1} \, \|u\|_{\mathrm{HS}} ,$$

so by the definition of the operator norm in $\mathcal{L}(H_1, H_2)$ as $\|u\| = \sup\{\|u(x)\|_{H_2} \mid \|x\|_{H_1} = 1\}$, we conclude that

$$\|u\| \leq \|u\|_{\mathrm{HS}} . \qquad (\dagger_1)$$

The previous proof in the case where $H_1 = H_2$ can also be adapted to prove that $\mathcal{L}_2(H_1, H_2)$ is a Banach space.

An important special case arises when the Hilbert space H is the space of functions $\mathrm{L}^2_\mu(X)$ for some measure space (X, μ). For simplicity of notation, we are omitting the σ-algebra $\mathcal{A}$ on which the positive measure μ is defined. Again, we assume that $\mathrm{L}^2_\mu(X)$ is separable. Then a classical ("folklore") result of functional analysis states that a linear map $T\colon \mathrm{L}^2_\mu(X) \to \mathrm{L}^2_\mu(X)$ is a Hilbert–Schmidt operator iff there is a kernel $\mathcal{K}\colon X \times X \to \mathbb{C}$, which is a function in $\mathrm{L}^2_{\mu\otimes\mu}(X \times X)$, such that

$$(T(f))(x) = \int_X \mathcal{K}(x, y) f(y) \, d\mu(y), \quad f \in \mathrm{L}^2_\mu(X),\ x \in X. \qquad \text{(HS1)}$$

Recall that to assert $\mathcal{K} \in \mathrm{L}^2_{\mu\otimes\mu}(X \times X)$ means that $\mathcal{K}$ is a $\mu \otimes \mu$-measurable map and

$$\|\mathcal{K}\|_2^2 = \int_X \int_X |\mathcal{K}(x, y)|^2 \, d\mu(x) \, d\mu(y) < \infty. \qquad \text{(HS2)}$$

It is worth explaining what is the significance of requiring that $\mathcal{K} \in \mathrm{L}^2_{\mu\otimes\mu}(X \times X)$. We follow the argument in Lang [Lang (1985)] (Chapter I, Section 3, Theorem 2).

Given any $f \in \mathrm{L}^2_\mu(X)$, by Fubini, for every fixed $x \in X$ (except for a subset of measure zero), the function $\mathcal{K}_x$ given by $\mathcal{K}_x(y) = \mathcal{K}(x, y)$ also belongs to $\mathrm{L}^2_\mu(X)$, so the function $y \mapsto \mathcal{K}_x(y)f(y)$ belongs to $\mathrm{L}^1_\mu(X)$, but in general *not to* $\mathrm{L}^2_\mu(X)$; see Lang [Lang (1996)] (Chapter VII, Theorem 1.1). By Cauchy–Schwarz, we have

$$\begin{aligned}|T(f)(x)|^2 = \left|\int_X \mathcal{K}_x(y)f(y)\,d\mu(y)\right|^2 &\leq \left(\int_X |\mathcal{K}_x(y)|\,|f(y)|\,d\mu(y)\right)^2\\ &\leq \|f\|_2^2 \int_X |\mathcal{K}_x(y)|^2\,d\mu(y),\end{aligned}$$

so by integrating, we obtain

$$\begin{aligned}\|T(f)\|_2^2 &= \int_X |T(f)(x)|^2\,d\mu(x) \leq \int_X \|f\|_2^2 \int_X |\mathcal{K}_x(y)|^2\,d\mu(y)\,d\mu(x)\\ &= \|f\|_2^2 \int_X \int_X |\mathcal{K}(x,y)|^2\,d\mu(y)\,d\mu(x) = \|f\|_2^2\,\|\mathcal{K}\|_2^2.\end{aligned}$$

Therefore, we have

$$\|T(f)\|_2^2 \leq \|f\|_2^2\,\|\mathcal{K}\|_2^2,$$

which implies that $T(f) \in \mathrm{L}^2_\mu(X)$, and if $\|T\|$ is the operator norm of T in $\mathcal{L}(\mathrm{L}^2_\mu(X))$, we have

$$\|T\| \leq \|\mathcal{K}\|_2, \tag{$\dagger_2$}$$

which shows that T is continuous.

We also sketch the rest of the proof given in Lang [Lang (1985)] (Chapter I, Section 3, Theorem 2). To prove that T is Hilbert–Schmidt, we express the kernel $\mathcal{K}$ in terms of the Hilbert basis of $\mathrm{L}^2_{\mu\otimes\mu}(X \times X)$ given by the family of functions $\varphi_i(x)\varphi_j(y)$ where (φ_i) is a Hilbert basis of $\mathrm{L}^2_\mu(X)$ (which is assumed to be separable). If

$$\mathcal{K}(x,y) = \sum_{ij} c_{ij}\varphi_i(x)\varphi_j(y), \quad c_{ij} \in \mathbb{C},$$

we consider the approximation kernels

$$\mathcal{K}_n(x,y) = \sum_{i,j=1}^{n} c_{ij}\varphi_i(x)\varphi_j(y)$$

and the corresponding linear maps T_n given by

$$T_n(f)(x) = \int_X \mathcal{K}_n(x,y) f(y)\, d\mu(y).$$

Then as in Lang [Lang (1985)] (Chapter I, Section 3, Theorem 2), we show that the T_n have finite rank and converge to T because the $\mathcal{K}_n$ converge to $\mathcal{K}$ in $\mathrm{L}^2_{\mu\otimes\mu}(X \times X)$ and because

$$\|T - T_n\| \leq \|\mathcal{K} - \mathcal{K}_n\|_2,$$

by ($\dagger_2$). This implies that T is Hilbert–Schmidt.

For the converse, if T is Hilbert–Schmidt, we use Lang [Lang (1996)] (Chapter XVII, Problem 9(d)). For each function φ_j in the Hilbert basis of $\mathrm{L}^2_\mu(X)$, each $T(\varphi_j)$ can be expressed as

$$T(\varphi_j) = \sum_j t_{ij}\varphi_i, \quad t_{ij} \in \mathbb{C}.$$

Then we prove that

$$\sum_{ij} |t_{ij}|^2 < \infty$$

and that the kernel

$$\mathcal{K}(x,y) = \sum_{ij} t_{ij}\varphi_i(x)\overline{\varphi_j(y)} \tag{$\mathcal{K}_1$}$$

works.

By the way, all this is Exercise 44 from Section 6, Chapter XI in Dunford and Schwartz [Dunford and Schwartz (1988)] without any hint!

In Chapter 8, we will need to consider continuous linear maps $T\colon \mathrm{L}^2_\mu(X, \mathcal{H}_1) \to \mathrm{L}^2_\mu(X, \mathcal{H}_2)$, where $\mathcal{H}_1$ and $\mathcal{H}_2$ are two finite-dimensional vector spaces equipped with Hermitian inner products. This leads us to consider kernels $\mathcal{K}\colon X \times X \to \mathrm{Hom}(\mathcal{H}_1, \mathcal{H}_2)$, which are functions in $\mathrm{L}^2_{\mu\otimes\mu}(X \times X, \mathrm{Hom}(\mathcal{H}_1, \mathcal{H}_2))$, such that

$$(T(f))(x) = \int_X \mathcal{K}(x,y)(f(y))\, d\mu(y), \quad f \in \mathrm{L}^2_\mu(X, \mathcal{H}_1),\ x \in X. \tag{HS3}$$

In this more general case, the inner product $\langle -, - \rangle_{\mu, \mathcal{H}_1}$ on $\mathrm{L}^2_\mu(X, \mathcal{H}_1)$ is given by

$$\langle f, g \rangle_{\mu, \mathcal{H}_1} = \int_X \langle f(x), g(x) \rangle_{\mathcal{H}_1}\, d\mu(x),$$

where $\langle -, - \rangle_{\mathcal{H}_1}$ is the inner product on $\mathcal{H}_1$, and the norm is given by

$$\|f\|_2^2 = \int_X \langle f(x), f(x) \rangle_{\mathcal{H}_1} \, d\mu(x) = \int_X \|f(x)\|_{\mathcal{H}_1}^2 \, d\mu(x).$$

The inner product $\langle -, - \rangle_{\mu, \mathcal{H}_2}$ on $\mathrm{L}^2_\mu(X, \mathcal{H}_2)$ is defined similarly.

The norm on the space $\mathrm{L}^2_{\mu\otimes\mu}(X \times X, \mathrm{Hom}(\mathcal{H}_1, \mathcal{H}_2))$ is given by

$$\|\mathcal{K}\|_2^2 = \int_X \int_X \|\mathcal{K}(x,y)\|^2_{\mathrm{Hom}(\mathcal{H}_1,\mathcal{H}_2)} \, d\mu(x) \, d\mu(y), \tag{$\dagger_3$}$$

where $\| \, \|_{\mathrm{Hom}(\mathcal{H}_1,\mathcal{H}_2)}$ is the norm induced by the inner product on $\mathrm{Hom}(\mathcal{H}_1, \mathcal{H}_2)$, which can be viewed as the Frobenius inner product; see Definition 4.15 in Section 4.5. As earlier, for any fixed $x \in X$, we define the function $\mathcal{K}_x$ by $\mathcal{K}_x(y) = \mathcal{K}(x,y)$. Then the function $y \mapsto \mathcal{K}_x(y)(f(y))$ (with $y \in X$) is in $\mathrm{L}^1_\mu(X, \mathcal{H}_2)$. Since the operator norm on $\mathrm{Hom}(\mathcal{H}_1, \mathcal{H}_2)$ has the property that

$$\|\mathcal{K}_x(y)(f(y))\|_{\mathcal{H}_2} \leq \|\mathcal{K}_x(y)\| \, \|f(y)\|_{\mathcal{H}_1}, \tag{$\dagger_4$}$$

where $\|\mathcal{K}_x(y)\|$ is the operator norm in $\mathrm{Hom}(\mathcal{H}_1, \mathcal{H}_2)$, by ($\dagger_4$) and Cauchy–Schwarz, we have

$$\begin{aligned}
\|T(f)(x)\|^2_{\mathcal{H}_2} &= \left\| \int_X \mathcal{K}_x(y)(f(y)) \, d\mu(y) \right\|^2_{\mathcal{H}_2} \\
&\leq \left(\int_X \|\mathcal{K}_x(y)(f(y))\|_{\mathcal{H}_2} \, d\mu(y) \right)^2 \\
&\leq \left(\int_X \|\mathcal{K}_x(y)\| \, \|f(y)\|_{\mathcal{H}_1} \, d\mu(y) \right)^2 \\
&\leq \int_X \|\mathcal{K}_x(y)\|^2 \, d\mu(y) \int_X \|f(y)\|^2_{\mathcal{H}_1} \, d\mu(y) \\
&= \|f\|_2^2 \int_X \|\mathcal{K}_x(y)\|^2 \, d\mu(y).
\end{aligned}$$

By integrating, we obtain

$$\begin{aligned}
\|T(f)\|_2^2 = \int_X \|T(f)(x)\|_2^2 \, d\mu(x) &\leq \int_X \|f\|_2^2 \int_X \|\mathcal{K}_x(y)\|^2 \, d\mu(y) \, d\mu(x) \\
&= \|f\|_2^2 \int_X \int_X \|\mathcal{K}(x,y)\|^2 \, d\mu(x) \, d\mu(y).
\end{aligned}$$

Therefore, if

$$\int_X \int_X \|\mathcal{K}(x,y)\|^2 \, d\mu(x) \, d\mu(y) < \infty,$$

then we have

$$\|T(f)\|_2^2 \leq \|f\|_2^2 \int_X \int_X \|\mathcal{K}(x,y)\|^2 \, d\mu(x) \, d\mu(y).$$

However, since the inner product on $\mathrm{Hom}(\mathcal{H}_1, \mathcal{H}_2)$ is the Frobenius inner-product, the corresponding norm is the Hilbert–Schmidt norm, so by ($\dagger_2$) we have

$$\|\mathcal{K}(x,y)\| \leq \|\mathcal{K}(x,y)\|_{\mathrm{Hom}(\mathcal{H}_1,\mathcal{H}_2)}$$

and the fact that $\mathcal{K} \in \mathrm{L}^2_{\mu\otimes\mu}(X \times X, \mathrm{Hom}(\mathcal{H}_1, \mathcal{H}_2))$ implies that

$$\int_X \int_X \|\mathcal{K}(x,y)\|^2_{\mathrm{Hom}(\mathcal{H}_1,\mathcal{H}_2)} \, d\mu(x) \, d\mu(y) < \infty,$$

so

$$\begin{aligned} \|T(f)\|_2^2 &\leq \|f\|_2^2 \int_X \int_X \|\mathcal{K}(x,y)\|^2_{\mathrm{Hom}(\mathcal{H}_1,\mathcal{H}_2)} \, d\mu(x) \, d\mu(y) \\ &= \|f\|_2^2 \, \|\mathcal{K}\|_2^2 < \infty, \end{aligned}$$

which implies that $T(f) \in \mathrm{L}^2_\mu(X, \mathcal{H}_2)$ and

$$\|T\| \leq \|\mathcal{K}\|_2 \, ,$$

which shows that T is continuous.

In order to prove that T is a Hilbert–Schmidt operator from $\mathrm{L}^2_\mu(X, \mathcal{H}_1)$ to $\mathrm{L}^2_\mu(X, \mathcal{H}_2)$, we need to express the kernel $\mathcal{K}$ over a Hilbert basis of $\mathrm{L}^2_{\mu\otimes\mu}(X \times X, \mathrm{Hom}(\mathcal{H}_1, \mathcal{H}_2))$. Such a basis can be obtained using duality.

Recall that there is an isomorphism between the finite-dimensional vector space $\mathcal{H}_1$ and its dual $\mathcal{H}_1^*$, the vector space of linear forms on $\mathcal{H}_1$ (which are continuous since $\mathcal{H}_1$ is finite-dimensional) defined such that for every $u \in \mathcal{H}_1$, the linear form $\flat(u)$ on $\mathcal{H}_1$ also denoted $u^\flat$ is defined by

$$\flat(u)(v) = u^\flat(v) = \langle v, u \rangle_{\mathcal{H}_1}, \quad v \in \mathcal{H}_1;$$

see Gallier and Quaintance [Gallier and Quaintance (2020c)] (Chapter 13, Section 2). Then given a Hilbert basis (φ_n) in $\mathrm{L}^2_\mu(X, \mathcal{H}_1)$ and a Hilbert basis (ψ_n) in $\mathrm{L}^2_\mu(X, \mathcal{H}_2)$, consider the family of functions $\varphi_j^\flat \psi_i$ from $X \times X$ to $\mathrm{Hom}(\mathcal{H}_1, \mathcal{H}_2)$ defined by

$$[(\varphi_j^\flat \psi_j)(x,y)](u) = \flat(\varphi_j(x))(u)\psi_i(y) = \langle u, \varphi_j(x) \rangle_{\mathcal{H}_1} \psi_i(y), \qquad (*_{\varphi_j^\flat \psi_j})$$

for all $x, y \in X$, $u \in \mathcal{H}_1$. For readers familiar with tensor products, note that we are implicitly using the isomorphism between $\mathrm{Hom}(\mathcal{H}_1, \mathcal{H}_2)$ and $\mathcal{H}_1^* \otimes \mathcal{H}_2$; see Gallier and Quaintance [Gallier and Quaintance (2020b)] (Chapter 2, Section 2.5) and Section 4.5.

Using the method described in Lang [Lang (1985)] (Chapter I, Section 3, Theorem 2), it can be shown that the functions $\varphi_j^\flat\psi_i$ form a Hilbert basis of $\mathrm{L}^2_{\mu\otimes\mu}(X \times X, \mathrm{Hom}(\mathcal{H}_1, \mathcal{H}_2))$. The details are quite technical and left as an exercise. Next, we express the kernel $\mathcal{K}$ in terms of this Hilbert basis as

$$\mathcal{K} = \sum_{ij} c_{ij}\varphi_j^\flat\psi_i, \quad c_{ij} \in \mathbb{C},$$

and we define the approximation kernels $\mathcal{K}_n$ by

$$\mathcal{K}_n = \sum_{ij=1}^{n} c_{ij}\varphi_j^\flat\psi_i,$$

and as before, we prove that T defined by (HS3) is Hilbert–Schmidt.

Conversely, given a Hilbert–Schmidt operator from $\mathrm{L}^2_\mu(X, \mathcal{H}_1)$ to $\mathrm{L}^2_\mu(X, \mathcal{H}_2)$, write

$$T\varphi_j = \sum_i t_{ij}\psi_i, \quad t_{ij} \in \mathbb{C},$$

and define the kernel $\mathcal{K}$ in $\mathrm{L}^2_{\mu\otimes\mu}(X \times X, \mathrm{Hom}(\mathcal{H}_1, \mathcal{H}_2))$ as

$$\mathcal{K}(x, y) = \sum_{ij} t_{ij}(\varphi_j^\flat\psi_i)(y, x), \quad t_{ij} \in \mathbb{C}. \tag{$\mathcal{K}_2$}$$

The permutation of the variables x and y on the right-hand side is necessary because the argument to $\mathcal{K}(x, y)$ in (HS3) is $f(y)$ and the output function is $T(f)(x)$, a function of x, but

$$[(\varphi_j^\flat\psi_j)(x, y)](\varphi_k(y)) = \langle\varphi_k(y), \varphi_j(x)\rangle_{\mathcal{H}_1}\psi_i(y),$$

which is not what we want, since x and y are switched. We check that $\mathcal{K}$ works by proving that $\sum_{ij}|t_{ij}|^2 < \infty$ (since T is Hilbert–Schmidt) and by verifying that the map S given by (HS3) agrees with T on the Hilbert bases. Since (φ_j) is an orthonormal basis, we have

$$\begin{aligned}
(S(\varphi_k))(x) &= \int_X \mathcal{K}(x, y)(\varphi_k(y))\, d\mu(y) \\
&= \int_X \sum_{ij} t_{ij}[(\varphi_j^\flat\psi_i)(y, x)](\varphi_k(y))\, d\mu(y) \\
&= \sum_{ij} t_{ij} \int_X \langle\varphi_k(y), \varphi_j(y)\rangle_{\mathcal{H}_1}\psi_i(x)\, d\mu(y) \\
&= \sum_{ij} t_{ij}\psi_i(x) \int_X \langle\varphi_k(y), \varphi_j(y)\rangle_{\mathcal{H}_1}\, d\mu(y) \\
&= \sum_{ij} t_{ij}\psi_i(x)\ \langle\varphi_k, \varphi_j\rangle_{\mu,\mathcal{H}_1} = \sum_i t_{ik}\psi_i(x) = (T(\varphi_k))(x),
\end{aligned}$$

as claimed.

The justification of the above computation is left as an exercise.

Note that we can immediately generalize all this to two spaces $\mathrm{L}^2_{\mu_X}(X, \mathcal{H}_1)$ and $\mathrm{L}^2_{\mu_Y}(Y, \mathcal{H}_2)$, where X and Y are different spaces and kernels $\mathcal{K} \in \mathrm{L}^2_{\mu_X \otimes \mu_Y}(X \times Y, \mathrm{Hom}(\mathcal{H}_1, \mathcal{H}_2))$ but then we need to define $T(f)$ as

$$(T(f))(y) = \int_X \mathcal{K}(x, y)(f(x))\, d\mu(x), \quad f \in \mathrm{L}^2_\mu(X, \mathcal{H}_1),\ y \in Y, \tag{HS4}$$

with $x \in X$ and $y \in Y$, which explains why we had to switch x and y in $(\mathcal{K}_2)$.

In retrospect, it might be better to use the formula

$$(T(f))(y) = \int_X \mathcal{K}(x, y)(f(x))\, d\mu(x), \quad f \in \mathrm{L}^2_\mu(X, \mathcal{H}_1),\ y \in Y$$

instead of the formula

$$(T(f))(x) = \int_X \mathcal{K}(x, y)(f(y))\, d\mu(y), \quad f \in \mathrm{L}^2_\mu(X, \mathcal{H}_1),\ x \in Y$$

used in (HS3) to define $T(f)$.

Remark: Lang [Lang (1996)] (Chapter XVII, Problem 9) deals with this technical detail by viewing (HS3) as defining a map from $\mathrm{L}^2_{\mu_Y}(Y, \mathcal{H}_2)$ to $\mathrm{L}^2_{\mu_X}(X, \mathcal{H}_1)$.

In summary, the fundamental representation of Hilbert–Schmidt operators in terms of kernels generalizes to operators from $\mathrm{L}^2_{\mu_X}(X, \mathcal{H}_1)$ to $\mathrm{L}^2_{\mu_Y}(Y, \mathcal{H}_2)$.

In the next sections, we consider the special case in which a Hilbert algebra is complete.

2.6 Complete Separable Hilbert Algebras

We now consider the case where the Hilbert algebra A is complete, which means that it is a Hilbert space. Although this is not obvious, as said in the previous section, it can be shown that the map $(x, y) \mapsto xy$ from $A \times A$ to A is continuous. Consequently, by Proposition 2.16, a complete Hilbert algebra is normable, and thus a Banach algebra.

Our main goal is to show that every separable complete Hilbert algebra is the (countable) Hilbert sum of two-sided ideals $\mathfrak{a}_k$, where each ideal $\mathfrak{a}_k$ is the (countable) Hilbert sum of minimal left ideals all isomorphic to a common left ideal (Theorem 2.33). This is a beautiful and powerful result which is one of the main steps in proving the Peter–Weyl theorem.

Following Dieudonné, we only deal with the case where the Hilbert algebras A are separable, because then only countable Hilbert sums are needed. This implies that when we consider the Hilbert algebra $A = \mathrm{L}^2(G)$, the group G is metrizable and compact. This is not a serious restriction, because every Lie group, being a second-countable manifold, is metrizable. The reader who wishes to see an exposition of the Peter–Weyl theorem in the general case of an arbitrary compact group is invited to consult Folland [Folland (1995)] (Chapter 5). Hilbert sums indexed by arbitrary (possibly uncountable) index sets arise.

We are led to the study of minimal left ideals and to irreducible self-adjoint idempotents which generate them. Recall that an element e of an algebra A is *idempotent* if $e^2 = e$, *self-adjoint* if $e = e^*$.[2]

Complete proofs are provided in Dieudonné [Dieudonné (2005)] (Chapter XV, Section 8). There are many tedious technical details so to make it easier on the reader, we decided to only state most results (except the most important ones) without proof.

First, observe that every closed self-adjoint subalgebra B (which means that $B^* = B$) of a complete Hilbert algebra A is a complete Hilbert algebra. Certain representations play a crucial role.

Proposition 2.20. *Let A be a complete Hilbert algebra. For any closed left ideal $\mathfrak{b}$ in A, let $U_{\mathfrak{b}}$ be the map from A to $\mathcal{L}(\mathfrak{b})$ given by*

$$U_{\mathfrak{b}}(x)(y) = xy, \qquad x \in A,\ y \in \mathfrak{b}.$$

Observe that $U_{\mathfrak{b}}(x)$ is left multiplication by $x \in A$. Then $U_{\mathfrak{b}}$ is a representation of A in $\mathfrak{b}$.

Proof. For every $x \in A$, the linear map $U_{\mathfrak{b}}(x)$ from $\mathfrak{b}$ to itself is continuous because of condition (U). The verification that $U_{\mathfrak{b}}(x_1x_2) = U_{\mathfrak{b}}(x_1) \circ U_{\mathfrak{b}}(x_2)$ is immediate. For all $y, z \in \mathfrak{b}$, by condition (2), we have

$$\begin{aligned}\langle U_{\mathfrak{b}}(x)^*(y), z\rangle &= \langle y, U_{\mathfrak{b}}(x)(z)\rangle \\ &= \langle y, xz\rangle \\ &= \langle x^*y, z\rangle \\ &= \langle U_{\mathfrak{b}}(x^*)(y), z\rangle,\end{aligned}$$

[2]Here, e does not denote the unit of A if A is unital. To avoid confusion, we denote the unit of A by $\mathbf{1}$ if A is unital.

which implies that $U_{\mathfrak{b}}(x)^* = U_{\mathfrak{b}}(x^*)$. If A has the unit element $\mathbf{1}$, then $U_{\mathfrak{b}}(\mathbf{1})$ is the identity transformation. □

Definition 2.17. Let A be a complete Hilbert algebra. For any closed left ideal $\mathfrak{b}$ in A, let $U_{\mathfrak{b}}\colon A \to \mathcal{L}(\mathfrak{b})$ be the representation given by

$$U_{\mathfrak{b}}(x)(y) = xy, \qquad x \in A,\ y \in \mathfrak{b}.$$

The representation $U_A\colon A \to \mathcal{L}(A)$ is called the *regular representation* of A.

When $\mathfrak{b} = A$, we usually write $U(x)$ instead of $U_A(x)$. By Proposition 2.17, the representation U_A is faithful. Moreover, since the map $(x, y) \mapsto xy$ from $A \times A$ to A is continuous, the map $x \mapsto U_A(x)$ from A to $\mathcal{L}(A)$ is also continuous.

Let A be a complete Hilbert algebra. For each left ideal $\mathfrak{b}$ of A, the set $\mathfrak{b}^* = \{x^* \mid x \in \mathfrak{b}\}$ is a right ideal in A. We now start a fairly long chain of definitions and results leading to our main result (Theorem 2.33).

In the sequel A denotes a complete Hilbert algebra. Two key concepts are the notion of an irreducible self-adjoint idempotent and a minimal left ideal.

Proposition 2.21. *For every left ideal $\mathfrak{l}$ of A, the orthogonal complement $\bar{\mathfrak{l}}^{\perp}$ of the closure of $\mathfrak{l}$ is a left ideal.*

Proposition 2.21 is proven in Dieudonné [Dieudonné (2005)] (Chapter XV, Proposition 15.8.2). The next proposition gives useful properties of idempotents.

Proposition 2.22. *Let $e \neq 0$ be an idempotent element in A ($e^2 = e$). Then the following properties hold:*

(1) $\|e\| \geq 1$;
(2) e^* *is idempotent;*
(3) The left ideal Ae is equal to the set $\{x \in A \mid x = xe\}$, and is closed in A.

Proposition 2.22 is proven in Dieudonné [Dieudonné (2005)] (Chapter XV, Proposition 15.8.3). The next proposition gives orthogonality properties of self-adjoint idempotents.

Proposition 2.23. *If e_1 and e_2 are self-adjoint idempotents in A, then the following properties are equivalent:*

(a) $\langle e_1, e_2\rangle = 0$;
(b) $e_1 e_2 = 0$;
(c) $e_2 e_1 = 0$.

The proof of Proposition 2.23 makes use of property (2) and property (2') of Definition 2.14; see Dieudonné [Dieudonné (2005)] (Chapter XV, Proposition 15.8.4).

The following proposition shows that there are plenty nonzero self-adjoint idempotents. Since Proposition 2.24 establishes a very important fact, we supply a proof.

Proposition 2.24. *Every left ideal $\mathfrak{l} \neq (0)$ in A contains a nonzero self-adjoint idempotent.*

Proof. Let x be any nonzero element in $\mathfrak{l}$. By Proposition 2.17, we have $x^*x \neq 0$. Let $z = x^*x$. Then z is a self-adjoint element of $\mathfrak{l}$, but in general it is not idempotent. Consider the representation U_A. By rescaling z, we may assume that $\|U_A(z)\| = 1$. Since $z = z^*$, the linear map $U_A(z)$ is self-adjoint, and we have $U_A(z^2) = U_A(z) \circ U_A(z) = (U_A(z))^2$. By Vol. I, Proposition D.11, we have

$$\left\|U_A(z^2)\right\| = \left\|(U_A(z))^2\right\| = \|U_A(z)\|^2 = 1. \qquad (*_1)$$

We claim that the sequence (z^{2k}) is a Cauchy sequence, whose limit e is a nonzero self-adjoint idempotent in $\mathfrak{l}$.

We are led to investigate bounds on $\left\|z^{2m} - z^{2n}\right\|$, with $m = n + p$, $n, p > 0$ and the proof uses the following steps.

Step 1. First, we prove that $\left\|U_A(z^k)\right\| = 1$ for all $k \geq 1$.

From Equation $(*_1)$, by induction we obtain

$$\left\|U_A(z^{2k})\right\| = 1, \quad k \geq 1. \qquad (*_2)$$

On the other hand, since $\|U_A(z)\| = 1$, we have

$$\left\|U_A(z^{k+1})\right\| = \left\|U_A(z) \circ U_A(z^k)\right\| \leq \|U_A(z)\| \left\|U_A(z^k)\right\| = \left\|U_A(z^k)\right\|.$$

Thus the sequence $(\left\|U_A(z^k)\right\|)$ is nonincreasing, and since it contains infinitely many terms (of the form $\left\|U_A(z^{2i})\right\|$) equal to 1, we must indeed have

$$\left\|U_A(z^k)\right\| = 1, \quad \text{for all } k \geq 1. \qquad (*_3)$$

Since we are using the operator norm, we deduce that

$$1 = \left\|U_A(z^k)\right\| \leq \|U_A\| \left\|z^k\right\|,$$

which implies that

$$\left\|z^k\right\| \geq 1/\left\|U_A\right\|, \quad k \geq 1. \tag{$*_4$}$$

Here, $\|U_A\|$ is the operator norm of U_A as a continuous linear map from A to $\mathcal{L}(A)$.

Step 2. Next, we show that the sequence $(\left\|z^{2k}\right\|^2)$ is nonincreasing, and since it is bounded from below by $1/\left\|U_A\right\|^2$, it has a limit $a > 0$.

Since z is self-adjoint and $U_A(x)(y) = xy$, using property (2) of Definition 2.14, we have (recall that $m = n + p$)

$$\begin{aligned}\langle z^{2m}, z^{2n}\rangle &= \langle z^p z^{p+2n}, z^{2n}\rangle = \langle z^{p+2n}, z^{p+2n}\rangle = \left\|U_A(z^p)(z^{2n})\right\|^2 \\ &\leq \left\|U_A(z^p)\right\|^2 \left\|z^{2n}\right\|^2 = \left\|z^{2n}\right\|^2 = \langle z^{2n}, z^{2n}\rangle,\end{aligned}$$

and

$$\begin{aligned}\langle z^{2m}, z^{2m}\rangle &= \left\|U(z^p)(z^{p+2n})\right\|^2 \leq \left\|U(z^p)\right\|^2 \left\|z^{p+2n}\right\|^2 = \left\|z^{p+2n}\right\|^2 \\ &= \langle z^{p+2n}, z^{p+2n}\rangle = \langle z^{2m}, z^{2n}\rangle,\end{aligned}$$

so that for all $m > n$, we have

$$1/\left\|U_A\right\|^2 \leq \langle z^{2m}, z^{2m}\rangle \leq \langle z^{2m}, z^{2n}\rangle \leq \langle z^{2n}, z^{2n}\rangle. \tag{$*_5$}$$

Then $(*_5)$ shows that the sequence $(\left\|z^{2k}\right\|^2)$ is nonincreasing, and since it is bounded from below by $1/\left\|U_A\right\|^2$, it has a limit $a > 0$.

Step 3. Using $(*_5)$, we also have

$$\begin{aligned}\left\|z^{2m} - z^{2n}\right\|^2 &= \langle z^{2m}, z^{2m}\rangle - 2\langle z^{2m}, z^{2n}\rangle + \langle z^{2n}, z^{2n}\rangle \\ &\leq \langle z^{2n}, z^{2n}\rangle - \langle z^{2m}, z^{2n}\rangle \\ &\leq \left\|z^{2n}\right\|^2 - a.\end{aligned}$$

Since the sequence $(\left\|z^{2k}\right\|^2)$ has limit a, the sequence (z^{2k}) is a Cauchy sequence, as asserted earlier.

Step 4. Let e be the limit of the Cauchy sequence (z^{2k}). By continuity,

$$e^2 = \lim_{k \mapsto \infty} z^{4k} = e,$$

and

$$e^* = \lim_{k \mapsto \infty} (z^*)^{4k} = \lim_{k \mapsto \infty} z^{4k} = e.$$

We also have

$$ez^2 = \lim_{k \mapsto \infty} z^{2k+2} = e,$$

and since $z \in \mathfrak{l}$ and $\mathfrak{l}$ is a left ideal, we see that $e \in \mathfrak{l}$. Finally, since by $(*_4)$ we have

$$\left\|z^k\right\| \geq 1/\left\|U_A\right\|, \quad k \geq 1,$$

and we deduce that $\|e\| > 0$, so $e \neq 0$. ☐

Definition 2.18. A self-adjoint idempotent $e \neq 0$ is *reducible* if there exist two orthogonal nonzero self-adjoint idempotents e_1 and e_2, such that $e = e_1 + e_2$. If a self-adjoint idempotent $e \neq 0$ is not reducible, we say that it is *irreducible*.

By Proposition 2.23, if $e = e_1 + e_2$ is reducible, then $ee_1 = e_1 e = e_1$ and $ee_2 = e_2 e = e_2$.

Proposition 2.25 shows that irreducible self-adjoint idempotents are the building blocks for self-adjoint idempotents.

Proposition 2.25. *Every self-adjoint idempotent $e \neq 0$ is the sum of a finite number of irreducible self-adjoint idempotents in Ae. Every left ideal $\mathfrak{l} \neq (0)$ contains an irreducible self-adjoint idempotent.*

Proof. Here is the proof of the first statement.

Let $e \neq 0$ be a self adjoint idempotent. Since $ee = e$, obviously $e \in Ae$. We claim that if $\|e\|^2 < 2$, then e is irreducible. Otherwise, $e = e_1 + e_2$ for two orthogonal nonzero self-adjoint idempotents e_1 and e_2, so

$$\|e\|^2 = \|e_1\|^2 + \|e_2\|^2 .$$

By Proposition 2.22, since $e_1, e_2 \neq 0$, we have $\|e_1\|, \|e_2\| \geq 1$, so $\|e\|^2 \geq 2$, a contradiction.

If $\|e\|^2 \geq 2$, we prove by complete induction on the smallest natural number n, such that $\|e\|^2 < n$, that e is the sum of a finite number of irreducible self-adjoint idempotents in Ae.

If e is reducible, then $e = e_1 + e_2$, where $e_1, e_2 \neq 0$ are orthogonal self-adjoint idempotents. By Proposition 2.23, we have $e_1 = e_1 e$ and $e_2 = e_2 e$, which implies that $e_1, e_2 \in Ae$. Since $\|e\|^2 = \|e_1\|^2 + \|e_2\|^2$, we have

$$\|e_1\|^2 = \|e\|^2 - \|e_2\|^2 \leq \|e\|^2 - 1 < n - 1,$$

and similarly,

$$\|e_2\|^2 < n - 1.$$

Therefore, we can apply the induction hypothesis to e_1 and e_2, and this finishes the proof. The second statement follows from Proposition 2.24 and what we just proved. $\square$

Definition 2.19. A left ideal $\mathfrak{l}$ is *minimal* if $\mathfrak{l} \neq (0)$ and if there exists no nonzero left ideal $\mathfrak{l}' \neq \mathfrak{l}$, such that $\mathfrak{l}' \subseteq \mathfrak{l}$. A similar definition applies to minimal right ideals.

Here is the first significant result which gives the structure of minimal left ideals in terms of irreducible self-adjoint idempotents.

Theorem 2.26. *A left ideal $\mathfrak{l}$ in A is minimal if and only if it is of the form $\mathfrak{l} = Ae$, where $e \neq 0$ and e is an irreducible self-adjoint idempotent.*

Proof. First, assume that $\mathfrak{l}$ is a minimal left ideal. By Proposition 2.25, the ideal $\mathfrak{l}$ contains an irreducible self-adjoint idempotent $e \neq 0$. Since $e \in \mathfrak{l}$ and $\mathfrak{l}$ is a left ideal, we have $Ae \subseteq \mathfrak{l}$ and $e = e^2 \in Ae$. This shows that Ae is a nonzero ideal contained in the minimal ideal $\mathfrak{l}$, which implies that $\mathfrak{l} = Ae$.

Conversely, let $e \neq 0$ be an irreducible self-adjoint idempotent. We need to prove that $\mathfrak{l} = Ae$ is a minimal ideal.

Suppose by contradiction that $\mathfrak{l}$ contains a left ideal $\mathfrak{l}' \neq (0)$, such that $\mathfrak{l}' \neq \mathfrak{l}$. By Proposition 2.24, there is some self-adjoint idempotent $e' \neq 0$ that belongs to $\mathfrak{l}'$.

If we let $e_1 = e - ee'$, and $e_2 = ee'$, then $e = e_1 + e_2$, and we claim that e_1 and e_2 are orthogonal nonzero self-adjoint idempotents. If so, this contradicts the fact that e is irreducible, and finishes the proof by contradiction.

Since $e' \in Ae$ and $ee = e$, we have

$$e' = e'e.$$

Consequently,

$$\begin{aligned} e_2^2 &= ee'ee' = ee'e' = ee' = e_2 \\ ee_2 &= eee' = ee' = e_2 \\ e_2e &= ee'e = ee' = e_2, \end{aligned}$$

and thus

$$\begin{aligned} e_1e_2 &= (e - e_2)e_2 = ee_2 - e_2^2 = e_2 - e_2 = 0 \\ e_2e_1 &= e_2(e - e_2) = e_2e - e_2^2 = e_2 - e_2 = 0 \\ e_1^2 &= (e - e_2)^2 = e^2 - ee_2 - e_2e + e_2^2 = e - e_2 - e_2 + e_2 = e - e_2 = e_1. \end{aligned}$$

We also have

$$e_2^* = (ee'e)^* = e^*(e')^*e^* = ee'e = e_2,$$

and so

$$e_1^* = (e - e_2)^* = e^* - e_2^* = e - e_2 = e_1.$$

In summary, we proved that e_1, e_2 are orthogonal self-adjoint idempotents. It remains to prove that $e_1 \neq 0$ and $e_2 \neq 0$.

Since $e = e_1 + e_2 = e_1 + ee'$, if $e_1 = 0$, then $e = ee' \in \mathfrak{l}'$ (since $e' \in \mathfrak{l}'$), hence $\mathfrak{l}' = \mathfrak{l}$, contradicting the hypothesis that $\mathfrak{l}' \neq \mathfrak{l}$. Since $e_2 = ee'$, we have

$$e'e_2 = e'ee' = e'e' = e' \neq 0,$$

which implies that $e_2 \neq 0$. Finally, we showed that $e = e_1 + e_2$ is a reducible decomposition of e, contradicting the hypothesis that e is irreducible. $\square$

Example 2.8. Consider the algebra $A = \mathrm{M}_n(\mathbb{C})$ from Example 2.1 consisting of $n \times n$ complex matrices with involution $X \mapsto X^*$ (the conjugate transpose). It is immediately verified that the map

$$\langle X, Y \rangle \mapsto \langle X, Y \rangle = \mathrm{tr}(Y^* X)$$

is a Hermitian inner product on A, which makes A into a complete Hilbert algebra, which is obviously separable. We immediately check that the $n \times n$ matrices E_i defined such that $(E_i)_{jk} = 1$ iff $j = k = i$, else $(E_i)_{jk} = 0$, are irreducible self-adjoint idempotents in A. Then the subspaces $\mathfrak{l}_i = AE_i$ $(1 \leq i \leq n)$ are minimal left ideals in A. Observe that $\mathfrak{l}_i = AE_i$ consists of those $n \times n$ matrices whose columns of index $1, \ldots, i-1, i+1, \ldots, n$ are zero columns. Observe that

$$A = \bigoplus_{j=1}^{n} \mathfrak{l}_j = \bigoplus_{j=1}^{n} AE_j.$$

Also note that

$$\mathfrak{l}_j = \bigoplus_{k=1}^{n} E_k A E_j,$$

where $E_k A E_j$ is the one-dimensional subspace of A consisting of the $n \times n$ matrices whose only nonzero entry, if any, is the entry of index (k, j). This example is an illustration of Theorem 2.34.

Theorem 2.26 and Proposition 2.23 immediately imply the following second significant result which shows that the minimal left ideals are the building blocks of left ideals.

Theorem 2.27. *Every left ideal in A contains a minimal left ideal. Every minimal left ideal is closed.*

The next two results are technical lemmas that are needed to prove Theorem 2.30. They build special kinds of orthogonal systems from self-adjoint idempotents.

Proposition 2.28. *The following properties hold.*

(1) If e and e' are two orthogonal self-adjoint idempotents, then the left ideals Ae and Ae' are orthogonal.
(2) Let $(e_i)_{1\leq i\leq n}$ be a finite family of pairwise orthogonal, self-adjoint idempotents. Then for every $x \in A$, the element $x - \sum_{i=1}^{n} xe_i$ is orthogonal to each Ae_j $(1 \leq j \leq n)$.

Proposition 2.28 is proven in Dieudonné [Dieudonné (2005)] (Chapter XV, Proposition 15.8.9). The proof is quite simple. On the other hand, the proof of Proposition 2.29 makes use of a fairly complicated inductive construction, which we omit. See Dieudonné [Dieudonné (2005)] (Chapter XV, Proposition 15.8.10) for complete details.

Proposition 2.29. *For every $x \in A$ there exists a finite or countably infinite sequence (e_n) of pairwise orthogonal irreducible self-adjoint idempotents belonging to the closure $\mathfrak{l}$ of the ideal Ax, such that $x = \sum_n xe_n$ (this series being convergent in A), and $\|x\|^2 = \sum_n \|xe_n\|^2$.*

From now on *we assume that the complete Hilbert algebra A is separable.* We have the following important result.

Theorem 2.30. *Suppose A is separable. Then every closed left ideal $\mathfrak{b}$ is the Hilbert sum of a finite or countably infinite sequence of minimal left ideals $\mathfrak{l}_n = Ae_n$, where e_n is an irreducible self-adjoint idempotent. For every $x \in \mathfrak{b}$, we have $x = \sum_n xe_n$, and for all $x, y \in \mathfrak{b}$ we have $\langle x, y\rangle = \sum_n \langle xe_n, ye_n\rangle$.*

Proof. We follow Dieudonné's proof from [Dieudonné (2005)] (Chapter XV, Section 8, Theorem 15.8.11). By Proposition 2.28 and by the properties of Hilbert sums (see Proposition 2.3), the vector xe_n is the orthogonal projection of x onto Ae_n, so the second and third assertions follow.

To prove the first assertion, let $(x_n)_{n\geq 1}$ be a dense sequence in $\mathfrak{b}$, which exists since A is separable. We define inductively, for each n, a finite or countably infinite sequence $(e_{n,i})_{i\in I_n}$ of irreducible self-adjoint idempotents as follows. For $n = 1$, by Proposition 2.29 applied to $x_1 \in \mathfrak{b}$, since $\overline{Ax_1} \subseteq \mathfrak{b}$, there is a finite or countably infinite sequence $(e_{1,i})_{i\in I_1}$ of pairwise orthogonal irreducible self-adjoint idempotents belonging to $\mathfrak{b}$, such that $x_1 = \sum_{i \in I_1} x_1 e_{1,i}$. Suppose that the $e_{m,i}$ have been defined for all $m \leq n$ in such a way that they are pairwise orthogonal and belong to $\mathfrak{b}$, and are such that the x_m with $m \leq n$ belong to the closure $\mathfrak{a}_n \subseteq \mathfrak{b}$ of the left ideal, which is the sum of the ideals $Ae_{m,i}$ for all $m \leq n$ and all $i \in I_m$,

$$\mathfrak{a}_n = \overline{\bigoplus_{m \leq n,\ i \in I_m} Ae_{m,i}}.$$

Let x'_{n+1} be the orthogonal projection of x_{n+1} onto $\mathfrak{a}_n^\perp \cap \mathfrak{b}$. Using Proposition 2.29 applied to $x'_{n+1} \in \mathfrak{a}_n^\perp \cap \mathfrak{b}$, since $\overline{Ax'_{n+1}} \subseteq \mathfrak{a}_n^\perp \cap \mathfrak{b}$, there is a finite or countably infinite sequence $(e_{n+1,i})_{i \in I_{n+1}}$ of pairwise orthogonal irreducible self-adjoint idempotents which belong to $\mathfrak{a}_n^\perp \cap \mathfrak{b}$ and are such that $x'_{n+1} = \sum_{i \in I_{n+1}} x'_{n+1} e_{n+1,i}$; see Figure 2.1. Since each family $(e_{n,i})_{i \in I_n}$ is countable and since there are countably many of these families, the union of these families is also countable, so it can be listed as a single sequence. Using Proposition 2.28 and the properties of Hilbert sums, we leave it as an exercise to check that this sequence has the desired properties. □

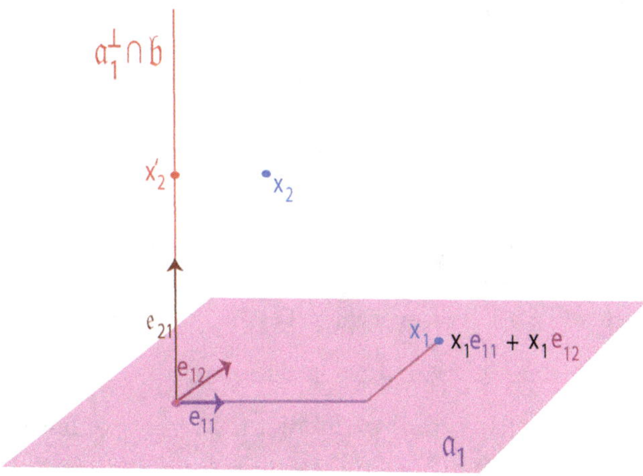

Fig. 2.1 A schematic illustration of the construction used in the proof of Theorem 2.30. Let $\mathfrak{a}_1 = \overline{Ae_{1,1} \oplus Ae_{1,2}}$ be the horizontal purple plane. The vertical red line is $\mathfrak{a}_1^\perp \cap \mathfrak{b}$. Then $\mathfrak{a}_2 = \overline{Ae_{1,1} \oplus Ae_{1,2} \oplus Ae_{2,1}}$.

Theorem 2.30 applies in particular when $\mathfrak{b} = A$. In this case, we get a decomposition of A as a Hilbert sum of minimal left ideals. However, in general, there are infinitely many such decompositions. More precisely, we have the following result.

Proposition 2.31. *Suppose A is separable, and let $\mathfrak{l}$ be a minimal left ideal of A. Then there exists a decomposition of A as a Hilbert sum of minimal left ideals $\mathfrak{l}_n$, such that $\mathfrak{l}_1 = \mathfrak{l}$.*

The following technical result is needed in the proof of the main theorem (Theorem 2.33).

Theorem 2.32. *Let e and e' be two irreducible self-adjoint idempotents, and let $\mathfrak{l} = Ae$, and $\mathfrak{l}' = Ae'$ be the corresponding minimal left ideals. The following properties hold.*

(1) Every homomorphism of the A-module $\mathfrak{l}$ into the A-module $\mathfrak{l}'$ is a map f_a of the form $f_a(x) = xa$ $(x \in \mathfrak{l})$, for some $a \in eAe' = eA \cap Ae'$; it is either zero or bijective, and the map $a \mapsto f_a$ is an isomorphism of the complex vector space eAe' onto $\mathrm{Hom}_A(\mathfrak{l}, \mathfrak{l}')$, such that $f_{ab} = f_b \circ f_a$.

(2) The $\mathbb{C}$-algebra eAe, isomorphic to $\mathrm{End}_A(\mathfrak{l}) = \mathrm{Hom}_A(\mathfrak{l}, \mathfrak{l})$ (the space of endomorphisms of $\mathfrak{l}$), is a field equal to $\mathbb{C}e$ (and therefore isomorphic to $\mathbb{C}$).

(3) If $\mathfrak{l}$ and $\mathfrak{l}'$ are not isomorphic as A-modules, then e and e' (and consequently $\mathfrak{l}$ and $\mathfrak{l}'$) are orthogonal, and $\mathfrak{l}\mathfrak{l}' = \mathfrak{l}'\mathfrak{l} = (0)$. If $\mathfrak{l}$ and $\mathfrak{l}'$ are isomorphic as A-modules, then eAe' is a complex vector space of dimension 1, and $\mathfrak{l}\mathfrak{l}' = \mathfrak{l}'$.

(4) If $x \in A$, then $\mathfrak{l}x$ is a left ideal which is either (0) or isomorphic (as A-module) to $\mathfrak{l}$.

Proof. We follow Dieudonné's proof from [Dieudonné (2005)] (Chapter XV, Section 8, Theorem 15.8.12). (1) Let $g\colon \mathfrak{l} \to \mathfrak{l}'$ be an A-module homomorphism, and let $a = g(e)$. For every $x \in \mathfrak{l} = Ae$, since $e^2 = e$, we have

$$g(x) = g(xe) = xg(e) = xa,$$

so $g = f_a$. Since $a \in g(\mathfrak{l}) \subseteq \mathfrak{l}'$, $\mathfrak{l}' = Ae'$ and $(e')^2 = e'$, we have $a = ae'$. On the other hand,

$$a = g(e) = g(e^2) = eg(e) = ea,$$

so $a = eae' \in eAe'$.

By definition of eAe', we have $eAe' \subseteq eA \cap Ae'$. Conversely, if $y \in eA \cap Ae'$, as $e^2 = e$ and $(e')^2 = e'$, we have $y = ye'$ and $y = ey$, so that $y = eye' \in eAe'$. Therefore, $eAe' = eA \cap Ae'$.

The image $g(\mathfrak{l})$ of g is a left ideal contained in $\mathfrak{l}'$, and since $\mathfrak{l}'$ is a minimal left ideal, either $g(\mathfrak{l}) = (0)$ or $g(\mathfrak{l}) = \mathfrak{l}'$. Likewise, the kernel $\mathrm{Ker}\, g$

of g is a left ideal contained in $\mathfrak{l}$, and since $\mathfrak{l}$ is a minimal left ideal, either $\mathrm{Ker}\, g = (0)$ or $\mathrm{Ker}\, g = \mathfrak{l}$. If $\mathrm{Ker}\, g = \mathfrak{l}$, then $g(\mathfrak{l}) = (0)$ and g is the zero map. If $\mathrm{Ker}\, g = (0)$, then $g(\mathfrak{l}) \neq (0)$, so we must have $g(\mathfrak{l}) = \mathfrak{l}'$, and g is bijective.

If $g = f_a = 0$, then $f_a(e) = ea = 0$; but $a \in eAe'$, so $ea = a$, and consequently $a = 0$, which shows that the map $a \mapsto f_a$ is an isomorphism.

(2) By Proposition 2.22(3), the $\mathbb{C}$-algebra eAe is a closed subalgebra of A. In (1), we saw that every element of $\mathrm{End}_A(\mathfrak{l})$ is either zero or invertible, so $\mathrm{End}_A(\mathfrak{l})$ is a (possibly noncommutative) field, and since eAe isomorphic to $\mathrm{End}_A(\mathfrak{l})$, it is also a field. Clearly, e is a unit in eAe, and since A is a complete Hilbert algebra, it is a Banach algebra, and since eAe is closed in A, it is also a Banach algebra. By the Gelfand–Mazur theorem (Vol. I, Theorem 9.14), $eAe = \mathbb{C}e \cong \mathbb{C}$.

(3) If $\mathfrak{l}$ and $\mathfrak{l}'$ are not isomorphic, by (1) we have $eAe' = (0)$, and in particular, $ee' = 0$. By Proposition 2.23, e and e' are orthogonal. By Proposition 2.28, the left ideals $\mathfrak{l}$ and $\mathfrak{l}'$ are orthogonal, and $\mathfrak{l}\mathfrak{l}' = AeAe' = A(0) = (0)$. Similarly, $e'Ae = (0)$, so $\mathfrak{l}'\mathfrak{l} = (0)$. If $\mathfrak{l}$ and $\mathfrak{l}'$ are isomorphic, and if $g\colon \mathfrak{l} \to \mathfrak{l}'$ is an isomorphism from $\mathfrak{l}$ to $\mathfrak{l}'$, then every homomorphism $h\colon \mathfrak{l} \to \mathfrak{l}'$ is of the form $h = g \circ u$, where $u \in \mathrm{End}_A(\mathfrak{l})$, which means that $\mathrm{End}_A(\mathfrak{l})$ and $\mathrm{Hom}_A(\mathfrak{l}, \mathfrak{l}')$ are isomorphic. By (2), the space $\mathrm{End}_A(\mathfrak{l})$ is one-dimensional, so $\mathrm{Hom}_A(\mathfrak{l}, \mathfrak{l}')$ is also one-dimensional, and by (1), the space $\mathrm{Hom}_A(\mathfrak{l}, \mathfrak{l}')$ is isomorphic to eAe', so we deduce that eAe' is complex vector space of dimension 1. The ideal $\mathfrak{l}\mathfrak{l}'$ is obviously contained in $\mathfrak{l}'$, and contains $eAe' \neq (0)$. Since $\mathfrak{l}'$ is a minimal left ideal, we must have $\mathfrak{l}\mathfrak{l}' = \mathfrak{l}'$.

(4) Since $\mathfrak{l}x$ is the image of $\mathfrak{l}$ under the homomorphism $\varphi\colon \mathfrak{l} \to A$ given by $\varphi(y) = yx$, it is a left ideal isomorphic to $\mathfrak{l}/\mathrm{Ker}\,\varphi$. But since $\mathrm{Ker}\,\varphi$ is a left ideal contained in the minimal left ideal $\mathfrak{l}$, either $\mathrm{Ker}\,\varphi = (0)$ or $\mathrm{Ker}\,\varphi = \mathfrak{l}$, which means that φ is either injective or zero, and so $\mathfrak{l}x$ is a left ideal, which is either (0) or isomorphic (as A-module) to $\mathfrak{l}$. $\square$

Finally, we come to the main theorems of this chapter.

2.7 The Structure of Complete Separable Hilbert Algebras

Theorem 2.33. *(Master decomposition theorem) Let A be a complete, separable Hilbert algebra. The following properties hold.*

(1) *There exists a finite or countably infinite sequence* $(\mathfrak{l}_k)_{k\in J}$ *of minimal left ideals, no pair of which are isomorphic, such that every minimal left ideal of* A *is isomorphic (as an* A*-module) to some* $\mathfrak{l}_k$.

(2) *For each index* $k \in J$, *the closure of the sum of all the minimal left ideals of* A *which are isomorphic to* $\mathfrak{l}_k$ *is a self-adjoint two-sided ideal* $\mathfrak{a}_k$. *Every minimal left ideal of the Hilbert algebra* $\mathfrak{a}_k$ *is a minimal left ideal of* A, *isomorphic to* $\mathfrak{l}_k$, *and the algebra* $\mathfrak{a}_k$ *contains no closed two-sided ideals other than* (0) *and* $\mathfrak{a}_k$.

(3) *Each of the algebras* $\mathfrak{a}_k$ *is a Hilbert sum*

$$\mathfrak{a}_k = \bigoplus_{j\in I_k} \mathfrak{l}'_j,$$

with I_k *finite or countably infinite, and where each* $\mathfrak{l}'_j$ *is a minimal left ideal isomorphic to* $\mathfrak{l}_k$. *The algebra* A *is the Hilbert sum*

$$A = \bigoplus_{k\in J} \mathfrak{a}_k,$$

and $\mathfrak{a}_h\mathfrak{a}_k = (0)$ *for all* $h \neq k$.

Proof. We reproduce Dieudonné's proof from [Dieudonné (2005)] (Chapter XV, Section 8, Theorem 15.8.13). By Theorem 2.30, we obtain A as the Hilbert sum

$$A = \bigoplus_{n\in L} \mathfrak{l}'_n$$

of minimal left ideals $\mathfrak{l}'_n$ of A, where L is finite or countably infinite. We define inductively the index set J and the sequence $(\mathfrak{l}_k)_{k\in J}$ of minimal left ideals $\mathfrak{l}_k$, no pair of which are isomorphic, as follows. Start with $\mathfrak{l}_1 = \mathfrak{l}'_1$. Having defined $\mathfrak{l}_1, \ldots, \mathfrak{l}_k$, let $\mathfrak{l}_{k+1}$ be the equal to $\mathfrak{l}'_m$, where m is the smallest integer, such that $\mathfrak{l}'_m$ is not isomorphic to any of the ideals $\mathfrak{l}_1, \ldots, \mathfrak{l}_k$. If all the $\mathfrak{l}'_n$ are isomorphic to one of the ideals $\mathfrak{l}_1, \ldots, \mathfrak{l}_k$, then stop. Let J be the finite or countably infinite sequence of indices k so obtained, and for every $k \in J$, let I_k be the sequence of integers n, such that $\mathfrak{l}'_n$ is isomorphic to $\mathfrak{l}_k$. If J is infinite, then each I_k is finite. Otherwise, $I_1, \ldots, I_{k-1}$ are finite, and I_k is finite or countably infinite. Define $\mathfrak{a}_k$ as the Hilbert sum

$$\mathfrak{a}_k = \bigoplus_{j\in I_k} \mathfrak{l}'_j.$$

By construction, it is clear that each $\mathfrak{a}_k$ is a left ideal and that H is the Hilbert sum

$$A = \bigoplus_{k\in J} \mathfrak{a}_k.$$

Let $\mathfrak{l}$ be any minimal left ideal in A. Then $\mathfrak{l}$ must be isomorphic to one of the $\mathfrak{l}_k$, for otherwise by Theorem 2.32(3), it would be orthogonal to all of the $\mathfrak{l}'_n$, and hence orthogonal to A itself, a contradiction. The same argument shows that $\mathfrak{l}$ is orthogonal to all the $\mathfrak{a}_h$ with $h \neq k$. Hence, since $\mathfrak{a}_k$ is the orthogonal complement of the Hilbert sum

$$\bigoplus_{h \in J - \{k\}} \mathfrak{a}_h,$$

we must have $\mathfrak{l} \subseteq \mathfrak{a}_k$. This implies that $\mathfrak{a}_k$ is the closure of the sum of all the minimal left ideals of A, which are isomorphic to $\mathfrak{l}_k$, and therefore, $\mathfrak{a}_k$ is independent of the decomposition of A as the Hilbert sum of the $\mathfrak{l}'_n$ from which we started. Moreover, for every $x \in A$, and every $n \in I_k$, by Theorem 2.32(4), we know that $\mathfrak{l}'_n x$ is a left ideal, which is either (0) or isomorphic to $\mathfrak{l}'_n$, hence contained in $\mathfrak{a}_k$. This proves that $\mathfrak{a}_k$ is a two-sided ideal. If $\mathfrak{l}'_n = Ae'_n$, where e'_n is an irreducible self-adjoint idempotent, then $(\mathfrak{l}'_n)^* = e'_n A$, hence $\mathfrak{a}_k^* = \mathfrak{a}_k$.

Let $\mathfrak{l}''$ be a minimal left ideal of the Hilbert algebra $\mathfrak{a}_k$. By Theorem 2.26, we have $\mathfrak{l}'' = \mathfrak{a}_k e''$, where e'' is a self-adjoint idempotent, and $e'_n e''$ cannot vanish for all $n \in I_k$, otherwise $\mathfrak{l}''$ would be orthogonal to all of the $\mathfrak{l}'_n$ with $n \in I_k$, and therefore to the closure of their sum, namely to $\mathfrak{a}_k$, which is impossible because $\mathfrak{l}'' \neq (0)$. Hence there exists at least one index $n \in I_k$, such that $\mathfrak{l}'_n \mathfrak{l}'' \neq (0)$; since $\mathfrak{l}'_n \mathfrak{l}''$ is a left ideal in $\mathfrak{a}_k$, we must have $\mathfrak{l}'_n \mathfrak{l}'' = \mathfrak{l}''$. By Theorem 2.32(3), $\mathfrak{l}''$ is a minimal left ideal of A necessarily isomorphic to $\mathfrak{l}'_n$, and therefore to $\mathfrak{l}_k$.

If $\mathfrak{b}$ is a nonzero two-sided ideal of the algebra $\mathfrak{a}_k$, by Theorem 2.27, it contains at least one minimal left ideal $\mathfrak{l}''$ of this algebra, hence also contains all the $\mathfrak{l}''\mathfrak{l}'_n$ $(n \in I_k)$. But by Theorem 2.32(3), we have $\mathfrak{l}''\mathfrak{l}'_n = \mathfrak{l}'_n$, and therefore $\mathfrak{b}$ contains the sum of all the $\mathfrak{l}'_n$ (with $n \in I_k$). If $\mathfrak{b}$ is closed, it follows that $\mathfrak{b} = \mathfrak{a}_k$. Finally, $\mathfrak{a}_h \mathfrak{a}_k \subseteq \mathfrak{a}_h \cap \mathfrak{a}_k = (0)$ if $h \neq k$, because $\mathfrak{a}_h$ and $\mathfrak{a}_k$ are two-sided ideals. □

The proof of the theorem shows that if J is infinite, then all the index sets I_k are finite. Also, observe that Theorem 2.32(3) plays a crucial role in proving that $\mathfrak{a}_h \mathfrak{a}_k = (0)$ for all $h \neq k$.

Roughly speaking, Theorem 2.33 says that if A is a complete separable Hilbert algebra, then there is an irredundant list $(\mathfrak{l}_k)_{k \in J}$ of the minimal left ideals of A, and A is the Hilbert sum of two-sided ideals $\mathfrak{a}_k$, where each $\mathfrak{a}_k$ is the Hilbert sum obtained by picking a certain number of copies of the minimal left ideal $\mathfrak{l}_k$ of A.

Example 2.9. As an aid to help the reader process the indexing for the master decomposition of Theorem 2.33, suppose A is a complete separable Hilbert algebra with the following finite Hilbert sum decomposition:

$$A = \mathfrak{l}'_1 \oplus \mathfrak{l}'_2 \oplus \mathfrak{l}'_3 \oplus \mathfrak{l}'_4 \oplus \mathfrak{l}'_5 \oplus \mathfrak{l}'_6,$$

where

$$\mathfrak{l}'_1 \cong \mathfrak{l}'_3 \cong \mathfrak{l}'_4, \qquad \mathfrak{l}'_2 \cong \mathfrak{l}_6.$$

Set

$$\mathfrak{l}_1 := \mathfrak{l}'_1, \qquad \mathfrak{l}_2 := \mathfrak{l}'_2, \qquad \mathfrak{l}_3 := \mathfrak{l}'_5.$$

Then $J = (1, 2, 3)$, and

$$I_1 = (1, 3, 4), \qquad I_2 = (2, 6), \qquad I_3 = (5).$$

Also, note that

$$\mathfrak{a}_1 := \bigoplus_{j \in I_1} \mathfrak{l}'_j = \mathfrak{l}'_1 \oplus \mathfrak{l}'_3 \oplus \mathfrak{l}'_4$$

$$\mathfrak{a}_2 := \bigoplus_{j \in I_2} \mathfrak{l}'_j = \mathfrak{l}'_2 \oplus \mathfrak{l}'_6$$

$$\mathfrak{a}_3 := \bigoplus_{j \in I_3} \mathfrak{l}'_j = \mathfrak{l}'_5,$$

which in turn implies

$$A = \mathfrak{a}_1 \oplus \mathfrak{a}_2 \oplus \mathfrak{a}_3 = (\mathfrak{l}'_1 \oplus \mathfrak{l}'_3 \oplus \mathfrak{l}'_4) \oplus (\mathfrak{l}'_2 \oplus \mathfrak{l}'_6) \oplus \mathfrak{l}'_5.$$

The ideals $\mathfrak{a}_k$ have properties analogous to those of simple algebras.

Definition 2.20. A complete Hilbert algebra A is *topologically simple* if it contains no *closed* two-sided ideal other than (0) and A.

Theorem 2.33 shows that the complete Hilbert algebras $\mathfrak{a}_k$ are topologically simple. Theorem 2.33 also shows that the study of complete separable Hilbert algebras reduces to the study of the topologically simple ones. We have the following theorem that gives us a sharper image of the structure of topologically simple complete separable Hilbert algebras.

Theorem 2.34. *(Structure of a topologically simple Hilbert algebra) Let A be topologically simple, complete, separable Hilbert algebra. For any minimal left ideal $\mathfrak{l}$ of A, the representation $U_{\mathfrak{l}}\colon A \to \mathcal{L}(\mathfrak{l})$ of A in the Hilbert space $\mathfrak{l}$ is faithful.*

If A is infinite-dimensional, then so is $\mathfrak{l}$. The image of A under $U_{\mathfrak{l}}$ is the algebra $\mathcal{L}_2(\mathfrak{l})$ of Hilbert–Schmidt operators on $\mathfrak{l}$, and there exists a constant $c > 0$, such that

$$c\langle x, y\rangle_A = \langle U_{\mathfrak{l}}(x), U_{\mathfrak{l}}(y)\rangle_{\mathrm{HS}}, \qquad \text{for all } x, y \in A. \qquad (*)$$

The inner product on the right-hand side is the inner product defined in Section 2.5.

If A is finite-dimensional, then the image of A under $U_{\mathfrak{l}}$ is the algebra $\mathrm{End}_{\mathbb{C}}(\mathfrak{l})$ of all endomorphisms of the vector space $\mathfrak{l}$, and $(*)$ remains valid (the inner product on the right-hand side is also the inner product defined in Example 2.5). In fact,

$$A = \bigoplus_{j=1}^{n} \mathfrak{l}_j, \qquad \mathfrak{l}_j = Ae_j = \bigoplus_{i=1}^{n} e_i A e_j,$$

where $e_1, \ldots, e_n$ are irreducible self-adjoint idempotents in A, the $\mathfrak{l}_j = Ae_j$ are isomorphic minimal left ideals, each space e_iAe_j is one-dimensional, and $e_1 + \cdots + e_n$ is the unit of A.

Proof. We follow Dieudonné's proof from [Dieudonné (2005)] (Chapter XV, Section 8, Theorem 15.8.14). By Proposition 2.31 and by Theorem 2.33, we may assume that A is the Hilbert sum of a finite or countably infinite sequence of minimal left ideals $\mathfrak{l}_n = Ae_n$, where $\mathfrak{l} = \mathfrak{l}_1$ and all the $\mathfrak{l}_n$ are isomorphic. We begin by observing that for any $x \in A$ such that $x \neq 0$, we have $Ax \neq (0)$. Indeed, if $x \neq 0$ and if $Ax = (0)$, then for $x^* \in A$ we have $x^*x = 0$, but by Proposition 2.17 this implies that $x = 0$, a contradiction.

First, we prove that the representation $U_{\mathfrak{l}}$ is faithful. If $U_{\mathfrak{l}}(x) = 0$ for some $x \neq 0$ in A, that is, if $x\mathfrak{l} = (0)$, then we should have $(Ax)\mathfrak{l} = (0)$. Since the ideal Ax is nonzero, by Theorem 2.27, the ideal Ax contains a minimal left ideal $\mathfrak{l}'$, and by Theorem 2.33, the left ideal $\mathfrak{l}'$ is isomorphic to $\mathfrak{l}$; hence $\mathfrak{l}'\mathfrak{l} = (0)$, contrary to Theorem 2.32(3). Therefore the representation $U_{\mathfrak{l}}$ is faithful.

Next, we claim that $\mathfrak{l}$ is the Hilbert sum of the subspaces e_nAe_1,

$$\mathfrak{l} = \bigoplus_{n} e_n A e_1,$$

where the spaces e_nAe_1 are one-dimensional.

Let $P_n = U_{\mathfrak{l}}(e_n)$. By Theorem 2.32(3), since the $\mathfrak{l}_i$ are isomorphic, the subspace e_nAe_1 is one-dimensional, and we shall show that P_n is the orthogonal projection of $\mathfrak{l} = Ae_1$ onto e_nAe_1. Since e_n is a self-adjoint idempotent and since an arbitrary element of Ae_1 is of the form xe_1 for some $x \in A$, and $U_\mathfrak{l}(e_n)(xe_1) = e_nxe_1$, we have

$$\langle xe_1 - e_nxe_1, e_nye_1\rangle = \langle e_nxe_1 - e_n^2xe_1, ye_1\rangle = \langle e_nxe_1 - e_nxe_1, ye_1\rangle = 0.$$

Since $e_me_n = 0$ whenever $m \neq n$, we have $P_mP_n = 0$, and therefore the subspaces e_nAe_1 are orthogonal in pairs. Moreover, $\mathfrak{l}$ is the Hilbert sum of these subspaces. Here, we use a standard fact of Hilbert space theory, which is that if (z_i) is an orthonormal family in $\mathfrak{l}$, and for every w, if w is orthogonal to all the z_i, then $w = 0$, then (z_i) is dense in $\mathfrak{l}$. Otherwise, $B = \overline{\bigoplus_i \mathbb{C}z_i}$ is a closed proper subspace of $\mathfrak{l}$, and thus its orthogonal complement $B^\perp$ is nonempty, so there is a nonzero $w \in B^\perp$ orthogonal to all the z_i. Since the subspaces e_iAe_1 are one-dimensional, we can choose z_i to be some nonzero vector in e_iAe_1. Now if xe_1 is orthogonal to all the subspaces e_nAe_1 (which are one-dimensional), then $P_n(xe_1) = 0$ for all n, so that $e_nxe_1 = 0$ for all n, and thus $y = xe_1$ belongs to the right annihilator of A. Proposition 2.17 implies that this right annihilator is equal to (0), because $y^* \in A$, so $y^*y = 0$, and by Proposition 2.17 we must have $y = xe_1 = 0$. A similar proof shows that $\mathfrak{l}_j$ is the Hilbert sum

$$\mathfrak{l}_j = \bigoplus_k e_kAe_j, \tag{†}$$

where the spaces e_kAe_j are one-dimensional.

Equation (†) shows that the sequence $(\mathfrak{l}_k)$ is finite iff $\mathfrak{l}$ (and thus each $\mathfrak{l}_k$) is finite-dimensional over $\mathbb{C}$, or equivalently iff A is finite-dimensional. If A is finite-dimensional, then

$$A = \bigoplus_{j=1}^n \mathfrak{l}_j, \qquad \mathfrak{l}_j = Ae_j = \bigoplus_{i=1}^n e_iAe_j,$$

where $e_1, \ldots, e_n$ are irreducible self-adjoint idempotents in A, the $\mathfrak{l}_j = Ae_j$ are isomorphic minimal left ideals, and each space e_iAe_j is one-dimensional. By Proposition 2.23, since the subspaces Ae_j are pairwise orthogonal, $e_ie_j = 0$ whenever $i \neq j$, so $e_1 + \cdots + e_n$ is a unit for e_iAe_j, and thus for A.

Let (a_n) be an orthonormal basis of $\mathfrak{l}$, such that $a_n \in e_nAe_1$ for each n. Since the e_n are self-adjoint idempotents, we have

$$a_ne_1 = a_n, \qquad e_na_n = a_n, \qquad a_na_n^* \in e_nAe_n, \qquad a_n^*a_n \in e_1Ae_1,$$

and by Theorem 2.32(2), we must have $a_n a_n^* = \alpha_n e_n$ for some nonzero $\alpha_n \in \mathbb{C}$. Similarly, $a_n^* a_n \in e_1 A e_1$, and we must have $a_n^* a_n = \alpha_n' e_1$ for some nonzero $\alpha_n' \in \mathbb{C}$.

We claim that $\alpha_n = \alpha_n'$ for all n.

On the one hand,

$$a_n a_n^* a_n a_n^* = \alpha_n e_n \alpha_n e_n = \alpha_n^2 e_n^2 = \alpha_n^2 e_n,$$

and on the other hand, since $a_n e_1 = a_n$, we have

$$a_n a_n^* a_n a_n^* = a_n \alpha_n' e_1 a_n^* = \alpha_n' a_n e_1 a_n^* = \alpha_n' a_n a_n^* = \alpha_n' \alpha_n e_n.$$

Consequently, $\alpha_n^2 e_n = \alpha_n' \alpha_n e_n$, and since $\alpha_n', \alpha_n \neq 0$, we conclude that $\alpha_n = \alpha_n'$ for all n.

We also have

$$\langle e_n, e_n \rangle = \langle e_1, e_1 \rangle, \qquad \text{for all } n \geq 1.$$

Indeed, we have

$$1 = \langle a_n, a_n \rangle = \langle a_n, e_n a_n \rangle = \langle a_n a_n^*, e_n \rangle = \alpha_n \langle e_n, e_n \rangle,$$

and

$$1 = \langle a_n, a_n \rangle = \langle a_n, a_n e_1 \rangle = \langle a_n^* a_n, e_1 \rangle = \alpha_n \langle e_1, e_1 \rangle,$$

and since $\alpha_n \neq 0$, we obtain $\langle e_n, e_n \rangle = \langle e_1, e_1 \rangle$ for all $n \geq 1$.

Let $c = 1/\langle e_1, e_1 \rangle = \alpha_n$. Then $a_n a_n^* = c e_n$, and for all $x, y \in A$, we have

$$\langle x a_n, y a_n \rangle = \langle y^* x, a_n a_n^* \rangle = \langle y^* x, c e_n \rangle = c \langle x e_n, y e_n \rangle.$$

By Theorem 2.30, the series with general term $\langle x e_n, y e_n \rangle$ is absolutely convergent with sum $\langle x, y \rangle$, and since $\langle x a_n, y a_n \rangle = c \langle x e_n, y e_n \rangle$, if A is infinite-dimensional, then $\sum_n \langle x a_n, x a_n \rangle = \sum_n \| U_{\mathfrak{l}}(x)(a_n) \|^2$ converges, so $U_{\mathfrak{l}}(x)$ is a Hilbert–Schmidt operator, and Equation $(*)$ holds. Since A is a Hilbert space, so is its image under $U_{\mathfrak{l}}$, and to show that this image is the whole of the Hilbert space $\mathcal{L}_2(\mathfrak{l})$ (see Example 2.5), it suffices to show that $U_{\mathfrak{l}}(A)$ is dense in $\mathcal{L}_2(\mathfrak{l})$.

Now for all m, n with $m \neq n$, by Theorem 2.32(3), we have

$$(e_m A e_n)(e_n A e_1) = e_m (A e_n)(A e_1) = e_m A e_1,$$

and since $e_n A e_1 = \mathbb{C} a_n$, $a_n \in e_n A e_1$, and $a_m \in e_m A e_1$, it follows that there exists $e_{mn} \in e_m A e_n$, such that $e_{mn} a_n = a_m$ (which implies that $e_{mn} = c^{-1} a_m a_n^*$, since $a_n a_n^* = c e_n$ and $e_{mn} \in e_m A e_n$), and clearly,

$e_{mn}a_p = 0$ if $p \neq n$. We conclude from this that $E_{mn} = U_{\mathfrak{l}}(e_{mn})$ is the continuous endomorphism of the Hilbert space $\mathfrak{l}$, such that $E_{mn}(a_n) = a_m$ and $E_{mn}(a_p) = 0$ if $p \neq n$. Our assertion now follows from the fact that it is not hard to show that the finite linear combinations of the E_{mn} are dense in $\mathcal{L}_2(\mathfrak{l})$.

We already took care of the case where A is finite-dimensional. □

Example 2.10. Let us apply the construction of Theorem 2.34 to $\mathfrak{a}_1$ of Example 2.9. In other words, let

$$A := \mathfrak{a}_1 = \mathfrak{l}_1' \oplus \mathfrak{l}_3' \oplus \mathfrak{l}_4' = \mathfrak{l}_1 \oplus \mathfrak{l}_2 \oplus \mathfrak{l}_3,$$

where

$$\mathfrak{l}_1 := \mathfrak{l}_1' \text{ and } \mathfrak{l}_1 = Ae_1, \quad \mathfrak{l}_2 := \mathfrak{l}_3' \text{ and } \mathfrak{l}_2 = Ae_2, \quad \mathfrak{l}_3 := \mathfrak{l}_4' \text{ and } \mathfrak{l}_3 = Ae_3.$$

Then

$$A = Ae_1 \oplus Ae_2 \oplus Ae_3.$$

Now set $\mathfrak{l} := \mathfrak{l}_1 = Ae_1$ and decompose each of the above $\mathfrak{l}_j$, where $1 \leq j \leq 3$, as

$$\begin{aligned}
\mathfrak{l} &= e_1Ae_1 \oplus e_2Ae_1 \oplus e_3Ae_1, \\
\mathfrak{l}_2 &= e_1Ae_2 \oplus e_2Ae_2 \oplus e_3Ae_2, \\
\mathfrak{l}_3 &= e_1Ae_3 \oplus e_2Ae_3 \oplus e_3Ae_3.
\end{aligned}$$

See Figure 2.2. We now scale the directions of e_1Ae_1, e_2Ae_1, and e_3Ae_1 to form the orthonormal basis $(a_i)_{1\leq i\leq 3}$ of $\mathfrak{l}$, where

$$a_1 \in e_1Ae_1, \qquad a_2 \in e_2Ae_1, \qquad a_3 \in e_3Ae_1.$$

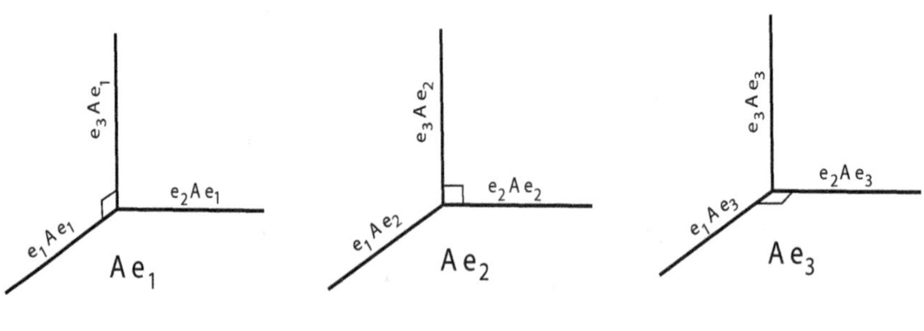

Fig. 2.2 An illustration of the one-dimensional perpendicular directions within each $\mathfrak{l}_i$; i.e. the decomposition $Ae_i = e_1Ae_i \oplus e_2Ae_i \oplus e_3Ae_i$, for $1 \leq i \leq 3$.

We also define six elements $e_{mn} \in e_m A e_n$, where $m, n \in \{1, 2, 3\}$ and $m \neq n$, i.e.

$$
\begin{array}{lll}
e_{21} \in e_2 A e_1, & e_{31} \in e_3 A e_1, & e_{12} \in e_1 A e_2, \\
e_{32} \in e_3 A e_2, & e_{13} \in e_1 A e_3, & e_{23} \in e_2 A e_3,
\end{array}
$$

such that

$$
\begin{array}{lll}
e_{21} a_1 = a_2, & e_{12} a_1 = a_1, & e_{13} a_3 = a_1 \\
e_{31} a_1 = a_3, & e_{32} a_2 = a_3, & e_{23} a_3 = a_2.
\end{array}
$$

See Figure 2.3.

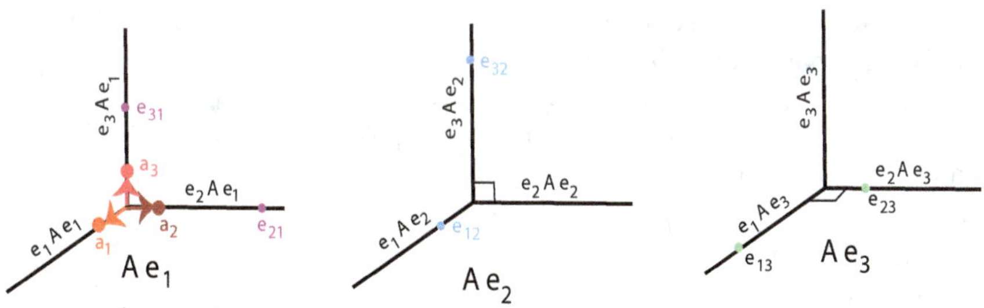

Fig. 2.3 The orthonormal basis (a_1, a_2, a_3) and the elements $e_{mn} \in e_m A e_n$ whenever $m \neq n$.

Theorem 2.34 implies that in the Hilbert sum

$$A = \bigoplus_{k \in J} \mathfrak{a}_k$$

given by Theorem 2.33, the Hilbert algebra $\mathfrak{a}_k$, which is a building block of the decomposition, is either isomorphic to the algebra $\mathcal{L}_2(\mathfrak{l}_k)$ of Hilbert–Schmidt operators on $\mathfrak{l}_k$, or to the finite-dimensional algebra $\mathrm{End}_{\mathbb{C}}(\mathfrak{l}_k)$ of all endomorphisms of the vector space $\mathfrak{l}_k$. If G is a metrizable compact group and $A = \mathrm{L}^2(G)$, then every $\mathfrak{a}_k$ in the Hilbert sum for A is isomorphic to the finite-dimensional algebra $\mathrm{End}_{\mathbb{C}}(\mathfrak{l}_k)$.

Observe that in all cases, the spaces $U_{\mathfrak{l}}(\mathfrak{l}_n)$ consist of endomorphisms of the form $U(x) \circ P_n$ (where the $P_n = U_{\mathfrak{l}}(e_n)$ are the orthogonal projections defined in the proof of Theorem 2.34), and therefore consist of endomorphisms of rank 1 of $\mathfrak{l}$.

The following result gives a sufficient condition for a Hilbert algebra satisfying the hypotheses of Theorem 2.34 to be finite. Recall that the center $Z(A)$ of an algebra A is the set $Z(A) = \{x \in A \mid xy = yx, \text{ for all } y \in A\}$.

Proposition 2.35. *Let A be topologically simple, complete, separable Hilbert algebra. If there is an element $c \neq 0$ in the center of A, then A is finite-dimensional. In that case, the center of A is $\mathbb{C}\mathbf{1}$, where $\mathbf{1}$ is the identity element of A.*

Proof. If c belongs to the center of A, then for any irreducible self-adjoint idempotent e in A (which exists by Theorem 2.26 and Theorem 2.27), $u = U_{\mathfrak{l}}(c)$ is an endomorphism of the A-module $\mathfrak{l} = Ae$, and by Proposition 2.32(1), it is a map of the form $x \mapsto \alpha x$ for all $x \in \mathfrak{l}$, with $\alpha \in \mathbb{C}$. However, such a map cannot be a Hilbert–Schmidt operator on an infinite-dimensional space unless it is the zero map ($\sum_{n=1}^{\infty} \|u(a_n)\|^2$ can't be finite, where (a_i) is a countable Hilbert basis of $\mathfrak{l}$). Therefore $\mathfrak{l}$ is finite-dimensional, so by Theorem 2.34, A is also finite-dimensional. In this case, by Theorem 2.34, the algebra A is isomorphic to an algebra of $n \times n$ matrices, and it is a well-known fact of linear algebra that the matrices that commute with all $n \times n$ matrices are of the form λI_n, with $\lambda \in \mathbb{C}$ (where I_n is the $n \times n$ identity matrix). ☐

Theorem 2.33 will be applied to the complete, separable, Hilbert algebra $\mathrm{L}^2(G)$ (G metrizable and compact), which is then a Hilbert sum of two-sided ideals $\mathfrak{a}_k$. It turns out that Proposition 2.35 applies to the Hilbert algebras $\mathfrak{a}_k$, so they are finite-dimensional. This will yield the first part of the Peter–Weyl theorem.

We conclude with a result about representations of complete Hilbert algebras that makes use of Theorem 2.33.

Theorem 2.36. *(Master decomposition for nondegenerate representations) Let A be a separable, complete Hilbert algebra and let $\mathfrak{a}_{k\in J}$ be the topologically simple Hilbert algebras which are the Hilbert summands as in Theorem 2.33(3). For every $k \in J$, let $\mathfrak{l}_k$ be a minimal left ideal of $\mathfrak{a}_k$. Let $V\colon A \to \mathcal{L}(H)$ be a nondegenerate representation of A in a separable Hilbert space H, such that $V\colon A \to \mathcal{L}(H)$ is continuous. The following properties hold.*

(1) *The Hilbert space H is the Hilbert sum of subspaces H_k ($k \in J$) invariant under V, such that if V_k is the restriction of V to H_k, we have*

$V_k(s) = 0$ for all $s \in \mathfrak{a}_h$ and for all h with $h \neq k$; thus V_k can be considered as a representation of $\mathfrak{a}_k$ in H_k.

(2) The representation V_k is the Hilbert sum of a finite or countably infinite sequence of irreducible representations, each equivalent to the representation $U_{\mathfrak{l}_k}$ of $\mathfrak{a}_k$ as in Theorem 2.34. If $\mathfrak{a}_k$ is finite-dimensional, then so is H_k.

Proof. We follow Dieudonné's proof from [Dieudonné (2005)] (Chapter XV, Section 8, Theorem 15.8.16). Let H_k be the closure of the subspace spanned by the set of vectors

$$\{V(s_k)(x) \mid s_k \in \mathfrak{a}_k,\ x \in H\}.$$

Since every $s \in A$ can be written as $s = \sum_k s_k$ for some $s_k \in \mathfrak{a}_k$ and since V is continuous, we have

$$V(s)(x) = \sum_k V(s_k)(x),$$

which implies that H is the closure of the sum of the H_k. Also, if $h \neq k$ and if $s_h \in \mathfrak{a}_h, s_k \in \mathfrak{a}_k$, we have

$$\begin{aligned}\langle V(s_h)(x), V(s_k)(y)\rangle &= \langle V^*(s_k)V(s_h)(x), y\rangle = \langle V(s_k^* s_h)(x), y\rangle \\ &= \langle V(0)(x), y\rangle = \langle 0, y\rangle = 0,\end{aligned}$$

because $\mathfrak{a}_k$ is self-adjoint and $\mathfrak{a}_k\mathfrak{a}_h = (0)$ since $h \neq k$, so $s_k^* s_h = 0$. Thus (1) holds.

Now assume that A is topologically simple and finite-dimensional over $\mathbb{C}$, and thus has a unit element. This means that A is equal to some $\mathfrak{a}_k$ in the master decomposition theorem (Theorem 2.33), V is equal to V_k, and H is equal to H_k. Our representation is $V_k\colon \mathfrak{a}_k \to \mathcal{L}(H_k)$, which we also denote by $V\colon A \to \mathcal{L}(H)$. The case where A is infinite-dimensional can also be handled but it is a bit more complicated; see Dieudonné's [Dieudonné (2005)] (Chapter XV, Section 8, Problem 1). If A is finite-dimensional, then $A = \bigoplus_{j=1}^n \mathfrak{l}_j$ is the sum of a finite number of isomorphic minimal left ideals $\mathfrak{l}_j$ in A. By Proposition 2.10, the representation $V\colon A \to \mathcal{L}(H)$ is the Hilbert sum (finite or countable) of topologically cyclic representations $V'_k\colon A \to \mathcal{L}(H'_k)$. Thus, it suffices to prove that each topologically cyclic representation $V'_k\colon A \to \mathcal{L}(H'_k)$ is a Hilbert sum (finite or countable) of irreducible representations $V'_{k,j}\colon A \to \mathcal{L}(H'_{k,j})$, each equivalent to the representation $U_{\mathfrak{l}}$, where $\mathfrak{l}$ is some minimal left ideal in A. Then $V\colon A \to \mathcal{L}(H)$ is the Hilbert sum of the family (finite or countable) of irreducible representations $V'_{k,j}\colon A \to \mathcal{L}(H'_{k,j})$.

For simplicity of notation, we may assume that $H = H'_k$, and let x_0 be a cyclic vector in H. The subspace of H spanned by the set of vectors $\{V(s)(x_0) \mid s \in A\}$ is finite-dimensional, and thus closed and dense in H, so it is equal to H. As a consequence, H is finite-dimensional and we can argue by induction on the dimension of H.

Since $A = \bigoplus_{j=1}^{n} \mathfrak{l}_j$ is the sum of a finite number of isomorphic minimal left ideals $\mathfrak{l}_j$, there is at least one minimal ideal, say $\mathfrak{l}$, such that the subspace $E = V(\mathfrak{l})(x_0) \subseteq H$ is nonzero. The surjection $\varphi\colon \mathfrak{l} \to E$ given by $\varphi(s) = V(s)(x_0)$ is then an A-module homomorphism, and since its kernel is a left ideal $\mathfrak{l}'$ contained in $\mathfrak{l}$ and distinct from $\mathfrak{l}$, since $\mathfrak{l}$ is minimal, we must have $\mathfrak{l}' = (0)$. Hence, $\varphi\colon \mathfrak{l} \to E$ is a linear isomorphism and E is a subspace of H invariant under V. By Theorem 2.34, the representation $U_{\mathfrak{l}}\colon A \to \mathcal{L}(\mathfrak{l})$ is faithful and it is an isomorphism between A and $\mathrm{End}_{\mathbb{C}}(\mathfrak{l})$, with $\mathfrak{l}$ finite-dimensional, so if $\mathfrak{l}'$ is a proper linear subspace of $\mathfrak{l}$ invariant under $U_{\mathfrak{l}}$, as $U_{\mathfrak{l}}$ remains a faithful representation in $\mathfrak{l}'$ (because any linear map on $\mathfrak{l}'$ can be extended by zero to a linear map on $\mathfrak{l}$), A would be isomorphic to $\mathrm{End}_{\mathbb{C}}(\mathfrak{l}')$ whose dimension is strictly smaller than the dimension of $\mathrm{End}_{\mathbb{C}}(\mathfrak{l})$, a contradiction. Therefore, the representation $U_{\mathfrak{l}}\colon A \to \mathcal{L}(\mathfrak{l})$ is irreducible, and $\varphi\colon \mathfrak{l} \to E$ establishes an equivalence between the representations $U_{\mathfrak{l}}\colon A \to \mathcal{L}(\mathfrak{l})$ and $V\colon A \to \mathcal{L}(E)$. Since the orthogonal complement $E^{\perp}$ of E in H is also invariant under V and has dimension strictly smaller than the dimension of H, we can apply the induction hypothesis to the representation $V\colon A \to \mathcal{L}(E^{\perp})$ to complete the proof. □

Example 2.11. Let us apply the proof techniques of Theorem 2.36 to $A = \mathfrak{a}_1 \oplus \mathfrak{a}_2 \oplus \mathfrak{a}_3$, where

$$\mathfrak{a}_1 = \mathfrak{l}_1 \oplus \mathfrak{l}_3 \oplus \mathfrak{l}_4, \qquad \mathfrak{a}_2 = \mathfrak{l}_2 \oplus \mathfrak{l}_6, \qquad \mathfrak{a}_3 = \mathfrak{l}_5;$$

see Example 2.9. Note that we have removed the primes so as to match the notation of the theorem. We are given a continuous representation $V\colon A \to \mathcal{L}(H)$. Define

$$\begin{aligned} H_1 &:= \overline{\mathrm{span}\{V(s_1)(x) \mid s_1 \in \mathfrak{a}_1,\ x \in H\}} \\ H_2 &:= \overline{\mathrm{span}\{V(s_2)(x) \mid s_2 \in \mathfrak{a}_2,\ x \in H\}} \\ H_3 &:= \overline{\mathrm{span}\{V(s_3)(x) \mid s_3 \in \mathfrak{a}_3,\ x \in H\}}. \end{aligned}$$

Then $H = H_1 \oplus H_2 \oplus H_3$, and since each H_i, for $1 \leq i \leq 3$, is invariant under V, we may define the restricted representations $V_1\colon \mathfrak{a}_1 \to \mathcal{L}(H_1)$, $V_2\colon \mathfrak{a}_2 \to \mathcal{L}(H_2)$, and $V_3\colon \mathfrak{a}_3 \to \mathcal{L}(H_3)$, such that if $s = s_1 + s_2 + s_3 \in \mathfrak{a}$, where $s_i \in \mathfrak{a}_i$ for $1 \le i \le 3$, and if $h = h_1 + h_2 + h_3 \in H$, where $h_i \in H_i$, for $1 \le i \le 3$, then

$$\begin{aligned}
V(s)(h) &= V(s_1 + s_2 + s_3)(h_1 + h_2 + h_3) \\
&= V(s_1 + s_2 + s_3)(h_1) + V(s_1 + s_2 + s_3)(h_2) \\
&\quad + V(s_1 + s_2 + s_3)(h_3) \\
&= V(s_1)(h_1) + V(s_2)(h_2) + V(s_3)(h_3) \\
&= V_1(s_1)(h_1) + V_2(s_2)(h_2) + V_3(s_3)(h_3).
\end{aligned}$$

The above decomposition is a schematic example of part 1 of Theorem 2.36.

To illustrate part 2 of Theorem 2.36, we now focus on $V_1 \colon \mathfrak{a}_1 \to \mathcal{L}(H_1)$. We also assume V_1 is a topologically cyclic representation with cyclic vector $x_0 \in H_1$, i.e.

$$\overline{\{V_1(s_1)(x_0) \mid s_1 \in \mathfrak{a}_1\}} = H_1.$$

Since $\mathfrak{a}_1 = \mathfrak{l}_1 \oplus \mathfrak{l}_3 \oplus \mathfrak{l}_4$, we will let $\mathfrak{l} := \mathfrak{l}_1$ be a minimal left ideal, such that $E := V_1(\mathfrak{l}_1)(x_0)$ is a nonzero subspace of H_1 invariant under V_1. This means $H_1 = E \oplus E^{\perp}$ and V_1 is the Hilbert sum of the two restricted representations $V_{1,1} \colon \mathfrak{a}_1 \to \mathcal{L}(E)$ and $V_{1,2} \colon \mathfrak{a}_1 \to \mathcal{L}(E^{\perp})$. In other words, given $h_1 \in H_1$, if $h_1 = h_{1,1} + h_{1,2}$ with $h_{1,1} \in E$ and $h_{1,2} \in E^{\perp}$, then

$$\begin{aligned}
V_1(s_1)(h_1) &= V_1(s_1)(h_{1,1} + h_{1,2}) \\
&= V_1(s_1)(h_{1,1}) + V_1(s_1)(h_{1,2}) = V_{1,1}(s_1)(h_1) + V_{1,2}(s_1)(h_{1,2}).
\end{aligned}$$

Furthermore, $V_{1,1}$ is an *irreducible* representation which is *equivalent* to $U_{\mathfrak{l}_1} : \mathfrak{a}_1 \to \mathcal{L}(\mathfrak{l}_1)$. The equivalence is provided by the isomorphism $\varphi \colon \mathfrak{l}_1 \to E$, where $\varphi(s_{1,1}) = V_1(s_{1,1})(x_0)$ with $s_{1,1} \in \mathfrak{l}_1 \subseteq \mathfrak{a}_1$.

Theorem 2.36 will be used to prove another part of the Peter–Weyl theorem. In fact, we will only use part (2) of Theorem 2.36 when $\mathfrak{a}_k$ is finite-dimensional, and we gave a proof in this case.

2.8 Positive Hilbert Forms And Representations

Our next goal is to state the Plancherel–Godement theorem, which will be used later in discussing harmonic analysis on the space induced by a Gelfand pair. A related theorem is the Bochner–Godement theorem. These theorems apply to a *commutative* Hilbert algebra (not necessarily complete) arising from the quotient of a commutative Hilbert algebra by a left ideal induced by a bitrace satisfying two additional conditions. Therefore, we go back to positive Hilbert forms (on an involutive algebra which is not necessarily commutative) to describe the construction of a certain representation. Commutativity is only required for the Plancherel–Godement theorem and the Bochner–Godement theorem. The first step is the following proposition.

The idea is that if g is a positive Hilbert form on an involutive (not necessarily commutative) algebra A, it almost defines an inner product, but in general it fails to be positive definite because there may be nonzero elements $s \in A$, such that $g(s,s) = 0$. However, if we take the quotient of A by the set $\mathfrak{n} = \{s \in A \mid g(s,s) = 0\}$, which is a left ideal because g is a positive Hilbert form, then we can define an inner product on the quotient vector space $A/\mathfrak{n}$. If g is a bitrace, then $A/\mathfrak{n}$ is an involutive algebra.

Proposition 2.37. *Let g be a positive Hilbert form on an involutive algebra A. The set*

$$\mathfrak{n} = \{s \in A \mid g(s,s) = 0\}$$

is a left ideal in A, and

$$\mathfrak{n} = \{s \in A \mid g(s,t) = 0 \quad \textit{for all } t \in A\}.$$

If $\pi\colon A \to A/\mathfrak{n}$ is the quotient map, then there exists a hermitian inner product $\langle -, - \rangle$ on $A/\mathfrak{n}$, such that

$$\langle \pi(s), \pi(t) \rangle = g(s,t), \qquad \textit{for all } s,t \in A.$$

Furthermore, if g is a bitrace, then $\mathfrak{n}$ is a self-adjoint two-sided ideal, so the quotient $A/\mathfrak{n}$ is an algebra. The involution $s \mapsto s^*$ on A induces an involution on $A/\mathfrak{n}$ given by $(\pi(s))^* = \pi(s^*)$, and the hermitian inner product $\langle -, - \rangle$ on $A/\mathfrak{n}$ is a bitrace on $A/\mathfrak{n}$.

Proof. By the Cauchy–Schwarz inequality

$$|g(s,t)|^2 \leq g(s,s)g(t,t),$$

we see that

$$\mathfrak{n} = \{s \in A \mid g(s,t) = 0 \quad \text{for all } t \in A\}. \tag{$*_1$}$$

Since g is a positive Hilbert form, we have

$$g(xy,z) = g(y,x^*z), \qquad \text{for all } x,y \in A,$$

so if $s \in \mathfrak{n}$, that is, $g(s,s) = 0$, then by $(*_1)$,

$$g(ts,ts) = g(s,t^*st) = 0,$$

so $\mathfrak{n}$ is a left ideal. Since g is hermitian, $g(t,s) = \overline{g(s,t)}$, and we also have

$$\mathfrak{n} = \{s \in A \mid g(t,s) = 0, \quad \text{for all } t \in A\}.$$

Consequently, if $s - s' \in \mathfrak{n}$ and if $t - t' \in \mathfrak{n}$ (that is, $\pi(s) = \pi(s')$ and $\pi(t) = \pi(t')$), since

$$g(s,t-t')+g(s-s',t') = g(s,t)-g(s,t')+g(s,t')-g(s',t') = g(s,t)-g(s',t')$$

and since $s, t' \in A$, $t - t', s - s' \in \mathfrak{n}$, we have $g(s, t - t') = g(s - s', t') = 0$, and thus $g(s,t) - g(s',t') = 0$, that is,

$$g(s,t) = g(s',t').$$

We can now define the function $\langle -, - \rangle$ on $A/\mathfrak{n}$ by

$$\langle \pi(s), \pi(t) \rangle = g(s,t), \qquad \text{for all } s, t \in A,$$

and it is well-defined since $\pi(s) = \pi(s')$ and $\pi(t) = \pi(t')$ imply that $g(s,t) = g(s',t')$. It is immediately verified that $\langle -, - \rangle$ is a hermitian inner product on $A/\mathfrak{n}$.

If g is a bitrace, then

$$g(y^*, x^*) = g(x,y), \qquad \text{for all } x, y \in A,$$

so

$$g(s^*, s^*) = g(s,s), \qquad \text{for all } s \in A,$$

which shows that $\mathfrak{n}^* = \mathfrak{n}$. If $s \in \mathfrak{n}$, then $s^* \in \mathfrak{n}$, and since $(st)^* = t^*s^*$ and $\mathfrak{n}$ is a left ideal,

$$g(st, st) = g((st)^*, (st)^*) = g(t^*s^*, t^*s^*) = 0,$$

which proves that $\mathfrak{n}$ is also a right ideal. It follows that $A/\mathfrak{n}$ is an algebra.

Since $\mathfrak{n} = \mathfrak{n}^*$, the map on $A/\mathfrak{n}$ given by

$$\pi(s) = s + \mathfrak{n} \mapsto (\pi(s))^* = s^* + \mathfrak{n}^* = s^* + \mathfrak{n} = \pi(s^*)$$

is an involution, and we have $(\pi(s))^* = \pi(s^*)$.

Finally, since by definition

$$\langle \pi(s), \pi(t) \rangle = g(s,t),$$

we get

$$\langle (\pi(t))^*, (\pi(s))^* \rangle = \langle \pi(t^*), \pi(s^*) \rangle = g(t^*, s^*) = g(s,t) = \langle \pi(s), \pi(t) \rangle,$$

which shows that $\langle -, - \rangle$ is a bitrace on $A/\mathfrak{n}$. $\square$

We are now going to show that if we add one more condition to a positive Hilbert form that insures that certain linear maps on $A/\mathfrak{n}$ are continuous, then we can define a representation of A into a Hilbert space, which is the completion of $A/\mathfrak{n}$.

However, let us first observe that if g arises from a positive linear form f_{x_0} induced by a topologically cyclic representation $U\colon A \to \mathcal{L}(H)$ with cyclic vector $x_0 \in H$, where

$$f_{x_0}(s) = \langle U(s)(x_0), x_0\rangle, \quad s \in A,$$

then the hermitian space $H_0 = \{U(s)(x_0) \mid s \in A\}$, which is dense in H, is determined by g_{x_0}, and in fact, the representation U is determined by g_{x_0}.

Indeed, since $g(s,t) = g_{x_0}(s,t) = f_{x_0}(t^* s)$, we have

$$g_{x_0}(s,t) = \langle U(s)(x_0), U(t)(x_0)\rangle, \tag{$*_2$}$$

and we see that

$$g_{x_0}(s,s) = \|U(s)(x_0)\|^2.$$

Hence, $\mathfrak{n}$ is the kernel of the linear map $h\colon A \to H_0$ given by

$$h(s) = U(s)(x_0).$$

Equation ($*_2$) shows that the quotient map $\widehat{h}\colon A/\mathfrak{n} \to H_0$ is a bijective isometry. This shows that H_0, and therefore the Hilbert space H (since H_0 is dense in H), is determined by g_{x_0} up to isomorphism.

Since $h = \widehat{h} \circ \pi$, for every $s \in A$, the map $U(s)$ is completely determined by $\widehat{h}$ on H_0, because

$$U(s)(U(t)(x_0)) = U(st)(x_0) = h(st) = \widehat{h}(\pi(st)).$$

Since H_0 is dense in H, the continuous map $U(s)$ extends uniquely to H. Therefore, the representation U is determined by g_{x_0}.

We also have the following uniqueness result up to equivalence.

Proposition 2.38. *If $U_1\colon A \to \mathcal{L}(H_1)$ and $U_2\colon A \to \mathcal{L}(H_2)$ are two topologically cyclic representations with respective cyclic vectors x_0 and x_0', and if $f_{x_0} = f_{x_0'}$, that is, $\langle U_1(s)(x_0), x_0\rangle = \langle U_2(s)(x_0'), x_0'\rangle$ for all $s \in A$, then the representations U_1 and U_2 are equivalent.*

Proof. We follow Dieudonné [Dieudonné (2005)] (Chapter XV, Section 6, Theorem 15.6.7). For all $s, t \in A$ we have

$$\begin{aligned}\langle U_1(s)(x_0), U_1(t)(x_0)\rangle &= \langle U_1(t^*s)(x_0), x_0\rangle \\ &= \langle U_2(t^*s)(x'_0), x'_0\rangle \\ &= \langle U_2(s)(x'_0), U_2(t)(x'_0)\rangle;\end{aligned}$$

that is,

$$\langle U_1(s)(x_0), U_1(t)(x_0)\rangle = \langle U_2(s)(x'_0), U_2(t)(x'_0)\rangle, \qquad \text{for all } s, t \in A. \quad (*_3)$$

Since the vectors $U_1(t)(x_0)$ (resp. $U_2(t)(x'_0)$) form a dense subset H_1^0 (resp. H_2^0) of H_1 (resp. H_2) and since the inner product is continuous in each argument, we deduce that $U_1(s)(x_0)$ is orthogonal to H_1 iff $U_2(s)(x'_0)$ is orthogonal to H_2, so $U_1(s)(x_0) = 0$ iff $U_2(s)(x'_0) = 0$ for all $s \in A$. As a consequence, for every $z \in H_1^0$ and for all $s, t \in A$, such that $U_1(s)(x_0) = U_1(t)(x_0) = z$, since $U_1(s-t)(x_0) = 0$ iff $U_2(s-t)(x'_0) = 0$, we have $U_2(s)(x'_0) = U_2(t)(x'_0)$, which means that the vector $U_2(s)(x'_0)$ has the same value $z' \in H_2^0$ for all $s \in A$, such that $U_1(s)(x_0) = z$, so we can define the map $T\colon H_1^0 \to H_2^0$ by $T(z) = z'$, or equivalently,

$$T(U_1(s)(x_0)) = U_2(s)(x'_0), \qquad s \in A.$$

It is immediately verified that T is a surjective linear map, and by $(*_3)$, it is an isometry of the hermitian space H_1^0 onto the hermitian space H_2^0. But then this isomorphism extends uniquely to an isomorphism, also denoted T, between the Hilbert spaces H_1 and H_2. It remains to show that T induces an equivalence of the representations U_1 and U_2. Since H_1^0 is dense in H_1 and H_2^0 is dense in H_2, it suffices to prove that

$$T(U_1(s)(z)) = U_2(s)(T(z))$$

for all z of the form $z = U_1(t)(x_0)$ $(t \in A)$. Since

$$U_1(s)(U_1(t)(x_0)) = U_1(st)(x_0),$$

by definition of T, we have

$$\begin{aligned}T(U_1(s)(z)) &= T(U_1(s)(U_1(t)(x_0))) \\ &= T(U_1(st)(x_0)) \\ &= U_2(st)(x'_0) \\ &= U_2(s)(U_2(t)(x'_0)) = U_2(s)(T(z)),\end{aligned}$$

as claimed. ∎

We now go back to our positive Hilbert form g, and we assume that it satisfies the analog of condition (U) of Definition 2.14: for every $s \in A$, there is some $M_s \geq 0$, such that

$$g(st, st) \leq M_s g(t, t), \quad \text{for all } t \in A, \tag{U}$$

If g arises from a representation $U \colon A \to \mathcal{L}(H)$ as

$$g(s,t) = \langle U(s)(x_0), U(t)(x_0) \rangle, \qquad s, t \in A,$$

then it is easy to see that g satisfies property (U).

If the positive Hilbert form g in Proposition 2.37 satisfies condition (U), then we check immediately that the inner product $\langle -, - \rangle$ on $A/\mathfrak{n}$ given by

$$\langle \pi(s), \pi(t) \rangle = g(s,t)$$

also satisfies property (U).

The following proposition shows how to construct a representation of A from g.

Proposition 2.39. *Let g be a positive Hilbert form on an involutive algebra A satisfying condition (U), let $\mathfrak{n}_g$ be the left ideal given by*

$$\mathfrak{n}_g = \{s \in A \mid g(s,s) = 0\},$$

and suppose that the hermitian space $A/\mathfrak{n}_g$ constructed in Proposition 2.37 is separable. If so, let H_g be the Hilbert space which is the completion of $A/\mathfrak{n}_g$, so that $A/\mathfrak{n}_g$ can be identified with a dense subspace H_0 of the separable Hilbert space H_g. If $\pi_g \colon A \to A/\mathfrak{n}_g$ denotes the quotient map, for every $s \in A$, the linear map $U_g(s) \colon A/\mathfrak{n}_g \to A/\mathfrak{n}_g$ given by

$$U_g(s)(\pi_g(t)) = \pi_g(st)$$

extends to a continuous linear map $U_g(s) \colon H_g \to H_g$, and the map $s \mapsto U_g(s)$ is a representation of A in H_g. If g is a bitrace, then $A/\mathfrak{n}_g$ is an involutive algebra, and the inner product $\langle -, - \rangle_g$ on $A/\mathfrak{n}_g$ given by

$$\langle \pi_g(s), \pi_g(t) \rangle_g = g(s,t)$$

is a bitrace that satisfies property (U).

Proof. If $\pi_g(t) = \pi_g(t')$, then $\pi_g(st) = \pi_g(st')$ because $\mathfrak{n}_g$ is a left ideal. Hence for any fixed $s \in A$, the endomorphism of H_0 given by $\pi_g(t) \mapsto \pi_g(st)$ is well-defined. The definition of the inner product $\langle -, - \rangle_g$ on $A/\mathfrak{n}_g$ and condition (U) ensure that this map is continuous. Since H_0 is dense in H_g

and the map $\pi_g(t) \mapsto U_g(s)(\pi_g(t)) = \pi_g(st)$ is continuous, it extends to a continuous map $U_g(s)\colon H_g \to H_g$.

Since $\pi_g((ss')t) = \pi_g(s(s't))$, we have $U_g(ss') = U_g(s) \circ U_g(s')$. Since g is a positive Hilbert form, we have

$$g(xy, z) = g(y, x^* z) \quad \text{for all } x, y, z \in A$$

and since g is hermitian, we have by Proposition 2.37 that

$$\begin{aligned}
\langle U_g(s^*)(\pi_g(t)), \pi_g(t')\rangle_g &= \langle \pi_g(s^*t), \pi_g(t')\rangle = g(s^*t, t') \\
&= g(t, st') \\
&= \overline{g(st', t)} = \overline{\langle \pi_g(st'), \pi_g(t)\rangle} \\
&= \overline{\langle U_g(s)(\pi_g(t')), \pi_g(t)\rangle_g} \\
&= \langle \pi_g(t), U_g(s)(\pi_g(t'))\rangle_g,
\end{aligned}$$

which shows that $U_g(s^*) = (U_g(s))^*$. If A has a unit element e then $U_g(e) = \mathrm{id}$. Therefore U_g is a representation of A in H_g.

When g is bitrace, the ideal $\mathfrak{n}_g$ is a two-sided ideal, in which case $A/\mathfrak{n}_g$ is an algebra and $\pi_g(s)\pi_g(t) = \pi_g(st)$. $\square$

In general, the representation $U_g\colon A \to \mathcal{L}(H_g)$ given by Proposition 2.39 may be degenerate. It is nondegenerate if and only if the following condition holds:

the subspace spanned by the set $\{\pi_g(st) \mid s, t \in A\}$ is dense in $A/\mathfrak{n}_g$. (N)

Observe that the above condition is the generalization of condition (N) of Definition 2.14 to $A/\mathfrak{n}_g$. In Definition 2.14, the bitrace g is an inner product so $\mathfrak{n}_g = (0)$. If A has a unit element, condition (N) holds trivially.

If the positive Hilbert form g in Proposition 2.37 satisfies condition (N), then we check immediately that the inner product $\langle -, -\rangle$ on $A/\mathfrak{n}$ given by

$$\langle \pi(s), \pi(t)\rangle = g(s, t)$$

also satisfies property (N).

It can be shown that if A is a unital Banach algebra with involution, then of course condition (N) is automatically satisfied, but condition (U) also holds. This follows from the following result proven in Dieudonné [Dieudonné (2005)] (Chapter XV, Section 7, Theorem 15.6.11).

Proposition 2.40. *Let A be a unital Banach algebra with involution and unit element $e \neq 0$. For any positive linear form f, the following properties hold.*

(1) f *is continuous and* $\|f\| = f(e)$.
(2) $|f(y^*xy)| \leq \|x\| f(y^*y)$, *for all* $x, y \in A$.

Recall that since A is unital, every positive Hilbert form g arises from the positive linear form f given by $f(s) = g(s, e)$. Then for every $s \in A$,

$$g(st, st) = f((st)^*st) = f(t^*s^*st) \leq \|s^*s\| f(t^*t) = \|s^*s\| g(t, t),$$

which is condition (U) with $M_s = \|s^*s\|$.

The proof of Proposition 2.40 makes use of the following result of independent interest also proven in Dieudonné [Dieudonné (2005)] (Chapter XV, Section 7, Theorem 15.6.11.1).

Proposition 2.41. *Let* A *be a unital Banach algebra with involution and unit element* $e \neq 0$. *If* $x \in A$ *is self-adjoint and* $\|x\| < 1$, *then there exists a self-adjoint element* $y \in A$ *such that* $y^2 = e + x$.

Here is a fairly general situation where a positive Hilbert form satisfies conditions (U) and (N).

Proposition 2.42. *Let* A *be a unital separable Banach algebra with involution and unit element* $e \neq 0$. *For any positive linear form* f *on* A, *let* g *be the corresponding positive Hilbert form given by* $g(x, y) = f(y^*x)$ *for all* $x, y \in A$. *Let* $\mathcal{A}_g$ *be the unital Banach algebra which is the closure of* $U_g(A)$ *in* $\mathcal{L}(H_g)$. *Then* g *satisfies property (U), and the hermitian space* $A/\mathfrak{n}_g$ *and the unital Banach algebra* $\mathcal{A}_g \subseteq \mathcal{L}(H_g)$ *are separable.*

Proof. By Proposition 2.40, the positive Hilbert form g induced by the positive linear form f satisfies property (U). By Proposition 2.39, we have

$$\|\pi_g(x)\|^2 = \langle \pi_g(x), \pi_g(x) \rangle_g = g(x, x) = f(x^*x).$$

By Proposition 2.40, since f is continuous we get

$$\|\pi_g(x)\|^2 = f(x^*x) \leq \|f\| \|x^*x\| \leq \|f\| \|x\|^2,$$

which shows that π_g is continuous. Since π_g is also surjective, it is easy to show that the image under π_g of a countable dense set in A is dense in $A/\mathfrak{n}_g$, so $A/\mathfrak{n}_g$ is separable. Since by Proposition 2.1, the map U_g is continuous, it sends a countable dense set into a countable dense set in $\mathcal{A}_g$, so $\mathcal{A}_g$ is also separable. $\square$

2.9 The Plancherel–Godement Theorem ⊛

After these preliminaries, we assume that A is a commutative (but not necessarily complete) involutive algebra equipped with a bitrace g satisfying conditions (U) and (N). We also assume that the hermitian space $A/\mathfrak{n}_g$ is separable, as in Proposition 2.39. Then by Proposition 2.39, the bitrace g induces a representation $U_g \colon A \to \mathcal{L}(H_g)$, where the separable Hilbert space H_g is the completion of $A/\mathfrak{n}_g$, so that $A/\mathfrak{n}_g$ is dense in the Hilbert space H_g. Since property (N) holds, the representation U_g is nondegenerate. The image of A under U_g is a commutative subalgebra of the C^*-algebra $\mathcal{L}(H_g)$. Let $\mathcal{A}_g$ *be the closure of* $U_g(A)$ *in* $\mathcal{L}(H_g)$, so that $\mathcal{A}_g$ is a commutative C^*-algebra (and thus, consists of normal operators). In general, $\mathcal{A}_g$ is not separable, *but we assume it is separable.*

A particular example of a trace f on A, which gives rise to a bitrace g (such that $g(x, y) = f(y^*x)$) is provided by the hermitian characters of A. These are the characters $\chi \in \mathsf{X}(A)$, such that

$$\chi(x^*) = \overline{\chi(x)}, \qquad \text{for all } x \in A.$$

We have $\chi(x^*x) = \chi(x^*)\chi(x) = |\chi(x)|^2$, and

$$g(x, y) = \chi(y^*x) = \chi(y^*)\chi(x) = \overline{\chi(y)}\chi(x).$$

Condition (U) holds because

$$g(st, st) = \overline{\chi(st)}\chi(st) = \overline{\chi(s)}\,\overline{\chi(t)}\chi(s)\chi(t) = |\chi(s)|^2|\chi(t)|^2 = |\chi(s)|^2 g(t, t).$$

The ideal $\mathfrak{n}_g$ is the kernel of χ, so by Vol. I, Proposition 9.12(1) it is a hyperplane in A, thus $A/\mathfrak{n}_g$ is isomorphic to $\mathbb{C}$, and condition (N) follows immediately from the fact that $\chi(x) \neq 0$ implies that $\chi(x^2) = (\chi(x))^2 \neq 0$. It can be shown that the corresponding representation is irreducible.

Let $\mathbf{H}(A)$ denote the set of hermitian characters in $\mathsf{X}(A)$. The set $\mathbf{H}(A)$ is a subset of the product space $\mathbb{C}^A$ and is closed under the product topology. We give $\mathbf{H}(A)$ the topology induced by the product topology (the topology of pointwise convergence). When A is a unital commutative Banach algebra with involution, the space $\mathbf{H}(A)$ is a compact subspace of $\mathsf{X}(A)$, since $\mathsf{X}(A)$ is compact, by Vol. I, Theorem 9.19. If A is also separable, then it can be shown that $\mathsf{X}(A)$ is metrizable; see Dieudonné [Dieudonné (2005)] (Chapter XV, Section 3, Theorem 15.3.2). In general, $\mathbf{H}(A) \neq \mathsf{X}(A)$, but if A is a unital C^*-algebra, then $\mathbf{H}(A) = \mathsf{X}(A)$, by Vol. I, Proposition 9.34.

The following theorem shows that the bitraces discussed in this section and the previous one all arise from a positive measure by a process of integration involving the hermitian characters.

Theorem 2.43. *(Plancherel–Godement theorem) Let A be a commutative involutive algebra, and let g be a bitrace on A satisfying conditions (U) and (N), such that the hermitian space $A/\mathfrak{n}_g$ and the C^*-algebra $\mathcal{A}_g \subseteq \mathcal{L}(H_g)$ are separable.*

(I) We can define canonically: (1) a subspace S_g of $\mathbf{H}(A)$ whose closure in $\mathbb{C}^A$ is either S_g or $S_g \cup \{0\}$ and is metrizable and compact (so that S_g is locally compact, metrizable, and separable); (2) a (positive) Radon measure m_g on S_g, with the following properties:

(i) For each $x \in A$, define the function $\widehat{x}\colon S_g \to \mathbb{C}$ by $\widehat{x}(\chi) = \chi(x)$. Then $\widehat{x} \in \mathcal{L}^2_{m_g}(S_g)$, and we have

$$g(x, y) = \int_{S_g} \chi(xy^*)\, dm_g(\chi) = \int_{S_g} \widehat{x}(\chi)\overline{\widehat{y}(\chi)}\, dm_g(\chi), \quad \text{for all } x, y \in A. \tag{†}$$

(ii) As x runs through A, the set of functions $\widehat{x}$ is contained in $\mathcal{C}_0(S_g; \mathbb{C})$ and is dense in this Banach space.

(iii) The map $x \mapsto \widehat{x}$ factors as $T_0 \circ \pi_g$ as illustrated below,

$$\begin{array}{ccccc} A & \xrightarrow{\ \pi_g\ } & A/\mathfrak{n}_g & \longrightarrow & H_g \\ & \overset{\frown}{\searrow} & \Big\downarrow{\scriptstyle T_0} & & \Big\downarrow{\scriptstyle T} \\ & & \mathrm{L}^2_{m_g}(S_g) \cap \mathcal{C}_0(S_g, \mathbb{C}) & \longrightarrow & \mathrm{L}^2_{m_g}(S_g), \end{array}$$

where the map $T_0\colon A/\mathfrak{n}_g \to \mathrm{L}^2_{m_g}(S_g) \cap \mathcal{C}_0(S_g; \mathbb{C})$ extends to an isomorphism $T\colon H_g \to \mathrm{L}^2_{m_g}(S_g)$, such that for all $x \in A$, we have $U_g(x) = T^{-1}M(\widehat{x})T$, where $M(\widehat{x})$ is multiplication by the class of $\widehat{x}$ in $\mathrm{L}^2_{m_g}(S_g)$, where U_g is the representation $U_g\colon A \to \mathcal{L}(H_g)$.

(iv) We have

$$\big\|\lambda \mathrm{id}_{H_g} + U_g(x)\big\| = \big\|\lambda \mathrm{id}_{S_g} + \widehat{x}\big\|$$

for all $\lambda \in \mathbb{C}$ and all $x \in A$.

(II) Conversely, let S be a subspace of $\mathbf{H}(A)$ such that $S \cup \{0\}$ is compact and metrizable, and let m be a positive measure on S, such that for all $x \in A$, the function $\widehat{x}\colon S \to \mathbb{C}$ given by $\widehat{x}(\chi) = \chi(x)$ belongs to

$$\mathcal{L}^2_m(S) \cap \mathcal{C}_0(S; \mathbb{C}).$$

Then the map g' given by

$$g'(x, y) = \int_S \widehat{x}(\chi)\overline{\widehat{y}(\chi)}\, dm(\chi), \quad \textit{for all } x, y \in A$$

is a bitrace on A satisfying conditions (U) and (N), such that $A/\mathfrak{n}_{g'}$ and $\mathcal{A}_{g'}$ are separable, and we have $S_{g'} = S$ and $m_{g'} = m$.

Note that for any $x \in A$, the map $\widehat{x}$ is the restriction of the Gelfand transform to either S_g or S.

Theorem 2.43 is proven in Dieudonné [Dieudonné (2005)] (Chapter XV, Section 9, Theorem 15.9.2). The proof is long and very technical. Among other results, it uses the Gelfand–Naimark theorem (Vol. I, Theorem 9.37). We simply describe the construction of S_g since it will be used in Theorem 9.21.

Let $\mathcal{A}'_g = \mathcal{A}_g \oplus \mathbb{C}\mathrm{id}_{H_g}$. This is a closed subalgebra of $\mathcal{L}(H_g)$, so it is a commutative C^* algebra with a unit element. By Vol. I, Proposition 9.34, every character $\xi' \in \mathsf{X}(\mathcal{A}'_g)$ is hermitian, and since $U_g(a^*) = (U_g(a))^*$, the map $\xi' \circ U_g$ is either the zero map or a hermitian character of A. This means that we have a map $\omega\colon \mathsf{X}(\mathcal{A}'_g) \to \mathbf{H}(A) \cup \{0\}$ given by $\omega(\xi') = \xi' \circ U_g$.

The map ω is injective because $\xi'(\mathrm{id}_{H_g}) = 1$ for all $\xi' \in \mathsf{X}(\mathcal{A}'_g)$, and since ξ' is continuous on $\mathcal{A}'_g$ and $U_g(A)$ is dense in $\mathcal{A}_g$, the restriction of ξ' to $U_g(A)$ has a unique extension to $\mathcal{A}_g$, so the character ξ' is uniquely determined. The map ω is also continuous with respect to the weak topologies on $\mathsf{X}(\mathcal{A}'_g)$ and $\mathbb{C}^A$. Since $\mathsf{X}(\mathcal{A}'_g)$ is metrizable and compact, the same is true of its image S'_g, with

$$S'_g = \omega(\mathsf{X}(\mathcal{A}'_g)) \subseteq \mathbf{H}(A) \cup \{0\},$$

and ω is a *homeomorphism* of $\mathsf{X}(\mathcal{A}'_g)$ onto S'_g. Then the space S_g is defined as follows.

(1) If $\mathrm{id}_{H_g} \in \mathcal{A}_g$, then $\mathcal{A}'_g = \mathcal{A}_g$ and S'_g does not contain the element $0 \in \mathbb{C}^A$. We set

$$S_g = S'_g.$$

(2) If $\mathrm{id}_{H_g} \notin \mathcal{A}_g$, then $\mathcal{A}_g$ is a closed hyperplane and an ideal in $\mathcal{A}'_g$, hence it is a maximal ideal. In fact, there is a nonzero character ξ'_0 of $\mathcal{A}'_g$, whose kernel is $\mathcal{A}_g$, so $\omega(\xi'_0) = 0 \in \mathbb{C}^A$. We set

$$S_g = S'_g - \{0\}.$$

In both cases, S_g is separable, metrizable, locally compact, and the complements in $S_g \cup \{0\}$ of the compact subsets of S_g are the open sets in $S_g \cup \{0\}$ that contain 0. The Gelfand–Naimark theorem (Vol. I, Theorem 9.37) also shows that the Gelfand transform is an isometry between $\mathcal{A}'_g$ and $\mathcal{C}_0(\mathsf{X}(\mathcal{A}'_g); \mathbb{C})$. Then (ii) follows quite easily.

The construction of the measure m_g is far more involved.

We should mention that Dieudonné uses a theory of integration in which positive Radon functionals are used instead of Borel measures. However, the version in Dieudonné also states that the support of the Radon functional is the whole of S_g, and by Vol. I, Proposition A.49, since S_g is locally compact, metrizable and separable, it is σ-compact, so there is no problem in obtaining the theorem for Radon measures by using Radon–Riesz II (Vol. I, Theorem 7.15).

If the bitrace g arises from a positive linear form f, which is a trace since A is commutative, the formula (†) leads us to ask whether we also have

$$f(x) = \int_{S_g} \widehat{x}(\chi)\, dm_g(\chi).$$

The Bochner–Godement theorem provides a partial answer to this question.

Theorem 2.44. *(Bochner–Godement theorem) Let A be a commutative involutive algebra A.*

(1) Let f be a positive linear form such that the bitrace g given by $g(x, y) = f(y^* x)$ satisfies the hypotheses of the Plancherel–Godement theorem (Theorem 2.43). Then if the formula

$$f(x) = \int_{S_g} \widehat{x}(\chi)\, dm_g(\chi) \tag{BG}$$

holds and if the (positive) Radon measure m_g is bounded, then f satisfies the following condition:

There is some $M > 0$, such that $|f(x)|^2 \leq M f(xx^*)$ for all $x \in A$. (B)

(2) Conversely, let f be positive linear form on A which satisfies condition (B), and suppose that the induced bitrace g given by $g(x, y) = f(y^* x)$ satisfies condition (U), and is such that the hermitian space $A/\mathfrak{n}_g$ and the C^*-algebra $\mathcal{A}_g$ are separable. Then g also satisfies condition (N), the (positive) Radon measure m_g is bounded, and formula (BG) holds.

Theorem 2.44 is proven in Dieudonné [Dieudonné (2005)] (Chapter XV, Section 9, Theorem 15.9.4). The proof is shorter than the proof of the Plancherel–Godement theorem and also uses the Gelfand–Naimark theorem.

If A is unital, then by Proposition 2.12(3), condition (B) is satisfied with $M = f(e)$. Also, by Proposition 2.42, if A is a unital separable Banach algebra with involution, then the hermitian space $A/\mathfrak{n}_g$ and the C^*-algebra $\mathcal{A}_g$ are separable. Therefore, if A is commutative unital separable Banach algebra with involution, then the Bochner–Godement theorem part 2 applies to *any* positive linear form on A.

Here is another situation in which the Bochner–Godement theorem part 2 applies. Let $U\colon A \to \mathcal{L}(H)$ be a representation of a commutative involutive algebra A into a *separable* Hilbert space H. For any $x_0 \in H$, let f_{x_0} be the positive linear form given by

$$f_{x_0}(s) = \langle U(s)(x_0), x_0 \rangle.$$

We claim that f_{x_0} satisfies condition (B). Indeed, by Cauchy–Schwarz, we have

$$|f_{x_0}(s)|^2 \leq \|U(s)(x_0)\|^2 \|x_0\|^2 = \|x_0\|^2 f_{x_0}(s^*s).$$

We know that the bitrace g induced by f_{x_0} (given by $g(x, y) = f_{x_0}(y^*x)$) satisfies condition (U), and because H is separable, the hermitian space $A/\mathfrak{n}_g$ and the C^*-algebra $\mathcal{A}_g$ are separable. *Therefore the Bochner–Godement theorem applies to the positive linear form f_{x_0}, and Equation (BG) shows that f_{x_0} is determined by the hermitian characters of* $\mathsf{X}(A)$.

As shown just after Proposition 2.37, if the topologically cyclic representation $U\colon A \to \mathcal{L}(H)$ has $x_0 \in H$ as cyclic vector, then it is completely determined by f_{x_0}. Also, by Proposition 2.10, if A is unital, then every (nondegenerate) representation of A in a separable Hilbert space is the countable Hilbert sum of topologically cyclic representations. Therefore, if A is commutative unital algebra, then every representation $U\colon A \to \mathcal{L}(H)$ of A in a separable Hilbert space H is completely determined by the hermitian characters of $\mathsf{X}(A)$.

As a nice application of both the Plancherel–Godement theorem and the Bochner–Godement theorem we obtain a characterization of the nondegenerate representation of the algebra $\mathcal{C}(K)$ of continuous functions on a compact metrizable space K.

2.10 Representations of Algebras of Continuous Functions

Let K be a compact metrizable space. Then $\mathcal{C}_{\mathbb{C}}(K)$ is a commutative unital C^*-algebra under pointwise multiplication, see Vol. I, Example 9.1(2) and Example 9.6(2). By Vol. I, Proposition 9.22, the space K is homeomorphic to the space $\mathsf{X}(\mathcal{C}_{\mathbb{C}}(K))$ of characters of $\mathcal{C}_{\mathbb{C}}(K)$ (which are Dirac measures on K), and the Gelfand transform from $\mathcal{C}_{\mathbb{C}}(K)$ to $\mathcal{C}_{\mathbb{C}}(\mathsf{X}(\mathcal{C}_{\mathbb{C}}(K))$ can be viewed as the identity. Furthermore, by Vol. I, Proposition 9.34, the characters in $\mathsf{X}(\mathcal{C}_{\mathbb{C}}(K))$ are hermitian (so $\mathbf{H}(\mathcal{C}_{\mathbb{C}}(K)) = \mathsf{X}(\mathcal{C}_{\mathbb{C}}(K)))$. Write $A = \mathcal{C}_{\mathbb{C}}(K)$, and let $U\colon A \to \mathcal{L}(H)$ be a topologically cyclic representation of the commutative C^*-algebra A into a *separable* Hilbert space H. As discussed at the end of Section 2.9, for any cyclic vector $x_0 \in H$, we have the positive linear form f_{x_0} given by

$$f_{x_0}(u) = \langle U(u)(x_0), x_0\rangle, \quad u \in A,$$

and by Bochner–Godement the positive linear form f_{x_0} is completely determined by the space $\mathsf{X}(A)$ of hermitian characters, the representation U is completely determined by f_{x_0}, and thus by $\mathsf{X}(A) \approx K$. If g is the bitrace associated with f_{x_0}, it can be shown (exercise left to the reader) that the subspace $S_g \subseteq \mathbf{H}(A)$ introduced in the Plancherel–Godement theorem (Theorem 2.43) is actually equal to $\mathbf{H}(A) = \mathsf{X}(A) \approx K$. Therefore we can apply Theorem 2.43(iii) to obtain the following remarkable result.

Theorem 2.45. *Let K be a compact metrizable space. Every topologically cyclic representation $U\colon \mathcal{C}_{\mathbb{C}}(K) \to \mathcal{L}(H)$ of the commutative unital C^*-algebra $\mathcal{C}_{\mathbb{C}}(K)$ in a separable Hilbert space H is equivalent to a representation $M_\mu\colon \mathcal{C}_{\mathbb{C}}(K) \to \mathcal{L}(\mathrm{L}^2_\mu(K;\mathbb{C}))$ obtained as follows: for some positive Radon measure μ on K, for every $u \in \mathcal{C}_{\mathbb{C}}(K)$, let $M_\mu(u)\colon \mathrm{L}^2_\mu(K;\mathbb{C}) \to \mathrm{L}^2_\mu(K;\mathbb{C})$ be the continuous linear map induced by multiplication by u; that is, for every $f \in \mathcal{L}^2_\mu(K;\mathbb{C})$, define $M_\mu(u)(\mathbf{f})$ as the equivalence class $\mathbf{uf}$ of uf in $\mathrm{L}^2_\mu(K;\mathbb{C})$. More precisely, there is some unitary map $W\colon H \to \mathcal{L}^2_\mu(K;\mathbb{C})$, such that*

$$WU(u)W^{-1} = M_\mu(u), \quad \textit{for all } u \in \mathcal{C}_{\mathbb{C}}(K).$$

It can be shown that

$$\|M_\mu(u)\| = \|u\|_\infty .$$

If the representation $U\colon \mathcal{C}_{\mathbb{C}}(K) \to \mathcal{L}(H)$ is not topologically cyclic (it is nondegenerate since $\mathcal{C}_{\mathbb{C}}(K)$ is unital), then by Proposition 2.10, the separable Hilbert space H ($H \neq (0)$) is the Hilbert sum of a sequence $(H_n)_{n\geq 1}$

of closed subspaces $H_n \neq (0)$ of H invariant under U, and such that the restriction U_n of U to each H_n is topologically cyclic. But then we can apply Theorem 2.45 to each topologically cyclic representation U_n, so there is a positive measure μ_n associated with H_n, such that H_n is isomorphic to $\mathrm{L}^2_{\mu_n}(K;\mathbb{C})$ and U_n is equivalent to M_{μ_n}.

A particularly interesting case for the space K arises if we consider a commutative unital C^*-subalgebra $\mathcal{A}$ of $\mathcal{L}(H)$. In this case, by the Gelfand–Naimark theorem (Vol. I, Theorem 9.37), the Gelfand transform $\mathcal{G}\colon \mathcal{A} \to \mathcal{C}_{\mathbb{C}}(\mathsf{X}(\mathcal{A}))$ is an isometric isomorphism between $\mathcal{A}$ and $\mathcal{C}_{\mathbb{C}}(\mathsf{X}(\mathcal{A}))$. Furthermore, $K = \mathsf{X}(\mathcal{A})$ is compact (see Vol. I, Theorem 9.19). But the inverse Gelfand transform $\mathcal{G}^{-1}\colon \mathcal{C}_{\mathbb{C}}(\mathsf{X}(\mathcal{A})) \to \mathcal{A}$ is a representation of $\mathcal{C}_{\mathbb{C}}(\mathsf{X}(\mathcal{A}))$ as continuous operators in $\mathcal{L}(H)$, so the results obtained above apply to the representation $U = \mathcal{G}^{-1}$. We obtain Theorem 2.47 which can be viewed as a generalization of the spectral theorem for normal linear maps. Note that since $\mathcal{A}$ is a unital C^*-subalgebra of $\mathcal{L}(H)$, the continuous linear maps in $\mathcal{A}$ are indeed normal. First, we need a technical result from the theory of Hilbert spaces.

Proposition 2.46. *Let E and F be two Hilbert spaces, where E is the Hilbert sum of a countable family $(E_n)_{n\in I}$ of closed subspaces of E and F is the Hilbert sum of a countable family $(F_n)_{n\in I}$ of closed subspaces of F (with the same index family I). For each n, let $T_n\colon E_n \to F_n$ be a continuous linear map, and assume that there is a uniform bound $b > 0$, such that $\|T_n\| \leq b$ for all n. Then there is a unique continuous linear map $T\colon E \to F$, whose restriction to E_n is equal to T_n. Furthermore, the restriction of T^* to E_n is equal to T_n^*. So if the T_n are normal, so is T, and if the T_n are unitary, so is T.*

A proof of Proposition 2.46 can be found in Dieudonné [Dieudonné (2005)] (Chapter XV, Section 10, Theorem 15.10.8.1).

Theorem 2.47. *(Spectral Theorem, I) Let $\mathcal{A}$ be a commutative unital C^*-subalgebra of $\mathcal{L}(H)$, with H a separable Hilbert space. There is a measure space $(\Omega, \mathcal{M}, \mu)$, a unitary map $W\colon H \to \mathrm{L}^2_\mu(\Omega, \mathcal{M};\mathbb{C})$, and an isometric algebra homomorphism $\varphi\colon \mathcal{A} \to \mathrm{L}^\infty_\mu(\Omega, \mathcal{M};\mathbb{C})$, such that*

$$WTW^{-1} = M_\mu(\varphi(T)), \quad \textit{for all } T \in \mathcal{A}\,.$$

Recall that $M_\mu(\varphi(T))(\mathbf{f})$ is the class of $\varphi(T)f$ in $\mathrm{L}^2_\mu(\Omega, \mathcal{M};\mathbb{C})$ for every $f \in \mathcal{L}^2_\mu(\Omega, \mathcal{M};\mathbb{C})$. Furthermore, Ω can be taken as a finite or countably infinite disjoint union of copies of $\mathsf{X}(\mathcal{A})$, in such a way that μ is a positive

Radon measure μ_n on each copy and $\varphi(T) = \mathcal{G}_T$ on each copy (where $\mathcal{G}_T$ is the Gelfand transform of T).

Proof. First, consider the case where $\mathcal{G}^{-1}$ is topologically cyclic. This means that there is some $x_0 \in H$, such that $\{\mathcal{G}^{-1}(u)(x_0) \mid u \in \mathcal{C}_{\mathbb{C}}(\mathsf{X}(\mathcal{A}))\}$ is dense in H, and since $\mathcal{G}$ is a bijection between $\mathcal{A}$ and $\mathcal{C}_{\mathbb{C}}(\mathsf{X}(\mathcal{A}))$, this is equivalent to saying that $\{Tx_0 \mid T \in \mathcal{A}\}$ is dense in H. In this case Theorem 2.45 applies, so there is a positive Radon measure μ on $\mathsf{X}(\mathcal{A})$ and a unitary map $W\colon H \to \mathrm{L}^2_{\mu}(\mathsf{X}(\mathcal{A}); \mathbb{C})$, such that

$$W\mathcal{G}^{-1}(u)W^{-1} = M_{\mu}(u), \quad \text{for all } u \in \mathcal{C}_{\mathbb{C}}(\mathsf{X}(\mathcal{A})). \tag{$\dagger_1$}$$

Since $\mathcal{G}$ is a bijection between $\mathcal{A}$ and $\mathcal{C}_{\mathbb{C}}(\mathsf{X}(\mathcal{A}))$, every $u \in \mathcal{C}_{\mathbb{C}}(\mathsf{X}(\mathcal{A}))$ is of the form $u = \mathcal{G}_T$ for a unique $T \in \mathcal{A}$, so ($\dagger_1$) is equivalent to

$$WTW^{-1} = M_{\mu}(\mathcal{G}_T), \quad \text{for all } T \in \mathcal{A}. \tag{$\dagger_2$}$$

If $\mathcal{G}^{-1}$ is not topologically cyclic (it is nondegenerate since $\mathcal{C}_{\mathbb{C}}(\mathsf{X}(\mathcal{A}))$ is unital), then by Proposition 2.10, the separable Hilbert space H ($H \neq (0)$) is the Hilbert sum of a finite or countably infinite sequence $(H_n)_{n\in I}$ of closed subspaces $H_n \neq (0)$ of H invariant under $\mathcal{G}^{-1}$, and such that the restriction U_n of $\mathcal{G}^{-1}$ to each H_n is topologically cyclic. The representation U_n is given by $U_n(u)(x) = \mathcal{G}^{-1}(u)(x)$ for every $u \in \mathcal{C}_{\mathbb{C}}(\mathsf{X}(\mathcal{A}))$ and every $x \in H_n$, but since $\mathcal{G}^{-1}(u) = T$ for a unique $T \in \mathcal{A}$, we see that $U_n(u)(x) = T(x) \in H_n$ for all $x \in H_n$ so the restriction of T to H_n is a continuous linear map $T_n\colon H_n \to H_n$. But then we can apply our previous result to the topologically cyclic representation U_n, so there is a positive Radon measure μ_n associated with H_n and a unitary map $W_n\colon H_n \to \mathcal{L}^2_{\mu_n}(\mathsf{X}(\mathcal{A}); \mathbb{C})$, such that

$$W_n T_n W_n^{-1} = M_{\mu_n}(\mathcal{G}_T), \quad \text{for all } T \in \mathcal{A}, \tag{$\dagger_3$}$$

where $T_n\colon H_n \to H_n$ is the restriction of T to H_n. Let $\Omega = \coprod_{n\in I} \mathsf{X}(\mathcal{A})$ be the disjoint union of copies of $\mathsf{X}(\mathcal{A})$, one for each index $n \in I$, and let us denote the nth copy as $\mathsf{X}(\mathcal{A})_n$. We can combine the measures μ_n and the unitary maps W_n to construct a measure μ and a unitary map W as follows. Let $\mathcal{M}$ be the σ-algebra consisting of all sets $E \subseteq \Omega$, such that $E \cap \mathsf{X}(\mathcal{A})_n$ is a Borel set in $\mathsf{X}(\mathcal{A})_n$, and define the measure μ on $(\Omega, \mathcal{M})$ by

$$\mu(E) = \sum_{n\in I} \mu_n(E \cap \mathsf{X}(\mathcal{A})_n).$$

It is easily verified that μ is a measure on Ω, and obviously it is finite on each copy $\mathsf{X}(\mathcal{A})_n$. It is easy to see that the Hilbert sum of the $\mathrm{L}^2_{\mu_n}(\mathsf{X}(\mathcal{A})_n; \mathbb{C})$ is isomorphic to the Hilbert space $\mathrm{L}^2_\mu(\Omega, \mathcal{M}; \mathbb{C})$. Using Proposition 2.46, since $\|W_n\| = 1$ because W_n is a unitary map, there is a unique unitary map W

$$W = \bigoplus_{n \in I} W_n \colon H \to \mathrm{L}^2_\mu(\Omega, \mathcal{M}; \mathbb{C}),$$

whose restriction to $\mathsf{X}(\mathcal{A})_n$ is equal to W_n. The maps $\mathcal{G}_T \colon \mathsf{X}(\mathcal{A}) \to \mathbb{C}$ yield a joint map $\coprod_{n\in I} \mathcal{G}_T \colon \Omega \to \mathbb{C}$, defined such that the restriction of $\coprod_{i\in I} \mathcal{G}_T$ to $\mathsf{X}(\mathcal{A})_i$ is equal to $\mathcal{G}_T$, so we obtain a map $\varphi\colon \mathcal{A} \to \mathrm{L}^\infty_\mu(\Omega, \mathcal{M}; \mathbb{C})$, given by $\varphi(T) = \coprod_{n\in I} \mathcal{G}_T$. By construction, for every $f \in \mathcal{L}^2_\mu(\Omega, \mathcal{M}; \mathbb{C})$, we have $M_\mu(\varphi(T))f = \varphi(T)f$, where on each copy $\mathsf{X}(\mathcal{A})_n$, the function $\varphi(T)f$ is equal to the pointwise product of $\mathcal{G}_T$ and f. It remains to verify that the map φ is an isometric algebra homomorphism. This technical fact is proven in Folland [Folland (1995)] (Section 1.4, Lemma 1.46). ☐

In the special case where the commutative unital C^*-algebra is generated by T, T^* and I, where T is a continuous normal linear map on a separable Hilbert space H, we can be more precise.

Let T be a normal continuous linear map on a Hilbert space H and let $\mathcal{A}_T$ be the subalgebra of $\mathcal{L}(H)$ generated by T, T^* and I. Since T and T^* commute, $\mathcal{A}_T$ is a commutative unital C^*-algebra. Volume I, Theorem 9.38 asserts that there is an isometric isomorphism $G\colon \mathcal{A}_T \to \mathcal{C}_\mathbb{C}(\sigma(T))$, such that

$$G(T) = \mathrm{id}_{\sigma(T)}.$$

As observed in the discussion following Vol. I, Theorem 9.38, the inverse $G^{-1}\colon \mathcal{C}_\mathbb{C}(\sigma(T)) \to \mathcal{A}_T$ of the isomorphism $G\colon \mathcal{A}_T \to \mathcal{C}_\mathbb{C}(\sigma(T))$ is a representation of $\mathcal{C}_\mathbb{C}(\sigma(T))$ in H. Here we should remind the reader that $\sigma(T)$ is the spectrum of T viewed as an element of the unital C^*-algebra $\mathcal{L}(H)$. As we noted just before stating Vol. I, Theorem 9.38, this spectrum is equal to the spectrum of T viewed as an element of the unital C^*-algebra $\mathcal{A}_T$.

Remark: The representation $G^{-1}\colon \mathcal{C}_\mathbb{C}(\sigma(T)) \to \mathcal{A}_T$ is often denoted $f \mapsto f(T)$; Dieudonné, Folland, Lang and Rudin use this notation.

The proof of Theorem 2.47 can be adapted to yield the following result (see Taylor [Taylor (1986)], Appendix B, for a proof using a different method).

Theorem 2.48. *(Spectral Theorem for Normal Bounded Operators, I) Let T be a normal continuous linear map on a separable Hilbert space H.*

(1) *If the representation* $G^{-1}\colon \mathcal{C}_{\mathbb{C}}(\sigma(T)) \to \mathcal{A}_T$ *is topologically cyclic, then there is a unitary map* $W\colon H \to \mathrm{L}^2_\mu(\sigma(T);\mathbb{C})$*, such that*

$$WTW^{-1} = M_\mu(\mathrm{id}_{\sigma(T)}).$$

(2) *If* G^{-1} *is not topologically cyclic, then* H *is the Hilbert sum of a family* $(H_n)_{n\in I}$ *of closed subspaces of* H *for some finite or countably infinite index set* I*, and if* $T_n\colon H_n \to H_n$ *is the restriction of* T *to* H_n*, then for each* $n \in I$ *there is a positive Radon measure* μ_n *on* $\sigma(T_n)$ *and a unitary map* $W_n\colon H_n \to \mathrm{L}^2_{\mu_n}(\sigma(T_n);\mathbb{C})$*, such that*

$$W_nT_nW_n^{-1} = M_{\mu_n}(\mathrm{id}_{\sigma(T_n)}).$$

If we let $\Omega = \coprod_{n\in I}\sigma(T_n)$ *be the disjoint union of the* $\sigma(T_n)$*, then we can define a* σ*-algebra* $\mathcal{M}$ *on* Ω*, a measure* μ *on* Ω *whose restriction to each* $\sigma(T_n)$ *is equal to* μ_n*, and a unitary map* $W\colon H \to \mathrm{L}^2_\mu(\Omega,\mathcal{M};\mathbb{C})$*, such that*

$$WTW^{-1} = M_\mu(\mathrm{id}_\Omega).$$

We simply indicate how to prove part (1) of Theorem 2.48, leaving the proof of part (2) as an exercise. Since we are assuming that the representation $G^{-1}\colon \mathcal{C}_{\mathbb{C}}(\sigma(T)) \to \mathcal{A}_T$ is topologically cyclic, Theorem 2.45 applies. Therefore, there is a positive Radon measure μ on $\sigma(T)$ and a unitary map $W\colon H \to \mathrm{L}^2_\mu(\sigma(T);\mathbb{C})$, such that

$$WG^{-1}(u)W^{-1} = M_\mu(u), \quad \text{for all } u \in \mathcal{C}_{\mathbb{C}}(\sigma(T)).$$

Since G is a bijection between $\mathcal{A}_T$ and $\mathcal{C}_{\mathbb{C}}(\sigma(T))$, such that $G(T) = \mathrm{id}_{\sigma(T)}$, for $u = G(T)$, we obtain

$$WTW^{-1} = M_\mu(\mathrm{id}_{\sigma(T)}).$$

The following fact is proven in Dieudonné [Dieudonné (2005)] (Chapter XV, Section 11, Proposition 15.11.5).

Proposition 2.49. *The spectrum* $\sigma(T)$ *of the normal continuous linear map* T *as above is the closure in* $\mathbb{C}$ *of the union* $\bigcup_{n\in I}\sigma(T_n)$.

The measures μ_n also determine which scalars $\lambda \in \sigma(T)$ are eigenvalues of T. Recall that a scalar $\lambda \in \mathbb{C}$ is an eigenvalue of T iff $\mathrm{Ker}\,(\lambda\mathrm{id} - T)$ is nontrivial, equivalently iff $\lambda\mathrm{id} - T$ is not injective. If λ is an eigenvalue of T, then $E(T,\lambda) = \mathrm{Ker}\,(\lambda\mathrm{id} - T)$ is called the eigenspace associated with λ, and the nonzero vectors in $E(T,\lambda)$ are the eigenvectors of T associated with λ. On the other hand, the spectrum of T consists of those $\lambda \in \mathbb{C}$, such that $\lambda \mathrm{id} - T$ is *not* invertible,

If H is finite-dimensional, a linear map is not invertible iff it is not injective, so in this case eigenvalues and spectral values coincide. But if H is infinite-dimensional, a linear map may be injective and yet not invertible because it is not surjective. As a consequence, if $\lambda \in \sigma(T)$, the set of eigenvectors associated with λ may be empty. There are continuous linear maps that have no eigenvalues. See Vol. I, Example 9.2.

The following result proven in Dieudonné [Dieudonné (2005)] (Chapter XV, Section 11, Proposition 15.11.6) gives a necessary and sufficient condition for a spectral value to be an eigenvalue. A similar result is proven in Rudin [Rudin (1991)] (Theorem 12.29) in the framework of projection-valued measures, which will be discussed in Section 2.12.

In what follows, we use the notation of Theorem 2.48. First, it is easy to see that $\lambda \in \sigma(T)$ is an eigenvalue of T if there is a nonempty subset $J \subseteq I$, such that $\lambda \in \sigma(T_n)$ is an eigenvalue of T_n for all $n \in J$.

Proposition 2.50. *Let T be a normal continuous linear map on a separable Hilbert space H. Using the notation of Theorem 2.48, a scalar $\lambda \in \sigma(T_n)$ is an eigenvalue of T_n iff $\mu_n(\{\lambda\}) \neq 0$. The space spanned by the eigenvectors of T_n associated with λ is a one-dimensional space D_n, which is an orthogonal projection of H_n.*

As a corollary of Proposition 2.50, if $\lambda \in \sigma(T)$ is an eigenvalue of T, then there is a finite or countably infinite index set J, such that the eigenspace $E(T, \lambda)$ is the Hilbert sum of the one-dimensional spaces D_n (with $n \in J$). Each D_n is spanned by the eigenvectors of T_n associated with λ. If λ and μ are two distinct eigenvalues of T, then $E(T, \lambda)$ and $E(T, \mu)$ are orthogonal.

We state a few more facts whose proof is left as an exercise. A normal continuous linear map is hermitian iff $\sigma(T) \subseteq \mathbb{R}$, unitary iff $\sigma(T) \subseteq \mathbf{U}(1)$.

Finally, a stronger result is obtained if T is a normal (continuous) linear map which is also compact. Recall that this means that the closure of $T(B)$ is compact if B is bounded. It can be shown that the spectrum $\sigma(T)$ of a compact operator is finite or countably infinite, and that the nonzero spectral values are eigenvalues of T. If H is infinite-dimensional, then $0 \in \sigma(T)$; see Lang [Lang (1996)] (Chapter XVII, Section 3). The spaces $E(T, \lambda_n)$, with $\lambda_n \in \sigma(T) - \{0\}$ are finite-dimensional, and together with $\operatorname{Ker} T$ form a Hilbert sum in H. These subspaces are all pairwise orthogonal; see Dieudonné [Dieudonné (2005)] (Chapter XV, Section 11, no. 15.11.14) and Folland [Folland (1995)] (Section 1.4, Theorem 1.52).

2.11 Extending Representations from $\mathcal{C}_{\mathbb{C}}(K)$ to $B(K)$

The next crucial step is to realize that a representation $U\colon \mathcal{C}_{\mathbb{C}}(K) \to \mathcal{L}(H)$ as above determines certain complex Radon measures $\mu_{u,v}$ on K, and that conversely these measures determine U. Then it is possible to extend the representation U of $\mathcal{C}_{\mathbb{C}}(K)$ to the larger commutative unital C^*-algebra $B(K)$ of bounded Borel measurable functions on K. What we gain in doing so is the fact that we can apply the extended representation U to the characteristic functions χ_E of Borel sets E (on K) (the functions χ_E are not continuous), and such operators $U(\chi_E)$ turn out to be orthogonal projections in $\mathcal{L}(H)$. These families of projections have properties that make them *projection-valued measures* (also called *spectral measures*), and such measures can be used to define representations of $B(K)$ that generalize the notion of integral.

For any fixed $u, v \in H$, consider the functional $\Phi_{u,v}$ on $\mathcal{C}_{\mathbb{C}}(K)$ given by

$$\Phi_{u,v}(f) = \langle U(f)(u), v\rangle, \quad f \in \mathcal{C}_{\mathbb{C}}(K).$$

Recall that from Proposition 2.1, we have $\|U(f)\| \leq \|f\|_\infty$, so by Cauchy–Schwarz, we have

$$|\langle U(f)(u), v\rangle| \leq \|U(f)(u)\| \, \|v\| \leq \|U(f)\| \, \|u\| \, \|v\| \leq \|f\|_\infty \, \|u\| \, \|v\| ,$$

so the functional $\Phi_{u,v}$ is bounded (continuous). By Radon–Riesz III, there is a unique complex Radon measure $\mu_{u,v}$ on K, such that

$$\langle U(f)(u), v\rangle = \int_K f \, d\mu_{u,v}, \quad f \in \mathcal{C}_{\mathbb{C}}(K). \tag{$*_1$}$$

The measure $\mu_{u,v}$ is often called a *spectral measure*; see Lang [Lang (1996)] (Chapter XX, Section 1). From the definition, we have

$$\|\mu_{u,v}\| \leq \|u\| \, \|v\| .$$

The following properties are easy to prove; see Folland [Folland (1995)] (Section 1.4, Proposition 1.34).

Proposition 2.51. *The map from $H \times H$ to $\mathbb{C}$ given by $(u, v) \mapsto \mu_{u,v}$ is sesquilinear. Moreover, $\mu_{v,u} = \overline{\mu_{u,v}}$, and $\mu_{u,u}$ is a positive measure.*

The next step is to extend U to the commutative unital C^*-algebra $B(K)$ of bounded Borel measurable functions on K. This can be done in two ways.

(1) By using Theorem 2.45 for a topologically cyclic representation and the decomposition of H as a Hilbert sum for an arbitrary representation, as explained above. This is the approach followed by Dieudonné [Dieudonné (2005)] (Chapter XV, Section 10).

(2) A faster way is to use the fact that every function $f \in B(K)$ is limit of a sequence of continuous functions f_n converging to f pointwise almost everywhere with respect to the measure $|\mu_{u,v}|$, such that $\|f_n\| \leq \|f\|_\infty$; see Lang [Lang (1996)] (Chapter XX, Section 1), and then to use the dominated convergence theorem (Vol. I, Theorem 5.34).

Using the second approach, we see that

$$\left| \int_K f \, d\mu_{u,v} \right| \leq \|f\|_\infty \, \|u\| \, \|v\| \,, \tag{$*_2$}$$

so for any fixed $f \in B(K)$ and any fixed u the map $v \mapsto \int_K f \, d\mu_{u,v}$ is a continuous semi-linear form on H, and by the Riesz representation theorem, there is a unique vector $\widetilde{U}(f)(u) \in H$, such that

$$\langle \widetilde{U}(f)(u), v \rangle = \int_K f \, d\mu_{u,v} \quad \text{for all } v \in H.$$

However, ($*_2$) shows that the map $u \mapsto \widetilde{U}(f)(u)$ is continuous, so indeed $\widetilde{U}(f) \in \mathcal{L}(H)$. Therefore, $\widetilde{U}(f) \in \mathcal{L}(H)$ is completely determined by the equation

$$\langle \widetilde{U}(f)(u), v \rangle = \int_K f \, d\mu_{u,v} \quad \text{for all } u, v \in H \text{ and for all } f \in B(K). \tag{$*_3$}$$

We also have

$$\left\| \widetilde{U}(f) \right\| \leq \|f\|_\infty \,.$$

It remains to prove that $\widetilde{U}$ is a representation of $B(K)$. The proof of Theorem 1.36 in Folland [Folland (1995)] applies immediately because all is needed is Proposition 2.51 and Equation ($*_3$).

Proposition 2.52. *The map $\widetilde{U} \colon B(K) \to \mathcal{L}(H)$ is a representation. Furthermore, if (f_n) is a uniformly bounded sequence of functions in $B(K)$ which converge pointwise to $f \in B(K)$, then*

$$\lim_{n \mapsto \infty} \langle \widetilde{U}(f_n)(u), v \rangle = \langle \widetilde{U}(f)(u), v \rangle \quad \textit{for all } u, v \in H.$$

We say that $\widetilde{U}(f_n)$ converges to $\widetilde{U}(f)$ in the weak operator topology.

For the sake of completeness, we define three notions of convergence on $\mathcal{L}(H)$, where H is a Hilbert space.

Definition 2.21. Let H be a Hilbert space with inner product $\langle -, - \rangle$ and corresponding norm $\|u\| = \sqrt{\langle u, u \rangle}$ ($u \in H$). As usual, we have the operator norm on $\mathcal{L}(H)$ defined such that for any $f \in \mathcal{L}(H)$,

$$\|f\| = \sup\{\|f(u)\| \mid \|u\| = 1, u \in H\}.$$

We have three notions of convergence corresponding to the following topologies on $\mathcal{L}(H)$:

(1) The *norm topology* on $\mathcal{L}(H)$ is the topology associated with the operator norm. A sequence (f_n) of continuous linear maps $f_n \in \mathcal{L}(H)$ converges to a continuous linear map $f \in \mathcal{L}(H)$ if $\lim_{n \mapsto \infty} \|f - f_n\| = 0$.
(2) The *topology of pointwise convergence* or *strong operator topology*, defined by the family of semi-norms $p_u(f) = \|f(u)\|$, for all $u \in H$, $f \in \mathcal{L}(H)$. A sequence (f_n) of maps $f_n \in \mathcal{L}(H)$ converges to a map $f \in \mathcal{L}(H)$ if

$$\lim_{n \mapsto \infty} \|f(u) - f_n(u)\| = 0 \quad \text{for all } u \in H.$$

This is *pointwise convergence*.
(3) The *weak operator topology*, defined by the family of semi-norms $p_{u,v}(f) = |\langle f(u), v \rangle|$, for all $u, v \in H$, $f \in \mathcal{L}(H)$. A sequence (f_n) of maps $f_n \in \mathcal{L}(H)$ converges to a map $f \in \mathcal{L}(H)$ if

$$\lim_{n \mapsto \infty} \langle f(u) - f_n(u), v \rangle = 0 \quad \text{for all } u, v \in H.$$

This is *weak pointwise convergence*.

From now on, to simplify notation we usually write U instead of $\widetilde{U}$. If we denote by μ the family of complex Radon measures $(\mu_{u,v})_{(u,v) \in H \times H}$, the usual convention is to write

$$U(f) = \int f \, d\mu.$$

Such integrals are often called *weak integrals*.

As we said just before Theorem 2.47, if $\mathcal{A}$ is a commutative unital C^*-subalgebra of $\mathcal{L}(H)$, then the inverse Gelfand transform $\mathcal{G}^{-1} \colon \mathcal{C}_{\mathbb{C}}(\mathsf{X}(\mathcal{A})) \to \mathcal{A}$ is a representation of $\mathcal{C}_{\mathbb{C}}(\mathsf{X}(\mathcal{A}))$ as continuous operators in $\mathcal{L}(H)$, and $K = \mathsf{X}(\mathcal{A})$ is compact. Thus, the results obtained above apply to the representation $U = \mathcal{G}^{-1}$. Since the Gelfand transform $\mathcal{G}_T$ belongs to

$\mathcal{C}_{\mathbb{C}}(\mathsf{X}(\mathcal{A})) \subseteq B(\mathsf{X}(\mathcal{A}))$ for every $T \in \mathcal{A}$, and since $U(\mathcal{G}_T) = \mathcal{G}^{-1}(\mathcal{G}_T) = T$, Equation $(*_1)$ says that

$$\langle T(u), v\rangle = \int_{\mathsf{X}(\mathcal{A})} \mathcal{G}_T \, d\mu_{u,v} \quad \text{for all } u, v \in H \text{ and for all } T \in \mathcal{A}, \qquad (*_4)$$

which is also written as

$$T = \int \mathcal{G}_T \, d\mu.$$

As a consequence, we obtain a preliminary version of another spectral theorem for a commutative unital C^*-subalgebra $\mathcal{A}$ of $\mathcal{L}(H)$.

Theorem 2.53. *Let $\mathcal{A}$ be a commutative unital C^*-subalgebra of $\mathcal{L}(H)$. The extension $U\colon B(\mathsf{X}(\mathcal{A})) \to \mathcal{L}(H)$ of $\mathcal{G}^{-1}\colon \mathcal{C}_{\mathbb{C}}(\mathsf{X}(\mathcal{A})) \to \mathcal{A}$ is a representation. There is a family of complex Radon measures $(\mu_{u,v})_{(u,v)\in H\times H}$ on $\mathsf{X}(\mathcal{A})$ satisfying the properties of Proposition 2.51, such that the following properties hold:*

$$\langle T(u), v\rangle = \int_{\mathsf{X}(\mathcal{A})} \mathcal{G}_T \, d\mu_{u,v} \quad \textit{for all } u, v \in H \textit{ and for all } T \in \mathcal{A},$$

and

$$\langle U(f)(u), v\rangle = \int_{\mathsf{X}(\mathcal{A})} f \, d\mu_{u,v} \quad \textit{for all } u, v \in H \textit{ and for all } f \in B(\mathsf{X}(\mathcal{A}))),$$

where U is the extension of $\mathcal{G}^{-1}$ to $B(\mathsf{X}(\mathcal{A}))$. In short, we write

$$T = \int \mathcal{G}_T \, d\mu, \quad U(f) = \int f \, d\mu.$$

Remark: For the sake of simplicity, we omitted to state another property that should be included in Theorem 2.53. This is the fact that for any $S \in \mathcal{L}(H)$, if S commutes with every $T \in \mathcal{A}$, then S commutes with $U(f)$ for every $f \in B(\mathsf{X}(\mathcal{A}))$; see Folland [Folland (1995)] (Proposition 1.36). This property is used to prove Schur's Lemma for irreducible unitary representations.

Remarkably, the families $(\mu_{u,v})_{(u,v)\in H\times H}$ of measures arise from families of projection-valued measures. Such projection-valued measures are defined by the operators $U(\chi_E)$, which are orthogonal projections in $\mathcal{L}(H)$, where E is a Borel set on K.

Proposition 2.54. *Consider the representation $\widetilde{U}\colon B(K) \to \mathcal{L}(H)$ that extends the representation $U\colon \mathcal{C}_{\mathbb{C}}(K) \to \mathcal{L}(H)$. The map P defined by $P(E) = \widetilde{U}(\chi_E)$, where E is any Borel set in K, has the following properties:*

(1) Each $P(E)$ is an orthogonal projection in $\mathcal{L}(H)$.
(2) $P(\emptyset) = 0$ and $P(K) = I$.
(3) $P(E \cap F) = P(E) \circ P(F)$.
(4) For any family $(E_i)_{i \geq 1}$ of pairwise disjoint Borel sets, we have

$$P\left(\bigcup_{i\geq 1} E_i\right) = \sum_{i\geq 1} P(E_i),$$

which means that if we define F_n and F as $F_n = \bigcup_{i=1}^n E_i$ and $F = \bigcup_{i\geq 1} E_i$, then $\lim_{n\mapsto\infty} \|P(F)(u) - P(F_n)(u)\| = 0$ for all $u \in H$ (convergence in the strong operator topology).

Proof. Since $\chi_E^2 = \chi_E = \overline{\chi_E}$, we have $P(E)^2 = P(E) = P(E)^*$. The equation $P(E)^2 = P(E)$ says that $P(E)$ is a projection, and since it is well-known from linear algebra that $\mathrm{Ker}(P(E))^\perp = \mathrm{Im}(P(E)^*)$, we have $\mathrm{Ker}(P(E))^\perp = \mathrm{Im}(P(E))$, which means that $\mathrm{Ker}(P(E))$ is the orthogonal complement of $\mathrm{Im}(P(E))$. Property (2) is obvious, and (3) follows from the fact that $\chi_{E\cap F} = \chi_E\chi_F$.

For a finite family $(E_i)_{i=1}^n$ of pairwise disjoint subsets, since

$$\chi_{\bigcup_{i=1}^n E_i} = \sum_{i=1}^n \chi_{E_i},$$

we have

$$P\left(\bigcup_{i=1}^n E_i\right) = \sum_{i=1}^n P(E_i).$$

Otherwise, if we write $F_n = \bigcup_{i=1}^n E_i$ and $F = \bigcup_{i\geq 1} E_i$ as above, since the sequence (χ_{F_n}) is uniformly bounded and converges pointwise to χ_F, by Proposition 2.52,

$$\sum_{i=1}^n P(E_i) = P(F_n) \quad \text{converges weakly to} \quad P(F).$$

In particular, $\lim_{n\mapsto\infty}\langle (P(F) - P(F_n))(u), u\rangle = 0$ for all $u \in H$. But F is the disjoint union $F = F_n \cup (F - F_n)$, so $P(F) = P(F_n) + P(F - F_n)$. Since $P(F - F_n)$ is an orthogonal projection, for every $u \in H$, we have

$$\begin{aligned}
\|(P(F) - P(F_n))(u)\|^2 &= \|(P(F - F_n)(u)\|^2 \\
&= \langle P(F - F_n)(u), P(F - F_n)(u)\rangle \\
&= \langle P(F - F_n)(u), u\rangle = \langle (P(F) - P(F_n))(u), u\rangle,
\end{aligned}$$

so weak convergence, which means that $\lim_{n\mapsto\infty}\langle (P(F) - P(F_n))(u), u\rangle = 0$, implies strong convergence, namely $\lim_{n\mapsto\infty} \|(P(F) - P(F_n))(u)\| = 0$. ☐

Proposition 2.55. *If E and F are two disjoint Borel sets, then the ranges of $P(E)$ and $P(F)$ are orthogonal.*

Proof. Since $P(E)$ and $P(F)$ are orthogonal projection, for any $u, v \in H$, we have

$$\begin{aligned}\langle P(E)(u), P(F)(v)\rangle &= \langle P(F)P(E)(u), v\rangle \\ &= \langle P(E \cap F)(u), v\rangle \\ &= \langle P(\emptyset)(u), v\rangle = \langle 0, v\rangle = 0,\end{aligned}$$

as claimed. ☐

We can use families of projections on a Hilbert space satisfying the properties of Proposition 2.54 to define families of measures similar to the $\mu_{u,v}$ introduced earlier, and using these measures, to also define representations.

2.12 Projection-Valued Measures and Representations

Let $(\Omega, \mathcal{M})$ be a measure space, where $\mathcal{M}$ is a σ-algebra on the set Ω (since we use the notation $\mathcal{A}$ to denote an algebra, to avoid a clash of notation we denote a σ-algebra by $\mathcal{M}$, departing from our earlier notation).

Definition 2.22. Given a measure space $(\Omega, \mathcal{M})$ and a Hilbert space H, a *projection-valued measure* is a map $P\colon \mathcal{M} \to \mathcal{L}(H)$ assigning an orthogonal projection $P(E)$ in $\mathcal{L}(H)$ to every set $E \in \mathcal{M}$, such that the properties of Proposition 2.54 hold, namely:

(1) Each $P(E)$ is an orthogonal projection in $\mathcal{L}(H)$ (which means that $P(E)^2 = P(E) = P(E)^*$).
(2) $P(\emptyset) = 0$ and $P(\Omega) = I$.
(3) $P(E \cap F) = P(E) \circ P(F)$.
(4) For any family $(E_i)_{i\geq 1}$ of pairwise disjoint sets in $\mathcal{M}$, we have

$$P\left(\bigcup_{i\geq 1} E_i\right) = \sum_{i\geq 1} P(E_i),$$

which means that if we define F_n and F as $F_n = \bigcup_{i=1}^n E_i$ and $F = \bigcup_{i\geq 1} E_i$, then $\lim_{n\mapsto\infty} \|P(F)(u) - P(F_n)(u)\| = 0$ for all $u \in H$ (convergence in the strong operator topology).

We can now define analogs of the measures $\mu_{u,v}$.

Definition 2.23. Let P be a projection-valued measure of a measure space $(\Omega, \mathcal{M})$ in a Hilbert space H. For all $u, v \in H$, define $P_{u,v}(E)$ as

$$P_{u,v}(E) = \langle P(E)(u), v\rangle$$

for all $E \in \mathcal{M}$.

Properties (2) and (4) of Definition 2.22 imply that each $P_{u,v}$ is a complex measure on Ω. The map $(u, v) \mapsto P_{u,v}$ is obviously sesquilinear. Since $P(E)^* = P(E)$, we have

$$P_{v,u}(E) = \langle P(E)(v), u\rangle = \langle v, P(E)(u)\rangle = \overline{\langle P(E)(u), v\rangle} = \overline{P_{u,v}(E)},$$

so $P_{v,u} = \overline{P_{u,v}}$. Since $P(E)^* = P(E) = P(E)^2$, we also have

$$\begin{aligned} P_{u,u}(E) &= \langle P(E)(u), u\rangle = \langle P(E)^2(u), u\rangle \\ &= \langle P(E)(u), P(E)^*(u)\rangle = \langle P(E)(u), P(E)(u)\rangle \geq 0, \end{aligned}$$

so each $P_{u,u}$ is a positive measure. Furthermore,

$$\|P_{u,u}\| = P_{u,u}(\Omega) = \langle P(\Omega)(u), u\rangle = \langle u, u\rangle = \|u\|^2 .$$

Finally, if $B(\Omega, \mathcal{M})$ denotes the space of bounded measurable functions on Ω, we will show that for all $u, v \in H$ and every $f \in B(\Omega, \mathcal{M})$, there is a unique continuous operator $U(f) \in \mathcal{L}(H)$, such that

$$\langle U(f)(u), v\rangle = \int_\Omega f\, dP_{u,v}, \quad \text{for all } u, v \in H \text{ and for all } f \in B(\Omega, \mathcal{M}).$$

Without any other assumptions, $B(\Omega, \mathcal{M})$ is simply a commutative unital algebra (under pointwise multiplication). We can think of $U(f)$ as a generalized integral,

$$U(f) = \int f\, dP.$$

Since $\|P_{u,u}\| = \|u\|^2$, for any $f \in B(\Omega, \mathcal{M})$, we have

$$\left| \int f\, dP_{u,u} \right| \leq \|f\|_\infty \|P_{u,u}\| = \|f\|_\infty \|u\|^2 .$$

By linear algebra, we have the polarization identity

$$\begin{aligned} \langle P(E)(u), v\rangle = \langle P(E)(u), P(E)(v)\rangle &= \frac{1}{4}(\|P(E)(u+v)\|^2 - \|P(E)(u-v)\|^2 \\ &\quad + i(\|P(E)(u+iv)\|^2 - \|P(E)(u-iv)\|^2)), \end{aligned}$$

so

$$\begin{aligned} P_{u,v}(E) = \frac{1}{4}(&P_{u+v,u+v}(E) - P_{u-v,u-v}(E) \\ &+ i(P_{u+iv,u+iv}(E) - P_{u-iv,u-iv}(E))). \end{aligned}$$

Note that all measures on the right-hand side are positive real measures. As a consequence,

$$\begin{aligned}\int f\,dP_{u,v} = \frac{1}{4}\Big(&\int f\,dP_{u+v,u+v} - \int f\,dP_{u-v,u-v}\\ &+ i\Big(\int f\,dP_{u+iv,u+iv} - \int f\,dP_{u-iv,u-iv}\Big)\Big),\end{aligned}$$

and so, for all $u, v \in H$, such that $\|u\| = \|v\| = 1$, we have

$$\begin{aligned}\left|\int f\,dP_{u,v}\right| &\leq \frac{1}{4}\,\|f\|_{\infty}\,(\|u+v\|^2 + \|u-v\|^2 + \|u-iv\|^2 + \|u-iv\|^2)\\ &\leq 4\,\|f\|_{\infty}\,.\end{aligned}$$

Replacing u by $u/\|u\|$ and v by $v/\|v\|$ $(u, v \neq 0)$, we obtain

$$\left|\int f\,dP_{u,v}\right| \leq 4\,\|f\|_{\infty}\,\|u\|\,\|v\|\,. \qquad (*_5)$$

As a consequence, for f and u fixed we obtain a continuous semi-linear form on H, so by the Riesz representation theorem there is a unique $U(f)(u) \in H$, such that

$$\langle U(f)(u), v\rangle = \int f\,dP_{u,v}, \quad \text{for all } u \in H,$$

but the map $u \mapsto U(f)(u)$ is also continuous, so we have a unique linear map $U(f) \in \mathcal{L}(H)$, such that

$$\langle U(f)(u), v\rangle = \int f\,dP_{u,v}, \quad \text{for all } u, v \in H \text{ and all } f \in B(\Omega, \mathcal{M}). \quad (*_6)$$

It is customary to write

$$U(f) = \int f\,dP.$$

Remark: If f is a step function $f = \sum_{i=1}^{n} c_j \chi_{E_i}$ (with $c_j \in \mathbb{C}$), then

$$\int f\,dP_{u,v} = \sum_{i=1}^{n} c_i P_{u,v}(E_i) = \sum_{i=1}^{n} c_i \langle P(E_i)(u), v\rangle = \left\langle \sum_{i=1}^{n} c_i P(E_i)(u), v\right\rangle,$$

so in this case, we have

$$\int f\,dP = \sum_{i=1}^{n} c_i P(E_i),$$

which is reassuring! Using the above fact and the definition of the integral using limits of step functions, the following result proven in Folland [Folland (1995)] (Theorem 1.43) shows that the map $f \mapsto U(f)$ as defined by $(*_6)$ is a representation of algebras. The preceding discussion and this last fact are combined in the following important theorem.

Theorem 2.56. *Let P be a projection-valued measure on a Hilbert space H. For every $f \in B(\Omega, \mathcal{M})$, there is a unique linear map $U(f) \in \mathcal{L}(H)$, such that*

$$\langle U(f)(u), v\rangle = \int f\, dP_{u,v}, \quad \textit{for all } u, v \in H \textit{ and all } f \in B(\Omega, \mathcal{M}).$$

For short, we write

$$U(f) = \int f\, dP.$$

The map $U\colon B(\Omega, \mathcal{M}) \to \mathcal{L}(H)$ is an isometric representation of algebras (it is linear and a multiplicative homomorphism). Moreover,

$$\|U(f)(u)\|^2 = \int |f|^2\, dP_{u,u} \quad \textit{for all } u \in H \textit{ and all } f \in B(\Omega, \mathcal{M}).$$

Remark: Rudin [Rudin (1991)] defines the notion of *resolution of the identity*, which is different from the notion of projection-valued measure, but essentially equivalent. A resolution of the identity need not satisfy condition (4) of Definition 2.22, but it is required that the $P_{u,v}$ as defined in Definition 2.23 are complex measures. Rudin also proves Theorem 2.56 in terms of resolutions of the identity; see Rudin [Rudin (1991)], Theorem 12.21.

Let us now return to the case of a representation $U\colon \mathcal{C}_{\mathbb{C}}(K) \to \mathcal{L}(H)$ and its extension $\widetilde{U}\colon B(K) \to \mathcal{L}(H)$ where K is compact, and let $\Omega = K$ and let $\mathcal{M}$ be the σ-algebra generated by the open sets in K. Since in this case $P(E) = \widetilde{U}(\chi_E)$, we have

$$P_{u,v}(E) = \langle P(E)(u), v\rangle = \langle \widetilde{U}(\chi_E)(u), v\rangle = \int_K \chi_E\, d\mu_{u,v} = \mu_{u,v}(E),$$

for all E, so we deduce that

$$P_{u,v} = \mu_{u,v}.$$

In particular, these are Radon measures, so they are regular. What this shows is that any representation $\widetilde{U}\colon B(K) \to \mathcal{L}(H)$ where K is compact arises from some regular projection-valued measure on a Hilbert space H.

As an application, if $\mathcal{A}$ is a commutative unital C^*-subalgebra of $\mathcal{L}(H)$, we showed that the inverse Gelfand transform $\mathcal{G}^{-1}\colon \mathcal{C}(\mathsf{X}(\mathcal{A}); \mathbb{C}) \to \mathcal{A}$ is a representation of $\mathcal{C}(\mathsf{X}(\mathcal{A}); \mathbb{C})$ as continuous operators in $\mathcal{L}(H)$. Here we have $K = \mathsf{X}(\mathcal{A})$. Thus we obtain a version of the spectral theorem for a commutative unital C^*-subalgebra of $\mathcal{L}(H)$.

Theorem 2.57. *(Spectral Theorem, II) Let $\mathcal{A}$ be a commutative unital C^*-subalgebra of $\mathcal{L}(H)$. There is a regular projection-valued measure P on $\mathsf{X}(\mathcal{A})$, such that the following properties hold:*

$$\langle T(u), v\rangle = \int_{\mathsf{X}(\mathcal{A})} \mathcal{G}_T \, dP_{u,v}, \quad \textit{for all } u, v \in H \textit{ and for all } T \in \mathcal{A},$$

and

$$\langle U(f)(u), v\rangle = \int_{\mathsf{X}(\mathcal{A})} f \, dP_{u,v}, \quad \textit{for all } u, v \in H \textit{ and for all } f \in B(\mathsf{X}(\mathcal{A})),$$

where U is the extension of $\mathcal{G}^{-1}$ to $B(\mathsf{X}(\mathcal{A}))$. In short, we write

$$T = \int \mathcal{G}_T \, dP, \quad U(f) = \int f \, dP.$$

In fact, it can be shown that the projection-valued measure P as above is unique; see Folland [Folland (1995)] (Theorem 1.44). Theorem 2.57 is also proven by Rudin in terms of resolutions of the identity; see Rudin [Rudin (1991)], Theorem 12.22.

As an application of Theorem 2.57, we obtain another version of the spectral theorem for normal continuous linear maps on a Hilbert space H.

Let T be a normal continuous linear map on a Hilbert space H and let $\mathcal{A}_T$ be the commutative unital C^*-algebra generated by T, T^* and I (which is a subalgebra of $\mathcal{L}(H)$). Recall that Vol. I, Theorem 9.38 asserts that there is an isometric isomorphism $G\colon \mathcal{A}_T \to \mathcal{C}_{\mathbb{C}}(\sigma(T))$, such that

$$G(T) = \mathrm{id}_{\sigma(T)}$$

and that the inverse $G^{-1}\colon \mathcal{C}_{\mathbb{C}}(\sigma(T)) \to \mathcal{A}_T$ of G is a representation of $\mathcal{C}_{\mathbb{C}}(\sigma(T))$ in H. By applying Theorem 2.57 to the representation G^{-1}, we obtain the following result.

Theorem 2.58. *(Spectral Theorem for Normal Bounded Operators, II) Let T be a continuous normal linear map on a Hilbert space H. There is a unique regular projection-valued measure P on $\sigma(T)$, such that*

$$\langle T(u), v\rangle = \int_{\sigma(T)} \mathrm{id} \, dP_{u,v}, \quad \textit{for all } u, v \in H.$$

In short, we write

$$T = \int \mathrm{id} \, dP.$$

Furthermore, for every $f \in B(\sigma(T))$, we have

$$U(f) = \int f \, dP,$$

where U is the extension of $G^{-1}\colon \mathcal{C}_{\mathbb{C}}(\sigma(T)) \to \mathcal{A}_T$ to $B(\sigma(T))$.

It is customary to denote the continuous linear operator $U(f) \in \mathcal{L}(H)$ given by $U(f) = \int f\, dP$ as $f(T)$; see Folland [Folland (1995)] (Chapter I, Equation 1.49). Some authors (Rudin [Rudin (1991)], Lax [Lax (2002)]) call the above result a *spectral decomposition* (or *resolution*) of T. Theorem 2.58 is proven by Rudin in terms of resolutions of the identity; see Rudin [Rudin (1991)], Theorem 12.23.

For the sake of simplicity, we omitted another property that should be included in Theorem 2.58. This property is used to prove Schur's Lemma for irreducible unitary representations.

Complement to Theorem 2.58. Let $T \in \mathcal{L}(H)$ be a continuous normal linear map. If $S \in \mathcal{L}(H)$ commutes with T and T^*, then S commutes with $f(T)$ for every $f \in B(\sigma(T))$.

The above fact is proven in Folland [Folland (1995)]; see Theorem 1.51(c). The proof uses the additional fact stated in the Remark after Theorem 2.53, which is also property (a) of Theorem 1.36 in Folland [Folland (1995)].

Theorems 2.48 and 2.58 constitute two ways of generalizing the spectral theorem for normal linear maps on a finite-dimensional Hilbert space. The careful reader will notice that Theorem 2.58 holds even if the Hilbert space is not separable. Theorem 2.48 can be generalized to nonseparable Hilbert spaces at the expense of using uncountable Hilbert sums. Also observe that the projection-valued measure P in Theorem 2.58 is uniquely determined by T, whereas the measure space $(\Omega, \mathcal{M}, \mu)$ and the unitary map W of Theorem 2.48 are not.

The usefulness of projection-valued measures becomes more apparent when we generalize Theorem 2.57 to representations of arbitrary commutative unital involutive Banach algebras.

Theorem 2.59. *(Spectral Theorem, III) Let $\mathcal{A}$ be any commutative unital involutive Banach algebra. For any representation $U\colon \mathcal{A} \to \mathcal{L}(H)$ of $\mathcal{A}$ in a Hilbert space H, there is a regular projection-valued measure P on $\mathsf{X}(\mathcal{A})$, such that*

$$\langle U(a)(u), v\rangle = \int_{\mathsf{X}(\mathcal{A})} \mathcal{G}_a\, dP_{u,v}, \quad \textit{for all } u, v \in H \textit{ and for all } a \in \mathcal{A},$$

which is abbreviated as

$$U(a) = \int \mathcal{G}_a\, dP, \quad a \in \mathcal{A}.$$

Sketch of proof. A more complete proof is given in Folland [Folland (1995)] (Section 1.5, Theorem 1.53). The key idea is to consider the closure $\mathcal{B}$ of $U(\mathcal{A})$ in $\mathcal{L}(H)$, because it is a commutative unital C^*-subalgebra of $\mathcal{L}(H)$, and so Theorem 2.57 applies to $\mathcal{B}$. Then there is a projection-valued measure P_0 on $\mathsf{X}(\mathcal{B})$ and we need to pull it back to $\mathsf{X}(\mathcal{A})$. Let us provide some details.

The map $U\colon \mathcal{A} \to \mathcal{B}$ induces a continuous map $U^*\colon \mathsf{X}(\mathcal{B}) \to \mathsf{X}(\mathcal{A})$ given by

$$U^*(h) = h \circ U, \quad \text{for all } h \in \mathsf{X}(\mathcal{B}).$$

Recall that for any $a \in \mathcal{A}$, we have $U(a) \in U(\mathcal{A}) \subseteq \mathcal{B} \subseteq \mathcal{L}(H)$ and $h\colon \mathcal{B} \to \mathbb{C}$, so we have $h \circ U\colon \mathcal{A} \to \mathbb{C}$. We claim that U^* is injective. Indeed, if $U^*(h_1) = U^*(h_2)$, then h_1 and h_2 (both in $\mathsf{X}(\mathcal{B})$) agree on $U(\mathcal{A}) \subseteq \mathcal{B}$, and since $\mathcal{B}$ is the closure of $U(\mathcal{A})$ in $\mathcal{L}(H)$, we must have $h_1 = h_2$. But $\mathsf{X}(\mathcal{B})$ is compact, so the injective continuous map U^* is a homeomorphism onto its image, which is compact in $\mathsf{X}(\mathcal{A})$.

As we said before, $\mathcal{B}$ is a commutative unital C^*-subalgebra of $\mathcal{L}(H)$, and so Theorem 2.57 applies to $\mathcal{B}$. By Theorem 2.57, there is a regular projection-valued measure P_0 on $\mathsf{X}(\mathcal{B})$, such that

$$T = \int \mathcal{G}_T \, dP_0, \quad \text{for all } T \in \mathcal{B}. \tag{$\dagger_1$}$$

We use U^* to define a regular projection-valued measure P on $\mathsf{X}(\mathcal{A})$ as follows: for every Borel set E on $\mathsf{X}(\mathcal{A})$, let

$$P(E) = P_0((U^*)^{-1}(E)),$$

where, $P_0((U^*)^{-1}(E))$ is really $P_0((U^*)^{-1}(E \cap U^*(\mathsf{X}(\mathcal{B}))))$. We leave it as an exercise to check that P is indeed a regular projection-valued measure on $\mathsf{X}(\mathcal{A})$.

Finally, observe that the Gelfand transforms of $\mathcal{B}$ and $\mathcal{A}$ are related as follows:

$$\mathcal{G}_{U(a)}(h) = \mathcal{G}_a(U^*(h)), \quad \text{for all } a \in \mathcal{A} \text{ and all } h \in \mathsf{X}(\mathcal{B}),$$

since

$$\mathcal{G}_{U(a)}(h) = h(U(a)) = U^*(h)(a) = \mathcal{G}_a(U^*(h)).$$

Then, since $U(a) \in \mathcal{B}$, by ($\dagger_1$), we have

$$U(a) = \int \mathcal{G}_{U(a)}(h) \, dP_0(h) = \int \mathcal{G}_a(U^*(h)) \, dP_0(h) = \int \mathcal{G}_a \, dP,$$

where the last equation is obtained by going back to the definitions of $\int \mathcal{G}_a(U^*(h)) \, dP_0(h)$ and $\int \mathcal{G}_a \, dP$ in terms of the inner product on H and using the definition of P in terms of P_0. ☐

The projection-valued measure in Theorem 2.59 is unique; see Folland [Folland (1995)] (Section 1.5). Theorem 2.59 can be promoted to nonunital commutative involutive Banach algebras as long as the representation U is nondegenerate; see Folland [Folland (1995)] (Section 1.5, Theorem 1.54).

Theorem 2.60. *(Spectral Theorem, IV) Let $\mathcal{A}$ be any commutative involutive Banach algebra. For any nondegenerate representation $U\colon \mathcal{A} \to \mathcal{L}(H)$ of $\mathcal{A}$ in a Hilbert space H, there is a unique regular projection-valued measure P on $\mathsf{X}(\mathcal{A})$, such that*

$$\langle U(a)(u), v\rangle = \int_{\mathsf{X}(\mathcal{A})} \mathcal{G}_a \, dP_{u,v}, \quad \text{for all } u, v \in H \text{ and for all } a \in \mathcal{A},$$

which is abbreviated as

$$U(a) = \int \mathcal{G}_a \, dP, \quad a \in \mathcal{A}.$$

The above theorem is crucial to the proof of Theorem 3.20 characterizing the unitary representations of an *abelian* locally compact group. Intuitively, the characters of G are glued by a suitable projection-valued measure. In turn, Theorem 3.20 is a key result used in Mackey's theory for constructing induced representations; see Chapter 7, Proposition 7.1.

For any locally compact space X, $\mathcal{C}_0(X;\mathbb{C})$ is a nonunital C^*-algebra, and since by Vol. I, Proposition 9.22, $\mathsf{X}(\mathcal{C}_0(X;\mathbb{C}))$ is homeomorphism to X itself, we can view the isometric isomorphism from $\mathcal{C}_0(X;\mathbb{C})$ to $\mathcal{C}_0(\mathsf{X}(\mathcal{C}_0(X;\mathbb{C}));\mathbb{C})$ provided by the Gelfand transform (by Gelfand–Naimark) as the identity. In other words, we can view the Gelfand transform on $\mathcal{C}_0(X;\mathbb{C})$ as the identity. Then we have the following corollary of Theorem 2.60.

Theorem 2.61. *Let X be a locally compact space, and let $U\colon \mathcal{C}_0(X;\mathbb{C}) \to \mathcal{L}(H)$ be a nondegenerate representation of $\mathcal{C}_0(X;\mathbb{C})$ in a Hilbert space H. There is a unique regular projection-valued measure P on X, such that*

$$\langle U(f)(u), v\rangle = \int_X f \, dP_{u,v}, \quad \text{for all } u, v \in H \text{ and for all } f \in \mathcal{C}_0(X;\mathbb{C}),$$

which is abbreviated as

$$U(f) = \int f \, dP, \quad f \in \mathcal{C}_0(X;\mathbb{C}).$$

The above theorem is used in Section 7.2 to give an alternate definition of a system of imprimitivity; see Definition 7.4.

2.13 Problems

Problem 2.1. Let $A = \mathrm{M}_n(\mathbb{C})$ be the algebra of $n \times n$ complex matrices with involution $X \mapsto X^*$ (the conjugate transpose); see Example 2.2. Define $\mathfrak{b}$ to be the subspace of A consisting of all $n \times n$ complex matrices whose last $n-1$ columns are zero.

(1) Show that $\mathfrak{b}$ is a minimal left ideal in A.
(2) Show that the subspace $\mathfrak{b}$ of $A = \mathrm{M}_n(\mathbb{C})$ is invariant under the representation $U \colon A \to \mathcal{L}(H)$ of Example 2.1, and $U_1 \colon A \to \mathcal{L}(\mathfrak{b})$ is a subrepresentation of A in $\mathfrak{b}$.

Problem 2.2. Prove that the representations $U_1 \colon A \to \mathcal{L}(\mathbb{C}^n)$ and $U_2 \colon A \to \mathcal{L}(\mathfrak{b})$ of Example 2.2 are equivalent.

Problem 2.3. Let $(H_\alpha, \langle -, - \rangle_\alpha)_{\alpha \in \Lambda}$ be a family of Hilbert spaces indexed by some index set Λ. Define the set H as the set of all sequences $(x_\alpha)_{\alpha\in\Lambda}$ with $x_\alpha \in H_\alpha$, such that $\sum_{\alpha\in\Lambda} \|x_\alpha\|_{H_\alpha}^2 < \infty$. Define a vector space structure on H by defining

$$\begin{aligned}(x_\alpha) + (y_\alpha) &= (x_\alpha + y_\alpha)\\ \lambda(x_\alpha) &= (\lambda x_\alpha),\end{aligned}$$

with $x_\alpha, y_\alpha \in H_\alpha$.

(1) Show that H, with the above operations, is a vector space.

    Define the inner product $\langle -, - \rangle$ on H by

$$\langle (x_\alpha), (y_\alpha) \rangle = \sum_{\alpha\in\Lambda} \langle x_\alpha, y_\alpha \rangle_\alpha.$$

(2) Show that $\langle -, - \rangle$ is a Hermitian inner product on H.
(3) Show that H is complete, so it is a Hilbert space.

Hint. See Dieudonné [Dieudonné (2010)] (Chapter VI, Section 4) and Schwartz [Schwartz (1980)] (Chapter XXIII, Theorem 1 and Theorem 2).

Problem 2.4. Prove Proposition 2.3.

Hint. See Dieudonné [Dieudonné (2010)] (Chapter VI, Section 4) and Schwartz [Schwartz (1980)] (Chapter XXIII, Theorem 4 and its corollaries).

Problem 2.5. Prove Proposition 2.4.

Hint. See Schwartz [Schwartz (1980)] (Chapter XXIII, Theorem 2, Corollary 2).

Problem 2.6. Prove part (1) of Proposition 2.5.

Hint. Use the continuity of $U(s)$; see Dieudonné [Dieudonné (2010)] (Chapter III, Section 11).

Problem 2.7. Prove Proposition 2.6.

Hint. See Dieudonné [Dieudonné (2005)] (Chapter XV, Section 5).

Problem 2.8. Prove Proposition 2.9.

Problem 2.9. Let H be a separable Hilbert space. Recall that this means H has a countable Hilbert basis, that is, a countable orthonormal basis $(a_i)_{i\geq 1}$, such that the subspace spanned by $(a_i)_{i\geq 1}$ is dense in H. A linear map $u \in \mathcal{L}(H)$ is a *Hilbert–Schmidt operator* if the series $\sum_{n=1}^{\infty} \|u(a_n)\|^2$ converges, that is, $\sum_{n=1}^{\infty} \|u(a_n)\|^2 < \infty$. The set of Hilbert–Schmidt operators is denoted by $\mathcal{L}_2(H)$. Then we define the map $u \mapsto \|u\|_{\text{HS}}$ on the set $\mathcal{L}_2(H)$ by

$$\|u\|_{\text{HS}}^2 = \sum_{n=1}^{\infty} \|u(a_n)\|^2 .$$

(1) Show that $\sum_{n=1}^{\infty} \|u(a_n)\|^2$ is independent of the Hilbert basis (a_n).
(2) Show that $\|u\|_{\text{HS}} = \|u^*\|_{\text{HS}}$.
(3) For any two Hilbert–Schmidt operators $u, v \in \mathcal{L}_2(H)$, show that

$$\|u \circ v\|_{\text{HS}} \leq \|u\|_{\infty} \|v\|_{\text{HS}} ,$$

and if $u \in \mathcal{L}_2(H)$, then

$$\|u\|_{\infty} \leq \|u\|_{\text{HS}} .$$

Problem 2.10. Using the norm $\| \, \|_{\text{HS}}$, show that the space $\mathcal{L}_2(H)$ of Hilbert–Schmidt operators is an involutive Banach algebra under composition, with the involution given by $u \mapsto u^*$ (where u^* is the adjoint of u).

Problem 2.11. Show that $\mathcal{L}_2(H)$ is a self-adjoint two-sided ideal in the involutive Banach algebra $\mathcal{L}(H)$.

Problem 2.12. If E and F are two normed vector spaces, a linear map $u \colon E \to F$ is a *compact operator* if the closure of $f(B)$ is compact for every bounded subset B of E. A compact operator is continuous. Prove that every Hilbert–Schmidt operator is a compact operator.

Hint. See Dieudonné [Dieudonné (2005)] (Chapter XV, Section 4).

Problem 2.13. Let $u, v \in \mathcal{L}_2(H)$.

(1) Show that

$$g(u,v) = \sum_{n \geq 1} \langle u(a_n), v(a_n) \rangle$$

is well-defined and independent of the Hilbert basis (a_n).

(2) Show that g is a hermitian inner product which is a bitrace, such that $g(u,u) = \|u\|_{\text{HS}}$; see Dieudonné [Dieudonné (2005)] (Chapter XV, Section 7).

(3) Show that g satisfies the properties (U) and (N).

Hint. See Dieudonné [Dieudonné (2005)] (Chapter XV, Section 7).

Problem 2.14. Complete the proof of Proposition 2.18.

Hint. See Dieudonné [Dieudonné (2005)] (Chapter XXI, Section 2).

Problem 2.15. Prove Proposition 2.21.

Hint. See Dieudonné [Dieudonné (2005)] (Chapter XV, Proposition 15.8.2).

Problem 2.16. Prove Proposition 2.22.

Hint. See Dieudonné [Dieudonné (2005)] (Chapter XV, Proposition 15.8.3).

Problem 2.17. Prove Proposition 2.28.

Hint. See Dieudonné [Dieudonné (2005)] (Chapter XV, Proposition 15.8.9).

Problem 2.18. (Advanced Exercise) Prove Proposition 2.29.

Hint. See Dieudonné [Dieudonné (2005)] (Chapter XV, Proposition 15.8.10).

Problem 2.19. Prove Proposition 2.40.

Hint. See Dieudonné [Dieudonné (2005)] (Chapter XV, Section 7, Theorem 15.6.11).

Problem 2.20. Prove Proposition 2.41.

Hint. See Dieudonné [Dieudonné (2005)] (Chapter XV, Section 7, Theorem 15.6.11.1).

Problem 2.21. Study the proof of the Plancherel–Godement theorem, Theorem 2.43, provided by Dieudonné [Dieudonné (2005)] (Chapter XV, Section 9, Theorem 15.9.2).

Problem 2.22. (Advanced Exercise) Prove the Bochner–Godement theorem, Theorem 2.44.

Hint. See Dieudonné [Dieudonné (2005)] (Chapter XV, Section 9, Theorem 15.9.4).

Problem 2.23. Prove Proposition 2.46.

Hint. See Dieudonné [Dieudonné (2005)] (Chapter XV, Section 10, Theorem 15.10.8.1).

Problem 2.24. Prove part (2) of Proposition 2.48.

Hint. Adapt the proof of Theorem 2.47 or see Taylor [Taylor (1986)], Appendix B.

Problem 2.25. Prove Proposition 2.49.

Hint. See Dieudonné [Dieudonné (2005)] (Chapter XV, Section 11, Proposition 15.11.5).

Problem 2.26. Prove Proposition 2.50.

Hint. See Dieudonné [Dieudonné (2005)] (Chapter XV, Section 11, Proposition 15.11.6), or see Rudin [Rudin (1991)] (Theorem 12.29).

Problem 2.27. Prove Proposition 2.51.

Hint. See Folland [Folland (1995)] (Section 1.4, Proposition 1.34).

Problem 2.28. Prove Theorem 2.56.

Hint. See Folland [Folland (1995)] (Theorem 1.43) or see Rudin [Rudin (1991)], Theorem 12.21.

Problem 2.29. Prove that the projection-valued measure of Theorem 2.57 is unique.

Hint. See Folland [Folland (1995)] (Theorem 1.44).

Problem 2.30. Let $T \in \mathcal{L}(H)$ be a continuous normal linear map. If $S \in \mathcal{L}(H)$ commutes with T and T^*, then S commutes with $f(T)$ for every $f \in B(\sigma(T))$.

Hint. See Folland [Folland (1995)] (Theorem 1.51(c)).

Problem 2.31. Prove that the projection-valued measure in Theorem 2.59 is unique.

Hint. See Folland [Folland (1995)] (Section 1.5).

Problem 2.32. (Advanced Exercise) Prove Theorem 2.60.

Hint. See Folland [Folland (1995)] (Section 1.5, Theorem 1.54).

Chapter 3

Unitary Representations of Locally Compact Groups

In this chapter, we discuss representations of locally compact groups. For simplicity, we begin with finite-dimensional representations, which are continuous group homomorphisms $\rho\colon G \to \mathbf{GL}(V)$, where V is a finite-dimensional complex vector space (see Section 3.1). Next we consider unitary representations, which are certain kinds of continuous homomorphisms $U\colon G \to \mathbf{U}(H)$, where H is a Hilbert space (typically separable), and $\mathbf{U}(H)$ is the group of unitary operators on H, that is, the continuous linear maps $f\colon H \to H$ that have a continuous inverse, and preserve the inner product; that is,

$$\langle f(x), f(y)\rangle = \langle x, y\rangle, \qquad \text{for all } x, y \in H.$$

Then a unitary operator is a continuous linear map $f\colon H \to H$, such that $f^{-1} = f^*$, where f^* is the adjoint of f, the unique continuous linear map determined by the equation

$$\langle f^*(x), y\rangle = \langle x, f(y)\rangle, \qquad \text{for all } x, y \in H.$$

The basic theory of unitary representations is discussed in Section 3.2.

There are three main results in this chapter.

The first main result (first shown by Naimark) is that every unitary representation $U\colon G \to \mathbf{U}(H)$ of a locally compact group G defines a nondegenerate representation $U_{\mathrm{ext}}\colon \mathrm{L}^1(G) \to \mathcal{L}(H)$ of the involutive Banach algebra $\mathrm{L}^1(G)$, and that conversely, for every nondegenerate representation $V\colon \mathrm{L}^1(G) \to \mathcal{L}(H)$ of $\mathrm{L}^1(G)$, there is a unique unitary representation $U\colon G \to \mathbf{U}(H)$ of the group G, such that $V = U_{\mathrm{ext}}$ (see Section 3.3, Theorem 3.17 and Theorem 3.18).

The bijection $U \mapsto U_{\mathrm{ext}}$ between unitary representations of a locally compact group G and nondegenerate representations of the algebra $\mathrm{L}^1(G)$

is a basic tool that allows the transfer of results about representations of algebras to representations of groups, and vice-versa. It will play a crucial role in the proof of the Peter–Weyl theorem.

The second main result (Theorem 3.20) is a characterization of the unitary representations $U\colon G \to \mathbf{U}(H)$ of a locally compact *abelian* group G in terms of projection-valued measures (as discussed in Section 2.12). This theorem plays a key role in the construction of induced representations using a method due to Mackey (the "Mackey machine"); see Chapter 7.

The third main result (Gelfand and Raikov, Godement) is that there is one-to-one correspondence between unitary cyclic representations of a locally compact group G and certain bounded continuous functions on G called functions of *positive type*.

Let $U\colon G \to \mathbf{U}(H)$ be a unitary representation of the locally compact group G in a Hilbert space H, let x_0 be any vector in H, and define the map $p = \psi_{U,x_0}$ by

$$p(s) = \psi_{U,x_0}(s) = \langle U(s)(x_0), x_0\rangle, \qquad s \in G.$$

It turns out that the function p is continuous and bounded and that it satisfies the following property:

$$\int (f^* * f)(s)p(s)\, d\lambda(s) \geq 0, \qquad \text{for all } f \in \mathcal{K}_{\mathbb{C}}(G),$$

where λ is a left Haar measure on G. Such functions are called functions of *positive type*. Remarkably, every continuous function p of positive type determines a unitary topologically cyclic representation U with a cyclic vector x_0, such that $p = \psi_{U,x_0}$ (see Theorem 3.22). The connection between cyclic unitary representations and functions of positive type is discussed in Section 3.5.

In Section 3.6, we present the Gelfand–Raikov theorem without proof (see Theorem 3.26). Informally, this theorem says that there is vast supply of irreducible unitary representations for any locally compact group. This is far from obvious a priori. For example, $\mathbf{SL}(2,\mathbb{R})$ does not have finite-dimensional unitary representations, and it is not that easy to find irreducible unitary representations.

Section 3.7 is devoted to measures of positive type, which generalize functions of positive type. A complex or σ-Radon measure μ is of positive type if

$$\int (f^* * f)(s)\, d\mu(s) = \int\!\!\int \overline{f(t)}f(ts)\, d\lambda(t)\, d\mu(s) \geq 0, \qquad \text{for all } f \in \mathcal{K}_{\mathbb{C}}(G).$$

The Dirac measure δ_e is a measure of positive type, and more generally, if ν is a complex measure, then the measure $\overline{\nu} * \nu$ is of positive type. The main point is that a measure μ of positive type defines a unitary representation U_μ of G in a separable Hilbert space H (see Theorem 3.30). This construction will be used in Section 9.9 to define the Plancherel transform.

Basically, all the material of this chapter is presented in a more condensed form in Dixmier's classical book Dixmier [Dixmier (1969)] (see also the English translation published by the AMS).

3.1 Finite-Dimensional Group Representations

For simplicity, we begin with finite-dimensional representations.

Definition 3.1. Given a locally compact group G and a normed vector space V of dimension n, a *continuous linear representation of G in V of dimension* (or *degree*) n is a group homomorphism $\rho\colon G \to \mathbf{GL}(V)$, where $\mathbf{GL}(V)$ denotes the group of invertible linear maps from V to itself, such that the following condition holds:

(C) The map $g \mapsto \rho(g)(u)$ is continuous for every $u \in V$.

The space V, called the *representation space*, may be a real or a complex vector space. If V has a Hermitian (resp. Euclidean) inner product $\langle -, - \rangle$, we say that $\rho\colon G \to \mathbf{GL}(V)$ is a *continuous unitary representation* if

(U) Every linear map $\rho(g)$ is an *isometry*, that is,

$$\langle \rho(g)(u), \rho(g)(v) \rangle = \langle u, v \rangle, \qquad \text{for all } g \in G \text{ and all } u, v \in V.$$

A unitary representation is denoted $\rho\colon G \to \mathbf{U}(V)$.

Thus, a continuous linear representation of G is a map $\rho\colon G \to \mathbf{GL}(V)$ satisfying condition (C) as well as the properties:

$$\begin{aligned} \rho(gh) &= \rho(g)\rho(h) \\ \rho(g^{-1}) &= \rho(g)^{-1} \\ \rho(1) &= \mathrm{id}_V \end{aligned}$$

for all $g, h \in G$. If ρ is a unitary representation, then we also have

$$(\rho(g))^{-1} = (\rho(g))^*.$$

If G is a finite group, the continuity requirement is omitted.

To avoid confusion when representations involving different groups arise we denote the space of the representation ρ by V_ρ, and so we denote a representation as $\rho\colon G \to \mathbf{GL}(V_\rho)$. To reduce the amount of parentheses, we often write $\rho_g(u)$ instead of $\rho(g)(u)$.

Note that a major difference with the notion of representation of an algebra, is that for a group representation $\rho\colon G \to \mathbf{GL}(V)$, the linear map $\rho(g)$ *must be invertible* for every $g \in G$. For an algebra representation $U\colon A \to \mathcal{L}(H)$ (where H is a Hilbert space), the linear maps $U(s)$ are generally *not invertible*.

For simplicity of language, we usually abbreviate *continuous linear (or unitary) representation* as *(unitary) representation*. The representation space V is also called a *G-module*, since the representation $\rho\colon G \to \mathbf{GL}(V)$ is equivalent to the left action $\cdot\colon G \times V \to V$, with $g \cdot v = \rho(g)(v)$. The representation, such that $\rho(g) = \mathrm{id}_V$ for all $g \in G$, is called the *trivial representation*.

It should be noted that because V is finite-dimensional, the condition that for every $u \in V$, the map $g \mapsto \rho(g)(u)$ is continuous, is actually equivalent to the fact that the map $g \mapsto \rho(g)$ from G to $\mathcal{L}(V)$ equipped with the operator norm induced by any norm on V is continuous.

Indeed, for any basis of V, the fact that the map $g \mapsto \rho(g)(u)$ is continuous implies that the matrix $(\rho_{ij}(g))$ representing $\rho(g)$ consists of continuous functions on G.

Since the space V of a representation $\rho\colon G \to \mathbf{GL}(V)$ is finite-dimensional, say n, it is often convenient to pick a basis $(e_1, \ldots, e_n)$ of V, and then every invertible linear map $\rho(g) \in \mathbf{GL}(V)$ is represented by an $n \times n$ matrix that we denote $M_\rho(g) = (\rho_{ij}(g))$.[1] We obtain a continuous map $M_\rho\colon G \to \mathbf{GL}(n, \mathbb{C})$ assigning an invertible $n \times n$ complex matrix $M_\rho(g) = (\rho_{ij}(g))$ to $g \in G$ satisfying the properties

$$\begin{aligned} M_\rho(gh) &= M_\rho(g)M_\rho(h) \\ M_\rho(g^{-1}) &= (M_\rho(g))^{-1} \\ M_\rho(1) &= I_n \end{aligned}$$

for all $g, h \in G$. The continuity of M_ρ is equivalent to the fact that the n^2 functions $g \mapsto \rho_{ij}(g)$ are continuous. If ρ is a unitary representation, then

[1]To be perfectly rigorous, the matrix M_ρ should be indexed by the basis $\mathcal{E} = (e_1, \ldots, e_n)$, say as $M_\rho^{\mathcal{E}}$, but this is just too much decoration.

we also have

$$(M_\rho(g))^{-1} = (M_\rho(g))^*.$$

If G is finite, we drop the continuity requirement. Conversely, we have the notion of representation in matrix form.

Definition 3.2. Given a locally compact group G a *continuous linear representation of G of dimension (or degree) n in matrix form* is a mapping $M_\rho\colon G \to \mathbf{GL}(n, \mathbb{C})$ assigning an invertible $n \times n$ complex matrix $M_\rho(g) = (\rho_{ij}(g))$ to $g \in G$ satisfying the properties

$$\begin{aligned} M_\rho(gh) &= M_\rho(g)M_\rho(h) \\ M_\rho(g^{-1}) &= (M_\rho(g))^{-1} \\ M_\rho(1) &= I_n \end{aligned}$$

for all $g, h \in G$, and such that the n^2 functions $g \mapsto \rho_{ij}(g)$ are continuous. If M_ρ is a unitary representation, then we also have

$$(M_\rho(g))^{-1} = (M_\rho(g))^*.$$

In this case, M_ρ is a homomorphism $M_\rho\colon G \to \mathbf{U}(n)$. If G is finite, we drop the continuity requirement.

A representation in matrix form $M_\rho\colon G \to \mathbf{GL}(n, \mathbb{C})$ (resp. $M_\rho\colon G \to \mathbf{U}(n)$) defines the representation $\rho\colon G \to \mathbf{GL}(\mathbb{C}^n)$ (resp. $\rho\colon G \to \mathbf{U}(\mathbb{C}^n)$) given by

$$(\rho(g))(z) = M_\rho(g)z, \quad z \in \mathbb{C}^n,\ g \in G.$$

Since the notation $M_\rho(g)$ is quite heavy, we often write $M(g)$ instead of $M_\rho(g)$. This is an abuse of notation since $M(g)$ is a linear map and $M_\rho(g)$ is a matrix representing it in some basis, and thus depends on this basis. We also often identify a matrix representation with the representation associated with it. The same issue arises in linear algebra and we hope that the reader is already familiar with it and will not be confused.

Given any basis $(e_1, \ldots, e_n)$ of V, we may think of the scalar functions $g \mapsto \rho_{ij}(g)$ as *special functions* on G. As explained in Dieudonné [Dieudonné (1980)] (see also Vilenkin [Vilenkin (1968)]), essentially all special functions (Legendre polynomials, ultraspherical polynomials, Bessel functions, *etc.*) arise in this way by choosing some suitable G and V.

Remark: In Chapter 6, we will consider the situation where G is a group not equipped with any topology, and V is a vector space, possibly infinite-dimensional, not equipped with any norm. Then a *linear representation* *of G in V* is simply a homomorphism $\rho\colon G \to \mathbf{GL}(V)$, which amounts to dropping condition (C) from Definition 3.1. However, in this chapter and the next, all representations satisfy condition (C).

Example 3.1. Consider the group $\mathfrak{S}_3$ of permutations on the set $\{1, 2, 3\}$. There are $3! = 6$ permutations

$$\begin{array}{lll} \pi_1 = (1,2,3) & \pi_2 = (1,3,2) & \pi_3 = (2,1,3) \\ \pi_4 = (2,3,1) & \pi_5 = (3,1,2) & \pi_6 = (3,2,1). \end{array}$$

The first permutation $\pi_1 = (1, 2, 3)$ is the identity; the permutations

$$\pi_2 = (1,3,2),\ \ \pi_3 = (2,1,3),\ \ \pi_6 = (3,2,1)$$

are transpositions and thus have negative signature, and the permutations

$$\pi_4 = (2,3,1),\ \ \pi_5 = (3,1,2)$$

are cyclic permutations and thus have positive signature. We obtain a representation $\rho_1\colon \mathfrak{S}_3 \to \mathbf{GL}(\mathbb{C}^3)$ as follows. If (e_1, e_2, e_3) is the canonical basis of $\mathbb{C}^3$, then $\rho_1(\pi_i)$ is the linear map given by

$$\rho_1(\pi_i)(e_j) = e_{\pi_i(j)}, \quad 1 \leq i, j \leq 3.$$

In the basis (e_1, e_2, e_3), the linear maps $\rho_1(\pi_i)$ are represented by the 3×3 matrices $M_1, \ldots, M_6$ given by

$$\begin{pmatrix} 1 & 0 & 0 \\ 0 & 1 & 0 \\ 0 & 0 & 1 \end{pmatrix}, \begin{pmatrix} 1 & 0 & 0 \\ 0 & 0 & 1 \\ 0 & 1 & 0 \end{pmatrix}, \begin{pmatrix} 0 & 1 & 0 \\ 1 & 0 & 0 \\ 0 & 0 & 1 \end{pmatrix}, \begin{pmatrix} 0 & 0 & 1 \\ 1 & 0 & 0 \\ 0 & 1 & 0 \end{pmatrix}, \begin{pmatrix} 0 & 1 & 0 \\ 0 & 0 & 1 \\ 1 & 0 & 0 \end{pmatrix}, \begin{pmatrix} 0 & 0 & 1 \\ 0 & 1 & 0 \\ 1 & 0 & 0 \end{pmatrix}.$$

Here is another representation of the group $\mathfrak{S}_3$ in $\mathbb{C}^6$.

Example 3.2. This time we define the representation $\rho_{\mathbf{R}}\colon \mathfrak{S}_3 \to \mathbf{GL}(\mathbb{C}^6)$ as follows. Let $(e_{\pi_1}, \ldots, e_{\pi_6})$ be the canonical basis of $\mathbb{C}^6$ indexed by the permutations π_i $(1 \leq i \leq 6)$, and set

$$\rho_{\mathbf{R}}(\pi_i)(e_{\pi_j}) = e_{\pi_i \circ \pi_j}, \quad 1 \leq i, j \leq 6.$$

Note that the 6×6 matrix representing $\rho_{\mathbf{R}}(\pi_i)$ in the basis $(e_{\pi_1}, \ldots, e_{\pi_6})$ consists of the permutation of the columns of the identity matrix I_6, whose indices are given by the ith row of the multiplication table of the group $\mathfrak{S}_3$. This multiplication table is given by

$$\begin{pmatrix} 1 & 2 & 3 & 4 & 5 & 6 \\ 2 & 1 & 5 & 6 & 3 & 4 \\ 3 & 4 & 1 & 2 & 6 & 5 \\ 4 & 3 & 6 & 5 & 1 & 2 \\ 5 & 6 & 2 & 1 & 4 & 3 \\ 6 & 5 & 4 & 3 & 2 & 1 \end{pmatrix},$$

where we denote π_i simply by i and where the (i,j) entry represents $\pi_i \circ \pi_j$. We obtain the following 6 matrices:

$$\begin{pmatrix} 1&0&0&0&0&0\\0&1&0&0&0&0\\0&0&1&0&0&0\\0&0&0&1&0&0\\0&0&0&0&1&0\\0&0&0&0&0&1 \end{pmatrix}, \quad \begin{pmatrix} 0&1&0&0&0&0\\1&0&0&0&0&0\\0&0&0&0&1&0\\0&0&0&0&0&1\\0&0&1&0&0&0\\0&0&0&1&0&0 \end{pmatrix}, \quad \begin{pmatrix} 0&0&1&0&0&0\\0&0&0&1&0&0\\1&0&0&0&0&0\\0&1&0&0&0&0\\0&0&0&0&0&1\\0&0&0&0&1&0 \end{pmatrix},$$

$$\begin{pmatrix} 0&0&0&0&1&0\\0&0&0&0&0&1\\0&1&0&0&0&0\\1&0&0&0&0&0\\0&0&0&1&0&0\\0&0&1&0&0&0 \end{pmatrix}, \quad \begin{pmatrix} 0&0&0&1&0&0\\0&0&1&0&0&0\\0&0&0&0&0&1\\0&0&0&0&1&0\\1&0&0&0&0&0\\0&1&0&0&0&0 \end{pmatrix}, \quad \begin{pmatrix} 0&0&0&0&0&1\\0&0&0&0&1&0\\0&0&0&1&0&0\\0&0&1&0&0&0\\0&1&0&0&0&0\\1&0&0&0&0&0 \end{pmatrix}.$$

The representation $\rho_{\mathbf{R}}$ is called the *regular representation* of $\mathfrak{S}_3$.

Example 3.3. For an example involving an infinite group, we describe a class of representations of $G = \mathbf{SL}(2,\mathbb{C})$, the group of complex matrices with determinant $+1$,

$$\begin{pmatrix} a & b \\ c & d \end{pmatrix}, \qquad ad - bc = 1.$$

Recall that $\mathcal{P}^{\mathbb{C}}_m(2)$ denotes the vector space of complex homogeneous polynomials of degree m in two variables (z_1, z_2). A complex homogeneous polynomials of degree m in two variables (z_1, z_2) is an expression of the form $P(z_1, z_2) = \sum_{i=0}^m c_i z_1^i z_2^{m-i}$, with $c_i \in \mathbb{C}$. For every matrix $A \in \mathbf{SL}(2,\mathbb{C})$, with

$$A = \begin{pmatrix} a & b \\ c & d \end{pmatrix},$$

for every homogeneous polynomial $P \in \mathcal{P}^{\mathbb{C}}_m(2)$, we define $U_m(A)(P(z_1, z_2))$ by

$$U_m(A)(P(z_1, z_2)) = P(dz_1 - bz_2, -cz_1 + az_2).$$

The reader may be puzzled by the fact that we departed from our implicit notational convention of using ρ for finite-dimensional representations. The reason is that U_m is also a representation of $\mathbf{SU}(2)$, and by defining a suitable inner product on $\mathbf{SU}(2)$, it become unitary. If we think of the homogeneous polynomial $Q(z_1, z_2)$ as a function $P\binom{z_1}{z_2}$ of the vector $\binom{z_1}{z_2}$, then

$$U_m(A)\left(P\begin{pmatrix} z_1 \\ z_2 \end{pmatrix}\right) = PA^{-1}\begin{pmatrix} z_1 \\ z_2 \end{pmatrix} = P\begin{pmatrix} d & -b \\ -c & a \end{pmatrix}\begin{pmatrix} z_1 \\ z_2 \end{pmatrix}.$$

The expression above makes it clear that

$$U_m(AB) = U_m(A)U_m(B)$$

for any two matrices $A, B \in \mathbf{SL}(2, \mathbb{C})$, so U_m is indeed a representation of $\mathbf{SL}(2, \mathbb{C})$ into $\mathcal{P}_m^{\mathbb{C}}(2)$. This is a left regular representation, as discussed later in Definition 3.6.

The representations U_m also yield representations of the subgroup $\mathbf{SU}(2)$ of $\mathbf{SL}(2, \mathbb{C})$. Recall that the group $\mathbf{SU}(2)$ consists of all 2×2 complex matrices

$$S = \begin{pmatrix} \alpha & \beta \\ -\overline{\beta} & \overline{\alpha} \end{pmatrix}, \quad \alpha\overline{\alpha} + \beta\overline{\beta} = 1.$$

As above, the representation $U_m \colon \mathbf{SU}(2) \to \mathbf{GL}(\mathcal{P}_m^{\mathbb{C}}(2))$ is given by

$$U_m(S)(P(z_1, z_2)) = P(\overline{\alpha}z_1 - \beta z_2, \overline{\beta}z_1 + \alpha z_2).$$

It can be shown that $\mathbf{SL}(2, \mathbb{C})$ has *no* nontrivial *unitary* finite-dimensional representations! This is because $\mathbf{SL}(2, \mathbb{C})$ is a connected simple noncompact Lie group with finite center; see Dieudonné [Dieudonné (2003a)] (Section 21.6, Problem 5).

Example 3.4. We define the representation $\rho_9 \colon \mathbf{SO}(3) \to \mathbf{GL}(\mathrm{M}_3(\mathbb{C}))$ as follows: for any 3×3 complex matrix $A \in \mathrm{M}_3(\mathbb{C})$, for any $Q \in \mathbf{SO}(3)$,

$$\rho_9(Q)(A) = QAQ^\top.$$

This is a representation in the vector space $\mathrm{M}_3(\mathbb{C})$, which has dimension 9. To obtain a version of ρ_9 as a matrix representation M_{ρ_9}, we need to pick a basis of $\mathrm{M}_3(\mathbb{C})$. Let us choose the canonical basis of nine matrices $E_{11}, E_{12}, E_{13}, E_{21}, E_{22}, E_{23}, E_{31}, E_{32}, E_{33}$, where E_{ij} contains 1 as the (i, j) entry and 0 otherwise. A matrix $M \in \mathrm{M}_3(\mathbb{C})$ is then written as the column vector

$$\mathrm{vec}(A) = (a_{11}, a_{12}, a_{13}, a_{21}, a_{22}, a_{23}, a_{31}, a_{32}, a_{33}).$$

It follows that over this basis, the matrix $M_{\rho_9}(Q)$ representing the linear map $\rho_9(Q)$ is given by

$$M_{\rho_9}(Q)(\mathrm{vec}(A)) = \mathrm{vec}(QAQ^\top).$$

However, it is a fact of linear algebra that for any $m \times m$ matrix A, any $n \times n$ matrix B, and $m \times n$ matrix Z, we have the identity

$$\mathrm{vec}(AZB) = (B^\top \otimes A)\mathrm{vec}(Z),$$

where $\otimes$ denotes the Kronecker product of matrices. Therefore, we deduce that

$$M_{\rho_9}(Q)(\mathrm{vec}(A)) = \mathrm{vec}(QAQ^\top) = (Q \otimes Q)\mathrm{vec}(A),$$

that is,

$$M_{\rho_9}(Q) = Q \otimes Q,$$

a 9×9-matrix. The definition of the representation ρ_9 as acting on the vector space $\mathrm{M}_3(\mathbb{C})$ is a lot more economical than its matrix version M_{ρ_9} acting on $\mathbb{C}^9$.

The representation ρ_9 is reducible (see Definition 3.4). Indeed, observe that both the subspace of symmetric matrices and the subspace of skew-symmetric matrices are invariant since $(QAQ^\top)^\top = QA^\top Q^\top$. The subspace of symmetric matrices A with $\mathrm{tr}(A) = 0$ is also invariant.

There is a natural and useful notion of equivalence of representations.

Definition 3.3. Given any two representations $\rho_1 \colon G \to \mathbf{GL}(V_1)$ and $\rho_2 \colon G \to \mathbf{GL}(V_2)$, a *$G$-map* (or *morphism of representations*) $\varphi \colon \rho_1 \to \rho_2$ is a linear map $\varphi \colon V_1 \to V_2$, which is *equivariant*, which means that the following diagram commutes for every $g \in G$:

$$\begin{array}{ccc} V_1 & \xrightarrow{\rho_1(g)} & V_1 \\ {\scriptstyle\varphi}\downarrow & & \downarrow{\scriptstyle\varphi} \\ V_2 & \xrightarrow[\rho_2(g)]{} & V_2, \end{array}$$

i.e.

$$\varphi \circ \rho_1(g) = \rho_2(g) \circ \varphi, \qquad g \in G.$$

The space of all G-maps between two representations as above is denoted $\mathrm{Hom}_G(\rho_1, \rho_2)$. Two representations $\rho_1 \colon G \to \mathbf{GL}(V_1)$ and $\rho_2 \colon G \to \mathbf{GL}(V_2)$ are *equivalent* iff $\varphi \colon V_1 \to V_2$ is an invertible linear map (which implies that $\dim V_1 = \dim V_2$). In matrix form, the representations $\rho_1 \colon G \to \mathbf{GL}(n, \mathbb{C})$ and $\rho_2 \colon G \to \mathbf{GL}(n, \mathbb{C})$ are equivalent iff there is some invertible $n \times n$ matrix P, so that

$$\rho_2(g) = P\rho_1(g)P^{-1}, \qquad g \in G.$$

If $W \subseteq V$ is a subspace of V, then in some cases, a representation $\rho\colon G \to \mathbf{GL}(V)$ yields a representation $\rho\colon G \to \mathbf{GL}(W)$. This is interesting because under certain conditions on G (*e.g.*, G is a compact group) every representation may be decomposed into a "sum" of so-called irreducible representations (defined below), and thus the study of all representations of G boils down to the study of irreducible representations of G; for instance, see Knapp [Knapp (2002)] (Chapter 4, Corollary 4.7), or Bröcker and tom Dieck [Bröcker and tom Dieck (1985)] (Chapter 2, Proposition 1.9).

Definition 3.4. Let $\rho\colon G \to \mathbf{GL}(V)$ be a representation of G. If $W \subseteq V$ is a subspace of V, then we say that W is *invariant* (or *stable*) under ρ iff $\rho(g)(w) \in W$, for all $g \in G$ and all $w \in W$. If W is invariant under ρ, then we have a homomorphism, $\rho\colon G \to \mathbf{GL}(W)$, called a *subrepresentation* of G. A representation $\rho\colon G \to \mathbf{GL}(V)$ with $V \neq (0)$ is *irreducible* iff it only has the two subrepresentations $\rho\colon G \to \mathbf{GL}(W)$ corresponding to $W = (0)$ or $W = V$.

Example 3.5. The representation $\rho_1\colon \mathfrak{S}_3 \to \mathbf{GL}(\mathbb{C}^3)$ of Example 3.1 is reducible. Indeed, the one-dimensional subspace V_1 spanned by $e_1+e_2+e_3$ is invariant under ρ_1 since each $\rho_1(\pi_i)$ permutes the indices $1, 2, 3$. The corresponding subrepresentation of $\mathfrak{S}_3$ in V_1 is equivalent to the irreducible trivial representation in $\mathbb{C}$, given by $\rho_{\mathrm{triv}}(\pi_i) = 1$ $(1 \leq i \leq 6)$. The orthogonal complement V_2 of V_1 is the plane of equation

$$x_1 + x_2 + x_3 = 0,$$

which has $(e_1 - e_2, e_2 - e_3)$ as a basis. It is easy to see that the subspace V_2 is also invariant under ρ_1. It is instructive to find an equivalent representation of ρ_1 in the basis (v_1, v_2, v_3) given by

$$\begin{aligned} v_1 &= (1/3)(e_1 + e_2 + e_3) \\ v_2 &= (1/3)(e_1 - e_2) \\ v_3 &= (1/3)(e_2 - e_3). \end{aligned}$$

The change of basis matrix P from the basis (e_1, e_2, e_3) to the basis (v_1, v_2, v_3) is

$$P = \begin{pmatrix} 1/3 & 1/3 & 0 \\ 1/3 & -1/3 & 1/3 \\ 1/3 & 0 & -1/3 \end{pmatrix},$$

whose inverse is

$$P^{-1} = \begin{pmatrix} 1 & 1 & 1 \\ 2 & -1 & -1 \\ 1 & 1 & -2 \end{pmatrix}.$$

Using the linear map φ from $\mathbb{C}^3$ to itself given by P^{-1} (which transforms the coordinates of a vector in $\mathbb{C}^3$ over the basis (e_1, e_2, e_3) to the coordinates of this vector over the basis (v_1, v_2, v_3)), we obtain the equivalent representation ρ_1' given by

$$\rho_1'(\pi_i) = \varphi\rho_1(\pi_i)\varphi^{-1},$$

and over the basis (v_1, v_2, v_3), the matrices representing the linear maps $\rho_1'(\pi_i)$ are the matrices $P^{-1}M_iP$ shown below:

$$\begin{pmatrix} 1 & 0 & 0 \\ 0 & 1 & 0 \\ 0 & 0 & 1 \end{pmatrix}, \quad \begin{pmatrix} 1 & 0 & 0 \\ 0 & 1 & 0 \\ 0 & 1 & -1 \end{pmatrix}, \quad \begin{pmatrix} 1 & 0 & 0 \\ 0 & -1 & 1 \\ 0 & 0 & 1 \end{pmatrix},$$
$$\begin{pmatrix} 1 & 0 & 0 \\ 0 & 0 & -1 \\ 0 & 1 & -1 \end{pmatrix}, \quad \begin{pmatrix} 1 & 0 & 0 \\ 0 & -1 & 1 \\ 0 & -1 & 0 \end{pmatrix}, \quad \begin{pmatrix} 1 & 0 & 0 \\ 0 & 0 & -1 \\ 0 & -1 & 0 \end{pmatrix}.$$

Some of the above matrices are not unitary. We can fix this by choosing an orthonormal basis (w_1, w_2, w_3) with $w_1 = (1/\sqrt{3})v_1$, a basis of V_1, and (w_2, w_3), a basis of V_2. For example, we can pick

$$\begin{aligned} w_1 &= (1/\sqrt{3})(e_1 + e_2 + e_3) \\ w_2 &= (1/\sqrt{2})(e_1 - e_2) \\ w_3 &= (1/\sqrt{6})(e_1 + e_2 - 2e_3). \end{aligned}$$

The change of basis matrix Q from the basis (e_1, e_2, e_3) to the basis (w_1, w_2, w_3) is

$$Q = \begin{pmatrix} 1/\sqrt{3} & 1/\sqrt{2} & 1/\sqrt{6} \\ 1/\sqrt{3} & -1/\sqrt{2} & 1/\sqrt{6} \\ 1/\sqrt{3} & 0 & -2/\sqrt{6} \end{pmatrix},$$

and $Q^{-1} = Q^\top$. We obtain an equivalent representation $\rho_1''(\pi_i)$ and over the basis (w_1, w_2, w_3), the unitary matrices representing the linear maps $\rho_1''(\pi_i)$ are the matrices $Q^{-1}M_iQ$ shown below:

$$\begin{pmatrix} 1 & 0 & 0 \\ 0 & 1 & 0 \\ 0 & 0 & 1 \end{pmatrix}, \quad \begin{pmatrix} 1 & 0 & 0 \\ 0 & 1/2 & \sqrt{3}/2 \\ 0 & \sqrt{3}/2 & -1/2 \end{pmatrix}, \quad \begin{pmatrix} 1 & 0 & 0 \\ 0 & -1 & 0 \\ 0 & 0 & 1 \end{pmatrix},$$
$$\begin{pmatrix} 1 & 0 & 0 \\ 0 & -1/2 & -\sqrt{3}/2 \\ 0 & \sqrt{3}/2 & -1/2 \end{pmatrix}, \quad \begin{pmatrix} 1 & 0 & 0 \\ 0 & -1/2 & \sqrt{3}/2 \\ 0 & -\sqrt{3}/2 & -1/2 \end{pmatrix}, \quad \begin{pmatrix} 1 & 0 & 0 \\ 0 & 1/2 & -\sqrt{3}/2 \\ 0 & -\sqrt{3}/2 & -1/2 \end{pmatrix}.$$

It is now clear that the subspace V_1 spanned by w_1 and the subspace V_2 spanned by w_2 and w_3 are invariant. It is not hard to show that the subrepresentation of ρ_1'' in V_2 is irreducible. This representation is usually called the *standard representation* of $\mathfrak{S}_3$; see Fulton and Harris [Fulton and Harris (1991)], Section 1.3. Thus, we have two irreducible representations of $\mathfrak{S}_3$, the second one being two-dimensional. It turns out that $\mathfrak{S}_3$ only has one more irreducible representation. How do we find it? The answer is, as a subrepresentation of the regular representation.

Recall the regular representation $\rho_{\mathbf{R}}\colon \mathfrak{S}_3 \to \mathbf{GL}(\mathbb{C}^6)$ of $\mathfrak{S}_3$ from Example 3.2. The notion of regular representation can be defined for any finite group.

Definition 3.5. Let G be a finite group with $g = |G|$ elements. We define the *regular representation* $\rho_{\mathbf{R}}\colon G \to \mathbf{GL}(\mathbb{C}^g)$ as follows. Let $(e_{s_1}, \dots, e_{s_g})$ be the canonical basis of $\mathbb{C}^g$ indexed by the g elements of G and set

$$\rho_{\mathbf{R}}(s_i)(e_{s_j}) = e_{s_i s_j}, \quad 1 \le i, j \le g.$$

The following facts about irreducible finite-dimensional representations of a finite group G can be shown.

(1) Every irreducible finite-dimensional representation $\rho_i\colon G \to \mathbf{GL}(\mathbb{C}^{n_i})$ of the finite group G is equivalent to a subrepresentation of the regular representation $\rho_{\mathbf{R}}\colon G \to \mathbf{GL}(\mathbb{C}^g)$ of G in $\mathbb{C}^g$ (where $g = |G|$).
(2) Every irreducible representation $\rho_i\colon G \to \mathbf{GL}(\mathbb{C}^{n_i})$ occurs n_i times in the regular representation; see Proposition 4.17.
(3) If there are h irreducible representations $\rho_i\colon G \to \mathbf{GL}(\mathbb{C}^{n_i})$ (up to equivalence), then

$$n_1^2 + \cdots + n_h^2 = g.$$

See Section 4.2, Example 4.2.
(4) The number h of irreducible representations of G (up to equivalence) is equal to the number of conjugacy classes of G; see Section 4.2, Example 4.2.

The proof of these standard facts of representation theory can be found in Serre [Serre (1977)], Fulton and Harris [Fulton and Harris (1991)], Simon [Simon (1996)], Hall [Hall (1976)], or any book on representation theory. We also prove these facts in Section 4.2 (Example 4.2) and in Section 4.3 (Proposition 4.17) as a special case of results applying to compact groups.

If G is finite of order $g = |G|$, if we write $G = \{s_1, \ldots, s_g\}$ and denote the canonical basis vectors of $\mathbb{C}^g$ as $(e_{s_1}, \ldots, e_{s_g})$, then there is an isomorphism between $\mathbb{C}^g$ and the vector space $\mathbb{C}^G$ of functions from G to $\mathbb{C}$ defined such that to every vector $x = z_{s_1}e_{s_1} + \cdots + z_{s_g}e_{s_g}$ in $\mathbb{C}^g$, we assign the function $f_x\colon G \to \mathbb{C}$ given by

$$f_x(s_i) = z_{s_i}, \quad 1 \leq i \leq g.$$

Now by the definition of the regular representation $\rho_{\mathbf{R}}$ of G, we have

$$\rho_{\mathbf{R}}(s_i)(x) = \rho_{\mathbf{R}}(s_i)\left(\sum_{j=1}^{g} z_{s_j}e_{s_j}\right) = \sum_{j=1}^{g} z_{s_j}e_{s_is_j}, \quad 1 \leq i \leq g. \qquad (*_1)$$

If we let $s_is_j = s_k$, then $s_j = s_i^{-1}s_k$, $z_{s_j}e_{s_is_j} = z_{s_i^{-1}s_k}e_{s_k}$, and the vector $y_i = \rho_{\mathbf{R}}(s_i)(x) = \sum_{k=1}^{g} z_{s_i^{-1}s_k}e_{s_k}$ corresponds the function f_{y_i} given by

$$f_{y_i}(s_k) = z_{s_i^{-1}s_k} = f_{y_i}(s_i^{-1}s_k), \quad 1 \leq k \leq g.$$

Therefore, $(*_1)$ induces the representation $\mathbf{R}\colon G \to \mathbf{GL}(\mathbb{C}^G)$ given by

$$(\mathbf{R}_{s_i}(f))(s_k) = f(s_i^{-1}s_k), \quad f \in \mathbb{C}^G, \ 1 \leq i, k \leq g.$$

Definition 3.6. Let G be a finite group with $g = |G|$ elements. The representation $\mathbf{R}$ given by

$$(\mathbf{R}_{s_i}(f))(s_k) = f(s_i^{-1}s_k), \quad f \in \mathbb{C}^G, \ 1 \leq i, k \leq g, \qquad (*_2)$$

is also called the *regular representation* of G in $\mathbb{C}^G$.

The representation of Definition 3.6 is a special case of the notion of regular representation defined in Definition 3.14 for locally compact groups. To be very precise, it is the *left regular representation* of G because it acts on the left on functions in $\mathbb{C}^G$ (recall that for two sets X and Y, the set of all functions $f\colon X \to Y$ is denoted Y^X). At first glance, the term $s_i^{-1}s_k$ may seem wrong, but it is necessary to use s_i^{-1} instead of s_i to insure that $\mathbf{R}$ is a left action on functions in $\mathbb{C}^G$. We already noticed this fact in Vol. I, Section 8.2, Definition 8.7. There is also a right regular representation defined by

$$(\mathbf{R}^r_{s_i}(f))(s_k) = f(s_ks_i), \quad f \in \mathbb{C}^G, \ 1 \leq i, k \leq g. \qquad (*_3)$$

Representations as given by $(*_2)$ are said to be representations by *left shifts*, and representations as given by $(*_3)$ are said to be representations by *right shifts*.

Obviously, the notion of left regular representation (and right regular representation) makes sense for any group G, finite or infinite, and any subspace $\mathcal{F}$ of the vector space all functions in $\mathbb{C}^G$, namely it is the representation $\mathbf{R}\colon G \to \mathbf{GL}(\mathcal{F})$ given by

$$(\mathbf{R}_s(f))(t) = f(s^{-1}t), \quad f \in \mathcal{F},\ s, t \in G. \tag{$*_4$}$$

If G is an infinite locally compact groups, it is necessary to replace the vector space $\mathbb{C}^G$ of the representation by a space of functions defined on G, namely $\mathrm{L}^2_\lambda(G; \mathbb{C})$ (where λ is a left Haar measure on G).

If V has a hermitian inner product, then we can prove that any irreducible linear representation $\rho\colon G \to \mathbf{GL}(V)$ of a group G, finite or infinite, where ρ is not assumed to satisfy condition (C), is equivalent to some (irreducible) subrepresentation $\widehat{\rho}\colon G \to \mathbf{GL}(\mathcal{F})$ of the left regular representation $\mathbf{R}\colon G \to \mathbf{GL}(\mathbb{C}^G)$. The key to the construction is the mapping $\varphi\colon V \to \mathbb{C}^G$ converting a vector $u \in V$ to a function $f_u \in \mathbb{C}^G$. This mapping is defined as follow.

Definition 3.7. Let V be a vector space with a hermitian inner product and let $\rho\colon G \to \mathbf{GL}(V)$ be a linear representation of a group G, finite or infinite, where ρ is not assumed to satisfy condition (C). Pick any nonzero vector $a \in V$ and define $\varphi(u) = f_u \in \mathbb{C}^G$ by

$$f_u(s) = \langle \rho(s^{-1})(u), a\rangle, \quad u \in V,\ s \in G. \tag{$*_5$}$$

The reason for using $\rho(s^{-1})$ is that we want the left regular representation. If we use $\rho(s)$, then we obtain the right regular representation, as in Vilenkin [Vilenkin (1968)] (Chapter I, Section 2.4). Since $\rho(s^{-1})$ is a linear map and the inner product is linear in its first argument, the function φ is linear. The trick is to see what is $f_{\rho(s)(u)}(t)$ $(s, t \in G)$. By definition,

$$\begin{aligned} f_{\rho(s)(u)}(t) &= \langle \rho(t^{-1})(\rho(s)(u)), a\rangle \\ &= \langle \rho(t^{-1}s)(u), a\rangle \\ &= \langle \rho((s^{-1}t)^{-1})(u), a\rangle \\ &= f_u(s^{-1}t), \end{aligned}$$

which we record as the equation

$$f_{\rho(s)(u)}(t) = f_u(s^{-1}t). \tag{$*_6$}$$

Also observe that

$$f_a(e) = \langle \rho(e^{-1})(a), a\rangle = \langle \rho(e)(a), a\rangle = \langle a, a\rangle,$$

so $f_a(e) \neq 0$ since $a \neq 0$. Using the above considerations we can prove the following result.

Proposition 3.1. *If V has a hermitian inner product, then any irreducible linear representation $\rho\colon G \to \mathbf{GL}(V)$ of a finite or infinite group G (where ρ is not assumed to satisfy condition (C)) is equivalent to some (irreducible) subrepresentation $\widehat{\rho}\colon G \to \mathbf{GL}(\mathcal{F})$ of the left regular representation $\mathbf{R}\colon G \to \mathbf{GL}(\mathbb{C}^G)$. The linear map $\varphi\colon V \to \mathbb{C}^G$ defined above is injective, $\mathcal{F} = \varphi(V)$, and $\varphi\colon V \to \mathcal{F}$ provides the equivalence between ρ and $\widehat{\rho}$.*

Proof. Since $\widehat{\rho}\colon G \to \mathbf{GL}(\mathcal{F})$ is a subrepresentation of the regular representation of G, it is given by

$$(\widehat{\rho}(s)(f))(t) = f(s^{-1}t), \quad f \in \mathcal{F},\ s, t \in G.$$

Let us first verify that the diagram

$$\begin{array}{ccc} V & \xrightarrow{\rho(s)} & V \\ {\scriptstyle\varphi}\downarrow & & \downarrow{\scriptstyle\varphi} \\ \mathcal{F} & \xrightarrow[\widehat{\rho}(s)]{} & \mathcal{F}, \end{array}$$

commutes, that is,

$$\varphi(\rho(s)(u)) = \widehat{\rho}(s)(\varphi(u)), \qquad s \in G, u \in V,$$

which means that

$$\varphi(\rho(s)(u))(t) = \widehat{\rho}(s)(\varphi(u))(t), \qquad s, t \in G, u \in V.$$

By definition of φ and $(*_6)$,

$$\varphi(\rho(s)(u))(t) = f_{\rho(s)(u)}(t) = f_u(s^{-1}t),$$

and

$$\widehat{\rho}(s)(\varphi(u))(t) = (\widehat{\rho}(s)(f_u))(t) = f_u(s^{-1}t),$$

which verifies the commutativity of the diagram. Consequently, as φ is surjective on $\mathcal{F}$ by definition, it suffices to prove that φ is injective to conclude that φ is an equivalence between ρ and $\widehat{\rho}$. Let $\mathcal{K}$ be the kernel of φ. We prove that $\mathcal{K}$ is invariant under ρ. For any $u \in V$, we have $u \in \mathcal{K}$ iff $f_u = 0$ iff $f_u(t) = 0$ for all $t \in G$ iff $f_u(s^{-1}t) = 0$ for all $t \in G$ and all $s \in G$ (since for fixed s, the map $t \mapsto s^{-1}t$ is a bijection of G), which by $(*_6)$ is equivalent to $f_{\rho(s)(u)}(t) = 0$ for all $t \in G$ iff $\rho(s)(u) \in \mathcal{K}$ for all $s \in G$. So $\mathcal{K}$ is indeed invariant. Since ρ is irreducible, either $\mathcal{K} = (0)$ of $\mathcal{K} = V$. But we observed earlier that $f_a(e) \neq 0$ so $\mathcal{K} = V$ is impossible since it means that $f_u(t) = 0$ for all $u \in V$ and all $t \in G$. Therefore, $\mathcal{K} = (0)$ and the map φ in injective. $\square$

If V is a Hilbert space, ρ is a unitary representation, all functions f_u belong to $\mathrm{L}^2(G)$, and the map $\varphi\colon V \to \mathrm{L}^2(G)$ is well-behaved, then $\widehat{\rho}$ is a unitary subrepresentation of the regular representation of G in $\mathrm{L}^2(G)$. This is the case for compact groups; see Proposition 4.17. Proposition 3.1 implies that if G is finite, since $\mathbb{C}^G$ is isomorphic to $\mathbb{C}^{|G|}$, then the dimension of the vector space V involved in an irreducible representation of G is at most the cardinality of G.

We now return to the regular representation of Example 3.2.

Example 3.6. It is easy to see that the symmetric group has three conjugacy classes, $\{\pi_1\}$, $\{\pi_2, \pi_3, \pi_6\}$ and $\{\pi_4, \pi_5\}$, so it has three irreducible representations. Going back to the regular representation $\rho_{\mathbf{R}}\colon \mathfrak{S}_3 \to \mathbf{GL}(\mathbb{C}^6)$, we see that the one-dimensional subspace V_1 spanned by $e_1 + e_2 + e_3 + e_4 + e_5 + e_6$ is invariant so the representation $\rho_{\mathbf{R}}$ is reducible. The subrepresentation of $\rho_{\mathbf{R}}$ in V_1 is equivalent to the trivial representation, which is irreducible. Although this is not obvious, there is another one-dimensional irreducible representation, which is the representation induced by the signature function ϵ on permutations. Recall that for any permutation π, its signature $\epsilon(\pi)$ is $+1$ if π is the composition of an even number of transpositions, -1 if it is the composition of an odd number of transpositions. The map $\epsilon\colon \mathfrak{S}_n \to \mathbb{C}$ is a homomorphism and it yields the irreducible representation $\rho_\epsilon\colon \mathfrak{S}_n \to \mathbf{U}(1)$ given by

$$(\rho_\epsilon(\pi))(z) = \epsilon(\pi)z, \quad z \in \mathbb{C}.$$

Then we see that the subspace V_2 spanned by the vector $e_1 - e_2 - e_3 + e_4 + e_5 - e_6$ (which corresponds to the signatures $+1, -1, -1, +1, +1, -1$ of the permutations $\pi_1, \ldots, \pi_6$) is invariant under $\rho_{\mathbf{R}}$, and the subrepresentation of $\rho_{\mathbf{R}}$ to V_2 is equivalent to the irreducible representation ρ_ϵ. The orthogonal complement V_3 of $V_1 \oplus V_2$ is the intersection of the two hyperplanes in $\mathbb{C}^6$ given by the equations

$$\begin{aligned} x_1 + x_2 + x_3 + x_4 + x_5 + x_6 &= 0 \\ x_1 - x_2 - x_3 + x_4 + x_5 - x_6 &= 0, \end{aligned}$$

a subspace of dimension 4. By adding and subtracting these equations, we see that the subspace V_3 is also defined by the equations

$$\begin{aligned} x_1 + x_4 + x_5 &= 0 \\ x_2 + x_3 + x_6 &= 0. \end{aligned}$$

We can prove directly that V_3 is invariant under $\rho_{\mathbf{R}}$, but since the representation $\rho_{\mathbf{R}}$ is actually unitary, we prefer using results from the next section.

An easy but crucial lemma about irreducible representations is "Schur's Lemma."

Lemma 3.2. *(Schur's Lemma) Let $\rho_1 \colon G \to \mathbf{GL}(V)$ and $\rho_2 \colon G \to \mathbf{GL}(W)$ be any two real or complex finite-dimensional representations of a group G. If ρ_1 and ρ_2 are irreducible, then the following properties hold:*

(i) Every G-map $\varphi \colon \rho_1 \to \rho_2$ in $\mathrm{Hom}_G(\rho_1, \rho_2)$ is either the zero map or an isomorphism.

(ii) If ρ_1 is a complex representation, then every G-map $\varphi \colon \rho_1 \to \rho_1$ in $\mathrm{Hom}_G(\rho_1, \rho_1)$ is of the form $\varphi = \lambda \mathrm{id}$, for some $\lambda \in \mathbb{C}$.

(iii) If $\rho_1 \colon G \to \mathbf{GL}(V)$ and $\rho_2 \colon G \to \mathbf{GL}(W)$ are real or complex irreducible representations, then ρ_1 and ρ_2 are equivalent iff $\mathrm{Hom}_G(\rho_1, \rho_2) \neq (0)$. If ρ_1 and ρ_2 are complex representations, then ρ_1 and ρ_2 are equivalent iff $\dim \mathrm{Hom}_G(\rho_1, \rho_2) = 1$.

Proof. (i) Observe that the kernel $\mathrm{Ker}\ \varphi \subseteq V$ of φ is invariant under ρ_1. Indeed, for every $v \in \mathrm{Ker}\ \varphi$ and every $g \in G$, we have

$$\varphi(\rho_1(g)(v)) = \rho_2(g)(\varphi(v)) = \rho_2(g)(0) = 0,$$

so $\rho_1(g)(v) \in \mathrm{Ker}\ \varphi$. Thus, $\rho_1 \colon G \to \mathbf{GL}(\mathrm{Ker}\ \varphi)$ is a subrepresentation of ρ_1, and as ρ_1 is irreducible, either $\mathrm{Ker}\ \varphi = (0)$ or $\mathrm{Ker}\ \varphi = V$. In the second case, $\varphi = 0$. If $\mathrm{Ker}\ \varphi = (0)$, then φ is injective. However, $\varphi(V) \subseteq W$ is invariant under ρ_2, since for every $v \in V$ and every $g \in G$,

$$\rho_2(g)(\varphi(v)) = \varphi(\rho_1(g)(v)) \in \varphi(V),$$

and as $\varphi(V) \neq (0)$ (as $V \neq (0)$ since ρ_1 is irreducible) and ρ_2 is irreducible, we must have $\varphi(V) = W$; that is, φ is an isomorphism. The proof also works for infinite-dimensional spaces.

(ii) Since V is a complex vector space of finite dimension, the linear map φ has some eigenvalue $\lambda \in \mathbb{C}$. Let $E_\lambda \subseteq V$ be the eigenspace associated with λ. The subspace E_λ is invariant under ρ_1, since for every $u \in E_\lambda$ and every $g \in G$, we have

$$\varphi(\rho_1(g)(u)) = \rho_1(g)(\varphi(u)) = \rho_1(g)(\lambda u) = \lambda \rho_1(g)(u),$$

so $\rho_1 \colon G \to \mathbf{GL}(E_\lambda)$ is a subrepresentation of ρ_1, and as ρ_1 is irreducible and $E_\lambda \neq (0)$, we must have $E_\lambda = V$.

(iii) If $\mathrm{Hom}_G(\rho_1, \rho_2) = (0)$, then ρ_1 and ρ_2 are inequivalent, since otherwise, $\mathrm{Hom}_G(\rho_1, \rho_2)$ would contain some isomorphism. Conversely, if $\mathrm{Hom}_G(\rho_1, \rho_2) \neq (0)$, then by (i), there is some isomorphism in

$\mathrm{Hom}_G(\rho_1, \rho_2)$, so ρ_1 and ρ_2 are equivalent. Therefore, ρ_1 and ρ_2 are inequivalent iff $\mathrm{Hom}_G(\rho_1, \rho_2) = (0)$, and ρ_1 and ρ_2 are equivalent iff $\mathrm{Hom}_G(\rho_1, \rho_2) \neq (0)$.

Suppose that ρ_1 and ρ_2 are complex equivalent representations and let φ_1 and φ_2 be two nonzero maps in $\mathrm{Hom}_G(\rho_1, \rho_2) \neq (0)$, which by (i), must be isomorphisms. It is immediately verified that $\varphi_2^{-1} \circ \varphi_1 \in \mathrm{Hom}_G(\rho_1, \rho_1)$ and $\varphi_2^{-1} \circ \varphi_1$ is an isomorphism, so by (ii), we have $\varphi_2^{-1} \circ \varphi_1 = \lambda \mathrm{id}$ for some nonzero $\lambda \in \mathbb{C}$, which means that $\varphi_1 = \lambda \varphi_2$. Therefore, $\dim \mathrm{Hom}_G(\rho_1, \rho_2) = 1$. Conversely, if $\dim \mathrm{Hom}_G(\rho_1, \rho_2) = 1$, by the first part, ρ_1 and ρ_2 are equivalent. ☐

Part (i) of Schur's Lemma also holds for infinite-dimensional representations as we noted in the proof.

An interesting corollary of Schur's Lemma is the following fact:

Proposition 3.3. *A complex irreducible finite-dimensional representation $\rho\colon G \to \mathbf{GL}(V)$ of a commutative group G is one-dimensional.*

Proof. Since G is abelian, we claim that for every $g \in G$, the map $\tau_g \colon V \to V$ given by $\tau_g(v) = \rho(g)(v)$ for all $v \in V$ is a G-map. This amounts to checking that the following diagram commutes

$$\begin{array}{ccc} V & \xrightarrow{\rho(g_1)} & V \\ {\scriptstyle \tau_g}\downarrow & & \downarrow{\scriptstyle \tau_g} \\ V & \xrightarrow[\rho(g_1)]{} & V \end{array}$$

for all $g, g_1 \in G$. This is equivalent to checking that

$$\begin{aligned} \tau_g(\rho(g_1)(v)) &= \rho(g)(\rho(g_1)(v)) = \rho(gg_1)(v) \\ &= \rho(g_1)(\tau_g(v)) = \rho(g_1)(\rho(g)(v)) = \rho(g_1 g)(v) \end{aligned}$$

for all $v \in V$, that is, $\rho(gg_1)(v) = \rho(g_1 g)(v)$, which holds since G is commutative (so $gg_1 = g_1 g$).

By Schur's Lemma (Lemma 3.2(ii)), $\tau_g = \lambda_g \mathrm{id}$ for some $\lambda_g \in \mathbb{C}$. It follows that any subspace of V is invariant. If the representation is irreducible, we must have $\dim(V) = 1$ since otherwise V would contain a one-dimensional invariant subspace, contradicting the assumption that ρ is irreducible. ☐

3.2 Unitary Group Representations

We now generalize representations to allow the representing space to be a *complex Hilbert space* (typically separable).

Definition 3.8. Given a locally compact group G and a complex Hilbert space H, a *unitary representation of G in H* is a group homomorphism $U\colon G \to \mathbf{U}(H)$, where $\mathbf{U}(H)$ is the group of unitary operators on H, such that:

(C) The map $g \mapsto U(g)(u)$ is continuous for every $u \in H$.
(U) Every linear map $U(g)$ is an isometry; that is,

$$\langle U(g)(u), U(g)(v)\rangle = \langle u, v\rangle, \qquad \text{for all } g \in G \text{ and all } u, v \in H.$$

In particular, $U(g)$ is continuous and

$$(U(g))^{-1} = (U(g))^{*}, \qquad \text{for all } g \in G.$$

As in Definition 3.1, to avoid confusion when representations involving different groups arise we denote the space of the representation U by H_U, and so we denote a representation as $U\colon G \to \mathbf{U}(H_U)$.

Remark: Sometimes, a unitary representation as in Definition 3.8 is called a *continuous* unitary representation. Note that if H is infinite-dimensional, the map $g \mapsto U(g)$ is *not necessarily continuous*. For a counter-example involving the regular representation of an infinite compact group G in $\mathrm{L}^2(G)$, see Dieudonné [Dieudonné (2003a)] (Chapter XXI, Section 1, Problem 3). However, the left action $U^a\colon G \times H \to H$ associated with U given by

$$U^a(s, x) = U(s)(x), \qquad \text{for all } s \in G \text{ and all } x \in H$$

is continuous. Indeed, since $U(s)$ is a unitary map, we have $\|U(s)(w)\| = \|w\|$ for all $w \in H$, so for all $s, t \in G$ and all $x, y \in H$, we have

$$\begin{aligned}\|U^a(s,x) - U^a(t,y)\| &\leq \|U(s)(x) - U(s)(y)\| + \|U(s)(y) - U(t)(y)\| \\ &= \|U(s)(x-y)\| + \|U(s)(y) - U(t)(y)\| \\ &= \|x - y\| + \|U(s)(y) - U(t)(y)\|,\end{aligned}$$

and since by hypothesis, for any fixed $y \in H$, the map $s \mapsto U(s)(y)$ is continuous, we see that the action U^a is continuous. Conversely, if the action $U^a\colon G \times H \to H$ is continuous, then obviously the map $s \mapsto U(s)(y)$ is continuous, so U is a unitary representation.

The notion of morphism of unitary representations and of equivalence is adapted as follows.

Definition 3.9. Given any two unitary representations $U_1\colon G \to \mathbf{U}(H_1)$ and $U_2\colon G \to \mathbf{U}(H_2)$, a *G-map* (or *morphism of representations*) $\varphi\colon U_1 \to U_2$ is a continuous linear map which is *equivariant*, which means that the following diagram commutes for every $g \in G$:

$$\begin{array}{ccc} H_1 & \xrightarrow{U_1(g)} & H_1 \\ {\scriptstyle\varphi}\downarrow & & \downarrow{\scriptstyle\varphi} \\ H_2 & \xrightarrow[U_2(g)]{} & H_2, \end{array}$$

i.e.

$$\varphi \circ U_1(g) = U_2(g) \circ \varphi, \qquad g \in G.$$

The space of all G-maps between two representations as above is denoted $\mathrm{Hom}_G(U_1, U_2)$. A G-map is also called an *intertwining operator*. Two unitary representations $U_1\colon G \to \mathbf{U}(H_1)$ and $U_2\colon G \to \mathbf{U}(H_2)$ are *equivalent* iff $\varphi\colon H_1 \to H_2$ is an invertible linear isometry whose inverse is also continuous; thus $U_2(g) = \varphi \circ U_1(g) \circ \varphi^{-1}$, for all $g \in G$.

When $U_1 = U_2$, the space of G-maps $\mathrm{Hom}_G(U, U)$ is a unital subalgebra of $\mathcal{L}(H)$ denoted by $\mathcal{C}(U)$ and is called the *commutant* or *centralizer* of U. Observe that

$$\mathcal{C}(U) = \{\varphi \in \mathcal{L}(H) \mid \varphi \circ U(g) = U(g) \circ \varphi \quad \text{for all } g \in G\}.$$

It is easy to show that the unital subalgebra $\mathcal{C}(U)$ of $\mathcal{L}(H)$ is actually a C^*-algebra and that it is closed in $\mathcal{L}(H)$ under weak limits (see Definition 2.21(3)). By the *von Neumann density theorem*, it is also closed in $\mathcal{L}(H)$ under strong limits (Definition 2.21(2)); see Folland [Folland (1995)], Section 1.6. Such a C^*-algebra is a *von Neumann algebra*.

Given a unitary representation $U\colon G \to \mathbf{U}(H)$, the definition of an invariant subspace $W \subseteq H$ is the same as in Definition 3.4. If $W \subseteq H$ is invariant under U, we say that the subrepresentation $U\colon G \to \mathbf{U}(W)$ is *closed* if W is closed in H. As in the case of unitary representations of algebras, the notion of subrepresentation is only well-defined for closed invariant subspaces of H. However, by Proposition 3.5, since the closure $\overline{W}$ of an invariant subspace W is closed, the notion of subrepresentation of G in $\overline{W}$ is well-defined.

In the definition of an *irreducible* unitary representation $U\colon G \to \mathbf{U}(H)$ $(H \neq (0))$, we require that the only *closed* subrepresentations $U\colon G \to$

$\mathbf{U}(W)$ of the representation $U\colon G \to \mathbf{U}(H)$ correspond to $W = (0)$ or $W = H$.

As for representations of algebras, we can define topologically cyclic representations and cyclic vectors.

Definition 3.10. Let $U\colon G \to \mathbf{U}(H)$ be a unitary representation of G in H. A vector $x_0 \in H$ is called a *totalizer*, or *totalizing vector*, or *cyclic vector* for the representation U if the subspace of H spanned by the set $\{U(s)(x_0) \mid s \in G\}$ is dense in H. Equivalently, if $\mathcal{M}_{x_0}$ denotes the closure of the set $\{U(s)(x_0) \mid s \in G\}$, called the *cyclic subspace* generated by x_0, which is invariant under U, then x_0 is a totalizer (a cyclic vector) if $\mathcal{M}_{x_0} = H$. A representation which admits a totalizer is said to be *topologically cyclic.*

The importance of totalizers stems from the following result which is the analog of Proposition 2.10 for group representations. In fact, the proof is essentially the same.

Proposition 3.4. *Let $U\colon G \to \mathbf{U}(H)$ be a unitary representation of G in H. Then H is the Hilbert sum of a sequence $(H_\alpha)_{\alpha\in\Lambda}$ of closed subspaces $H_\alpha \neq (0)$ of H invariant under U, and such that the restriction of U to each H_α is topologically cyclic. If H is separable, the family Λ is countable (possibly finite).*

Proposition 3.4 is proven in the separable case in Dieudonné [Dieudonné (2005)] (Chapter XV, Section 5), and in general, using Zorn's Lemma; see Folland [Folland (1995)] (Chapter 3, Proposition 3.3).

Hilbert sums of unitary representations of a locally compact group are defined just as in the case of an algebra; see Definition 2.6. We also have the following version of Proposition 2.5 for group representations.

Proposition 3.5. *Let $U\colon G \to \mathbf{U}(H)$ be a unitary representation of G in H.*

(1) If the subspace E of H is invariant under U, then its closure $\overline{E}$ is also invariant under U.
(2) Let E be a closed subspace of H invariant under U. If $E^{\perp}$ is the orthogonal complement of E in H, then $E^{\perp}$ is invariant under U. If $U_1(s)$ and $U_2(s)$ are the restrictions of $U(s)$ to E and $E^{\perp}$, then $H = E \oplus E^{\perp}$ (the algebraic direct sum), and the representation U is the Hilbert sum of the representations U_1 and U_2.

Proof. Part (1) is easy to prove and follows from the continuity of $U(s)$; see Dieudonné [Dieudonné (2010)], (Chapter III, Section 11). For part (2), let $x \in E$ and $y \in E^{\perp}$. For any $s \in G$, we have

$$\langle x, U(s)(y)\rangle = \langle (U(s))^{*}(x), y\rangle = \langle (U(s))^{-1}(x), y\rangle = \langle U(s^{-1})(x), y\rangle = 0,$$

since E is invariant under U, so $U(s^{-1})(x) \in E$, and since $E^{\perp}$ is the orthogonal complement of E and $y \in E^{\perp}$. Then $U(s)(y)$ is orthogonal to all $x \in E$, which means that $U(s)(y) \in E^{\perp}$, so $E^{\perp}$ is invariant under U. The last property is obvious. $\square$

One should realize that property (2) of Proposition 3.5 fails for nonunitary representations. For example, the map

$$U \colon x \mapsto \begin{pmatrix} 1 & x \\ 0 & 1 \end{pmatrix}$$

is a representation of $\mathbb{R}$ in $\mathbb{C}^2$, but the only nontrivial invariant subspace is the subspace spanned by $(1, 0)$, which is one-dimensional. The problem is that because $\mathbb{R}$ is not compact, there is no way to define an inner product on $\mathbb{C}^2$ invariant under U.

However, using the Haar measure, Vol. I, Theorem 8.36 shows that if H is a finite-dimensional hermitian space, then there is an inner-product on H for which the linear maps $U(s)$ are unitary.

Theorem 3.6. *(Complete Reducibility) Let $U \colon G \to \mathbf{GL}(H)$ be a linear representation of a compact group G in a complex space H of dimension $n \geq 1$. There is a hermitian inner product $\langle -, -\rangle$ on H, such that $U \colon G \to \mathbf{U}(H)$ is a unitary representation of G in the hermitian space $(H, \langle -, -\rangle)$. The representation U is the direct sum of a finite number of irreducible unitary representations.*

Proof. As we noted in the discussion following Definition 3.1, the representation $U \colon G \to \mathbf{GL}(H)$ is a continuous linear map $g \mapsto U(g)$ from G to $\mathcal{L}(H)$ equipped with any norm. Since H is finite-dimensional we can define a hermitian inner product on it, for example by picking a basis and considering the hermitian inner product for which this basis is orthonormal. Since G is compact and H is a finite-dimensional hermitian space, Vol. I, Theorem 8.36 yields an inner product on H which is invariant under U.

We proceed by complete induction on the dimension $n \geq 1$ of H. When $n = 1$, the representation is automatically irreducible. If $n > 1$ and the representation is not irreducible, then it has some invariant subspace H_1

of dimension n_1 with $1 \leq n_1 < n$. By Proposition 3.5, the orthogonal complement $H_2 = H_1^{\perp}$ of H_1 is also invariant under U, and its dimension n_2 satisfies $n_2 \geq 1$ and $n_1 + n_2 = n$, with $n > 1$ and $1 \leq n_1 < n$, so we also have $1 \leq n_2 < n$. We can apply the induction hypothesis to the subrepresentations $U \colon G \to \mathbf{U}(H_1)$ and $U \colon G \to \mathbf{U}(H_2)$, with $H = H_1 \oplus H_2$, and we obtain a collection of irreducible representations of G whose direct sum is U. ☐

Theorem 3.6 is very significant because it shows that the study of *arbitrary finite-dimensional* representations of a compact group G reduces to the study of the *irreducible unitary (finite-dimensional) representations* of G.

Example 3.7. The regular representation $\rho_{\mathbf{R}} \colon \mathfrak{S}_3 \to \mathbf{GL}(\mathbb{C}^6)$ of $\mathfrak{S}_3$ from Example 3.2 is obviously unitary. Theorem 3.6 tells us that $\rho_{\mathbf{R}}$ is the direct sum of irreducible representations, and in Example 3.6 we already found two irreducible representations which are one-dimensional. The discussion before Example 3.6 also shows that the standard representation (see Example 3.5) must occur in the representation $\rho_{\mathbf{R}}$, and if there are h irreducible representations, the equation $n_1^2 + \cdots + n_h^2 = g = 6$ implies that $1 + 1 + 2^2 + \cdots + n_h^2 = 6$, so $h = 3$ and the standard representation occurs twice. Therefore, the orthogonal complement V_3 of the direct sum $V_1 \oplus V_2$ given by the equations

$$
\begin{aligned}
x_1 + x_4 + x_5 &= 0 \\
x_2 + x_3 + x_6 &= 0
\end{aligned}
$$

must be the direct sum of 2 two-dimensional invariant subspaces. With a little help from `Matlab`, we find that the subspace V_1^3 spanned by the vectors

$$
e_1 + e_2 - e_3 - e_4, \;\; e_3 + e_4 - e_5 - e_6
$$

is invariant under $\rho_{\mathbf{R}}$, the subspace V_2^3 spanned by the vectors

$$
e_1 - e_3 - e_4 + e_6, \;\; e_2 + e_4 - e_5 - e_6,
$$

is also invariant under $\rho_{\mathbf{R}}$, both V_1^3 and V_2^3 are orthogonal to $V_1 \oplus V_2$, and

$$
\mathbb{C}^6 = V_1 \oplus V_2 \oplus V_1^3 \oplus V_2^3.
$$

To show that V_1^3 is invariant, we observe that V_1^3 is also spanned by

$$
e_1 + e_2 - e_3 - e_4, \;\; e_3 + e_4 - e_5 - e_6, \;\; e_1 + e_2 - e_5 - e_6,
$$

and the action of $\rho_{\mathbf{R}}(\pi_i)$ is to permute these vectors, possibly flipping signs, and similarly V_2^3 is also spanned by

$$e_1 - e_3 - e_4 + e_6,\ e_2 + e_4 - e_5 - e_6,\ e_1 + e_2 - e_3 - e_5,$$

and the action of $\rho_{\mathbf{R}}(\pi_i)$ is also to permute these vectors, possibly flipping signs. According to our previous discussion, these two sub-representations of $\mathfrak{S}_3$ in V_1^3 and V_2^3 are equivalent to the standard representation given in Example 3.5. Thus, we identified explicitly the three irreducible representations of $\mathfrak{S}_3$ as subrepresentations of the regular representation.

The analog of Proposition 2.7 holds for unitary group representations.

Proposition 3.7. *Let $U\colon G \to \mathbf{U}(H)$ be a unitary representation of G in H. A closed subspace E of H is invariant under U iff $P_E U(g) = U(g) P_E$ for all $g \in G$, in other words, $P_E \in \mathcal{C}(U) = \mathrm{Hom}_G(U, U)$, where $P_E\colon H \to E$ is the orthogonal projection of H onto E.*

Proof. Assume that $P_E \in \mathrm{Hom}_G(U, U)$, so that the following diagram commutes:

$$\begin{array}{ccc} H & \xrightarrow{U(g)} & H \\ {\scriptstyle P_E}\downarrow & & \downarrow{\scriptstyle P_E} \\ H & \xrightarrow[U(g)]{} & H. \end{array}$$

For any $x \in E$, since P_E is the orthogonal projection of H onto E, we have $P_E(x) = x$, so

$$P_E(U(g)(x)) = U(g)(P_E(x)) = U(g)(x),$$

which shows that $U(g)(x) \in E$, and thus E is invariant under U.

Conversely, assume that E is invariant under U. Since E is closed, by a well-known result of Hilbert space theory, we have $H = E \oplus E^\perp$, an algebraic direct sum. For any $x \in E$, since E is invariant under U, we have $U(g)(x) \in E$ for all $g \in G$, and since P_E is a projection onto E, we have

$$U(g)(P_E(x)) = U(g)(x) = P_E(U(g)(x)), \qquad \text{for all } x \in E.$$

By Proposition 3.5, the subspace $E^\perp$ is also invariant under U. For any $x \in E^\perp$, we have $U(g)(x) \in E^\perp$, so $P_E(x) = P_E(U(g)(x)) = 0$, and we have

$$U(g)(P_E(x)) = 0 = P_E(U(g)(x)), \qquad \text{for all } x \in E^\perp.$$

Since $U = E \oplus E^\perp$, we have

$$U(g)(P_E(x)) = P_E(U(g)(x)), \qquad \text{for all } x \in H,$$

namely, $P_E \in \mathrm{Hom}_G(U, U)$. ∎

Proposition 3.7 yields a method for proving that a unitary representation $U\colon G \to \mathbf{U}(H)$ is irreducible. Indeed, if U is reducible, then there is some nonzero G-map $\varphi \in \mathrm{Hom}_G(U,U)$, which is *not invertible*. Thus, *if every nonzero G-map in* $\mathrm{Hom}_G(U,U)$ *is invertible, then U must be irreducible*. This technique is illustrated in the next example.

Example 3.8. Recall the representations $U_m\colon \mathbf{SU}(2) \to \mathbf{GL}(\mathcal{P}_m^{\mathbb{C}}(2))$ from Example 3.3, where $\mathcal{P}_m^{\mathbb{C}}(2)$ denotes the vector space of complex homogeneous polynomials $P(z_1,z_2) = \sum_{k=0}^{m} c_k z_1^k z_2^{m-k}$ of degree m ($c_i \in \mathbb{C}$). The $m+1$ monomials $P_k = z_1^k z_2^{m-k}$ ($0 \le k \le m$) form a basis of $\mathcal{P}_m^{\mathbb{C}}(2)$. In the physics literature, it is customary to index homogeneous polynomials in terms of $\ell = m/2$, which is an integer when m is even but a half integer when m is odd. In this context, the number $\ell = m/2$ is the *spin* of a particle. In terms of $\ell = m/2$, a homogeneous polynomial is written as

$$P(z_1,z_2) = \sum_{k=-\ell}^{\ell} c_k z_1^{\ell-k} z_2^{\ell+k},$$

where it is assumed that $\ell + k = j$, where j takes the *integral* values $j = 0,1,\ldots,2\ell = m$, so that $\ell - k = 2\ell - (\ell+k) = 2\ell - j$ takes the values $2\ell, 2\ell-1,\ldots,0$. Note that $k = j - \ell = j - m/2$ with $j = 0,1,\ldots,2\ell = m$, so k is an integer only if m is even. The physics notation makes it easier to make the connection between the matrix expression of the representations U_m (renamed as U_ℓ) and the special functions expressed in terms of Jacobi polynomials; see Vilenkin [Vilenkin (1968)] (Chapter III, Sections 2 and 3).

For every matrix $S \in \mathbf{SU}(2)$, with

$$S = \begin{pmatrix} \alpha & \beta \\ -\overline{\beta} & \overline{\alpha} \end{pmatrix}, \quad \alpha\overline{\alpha} + \beta\overline{\beta} = 1,$$

for every homogeneous polynomial $P \in \mathcal{P}_m^{\mathbb{C}}(2)$, $U_m(S)(P(z_1,z_2))$ is defined by

$$U_m(S)(P(z_1,z_2)) = P(\overline{\alpha}z_1 - \beta z_2, \overline{\beta}z_1 + \alpha z_2). \tag{U_m}$$

As defined, the representations U_m are not unitary, but since $\mathbf{SU}(2)$ is compact, we can apply Theorem 3.6 to find an invariant inner product on $\mathcal{P}_m^{\mathbb{C}}(2)$. This can actually be done quite explicitly; we will come back to this point later.

Proposition 3.8. *The representations* $U_m\colon \mathbf{SU}(2) \to \mathbf{GL}(\mathcal{P}_m^{\mathbb{C}}(2))$ *are irreducible.*

Proof. To prove that the representations U_m are irreducible, it suffices to prove that every nonzero equivariant map A in $\mathrm{Hom}_{\mathbf{SU}(2)}(U_m, U_m)$ is invertible. Actually, we will prove that $A = \lambda \mathrm{id}$, with $\lambda \in \mathbb{C}, \lambda \neq 0$. A nice and rather short proof is given in Bröcker and tom Dieck [Bröcker and tom Dieck (1985)], Chapter 2, Proposition 5.1. The trick is to consider the matrices

$$r_x(\varphi) = \begin{pmatrix} e^{i\varphi} & 0 \\ 0 & e^{-i\varphi} \end{pmatrix}, \quad 0 < \varphi < \pi.$$

Plugging the matrix $r_x(\varphi)$ and $P = P_k = z_1^k z_2^{m-k}$ in Equation (U_m) yields $U_m(r_x(\varphi))(P_k) = (e^{-i\varphi} z_1)^k (e^{i\varphi} z_2)^{m-k} = e^{i(m-2k)\varphi} z_1^k z_2^{m-k} = e^{i(m-2k)\varphi} P_k$. Therefore, $(P_0, \ldots, P_m)$ is a basis (in fact, orthogonal) of eigenvectors of $U_m(r_x(\varphi))$ for the eigenvalues $(e^{im\varphi}, e^{i(m-2)\varphi}, \ldots, e^{-im\varphi})$. We can pick φ such that these eigenvalues are all distinct, for example $\varphi = 2\pi/m$. Now if $A \in \mathrm{Hom}_{\mathbf{SU}(2)}(U_m, U_m)$ is equivariant, then $U_m(r_x(\varphi))A = AU_m(r_x(\varphi))$, so for $k = 0, \ldots, m$ we have

$$U_m(r_x(\varphi))AP_k = AU_m(r_x(\varphi))P_k = Ae^{i(m-2k)\varphi} P_k = e^{i(m-2k)\varphi} AP_k.$$

The above implies that either $AP_k = 0$ or AP_k is an eigenvector of $U_m(r_x(\varphi))$ for the eigenvalue $e^{i(m-2k)\varphi}$. Since φ was chosen so that the eigenvalues $(e^{im\varphi}, \ldots, e^{i(m-2)\varphi}, \ldots, e^{-im\varphi})$ are all distinct, each eigenspace is one-dimensional, so $AP_k = c_k P_k$ for some $c_k \in \mathbb{C}$, $c_k \neq 0$. In either case,

$$AP_k = c_k P_k$$

for some $c_k \in \mathbb{C}$. We will now prove that $c_0 = c_1 = \cdots = c_m$. This shows that $A = c_0 \mathrm{id}_{m+1}$, and since A is not the zero map, $c_0 \neq 0$, so A is invertible, as desired.

To prove that the c_k have the same value, we use the matrices

$$r_y(t) = \begin{pmatrix} \cos t & -\sin t \\ \sin t & \cos t \end{pmatrix}, \quad t \in \mathbb{R}.$$

Since A is equivariant, $AU_m(r_y(t)) = U_m(r_y(t))A$, so we need to compute $AU_m(r_y(t))P_m$ and $U_m(r_y(t))AP_m$. Since $P_m = z_1^m$ and $AP_k = c_k P_k$, using Equation (U_m) we have

$$\begin{aligned}
AU_m(r_y(t))P_m &= A(z_1 \cos t + z_2 \sin t)^m \\
&= A \sum_{k=1}^{m} \binom{m}{k} (\cos t)^k (\sin t)^{m-k} z_1^k z_2^{m-k} \\
&= \sum_{k=1}^{m} \binom{m}{k} (\cos t)^k (\sin t)^{m-k} AP_k \\
&= \sum_{k=1}^{m} \binom{m}{k} (\cos t)^k (\sin t)^{m-k} c_k P_k.
\end{aligned}$$

We also have

$$\begin{aligned}U_m(r_y(t))AP_m &= U_m(r_y(t))c_mP_m = c_mU_m(r_y(t))P_m\\&= c_m(z_1\cos t + z_2\sin t)^m\\&= \sum_{k=1}^{m}\binom{m}{k}(\cos t)^k(\sin t)^{m-k}c_mP_k.\end{aligned}$$

Since $AU_m(r_y(t))P_m = U_m(r_y(t))AP_m$, comparing coefficients (since these equations hold for all $t \in \mathbb{R}$) we obtain

$$c_k = c_m, \quad 0 \le k \le m.$$

Therefore, on the basis $(P_0, \ldots, P_m)$ we have $AP_k = c_0P_k$, which means that $A = c_0\mathrm{id}_{m+1}$, as claimed. □

Therefore, the representations $U_m\colon \mathbf{SU}(2) \to \mathbf{GL}(\mathcal{P}_m^{\mathbb{C}}(2))$ are irreducible unitary representations of $\mathbf{SU}(2)$. In fact, they constitute all of them up to equivalence, but this is harder to prove. A good strategy is to use properties of the characters of compact groups; see Section 4.2.

The groups $\mathbf{SU}(2)$ and $\mathbf{SO}(3)$ are intimately related by the adjoint representation that we review next. Details can be found in Gallier and Quaintance [Gallier and Quaintance (2020b)] (Chapter 15) and Gallier [Gallier (2011)] (Chapter 9). The group $\mathbf{SU}(2)$ turns out to be the group of unit quaternions but all we need here is Theorem 3.9.

The group $\mathbf{SU}(2)$ is the group of 2×2 complex matrices q of the form

$$q = \begin{pmatrix} a+ib & c+id\\ -(c-id) & a-ib\end{pmatrix}, \quad a,b,c,d \in \mathbb{R},\ a^2+b^2+c^2+d^2 = 1.$$

If we get rid of the condition $a^2+b^2+c^2+d^2 = 1$, the set of *all* matrices X of the form

$$X = \begin{pmatrix} a+ib & c+id\\ -(c-id) & a-ib\end{pmatrix}, \quad a,b,c,d \in \mathbb{R}$$

is a real vector space, which turns out to be closed under multiplication and in which every nonzero element has a multiplicative inverse. It is the skew-field of *quaternions*, denoted $\mathbb{H}$.

If we write $\alpha = a+ib$ and $\beta = c+id$, then a matrix $q \in \mathbf{SU}(2)$ can be written as

$$q = \begin{pmatrix} \alpha & \beta\\ -\overline{\beta} & \overline{\alpha}\end{pmatrix}, \quad \text{with} \quad |\alpha|^2 + |\beta|^2 = 1.$$

Since the matrices in $\mathbf{SU}(2)$ are unitary, the inverse of q is q^*, given by

$$q^* = \begin{pmatrix} \overline{\alpha} & -\beta \\ \overline{\beta} & \alpha \end{pmatrix}.$$

The group $\mathbf{SU}(2)$ is a Lie group, whose Lie algebra $\mathfrak{su}(2)$ is defined as follows.

Definition 3.11. The (real) vector space $\mathfrak{su}(2)$ of 2×2 *skew Hermitian matrices with zero trace* is given by

$$\mathfrak{su}(2) = \left\{ \begin{pmatrix} ix & y + iz \\ -y + iz & -ix \end{pmatrix} \;\middle|\; (x, y, z) \in \mathbb{R}^3 \right\}.$$

Observe that for every matrix $A \in \mathfrak{su}(2)$, we have $A^* = -A$, that is, A is skew Hermitian, and that $\mathrm{tr}(A) = 0$. Also note that $\mathfrak{su}(2) \subseteq \mathbb{H}$. The quaternions in $\mathfrak{su}(2)$ are also called *pure quaternions* (they have no "real part" a).

Definition 3.12. The *adjoint representation* of the group $\mathbf{SU}(2)$ is the group homomorphism
$\mathrm{Ad}\colon \mathbf{SU}(2) \to \mathbf{GL}(\mathfrak{su}(2))$ defined such that for every $q \in \mathbf{SU}(2)$, with

$$q = \begin{pmatrix} \alpha & \beta \\ -\overline{\beta} & \overline{\alpha} \end{pmatrix} \in \mathbf{SU}(2),$$

we have

$$\mathrm{Ad}_q(A) = qAq^*, \quad A \in \mathfrak{su}(2),$$

where q^* is the inverse of q.

One needs to verify that the map Ad_q is an invertible linear map from $\mathfrak{su}(2)$ to itself, and that Ad is a group homomorphism, which is easy to do.

In order to associate a rotation ρ_q (in $\mathbf{SO}(3)$) to q, we need to embed $\mathbb{R}^3$ into $\mathfrak{su}(2) \subseteq \mathbb{H}$ as the pure quaternions, by

$$\mathrm{su}(x, y, z) = \begin{pmatrix} ix & y + iz \\ -y + iz & -ix \end{pmatrix}, \quad (x, y, z) \in \mathbb{R}^3.$$

Then q defines the rotation $\rho_q \in \mathbf{SO}(3)$ given by

$$\rho_q(x, y, z) = \mathrm{su}^{-1}(q\,\mathrm{su}(x, y, z)q^*).$$

Therefore, modulo the isomorphism su, the linear map ρ_q is the linear isomorphism Ad_q. The reason why this is interesting is summarized in the following result proven in Gallier [Gallier (2011)] (Chapter 9).

Theorem 3.9. *Let $\rho\colon \mathbf{SU}(2) \to \mathbf{SO}(3)$ be the map given by*

$$\rho_q(x, y, z) = \mathrm{su}^{-1}(q\,\mathrm{su}(x, y, z)q^*), \quad q \in \mathbf{SU}(2),\ (x, y, z) \in \mathbb{R}^3.$$

The map $\rho\colon \mathbf{SU}(2) \to \mathbf{SO}(3)$ is a surjective homomorphism whose kernel is $\{I, -I\}$. If

$$q = \begin{pmatrix} a + ib & c + id \\ -(c - id) & a - ib \end{pmatrix}, \quad a, b, c, d \in \mathbb{R},\ a^2 + b^2 + c^2 + d^2 = 1,$$

let $u = (b, c, d)$. We have $\rho_q = I_3$ iff $u = (b, c, d) = 0$ iff $|a| = 1$. If $u \neq 0$, then either $a = 0$ and ρ_q is a rotation by π around the axis of rotation determined by the vector $u = (b, c, d)$, or $0 < |a| < 1$ and ρ_q is the rotation around the axis of rotation determined by the vector $u = (b, c, d)$ and the angle of rotation $\theta \neq \pi$ with $0 < \theta < 2\pi$, is given by

$$\tan(\theta/2) = \frac{\|u\|}{a}.$$

Here we are assuming that a basis (w_1, w_2) has been chosen in the plane orthogonal to $u = (b, c, d)$, such that (w_1, w_2, u) is positively oriented, that is, $\det(w_1, w_2, u) > 0$ *(where w_1, w_2, u are expressed over the canonical basis (e_1, e_2, e_3), which is chosen to define positive orientation).*

Remark: Under the orientation defined above, we have

$$\cos(\theta/2) = a, \quad 0 < \theta < 2\pi.$$

Note that the condition $0 < \theta < 2\pi$ implies that θ is uniquely determined by the above equation. This is not the case if we choose π such that $-\pi < \theta < \pi$ since both θ and $-\theta$ satisfy the equation, and this shows why the condition $0 < \theta < 2\pi$ is preferable. If $0 < a < 1$, then $0 < \theta < \pi$, and if $-1 < a < 0$, then $\pi < \theta < 2\pi$. In the second case, ρ_q is also the rotation of axis $-u$ and of angle $-(2\pi - \theta) = \theta - 2\pi$ with $0 < 2\pi - \theta < \pi$, but this time the orientation of the plane orthogonal to $-u = (b, c, d)$ is the opposite orientation from before. This orientation is given by (w_2, w_1), so that $(w_2, w_1, -u)$ has positive orientation. Since the quaternions q and $-q$ define the same rotation, we may assume that $a > 0$, in which case $0 < \theta < \pi$, but we have to remember that if $a < 0$ and if we pick $-q$ instead of q, the vector defining the axis of rotation becomes $-u$, which amounts to flipping the orientation of the plane orthogonal to the axis of rotation.

Because there is a surjective homomorphism $\rho\colon \mathbf{SU}(2) \to \mathbf{SO}(3)$ whose kernel is $\{-I, I\}$, the irreducible representations of $\mathbf{SO}(3)$ can also be determined (up to equivalence).

Example 3.9. If $U\colon \mathbf{SO}(3) \to \mathbf{U}(H)$ is an irreducible unitary representation of $\mathbf{SO}(3)$, then $V = U \circ \rho$ is a unitary representation $V\colon \mathbf{SU}(2) \to \mathbf{U}(H)$ of $\mathbf{SU}(2)$ which must be irreducible, and $V(-I)$ is the identity. Conversely, an irreducible unitary representation $V\colon \mathbf{SU}(2) \to \mathbf{U}(H)$ of $\mathbf{SU}(2)$ descends to an irreducible unitary representation $U\colon \mathbf{SO}(3) \to \mathbf{U}(H)$ iff $V(-I) = \mathrm{id}$. Now by definition of U_m,

$$U_m(-I)(P_k) = (-z_1)^k(-z_2)^{m-k} = (-1)^m z_1^k z_2^{m-k} = (-1)^m P_k.$$

Therefore, $U_m(-I) = \mathrm{id}_{m+1}$ iff $(-1)^m = 1$ iff m is even. In summary, we obtained the following result.

Proposition 3.10. *The representations $W_\ell\colon \mathbf{SO}(3) \to \mathbf{GL}(\mathcal{P}^{\mathbb{C}}_{2\ell}(2))$ given by*

$$W_\ell(\rho_q) = U_{2\ell}(q) \quad q \in \mathbf{SU}(2), \quad \ell \geq 0$$

are irreducible. Observe that $\mathcal{P}^{\mathbb{C}}_{2\ell}(2)$ has odd dimension $2\ell + 1$.

We will prove later that every irreducible unitary representation of $\mathbf{SU}(2)$ is equivalent to some representation U_m, and that every irreducible unitary representation of $\mathbf{SO}(3)$ is equivalent to some representation W_ℓ; see Proposition 5.1. We will also present a more pleasant description of the irreducible unitary representation of $\mathbf{SO}(3)$ in terms of spaces of harmonic polynomials.

Remark: The representations $U_m\colon \mathbf{SL}(2,\mathbb{C}) \to \mathbf{GL}(\mathcal{P}^{\mathbb{C}}_m(2))$ are not unitary, but they are irreducible. If some nontrivial proper subspace F of $\mathcal{P}^{\mathbb{C}}_m(2)$ was invariant under U_s for all $s \in \mathbf{SL}(2,\mathbb{C})$, then F would also be invariant under U_s for all $s \in \mathbf{SU}(2)$, contradicting the irreducibility of $U_m\colon \mathbf{SU}(2) \to \mathbf{GL}(\mathcal{P}^{\mathbb{C}}_m(2))$. The group $\mathbf{SL}(2,\mathbb{C})$ is the complexification of the group $\mathbf{SU}(2)$.

There is a generalization of Schur's Lemma to (complex) unitary representations, which says that if a unitary representation $U\colon G \to \mathbf{U}(H)$ is irreducible, then every G-map in $\mathrm{Hom}_G(U,U)$ is of the form $\alpha\,\mathrm{id}_H$, for some $\alpha \in \mathbb{C}$.

The proof requires much more machinery because a linear map on an infinite-dimensional vector space may not have eigenvectors! It uses some results from the spectral theory of algebras, in particular, the complement to Theorem 2.58.

Theorem 3.11. *(Schur's Lemma for unitary representations) The following properties hold.*

(1) A (complex) unitary representation $U: G \to \mathbf{U}(H)$ *is irreducible iff every G-map in* $\mathcal{C}(U) = \mathrm{Hom}_G(U, U)$ *is of the form* $\alpha\, \mathrm{id}_H$*, for some* $\alpha \in \mathbb{C}$.

(2) Let $U_1: G \to \mathbf{U}(H_1)$ *and* $U_2: G \to \mathbf{U}(H_2)$ *be two complex irreducible unitary representations. If* U_1 *and* U_2 *are equivalent, then* $\mathrm{Hom}_G(U_1, U_2)$ *is one-dimensional; otherwise we have* $\mathrm{Hom}_G(U_1, U_2) = (0)$.

Proof. We follow Folland's proof; see Chapter 3, Proposition 3.5.

(a) If U is reducible, then by Proposition 3.7, $\mathcal{C}(U)$ contains a nontrivial projection.

Conversely, assume that there is some $T \in \mathcal{C}(U)$, such that T is not a scalar multiple of the identity. We also have $T^* \in \mathcal{C}(U)$ because for all $s \in G$, we have

$$T^* \circ U(s) = (U(s^{-1}) \circ T)^* = (T \circ U(s^{-1}))^* = U(s) \circ T^*.$$

Then $A_1 = (1/2)(T + T^*)$ and $A_2 = (1/2)(T - T^*)$ belong to $\mathcal{C}(U)$, and they can't be both scalar multiples of the identity, because if $A_1 = (1/2)(T + T^*) = \lambda_1 \mathrm{id}$ and $A_2 = (1/2)(T - T^*) = \lambda_2 \mathrm{id}$, then $T = A_1 + A_2 = (\lambda_1 + \lambda_2)\mathrm{id}$. By definition, $A_1^* = A_1$ and $A_2^* = -A_2$. We may assume that A_1 is not a scalar multiple of the identity since the case where A_2 is not a scalar multiple of the identity is similar. By Theorem 2.58, since A_1 is a normal (continuous) operator on H, there is a projection-valued measure P on $\sigma(A_1)$, such that

$$A_1 = \int \mathrm{id}\, dP,$$

and for every $f \in B(\sigma(A_1))$ we define $f(A_1)$ as the linear bounded operator on H given by

$$f(A_1) = \int f\, dP.$$

Now by the complement to Theorem 2.58, any $S \in \mathcal{L}(H)$ which commutes with A_1 also commutes with $A_1^* = A_1$ (if we use A_2, then if S commutes with A_2, it also commutes with $A_2^* = -A_2$), so S commutes with $f(A_1)$ for all $f \in B(\sigma(A_1))$. In particular, for $f = \chi_E$ with $E \subseteq \sigma(A_1)$, we have the projections $\chi_E(A_1)$. If follows that every $S \in \mathcal{L}(H)$ of the form $U(s)$ commutes with all the projections $\chi_E(A_1)$, thus $\mathcal{C}(U)$ contains nontrivial projections, and by Proposition 3.7, U is reducible.

(b) If $T \in \mathrm{Hom}_G(U_1, U_2)$ and $T \neq 0$, then we also have $T^* \in \mathrm{Hom}_G(U_2, U_1)$, because

$$T^* \circ U_2(s) = (U_2(s^{-1}) \circ T)^* = (T \circ U_1(s^{-1}))^* = U_1(s) \circ T^*.$$

It follows that $T^* \circ T \in \mathcal{C}(U_1)$ and $T \circ T^* \in \mathcal{C}(U_2)$, and since U_1 and U_2 are irreducible, by part (a), we have $T^* \circ T = \lambda_1 \mathrm{id}$ and $T \circ T^* = \lambda_2 \mathrm{id}$ for some $\lambda_1, \lambda_2 \in \mathbb{C}$. Then

$$\lambda_1 T = T \circ T^* \circ T = \lambda_2 T.$$

Since $T \neq 0$, we must have $\lambda_1 = \lambda_2 = \lambda$. Actually, λ is real and positive because $T^* \circ T$ is positive semi-definite. Indeed, for all $x \in H$, we have

$$\langle (T^* \circ T)(x), x \rangle = \langle T(x), T(x) \rangle \geq 0,$$

so for $x \neq 0$, we have

$$\langle (T^* \circ T)(x), x \rangle = \langle \lambda x, x \rangle = \lambda \langle x, x \rangle = \langle T(x), T(x) \rangle,$$

which implies that $\lambda > 0$. Since $\lambda > 0$, the map $\lambda^{-1/2} T$ is unitary, so U_1 and U_2 are equivalent. Consequently, $\mathrm{Hom}_G(U_1, U_2) = \{0\}$ iff U_1 and U_2 are not equivalent. If U_1 and U_2 are equivalent and T_1, T_2 are nonzero G-maps in $\mathrm{Hom}_G(U_1, U_2)$, then T_1 and T_2 are unitary, so

$$T_2^{-1} \circ T_1 = T_2^* \circ T_1 \in \mathcal{C}(U_1),$$

so by part (a), $T_2^{-1} \circ T_1 = \lambda I$ for some $\lambda \in \mathbb{C}$, namely $T_1 = \lambda T_2$, which implies that $\mathrm{Hom}_G(U_1, U_2)$ is one-dimensional. □

As in the case of representations in finite dimensional vector spaces, an important corollary of Theorem 3.11 is the following result.

Proposition 3.12. *Every complex irreducible unitary representation $U \colon G \to \mathbf{U}(H)$ of a locally compact abelian group G in a Hilbert space H is one-dimensional.*

Proof. If G is abelian, then $U(s) \circ U(t) = U(t) \circ U(s)$ for all $s, t \in G$, which implies that $U(s) \in \mathcal{C}(U)$ for all $s \in G$. If U is irreducible, then by part (1) of Schur's Lemma, we have $U(s) = \alpha_s \mathrm{id}$ for some $\alpha_s \in \mathbb{C}$. It follows that every one-dimensional subspace of H is invariant, so H itself is one-dimensional. □

If the locally compact group G is abelian, then the following result shows that every irreducible unitary representation of G is uniquely defined by a *character* of G, as introduced in Vol. I, Definition 10.1.

Proposition 3.13. *Let G be a locally compact abelian group. Every irreducible unitary representation $U\colon G \to \mathbf{U}(1)$ of G is of the form*

$$U(s)(z) = \chi(s)z, \quad \textit{for all } s \in G \textit{ and all } z \in \mathbb{C}$$

for a unique character $\chi \in \widehat{G}$.

Proof. If $U\colon G \to \mathbf{U}(1)$ is a unitary representation of G, then $U(s)$ is a unitary map of $\mathbb{C}$ for every $s \in G$, which means that there is a complex number of unit length, say $\chi(s) \in \mathbb{T}$, such that

$$U(s)(z) = \chi(s)z, \quad \text{for all } z \in \mathbb{C},$$

and for all $s_1, s_2 \in G$, we have

$$\chi(s_1 s_2)z = U(s_1 s_2)(z) = U(s_1)(U(s_2)(z)) = \chi(s_1)\chi(s_2)z \quad \text{for all } z \in \mathbb{C},$$

which implies that

$$\chi(s_1 s_2) = \chi(s_1)\chi(s_2).$$

But then $\chi\colon G \to \mathbb{T}$ is a character of G, and so every unitary representation $U\colon G \to \mathbf{U}(1)$ of G is of the form

$$U(s)(z) = \chi(s)z, \quad \text{for all } s \in G \text{ and all } z \in \mathbb{C}$$

for a unique character $\chi \in \widehat{G}$. ☐

As an application of Theorem 3.6, Proposition 3.12 and Proposition 3.13, we describe *all finite-dimensional unitary representations of* $\mathbf{SO}(2) \simeq \mathbf{U}(1)$. Here we use the isomorphism between $\mathbf{SO}(2)$ and $\mathbf{U}(1)$ given by

$$\begin{pmatrix} \cos\theta & -\sin\theta \\ \sin\theta & \cos\theta \end{pmatrix} \mapsto e^{i\theta}, \ \theta \in [0, 2\pi).$$

Proposition 3.14. *Every finite-dimensional unitary representation $U\colon \mathbf{SO}(2) \to \mathbf{U}(n)$ of $\mathbf{SO}(2) \simeq \mathbf{U}(1)$ $(n \geq 1)$ is of the form*

$$U(e^{i\theta})(z) = \begin{pmatrix} e^{ik_1\theta} & 0 & \dots & 0 \\ 0 & e^{ik_2\theta} & \dots & 0 \\ \vdots & \vdots & \ddots & \vdots \\ 0 & \dots & \dots & e^{ik_n\theta} \end{pmatrix} \begin{pmatrix} z_1 \\ z_2 \\ \vdots \\ z_n \end{pmatrix}, \quad z \in \mathbb{C}^n, 0 \leq \theta < 2\pi,$$

for some $k_1, \dots, k_n \in \mathbb{Z}$.

Proof. Since $\mathbf{SO}(2) \simeq \mathbf{U}(1)$ is compact and abelian, by Proposition 3.12, every irreducible unitary representation of $\mathbf{SO}(2) \simeq \mathbf{U}(1)$ is one-dimensional. By Proposition 3.13, the irreducible unitary representations of $\mathbf{SO}(2) \simeq \mathbf{U}(1)$ are determined by the characters of $\mathbf{U}(1) = \mathbb{T}$. By Vol. I, Proposition 10.9(2), the characters of $\mathbf{U}(1) = \mathbb{T}$ are of the form

$$\chi_k(e^{i\theta}) = e^{ik\theta},$$

for some $k \in \mathbb{Z}$. Since $\mathbf{SO}(2) \simeq \mathbf{U}(1)$ is compact, by Theorem 3.6, every finite-dimensional unitary representation $U\colon \mathbf{SO}(2) \to \mathbf{U}(n)$ of $\mathbf{SO}(2)$ is the direct sum of n one-dimensional unitary representations $U_j\colon \mathbf{SO}(2) \to \mathbf{U}(1)$. But each representation $U_j\colon \mathbf{SO}(2) \to \mathbf{U}(1)$ arises from a character of $\mathbf{U}(1)$, and so is of the form

$$U_j(e^{i\theta})(y) = e^{ik_j\theta}y, \quad y \in \mathbb{C},$$

for some $k_j \in \mathbb{Z}$. The direct sum U of the representation $U_j\colon \mathbf{SO}(2) \to \mathbf{U}(1)$ acts on $\mathbb{C}^n$ as multiplication by a unitary matrix, namely

$$U(e^{i\theta}) = \begin{pmatrix} e^{ik_1\theta} & 0 & \dots & 0 \\ 0 & e^{ik_2\theta} & \dots & 0 \\ \vdots & \vdots & \ddots & \vdots \\ 0 & \dots & \dots & e^{ik_n\theta} \end{pmatrix},$$

as claimed. $\square$

Remark: Let Q be the $n \times n$ matrix given by

$$Q = \begin{pmatrix} ik_1 & 0 & \dots & 0 \\ 0 & ik_2 & \dots & 0 \\ \vdots & \vdots & \ddots & \vdots \\ 0 & \dots & \dots & ik_n \end{pmatrix}.$$

Observe that Q is skew-symmetric, so that $Q \in \mathfrak{u}(n)$, and we have

$$U(e^{i\theta}) = e^{\theta Q}.$$

3.3 Unitary Representations of G and $\mathrm{L}^1(G)$

In this section, we discuss the crucial fact that every unitary representation $U\colon G \to \mathbf{U}(H)$ of a locally compact group G defines a nondegenerate representation $U_{\mathrm{ext}}\colon \mathrm{L}^1(G) \to \mathcal{L}(H)$ of the involutive Banach algebra $\mathrm{L}^1(G)$, and that conversely, for every nondegenerate representation $V\colon \mathrm{L}^1(G) \to \mathcal{L}(H)$ of $\mathrm{L}^1(G)$, there is a unique unitary representation $U\colon G \to \mathbf{U}(H)$ of the group G, such that $V = U_{\text{ext}}$. These results hold for any Hilbert space H, but when dealing with Hilbert sums H is assumed to be separable.

Dieudonné [Dieudonné (2003a)] (Chapter XXI, Section 1) proves the above results under the simplifying assumption that G is metrizable, separable, and unimodular (and of course locally compact). One of the reasons is that Dieudonné only shows the existence of the Haar measure for a metrizable, separable, locally compact group. We prove it for metrizable locally compact groups.

The bijection holds for any locally compact group, not necessarily unimodular, and is proven in Folland [Folland (1995)] (Chapter 3). Since the technical details are not particularly illuminating, we will give an outline of the constructions and proofs, using the simplifying assumption that G is metrizable. This includes the case of Lie groups. The involution $f \mapsto f^*$ in $\mathrm{L}^1(G)$ is given by $f^*(s) = \Delta(s^{-1})\overline{f(s^{-1})}$, but not much simplification is afforded if we assume that G is unimodular.

First, we show that representations of $\mathrm{L}^1(G)$ are continuous.

Proposition 3.15. *Let G be a locally compact group and let $V\colon \mathrm{L}^1(G) \to \mathcal{L}(H)$ be a representation. We have*

$$\|V(f)\| \le \|f\|_1, \qquad \text{for all } f \in \mathcal{L}^1(G), \tag{$*$}$$

and thus V is continuous.

Proof. If G is discrete, this follows by Proposition 2.1. Otherwise, we can extend V to a representation of the unital subalgebra $\mathrm{L}^1(G) \oplus \mathbb{C}\delta_e$ of $\mathcal{M}^1(G)$ by setting $V(f\,d\lambda + \alpha\delta_e) = V(f) + \alpha\,\mathrm{id}_H$, and then we apply Proposition 2.1 to this representation. ☐

Our goal is to construct a nondegenerate representation of the algebra $\mathrm{L}^1(G)$ in H from a continuous unitary representation $U\colon G \to \mathbf{U}(H)$ of G. Technically, it is more advantageous to construct a nondegenerate representation of the algebra $\mathcal{M}^1(G)$ of complex regular Borel measures (see Vol. I, Definition 7.22) from a continuous unitary representation $U\colon G \to \mathbf{U}(H)$ of G but to motivate the construction let us stick with $\mathrm{L}^1(G)$. We need to define a map $\widetilde{U}\colon \mathrm{L}^1(G) \to \mathcal{L}(H)$, which is an algebra homomorphism. For every function $f \in \mathrm{L}^1(G)$, an obvious candidate $\widetilde{U}(f)$ for a continuous linear map from H to itself is

$$\widetilde{U}(f)(x) = \int f(s)U(s)(x)\,d\lambda(s), \qquad f \in \mathrm{L}^1(G),\ x \in H, \tag{1}$$

where λ is a left Haar measure on G. However, the right-hand side is an integral over the vector-valued function $s \mapsto f(s)U(s)(x)$ from G to H (in general, an infinite-dimensional vector space) so the theory of integration that we have presented does not apply. We will see how to circumvent this difficulty using weak integrals a little later, but since this method works if G is a *finite* group, let us assume temporarily that G is a finite group.

Let G be a finite group of order $|G|$. In this case, the algebras $\mathrm{L}^1(G)$ and $\mathrm{L}^2(G)$ are the same and equal to the space $[G \to \mathbb{C}]$ of functions from G to $\mathbb{C}$. Convolution of two functions $f, h\colon G \to \mathbb{C}$ is given by

$$(f * h)(s) = \frac{1}{|G|} \sum_{s_1 s_2 = s} f(s_1)h(s_2) = \frac{1}{|G|} \sum_{t \in G} f(t)h(t^{-1}s). \tag{2}$$

Recall that for every $s \in G$, the function $\delta_s\colon G \to \mathbb{C}$ is given by

$$\delta_s(t) = \begin{cases} 1 & \text{if } t = s \\ 0 & \text{if } t \neq s. \end{cases}$$

We define an involution $f \mapsto f^*$ on $\mathrm{L}^1(G)$ by $f^*(s) = \overline{f(s^{-1})}$. Then $\mathrm{L}^1(G) = [G \to \mathbb{C}]$ is a unital involutive algebra under convolution with unit δ_e (where e is the identity element of G).

Using the discrete analog of (1) where the integral is replaced by a sum, given a unitary representation $U\colon G \to \mathbf{U}(H)$ of G where H is finite-dimensional, define $\widetilde{U}(f)(x)$ by

$$\widetilde{U}(f)(x) = \frac{1}{|G|} \sum_{s \in G} f(s)U(s)(x), \quad x \in H,\ f \in \mathrm{L}^1(G). \tag{3}$$

It is not hard to prove that $\widetilde{U}\colon \mathrm{L}^1(G) \to \mathcal{L}(H)$ is an algebra representation; for details, see Simon [Simon (1996)] (Chapter II, Section 3). For instance, it is instructive to verify that

$$\widetilde{U}(f * h) = \widetilde{U}(f) \circ \widetilde{U}(h).$$

This result for finite groups is generalized to locally compact metrizable groups in Theorem 3.17.

Conversely, let $V\colon \mathrm{L}^1(G) \to \mathcal{L}(H)$ be an algebra representation. Then we can construct a unitary group representation $U\colon G \to \mathbf{U}(H)$ from V, such that $\widetilde{U} = V$.

If we define $U\colon G \to \mathbf{U}(H)$ by

$$U(s) = V(\delta_s), \quad s \in G, \tag{4}$$

we can verify that U is a unitary representation, such that $\widetilde{U} = V$; for details, see Simon [Simon (1996)] (Chapter II, Section 3). This result for finite groups is generalized to locally compact metrizable groups in Theorem 3.18.

As an application of the first construction going from a representation of G to a representation of $L^1(G)$, consider the left regular representation of G. The space $L^2(G) = L^1(G)$ has the hermitian inner product given by

$$\langle f, g\rangle = \frac{1}{|G|}\sum_{s\in G} f(s)\overline{g(s)}, \quad f, g \in L^2(G). \tag{5}$$

For any $s \in G$, define $\mathbf{R}_s \colon L^2(G) \to L^2(G)$ by

$$(\mathbf{R}_s(f))(t) = f(s^{-1}t), \quad s, t \in G, \ L^2(G). \tag{6}$$

It is easily verified that the map $s \mapsto \mathbf{R}_s$ is a linear representation of G in $L^2(G)$, and since the inner product is left and right invariant under G, each $\mathbf{R}_s$ is unitary, so the map $\mathbf{R}\colon G \to \mathbf{U}(L^2(G))$ is a unitary representation of G called the *left regular representation of* G. The left regular representation is generalized to locally compact metrizable groups in Definition 3.14.

We leave it as an exercise to prove that if we apply (3) to define $\widetilde{\mathbf{R}}(f)$ we find that

$$(\widetilde{\mathbf{R}}(f))(g) = f * g, \quad f, g \in L^2(G). \tag{7}$$

The algebra representation $\widetilde{\mathbf{R}}\colon L^1(G) \to \mathcal{L}(L^2(G))$ is generalized to locally compact metrizable groups in Definition 3.15. If G is a finite group, then $L^1(G) = L^2(G)$, but for infinite groups this is generally false.

We now return to the situation when G is a locally compact metrizable group and H is Hilbert space (for simplicity, we may assume that H is separable). Recall that if λ is a left Haar measure on G, then we have an embedding of $L^1(G)$ into $\mathcal{M}^1(G)$ given by $f \mapsto f\,d\lambda$.

As we stated earlier, given a unitary representation $U\colon G \to \mathbf{U}(H)$ we need to construct an algebra representation of $L^1(G)$ but technically it is preferable to construct an algebra representation (an algebra homomorphism) $\widetilde{U}\colon \mathcal{M}^1(G) \to \mathcal{L}(H)$ of the measure algebra $\mathcal{M}^1(G)$.

Pick any complex regular Borel measure $\mu \in \mathcal{M}^1(G)$. We need to define $\widetilde{U}(\mu)$ as a continuous linear map from H to itself. An obvious candidate is

$$\widetilde{U}(\mu)(x) = \int U(s)(x)\,d\mu(s), \qquad x \in H$$

but the right-hand side is an integral over the vector-valued function $s \mapsto U(s)(x)$ from G to H, and μ is generally not a positive measure, so the theory of integration that we have presented does not apply. The theory does extend to complex measures (see Schwartz [Schwartz (1993)]), but we do not know whether this type of integral has the properties needed to obtain the desired results, so instead we will resort to a so-called weak integral. The idea is to use the duality between the Hilbert space H and its dual H', the space of continuous linear forms. Technically, we use the Riesz representation theorem for Hilbert spaces (see Vol. I, Theorem D.9). For the reader's convenience, we briefly review the Riesz representation theorem.

If H is a Hilbert space, its *dual* H' is the vector space of *continuous linear forms* $\varphi\colon H \to \mathbb{C}$. We also define $\overline{H}'$ as the space of *continuous semilinear forms* $\varphi\colon H \to \mathbb{C}$, which are the continuous functions such that for all $x, y \in H$ and all $\lambda \in \mathbb{C}$, we have

$$\begin{aligned} \varphi(x+y) &= \varphi(x) + \varphi(y) \\ \varphi(\lambda x) &= \overline{\lambda}\varphi(x). \end{aligned}$$

Theorem 3.16. *(Riesz representation theorem) Let H be a Hilbert space.*

(1) The mapping $\flat\colon H \to H'$ defined such that for every $x \in H$, the linear form $\flat(x)$ is given by

$$\flat(x)(y) = \langle y, x\rangle, \quad y \in H,$$

is a semilinear continuous bijection. Thus for every continuous linear form $\varphi \in H'$, there is a unique $x \in H$, such that

$$\varphi(y) = \flat(x)(y) = \langle y, x\rangle, \quad y \in H.$$

(2) The mapping $\flat^l\colon H \to \overline{H}'$ defined such that for every $x \in H$, the semilinear form $\flat^l(x)$ is given by

$$\flat^l(x)(y) = \langle x, y\rangle, \quad y \in H,$$

is a continuous bijection. Thus for every continuous semilinear linear form $\varphi \in \overline{H}'$, there is a unique $x \in H$, such that

$$\varphi(y) = \flat^l(x)(y) = \langle x, y\rangle, \quad y \in H.$$

Returning to our unitary representation $U\colon G \to \mathbf{U}(H)$, for $x \in H$ fixed, we define the semilinear form $\Phi_{\mu,x}\colon H \to \mathbb{C}$ given by

$$\Phi_{\mu,x}(y) = \int \langle U(s)(x), y\rangle \, d\mu(s), \qquad y \in H;$$

this form is semilinear because $\Phi_{\mu,x}(y_1 + y_2) = \Phi_{\mu,x}(y_1) + \Phi_{\mu,x}(y_2)$, but $\Phi_{\mu,x}(\lambda y) = \overline{\lambda}\Phi_{\mu,x}(y)$. The function $s \mapsto \langle U(s)(x), y\rangle$ is continuous and bounded because $\|U(s)(x)\| = \|x\|$ since $U(s)$ is unitary, so it is μ-integrable (recall that $|\mu|(X)$ is finite). Using the Cauchy–Schwarz inequality we also have

$$|\Phi_{\mu,x}(y)| = \left|\int \langle U(s)(x), y\rangle\, d\mu(s)\right| \leq \|\mu\|\,\|x\|\,\|y\|\,,$$

so the semilinear form $\Phi_{\mu,x}$ is continuous. By the Riesz representation theorem (Theorem 3.16(2)), there is a unique vector $\widetilde{U}(\mu)(x) \in H$, such that

$$\langle \widetilde{U}(\mu)(x), y\rangle = \Phi_{\mu,x}(y), \qquad \text{for all } y \in H.$$

If we let $y = \widetilde{U}(\mu)(x)$ in the inequality

$$|\langle \widetilde{U}(\mu)(x), y\rangle| \leq \|\mu\|\,\|x\|\,\|y\|\,,$$

we get

$$\left\|\widetilde{U}(\mu)(x)\right\| \leq \|\mu\|\,\|x\|\,,$$

and so

$$\left\|\widetilde{U}(\mu)\right\| \leq \|\mu\|\,. \tag{C}$$

This shows that $\widetilde{U}(\mu)$ is a continuous linear map (in $\mathcal{L}(H)$, we use the operator norm induced by the Hermitian norm on H).

Definition 3.13. Given any complex regular Borel measure $\mu \in \mathcal{M}^1(G)$, for every $x \in H$, let $\Phi_{\mu,x} \colon H \to \mathbb{C}$ be the continuous semilinear form given by

$$\Phi_{\mu,x}(y) = \int \langle U(s)(x), y\rangle\, d\mu(s), \qquad y \in H.$$

The unique vector $\widetilde{U}(\mu)(x) \in H$, such that

$$\langle \widetilde{U}(\mu)(x), y\rangle = \Phi_{\mu,x}(y) = \int \langle U(s)(x), y\rangle\, d\mu(s), \qquad \text{for all } y \in H$$

is called the *weak integral* of the function $s \mapsto U(s)(x)$ from G to H with respect to μ, and is denoted by

$$\int U(s)(x)\, d\mu(s) = \widetilde{U}(\mu)(x).$$

Observe that

$$\widetilde{U}(\delta_s) = U(s), \qquad \text{for all } s \in G, \qquad (\widetilde{U}(\delta_s))$$

where δ_s is the Dirac measure at s. Also, when $\mu = f\, d\lambda$ with $f \in \mathrm{L}^1(G)$, we have

$$\langle \widetilde{U}(f\, d\lambda)(x), y\rangle = \int f(s)\langle U(s)(x), y\rangle\, d\lambda(s), \qquad \text{for all } y \in H.$$

For simplicity of notation, we also write $\widetilde{U}(f)$ instead of $\widetilde{U}(f\, d\lambda)$ and we write

$$\widetilde{U}(f)(x) = \int f(s)U(s)(x)\, d\lambda(s).$$

The next step is to show that the map $\mu \mapsto \widetilde{U}(\mu)$ is a representation of the unital involutive Banach algebra $\mathcal{M}^1(G)$.

Theorem 3.17. *Let G be a metrizable locally compact group, and let $U\colon G \to \mathbf{U}(H)$ be a unitary representation of G in H. The map $\widetilde{U}\colon \mathcal{M}^1(G) \to \mathcal{L}(H)$ defined above is a representation of the unital involutive Banach algebra $\mathcal{M}^1(G)$. The restriction $U_{\mathrm{ext}}\colon \mathrm{L}^1(G) \to \mathcal{L}(H)$ of $\widetilde{U}$ to the involutive Banach algebra $\mathrm{L}^1(G)$ is nondegenerate. The theorem also holds for any arbitrary locally compact group G.*

Proof. Theorem 3.17 is proven in Dieudonné [Dieudonné (2003a)] (Chapter XXI, Section 1, Theorem 21.1.6). We leave the verification that $\widetilde{U}(\mu)$ is linear as exercise. Let us verify that $\widetilde{U}(\mu * \nu) = \widetilde{U}(\mu) \circ \widetilde{U}(\nu)$. Recall the definition of the convolution of measures, Vol. I, Definition 8.21. For all $x, y \in H$ and all $s \in G$, we have

$$\begin{aligned}
\langle \widetilde{U}(\mu * \nu)(x), y\rangle &= \int \langle U(s)(x), y\rangle\, d(\mu * \nu) = \int\!\!\int \langle U(st)(x), y\rangle\, d\mu(s)\, d\nu(t)\\
&= \int\!\!\int \langle U(t)(x), U(s)^*(y)\rangle\, d\nu(t)\, d\mu(s)\\
&= \int \langle \widetilde{U}(\nu)(x), U(s)^*(y)\rangle\, d\mu(s)\\
&= \int \langle U(s)(\widetilde{U}(\nu)(x)), y\rangle\, d\mu(s) = \langle \widetilde{U}(\mu)(\widetilde{U}(\nu)(x)), y\rangle,
\end{aligned}$$

which proves that $\widetilde{U}(\mu * \nu) = \widetilde{U}(\mu) \circ \widetilde{U}(\nu)$.

Next, recall that

$$\int \varphi(s) d\overline{\mu}(s) = \overline{\int \overline{\varphi(s)}\, d\mu(s)},$$

and

$$\int \varphi(s) d\check{\mu}(s) = \int \varphi(s^{-1})\, d\mu(s),$$

see Vol. I, Proposition 7.24 and Proposition 8.45. Then using the fact that since U is a unitary representation, we have $(U(s))^* = U(s^{-1})$, and we have

$$\begin{aligned}
\langle (\widetilde{U}(\mu))^*(x), y\rangle &= \langle x, \widetilde{U}(\mu)(y)\rangle = \overline{\langle \widetilde{U}(\mu)(y), x\rangle} \\
&= \overline{\int \langle U(s)(y), x\rangle\, d\mu(s)} \\
&= \int \overline{\langle U(s)(y), x\rangle}\, d\overline{\mu}(s) = \int \langle x, U(s)(y)\rangle\, d\overline{\mu}(s) \\
&= \int \langle (U(s))^*(x), y\rangle\, d\overline{\mu}(s) = \int \langle U(s^{-1})(x), y\rangle\, d\overline{\mu}(s) \\
&= \int \langle U(s)(x), y\rangle\, d\check{\overline{\mu}}(s) = \langle \widetilde{U}(\check{\overline{\mu}})(x), y\rangle,
\end{aligned}$$

which proves that $(\widetilde{U}(\mu))^* = \widetilde{U}(\check{\overline{\mu}})$.

Recall from Definition 2.8 that the algebra representation $\widetilde{U}$ is nondegenerate iff $\widetilde{U}(f\, d\lambda)(x) = 0$ for all $f \in L^1(G)$ implies that $x = 0$. To prove that the restriction of $\widetilde{U}$ to $L^1(G)$ is nondegenerate, since G is metrizable, we can find a neighborhood base of e (the identity element of G) consisting of a sequence (V_n) of open neighborhoods of e, such that $V_{n+1} \subset V_n$ for all n.[2] Fix $s \in G$. Using Vol. I, Proposition A.39, for every $n \geq 1$ we can define a positive function $u_n \in \mathcal{K}_{\mathbb{R}}(G)$ of compact support contained in sV_n, such that $\int u_n\, d\lambda = 1$. Since for any fixed $x \in H$ the map $s \mapsto U(s)(x)$ is continuous (condition (2) of Definition 3.8), for every $x \in H$ and every $\epsilon > 0$, there is some $n > 0$, such that

$$\|U(t)(x) - U(s)(x)\| \leq \epsilon, \qquad \text{for all } t \in sV_n.$$

For all $y \in H$, since $\int u_n\, d\lambda = 1$, we have

$$\langle \widetilde{U}(u_n\, d\lambda)(x) - U(s)(x), y\rangle = \int \langle U(t)(x) - U(s)(x), y\rangle u_n(t)\, d\lambda(t),$$

so if we choose $y = \widetilde{U}(u_n\, d\lambda)(x) - U(s)(x)$, we get

$$\begin{aligned}
&\left\|\widetilde{U}(u_n\, d\lambda)(x) - U(s)(x)\right\|^2 \\
&\qquad = \int \langle U(t)(x) - U(s)(x), \widetilde{U}(u_n\, d\lambda)(x) - U(s)(x)\rangle u_n(t)\, d\lambda(t).
\end{aligned}$$

[2]This is where the assumption that G is metrizable is used. Otherwise, we may have to use an uncountable family.

Using the Cauchy–Schwarz inequality, we have

$$
\begin{aligned}
&\left\|\widetilde{U}(u_n\,d\lambda)(x) - U(s)(x)\right\|^2 \\
&\quad = \int \langle U(t)(x) - U(s)(x), \widetilde{U}(u_n\,d\lambda)(x) - U(s)(x)\rangle u_n(t)\,d\lambda(t) \\
&\quad \leq \int \left|\langle U(t)(x) - U(s)(x), \widetilde{U}(u_n\,d\lambda)(x) - U(s)(x)\rangle u_n(t)\right|\,d\lambda(t) \\
&\quad \leq \int \|U(t)(x) - U(s)(x)\|\,\|\widetilde{U}(u_n\,d\lambda)(x) - U(s)(x)\|u_n(t)\,d\lambda(t) \\
&\leq \|\widetilde{U}(u_n\,d\lambda)(x) - U(s)(x)\|\epsilon\int u_n(t)\,d\lambda(t) = \|\widetilde{U}(u_n\,d\lambda)(x) - U(s)(x)\|\epsilon,
\end{aligned}
$$

so we deduce that

$$\left\|\widetilde{U}(u_n\,d\lambda)(x) - U(s)(x)\right\| \leq \epsilon.$$

If there was some $x \neq 0$, such that $\widetilde{U}(f\,d\lambda)(x) = 0$ for all $f \in \mathrm{L}^1(G)$, then for $f = u_n$ we would have $U(s)(x) = 0$ for all $s \in G$, which is absurd for $s = e$ (since $U(e) = \mathrm{id}$). Therefore, the restriction of $\widetilde{U}$ to $\mathrm{L}^1(G)$ is nondegenerate.

If G is not metrizable, we have to use a more general neighborhood base and a filter argument ☐

For simplicity of notation, we write $U_{\mathrm{ext}}(f)$ instead of $U_{\mathrm{ext}}(f\,d\lambda)$. The following converse holds.

Theorem 3.18. *Let G be a metrizable and locally compact group. For every nondegenerate representation $V\colon \mathrm{L}^1(G) \to \mathcal{L}(H)$ of $\mathrm{L}^1(G)$, there is a unique unitary representation $U\colon G \to \mathbf{U}(H)$ of the group G, such that $V = U_{\mathrm{ext}}$. Consequently, the map $U \mapsto U_{\mathrm{ext}}$ is a bijection between the set of unitary representations of the group G and the set of nondegenerate representations of the involutive Banach algebra $\mathrm{L}^1(G)$. Furthermore, a closed subspace E of H is invariant under the linear map $U(s)$ for every $s \in G$ if and only if it is invariant under the linear map $V(f)$ for every $f \in \mathrm{L}^1(G)$ (in fact, since $\mathcal{K}_{\mathbb{R}}(G)$ is dense in $\mathrm{L}^1(G)$, for every $f \in \mathcal{K}_{\mathbb{R}}(G)$). Consequently, the map $U \mapsto U_{\mathrm{ext}}$ is a bijection between the set of irreducible unitary representations of G and the set of nondegenerate topologically irreducible representations of $\mathrm{L}^1(G)$. If $H = \bigoplus_n H_n$ is a Hilbert sum, then there is a bijection between the Hilbert sum $U = \bigoplus_n U_n$ of the unitary representations $U_n\colon G \to \mathbf{U}(H_n)$ and the Hilbert sum $U_{\mathrm{ext}} = \bigoplus_n (U_n)_{\mathrm{ext}}$ of the unitary nondegenerate representations $(U_n)_{\mathrm{ext}}\colon \mathrm{L}^1(G) \to \mathcal{L}(H_n)$. The theorem also holds for any arbitrary locally compact group G.*

Proof. Theorem 3.18 is proven in Dieudonné [Dieudonné (2003a)] (Chapter XXI, Section 1, Theorem 21.1.7). Folland [Folland (1995)] (Chapter, Theorem 3.11) gives a different proof that applies to any locally compact group.

The proof of Theorem 3.17 shows that $U(s)(x)$ is the limit of a sequence $\widetilde{U}(u_n\, d\lambda)(x)$, with $u_n \in \mathcal{K}_{\mathbb{R}}(G)$, which shows that the map $U \mapsto U_{\text{ext}}$ is injective. It also shows that if the closed subspace E is invariant under the $\widetilde{U}(f)$ with $f \in \mathcal{K}_{\mathbb{R}}(G)$, then it is invariant under the maps $U(s)$ for all $s \in G$. Conversely, by definition of $\widetilde{U}(\mu)$, it is immediate that if the closed subspace E is invariant under the maps $U(s)$ for all $s \in G$, then it is invariant under the $\widetilde{U}(f)$ with $f \in \mathcal{K}_{\mathbb{R}}(G)$,

Given a nondegenerate representation $V\colon L^1(G) \to \mathcal{L}(H)$ of $L^1(G)$, we need to construct a unitary representation $U\colon G \to \mathbf{U}(H)$ of the group G such that $V = U_{\text{ext}}$. If G is a finite group, then we can use (4) to define U by $U(s) = V(\delta_s)$. Unfortunately, if G is infinite then $\delta_s \notin L^1(G)$, so we have to proceed differently.

The idea is that $U(s)(y)$ is the limit of a sequence $V(u_n)(y)$ for a sequence (u_n) of functions that tends to the Dirac delta function at s. To make this rigorous, we proceed as follows.

Consider the subspace E of H spanned by the set

$$\{V(f)(x) \mid f \in \mathcal{L}^1(G),\ x \in H\}.$$

Since V is nondegenerate, by Proposition 2.9, E is dense in H. Pick $s \in G$ and define a neighborhood base of e consisting of a sequence (V_n) of open neighborhoods of e, such that $V_{n+1} \subset V_n$ for all n and a sequence (u_n) of functions $u_n \in \mathcal{K}_{\mathbb{R}}(G)$ of compact support contained in sV_n, as in the proof of Theorem 3.17, so that $\int u_n\, d\lambda = 1$. Since the Haar measure is left-invariant, $\int (\delta_{s^{-1}} * u_n)\, d\lambda = \int u_n\, d\lambda$, and since the function u_n has support contained in sV_n, the function $\delta_{s^{-1}} * u_n$ has support contained in V_n. For any open subset W containing e, since the V_n a form a neighborhood base of e, we have $V_n \subseteq W$ for n large enough, so $G - W \subseteq G - V_n$ and as a result, since $\delta_{s^{-1}} * u_n$ has support contained in V_n, $\int_{G-W} (\delta_{s^{-1}} * u_n)\, d\lambda = 0$ as n tends to infinity. Then Vol. I, Proposition 8.50 shows that $\lim_{n \mapsto \infty} \|\delta_{s^{-1}} * u_n * f - f\|_1 = 0$, and since

$$\begin{aligned} \|u_n * f - \delta_s * f\|_1 &= \|\delta_s * \delta_{s^{-1}} * u_n * f - \delta_s * f\|_1 \\ &\leq \|\delta_s\|\, \|\delta_{s^{-1}} * u_n * f - f\|_1, \end{aligned}$$

we deduce that $\lim_{n\mapsto\infty} \|u_n * f - \delta_s * f\|_1 = 0$. Alternatively, we can prove this by going back to the proof of Vol. I, Proposition 8.50.

Remark: In [Folland (1995)] (Theorem 3.11), Folland defines functions ψ_{V_n} with compact support contained in V_n, so that $\lim_{n\mapsto\infty} \|\psi_{V_n} * f - f\|_1 = 0$. The connection with our u_n is that $u_n = \delta_s * \psi_{V_n}$ and that $\lim_{n\mapsto\infty} \|(\delta_s * \psi_{V_n}) * f - \delta_s * f\|_1 = 0$.

By Proposition 3.15, we have

$$\|V(f)\| \leq \|f\|_1, \qquad \text{for all } f \in \mathcal{L}^1(G). \tag{$*_1$}$$

Applying V to $u_n * f - \delta_s * f$, using the above inequality, we get

$$\lim_{n\mapsto\infty} \|V(u_n) \circ V(f) - V(\delta_s * f)\| = 0.$$

The above proves that for every linear combination $y = \sum_k V(f_k)(x_k) \in E$, with $f_k \in \mathcal{L}^1(G)$ and $x_k \in H$, the sequence $(V(u_n)(y))$ has a limit in H equal to $\sum_k V(\delta_s * f_k)(x_k)$, because

$$\begin{aligned}\left\| V(u_n)\Big(\sum_k V(f_k)(x_k)\Big) - \sum_k V(\delta_s * f_k)(x_k) \right\| &\\ \leq \sum_k \|V(u_n)(V(f_k)(x_k)) - V(\delta_s * f_k)(x_k)\| &\\ \leq \sum_k \|V(u_n) \circ V(f_k) - V(\delta_s * f_k)\| \, \|x_k\|, &\end{aligned}$$

and since

$$\lim_{n\mapsto\infty} \|V(u_n) \circ V(f_k) - V(\delta_s * f_k)\| = 0,$$

we also have

$$\lim_{n\mapsto\infty} \left\| V(u_n)\Big(\sum_k V(f_k)(x_k)\Big) - \sum_k V(\delta_s * f_k)(x_k) \right\| = 0. \tag{$*_2$}$$

Therefore, for $y = \sum_k V(f_k)(x_k)$, we define $U(s)(y)$ by

$$U(s)(y) = U(s)\Big(\sum_k V(f_k)(x_k)\Big) = \sum_k V(\delta_s * f_k)(x_k). \tag{$*_3$}$$

We obtain a linear map $U(s)$ from E to H, such that

$$U(s) \circ V(f) = V(\delta_s * f), \qquad \text{for all } f \in \mathcal{L}^1(G), \tag{$\dagger$}$$

which shows that $U(s)$ maps E into itself. Note that $(*_2)$ says that

$$\lim_{n\mapsto\infty} \|V(u_n)(y) - U(s)(y)\| = 0, \quad \text{for all } y \in E, \tag{$*_4$}$$

which means that $V(u_n)$ converges strongly to $U(s)$ on E. By $(*_1)$, we have $\|V(u_n)\| \leq \|u_n\|_1 = 1$, so by $(*_4)$

$$\|U(s)(y)\| \leq \|y\|, \qquad \text{for all } y \in E \text{ and all } s \in G,$$

and $U(s)$ extends uniquely to a continuous map on H, also denoted $U(s)$. What we just did also shows that

$$\|U(s)\| \leq 1, \qquad \text{for all } s \in G. \tag{††}$$

It remains to prove that the map $s \mapsto U(s)$ is a unitary representation of G and that $V = U_{\mathrm{ext}}$.

By definition, for any $y = \sum_k V(f_k)(x_k) \in E$,

$$U(s)(y) = \sum_k V(\delta_s * f_k)(x_k),$$

and the above expression is continuous in s.

For all $s, t \in G$ and all $f \in \mathcal{L}^1(G)$, using (†) we have

$$\begin{aligned} U(st) \circ V(f) &= V(\delta_{st} * f) \\ &= V(\delta_s * (\delta_t * f)) \\ &= U(s) \circ V(\delta_t * f) \\ &= U(s) \circ U(t) \circ V(f), \end{aligned}$$

which implies that $U(st)(y) = U(s)(U(t)(y))$ for all $y \in E$, and then by continuity $U(st) = U(s) \circ U(t)$ in $\mathcal{L}(H)$. By (†), we also have $U(e) = \mathrm{id}_H$.

By (††), we have $\|U(s)(x)\| \leq \|x\|$ for all $s \in G$ and all $x \in H$, so $\left\|U(s^{-1})(x)\right\| \leq \|x\|$, and then $\|x\| = \left\|U(s^{-1}s)(x)\right\| = \left\|U(s^{-1})(U(s)(x))\right\| \leq \|U(s)(x)\|$, so $\|U(s)(x)\| = \|x\|$, and since $U(s)$ is linear, by the polarization identity for a hermitian inner product, $U(s)$ is a continuous isometry. Therefore, U is a unitary representation of G in H.

To prove that $V = U_{\mathrm{ext}}$, we use the fact that by Vol. I, Theorem 5.51, the dual $\mathrm{L}^1(G)'$ of $\mathrm{L}^1(G)$ is isomorphic to $\mathrm{L}^\infty(G)$. This means that for every continuous form $\Phi \in \mathrm{L}^1(G)'$, there is a unique function $h \in \mathrm{L}^\infty(G)$, such that

$$\Phi(f) = \int f(s)h(s)\, d\lambda(s), \qquad \text{for all } f \in \mathrm{L}^1(G).$$

With some abuse of notation, we write $\Phi(f) = (h, f) = \int f(s)h(s)\, d\lambda(s)$.

We use the following trick (see Dieudonné [Dieudonné (2003a)] (Chapter XXI, Section 1, Theorem 21.1.7). For all $f, g \in \mathcal{L}^1(G)$ and all $h \in \mathcal{L}^\infty(G)$,

$$(h, f * g) = \int f(s)(h, \delta_s * g)\, d\lambda(s). \qquad (**)$$

Indeed, using the fact that $(\delta_s * g)(t) = g(s^{-1}t)$ and Fubini's theorem, we have

$$\begin{aligned}
(h, f * g) &= \int (f * g)(t)h(t)\, d\lambda(t) \\
&= \int \left(\int f(s)g(s^{-1}t)\, d\lambda(s) \right) h(t)\, d\lambda(t) \\
&= \int \left(\int f(s)(\delta_s * g)(t)\, d\lambda(s) \right) h(t)\, d\lambda(t) \\
&= \int f(s) \left(\int (\delta_s * g)(t)h(t)\, d\lambda(t) \right) d\lambda(s) \\
&= \int f(s)(h, \delta_s * g)\, d\lambda(s).
\end{aligned}$$

For any fixed pair $x, y \in H$, the map $f \mapsto \langle V(f)(x), y\rangle$ is a continuous linear form on $\mathcal{L}^1(G)$, so there is a unique $h \in \mathcal{L}^\infty(G)$, such that $(h, f) = \langle V(f)(x), y\rangle$, for all $f \in \mathrm{L}^1(G)$, and we get

$$\begin{aligned}
\langle V(f)(V(g)(x)), y\rangle &= \langle V(f * g)(x), y\rangle \\
&= (h, f * g) = \int f(s)(h, \delta_s * g)\, d\lambda(s) \quad \text{by } (**) \\
&= \int f(s)\langle V(\delta_s * g)(s), y\rangle\, d\lambda(s) \\
&= \int \langle U(s)(V(g)(x)), y\rangle f(s)\, d\lambda(s) \qquad \text{by } (\dagger) \\
&= \langle \widetilde{U}(f)(V(g)(x)), y\rangle \qquad \text{by definition.}
\end{aligned}$$

This proves that $\langle \widetilde{U}(f)(z), y\rangle = \langle V(f)(z), y\rangle$ for all $z \in E$ and all $y \in H$, and since E is dense in H, since U_{ext} is the restriction of $\widetilde{U}$ to $\mathrm{L}^1(G)$, we conclude that $U_{\mathrm{ext}}(f) = V(f)$.

If G is not metrizable, we have to use a more general neighborhood base and a filter argument. ☐

Since the preceding proof involves many technical details, a summary of the proof focusing on the main points should be helpful.

First, pick a neighborhood base of e (the identity element of G) consisting of a sequence (V_n) of open neighborhoods of e, such that $V_{n+1} \subset V_n$ for all n. Second, for every $s \in G$, for every $n \geq 1$ define a positive function $u_n \in \mathcal{K}_{\mathbb{R}}(G)$ of support contained in sV_n, such that $\int u_n \, d\lambda = 1$. Then we proved that $V(u_n)(V(f)(x))$ converges to $V(\delta_s * f)(x)$, for any $f \in \mathrm{L}^1(G)$ and any $x \in H$, and since by definition $U(s)(V(f)(x)) = V(\delta_s * f)(x)$, actually $V(u_n)(V(f)(x))$ converges to $U(s)(V(f)(x))$. But the set of linear combinations of terms of the form $V(f)(x)$ is dense in H, so we proved that $V(u_n)(y)$ converges to $U(s)(y)$ for all $y \in H$, which is strong convergence of $V(u_n)$ to $U(s)$.

As an application of Theorem 3.17, we obtain an injective representation of $\mathrm{L}^1(G)$ into $\mathrm{L}^2(G)$, which will be needed in the proof of the Peter–Weyl theorem. It is shown in Dieudonné [Dieudonné (2005)] (Chapter XIV, Section 9, Theorem 14.9.2) that for every $s \in G$, for any $f \in \mathrm{L}^2(G)$, we have $\delta_s * f = \lambda_s(f) \in \mathrm{L}^2(G)$. By left-invariance of the (left) Haar measure, we have $\|\delta_s * f\|_2 = \|f\|_2$. Consequently, the map $f \mapsto \delta_s * f = \lambda_s(f)$, denoted $\mathbf{R}(s)$, is a unitary operator on $\mathrm{L}^2(G)$. Furthermore, if G is unimodular, by Theorem 14.10.6.3 of Dieudonné [Dieudonné (2005)] (Chapter XIV, Section 10), the map $s \mapsto \mathbf{R}(s)$ is continuous and so it is a unitary representation of G in $\mathrm{L}^2(G)$. If G is not unimodular, then the continuity follows from the argument in Proposition 2.41 of Folland [Folland (1995)].

Definition 3.14. The representation $\mathbf{R}\colon G \to \mathbf{U}(\mathrm{L}^2(G))$ given by

$$(\mathbf{R}(s)(f))(t) = \lambda_s(f)(t) = f(s^{-1}t), \qquad f \in \mathrm{L}^2(G),\ s, t \in G,$$

is called the *left regular representation* of G in $\mathrm{L}^2(G)$.

By Theorem 3.17, we obtain a representation $\mathbf{R}_{\mathrm{ext}}$ of $\mathrm{L}^1(G)$ in $\mathrm{L}^2(G)$ (a homomorphism from $\mathrm{L}^1(G)$ to $\mathcal{L}(\mathrm{L}^2(G))$). Going back to Definition 3.13 of a weak integral,

$$\langle \widetilde{U}(f)(x), y\rangle = \int f(s)\langle U(s)(x), y\rangle \, d\lambda(s), \qquad \text{for all } y \in H,$$

it is not hard to prove that

$$\mathbf{R}_{\mathrm{ext}}(f)(g) = f * g,$$

with $f \in \mathrm{L}^1(G)$ and $g \in \mathrm{L}^2(G)$ (in the equation defining $\widetilde{U}(f) = \mathbf{R}_{\mathrm{ext}}(f)$, x is the function g and y is a function h). Using Vol. I, Proposition 8.50, it can be shown that $\mathbf{R}_{\mathrm{ext}}$ is injective, because if $f * g$ is zero almost everywhere for all $g \in \mathrm{L}^2(G)$, then $f = 0$ almost everywhere.

Definition 3.15. The representation $\mathbf{R}_{\text{ext}}\colon \mathrm{L}^1(G) \to \mathcal{L}(\mathrm{L}^2(G))$ given by

$$(\mathbf{R}_{\text{ext}}(f))(g) = f * g, \qquad f \in \mathrm{L}^1(G),\ g \in \mathrm{L}^2(G),$$

is called the *left regular representation* of $\mathrm{L}^1(G)$ in $\mathrm{L}^2(G)$.

3.4 Unitary Representations of LCA Groups

We know from Proposition 3.12 that every irreducible unitary representation $U\colon G \to \mathbf{U}(H)$ of a locally compact abelian group G has dimension one, and by Proposition 3.13, every irreducible unitary representation of G is uniquely defined by a *character* of G, as introduced in Vol. I, Definition 10.1.

It is remarkable that *any* unitary representation $U\colon G \to \mathbf{U}(H)$ of a locally compact abelian group G can be expressed in terms of a projection-valued measure, as discussed in Section 2.12. Intuitively, the projection-valued measure glues the characters in the dual group $\widehat{G}$.

In order to state our theorem, we need to recall the fundamental fact that for a locally compact abelian group G, the dual group $\widehat{G}$ and the space $\mathsf{X}(\mathrm{L}^1(G))$ of algebra characters of $\mathrm{L}^1(G)$ are homeomorphic; see Vol. I, Theorem 10.6. More precisely, the map $j\colon \widehat{G} \to \mathsf{X}(\mathrm{L}^1(G))$ given by

$$j(\chi)(f) = \int_G \chi(a) f(a)\, d\lambda(a), \qquad \chi \in \widehat{G},\ f \in \mathrm{L}^1(G),$$

is a homeomorphism (where λ is a Haar measure on G). As a matter of notation, we denote the group characters in $\widehat{G}$ by χ and the algebra characters in $\mathsf{X}(\mathrm{L}^1(G))$ by ζ. Then our map is also expressed by $\chi \mapsto \zeta_\chi = j(\chi)$, with

$$\zeta_\chi(f) = \int_G \chi(a) f(a)\, d\lambda(a).$$

Also recall that the Gelfand map from $\mathrm{L}^1(G)$ to $\mathsf{X}(\mathrm{L}^1(G))$ is given by $\mathcal{G}_f(\zeta) = \zeta(f)$ and that

$$\zeta_\chi(f) = \mathcal{G}_f(\zeta_\chi) = \overline{\mathcal{F}}(f)(\chi),$$

where $\overline{\mathcal{F}}(f)$ is the Fourier co-transform of f.

The other fact that we need to recall is that every unitary representation $U\colon G \to \mathbf{U}(H)$ of G induces a nondegenerate representation $U_{\text{ext}}\colon \mathrm{L}^1(G) \to \mathcal{L}(H)$ of the involutive Banach algebra $\mathrm{L}^1(G)$; see Theorem 3.17. Before stating our next theorem, we need to address a notational issue, which is to make sense of the "integrals" $\int_{\widehat{G}} \chi(s)\, dP(\chi)$ and $\int_{\widehat{G}} \zeta_\chi(f)\, dP(\chi)$, for $s \in G$ and $f \in \mathrm{L}^1(G)$.

Recall that $\int_{\widehat{G}} \chi(s)\, dP(\chi)$ is the unique continuous linear map T in $\mathcal{L}(H)$, such that

$$\langle T(u), v\rangle = \int_{\widehat{G}} \chi(s)\, dP_{u,v}(\chi), \quad \text{for all } u, v \in H,$$

where for any Borel set E on $\widehat{G}$, the finite Radon measure $P_{u,v}$ is defined by

$$P_{u,v}(E) = \langle P(E)(u), v\rangle.$$

We are actually a bit sloppy because the integrand should be the function $\chi \mapsto \chi(s)$, evaluation at s. It would be more rigorous to introduce for every $s \in G$ the evaluation map $\mathrm{eval}_s^{\widehat{G}} \colon \widehat{G} \to \mathbb{C}$ given by

$$\mathrm{eval}_s^{\widehat{G}}(\chi) = \chi(s), \quad \chi \in \widehat{G}.$$

Then, the integral $\int_{\widehat{G}} \chi(s)\, dP_{u,v}(\chi)$ is really

$$\int_{\widehat{G}} \mathrm{eval}_s^{\widehat{G}}\, dP_{u,v}.$$

Similarly, $\int_{\widehat{G}} \zeta_\chi(f)\, dP(\chi)$ is the unique continuous linear map S in $\mathcal{L}(H)$, such that

$$\langle S(u), v\rangle = \int_{\widehat{G}} \zeta_\chi(f)\, dP_{u,v}(\chi), \quad \text{for all } u, v \in H.$$

This time, for every $f \in \mathsf{X}(\mathrm{L}^1(G))$, we have the evaluation map $\mathrm{eval}_f^{\mathsf{X}(\mathrm{L}^1(G))} \colon \mathsf{X}(\mathrm{L}^1(G)) \to \mathbb{C}$ given by

$$\mathrm{eval}_f^{\mathsf{X}(\mathrm{L}^1(G))}(\zeta) = \zeta(f), \quad f \in \mathsf{X}(\mathrm{L}^1(G)).$$

But note that $\mathrm{eval}_f^{\mathsf{X}(\mathrm{L}^1(G))} = \mathcal{G}_f$, where $\mathcal{G}$ is the Gelfand map from $\mathrm{L}^1(G)$ to $\mathsf{X}(\mathrm{L}^1(G))$! Then

$$\zeta_\chi(f) = \mathcal{G}_f(\zeta_\chi) = \mathcal{G}_f(\jmath(\chi)) = (\mathcal{G}_f \circ \jmath)(\chi),$$

so the second integral $\int_{\widehat{G}} \zeta_\chi(f)\, dP_{u,v}(\chi)$ is really

$$\int_{\widehat{G}} (\mathcal{G}_f \circ \jmath)\, dP_{u,v}.$$

Another technical point comes up in the proof of Theorem 3.20, which is the fact that we use the notion of direct image of a measure.

Definition 3.16. Let $(X, \mathcal{A})$ and $(Y, \mathcal{B})$ be two measure spaces, and let $\varphi\colon X \to Y$ be a map such that $\varphi^{-1}(B) \in \mathcal{A}$ for all $B \in \mathcal{B}$ (φ is a measurable map; see Vol. I, Definition 5.1). If μ is a (positive) measure on $(X, \mathcal{A})$, we define the *direct image* $\varphi_*\mu$ of μ as the measure on $(Y, \mathcal{B})$ given by

$$\varphi_*\mu(B) = \mu(\varphi^{-1}(B)), \quad B \in \mathcal{B}.$$

We leave it as an exercise to prove that $\varphi_*\mu$ is a measure. Then we have the following result.

Proposition 3.19. *With the notations of Definition 3.16, if* $g \in \mathcal{L}^1_{\varphi_*\mu}(Y, \mathcal{B}, \mathbb{C})$*, then* $g \circ \varphi \in \mathcal{L}^1_\mu(X, \mathcal{A}, \mathbb{C})$ *and*

$$\int_X (g \circ \varphi)\, d\mu = \int_Y g\, d(\varphi_*\mu).$$

The proof is not difficult but if you get stuck, see Folland [Folland (1995)] (Proposition 10.1) or Lang [Lang (1996)] (Chapter VI, Exercise 8). Proposition 3.19 can be extended to complex Radon measures on locally compact spaces, where $\mathcal{A}$ and $\mathcal{B}$ are the Borel σ-algebras on X and Y, respectively.

Now that we have given a precise meaning to our generalized integrals, we can state the following important result.

Theorem 3.20. *Let G be a locally compact abelian group with Haar measure λ. For every unitary representation $U\colon G \to \mathbf{U}(H)$ of G there is a unique regular projection-valued measure P on the dual group $\widehat{G}$ such that*

$$U(s) = \int_{\widehat{G}} \chi(s)\, dP(\chi), \quad s \in G$$

$$U_{\mathrm{ext}}(f) = \int_{\widehat{G}} \zeta_\chi(f)\, dP(\chi), \quad f \in \mathrm{L}^1(G).$$

According to the preceding remarks, a more rigorous statement of the above equations is

$$U(s) = \int_{\widehat{G}} \mathrm{eval}_s^{\widehat{G}}\, dP, \quad s \in G$$

$$U_{\mathrm{ext}}(f) = \int_{\widehat{G}} (\mathcal{G}_f \circ j)\, dP, \quad f \in \mathrm{L}^1(G),$$

where j is the homeomorphism $j\colon \widehat{G} \to \mathsf{X}(\mathrm{L}^1(G))$. Moreover, a continuous linear map $T \in \mathcal{L}(H)$ belongs to $\mathcal{C}(U)$ iff T commutes with $P(E)$ for every Borel set $E \subseteq \widehat{G}$.

Proof sketch. The statement about commuting operators is proven in Folland [Folland (1995)]; see Theorem 1.54 and Theorem 3.12(b). The second equation in Theorem 3.20 follows from Theorem 2.60 (Spectral Theorem IV). Indeed, this theorem says that there is a unique regular projection-valued measure P^L on $\mathsf{X}(\mathrm{L}^1(G))$, such that

$$U_{\mathrm{ext}}(f) = \int_{\mathsf{X}(\mathrm{L}^1(G))} \mathcal{G}_f \, dP^L = \int_{\mathsf{X}(\mathrm{L}^1(G))} \mathcal{G}_f(\zeta) \, dP^L(\zeta), \quad f \in \mathrm{L}^1(G).$$

Using the homeomorphism $j \colon \widehat{G} \to \mathsf{X}(\mathrm{L}^1(G))$, we define the projection-valued measure P on $\widehat{G}$ given by

$$P(E) = P^L(j(E))$$

for every Borel set E on $\widehat{G}$. Because j is a homeomorphism, P is the direct image of the measure P^L by j^{-1} and P^L is the direct image of the measure P by j, so by Proposition 3.19, we have

$$\int_{\widehat{G}} (\mathcal{G}_f \circ j) \, dP = \int_{\mathsf{X}(\mathrm{L}^1(G))} \mathcal{G}_f \, dP^L.$$

Thus, we proved that

$$U_{\mathrm{ext}}(f) = \int_{\widehat{G}} (\mathcal{G}_f \circ j) \, dP = \int_{\widehat{G}} \zeta_\chi(f) \, dP(\chi), \quad f \in \mathrm{L}^1(G),$$

as claimed.

The first equation follows from the second equation but the proof is more involved. The argument uses the technique from the proof of Theorem 3.18. To simplify notation, write $V = U_{\mathrm{ext}}$. We need to recover U from V. The idea is that $U(s)(y)$ $(s \in G, y \in H)$ is the limit of the sequence $V(u_n)(y)$ for a sequence (u_n) of functions that tends to the Dirac delta function at s. If G is metrizable, we can use the proof method of Theorem 3.18. In this case, we introduced a neighborhood base of e consisting of a sequence (V_n) of open neighborhoods of e, such that $V_{n+1} \subset V_n$ for all n and a sequence of positive functions $u_n \in \mathcal{K}_{\mathbb{R}}(G)$ of support contained in sV_n, such that $\int u_n \, d\lambda = 1$. We proved that $V(u_n)(V(f)(x))$ converges to $V(\delta_s * f)(x)$, for any $f \in \mathrm{L}^1(G)$ and any $x \in H$, and since by definition $U(s)(V(f)(x)) = V(\delta_s * f)(x)$, actually $V(u_n)(V(f)(x))$ converges to $U(s)(V(f)(x))$. But the set of linear combinations of terms of the form $V(f)(x)$ is dense in H, so we proved that $V(u_n)(y)$ converges to $U(s)(y)$ for all $y \in H$, which is strong convergence of $V(u_n)$ to $U(s)$. Let us take a closer look at

$$V(u_n) = \int_{\widehat{G}} \zeta_\chi(u_n) \, dP(\chi).$$

The u_n have support in sV_n, where the V_n are neighborhoods of e, so the functions $\delta_{s^{-1}} * u_n$ have support in V_n, and $\int(\delta_{s^{-1}} * u_n)\, d\lambda = 1$. But since ζ is an algebra homomorphism with respect to convolution, and since by Vol. I, Proposition 10.19,

$$\zeta_\chi(\delta_s) = \overline{\mathcal{F}}(\delta_s)(\chi) = \chi(s),$$

we obtain

$$\begin{aligned} V(u_n) &= \int_{\widehat{G}} \zeta_\chi(u_n)\, dP(\chi) \\ &= \int_{\widehat{G}} \zeta_\chi(\delta_s * (\delta_{s^{-1}} * u_n))\, dP(\chi) \\ &= \int_{\widehat{G}} \zeta_\chi(\delta_s)\zeta_\chi(\delta_{s^{-1}} * u_n)\, dP(\chi) \\ &= \int_{\widehat{G}} \chi(s)\zeta_\chi(\delta_{s^{-1}} * u_n)\, dP(\chi). \end{aligned}$$

For every $\epsilon > 0$, for every compact subset K of $\widehat{G}$, consider the set

$$W_{K,\epsilon} = \{a \in G \mid |\chi(a) - 1| < \epsilon, \text{ for all } \chi \in K\}.$$

It is easily checked that $W_{K,\epsilon}$ is a neighborhood of e. For $V_n \subseteq W_{K,\epsilon}$, for all $\chi \in K$, since $\int(\delta_{s^{-1}} * u_n)\, d\lambda = 1$, we have

$$|\zeta_\chi(\delta_{s^{-1}} * u_n) - 1| = \left| \int_{W_{K,\epsilon}} (\chi(a) - 1)(\delta_{s^{-1}} * u_n)\, d\lambda \right| < \epsilon. \tag{$*_1$}$$

For every $\epsilon > 0$, since $P_{u,v}$ is a finite Radon measure, there is a compact $K \subseteq \widehat{G}$ such that $|\mu_{u,v}|(\widehat{G} - K) < \epsilon$. Since

$$V(u_n) = \int_{\widehat{G}} \chi(s)\zeta_\chi(\delta_{s^{-1}} * u_n)\, dP(\chi),$$

we also have

$$\left\langle \left(V(u_n) - \int_{\widehat{G}} \chi(s)\, dP(\chi) \right)(u), v \right\rangle = \int_{\widehat{G}} \chi(s)(\zeta_\chi(\delta_{s^{-1}} * u_n) - 1)\, dP_{u,v}.$$

The integral on the right can be written as

$$\int_K \chi(s)(\zeta_\chi(\delta_{s^{-1}} * u_n) - 1)\, dP_{u,v} + \int_{\widehat{G}-K} \chi(s)(\zeta_\chi(\delta_{s^{-1}} * u_n) - 1)\, dP_{u,v}.$$

For all n, such that $V_n \subseteq W_{K,\epsilon}$, by $(*_1)$ the first integral is bounded by ϵ. Since $|\chi(a)| = 1$, we have $|\chi(a) - 1| \leq 2$, so for all $\chi \in \widehat{G} - K$, since $\int |\delta_{s^{-1}} * u_n|\, d\lambda = 1$, we have

$$|\zeta_\chi(\delta_{s^{-1}} * u_n) - 1| = \left| \int_{\widehat{G}-K} (\chi(a) - 1)(\delta_{s^{-1}} * u_n)\, d\lambda \right| \leq 2, \tag{$*_2$}$$

and since $|\mu_{u,v}|(\widehat{G} - K) < \epsilon$ and $|\chi(a)| = 1$, the second integral

$$\int_{\widehat{G}-K} \chi(s)(\zeta_\chi(\delta_{s^{-1}} * u_n) - 1)\, dP_{u,v}$$

is bounded by 2ϵ. Finally, the above argument shows that

$$\langle U(s)(u), v\rangle = \lim_{n\mapsto\infty} \langle V(u_n)(u), v\rangle = \int_{\widehat{G}} \chi(s)\, dP_{u,v}(\chi),$$

as claimed.

Otherwise, we need to use a more general neighborhood base and a filter (or net) argument. Technically, this is achieved by Theorem 3.11 of Folland [Folland (1995)], which relies on Proposition 2.42. Then another limit argument very similar to the one we gave above shows that the equation

$$U(s) = \int_{\widehat{G}} \chi(s)\, dP(\chi), \quad s \in G$$

follows from the equation

$$U_{\mathrm{ext}}(f) = \int_{\widehat{G}} \zeta_\chi(f)\, dP(\chi), \quad f \in \mathrm{L}^1(G).$$

The details of this proof are worked out in Folland [Folland (1995)] after Lemma 4.46. ☐

Theorem 3.20 plays a crucial role in Mackey's theory for constructing induced representations; see Chapter 7, Proposition 7.1.

As a corollary of Theorem 3.20, since by Vol. I, Corollary 10.93 the characters of $\mathbb{R}^n$ are the homomorphisms

$$x \mapsto e^{iy\cdot x}, \qquad x, y \in \mathbb{R}^n,$$

where $y \cdot x$ is the Euclidean product in $\mathbb{R}^n$, we obtain the following result due to Stone.

Theorem 3.21. *(Stone) For every unitary representation* $U\colon \mathbb{R}^n \to \mathbf{U}(H)$ *of* $\mathbb{R}^n$*, there is a unique projection measure* P *on* $\mathbb{R}^n$*, such that*

$$U(x) = \int_{\mathbb{R}^n} e^{iy\cdot x}\, dP(y), \quad x \in \mathbb{R}^n.$$

3.5 Functions of Positive Type and Unitary Representations

There is a deep and fruitful connection between topologically cyclic unitary representations $U\colon G \to \mathbf{U}(H)$ and certain kinds of continuous functions $p \in \mathcal{C}(G;\mathbb{C})$ called functions of positive type.

Let $U\colon G \to \mathbf{U}(H)$ be a unitary representation of the locally compact group G in a Hilbert space H, let x_0 be any vector in H, and define the map $p = \psi_{U,x_0}$ by

$$p(s) = \psi_{U,x_0}(s) = \langle U(s)(x_0), x_0\rangle, \qquad s \in G.$$

Note that this definition is analogous to Definition 2.11, which involves representations of algebras, but here we are dealing with a group representation. By its very definition, the function ψ_{U,x_0} is continuous, but it is also bounded, because $U(s)$ is unitary for every $s \in G$, so $\|U(s)(x_0)\| = \|x_0\|$, which implies by Cauchy–Schwarz that

$$|\psi_{U,x_0}(s)| = |\langle U(s)(x_0), x_0\rangle| \leq \|U(s)(x_0)\|\,\|x_0\| = \|x_0\|^2 = \psi_{U,x_0}(e),$$

for all $s \in G$. Consequently,

$$\|p\|_\infty = p(e),$$

and $p = \psi_{U,x_0} \in \mathcal{L}^\infty(G;\mathbb{C})$, so we obtain a continuous linear form $\omega\colon \mathcal{L}^1(G;\mathbb{C}) \to \mathbb{C}$ given by

$$\omega(f) = \int f(s)p(s)\,d\lambda(s) = \int f(s)\langle U(s)(x_0), x_0\rangle\,d\lambda(s),$$

for all $f \in \mathcal{L}^1(G;\mathbb{C})$. We recognize above the weak integral $U_{\text{ext}}(f)$, so we have

$$\omega(f) = \int f(s)\langle U(s)(x_0), x_0\rangle\,d\lambda(s) = \langle U_{\text{ext}}(f)(x_0), x_0\rangle.$$

The term on the right-hand side is exactly the term of Definition 2.11, so by Proposition 2.11, ω is a positive linear form, which means that $\omega(f^* * f) \geq 0$ for all $f \in \mathcal{L}^1(G;\mathbb{C})$, that is,

$$\int (f^* * f)(s)p(s)\,d\lambda(s) \geq 0, \qquad \text{for all } f \in \mathcal{L}^1(G;\mathbb{C}).$$

But $f^*(s) = \Delta(s^{-1})\overline{f(s^{-1})}$, so by changing t to t^{-1}, by Fubini, the left invariance of the Haar measure, and Vol. I, Proposition 8.27,

$$\begin{aligned}\int (f^* * f)(s)p(s)\,d\lambda(s) &= \int\int \Delta(t^{-1})\overline{f(t^{-1})}f(t^{-1}s)p(s)\,d\lambda(t)\,d\lambda(s)\\ &= \int\int \overline{f(t)}f(ts)p(s)\,d\lambda(t)\,d\lambda(s)\\ &= \int\int \overline{f(t)}f(ts)p(s)\,d\lambda(s)\,d\lambda(t)\\ &= \int\int p(t^{-1}s)\overline{f(t)}f(s)\,d\lambda(s)\,d\lambda(t).\end{aligned}$$

Observe that we also have

$$\psi_{U,x_0}(s^{-1}) = \overline{\psi_{U,x_0}(s)},$$

because

$$\begin{aligned}\psi_{U,x_0}(s^{-1}) &= \langle U(s^{-1})(x_0), x_0\rangle = \langle (U(s))^*(x_0), x_0\rangle\\ &= \langle x_0, U(s)(x_0)\rangle = \overline{\psi_{U,x_0}(s)}.\end{aligned}$$

Definition 3.17. If G is a locally compact group, then a continuous function $p \in \mathcal{C}(G;\mathbb{C})$ is of *positive type*, if

$$\int (f^* * f)(s)p(s)\,d\lambda(s) \geq 0, \qquad \text{for all } f \in \mathcal{K}_{\mathbb{C}}(G),$$

or equivalently, if

$$\int\int p(t^{-1}s)\overline{f(t)}f(s)\,d\lambda(s)\,d\lambda(t) \geq 0, \qquad \text{for all } f \in \mathcal{K}_{\mathbb{C}}(G).$$

The set of functions of positive type is denoted by $\mathcal{P}$.

Observe that $\mathcal{P} \subseteq \mathcal{C}(G;\mathbb{C}) \cap \mathcal{L}^1(G;\mathbb{C})$ and $\|p\|_\infty = p(e)$ for all $p \in \mathcal{P}$.

Remark: If p is of positive type, then $\int (f^* * f)(s)p(s)\,d\lambda(s) \geq 0$ for all $f \in \mathrm{L}^1(G)$. Indeed, $\mathcal{K}_{\mathbb{C}}(G)$ is dense in $\mathrm{L}^1(G)$, and for any sequence (f_n) with $f_n \in \mathcal{K}_{\mathbb{C}}(G)$ converging to f in $\mathrm{L}^1(G)$, the sequence $f_n^* * f_n$ converges to $f^* * f$ in $\mathrm{L}^1(G)$, and this implies that the sequence $\int (f_n^* * f_n)(s)p(s)\,d\lambda(s) \geq 0$ converges to $\int (f^* * f)(s)p(s)\,d\lambda(s) \geq 0$.

Every constant function with a nonnegative value is of positive type. We see this using the fact that the Haar measure is a positive measure and by applying Vol. I, Proposition 7.24 to the integral $\int\int \overline{f(t)}f(s)\,d\lambda(s)\,d\lambda(t)$.

For every $f \in \mathcal{L}^2(G;\mathbb{C})$, we have the left regular representation of G in $\mathrm{L}^2(G)$ with $(\mathbf{R}(s))(f) = \lambda_s f$ (see Definition 3.14), and we have

$$\langle (\mathbf{R}(s))(\overline{f}), \overline{f}\rangle = \int \overline{f(s^{-1}t)} f(t)\, d\lambda(t) = \int \check{\overline{f}}(t^{-1}s) f(t)\, d\lambda(t) = (f * \check{\overline{f}})(s),$$

so as a special case of a function p of the form ψ_{U,x_0}, $f * \check{\overline{f}} = \psi_{\mathbf{R},\overline{f}}$ is of positive type.

We showed that the functions of the form ψ_{U,x_0} are of positive type. Remarkably, every continuous function p of positive type determines a unitary topologically cyclic representation U with a cyclic vector x_0, such that $p = \psi_{U,x_0}$. Before stating our next theorem, we need to recall that by Vol. I, Theorem 8.34, if G is a metrizable, separable, locally compact group, then $\mathrm{L}^1(G)$ is separable.

Theorem 3.22. *Let G be a metrizable, separable, locally compact group. For any continuous function $p \in \mathcal{C}(G;\mathbb{C})$, the following properties are equivalent:*

(a) There is a unitary representation $U\colon G \to \mathbf{U}(H)$ of G in a separable Hilbert space H and a vector $x_0 \in H$, such that $p = \psi_{U,x_0}$.
(b) The function p is of positive type, that is,

$$\int\!\!\int p(t^{-1}s)\overline{f(t)} f(s)\, d\lambda(s)\, d\lambda(t) \geq 0, \qquad \text{for all } f \in \mathcal{K}_{\mathbb{C}}(G).$$

(c) The function p is bounded by $p(e) \geq 0$, $\check{\overline{p}} = p$, and for every complex measure $\mu \in \mathcal{M}^1(G)$, we have

$$\int p(s)\, d(\check{\overline{\mu}} * \mu)(s) = \int\!\!\int p(t^{-1}s)\, d\overline{\mu}(t)\, d\mu(s) \geq 0.$$

If p satisfies the above conditions, then there exists a topologically cyclic unitary representation V_1 of G in a separable Hilbert space H_1 and a cyclic vector x_1, such that $p = \psi_{V_1,x_1}$. The topologically cyclic representation is unique up to equivalence, in the sense that if V_2 is another topologically cyclic representation in a separable Hilbert H_2 and if x_2 is a cyclic vector for V_2, such that $p = \psi_{V_2,x_2}$, then there is an isomorphism $T\colon H_1 \to H_2$, such that $T(x_1) = x_2$ and $V_2 = TV_1T^{-1}$.

Proof. We follow Dieudonné [Dieudonné (2003b)] (Chapter XXII, Section 1, Theorem 22.1.3). We already proved that (a) implies (b). Let us prove that (b) implies (c).

By Vol. I, Proposition 8.45, we have $\int \varphi(t)d\check{\mu}(t) = \int \varphi(t^{-1})\,d\mu(t)$. By the definition of the convolution of measures, we obtain

$$\int p(z)\,d(\check{\overline{\mu}}*\mu)(z) = \int\!\!\int p(ts)\,d\check{\overline{\mu}}(t)\,d\mu(s) = \int\!\!\int p(t^{-1}s)\,d\overline{\mu}(t)\,d\mu(s). \quad (*_1)$$

For every complex measure μ, the union of all the open sets A of measure zero (that is, $\mu(A) = 0$) has measure zero, so there is a largest open set of measure zero.

Definition 3.18. The *support* $\mathrm{supp}(\mu)$ of the measure μ is the complement of the largest open set of measure zero.

The support of the measure μ has the property that for every $x \in \mathrm{supp}(\mu)$, for every neighborhood V of x, there is a continuous function f with compact support contained in V, such that $\int f(s)\,d\mu(s) \neq 0$; see Dieudonné [Dieudonné (2005)] (Chapter XIII, Section 19).

Let us first assume that μ has compact support. In this case, for any $f \in \mathcal{K}_{\mathbb{C}}(G)$, we have $\mu * f \in \mathcal{K}_{\mathbb{C}}(G)$ (see Dieudonné [Dieudonné (2005)] (Chapter XIV, Sections 14, 14.5.4 and 14.9.2), and it follows that

$$0 \leq \int\!\!\int p(t^{-1}s)\overline{(\mu * f)(t)}(\mu * f)(s)\,d\lambda(s)\,d\lambda(t),$$

and since by Vol. I, Definition 8.25,

$$(\mu * f)(t) = \int f(x^{-1}t)\,d\mu(x),$$

using Proposition 7.24 and Proposition 8.45, we have

$$\begin{aligned}
&\int\!\!\int p(t^{-1}s)\overline{(\mu * f)(t)}(\mu * f)(s)\,d\lambda(s)\,d\lambda(t)\\
&= \int\!\!\int p(t^{-1}s)\left(\overline{\int f(x^{-1}t)\,d\mu(x)}\right)\left(\int f(y^{-1}s)\,d\mu(y)\right)d\lambda(s)\,d\lambda(t)\\
&= \int\!\!\int p(t^{-1}s)\left(\int \overline{f(x^{-1}t)}\,d\overline{\mu}(x)\right)\left(\int f(y^{-1}s)\,d\mu(y)\right)d\lambda(s)\,d\lambda(t)\\
&= \int\!\!\int p(t^{-1}s)\left(\int \overline{f(xt)}\,d\check{\overline{\mu}}(x)\right)\left(\int f(ys)\,d\check{\mu}(y)\right)d\lambda(s)\,d\lambda(t)\\
&= \int\!\!\int \left(\int\!\!\int p(t^{-1}xy^{-1}s)\overline{f(t)}\,d\lambda(t) f(s)\,d\lambda(s)\right)d\check{\overline{\mu}}(x)\,d\check{\mu}(y)\\
&= \int\!\!\int \left(\int\!\!\int \overline{f}(t)f(s)p(t^{-1}x(s^{-1}y)^{-1})\,d\lambda(t)\,d\lambda(s)\right)d\check{\overline{\mu}}(x)\,d\check{\mu}(y).
\end{aligned}$$

Define the group $G \times G$ as the Cartesian product $G \times G$ with the multiplication

$$(s_1, t_1)(s_2, t_2) = (s_1 s_2, t_1 t_2).$$

Then the convolution of the functions $(t, s) \mapsto F(t, s) = \overline{f}(t) f(s)$ and $(x, y) \mapsto \Pi(x, y) = p(xy^{-1})$ is given by

$$\begin{aligned}(x, y) \mapsto &\iint F(t, s)\Pi((t^{-1}, s^{-1})(x, y))\, d\lambda(t)\, d\lambda(s) \\ &= \iint F(t, s)\Pi(t^{-1}x, s^{-1}y)\, d\lambda(t)\, d\lambda(s) \\ &= \iint \overline{f}(t) f(s) p(t^{-1}xy^{-1}s)\, d\lambda(t)\, d\lambda(s).\end{aligned}$$

This suggests using the regularization method (Vol. I, Proposition 8.50). Let (V_n) be a fundamental system of compact neighborhoods of e, such that $V_{n+1} \subseteq V_n$ for all n, and let f_n be a continuous function $f_n \geq 0$ with compact support contained in V_n and such that $\int f_n \, d\lambda = 1$. Since f_n is real, $\overline{f_n} = f_n$. Then if we let $F_n(t, s) = \overline{f}_n(t) f_n(s) = f_n(t) f_n(s)$, we have $\iint F_n(t, s)\, d\lambda(t)\, d\lambda(s) = \iint f_n(t) f_n(s)\, d\lambda(t)\, d\lambda(s) = 1$, and by Vol. I, Proposition 8.50, the sequence of functions

$$(F_n * \Pi)(x, y)$$

converges uniformly to the function $(x, y) \mapsto p(xy^{-1})$ on every compact subset. By passing to the limit, using Proposition 13.19.3 of Dieudonné [Dieudonné (2005)] (Chapter XIII, Section 19) which says that on a compact subset we can interchange the integral and the limit and $(*_1)$, we obtain the inequality

$$\iint p(xy^{-1})\, d\check{\overline{\mu}}(x)\, d\check{\mu}(y) = \iint p(xy)\, d\check{\overline{\mu}}(x)\, d\mu(y) = \int p(s) d(\check{\overline{\mu}} * \mu)(s) \geq 0.$$

Now we show that p is bounded by $p(e)$. For any finite subset $\{s_1, \ldots, s_n\}$ of G and for any complex numbers $\xi_1, \ldots, \xi_n$, the linear functional α on $\mathcal{K}_{\mathbb{C}}(G)$ given by

$$\alpha(f) = \sum_{j=1}^{n} \xi_j f(s_j)$$

is continuous, so by Radon–Riesz III (Vol. I, Theorem 7.30), there is a unique complex measure μ corresponding to α, called an *atomic measure*. For the measure μ, the inequality in (c) becomes

$$\sum_{j,k} p(s_j^{-1} s_k) \overline{\xi}_j \xi_k \geq 0.$$

This means that the sesquilinear form Φ defined by $\Phi(x, y) = \sum_{i,j=1}^{n} p(s_i^{-1}s_j)x_i\overline{y}_j$ must satisfy the property $\Phi(x, x) \geq 0$ for all $x \in \mathbb{C}^n$. Since

$$\Phi(x + y, x + y) = \Phi(x, x) + \Phi(x, y) + \Phi(y, x) + \Phi(y, y),$$

we see that $\Phi(x, y) + \Phi(y, x)$ must be real. By replacing x by ix, we see that $i\Phi(x, y) - i\Phi(y, x)$ must be real, so we must have

$$\Phi(y, x) = \overline{\Phi(x, y)}.$$

Therefore, the matrix $(p(s_j^{-1}s_k))$ is hermitian positive semidefinite. In particular, when $n = 2$ and with the set $\{e, s\}$, the matrix

$$\begin{pmatrix} p(e) & p(s) \\ p(s^{-1}) & p(e) \end{pmatrix}$$

must be hermitian positive semidefinite, which implies that $p(e) \geq 0$,

$$p(s^{-1}) = \overline{p(s)},$$

so $\overline{\check{p}} = p$, and

$$p(e)^2 - p(s)p(s^{-1}) = p(e)^2 - p(s)\overline{p(s)} \geq 0,$$

and thus

$$|p(s)| \leq p(e), \qquad \text{for all } s \in G,$$

namely, p is bounded (by $(p(e))$.

Let us now consider an arbitrary complex measure μ. By Vol. I, Proposition A.49, since G is locally compact and metrizable, there is a sequence (K_n) of compact subsets of G, such that $K_n \subseteq K_{n+1}$ and $G = \bigcup_n K_n$. Then it can be shown that $\lim_{n \mapsto \infty} |\mu|(G - K_n) = 0$ (see Dieudonné [Dieudonné (2005)], Chapter XIII, Section 8, Proposition 13.8.7). By Radon–Riesz III, the continuous linear functional $f \mapsto \int \chi_{K_n} f\, d\mu$ (with $f \in \mathcal{K}_{\mathbb{C}}(G)$) corresponds to a measure μ_n of compact support K_n. Then it can be shown that $\lim_{n \mapsto \infty} \|\mu - \mu_n\| = 0$, and also $\lim_{n \mapsto \infty} \left\|\overline{\check{\mu}} * \mu - \overline{\check{\mu}}_n * \mu_n\right\| = 0$ (see Dieudonné [Dieudonné (2005)], Chapter XIV, Section 6, Proposition 14.6.2). Since μ_n has compact support, by our previous result $\int p(s) d(\overline{\check{\mu}}_n * \mu_n)(s) \geq 0$, but p is bounded and by the dominated convergence theorem, $\int p(s) d(\overline{\check{\mu}}_n * \mu_n)(s)$ tends to $\int p(s) d(\overline{\check{\mu}} * \mu)(s)$, and thus $\int p(s) d(\overline{\check{\mu}} * \mu)(s) \geq 0$.

Finally, we prove that (c) implies (a). Consider the linear form φ_p defined on the unital involutive Banach algebra $\mathcal{M}^1(G)$ given by

$$\varphi_p(\mu) = \int p(s)\, d\mu(s), \qquad \mu \in \mathcal{M}^1(G),$$

where p is a bounded continuous function satisfying (c), and with the involution of $\mathcal{M}^1(G)$ being given by $\mu^* = \check{\overline{\mu}}$. Observe that $\varphi_p(\check{\overline{\mu}} * \mu) = \int p(s)\, d(\check{\overline{\mu}} * \mu)(s) \geq 0$, by (c). Therefore, φ_p is a positive linear form on $\mathcal{M}^1(G)$, according to Definition 2.10. Recall that $\mathrm{L}^1(G) \oplus \mathbb{C}\delta_e$ is also a unital involutive Banach algebra, and the restriction of φ_p to $\mathrm{L}^1(G) \oplus \mathbb{C}\delta_e$ is also a positive linear form. By Proposition 2.40(1), the linear form φ_p is continuous.

Recall from Proposition 2.12 that φ_p induces a positive Hilbert form γ given by $\gamma(\mu, \nu) = \varphi_p(\nu^* * \mu)$. Then we are almost in the position of applying Proposition 2.39 to obtain a representation of the algebra $\mathrm{L}^1(G) \oplus \mathbb{C}\delta_e$, but it is not clear that condition (U) is satisfied so we proceed directly.

Proposition 2.37 applies to the positive Hilbert form γ. To simplify notation, write $A = \mathrm{L}^1(G) \oplus \mathbb{C}\delta_e$. If

$$\mathfrak{n} = \{\mu \in A \mid \gamma(\mu, \mu) = \varphi_p(\mu^* * \mu) = 0\},$$

then $\mathfrak{n}$ is a left ideal in A and $A/\mathfrak{n} = H_0$ is a hermitian space with the inner product given by

$$\langle \pi(\mu), \pi(\nu) \rangle = \gamma(\mu, \nu) = \varphi_p(\nu^* * \mu),$$

where $\pi\colon A \to A/\mathfrak{n} = H_0$ is the quotient map. Observe that by Proposition 2.40, φ_p is a continuous linear form, such that $\|\varphi_p\| = \varphi_p(e)$, so we have

$$\|\pi(\mu)\|^2 = \gamma(\mu, \mu) = \varphi_p(\mu^* * \mu) \leq \varphi_p(e) \|\mu^* * \mu\| \leq \varphi_p(e) \|\mu\|^2,$$

so π is continuous. Since A is separable, so is H_0. The completion H of H_0 is a separable Hilbert space. As in the proof of Proposition 2.39, the endomorphism $V(\mu)$ given by

$$V(\mu)(\pi(\nu)) = \pi(\mu * \nu)$$

extends to a continuous map $V(\mu)\colon H \to H$, which is a representation of A (left multiplication). Since A has a unit element δ_e, we see that

$$V(\mu)(\pi(\delta_e)) = \pi(\mu * \delta_e) = \pi(\mu),$$

so $x_0 = \pi(\delta_e)$ is a cyclic vector for V. Since

$$V(\mu)(x_0) = \pi(\mu),$$

we have

$$\langle V(\mu)(x_0), x_0 \rangle = \langle \pi(\mu), \pi(\delta_e) \rangle = \gamma(\mu, \delta_e) = \varphi_p(\delta_e^* * \mu) = \varphi_p(\mu).$$

We also claim that the representation V is nondegenerate. It suffices to prove that the set of elements of the form $f * g$ with $f, g \in \mathcal{L}^1(G)$ is dense in $\mathcal{L}^1(G)$ (property (N)). But this follows immediately by regularization (Vol. I, Proposition 8.50).

We can now apply Theorem 3.18, to obtain a unitary representation $U\colon G \to \mathbf{U}(H)$, topologically cyclic, and such that $U_{\text{ext}} = V$. We know from (†) in the proof of Theorem 3.18 that U is given by

$$U(s) \circ V(\nu) = V(\delta_s * \nu),$$

and since $V(\mu)(x_0) = \pi(\mu)$, this means that

$$U(s)(\pi(\nu)) = \pi(\delta_s * \nu).$$

In particular, $U(s)(x_0) = \pi(\delta_s)$. Since $H_U = \overline{\{U(s)(x_0) \mid s \in G\}}$ is invariant under $U(s)$ for every $s \in G$, by Theorem 3.18, the closed subset H_U is also invariant under $V(\mu)$ for all $\mu \in A$, but $H_0 = \{V(\mu)(x_0) \mid \mu \in A\}$ and $x_0 \in H_U$, so we must have $H_U = H_0$, and $\{U(s)(x_0) \mid s \in G\}$ is dense in H. Therefore, x_0 is a cyclic vector for U, which means that the set $\{\pi(\delta_s) \mid s \in G\}$ is dense in H.

We have

$$\begin{aligned}\langle U(s)(x_0), x_0\rangle &= \langle \pi(\delta_s), \pi(\delta_e)\rangle = \gamma(\delta_s, \delta_e)\\ &= \varphi_p(\delta_e^* * \delta_s) = \varphi_p(\delta_s) = \int p(t)\, d\delta_s(t) = p(s),\end{aligned}$$

so $p = \psi_{U,x_0}$, as desired. The uniqueness of U up to equivalence follows from Proposition 2.38. □

In the next section, we present the Gelfand–Raikov theorem.

3.6 The Gelfand–Raikov Theorem

We will not prove the Gelfand–Raikov theorem but we will prove several technical propositions needed for its proof that are of independent interest.

Proposition 3.23. *Let p be a function of positive type on G. For all $s, t \in G$, we have*

$$|p(s) - p(t)|^2 \leq 2p(e)(p(e) - \Re(p(s^{-1}t))).$$

Proof. By Theorem 3.22, we may assume that there is cyclic unitary representation U and a cyclic vector x_0, such that

$$p(s) = \langle U(s)(x_0), x_0\rangle.$$

This immediately implies that $p(e) = \langle U(e)(x_0), x_0\rangle = \langle x_0, x_0\rangle = \|x_0\|^2$. By Cauchy–Schwarz and the fact that $p(s^{-1}) = \overline{p(s)}$, we have

$$\begin{aligned}|p(s) - p(t)|^2 &= |\langle (U(s) - U(t))(x_0), x_0\rangle|^2 \leq \|x_0\|^2 \, \|(U(s) - U(t))(x_0)\|^2 \\ &= p(e)(\|U(s)(x_0)\|^2 + \|U(t)(x_0)\|^2 \\ &\quad - 2\Re(\langle U(s)(x_0), U(t)(x_0)\rangle)) \\ &= p(e)(2\,\|x_0\|^2 - 2\Re(\langle U(t^{-1}s)(x_0), x_0\rangle)) \\ &= 2p(e)(p(e) - \Re(p(t^{-1}s))) = 2p(e)(p(e) - \Re(\overline{p(s^{-1}t)})) \\ &= 2p(e)(p(e) - \Re(p(s^{-1}t))),\end{aligned}$$

as claimed. ☐

Let $p = \psi_{U,x_0}$ be a function of positive type given by a cyclic unitary representation $U\colon G \to \mathbf{U}(H)$ with cyclic vector x_0. If (a_n) is a Hilbert basis of H (recall that the Hilbert space H is separable), then we can write

$$U(s)(x_0) = \sum_n p_n(s) a_n,$$

where each function p_n is continuous, and we have

$$\sum_n |p_n(s)|^2 = \|x_0\|^2, \qquad \text{for all } s \in G.$$

We deduce that

$$p(s^{-1}t) = \langle U(t)(x_0), U(s)(x_0)\rangle = \sum_n \overline{p_n(s)} p_n(t), \tag{$*_2$}$$

with $\sum_n |p_n(s)p_n(t)| \leq \|x_0\|^2$.

Proposition 3.24. *The product pq of two functions p and q of positive type on G is a function of positive type.*

Proof. Using a Hilbert basis as above, $(*_2)$, and a corollary of the dominated convergence theorem (Vol. I, Proposition 5.37), for every $f \in \mathcal{K}_{\mathbb{C}}(G)$, we have

$$\iint p(s^{-1}t)q(s^{-1}t)\overline{f(s)}f(t)\, d\lambda(s)\, d\lambda(t) \\ = \sum_n \iint q(s^{-1}t)\overline{p_n(s)f(s)}p_n(t)f(t)\, d\lambda(s)\, d\lambda(t),$$

but since q is also of positive type, we have

$$\iint q(s^{-1}t)\overline{p_n(s)f(s)}p_n(t)f(t)\, d\lambda(s)\, d\lambda(t) \geq 0, \qquad \text{for all } n,$$

so

$$\iint p(s^{-1}t)q(s^{-1}t)\overline{f(s)}f(t)\,d\lambda(s)\,d\lambda(t) \geq 0,$$

that is, pq is of positive type. □

Two subsets of the set $\mathcal{P}$ of continuous functions of positive type on G come up in the proof of the Gelfand–Raikov theorem and are of particular interest:

$$\mathcal{P}_1 = \{f \in \mathcal{P} \mid f(e) = 1\} = \{f \in \mathcal{P} \mid \|f\|_\infty = 1\}$$
$$\mathcal{P}_0 = \{f \in \mathcal{P} \mid 0 \leq f(e) \leq 1\} = \{f \in \mathcal{P} \mid \|f\|_\infty \leq 1\}.$$

Since $\mathcal{P} \subseteq \mathcal{C}(G;\mathbb{C}) \cap \mathcal{L}^1(G;\mathbb{C})$, the space $\mathcal{P}$ can be given several topologies. The subspace $\mathcal{P}_1$ is particularly important because of its role in the proof of the Gelfand–Raikov theorem and remarkably, three natural topologies on $\mathcal{P}_1$ coincide.

Remark: The above notation is from Folland [Folland (1995)] (Chapter 3). Unfortunately, Dieudonné denotes $\mathcal{P}_1$ as $\mathcal{P}_0$.

The sets $\mathcal{P}_0$ and $\mathcal{P}_1$ are convex and bounded ($\mathcal{P}$ itself is a convex cone). Recall the definition of an *extreme point*. Given a nonempty convex set S, a point a of the boundary of S is *extreme* (or *extremal*) if $S - \{a\}$ is still convex. Equivalently, there does not exist two distinct points $x, y \in S$, such that $a = (1-\lambda)x + \lambda y$, with $0 < \lambda < 1$.

Let $\mathcal{E}(\mathcal{P}_0)$ (resp. $\mathcal{E}(\mathcal{P}_1)$) be the set of extreme points of $\mathcal{P}_0$ (resp. $\mathcal{P}_1$). The following results are shown in Folland [Folland (1995)] (Chapter 3, Theorem 3.25 and Lemma 3.26).

Theorem 3.25. *If $p \in \mathcal{P}_1$, then the cyclic unitary representation U associated with p given by Theorem 3.22 is irreducible iff $p \in \mathcal{E}(\mathcal{P}_1)$. We have $\mathcal{E}(\mathcal{P}_0) = \mathcal{E}(\mathcal{P}_1) \cup \{0\}$.*

In order to state the next results, we need to define the weak *-topology on $\mathrm{L}^\infty(G)$. Recall from Vol. I, Theorem 5.51 that $\mathrm{L}^\infty(G)$ is isomorphic to the dual $(\mathrm{L}^1(G))'$ of $\mathrm{L}^1(G)$ under the pairing $(-,-)\colon \mathrm{L}^1(G) \times \mathrm{L}^\infty(G) \to \mathbb{C}$ given by

$$(f, g) = \int f(s)g(s)\,d\lambda(s).$$

Every function $g \in \mathrm{L}^\infty(G)$ defines the continuous linear form in $(\mathrm{L}^1(G))'$ given by $f \mapsto (f, g)$, for every $f \in \mathrm{L}^1(G)$, and every linear form in $(\mathrm{L}^1(G))'$ arises in this fashion for a unique function $g \in \mathrm{L}^\infty(G)$.

Definition 3.19. The weak*-topology on $\mathrm{L}^\infty(G)$ is the topology of pointwise convergence on $(\mathrm{L}^1(G))'$. This topology is defined directly on $\mathrm{L}^\infty(G)$ by the family $(p_f)_{f\in \mathrm{L}^1(G)}$ of semi-norms indexed by the set of functions $\mathrm{L}^1(G)$, such that for every $f \in \mathrm{L}^1(G)$,

$$p_f(g) = |(f, g)|, \qquad \text{for every } g \in \mathrm{L}^\infty(G).$$

(See Vol. I, Section 2.7 and Dieudonné [Dieudonné (2005)] (Chapter XII, Section 15).)

It is proven in Folland [Folland (1995)] (Chapter 3, Theorem 3.31) that the topology induced on $\mathcal{P}_1 \subseteq \mathrm{L}^\infty(G)$ by the weak*-topology of $\mathrm{L}^\infty(G)$ coincides with the topology induced on $\mathcal{P}_1$ by the topology of compact convergence in $\mathbb{C}^G$ (see Vol. I, Definition 2.9). This result is one of the key facts in the proof of the Gelfand–Raikov theorem.

In Dieudonné [Dieudonné (2003b)] (Chapter XXII, Section 1, Theorem 2.1.11), it is shown that the topology induced on $\mathcal{P}_1 \subseteq \mathcal{C}(G;\mathbb{C})$ by the topology of Fréchet space of $\mathcal{C}(G;\mathbb{C})$ (see Vol. I, Section 2.7) and the topology induced on $\mathcal{P}_1$ by the weak*-topology of $\mathrm{L}^\infty(G)$ coincide.

An important theorem due to Gelfand and Raikov shows that there is vast supply of irreducible unitary representations for any locally compact group. This is far from obvious a priori. For example, $\mathbf{SL}(2,\mathbb{R})$ does not have finite-dimensional unitary representations, and it is not that easy to find irreducible unitary representations.

Theorem 3.26. *(Gelfand–Raikov) If G is a locally compact group, then the irreducible unitary representations of G separate points. This means that for any $s, t \in G$, if $s \neq t$, then there is an irreducible representation U, such that $U(s) \neq U(t)$.*

Theorem 3.26 is proven in Folland [Folland (1995)] (Chapter 3, Theorem 3.34).

The notion of function of positive type is closely related to the notion of positive semidefinite function defined below, which came up during the proof of Theorem 3.22.

Definition 3.20. A function $p\colon G \to \mathbb{C}$ (not necessarily continuous) is *positive semidefinite* if for all $s_1, \ldots, s_n \in G$ and all $\xi_1, \ldots, \xi_n \in \mathbb{C}$, we have

$$\sum_{j,k=1}^{n} p(s_j^{-1} s_k)\xi_k \overline{\xi_j} \geq 0.$$

As we showed during the proof of Theorem 3.22, the matrix $(p(s_j^{-1}s_k))$ is hermitian positive semidefinite. We also have $p(s^{-1}) = \overline{p(s)}$ and $|p(s)| \leq p(e)$, so p is bounded, but examples of discontinuous or even nonmeasurable positive semidefinite p can be given. However, if p is continuous, then p is actually a function of positive type. The following result is shown in Folland [Folland (1995)] (Chapter 3, Proposition 3.35).

Proposition 3.27. *Let G be a locally compact group. For any bounded continuous function $p\colon G \to \mathbb{C}$, the following are equivalent:*

(1) The function p is of positive type.
(2) The function p is positive semidefinite.
(3) We have $\int (f^* * f)(s)p(s)\, d\lambda(s) \geq 0$ for all $f \in \mathcal{K}_{\mathbb{C}}(G)$.

In Section 9.9, we will need the notion of measure of positive type, a natural generalization of the notion of function of positive type.

3.7 Measures of Positive Type and Unitary Representations

As in the previous section, we assume that G is a separable, metrizable, locally compact group.

For any complex or σ-Radon measure μ, and any function $f \in \mathcal{K}_{\mathbb{C}}(G)$, we have

$$\begin{aligned}\int (f^* * f)(s)\, d\mu(s) &= \iint \Delta(t^{-1})\overline{f(t^{-1})}f(t^{-1}s)\, d\lambda(t)\, d\mu(s)\\ &= \iint \overline{f(t)}f(ts)\, d\lambda(t)\, d\mu(s).\end{aligned}$$

This suggests defining a measure of positive type as follows.

Definition 3.21. A complex or σ-Radon measure μ is of *positive type* if

$$\int (f^* * f)(s)\, d\mu(s) = \iint \overline{f(t)}f(ts)\, d\lambda(t)\, d\mu(s) \geq 0, \quad \text{for all } f \in \mathcal{K}_{\mathbb{C}}(G).$$

Observe that if $\mu = p\, d\lambda$ for some $p \in \mathrm{L}^1(G)$, then

$$\begin{aligned}\int (f^* * f)(s)\, d\mu(s) &= \iint \overline{f(t)}f(ts)p(s)\, d\lambda(t)\, d\lambda(s)\\ &= \iint \overline{f(t)}f(s)p(t^{-1}s)\, d\lambda(t)\, d\lambda(s),\end{aligned}$$

which is exactly the expression defining a function of positive type in Definition 3.17. But here $p \in \mathrm{L}^1(G)$ is not necessarily continuous, so Definition 3.21 yields a generalization of the notion of function of positive type.

Proposition 3.28. *For every complex measure $\nu \in \mathcal{M}^1(G)$, the measure $\mu = \check{\overline{\nu}} * \nu$ is of positive type on G.*

Proof. For every function $f \in \mathcal{K}_{\mathbb{C}}(G)$, we have

$$\begin{aligned}
\int (f^* * f)(s) d(\check{\overline{\nu}} * \nu)(s) &= \int \Big(\int \overline{f(t)} f(ts)\, d\lambda(t) \Big)\, d(\check{\overline{\nu}} * \nu)(s) \\
&= \int\!\!\int\!\!\int \overline{f(t)} f(tyz)\, d\lambda(t)\, d\check{\overline{\nu}}(y)\, d\nu(z) \\
&= \int\!\!\int\!\!\int \overline{f(t)} f(ty^{-1}z)\, d\lambda(t)\, d\overline{\nu}(y)\, d\nu(z) \\
&= \int\!\!\int\!\!\int \Delta(t^{-1}) \overline{f(t^{-1})} f(t^{-1}y^{-1}z)\, d\lambda(t)\, d\overline{\nu}(y)\, d\nu(z) \\
&= \int\!\!\int\!\!\int \Delta(t^{-1}y) \overline{f(t^{-1}y)} f(t^{-1}z)\, d\overline{\nu}(y)\, d\nu(z)\, d\lambda(t) \\
&= \int \Delta(t^{-1}y) \left| \int f(t^{-1}x)\, d\nu(x) \right|^2 d\lambda(t) \geq 0,
\end{aligned}$$

since the modular function Δ is strictly positive. □

As a corollary of Proposition 3.28, since $\delta_e = \check{\overline{\delta_e}} = \check{\overline{\delta_e}} * \delta_e$, we see that the Dirac measure δ_e is of positive type.

In the special case where $\nu = f\, d\lambda$, with $f \in \mathcal{L}^1(G)$, we see that $f^* * f$ is of positive type. It should be noted that if $f \in \mathcal{L}^1(G)$, the function of positive type $f^* * f \in \mathcal{L}^1(G)$ may not be bounded.

Example 3.10. For example, if $G = \mathbb{R}$ and if $f(x) = x^{-1/2}$ for $0 < x < 1$ and $f(x) = 0$ otherwise, then $f^* * f$ is not bounded.

Observe that $f^*(x) = f(x^{-1}) = \sqrt{x}$ if $x > 1$ and $f^*(x) = 0$ otherwise. See Figure 3.1.

We have

$$g(x) = (f^* * f)(x) = \int_{\mathbb{R}} f^*(t) f(x - t)\, dt,$$

and this integral is not zero if

$$t > 1, \qquad x - 1 < t < x.$$

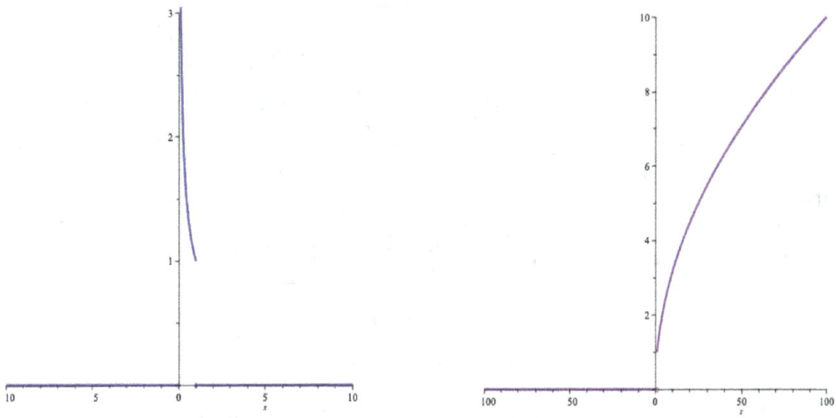

Fig. 3.1 The left figure is the graph of $f(x) = x,^{-1/2}$ for $0 < x < 1$, while the right figure is the graph of $f^*(x) = f(x^{-1}) = \sqrt{x}$ if $x > 1$.

If $x \leq 1$, then $g(x) = 0$. If $1 < x \leq 2$, then

$$g(x) = (f^* * f)(x) = \int_1^x \frac{\sqrt{t}}{\sqrt{x - t}}\, dt,$$

and if $x > 2$, then

$$g(x) = (f^* * f)(x) = \int_{x-1}^x \frac{\sqrt{t}}{\sqrt{x - t}}\, dt.$$

Let us compute the integral

$$\int_{x-1}^x \frac{\sqrt{t}}{\sqrt{x - t}} dt = \int_{x-1}^x \frac{1}{\sqrt{\frac{x}{t} - 1}} dt.$$

If we do the change of variable

$$u = \frac{x}{t} - 1,$$

we get

$$t = \frac{x}{u + 1}, \qquad dt = -\frac{x du}{(u + 1)^2},$$

so

$$\int_{x-1}^x \frac{dt}{\sqrt{\frac{x}{t} - 1}} = \int_0^{\frac{1}{x-1}} \frac{x du}{\sqrt{u}(u + 1)^2}.$$

Next, we make the change of variable

$$u = w^2,$$

so we have

$$w = \sqrt{u}, \qquad du = 2wdw,$$

and we get

$$\int_0^{\frac{1}{x-1}} \frac{xdu}{\sqrt{u}(u+1)^2} = 2x \int_0^{\frac{1}{\sqrt{x-1}}} \frac{dw}{(w^2+1)^2}.$$

But

$$\begin{aligned}\int \frac{dw}{(w^2+1)^2} &= \int \frac{w^2+1-w^2}{(w^2+1)^2} dw = \int \frac{dw}{(w^2+1)} - \int \frac{w^2 dw}{(w^2+1)^2} \\ &= \arctan w - \int w \frac{wdw}{(w^2+1)^2},\end{aligned}$$

and by integrating the second term by parts, we get

$$\int \frac{dw}{(w^2+1)^2} = \frac{w}{2(w^2+1)} + \frac{1}{2} \arctan w.$$

We finally obtain

$$\begin{aligned}g(x) &= 2x \left[\frac{w}{2(w^2+1)} + \frac{1}{2} \arctan w \right]_0^{\frac{1}{\sqrt{x-1}}} \\ &= \sqrt{x-1} + x \arctan\left(\frac{1}{\sqrt{x-1}} \right), \qquad x > 2.\end{aligned}$$

When $x > 2$ goes to infinity, the second term remains positive (in fact, goes to infinity, as we can see by using the power series for $\arctan y$ with $|y| < 1$), and the first term goes to infinity.

For the sake of completeness, if $1 < x \leq 2$, we have

$$\int_1^x \frac{dt}{\sqrt{\frac{x}{t} - 1}} = \int_0^{x-1} \frac{xdu}{\sqrt{u}(u+1)^2},$$

and

$$\int_0^{x-1} \frac{xdu}{\sqrt{u}(u+1)^2} = 2x \int_0^{\sqrt{x-1}} \frac{dw}{(w^2+1)^2}.$$

It follows that for $1 < x \leq 2$, we have

$$\begin{aligned}g(x) &= 2x \left[\frac{w}{2(w^2+1)} + \frac{1}{2} \arctan w \right]_0^{\sqrt{x-1}} \\ &= \sqrt{x-1} + x \arctan\left(\sqrt{x-1}\right).\end{aligned}$$

Therefore, the function $g = f^* * f$ is given by

$$g(x) = \begin{cases} 0 & \text{if } x \leq 1 \\ \sqrt{x-1} + x \arctan\left(\sqrt{x-1}\right) & \text{if } 1 < x \leq 2 \\ \sqrt{x-1} + x \arctan\left(\frac{1}{\sqrt{x-1}}\right) & \text{if } x > 2. \end{cases}$$

See Figure 3.2.

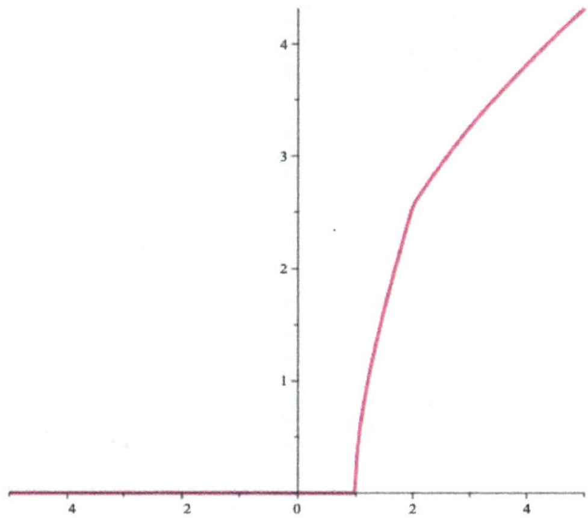

Fig. 3.2 The graph of $g = f^* * f$.

We showed that a continuous function p of positive type satisfies the property $p(s^{-1}) = \overline{p(s)}$, equivalently, $\check{\overline{p}} = p$. This property generalizes to measures of positive type.

Proposition 3.29. *For every measure μ of positive type on G, we have $\check{\overline{\mu}} = \mu$.*

Proof. Observe that if we can prove for all $f, g \in \mathcal{K}_{\mathbb{C}}(G)$ that

$$\int (f^* * g)(s)\, d\mu(s) = \int (f^* * g)(s)\, d\check{\overline{\mu}}(s), \tag{$*_3$}$$

then by regularization (Vol. I, Proposition 8.50), we will have

$$\int h(s)\, d\mu(s) = \int h(s)\, d\check{\overline{\mu}}(s), \qquad \text{for all } h \in \mathcal{K}_{\mathbb{C}}(G),$$

which proves that $\check{\overline{\mu}} = \mu$. By polarization, to prove ($*_3$), it suffices to prove it for $g = f$, since

$$\begin{aligned} 4g^* * f = (f+g)^* * (f+g) - (f-g)^* * (f-g) \\ + i(f+ig)^* * (f+ig) - i(f-ig)^* * (f-ig). \end{aligned}$$

If $f \in \mathcal{K}_{\mathbb{C}}(G)$, with $\nu = f\, d\lambda$, by Proposition 3.28, we see that $f^* * f \in \mathcal{K}_{\mathbb{C}}(G)$ is of positive type, so by Theorem 3.22(c),

$$\overline{(f^* * f)(s^{-1})} = (f^* * f)(s),$$

and since μ is of positive type, we have $\int (f^* * f)(s)\, d\mu(s) \geq 0$. By $(*)$ just before Vol. I, Proposition 8.46,

$$\int g(s)\, d\check{\overline{\mu}}(s) = \overline{\int \check{\overline{g}}(s)\, d\mu(s)},$$

and we obtain

$$\begin{aligned}\int (f^* * f)(s)\, d\check{\overline{\mu}}(s) &= \overline{\int \overline{(f^* * f)(s^{-1})}\, d\mu(s)}\\ &= \overline{\int (f^* * f)(s)\, d\mu(s)}\\ &= \int (f^* * f)(s)\, d\mu(s),\end{aligned}$$

as claimed. □

We conclude this section by showing that a measure μ of positive type defines a unitary representation U_μ of G. This construction will be used in Section 9.9 to define the Plancherel transform.

The vector space $\mathcal{K}_{\mathbb{C}}(G)$ is a nonunital algebra under convolution with involution $f \mapsto f^*$, with $f^*(s) = \Delta(s^{-1})\check{\overline{f}}(s)$. Because μ is of positive type, the linear form $\varphi_\mu \colon \mathcal{K}_{\mathbb{C}}(G) \to \mathbb{C}$ given by

$$\varphi_\mu(f) = \int f(s)\, d\mu(s)$$

is a positive linear form in the sense of Definition 2.10. As in Section 3.5, the set

$$\mathfrak{n} = \{f \in \mathcal{K}_{\mathbb{C}}(G) \mid \varphi_\mu(f^* * f) = 0\}$$

is a left ideal in $\mathcal{K}_{\mathbb{C}}(G)$, and $H_0 = \mathcal{K}_{\mathbb{C}}(G)/\mathfrak{n}$ is a hermitian space with the hermitian inner product

$$\langle \pi(f), \pi(g)\rangle_\mu = \varphi_\mu(g^* * f) = \int (g^* * f)(s)\, d\mu(s), \tag{$\dagger_1$}$$

where $\pi \colon \mathcal{K}_{\mathbb{C}}(G) \to \mathcal{K}_{\mathbb{C}}(G)/\mathfrak{n}$ is the quotient map. Since

$$\begin{aligned}\int (g^* * f)(s)\, d\mu(s) &= \iint \Delta(t^{-1})\overline{g(t^{-1})} f(t^{-1}s)\, d\lambda(t)\, d\mu(s)\\ &= \iint \overline{g(t)} f(ts)\, d\lambda(t)\, d\mu(s),\end{aligned}$$

we have

$$\langle \pi(f), \pi(g)\rangle_\mu = \varphi_\mu(g^* * f) = \iint \overline{g(t)} f(ts)\, d\lambda(t)\, d\mu(s). \tag{$\dagger_2$}$$

We claim that $H_0 = \mathcal{K}_{\mathbb{C}}(G)/\mathfrak{n}$ is separable. Recall from Vol. I, Proposition 2.16 that since G is a locally compact separable metric space, the space $\mathcal{K}_{\mathbb{C}}(G)$ is separable. If (f_n) is a sequence of functions in $\mathcal{K}_{\mathbb{C}}(G)$ converging uniformly to a function $f \in \mathcal{K}_{\mathbb{C}}(G)$, with the supports of the f_n remaining within some fixed compact subset, then $f_n^* * f_n$ converges uniformly to $f^* * f$, the supports of the $f_n^* * f_n$ remaining with some fixed compact set, thus $\|\pi(f_n) - \pi(f)\|_\mu$ tends to zero as n tends to infinity. This shows that H_0 is separable, and we let H (or H_μ) denote the separable Hilbert space which is its completion.

Instead of first defining a nondegenerate representation V of the algebra $\mathcal{K}_{\mathbb{C}}(G)$ and then the unitary representation U of G, such that $U_{\text{ext}} = V$, we define $U_\mu \colon G \to \mathbf{GL}(H_0)$ as follows:

$$U_\mu(s)(\pi(f)) = \pi(\delta_s * f), \qquad \text{for all } s \in G \text{ and all } f \in \mathcal{K}_{\mathbb{C}}(G). \qquad (*_{U_\mu})$$

Recall $(\delta_s * f)(t) = f(s^{-1}t)$, and that if $f \in \mathcal{K}_{\mathbb{C}}(G)$, then $\delta_s * f \in \mathcal{K}_{\mathbb{C}}(G)$.

Theorem 3.30. *For any measure μ of positive type, with the notation as above, if $U_\mu \colon G \to \mathbf{GL}(H_0)$ is the map defined by*

$$U_\mu(s)(\pi(f)) = \pi(\delta_s * f), \qquad \text{for all } s \in G \text{ and all } f \in \mathcal{K}_{\mathbb{C}}(G),$$

then each linear map $U_\mu(s)$ from H_0 to itself is continuous and unitary, thus the map $U_\mu(s)$ extends to a unitary map of H, and we obtain a homomorphism $U_\mu \colon G \to \mathbf{U}(H)$. For each $x \in H$, the map $s \mapsto U_\mu(s)(x)$ is continuous, therefore, $U_\mu \colon G \to \mathbf{U}(H)$ is a unitary representation of G in H.

Proof. By $(\dagger_2)$, we have

$$\begin{aligned}
\|U_\mu(s)(\pi(f))\|_\mu^2 &= \langle U_\mu(s)(\pi(f)), U_\mu(s)(\pi(f))\rangle_\mu = \langle \pi(\delta_s * f), \pi(\delta_s * f)\rangle_\mu \\
&= \int\!\!\int \overline{f(s^{-1}t)}\, f(s^{-1}tu)\, d\lambda(t)\, d\mu(u) \\
&= \int\!\!\int \overline{f(t)}\, f(tu)\, d\lambda(t)\, d\mu(u) \\
&= \langle \pi(f), \pi(f)\rangle_\mu = \|\pi(f)\|_\mu^2 .
\end{aligned}$$

Thus, $U_\mu(s)$ is unitary and continuous.

For any $f \in \mathcal{K}_{\mathbb{C}}(G)$, and $s \in G$, and any sequence (s_n) in G converging to $s \in G$, the sequence $(\lambda_{s_n} f)$ converges uniformly to $\lambda_s f$ (recall that $(\lambda_s f)(t) = f(s^{-1}t)$), the support of each $\lambda_s f$ remaining in a fixed compact set, so as before, the sequence $(\delta_{s_n} * f)$ converges to $\delta_s * f$, and since

$U_\mu(s_n)(\pi(f)) = \pi(\delta_{s_n} * f) = \pi(\lambda_{s_n} f)$, the sequence $(U_\mu(s_n)(\pi(f)))$ converges to $U_\mu(s)(\pi(f)) \in H$. Since H_0 is dense in H, and since the set of maps $\{U_\mu(s) \mid s \in G\}$ from H to itself is equicontinuous (see Vol. I, Proposition 2.13 or Dieudonné [Dieudonné (2005)], Chapter XII, Section 15, Theorem 12.15.7.1), as a consequence, each map $s \mapsto U_\mu(s)(x)$ is continuous (see Vol. I, Proposition 2.12 or Dieudonné [Dieudonné (2010)], Chapter VII, Section 5, Theorem 7.5.5). $\square$

Remark: According to Definition 3.13, the algebra representation $(U_\mu)_{\mathrm{ext}} \colon \mathrm{L}^1(G) \to \mathcal{L}(H)$ is defined such that for all $f \in \mathrm{L}^1(G)$ and all $g \in \mathcal{K}_\mathbb{C}(G)$, the vector $(U_\mu)_{\mathrm{ext}}(f)(\pi(g))$ is uniquely determined by the equation

$$\langle (U_\mu)_{\mathrm{ext}}(f)(\pi(g)), \pi(h)\rangle_\mu = \int f(s)\langle U_\mu(s)(\pi(g)), \pi(h)\rangle_\mu \, d\lambda(s)$$

for all $h \in \mathcal{K}_\mathbb{C}(G)$, and since

$$U_\mu(s)(\pi(g)) = \pi(\delta_s * g),$$

and by $(\dagger_2)$,

$$\begin{aligned}\langle U_\mu(s)(\pi(g)), \pi(h)\rangle_\mu &= \langle \pi(\delta_s * g), \pi(h)\rangle_\mu \\ &= \int\int \overline{h(u)}(\delta_s * g)(ut)\, d\lambda(u)\, d\mu(t) \\ &= \int\int g(s^{-1}ut)\overline{h(u)}\, d\lambda(u)\, d\mu(t),\end{aligned}$$

we obtain

$$\begin{aligned}\langle (U_\mu)_{\mathrm{ext}}(f)(\pi(g)), \pi(h)\rangle_\mu &= \int\int\int f(s)g(s^{-1}ut)\overline{h(u)}\, d\lambda(u)\, d\mu(t)\, d\lambda(s) \\ &= \int\int\int f(s)g(s^{-1}ut)\overline{h(u)}\, d\lambda(s)\, d\lambda(u)\, d\mu(t),\end{aligned}$$

as in Dieudonné [Dieudonné (2003b)] (Chapter XXII, Section 7, no. 22.7.2.1), except that in Dieudonné [Dieudonné (2003b)], u^{-1} occurs instead of u. But Dieudonné assumes that G is unimodular, so this does not make any difference. To deal with the case where G is not unimodular, we need to replace $\overline{\check{f}}$ by $f^* = \Delta^{-1}\overline{\check{f}}$, as we did, following Folland [Folland (1995)].

If $f \in \mathcal{K}_\mathbb{C}(G)$, then for all $g \in \mathcal{K}_\mathbb{C}(G)$, we have the simpler expression

$$(U_\mu)_{\mathrm{ext}}(f)(\pi(g)) = \pi(f * g),$$

as in Proposition 2.39.

Representation theory is a vast area of mathematics and we will only give a few references. A classic on the general theory is Kirillov [Kirillov (1976)]. Kirillov's survey [Kirillov (1994)] gives an excellent panorama of the field. Another encyclopedic source that covers a lot of the general theory is Hewitt and Ross [Hewitt and Ross (1970)]. A good source for the general theory is Folland [Folland (1995)]. Bröcker and tom Dieck [Bröcker and tom Dieck (1985)], Dieudonné [Dieudonné (2003a)], and Knapp [Knapp (2002)] cover the representation theory of compact groups in great depth. Fulton and Harris [Fulton and Harris (1991)], Humphreys [Humphreys (1972)], Knapp [Knapp (2001)], Taylor [Taylor (1986)], Varadarajan [Varadarajan (1984, 1999)], and Vilenkin [Vilenkin (1968)] cover the representation theory of Lie groups.

We are now ready to prove the famous Peter–Weyl theorem.

3.8 Problems

Problem 3.1. Define the representation $\rho_9\colon \mathbf{SO}(3) \to \mathbf{GL}(\mathrm{M}_3(\mathbb{C}))$ as follows: for any 3×3 complex matrix $A \in \mathrm{M}_3(\mathbb{C})$, for any $Q \in \mathbf{SO}(3)$,

$$\rho_9(Q)(A) = QAQ^\top.$$

To obtain a version of ρ_9 as a matrix representation M_{ρ_9}, we choose the canonical basis of nine matrices $E_{11}, E_{12}, E_{13}, E_{21}, E_{22}, E_{23}, E_{31}, E_{32}, E_{33}$, where E_{ij} contains 1 as the (i,j) entry and 0 otherwise. A matrix $M \in \mathrm{M}_3(\mathbb{C})$ is then written as the column vector

$$\mathrm{vec}(A) = (a_{11}, a_{12}, a_{13}, a_{21}, a_{22}, a_{23}, a_{31}, a_{32}, a_{33}).$$

(1) Show that over this basis, the matrix $M_{\rho_9}(Q)$ representing the linear map $\rho_9(Q)$ is given by

$$M_{\rho_9}(Q)(\mathrm{vec}(A)) = \mathrm{vec}(QAQ^\top).$$

(2) Prove the identity

$$\mathrm{vec}(AZB) = (B^\top \otimes A)\mathrm{vec}(Z),$$

where $\otimes$ denotes the Kronecker product of matrices.

(3) Prove that

$$M_{\rho_9}(Q)(\mathrm{vec}(A)) = \mathrm{vec}(QAQ^\top) = (Q \otimes Q)\mathrm{vec}(A).$$

(4) Prove that the subspace of symmetric matrices and the subspace of skew-symmetric matrices are invariant under $\rho_9\colon \mathbf{SO}(3) \to \mathbf{GL}(\mathrm{M}_3(\mathbb{C}))$.

(5) Prove that the subspace of symmetric matrices A with $\mathrm{tr}(A) = 0$ is also invariant.

Problem 3.2. Let G be a finite group with $g = |G|$ elements. Recall that the regular representation of G, $\rho_{\mathbf{R}} \colon G \to \mathbf{GL}(\mathbb{C}^g)$, is defined as

$$\rho_{\mathbf{R}}(s_i)(e_{s_j}) = e_{s_i s_j}, \quad 1 \leq i, j \leq g,$$

where $(e_{s_1}, \ldots, e_{s_g})$ is the canonical basis of $\mathbb{C}^g$ indexed by the g elements of G. Prove the following facts about irreducible finite-dimensional representations of a finite group G.

(1) Every irreducible finite-dimensional representation $\rho_i \colon G \to \mathbf{GL}(\mathbb{C}^{n_i})$ of the finite group G is equivalent to a subrepresentation of the regular representation $\rho_{\mathbf{R}} \colon G \to \mathbf{GL}(\mathbb{C}^g)$ of G in $\mathbb{C}^g$ (where $g = |G|$).
(2) Every irreducible representation $\rho_i \colon G \to \mathbf{GL}(\mathbb{C}^{n_i})$ occurs n_i times in the regular representation.
(3) If there are h irreducible representations $\rho_i \colon G \to \mathbf{GL}(\mathbb{C}^{n_i})$ (up to equivalence), then

$$n_1^2 + \cdots + n_h^2 = g.$$

(4) The number h of irreducible representations of G (up to equivalence) is equal to the number of conjugacy classes of G.

Hint. See Section 4.2 (Example 4.2), Proposition 4.17, and Serre [Serre (1977)].

Problem 3.3. Recall that

$$\mathcal{C}(U) = \{\varphi \in \mathcal{L}(H) \mid \varphi \circ U(g) = U(g) \circ \varphi \quad \text{for all } g \in G\}.$$

Show that $\mathcal{C}(U)$ is a unital C^*-algebra that is closed in $\mathcal{L}(H)$ under weak limits (see Definition 2.21(3)).

Hint. See Folland [Folland (1995)], Section 1.6.

Problem 3.4. Prove Proposition 3.4.

Hint. See Dieudonné [Dieudonné (2005)] (Chapter XV, Section 5) or Folland [Folland (1995)] (Chapter 3, Proposition 3.3).

Problem 3.5. Let $\alpha = a + ib$ and $\beta = c + id$. A matrix $q \in \mathbf{SU}(2)$ can be written as

$$q = \begin{pmatrix} \alpha & \beta \\ -\overline{\beta} & \overline{\alpha} \end{pmatrix}, \quad \text{with} \quad |\alpha|^2 + |\beta|^2 = 1.$$

The group $\mathbf{SU}(2)$ is a Lie group, whose (real) Lie algebra $\mathfrak{su}(2)$ is defined as

$$\mathfrak{su}(2) = \left\{ \begin{pmatrix} ix & y+iz \\ -y+iz & -ix \end{pmatrix} \;\middle|\; (x,y,z) \in \mathbb{R}^3 \right\}.$$

The adjoint representation $\mathbf{SU}(2)$ Ad: $\mathbf{SU}(2) \to \mathbf{GL}(\mathfrak{su}(2))$ is defined such that for every $q \in \mathbf{SU}(2)$, we have

$$\mathrm{Ad}_q(A) = qAq^*, \quad A \in \mathfrak{su}(2),$$

where q^* is the inverse of q. Prove that Ad_q is an invertible linear map from $\mathfrak{su}(2)$ to itself, and that Ad is a group homomorphism.

Problem 3.6. Prove Theorem 3.9.

Hint. See Gallier [Gallier (2011)] (Chapter 9).

Problem 3.7. Let G be a finite group of order $|G|$. The algebras $\mathrm{L}^1(G)$ and $\mathrm{L}^2(G)$ are the same and equal to the space $[G \to \mathbb{C}]$ of functions from G to $\mathbb{C}$. Convolution of two functions $f, h\colon G \to \mathbb{C}$ is given by

$$(f * h)(s) = \frac{1}{|G|} \sum_{s_1 s_2 = s} f(s_1)h(s_2) = \frac{1}{|G|} \sum_{t \in G} f(t)h(t^{-1}s).$$

Recall that for every $s \in G$, the function $\delta_s\colon G \to \mathbb{C}$ is given by

$$\delta_s(t) = \begin{cases} 1, & \text{if } t = s \\ 0, & \text{if } t \neq s. \end{cases}$$

We define an involution $f \mapsto f^*$ on $\mathrm{L}^1(G)$ by $f^*(s) = \overline{f(s^{-1})}$. Then $\mathrm{L}^1(G) = [G \to \mathbb{C}]$ is a unital involutive algebra under convolution with unit δ_e (where e is the identity element of G).

(1) Given a unitary representation $U\colon G \to \mathbf{U}(H)$ of G, where H is finite-dimensional, define $\widetilde{U}(f)(x)$ by

$$\widetilde{U}(f)(x) = \frac{1}{|G|} \sum_{s \in G} f(s)U(s)(x), \quad x \in H,\ f \in \mathrm{L}^1(G). \tag{(1)}$$

Prove that $\widetilde{U}\colon \mathrm{L}^1(G) \to \mathcal{L}(H)$ is an algebra representation.

Hint. See Simon [Simon (1996)] (Chapter II, Section 3).

(2) Conversely, let $V\colon \mathrm{L}^1(G) \to \mathcal{L}(H)$ be an algebra representation. Define $U\colon G \to \mathbf{U}(H)$ by

$$U(s) = V(\delta_s), \quad s \in G.$$

Verify that U is a unitary representation, such that $\widetilde{U} = V$.

Hint. See Simon [Simon (1996)] (Chapter II, Section 3).

Problem 3.8. (A continuation of Problem 3.7. Note that G is still a finite group). Consider the left regular representation of G. The space $\mathrm{L}^2(G) = \mathrm{L}^1(G)$ has the hermitian inner product given by

$$\langle f, g\rangle = \frac{1}{|G|}\sum_{s\in G} f(s)\overline{g(s)}, \quad f, g \in \mathrm{L}^2(G).$$

For any $s \in G$, define $\mathbf{R}_s\colon \mathrm{L}^2(G) \to \mathrm{L}^2(G)$ by

$$(\mathbf{R}_s(f))(t) = f(s^{-1}t), \quad s, t \in G,\ \mathrm{L}^2(G).$$

Prove that if we apply (1) to define $\widetilde{\mathbf{R}}(f)$, we find that

$$(\widetilde{\mathbf{R}}(f))(g) = f * g, \quad f, g \in \mathrm{L}^2(G).$$

Problem 3.9. Let G be a locally compact group. Recall that the representation $\mathbf{R}\colon G \to \mathbf{U}(\mathrm{L}^2(G))$ given by

$$(\mathbf{R}(s)(f))(t) = \lambda_s(f)(t) = f(s^{-1}t), \qquad f \in \mathrm{L}^2(G),\ s, t \in G,$$

is the left regular representation of G in $\mathrm{L}^2(G)$. By Theorem 3.17, we obtain $\mathbf{R}_{\mathrm{ext}}$ of $\mathrm{L}^1(G)$ in $\mathrm{L}^2(G)$.

(1) Prove that
$$\mathbf{R}_{\mathrm{ext}}(f)(g) = f * g,$$
with $f \in \mathrm{L}^1(G)$ and $g \in \mathrm{L}^2(G)$ (in the equation defining $\widetilde{U}(f) = \mathbf{R}_{\mathrm{ext}}(f)$, x is the function g and y is a function h).
(2) Using Vol. I, Proposition 8.50, prove that $\mathbf{R}_{\mathrm{ext}}$ is injective.

Problem 3.10. Let $(X, \mathcal{A})$ and $(Y, \mathcal{B})$ be two measure spaces, and let $\varphi\colon X \to Y$ be a measurable map. Let μ be a (positive) measure on $(X, \mathcal{A})$ and recall that $\varphi_*\mu$ of μ is defined as

$$\varphi_*\mu(B) = \mu(\varphi^{-1}(B)), \quad B \in \mathcal{B}.$$

(1) Prove that $\varphi_*\mu$ is a measure on $(Y, \mathcal{B})$.
(2) Prove Proposition 3.19.

Hint. See Folland [Folland (1995)] (Proposition 10.1) or Lang [Lang (1996)] (Chapter VI, Exercise 8).

Problem 3.11. Prove that the support of a measure μ satisfies the following property: For every $x \in \mathrm{supp}(\mu)$, for every neighborhood V of x, there is a continuous function f with compact support contained in V, such that $\int f(s)\,d\mu(s) \neq 0$.

Hint. See Dieudonné [Dieudonné (2005)] (Chapter XIII, Section 19).

Problem 3.12. (Advanced Exercise for Theorem 3.22) Complete the proof of (b) implies (c) when μ is an arbitrary complex measure.

Hint. See Dieudonné [Dieudonné (2005)] (Chapter XIII, Section 8, Proposition 13.8.7).

Problem 3.13. (Advanced Exercise) Prove Theorem 3.25.

Hint. See Folland [Folland (1995)] (Chapter 3, Theorem 3.25 and Lemma 3.26).

Problem 3.14. Prove Proposition 3.27.

Hint. See Folland [Folland (1995)] (Chapter 3, Proposition 3.35).

Chapter 4

Analysis on Compact Groups and Representations

Volume I, Chapter 10 is devoted to harmonic analysis on *abelian* locally compact (not necessarily compact) groups. In this chapter, we consider the case of a compact *not necessarily abelian* group G. Noncommutativity causes trouble. In particular, the characters no longer form a group. Irreducible representations of the group G become a substitute for the group characters of abelian groups. Fortunately, compactness also has a positive influence.

The structure of the algebra $\mathrm{L}^2(G)$ is described by a Hilbert sum of *finite-dimensional* matrix algebras, which determine irreducible unitary representations of G (in fact, up to equivalence, all of them). These results constitute a deep and beautiful theorem due to Peter and Weyl, and most of this chapter is devoted to its proof. To avoid technical complications (namely to avoid uncountable Hilbert sums and not use filters to deal with convergence issues), in this chapter *we assume that all locally compact groups are metrizable and separable and that all compact groups are metrizable*. This is not really a restriction since most groups that we will consider are Lie groups, which are metrizable and separable. As observed just before Proposition 2.18, if a topological space is metrizable and compact, then it is separable, so a metrizable compact group is also separable.

If the (metrizable) group G is compact, then some remarkable things happen:

(1) The involutive algebra (under convolution) $\mathrm{L}^2(G)$ is a complete Hilbert algebra, and as a consequence of Theorem 2.33, the algebra $\mathrm{L}^2(G)$ is a finite or countably infinite Hilbert sum $\bigoplus_{\rho \in R} \mathfrak{a}_\rho$ of topologically simple algebras, but because G is compact, each $\mathfrak{a}_\rho$ is isomorphic to a *finite-dimensional matrix algebra* $\mathrm{M}_{n_\rho}(\mathbb{C})$. The elements of $\mathfrak{a}_\rho$ are continuous

functions on G. This is the first half of the first part of a theorem due to Peter and Weyl (1927); see Theorem 4.2.

Since each minimal two-sided ideal $\mathfrak{a}_\rho$ is finite-dimensional, it can be expressed as a finite direct sum

$$\mathfrak{a}_\rho = \bigoplus_{1 \le j \le n_\rho} \mathfrak{a}_\rho * m_j,$$

of orthogonal minimal left ideals, where the m_j are self-adjoint irreducible idempotents. We can pick a Hilbert basis $(a_j)_{1 \le j \le n_\rho}$ in $\mathfrak{l}_1 = \mathfrak{a}_\rho * m_1$, such that $a_j \in m_j * \mathfrak{a}_\rho * m_1$, and then it turns out that there is some $\gamma > 0$, such that

$$a_j * \check{\bar{a}}_j = \gamma m_j, \qquad \check{\bar{a}}_j * a_j = \gamma m_1, \qquad 1 \le j \le n_\rho.$$

In fact, we will show that $\gamma = n_\rho^{-1}$. Finally, for all j, k with $1 \le j, k \le n_\rho$, let

$$m_{jk} = \gamma^{-1} a_j * \check{\bar{a}}_k,$$

which we also denote by $m_{jk}^{(\rho)}$. We have $m_{jj} = m_j$.

Then the family

$$\left(\frac{1}{\sqrt{n_\rho}} m_{ij}^{(\rho)} \right)_{1 \le i,j \le n_\rho,\ \rho \in R}$$

is a Hilbert basis of $\mathrm{L}^2(G)$, and for ρ fixed, it is an orthonormal basis of $\mathfrak{a}_\rho$.

Furthermore, for every $s \in G$, if we define the $n_\rho \times n_\rho$ matrix $M_\rho(s)$ by

$$M_\rho(s) = \left(\frac{1}{n_\rho} m_{ij}(s) \right),$$

then these matrices are invertible and satisfy the equations

$$M_\rho(st) = M_\rho(s) M_\rho(t) \qquad \text{and} \qquad M_\rho(s^{-1}) = (M_\rho(s))^*.$$

Thus, the map $s \mapsto M_\rho(s)$ is a continuous unitary representation in matrix form $M_\rho \colon G \to \mathbf{U}(n_\rho)$ of G in $\mathbb{C}^{n_\rho}$.

The above results are parts of Theorem 4.6, which constitutes the second half of the part of the Peter–Weyl theorem dealing with the structure of $\mathrm{L}^2(G)$ as Hilbert sum of finite-dimensional matrix algebras. But already, representations show their nose.

The unit of every two-sided ideal $\mathfrak{a}_\rho$ is $u_\rho = \sum_{j=1}^{n_\rho} m_{jj}$, and we show that the center of the Hilbert algebra $\mathrm{L}^2(G)$ is the Hilbert sum of the one-dimensional spaces $\mathbb{C}u_\rho$. The above results are shown in Section 4.1.

(2) Besides characters of groups and characters of algebras, there is one more kind of characters, namely, characters of finite-dimensional representations. For every $\rho \in R$, define the *character* χ_ρ *of* G *associated with the ideal* $\mathfrak{a}_\rho$ as the function given by

$$\chi_\rho(s) = \frac{1}{n_\rho}\, u_\rho(s) = \mathrm{tr}(M_\rho(s)), \qquad \text{for all } s \in G.$$

The character χ_{ρ_0} associated with $\mathfrak{a}_{\rho_0}$ is the constant function $\chi_{\rho_0}(s) = 1$ for all $s \in G$, called the *trivial character* of G. One of the main properties of the characters is that the family of characters $(\chi_\rho)_{\rho\in R}$ forms a Hilbert basis of the center of $\mathrm{L}^2(G)$; see Proposition 4.10. Other properties of the characters χ_ρ are shown in Section 4.2. In particular, if G is compact and abelian, then the characters are continuous homomorphisms of G to $\mathbf{U}(1)$, and they form a Hilbert basis for $\mathrm{L}^2(G)$.

(3) The second part of the Peter–Weyl theorem (Theorem 4.16) deals with unitary representations and is discussed in Section 4.3. This theorem asserts the following facts. Let $V\colon G \to \mathbf{U}(H)$ be a unitary representation of G in a separable Hilbert space H. Then H is a Hilbert sum of subspaces E_ρ invariant under V, and each nontrivial E_ρ is the Hilbert sum of invariant subspaces corresponding to irreducible representations of G. More precisely:

 (1) For every $\rho \in R$, there is an orthogonal projection of H onto a closed subspace E_ρ (which may be reduced to (0)), and H is the Hilbert sum of the $E_\rho \neq (0)$.
 (2) Every subspace $E_\rho \neq (0)$ is invariant under V, and the restriction V_ρ of V to E_ρ is a finite or countably infinite Hilbert sum of irreducible representations, all equivalent to M_ρ.

In particular, all the representations $M_\rho\colon G \to \mathbf{U}(n_\rho)$ occurring in Peter–Weyl I *are irreducible*, and since every unitary irreducible representation is equivalent to some representation of the form M_ρ, the index set R corresponds to a complete set of unitary representations of G. Now, because G is compact, there is a normalized Haar measure λ_G on G, such that $\lambda_G(G) = 1$, and it can be shown that for *any finite-dimensional* representation $V\colon G \to \mathbf{GL}(H)$, there is an inner product on H, such that V becomes a unitary representation. Then we define the *character* χ_V *of the representation* V by

$$\chi_V(s) = \mathrm{tr}(V(s)), \qquad s \in G.$$

Theorem 4.19 shows that two finite-dimensional unitary representations $V_1\colon G \to \mathbf{U}(H_2)$ and $V_2\colon G \to \mathbf{U}(H_2)$ of G are equivalent if and only if $\chi_{V_1} = \chi_{V_2}$. This confirms the importance of the characters; they determine the equivalence classes of finite-dimensional representations of a (metrizable) compact group.

In Section 4.4, we discuss tensor products of finite-dimensional representations. We begin with the definition of the tensor product representation $U_1 \otimes U_2 \colon G \to \mathbf{U}(H_1 \otimes H_2)$ of two finite-dimensional unitary representations $U_1 \colon G \to \mathbf{U}(H_1)$ and $U_2 \colon G \to \mathbf{U}(H_2)$ of the same locally compact (metrizable, separable) group G. In general, if U_1 and U_2 are irreducible, then the tensor product representation $U_1 \otimes U_2$ is *not* irreducible. If G is compact, the representation $U_1 \otimes U_2$ splits as a sum of irreducible representations of G, but finding this decomposition is generally very difficult. In the special case $G = \mathbf{SU}(2)$, this can be done. This is an important result of quantum physics; see Section 5.17 on the Clebsch–Gordan coefficients.

Next, we define the tensor product representation $U_1 \otimes U_2 \colon G_1 \times G_2 \to \mathbf{U}(H_1 \otimes H_2)$ of the finite-dimensional unitary representations $U_1 \colon G_1 \to \mathbf{U}(H_1)$ and $U_2 \colon G_2 \to \mathbf{U}(H_2)$ of two locally compact groups G_1 and G_2. This time it turns out that $U_1 \otimes U_2$ is irreducible iff U_1 and U_2 are irreducible. We prove this result when G is compact. Furthermore, if G_1 and G_2 are compact, then every finite-dimensional irreducible unitary representation $U \colon G_1 \times G_2 \to \mathbf{U}(H)$ is equivalent to the tensor product $U_1 \otimes U_2$ of two finite-dimensional irreducible unitary representations $U_1 \colon G_1 \to \mathbf{U}(H_1)$ and $U_2 \colon G_2 \to \mathbf{U}(H_2)$. This fact can be used to determine the irreducible representations of the compact groups $\mathbf{O}(2m+1)$ and $\mathbf{U}(2)$. The case of $\mathbf{O}(2m)$ is more difficult because $\mathbf{O}(2m)$ is not isomorphic to the direct product of $\mathbf{SO}(2m)$ with some other subgroup, but instead a semi-direct product (see Section 7.4). For $m = 1$, the group $\mathbf{SO}(2)$ is abelian so the method of Section 7.4 involving Mackey's little group method can be used to determine the irreducible representations of $\mathbf{O}(2)$.

In Section 4.5, we define the notion of *contragredient representation* $\overline{U} \colon G \to \mathbf{GL}(H^*)$ of a representation $U \colon G \to \mathbf{GL}(H)$ and the notion of Hom *representation* $\mathrm{Hom}(U_1, U_2) \colon G \to \mathbf{GL}(\mathrm{Hom}(H_1, H_2))$ of two representations $U_1 \colon G \to \mathbf{GL}(H_1)$ and $U_2 \colon G \to \mathbf{GL}(H_2)$. These notions will be needed in Chapter 8. The main result is that if H_1 and H_2 are finite-dimensional vector spaces then the representations $\overline{U_1} \otimes U_2$ and $\mathrm{Hom}(U_1, U_2)$ are equivalent; see Proposition 4.23.

The Fourier transform and Fourier cotransform can also be generalized, but they involve the unitary irreducible representations of G, which are usually very difficult to determine, so they are generally not so useful in practice. The groups $\mathbf{SO}(2)$, $\mathbf{SO}(3)$, $\mathbf{SU}(2)$ are an exception, their irreducible representations are determined explicitly; see Chapter 5.

In Section 4.6, we define a notion of Fourier transform and Fourier cotransform for a (metrizable) compact group G. Since for a nonabelian compact group the set of characters is not a group, the definition of the spaces $\mathrm{L}^p(\widehat{G})$ is more complicated. The *Fourier transform* $\mathcal{F}f$ of a function $f \in \mathrm{L}^1(G)$ is now a function with domain R, a complete set of irreducible unitary representations of G, such that for every $\rho \in R$,

$$\mathcal{F}(f)(\rho) = \int f(t)(M_\rho(t))^* \, d\lambda_g(t).$$

The Fourier transform defined above is the natural generalization of the definition of the Fourier transform when G is an abelian compact group (Vol. I, Definition 10.3),

$$\mathcal{F}(f)(\chi) = \int f(s)\overline{\chi(s)} \, d\lambda(s) = \int f(s)\chi(s^{-1}) \, d\lambda(s);$$

the character χ is replaced by the irreducible representation M_ρ.

The definition of $\mathcal{F}(f)(\rho)$ implies that $\mathcal{F}(f)(\rho)$ is a linear map from $\mathbb{C}^{n_\rho}$ to itself (since $(M_\rho(t))^*$ is a matrix). Thus, $\mathcal{F}(f) \in \prod_{\rho \in R} \mathrm{M}_{n_\rho}(\mathbb{C})$. Every element $F \in \prod_{\rho \in R} \mathrm{M}_{n_\rho}(\mathbb{C})$ is an R-indexed sequence $F = (F(\rho))_{\rho \in R}$ of $n_\rho \times n_\rho$ matrices $F(\rho)$. These sequences can be added and rescaled componentwise, so we obtain a vector space.

It is natural to define $\widehat{G}$ as R, but the vector space $\prod_{\rho \in R} \mathrm{M}_{n_\rho}(\mathbb{C})$ is too big. Thus, we define some normed vector spaces $\mathrm{L}^p(\widehat{G})$, which are subspaces of $\prod_{\rho \in R} \mathrm{M}_{n_\rho}(\mathbb{C})$. For this, we need to define some norms due to von Neumann; see Section 4.7.

We obtain some Banach spaces $\mathrm{L}^1(\widehat{G})$, $\mathrm{L}^2(\widehat{G})$, and $\mathrm{L}^\infty(\widehat{G})$; the space $\mathrm{L}^2(\widehat{G})$ is a Hilbert space. The following result is obtained (Theorem 4.29). Let G be a compact group.

(1) The map $f \mapsto \mathcal{F}(f)$ is a non norm-increasing injective involutive algebra homomorphism from $\mathrm{L}^1(G)$ into $\mathrm{L}^\infty(\widehat{G})$. In particular, for all $f, g \in \mathrm{L}^1(G)$, for all $\rho \in R$, we have

$$(\mathcal{F}(f * g))(\rho) = \mathcal{F}(g)(\rho) \circ \mathcal{F}(f)(\rho).$$

(2) For every $\rho \in R$, the map $f \mapsto \mathcal{F}(f)(\rho)$ is an algebra representation of $\mathrm{L}^1(G)$ in $\mathbb{C}^{n_\rho}$.

We also have a version of Plancherel's theorem (see Theorem 4.32). If G is a compact group, then the map $f \mapsto \mathcal{F}(f)$ is an isometric isomorphism between the Hilbert space $\mathrm{L}^2(G)$ and the Hilbert space $\mathrm{L}^2(\widehat{G})$.

We can also define a notion of Fourier cotransform and there are versions of Fourier inversion; see Section 4.8. For any $F \in \prod_{\rho\in R} \mathrm{M}_{n_\rho}(\mathbb{C})$, the *Fourier cotransform* $\overline{\mathcal{F}}(F)$ of F is the function on G given by

$$\overline{\mathcal{F}}(F)(s) = \sum_{\rho\in R} n_\rho \operatorname{tr}(F(\rho)M_\rho(s)), \qquad s \in G.$$

Of course, there are convergence issues. It can be shown (Theorem 4.33) that if $F \in \mathrm{L}^1(\widehat{G})$, then the map

$$s \mapsto (\overline{\mathcal{F}}(F))(s) = \sum_{\rho\in R} n_\rho \operatorname{tr}(F(\rho)M_\rho(s))$$

converges uniformly to a continuous function f. Furthermore, we have the Fourier inversion formula

$$(\overline{\mathcal{F}}(\mathcal{F}(f)))(s) = \sum_{\rho\in R} n_\rho \operatorname{tr}(\mathcal{F}(f)(\rho)M_\rho(s)), \qquad s \in G.$$

Also, Fourier inversion holds for $\mathrm{L}^2(G)$ (see Theorem 4.35). The Fourier cotransform $\overline{\mathcal{F}}(F) \in \mathrm{L}^2(G)$ of any $F \in \mathrm{L}^2(\widehat{G})$ converges in the L^2-norm, and for every $f \in \mathrm{L}^2(G)$, we have

$$f(s) = (\overline{\mathcal{F}}(\mathcal{F}(f)))(s) = \sum_{\rho\in R} n_\rho \operatorname{tr}\left(\mathcal{F}(f)(\rho)M_\rho(s)\right), \qquad s \in G,$$

in the L^2-norm.

4.1 The Peter–Weyl Theorem, Part I

The theorem below is the first of several theorems describing the structure of the involutive Banach algebra $\mathrm{L}^2(G)$, where G is a (metrizable) compact group. By Proposition 2.18, the Banach algebra $\mathrm{L}^2(G)$ is a complete separable Hilbert algebra, so Theorem 2.33 is applicable and yields most of a deep theorem first proved by Peter and Weyl (1927); see Theorem 4.2.

No matter how it is approached, the proof of the Peter–Weyl theorem (Theorem 4.2) is hard. We follow Dieudonné's exposition [Dieudonné (2003a)] (Sections 1–4). The disadvantage in doing so is that it requires some material on Hilbert algebras from Chapter 2, in particular, Theorem 2.33, whose proof is long. The advantage is that we obtain a sharper and more informative version of the Peter–Weyl theorem.

Since G is compact, it is unimodular, and so it has a Haar measure λ which is both left and right invariant. We also assume that λ is normalized so that $\lambda(G) = 1$.

When we describe operations on elements of $\mathrm{L}^2(G)$, such as $f * g$ for the convolution $[f] * [g]$ of $[f]$ and $[g]$ in $\mathrm{L}^2(G)$, we mean the equivalence class $[f * g]$ of $f * g$, where $f, g \in \mathcal{L}^2(G)$ are representatives in the equivalence classes $[f], [g] \in \mathrm{L}^2(G)$ (where two functions are equivalent iff they are equal almost everywhere). To be perfectly rigorous, we should check that these constructions do not depend on the representatives chosen in these equivalence classes, but we will not inflict such verifications on the reader.

Recall that by Vol. I, Proposition 8.49, if $f, g \in \mathrm{L}^2(G)$, then $f * g \in \mathcal{C}_0(G; \mathbb{C})$. In particular, $f * g$ is continuous.

Definition 4.1. A function $h \in \mathcal{L}^2(G)$ is *central* if its class in $\mathrm{L}^2(G)$ belongs to the center of $\mathrm{L}^2(G)$. This means that for every $f \in \mathcal{L}^2(G)$, we have $f * h = h * f$ almost everywhere.

The following auxiliary result is needed.

Proposition 4.1. *Let G be a compact group. A continuous function $h \in \mathcal{L}^2(G)$ is central if and only if $h(sts^{-1}) = h(t)$ for all $s, t \in G$. The class of every central function $f \in \mathcal{L}^2(G)$ belongs to the center of $\mathcal{M}^1(G)$.*

Proof. If $f*h = h*f$ almost everywhere, since $f*h$ and $h*f$ are continuous, we must have $f*h = h*f$ everywhere. Since G is compact, it is unimodular, so have

$$f * h(s) = \int f(t)h(t^{-1}s)\, d\lambda(t) = \int f(t^{-1})h(ts)\, d\lambda(t),$$

and

$$h * f(s) = \int h(t)f(t^{-1}s)\, d\lambda(t) = \int h(st)f(t^{-1})\, d\lambda(t).$$

Thus, for every $s \in G$, we have

$$\int_G f(t^{-1})(h(st) - h(ts))\, d\lambda(t) = 0.$$

The above implies that $h(st) = h(ts)$ for all t in the complement of a set of measure zero (depending on s), but since h is continuous, this subset must be empty. It follows that $h(st) = h(ts)$ for all $s, t \in G$, and if we replace t by ts^{-1}, we obtain $h(sts^{-1}) = h(t)$ for all $s, t \in G$.

By the formula

$$(\mu * f)(s) = \int f(t^{-1}s)\, d\mu(t)$$

in Vol. I, Definition 8.25, and the formula

$$(f * \mu)(s) = \int f(st^{-1})\Delta(t^{-1})\, d\mu(t)$$

in Vol. I, Definition 8.26, since G is compact, it is unimodular, so $\Delta(t^{-1}) = 1$, and if f is a central function, then it is easy to show that $\mu * (fd\lambda) = (fd\lambda) * \mu$, which shows that $fd\lambda$ is in the center of $\mathcal{M}^1(G)$. Recall that $\mathcal{L}^1(G)$ is embedded in $\mathcal{M}^1(G)$ by mapping f to the measure $fd\lambda$ and we usually identify f and $fd\lambda$. ☐

Theorem 4.2. *(Peter–Weyl theorem, I) Let G be a (metrizable) compact group. The complete Hilbert algebra $\mathrm{L}^2(G)$ is the Hilbert sum*

$$\mathrm{L}^2(G) = \bigoplus_{\rho \in R} \mathfrak{a}_\rho$$

of a finite or countably infinite family $(\mathfrak{a}_\rho)_{\rho\in R}$ of topologically simple Hilbert algebras $\mathfrak{a}_\rho$ of finite dimension n_ρ^2, where each $\mathfrak{a}_\rho$ is a minimal two-sided ideal of $\mathrm{L}^2(G)$ isomorphic to the matrix algebra $\mathrm{M}_{n_\rho}(\mathbb{C})$, and $\mathfrak{a}_h\mathfrak{a}_k = (0)$ for all $h \neq k$ $(h, k \in R)$. What this means is that each ideal $\mathfrak{a}_\rho$ has an orthogonal basis of functions $\left(m_{ij}^{(\rho)}\right)_{1\leq i,j\leq n_\rho}$ satisfying certain properties stated in Theorem 4.6, so that the map from $\mathfrak{a}_\rho$ to $\mathrm{M}_{n_\rho}(\mathbb{C})$ given by $\sum_{i,j} \lambda_{ij} m_{ij}^{(\rho)} \mapsto (\lambda_{ij})_{1\leq i,j\leq n_\rho}$ is an algebra isomorphism. The elements of $\mathfrak{a}_\rho$ are classes of continuous functions on G; the unit element of $\mathfrak{a}_\rho$ is the class of a continuous function u_ρ, such that $\check{\overline{u}}_\rho = u_\rho$, and the orthogonal projection of $\mathrm{L}^2(G)$ onto $\mathfrak{a}_\rho$ is the map $f \mapsto f * u_\rho = u_\rho * f$, for every $f \in \mathcal{L}^2(G)$. Furthermore, for every $f \in \mathcal{L}^2(G)$, we have

$$f = \sum_{\rho\in R} f * u_\rho,$$

where the series on the right-hand side is commutatively convergent.

Proof. We follow Dieudonné's proof [Dieudonné (2003a)] (Chapter XXI, Section 2, Theorem 21.2.3). Since by Proposition 2.18, the Banach algebra $A = \mathrm{L}^2(G)$ is a complete separable Hilbert algebra, Theorem 2.33 shows that $\mathrm{L}^2(G)$ is the Hilbert sum of a finite or countably infinite family $(\mathfrak{a}_\rho)_{\rho\in R}$ of two-sided ideals, which are topologically simple Hilbert algebras, and $\mathfrak{a}_h\mathfrak{a}_k = (0)$ for all $h \neq k$. If we can prove that every $\mathfrak{a}_\rho$ is finite-dimensional, we will be done because then, by Theorem 2.34, each $\mathfrak{a}_\rho$ will be a finite Hilbert sum of isomorphic minimal left ideals $\mathfrak{l}_j$, each generated by a self-adjoint irreducible idempotent e_j, and the sum of these idempotents will be the unit $\mathbf{1}_\rho$ of the algebra $\mathfrak{a}_\rho$. If u_ρ is a function whose class is $\mathbf{1}_\rho$, every element of $\mathfrak{a}_\rho$ will be the class of a function of the form $f * u_\rho$, with $f \in \mathcal{L}^2(G)$, which is a continuous function by Vol. I, Proposition 8.49. The other assertions of the theorem follow from Theorem 2.30, since the orthogonal projection of f onto $\mathfrak{a}_\rho$ is of the form $\sum_{j=1}^{n_\rho} f * e_j = f * u_\rho$, since $u_\rho = \sum_{j=1}^{n_\rho} e_j$.

By Proposition 2.35, if we can prove that there is a nonzero element in the center of $\mathfrak{a}_\rho$, then $\mathfrak{a}_\rho$ will be finite-dimensional. This is a consequence of the following proposition.

Proposition 4.3. *For every closed two-sided ideal $\mathfrak{b} \neq (0)$ in $\mathrm{L}^2(G)$, there is some nonzero element $c \in \mathfrak{b}$ in the center of $\mathrm{L}^2(G)$.*

Proof. The proof of Proposition 4.3 makes use of the following result.

Proposition 4.4. *Let $\mathfrak{b}$ be a closed subspace (as a vector space) of $\mathrm{L}^2(G)$. Then the following conditions are equivalent.*

(1) $\mathfrak{b}$ is a left ideal in $\mathrm{L}^2(G)$.
(2) $\mathfrak{b}$ is invariant under the regular representation $\mathbf{R}_{\mathrm{ext}}$ of $\mathrm{L}^1(G)$ in $\mathrm{L}^2(G)$ (see Definition 3.15).
(3) For every function f whose class is in $\mathfrak{b}$ and for all $s \in G$, the class of $\delta_s * f = \lambda_s(f)$ is in $\mathfrak{b}$.

Proof. The equivalence of (2) and (3) follows from Theorem 3.18 applied to the regular representation $\mathbf{R}_{\mathrm{ext}}$. It is clear that (2) implies (1). On the other hand, Vol. I, Theorem 7.10 implies that $\mathcal{L}^2(G)$ is dense in $\mathcal{L}^1(G)$, and by Vol. I, Proposition 8.48, the map $f \mapsto f * g$ from $\mathcal{L}^1(G)$ to $\mathcal{L}^2(G)$ is continuous for every $g \in \mathcal{L}^2(G)$, so (1) implies (2). □

Proposition 4.4 also applies to right ideals in (1) and to $f * \delta_s = \rho_{s^{-1}}(f)$ in (3).

We can now prove Proposition 4.3. First, let us prove that $\mathfrak{b}$ contains the class of a continuous function f, such that f is not the zero function (we need a continuous function, because we want to construct a central function, and to apply Proposition 4.1, such a function must be continuous). Indeed, let $g \in \mathfrak{b}$ be a function not zero almost everywhere. Then the class of the function $g * \check{\overline{g}}$ also belongs to $\mathfrak{b}$. But $g * \check{\overline{g}}$ is continuous (by Vol. I, Proposition 8.49), and since the definition of the convolution of functions implies that

$$(g * \check{\overline{g}})(e) = \int g(s)\check{\overline{g}}(s^{-1})\, d\lambda(s) = \int g(s)\overline{g}(s)\, d\lambda(s) = \|g\|_2^2 > 0,$$

we can pick $f = g * \check{\overline{g}}$ in $\mathfrak{b}$. Consider the function h given by

$$h(t) = \int_G f(sts^{-1})\, d\lambda(s).$$

Since the function $(x, y, z) \mapsto f(xyz)$ is uniformly continuous on $G \times G \times G$, we see that h is continuous, and since $h(e) = f(e) \neq 0$, it is not the zero function. For all $s \in G$, we have

$$h(xtx^{-1}) = \int_G f((sx)t(sx)^{-1})\, d\lambda(s) = h(t),$$

since the Haar measure on a compact group is left and right invariant. By Proposition 4.1, the function h belongs to the center of $\mathrm{L}^2(G)$. It remains to show that the class of h belongs to $\mathfrak{b}$. Since $\mathrm{L}^2(G) = \mathfrak{b} \oplus \mathfrak{b}^{\perp}$ as a Hilbert sum, and $\mathfrak{b}^{\perp}$ is also a two-sided ideal by Proposition 2.21, it suffices to prove that $\langle h, w\rangle = 0$ for all $w \in \mathfrak{b}^{\perp}$ (we are abusing notation, h and w should be equivalence classes). Using the fact that the Haar measure is left and right invariant, and Fubini's theorem, we have

$$\begin{aligned}\langle h, w\rangle &= \int \overline{w(t)} \int f(sts^{-1})\, d\lambda(s)\, d\lambda(t)\\ &= \int \left(\int \overline{w(t)} f(sts^{-1})\, d\lambda(t)\right) d\lambda(s)\\ &= \int \left(\int \overline{w(s^{-1}ts)} f(t)\, d\lambda(t)\right) d\lambda(s).\end{aligned}$$

Since $w \in \mathfrak{b}^{\perp}$, by Proposition 4.4 and its version for right ideals, the class of $\delta_s * w * \delta_{s^{-1}}$ also belongs to $\mathfrak{b}^{\perp}$. Since G is unimodular (see after Vol. I, Definition 8.25 and Definition 8.26), we have

$$(\delta_s * w * \delta_{s^{-1}})(t) = w(s^{-1}ts),$$

and since $f \in \mathfrak{b}$ and $\delta_s * w * \delta_{s^{-1}} \in \mathfrak{b}^{\perp}$,

$$\int \overline{w(s^{-1}ts)} f(t)\, d\lambda(t) = 0,$$

which concludes the proof of Proposition 4.3. □

This also concludes the proof of Theorem 4.2. □

We will identify every element of $\mathfrak{a}_\rho$ with the unique continuous function belonging to this class.

Our next goal is to get a better understanding of the structure of the algebras $\mathfrak{a}_\rho$ by decomposing them as finite Hilbert sums of minimal left ideals, and by choosing some Hilbert bases in these ideals.

For every $\rho \in R$, we assume that we have chosen a decomposition of $\mathfrak{a}_\rho$ as a finite Hilbert sum of n_ρ minimal left ideals $\mathfrak{l}_j = \mathfrak{a}_\rho * m_j$,

$$\mathfrak{a}_\rho = \bigoplus_{j=1}^{n_\rho} \mathfrak{l}_j = \bigoplus_{1 \leq j \leq n_\rho} \mathfrak{a}_\rho * m_j,$$

where the $\mathfrak{a}_\rho * m_j$, also denoted $\mathfrak{l}_j^{(\rho)}$, are pairwise isomorphic and orthogonal, and where m_j is a self-adjoint irreducible idempotent ($1 \leq j \leq n_\rho$), so that the unit of $\mathfrak{a}_\rho$ is

$$u_\rho = \sum_{j=1}^{n_\rho} m_j; \qquad (u_\rho)$$

see Theorem 2.34.[1] By Proposition 2.23, since m_j and m_j are orthogonal when $i \neq j$, we have $m_i * m_j = 0$ if $i \neq j$. Let $(a_j)_{1 \leq j \leq n_\rho}$ be a Hilbert basis of $\mathfrak{l}_1 = \mathfrak{a}_\rho * m_1$, such that $a_j \in m_j * \mathfrak{a}_\rho * m_1$.

Since $a_j \in m_j * \mathfrak{a}_\rho * m_1$, we have

$$a_j * m_1 = a_j, \qquad m_j * a_j = a_j, \quad 1 \leq j \leq n_\rho.$$

We have the following proposition, which is in fact part of the proof of Theorem 2.34. To simplify notation, write $a^* = \breve{\bar{a}}$.

Proposition 4.5. *The inner products $\langle m_j, m_j \rangle$ have the same value $\gamma > 0$, and we have*

$$a_j * a_j^* = \gamma m_j, \qquad a_j^* * a_j = \gamma m_1, \qquad 1 \leq j \leq n_\rho.$$

Proof. Since each $\mathfrak{a}_\rho$ is a Hilbert algebra, $\mathfrak{a}_\rho = \mathfrak{a}_\rho^*$ and $\mathfrak{a}_\rho \mathfrak{a}_\rho = \mathfrak{a}_\rho$. Since the m_j are self-adjoint idempotent ($m_j * m_j = m_j$ and $m_j^* = m_j$) and $(a * b)^* = b^* * a^*$, as $a_j \in m_j * \mathfrak{a}_\rho * m_1$, we have $a_j * a_j^* \in m_j * \mathfrak{a}_\rho * m_j$. By Theorem 2.32(2), we must have $a_j * a_j^* = \lambda_j m_j$, for some $\lambda_j \in \mathbb{C}$, with $\lambda_j \neq 0$. Similarly, $a_j^* * a_j \in m_1 * \mathfrak{a}_\rho * m_1$, so $a_j^* * a_j = \lambda_j' m_1$ for some $\lambda_j' \in \mathbb{C}$, with $\lambda_j' \neq 0$. We claim that $\lambda_j = \lambda_j'$.

[1] Note that the m_j are the e_j used in the proof of Theorem 2.34.

First, we have

$$a_j * a_j^* * a_j * a_j^* = \lambda_j m_j * \lambda_j m_j = \lambda_j^2 m_j * m_j = \lambda_j^2 m_j,$$

and second

$$a_j * a_j^* * a_j * a_j^* = a_j * \lambda_j' m_1 * a_j^* = \lambda_j' a_j * m_1 * a_j^* = \lambda_j' a_j * a_j^* = \lambda_j' \lambda_j m_j,$$

since $a_j * m_1 = a_j$. Therefore, $\lambda_j^2 = \lambda_j' \lambda_j$, and since λ_j and λ_j' are nonzero, we deduce that $\lambda_j = \lambda_j'$, for $j = 1, \ldots, n_\rho$.

We also have

$$1 = \langle a_j, a_j \rangle = \langle a_j, m_j * a_j \rangle = \langle a_j * a_j^*, m_j \rangle = \lambda_j \langle m_j, m_j \rangle,$$

and

$$1 = \langle a_j, a_j \rangle = \langle a_j, a_j * m_1 \rangle = \langle a_j^* * a_j, m_1 \rangle = \lambda_j \langle m_1, m_1 \rangle.$$

Since $\lambda_j \neq 0$, we deduce that

$$\langle m_j, m_j \rangle = \langle m_1, m_1 \rangle, \qquad 1 \leq j \leq n_\rho,$$

and so

$$\lambda_j = \langle m_1, m_1 \rangle^{-1} = \gamma, \qquad 1 \leq j \leq n_\rho. \qquad \square$$

Thus, there is some $\gamma > 0$, such that

$$a_j * \breve{\bar{a}}_j = \gamma m_j, \qquad \breve{\bar{a}}_j * a_j = \gamma m_1, \qquad 1 \leq j \leq n_\rho.$$

Since $a_i \in m_i * \mathfrak{a}_\rho * m_1$ and $\breve{\bar{a}}_j \in m_1 * \mathfrak{a}_\rho * m_j$, we have $a_i * \breve{\bar{a}}_j \in m_i * \mathfrak{a}_\rho * m_j$ and since $\mathfrak{l}_j = \mathfrak{a}_\rho * m_j$ is a left ideal, $a_i * \breve{\bar{a}}_j \in \mathfrak{l}_j$.

Definition 4.2. For all j, k with $1 \leq j, k \leq n_\rho$, let

$$m_{jk} = \gamma^{-1} a_j * \breve{\bar{a}}_k \in \mathfrak{l}_k.$$

In particular, $m_{jj} = m_j$.

Since $a_h \in m_h * \mathfrak{a}_\rho * m_1$, $\breve{\bar{a}}_k \in m_1 * \mathfrak{a}_\rho * m_k$, and $m_k * m_h = 0$ whenever $h \neq k$, we have $\breve{\bar{a}}_k * a_h = 0$ whenever $h \neq k$, so

$$m_{jk} * a_h = \delta_{kh} a_j. \qquad (*)$$

Remark: Observe that the m_{ij} are the e_{mn} introduced during the proof of Theorem 2.34.

We will also write $m_{ij}^{(\rho)}$ instead of m_{ij}. The following result reveals that some representations are hidden in the Hilbert sum of the $\mathfrak{a}_\rho$.

Theorem 4.6. *With the above notation, the following properties hold.*

(1) *For every j with $1 \leq j \leq n_\rho$, the $(m_{ij})_{1\leq i\leq n_\rho}$ form an orthogonal basis of $\mathfrak{l}_j$, and the $(m_{ij})_{1\leq i,j\leq n_\rho}$ form an orthogonal basis of $\mathfrak{a}_\rho = \bigoplus_{j=1}^{n_\rho} \mathfrak{l}_j$.*
(2) *We have $m_{ji} = \overline{\check{m}}_{ij}$ and $m_{ij} * m_{hk} = \delta_{jh}\, m_{ik}$.*
(3) *We have $\langle m_{ij}, m_{ij}\rangle = n_\rho$, $m_{ij}(e) = n_\rho\, \delta_{ij}$, for all i, j with $1 \leq i, j \leq n_\rho$ (in other words, $\gamma = (n_\rho)^{-1}$), and $u_\rho = \sum_{j=1}^{n_\rho} m_{jj}$. Thus the family of functions*

$$\left(\frac{1}{\sqrt{n_\rho}} m_{ij}^{(\rho)}\right)_{1\leq i,j\leq n_\rho,\ \rho\in R}$$

is a Hilbert basis of $\mathrm{L}^2(G)$.
(4) *For every $s \in G$, if we define the $n_\rho \times n_\rho$ matrix $M_\rho(s)$ by*

$$M_\rho(s) = \left(\frac{1}{n_\rho} m_{ij}^{(\rho)}(s)\right),$$

then these matrices are invertible and satisfy the equations

$$M_\rho(st) = M_\rho(s) M_\rho(t) \qquad \text{and} \qquad M_\rho(s^{-1}) = (M_\rho(s))^*.$$

Thus, the map $s \mapsto M_\rho(s)$ is a continuous unitary representation in matrix form $M_\rho\colon G \to \mathbf{U}(n_\rho)$ of G in $\mathbb{C}^{n_\rho}$, for the standard hermitian inner product $\sum_{j=1}^{n_\rho} \alpha_j \overline{\beta}_j$.

Proof. (1) Since $m_{ij} = \gamma^{-1} a_i * \check{\overline{a}}_j = \gamma^{-1} a_i * a_j^*$, it suffices to prove that for any fixed j,

$$\langle a_i * a_j^*, a_k * a_j^*\rangle = 0, \qquad \text{for all } i \neq k.$$

Since $a_j^* * a_j = \gamma m_1$ (by Proposition 4.5), $m_1^* = m_1$, and $a_i * m_1 = a_i$, we have

$$\begin{aligned}
\langle a_i * a_j^*, a_k * a_j^*\rangle &= \langle a_i * a_j^* * a_j, a_k\rangle \\
&= \langle a_i * \gamma m_1, a_k^*\rangle \\
&= \gamma\langle a_i, a_k\rangle = 0,
\end{aligned}$$

since the a_i are pairwise orthogonal.

(2) We have

$$\begin{aligned}
m_{ji}(s) &= \gamma^{-1}(a_j * \check{\bar{a}}_i)(s) \\
&= \gamma^{-1}\int a_j(t)\check{\bar{a}}_i(t^{-1}s)\,d\lambda(t) \\
&= \gamma^{-1}\overline{\int \overline{a_j(t)}a_i(s^{-1}t)\,d\lambda(t)} \\
&= \gamma^{-1}\overline{\int a_i(t)\overline{a_j(st)}\,d\lambda(t)} \\
&= \gamma^{-1}\overline{\int a_i(t)\check{\bar{a}}_j(t^{-1}s^{-1})\,d\lambda(t)} \\
&= \gamma^{-1}\overline{(a_i * \check{\bar{a}}_j)(s^{-1})} \\
&= \overline{m}_{ij}(s^{-1}) = \check{\overline{m}}_{ij}(s),
\end{aligned}$$

where Vol. I, Proposition 7.24 was used to derive the third equation. Since $a_i \in m_i * \mathfrak{a}_\rho * m_1$, we have $a_i^* * a_j \in m_1 * \mathfrak{a}_\rho * m_i * m_j * \mathfrak{a}_\rho * m_1 = 0$ whenever $i \neq j$, since the m_i are pairwise orthogonal self-adjoint irreducible idempotents, and thus $m_i * m_j = 0$ whenever $i \neq j$. Consequently

$$a_j^* * a_h = \delta_{jh}\gamma m_1,$$

and thus

$$\begin{aligned}
m_{ij} * m_{hk} &= \gamma^{-1}a_i * a_j^* * \gamma^{-1}a_h * a_k^* = \gamma^{-2}a_i * \delta_{jh}\gamma m_1 * a_k^* \\
&= \delta_{jh}\gamma^{-1}a_i * a_k^* = \delta_{jh}m_{ik}.
\end{aligned}$$

(3) Since $\mathfrak{a}_\rho$ is a Hilbert algebra, by (2) and (2') (see Definition 2.14), we have

$$\langle m_{ij}, m_{ij}\rangle = \gamma^{-2}\langle a_i * \check{\bar{a}}_j, a_i * \check{\bar{a}}_j\rangle = \gamma^{-2}\langle \check{\bar{a}}_i * a_i, \check{\bar{a}}_j * a_j\rangle = \langle m_1, m_1\rangle. \quad (*_1)$$

To compute this value, observe that for every k, by Proposition 4.4, the function $t \mapsto m_{ik}(st)$ belongs to $\mathfrak{l}_k$ for all $s \in G$. Thus we can write

$$m_{ik}(st) = \sum_{j=1}^{n_\rho} c_{ij}(s)m_{jk}(t). \quad (*_2)$$

On the other hand, using the fact that $m_{1k} = \check{\overline{m}}_{k1}$ and (2), we have

$$\begin{aligned}
m_{jk}(t) = (m_{j1} * m_{1k})(t) &= \int_G m_{j1}(tx)m_{1k}(x^{-1})\,d\lambda(x) \\
&= \int_G m_{j1}(tx)\overline{m_{k1}(x)}\,d\lambda(x),
\end{aligned}$$

which yields

$$m_{jk}(e) = \langle m_{j1}, m_{k1} \rangle, \tag{$*_3$}$$

and if we let $t = e$ in ($*_2$), using the orthogonality properties of the m_{ij} and the fact that $m_{jj} = m_j$ and $\langle m_j, m_j \rangle = \langle m_1, m_1 \rangle$, we get

$$m_{ik}(s) = \langle m_1, m_1 \rangle c_{ik}(s). \tag{$*_4$}$$

If we let $s = t^{-1}$ and $i = k = 1$ in ($*_2$) and $j = k = 1$ in $m_{jk}(e) = \langle m_{j1}, m_{k1} \rangle$, we get

$$\langle m_1, m_1 \rangle = m_1(e) = \sum_{j=1}^{n_\rho} c_{1j}(s) m_{j1}(s^{-1}) = \sum_{j=1}^{n_\rho} c_{1j}(s) \overline{m_{1j}(s)},$$

and using ($*_4$),

$$c_{1j} = \frac{1}{\langle m_1, m_1 \rangle} m_{1j},$$

we obtain

$$\sum_{j=1}^{n_\rho} m_{1j}(s) \overline{m_{1j}(s)} = \langle m_1, m_1 \rangle^2. \tag{$*_5$}$$

Since by ($*_1$) we have

$$\langle m_1, m_1 \rangle = \langle m_{1j}, m_{1j} \rangle = \int m_{1j}(s) \overline{m_{1j}(s)}\, d\lambda(s), \tag{$*_6$}$$

and since $\langle m_1, m_1 \rangle^2$ is a constant and λ is the normalized Haar measure, if we integrate both sides of ($*_5$), we obtain

$$\begin{aligned} \sum_{j=1}^{n_\rho} \int_G m_{1j}(s) \overline{m_{1j}(s)}\, d\lambda(s) &= \int_G \langle m_1, m_1 \rangle^2 \, d\lambda(s) \\ &= \langle m_1, m_1 \rangle^2 \int_G d\lambda(s) = \langle m_1, m_1 \rangle^2, \end{aligned}$$

and by ($*_6$) we have

$$\sum_{j=1}^{n_\rho} \langle m_1, m_1 \rangle = n_\rho \langle m_1, m_1 \rangle = \langle m_1, m_1 \rangle^2,$$

so $\langle m_{ij}, m_{ij} \rangle = \langle m_1, m_1 \rangle = n_\rho$, which proves (3). The equations in (4) follow immediately from (2), ($*_2$), and ($*_4$). □

As in Definition 3.2, the unitary matrix representation $M_\rho\colon G \to \mathbf{U}(n_\rho)$ defines (with a small abuse of notation) the representation $M_\rho\colon G \to \mathbf{U}(\mathbb{C}^{n_\rho})$ given by

$$(M_\rho(s))(z) = M_\rho(s)z, \quad z \in \mathbb{C}^{n_\rho},\ s \in G.$$

We usually identify these two variants. We will see later that the representations M_ρ are irreducible. The center of $\mathrm{L}^2(G)$ is characterized as follows.

Proposition 4.7. *Let G be a (metrizable) compact group. The center of the Hilbert algebra $\mathrm{L}^2(G)$ is the Hilbert sum of the one-dimensional spaces $\mathbb{C}u_\rho$ (with $\rho \in R$). In particular, if G is commutative, then every ideal $\mathfrak{a}_\rho$ is one-dimensional ($n_\rho = 1$).*

Proof. Since u_ρ is the unit element of $\mathfrak{a}_\rho$ and since $\mathfrak{a}_\rho * \mathfrak{a}_{\rho'} = (0)$ whenever $\rho \neq \rho'$, we see that u_ρ belongs to the center of $\mathrm{L}^2(G)$. If the class of a function $f \in \mathcal{L}^2(G)$ belongs to the center of $\mathrm{L}^2(G)$, since u_ρ also belongs to this center, we deduce that the class of $f * u_\rho \in \mathfrak{a}_\rho$ belongs to the center $\mathrm{L}^2(G)$, but since $\mathfrak{a}_\rho$ is topologically simple, complete, separable algebra, by Proposition 2.35, the center of $\mathfrak{a}_\rho$ is one-dimensional, so $f * u_\rho = c_\rho u_\rho$ for some $c_\rho \in \mathbb{C}$. Since by Theorem 4.2, we have

$$f = \sum_{\rho \in R} f * u_\rho$$

for every $f \in \mathcal{L}^2(G)$, we must have $f = \sum_{\rho \in R} c_\rho u_\rho$, as claimed. $\square$

Since the group G is compact, for every $f \in \mathrm{L}^2(G)$ and every constant function α, we have

$$f * \alpha = \alpha * f = \alpha \left(\int_G f(s)\, d\lambda(s) \right).$$

Therefore the (complex) constant functions form a two-sided ideal in $\mathrm{L}^2(G)$, and thus must be an ideal of the form $\mathfrak{a}_{\rho_0}$.

Definition 4.3. The ideal $\mathfrak{a}_{\rho_0}$ is called the *trivial ideal.*

The corresponding representation M_{ρ_0} is one-dimensional, and $M_{\rho_0}(s) = 1$ for all $s \in G$. In other words, M_{ρ_0} is the trivial representation of G. For all $\rho \neq \rho_0$, since the spaces $\mathfrak{a}_\rho$ and $\mathfrak{a}_{\rho_0}$ are orthogonal, we have

$$\int_G m_{ij}^{(\rho)}(s)\, d\lambda(s) = \int_G m_{ij}^{(\rho)}(s)\bar{1}\, d\lambda(s) = 0.$$

Therefore, for every $\rho \neq \rho_0$, we have

$$\int_G m_{ij}^{(\rho)}(s)\, d\lambda(s) = 0. \qquad (*_{\rho\neq\rho_0})$$

We also have the following results.

Proposition 4.8. *Let G be a (metrizable) compact group. With the notation as above, the following properties hold.*

(1) If f and g are two continuous functions in $\mathcal{C}(G;\mathbb{C})$, then we have

$$f * g = \sum_{\rho\in R}\left(\sum_{1\leq i,j\leq n_\rho} \frac{1}{n_\rho}\langle g, m_{ij}^{(\rho)}\rangle (f * m_{ij}^{(\rho)})\right),$$

where the family on the right-hand side converges for the topology of uniform convergence.

(2) The family of continuous functions $\{m_{ij}^{(\rho)} \mid \rho \in R, 1 \leq i,j \leq n_\rho\}$ is an orthogonal system that is dense in $\mathcal{C}(G;\mathbb{C})$ for the topology of uniform convergence.

Proof. (1) From Theorem 4.6, which says that the family of functions

$$\left(\frac{1}{\sqrt{n_\rho}} m_{ij}^{(\rho)}\right)_{1\leq i,j\leq n_\rho,\ \rho\in R}$$

is a Hilbert basis of $\mathrm{L}^2(G)$, we have

$$g = \sum_{\rho\in R}\sum_{1\leq i,j\leq n_\rho} \frac{1}{n_\rho}\langle g, m_{ij}^{(\rho)}\rangle\, m_{ij}^{(\rho)}.$$

Since the map $h \mapsto f * h$ is a continuous map from $\mathrm{L}^2(G)$ to $\mathcal{C}(G;\mathbb{C})$, (since by Vol. I, Proposition 8.49, $\|f * h\|_\infty \leq \|f\|_2\,\|h\|_2$), we can apply convolution to both sides, and we get the equation in (1).

(2) By Vol. I, Proposition 8.50, for every continuous function $g \in \mathcal{C}(G;\mathbb{C})$, there is some continuous function f, such that $\|f * g - g\|_\infty$ can be made arbitrarily small. But, for every $\rho \in R$, the functions $f * m_{ij}^{(\rho)}$ belong to the two-sided ideal $\mathfrak{a}_\rho$, and so they are (complex) linear combinations of the $m_{hk}^{(\rho)}$ with $1 \leq h,k \leq n_\rho$. Therefore, the formula of (1) for $f * g$ shows that $f * g$ can be expressed in terms of the $m_{hk}^{(\rho)}$, which proves (2). ☐

In Section 6.9, we will need the following result.

Proposition 4.9. *For any unitary $n_\rho \times n_\rho$ matrix P, for every $s \in G$, let $Q_\rho(s) = P^* M_\rho(s) P$. The matrices $Q_\rho(s) = (q_{ij}(s))$ define n_ρ^2 functions $q_{ij} \in \mathfrak{a}_\rho$ which are linear combinations of the functions m_{ij}, where $m_{ij}(s) \in M_\rho(s)$, and satisfy the following properties:*

(1) The $(q_{ij})_{1\leq i,j\leq n_\rho}$ form an orthogonal basis of $\mathfrak{a}_\rho$.
(2) We have $q_{ji} = \check{\overline{q}}_{ij}$ and $q_{ij} * q_{hk} = \delta_{jh}\, q_{ik}$.
(3) We have $\langle q_{ij}, q_{ij}\rangle = n_\rho$ and $q_{ij}(e) = n_\rho\, \delta_{ij}$, for all i, j with $1 \leq i, j \leq n_\rho$.
(4) The map $s \mapsto Q_\rho(s)$ is unitary representation in matrix form $Q_\rho \colon G \to \mathbf{U}(n_\rho)$ of G in $\mathbb{C}^{n_\rho}$, equivalent to the unitary representation $M_\rho \colon G \to \mathbf{U}(n_\rho)$.
(5) If $\mathfrak{l}_j$ is the minimal left ideal of $\mathfrak{a}_\rho$ spanned by the jth column M_ρ^j of M_ρ,

$$\mathfrak{l}_j = \bigoplus_{i=1}^{n_\rho} \mathbb{C} m_{ij}^{(\rho)},$$

then the jth column of $Q_\rho = P^* M_\rho P$ spans a minimal ideal $\mathfrak{l}_j^Q$ of $\mathfrak{a}_\rho$ (of dimension n_ρ) given by

$$\mathfrak{l}_j^Q = \left\{ \sum_{h=1}^{n_\rho} p_{hj} \left(\sum_{k=1}^{n_\rho} \mu_k m_{kh} \right) \;\middle|\; \mu = (\mu_1, \ldots, \mu_{n_\rho}) \in \mathbb{C}^{n_\rho} \right\},$$

where every $\sum_{k=1}^{n_\rho} \mu_k m_{kh} \in \mathfrak{l}_h$ is a linear combination of the entries of the hth column of M_ρ involving the same scalars $(\mu_1, \ldots, \mu_{n_\rho})$ for all $h = 1, \ldots, n_\rho$.

Proof. The (i, h) entry of the matrix $P^* M_\rho(s)$ is

$$\sum_{k=1}^{n_\rho} \overline{p_{ki}}\, m_{kh}(s),$$

and $q_{ij}(s)$ (the (i, j) entry in $Q_\rho(s) = P^* M_\rho(s) P$) is given by

$$q_{ij}(s) = \sum_{h,k=1}^{n_\rho} \overline{p_{ki}}\, p_{hj}\, m_{kh}(s). \tag{q_{ij}}$$

The $q_{ij}(s)$ are indeed linear combinations of the $m_{ij}(s)$. Let us compute the inner product

$$\langle q_{i_1 j_1}, q_{i_2 j_2}\rangle = \int_G q_{i_1 j_1}(s) \overline{q_{i_2 j_2}(s)}\, d\lambda(s).$$

We have

$$q_{i_1j_1}\overline{q_{i_2j_2}} = \sum_{h,k=1}^{n_\rho}\sum_{h',k'=1}^{n_\rho} \overline{p_{ki_1}}\, p_{hj_1} m_{kh}(s) p_{k'i_2}\overline{p_{h'j_2}}\,\overline{m_{k'h'}(s)},$$

and so

$$\begin{aligned}\langle q_{i_1j_1}, q_{i_2j_2}\rangle &= \int_G q_{i_1j_1}(s)\overline{q_{i_2j_2}(s)}\, d\lambda(s)\\ &= \sum_{h,k=1}^{n_\rho}\sum_{h',k'=1}^{n_\rho} \overline{p_{ki_1}}\, p_{hj_1} p_{k'i_2}\overline{p_{h'j_2}} \int_G m_{kh}(s)\overline{m_{k'h'}(s)}\, d\lambda(s)\\ &= \sum_{h,k=1}^{n_\rho}\sum_{h',k'=1}^{n_\rho} p_{k'i_2}\overline{p_{ki_1}}\, p_{hj_1}\overline{p_{h'j_2}}\, \langle m_{kh}, m_{k'h'}\rangle.\end{aligned}$$

Since the m_{ij} form an orthogonal family and $\langle m_{kh}, m_{kh}\rangle = n_\rho$, we obtain

$$\langle q_{i_1j_1}, q_{i_2j_2}\rangle = \sum_{h,k=1}^{n_\rho} p_{ki_2}\overline{p_{ki_1}}\, p_{hj_1}\overline{p_{hj_2}}\, n_\rho = n_\rho \sum_{k=1}^{n_\rho} p_{ki_2}\overline{p_{ki_1}} \sum_{h=1}^{n_\rho} p_{hj_1}\overline{p_{hj_2}}.$$

If $i_1 \neq i_2$, since P is a unitary matrix the columns of index i_1 and i_2 are orthogonal and so $\sum_{k=1}^{n_\rho} p_{ki_2}\overline{p_{ki_1}} = 0$, and similarly, if $j_1 \neq j_2$, the columns of index j_1 and j_2 are orthogonal and $\sum_{h=1}^{n_\rho} p_{hj_1}\overline{p_{hj_2}} = 0$. Thus $q_{i_1j_1}$ and $q_{i_2j_2}$ are orthogonal if $(i_1, j_1) \neq (i_2, j_2)$.

If $i_1 = i_2$ and $j_1 = j_2$, since P is unitary, its columns are unit vectors, so $\sum_{k=1}^{n_\rho} p_{ki_1}\overline{p_{ki_1}} = 1$ and $\sum_{h=1}^{n_\rho} p_{hj_1}\overline{p_{hj_1}} = 1$, and thus

$$\langle q_{i_1j_1}, q_{i_1j_1}\rangle = n_\rho.$$

This concludes the proof of (1) and part of (3).

For $s = e$, since $m_{kh}(e) = n_\rho\, \delta_{kh}$, we have

$$q_{ij}(e) = \sum_{h,k=1}^{n_\rho} \overline{p_{ki}}\, p_{hj}\, m_{kh}(e) = n_\rho \sum_{h,k=1}^{n_\rho} p_{kj}\overline{p_{ki}}\, \delta_{kh} = n_\rho \sum_{k=1}^{n_\rho} p_{kj}\overline{p_{ki}}.$$

Since P is a unitary matrix, if $i \neq j$, then $\sum_{k=1}^{n_\rho} p_{kj}\overline{p_{ki}} = 0$, and if $i = j$, then $\sum_{k=1}^{n_\rho} p_{ki}\overline{p_{ki}} = 1$, so we have

$$q_{ij}(e) = n_\rho \delta_{ij}.$$

This finishes the proof of (3).

Since $m_{kh}(s) = \overline{m_{hk}(s^{-1})}$, using (q_{ij}) twice, we have

$$\begin{aligned} q_{ji}(s) &= \sum_{h,k=1}^{n_\rho} \overline{p_{kj}}\, p_{hi}\, m_{kh}(s) \\ &= \sum_{h,k=1}^{n_\rho} \overline{\overline{p_{hi}}\, p_{kj}}\; \overline{m_{hk}(s^{-1})} \\ &= \overline{q_{ij}(s^{-1})}. \end{aligned}$$

Since $m_{kh} * m_{k'h'} = \delta_{hk'} m_{kh'}$, we have

$$\begin{aligned} q_{i_1j_1} * q_{i_2j_2} &= \sum_{h,k=1}^{n_\rho} \sum_{h',k'=1}^{n_\rho} \overline{p_{ki_1}}\, p_{hj_1} \overline{p_{k'i_2}}\, p_{h'j_2}\, m_{kh} * m_{k'h'} \\ &= \sum_{h,k=1}^{n_\rho} \sum_{h',k'=1}^{n_\rho} \overline{p_{ki_1}}\, p_{hj_1} \overline{p_{k'i_2}}\, p_{h'j_2}\, \delta_{hk'} m_{kh'} \\ &= \sum_{h=1}^{n_\rho} p_{hj_1} \overline{p_{hi_2}} \sum_{k,h'=1}^{n_\rho} \overline{p_{ki_1}}\, p_{h'j_2} m_{kh'}. \end{aligned}$$

If $j_1 \neq i_2$, since P is unitary we have $\sum_{h=1}^{n_\rho} p_{hj_1}\overline{p_{hi_2}} = 0$ and then $q_{i_1j_1} * q_{i_2j_2} = 0$. If $j_1 = i_2$, since P is unitary we have $\sum_{h=1}^{n_\rho} p_{hj_1}\overline{p_{hj_1}} = 1$, in which case, using (q_{ij}),

$$q_{i_1j_1} * q_{j_1j_2} = \sum_{k,h'=1}^{n_\rho} \overline{p_{ki_1}}\, p_{h'j_2}\, m_{kh'} = q_{i_1j_2},$$

which concludes the proof of (2).

Since $Q_\rho(s) = P^* M_\rho(s) P$, part (4) is trivial.

Since the (i,j) entry q_{ij} of $Q_\rho = P^* M_\rho P$ is given by

$$q_{ij}(s) = \sum_{h,k=1}^{n_\rho} \overline{p_{ki}}\, p_{hj}\, m_{kh}(s),$$

any linear combination of the entries of the jth column of Q_ρ is of the form

$$\begin{aligned} \sum_{i=1}^{n_\rho} \lambda_i \sum_{h,k=1}^{n_\rho} \overline{p_{ki}}\, p_{hj}\, m_{kh}(s) &= \sum_{h=1}^{n_\rho} p_{hj} \left(\sum_{k=1}^{n_\rho} \sum_{i=1}^{n_\rho} \overline{p_{ki}}\, \lambda_i m_{kh} \right) \\ &= \sum_{h=1}^{n_\rho} p_{hj} \left(\sum_{k=1}^{n_\rho} (P^*\lambda)_k m_{kh} \right), \end{aligned}$$

with $\lambda = (\lambda_1, \dots, \lambda_{n_\rho}) \in \mathbb{C}^{n_\rho}$ and where $(P^*\lambda)_k$ is the kth component of the vector $P^*\lambda$, and since P^* is invertible, we deduce that the subspace $\mathfrak{l}_j^Q$ of all linear combinations of entries in the jth column of Q_ρ is given by

$$\mathfrak{l}_j^Q = \left\{ \sum_{h=1}^{n_\rho} p_{hj} \left(\sum_{k=1}^{n_\rho} \mu_k m_{kh} \right) \;\middle|\; \mu = (\mu_1, \dots, \mu_{n_\rho}) \in \mathbb{C}^{n_\rho} \right\}.$$

To check that $\mathfrak{l}_j^Q$ is a left ideal, we need to check that $m_{ij'} * \mathfrak{l}_j^Q \subseteq \mathfrak{l}_j^Q$ for all $m_{ij'} \in \mathfrak{a}_\rho$. Since $m_{ij'} * m_{kh} = \delta_{j'k} m_{ih}$, we have

$$\begin{aligned} m_{ij'} * \mathfrak{l}_j^Q &= \left\{ \sum_{h=1}^{n_\rho} p_{hj} \left(\sum_{k=1}^{n_\rho} \mu_k m_{ij'} * m_{kh} \right) \;\middle|\; \mu = (\mu_1, \dots, \mu_{n_\rho}) \in \mathbb{C}^{n_\rho} \right\} \\ &= \left\{ \sum_{h=1}^{n_\rho} p_{hj} \left(\sum_{k=1}^{n_\rho} \mu_k \delta_{j'k} m_{ih} \right) \;\middle|\; \mu = (\mu_1, \dots, \mu_{n_\rho}) \in \mathbb{C}^{n_\rho} \right\} \\ &= \left\{ \sum_{h=1}^{n_\rho} p_{hj} (\mu_{j'} m_{ih}) \;\middle|\; \mu_{j'} \in \mathbb{C} \right\} \subseteq \mathfrak{l}_j^Q. \end{aligned}$$

Thus $\mathfrak{l}_j^Q$ is indeed a left ideal. This left ideal has dimension n_ρ since $(q_{1j}, \dots, q_{n_\rho j})$ is a basis for it, so it is a minimal ideal because all minimal ideals of $\mathfrak{a}_\rho$ are isomorphic to $\mathfrak{l}_1$, which has dimension n_ρ. ☐

In summary, the functions q_{ij} in the matrix $Q_\rho = P^* M_\rho P$ provide another isomorphism of the minimal two-sided ideal $\mathfrak{a}_\rho$ of $\mathrm{L}^2(G)$ with the matrix algebra $\mathrm{M}_{n_\rho}(\mathbb{C})$ and the family of functions

$$\left(\frac{1}{\sqrt{n_\rho}} q_{ij} \right)_{1 \le i,j \le n_\rho}$$

is an orthonormal basis of $\mathfrak{a}_\rho$.

4.2 Characters of Compact Groups

Besides characters of groups and characters of algebras, there is one more kind of characters, namely, characters of finite-dimensional representations. As in the previous section, we work with metrizable compact groups. Since we have the Peter–Weyl theorem and Theorem 4.6 at our disposal, it will be fairly easy to prove the properties of characters of these groups.

Definition 4.4. Let G be a (metrizable) compact group. With the notations of Section 4.2, for every $\rho \in R$, define the *character* χ_ρ *of* G *associated* *with the ideal* $\mathfrak{a}_\rho$ as the function given by

$$\chi_\rho(s) = \frac{1}{n_\rho}\, u_\rho(s) = \frac{1}{n_\rho}\sum_{j=1}^{n_\rho} m_{jj}^{(\rho)}(s) = \mathrm{tr}(M_\rho(s)), \qquad \text{for all } s \in G.$$

The character χ_{ρ_0} associated with $\mathfrak{a}_{\rho_0}$ is the constant function $\chi_{\rho_0}(s) = 1$ for all $s \in G$, called the *trivial character* of G.

The properties stated in the following proposition are immediate consequences of Theorem 4.2 and Theorem 4.6.

Proposition 4.10. *The following properties hold.*

(1) Every character χ_ρ *is a continuous central function, which means that*

$$\chi_\rho(sts^{-1}) = \chi_\rho(t), \qquad \textit{for all } s, t \in G.$$

(2) We have

$$\chi_\rho(s^{-1}) = \overline{\chi_\rho(s)}, \qquad \textit{for all } s \in G.$$

(3) We have

$$\chi_\rho * \chi_{\rho'} = 0 \quad \textit{whenever } \rho \neq \rho', \quad \textit{and} \quad \chi_\rho * \chi_\rho = \frac{1}{n_\rho}\,\chi_\rho.$$

(4) The family of characters $(\chi_\rho)_{\rho \in R}$ *forms a Hilbert basis of the center of* $\mathrm{L}^2(G)$*, which means that:*

(a) We have

$$\langle \chi_\rho, \chi_{\rho'} \rangle = \int \chi_\rho(s)\overline{\chi_{\rho'}(s)}\, d\lambda(s) = 0, \qquad \textit{whenever } \rho \neq \rho'$$

$$\langle \chi_\rho, \chi_\rho \rangle = \int |\chi_\rho(s)|^2\, d\lambda(s) = 1.$$

(b) For every function $f \in \mathrm{L}^2(G)$*, we have*

$$f = \sum_{\rho \in R} n_\rho\, (f * \chi_\rho),$$

and for every central function $f \in \mathrm{L}^2(G)$*, we have*

$$f = \sum_{\rho \in R} \langle f, \chi_\rho \rangle \chi_\rho.$$

(c) We have

$$\int \chi_\rho(s)\, d\lambda(s) = 0, \qquad \textit{for all } \rho \neq \rho_0.$$

(d) *For all* $s \in G$, *we have*

$$\chi_\rho(s) = \mathrm{tr}(M_\rho(s)),$$

and

$$\chi_\rho(e) = n_\rho.$$

The only nontrivial proof is the proof of property (b). By Theorem 4.2 and the fact that $u_\rho(s) = n_\rho \chi_\rho(s)$, we have

$$f = \sum_{\rho \in R} f * u_\rho = \sum_{\rho \in R} n_\rho (f * \chi_\rho).$$

By Proposition 4.7, the center of the Hilbert algebra $\mathrm{L}^2(G)$ is the Hilbert sum of the one-dimensional spaces $\mathbb{C}u_\rho$ (with $\rho \in R$). Since $u_\rho = n_\rho \chi_\rho$, by (1) and (3), the family of characters $(\chi_\rho)_{\rho \in R}$ forms a Hilbert basis of the center of $\mathrm{L}^2(G)$. It follows that for every central function $f \in \mathrm{L}^2(G)$, we have

$$f = \sum_{\rho \in R} \langle f, \chi_\rho \rangle \chi_\rho,$$

Observe that unlike the characters of a locally compact *abelian* group G, which take their values in $\mathbf{U}(1) \cong \mathbf{T}$, the characters χ_ρ of a compact *not necessarily abelian* group G take their values in $\mathbb{C}$. For instance $\chi_\rho(e) = n_\rho$, and in general, $n_\rho > 1$. Also, the characters χ_ρ are *not* homomorphisms. In general, $\chi_\rho(st) \neq \chi_\rho(s)\chi_\rho(t)$.

The next proposition is needed to prove Theorem 4.12.

Proposition 4.11. *For any two continuous central functions* f, g *in* $\mathcal{C}(G; \mathbb{C})$, *we have*

$$f * g = \sum_{\rho \in R} \langle g, \chi_\rho \rangle (f * \chi_\rho),$$

where the family on the right-hand side converges in the topology of uniform convergence.

Proof. This follows from the fact shown in Proposition 4.10(4) that the family of characters $(\chi_\rho)_{\rho \in R}$ is a Hilbert basis of the center of $\mathrm{L}^2(G)$, and the fact that $\|f * g\|_\infty \leq \|f\|_2 \, \|g\|_2$. □

The next theorem will require an auxiliary proposition.

Theorem 4.12. *The family of continuous central functions* $(\chi_\rho)_{\rho \in R}$ *constitutes an orthonormal system, which is dense in the space of continuous central functions in* $\mathcal{C}(G; \mathbb{C})$ *for the topology of uniform convergence.*

Proof. For every continuous central function f, since $\chi_\rho = \frac{1}{n_\rho} u_\rho$, by Proposition 4.7, the function $f * \chi_\rho$ is a scalar multiple of χ_ρ. In view of the formula of Proposition 4.11, it suffices to show that for every continuous central function g, there exists a continuous central function f, such that $\|f * g - g\|_\infty$ is arbitrarily small. Recall that for any element $t \in G$ the *inner automorphism* C_t is defined by $C_t(s) = tst^{-1}$, for all $s \in G$. The following result is needed.

Proposition 4.13. *The following properties hold.*

(1) Let G be a (metrizable) topological group, and let K be a compact subset of G. For every neighborhood U of the identity element e of G, there is a neighborhood $V \subseteq U$ of e, such that $tVt^{-1} \subseteq U$ for all $t \in K$.

(2) In a (metrizable) compact group G, there exists a neighborhood base of neighborhoods of e invariant under all inner automorphisms of G. For such a neighborhood T, there exists a continuous central function $h \geq 0$, of support contained in T, such that $\int_G h(s)\, d\lambda(s) = 1$.

Proof. (1) Using the technique used to prove Vol. I, Proposition 8.2 applied to the continuous map $(g_1, g_2, g_3) \mapsto g_1 g_2 g_3$, we can find a neighborhood U_0 of e, such that $U_0^3 \subseteq U$. Using this technique again, for every $s \in G$, by continuity of the map $g \mapsto sgs^{-1}$, there is a neighborhood V_s of e in G, such that $V_s^{-1} = V_s$ and $sV_s s^{-1} \subseteq U_0$. Note that

$$sV_s^3 s^{-1} = sV_s s^{-1} sV_s s^{-1} sV_s s^{-1} \subseteq U_0^3 \subseteq U.$$

Since

$$tV_s t^{-1} = ss^{-1}tV_s t^{-1} ss^{-1},$$

for any $t \in G$, if $s^{-1}t \in V_s$, which implies that $t^{-1}s \in V_s^{-1} = V_s$, then $s^{-1}tV_s t^{-1}s \in V_s^3$, so

$$tV_s t^{-1} = ss^{-1}tV_s t^{-1} ss^{-1} \in sV_s^3 s^{-1} \subseteq U.$$

Therefore if we let $W_s = sV_s$, then for all $t \in W_s = sV_s$, we have $s^{-1}t \in V_s$ and so $tV_s t^{-1} \subseteq U$. Since K is compact, there exists a finite number of m elements $s_j \in K$, such that $W_{s_1} \cup \cdots \cup W_{s_m}$ covers K. If we let $V = \bigcap_{j=1}^m V_{s_j}$, we have $tVt^{-1} \subseteq U$ for all $t \in K$.

(2) Apply (1) with $K = G$. Then

$$T = \bigcup_{t \in G} tVt^{-1}$$

is a neighborhood of e contained in U, obviously invariant under the inner automorphisms of G. To define h, start with a continuous function $f \geq 0$ of support contained in T, such that $f(e) > 0$. Let

$$h(t) = c \int_G f(sts^{-1})\, d\lambda(s),$$

where the constant $c > 0$ is chosen in a suitable way, and then the proof that h works is the same as in the proof of Theorem 4.2. This finishes the proof of Proposition 4.13. □

In view of Proposition 4.13(2), Vol. I, Proposition 8.50(i) shows that for every continuous central function g, there is some continuous central function h, such that $\|g * h - g\|_\infty$ can be made arbitrarily small, which finishes the proof of Theorem 4.12. □

The following result shows certain independence results.

Theorem 4.14. *The following properties hold.*

(1) For every $s \in G$, if $s \neq e$, then there is some $\rho \in R$, such that $\chi_\rho(s) \neq \chi_\rho(e)$.
(2) We have

$$\bigcap_{\rho \in R} N_\rho = \{e\},$$

where N_ρ is the kernel of the (group) homomorphism $s \mapsto M_\rho(s)$.

Proof. (1) If there is some $s \neq e$, such that $\chi_\rho(s) = \chi_\rho(e)$ for all $\rho \in R$, then by Theorem 4.12, we would have $f(s) = f(e)$ for all continuous central functions, but this contradicts Proposition 4.13(2), since we can find a continuous central function h with $h(e) \neq 0$, whose support T does not contain $s \neq 0$.

(2) If $s \in N_\rho$, then $M_\rho(s) = I_{n_\rho}$, and since $\chi_\rho(s) = \operatorname{tr}(M_\rho(s)) = \operatorname{tr}(I_{n_\rho}) = n_\rho$, by Proposition 4.10(4)(d), we have $\chi_\rho(s) = \chi_\rho(e) = n_\rho$ and (1) implies that

$$\bigcap_{\rho \in R} N_\rho = \{e\},$$

as claimed. □

The following is a product formula for the characters.

Proposition 4.15. *For every character χ of G, we have*

$$\chi(s)\chi(t) = \chi(e)\int_G \chi(usu^{-1}t)\,d\lambda(u).$$

Proof. By definition and since $M_\rho(st) = M_\rho(s)M_\rho(t)$ for all $s,t \in G$, we have

$$\begin{aligned}\chi_\rho(usu^{-1}t) &= \frac{1}{n_\rho}\sum_i m_{ii}^{(\rho)}(usu^{-1}t)\\ &= \frac{1}{n_\rho^4}\sum_{i,j,h,k} m_{ij}(u)m_{jh}(s)m_{hk}(u^{-1})m_{ki}(t).\end{aligned}$$

Using Theorem 4.6(2) and (3) (convolution evaluated at e), it follows that

$$\begin{aligned}\int_G \chi_\rho(usu^{-1}t)\,d\lambda(u) &= \frac{1}{n_\rho^4}\sum_{i,j,h,k} m_{jh}(s)m_{ki}(t)\int m_{ij}(u)m_{hk}(u^{-1})\,d\lambda(u)\\ &= \frac{1}{n_\rho^3}\sum_{i,j,h,k}\delta_{jh}\delta_{ik}m_{jh}(s)m_{ki}(t)\\ &= \frac{1}{n_\rho^3}\sum_{i,j} m_{jj}(s)m_{ii}(t)\\ &= \frac{1}{n_\rho}\chi_\rho(s)\chi_\rho(t),\end{aligned}$$

as claimed. □

Since by Vol. I, Proposition 8.47, $\overline{f*g} = \overline{f}*\overline{g}$, the function which maps the class of a function $f \in \mathcal{L}^2(G)$ to the class of its complex conjugate $\overline{f}$ is a semilinear bijection of $\mathrm{L}^2(G)$, and an automorphism of its ring structure (under convolution).

Definition 4.5. The above automorphism of $\mathrm{L}^2(G)$ maps every ideal $\mathfrak{a}_\rho$ into the minimal two-sided ideal $\overline{\mathfrak{a}_\rho} = \{\overline{f} \mid f \in \mathfrak{a}_\rho\}$ that we denote by $\mathfrak{a}_{\overline{\rho}}$.

The map $\mathfrak{a}_\rho \mapsto \mathfrak{a}_{\overline{\rho}}$ permutes the indices of R but leaves the Hilbert sum unchanged, namely $\mathrm{L}^2(G)$ is the Hilbert sum of both families $(\mathfrak{a}_\rho)_{\rho\in R}$ and $(\mathfrak{a}_{\overline{\rho}})_{\rho\in R} = (\overline{\mathfrak{a}_\rho})_{\rho\in R}$.

If (as usual), given a complex matrix $X = (x_{ij})$, we denote by $\overline{X}$ the matrix $(\overline{x}_{ij})$, then we have

$$\overline{M_\rho(s)} = M_{\overline{\rho}}(s), \qquad \text{for all } s \in G,$$

and as a consequence, since $u_\rho(s) = n_\rho\mathrm{tr}(M_\rho(s)) = n_\rho\chi_\rho(s)$, we have

$$\overline{u_\rho} = u_{\overline{\rho}} \quad\text{and}\quad \overline{\chi_\rho} = \chi_{\overline{\rho}}.$$

Thus, the equation $\mathfrak{a}_\rho = \mathfrak{a}_{\overline{\rho}}$ is equivalent to saying that the character χ_ρ only takes *real values*.

Let us now consider the special cases where either G is compact and abelian or G is finite.

Example 4.1. We will consider the case where G is a compact (metrizable) *abelian* group, but before doing this, let G be a not necessarily abelian compact group. Let $f \in \mathcal{L}^2(G)$ be a function not zero everywhere, such that for every $s \in G$,

$$f(st) = f(s)f(t), \qquad \text{for almost all } t \in G.$$

Then

$$f(s^{-1}t) = f(s^{-1})f(t),$$

so the subgroup $\mathbb{C}f$ of $\mathrm{L}^2(G)$ is invariant under the map $g \mapsto \lambda_s(g)$ for every $s \in G$, and by Proposition 4.4, this subgroup is a closed minimal left ideal of dimension 1. This is only possible if this left ideal is one of the $\mathfrak{a}_\rho$ for which $n_\rho = 1$, and then f is equal to the character χ_ρ almost everywhere. Such characters are called *abelian characters*. The above reasoning shows that these are the only continuous homomorphisms of G into $\mathbb{C}^*$. Since the image of G by such a character χ_ρ is a compact subgroup of $\mathbb{C}^*$, it must be contained in $\mathbf{U}(1)$.

If G is compact *and abelian*, then every character is obviously abelian, since the algebras $\mathfrak{a}_\rho$ are commutative. Then by Theorem 4.6(3), the characters of G form a Hilbert basis of $\mathrm{L}^2(G)$, and every continuous function is a uniform limit of linear combinations of characters (by Theorem 4.12). We have determined the characters of several compact abelian groups, such as $\mathbb{Z}/n\mathbb{Z}$ and $\mathbb{T}^n$, in Vol. I, Proposition 10.9 and Corollary 10.11.

Example 4.2. Let G be a *finite* (not necessarily abelian) group of order $|G| = g$. In this case, the algebras $\mathrm{L}^1(G)$ and $\mathrm{L}^2(G)$ are the same and equal to the group algebra $\mathbb{C}[G] = [G \to \mathbb{C}]$ of formal linear combinations $\sum_{s \in G} x_s\, s$ with $x_s \in \mathbb{C}$, with the convolution

$$\begin{aligned}\frac{1}{g}\left(\sum_{s_1 \in G} x_{s_1}\, s_1\right) * \left(\sum_{s_2 \in G} y_{s_2}\, s_2\right) &= \frac{1}{g}\sum_{s \in G}\left(\sum_{s_1 s_2 = s} x_{s_1} y_{s_2}\right) s \\ &= \frac{1}{g}\sum_{s \in G}\left(\sum_{t \in G} x_t y_{t^{-1}s}\right) s.\end{aligned}$$

Recall that two elements $a, b \in G$ are *conjugate* if $b = sas^{-1}$ for some $s \in G$. Conjugation is an equivalence relation in G, and its classes, called the *conjugacy classes* of G, are the sets

$$C_a = \{sas^{-1} \mid s \in G\}.$$

Since G is finite, it has finite number r of conjugacy classes $C_1, \ldots, C_r$, and we assume that $C_1 = \{e\}$.

The central functions (also called *class functions*) are constant on the conjugacy classes. Also, since G is finite, by Theorem 4.12, every central function is a linear combination of characters, and since they are linearly independent, the number r of conjugacy classes is equal to the number of characters, and to the dimension of the center of the algebra $Z(\mathbb{C}[G])$ of $\mathbb{C}[G]$.

Let $R = \{\rho_1, \ldots, \rho_r\}$, and write χ_{ij} for the value of the character χ_i on the conjugacy class C_j $(1 \leq i, j \leq r)$. If g is the order of the group G and h_j is the number of elements in the conjugacy class C_j, then the orthogonality relations in Proposition 4.10(4) become

$$\frac{1}{g}\sum_{k=1}^{r} h_k \chi_{ik}\overline{\chi_{jk}} = \delta_{ij}, \qquad (1 \leq i, j \leq r).$$

In other words, the matrix

$$\left(\frac{h_k}{g}\chi_{ik}\right)_{1\leq i,k\leq r}$$

is unitary. We get more identities by expressing that the transpose of the above matrix is unitary, namely:

$$\sum_{i=1}^{r} \chi_{ik}\overline{\chi_{il}} = 0, \qquad \text{if } k \neq l$$

and

$$\sum_{i=1}^{r} |\chi_{ik}|^2 = \frac{g}{h_k}.$$

These formulae can also be written as

$$\sum_{\rho\in R} \chi_\rho(s)\chi_\rho(t^{-1}) = 0 \tag{$\dagger_1$}$$

if s and t are not conjugate in G, and

$$\sum_{\rho\in R} |\chi_\rho(s)|^2 = \frac{g}{h_k}, \qquad \text{if } s \in C_k. \tag{$\dagger_2$}$$

Since e is not conjugate to any other element in G, if we let $t = e$ in $(\dagger_1)$, using the fact that $\chi_\rho(e) = n_\rho$, we obtain

$$\sum_{\rho\in R} n_\rho \chi_\rho(s) = 0, \qquad \text{if } s \neq e.$$

If we let $s = e$ in $(\dagger_2)$, we obtain the relation

$$\sum_{\rho\in R} n_\rho^2 = g. \tag{$\dagger_3$}$$

The above equation confirms that $\mathrm{L}^2(G)$ is the direct sum of the $\mathfrak{a}_\rho$.

4.3 The Peter–Weyl Theorem, Part II

In this section, we prove the second part of the Peter–Weyl theorem which has to do with unitary representations. In particular, we prove the important result that the unitary representations $s \mapsto M_\rho(s)$ of G discussed in Theorem 4.6 are irreducible (in fact, all of them, up to equivalence).

Given a unitary representation $V\colon G \to \mathbf{U}(H)$, recall from Definition 3.13 specialized to measures of the form $f\,d\lambda$, where $f \in \mathrm{L}^1(G)$, that for every $x \in H$, the unique vector $\widetilde{V}(f\,d\lambda)(x) \in H$, such that

$$\langle \widetilde{V}(f\,d\lambda)(x), y\rangle = \int f(s)\langle V(s)(x), y\rangle\, d\lambda(s), \qquad \text{for all } y \in H,$$

is called the *weak integral* of the function $s \mapsto V(s)(x)$ from G to H with respect to $fd\lambda$, and we write

$$\widetilde{V}(f\,d\lambda)(x) = \int_G f(s)V(s)(x)\, d\lambda(s).$$

We also denote $\widetilde{V}(f\,d\lambda)$ by $V_{\mathrm{ext}}(f)$. We know from by Proposition 3.15 that $\|V_{\mathrm{ext}}(f)\| \leq \|f\|_1$. Recall from Definition 4.4 that $u_\rho(s) = n_\rho\chi_\rho(s)$.

Theorem 4.16. *(Peter–Weyl theorem, II) Let G be a (metrizable) compact group, and let $V\colon G \to \mathbf{U}(H)$ be a unitary representation of G in a separable Hilbert space H.*

(1) For every $\rho \in R$, the map $V_{\mathrm{ext}}(\overline{u_\rho})$ given by

$$\begin{aligned} V_{\mathrm{ext}}(\overline{u_\rho})(x) &= \int_G \overline{u_\rho}(s)V(s)(x)\, d\lambda(s) \\ &= n_\rho \int_G \overline{\chi_\rho}(s)V(s)(x)\, d\lambda(s), \;\; x \in H, \end{aligned} \tag{proj}$$

is an orthogonal projection of H onto a closed subspace E_ρ (which may be reduced to (0)), and H is the Hilbert sum

$$H = \bigoplus_{\rho \in R,\, E_\rho \neq (0)} E_\rho$$

of the $E_\rho \neq (0)$.

(2) *Every subspace $E_\rho \neq (0)$ is invariant under V, and the restriction V_ρ of V to E_ρ is a finite or countably infinite Hilbert sum of irreducible representations, all equivalent to M_ρ, viewed as a representation $M_\rho\colon G \to \mathbf{U}(\mathbb{C}^{n_\rho})$. Thus E_ρ is a finite or countably infinite Hilbert sum of d_ρ finite-dimensional subspaces $E_\rho^{k_\rho}$ (where $d_\rho = \infty$ is possible),*

$$E_\rho = \bigoplus_{k_\rho=1}^{d_\rho} E_\rho^{k_\rho},$$

and each $E_\rho^{k_\rho}$ is isomorphic to $\mathbb{C}^{n_\rho}$. More precisely, each subrepresentation $V_\rho^{k_\rho}: G \to \mathbf{U}(E_\rho^{k_\rho})$ is equivalent to the irreducible representation $M_\rho\colon G \to \mathbf{U}(\mathbb{C}^{n_\rho})$.

Proof. To help the reader navigate through the flow of this proof, we provide the following proof outline. By Theorem 4.2, we have the Hilbert sum

$$\mathrm{L}^2(G) = \bigoplus_{\rho \in R} \mathfrak{a}_\rho = \bigoplus_{\rho \in R} \mathfrak{a}_{\overline{\rho}},$$

with

$$\mathfrak{a}_\rho = \mathfrak{l}_1 \oplus \cdots \oplus \mathfrak{l}_{n_\rho}.$$

We are given a unitary representation $V\colon G \to \mathbf{U}(H)$ of G, and we use Theorem 3.17 to form the algebra representation $V_{\text{ext}}\colon \mathrm{L}^1(G) \to \mathcal{L}(H)$ of $\mathrm{L}^1(G)$. For part 1, we define

$$E_\rho = \{V_{\text{ext}}(\overline{u_\rho})(x) \mid x \in H\},$$

and we show that H is the Hilbert sum

$$H = \bigoplus_{\rho \in R} E_\rho.$$

In part (2), step 1, we prove that E_ρ is invariant under V. For step 2 we consider the restriction $V_\rho\colon G \to \mathbf{U}(E_\rho)$ of V to E_ρ and its extension $(V_\rho)_{\text{ext}}\colon \mathrm{L}^1(G) \to \mathcal{L}(E_\rho)$ to $\mathrm{L}^1(G)$. Since G is compact, $L^2(G) \subseteq L^1(G)$, and we can show that $(V_\rho)_{\text{ext}}$ is zero on every $\mathfrak{a}_{\overline{\rho'}}$ with $\rho' \neq \rho$. Consequently, the restriction of $(V_\rho)_{\mathrm{ext}}$ to $\mathrm{L}^2(G)$ can be viewed as a nondegenerate representation

$$(V_\rho)_{\mathrm{ext}} \colon \mathfrak{a}_{\overline{\rho}} \to \mathcal{L}(E_\rho)$$

of the topologically simple algebra $\mathfrak{a}_{\overline{\rho}}$ in E_ρ. This allows us to use Theorem 2.36(2) to obtain a finite or countably infinite Hilbert sum

$$E_\rho = \bigoplus_k E_\rho^{(k)},$$

such that every representation $(V_\rho)_{\mathrm{ext}}^{(k)} \colon \mathfrak{a}_{\overline{\rho}} \to \mathcal{L}(E_\rho^{(k)})$ is equivalent to the irreducible representation $U_{\overline{\mathfrak{l}_1}} \colon \mathfrak{a}_{\overline{\rho}} \to \mathcal{L}(\overline{\mathfrak{l}_1})$.

In step 3, we observe that $U_{\overline{\mathfrak{l}_1}}$ is the restriction of $\mathbf{R}_{\mathrm{ext}} \colon \mathrm{L}^1(G) \to \mathcal{L}(\mathrm{L}^2(G))$ to $\mathfrak{a}_{\overline{\rho}}$, where $\mathbf{R} \colon G \to \mathbf{U}(\mathrm{L}^2(G))$ is the left regular representation of G in $\mathrm{L}^2(G)$. Thus we can view $U_{\overline{\mathfrak{l}_1}}$ as the nondegenerate topologically irreducible representation $\widetilde{U_{\overline{\mathfrak{l}_1}}} \colon \mathrm{L}^1(G) \to \mathcal{L}(\overline{\mathfrak{l}_1})$ obtained by extending the nondegenerate representation $U_{\overline{\mathfrak{l}_1}} \colon \mathfrak{a}_{\overline{\rho}} \to \mathcal{L}(\overline{\mathfrak{l}_1})$ to $\mathrm{L}^1(G)$ (we set $\widetilde{U_{\overline{\mathfrak{l}_1}}}$ to zero on the orthogonal complement of $\mathfrak{a}_{\overline{\rho}}$), which is equal to $\mathbf{R}_{\mathrm{ext}}$ on $\mathfrak{a}_{\overline{\rho}}$. The corresponding representation of G is an irreducible unitary representation of G in $\overline{\mathfrak{l}_1}$ that agrees with $\mathbf{R}$, so we compute the matrix of $\mathbf{R}(s)$ in the basis of $\overline{\mathfrak{l}_1}$ consisting of the vectors $(\frac{1}{n_\rho}\overline{m_{i1}})_{1\leq i\leq n_\rho}$, and we find M_ρ.

And now comes the detailed proof. (1) By the Peter–Weyl theorem (Theorem 4.2), $\mathrm{L}^2(G)$ is the Hilbert sum of both families $(\mathfrak{a}_\rho)_{\rho\in R}$ and $(\mathfrak{a}_{\overline{\rho}})_{\rho\in R} = (\overline{\mathfrak{a}_\rho})_{\rho\in R}$, but to obtain the matrix representations M_ρ we need to use the Hilbert sum $(\mathfrak{a}_{\overline{\rho}})_{\rho\in R}$. We observed just after Definition 4.5 that

$$\overline{M_\rho(s)} = M_{\overline{\rho}}(s), \quad \overline{u_\rho} = u_{\overline{\rho}}, \quad \overline{\chi_\rho} = \chi_{\overline{\rho}}.$$

The first equation implies that if $m_{ij}^{(\rho)}(s)$ are the elements of the matrix $M_\rho(s)$, then the elements $m_{ij}^{(\overline{\rho})}(s)$ of the matrix $M_{\overline{\rho}}(s)$ are given by $m_{ij}^{(\overline{\rho})}(s) = \overline{m_{i,j}^{(\rho)}}$. The $u_{\overline{\rho}}$ are self-adjoint idempotents, that is, $u_{\overline{\rho}} * u_{\overline{\rho}} = u_{\overline{\rho}}$ and $u_{\overline{\rho}}^* = \check{\overline{u_{\overline{\rho}}}} = u_{\overline{\rho}}$, and since $V_{\mathrm{ext}} \colon \mathrm{L}^1(G) \to \mathcal{L}(H)$ is an algebra homomorphism, we have

$$\begin{aligned} V_{\mathrm{ext}}(u_{\overline{\rho}}) &= V_{\mathrm{ext}}(u_{\overline{\rho}} * u_{\overline{\rho}}) \\ &= V_{\mathrm{ext}}(u_{\overline{\rho}}) \circ V_{\mathrm{ext}}(u_{\overline{\rho}}), \end{aligned}$$

and

$$\begin{aligned} V_{\mathrm{ext}}(u_{\overline{\rho}}) &= V_{\mathrm{ext}}(u_{\overline{\rho}}^*) \\ &= (V_{\mathrm{ext}}(u_{\overline{\rho}}))^*, \end{aligned}$$

so the continuous linear map $V_{\rm ext}(u_{\overline{\rho}})$ is idempotent and hermitian, and by Proposition 2.6, it is an orthogonal projection. Since $u_{\overline{\rho}} * u_{\overline{\rho'}} = 0$ if $\rho \neq \rho'$, we have $V_{\rm ext}(u_{\overline{\rho}}) \circ V_{\rm ext}(u_{\overline{\rho'}}) = 0$, which implies that the images E_ρ of the projections $V_{\rm ext}(u_{\overline{\rho}})$ are closed subspaces that are pairwise orthogonal. To prove that H is the Hilbert sum of the family $(E_\rho)_{\rho\in R}$, we need to show that the algebraic direct sum $\bigoplus_{\rho\in R} E_\rho$ is dense in H. We know from the proof of Theorem 3.18 that the linear span E of the set $\{V_{\rm ext}(f)(x) \mid f \in \mathcal{C}(G;\mathbb{C}), x \in H\}$ is dense in H, and by Proposition 3.15, we have $\|V_{\rm ext}(f)\| \leq \|f\|_1$. By Proposition 4.8(2), if f is continuous, then for every $\epsilon > 0$, there is a (finite) linear combination $\sum_{i,j,\rho} c_{ij}^{(\overline{\rho})} m_{ij}^{(\overline{\rho})}$ such that

$$\left\| f - \sum_{i,j,\rho} c_{ij}^{(\overline{\rho})} m_{ij}^{(\overline{\rho})} \right\|_\infty \leq \epsilon,$$

and since $\|V_{\rm ext}(f)\| \leq \|f\|_1$, G is compact, and $\lambda(G) = 1$, we have $\|g\|_1 \leq \|g\|_\infty$ for any $g \in \mathrm{L}^1(G)$, and this implies that

$$\begin{aligned}\left\| V_{\rm ext}(f) - \sum_{i,j,\rho} c_{ij}^{(\overline{\rho})} V_{\rm ext}(m_{ij}^{(\overline{\rho})}) \right\| &\leq \left\| f - \sum_{i,j,\rho} c_{ij}^{(\overline{\rho})} m_{ij}^{(\overline{\rho})} \right\|_1 \\ &\leq \left\| f - \sum_{i,j,\rho} c_{ij}^{(\overline{\rho})} m_{ij}^{(\overline{\rho})} \right\|_\infty \leq \epsilon.\end{aligned}$$

Since $m_{ij}^{(\overline{\rho})} = u_{\overline{\rho}} * m_{ij}^{(\overline{\rho})}$, we have $V_{\rm ext}(m_{ij}^{(\overline{\rho})}) = V_{\rm ext}(u_{\overline{\rho}}) \circ V_{\rm ext}(m_{ij}^{(\overline{\rho})})$, thus the vector $\left(\sum_{i,j,\rho} c_{ij}^{(\overline{\rho})} V_{\rm ext}(m_{ij}^{(\overline{\rho})})\right)(x)$ belongs to the sum of the E_ρ, which proves that E is dense in H. We delete the summands E_ρ, such that $E_\rho = (0)$.

(2) *Step 1*. We prove that the subspaces E_ρ which are invariant under $V_{\rm ext}$ are also invariant under V. Since by Proposition 4.7, the $u_{\overline{\rho}}$ belong to the center of $\mathrm{L}^2(G)$, by Proposition 4.1, the $u_{\overline{\rho}}$ belong to the center of $\mathcal{M}^1(G)$. Recall that Theorem 3.17 actually yields an algebra representation $\widetilde{V}\colon \mathcal{M}^1(G) \to \mathcal{L}(H)$ of the unital involutive Banach algebra $\mathcal{M}^1(G)$ and that $V_{\rm ext}$ is the restriction of $\widetilde{V}$ to $\mathrm{L}^1(G)$. In particular, even though $\delta_s \notin \mathrm{L}^1(G)$ (unless G is discrete), $\widetilde{V}(\delta_s)$ makes sense, and

$$\widetilde{V}(\delta_s) = V(s),$$

as stated in $(\widetilde{U}(\delta_s))$ just after Definition 3.13. Also recall (†) from the proof of Theorem 3.18,

$$V(s)(V_{\rm ext}(f)(x)) = V_{\rm ext}(\delta_s * f)(x), \text{ for all } f \in \mathcal{L}^1(G), \text{ and all } x \in H. \quad (\dagger)$$

Then as any $E_\rho \neq (0)$ is the image of $V_{\text{ext}}(u_{\overline{\rho}})$, we have

$$\begin{aligned}V(s)(V_{\text{ext}}(u_{\overline{\rho}})(x)) &= V_{\text{ext}}(\delta_s * u_{\overline{\rho}})(x)\\ &= \widetilde{V}(\delta_s * u_{\overline{\rho}})(x)\\ &= \widetilde{V}(u_{\overline{\rho}} * \delta_s)(x)\\ &= \widetilde{V}(u_{\overline{\rho}})(\widetilde{V}(\delta_s)(x))\\ &= V_{\text{ext}}(u_{\overline{\rho}})(V(s)(x)),\end{aligned}$$

we conclude that the E_ρ are invariant under V.

Step 2. We prove that if V_ρ is the restriction of the representation V of G to E_ρ we obtain a Hilbert sum decomposition into irreducible representations for the restriction of the nondegenerate representation $(V_\rho)_{\text{ext}}$ to $\mathfrak{a}_{\overline{\rho}}$ in E_ρ.

If V_ρ is the restriction of the representation V of G to E_ρ, then $(V_\rho)_{\text{ext}}(u_{\overline{\rho'}}) = 0$ for all $\rho' \neq \rho$, since $u_{\overline{\rho}} * u_{\overline{\rho'}} = 0$. The representation $(V_\rho)_{\text{ext}}$ is a representation of $\mathrm{L}^1(G)$ in E_ρ, and since $\mathrm{L}^2(G) \subseteq \mathrm{L}^1(G)$ is the Hilbert sum $\mathrm{L}^2(G) = \bigoplus_{\rho \in R} \mathfrak{a}_{\overline{\rho}}$, and the projection of $\mathrm{L}^2(G)$ onto $\mathfrak{a}_{\overline{\rho}}$ is the map $f \mapsto f * u_{\overline{\rho}}$, we have $(V_\rho)_{\text{ext}}(f * u_{\overline{\rho'}}) = (V_\rho)_{\text{ext}}(f) \circ (V_\rho)_{\text{ext}}(u_{\overline{\rho'}}) = 0$, which means that $(V_\rho)_{\text{ext}}$ is zero on every $\mathfrak{a}_{\overline{\rho'}}$ with $\rho' \neq \rho$. Consequently, the restriction of $(V_\rho)_{\text{ext}}$ to $\mathrm{L}^2(G)$ can be viewed as a nondegenerate representation of the topologically simple algebra $\mathfrak{a}_{\overline{\rho}}$ in E_ρ. Since by Proposition 3.15, the representation $(V_\rho)_{\text{ext}}$ is continuous, by Theorem 2.36(2), the nondegenerate representation $(V_\rho)_{\text{ext}}$ of $\mathfrak{a}_{\overline{\rho}}$ in E_ρ is a finite or countably infinite (if E_ρ is infinite dimensional) Hilbert sum of topologically irreducible representations all equivalent to the representation $U_{\overline{\mathfrak{l}_1}}\colon \mathfrak{a}_{\overline{\rho}} \to \mathcal{L}(\overline{\mathfrak{l}_1})$.

Step 3. We observe that $U_{\overline{\mathfrak{l}_1}}$ is the restriction of $\mathbf{R}_{\text{ext}}\colon \mathrm{L}^1(G) \to \mathcal{L}(\mathrm{L}^2(G))$ to $\mathfrak{a}_{\overline{\rho}}$, where $\mathbf{R}\colon G \to \mathbf{U}(\mathrm{L}^2(G))$ is the left regular representation of G in $\mathrm{L}^2(G)$ given by

$$(\mathbf{R}(s)(f))(t) = \lambda_s(f)(t) = f(s^{-1}t), \qquad f \in \mathrm{L}^2(G),\ s, t \in G;$$

see Definition 3.14. This is because by definition of $U_{\overline{\mathfrak{l}_1}}$ in Proposition 2.20,

$$U_{\overline{\mathfrak{l}_1}}(f)(g) = f * g, \qquad f \in \mathfrak{a}_{\overline{\rho}},\ g \in \overline{\mathfrak{l}_1},$$

and Definition 3.15 of the left regular representation $\mathbf{R}_{\text{ext}}$ of $\mathrm{L}^1(G)$ in $\mathrm{L}^2(G)$,

$$(\mathbf{R}_{\text{ext}}(f))(g) = f * g, \qquad f \in \mathrm{L}^1(G),\ g \in \mathrm{L}^2(G),$$

so $U_{\overline{\mathfrak{l}_1}}$ is the restriction of $\mathbf{R}_{\text{ext}}$ to $\mathfrak{a}_{\overline{\rho}}$. Thus, we can view $U_{\overline{\mathfrak{l}_1}}$ as the nondegenerate topologically irreducible representation $\widetilde{U_{\overline{\mathfrak{l}_1}}}\colon \mathrm{L}^1(G) \to \mathcal{L}(\overline{\mathfrak{l}_1})$ obtained by extending the nondegenerate representation $U_{\overline{\mathfrak{l}_1}}\colon \mathfrak{a}_{\overline{\rho}} \to \mathcal{L}(\overline{\mathfrak{l}_1})$ to

$L^1(G)$ (we set $\widetilde{U_{\overline{\mathfrak{l}_1}}}$ to zero on the orthogonal complement of $\mathfrak{a}_{\overline{\rho}}$), which is equal to $\mathbf{R}_{\text{ext}}$ on $\mathfrak{a}_{\overline{\rho}}$. The corresponding representation of G is an irreducible unitary representation of G in $\overline{\mathfrak{l}_1}$ that agrees with $\mathbf{R}$, so we compute the matrix of $\mathbf{R}(s)$ in the basis of $\overline{\mathfrak{l}_1}$ consisting of the vectors $(\frac{1}{n_\rho}\overline{m_{i1}})_{1\leq i\leq n_\rho}$.

Using Theorem 4.6(4), we have

$$\mathbf{R}(s)(\overline{m_{j1}}) = \overline{m_{j1}}(s^{-1}t) = \frac{1}{n_\rho}\sum_{i=1}^{n_\rho} \overline{m_{ji}}(s^{-1})\overline{m_{i1}}(t),$$

and since by Theorem 4.6(2), $\overline{m_{ji}}(s^{-1}) = m_{ij}(s)$, we recognize that the matrix of $\mathbf{R}(s)$ is $M_\rho(s)$, as claimed. □

The above proof is an adaptation of Dieudonné's proof [Dieudonné (2003a)] (Section 4, Theorem 21.4.1). Dieudonné's proof uses the projection $V_{\text{ext}}(u_\rho)$ instead of the projection $V_{\text{ext}}(\overline{u_\rho})$. The second option is the projection used by Serre in his short section on the representation of compact groups and also in Hewitt and Ross; see Serre [Serre (1977)] (Section 4.3) and Hewitt and Ross [Hewitt and Ross (1970)] (Chapter VII, Theorem 27.44). The advantage of Dieudonné's choice is that we avoid a plethora of indices $\overline{\rho}$, but the disadvantage is that the irreducible representations that occur in a given representation are the $\overline{M_\rho} = M_{\overline{\rho}}$. With the second option (as in Serre and Hewitt and Ross), the irreducible representations that occur are the M_ρ; no conjugation needed. Even though using the second option causes an additional notational burden in the proof (most indices are $\overline{\rho}$ instead of ρ), in the long-term this simplifies matters because the representations that occur are the M_ρs.

Let us emphasize that Theorem 4.16 proves that *every* representation M_ρ is irreducible, which is not at all obvious from their definition. Theorem 4.16 also shows that every irreducible unitary representation of G is equivalent to some representation of the form M_ρ, and M_ρ is not equivalent to $M_{\rho'}$ for $\rho \neq \rho'$.

Definition 4.6. Let G be a locally compact (metrizable, separable) group. A sequence of unitary representations $(U_\rho\colon G \to \mathbf{U}(H_\rho))_{\rho\in R}$ of G, where R is some index set (possibly infinite) is called a *complete set of irreducible unitary representations of* G if

(1) Each unitary representation $U_\rho\colon G \to \mathbf{U}(H_\rho)$ is irreducible.
(2) Any two representations U_ρ and $U_{\rho'}$ with $\rho \neq \rho'$ are inequivalent.

(3) Every irreducible unitary representation $V\colon G \to \mathbf{U}(H)$ of G is equivalent to some representation U_ρ (necessarily unique).

Consequently, $(M_\rho)_{\rho\in R}$ is a complete set of unitary irreducible representations of G in a separable Hilbert space. When we deal with more than one group G (say also a closed subgroup of G), we use the notation $R(G)$ instead of R.

Remark: It would be tempting to say that each ρ corresponds to an equivalence class of unitary representations (under equivalence) but there is a set-theoretic difficulty since the collection of unitary representations is not a set. This sticky point appears to be ignored by most authors, who do not hesitate to refer to the "set of equivalence classes" of irreducible representations of a group G, and even to the set of *all* representations of G. Some authors are more careful and avoid the term "equivalence classes of irreducible representations." The only source we are aware of that brings up this issue is Hewitt and Ross [Hewitt and Ross (1970)] (Chapter VII, second footnote on page 2). They suggest that a way to circumvent this set-theoretic difficulty is to observe that for a given group G, the cardinality of the vector spaces involved in irreducible representations of G is bounded. In fact, by Proposition 3.1, it is bounded by $\aleph_1^{|G|}$, where $|G|$ denotes the cardinality of G. Then by Riesz–Fischer (Vol. I, Theorem D.19), we can pick representatives for the Hilbert spaces of cardinality at most $\aleph_1^{|G|}$ among $\ell^p(K)$-spaces with K of cardinality bounded by $\aleph_1^{|G|}$. For compact groups, we just showed that the irreducible unitary representations are finite-dimensional so we can pick these vector spaces as the spaces $\mathbb{C}^n$ (countably many). Hewitt and Ross's footnote ends with the sentence: "The exact details are of little interest for the purposes of the present book." We tend to agree! Definition 4.6 is designed to avoid set-theoretic difficulties. With a small abuse of language, we may still say that the unitary representations equivalent to the representation M_ρ are *of class* ρ.

If the compact group G is abelian, then every algebra $\mathfrak{a}_\rho$ is abelian, and since it is simple, it must be one-dimensional. Therefore, every unitary representation of a (metrizable) compact abelian group is a finite or a countably infinite Hilbert sum of *one-dimensional* representations.

It is customary to introduce the following terminology.

Definition 4.7. With the notations of Theorem 4.16, if $V\colon G \to \mathbf{U}(H)$ is a unitary representation of G in a separable Hilbert space H, and if

$H = \bigoplus_{\rho \in R, E_\rho \neq (0)} E_\rho$ is the Hilbert sum induced by the projections $\pi_\rho^V = V_{\text{ext}}(\overline{u_\rho})$, with

$$\pi_\rho^V(x) = n_\rho \int_G \overline{\chi_\rho}(s) V(s)(x)\, d\lambda(s), \quad x \in H,$$

whenever $E_\rho \neq (0)$ and $V_\rho \colon G \to \mathbf{U}(E_\rho)$ is the corresponding representation, we say that the irreducible representation M_ρ is *contained* in the representation V. If

$$E_\rho = \bigoplus_{k_\rho = 1}^{d_\rho} E_\rho^{k_\rho}$$

is finite-dimensional of dimension $d_\rho n_\rho > 0$ (recall that each subspace $E_\rho^{k_\rho}$ is isomorphic to $\mathbb{C}^{n_\rho}$), we say that M_ρ is *contained d_ρ times* in V (or *infinitely many times* if E_ρ is infinite-dimensional). We also call d_ρ the *multiplicity* of M_ρ in V_ρ. The representations M_ρ such that $d_\rho > 0$ are called the *irreducible components* of the representation V.

If we consider the left regular representation $\mathbf{R}$ of G in $\mathrm{L}^2(G)$, then the projection $\pi_\rho^{\mathbf{R}}$ is given by

$$\pi_\rho^{\mathbf{R}}(f) = \int \overline{u_\rho}(s) \mathbf{R}_s(f)\, d\lambda(s) = \int \overline{u_\rho}(s) \lambda_s(f)\, d\lambda(s) = \overline{u_\rho} * f = \overline{u_\rho * \overline{f}},$$

so $E_\rho = \mathfrak{a}_{\overline{\rho}} = \overline{\mathfrak{a}_\rho}$ for all $\rho \in R$, and Theorem 4.16 says that on $\mathfrak{a}_{\overline{\rho}}$, the representation $\mathbf{R}$ splits into n_ρ irreducible representations all equivalent to M_ρ. We can view these representation as acting on the columns of $M_{\overline{\rho}} = \overline{M_\rho}$, which span n_ρ minimal left ideals $\mathfrak{l}_j^{(\overline{\rho})}$ of $\mathfrak{a}_{\overline{\rho}}$; that is,

$$\mathfrak{a}_{\overline{\rho}} = \bigoplus_{j=1}^{n_\rho} \mathfrak{l}_j^{(\overline{\rho})} \quad \text{and} \quad \mathfrak{l}_j^{(\overline{\rho})} = \bigoplus_{k=1}^{n_{\overline{\rho}}} \mathbb{C} m_{kj}^{(\overline{\rho})}.$$

Remark: The statement $E_\rho = \mathfrak{a}_{\overline{\rho}} = \overline{\mathfrak{a}_\rho}$ may seem wrong, but it is correct. It is a consequence of the definition of the projection π_ρ^V. The exact same fact is noted in Hewitt and Ross [Hewitt and Ross (1970)] (Chapter VII, Section 27.49).

The above fact is worth recording as a proposition.

Proposition 4.17. *The left regular representation $\mathbf{R} \colon G \to \mathbf{U}(\mathrm{L}^2(G))$ of a compact (metrizable) group G in $\mathrm{L}^2(G)$ contains every irreducible unitary representation M_ρ of G, and each one is contained n_ρ times, where n_ρ is the dimension of the space of the representation.*

Proposition 4.17 is a generalization to compact groups of a property holding for finite groups for which the proof is much easier; see Serre [Serre (1977)] (Section 2.4).

If V is a finite-dimensional unitary representation, then the trace of the linear map $V(s)$ plays a crucial role. In fact, it determines this representation up to equivalence.

Proposition 4.18. *Let G be a (metrizable) compact group. For any unitary representation $V\colon G \to \mathbf{U}(H)$ of G in a finite-dimensional hermitian space H of dimension d, assume that for every $\rho \in R$, the irreducible representation M_ρ is contained d_ρ times in V, so that*

$$d = \sum_{\rho \in R} d_\rho n_\rho,$$

where $d_\rho \neq 0$ for only finitely many $\rho \in R$. Then we have

$$\mathrm{tr}(V(s)) = \sum_{\rho \in R} d_\rho \chi_\rho(s), \qquad \textit{for all } s \in G.$$

Proof. We can write H as the direct sum of finite-dimensional spaces, and by picking bases, we can express $V(s)$ as a sum of matrices similar to some of the $M_\rho(s)$. Then the above formula follows from the fact that $\chi_\rho(s) = \mathrm{tr}(M_\rho(s))$ and the fact that the trace is invariant under conjugation, $\mathrm{tr}(PVP^{-1}) = \mathrm{tr}(V)$. ☐

Theorem 4.19. *Let G be a (metrizable) compact group. Two unitary representations $V_1\colon G \to \mathbf{U}(H_2)$ and $V_2\colon G \to \mathbf{U}(H_2)$ of G in finite-dimensional hermitian spaces H_1 and H_2 of dimensions d_1 and d_2 are equivalent if and only if*

$$\mathrm{tr}(V_1(s)) = \mathrm{tr}(V_2(s)), \qquad \textit{for all } s \in G.$$

In particular, if V_1 and V_2 are equivalent, then $d_1 = d_2$. Moreover, if V_1 and V_2 are any two equivalent irreducible unitary representations, then

$$\mathrm{tr}(V_1(s)) = \mathrm{tr}(V_2(s)) = \chi_\rho(s), \quad s \in G,$$

where M_ρ is the irreducible representation from Theorem 4.16 to which V_2 and V_2 are equivalent.

Proof. Clearly, if V_1 and V_2 are equivalent, the formula of the theorem holds. Conversely, by Proposition 4.18, since

$$\mathrm{tr}(V_1(s)) = \sum_{\rho \in R_1} d_\rho \chi_\rho(s) \quad \text{and} \quad \mathrm{tr}(V_2(s)) = \sum_{\rho \in R_2} d_\rho \chi_\rho(s)$$

for some finite subsets R_1 and R_2 of R, and since by Theorem 4.6(3) the characters are linearly independent, we must have $R_1 = R_2$ and $d_1 = d_2$. $\square$

Theorem 4.19 suggests the following (standard) definition.

Definition 4.8. Let G be a (metrizable) compact group. For any unitary representation $V: G \to \mathbf{U}(H)$ of G in a finite-dimensional hermitian space H of dimension d, we define the *character* χ_V *of the representation* V as the map $\chi_V: G \to \mathbb{C}$ given by

$$\chi_V(s) = \mathrm{tr}(V(s)), \qquad \text{for all } s \in G.$$

The characters of a finite-dimensional unitary representation are central functions, and in view of Proposition 4.18, they have many of the properties of the characters χ_ρ.

By definition, the character χ_ρ of the compact group G is identical to the character χ_{M_ρ} of the special representation M_ρ, which is irreducible by Peter–Weyl II, and by Theorem 4.19, it is also the character of *all* equivalent irreducible unitary representations of G equivalent to M_ρ. Thus the set $(\chi_\rho)_{\rho \in R}$ is the set of characters of *all* irreducible unitary representations of G. If we have some complete set of irreducible unitary representations for G, we can determine the characters of G. If the group G is finite, then there are finitely many irreducible representations up to equivalence, so this method can be used practically.

Example 4.3. Let G be a finite group and assume that $\{\rho_1, \ldots, \rho_r\}$ is a complete set of irreducible unitary representations $\rho_i: G \to \mathbf{U}(W_i)$ of G (where r is the number of conjugacy classes of G) so that $R = \{\rho_1, \ldots, \rho_r\}$, write $n_i = \dim(W_i)$, and let $\chi_1, \ldots, \chi_r$ be the characters of G (which are equal to the characters of the ρ_i). If $U: G \to \mathbf{U}(E)$ is any unitary representation of G (where E is finite-dimensional), then by Peter–Weyl II, we have a direct sum

$$E = E_{i_1} \oplus \cdots \oplus E_{i_h} \tag{$\dagger_1$}$$

for some subset $\{\rho_{i_1}, \ldots, \rho_{i_h}\}$ of R ($h \leq r$), and each E_{i_j} ($1 \leq j \leq h$) is a direct sum

$$E_{i_j} = E^1_{i_j} \oplus \cdots \oplus E^{d_j}_{i_j} \quad (d_j \geq 1), \tag{$\dagger_2$}$$

such that for $k = 1, \ldots, d_j$, each representation $U: G \to \mathbf{U}(E^k_{i_j})$ is equivalent to the irreducible representation $\rho_{i_j}: G \to \mathbf{U}(W_{i_j})$. Each subspace E_{i_j}

is the projection of E by the projection $\pi^U_{i_j}$ given by

$$\pi^U_{i_j}(x) = \frac{n_{i_j}}{|G|} \sum_{s \in G} \overline{\chi_{i_j}}(s) U(s)(x) \quad x \in E. \tag{$\dagger_3$}$$

The E_{i_j} in ($\dagger_1$) are uniquely determined by U (in terms of the projections $\pi^U_{i_j}$), but the splitting of E_{i_j} as a direct sum as above in ($\dagger_2$) is not.

The decomposition of $U: G \to \mathbf{U}(E)$ into the h unitary representations $U: G \to \mathbf{U}(E_{i_j})$ $(1 \leq j \leq h)$ is called the *canonical decomposition of U*. For finite groups, these results can be obtained more directly; see Serre [Serre (1977)] (Section 2.6, in particular, Theorem 8). Each representation $U: G \to \mathbf{U}(E_{i_j})$ $(1 \leq j \leq h)$ contains the irreducible representation $\rho_{i_j}: G \to \mathbf{U}(W_{i_j})$ d_j times, so it is not irreducible unless $d_j = 1$. It is actually possible to obtain a specific decomposition of each E_{i_j} into some subspaces $E^k_{i_j}$ as in ($\dagger_2$) given by projections expressed in terms of matrix representations for the irreducible representations $\rho_{i_j}: G \to \mathbf{U}(W_{i_j})$; see Serre [Serre (1977)] (Section 2.7).

Example 4.4. Recall from Example 3.1 that the group $\mathfrak{S}_3$ consists of the permutations on the set $\{1, 2, 3\}$. There are $3! = 6$ permutations

$$\begin{aligned} \sigma_1 &= (1,2,3), \;\; \sigma_2 = (1,3,2), \;\; \sigma_3 = (2,1,3), \\ \sigma_4 &= (2,3,1), \;\; \sigma_5 = (3,1,2), \;\; \sigma_6 = (3,2,1), \end{aligned}$$

three conjugacy classes, $C_1 = \{\sigma_1\}$, $C_2 = \{\sigma_2, \sigma_3, \sigma_6\}$, $C_3 = \{\sigma_4, \sigma_5\}$, and three irreducible representations (up to equivalence). The two one-dimensional irreducible representations are the trivial representation $\rho_1: \mathfrak{S}_3 \to \mathbf{U}(1)$ with

$$\rho_1(\sigma_i) = 1, \quad i = 1, \ldots, 6,$$

and the signature representation $\rho_2: \mathfrak{S}_3 \to \mathbf{U}(1)$ from Example 3.6, with

$$\rho_2(\sigma_i) = \begin{cases} +1, & \sigma_i \in C_1 \\ -1, & \sigma_i \in C_2 \\ +1, & \sigma_i \in C_3. \end{cases}$$

The third irreducible representation ρ_3 is two-dimensional and is obtained from Example 3.5. We obtained the matrix representation of $\rho_3: \mathfrak{S}_3 \to \mathbf{U}(2)$ by 3×3 matrices with respect to the basis (w_1, w_2, w_3) and we just have to consider the 2×2 matrices obtained by deleting the first row and the first column since w_1 is invariant. We get

$$\begin{pmatrix}1 & 0\\ 0 & 1\end{pmatrix}, \quad \begin{pmatrix}1/2 & \sqrt{3}/2\\ \sqrt{3}/2 & -1/2\end{pmatrix}, \quad \begin{pmatrix}-1 & 0\\ 0 & 1\end{pmatrix}, \quad \begin{pmatrix}-1/2 & -\sqrt{3}/2\\ \sqrt{3}/2 & -1/2\end{pmatrix},$$
$$\begin{pmatrix}-1/2 & \sqrt{3}/2\\ -\sqrt{3}/2 & -1/2\end{pmatrix}, \quad \begin{pmatrix}1/2 & -\sqrt{3}/2\\ -\sqrt{3}/2 & -1/2\end{pmatrix}.$$

In Example 3.7, we found the canonical decomposition of the regular representation $\rho_{\mathbf{R}}\colon \mathfrak{S}_3 \to \mathbf{GL}(\mathbb{C}^6)$ of $\mathfrak{S}_3$ from Example 3.2. We have

$$\mathbb{C}^6 = V_1 \oplus V_2 \oplus V_3 = V_1 \oplus V_2 \oplus V_1^3 \oplus V_2^3,$$

where V_1 is spanned by the vector $e_1+e_2+e_3+e_4+e_5+e_6$, V_2 is spanned by the vector $e_1-e_2-e_3+e_4+e_5-e_6$, V_1^3 is spanned by the vectors $e_1+e_2-e_3-e_4$, $e_3+e_4-e_5-e_6$, and V_2^3 is spanned by the vectors $e_1-e_3-e_4+e_6$, $e_2+e_4-e_5-e_6$. The representation $\rho_{\mathbf{R}}\colon \mathfrak{S}_3 \to \mathbf{GL}(V_1)$ is equivalent to the irreducible representation $\rho_1\colon \mathfrak{S}_3 \to \mathbf{U}(1)$, the representation $\rho_{\mathbf{R}}\colon \mathfrak{S}_3 \to \mathbf{GL}(V_2)$ is equivalent to the irreducible representation $\rho_2\colon \mathfrak{S}_3 \to \mathbf{U}(1)$, and both representations $\rho_{\mathbf{R}}\colon \mathfrak{S}_3 \to \mathbf{GL}(V_k^3)$, $k = 1, 2$, are equivalent to the irreducible representation $\rho_3\colon \mathfrak{S}_3 \to \mathbf{U}(2)$.

Theorem 4.19 shows that two finite-dimensional unitary representations $V_1\colon G \to \mathbf{U}(H_2)$ and $V_2\colon G \to \mathbf{U}(H_2)$ of G are equivalent if and only if $\chi_{V_1} = \chi_{V_2}$. This confirms the importance of the characters; they determine the equivalence classes of finite-dimensional unitary representations of a (metrizable) compact group.

Observe that the definition of the character of a representation makes sense even if the representation is not unitary (it only needs to be finite-dimensional). In view of Theorem 3.6 and the discussion following it, every *finite-dimensional* representation of G (not necessarily unitary) can be viewed as a unitary representation for some suitable hermitian inner product, so Proposition 4.18 and Theorem 4.19 also apply to such representations. Consequently, the characters also determine the equivalence classes of all finite-dimensional, not necessarily unitary, representations of a (metrizable) compact group.

If G is finite, it may be possible to build a character table for G by determining a complete set of irreducible representations of G (in view of the above remarks, not necessarily unitary). In general, this is difficult. It should be noted that for finite groups, using Peter–Weyl II to introduce characters and obtain some of their properties is a very heavy-handed method. A more gentle (and standard) approach is to define the characters of finite-dimensional representations and to derive their properties directly, singling out the role played by the characters of irreducible representations. Such an approach is presented in the excellent texts of Serre [Serre (1977)] and Simon [Simon (1996)]. Here is an example of the computation of the character table of the symmetric group $\mathfrak{S}_3$. Since we determined the irreducible representations of the symmetric group $\mathfrak{S}_3$ in Section 3.1, we can build its table of characters.

Example 4.5. In Example 4.4, we found the three irreducible unitary representations (up to equivalence) of the group $\mathfrak{S}_3$ consisting of the permutations on the set $\{1, 2, 3\}$. Recall that there are $3! = 6$ permutations

$$\sigma_1 = (1,2,3),\ \ \sigma_2 = (1,3,2),\ \ \sigma_3 = (2,1,3),$$
$$\sigma_4 = (2,3,1),\ \ \sigma_5 = (3,1,2),\ \ \sigma_6 = (3,2,1),$$

and three conjugacy classes, $C_1 = \{\sigma_1\}$, $C_2 = \{\sigma_2, \sigma_3, \sigma_6\}$, $C_3 = \{\sigma_4, \sigma_5\}$. The two one-dimensional irreducible unitary representations are the trivial representation ρ_1 with

$$\rho_1(\sigma_i) = 1, \quad i = 1, \ldots, 6,$$

and the signature representation ρ_2 from Example 3.6, with

$$\rho_2(\sigma_i) = \begin{cases} +1, & \sigma_i \in C_1 \\ -1, & \sigma_i \in C_2 \\ +1, & \sigma_i \in C_3. \end{cases}$$

The corresponding characters χ_1, χ_2 are the central functions obtained by taking traces, in this scalar case the identity, so we get

$$\chi_1(\sigma_i) = \begin{cases} 1, & \sigma_i \in C_1 \\ 1, & \sigma_i \in C_2 \\ 1, & \sigma_i \in C_3 \end{cases} \quad \chi_2(\sigma_i) = \begin{cases} +1, & \sigma_i \in C_1 \\ -1, & \sigma_i \in C_2 \\ +1, & \sigma_i \in C_3. \end{cases}$$

The third irreducible unitary representation $\rho_3 \colon \mathfrak{S}_3 \to \mathbf{U}(2)$ is given by the matrices

$$\begin{pmatrix} 1 & 0 \\ 0 & 1 \end{pmatrix}, \quad \begin{pmatrix} 1/2 & \sqrt{3}/2 \\ \sqrt{3}/2 & -1/2 \end{pmatrix}, \quad \begin{pmatrix} -1 & 0 \\ 0 & 1 \end{pmatrix}, \quad \begin{pmatrix} -1/2 & -\sqrt{3}/2 \\ \sqrt{3}/2 & -1/2 \end{pmatrix},$$
$$\begin{pmatrix} -1/2 & \sqrt{3}/2 \\ -\sqrt{3}/2 & -1/2 \end{pmatrix}, \quad \begin{pmatrix} 1/2 & -\sqrt{3}/2 \\ -\sqrt{3}/2 & -1/2 \end{pmatrix}.$$

By taking traces, we obtain

$$\chi_3(\sigma_i) = \begin{cases} 2, & \sigma_i \in C_1 \\ 0, & \sigma_i \in C_2 \\ -1, & \sigma_i \in C_3. \end{cases}$$

Thus, we obtain the following character table for $\mathfrak{S}_3$.

	C_1	C_2	C_3
χ_1	1	1	1
χ_2	1	-1	1
χ_3	2	0	-1

For much more about the representation of finite groups, see Serre [Serre (1977)], Simon [Simon (1996)] and Fulton and Harris [Fulton and Harris (1991)]. The conjugacy classes and the characters of the symmetric group are discussed in Fulton and Harris [Fulton and Harris (1991)] and Simon [Simon (1996)]. This beautiful theory makes use of Young tableaux.

Operations on (finite-dimensional) vector space induce operations on finite-dimensional, not necessarily unitary, representations, which in turn induce operations on their characters.

Given two finite-dimensional representations $U_1 \colon G \to \mathbf{U}(H_2)$ and $U_2 \colon G \to \mathbf{U}(H_2)$, with $d_1 = \dim(H_1)$ and $d_2 = \dim(H_2)$, we already defined their *direct sum* as the representation $U_1 \oplus U_2$ of G in $H_1 \oplus H_2$ given by

$$(U_1 \oplus U_2)(s)(x_1 + x_2) = U_1(s)(x_1) + U_2(s)(x_2), \quad s \in G,\, x_1 \in H_1,\, x_2 \in H_2.$$

The tensor product of representations is also useful because it can be used to characterize the irreducible finite-dimensional representations of products of compact groups.

4.4 Tensor Products of Finite-Dimensional Representations

If H_1 and H_2 are two finite-dimensional vector spaces, following Serre, the tensor product of H_1 and H_2 can be defined in a way that avoids the rather abstract universal mapping property.

Definition 4.9. If H_1 and H_2 are two finite-dimensional (real or complex) vector spaces, a tensor product $H_1 \otimes H_2$ of H_1 and H_2 is a (real or complex) vector space together with a map $\iota_\otimes \colon H_1 \times H_1 \to H_1 \otimes H_2$, such that the following two conditions hold:

(1) The map $\iota_\otimes \colon H_1 \times H_1 \to H_1 \otimes H_2$ is bilinear. For any $u \in H_1$ and any $v \in H_2$, we denote $\iota_\otimes(u, v)$ by $u \otimes v$.
(2) For any basis $(u_1, \ldots, u_m)$ of H_1 and any basis $(v_1, \ldots, v_n)$ of H_2, the $m \times n$ vectors $u_i \otimes v_j$ form a basis of $H_1 \otimes H_2$.

By standard methods of linear algebra, it can shown that such a space $H_1 \otimes H_2$ exists and is unique up to isomorphism; for example, see Gallier and Quaintance [Gallier and Quaintance (2020b)] (Chapter 2). The tensor product $H_1 \otimes H_2$ has the following *universal mapping property*: for every vector space F and every bilinear map $f\colon H_1 \times H_2 \to F$, there is a *unique linear map* $f_\otimes\colon H_1 \otimes H_2 \to F$, such that

$$f = f_\otimes \circ \iota_\otimes,$$

as illustrated in the following diagram:

$$\begin{array}{ccc} H_1 \times H_2 & \xrightarrow{\iota_\otimes} & H_1 \otimes H_2 \\ & \searrow^{f} & \downarrow f_\otimes \\ & & F. \end{array}$$

Given two linear maps $f\colon E \to E'$ and $g\colon F \to F'$, there is a unique linear map

$$f \otimes g\colon E \otimes F \to E' \otimes F',$$

such that

$$(f \otimes g)(u \otimes v) = f(u) \otimes g(v), \quad \text{for all } u \in E \text{ and all } v \in F. \qquad (f \otimes g)$$

This is because we can define $h\colon E \times F \to E' \otimes F'$ by

$$h(u,\, v) = f(u) \otimes g(v).$$

It is immediately verified that h is bilinear, and thus by the universal mapping property it induces a unique linear map

$$f \otimes g\colon E \otimes F \to E' \otimes F'$$

making the following diagram commute

$$\begin{array}{ccc} E \times F & \xrightarrow{\iota_\otimes} & E \otimes F \\ & \searrow^{h} & \downarrow f\otimes g \\ & & E' \otimes F', \end{array}$$

such that $(f \otimes g)(u \otimes v) = f(u) \otimes g(v)$, for all $u \in E$ and all $v \in F$. For proofs of the above facts (in a more general framework) see Gallier and Quaintance [Gallier and Quaintance (2020b)] (Chapter 2).

In terms of matrices, given a basis $(u_1, \ldots, u_{d_1})$ of H_1 and a basis $(v_1, \ldots, v_{d_2})$ of H_2, assume f_1 is represented by the matrix A_1 and f_2 is represented by the matrix A_2. Then with respect to the basis

$(u_i \otimes v_j)_{1\leq i\leq d_1, 1\leq j\leq d_2}$, the linear map $f_1 \otimes f_2$ is defined by a $(d_1 d_2)\times(d_1 d_2)$ matrix; as a block matrix, it is the $d_1 \times d_1$ matrix of $d_2 \times d_2$ blocks where the (i,j) block is the matrix $(A_1)_{ij} A_2$ $(1 \leq i,j \leq d_1)$. This matrix is called the *Kronecker product* of A_1 and A_2.

Given a complex vector space H, recall that $\overline{H}$ is the complex vector space with the same additive operation $+$ but with multiplication by a scalar defined by

$$(\lambda, u) \mapsto \overline{\lambda} u, \quad u \in H, \lambda \in \mathbb{C}.$$

Then a map $f\colon H \to \mathbb{C}$ is *semilinear* iff $f\colon \overline{H} \to \mathbb{C}$ is linear, which means that

$$\begin{aligned} f(u+v) &= f(u) + f(v) \\ f(\lambda u) &= \overline{\lambda} u, \end{aligned}$$

for all $u, v \in H$ and all $\lambda \in \mathbb{C}$. Observe that a map $\varphi\colon H \times H \to \mathbb{C}$ is *sesquilinear*, which means linear in its first argument and semilinear in its second argument, iff $\varphi\colon H \times \overline{H} \to \mathbb{C}$ is bilinear.

We define a hermitian inner product on the tensor product $H_1 \otimes H_2$ of two finite-dimensional complex vector spaces H_1 and H_2 each equipped with a hermitian inner product $\langle -,- \rangle_i$ $(i = 1, 2)$ following Bourbaki [Bourbaki (1968)] (Chapter 9, Section 1.9), which considers a more general situation. The map $\langle -,- \rangle\colon (H_1 \times H_2) \times (\overline{H_1} \times \overline{H_2}) \to \mathbb{C}$ is defined as follows: for all $u_1, u_2 \in H_1$ and all $v_1, v_2 \in H_2$,

$$\langle (u_1, v_1), (u_2, v_2) \rangle = \langle u_1, u_2 \rangle_1 \langle v_1, v_2 \rangle_2. \tag{$\langle\,\rangle$}$$

It is immediately verified that this map is linear in each of its arguments. By the universal mapping property, the above map extends to a unique bilinear map $\langle -,- \rangle_\otimes\colon (H_1 \otimes H_2) \times (\overline{H_1} \otimes \overline{H_2}) \to \mathbb{C}$, such that

$$\langle u_1 \otimes v_1, u_2 \otimes v_2 \rangle_\otimes = \langle u_1, u_2 \rangle_1 \langle v_1, v_2 \rangle_2$$

for all $u_1, u_2 \in H_1$ and all $v_1, v_2 \in H_2$. However, $\overline{H_1} \otimes \overline{H_2}$ is isomorphic to $\overline{H_1 \otimes H_2}$, so we obtain a sesquilinear map $\langle -,- \rangle\colon (H_1 \otimes H_2) \times (H_1 \otimes H_2) \to \mathbb{C}$. Since

$$\begin{aligned} \langle u_2 \otimes v_2, u_1 \otimes v_1 \rangle_\otimes &= \langle u_2, u_1 \rangle_1 \langle v_2, v_1 \rangle_2 = \overline{\langle u_1, u_2 \rangle_1}\, \overline{\langle v_1, v_2 \rangle_2} \\ &= \overline{\langle u_1 \otimes v_1, u_2 \otimes v_2 \rangle_\otimes}, \end{aligned}$$

the sesquilinear form $\langle -, - \rangle\colon (H_1 \otimes H_2) \times (H_1 \otimes H_2) \to \mathbb{C}$ is hermitian. Finally, observe that

$$\langle u_1 \otimes v_1, u_1 \otimes v_1 \rangle_\otimes = \langle u_1, u_1 \rangle_1 \langle v_1, v_1 \rangle_2,$$

and $\langle u_1, u_1 \rangle_1 \langle v_1, v_1 \rangle_2 > 0$ iff $\langle u_1, u_1 \rangle_1 > 0$ and $\langle v_1, v_1 \rangle_2 > 0$ iff $u_1 \neq 0$ and $v_1 \neq 0$, which means that our inner product is positive definite. Therefore the map $\langle -, - \rangle\colon (H_1 \otimes H_2) \times (H_1 \otimes H_2) \to \mathbb{C}$ uniquely defined by

$$\langle u_1 \otimes v_1, u_2 \otimes v_2 \rangle_\otimes = \langle u_1, u_2 \rangle_1 \langle v_1, v_2 \rangle_2$$

for all $u_1, u_2 \in H_1$ and all $v_1, v_2 \in H_2$, is a hermitian inner product on $H_1 \otimes H_2$.

Definition 4.10. If $(H_1, \langle -, - \rangle_1)$ and $(H_2, \langle -, - \rangle_2)$ are two finite-dimensional complex vector spaces each equipped with a hermitian inner product $\langle -, - \rangle_i$ $(i = 1, 2)$, there is a unique hermitian inner product $\langle -, - \rangle_\otimes\colon (H_1 \otimes H_2) \times (H_1 \otimes H_2) \to \mathbb{C}$ on $H_1 \otimes H_2$ satisfying the equation

$$\langle u_1 \otimes v_1, u_2 \otimes v_2 \rangle_\otimes = \langle u_1, u_2 \rangle_1 \langle v_1, v_2 \rangle_2 \tag{$\langle\,\rangle_\otimes$}$$

for all $u_1, u_2 \in H_1$ and all $v_1, v_2 \in H_2$.

Observe that if $(u_1, \ldots, u_{d_1})$ is an orthonormal basis of H_1 and $(v_1, \ldots, v_{d_2})$ is an orthonormal basis of H_2, then $(u_i \otimes v_j)_{1 \leq i \leq d_1, 1 \leq j \leq d_2}$ is an orthonormal basis of $H_1 \otimes H_2$ with respect to the inner product $\langle -, - \rangle_\otimes$.

If $f_1\colon H_1 \to H_1$ and $f_2\colon H_2 \to H_2$ are unitary linear maps, then for all $u_1, u_2 \in H_1$ and all $v_1, v_2 \in H_2$, we have

$$\begin{aligned}
&\langle (f_1 \otimes f_2)(u_1 \otimes v_1), (f_1 \otimes f_2)(u_2 \otimes v_2) \rangle_\otimes \\
&\quad = \langle f_1(u_1) \otimes f_2(v_1), f_1(u_2) \otimes f_2(v_2) \rangle \\
&\quad = \langle f_1(u_1), f_1(u_2) \rangle_1 \langle f_2(v_1), f_2(v_2) \rangle_2 \\
&\quad = \langle u_1, u_2 \rangle_1 \langle v_1, v_2 \rangle_2 = \langle u_1 \otimes v_1, u_2 \otimes v_2 \rangle_\otimes,
\end{aligned}$$

which proves that $f_1 \otimes f_2\colon H_1 \otimes H_2 \to H_1 \otimes H_2$ is unitary for the hermitian inner product $\langle -, - \rangle_\otimes$ on $H_1 \otimes H_2$. As a consequence of all this, we can make the following definition.

Definition 4.11. Given two finite-dimensional unitary representations $U_1\colon G \to \mathbf{U}(H_1)$ and $U_2\colon G \to \mathbf{U}(H_2)$ of the locally compact (metrizable, separable) group G, we define the *tensor product* $U_1 \otimes U_2$ of U_1 and U_2 as the unitary representation $U_1 \otimes U_2\colon G \to \mathbf{U}(H_1 \otimes H_2)$ of G in $H_1 \otimes H_2$ (with the hermitian inner product $\langle -, - \rangle_\otimes$ on $H_1 \otimes H_2$ defined in $(\langle\,\rangle_\otimes)$) given by

$$(U_1 \otimes U_2)(s) = U_1(s) \otimes U_2(s), \quad s \in G,$$

where $U_1(s) \otimes U_2(s)$ is the tensor product linear map given by

$$(U_1(s) \otimes U_2(s))(x_1 \otimes x_2) = U(s)(x_1) \otimes U(s)(x_2), \text{ for all } x_1 \in H_1, x_2 \in H_2.$$

In terms of matrices, the linear map $U_1(s) \otimes U_2(s)$ is defined by a $(d_1d_2) \times (d_1d_2)$ matrix, namely the Kronecker product of $U_1(s)$ and $U_2(s)$. As a block matrix, it is the $d_1 \times d_1$ matrix of $d_2 \times d_2$ blocks where the (i,j) block is the matrix $U_1(s)_{ij}U_2(s)$ $(1 \leq i,j \leq d_1)$.

It is well-known that

$$\begin{aligned}\mathrm{tr}(U_1(s) \oplus U_2(s)) &= \mathrm{tr}(U_1(s)) + \mathrm{tr}(U_2(s))\\ \mathrm{tr}(U_1(s) \otimes U_2(s)) &= \mathrm{tr}(U_1(s))\mathrm{tr}(U_2(s)).\end{aligned}$$

Let us now assume that G is compact until Definition 4.12. If $U_1 = M_{\rho'}$ and $U_2 = M_{\rho''}$ are two irreducible representations of G, then since $\chi_{\rho'}\chi_{\rho''} = \mathrm{tr}(M_{\rho'} \otimes M_{\rho''})$ and $M_{\rho'} \otimes M_{\rho''}$ is finite-dimensional, Proposition 4.18 implies that

$$\chi_{\rho'}\chi_{\rho''} = \sum_{\rho \in R} c^{\rho}_{\rho',\rho''}\chi_{\rho}, \tag{$\otimes$}$$

where $c^{\rho}_{\rho',\rho''} \geq 0$ is an integer, the number of times that the representations M_ρ is contained in $M_{\rho'} \otimes M_{\rho''}$ (this is d_ρ). The integers $c^{\rho}_{\rho',\rho''}$ are often called *Clebsch–Gordan coefficients*.

The determination of the $c^{\rho}_{\rho',\rho''}$ is usually very difficult. When $G = \mathbf{SU}(2)$, the irreducible representations can be completely determined and the $c^{\rho}_{\rho',\rho''}$ turn out to be either 1 or 0; see Chapter 5, Section 5.17. They play an important role in physics.

Since the characters are linearly independent, we see that they form a subring of $\mathcal{C}(G;\mathbb{C})$ spanned by the characters, which is a $\mathbb{Z}$-algebra having the trivial character as identity, where the characters form a basis over $\mathbb{Z}$, and whose multiplication table is given as above.

Remark: For every $\rho \in R$, the trivial representation is contained in $M_\rho \otimes M_{\overline{\rho}} = M_\rho \otimes \overline{M}_\rho$. Otherwise, by Proposition 4.10(4)(a,c) and by $(\otimes)$, we would have

$$0 = \sum_{\rho'} c^{\rho'}_{\rho,\overline{\rho}} \int_G \chi_{\rho'}\, d\lambda(s) = \int \chi_\rho(s)\overline{\chi_\rho(s)}\, d\lambda(s) = \int |\chi_\rho(s)|^2\, d\lambda(s),$$

which is absurd.

Since any irreducible representation V of G is equivalent to a unique representation M_ρ, we call ρ the *class* of V and we write $\rho = \mathrm{cl}(V)$. Any finite-dimensional representation V of G corresponds uniquely to the formal linear combinations $\mathrm{cl}(V) = \sum_{\rho\in R} d_\rho \rho$, over those ρ for which M_ρ occurs d_ρ times. The $\mathbb{Z}$-module $\mathbb{Z}^{(R)}$ of formal linear combinations $\sum_{\rho\in R_1} m_\rho \rho$, with $m_\rho \in \mathbb{Z}$ and R_1 a finite subset of R, is isomorphic to the subring of $\mathcal{C}(G;\mathbb{C})$ spanned by the characters, and we can give it a multiplication operation using formula ($\otimes$). With this multiplication, we have

$$\mathrm{cl}(U_1 \otimes U_2) = \mathrm{cl}(U_1)\mathrm{cl}(U_2).$$

This ring is the *ring of linear representations of* G. It is a substitute for the group of characters $\widehat{G}$, when G is abelian.

Definition 4.11 is a special case of the notion of the tensor product of finite-dimensional unitary representations of two locally compact groups.

Definition 4.12. Given two finite-dimensional unitary representations $U_1 \colon G_1 \to \mathbf{U}(H_1)$ and $U_2 \colon G_2 \to \mathbf{U}(H_2)$ of the locally compact (metrizable, separable) groups G_1 and G_2, we define the *tensor product* $U_1 \otimes U_2$ of U_1 and U_2 as the unitary representation $U_1 \otimes U_2 \colon G_1 \times G_2 \to \mathbf{U}(H_1 \otimes H_2)$ of $G_1 \times G_2$ in $H_1 \otimes H_2$ (with the hermitian inner product $\langle -, -\rangle_\otimes$ on $H_1 \otimes H_2$ defined in ($\langle\,\rangle_\otimes$)) given by

$$(U_1 \otimes U_2)(s_1, s_2) = U_1(s_1) \otimes U_2(s_2), \quad s_1 \in G_1,\ s_2 \in G_2,$$

where $U_1(s_1) \otimes U_2(s_2)$ is the tensor product linear map given by

$$(U_1(s_1) \otimes U_2(s_2))(x_1 \otimes x_2) = U(s_1)(x_1) \otimes U(s_2)(x_2),$$

for all $x_1 \in H_1$ and all $x_2 \in H_2$. As earlier, in terms of matrices, the linear map $U_1(s_1) \otimes U_2(s_2)$ is defined by a $(d_1d_2) \times (d_1d_2)$ matrix, namely the Kronecker product of $U_1(s_1)$ and $U_2(s_2)$. As a block matrix, it is the $d_1 \times d_1$ matrix of $d_2 \times d_2$ blocks where the (i, j) block is the matrix $U_1(s_1)_{ij}U_2(s_2)$ $(1 \leq i, j \leq d_1)$.

This time, if $U_1 \colon G_1 \to \mathbf{U}(H_1)$ and $U_2 \colon G_2 \to \mathbf{U}(H_2)$ are irreducible, then the representation $U_1 \otimes U_2 \colon G_1 \times G_2 \to \mathbf{U}(H_1 \otimes H_2)$ is also irreducible. If G_1 and G_2 are compact, this can be easily proven using the characters.

Proposition 4.20. *If G_1 and G_2 are two compact groups and if $U_1 \colon G_1 \to \mathbf{U}(H_1)$ and $U_2 \colon G_2 \to \mathbf{U}(H_2)$ are irreducible unitary representations, then the unitary representation $U_1 \otimes U_2 \colon G_1 \times G_2 \to \mathbf{U}(H_1 \otimes H_2)$ is also irreducible.*

Proof. First, observe that $G_1 \times G_2$ is compact since G_1 and G_2 are compact, and that since H_1 and H_2 must be finite-dimensional (since U_1 and U_2 are irreducible and G_1 and G_2 are compact), then $H_1 \otimes H_2$ is finite-dimensional. Recall that if V is a finite-dimensional representation of a compact group G, by Proposition 4.18 and Definition 4.8,

$$\chi_V(s) = \mathrm{tr}(V(s)) = \sum_{\rho \in R} d_\rho \chi_\rho(s), \qquad \text{for all } s \in G.$$

If V is irreducible, then V is equivalent to one of the irreducible representations M_ρ, so by Proposition 4.10(4)(a), we have $\langle \chi_V, \chi_V \rangle = \langle \chi_\rho, \chi_\rho \rangle = 1$. Conversely, if $\langle \chi_V, \chi_V \rangle = 1$, the representation V must be irreducible. Indeed, since the characters of a compact group form an orthogonal system, we have

$$\langle \chi_V, \chi_V \rangle = \left\langle \sum_{\rho \in R} d_\rho \chi_\rho, \sum_{\rho \in R} d_\rho \chi_\rho \right\rangle = \sum_{\rho \in R} d_\rho^2.$$

If $\langle \chi_V, \chi_V \rangle = 1$, then V must be equivalent to one of the irreducible representations M_ρ. We now apply the above criterion to the representation $V = U_1 \otimes U_2$ of the compact group $G = G_1 \times G_2$. Let us compute $\langle \chi_V, \chi_V \rangle$. We have

$$\begin{aligned}
\langle \chi_V, \chi_V \rangle &= \int_{G_1 \times G_2} \chi_{U_1 \otimes U_2} \overline{\chi_{U_1 \otimes U_2}}\, d\lambda_{G_1 \times G_2} \\
&= \int_{G_1 \times G_2} \chi_{U_1}(s_1)\chi_{U_2}(s_2)\overline{\chi_{U_1}(s_1)}\,\overline{\chi_{U_2}(s_2)}\, d\lambda_{G_1}(s_1)\, d\lambda_{G_2}(s_2) \\
&= \int_{G_1} \chi_{U_1}(s_1)\overline{\chi_{U_1}(s_1)}\, d\lambda_{G_1}(s_1) \int_{G_2} \chi_{U_2}(s_2)\overline{\chi_{U_2}(s_2)}\, d\lambda_{G_2}(s_2) \\
&= 1 \cdot 1 = 1.
\end{aligned}$$

Therefore, $V = U_1 \otimes U_2$ is indeed irreducible.

In the above derivation, $\lambda_{G_1 \times G_2}$ is the product of the Radon measures λ_{G_1} on G_1 and λ_{G_2} on G_2; see Folland [Folland (1999)] (Chapter 7, Section 7.4). Since λ_{G_1} and G_1 and λ_{G_2} are Haar measures, so is $\lambda_{G_1 \times G_2}$; see [Folland (1995)] (Chapter 2, Section 2.2). If G_1 and G_2 are second-countable, which is the case if they are compact, then $\lambda_{G_1 \times G_2}$ agrees with the product measure $\lambda_{G_1} \otimes \lambda_{G_2}$ (see Vol. I, Section 5.12), as shown in Folland [Folland (1999)] (Chapter 7, Section 7.4). We are also using Fubini's theorem; see Folland [Folland (1999)] (Chapter 7, Theorem 7.27). □

Remark: Observe that if $U_1 \colon G \to \mathbf{U}(H_1)$ and $U_2 \colon G \to \mathbf{U}(H_2)$ are two unitary finite-dimensional representation of G, we actually have two versions of tensor products, namely the first version which is a representation $U_1 \otimes U_2 \colon G \to \mathbf{U}(H_1 \otimes H_2)$ of G, and the second version $U_1 \otimes U_2 \colon G \times G \to \mathbf{U}(H_1 \otimes H_2)$ which is a representation of $G \times G$. This confusion could be avoided by using a different notation for the two kinds of tensor products, but in most cases it is clear which one is used. This also explains the apparent contradiction that if U_1 and U_2 are irreducible, then $U_1 \otimes U_2 \colon G \to \mathbf{U}(H_1 \otimes H_2)$ is not necessarily irreducible, while $U_1 \otimes U_2 \colon G \times G \to \mathbf{U}(H_1 \otimes H_2)$ is irreducible.

Actually, the converse of Proposition 4.20 holds.

Theorem 4.21. *Let G_1 and G_2 be two compact groups. The finite-dimensional unitary representations $U_1 \colon G_1 \to \mathbf{U}(H_1)$ and $U_2 \colon G_2 \to \mathbf{U}(H_2)$ are irreducible iff the finite-dimensional unitary representation $U_1 \otimes U_2 \colon G_1 \times G_2 \to \mathbf{U}(H_1 \otimes H_2)$ is irreducible.*

Proof. Half of the theorem was proven in Proposition 4.20. For the converse, we need to prove that if $U_1 \otimes U_2 \colon G_1 \times G_2 \to \mathbf{U}(H_1 \otimes H_2)$ is irreducible, then so are $U_1 \colon G_1 \to \mathbf{U}(H_1)$ and $U_2 \colon G_2 \to \mathbf{U}(H_2)$. Equivalently, we need to prove that if $U_1 \colon G_1 \to \mathbf{U}(H_1)$ or $U_2 \colon G_2 \to \mathbf{U}(H_2)$ is reducible, then $U_1 \otimes U_2 \colon G_1 \times G_2 \to \mathbf{U}(H_1 \otimes H_2)$ is reducible. But it is easy to see that if M_2 is an invariant subspace of H_2 for U_2, then $H_1 \otimes M_2$ is invariant for $U_1 \otimes U_2$. Similarly, it is easy to see that if M_1 is an invariant subspace of H_1 for U_1, then $M_1 \otimes H_2$ is invariant for $U_1 \otimes U_2$. $\square$

Observe that the converse of Proposition 4.20 actually holds for any locally compact groups. In fact, Theorem 4.21 holds for any locally compact (metrizable) groups. This stronger version of Theorem 4.21 is proven in Folland [Folland (1995)] (Chapter 7, Theorem 7.20). Folland also defines tensor products of Hilbert spaces and proves a version of Theorem 4.21 for unitary representations in Hilbert spaces.

If G_1 and G_2 are compact, then we have another very useful result.

Proposition 4.22. *If G_1 and G_2 are compact, then every finite-dimensional irreducible unitary representation $U \colon G_1 \times G_2 \to \mathbf{U}(H)$ is equivalent to the tensor product $U_1 \otimes U_2$ of two finite-dimensional irreducible unitary representations $U_1 \colon G_1 \to \mathbf{U}(H_1)$ and $U_2 \colon G_2 \to \mathbf{U}(H_2)$.*

Proposition 4.22 is proven in Bröcker and tom Dieck [Bröcker and tom Dieck (1985)] (Chapter 2, Section 4, Proposition 4.14) and Folland [Folland (1995)] (Chapter 7, Theorem 7.25). The proof in Bröcker and tom Dieck uses the fact that if G is a compact group and if $U\colon G \to \mathbf{U}(H)$ is a finite-dimensional representation, then there is an isomorphism

$$d\colon \bigoplus_{W} \mathrm{Hom}_G(W, U) \otimes_{\mathbb{C}} H_W \to H,$$

where $\mathrm{Hom}_G(W, U)$ is the set of G-maps between W and U (see Definition 3.3) and $W\colon G \to \mathbf{U}(H_W)$ ranges over all irreducible representations of G; only finitely many summands $\mathrm{Hom}_G(W, U)$ are not reduced to (0). The map d is the direct sum of the maps

$$d_W\colon \mathrm{Hom}_G(W, U) \otimes_{\mathbb{C}} H_W \to H$$

given by

$$d_W(\varphi \otimes w) = \varphi(w), \quad \varphi \in \mathrm{Hom}_G(W, U),\ w \in H_W.$$

See Bröcker and tom Dieck [Bröcker and tom Dieck (1985)] (Chapter 2, Section 1, Proposition 1.14).

Folland's result is more general because it applies to locally compact groups that are not necessarily compact, but of type I. The definition of a *group of type I* is given in Folland [Folland (1995)], Chapter 7, Section 7.2, page 206, and involves the notion of primary representations. A representation U is *primary* if the center of $\mathcal{C}(U)$ consists of scalar multiples of the identity (see Definition 3.9 for the definition of $\mathcal{C}(U)$). A locally compact group G *is of type I* if every primary representation of G is a direct sum of copies of some irreducible representation of G. Note that *locally compact abelian groups and compact groups are of type I*; see Folland [Folland (1995)], Chapter 7, Section 7.2, page 206. Folland proves that Proposition 4.22 holds if either G_1 or G_2 is of type I. For a further discussion regarding the characterization of groups of type I, see Folland [Folland (1995)], Chapter 7.

Folland's proof ([Folland (1995)], Chapter 7, Theorem 7.25) makes use of the following observation. For every finite-dimensional irreducible unitary representation $U\colon G_1 \times G_2 \to \mathbf{U}(H)$, by definition of the product operation in a direct product of groups, we have

$$(s_1, s_2) = (s_1, e)(e, s_2) = (e, s_2)(s_1, e), \quad s_1 \in G_1,\ s_2 \in G_2.$$

We can then define the finite-dimensional unitary representation $U^1\colon G_1 \to \mathbf{U}(H)$ given by $U^1(s_1) = U(s_1, e)$ and the finite-dimensional unitary representation $U^2\colon G_2 \to \mathbf{U}(H)$ given by $U^2(s_2) = U(e, s_2)$. The next step is to prove that U^1 is primary, which is not difficult. Since we are assuming that

G_1 is of type I, it can be shown that H is isomorphic to a tensor product $H_1 \otimes H_2$ and that $U^1 = U_1 \otimes I_{H_2}$, for some irreducible unitary representation $U_1 \colon G_1 \to \mathbf{U}(H_1)$ (here, I_{H_2} denotes the trivial representation of G_2 in H_2, namely $I_{H_2}(s_2) = \mathrm{Id}_{H_2}$ for all $s_2 \in G_2$). Using Schur's Lemma, it can be shown that $U^2(s_2) = I_{H_1} \otimes U_2(s_2)$ for some $U_2(s_2) \in \mathcal{L}(H_2)$, and that $U_2 \colon G_2 \to \mathbf{U}(H_2)$ is a unitary representation of G_2, necessarily irreducible. Finally, it is easy to show that U is equivalent to $U_1 \otimes U_2$.

Example 4.6. Theorem 4.21 and Proposition 4.22 can be used to determine the irreducible representations of $\mathbf{O}(2m+1)$ in terms of the irreducible representations of $\mathbf{SO}(2m+1)$. This is because if $Q \in \mathbf{O}(2m+1)$ and $\det(Q) = -1$, since $\det(-I_{2m+1}) = (-1)^{2m+1} = -1$, then $Q(-I) \in \mathbf{SO}(2m+1)$, and so the direct product $\mathbf{SO}(2m+1) \times \{I_{2m+1}, -I_{2m+1}\}$ is isomorphic to $\mathbf{O}(2m+1)$ under the isomorphism

$$(Q, X) \mapsto QX, \quad Q \in \mathbf{SO}(2m+1),\ X \in \{I_{2m+1}, -I_{2m+1}\}.$$

The reason why the above map is a homomorphism is that Q and $-I_{2m+1}$ commute for all $Q \in \mathbf{SO}(2m+1)$. It follows that the irreducible representations of $\mathbf{O}(2m+1)$ are the tensor product representations of irreducible representations of $\mathbf{SO}(2m+1)$ and irreducible representations of the finite abelian group $\{I_{2m+1}, -I_{2m+1}\} \simeq \mathbb{Z}/2\mathbb{Z}$, which are determined by their group of characters. These are the trivial character ρ_0 given by $\rho_0(I_{2m+1}) = \rho_0(-I_{2m+1}) = 1$ and the character ρ_1 given by $\rho_1(I_{2m+1}) = 1$ and $\rho_1(-I_{2m+1}) = -1$. Observe that ρ_1 is the determinant map. When $m = 1$, the irreducible representations of $\mathbf{SO}(3)$ can be described in terms of harmonic polynomials (see Section 5.2, Proposition 5.3), so we have a complete description of the irreducible representation of $\mathbf{O}(3)$. The irreducible representations of $\mathbf{O}(3)$ are of the form $\mathbf{R}_n \otimes \rho_k$, with $k \in \{0,1\}$ and $n \in \mathbb{N}$, or more explicitly

$$(\mathbf{R}_n \otimes \rho_k)(Q, X) = \rho_k(X)\mathbf{R}_n(Q), \quad Q \in \mathbf{SO}(3), \quad X \in \{I_{2m+1}, -I_{2m+1}\},$$
$$k \in \{0,1\},\ n \in \mathbb{N}.$$

The case of $\mathbf{O}(2m)$ is more delicate. The problem is that $-I_{2m}$ is no longer a reflection since $\det(-I_{2m}) = (-1)^{2m} = +1$. We need to use a hyperplane reflection, such as the $(2m) \times (2m)$-matrix $J = \mathrm{diag}(-1, 1, \ldots, 1)$. If $Q \in \mathbf{O}(2m)$, then $QJ \in \mathbf{SO}(2m)$. However, J does *not* commute with all matrices in $\mathbf{SO}(2m)$, so this time we have an isomorphism between the *semi-direct* product $\mathbf{SO}(2m) \rtimes \{I_{2m}, J\}$ and $\mathbf{O}(2m)$; see Section 7.4 (note that $J^2 = I_{2m}$). Unfortunately, the normal subgroup $\mathbf{SO}(2m)$ of $\mathbf{O}(2m)$ is *not* abelian for $m > 1$, which complicates matters. For $m = 1$, the group $\mathbf{SO}(2)$ is abelian, so Mackey's little group method can be used to determine the irreducible representations of $\mathbf{O}(2)$; see Section 7.4.

The irreducible representations of $\mathbf{U}(2)$ can also be determined using the following trick. The trick is that $\mathbf{U}(2)$ is isomorphic to the quotient $(\mathbf{U}(1) \times \mathbf{SU}(2))/(\{(1, I_2), -(1, I_2)\})$. But $\mathbf{U}(1) \simeq \mathbb{T}$ is a locally compact abelian group and its irreducible representations are determined by its characters χ_m, which are given by

$$e^{i\theta} \mapsto e^{im\theta}, \ m \in \mathbb{Z}, \ \theta \in \mathbb{R}/2\pi.$$

The irreducible representations of $\mathbf{SU}(2)$ are determined in Chapter 5; in particular, we have the irreducible representations U_n, with $n \in \mathbb{N}$; see Section 5.1. The fact that we need to mod out by the subgroup $\{(1, I_2), -(1, I_2)\}$ implies that the irreducible representations of $\mathbf{U}(2)$ are of the form $\chi_m \otimes U_n$, with $m + n$ even. More explicitly,

$$(\chi_m \otimes U_n)(e^{i\theta}T) = e^{im\theta}U_n(T), \ \theta \in \mathbb{R}/2\pi, \ T \in \mathbf{SU}(2),$$

with $m \in \mathbb{Z}, n \in \mathbb{N}$, and $m + n$ even. Details can be found in Bröcker and tom Dieck [Bröcker and tom Dieck (1985)] (Chapter 2, Section 5, page 87) and Folland [Folland (1995)] (Chapter 5, Section 5.4).

4.5 Contragredient Representations and Hom Representations

Later on in Chapter 8, we will need to consider the Hom representation defined by two representations $U_1 \colon G \to \mathbf{U}(H_1)$ and $U_2 \colon G \to \mathbf{U}(H_2)$. In order to promote the isomorphism between the tensor product $E_1^* \otimes E_2$ and the space of linear maps $\mathrm{Hom}(E_1, E_2)$ (where E_1 and E_2 are finite-dimensional) to representations, given a representation $U \colon G \to \mathbf{U}(H)$, we first need to define a representation $\overline{U} \colon G \to \mathbf{U}(H^*)$ defined on the dual of H^*, namely the space of linear forms on H.

We begin by reviewing the duality between a finite-dimensional hermitian vector space E and its dual E^*; for a complete exposition see Gallier and Quaintance [Gallier and Quaintance (2020c)] (Chapter 13, Section 2). For any $u \in E$, define the linear form $\varphi_u \in E^*$ by

$$\varphi_u(v) = \langle v, u \rangle, \quad v \in E.$$

Then the map $\flat$ from E to E^* given by $\flat(u) = \varphi_u$ is a semi-linear isomorphism (semi-linear means that $\flat(\lambda u) = \overline{\lambda}\, \flat(u)$, for $\lambda \in \mathbb{C}$).

Definition 4.13. Given a finite-dimensional hermitian space E, we give E^* the hermitian inner product induced by $\flat^{-1}\colon E^* \to E$, namely

$$\langle \varphi_1, \varphi_2 \rangle_{E^*} = \overline{\langle \flat^{-1}(\varphi_1), \flat^{-1}(\varphi_2) \rangle}, \quad \varphi_1, \varphi_2 \in E^*.$$

Definition 4.13 is a special case of a definition given in Bourbaki [Bourbaki (1968)] (Chapter 9, Section 1.7, Definition 9) which considers a more general situation. As observed in Bourbaki, without conjugation we obtain a left-sesquilinear form. The conjugation on the right-hand side is necessary to make $\langle -, - \rangle_{E^*}$ a right-sesquilinear form, which means that it is linear in the first argument and semi-linear in the second argument since $\flat^{-1}$ is only semi-linear. Observe that for all $u, v \in E$, we have

$$\langle \flat(u), \flat(v) \rangle_{E^*} = \overline{\langle \flat^{-1}(\flat(u)), \flat^{-1}(\flat(v)) \rangle} = \overline{\langle u, v \rangle}.$$

Also observe that if $f\colon E \to E$ is a linear map, then

$$(\varphi_u \circ f)(v) = \varphi_u(f(v)) = \langle f(v), u \rangle = \langle v, f^*(u) \rangle,$$

where f^* is the adjoint of f, which shows that

$$\varphi_u \circ f = \varphi_{f^*(u)}. \tag{$\dagger_4$}$$

If $(u_1, \ldots, u_n)$ is an orthonormal basis of E, then the definition of the hermitian inner product on E^* immediately implies that $(\varphi_{u_1}, \ldots, \varphi_{u_n})$ is an orthonormal basis of E^*. Also, we have

$$\varphi_{u_i}(u_j) = \langle u_j, u_i \rangle = \delta_{i\,j},$$

which shows that φ_{u_i} is the ith coordinate function over the basis $(u_1, \ldots, u_n)$.

Definition 4.14. Given any complex representation $U\colon G \to \mathbf{GL}(H)$ in a finite-dimensional vector space H, the *contragredient representation* $\overline{U}\colon G \to \mathbf{GL}(H^*)$ *of* $U\colon G \to \mathbf{GL}(H)$ is given by

$$\overline{U}_g(\psi) = \psi \circ U_{g^{-1}}, \quad \psi \in H^*,\ g \in G.$$

Observe that $\overline{U}_g = (U_{g^{-1}})^\top$, the transpose of the linear map $U_{g^{-1}}$.

If H a finite-dimensional vector space with a hermitian inner product and U is a unitary representation $U\colon G \to \mathbf{U}(H)$, in terms of matrices, since U_g is a unitary matrix, $\overline{U}_g$ is the conjugate of the matrix U_g. We need to check that $\overline{U}_g$ is a unitary map on H^*. For any $\psi_1, \psi_2 \in H^*$, since $\flat\colon H \to H^*$ is a bijection there are unique vectors $u_1, u_2 \in H$, such that

$\psi_1 = \varphi_{u_1}$ and $\psi_2 = \varphi_{u_2}$, and by definition of $\overline{U}$ and of the inner product on H^*, by ($\dagger_4$) and since U_g is unitary, we have

$$\begin{aligned}
\langle \overline{U}_g(\psi_1), \overline{U}_g(\psi_2)\rangle_{E^*} &= \langle \psi_1 \circ U_{g^{-1}}, \psi_2 \circ U_{g^{-1}}\rangle_{E^*} \\
&= \langle \varphi_{u_1} \circ U_{g^{-1}}, \varphi_{u_2} \circ U_{g^{-1}}\rangle_{E^*} \\
&= \langle \varphi_{U^*_{g^{-1}}(u_1)}, \varphi_{U^*_{g^{-1}}(u_2)}\rangle_{E^*} = \overline{\langle U^*_{g^{-1}}(u_1), U^*_{g^{-1}}(u_2)\rangle} \\
&= \overline{\langle U_g(u_1), U_g(u_2)\rangle} = \overline{\langle u_1, u_2\rangle} \\
&= \langle \varphi_{u_1}, \varphi_{u_2}\rangle_{E^*} = \langle \psi_1, \psi_2\rangle_{E^*}.
\end{aligned}$$

Therefore, $\overline{U}\colon G \to \mathbf{U}(H^*)$ is indeed a unitary representation.

Remark: If H is a Hilbert space (of infinite dimension), then the dual of H is the space H' of continuous linear forms on H, and by the Riesz representation theorem, $\flat\colon H \to H'$ is a bijection, so the above calculations go through and Definition 4.14 yields a unitary representation $\overline{U}\colon G \to \mathbf{U}(H')$.

We now review the relationship between $E^* \otimes F$ and $\mathrm{Hom}(E, F)$. For a complete exposition see Gallier and Quaintance [Gallier and Quaintance (2020b)] (Chapter 2, Section 2.5).

Let E and F be two vector spaces and let $\Psi\colon E^* \times F \to \mathrm{Hom}(E, F)$ be the map defined, such that

$$\Psi(u^*, f)(x) = u^*(x)f,$$

for all $u^* \in E^*$, $f \in F$, and $x \in E$. This map is clearly bilinear, and thus it induces a linear map $\Psi_\otimes\colon E^* \otimes F \to \mathrm{Hom}(E, F)$, making the following diagram commute

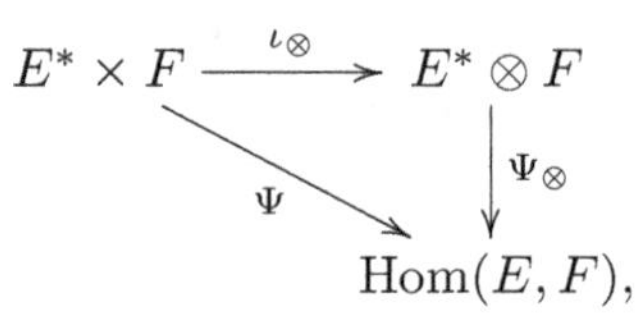

such that

$$\Psi_\otimes(u^* \otimes f)(x) = u^*(x)f. \qquad (\dagger_5)$$

Then Proposition 2.7 in Gallier and Quaintance [Gallier and Quaintance (2020b)] tells us that

(1) The linear map $\Psi_\otimes\colon E^* \otimes F \to \mathrm{Hom}(E, F)$ is injective.

(2) If E or if F is finite-dimensional, then $\Psi_\otimes \colon E^* \otimes F \to \mathrm{Hom}(E, F)$ is an isomorphism.

If E and F are finite-dimensional and if each of them has a hermitian inner product, the isomorphism $\Psi_\otimes$ can be made more concrete by picking bases. If $(u_1, \ldots, u_n)$ is an orthonormal basis of E and $(v_1, \ldots, v_m)$ is an orthonormal basis of F, then $(\varphi_{u_1}, \ldots, \varphi_{n_n})$ is an orthonormal basis of E^* (with the inner product on E^* induced by the inner product on E of Definition 4.13). Then the $m \times n$ tensors $\varphi_{u_j} \otimes v_i$ form a basis of $E^* \otimes F$, so any tensor $T \in E^* \otimes F$ can be expressed in terms of an $m \times n$ matrix $A = (a_{ij})$ $(a_{ij} \in \mathbb{C})$ as

$$T = \sum_{i=1}^{m} \sum_{j=1}^{n} a_{ij}\, \varphi_{u_j} \otimes v_i.$$

If we denote the linear map $\Psi_\otimes(\varphi_{u_j} \otimes v_i)$ from E to F as $\varphi_{u_j} v_i$, (this is the linear map such that $(\varphi_{u_j} v_i)(x) = \varphi_{u_j}(x) v_i$ for all $x \in E$), then the linear map $\Psi_\otimes(T)$ is expressed as

$$\Psi_\otimes(T) = \sum_{i=1}^{m} \sum_{j=1}^{n} a_{ij}\, \varphi_{u_j} v_i.$$

Because $\Psi_\otimes$ is an isomorphism, the linear maps $\varphi_{u_j} v_i$ are linearly independent and form a basis of $\mathrm{Hom}(E, F)$. The matrix representing the linear map $\Psi_\otimes(T)$ with respect to the bases $(u_1, \ldots, u_n)$ and $(v_1, \ldots, v_m)$ has for its j-column the coordinates of the vector

$$\Psi_\otimes(T)(u_j) = \sum_{i=1}^{m} \sum_{k=1}^{n} a_{ik}\, \varphi_{u_k}(u_j) v_i = \sum_{i=1}^{m} \sum_{k=1}^{n} a_{ik}\, \delta_{k\,j} v_i = \sum_{i=1}^{m} a_{ij} v_i,$$

and so it is also the matrix A.

We can use the isomorphism $\Psi_\otimes$ to transfer the hermitian inner product on $E^* \otimes F$ (see Definition 4.10 and Definition 4.13) to a hermitian inner product on $\mathrm{Hom}(E, F)$ so that $\Psi_\otimes$ becomes unitary.

Definition 4.15. Given two finite-dimensional hermitian spaces E and F, the inner product $\langle -, - \rangle_{\mathrm{Hom}}$ on $\mathrm{Hom}(E, F)$ is given by

$$\langle h_1, h_2 \rangle_{\mathrm{Hom}} = \langle \Psi_\otimes^{-1}(h_1), \Psi_\otimes^{-1}(h_2) \rangle_{E^* \otimes F}, \quad h_1, h_2 \in \mathrm{Hom}(E, F).$$

Observe that

$$\begin{aligned} &\langle \Psi_\otimes(u_1^* \otimes f_1), \Psi_\otimes(u_2^* \otimes f_2) \rangle_{\mathrm{Hom}} \\ &\qquad = \langle \Psi_\otimes^{-1}(\Psi_\otimes(u_1^* \otimes f_1)), \Psi_\otimes^{-1}(\Psi_\otimes(u_2^* \otimes f_2)) \rangle_{E^* \otimes F} \\ &\qquad = \langle u_1^* \otimes f_1, u_2^* \otimes f_2 \rangle_{E^* \otimes F} = \langle u_1^*, u_2^* \rangle_{E^*} \langle f_1, f_2 \rangle_F, \end{aligned}$$

so the inner product $\langle -, - \rangle_{\mathrm{Hom}}$ on $\mathrm{Hom}(E, F)$ is the inner product that makes the linear map $\Psi_{\otimes} \colon E^* \otimes F \to \mathrm{Hom}(E, F)$ an isometry.

In terms of the orthonormal bases, the tensors $\varphi_{u_j} \otimes v_i$ form an orthonormal basis of $E^* \otimes F$, so the hermitian inner product on $\mathrm{Hom}(E, F)$ is the one that makes the basis $(\varphi_{u_j} v_i)_{1 \leq i \leq m, 1 \leq j \leq n}$ orthonormal in $\mathrm{Hom}(E, F)$. Then if $h_1 \colon E \to F$ and $h_2 \colon E \to F$ are two linear maps given by the matrices A and B with respect to the bases $(u_1, \ldots, u_n)$ and $(v_1, \ldots, v_m)$, a simple computation shows that the inner product of h_1 and h_2 is given by

$$\langle h_1, h_2 \rangle_{\mathrm{Hom}} = \mathrm{tr}(B^* A) = \mathrm{tr}(A^* B),$$

the Frobenius inner product of complex matrices!

We now define the Hom-representation, first for arbitrary vector spaces not necessarily equipped with an inner product.

Definition 4.16. Let $U_1 \colon G \to \mathbf{GL}(H_1)$ and $U_2 \colon G \to \mathbf{GL}(H_2)$ be two representations. The representation $\mathrm{Hom}(U_1, U_2) \colon G \to \mathbf{GL}(\mathrm{Hom}(H_1, H_2))$ is given by

$$[\mathrm{Hom}(U_1, U_2)(g)](f) = U_2(g) \circ f \circ U_1(g^{-1}), \quad f \in \mathrm{Hom}(H_1, H_2),\ g \in G.$$

Working through the definitions, we prove the following result.

Proposition 4.23. *If $U_1 \colon G \to \mathbf{GL}(H_1)$ and $U_2 \colon G \to \mathbf{GL}(H_2)$ are two finite-dimensional representations, then the linear map $\Psi_{\otimes} \colon H_1^* \otimes H_2 \to \mathrm{Hom}(H_1, H_2)$ is an equivalence between the representations $\overline{U_1} \otimes U_2$ and $\mathrm{Hom}(U_1, U_2)$; that is, the diagram*

$$\begin{array}{ccc} H_1^* \otimes H_2 & \xrightarrow{(\overline{U_1} \otimes U_2)(g)} & H_1^* \otimes H_2 \\ \Psi_{\otimes} \downarrow & & \downarrow \Psi_{\otimes} \\ \mathrm{Hom}(H_1, H_2) & \xrightarrow[\mathrm{Hom}(U_1, U_2)(g)]{} & \mathrm{Hom}(H_1, H_2) \end{array}$$

commutes for all $g \in G$.

Proof. It suffices to prove that the maps $\Psi_{\otimes} \circ (\overline{U_1} \otimes U_2)(g)$ and $\mathrm{Hom}(U_1, U_2)(g) \circ \Psi_{\otimes}$ agree on generators. For any $h_1^* \in H_1^*$, any $x \in H_1$ and any $h_2 \in H_2$, using $(\dagger_5)$, we have

$$\begin{aligned} \Psi_{\otimes}\left((\overline{U_1} \otimes U_2)(g)(h_1^* \otimes h_2)\right)(x) &= \Psi_{\otimes}\left(\overline{U_1}(g)(h_1^*) \otimes U_2(g)(h_2)\right)(x) \\ &= \Psi_{\otimes}\left((h_1^* \circ U_1(g^{-1})) \otimes U_2(g)(h_2)\right)(x) \\ &= (h_1^* \circ U_1(g^{-1}))(x)[U_2(g)(h_2)] \\ &= (h_1^*(U_1(g^{-1})(x))[U_2(g)(h_2)]. \end{aligned}$$

We also have

$$\begin{aligned}
&[\mathrm{Hom}(U_1, U_2)(g)\,(\Psi_\otimes(h_1^* \otimes h_2))](x)\\
&\quad = [U_2(g) \circ \Psi_\otimes(h_1^* \otimes h_2) \circ U_1(g^{-1})](x)\\
&\quad = U_2(g)[\Psi_\otimes(h_1^* \otimes h_2)(U_1(g^{-1})(x))]\\
&\quad = U_2(g)[h_1^*(U_1(g^{-1})(x))h_2]\\
&\quad = h_1^*(U_1(g^{-1})(x))[U_2(g)(h_2)],
\end{aligned}$$

since $h_1^*(U_1(g^{-1})(x)) \in \mathbb{C}$ and $U_2(g)$ is linear. Thus

$$\Psi_\otimes\left((\overline{U_1} \otimes U_2)(g)(h_1^* \otimes h_2)\right)(x) = [\mathrm{Hom}(U_1, U_2)(g)\,(\Psi_\otimes(h_1^* \otimes h_2))](x),$$

as claimed. □

If H_1 and H_2 are finite-dimensional and each one has a hermitian inner product so that $U_1\colon G \to \mathbf{U}(H_1)$ and $U_2\colon G \to \mathbf{U}(H_2)$ are unitary representations, then the representation of Definition 4.16 becomes a unitary representation $\mathrm{Hom}(U_1, U_2)\colon G \to \mathbf{U}(\mathrm{Hom}(H_1, H_2))$ for the hermitian inner product on $\mathrm{Hom}(H_1, H_2)$ given in Definition 4.15 making $\Psi_\otimes$ unitary.

4.6 The Fourier Transform for Compact Groups

First, we need to discuss the notion of weak integral a bit more. The reasoning used in Section 3.3 can be immediately adapted to show the following fact. Let G be a locally compact group equipped with a Haar measure λ and let $A\colon G \to \mathbf{U}(H)$ be a map such that $s \mapsto A(s)(x)$ is continuous for any fixed $x \in H$, where H is a Hilbert space. For any function $f \in \mathrm{L}^1(G)$, for all $x, y \in H$, the function $s \mapsto f(s)\langle A(s)(x), y\rangle d\lambda(s)$ is integrable and the functional Φ_x given by

$$\Phi_x(y) = \int f(s)\langle A(s)(x), y\rangle d\lambda(s)$$

is a bounded linear functional on H, so by the Riesz representation theorem, there is a unique vector in H, called a weak integral and denoted $\int f(s)A(s)(x)d\lambda(s)$ (or even $A(f)(x)$), such that

$$\left\langle \int f(s)A(s)(x)d\lambda(s), y \right\rangle = \int f(s)\langle A(s)(x), y\rangle d\lambda(s), \quad \text{for all } x, y \in H.$$

Note that here, A is not necessarily a representation of G.

In the special case where H is a finite-dimensional space of dimensional n, we can pick an orthonormal basis in H and we can view $A(s)$ as an $n \times n$

matrix whose entry $a(s)_{ij}$ is a function on G. In this case, x and y are vectors of dimension n, so we have

$$\begin{aligned}\int f(s)\langle A(s)(x), y\rangle d\lambda(s) &= \int \sum_{i,j}^{n} f(s)a(s)_{ij}x_j\overline{y_i}d\lambda(s)\\ &= \sum_{i,j}^{n} \int f(s)a(s)_{ij}d\lambda(s)x_j\overline{y_i}\\ &\left\langle \left(\int f(s)a(s)_{ij}d\lambda(s)\right) x, y \right\rangle.\end{aligned}$$

The above shows that the weak integral $\int f(s)A(s)(x)d\lambda(s)$ is equal to the product by x of the $n \times n$ matrix $\left(\int f(s)a(s)_{ij}d\lambda(s)\right)$ obtained by integrating every entry in the matrix $f(s)A(s)$. We also denote the matrix $\left(\int f(s)a(s)_{ij}d\lambda(s)\right)$ by $\int f(s)A(s)d\lambda(s)$, or even $A(f)$.

Remark: More generally, let $\mu \in \mathcal{M}^1(G)$ be a complex regular Borel measure and let $h\colon G \to H$ be a function from G to a Hilbert space H, such that:

(1) For every $y \in H$, the map $s \mapsto \langle h(s), y\rangle$ belongs to $\mathrm{L}^1(G)$.
(2) The map $s \mapsto \|h(s)\|$ belongs to $\mathrm{L}^1(G)$.

Then there is a unique vector in H, denoted $\int h(s)d\mu$, such that

$$\left\langle \int h(s)d\mu, y \right\rangle = \int \langle h(s), y\rangle d\mu, \quad \text{for all } y \in H;$$

see Dieudonné [Dieudonné (2005)] (Chapter XIII, Section 10). The quantity $\int h(s)d\mu$ is called the weak integral of h. If we have a map $A\colon G \to \mathbf{U}(H)$ as before, for every fixed $x \in H$, if we let $h(s) = A(s)(x)$ and $\mu = fd\lambda$, we obtain the weak integral $\int f(s)A(s)(x)d\lambda(s)$ as a special case. Even more general notions of weak integrals are discussed in Folland [Folland (1995)] (Appendix 3).

We now return to the case where G is a compact group. Recall that

$$M_\rho(s) = \left(\frac{1}{n_\rho}m_{ij}(s)\right),$$

We now apply the above discussion to the matrix

$$A(t) = M_\rho(t^{-1}s).$$

Note that due to the presence of t^{-1}, for s fixed, the map $t \mapsto M_\rho(t^{-1}s)$ is not a representation.

Using the notations introduced just after Theorem 4.2, the formula

$$f = \sum_{\rho \in R} f * u_\rho, \qquad f \in \mathcal{L}^2(G),$$

given by this theorem can be written as

$$f = \sum_{\rho \in R} \left(\sum_{j=1}^{n_\rho} (f * m_{jj}^{(\rho)}) \right). \tag{$*_1$}$$

But by definition,

$$(f * m_{jj}^{(\rho)})(s) = \int f(t) m_{jj}^{(\rho)}(t^{-1}s)\, d\lambda(t),$$

so we get

$$\sum_{j=1}^{n_\rho} (f * m_{jj}^{(\rho)})(s) = n_\rho \operatorname{tr}\left(\int f(t) M_\rho(t^{-1}s)\, d\lambda(t) \right),$$

and so

$$f(s) = \sum_{\rho \in R} n_\rho \operatorname{tr}\left(\int f(t) M_\rho(t^{-1}s)\, d\lambda(t) \right). \tag{$*_2$}$$

However, we also have

$$M_\rho(\check{f}) = \int f(t) M_\rho(t^{-1})\, d\lambda(t),$$

(where again we integrate term by term), because for every $x \in \mathbb{C}^{n_g}$, by definition the vector $M_\rho(\check{f})(x)$ is the unique vector $\Phi(x) \in \mathbb{C}^{n_\rho}$, such that

$$\langle \Phi(x), y \rangle = \int f(t^{-1}) \langle M_\rho(t)(x), y \rangle\, d\lambda(t), \qquad \text{for all } y \in \mathbb{C}^{n_\rho},$$

and since G is unimodular,

$$\langle \Phi(x), y \rangle = \int f(t^{-1}) \langle M_\rho(t)(x), y \rangle\, d\lambda(t) = \int f(t) \langle M_\rho(t^{-1})(x), y \rangle\, d\lambda(t),$$

so by definition of the weak integral,

$$\Phi(x) = \int f(t) M_\rho(t^{-1})(x)\, d\lambda(t).$$

Recall that we also have

$$M_\rho(t^{-1}s) = M_\rho(t^{-1}) M_\rho(s).$$

Therefore,

$$\begin{aligned}\operatorname{tr}\left(\int f(t)M_\rho(t^{-1}s)\,d\lambda(t)\right) &= \operatorname{tr}\left(\int f(t)M_\rho(t^{-1})M_\rho(s)\,d\lambda(t)\right)\\ &= \operatorname{tr}\left(\left(\int f(t)M_\rho(t^{-1})\,d\lambda(t)\right)M_\rho(s)\right)\\ &= \operatorname{tr}\left(M_\rho(\check{f})M_\rho(s)\right),\end{aligned}$$

and by substituting this result in $(*_2)$, we obtain

$$f(s) = \sum_{\rho\in R} n_\rho \operatorname{tr}\left(M_\rho(\check{f})M_\rho(s)\right), \qquad f \in \mathrm{L}^2(G),\ s\in G. \tag{FI$_1$}$$

The above suggests the following definition for the generalization of the Fourier transform to compact groups.

Definition 4.17. Let G be a compact group. For any function $f \in \mathrm{L}^1(G)$, the *Fourier transform* $\mathcal{F}(f)$ of f is the map with domain R given by

$$\mathcal{F}(f)(\rho) = M_\rho(\check{f}) = \int f(t)M_\rho(t^{-1})\,d\lambda(t) = \int f(t)(M_\rho(t))^*\,d\lambda(t),\ \rho\in R.$$

We can view $\mathcal{F}(f)(\rho)$ as being defined as a weak integral, or in view of the discussion at the beginning of this section as the result of integrating term by term the matrix $f(t)(M_\rho(t))^*$.

Observe that $\mathcal{F}(f)(\rho) \in \mathrm{M}_{n_\rho}(\mathbb{C}) \cong \mathrm{Hom}_{\mathbb{C}}(\mathbb{C}^{n_\rho}, \mathbb{C}^{n_\rho})$. The Fourier transform of Definition 4.17 is the natural generalization of the definition of the Fourier transform when G is an abelian compact group (Vol. I, Definition 10.3),

$$\mathcal{F}(f)(\chi) = \int f(s)\overline{\chi(s)}\,d\lambda(s) = \int f(s)\chi(s^{-1})\,d\lambda(s);$$

the character χ is replaced by the irreducible representation M_ρ.

Remark: The Fourier transform of Definition 4.17 is related to the Fourier transform $\mathcal{F}_2$ defined by Hewitt and Ross [Hewitt and Ross (1970)] (Chapter VII, Definition 28.34) as the map

$$\mathcal{F}_2(f)(\rho) = \int f(t)\overline{M_\rho}(t)\,d\lambda(t) = \overline{M_\rho}(f), \qquad f\in \mathrm{L}^1(G),\ \rho\in R.$$

Let $D_\rho\colon \mathbb{C}^{n_\rho} \to \mathbb{C}^{n_\rho}$ be the semilinear map given by

$$D_\rho\left(\sum_{k=1}^{n_\rho} \alpha_k e_k\right) = \sum_{k=1}^{n_\rho} \overline{\alpha_k}\, e_k, \qquad \alpha_k\in\mathbb{C},$$

where $(e_1, \ldots, e_{n_\rho})$ is the canonical basis of $\mathbb{C}^{n_\rho}$. It is immediately verified that

$$\langle D_\rho x, D_\rho y\rangle = \langle y, x\rangle = \overline{\langle x, y\rangle},$$

and

$$D_\rho^2 = \mathrm{id}.$$

Then it is not hard to show (see Hewitt and Ross [Hewitt and Ross (1970)], Chapter IX, Lemma 34.1) that

$$\mathcal{F}(f)(\rho) = D_\rho \circ (\mathcal{F}_2(f)(\rho))^* \circ D_\rho.$$

The definition of the Fourier transform $\mathcal{F}$ given in Definition 4.17 is identical to the definition given by Kirillov; see [Kirillov (1994)] (Section 2.3) and by Folland [Folland (1995)] (Chapter 5, Section 5.3). It has the advantage that the Fourier cotransform has a simpler formulation, and since for p with $1 \leq p \leq \infty$, the spaces $\mathrm{L}^p(\widehat{G})$ are closed under adjunction and conjugation, all the results proved for the Fourier transform $\mathcal{F}_2$ in Hewitt and Ross [Hewitt and Ross (1970)] also hold for the Fourier transform $\mathcal{F}$.

Equation (FI_1) can also be written as

$$f(s) = \sum_{\rho\in R} n_\rho \,\mathrm{tr}\left(\mathcal{F}(f)(\rho) M_\rho(s)\right), \qquad f \in \mathrm{L}^2(G),\ s \in G. \tag{FI}$$

The Fourier transform $\mathcal{F}(f)$ is a function with domain R, the set of "equivalence classes" of irreducible representations of G, which plays the analog of $\widehat{G}$, to the space $\prod_{\rho\in R} \mathrm{M}_{n_\rho}(\mathbb{C})$, where $\mathrm{M}_{n_\rho}(\mathbb{C})$ is the algebra of $n_\rho \times n_\rho$ complex matrices. Every element $F \in \prod_{\rho\in R} \mathrm{M}_{n_\rho}(\mathbb{C})$ is an R-indexed sequence $F = (F(\rho))_{\rho\in R}$ of $n_\rho \times n_\rho$ matrices $F(\rho)$. Sequences in $\prod_{\rho\in R} \mathrm{M}_{n_\rho}(\mathbb{C})$ are added, multiplied, and multiplied by a scalar, componentwise. Thus $\prod_{\rho\in R} \mathrm{M}_{n_\rho}(\mathbb{C})$ is a (complex) algebra. Given $F \in \prod_{\rho\in R} \mathrm{M}_{n_\rho}(\mathbb{C})$, the adjoint F^* of F is defined componentwise by $F^* = (F_\rho^*)_{\rho\in R}$.

Definition 4.18. We define $\widehat{G}$ as $R(G)$ the set of indices of a complete set of unitary irreducible representations of G (see the comment just after Theorem 4.16).

Note the analogy to the situation where $G = \mathbb{T}$ and $\widehat{G} = \widehat{\mathbb{T}} = \mathbb{Z}$, except that, $\mathrm{L}^1(\widehat{\mathbf{T}}) = l^1(\mathbb{Z})$ consists of $\mathbb{Z}$-indexed sequences of complex numbers, but the $F(\rho)$ are matrices. By analogy with the case $G = \mathbb{T}$ and $\widehat{\mathbb{T}} = \mathbb{Z}$, where the numbers $\mathcal{F}(f)(m) = \widehat{f}(m)$ are the Fourier coefficients of $f \in \mathrm{L}^1(\mathbb{T})$, the endomorphisms $\mathcal{F}(f)(\rho) \in \mathrm{Hom}_{\mathbb{C}}(\mathbb{C}^{n_\rho}, \mathbb{C}^{n_\rho})$, represented by matrices in $\mathrm{M}_{n_\rho}(\mathbb{C})$, can be viewed as *generalized Fourier coefficients* of $f \in \mathrm{L}^1(G)$, where G is a compact group.

Equation (FI) is a kind of Fourier inversion formula. Explicit examples of the Fourier transform and Fourier inversion formula (FI) will be given in Section 5.15 for the groups $\mathbf{SU}(2)$ and $\mathbf{SO}(3)$.

We can define the Fourier cotransform $\overline{\mathcal{F}}$, defined on $\prod_{\rho\in R} \mathrm{M}_{n_\rho}(\mathbb{C})$, with input in G, by

$$\overline{\mathcal{F}}(F)(s) = \sum_{\rho \in R} n_\rho \operatorname{tr}(F(\rho)M_\rho(s)), \qquad F \in \prod_{\rho\in R} \mathrm{M}_{n_\rho}(\mathbb{C}),\ s \in G. \tag{FC}$$

Of course, there is an issue of convergence with (FC). The space $\prod_{\rho\in R} \mathrm{M}_{n_\rho}(\mathbb{C})$ is just too big, so following Hewitt and Ross [Hewitt and Ross (1970)] (Chapter VII, Section 28.24), we define normed subspaces $\mathrm{L}^p(\widehat{G})$ as follows.

First, we need to define some norms on $n \times n$ matrices introduced by von Neumann.

4.7 von Neumann Norms and the Algebras $\mathbf{L}^p(\widehat{G})$

Definition 4.19. Let $A \in \mathrm{M}_n(\mathbb{C})$ be any complex $n \times n$ matrix, and let $(\sigma_1, \ldots, \sigma_n)$ be the sequence of nonnegative square roots of the eigenvalues of A^*A listed in any order (the positive square roots are the *singular values* of A). For any p, $1 \le p < \infty$, define the *von Neumann norm* $\|A\|_{\varphi_p}$ of A by

$$\|A\|_{\varphi_p} = \left(\sum_{k=1}^{n} \sigma_k^p\right)^{1/p},$$

and $\|A\|_{\varphi_\infty}$ by

$$\|A\|_{\varphi_\infty} = \max_{1\le k\le n} \sigma_k.$$

It is not obvious that the functions defined in Definition 4.19 are matrix norms, but this is proven in Hewitt and Ross, see [Hewitt and Ross (1970)] (Appendix D, Theorem D40).

Since $(\sigma_1^2, \ldots, \sigma_n^2)$ are the eigenvalues of A^*A, we see that

$$\|A\|_{\varphi_2}^2 = \sum_{k=1}^{n} \sigma_k^2 = \operatorname{tr}(A^*A) = \|A\|_{\mathrm{HS}}^2,$$

where $\|A\|_{\mathrm{HS}}$ is a *Hilbert–Schmidt norm*, also known as *Frobenius norm*, of A (see Vol. I, Definition B.6). We also have

$$\|A\|_{\varphi_1} = \sum_{k=1}^{n} \sigma_k,$$

and

$$\|A\|_{\varphi_\infty} = \max_{1 \leq k \leq n} \sigma_k = \|A\|_2\,,$$

where $\|A\|_2$ is the *operator norm* induced by the 2-norm; see Vol. I, Definition B.7 and Proposition B.8.

Next, we use the norms of Definition 4.19 to define norms on $\prod_{\rho \in R} \mathrm{M}_{n_\rho}(\mathbb{C})$.

Definition 4.20. For any fixed sequence $(a_\rho)_{\rho \in R}$ of reals $a_\rho \geq 1$, for any $F \in \prod_{\rho \in R} \mathrm{M}_{n_\rho}(\mathbb{C})$, if $1 \leq p < \infty$, define $\|F\|_p$ by

$$\|F\|_p = \left(\sum_{\rho \in R} a_\rho \|F(\rho)\|_{\varphi_p}^p\right)^{1/p},$$

and for $p = \infty$, let

$$\|F\|_\infty = \sup_{\rho \in R} \|F(\rho)\|_{\varphi_\infty}\,,$$

where $\|F(\rho)\|_{\varphi_p}$ is the von Neumann p-norm of the matrix $F(\rho)$. Observe that for $p = 2$, we have

$$\|F\|_2 = \left(\sum_{\rho \in R} a_\rho \|F(\rho)\|_{\mathrm{HS}}^2\right)^{1/2} = \left(\sum_{\rho \in R} a_\rho \operatorname{tr}\Big(F(\rho)^* F(\rho)\Big)\right)^{1/2}.$$

Following Hewitt and Ross [Hewitt and Ross (1970)] (Chapter VII, Section 28.24), we make the following definitions.

Definition 4.21. Denote $\prod_{\rho \in R} \mathrm{M}_{n_\rho}(\mathbb{C})$ by $\mathfrak{E}(\widehat{G})$. Pick a fixed sequence $(a_\rho)_{\rho \in R}$ of reals $a_\rho \geq 1$. Let $\mathfrak{E}(\widehat{G})_{0,0}$ be the subspace of $\mathfrak{E}(\widehat{G})$ consisting of all sequences $F = (F(\rho))_{\rho \in R}$, such that the set $\{\rho \in R \mid F(\rho) \neq 0\}$ is finite, and let $\mathfrak{E}(\widehat{G})_0$ be the subspace of $\mathfrak{E}(\widehat{G})$ consisting of all sequences $F = (F(\rho))_{\rho \in R}$, such that the set $\{\rho \in R \mid \|F(\rho)\|_{\varphi_\infty} \geq \epsilon\}$ is finite for all $\epsilon > 0$.

For any p with $1 \leq p \leq \infty$, we define $\mathrm{L}^p(R) = \mathrm{L}^p(\widehat{G})$ as

$$\mathrm{L}^p(\widehat{G}) = \left\{F \in \prod_{\rho \in R} \mathrm{M}_{n_\rho}(\mathbb{C}) \mid \|F\|_p < \infty\right\} = \left\{F \in \mathfrak{E}(\widehat{G}) \mid \|F\|_p < \infty\right\}.$$

The following results are shown in Hewitt and Ross [Hewitt and Ross (1970)] (Theorem 28.25 and Theorem 28.26).

Proposition 4.24. *Let G be a compact group. For any fixed sequence $(a_\rho)_{\rho\in R}$ of reals $a_\rho \geq 1$, for any p, such that $1 \leq p \leq \infty$, the space $\mathrm{L}^p(\widehat{G})$ is a Banach space. For any $F \in \mathrm{L}^p(\widehat{G})$, we have $F^* \in \mathrm{L}^p(\widehat{G})$ and $\|F^*\|_p = \|F\|_p$. The space $\mathrm{L}^\infty(\widehat{G})$ is a Banach algebra under component-wise multiplication, and $\|FF^*\|_\infty = \|F\|_\infty^2$ for any $F \in \mathrm{L}^\infty(\widehat{G})$.*

The following result is also shown in Hewitt and Ross [Hewitt and Ross (1970)] (Theorem 28.27).

Proposition 4.25. *Let G be a compact group, and let $(a_\rho)_{\rho\in R}$ be any fixed sequence of reals $a_\rho \geq 1$. With the norm $\| \ \|_\infty$, the space $\mathfrak{E}(\widehat{G})_0$ is a closed two-sided ideal of $\mathrm{L}^\infty(\widehat{G})$. For any p, such that $1 \leq p < \infty$, the space $\mathfrak{E}(\widehat{G})_{0,0}$ is a dense two-sided ideal of $\mathfrak{E}(\widehat{G})_0$, and a dense two-sided ideal of $\mathrm{L}^p(\widehat{G})$. Both $\mathfrak{E}(\widehat{G})_{0,0}$ and $\mathfrak{E}(\widehat{G})_0$ are closed under adjunction ($F \mapsto F^*$).*

It is also possible to define an inner product on $\mathrm{L}^p(\widehat{G})$ based on the following proposition shown in Hewitt and Ross [Hewitt and Ross (1970)] (Lemma 28.28).

Proposition 4.26. *Let G be a compact group, and let $(a_\rho)_{\rho\in R}$ be any fixed sequence of reals $a_\rho \geq 1$. For any p, $1 \leq p \leq \infty$, if q is defined such that $\frac{1}{p} + \frac{1}{q} = 1$, then for all $E \in \mathrm{L}^p(\widehat{G})$ and all $F \in \mathrm{L}^q(\widehat{G})$, the following facts hold:*

(1) The number

$$\langle E, F\rangle = \sum_{\rho\in R} a_\rho \mathrm{tr}(F_\rho^* E_\rho)$$

is well-defined (the series converges absolutely).

(2) We have

$$\langle F, E\rangle = \overline{\langle E, F\rangle}.$$

(3) (Hölder's inequality)

$$|\langle E, F\rangle| \leq \|E\|_p \|F\|_q \,.$$

Then we have the following theorem shown in Hewitt and Ross [Hewitt and Ross (1970)] (Theorem 28.30).

Theorem 4.27. *Let G be a compact group, and let $(a_\rho)_{\rho\in R}$ be any fixed sequence of reals $a_\rho \geq 1$. The space $\mathrm{L}^2(\widehat{G})$ is a Hilbert space with the inner product*

$$\langle E, F\rangle = \sum_{\rho\in R} a_\rho \mathrm{tr}(F_\rho^* E_\rho),$$

and we have

$$\|E\|_2^2 = \langle E, E\rangle.$$

We also have the following result shown in Hewitt and Ross [Hewitt and Ross (1970)] (Theorem 28.32).

Proposition 4.28. *Let G be a compact group, and let $(a_\rho)_{\rho\in R}$ be any fixed sequence of reals $a_\rho \geq 1$.*

(1) For any p, such that $1 \leq p \leq \infty$, if q is such that $\frac{1}{p} + \frac{1}{q} = 1$, for any $E \in \mathrm{L}^p(\widehat{G})$ and $F \in \mathrm{L}^q(\widehat{G})$, we have $EF \in \mathrm{L}^1(\widehat{G})$, and

$$\|EF\|_1 \leq \|E\|_p \|F\|_q .$$

(2) For any p, q, such that $1 \leq p < q \leq \infty$, we have

$$\mathrm{L}^p(\widehat{G}) \subseteq \mathrm{L}^q(\widehat{G}),$$

and for every $E \in \mathrm{L}^p(\widehat{G})$,

$$\|E\|_q \leq \|E\|_p .$$

(3) For any p, such that $1 \leq p \leq \infty$, for all $E, F \in \mathrm{L}^p(\widehat{G})$, we have $EF \in \mathrm{L}^p(\widehat{G})$, and

$$\|EF\|_p \leq \|E\|_p \|F\|_p .$$

We now have the following results about the Fourier transform on a compact group, generalizing similar results about the Fourier transform on $\mathbb{T}$. *From now on, we assume that the sequence $(a_\rho)_{\rho\in R}$ of reals $a_\rho \geq 1$ is the sequence of positive integers $(n_\rho)_{\rho\in R}$.*

Theorem 4.29. *Let G be a compact group.*

(1) If we define the multiplication on $\mathrm{L}^\infty(\widehat{G})$ as $(F_1 \cdot F_2)(\rho) = F_2(\rho)F_1(\rho)$, then the map $f \mapsto \mathcal{F}(f)$ is a non norm-increasing injective involutive algebra homomorphism from $\mathrm{L}^1(G)$ into $\mathrm{L}^\infty(\widehat{G})$. In particular, for all $f, g \in \mathrm{L}^1(G)$, for all $\rho \in R$, we have

$$(\mathcal{F}(f * g))(\rho) = \mathcal{F}(g)(\rho) \circ \mathcal{F}(f)(\rho).$$

(2) For every $\rho \in R$, the map $f \mapsto \mathcal{F}(f)(\rho)$ is an algebra representation of $\mathrm{L}^1(G)$ *in* $\mathbb{C}^{n_\rho}$.

Proof sketch. Theorem 4.29 is proven in Hewitt and Ross [Hewitt and Ross (1970)] (Theorem 28.36); see also Folland [Folland (1995)] (Section 5.3, Equations 5.17, 5.18). It is instructive to prove that

$$(\mathcal{F}(f*g))(\rho) = \mathcal{F}(g)(\rho) \circ \mathcal{F}(f)(\rho).$$

By definition as a weak integral, $(\mathcal{F}(f*g)(\rho))(x)$ is the unique vector, such that

$$\langle (\mathcal{F}(f*g)(\rho))(x), y\rangle = \int \langle M_\rho^*(s)(x), y\rangle (f*g)(s)\, d\lambda(s), \quad \text{for all } x, y \in \mathbb{C}^{n_\rho},$$

and using the fact $(\mathcal{F}(g)(\rho))(z)$ is the unique vector, such that

$$\langle (\mathcal{F}(g)(\rho))(z), y\rangle = \int \langle M_\rho^*(s)(z), y\rangle g(s)\, d\lambda(s), \qquad \text{for all } y, z \in \mathbb{C}^{n_\rho},$$

and $(\mathcal{F}(f)(\rho))(x)$ is the unique vector, such that

$$\langle (\mathcal{F}(f)(\rho))(x), y\rangle = \int \langle M_\rho^*(s)(x), y\rangle f(s)\, d\lambda(s), \qquad \text{for all } x, y \in \mathbb{C}^{n_\rho},$$

we have

$$\begin{aligned}
&\int \langle M_\rho^*(s)(x), y\rangle (f*g)(s)\, d\lambda(s)\\
&\qquad = \int \langle M_\rho^*(s)(x), y\rangle \int f(t) g(t^{-1}s)\, d\lambda(t)\, d\lambda(s)\\
&\qquad = \int \langle M_\rho^*(t^{-1}s) M_\rho^*(t)(x), y\rangle \int f(t) g(t^{-1}s)\, d\lambda(t)\, d\lambda(s)\\
&\qquad = \int \left(\int \langle M_\rho^*(t^{-1}s) M_\rho^*(t)(x), y\rangle g(t^{-1}s)\, d\lambda(s)\right) f(t)\, d\lambda(t)\\
&\qquad = \int \left(\int \langle M_\rho^*(s) M_\rho^*(t)(x), y\rangle g(s)\, d\lambda(s)\right) f(t)\, d\lambda(t)\\
&\qquad = \int \langle (\mathcal{F}(g)(\rho))(M_\rho^*(t)(x)), y\rangle f(t)\, d\lambda(t)\\
&\qquad = \int \langle M_\rho^*(t)(x), (\mathcal{F}(g)(\rho))^*(y)\rangle f(t)\, d\lambda(t)\\
&\qquad = \langle (\mathcal{F}(f)(\rho))(x), (\mathcal{F}(g)(\rho))^*(y)\rangle\\
&\qquad = \langle (\mathcal{F}(g)(\rho))((\mathcal{F}(f)(\rho))(x)), y\rangle,
\end{aligned}$$

which proves that

$$(\mathcal{F}(f*g))(\rho) = \mathcal{F}(g)(\rho) \circ \mathcal{F}(f)(\rho),$$

as claimed. ☐

Observe that part (1) of Theorem 4.29 implies that the map $f \mapsto \mathcal{F}(f)$ is continuous since the operator norm of this map is bounded by 1.

Remark: Notice that the order of f and g is switched on the right-hand side. This is the reason why, if we want $\mathcal{F}$ to be a homomorphism, that we have to switch the order of the arguments in the multiplication on $\mathrm{L}^\infty(\widehat{G})$. If we use the Fourier transform $\mathcal{F}_2$ instead of the Fourier transform $\mathcal{F}$, then we get

$$(\mathcal{F}_2(f * g))(\rho) = \mathcal{F}_2(f)(\rho) \circ \mathcal{F}_2(g)(\rho),$$

as in Hewitt and Ross [Hewitt and Ross (1970)].

Definition 4.22. Let G be a compact group. For any $\rho \in R$, let $\mathcal{T}_\rho$ be the space of functions from G to $\mathbb{C}$ spanned by the set of functions

$$s \mapsto \langle M_\rho(s)(x), y\rangle, \qquad x, y \in \mathbb{C}^{n_\rho},$$

called *matrix coefficients*. Let $\mathcal{T}(G)$ be the space of functions spanned by the set

$$\bigcup_{\rho \in R} \mathcal{T}_\rho(G).$$

Since the M_ρ are representations, we have $\mathcal{T}(G) \subseteq \mathcal{C}(G; \mathbb{C})$.

Theorem 4.30. *Let G be a compact group.*

(1) For every $\rho \in R$, we have

$$\{\mathcal{F}(f)(\rho) \mid f \in \mathcal{T}_\rho(G)\} = \mathrm{Hom}_{\mathbb{C}}(\mathbb{C}^{n_\rho}, \mathbb{C}^{n_\rho}).$$

(2) We have

$$\{\mathcal{F}(f) \mid f \in \mathcal{T}(G)\} = \mathfrak{C}_{0,0}(\widehat{G}).$$

Theorem 4.30 is proven in Hewitt and Ross [Hewitt and Ross (1970)] (Theorem 28.39).

Theorem 4.31. *Let G be a compact group. The map $f \mapsto \mathcal{F}(f)$ is a non norm-increasing involutive isomorphism of $\mathrm{L}^1(G)$ onto a dense subalgebra of $\mathfrak{C}_0(\widehat{G}) \subseteq \mathrm{L}^\infty(\widehat{G})$. In particular, the map $f \mapsto \mathcal{F}(f)$ is continuous.*

Theorem 4.31 is proven in Hewitt and Ross [Hewitt and Ross (1970)] (Theorem 28.40). It is a version of the Riemann–Lebesgue lemma for compact groups; indeed, since $\widehat{G} = R$ is discrete, by definition of $\mathfrak{C}_0(\widehat{G})$, we can view the vectors in $\mathfrak{C}_0(\widehat{G})$ as functions of $\rho \in R$ that tend to zero at infinity. See Vol. I, Proposition 10.18 in the case of abelian locally compact groups.

Finally, we have the following version of Plancherel's theorem.

Theorem 4.32. *(Plancherel) Let G be a compact group. The map $f \mapsto \mathcal{F}(f)$ is an isometric isomorphism between the Hilbert space $\mathrm{L}^2(G)$ and the Hilbert space $\mathrm{L}^2(\widehat{G})$. In particular, the map $f \mapsto \mathcal{F}(f)$ is continuous. If we pick any orthonormal basis $(e_1^\rho, \ldots, e_{n_\rho}^\rho)$ in $\mathbb{C}^{n_\rho}$, then for every $f \in \mathrm{L}^2(G)$, we have*

$$f = \sum_{\rho \in R} n_\rho \sum_{j,k=1}^{n_\rho} \langle (\mathcal{F}(f)(\rho))(e_k^\rho), e_j^\rho \rangle u_{jk}^\rho,$$

where u_{jk}^ρ is the function on G given by

$$u_{jk}^\rho(s) = \langle M_\rho(s)(e_k^\rho), e_j^\rho \rangle, \qquad s \in G,\ 1 \leq j, k \leq n_\rho.$$

The functions u_{jk}^ρ are called the *coordinate functions for M_ρ and the basis $(e_1^\rho, \ldots, e_{n_\rho}^\rho)$*. Theorem 4.32 is proven in Hewitt and Ross [Hewitt and Ross (1970)] (Theorem 28.43).

We now return to the Fourier cotransform.

4.8 Fourier Inversion for Compact Groups

Definition 4.23. Let G be a compact group. For any $F \in \mathfrak{E}(\widehat{G}) = \prod_{\rho \in R} \mathrm{M}_{n_\rho}(\mathbb{C})$, the *Fourier cotransform* $\overline{\mathcal{F}}(F)$ of F is the function on G given by

$$\overline{\mathcal{F}}(F)(s) = \sum_{\rho \in R} n_\rho \operatorname{tr}(F(\rho) M_\rho(s)), \qquad s \in G.$$

In the above definition, the infinite sum should be viewed as a formal expression. We will give below sufficient conditions that guarantee convergence.

Remark: Since Hewitt and Ross use the Fourier transform $\mathcal{F}_2$, related to the Fourier transform $\mathcal{F}$ by the equation

$$\mathcal{F}(f)(\rho) = D_\rho \circ (\mathcal{F}_2(f)(\rho))^* \circ D_\rho,$$

the definition of the Fourier cotransform (called inverse Fourier transform) given by Hewitt and Ross (Chapter IX, Section 34.47) is

$$\overline{\mathcal{F}}_2(F)(s) = \sum_{\rho \in R} n_\rho \operatorname{tr}(D_\rho F(\rho)^* D_\rho M_\rho(s)), \qquad s \in G.$$

Following Hewitt and Ross, it is natural to make the following definition (see [Hewitt and Ross (1970)], Definition 34.4).

Definition 4.24. Let G be a compact group. The subspace $\mathfrak{R}(G)$ of $\mathrm{L}^1(G)$ is defined by

$$\begin{aligned}\mathfrak{R}(G) &= \{f \in \mathrm{L}^1(G) \mid \|\mathcal{F}(f)\|_1 < \infty\}\\ &= \left\{ f \in \mathrm{L}^1(G) \;\middle|\; \sum_{\rho \in R} n_\rho \|\mathcal{F}(f)(\rho)\|_{\varphi_1} < \infty \right\}.\end{aligned}$$

The subspace $\mathfrak{R}(G)$ is called the space of *absolutely convergent Fourier series*. We define $\|f\|_{\varphi_1}$ by

$$\|f\|_{\varphi_1} = \|\mathcal{F}(f)\|_1 .$$

For any function $f \in \mathrm{L}^1(G)$, the formal expression

$$(\overline{\mathcal{F}}(\mathcal{F}(f)))(s) = \sum_{\rho \in R} n_\rho \operatorname{tr}(\mathcal{F}(f)(\rho) M_\rho(s))$$

is called the *Fourier series of f*.

Observe that Definition 4.24 is the generalization of the case $G = \mathbb{T}$ and $\widehat{\mathbb{T}} = \mathbb{Z}$, where for every $f \in \mathrm{L}^1(\mathbb{T})$ we define the Fourier series of f as the map

$$\theta \mapsto \sum_{m \in \mathbb{Z}} \widehat{f}(m) e^{im\theta} = (\overline{\mathcal{F}}(\widehat{f}))(\theta).$$

Here the character $\theta \mapsto e^{im\theta}$ is replaced by the irreducible representation M_ρ, and the trace function is needed to convert the matrix $\mathcal{F}(f)(\rho)M_\rho(s)$ to a number (and the dimensions n_ρ must be accounted for).

The following results are shown in Hewitt and Ross [Hewitt and Ross (1970)] (Theorem 34.5).

Theorem 4.33. *Let G be a compact group.*

(1) If $F \in \mathrm{L}^1(\widehat{G})$, then the map

$$s \mapsto \sum_{\rho \in R} n_\rho \left|\operatorname{tr}(F(\rho) M_\rho(s))\right|$$

is uniformly convergent.

(2) The map

$$s \mapsto (\overline{\mathcal{F}}(F))(s) = \sum_{\rho \in R} n_\rho \operatorname{tr}(F(\rho) M_\rho(s))$$

converges uniformly to a continuous function f. Furthermore, we have the Fourier inversion formula

$$(\overline{\mathcal{F}}(\mathcal{F}(f)))(s) = \sum_{\rho \in R} n_\rho \operatorname{tr}(\mathcal{F}(f)(\rho) M_\rho(s)), \qquad s \in G,$$

where $(\overline{\mathcal{F}}(\mathcal{F}(f)))(s)$ is the Fourier series of f, so $f \in \mathfrak{R}(G)$.

(3) *We have*

$$\|f\|_\infty \leq \|f\|_{\varphi_1} = \|\mathcal{F}(f)\|_1 ,$$

where $\|f\|_\infty$ *is the sup norm on* $\mathcal{C}(G;\mathbb{C})$.

The Fourier series of f is not necessarily convergent, but we have the following results; see Hewitt and Ross [Hewitt and Ross (1970)] (Corollary 34.6 and Corollary 34.7).

Theorem 4.34. *Let* G *be a compact group.*

(1) *For any function* $f \in \mathfrak{R}(G)$, *the Fourier series of* f *converges uniformly and*

$$f = (\overline{\mathcal{F}}(\mathcal{F}(f)))(s) = \sum_{\rho\in R} n_\rho \operatorname{tr}(\mathcal{F}(f)(\rho)M_\rho(s)),$$

for almost all $s \in G$. *We have*

$$\|f\|_\infty \leq \|f\|_{\varphi_1} .$$

(2) *The map* $f \mapsto \mathcal{F}(f)$ *is a norm-preserving linear isomorphism from* $\mathfrak{R}(G)$ *onto* $\mathrm{L}^1(\widehat{G})$, *so* $\mathfrak{R}(G)$ *is a Banach space.*

For the record, in view of Theorem 4.32 and (FI), we have the following result (see also Hewitt and Ross [Hewitt and Ross (1970)], Chapter IX, Section 34.47(a)).

Theorem 4.35. *(Fourier inversion for* $\mathrm{L}^2(G)$*) Let* G *be a compact group. The Fourier cotransform* $\overline{\mathcal{F}}(F) \in \mathrm{L}^2(G)$ *of any* $F \in \mathrm{L}^2(\widehat{G})$ *converges as a series in the* L^2*-norm, and for every* $f \in \mathrm{L}^2(G)$, *we have*

$$f(s) = (\overline{\mathcal{F}}(\mathcal{F}(f)))(s) = \sum_{\rho\in R} n_\rho \operatorname{tr}\left(\mathcal{F}(f)(\rho)M_\rho(s)\right), \qquad s \in G,$$

where the series converges in the L^2*-norm.*

Remark: There is another way to reprove Proposition 4.10(4)(b). Indeed, we have

$$\operatorname{tr}\left(\int f(t)M_\rho(t^{-1}s)\,d\lambda(t)\right) = \int f(t)\operatorname{tr}\left(M_\rho(t^{-1}s)\right)\,d\lambda(t) = (f * \chi_\rho)(s).$$

Therefore, by $(*_2)$, we have

$$f = \sum_{\rho\in R} n_\rho\, f * \chi_\rho, \qquad f \in \mathrm{L}^2(G). \tag{FI'}$$

See also Hewitt and Ross [Hewitt and Ross (1970)] (Chapter XII, Theorem 27.41).

Example 4.7. If G is a finite group, then $\widehat{G} = \{\rho_1, \dots, \rho_r\}$, where r is the number of conjugacy classes of G. If we give G the counting measure normalized so that G has measure 1, then the Fourier transform of any function $f \in \mathrm{L}^2(G)$ is given by

$$\mathcal{F}(f)(\rho) = \frac{1}{|G|} \sum_{s \in G} f(s)(M_\rho(s))^*,$$

where $M_{\rho_1}, \dots, M_{\rho_r}$ are the irreducible representations of G (up to equivalence). For every $F \in \mathrm{L}^2(\widehat{G})$, the Fourier cotransform of F is given by

$$\overline{\mathcal{F}}(F)(s) = \sum_{k=1}^{r} n_{\rho_k} \mathrm{tr}(F(\rho_k) M_{\rho_k}(s)), \qquad s \in G,$$

and the Fourier inversion formula is given by

$$f = \sum_{k=1}^{r} n_{\rho_k} \mathrm{tr}((\mathcal{F}(f))(\rho_k) M_{\rho_k}(s)).$$

The fact that $\mathcal{F}$ is an isometry is expressed by the equation

$$\langle f_1, f_2 \rangle = \frac{1}{|G|} \sum_{s \in G} f_1(s)\overline{f_2(s)} = \langle \mathcal{F}(f_1), \mathcal{F}(f_2) \rangle = \sum_{k=1}^{r} n_{\rho_k} \mathrm{tr}((\mathcal{F}(f_2))^* \mathcal{F}(f_1)),$$

for all $f_1, f_2 \in \mathrm{L}^2(G)$.

For all $f, g \in \mathrm{L}^2(G)$, the convolution $f * g$ is given by

$$(f * g)(s) = \frac{1}{|G|} \sum_{s_1 s_2 = s} f(s_1) g(s_2) = \frac{1}{|G|} \sum_{t \in G} f(t) g(t^{-1} s),$$

and we can write explicitly the equation

$$(\mathcal{F}(f * g))(\rho) = (\mathcal{F}(g))(\rho) \circ (\mathcal{F}(f))(\rho).$$

We leave it to the diligence of the reader to check that it holds.

4.9 Problems

Problem 4.1. Prove Proposition 4.10.

Hint. Use Theorem 4.2 and Theorem 4.6.

Problem 4.2. Show that the function which maps the class of a function $f \in \mathcal{L}^2(G)$ to the class of its complex conjugate $\overline{f}$ is a semilinear bijection of $\mathrm{L}^2(G)$ and an automorphism of its ring structure (under convolution).

Hint. Use the fact that $\overline{f * g} = \overline{f} * \overline{g}$.

Problem 4.3. Given two linear maps $f\colon E \to E'$ and $g\colon F \to F'$, recall that $f \otimes g\colon E \otimes F \to E' \otimes F'$ is the unique linear map, such that

$$(f \otimes g)(u \otimes v) = f(u) \otimes g(v), \quad \text{for all } u \in E \text{ and all } v \in F. \qquad (f \otimes g)$$

Given a basis $(u_1, \ldots, u_{d_1})$ of H_1 and a basis $(v_1, \ldots, v_{d_2})$ of H_2, assume f_1 is represented by the matrix A_1 and f_2 is represented by the matrix A_2. Show that with respect to the basis $(u_i \otimes v_j)_{1 \leq i \leq d_1, 1 \leq j \leq d_2}$, the linear map $f_1 \otimes f_2$ is defined by a $(d_1 d_2) \times (d_1 d_2)$ matrix; as a block matrix, it is the $d_1 \times d_1$ matrix of $d_2 \times d_2$ blocks where the (i, j) block is the matrix $(A_1)_{ij} A_2$ $(1 \leq i, j \leq d_1)$.

Problem 4.4. Let $(u_1, \ldots, u_{d_1})$ be an orthonormal basis of H_1, and let $(v_1, \ldots, v_{d_2})$ be an orthonormal basis of H_2. Show that $(u_i \otimes v_j)_{1 \leq i \leq d_1, 1 \leq j \leq d_2}$ is an orthonormal basis of $H_1 \otimes H_2$ with respect to the inner product $\langle -, - \rangle_{\otimes}$.

Problem 4.5.

(1) Show that the linear map $U_1(s) \otimes U_2(s)$ is given by the $(d_1 d_2) \times (d_1 d_2)$ matrix, which is the Kronecker product of $U_1(s)$ and $U_2(s)$.
(2) Prove the following two identities:

$$\begin{aligned}
\operatorname{tr}(U_1(s) \oplus U_2(s)) &= \operatorname{tr}(U_1(s)) + \operatorname{tr}(U_2(s)) \\
\operatorname{tr}(U_1(s) \otimes U_2(s)) &= \operatorname{tr}(U_1(s)) \operatorname{tr}(U_2(s)).
\end{aligned}$$

Problem 4.6. Let R be a complete set of irreducible representations of a compact group G. Show that the $\mathbb{Z}$-module $\mathbb{Z}^{(R)}$ of formal linear combinations $\sum_{\rho \in R_1} m_\rho \rho$, with $m_\rho \in \mathbb{Z}$ and R_1 a finite subset of R, is isomorphic to the subring of $\mathcal{C}(G; \mathbb{C})$ spanned by the characters.

Hint. To define a multiplication structure on $\mathbb{Z}^{(R)}$ see formula $(\otimes)$.

Problem 4.7. (Advanced Exercise) Prove Proposition 4.22.

Hint. See Bröcker and tom Dieck [Bröcker and tom Dieck (1985)] (Chapter 2, Section 4, Proposition 4.14) or Folland [Folland (1995)] (Chapter 7, Theorem 7.25).

Problem 4.8. If $(u_1, \ldots, u_n)$ is an orthonormal basis of E, use Definition 4.13 to show that $(\varphi_{u_1}, \ldots, \varphi_{u_n})$ is an orthonormal basis of E^*.

Problem 4.9. If $h_1\colon E \to F$ and $h_2\colon E \to F$ are two linear maps given by the matrices A and B with respect to the bases $(u_1, \ldots, u_n)$ and $(v_1, \ldots, v_m)$, show that the inner product of h_1 and h_2 is given by

$$\langle h_1, h_2 \rangle_{\mathrm{Hom}} = \operatorname{tr}(B^* A) = \operatorname{tr}(A^* B).$$

Problem 4.10. Let H_1 and H_2 be finite-dimensional vector spaces and each one has a hermitian inner product so that $U_1 \colon G \to \mathbf{U}(H_1)$ and $U_2 \colon G \to \mathbf{U}(H_2)$ are unitary representations. Show that the representation of Definition 4.16 becomes a unitary representation $\mathrm{Hom}(U_1, U_2) \colon G \to \mathbf{U}(\mathrm{Hom}(H_1, H_2))$.

Hint. Use the hermitian inner product on $\mathrm{Hom}(H_1, H_2)$ given in Definition 4.15.

Problem 4.11. The Fourier transform of Definition 4.17 is related to the Fourier transform $\mathcal{F}_2$ defined by

$$\mathcal{F}_2(f)(\rho) = \int f(t)\overline{M_\rho}(t)\, d\lambda(t) = \overline{M_\rho}(f), \qquad f \in \mathrm{L}^1(G),\ \rho \in R.$$

Let $D_\rho \colon \mathbb{C}^{n_\rho} \to \mathbb{C}^{n_\rho}$ be the semilinear map given by

$$D_\rho\left(\sum_{k=1}^{n_\rho} \alpha_k e_k\right) = \sum_{k=1}^{n_\rho} \overline{\alpha_k}\, e_k, \qquad \alpha_k \in \mathbb{C},$$

where $(e_1, \ldots, e_{n_\rho})$ is the canonical basis of $\mathbb{C}^{n_\rho}$.

(1) Show that

$$\langle D_\rho x, D_\rho y\rangle = \langle y, x\rangle = \overline{\langle x, y\rangle},$$

and that

$$D_\rho^2 = \mathrm{id}.$$

(2) Show that

$$\mathcal{F}(f)(\rho) = D_\rho \circ (\mathcal{F}_2(f)(\rho))^* \circ D_\rho.$$

Hint. See Hewitt and Ross [Hewitt and Ross (1970)], Chapter IX, Lemma 34.1.

Problem 4.12. Let $A \in \mathrm{M}_n(\mathbb{C})$ be any complex $n \times n$ matrix, and let $(\sigma_1, \ldots, \sigma_n)$ be the sequence of nonnegative square roots of the eigenvalues of A^*A.

(1) For any p, $1 \leq p < \infty$, show that

$$\|A\|_{\varphi_p} = \left(\sum_{k=1}^{n} \sigma_k^p\right)^{1/p}$$

is a matrix norm.

(2) Show that

$$\|A\|_{\varphi_\infty} = \max_{1 \le k \le n} \sigma_k$$

is a matrix norm.

Hint. See Hewitt and Ross [Hewitt and Ross (1970)] (Appendix D, Theorem D40).

Problem 4.13. For any fixed sequence $(a_\rho)_{\rho \in R}$ of reals $a_\rho \geq 1$, for any $F \in \prod_{\rho \in R} \mathrm{M}_{n_\rho}(\mathbb{C})$, if $1 \leq p < \infty$, show that

$$\|F\|_p = \left(\sum_{\rho \in R} a_\rho \|F(\rho)\|_{\varphi_p}^p \right)^{1/p}$$

is a norm on $\prod_{\rho \in R} \mathrm{M}_{n_\rho}(\mathbb{C})$. Also show that

$$\|F\|_\infty = \sup_{\rho \in R} \|F(\rho)\|_{\varphi_\infty}$$

is a norm on $\prod_{\rho \in R} \mathrm{M}_{n_\rho}(\mathbb{C})$.

Problem 4.14. (Advanced Exercise) Prove Proposition 4.24.

Hint. See Hewitt and Ross [Hewitt and Ross (1970)], Theorem 28.25 and Theorem 28.26.

Problem 4.15. (Advanced Exercise) Prove Proposition 4.25.

Hint. See Hewitt and Ross [Hewitt and Ross (1970)], Theorem 28.27.

Problem 4.16. Prove Proposition 4.26.

Hint. See Hewitt and Ross [Hewitt and Ross (1970)], Lemma 28.28.

Problem 4.17. Prove Theorem 4.27.

Hint. See Hewitt and Ross [Hewitt and Ross (1970)], Theorem 28.30.

Problem 4.18. Prove Proposition 4.28.

Hint. See Hewitt and Ross [Hewitt and Ross (1970)], Theorem 28.32.

Problem 4.19. (Advanced Exercise) Complete the proof of Theorem 4.29.

Hint. See Hewitt and Ross, Theorem 28.36 or Folland, Section 5.3, Equations 5.17, 5.18.

Problem 4.20. Prove Theorem 4.30.

Hint. See Hewitt and Ross [Hewitt and Ross (1970)], Theorem 28.39.

Problem 4.21. (Advanced Exercise) Prove Theorem 4.31.

Hint. See Hewitt and Ross [Hewitt and Ross (1970)], Theorem 28.40.

Problem 4.22. (Advanced Exercise) Prove Theorem 4.33.

Hint. See Hewitt and Ross [Hewitt and Ross (1970)], Theorem 34.5.

Problem 4.23. Prove Theorem 4.34.

Hint. See Hewitt and Ross [Hewitt and Ross (1970)], Corollary 34.6 and Corollary 34.7.

Problem 4.24. Prove via direct computation that

$$(\mathcal{F}(f * g))(\rho) = (\mathcal{F}(g))(\rho) \circ (\mathcal{F}(f))(\rho)$$

is true when G is a finite group; see Example 4.7.

Chapter 5

Matrix Representations of $\mathbf{SL}(2,\mathbb{C})$, $\mathbf{SU}(2)$ and $\mathbf{SO}(3)$

This chapter deals with explicit matrix descriptions of the irreducible representations of the groups $\mathbf{SL}(2,\mathbb{C})$, $\mathbf{SU}(2)$ and $\mathbf{SO}(3)$ (unitary representation in the last two cases). Our presentation (except for Section 5.7) relies heavily on Vilenkin's exposition [Vilenkin (1968)], especially Chapter III. To the best of our knowledge, Vilenkin contains the most detailed presentation of this type material.

We begin by proving (Section 5.1) that the representations $U_m\colon \mathbf{SU}(2) \to \mathbf{GL}(\mathcal{P}^{\mathbb{C}}_m(2))$ and $W_\ell\colon \mathbf{SO}(3) \to \mathbf{GL}(\mathcal{P}^{\mathbb{C}}_{2\ell}(2))$, which were shown to be irreducible in Example 3.8 and Example 3.9, form complete sets of set of irreducible (unitary) representations. The proof involves computing the value of the character χ_{U_m} on the matrices $r_x(\varphi)$ given by

$$r_x(\varphi) = \begin{pmatrix} e^{i\varphi} & 0 \\ 0 & e^{-i\varphi} \end{pmatrix}.$$

Namely, we have

$$\chi_{U_m}(r_x(\varphi)) = \mathrm{tr}(U_m(r_x(\varphi))) = \frac{\sin((m+1)\varphi)}{\sin\varphi}.$$

Here, $\mathcal{P}^{\mathbb{C}}_m(2)$ is the vector space of complex homogeneous polynomials of degree m in two variables (z_1 and z_2).

In Section 5.2, we give a more pleasant description of the irreducible unitary representations of $\mathbf{SO}(3)$ in terms of the spaces $\mathcal{H}^{\mathbb{C}}_k(3)$ of complex homogeneous harmonic polynomials in three variables. In fact, we show that the regular representation $\mathbf{R}_k\colon \mathbf{SO}(3) \to \mathbf{U}(\mathcal{H}^{\mathbb{C}}_k(3))$ and the representations $W_k\colon \mathbf{SO}(3) \to \mathbf{U}(\mathcal{P}^{\mathbb{C}}_{2k}(2))$ are equivalent, where

$$((\mathbf{R}_k)_Q(P))(x) = P(Q^{-1}x), \quad Q \in \mathbf{SO}(3),\ P \in \mathcal{P}^{\mathbb{C}}_k(3),\ x \in \mathbb{R}^3.$$

It turns out that to obtain the most explicit matrix descriptions of the representations of $\mathbf{SU}(2)$ and $\mathbf{SO}(3)$, it is crucial to factor a unit quaternion q as the product of three types of unit quaternions $r_x(\varphi/2), r_y(\psi/2), r_z(\theta/2)$, which happen to induce the well-known rotations of $\mathbb{R}^3$ associated with the Euler angles. For example, we have the factorizations $q = r_x(\varphi/2)r_z(\theta/2)r_x(\psi/2)$ and $q = r_x(-\varphi/2)r_y(\theta/2)r_x(-\psi/2)$. This matter is treated in great detail in Section 5.3. This is a standard topic in quantum mechanics but it is also a source of confusion because different formulae are obtained depending on the method chosen for defining the rotation in $\mathbf{SO}(3)$ induced by a unit quaternion q in $\mathbf{SU}(2)$. The main difference has to do with the way $\mathbb{R}^3$ is represented; mathematicians tend to use 2×2 skew-hermitian matrices, but physicists seem to prefer 2×2 hermitian matrices. So Dieudonné, Vilenkin and Wigner obtain different formulae! We thoroughly discuss this issue.

Until now, the representations $U_m\colon \mathbf{SU}(2) \to \mathbf{GL}(\mathcal{P}^{\mathbb{C}}_m(2))$, which are also representations of $\mathbf{SL}(2)$, act on the vector space $\mathcal{P}^{\mathbb{C}}_m(2)$ of complex homogeneous polynomials of degree m in two variables. In quantum mechanics, it is preferable to use the integer or half-integer index $\ell = m/2$. The space $\mathcal{P}^{\mathbb{C}}_m(2) = \mathcal{P}^{\mathbb{C}}_{2\ell}(2)$ then has dimension $2\ell + 1$ and the monomials $c_k z_1^{\ell-k} z_2^{\ell+k}$ of a polynomial $P(z_1, z_2)$ are indexed by the index k which ranges from $-\ell$ to ℓ. It is actually preferable to use the "dehomogenized" polynomial $Q(z) = P(z, 1)$ in the single variable z. The vector space of such polynomials (of degree $2\ell + 1$) is denoted $\mathcal{P}^{\mathbb{C}}_\ell$, and we define the representation $T_\ell\colon \mathbf{SL}(2,\mathbb{C}) \to \mathbf{GL}(\mathcal{P}^{\mathbb{C}}_\ell)$, which yields a representation $T_\ell\colon \mathbf{SU}(2) \to \mathbf{GL}(\mathcal{P}^{\mathbb{C}}_\ell)$ when restricted to the subgroup $\mathbf{SU}(2)$ of $\mathbf{SL}(2,\mathbb{C})$; see Section 5.5, Definition 5.3. We caution the reader that the formula for $T_\ell(A)$ ($A \in \mathbf{SL}(2)$) is *not* what we would obtain directly from the representation U_ℓ. We are using Vilenkin's formula to facilitate comparison with his exposition; see Vilenkin [Vilenkin (1968)] (Chapter III, Section 2.1). The representation $U_{2\ell}\colon \mathbf{SL}(2,\mathbb{C}) \to \mathbf{GL}(\mathcal{P}^{\mathbb{C}}_{2\ell}(2))$ is equivalent to the representation $T_\ell\colon \mathbf{SL}(2,\mathbb{C}) \to \mathbf{GL}(\mathcal{P}^{\mathbb{C}}_\ell)$ and similarly, the representation $U_{2\ell}\colon \mathbf{SU}(2) \to \mathbf{GL}(\mathcal{P}^{\mathbb{C}}_{2\ell}(2))$ is equivalent to the representation $T_\ell\colon \mathbf{SU}(2) \to \mathbf{GL}(\mathcal{P}^{\mathbb{C}}_\ell)$. In particular, the representations $T_\ell\colon \mathbf{SU}(2) \to \mathbf{GL}(\mathcal{P}^{\mathbb{C}}_\ell)$ form a complete set of irreducible representations of $\mathbf{SU}(2)$.

We will need to define an $\mathbf{SU}(2)$-invariant hermitian inner product on each space $\mathcal{P}^{\mathbb{C}}_\ell$, and for this, it is useful to figure out what is the derivative of the representation $T_\ell\colon \mathbf{SL}(2,\mathbb{C}) \to \mathbf{GL}(\mathcal{P}^{\mathbb{C}}_\ell)$ at the identity. This yields a representation $\mathfrak{t}_\ell \colon \mathfrak{sl}(2,\mathbb{C}) \to \mathrm{Hom}(\mathcal{P}_\ell^{\mathbb{C}}, \mathcal{P}_\ell^{\mathbb{C}})$, which is a representation of Lie algebras. This topic is discussed in Section 5.6. In particular, we need to obtain formulae for the action of $\mathfrak{t}_\ell$ on a specific basis (ξ_1, ξ_2, ξ_3) of the Lie algebra $\mathfrak{sl}(2,\mathbb{C})$ (which is also a basis of $\mathfrak{su}(2)$ over $\mathbb{R}$; see Definition 5.4).

In Section 5.7, we determine all the irreducible Lie algebra representations of $\mathfrak{sl}(2,\mathbb{C})$ (and again, of $\mathfrak{su}(2)$). The Lie algebra $\mathfrak{sl}(2,\mathbb{C})$ is a (complex) *simple* Lie algebra, which means that it is not abelian and that its only ideals are $\{0\}$ and $\mathfrak{sl}(2,\mathbb{C})$ itself. One of the most beautiful results of Lie theory is that the complex simple(!) Lie algebras fall into four infinite families plus five exceptional simple Lie algebras. Furthermore, the irreducible representations of the complex simple Lie algebras can be completely determined. These results are presented in Fulton and Harris [Fulton and Harris (1991)] and Knapp [Knapp (2002)], among other sources. The determination of the irreducible Lie algebra representations of $\mathfrak{sl}(2,\mathbb{C})$ is a "miniature" case. We also state H. Weyl's famous complete reducibility result for the finite-dimensional unitary representations of $\mathfrak{sl}(2,\mathbb{C})$. This section presents key results of representation theory which occur in every book in representation theory. We follow Serre's exposition [Serre (2000)].

We return to our goal of finding explicit formulae for the matrix representations of $\mathbf{SL}(2,\mathbb{C})$, $\mathbf{SU}(2)$, and $\mathbf{SO}(3)$. In Section 5.8, we prove that if we consider the polynomials $\psi_k(z)$ given by

$$\psi_k(z) = \frac{z^{\ell-k}}{\sqrt{(\ell-k)!(\ell+k)!}}, \quad -\ell \leq k \leq \ell,$$

then the hermitian inner product on $\mathcal{P}_\ell^{\mathbb{C}}$ making the basis (ψ_k) orthonormal is $\mathbf{SU}(2)$-invariant (see Proposition 5.20). A key technical result used in this section is the fact that in the basis $(z^{\ell-k})_{-\ell\leq k\leq \ell}$, the matrix of $T_\ell(r_x(\varphi/2))$ is the diagonal matrix

$$\begin{pmatrix} e^{i\ell\varphi} & & & & \\ & e^{i(\ell-1)\varphi} & & & \\ & & \ddots & & \\ & & & e^{-i(\ell-1)\varphi} & \\ & & & & e^{-i\ell\varphi} \end{pmatrix}.$$

This result already appears in Wigner [Wigner (1959)] (Formula 15.6, page 155, with $-\alpha$ instead of φ).

In Section 5.9, we give $\mathcal{P}_\ell^{\mathbb{C}}$ the hermitian inner product making (ψ_k) an orthonormal basis and we give various expressions for the matrix entries of the matrix $t^{(\ell)}(A)$ representing $T_\ell(A)$ in this basis.

In Section 5.10, we restrict our attention to matrices in the group $\mathbf{SU}(2)$, in which case the hermitian inner product on $\mathcal{P}_\ell^{\mathbb{C}}$ making the basis (ψ_k) orthonormal is $\mathbf{SU}(2)$-invariant (see Proposition 5.20). Using the Euler angles representation of Section 5.3, we prove the important fact (see Proposition 5.23) that for any matrix $q \in \mathbf{SU}(2)$ expressed in terms of the Euler angles as $q = u(\varphi, \theta, \psi) = r_x(\varphi/2)r_z(\theta/2)r_x(\psi/2)$, with respect to the orthonormal basis (ψ_k) of $\mathcal{P}_\ell^{\mathbb{C}}$, we have

$$t_{jk}^{(\ell)}(q) = e^{-i(j\varphi+k\psi)}\, t_{jk}^{(\ell)}(r_z(\theta/2)), \quad -\ell \leq j, k \leq \ell,$$

Thus, we are left with finding an explicit expression for the matrix $t^{(\ell)}(r_z(\theta/2))$, which we denote as $t^{(\ell)}(\theta)$ (see Definition 5.11). Such a formula is given in Proposition 5.24.

Since $\mathbf{SU}(2)$ is the universal cover of $\mathbf{SO}(3)$, we obtain a formula for the matrix $w^{(\ell)}(R)$ of the unitary map $W_\ell(R)$ associated with the irreducible representation $W_\ell\colon \mathbf{SO}(3) \to \mathbf{U}(\mathcal{P}_\ell^{\mathbb{C}})$, where $R \in \mathbf{SO}(3)$ is expressed in terms of the Euler angles as $R = R_x(\varphi)R_z(\theta)R_x(\psi)$. With respect to the orthonormal basis (ψ_k) of $\mathcal{P}_\ell^{\mathbb{C}}$, the matrix $w^{(\ell)}(R)$ is given by

$$w_{jk}^{(\ell)}(R) = e^{-i(j\varphi+k\psi)}\, t_{jk}^{(\ell)}(\theta), \quad \ell \in \mathbb{N}.$$

We also discuss the famous Wigner d-matrices and $\mathcal{D}$-matrices.

There is one more method for computing the matrix elements $t_{jk}^{(\ell)}(A)$ (with $A \in \mathbf{SL}(2,\mathbb{C})$) based on integration. The idea is to use another representing space for the representation T_ℓ, namely the vector space $\mathfrak{F}_\ell$ (of dimension $2\ell+1$) of finite Fourier series

$$\Phi(e^{i\varphi}) = \sum_{k=-\ell}^{\ell} c_k e^{-ik\varphi},$$

with $c_k \in \mathbb{C}$. In Section 5.11, we define the (irreducible) representations $\mathcal{T}_\ell\colon \mathbf{SL}(2,\mathbb{C}) \to \mathbf{GL}(\mathfrak{F}_\ell)$. In Proposition 5.26, we obtain an integral formula for the matrix elements $t_{jk}^{(\ell)}(A)$. By specializing to the matrices $A = r_z(\theta/2)$, we obtain an integral formula for computing the matrix elements $t_{jk}^{(\ell)}(\theta)$ (see Proposition 5.27). For small values of ℓ, this equation is quite practical.

In Section 5.12, we show that the matrix elements $t_{jk}^{(\ell)}(\theta)$ can be expressed in terms of certain polynomials known as the *Jacobi polynomials*.

Indeed, if the unit quaternion q is expressed in terms of the Euler angles, there is a function $P^{\ell}_{jk}(z)$, such that

$$t^{(\ell)}_{jk}(\theta) = P^{\ell}_{jk}(\cos\theta), \quad 0 \leq \theta < \pi.$$

Various formulae for the functions $P^{\ell}_{jk}(z)$ are obtained. The Jacobi polynomials $P^{\lambda,\mu}_{h}(z)$ are defined in Definition 5.14, and the relationship between the functions $P^{\ell}_{jk}(z)$ and the Jacobi polynomials $P^{\lambda,\mu}_{h}(z)$ is shown in Proposition 5.30.

In the special case where $k = 0$, in which case the function

$$t^{(\ell)}_{j0}(q) = e^{-ij\varphi}\, P^{\ell}_{j0}(\cos\theta)$$

is independent of the angle ψ, the function $P^{\ell}_{j0}(z)$ is a rescaling of the *associated Legendre function* $P^{j}_{\ell}(z)$. The function $t^{(\ell)}_{j0}(q)$ (with $q = r_x(\varphi/2)r_z(\theta/2)$) can be viewed as a function on the sphere S^2 and is denoted $Y_{\ell j}(\varphi, \theta)$, with $0 \leq \varphi < 2\pi$ and $0 \leq \theta < \pi$. The function $Y_{\ell j}(\varphi, \theta)$ is called a *spherical function*. Up to a constant, $Y_{\ell j}(\varphi, \theta)$ is the classical spherical harmonic (unfortunately) denoted $Y^{j}_{\ell}(\theta, \varphi)$ and called the *Laplace spherical harmonic* by Dieudonné.

In Section 5.14, we derive explicit formulae for the normalized Haar measures on **SU**(2) and **SO**(3) when these groups are parametrized by the Euler angles. Technically, these parametrizations are injective only on open subsets of **SU**(2) and **SO**(3), but the complements of these open sets have measure zero so from the point of view integration we obtain formulae for integrating all functions in $\mathrm{L}^2(\mathbf{SU}(2))$ and all functions in $\mathrm{L}^2(\mathbf{SO}(3))$ (respectively equipped with these left and right invariant Haar measures).

As a first step, we will need to derive a formula for an **SU**(2)-invariant volume form on **SU**(2) as a pull-back of the **SO**(4)-invariant volume form ω_{S^3} on S^3. We define the bijection $\Sigma\colon \mathbb{H} \to \mathbb{R}^4$ from the space $\mathbb{H}$ of quaternions to $\mathbb{R}^4$ as follows: for every quaternion $A \in \mathbb{H}$, with

$$A = \begin{pmatrix} a + ib & c + id \\ -c + id & a - ib \end{pmatrix}, \quad a, b, c, d \in \mathbb{R},$$

we have

$$\Sigma(A) = (a, b, c, d),$$

where, as usual, we view (a, b, c, d) as a column vector. The bijection Σ restricts to a bijection $\Sigma\colon \mathbf{SU}(2) \to S^3$ from **SU**(2) to the sphere S^3 (in $\mathbb{R}^4$). The *volume form* ω *on* **SU**(2) is defined as the pull-back $\omega = \Sigma^*(\omega_{S^3})$, where ω_{S^3} is the standard $\mathbf{SO}(4)$-invariant volume form on S^3; that is, for all $A \in \mathbf{SU}(2)$ and all $Y \in T_A\mathbf{SU}(2)$, we have

$$\omega_A(Y) = (\omega_{S^3})_{\Sigma(A)}(\Sigma(Y)).$$

Then, we prove that the volume form ω is left and right invariant; see Proposition 5.35. The proof uses an old fact about the quaternions, namely that if

$$A = \begin{pmatrix} a+ib & c+id \\ -c+id & a-ib \end{pmatrix},$$

is a unit quaternion (resp. A' is a unit quaternion), then

$$M(L_A) = \begin{pmatrix} a & -b & -c & -d \\ b & a & -d & c \\ c & d & a & -b \\ d & -c & b & a \end{pmatrix}, \quad M(R_{A'}) = \begin{pmatrix} a' & -b' & -c' & -d' \\ b' & a' & d' & -c' \\ c' & -d' & a' & b' \\ d' & c' & -b' & a' \end{pmatrix},$$

belong to $\mathbf{SO}(4)$.

Let $\Omega \subseteq \mathbb{R}^3$ be the open subset

$$\Omega = (0, 2\pi) \times (0, \pi) \times (-2\pi, 2\pi).$$

By Proposition 5.4, the map $u \colon \Omega \to \mathbf{SU}(2)$ given by $u(\varphi, \theta, \psi) = r_x(\varphi/2)r_z(\theta/2)r_x(\psi/2)$ is a diffeomorphism, and let $\Phi \colon \Omega \to S^3$ be the composed map $\Phi = \Sigma \circ u$ from Ω onto an open subset of S^3. Let ω_Ω be the pull-back form $\omega_\Omega = \Phi^*\omega_{S^3}$. Then we can prove that the volume form ω_Ω is given by

$$\omega_\Omega = \frac{1}{8}\sin\theta\, d\theta \wedge d\varphi \wedge d\psi.$$

See Proposition 5.36.

Similarly, let $\Omega_0 \subseteq \mathbb{R}^3$ be the open subset

$$\Omega = (0, 2\pi) \times (0, \pi) \times (0, 2\pi),$$

let $u_0 \colon \Omega_0 \to \mathbf{SU}(2)$ be the restriction of u to Ω_0, and let $\Phi_0 \colon \Omega_0 \to S^3$ be the composed map $\Phi_0 = \Sigma \circ u_0$ from Ω_0 onto an open subset of S^3. Let ω_{Ω_0} be the pull-back form $\omega_{\Omega_0} = \Phi_0^*\omega_{S^3}$. Then we can prove that the volume form ω_{Ω_0} is also given by

$$\omega_{\Omega_0} = \frac{1}{8}\sin\theta\, d\theta \wedge d\varphi \wedge d\psi.$$

See Proposition 5.40. Together with the injection $R_0 \colon \Omega_0 \to \mathbf{SO}(3)$ given by

$$R_0(\varphi, \theta, \psi) = R_x(\varphi)R_z(\theta)R_x(\psi),$$

we obtain a formula for integrating functions in $\mathrm{L}^2(\mathbf{SO}(3))$; see Proposition 5.40.

Combining results from Section 5.14 and the previous sections, in Section 5.15, we obtain explicit Fourier series expansions for the functions in $\mathrm{L}^2(\mathbf{SU}(2))$ and $\mathrm{L}^2(\mathbf{SO}(3))$ in terms of the matrix elements $t^{(\ell)}_{jk}$. The reason is that by Peter–Weyl the family of functions

$$\left(\sqrt{2\ell+1}\, t^{(\ell)}_{ij}\right)_{-\ell\leq i,j\leq \ell,\, \ell\in R},$$

with $R = \{0, 1/2, 1, 3/2, 2, \ldots\}$, is a Hilbert basis of $\mathrm{L}^2(\mathbf{SU}(2))$. Actually, we obtain explicit formulae for the Fourier transform and the Fourier cotransform (discussed in Section 4.6) on $\mathrm{L}^2(\mathbf{SU}(2))$. Similarly, the family of functions

$$\left(\sqrt{2\ell+1}\, w^{(\ell)}_{ij}\right)_{-\ell\leq i,j\leq \ell,\, \ell\in \mathbb{N}}$$

is a Hilbert basis of $\mathrm{L}^2(\mathbf{SO}(3))$. This yields another explicit example of the Fourier transform and Fourier cotransform on $\mathrm{L}^2(\mathbf{SO}(3))$. If the functions are expressed in terms of the Euler angles, then we obtain formulae that are practically computable.

We also provide Fourier series expansions for two subspaces $\mathfrak{L}^2_k$ and ${}_j\mathfrak{L}^2$ of $\mathrm{L}^2(\mathbf{SU}(2))$ defined by Vilenkin. A special case of $\mathfrak{L}^2_k$ yields another derivation of the well-known series expansion of functions in $\mathrm{L}^2(S^2)$ in terms of spherical harmonics.

In Section 5.16, following Vilenkin, we show how to decompose not only scalar-valued but also vector-valued functions on the sphere S^2 into Fourier series that behave nicely under rotations of the sphere.

To simplify notation, we will write $\mathcal{P}_\ell$ instead of $\mathcal{P}^{\mathbb{C}}_\ell$. Let $T_\ell \colon \mathbf{SU}(2) \to \mathbf{U}(\mathcal{P}_\ell)$ be the irreducible representation of $\mathbf{SU}(2)$ associated with $\ell \in R = \{0, 1/2, 1, 3/2, 2, 5/2, 3, \ldots\}$. We wish to consider the Hilbert space $\mathfrak{F}^S_\ell$ of functions $f \colon S^2 \to \mathcal{P}_\ell$ defined by the isomorphism

$$\mathfrak{F}^S_\ell \simeq \bigoplus_{j=-\ell}^{\ell} \mathrm{L}^2(S^2)\psi_j,$$

where the ψ_j constitute an orthonormal basis of $\mathcal{P}_\ell$ for an $\mathbf{SU}(2)$-invariant hermitian inner product defined in Section 5.8 ($\mathcal{P}_\ell$ is a complex vector space of dimension $2\ell+1$). Vilenkin calls the functions in $\mathfrak{F}^S_\ell$ *fields of quantities on the sphere transforming according to the irreducible representation* T_ℓ.

For example, for $\ell = 1$, since $2\ell+1 = 3$, we get a vector field on the sphere. We show how to decompose the functions in $\mathfrak{F}_\ell^S$ in terms of certain representations $V_\ell\colon \mathbf{SU}(2) \to \mathbf{GL}(\mathfrak{F}_\ell^S)$ defined in terms of the representations T_ℓ.

The last section of this chapter (Section 5.16) deals with the *Clebsch–Gordan coefficients*, a standard topic in quantum mechanics. In general, the tensor product $T_{\ell_1} \otimes T_{\ell_2}$ of two irreducible representations T_{ℓ_1} and T_{ℓ_2} of $\mathbf{SU}(2)$ is not irreducible, so according to the Peter–Weyl theorem (Theorem 4.16) it splits as a direct sum of irreducible representations. Since the character associated with the representation $T_{\ell_1} \otimes T_{\ell_2}$ is equal to the product $\chi_{T_{\ell_1}}\chi_{T_{\ell_2}}$ of the characters $\chi_{T_{\ell_1}}$ and $\chi_{T_{\ell_2}}$ associated with T_{ℓ_1} and T_{ℓ_2}, it turns out that the following famous result (known to H. Weyl and E. Wigner) can be obtained (see Proposition 5.46). For any two irreducible representations T_{ℓ_1} and T_{ℓ_2} of $\mathbf{SU}(2)$, we have

$$\chi_{T_{\ell_1}}(q)\chi_{T_{\ell_2}}(q) = \sum_{\ell=|\ell_1-\ell_2|}^{\ell_1+\ell_2} \chi_{T_\ell}(q), \quad q \in \mathbf{SU}(2).$$

As a consequence, we also have an isomorphism

$$\mathcal{P}_{\ell_1} \otimes \mathcal{P}_{\ell_2} \simeq \bigoplus_{\ell=|\ell_1-\ell_2|}^{\ell_1+\ell_2} \mathcal{P}_\ell.$$

The space $\mathcal{P}_{\ell_1} \otimes \mathcal{P}_{\ell_2}$ has dimension $(2\ell_1+1)(2\ell_2+1)$ and each summand $\mathcal{P}_\ell$ has dimension $2\ell+1$.

By Proposition 5.16, each vector space $\mathcal{P}_\ell$ has an orthonormal basis (ψ_k) $(-\ell \le k \le \ell)$ invariant under the action of $\mathbf{SU}(2)$. Following Vilenkin [Vilenkin (1968)] (Chapter III, Section 8.2), we denote the basis of $\mathcal{P}_{\ell_1}$ as $(\mathbf{f}_j)$ $(-\ell_1 \le j \le \ell_1)$ and the basis of $\mathcal{P}_{\ell_2}$ as $(\mathbf{h}_k)$ $(-\ell_2 \le k \le \ell_2)$. Then the family of tensor products

$$\mathbf{f}_j \otimes \mathbf{h}_k, \quad -\ell_1 \le j \le \ell_1,\ -\ell_2 \le k \le \ell_2,$$

is a basis of $\mathcal{P}_{\ell_1} \otimes \mathcal{P}_{\ell_2}$. If we give $\mathcal{P}_{\ell_1} \otimes \mathcal{P}_{\ell_2}$ the inner product defined in Definition 4.10 induced by the inner products associated with the bases $(\mathbf{f}_j)$ and $(\mathbf{h}_k)$, then the vectors $(\mathbf{f}_j \otimes \mathbf{h}_k)$ form an orthonormal basis of $\mathcal{P}_{\ell_1} \otimes \mathcal{P}_{\ell_2}$.

Since we have the direct sum

$$\mathcal{P}_{\ell_1} \otimes \mathcal{P}_{\ell_2} \simeq \bigoplus_{\ell=|\ell_1-\ell_2|}^{\ell_1+\ell_2} \mathcal{P}_\ell,$$

we also have a basis of $\mathcal{P}_{\ell_1} \otimes \mathcal{P}_{\ell_2}$ consisting of the union of the bases associated with each of the summand $\mathcal{P}_l$, which Vilenkin denotes by

$$\mathbf{a}^{\ell}_{m}, \quad |\ell_1 - \ell_2| \leq \ell \leq \ell_1 + \ell_2, \ -\ell \leq m \leq \ell,$$

where for ℓ fixed, $(\mathbf{a}^{\ell}_{m})$ $(-\ell \leq m \leq \ell)$ is the basis of $\mathcal{P}_\ell$. Since both bases are orthonormal bases of $\mathcal{P}_{\ell_1} \otimes \mathcal{P}_{\ell_2}$, there is a unitary matrix C expressing the basis $(\mathbf{f}_j \otimes \mathbf{h}_k)$ in terms of the basis $(\mathbf{a}^{\ell}_{m})$, and the entries of the matrix C are called the *Clebsch–Gordan coefficients*. More precisely, the change of basis matrix $C = (C_{(\ell m),(jk)})$ is the unitary matrix defined such that the (jk)th column of C consists of the coefficients of $\mathbf{f}_j \otimes \mathbf{h}_k$ over the basis $(\mathbf{a}^{\ell}_{m})$, namely

$$\mathbf{f}_j \otimes \mathbf{h}_k = \sum_{\ell=|\ell_1-\ell_2|}^{\ell_1+\ell_2} \sum_{m=-\ell}^{\ell} C_{(\ell m),(jk)} \mathbf{a}^{\ell}_{m},$$

with $-\ell_1 \leq j \leq \ell_1$, $-\ell_2 \leq k \leq \ell_2$.

Amazingly, the coefficients $C_{(\ell m),(jk)}$ can be computed explicitly, but the formulae are very complicated and the technical details of the computations are quite involved. Complete details can be found in Vilenkin [Vilenkin (1968)] (Chapter III, Section 8). In this section, we will content ourselves by providing an outline of these computations.

5.1 Irreducible Representations of SU(2) and SO(3)

In Example 3.8, it was proven that the representations $U_m \colon \mathbf{SU}(2) \to \mathbf{GL}(\mathcal{P}^{\mathbb{C}}_m(2))$ are irreducible. In Example 3.9, it was proven that the representations $W_\ell \colon \mathbf{SO}(3) \to \mathbf{GL}(\mathcal{P}^{\mathbb{C}}_{2\ell}(2))$ are irreducible. Recall that since $\mathbf{SU}(2)$ is compact and $\mathcal{P}^{\mathbb{C}}_m(2)$ is finite-dimensional, there is an invariant inner product on $\mathcal{P}^{\mathbb{C}}_m(2)$ so we may assume that these representations are unitary.

Let us now prove that the representations U_m form a complete set of irreducible unitary representations.

Proposition 5.1. *Every irreducible unitary representation of* $\mathbf{SU}(2)$ *is equivalent to one of the irreducible unitary representations* $U_m \colon \mathbf{SU}(2) \to \mathbf{U}(\mathcal{P}^{\mathbb{C}}_m(2))$. *Furthermore, every irreducible unitary representation of* $\mathbf{SO}(3)$ *is equivalent to one of the irreducible unitary representations* $W_m \colon \mathbf{SO}(3) \to \mathbf{U}(\mathcal{P}^{\mathbb{C}}_{2m}(2))$.

Proof. The key point is to figure out what are the characters χ_{U_m} of the irreducible unitary representations U_m. Every unitary matrix $q \in \mathbf{SU}(2)$ is diagonalizable as

$$q = Rr_x(\varphi)R^*$$

for some unitary matrix $R \in \mathbf{U}(2)$, where

$$r_x(\varphi) = \begin{pmatrix} e^{i\varphi} & 0 \\ 0 & e^{-i\varphi} \end{pmatrix}.$$

If $\det(R) = e^{i\omega} \neq 1$, we replace R by $e^{-i\omega/2}R$, which is unitary $((e^{-i\omega/2}R)(e^{-i\omega/2}R)^* = e^{-i\omega/2}Re^{i\omega/2}R^* = I_2)$, so that $\det(e^{-i\omega/2}R) = (e^{-i\omega/2})^2\det(R) = e^{-i\omega}e^{i\omega} = +1$, and then

$$(e^{-i\omega/2}R)r_x(\varphi)(e^{-i\omega/2}R)^* = e^{-i\omega/2}Rr_x(\varphi)e^{i\omega/2}R^* = Rr_x(\varphi)R^* = q.$$

Therefore, we may assume that $R \in \mathbf{SU}(2)$. Here $-\pi \leq \varphi \leq \pi$, but if $-\pi \leq \varphi < 0$, we can replace R by

$$R\begin{pmatrix} 0 & e^{i\pi/2} \\ e^{i\pi/2} & 0 \end{pmatrix} \in \mathbf{SU}(2),$$

because then

$$\begin{aligned} &\begin{pmatrix} 0 & e^{i\pi/2} \\ e^{i\pi/2} & 0 \end{pmatrix}\begin{pmatrix} e^{i\varphi} & 0 \\ 0 & e^{-i\varphi} \end{pmatrix}\begin{pmatrix} 0 & e^{-i\pi/2} \\ e^{-\pi/2} & 0 \end{pmatrix} \\ &= \begin{pmatrix} 0 & e^{-i(\varphi-\pi/2)} \\ e^{i(\varphi+\pi/2)} & 0 \end{pmatrix}\begin{pmatrix} 0 & e^{-i\pi/2} \\ e^{-i\pi/2} & 0 \end{pmatrix} = \begin{pmatrix} e^{-i\varphi} & 0 \\ 0 & e^{i\varphi} \end{pmatrix}. \end{aligned}$$

Thus, we may assume that $0 \leq \varphi \leq \pi$. *Therefore we proved that every matrix in* $\mathbf{SU}(2)$ *is conjugate to a unique matrix* $r_x(\varphi)$, *with* $0 \leq \varphi \leq \pi$.

Since the characters are central functions (Proposition 4.10(1)), it suffices to compute the value of the character χ_{U_m} on $r_x(\varphi)$. Also observe that $r_x(\varphi)$ and $r_x(\theta)$ are conjugate iff they have the same eigenvalues iff $\varphi = \pm\theta \pmod{2\pi}$. But then we obtain a bijection between the space of central functions of $\mathrm{L}^2(\mathbf{SU}(2))$ and the space of even continuous 2π-periodic functions from $\mathbb{R}$ to $\mathbb{C}$ given by $f \mapsto S(f)$, with

$$S(f)(\varphi) = f(r_x(\varphi)).$$

We will compute the values $\chi_{U_m}(r_x(\varphi))$ and prove that the characters χ_{U_m} are dense in the space of central functions of $\mathrm{L}^2(\mathbf{SU}(2))$.

We proved that the eigenvalues of $U_m(r_x(\varphi))$ are $(e^{im\varphi}, e^{i(m-2)\varphi}, \ldots, e^{-im\varphi})$ in Example 3.8. Therefore,

$$\chi_{U_m}(r_x(\varphi)) = \mathrm{tr}(U_m(r_x(\varphi))) = \sum_{k=0}^{m} e^{i(m-2k)\varphi}.$$

We have

$$\begin{aligned}\sum_{k=0}^{m} e^{i(m-2k)\varphi} &= e^{im\varphi}\sum_{k=0}^{m}(e^{-i2\varphi})^k = e^{im\varphi}\frac{(1-(e^{-i2\varphi})^{m+1})}{1-e^{-i2\varphi}} \\ &= e^{im\varphi}\frac{e^{i\varphi}(1-(e^{-i2\varphi})^{m+1})}{e^{i\varphi}(1-e^{-i2\varphi})} = e^{i(m+1)\varphi}\frac{(1-e^{-i2(m+1)\varphi})}{e^{i\varphi}-e^{-i\varphi}} \\ &= \frac{(e^{i(m+1)\varphi}-e^{-i(m+1)\varphi})}{e^{i\varphi}-e^{-i\varphi}} = \frac{\sin((m+1)\varphi)}{\sin\varphi}.\end{aligned}$$

We also easily check that

$$\chi_{U_m}(r_x(k\pi)) = e^{imk\pi}(m+1) = (-1)^{mk}(m+1).$$

In summary, we obtained the following result.

For every $q \in \mathbf{SU}(2)$, if $r_x(\varphi)$ is the unique diagonal matrix conjugate to q with $0 \leq \varphi \leq \pi$, then $\chi_{U_m}(q)$ is given by

$$\chi_{U_m}(q) = \chi_{U_m}(r_x(\varphi)) = \frac{\sin((m+1)\varphi)}{\sin\varphi}.$$

If we write

$$\kappa_m(\varphi) = \chi_{U_m}(r_x(\varphi)) = \frac{\sin((m+1)\varphi)}{\sin\varphi},$$

then for $m \geq 1$, we get

$$\begin{aligned}\kappa_m(\varphi) &= \frac{\sin((m+1)\varphi)}{\sin\varphi} = \frac{\sin(m\varphi)\cos\varphi + \cos(m\varphi)\sin\varphi}{\sin\varphi} \\ &= \cos(m\varphi) + \kappa_{m-1}(\varphi)\cos\varphi.\end{aligned}$$

Note that $\kappa_0(\varphi) = 1$. The formula for $\kappa_m(\varphi)$ still holds for $\varphi = k\pi$. In summary,

$$\kappa_m(\varphi) = \cos(m\varphi) + \kappa_{m-1}(\varphi)\cos\varphi, \quad m \geq 1, \quad \kappa_0(\varphi) = 1. \qquad (\kappa)$$

The above equation shows that $\kappa_0(\varphi), \kappa_1(\varphi), \ldots, \kappa_m(\varphi)$ generates the same vector space as $1, \cos\varphi, \ldots, \cos m\varphi$.

It is known from Fourier analysis that the space generated by the family of functions $(\cos m\varphi))_{m\geq 0}$ is dense in the space of even 2π-periodic continuous functions from $\mathbb{R}$ to $\mathbb{C}$; for example, see Folland [Folland (1999, 1992)].

Consequently, the family $(\kappa_m)_{m\geq 0}$ is also dense in the space of even continuous 2π-periodic functions from $\mathbb{R}$ to $\mathbb{C}$. Since the map $f \mapsto S(f)$ is a bijection between the space of central functions of $\mathrm{L}^2(\mathbf{SU}(2))$ and the space of even continuous 2π-periodic functions from $\mathbb{R}$ to $\mathbb{C}$, we conclude that the family of characters $(\chi_{U_m})_{m\geq 0}$ is dense in the space of central functions of $\mathrm{L}^2(\mathbf{SU}(2))$.

To finish the proof, we use Proposition 4.10(4) which says that the characters χ_m of $\mathbf{SU}(2)$ form a Hilbert basis of the space of central functions of $\mathrm{L}^2(\mathbf{SU}(2))$. Since the χ_{U_m} are characters of irreducible unitary representations, they are equal to some of the characters χ_ρ of $\mathbf{SU}(2)$, and they are not equivalent since the dimensions $m+1$ of the representing spaces are different. If some character χ_ρ is not equivalent to one of the χ_{U_m}, then by Proposition 4.10(4)(a),

$$\langle \chi_{U_m}, \chi_\rho\rangle = 0, \quad \text{for all} \quad m \geq 0,$$

but since the χ_{U_m} are dense in the space of central functions of $\mathrm{L}^2(\mathbf{SU}(2))$, this implies that $\chi_\rho = 0$, a contradiction.

The second statement follows from the fact that the unitary representations $W_\ell \colon \mathbf{SO}(3) \to \mathbf{U}(\mathcal{P}^{\mathbb{C}}_{2\ell}(2))$ are given by

$$W_\ell(\rho_q) = U_{2\ell}(q) \quad q \in \mathbf{SU}(2), \quad \ell \geq 0. \qquad \square$$

We now give a more pleasant description of the irreducible representations of $\mathbf{SO}(3)$ in terms of harmonic polynomials.

5.2 Irreducible Representations of SO(3); Harmonics

Recall that the Laplacian in $\mathbb{R}^n$ is given by

$$\Delta f = \frac{\partial^2 f}{\partial x_1^2} + \frac{\partial^2 f}{\partial x_2^2} + \cdots + \frac{\partial^2 f}{\partial x_n^2},$$

where $f \colon \mathbb{R}^n \to \mathbb{C}$ is twice differentiable. The n-sphere $S^n \subseteq \mathbb{R}^{n+1}$ is given by

$$S^n = \{(x_1, \ldots, x_{n+1}) \in \mathbb{R}^{n+1} \mid x_1^2 + \cdots + x_{n+1}^2 = 1\}.$$

Definition 5.1. Let $\mathcal{P}^{\mathbb{C}}_k(n+1)$ denote the space of homogeneous polynomials of degree k in $n+1 \geq 2$ variables with complex coefficients, and let $\mathcal{P}^{\mathbb{C}}_k(S^n)$ denote the restrictions of homogeneous polynomials in $\mathcal{P}^{\mathbb{C}}_k(n+1)$

to S^n. Let $\mathcal{H}_k^{\mathbb{C}}(n+1)$ denote the space of *complex harmonic polynomials*, with

$$\mathcal{H}_k^{\mathbb{C}}(n+1) = \{P \in \mathcal{P}_k^{\mathbb{C}}(n+1) \mid \Delta P = 0\};$$

in the above equation, we view P as a function on $\mathbb{R}^{n+1}$. Harmonic polynomials are sometimes called *solid harmonics*. Finally, let $\mathcal{H}_k^{\mathbb{C}}(S^n)$ denote the space of *complex spherical harmonics* as the set of restrictions of harmonic polynomials in $\mathcal{H}_k^{\mathbb{C}}(n+1)$ to S^n.

It is not hard to prove that the restriction map from $\mathcal{H}_k^{\mathbb{C}}(n+1)$ to $\mathcal{H}_k^{\mathbb{C}}(S^n)$ is a bijection, and thus a linear isomorphism; see Gallier and Quaintance [Gallier and Quaintance (2020b)] (Section 7.5). The functions in $\mathcal{H}_k^{\mathbb{C}}(S^n)$, the spherical harmonics, have been studied extensively. They are the eigenspaces of the Laplacian on the sphere S^n; see Gallier and Quaintance [Gallier and Quaintance (2020b)] (Chapter 7). We will return to these functions later.

The group $\mathbf{SO}(n+1)$ acts on $\mathcal{P}_k^{\mathbb{C}}(n+1)$ by the (left regular) action

$$(\mathbf{R}_Q(P))(x) = P(Q^{-1}x), \quad Q \in \mathbf{SO}(n+1),\ P \in \mathcal{P}_k^{\mathbb{C}}(n+1),\ x \in \mathbb{R}^{n+1}.$$

Note that the above formula shows that $\mathbf{R}$ is also an action of $\mathbf{SO}(n+1)$ on smooth functions on $\mathbb{R}^{n+1}$.

The action $\mathbf{R}$ on $\mathcal{P}_k^{\mathbb{C}}(n+1)$ is reducible for $k \geq 2$. For example, we easily check that the subspace of $\mathcal{P}_2^{\mathbb{C}}(n+1)$ generated by the polynomial $x_1^2+\cdots+x_{n+1}^2$ is invariant. However, this action turns out to be irreducible on $\mathcal{H}_k^{\mathbb{C}}(n+1)$. This will be shown in Section 6.10. But first we need to prove that the action of the Laplacian on smooth functions on $\mathbb{R}^{n+1}$ commutes with the action $\mathbf{R}$. Recall that $\lambda_Q f$ is the function given by $(\lambda_Q f)(x) = f(Q^{-1}x)$.

Proposition 5.2. *The action of the Laplacian on smooth functions on $\mathbb{R}^{n+1}$ commutes with the action $\mathbf{R}$; that is, for every smooth function f on $\mathbb{R}^{n+1}$, for every $Q \in \mathbf{SO}(n+1)$, for all $u \in \mathbb{R}^{n+1}$, we have*

$$\Delta(\lambda_Q f)(u) = (\Delta f)(Q^{-1}u).$$

Proof. For simplicity of notation, write $A = Q^{-1} = Q^\top$. The proof makes heavy use of the chain rule. If we let h be the function given by $h(x) = Ax$ and view f as a function $y \mapsto f(y)$ of the variable y, if we write $g = f \circ h$ so that $g(x) = f(Ax)$, then we need to compute $(\partial g/\partial x_j)(u)$ $(x, u \in \mathbb{R}^{n+1})$, which by the chain rule is given by

$$\frac{\partial g}{\partial x_j}(u) = df_{h(u)}\left(\frac{\partial h}{\partial x_j}(u)\right).$$

Since

$$h(x) = Ax = \left(\sum_{j=1}^{n+1} a_{1j} x_j, \ldots, \sum_{j=1}^{n+1} a_{n+1j} x_j \right),$$

we have

$$\frac{\partial h}{\partial x_j}(u) = (a_{1j}, \ldots, a_{n+1j})$$

(independently of u), and since

$$df_{Au}(w) = \sum_{i=1}^{n+1} w_i \frac{\partial f}{\partial y_i}(Au),$$

we obtain

$$\frac{\partial g}{\partial x_j}(u) = \sum_{i=1}^{n+1} a_{ij} \frac{\partial f}{\partial y_i}(Au).$$

To compute $\frac{\partial^2 g}{\partial x_j^2}(u)$, we view the function $y \mapsto \frac{\partial f}{\partial y_i}(y)$ as the function f, so we obtain

$$\frac{\partial^2 g}{\partial x_j^2}(u) = \sum_{i=1}^{n+1} a_{ij} \sum_{k=1}^{n+1} a_{kj} \frac{\partial^2 f}{\partial y_i \partial y_k}(Au),$$

and thus the Laplacian is given by

$$\Delta g(u) = \sum_{i=1}^{n+1} \sum_{j=1}^{n+1} \sum_{k=1}^{n+1} a_{ij} a_{kj} \frac{\partial^2 f}{\partial y_i \partial y_k}(Au).$$

The right-hand side can be rewritten as

$$\begin{aligned}
\Delta g(u) &= \sum_{i=1}^{n+1} \sum_{j=1}^{n+1} \sum_{k=1}^{n+1} a_{ij} a_{kj} \frac{\partial^2 f}{\partial y_i \partial y_k}(Au) \\
&= \sum_{i=1}^{n+1} \left(\sum_{j=1}^{n+1} a_{ij}^2 \right) \frac{\partial^2 f}{\partial y_i^2}(Au) + 2 \sum_{i<k}^{n+1} \left(\sum_{j=1}^{n+1} a_{ij} a_{kj} \right) \frac{\partial^2 f}{\partial y_i \partial y_k}(Au),
\end{aligned}$$

and since A is an orthogonal matrix, its rows have unit length and are pairwise orthogonal, which means that

$$\sum_{j=1}^{n+1} a_{ij}^2 = 1, \quad 1 \leq i \leq n+1$$

$$\sum_{j=1}^{n+1} a_{ij} a_{kj} = 0, \quad i < k,$$

so we obtain

$$\Delta g(u) = (\Delta f)(Au),$$

which means that $\Delta(\lambda_{A^{-1}} f)(u) = (\Delta f)(Au)$, as claimed. □

As a corollary of Proposition 5.2, the vector space $\mathcal{H}_k^{\mathbb{C}}(n+1)$ is invariant under $\mathbf{R}$, and so $\mathbf{R}\colon \mathbf{SO}(n+1) \to \mathbf{GL}(\mathcal{H}_k^{\mathbb{C}}(n+1))$ is a representation. Since $\mathbf{SO}(n+1)$ is compact and $\mathcal{H}_k^{\mathbb{C}}(n+1)$ is finite-dimensional, we may assume that $\mathbf{R}$ is unitary.

It is shown in Gallier and Quaintance [Gallier and Quaintance (2020b)] (Section 7.5) that $\mathcal{H}_k^{\mathbb{C}}(n+1)$ has dimension

$$a_{k,n+1} = \binom{n+k}{k} - \binom{n+k-2}{k-2}$$

if $n \geq 1$, $k \geq 2$, with $a_{0,n+1} = 1$ and $a_{1,n+1} = n$, For $n = 2$, we get $a_{k,3} = 2k+1$. Here is a list of bases of the homogeneous harmonic polynomials of degree k in three variables up to $k = 4$.

$$\begin{array}{ll}
k = 0 & 1 \\
k = 1 & x,\ y,\ z \\
k = 2 & x^2 - y^2,\ x^2 - z^2,\ xy,\ xz,\ yz \\
k = 3 & x^3 - 3xy^2,\ 3x^2y - y^3,\ x^3 - 3xz^2,\ 3x^2z - z^3, \\
 & y^3 - 3yz^2,\ 3y^2z - z^3,\ xyz \\
k = 4 & x^4 - 6x^2y^2 + y^4,\ x^4 - 6x^2z^2 + z^4,\ y^4 - 6y^2z^2 + z^4, \\
 & x^3y - xy^3,\ x^3z - xz^3,\ y^3z - yz^3, \\
 & 3x^2yz - yz^3,\ 3xy^2z - xz^3,\ 3xyz^2 - x^3y.
\end{array}$$

To prove that the representations $\mathbf{R}\colon \mathbf{SO}(n+1) \to \mathbf{U}(\mathcal{H}_k^{\mathbb{C}}(n+1))$ are irreducible we restrict ourselves to the case where $n = 2$. In order to deal with the case where $n > 2$, we need results from the next chapter. Since these regular representations map to different spaces, for clarity we index them by k, that is, we write $\mathbf{R}_k\colon \mathbf{SO}(n+1) \to \mathbf{U}(\mathcal{H}_k^{\mathbb{C}}(n+1))$.

Proposition 5.3. *The representations $\mathbf{R}_k\colon \mathbf{SO}(3) \to \mathbf{U}(\mathcal{H}_k^{\mathbb{C}}(3))$ are irreducible. In fact, the representations $\mathbf{R}_k\colon \mathbf{SO}(3) \to \mathbf{U}(\mathcal{H}_k^{\mathbb{C}}(3))$ and $W_k\colon \mathbf{SO}(3) \to \mathbf{U}(\mathcal{P}_{2k}^{\mathbb{C}}(2))$ are equivalent.*

Proof. By Peter–Weyl II (Theorem 4.16), the representation $\mathbf{R}_k\colon \mathbf{SO}(3) \to \mathbf{U}(\mathcal{H}_k^{\mathbb{C}}(3))$ is equivalent to the direct sum of a finite number of irreducible representations $W_{\ell_j}\colon \mathbf{SO}(3) \to \mathbf{U}(\mathcal{P}_{2\ell_j}^{\mathbb{C}}(2))$, so that we have an isomorphism

$$\mathbf{R}_k \approx \bigoplus_{j=1}^{p} W_{\ell_j}, \quad \ell_j \leq \ell_{j+1}.$$

Since $\dim(\mathcal{H}_k^{\mathbb{C}}(3)) = 2k+1$ and $\dim(\mathcal{P}_{2\ell_j}^{\mathbb{C}}(2)) = 2\ell_j+1$, if we can prove that $\ell_j \geq k$ for some j, then $2k+1 = \sum_{j=1}^{p}(2\ell_j+1)$ implies that $p = 1$ and $k = \ell_1$, and so $\mathbf{R}_k\colon \mathbf{SO}(3) \to \mathbf{U}(\mathcal{H}_k^{\mathbb{C}}(3))$ is equivalent to $W_k\colon \mathbf{SO}(3) \to \mathbf{U}(\mathcal{P}_{2k}^{\mathbb{C}}(2))$.

The key point is to figure out what is the value of the character χ_{W_ℓ} on the rotation of angle φ and axis Ox given by

$$R_x(\varphi) = \begin{pmatrix} 1 & 0 & 0 \\ 0 & \cos\varphi & -\sin\varphi \\ 0 & \sin\varphi & \cos\varphi \end{pmatrix}.$$

The trick is that $R_x(\varphi)$ is the rotation in $\mathbf{SO}(3)$ corresponding to the (familiar) quaternion

$$r_x(\varphi/2) = \begin{pmatrix} e^{i\varphi/2} & 0 \\ 0 & e^{-i\varphi/2} \end{pmatrix}.$$

This fact is easily verified by direct computation. But remember that W_ℓ is given by

$$W_\ell(\rho_q) = U_{2\ell}(q) \quad q \in \mathbf{SU}(2), \quad \ell \geq 0,$$

which proves that

$$\chi_{W_\ell}(\rho_q) = \mathrm{tr}(W_\ell(\rho_q)) = \mathrm{tr}(U_{2\ell}(q)) = \chi_{U_{2\ell}}(q).$$

If we apply the above equation to $q = r_x(\varphi/2)$ and $R_x(\varphi)$, we obtain

$$\chi_{W_\ell}(R_x(\varphi)) = \chi_{U_{2\ell}}(r_x(\varphi/2)) = \sum_{j=0}^{2\ell} e^{i(2\ell-2j)\varphi/2} = \sum_{j=0}^{2\ell} e^{i(\ell-j)\varphi}.$$

Since $\mathbf{R}_k\colon \mathbf{SO}(3) \to \mathbf{U}(\mathcal{H}_k^{\mathbb{C}}(3))$ is equivalent to a finite direct sum of p irreducible representations $W_{\ell_j}\colon \mathbf{SO}(3) \to \mathbf{U}(\mathcal{P}_{2\ell_j}^{\mathbb{C}}(2))$, by Proposition 4.18, the value of the character $\chi_{\mathbf{R}_k}$ on $R_x(\varphi)$ is the sum of the values of the characters $\chi_{W_{\ell_j}}$ on $R_x(\varphi)$, and by the above equation, it is an integral combination of terms of the form $e^{ij\varphi}$, with $|j| \leq \ell_p$. Consequently, if we find an eigenvector of $\mathbf{R}_k(R_x(\varphi))$ for the eigenvalue $e^{-ik\varphi}$, the representation $W_k\colon \mathbf{SO}(3) \to \mathbf{U}(\mathcal{P}_{2k}^{\mathbb{C}}(2))$ must occur. Consider $P(x) = P(x_1, x_2, x_3) = (x_2 + ix_3)^k$. We immediately check that $\Delta P = 0$, and since

$$R_x(\varphi)^{-1} = R_x(-\varphi) = \begin{pmatrix} 1 & 0 & 0 \\ 0 & \cos\varphi & \sin\varphi \\ 0 & -\sin\varphi & \cos\varphi \end{pmatrix},$$

we obtain

$$\begin{aligned}(\mathbf{R}_k(R_x(\varphi)))(P) &= P(R_x(-\varphi)) \\ &= P((\cos\varphi)x_2 + (\sin\varphi)x_3, -(\sin\varphi)x_2 + (\cos\theta)x_3) \\ &= ((\cos\varphi)x_2 + (\sin\varphi)x_3 + i(-(\sin\varphi)x_2 + (\cos\theta)x_3))^k \\ &= (\cos\varphi - i\sin\varphi)x_2 + i(\cos\varphi - i\sin\varphi)x_3)^k \\ &= e^{-ik\varphi}(x_2 + ix_3)^k = e^{-ik\varphi}P(x),\end{aligned}$$

as desired. □

Proposition 5.3 also shows that the representations $\mathbf{R}_k \colon \mathbf{SO}(3) \to \mathbf{U}(\mathcal{H}_k^{\mathbb{C}}(3))$ form a complete set of irreducible representations of $\mathbf{SO}(3)$.

5.3 Factorization of the Unit Quaternions Using Euler Angles

In order to obtain formulae for the matrix elements of the representations of $\mathbf{SU}(2)$ in terms of special functions, the Jacobi polynomials, it is necessary to understand how to express the unit quaternions in terms of Euler angles. The key fact is that there are three types of unit quaternions, $r_x(\varphi)$, $r_y(\psi)$, $r_z(\theta)$ that define rotations around the x-axis, y-axis, and z-axis, respectively, namely

$$r_x(\varphi/2) = \begin{pmatrix} e^{\frac{i\varphi}{2}} & 0 \\ 0 & e^{-\frac{i\varphi}{2}} \end{pmatrix}, \quad r_y(\psi/2) = \begin{pmatrix} \cos\frac{\psi}{2} & -\sin\frac{\psi}{2} \\ \sin\frac{\psi}{2} & \cos\frac{\psi}{2} \end{pmatrix},$$

$$r_z(\theta/2) = \begin{pmatrix} \cos\frac{\theta}{2} & i\sin\frac{\theta}{2} \\ i\sin\frac{\theta}{2} & \cos\frac{\theta}{2} \end{pmatrix}.$$

We immediately check that the rotations corresponding to $r_x(\varphi/2)$, $r_y(\psi/2)$, $r_z(\theta/2)$ under the homomorphism $\rho\colon \mathbf{SU}(2) \to \mathbf{SO}(3)$ (see Theorem 3.9) are given by the matrices

$$R_x(\varphi) = \begin{pmatrix} 1 & 0 & 0 \\ 0 & \cos\varphi & -\sin\varphi \\ 0 & \sin\varphi & \cos\varphi \end{pmatrix}, \quad R_y(\psi) = \begin{pmatrix} \cos\psi & 0 & -\sin\psi \\ 0 & 1 & 0 \\ \sin\psi & 0 & \cos\psi \end{pmatrix},$$

$$R_z(\theta) = \begin{pmatrix} \cos\theta & -\sin\theta & 0 \\ \sin\theta & \cos\theta & 0 \\ 0 & 0 & 1 \end{pmatrix}.$$

So $R_x(\varphi)$ is a rotation by φ around the x-axis (with the plane orthogonal to the x-axis oriented by (e_2, e_3, e_1)), $R_y(\psi)$ is a rotation by ψ around the $-y$-axis (with the plane orthogonal to the $-y$-axis oriented by $(e_1, e_3, -e_2)$, or equivalently a rotation by $-\psi$ around the y-axis with the plane orthogonal to the y-axis oriented by (e_3, e_1, e_2)), and $R_z(\theta)$ is a rotation by θ around the z-axis (with the plane orthogonal to the z-axis oriented by (e_1, e_2, e_3)).

Remark: Beware that a number of authors switch the roles of x and z, in particular Vilenkin [Vilenkin (1968)], and most books on quantum mechanics. As a consequence, the orientation of the plane normal to the y-axis is flipped. In this case, $R_x(\varphi)$ and $R_z(\varphi)$ are swapped, but $R_y(\psi)$ becomes $R_y(-\psi)$, which is a rotation by ψ around the y-axis (with the plane orthogonal to the y-axis oriented by (e_3, e_1, e_2)). Vilenkin denotes our matrices r_x, r_y, r_z as ω_3, ω_2, ω_1.

The issue of deciding exactly how a quaternion acts on $\mathbb{R}^3$ as a rotation is quite confusing, and we feel that some clarifications are in order. First, we need to decide whether a vector $(x, y, z) \in \mathbb{R}^3$ is represented as a skew-hermitian matrix (a matrix in $\mathfrak{su}(2)$) or as a hermitian matrix. The first option seems to be followed by most mathematicians and by the computer graphics community. On the other hand, physicists seem to prefer hermitian matrices to skew-hermitian matrices. Of course, if S is a skew-hermitian matrix, then iS is a hermitian matrix, and this is the method used to make the conversion, although sometimes $(-i)S$ is used instead.

In the first method, we embed $\mathbb{R}^3$ into $\mathfrak{su}(2) \subseteq \mathbb{H}$ using the map

$$\mathrm{su}(x, y, z) = \begin{pmatrix} ix & y + iz \\ -y + iz & -ix \end{pmatrix}, \quad (x, y, z) \in \mathbb{R}^3.$$

Then $q \in \mathbf{SU}(2)$ defines the map ρ_q (on $\mathbb{R}^3$) given by

$$\rho_q(x, y, z) = \mathrm{su}^{-1}(q\, \mathrm{su}(x, y, z) q^*).$$

This is the method used in *this book* and in Gallier and Quaintance [Gallier and Quaintance (2020b)] (Chapter 15). It is possible to derive an explicit orthogonal matrix corresponding to ρ_q; see Proposition 15.5.

The representation of $\mathbb{R}^3$ as the space of hermitian matrices has several variations, and this is the source of the confusion. One option is to represent $(x, y, z) \in \mathbb{R}^3$ by the hermitian matrix

$$(-i)\mathrm{su}(x, y, z) = \begin{pmatrix} x & z - iy \\ z + iy & -x \end{pmatrix}.$$

A nice feature of this representation is that

$$\begin{pmatrix} x & z - iy \\ z + iy & -x \end{pmatrix} = x\sigma_3 + y\sigma_2 + z\sigma_1,$$

where σ_1, σ_2, σ_3 are the *Pauli spin matrices*, where

$$\sigma_1 = \begin{pmatrix} 0 & 1 \\ 1 & 0 \end{pmatrix}, \quad \sigma_2 = \begin{pmatrix} 0 & -i \\ i & 0 \end{pmatrix}, \quad \sigma_3 = \begin{pmatrix} 1 & 0 \\ 0 & -1 \end{pmatrix}.$$

This representation is equivalent to the representation using su and yields the *exact same* rotation ρ_q. See Gallier and Quaintance [Gallier and Quaintance (2020b)] (Chapter 15).

The second option, apparently adopted in most of the quantum mechanics literature is to use a version of $i$su, *except that x and z are swapped and y becomes $-y$*. Vilenkin [Vilenkin (1968)] (Chapter II, Section 1) uses the map

$$(x_1, y_1, z_1) \mapsto \begin{pmatrix} z_1 & x_1 + iy_1 \\ x_1 - iy_1 & -z_1 \end{pmatrix},$$

so in terms of our embedding,

$$z_1 = x, \quad x_1 = z, \quad y_1 = -y.$$

We can check that the unit quaternions $r_x(\varphi/2)$, $r_y(\psi/2)$, $r_z(\theta/2)$ induce the rotations $R_z(\varphi)$, $R_y(-\psi)$, and $R_x(\theta)$, namely

$$R_z(\varphi) = \begin{pmatrix} \cos\varphi & -\sin\varphi & 0 \\ \sin\varphi & \cos\varphi & 0 \\ 0 & 0 & 1 \end{pmatrix}, \quad R_y(-\psi) = \begin{pmatrix} \cos\psi & 0 & \sin\psi \\ 0 & 1 & 0 \\ -\sin\psi & 0 & \cos\psi \end{pmatrix},$$

$$R_x(\theta) = \begin{pmatrix} 1 & 0 & 0 \\ 0 & \cos\theta & -\sin\theta \\ 0 & \sin\theta & \cos\theta \end{pmatrix}.$$

These are the rotation matrices used in most books on quantum mechanics, including Sakurai and Napolitano [Sakurai and Napolitano (2021)]. Using our notation, Vilenkin factors a unit quaternion as

$$q = r_x(\varphi/2) r_z(\theta/2) r_x(\psi/2);$$

see page 99 of Vilenkin [Vilenkin (1968)]. This quaternion induces the rotation $R_z(\varphi)R_x(\theta)R_z(\psi)$.

Wigner [Wigner (1959)] (page 158) uses the map

$$(x_1, y_1, z_1) \mapsto \begin{pmatrix} -z_1 & x_1 + iy_1 \\ x_1 - iy_1 & z_1 \end{pmatrix},$$

so in terms of our embedding,

$$z_1 = -x, \quad x_1 = z, \quad y_1 = -y.$$

With Wigner's map, we have

$$\begin{pmatrix} -z_1 & x_1 + iy_1 \\ x_1 - iy_1 & z_1 \end{pmatrix} = x_1\sigma_1 - y_1\sigma_2 - z_1\sigma_3,$$

and Wigner calls $\sigma_1, -i\sigma_2, -i\sigma_3$ the *Pauli matrices*! These days, most books on quantum mechanics seem to be using the definition of the Pauli matrices $\sigma_1, \sigma_2, \sigma_3$ that we gave above. For instance, this is the definition given in Sakurai and Napolitano [Sakurai and Napolitano (2021)] (see page 160). We can check that the unit quaternions $r_x(\varphi/2), r_y(\psi/2), r_z(\theta/2)$ induce the rotations $R_z(\varphi)$, $R_y(\psi)$, and $R_x(-\theta)$, namely

$$R_z(\varphi) = \begin{pmatrix} \cos\varphi & -\sin\varphi & 0 \\ \sin\varphi & \cos\varphi & 0 \\ 0 & 0 & 1 \end{pmatrix}, \quad R_y(\psi) = \begin{pmatrix} \cos\psi & 0 & -\sin\psi \\ 0 & 1 & 0 \\ \sin\psi & 0 & \cos\psi \end{pmatrix},$$

$$R_x(-\theta) = \begin{pmatrix} 1 & 0 & 0 \\ 0 & \cos\theta & \sin\theta \\ 0 & -\sin\theta & \cos\theta \end{pmatrix}.$$

Using our notation, Wigner [Wigner (1959)] factors a unit quaternion q as

$$q = r_x(-\varphi/2)r_y(\theta/2)r_x(-\psi/2);$$

see Formula (15.15) with φ, θ, ψ replaced by α, β, γ. This quaternion induces the rotation $R_z(-\varphi)R_y(\theta)R_z(-\psi)$.

One final word of caution. In quantum mechanics, it is customary to express $r_x(-\varphi/2)$, $r_y(\psi/2)$, $r_z(-\theta/2)$ in terms of the Pauli spin matrices as

$$r_x(-\varphi/2) = e^{-i\frac{\varphi\sigma_3}{2}}, \quad r_y(\psi/2) = e^{-i\frac{\psi\sigma_2}{2}}, \quad r_z(-\theta/2) = e^{-i\frac{\theta\sigma_1}{2}};$$

beware of the different sign in $r_y(\psi/2)$. See Formula 3.91 on page 168 of Sakurai and Napolitano [Sakurai and Napolitano (2021)], and remember that x and z are swapped.

Analogously to the factorization of rotation matrices in terms of the Euler angles, we will prove that every unit quaternion q can be written in the form

$$q = u(\varphi, \theta, \psi) = r_x(\varphi/2)r_z(\theta/2)r_x(\psi/2).$$

Multiplying out the above matrices, we get

$$u(\varphi, \theta, \psi) = \begin{pmatrix} e^{\frac{i\varphi}{2}} & 0 \\ 0 & e^{-\frac{i\varphi}{2}} \end{pmatrix} \begin{pmatrix} \cos\frac{\theta}{2} & i\sin\frac{\theta}{2} \\ i\sin\frac{\theta}{2} & \cos\frac{\theta}{2} \end{pmatrix} \begin{pmatrix} e^{\frac{i\psi}{2}} & 0 \\ 0 & e^{-\frac{i\psi}{2}} \end{pmatrix}$$

$$= \begin{pmatrix} \cos\frac{\theta}{2}\, e^{\frac{i(\varphi+\psi)}{2}} & i\sin\frac{\theta}{2}\, e^{\frac{i(\varphi-\psi)}{2}} \\ i\sin\frac{\theta}{2}\, e^{-\frac{i(\varphi-\psi)}{2}} & \cos\frac{\theta}{2}\, e^{-\frac{i(\varphi+\psi)}{2}} \end{pmatrix}.$$

The reader can reconfirm by inspection that $u(\varphi, \theta, \psi)^{-1} = u(\varphi, \theta, \psi)^*$.

Proposition 5.4. *Every unit quaternion*

$$q = \begin{pmatrix} \alpha & \beta \\ -\overline{\beta} & \overline{\alpha} \end{pmatrix}, \quad \alpha, \beta \in \mathbb{C},\ |\alpha|^2 + |\beta|^2 = 1$$

can be expressed as

$$q = u(\varphi, \theta, \psi) = r_x(\varphi/2) r_z(\theta/2) r_x(\psi/2)$$

$$= \begin{pmatrix} e^{\frac{i\varphi}{2}} & 0 \\ 0 & e^{-\frac{i\varphi}{2}} \end{pmatrix} \begin{pmatrix} \cos\frac{\theta}{2} & i\sin\frac{\theta}{2} \\ i\sin\frac{\theta}{2} & \cos\frac{\theta}{2} \end{pmatrix} \begin{pmatrix} e^{\frac{i\psi}{2}} & 0 \\ 0 & e^{-\frac{i\psi}{2}} \end{pmatrix}.$$

If $\beta = 0$, we can pick $\theta = 0$ and φ and ψ, such that

$$\alpha = e^{i\frac{(\varphi+\psi)}{2}},$$

and in particular, $\psi = 0$. If $\alpha = 0$, we can pick $\theta = \pi$ and φ and ψ, such that

$$\beta = e^{i\frac{(\varphi-\psi+\pi)}{2}},$$

and in particular, $\psi = \pi$. If $\alpha\beta \neq 0$ and if we require that

$$0 \leq \varphi < 2\pi, \quad 0 < \theta < \pi, \quad -2\pi \leq \psi < 2\pi,$$

then φ and ψ are unique. In this case,

$$\cos\theta = 2|\alpha|^2 - 1, \quad e^{i\varphi} = -\frac{\alpha\beta i}{|\alpha||\beta|}, \quad e^{\frac{i\psi}{2}} = \frac{\alpha}{|\alpha|} e^{-\frac{i\varphi}{2}}.$$

Proof. Since $|\alpha|^2 + |\beta|^2 = 1$, we can write $\alpha = re^{i\omega}$ and $\beta = \sqrt{1-r^2}e^{i\sigma}$, with $0 \leq r \leq 1$ and where ω and σ are defined modulo 2π. We will see shortly that it is convenient to assume that $0 \leq \omega < 2\pi$ and $\frac{\pi}{2} \leq \sigma < \frac{5\pi}{2}$. The equation $q = u(\varphi, \theta, \psi)$ is equivalent to the two equations

$$re^{i\omega} = \cos\frac{\theta}{2}\, e^{\frac{i(\varphi+\psi)}{2}}$$

$$\sqrt{1-r^2}e^{i\sigma} = i\sin\frac{\theta}{2}\, e^{\frac{i(\varphi-\psi)}{2}} = \sin\frac{\theta}{2}\, e^{\frac{i(\varphi-\psi+\pi)}{2}},$$

since $i = e^{\frac{i\pi}{2}}$. If $r = 1$, we pick $\theta = 0$ and then $e^{i\omega} = e^{\frac{i(\varphi+\psi)}{2}}$, so we choose φ, ψ so that $2\omega = \varphi + \psi$. If $r = 0$, we pick $\theta = \pi$ and then $e^{i\sigma} = e^{\frac{i(\varphi-\psi+\pi)}{2}}$, so we choose φ, ψ so that $2\sigma = \varphi - \psi + \pi$. If $0 < r < 1$, namely $\alpha\beta \neq 0$, then there is a unique θ, such that $0 < \theta < \pi$ and $r = \cos\frac{\theta}{2}$, $\sqrt{1-r^2} = \sin\frac{\theta}{2}$. The angles φ and ψ must satisfy the equations

$$\begin{aligned} \omega + k_1 2\pi &= \frac{(\varphi+\psi)}{2} \\ \sigma + k_2 2\pi &= \frac{(\varphi-\psi+\pi)}{2}, \end{aligned}$$

with $k_1, k_2 \in \mathbb{Z}$, and these are equivalent to the equations

$$\begin{aligned} \varphi &= \omega + \sigma - \frac{\pi}{2} + (k_1 + k_2)2\pi \\ \psi &= \omega - \sigma + \frac{\pi}{2} + (k_1 - k_2)2\pi, \end{aligned}$$

with $k_1, k_2 \in \mathbb{Z}$. These equations always have solutions, but we would like to show that if we require that $0 \leq \varphi < 2\pi$ and $-2\pi \leq \psi < 2\pi$, then φ and ψ are unique.

First, since $-\sigma + \frac{\pi}{2} = -(\sigma - \frac{\pi}{2})$, we let $\delta = \sigma - \frac{\pi}{2}$ so that the above equations become

$$\begin{aligned} \varphi &= \omega + \delta + (k_1 + k_2)2\pi \\ \psi &= \omega - \delta + (k_1 - k_2)2\pi, \end{aligned}$$

with $k_1, k_2 \in \mathbb{Z}$, and we may assume that $0 \leq \omega < 2\pi$, $0 \leq \delta < 2\pi$. Since $0 \leq \omega, \delta < 2\pi$, we have $0 \leq \omega + \delta < 4\pi$.

If $\omega + \delta < 2\pi$, since $\omega, \delta \geq 0$, we have $-2\pi < \omega - \delta < 2\pi$, so we must pick $k_1 = 0$ and $k_2 = 0$ to make sure that $0 \leq \varphi < 2\pi$ and $-2\pi \leq \psi < 2\pi$.

Let us now assume that $2\pi \leq \omega + \delta < 4\pi$. Since $0 \leq \omega < 2\pi$, $0 \leq \delta < 2\pi$, we have $-2\pi < \omega - \delta < 2\pi$.

Case 1. $\omega - \delta \geq 0$. Since $2\pi \leq \omega + \delta < 4\pi$, by subtracting 2π we get $0 \leq \omega + \delta - 2\pi < 2\pi$. This can be achieved by setting $k_1 = -1$, $k_2 = 0$. Then, since $\omega - \delta \geq 0$, we have $\omega - \delta - 2\pi = \omega - \delta - 2\pi \geq -2\pi$. Consequently,

$$\begin{aligned} \varphi &= \omega + \delta - 2\pi \\ \psi &= \omega - \delta - 2\pi \end{aligned}$$

satisfy the required conditions $0 \leq \varphi < 2\pi$ and $-2\pi \leq \psi < 2\pi$.

Case 2. $\omega - \delta < 0$. Since $2\pi \leq \omega + \delta < 4\pi$, by subtracting 2π we get $0 \leq \omega + \delta - 2\pi < 2\pi$. This can be achieved by setting $k_1 = 0$, $k_2 = -1$. Then, since $\omega - \delta < 0$, we have

$$\omega - \delta + 2\pi = \omega - \delta + 2\pi < 2\pi.$$

Consequently,

$$\begin{aligned}\varphi &= \omega + \delta - 2\pi \\ \psi &= \omega - \delta + 2\pi\end{aligned}$$

satisfy the required conditions $0 \leq \varphi < 2\pi$ and $-2\pi \leq \psi < 2\pi$.

The last part is immediately verified. □

An interesting corollary of Proposition 5.4 is the fact that every rotation matrix $Q \in \mathbf{SO}(3)$ can be written in the terms of the Euler angles as a product

$$Q = R_x(\varphi)R_z(\theta)R_x(\psi),$$

namely

$$Q = \begin{pmatrix} 1 & 0 & 0 \\ 0 & \cos\varphi & -\sin\varphi \\ 0 & \sin\varphi & \cos\varphi \end{pmatrix} \begin{pmatrix} \cos\theta & -\sin\theta & 0 \\ \sin\theta & \cos\theta & 0 \\ 0 & 0 & 1 \end{pmatrix} \begin{pmatrix} 1 & 0 & 0 \\ 0 & \cos\psi & -\sin\psi \\ 0 & \sin\psi & \cos\psi \end{pmatrix}.$$

But in this case, we may assume that $0 \leq \psi < 2\pi$. This is because both q and $-q$ define the same rotation ρ_q, but since $e^{i\pi} = e^{-i\pi} = -1$, we have $-r_x(\psi/2) = r_x(\frac{\psi+2\pi}{2})$, so if $-2\pi \leq \psi < 0$, then $0 \leq \psi + 2\pi < 2\pi$ and $Q = R_x(\varphi)R_z(\theta)R_x(\psi + 2\pi)$.

One might wonder what happens if we make the bold move of replacing the *real* angle parameters φ, θ, ψ by *arbitrary* complex numbers? This certainly makes sense since the complex power series $z \mapsto e^z$, $z \mapsto \cos z$, $\mapsto \sin z$ are perfectly well-defined. We see immediately that $\det(u(\varphi, \theta, \psi)) = 1$, so these complex matrices belong to $\mathbf{SL}(2, \mathbb{C})$. Remarkably, *every* matrix $A \in \mathbf{SL}(2, \mathbb{C})$ can be expressed as $A = u(\varphi, \theta, \psi)$ for some choice of complex numbers φ, θ, ψ. We also have uniqueness of the representation if φ, θ, $\psi \in \mathbb{C}$ satisfy the conditions

$$0 < \Re(\theta) < \pi, \quad 0 \leq \Re(\varphi) < 2\pi, \quad -2\pi \leq \Re(\psi) < 2\pi.$$

See Vilenkin [Vilenkin (1968)] (Chapter III, Section 1.4). In some sense, the above fact illustrates the fact that $\mathbf{SL}(2, \mathbb{C})$ is the complexification of $\mathbf{SU}(2)$.

5.4 Multiplication of Quaternions in Terms of Euler Angles

Let q_1 and q_2 be two unit quaternions ($q_1, q_2 \in \mathbf{SU}(2)$) expressed in terms of Euler angles as $q_1 = r_x(\varphi_1/2)r_z(\theta_1/2)r_x(\psi_1/2)$ and $q_2 = r_x(\varphi_2/2)r_z(\theta_2/2)r_x(\psi_2/2)$. It is possible, although somewhat complicated, to express the product $q = q_1q_2$ as $q = r_x(\varphi/2)r_z(\theta/2)r_x(\psi/2)$ for some Euler angles φ, θ, ψ which can be computed from φ_1, θ_1, ψ_1, φ_2, θ_2, ψ_2. Details of this computation are given in Vilenkin [Vilenkin (1968)] (Chapter III, Section 1.2).

The key point is that

$$\begin{aligned} q = q_1q_2 &= r_x(\varphi_1/2)r_z(\theta_1/2)r_x(\psi_1/2)r_x(\varphi_2/2)r_z(\theta_2/2)r_x(\psi_2/2) \\ &= r_x(\varphi_1/2)r_z(\theta_1/2)r_x((\varphi_2+\psi_1)/2)r_z(\theta_2/2)r_x(\psi_2/2), \end{aligned}$$

so if we can find α, β, γ, such that

$$r_x(\alpha/2)r_z(\beta/2)r_x(\gamma/2) = r_z(\theta_1/2)r_x((\varphi_2+\psi_1)/2)r_z(\theta_2/2),$$

then

$$\begin{aligned} q &= r_x(\varphi_1/2)r_x(\alpha/2)r_z(\beta/2)r_x(\gamma/2)r_x(\psi_2/2) \\ &= r_x((\varphi_1+\alpha)/2)r_z(\beta/2)r_x((\psi_2+\gamma)/2), \end{aligned}$$

and so

$$q = q_1q_2 = r_x(\varphi/2)r_z(\theta/2)r_x(\psi/2), \text{ with } \varphi = \varphi_1+\alpha,\ \theta = \beta,\ \psi = \psi_2+\gamma.$$

The problem reduces to finding some Euler angles α, β, γ, such that

$$r_x(\alpha/2)r_z(\beta/2)r_x(\gamma/2) = q = r_z(\theta_1/2)r_x(\varphi_2/2)r_z(\theta_2/2),$$

where for simplicity of notation, we use the variable φ_2 instead of $\varphi_2+\psi_1$. From Section 5.3, we have

$$r_x(\varphi_2/2)r_z(\theta_2/2) = \begin{pmatrix} \cos\frac{\theta_2}{2}\, e^{\frac{i\varphi_2}{2}} & i\sin\frac{\theta_2}{2}\, e^{\frac{i\varphi_2}{2}} \\ i\sin\frac{\theta_2}{2}\, e^{-\frac{i\varphi_2}{2}} & \cos\frac{\theta_2}{2}\, e^{-\frac{i\varphi_2}{2}} \end{pmatrix},$$

so

$$q = \begin{pmatrix} \cos\frac{\theta_1}{2} & i\sin\frac{\theta_1}{2} \\ i\sin\frac{\theta_1}{2} & \cos\frac{\theta_1}{2} \end{pmatrix} \begin{pmatrix} \cos\frac{\theta_2}{2}\, e^{\frac{i\varphi_2}{2}} & i\sin\frac{\theta_2}{2}\, e^{\frac{i\varphi_2}{2}} \\ i\sin\frac{\theta_2}{2}\, e^{-\frac{i\varphi_2}{2}} & \cos\frac{\theta_2}{2}\, e^{-\frac{i\varphi_2}{2}} \end{pmatrix}.$$

Multiplying out and using Proposition 5.4, we find that α, β, γ must satisfy the equations

$$\begin{aligned} \cos\beta &= \cos\theta_1\cos\theta_2 - \sin\theta_1\sin\theta_2\cos\varphi_2 \\ e^{i\alpha} &= \frac{\sin\theta_1\cos\theta_2 + \cos\theta_1\sin\theta_2\cos\varphi_2 + i\sin\theta_2\sin\varphi_2}{\sin\beta} \\ e^{\frac{i(\alpha+\gamma)}{2}} &= \frac{\cos\frac{\theta_1}{2}\cos\frac{\theta_2}{2}e^{\frac{i\varphi_2}{2}} - \sin\frac{\theta_1}{2}\sin\frac{\theta_2}{2}e^{\frac{-i\varphi_2}{2}}}{\cos\frac{\beta}{2}}. \end{aligned}$$

It is shown in Vilenkin that we obtain

$$\tan\alpha = \frac{\sin\theta_2 \sin\varphi_2}{\cos\theta_1 \sin\theta_2 \cos\varphi_2 + \sin\theta_1 \cos\theta_2}$$
$$\tan\gamma = \frac{\sin\theta_1 \sin\varphi_2}{\sin\theta_1 \cos\theta_2 \cos\varphi_2 + \cos\theta_1 \sin\theta_2},$$

and β is determined by the equation

$$\cos\beta = \cos\theta_1 \cos\theta_2 - \sin\theta_1 \sin\theta_2 \cos\varphi_2,$$

where $0 \leq \theta \leq \pi$. Then φ_2 is replaced by $\varphi_2 + \psi_1$ in the equations above and we have

$$\varphi = \varphi_1 + \alpha, \qquad \theta = \beta, \qquad \psi = \psi_2 + \gamma.$$

It should be noted that the equations for α and γ do not determine these angles uniquely when $\alpha \in [0, 2\pi)$ and $\gamma \in [-2\pi, 2\pi)$.

Since the rotation matrix $R = R_x(\varphi)R_z(\theta)R_x(\psi)$ corresponds to the quaternion $q = r_x(\varphi/2)r_z(\theta/2)r_x(\psi/2)$ under the homomorphism $\rho\colon \mathbf{SU}(2) \to \mathbf{SO}(3)$, with $R = \rho_q$ (see Theorem 3.9), exactly the same formulae as above can be used to determine φ, θ, ψ, so that

$$R_x(\varphi)R_z(\theta)R_x(\psi) = R_x(\varphi_1)R_z(\theta_1)R_x(\psi_1)R_x(\varphi_2)R_z(\theta_2)R_x(\psi_2),$$

except that here $\gamma \in [0, 2\pi)$.

5.5 Dehomogenized Representations of SL(2, ℂ) and SU(2)

In Example 3.8, we defined the irreducible representations $U_m\colon \mathbf{SU}(2) \to \mathbf{GL}(\mathcal{P}_m^{\mathbb{C}}(2))$ of $\mathbf{SU}(2)$, whose representing spaces are the vector spaces $\mathcal{P}_m^{\mathbb{C}}(2)$ of homogeneous polynomials in two variables. We also said that it is customary, especially in the physics literature, to index homogeneous polynomials in terms of $\ell = m/2$, which is an integer when m is even but a half integer when m is odd. In this context, in terms of $\ell = m/2$, a homogeneous polynomial is written as

$$P(z_1, z_2) = \sum_{k=-\ell}^{\ell} c_k z_1^{\ell-k} z_2^{\ell+k},$$

where it is assumed that $\ell + k = j$, where j takes the *integral* values $j = 0, 1, \ldots, 2\ell = m$, so that $\ell - k = 2\ell - (\ell + k) = 2\ell - j$ takes the values $2\ell, 2\ell - 1, \ldots, 0$. Note that $k = j - \ell = j - m/2$ with $j = 0, 1, \ldots, 2\ell = m$, so k is an integer only if m is even. If m is odd, say $m = 2h+1$, then $\ell = h + \frac{1}{2}$ and k takes the $2\ell + 1 = m + 1$ values

$$-h - \frac{1}{2},\ -(h-1) - \frac{1}{2}, \ldots, -\frac{1}{2},\ \frac{1}{2},\ 1 + \frac{1}{2}, \ldots, h + \frac{1}{2},$$

and so $k \neq 0$. If m is even, say $m = 2h$, then $\ell = h$ and k takes the $2\ell + 1 = m + 1$ values

$$-h,\ -(h-1), \ldots, -1,\ 0,\ 1, \ldots, h-1, h.$$

For example, if $\ell = \frac{3}{2}$, then k takes the four values

$$-\frac{3}{2},\ -\frac{1}{2},\ \frac{1}{2},\ \frac{3}{2},$$

and if $\ell = 2$, then k takes the five values

$$-2,\ -1,\ 0,\ 1,\ 2.$$

The representing space is then $\mathcal{P}^{\mathbb{C}}_{2\ell}(2)$ and it has dimension $2\ell + 1$. Using the standard technique of "dehomogenizing" and "homogenizing," we can use the space of complex polynomials of degree $2\ell + 1$ in *one* variable z instead of the space $\mathcal{P}^{\mathbb{C}}_{2\ell}(2)$ of homogeneous polynomials in two variables z_1, z_2. Given a homogeneous polynomial $P(z_1, z_2)$ of degree $m = 2\ell$, by dehomogenizing we obtain the polynomial $Q(z)$ of degree $m = 2\ell$ given by

$$Q(z) = P(z, 1). \qquad \text{(dehomog)}$$

So given

$$P(z_1, z_2) = \sum_{k=-\ell}^{\ell} c_k z_1^{\ell-k} z_2^{\ell+k},$$

we obtain

$$Q(z) = \sum_{k=-\ell}^{\ell} c_k z^{\ell-k}. \qquad (Q)$$

Observe that due to our indexing scheme, the coefficients of Q have "funny" indices. For example, for $\ell = 2$, so that $m = 2\ell = 4$,

$$Q(z) = c_{-2}z^4 + c_{-1}z^3 + c_0 z^2 + c_1 z + c_2,$$

and when $\ell = 5/2$, so that $m = 2\ell = 5$, we have

$$Q(z) = c_{-5/2}z^5 + c_{-3/2}z^4 + c_{-1/2}z^3 + c_{1/2}z^2 + c_{3/2}z + c_{5/2}.$$

Conversely, given a polynomial $Q(z)$ of degree $m = 2\ell$, by homogenizing we obtain the homogeneous polynomial $P(z_1, z_2)$ of degree $m = 2\ell$ given by

$$P(z_1, z_2) = z_2^{2\ell} Q\left(\frac{z_1}{z_2}\right). \qquad \text{(homog)}$$

Definition 5.2. Following Vilenkin, we denote the space of polynomials of degree $m = 2\ell$ with complex coefficients in one variable by $\mathcal{P}_\ell^{\mathbb{C}}$.

Note that the "funny" index ℓ is a half integer when m is odd. We can convert our representations $U_m\colon \mathbf{SU}(2) \to \mathbf{GL}(\mathcal{P}_m^{\mathbb{C}}(2))$ to representations in the spaces $\mathcal{P}_\ell^{\mathbb{C}}$. Actually, until we use the fact that $\mathbf{SU}(2)$ is compact, we consider representations of $\mathbf{SL}(2,\mathbb{C})$.

Definition 5.3. Given any matrix

$$A = \begin{pmatrix} a & b \\ c & d \end{pmatrix}, \quad a,b,c,d \in \mathbb{C},\ ad - bc = 1,$$

in $\mathbf{SL}(2,\mathbb{C})$, for any polynomial $Q \in \mathcal{P}_\ell^{\mathbb{C}}$, define $T_\ell(A)(Q(z))$ by

$$T_\ell(A)(Q(z)) = (bz+d)^{2\ell} Q\left(\frac{az+c}{bz+d}\right). \tag{T_ℓ}$$

It is immediately verified that the above formula yields a representation $T_\ell\colon \mathbf{SL}(2,\mathbb{C}) \to \mathbf{GL}(\mathcal{P}_\ell^{\mathbb{C}})$, which yields a representation $T_\ell\colon \mathbf{SU}(2) \to \mathbf{GL}(\mathcal{P}_\ell^{\mathbb{C}})$ when restricted to the subgroup $\mathbf{SU}(2)$ of $\mathbf{SL}(2,\mathbb{C})$.

Note that the above formula for $T_\ell(A)(Q(z))$ is *not* what we would obtain directly from the representation U_ℓ. We are using Vilenkin's formula to facilitate comparison with his exposition; see Vilenkin [Vilenkin (1968)] (Chapter III, Section 2.1) and Kosmann-Schwarzbach [Kosmann-Schwarzbach (2010)]. With our version, we define the representations T_ℓ as

$$T_\ell(A)(Q(z)) = (-cz+a)^{2\ell} Q\left(\frac{dz-b}{-cz+a}\right).$$

In its homogeneous form, Vilenkin's version of the representation U_ℓ is

$$U_\ell^v(A)(Q(z_1,z_2)) = Q(az_1 + cz_2, bz_1 + dz_2).$$

Observe that

$$\begin{pmatrix} az_1 + cz_2 \\ bz_1 + dz_2 \end{pmatrix} = \begin{pmatrix} a & c \\ b & d \end{pmatrix} \begin{pmatrix} z_1 \\ z_2 \end{pmatrix} = A^\top \begin{pmatrix} z_1 \\ z_2 \end{pmatrix},$$

but in our case

$$\begin{pmatrix} dz_1 - bz_2 \\ -cz_1 + az_2 \end{pmatrix} = \begin{pmatrix} d & -b \\ -c & a \end{pmatrix} \begin{pmatrix} z_1 \\ z_2 \end{pmatrix} = A^{-1} \begin{pmatrix} z_1 \\ z_2 \end{pmatrix}.$$

We immediately check that if

$$Y = \begin{pmatrix} b & d \\ -a & -c \end{pmatrix},$$

then

$$YA^\top = A^{-1}Y,$$

and $\det(Y) = ad - bc = 1$. Then Y defines a linear isomorphism of $\mathcal{P}_{2\ell}^{\mathbb{C}}(2)$ given by $Q(z_1, z_2) \mapsto Q(bz_1 + dz_2, -az_1 - cz_2)$, and this map is an equivalence between the representations U_ℓ and U_ℓ^v (we leave the details as an exercise). We also leave it as an exercise (using the dehomogenization and the homogenization maps, which are linear isomorphisms) to check that the representation $U_{2\ell}\colon \mathbf{SL}(2,\mathbb{C}) \to \mathbf{GL}(\mathcal{P}_{2\ell}^{\mathbb{C}}(2))$ is equivalent to the representation $T_\ell\colon \mathbf{SL}(2,\mathbb{C}) \to \mathbf{GL}(\mathcal{P}_\ell^{\mathbb{C}})$ and similarly, the representation $U_{2\ell}\colon \mathbf{SU}(2) \to \mathbf{GL}(\mathcal{P}_{2\ell}^{\mathbb{C}}(2))$ is equivalent to the representation $T_\ell\colon \mathbf{SU}(2) \to \mathbf{GL}(\mathcal{P}_\ell^{\mathbb{C}})$. In particular, the representations $T_\ell\colon \mathbf{SU}(2) \to \mathbf{GL}(\mathcal{P}_\ell^{\mathbb{C}})$ form a complete set of irreducible representations of $\mathbf{SU}(2)$.

5.6 The Lie Algebra Representation Associated with T_ℓ

We will need to define an $\mathbf{SU}(2)$-invariant hermitian inner product on each space $\mathcal{P}_\ell^{\mathbb{C}}$, and for this, it is useful to figure out what is the derivative of the representation $T_\ell\colon \mathbf{SL}(2,\mathbb{C}) \to \mathbf{GL}(\mathcal{P}_\ell^{\mathbb{C}})$ at the identity. This yields a representation $\mathfrak{t}_\ell\colon \mathfrak{sl}(2,\mathbb{C}) \to \mathrm{Hom}(\mathcal{P}_\ell^{\mathbb{C}}, \mathcal{P}_\ell^{\mathbb{C}})$, which is a representation of Lie algebras! Following Kosmann-Schwarzbach [Kosmann-Schwarzbach (2010)] (Problem 9), we use the standard technique of "passing a curve" though the identity whose tangent vector for $t = 0$ is a vector in the tangent space. So for any tangent vector $X \in \mathfrak{sl}(2,\mathbb{C})$,

$$X = \begin{pmatrix} \alpha & \beta \\ \gamma & \delta \end{pmatrix}, \quad \alpha + \delta = 0,$$

we consider the curve

$$C(t) = e^{tX} = \begin{pmatrix} a(t) & b(t) \\ c(t) & d(t) \end{pmatrix}$$

through I_2, such that $C'(0) = X$, and by the chain rule, we have

$$\begin{aligned}(\mathfrak{t}_\ell(X))(Q(z)) = (d(T_\ell)_I(X))(Q(z)) &= \frac{d}{dt}\left(T_\ell(C(t))(Q(z))\right)\Big|_{t=0} \\ &= \frac{d}{dt}\left((b(t)z + d(t))^{2\ell} Q\left(\frac{a(t)z + c(t)}{b(t)z + d(t)}\right)\right)\Bigg|_{t=0}.\end{aligned}$$

We have

$$\begin{aligned}&\frac{d}{dt}\left(\frac{a(t)z + c(t)}{b(t)z + d(t)}\right) \\ &\quad = \frac{(a'(t)z + c'(t))(b(t)z + d(t)) - (a(t)z + c(t))(b'(t)z + d'(t))}{(b(t)z + d(t))^2},\end{aligned}$$

and since $a(0) = d(0) = 1$, $b(0) = c(0) = 0$, $a'(0) = \alpha$, $b'(0) = \beta$, $c'(0) = \gamma$, $d'(0) = \delta$,

$$\frac{d}{dt}\left(\frac{a(t)z + c(t)}{b(t)z + d(t)}\right)\bigg|_{t=0} = \alpha z + \gamma - z(\beta z + \delta) = -\beta z^2 + (\alpha - \delta)z + \gamma,$$

so we obtain

$$(\mathfrak{t}_\ell(X))(Q(z)) = 2\ell(\beta z + \delta)Q(z) + (-\beta z^2 + (\alpha - \delta)z + \gamma)\frac{d}{dz}(Q(z)),$$

which can be written as

$$\mathfrak{t}_\ell(X) = 2\ell(\beta z + \delta) + (-\beta z^2 + (\alpha - \delta)z + \gamma)\frac{d}{dz},$$

viewed as a differential operator on polynomials $Q(z)$ in z. In summary, we obtained the following result.

Proposition 5.5. *For any representation* $T_\ell\colon \mathbf{SL}(2,\mathbb{C}) \to \mathbf{GL}(\mathcal{P}_\ell^{\mathbb{C}})$, *the derivative* $\mathfrak{t}_\ell = d(T_\ell)_I$ *of* T_ℓ *at the identity is the representation* $\mathfrak{t}_\ell\colon \mathfrak{sl}(2,\mathbb{C}) \to \mathrm{Hom}(\mathcal{P}_\ell^{\mathbb{C}}, \mathcal{P}_\ell^{\mathbb{C}})$ *given by*

$$\mathfrak{t}_\ell(X) = 2\ell(\beta z + \delta) + (-\beta z^2 + (\alpha - \delta)z + \gamma)\frac{d}{dz} \tag{$\mathfrak{t}_\ell$}$$

viewed as a differential operator on polynomials $Q(z)$ *in* z, *for any*

$$X = \begin{pmatrix} \alpha & \beta \\ \gamma & \delta \end{pmatrix} \in \mathfrak{sl}(2,\mathbb{C}).$$

Now $\mathfrak{su}(2)$ is the *real* vector space consisting of skew-hermitian matrices with zero trace, which are of the form

$$X = \begin{pmatrix} iu_1 & u_2 + iu_3 \\ -u_2 + iu_3 & -iu_1 \end{pmatrix}, \quad u_1, u_2, u_3 \in \mathbb{R},$$

and a basis (of course, over $\mathbb{R}$) is given by the matrices

$$\begin{pmatrix} i & 0 \\ 0 & -i \end{pmatrix}, \quad \begin{pmatrix} 0 & 1 \\ -1 & 0 \end{pmatrix}, \quad \begin{pmatrix} 0 & i \\ i & 0 \end{pmatrix}.$$

Observe that these are the matrices $i\sigma_3$, $i\sigma_2$, $i\sigma_1$, where σ_1, σ_2, σ_3 are the Pauli spin matrices. The exponential map $\exp\colon \mathfrak{su}(2) \to \mathbf{SU}(2)$ is very nicely expressed. For any $X \in \mathfrak{su}(2)$ given by

$$X = \begin{pmatrix} iu_1 & u_2 + iu_3 \\ -u_2 + iu_3 & -iu_1 \end{pmatrix},$$

if we write $\theta = \sqrt{u_1^2 + u_2^2 + u_3^2}$, then

$$e^X = \cos\theta I + \frac{\sin\theta}{\theta}X, \quad \theta \neq 0,$$

and $e^0 = I$. It is not hard to prove that the map $\exp\colon \mathfrak{su}(2) \to \mathbf{SU}(2)$ is surjective. See Gallier and Quaintance [Gallier and Quaintance (2020c)] (Section 15.5).

The Lie algebra $\mathfrak{sl}(2,\mathbb{C})$ is the complex vector space consisting of all complex 2×2 matrices with zero trace, and since the above three matrices are linearly independent over $\mathbb{C}$, they also form a basis of $\mathfrak{sl}(2,\mathbb{C})$. However, for the sake of consistency with other sources, especially Kosmann-Schwarzbach and Vilenkin, it is preferable to use the basis denoted (ξ_1, ξ_2, ξ_3) in Kosmann-Schwarzbach [Kosmann-Schwarzbach (2010)] (Chapter 5, Section 1).

Definition 5.4. The *basis* (ξ_1, ξ_2, ξ_3) of $\mathfrak{sl}(2,\mathbb{C})$ (over $\mathbb{C}$), which is also a basis of $\mathfrak{su}(2)$ (over $\mathbb{R}$), is given by

$$\xi_1 = \frac{1}{2}\begin{pmatrix} 0 & i \\ i & 0 \end{pmatrix}, \quad \xi_2 = \frac{1}{2}\begin{pmatrix} 0 & -1 \\ 1 & 0 \end{pmatrix}, \quad \xi_3 = \frac{1}{2}\begin{pmatrix} i & 0 \\ 0 & -i \end{pmatrix}.$$

It is denoted (a_1, a_2, a_3) in Vilenkin [Vilenkin (1968)] (Chapter III, Section 1.3). Note that these are the matrices $(i/2)\sigma_1$, $(-i/2)\sigma_2$, $(i/2)\sigma_3$, where σ_1, σ_2, σ_3 are the Pauli spin matrices.

It is easily verified that we have the following nice cyclic equations regarding Lie brackets:

$$[\xi_1, \xi_2] = \xi_3, \quad [\xi_2, \xi_3] = \xi_1, \quad [\xi_3, \xi_1] = \xi_2.$$

The basis (ξ_1, ξ_2, ξ_3) also has the advantage that

$$e^{\varphi\xi_3} = r_x(\varphi/2), \quad e^{\theta\xi_2} = r_y(\theta/2), \quad e^{\psi\xi_1} = r_z(\psi/2).$$

If we pick the following basis for $\mathfrak{so}(3)$,

$$E_1 = \begin{pmatrix} 0 & 0 & 0 \\ 0 & 0 & -1 \\ 0 & 1 & 0 \end{pmatrix}, \quad E_2 = \begin{pmatrix} 0 & 0 & 1 \\ 0 & 0 & 0 \\ -1 & 0 & 0 \end{pmatrix}, \quad E_3 = \begin{pmatrix} 0 & -1 & 0 \\ 1 & 0 & 0 \\ 0 & 0 & 0 \end{pmatrix},$$

then we easily check that

$$e^{\varphi E_1} = R_x(\varphi), \quad e^{-\theta E_2} = R_y(\theta), \quad e^{\psi E_3} = R_z(\psi).$$

The basis (E_1, E_2, E_3) also satisfies the cyclic equations regarding Lie brackets:

$$[E_1, E_2] = E_3, \quad [E_2, E_3] = E_1, \quad [E_3, E_1] = E_2.$$

Observe that the derivative $d\rho_I\colon \mathfrak{su}(2) \to \mathfrak{so}(3)$ of the group homomorphism $\rho\colon \mathbf{SU}(2) \to \mathbf{SO}(3)$ is a Lie algebra isomorphism given by

$$d\rho_I(\xi_1) = E_3, \quad d\rho_I(\xi_2) = -E_2, \quad d\rho_I(\xi_3) = E_1.$$

The fact that E_1 and E_3 are permuted is compensated by the fact that ξ_2 is mapped to $-E_2$.

Remark: The swap between ξ_1 and ξ_3 has to do with fact that Vilenkin and Kosmann-Schwarzbach swap x and z.

Definition 5.5. In quantum physics, it is customary to define the *hermitian* matrices J_x, J_y, J_z given by

$$J_x = iE_1, \quad J_y = iE_2, \quad J_z = iE_3,$$

and under the conventions used by physicists the rotations $R_x(\alpha)$, $R_y(-\beta)$, $R_z(\gamma)$ are expressed as

$$R_x(\alpha) = e^{-i\alpha J_x}, \quad R_y(-\beta) = e^{-i\beta J_y}, \quad R_z(\gamma) = e^{-i\gamma J_z}.$$

Observe that according to the quantum physics notation, $e^{-i\beta J_y} = R_y(-\beta)$ using our definition of $R_y(\beta)$, namely

$$R_y(-\beta) = \begin{pmatrix} \cos\beta & 0 & \sin\beta \\ 0 & 1 & 0 \\ -\sin\beta & 0 & \cos\beta \end{pmatrix}.$$

See the discussion before Proposition 5.4.

It is useful to obtain formulae for the action of $\mathfrak{t}_\ell$ on the basis (ξ_1, ξ_2, ξ_3) of $\mathfrak{sl}(2,\mathbb{C})$. Using Formula ($\mathfrak{t}_\ell$), we obtain

$$\mathfrak{t}_\ell(\xi_1) = i\ell z + \frac{i}{2}(1 - z^2)\frac{d}{dz} \tag{t1}$$

$$\mathfrak{t}_\ell(\xi_2) = -\ell z + \frac{1}{2}(1 + z^2)\frac{d}{dz} \tag{t2}$$

$$\mathfrak{t}_\ell(\xi_3) = i\left(z\frac{d}{dz} - \ell\right). \tag{t3}$$

It is instructive to see what is the action of the above operators on the basis of $\mathcal{P}_\ell^{\mathbb{C}}$ consisting of the $2\ell + 1$ polynomials $z^{\ell-k}$, $k = -\ell, -\ell+1, \ldots, +\ell$. We obtain

$$\mathfrak{t}_\ell(\xi_1)z^{\ell-k} = \frac{i}{2}(\ell - k)z^{\ell-k-1} + \frac{i}{2}(\ell + k)z^{\ell-k+1} \tag{t4}$$

$$\mathfrak{t}_\ell(\xi_2)z^{\ell-k} = \frac{1}{2}(\ell - k)z^{\ell-k-1} - \frac{1}{2}(\ell + k)z^{\ell-k+1} \tag{t5}$$

$$\mathfrak{t}_\ell(\xi_3)z^{\ell-k} = -ikz^{\ell-k}. \tag{t6}$$

These formulae can be made more revealing by introducing the linear maps H_+, H_-, H_3 on $\mathfrak{sl}(2,\mathbb{C})$ given by

$$H_+ = i\mathfrak{t}_\ell(\xi_1) - \mathfrak{t}_\ell(\xi_2) = -\frac{d}{dz} \tag{t7}$$

$$H_- = i\mathfrak{t}_\ell(\xi_1) + \mathfrak{t}_\ell(\xi_2) = -2\ell z + z^2\frac{d}{dz} \tag{t8}$$

$$H_3 = i\mathfrak{t}_\ell(\xi_3) = \ell - z\frac{d}{dz}. \tag{t9}$$

Remark: Kosmann-Schwarzbach defines J_+, J_-, J_3 as

$$J_+ = i\xi_1 - \xi_2 = \begin{pmatrix} 0 & 0 \\ -1 & 0 \end{pmatrix}, J_- = i\xi_1 + \xi_2 = \begin{pmatrix} 0 & -1 \\ 0 & 0 \end{pmatrix}, J_3 = i\xi_3 = \frac{1}{2}\begin{pmatrix} -1 & 0 \\ 0 & 1 \end{pmatrix},$$

and so $H_+ = \mathfrak{t}_\ell(J_+)$, $H_- = \mathfrak{t}_\ell(J_-)$, $H_3 = \mathfrak{t}_\ell(J_3)$. In quantum physics, the linear operator H_3 on $\mathfrak{sl}(2,\mathbb{C})$ is an observable. Another notation found in the literature, for example Dieudonné [Dieudonné (2003a)] (Chapter XXI, Section 9) is $X_+ = -J_-$, $X_- = -J_+$, $H = -2J_3$, that is,

$$X_- = \begin{pmatrix} 0 & 0 \\ 1 & 0 \end{pmatrix}, \quad X_+ = \begin{pmatrix} 0 & 1 \\ 0 & 0 \end{pmatrix}, \quad H = \begin{pmatrix} 1 & 0 \\ 0 & -1 \end{pmatrix}.$$

In Serre [Serre (2000)] (Chapter IV), X_+ is denoted X and X_- is denoted Y.

Using the formulae (t7), (t8), (t9), we obtain

$$H_+ z^{\ell-k} = -(\ell-k)z^{\ell-k-1} \tag{H_+}$$

$$H_- z^{\ell-k} = -(\ell+k)z^{\ell-k+1} \tag{H_-}$$

$$H_3 z^{\ell-k} = kz^{\ell-k}. \tag{H_3}$$

In all of the above formulae, recall that $k = -\ell, -\ell+1, \ldots, +\ell$.

The above formulae show the following interesting facts:

(1) The polynomial $z^{\ell-k}$ is an eigenvector of H_3 for the eigenvalue k.
(2) The linear map H_+ sends $z^{\ell-k}$ to an eigenvector of H_3 for the eigenvalue $k+1$. In particular, when $k=\ell$, $H_+(1)$ is the zero polynomial.
(3) The linear map H_- sends $z^{\ell-k}$ to an eigenvector of H_3 for the eigenvalue $k-1$. In particular, when $k=-\ell$, $H_-(z^{2\ell})$ is the zero polynomial.

The above facts can be used to prove that the representation $\mathfrak{t}_\ell$ of $\mathfrak{sl}(2,\mathbb{C})$ is irreducible; see Section 5.7. Then it can be shown that the representation T_ℓ of $\mathbf{SL}(2,\mathbb{C})$ and its subgroup $\mathbf{SU}(2)$ is also irreducible.

Remark: Another interesting linear operator on $\mathfrak{sl}(2,\mathbb{C})$ is the operator traditionally denoted J^2 given by

$$J^2 = \mathfrak{t}_\ell(i\xi_1)^2 + \mathfrak{t}_\ell(i\xi_2)^2 + \mathfrak{t}_\ell(i\xi_3)^2 = -(\mathfrak{t}_\ell(\xi_1)^2 + \mathfrak{t}_\ell(\xi_2)^2 + \mathfrak{t}_\ell(\xi_3)^2).$$

It is easy to see that

$$J^2 = H_+H_- + H_3(H_3 - I) = H_-H_+ + H_3(H_3 + I).$$

Using the formulae above, we can check that

$$J^2(z^{\ell-k}) = \ell(\ell+1)z^{\ell-k}.$$

Thus $\ell(\ell+1)$ is a common eigenvalue for all basis vectors $z^{\ell-k}$. In some sense the operator J^2 behave like a Laplacian. It is called the *Casimir operator* of the representation $\mathfrak{t}_\ell$. In quantum physics, it an observable of the angular momentum.

5.7 Irreducible Lie Algebra Representations of $\mathfrak{sl}(2,\mathbb{C})$ and $\mathfrak{su}(2)$

This section assumes some background of Lie algebras and Lie groups. Elementary presentations are found in Carter, Segal and Macdonald [Carter *et al.* (1995)], Hall [Hall (2003)], and Gallier and Quaintance [Gallier and Quaintance (2020a)]. More advanced treatments are given in Dieudonné [Dieudonné (2003a)], Duistermaat and Kolk [Duistermaat and Kolk (2000)], Fulton and Harris [Fulton and Harris (1991)], Hall [Hall (2003)], Helgason [Helgason (2001)], Humphreys [Humphreys (1972)], Knapp [Knapp (2002, 2001)], Samelson [Samelson (1990)], Serre [Serre (2000, 1992)], and Varadarajan [Varadarajan (1984)].

In this section, we determine all the irreducible Lie algebra representations of $\mathfrak{sl}(2,\mathbb{C})$. The Lie algebra $\mathfrak{sl}(2,\mathbb{C})$ is a (complex) *simple* Lie algebra, which means that it is not abelian and that its only ideals are $\{0\}$ and $\mathfrak{sl}(2,\mathbb{C})$ itself. One of the most beautiful result of Lie theory is that the complex simple(!) Lie algebras fall into four infinite families plus five exceptional simple Lie algebras. Furthermore, the irreducible representations of the complex simple Lie algebras can be completely determined. These results are presented in Fulton and Harris [Fulton and Harris (1991)] and Knapp [Knapp (2002)], among other sources. The determination of the irreducible Lie algebra representations of $\mathfrak{sl}(2,\mathbb{C})$ is a "miniature" case.

As a basis of $\mathfrak{sl}(2, \mathbb{C})$, it is convenient to use the basis (X, Y, H) of Section 5.6, namely

$$X = \begin{pmatrix} 0 & 1 \\ 0 & 0 \end{pmatrix}, \quad Y = \begin{pmatrix} 0 & 0 \\ 1 & 0 \end{pmatrix}, \quad H = \begin{pmatrix} 1 & 0 \\ 0 & -1 \end{pmatrix}.$$

We immediately find the equations

$$[X, Y] = H, \quad [H, X] = 2X, \quad [H, Y] = -2Y.$$

(The Lie bracket $[A, B]$ of two square matrices A and B is defined as $[A, B] = AB - BA$.)

Since we never actually defined Lie algebra representations, we recall the definition below.

Definition 5.6. Let K denote the field $K = \mathbb{R}$ or $K = \mathbb{C}$ and let $\mathfrak{g}$ be a Lie algebra. If $\mathfrak{g}$ is a real Lie algebra, then a *Lie algebra representation* of $\mathfrak{g}$ in a K-vector space V is a $\mathbb{R}$-linear map $\rho\colon \mathfrak{g} \to \mathrm{Hom}(V, V)$, which means that $\rho(\lambda X) = \lambda\rho(X)$ for all $\lambda \in \mathbb{R}$ and all $X \in \mathfrak{g}$. If $\mathfrak{g}$ is a complex Lie algebra, then a *Lie algebra representation* of $\mathfrak{g}$ in a $\mathbb{C}$-vector space V is a $\mathbb{C}$-linear map $\rho\colon \mathfrak{g} \to \mathrm{Hom}(V, V)$, which means that $\rho(\lambda X) = \lambda\rho(X)$ for all $\lambda \in \mathbb{C}$ and all $X \in \mathfrak{g}$. In both cases, ρ also has the property

$$\rho([X, Y])(v) = \rho(X)(\rho(Y)(v)) - \rho(Y)(\rho(X)(v)), \quad X, Y \in \mathfrak{g},\ v \in V. \tag{$[-,-]$}$$

When no confusion arises, $\rho(X)(v)$ is abbreviated as $X \cdot v$. With this convention, the above equation is written

$$[X, Y] \cdot v = X \cdot (Y \cdot v) - Y \cdot (X \cdot v), \quad X, Y \in \mathfrak{g},\ v \in V. \tag{$[-,-]$}$$

It should be noted that if $\mathfrak{g}$ is a real Lie algebra and if V is a complex vector space, then the linear maps $\rho(X)\colon V \to V$ are $\mathbb{C}$-linear.

Definition 5.7. A representation $\rho\colon \mathfrak{g} \to \mathrm{Hom}(V, V)$ is *irreducible* if $V \neq \{0\}$ and if the only subspaces W of V invariant under $\rho(X)$ for all $X \in \mathfrak{g}$ are $W = \{0\}$, and $W = V$. Note that if V is a *complex* space, then W is also a *complex* subspace of V.

The notion of map of Lie algebra representations is essentially the same as in the case of groups (see Definition 3.3).

Definition 5.8. Given two representations $\rho_1\colon \mathfrak{g} \to \mathrm{Hom}(V_1, V_1)$ and $\rho_2\colon \mathfrak{g} \to \mathrm{Hom}(V_2, V_2)$ of a Lie algebra $\mathfrak{g}$, a *map (or morphism) of representations* $\varphi\colon \rho_1 \to \rho_2$ is a linear map $\varphi\colon V_1 \to V_2$ which is *equivariant*, which means that the following diagram commutes for every $X \in \mathfrak{g}$:

$$\begin{array}{ccc} V_1 & \xrightarrow{\rho_1(X)} & V_1 \\ {\scriptstyle\varphi}\big\downarrow & & \big\downarrow{\scriptstyle\varphi} \\ V_2 & \xrightarrow[\rho_2(X)]{} & V_2, \end{array}$$

i.e.

$$\varphi \circ \rho_1(X) = \rho_2(X) \circ \varphi, \qquad X \in \mathfrak{g}.$$

The space of all maps between two representations as above is denoted $\mathrm{Hom}_{\mathfrak{g}}(\rho_1, \rho_2)$. Two representations $\rho_1 \colon \mathfrak{g} \to \mathrm{Hom}(V_1, V_1)$ and $\rho_2 \colon \mathfrak{g} \to \mathrm{Hom}(V_2, V_2)$ are *equivalent* iff $\varphi \colon V_1 \to V_2$ is an invertible linear map.

It should be noted that the map $\varphi \colon V_1 \to V_2$ is $\mathbb{R}$-linear if both V_1 and V_2 are real vector spaces (in which case $\mathfrak{g}$ is a real Lie algebra), and $\mathbb{C}$-linear if both V_1 and V_2 are complex vector spaces (in which case $\mathfrak{g}$ is a real or a complex Lie algebra).

As in Section 9.3, given a real Lie algebra $\mathfrak{g}$ we can construct its complexification $\mathfrak{g}_{\mathbb{C}}$, which is the complex Lie algebra whose carrier is the complex vector space $\mathfrak{g} \oplus i\mathfrak{g}$ as a direct sum of real subspaces (technically, $(\mathfrak{g}_{\mathbb{C}})|_{\mathbb{R}} = \mathfrak{g} \oplus i\mathfrak{g}$, see the beginning of Section 9.3) with the Lie bracket given by

$$[u + iv, x + iy]_{\mathbb{C}} = [u, x] - [v, y] + i([u, y] + [v, x]).$$

Then for any representation $\rho \colon \mathfrak{g} \to \mathrm{Hom}(V, V)$ with V a complex vector space, we obtain the complex representation $\rho_{\mathbb{C}} \colon \mathfrak{g}_{\mathbb{C}} \to \mathrm{Hom}(V, V)$ given by

$$\rho_{\mathbb{C}}(X + iY) = \rho(X) + i\rho(Y), \quad X, Y \in \mathfrak{g}.$$

Since $\rho(X) \colon V \to V$ and $\rho(Y) \colon V \to V$ are $\mathbb{C}$-linear maps of the complex vector space V, $i\rho(Y) \colon V \to V$ is also a $\mathbb{C}$-linear map, and so $\rho_{\mathbb{C}}(X + iY)$ makes sense. Observe that the restriction of $\rho_{\mathbb{C}}$ to $\mathfrak{g}$ is the original representation $\rho \colon \mathfrak{g} \to \mathrm{Hom}(V, V)$.

We have the following useful result which shows that for a real Lie algebra $\mathfrak{g}$ and a complex vector space V, the study of the representations $\rho \colon \mathfrak{g} \to \mathrm{Hom}(V, V)$ is equivalent to the study of the complex representations $\rho_{\mathbb{C}} \colon \mathfrak{g}_{\mathbb{C}} \to \mathrm{Hom}(V, V)$.

Proposition 5.6. *Let $\mathfrak{g}$ be a real Lie algebra and let V be a complex vector space V. There is a bijection between the set of representations $\rho\colon \mathfrak{g} \to \mathrm{Hom}(V,V)$ of $\mathfrak{g}$ in V and the set of representations $\rho'\colon \mathfrak{g}_{\mathbb{C}} \to \mathrm{Hom}(V,V)$ of $\mathfrak{g}_{\mathbb{C}}$ in V given by the map $\rho \mapsto \rho_{\mathbb{C}}$, whose inverse is the restriction of ρ' to $\mathfrak{g}$. The representation $\rho\colon \mathfrak{g} \to \mathrm{Hom}(V,V)$ is irreducible iff the representation $\rho_{\mathbb{C}}\colon \mathfrak{g}_{\mathbb{C}} \to \mathrm{Hom}(V,V)$ is irreducible.*

Proof. We already explained the reason for the bijection. Suppose that $\rho\colon \mathfrak{g} \to \mathrm{Hom}(V,V)$ is irreducible, and let W be any subspace of V invariant under $\rho_{\mathbb{C}}(X+iY)$ for all $X,Y \in \mathfrak{g}$. Then by setting $Y=0$, the subspace W is invariant under $\rho(X)$ for all $X \in \mathfrak{g}$, and since $\rho\colon \mathfrak{g} \to \mathrm{Hom}(V,V)$ is irreducible, we must have $W=\{0\}$ or $W=V$, so $\rho_{\mathbb{C}}$ is also irreducible.

Let us now assume that $\rho_{\mathbb{C}}\colon \mathfrak{g}_{\mathbb{C}} \to \mathrm{Hom}(V,V)$ is irreducible and let W be any subspace of V invariant under $\rho(X)$ for all $X \in \mathfrak{g}$. Since W is a complex subspace, we have $\rho_{\mathbb{C}}(X+iY) = \rho(X) + i\rho(Y) \in W$ for all $X,Y \in \mathfrak{g}$, and since $\rho_{\mathbb{C}}\colon \mathfrak{g}_{\mathbb{C}} \to \mathrm{Hom}(V,V)$ is irreducible, we must have $W=\{0\}$ or $W=V$, so ρ is also irreducible. $\square$

Proposition 5.6 applies to the real Lie algebra $\mathfrak{su}(2)$. Indeed, the matrices

$$\begin{pmatrix} i & 0 \\ 0 & -i \end{pmatrix}, \quad \begin{pmatrix} 0 & 1 \\ -1 & 0 \end{pmatrix}, \quad \begin{pmatrix} 0 & i \\ i & 0 \end{pmatrix}$$

are a basis (over $\mathbb{R}$) of $\mathfrak{su}(2)$ and also a basis (over $\mathbb{C}$) of $\mathfrak{sl}(2,\mathbb{C})$, and it can be shown that

$$\mathfrak{sl}(2,\mathbb{C}) = \mathfrak{su}(2) \oplus i\mathfrak{su}(2)$$

as a direct sum of real subspaces; see Example 9.1 for details. Therefore $\mathfrak{sl}(2,\mathbb{C})$ is the complexification of $\mathfrak{su}(2)$, and by Proposition 5.6, the irreducible representations of $\mathfrak{su}(2)$ in a complex vector space V are in bijection with the irreducible representations of $\mathfrak{sl}(2,\mathbb{C})$ in V.

We now consider the complex Lie algebra $\mathfrak{sl}(2,\mathbb{C})$. Let $\rho\colon \mathfrak{sl}(2,\mathbb{C}) \to \mathrm{Hom}(V,V)$ be any representation of $\mathfrak{sl}(2,\mathbb{C})$ with V of finite dimension $m+1$. Since $\rho(H)$ is a linear map over a complex vector space of finite dimension $m+1$, it has $m+1$ complex eigenvalues (counted with their multiplicities). For every eigenvalue λ of $\rho(H)$, let V^{λ} be the corresponding eigenspace. In the context of Lie algebras, λ is called a *weight*. It turns out that $\rho(H)$ is diagonalizable, but we will not need this fact to characterize the irreducible representations of $\mathfrak{sl}(2,\mathbb{C})$.

The first important property is this.

Proposition 5.7. *For any complex representation $\rho\colon \mathfrak{sl}(2,\mathbb{C}) \to \mathrm{Hom}(V,V)$ with V finite-dimensional, for any eigenvalue λ of $\rho(H)$ and any vector $v \in V^{\lambda}$, we have*

$$H \cdot (X \cdot v) = (\lambda+2)X \cdot v, \quad H \cdot (Y \cdot v) = (\lambda-2)Y \cdot v. \tag{V1}$$

Consequently, $X\colon V^{\lambda} \to V^{\lambda+2}$ and $Y\colon V^{\lambda} \to V^{\lambda-2}$.

Proof. Since $HX - XH = [H,X] = 2X$ and v is an eigenvector of $\rho(H)$ for λ, we get

$$H \cdot (X \cdot v) = [H,X] \cdot v + X \cdot (H \cdot v) = 2X \cdot v + X \cdot \lambda v = (\lambda+2)X \cdot v.$$

Similarly, since $HY - YH = [H,Y] = -2Y$,

$$H \cdot (Y \cdot v) = [H,Y] \cdot v + Y \cdot (H \cdot v) = -2Y \cdot v + Y \cdot \lambda v = (\lambda-2)Y \cdot v,$$

as claimed. ∎

Now let $z \neq 0$ be some vector $z \in V^{\lambda}$, for some eigenvalue λ of $\rho(H)$. Consider the sequence

$$z,\ X \cdot z,\ X^2 \cdot z, \ldots, X^n \cdot z, \ldots.$$

By Proposition 5.7, $X^n \cdot z \in V^{\lambda+2n}$. The nonzero vectors of the form $X^n \cdot z$ correspond to distinct eigenvalues $\lambda + 2n$ of $\rho(H)$ so they are linearly independent. But V is finite-dimensional, so there is a smallest n, such that $X^{n+1} \cdot z = 0$, and if we let $x = X^n \cdot z$, then $x \neq 0$, $X \cdot x = 0$, and $Hx = (\lambda + 2n)x$. This suggests the following definition.

Definition 5.9. Let $\rho\colon \mathfrak{sl}(2,\mathbb{C}) \to \mathrm{Hom}(V,V)$ be complex representation of $\mathfrak{g}$ with V finite-dimensional. A *nonzero* vector $e \in V$ is *primitive of weight* $\lambda \in \mathbb{C}$ if

$$Xe = 0, \quad He = \lambda e. \tag{V2}$$

The argument just before Definition 5.9 proved the following result.

Proposition 5.8. *Given any complex representation $\rho\colon \mathfrak{sl}(2,\mathbb{C}) \to \mathrm{Hom}(V,V)$ with V finite-dimensional, there is some primitive element $e \in V$ for some weight λ.*

A priori, λ is a complex number, but in fact we will prove that it is a nonnegative integer. The next proposition is the key result.

Proposition 5.9. *Given any complex representation $\rho\colon \mathfrak{sl}(2,\mathbb{C}) \to \mathrm{Hom}(V,V)$ with V finite-dimensional, consider any primitive element $e \in V$ of weight λ. Define the sequence $(e_n)_{n\geq -1}$ as follows:*

$$e_n = (1/n!)Y^n \cdot e, \quad n \geq 0, \tag{V3}$$

with $e_{-1} = 0$. Then the following properties hold:

$$H \cdot e_n = (\lambda - 2n)e_n \tag{V4}$$

$$Y \cdot e_n = (n+1)e_{n+1} \tag{V5}$$

$$X \cdot e_n = (\lambda - n + 1)e_{n-1}. \tag{V6}$$

There is a smallest $m \geq 0$, such that $e_{m+1} = 0$ and $(e_0, \ldots, e_m)$ are linearly independent. (We also have $e_n = 0$ for all $n \geq m+1$.) The weight λ is a nonnegative integer, namely $\lambda = m$, and $e_n \in V^{m-2n}$ for $n = 0, \ldots, m$. See the diagram below.

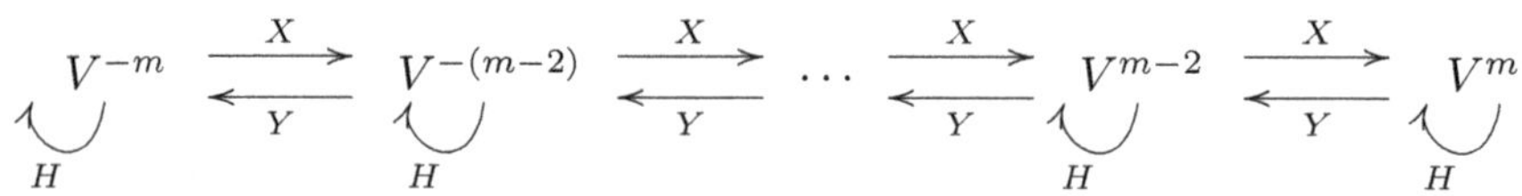

Proof. Equation (V5) follows by the definition of e_n since

$$Y \cdot e_n = (1/n!)Y \cdot Y^n \cdot e = (n+1)(1/((n+1)!)Y^{n+1} \cdot e = (n+1)e_{n+1}.$$

Equation (V4) is proven by induction. For the base case $n = 0$, since $e_0 = e$ is a primitive element of weight λ, we have $H \cdot e_0 = H \cdot e = \lambda e = \lambda e_0$.

For the induction step, since by the induction hypothesis, $H \cdot e_n = (\lambda - 2n)e_n$, from the second equation of Proposition 5.7 with $\lambda - 2n$ instead of λ, we get

$$H \cdot (Y \cdot e_n) = (\lambda - 2n - 2) \cdot (Y \cdot e_n),$$

and by (V5), we obtain

$$H \cdot e_{n+1} = (\lambda - 2(n+1)) \cdot e_{n+1}.$$

Equation (V6) is proven by induction. The base case is trivial since we set $e_{-1} = 0$. For the induction step, since $XY - YX = [X,Y] = H$ and $ne_n = Y \cdot e_{n-1}$, we have

$$\begin{aligned} nX \cdot e_n &= XY \cdot e_{n-1} \\ &= [X,Y] \cdot e_{n-1} + YX \cdot e_{n-1} \\ &= H \cdot e_{n-1} + (\lambda - n + 2)Y \cdot e_{n-2} \\ &= (\lambda - 2n + 2)e_{n-1} + (\lambda - n + 2)(n-1)e_{n-1} \\ &= n(\lambda - n + 1)e_{n-1}, \end{aligned}$$

finishing the induction step.

By (V4), the nonzero e_ns correspond to distinct eigenvalues $\lambda - 2n$, so they are linearly independent, and since V is finite-dimensional, there is some smallest $m \geq 0$, such that $e_{m+1} = 0$. If we apply (V6) with $n = m+1$, we get

$$0 = X \cdot 0 = X \cdot e_{m+1} = (\lambda - m)e_m$$

with $e_m \neq 0$, so $\lambda = m$. $\square$

We deduce the following theorem.

Theorem 5.10. *Given any complex representation* $\rho\colon \mathfrak{sl}(2,\mathbb{C}) \to \mathrm{Hom}(V,V)$ *with* V *finite-dimensional, for any primitive element* $e \in V$ *of weight* $m \in \mathbb{N}$*, the subspace* W *of* V *with basis* $(e_0,\ldots,e_m)$ *as in Proposition 5.9 is invariant under* ρ *and the restriction* $\rho\colon \mathfrak{sl}(2,\mathbb{C}) \to \mathrm{Hom}(W,W)$ *of* ρ *to* W *is irreducible.*

Proof. Equations (V4), (V5), (V6) show that W is invariant under ρ. By equation (V4), the $m+1$ eigenvalues of the restriction of $\rho(H)$ to W are $m, m-2, m-4, \ldots, -(m-2), -m$ and have multiplicity 1 (since W has dimension $m+1$). Suppose W' is a nonzero subspace of W invariant under ρ. Since $(e_0,\ldots,e_m)$ is a basis of W, one of the e_i must belong to W'. Since W' is invariant under ρ, we can apply (V6) to e_i several times and since $m - j + 1 \neq 0$ if $0 \leq j \leq i \leq m$, we see that $e_i, e_{i-1}, \ldots, e_0 = e$ belongs to W'. By applying (V5) to e_i, we see that $e_i, e_{i+1}, \ldots, e_m$ all belong to W', so $W' = W$, that is, W is irreducible. $\square$

The nonnegative integer m is called the *highest weight* of the irreducible representation ρ in W.

Remark: For *any* representation $\rho\colon \mathfrak{sl}(2,\mathbb{C}) \to \mathrm{Hom}(V,V)$ with V finite-dimensional, $\rho(H)$ is diagonalizable so we have a direct sum

$$V = \bigoplus_{\lambda} V^{\lambda}.$$

This is because $\mathfrak{sl}(2,\mathbb{C})$ is a semisimple Lie algebra (in fact, a simple Lie algebra) and $\mathrm{ad}(H)$ is diagonalizable since it has the three distinct eigenvalues $2, 0, -2$ (recall that $\mathrm{ad}(H)(Z) = [H,Z]$, and that $[H,X] = 2X$, $[H,H] = 0$, $[H,Y] = -2Y$). Then for any complex representation $\rho\colon \mathfrak{sl}(2,\mathbb{C}) \to \mathrm{Hom}(V,V)$ with V finite-dimensional, $\rho(H)$ is diagonalizable. This is a special case of results about semisimple Lie algebras found in Fulton and Harris [Fulton and Harris (1991)] (Appendix C, Section C.2) and Serre [Serre (1992)] (Part I, Chapter VI, Theorem 5.7).

We can now characterize all the irreducible (complex) representations of $\mathfrak{sl}(2, \mathbb{C})$.

Definition 5.10. Let $m \geq 0$ be any natural number, and let W_m be a complex vector space of dimension $m+1$ with basis $(e_0, \ldots, e_m)$. Define the endormorphisms X^{W_m}, Y^{W_m}, Z^{W_m} of W_m as follows (by convention, $e_{-1} = e_{m+1} = 0$).

$$H^{W_m} e_n = (m - 2n)e_n \tag{V7}$$
$$Y^{W_m} e_n = (n+1)e_{n+1} \tag{V8}$$
$$X^{W_m} e_n = (m - n + 1)e_{n-1}. \tag{V9}$$

We define the homomorphism $\rho_m \colon \mathfrak{sl}(2, \mathbb{C}) \to \mathrm{Hom}(W_m, W_m)$ by

$$\rho_m(H) = H^{W_m}, \quad \rho_m(X) = X^{W_m}, \quad \rho_m(Y) = Y^{W_m}. \tag{V10}$$

Theorem 5.11. *The homomorphism* $\rho_m \colon \mathfrak{sl}(2, \mathbb{C}) \to \mathrm{Hom}(W_m, W_m)$ *is an irreducible representation of* $\mathfrak{sl}(2, \mathbb{C})$. *Every irreducible complex representation* $\rho \colon \mathfrak{sl}(2, \mathbb{C}) \to \mathrm{Hom}(V, V)$ *with* $\dim(V) = m + 1$ *is equivalent to the representation* $\rho_m \colon \mathfrak{sl}(2, \mathbb{C}) \to \mathrm{Hom}(W_m, W_m)$.

Proof. It is easy to check that the formulae defining $X^{W_m}, Y^{W_m}, H^{W_m}$ imply that

$$\begin{aligned}
H^{W_m} X^{W_m}(e_n) - X^{W_m} H^{W_m}(e_n) &= 2X^{W_m}(e_n) \\
H^{W_m} Y^{W_m}(e_n) - Y^{W_m} H^{W_m}(e_n) &= -2Y^{W_m}(e_n) \\
X^{W_m} Y^{W_m}(e_n) - Y^{W_m} X^{W_m}(e_n) &= H^{W_m}(e_n),
\end{aligned}$$

so ρ_m is a representation. Observe that by construction $e = e_0$ is a primitive element of weight m and that the $Y^n e$ span W_m. Theorem 5.10 implies that ρ_m is irreducible.

Let $\rho \colon \mathfrak{sl}(2, \mathbb{C}) \to \mathrm{Hom}(V, V)$ be any irreducible representation of $\mathfrak{sl}(2, \mathbb{C})$ of dimension $m + 1$. By Proposition 5.8 and Proposition 5.9, V contains some primitive element e' of weight m', for some natural number m'. By Proposition 5.9, e' generates a subspace W of V invariant under ρ that has dimension $m' + 1$. Since ρ is irreducible, we must have $W = V$. It follows that V has $(e'_0, e'_1, \ldots, e'_m)$ as a basis, with $e'_n = (1/n!)Y^n \cdot e'$. Define the linear isomorphism $\varphi \colon W_m \to V$ by $\varphi(e_n) = e'_n$, for $n = 0, \ldots, m$. Since by Proposition 5.9 the e'_n satisfy equations (V4), (V5), (V6), and by construction the e_n satisfy equations (V7), (V8), (V9), it is immediately verified that φ is an equivalence between the representations $\rho_m\colon \mathfrak{sl}(2,\mathbb{C}) \to \mathrm{Hom}(W_m, W_m)$ and $\rho\colon \mathfrak{sl}(2,\mathbb{C}) \to \mathrm{Hom}(V,V)$. $\square$

Since $\mathfrak{sl}(2,\mathbb{C})$ is the complexification of $\mathfrak{su}(2)$, by Proposition 5.6 and Theorem 5.11, we obtain the following result.

Theorem 5.12. *The irreducible representation $\rho_m\colon \mathfrak{sl}(2,\mathbb{C}) \to \mathrm{Hom}(W_m, W_m)$ induces by restriction an irreducible representation $\rho_m\colon \mathfrak{su}(2) \to \mathrm{Hom}(W_m, W_m)$. Every irreducible representation $\rho\colon \mathfrak{su}(2) \to \mathrm{Hom}(V,V)$ with V a complex vector space of dimension $m+1$ is equivalent to the irreducible representation $\rho_m\colon \mathfrak{su}(2) \to \mathrm{Hom}(W_m, W_m)$.*

For $m = 0$, the space W_0, is one-dimensional space isomorphic to $\mathbb{C}$, in which case H, X, Y are the zero map on W_0; ρ_0 is the trivial representation.

For $m = 1$, the space $W_1 \simeq \mathbb{C}^2$ has the basis (e_0, e_1), and $H \cdot e_0 = e_0$, $H \cdot e_1 = -e_1$, $X \cdot e_0 = 0$, $X \cdot e_1 = e_0$, $Y \cdot e_0 = e_1$, $Y \cdot e_1 = 0$, so $W_1 = W_1^{-1} \oplus W_1^1$ where W_1^{-1} is the eigenspace spanned by e_1 associated with the eigenvalue -1, W_1^1 is the eigenspace spanned by e_0 associated with the eigenvalue 1, and ρ_1 is the standard representation on $\mathbb{C}^2$.

Remark: It turns out that ρ_m is equivalent to the representation induced by ρ_1 on the symmetric tensor power $\mathrm{Sym}^m W_1$, but given a Lie algebra representation $\rho\colon \mathfrak{g} \to \mathrm{Hom}(V,V)$, one has to define the representation $\rho^{m\odot}\colon \mathfrak{g} \to \mathrm{Hom}(\mathrm{Sym}^m V, \mathrm{Sym}^m V)$. This can be done as follows. First, given two representations $\rho_1\colon \mathfrak{g} \to \mathrm{Hom}(V,V)$ and $\rho_2\colon \mathfrak{g} \to \mathrm{Hom}(W,W)$, we define the tensor product representation $\rho_1 \otimes \rho_2\colon \mathfrak{g} \to \mathrm{Hom}(V \otimes W, V \otimes W)$ by

$$[(\rho_1 \otimes \rho_2)(X)](v \otimes w) = [\rho_1(X)](v) \otimes w + v \otimes [\rho_2(X)](w),$$
$$X \in \mathfrak{g},\ v \in V,\ w \in W.$$

Taking inspiration from the above equation, since $\mathrm{Sym}^m W_1$ is generated by the m-fold powers $v_1 \odot \cdots \odot v_m$ with $v_1, \ldots, v_m \in V$, we define $\rho^{m\odot}$ recursively by

$$\begin{aligned}
&[(\rho^{m\odot})(X)](v_1 \odot \cdots \odot v_m) \\
&\quad = [\rho(X)](v_1) \odot v_2 \cdots \odot v_m + v_1 \odot [\rho^{(m-1)}(X)](v_2 \odot \cdots \odot v_m)
\end{aligned}$$

for $m \geq 2$, with $\rho^{\odot} = \rho$. Since W_1 has the basis (e_0, e_1), it is a fact of linear algebra that $\mathrm{Sym}^m W_1$ has the basis

$$(e_0^m, \ldots, e_0^{m-n} e_1^n, \ldots, e_1^m), \quad 0 \leq n \leq m,$$

where for notational simplicity, we suppress the symbol $\odot$, so we can find $[\rho_1^{m\odot}(H)](e_0^{m-n}e_1^n)$. We can show by induction that

$$\rho_1^{m\odot}(H)(e_0^{m-n}e_1^n) = (m-n)e_0^{m-n-1}e_1^n\rho_1(H)\cdot e_0 + ne_0^{m-n}e_1^{n-1}\rho_1(H)\cdot e_1,$$

and since $\rho_1(H)\cdot e_0 = e_0$ and $\rho_1(H)\cdot e_1 = -e_1$, we get

$$\rho_1^{m\odot}(H)(e_0^{m-n}e_1^n) = (m-2n)e_0^{m-n}e_1^n.$$

Thus the eigenvalues of $\rho_1^{m\odot}(H)$ are the $m+1$ integers $m, m-2, \ldots, -(m-2), -m$, and this implies that $\rho_1^{m\odot}$ is equivalent to ρ_m. Since $\mathrm{Sym}^m W_1$ is isomorphic to the space of homogeneous polynomials of degree m in two variables, we have an "a posteriori" explanation of the fact that the spaces of the irreducible representations of $\mathbf{SL}(2,\mathbb{C})$ (and $\mathbf{SU}(2)$) are these spaces of homogeneous polynomials. See Fulton and Harris [Fulton and Harris (1991)] (Chapter 11, Section 11.1).

As in the case of the representations of compact groups, we have the following result but its proof is far from immediate.

Theorem 5.13. *Every representation* $\rho_m\colon \mathfrak{sl}(2,\mathbb{C}) \to \mathrm{Hom}(V,V)$ *of* $\mathfrak{sl}(2,\mathbb{C})$ *with* V *of finite dimension splits as a direct sum of irreducible representation* $\rho_m\colon \mathfrak{sl}(2,\mathbb{C}) \to \mathrm{Hom}(W_m, W_m)$. *The number of irreducible factors isomorphic to* ρ_m *is the sum of the multiplicities of* 0 *and* 1 *as eigenvalues of* $\rho(H)$.

Theorem 5.13 is known as *complete reducibility* and is usually attributed to H. Weyl. A fascinating account of the history of its proof, starting in the mid 1890's with proofs of E. Cartan and G. Fano, can be found in Borel [Borel (2001)].

For a proof of Theorem 5.13, see Fulton and Harris [Fulton and Harris (1991)] (Appendix C, Section C.2), Serre [Serre (1992)] (Part I, Chapter VI, Section 3), Humphreys [Humphreys (1972)] (Chapter II, Section 6.3) and Samelson [Samelson (1990)] (Chapter 1, Section 1.12). See also Fulton and Harris [Fulton and Harris (1991)] (Chapter 9, Section 3) for a sketch of a proof using "Weyl's unitary trick."

Weyl's unitary trick (actually called "unitarian trick" by Weyl himself) is discussed in Serre [Serre (2000)] (Chapter IV, Theorem 6) in the special case of $\mathbf{SU}(2), \mathbf{SL}(2,\mathbb{C})$, $\mathfrak{su}(2)$, and $\mathfrak{sl}(2,\mathbb{C})$.

Let G be a complex Lie group and let $\mathfrak{g}$ be its (complex) Lie algebra. The trick works for the following reasons:

(1) The complex Lie algebra $\mathfrak{sl}(2,\mathbb{C})$ is the complexification of the real Lie algebra $\mathfrak{su}(2)$. It follows by (an easy adaptation of) Proposition 5.6 that there is a bijection d between the set $\mathrm{Hom}_{\mathbb{C}}(\mathfrak{sl}(2,\mathbb{C}),\mathfrak{g})$ of $\mathbb{C}$-homomorphisms of the Lie algebras $\mathfrak{sl}(2,\mathbb{C})$ and $\mathfrak{g}$ and the set $\mathrm{Hom}_{\mathbb{R}}(\mathfrak{su}(2),\mathfrak{g})$ of $\mathbb{R}$-homomorphisms of the Lie algebras $\mathfrak{su}(2)$ and $\mathfrak{g}$.

(2) The Lie groups $\mathbf{SU}(2)$ and $\mathbf{SL}(2,\mathbb{C})$ are connected and simply-connected.

(3) It follows from (2) (see Gallier and Quaintance [Gallier and Quaintance (2020a)] (Theorem 19.20), Fulton and Harris [Fulton and Harris (1991)] (Chapter 8, Section 3), Warner [Warner (1983)] (Chapter 3, Theorem 3.27) that there is a bijection b between the set $\mathrm{Hom}_{\mathbb{C}}(\mathbf{SL}(2,\mathbb{C}),G)$ of $\mathbb{C}$-homomorphisms (holomorphic maps) of the Lie groups $\mathbf{SL}(2,\mathbb{C})$ and G and the set $\mathrm{Hom}_{\mathbb{C}}(\mathfrak{sl}(2,\mathbb{C}),\mathfrak{g})$ of $\mathbb{C}$-homomorphisms of the Lie algebras $\mathfrak{sl}(2,\mathbb{C})$ and $\mathfrak{g}$, and a bijection c between the set $\mathrm{Hom}_{\mathbb{R}}(\mathbf{SU}(2),G)$ of $\mathbb{R}$-homomorphisms of the Lie groups $\mathbf{SU}(2)$ and G and the set $\mathrm{Hom}_{\mathbb{R}}(\mathfrak{su}(2),\mathfrak{g})$ of $\mathbb{R}$-homomorphisms of the Lie algebras $\mathfrak{su}(2)$ and $\mathfrak{g}$.

As a consequence, we obtain the following result.

Theorem 5.14. *(Weyl's Unitarian Trick) Let G be a complex Lie group and let $\mathfrak{g}$ be its (complex) Lie algebra. The following diagram commutes and all maps in it are bijections.*

$$
\begin{array}{ccc}
\mathrm{Hom}_{\mathbb{C}}(\mathbf{SL}(2,\mathbb{C}),G) & \xrightarrow{\ a\ } & \mathrm{Hom}_{\mathbb{R}}(\mathbf{SU}(2),G) \\
\Big\downarrow b & & \Big\downarrow c \\
\mathrm{Hom}_{\mathbb{C}}(\mathfrak{sl}(2,\mathbb{C}),\mathfrak{g}) & \xrightarrow[\ d\ ]{} & \mathrm{Hom}_{\mathbb{R}}(\mathfrak{su}(2),\mathfrak{g}).
\end{array}
$$

The only nonobvious map is a, which is the composition $c^{-1}\circ d\circ b$.

If we apply Theorem 5.14 to $G=\mathbf{GL}(V)$ and $\mathfrak{g}=\mathrm{Hom}(V,V)$ where V is a complex vector space, since $\mathbf{SU}(2)$ is a compact Lie group, by Peter–Weyl II we obtain complete reducibility, the fact that the representations of $\mathfrak{sl}(2,\mathbb{C})$, $\mathfrak{su}(2)$ and $\mathbf{SL}(2,\mathbb{C})$ (and of course $\mathbf{SU}(2)$) split as direct sums of irreducible representations whose representing spaces are all described by Definition 5.10. Using the isomorphism (c), we also rediscover the structure of the irreducible representations of $\mathbf{SU}(2)$.

5.8 SU(2)-Invariant Hermitian Inner Product on $\mathcal{P}_\ell^{\mathbb{C}}$

We now restrict our attention to the representations T_ℓ of $\mathbf{SU}(2)$. Our goal is to find explicitly an $\mathbf{SU}(2)$-invariant hermitian inner product on $\mathcal{P}_\ell^{\mathbb{C}}$. Because $\mathbf{SU}(2)$ is compact, such an inner product must exist. If such an invariant hermitian inner product $\langle -,-\rangle$ exists, in particular, it must be invariant for the matrices $T_\ell(r_x(\varphi/2))$, $T_\ell(r_y(\theta/2))$ and $T_\ell(r_z(\psi/2))$, so we assert such invariance and deduce consequences by taking derivatives. In fact, the proof shows that is suffices to assert invariance for the matrices $T_\ell(r_x(\varphi/2))$ and $T_\ell(r_y(\theta/2))$.

First, we need to figure out what is $T_\ell(r_x(\varphi/2))(z^{\ell-k})$. Since

$$r_x(\varphi/2) = \begin{pmatrix} e^{i\frac{\varphi}{2}} & 0 \\ 0 & e^{-i\frac{\varphi}{2}} \end{pmatrix},$$

with $a = e^{i\frac{\varphi}{2}}$, $b = c = 0$, and $d = e^{-i\frac{\varphi}{2}}$, the formula

$$T_\ell(A)(Q(z)) = (bz+d)^{2\ell} Q\left(\frac{az+c}{bz+d}\right)$$

yields

$$T_\ell(r_x(\varphi/2))(z^{\ell-k}) = e^{-i\ell\varphi}\left(\frac{e^{i\frac{\varphi}{2}}z}{e^{-i\frac{\varphi}{2}}}\right)^{\ell-k} = e^{-i\ell\varphi}e^{i(\ell-k)\varphi}z^{\ell-k} = e^{-ik\varphi}z^{\ell-k},$$

that is,

$$T_\ell(r_x(\varphi/2))(z^{\ell-k}) = e^{-ik\varphi}z^{\ell-k}.$$

The above equation is important enough to be recorded as a proposition.

Proposition 5.15. *Each polynomial $z^{\ell-k}$ is an eigenvector of $T_\ell(r_x(\varphi/2))$ for the eigenvalue $e^{-ik\varphi}$, that is,*

$$T_\ell(r_x(\varphi/2))(z^{\ell-k}) = e^{-ik\varphi}z^{\ell-k}. \qquad (*_1)$$

Thus, in the basis $(z^{\ell-k})_{-\ell\le k\le\ell}$, the matrix of $T_\ell(r_x(\varphi/2))$ is the diagonal matrix

$$\begin{pmatrix} e^{i\ell\varphi} & & & & \\ & e^{i(\ell-1)\varphi} & & & \\ & & \ddots & & \\ & & & e^{-i(\ell-1)\varphi} & \\ & & & & e^{-i\ell\varphi} \end{pmatrix}.$$

The invariance of the inner product for $T_\ell(r_x(\varphi/2))$ is stated as

$$\langle T_\ell(r_x(\varphi/2))(z^{\ell-j}),\, T_\ell(r_x(\varphi/2))(z^{\ell-k})\rangle = \langle z^{\ell-j},\, z^{\ell-k}\rangle \qquad (*_2)$$

for all j, k with $-\ell \leq j, k \leq \ell$, and since

$$T_\ell(r_x(\varphi/2))(z^{\ell-j}) = e^{-ij\varphi} z^{\ell-j} \quad \text{and} \quad T_\ell(r_x(\varphi/2))(z^{\ell-k}) = e^{-ik\varphi} z^{\ell-k}$$

(remembering that the hermitian inner product is semilinear on the second argument!), we obtain

$$\langle T_\ell(r_x(\varphi/2))(z^{\ell-j}),\, T_\ell(r_x(\varphi/2))(z^{\ell-k})\rangle = e^{-i(j-k)\varphi}\langle z^{\ell-j}, z^{\ell-k}\rangle. \qquad (*_3)$$

Equations $(*_2)$ and $(*_3)$ yield

$$e^{-i(j-k)\varphi}\langle z^{\ell-j}, z^{\ell-k}\rangle = \langle z^{\ell-j}, z^{\ell-k}\rangle,$$

and these equations show that

$$\langle z^{\ell-j},\, z^{\ell-k}\rangle = 0, \quad \text{for all } j \neq k. \qquad (*_4)$$

Next, we need to compute $\langle z^{\ell-k}, z^{\ell-k}\rangle$ to find the normalization factors. Here we assert invariance of the inner product for $T_\ell(r_y(\theta/2))$ for $z^{\ell-k}$ and $z^{\ell-k+1}$, which is stated as

$$\langle T_\ell(r_y(\theta/2))(z^{\ell-k}),\, T_\ell(r_y(\theta/2))(z^{\ell-k+1})\rangle = \langle z^{\ell-k},\, z^{\ell-k+1}\rangle \qquad (*_5)$$

for all k with $-\ell \leq k \leq \ell$. The trick is to differentiate the above equation at $\theta = 0$. Since $r_y(\theta/2) = e^{\theta\xi_2}$, we obtain

$$\langle \mathfrak{t}_\ell(\xi_2)(z^{\ell-k}),\, z^{\ell-k+1}\rangle + \langle z^{\ell-k},\, \mathfrak{t}_\ell(\xi_2)(z^{\ell-k+1})\rangle = 0. \qquad (*_6)$$

Using equation (t5), we obtain

$$-(\ell+k)\langle z^{\ell-k+1},\, z^{\ell-k+1}\rangle + (\ell-k+1)\langle z^{\ell-k},\, z^{\ell-k}\rangle = 0. \qquad (*_7)$$

By changing k to $k+1$, we obtain

$$(\ell+k+1)\langle z^{\ell-k},\, z^{\ell-k}\rangle = (\ell-k)\langle z^{\ell-k-1},\, z^{\ell-k-1}\rangle, \qquad (*_8)$$

and this recurrence equation yields

$$\langle z^{\ell-k},\, z^{\ell-k}\rangle = \frac{(\ell-k)!}{(\ell+k+1)\cdots(2\ell)}\langle 1,\, 1\rangle = \frac{(\ell-k)!(\ell+k)!}{(2\ell)!}\langle 1,\, 1\rangle.$$

It is natural to pick

$$\langle 1, 1\rangle = (2\ell)!,$$

and so we obtain

$$\langle z^{\ell-k},\, z^{\ell-k}\rangle = (\ell-k)!(\ell+k)!, \quad -\ell \leq k \leq \ell. \qquad (*_9)$$

Equations $(*_4)$ and $(*_9)$ show that the $2\ell+1$ polynomials

$$\frac{z^{\ell-k}}{\sqrt{(\ell-k)!(\ell+k)!}},$$

form an orthonormal basis of $\mathcal{P}_\ell^{\mathbb{C}}$ for an invariant hermitian inner product on $\mathbf{SU}(2)$ which is uniquely determined by setting $\langle 1, 1\rangle = (2\ell)!$. This is an important result that we record below.

Proposition 5.16. *In Vilenkin's notation, the polynomials*

$$\psi_k(z) = \frac{z^{\ell-k}}{\sqrt{(\ell-k)!(\ell+k)!}}, \quad -\ell \leq k \leq \ell, \tag{$*_{10}$}$$

form an orthonormal basis of $\mathcal{P}_\ell^{\mathbb{C}}$ *for a unique invariant hermitian inner product on* $\mathbf{SU}(2)$. *The* ψ_k *are the unit-length eigenvectors of the linear map* $T_\ell(r_x(\varphi/2))$.

Also note that the formulae (H_+), (H_-), (H_3) become

$$H_+\psi_k(z) = -\sqrt{(\ell-k)(\ell+k+1)}\,\psi_{k+1}(z) \tag{H'_+}$$

$$H_-\psi_k(z) = -\sqrt{(\ell+k)(\ell-k+1)}\,\psi_{k-1}(z) \tag{H'_-}$$

$$H_3\psi_k(z) = k\,\psi_k(z). \tag{H'_3}$$

Actually, it is remarkable that if we define a hermitian inner product on $\mathcal{P}_\ell^{\mathbb{C}}$ by requiring that the polynomials ψ_k form an orthonormal basis, then this inner product is $\mathbf{SU}(2)$ invariant. The proof of this fact relies on two standard facts of Lie group theory about the relationship between a representation and its derivative.

First, recall that if $f\colon G \to H$ is a homomorphism of Lie groups, then the derivative df_e of f at the identity element e of G is a Lie algebra homomorphism $df_e\colon \mathfrak{g} \to \mathfrak{h}$; see Gallier and Quaintance [Gallier and Quaintance (2020a)] (Chapter 19). In particular, if $H = \mathbf{GL}(E)$, where E is a finite-dimensional (complex) vector space, then f is a representation, and since the Lie algebra of the group $\mathbf{GL}(E)$ is $\mathfrak{gl}(E) = \mathrm{Hom}(E,E)$, the space of all linear maps from E to itself, $df_e\colon \mathfrak{g} \to \mathrm{Hom}(E,E)$ is what is called a *Lie algebra representation.*

If E has a hermitian inner product $\langle -, -\rangle$ and if $H = \mathbf{U}(E)$, the group of unitary linear maps with respect to the hermitian inner product $\langle -, -\rangle$, we claim that the Lie algebra $\mathfrak{u}(E)$ of $\mathbf{U}(E)$ consists of the linear maps $Z\colon E \to E$, such that

$$\langle Z(u), v\rangle + \langle u, Z(v)\rangle = 0, \quad \text{for all } u, v \in E, \tag{skew$_1$}$$

or equivalently

$$Z^* = -Z, \tag{skew$_2$}$$

where Z^* is the adjoint of Z with respect to the hermitian inner product $\langle -, -\rangle$, which is the unique linear map Z^* defined by the property that

$$\langle Z(u), v\rangle = \langle u, Z^*(v)\rangle, \quad \text{for all } u, v \in E.$$

Linear maps $Z\colon E \to E$ satisfying property (skew_1) (equivalently (skew_2)) are called *skew-hermitian* with respect to the hermitian inner product $\langle -, - \rangle$. First, since $\mathfrak{u}(E)$ is the tangent space to $\mathbf{U}(E)$ at id, by definition $Z = C'(0)$ for any smooth curve $C\colon (-\epsilon, \epsilon) \to \mathbf{U}(E)$, such that $C(0) = \text{id}$, and since each $C(t)$ is unitary, we have

$$\langle C(t)(u),\, C(t)(v)\rangle = \langle u, v\rangle$$

for all $t \in (-\epsilon, \epsilon)$ and all $u, v \in E$, so by differentiating at $t = 0$ we get

$$\langle C'(0)(u),\, C(0)(v)\rangle + \langle C(0)(u),\, C'(0)(v)\rangle = 0,$$

which, since $C'(0) = Z$ and $C(0) = \text{id}$, yields

$$\langle Z(u),\, v\rangle + \langle u,\, Z(v)\rangle = 0,$$

which is equation (skew_1). To show that all skew-hermitian linear maps belong to $\mathfrak{u}(E)$, we use standard properties of the exponential map, namely that if Z is skew-hermitian, then $(e^Z)^* = e^{(Z^*)} = e^{-Z}$, and so e^Z is unitary (all with respect to the hermitian inner product $\langle -, - \rangle$ on E). Since E is finite-dimensional, we can pick an orthonormal basis of E with respect to $\langle -, - \rangle$, and work with matrices. As a corollary, we have the following result.

Proposition 5.17. *Define a hermitian inner product* $\langle -, - \rangle$ *on* $\mathcal{P}_\ell^{\mathbb{C}}$ *by requiring that the polynomials* ψ_k *form an orthonormal basis. For any unitary representation* $T_\ell\colon \mathbf{SU}(2) \to \mathbf{U}(\mathcal{P}_\ell^{\mathbb{C}})$ *(with respect to* $\langle -, - \rangle$*) we obtain the Lie algebra representation* $\mathfrak{t}_\ell\colon \mathfrak{su}(2) \to \mathfrak{u}(\mathcal{P}_\ell^{\mathbb{C}})$*, where* $\mathfrak{t}_\ell = d(T_\ell)_I$*. Thus*

$$\mathfrak{t}_\ell(X)^* = -\mathfrak{t}_\ell(X), \quad X \in \mathfrak{su}(2), \tag{$*_{15}$}$$

namely, $\mathfrak{t}_\ell(X)$ *is skew-hermitian with respect to the hermitian inner product* $\langle -, - \rangle$.

The converse is true.

Proposition 5.18. *For any representation* $T_\ell\colon \mathbf{SU}(2) \to \mathbf{GL}(\mathcal{P}_\ell^{\mathbb{C}})$*, let* $\mathfrak{t}_\ell = d(T_\ell)_I$*, so that* $\mathfrak{t}_\ell\colon \mathfrak{su}(2) \to \text{Hom}(\mathcal{P}_\ell^{\mathbb{C}}, \mathcal{P}_\ell^{\mathbb{C}})$ *is the corresponding Lie algebra representation. If for every* $X \in \mathfrak{su}(2)$ *the linear map* $\mathfrak{t}_\ell(X)\colon \mathcal{P}_\ell^{\mathbb{C}} \to \mathcal{P}_\ell^{\mathbb{C}}$ *is skew-hermitian with respect to the hermitian inner product* $\langle -, - \rangle$ *on* $\mathcal{P}_\ell^{\mathbb{C}}$ *making the basis* (ψ_k) *orthonormal, then* $T_\ell(A)$ *is unitary with respect to* $\langle -, - \rangle$ *for all* $A \in \mathbf{SU}(2)$*; in other words,* T_ℓ *is a unitary representation* $T_\ell\colon \mathbf{SU}(2) \to \mathbf{U}(\mathcal{P}_\ell^{\mathbb{C}})$.

Proof. This result is actually true for any representation $U\colon G \to \mathbf{GL}(E)$ where G is a connected Lie group and E is finite-dimensional and equipped with a hermitian inner product, but since the exponential map $\exp\colon \mathfrak{su}(2) \to \mathbf{SU}(2)$ is surjective (Gallier and Quaintance [Gallier and Quaintance (2020c)], Section 15.5) we can give a simpler proof. Since every $q \in \mathbf{SU}(2)$ can be written as $q = e^X$ for some $X \in \mathfrak{su}(2)$, consider the function $F\colon \mathbb{R} \to \mathbb{C}$ given by

$$F(t) = \langle T_\ell(e^{tX})(\psi_j),\, T_\ell(e^{tX})(\psi_k)\rangle, \tag{$*_{16}$}$$

which has the property that $F(0) = \langle \psi_j, \psi_k\rangle$. We prove that F is constant by showing that its derivative is zero for all t. If so, since $\mathbf{SU}(2)$ is connected, F must be constant, and since its value at $t = 0$ is $\langle \psi_j, \psi_k\rangle$, for $t = 1$, we obtain

$$\langle T_\ell(q)(\psi_j),\, T_\ell(q)(\psi_k)\rangle = \langle \psi_j, \psi_k\rangle,$$

which proves that $T_\ell(q)$ is unitary with respect to $\langle -, -\rangle$. Because the map $t \mapsto h(t) = T_\ell(e^{tX})$ is a one-parameter group and $h'(0) = d(T_\ell)_I(X) = \mathfrak{t}_\ell(X)$, by Lie group theory,

$$T_\ell(e^{tX}) = e^{t\mathfrak{t}_\ell(X)};$$

see Gallier and Quaintance [Gallier and Quaintance (2020a)] (Proposition 4.13). By the chain rule

$$d(T_\ell(e^{tX}))_s = d(e^{t\mathfrak{t}_\ell(X)})_s = \mathfrak{t}_\ell(X) \circ e^{s\mathfrak{t}_\ell(X)} = \mathfrak{t}_\ell(X) \circ T_\ell(e^{sX}).$$

If we take the derivative of equation ($*_{16}$) (at any $t = s$), we get

$$\begin{aligned} F'(s) &= \langle \mathfrak{t}_\ell(X)(T_\ell(e^{sX})(\psi_j)),\, T_\ell(e^{sX})(\psi_k)\rangle \\ &\quad + \langle T_\ell(e^{sX})(\psi_j),\, \mathfrak{t}_\ell(X)(T_\ell(e^{sX})(\psi_k))\rangle. \end{aligned} \tag{$*_{17}$}$$

Since $\mathfrak{t}_\ell(X)$ is skew-hermitian by hypothesis, we conclude that $F'(s) = 0$ for all s. □

According to Proposition 5.18, to prove that the hermitian inner product $\langle -, -\rangle$ on $\mathcal{P}_\ell^{\mathbb{C}}$ making the basis (ψ_k) orthonormal is $\mathbf{SU}(2)$-invariant, it suffices to prove that the linear maps $\mathfrak{t}_\ell(X)$ are skew-hermitian with respect to $\langle -, -\rangle$ for all $X \in \mathfrak{su}(2)$. Since (ξ_1, ξ_2, ξ_3) is a basis of $\mathfrak{su}(2)$, we need to prove that $\mathfrak{t}_\ell(\xi_i)$ is skew-hermitian for $i = 1, 2, 3$.

Proposition 5.19. *The linear maps* $\mathfrak{t}_\ell(\xi_i)$ $(1 \le i \le 3)$ *are skew-hermitian for the hermitian inner product* $\langle -, -\rangle$ *on* $\mathcal{P}_\ell^{\mathbb{C}}$ *making the basis* (ψ_k) *orthonormal.*

Proof. First, we prove that $\mathrm{t}_\ell(\xi_1)$ is skew-hermitian using equation (t4),

$$\mathrm{t}_\ell(\xi_1)z^{\ell-k} = \frac{i}{2}(\ell-k)z^{\ell-k-1} + \frac{i}{2}(\ell+k)z^{\ell-k+1},$$

which is expressed in terms of the basis $(z^{\ell-k})$, and thus needs some adjustment. We divide both sides by $\sqrt{(\ell-k)!(\ell+k)!}$, which yields

$$\begin{aligned}\mathrm{t}_\ell(\xi_1)&\frac{z^{\ell-k}}{\sqrt{(\ell-k)!(\ell+k)!}}\\ &= \frac{i}{2}(\ell-k)\frac{z^{\ell-k-1}}{\sqrt{(\ell-k)!(\ell+k)!}} + \frac{i}{2}(\ell+k)\frac{z^{\ell-k+1}}{\sqrt{(\ell-k)!(\ell+k)!}}.\end{aligned}$$

Since

$$\begin{aligned}\psi_{k+1}(z) &= \frac{z^{\ell-k-1}}{\sqrt{(\ell-k-1)!(\ell+k+1)!}}\\ \psi_{k-1}(z) &= \frac{z^{\ell-k+1}}{\sqrt{(\ell-k+1)!(\ell+k-1)!}},\end{aligned}$$

we need to compute

$$\begin{aligned}&\frac{(\ell-k)\sqrt{(\ell-k-1)!(\ell+k+1)!}}{\sqrt{(\ell-k)!(\ell+k)!}}\\ &\qquad= \sqrt{(\ell-k)(\ell+k+1)}\frac{\sqrt{(\ell-k-1)!(\ell-k)(\ell+k)!}}{\sqrt{(\ell-k)!(\ell+k)!}}\\ &\qquad= \sqrt{(\ell-k)(\ell+k+1)},\end{aligned}$$

and

$$\begin{aligned}&\frac{(\ell+k)\sqrt{(\ell-k+1)!(\ell+k-1)!}}{\sqrt{(\ell-k)!(\ell+k)!}}\\ &\qquad= \sqrt{(\ell+k)(\ell-k+1)}\frac{\sqrt{(\ell-k)!(\ell+k-1)!(\ell+k)}}{\sqrt{(\ell-k)!(\ell+k)!}}\\ &\qquad= \sqrt{(\ell+k)(\ell-k+1)}.\end{aligned}$$

Consequently, we obtain the equation

$$\begin{aligned}\mathrm{t}_\ell(\xi_1)\psi_k(z) &= \frac{i}{2}\sqrt{(\ell-k)(\ell+k+1)}\psi_{k+1}(z)\\ &\quad+ \frac{i}{2}\sqrt{(\ell+k)(\ell-k+1)}\psi_{k-1}(z). \qquad (*_{18})\end{aligned}$$

Since (ψ_k) is an orthonormal basis, the (j,k) entry of the matrix $\mathbf{t}^{(1)}$ representing $\mathfrak{t}_\ell(\xi_1)$ is

$$\begin{aligned}\mathbf{t}^{(1)}_{jk} &= \langle \mathfrak{t}_\ell(\xi_1)(\psi_k(z)),\, \psi_j(z)\rangle = \frac{i}{2}\sqrt{(\ell-k)(\ell+k+1)}\langle \psi_{k+1}(z),\, \psi_j(z)\rangle \\ &\quad + \frac{i}{2}\sqrt{(\ell+k)(\ell-k+1)}\langle \psi_{k-1}(z),\, \psi_j(z)\rangle,\end{aligned}$$

and so the only nonzero entries are

$$\begin{aligned}\mathbf{t}^{(1)}_{k+1k} &= \frac{i}{2}\sqrt{(\ell-k)(\ell+k+1)} \\ \mathbf{t}^{(1)}_{k-1k} &= \frac{i}{2}\sqrt{(\ell+k)(\ell-k+1)},\end{aligned}$$

and by changing k to $k+1$

$$\mathbf{t}^{(1)}_{kk+1} = \frac{i}{2}\sqrt{(\ell+k+1)(\ell-k)},$$

and finally

$$\mathbf{t}^{(1)}_{kk+1} = \mathbf{t}^{(1)}_{k+1k} = \frac{i}{2}\sqrt{(\ell-k)(\ell+k+1)}.$$

It follows that $\mathbf{t}^{(1)}$ is a pure imaginary matrix, such that $-\overline{\mathbf{t}^{(1)}_{kk+1}} = \mathbf{t}^{(1)}_{kk+1} = \mathbf{t}^{(1)}_{k+1k}$, which proves that $\mathfrak{t}_\ell(\xi_1)$ is skew-hermitian.

To prove that $\mathfrak{t}_\ell(\xi_2)$ is skew-hermitian, we use equation (t5),

$$\mathfrak{t}_\ell(\xi_2)z^{\ell-k} = \frac{1}{2}(\ell-k)z^{\ell-k-1} - \frac{1}{2}(\ell+k)z^{\ell-k+1},$$

which only differs by the absence of i and the fact that the sign in front of the second term is -1 instead of 1. This time we find that

$$\begin{aligned}&\mathfrak{t}_\ell(\xi_2)\psi_k(z) \\ &= \frac{1}{2}\sqrt{(\ell-k)(\ell+k+1)}\psi_{k+1}(z) - \frac{1}{2}\sqrt{(\ell+k)(\ell-k+1)}\psi_{k-1}(z). \quad (*_{19})\end{aligned}$$

The (j,k) entry of the matrix $\mathbf{t}^{(2)}$ representing $\mathfrak{t}_\ell(\xi_2)$ is nonzero iff

$$\begin{aligned}\mathbf{t}^{(2)}_{kk+1} &= \frac{1}{2}\sqrt{(\ell-k)(\ell+k+1)} \\ \mathbf{t}^{(2)}_{k+1k} &= -\frac{1}{2}\sqrt{(\ell+k+1)(\ell-k)}.\end{aligned}$$

It follows that $\mathbf{t}^{(2)}$ is a real matrix, such that $-\overline{\mathbf{t}^{(2)}_{kk+1}} = \mathbf{t}^{(2)}_{k+1k}$, which proves that $\mathfrak{t}_\ell(\xi_2)$ is skew-hermitian.

To prove that $\mathfrak{t}_\ell(\xi_3)$ is skew-hermitian, we use equation (t6),

$$\mathfrak{t}_\ell(\xi_3) z^{\ell-k} = -ikz^{\ell-k}.$$

Since $z^{\ell-k}$ is an eigenvector, this is simpler. By dividing both sides by $\sqrt{(\ell-k)!(\ell+k)!}$, we obtain

$$\mathfrak{t}_\ell(\xi_3)\psi_k(z) = -ik\psi_k(z). \qquad (*_{20})$$

It follows that $\mathbf{t}^{(3)}$ is a pure imaginary diagonal matrix with diagonal elements

$$\mathbf{t}^{(3)}_{kk} = -ik,$$

and so $\mathfrak{t}_\ell(\xi_3)$ is skew-hermitian. Having verified that the three linear maps $\mathfrak{t}_\ell(\xi_i)$ are skew-hermitian, we conclude as we said earlier that the hermitian inner product defined by requiring that the (ψ_k) form an orthonormal basis is $\mathbf{SU}(2)$-invariant. □

In summary, we proved the following result.

Proposition 5.20. *The hermitian inner product on $\mathcal{P}^{\mathbb{C}}_\ell$ making the basis (ψ_k) orthonormal is $\mathbf{SU}(2)$-invariant.*

Note that this inner product is *not* invariant with respect to $\mathbf{SL}(2, \mathbb{C})$, because as before the linear maps $\mathfrak{t}_\ell(X)$ are skew-hermitian for $X \in \mathfrak{su}(2)$, but are *hermitian* for $X \in i\mathfrak{su}(2)$ (recall that $\mathfrak{sl}(2, \mathbb{C}) = \mathfrak{su}(2) \oplus i\mathfrak{su}(2)$, as a *real* vector space).

5.9 Matrices of the Irreducible Representations of $\mathbf{SL}(2, \mathbb{C})$

We now use the basis (ψ_k) to find various expressions for the matrix entries of the matrix $t^{(\ell)}(A)$ representing $T_\ell(A)$ in this basis. We give $\mathcal{P}^{\mathbb{C}}_\ell$ the hermitian inner product making (ψ_k) an orthonormal basis. In this section, we consider an arbitrary matrix

$$A = \begin{pmatrix} \alpha & \beta \\ \gamma & \delta \end{pmatrix}, \quad \alpha, \beta, \gamma, \delta \in \mathbb{C},\ \alpha\delta - \beta\gamma = 1$$

in $\mathbf{SL}(2, \mathbb{C})$. The special case of $\mathbf{SU}(2)$ is considered in later sections. In this latter case, these matrices are unitary. We use $\alpha, \beta, \gamma, \delta$ instead of a, b, c, d to make it easier to follow Vilenkin's exposition. Since the ψ_k form an orthonormal basis, we have

$$t^{(\ell)}_{jk}(A) = \langle T_\ell(A)(\psi_k),\, \psi_j\rangle = \frac{\langle T_\ell(A)(z^{\ell-k}),\, z^{\ell-j}\rangle}{\sqrt{(\ell-j)!(\ell+j)!(\ell-k)!(\ell+k)!}}. \qquad (*_{21})$$

By (T_ℓ) we have

$$T_\ell(A)(z^{\ell-k}) = (\beta z + \delta)^{2\ell}\left(\frac{\alpha z + \gamma}{\beta z + \delta}\right)^{\ell-k} = (\alpha z + \gamma)^{\ell-k}(\beta z + \delta)^{\ell+k},$$

so we obtain

$$t^{(\ell)}_{jk}(A) = \frac{\langle(\alpha z + \gamma)^{\ell-k}(\beta z + \delta)^{\ell+k},\, z^{\ell-j}\rangle}{\sqrt{(\ell-j)!(\ell+j)!(\ell-k)!(\ell+k)!}}. \qquad (*_{22})$$

The expression on the right-hand side can be "doctored on" in various ways.

The first brute-force method is to use the binomial formula together with the orthogonality of $z^{\ell-j}$ and $z^{\ell-k}$ for $j \neq k$ and the formulae

$$\langle z^{\ell-k},\, z^{\ell-k}\rangle = (\ell-k)!(\ell+k)!, \quad -\ell \leq k \leq \ell.$$

We get

$$t^{(\ell)}_{jk}(A) = \sqrt{\frac{(\ell-j)!(\ell+j)!}{(\ell-k)!(\ell+k)!}} \times \sum_{h=M}^{N} \binom{\ell-k}{\ell-j-h}\binom{\ell+k}{h} \alpha^{\ell-j-h}\beta^{h}\gamma^{j+h-k}\delta^{\ell+k-h} \qquad (*_{23})$$

with $M = \max(0, k-j)$, $N = \min(\ell-j, \ell+k)$, which can be somewhat simplified as

$$t^{(\ell)}_{jk}(A) = \sqrt{(\ell-j)!(\ell+j)!(\ell-k)!(\ell+k)!} \times \left[\sum_{h=M}^{N} (h!(\ell-j-h)!(\ell+k-h)!(j-k+h)!)^{-1} \alpha^{\ell-j-h}\beta^{h}\gamma^{j+h-k}\delta^{\ell+k-h}\right], \qquad (*_{24})$$

also with $M = \max(0, k-j)$, $N = \min(\ell-j, \ell+k)$. It is understood that if any of $\alpha, \beta, \gamma, \delta$ is zero, then the corresponding exponent must be zero. Of course, since $\alpha\delta - \beta\gamma = 1$, at most two of these coefficients must be nonzero.

Using the factorization of A as the product of an upper triangular matrix and a lower triangular matrix, Vilenkin obtains simpler formulae; see Vilenkin [Vilenkin (1968)] (Chapter III, Section 3.2). Suppose $\delta \neq 0$. Then we check immediately that if $\alpha\delta - \beta\gamma = 1$, then $\alpha = \delta^{-1} + (\beta\gamma)/\delta$, and so

$$A = \begin{pmatrix} \alpha & \beta \\ \gamma & \delta \end{pmatrix} = \begin{pmatrix} \delta^{-1} & \beta \\ 0 & \delta \end{pmatrix}\begin{pmatrix} 1 & 0 \\ \gamma/\delta & 1 \end{pmatrix}.$$

If we denote the first of the two matrices on the right-hand side by B and the second matrix by C, we have $A = BC$, and since T_ℓ is a representation,

$$T_\ell(A) = T_\ell(B)T_\ell(C),$$

which in terms of the matrices $t^{(\ell)}(A)$, $t^{(\ell)}(B)$, $t^{(\ell)}(C)$ means that

$$t^{(\ell)}(A) = t^{(\ell)}(B)t^{(\ell)}(C).$$

Therefore, if we compute the matrices $t^{(\ell)}(B)$ and $t^{(\ell)}(C)$, then $t^{(\ell)}_{jk}(A)$ will be given by

$$t^{(\ell)}_{jk}(A) = \sum_{h=-\ell}^{\ell} t^{(\ell)}_{jh}(B)t^{(\ell)}_{hk}(C).$$

To compute $t^{(\ell)}(B)$, we set $\gamma = 0$ and $\alpha = \delta^{-1}$ in Formula $(*_{24})$. The only nonzero term is obtained for $h = k - j$, and since h must be a nonnegative integer, we must have $j \leq k$. We obtain the formula

$$t^{(\ell)}_{jk}(B) = \begin{cases} 0, & \text{if } j > k \\ \sqrt{\dfrac{(\ell-j)!(\ell+k)!}{(\ell+j)!(\ell-k)!}}\, \dfrac{\beta^{k-j}\delta^{j+k}}{(k-j)!}, & \text{if } j \leq k. \end{cases}$$

To compute $t^{(\ell)}(C)$, we set $\beta = 0$, $\alpha = \delta = 1$, and substitute γ/δ for γ in Formula $(*_{24})$. The only nonzero term is obtained for $h = 0$, and for $(j-k)!$ to make sense we must have $j \geq k$. We obtain the formula

$$t^{(\ell)}_{jk}(C) = \begin{cases} 0, & \text{if } j < k \\ \sqrt{\dfrac{(\ell+j)!(\ell-k)!}{(\ell-j)!(\ell+k)!}}\, \dfrac{\gamma^{j-k}\delta^{k-j}}{(j-k)!}, & \text{if } j \geq k. \end{cases}$$

The $\beta\delta$ term in $t^{(\ell)}_{jh}(B)$ is $\beta^{h-j}\delta^{j+h}$ and the $\gamma\delta$ term in $t^{(\ell)}_{hk}(B)$ is $\gamma^{h-k}\delta^{k-h}$, so the $\beta\gamma\delta$ term in $t^{(\ell)}_{jh}(B)t^{(\ell)}_{hk}(C)$ is

$$\beta^{h-j}\delta^{j+h}\gamma^{h-k}\delta^{k-h} = \beta^{h-j}\gamma^{h-k}\delta^{j+k}.$$

Since $t^{(\ell)}_{jh}(B) = 0$ if $j > h$ and $t^{(\ell)}_{hj}(C) = 0$ if $h < k$, the only nonzero terms occur for $h \geq \max(j,k)$. In summary, we proved the following result.

Proposition 5.21. *With respect to the orthonormal basis* (ψ_k) *of* $\mathcal{P}^{\mathbb{C}}_\ell$, *the entries in the matrix* $t^{(\ell)}(A)$ *are given by the formulae below.*

(1) If $\delta \neq 0$, *then*

$$t^{(\ell)}_{jk}(A) = \sqrt{\frac{(\ell-j)!(\ell-k)!}{(\ell+j)!(\ell+k)!}} \times \sum_{h=\max(j,k)}^{\ell} \frac{(\ell+h)!}{(\ell-h)!(h-j)!(h-k)!}\, \beta^{h-j}\gamma^{h-k}\delta^{j+k}. \qquad (*_{25})$$

In particular, if $\beta = \gamma = 0$, *then* $\alpha\delta = 1$, $A = \begin{pmatrix} \alpha & 0 \\ 0 & 1/\alpha \end{pmatrix}$, *and* $t^{(\ell)}_{jk}(A)$ *is the diagonal matrix with*

$$t^{(\ell)}_{kk}(A) = \alpha^{-2k} = \delta^{2k}.$$

(2) If $\delta = 0$ *and* $\alpha \neq 0$, *then*

$$t^{(\ell)}_{jk}(A) = \begin{cases} 0, & \text{if } j+k>0 \\ \sqrt{\dfrac{(\ell-j)!(\ell-k)!}{(\ell+j)!(\ell+k)!}}\, \dfrac{(-1)^{\ell+j}\beta^{k-j}}{(-(j+k))!\,\alpha^{j+k}}, & \text{if } j+k \leq 0. \end{cases} \qquad (*_{26})$$

(3) If $\alpha = \delta = 0$, *then we obtain an anti-diagonal matrix*

$$t^{(\ell)}_{jk}(A) = \begin{cases} 0, & \text{if } j+k \neq 0 \\ (-1)^{\ell-j}\gamma^{2j}, & \text{if } k = -j. \end{cases} \qquad (*_{27})$$

In particular, if $A = r_z(\pi/2) = \begin{pmatrix} 0 & i \\ i & 0 \end{pmatrix}$, *then* $t^{(\ell)}_{jk}(A) = 0$ *if* $j \neq k$ *and* $t^{(\ell)}_{j\,-j}(A) = i^{2\ell}$.

In Proposition 5.21, we should remember that if ℓ is a half integer, then in $(*_{25})$ h is also a half integer. Of course, if ℓ is a half integer, then so are j, k.

Observe that

$$\begin{aligned} &\frac{(\ell+h)!}{(\ell-h)!(h-j)!(h-k)!} \\ &\quad = \frac{(\ell+h)!}{(\ell-h)!(2h)!}\frac{(2h)!}{(h-k)!(h+k)!}\frac{(h+k)!}{(h-j)!(j+k)!}(j+k)! \\ &\quad = \binom{\ell+h}{2h}\binom{2h}{h-k}\binom{h+k}{h-j}(j+k)!, \end{aligned}$$

with $\max(j,k) \leq h \leq \ell$. In particular, if $j = -k$, then

$$\frac{(\ell+h)!}{(\ell-h)!(h+k)!(h-k)!} = \binom{\ell+h}{2h}\binom{2h}{h-k}. \qquad (\dagger)$$

Another strategy is to use Taylor's formula. Recall that for polynomial $P(z)$ of degree m, we have

$$P(z) = \sum_{j=0}^{m} \frac{P^{(j)}(0)}{j!} z^j,$$

where $P^{(k)}(0)$ is the value of the kth derivative of P at $z = 0$. Now by definition the kth column of the matrix $t^{(\ell)}(A)$ consists of the coordinates $t^{(\ell)}_{jk}(A)$ of

$$T_\ell(A)(\psi_k(z)) = \sum_{j=-\ell}^{\ell} t^{(\ell)}_{jk}(A)\psi_j(z) = \sum_{j=-\ell}^{\ell} \frac{t^{(\ell)}_{jk}(A)}{\sqrt{(\ell-j)!(\ell+j)!}} z^{\ell-j},$$

and since

$$T_\ell(A)(\psi_k(z)) = \frac{(\alpha z+\gamma)^{\ell-k}(\beta z+\delta)^{\ell+k}}{\sqrt{(\ell-k)!(\ell+k)!}}, \tag{$*_{28}$}$$

we deduce that $t^{(\ell)}_{jk}(A)/\sqrt{(\ell-j)!(\ell+j)!}$ is the coefficient of $z^{\ell-j}$ in the expansion of ($*_{28}$) in powers of z. Using Taylor's formula, we obtain

$$t^{(\ell)}_{jk}(A) = \sqrt{\frac{(\ell+j)!}{(\ell-k)!(\ell+k)!(\ell-j)!}} \frac{d^{\ell-j}}{z^{\ell-j}} [(\alpha z+\gamma)^{\ell-k}(\beta z+\delta)^{\ell+k}]_{z=0}.$$

If $\alpha = 0$ or $\beta = 0$, the above formula simplifies. If $\alpha\beta \neq 0$, we substitute $y = \alpha(\beta z + \delta)$, then from $\alpha\delta - \beta\gamma = 1$ we get $\alpha z + \gamma = (y-1)/\beta$, so we obtain

$$t^{(\ell)}_{jk}(A) = \sqrt{\frac{(\ell+j)!}{(\ell-k)!(\ell+k)!(\ell-j)!}} \frac{\beta^{k-j}}{\alpha^{k+j}} \frac{d^{\ell-j}}{dy^{\ell-j}} [y^{\ell+k}(y-1)^{\ell-k}]_{y=\alpha\delta}.$$

Finally, we let $z = y - 1$, and since $\alpha\delta - 1 = \beta\gamma$, we obtain

$$t^{(\ell)}_{jk}(A) = \sqrt{\frac{(\ell+j)!}{(\ell-k)!(\ell+k)!(\ell-j)!}} \frac{\beta^{k-j}}{\alpha^{k+j}} \frac{d^{\ell-j}}{dz^{\ell-j}} [z^{\ell-k}(z+1)^{\ell+k}]_{z=\beta\gamma}.$$

In summary, we have the following result.

Proposition 5.22. *With respect to the orthonormal basis* (ψ_k) *of* $\mathcal{P}^{\mathbb{C}}_\ell$, *the entries in the matrix* $t^{(\ell)}(A)$ *are given by the formulae below.*

$$t^{(\ell)}_{jk}(A) = \sqrt{\frac{(\ell+j)!}{(\ell-k)!(\ell+k)!(\ell-j)!}} \frac{d^{\ell-j}}{z^{\ell-j}} [(\alpha z+\gamma)^{\ell-k}(\beta z+\delta)^{\ell+k}]_{z=0}. \tag{$*_{29}$}$$

If $\alpha\beta \neq 0$, *then*

$$t^{(\ell)}_{jk}(A) = \sqrt{\frac{(\ell+j)!}{(\ell-k)!(\ell+k)!(\ell-j)!}} \frac{\beta^{k-j}}{\alpha^{k+j}} \frac{d^{\ell-j}}{dz^{\ell-j}} [z^{\ell-k}(z+1)^{\ell+k}]_{z=\beta\gamma}. \tag{$*_{30}$}$$

5.10 Euler Angles Matrix Representations of T_ℓ

The "best" formula is obtained by using the Euler angles. We now restrict ourselves to $\mathbf{SU}(2)$, although it is possible to handle the more general case; see Vilenkin [Vilenkin (1968)] (Chapter III, Sections 3.3–3.9).

By Proposition 5.4, every matrix $q \in \mathbf{SU}(2)$, where

$$q = \begin{pmatrix} \alpha & \beta \\ -\overline{\beta} & \overline{\alpha} \end{pmatrix}, \quad |\alpha|^2 + |\beta|^2 = 1,$$

can be expressed as

$$\begin{aligned} q = u(\varphi, \theta, \psi) &= r_x(\varphi/2) r_z(\theta/2) r_x(\psi/2) \\ &= \begin{pmatrix} e^{\frac{i\varphi}{2}} & 0 \\ 0 & e^{-\frac{i\varphi}{2}} \end{pmatrix} \begin{pmatrix} \cos\frac{\theta}{2} & i\sin\frac{\theta}{2} \\ i\sin\frac{\theta}{2} & \cos\frac{\theta}{2} \end{pmatrix} \begin{pmatrix} e^{\frac{i\psi}{2}} & 0 \\ 0 & e^{-\frac{i\psi}{2}} \end{pmatrix}, \end{aligned}$$

with

$$0 \leq \varphi < 2\pi, \quad 0 \leq \theta \leq \pi, \quad -2\pi \leq \psi < 2\pi.$$

Furthermore, if $\alpha\beta \neq 0$ and if we require that $0 < \theta < \pi$, then φ, θ, ψ are unique. Since T_ℓ is a representation, we have

$$T_\ell(q) = T_\ell(r_x(\varphi/2)) T_\ell(r_z(\theta/2)) T_\ell(r_x(\psi/2)).$$

We also proved that the polynomials in the basis $(\psi_k(z))$ are eigenvectors of $T_\ell(r_x(\varphi/2))$ and $T_\ell(r_x(\psi/2))$, namely (by $(*_1)$)

$$\begin{aligned} T_\ell(r_x(\varphi/2))\psi_k(z) &= e^{-ik\varphi}\psi_k(z) \\ T_\ell(r_x(\psi/2))\psi_k(z) &= e^{-ik\psi}\psi_k(z). \end{aligned}$$

Thus $T_\ell(r_x(\varphi/2))$ is represented by the diagonal matrix $t^{(\ell)}(r_x(\varphi/2))$ with $t^{(\ell)}_{kk}(r_x(\varphi/2)) = e^{-ik\varphi}$, and $T_\ell(r_x(\psi/2))$ is represented by the diagonal matrix $t^{(\ell)}(r_x(\psi/2))$ with $t^{(\ell)}_{kk}(r_x(\psi/2)) = e^{-ik\psi}$. Since

$$T_\ell(q) = T_\ell(r_x(\varphi/2)) T_\ell(r_z(\theta/2)) T_\ell(r_x(\psi/2)),$$

we have

$$t^{(\ell)}(q) = t^{(\ell)}(r_x(\varphi/2)) t^{(\ell)}(r_z(\theta/2)) t^{(\ell)}(r_x(\psi/2)),$$

and since $t^{(\ell)}(r_x(\varphi/2))$ and $t^{(\ell)}(r_x(\psi/2))$ are diagonal matrices, the (j,k) entry of the matrix $t^{(\ell)}(q)$ is

$$\begin{aligned} t^{(\ell)}_{jk}(q) &= t^{(\ell)}_{jj}(r_x(\varphi/2)) t^{(\ell)}_{jk}(r_z(\theta/2)) t^{(\ell)}_{kk}(r_x(\psi/2)) \\ &= e^{-ij\varphi}\, t^{(\ell)}_{jk}(r_z(\theta/2))\, e^{-ik\psi} = e^{-i(j\varphi + k\psi)}\, t^{(\ell)}\ell_{jk}(r_z(\theta/2)), \end{aligned}$$

that is,

$$t^{(\ell)}_{jk}(q) = e^{-i(j\varphi + k\psi)}\, t^{(\ell)}_{jk}(r_z(\theta/2)).$$

We record this important result below.

Proposition 5.23. *For any matrix $q \in \mathbf{SU}(2)$ expressed in terms of the Euler angles as $q = u(\varphi, \theta, \psi) = r_x(\varphi/2) r_z(\theta/2) r_x(\psi/2)$, with respect to the orthonormal basis (ψ_k) of $\mathcal{P}_\ell^{\mathbb{C}}$, we have*

$$t_{jk}^{(\ell)}(q) = e^{-i(j\varphi + k\psi)}\, t_{jk}^{(\ell)}(r_z(\theta/2)). \tag{$*_{31}$}$$

Thus, we are left with finding an explicit expression for the matrix $t^{(\ell)}(r_z(\theta/2))$.

Definition 5.11. Define the matrix $t^{(\ell)}(\theta)$ as $t^{(\ell)}(\theta) = t^{(\ell)}(r_z(\theta/2))$, with

$$r_z(\theta/2) = \begin{pmatrix} \cos\frac{\theta}{2} & i\sin\frac{\theta}{2} \\ i\sin\frac{\theta}{2} & \cos\frac{\theta}{2} \end{pmatrix}.$$

If $\theta = \pi$, then $r_z(\pi/2) = \begin{pmatrix} 0 & i \\ i & 0 \end{pmatrix}$, and by $(*_{27})$ we know that $t^{(\ell)}(\pi)$ is the anti-diagonal matrix with $t_{jk}^{(\ell)}(\pi) = 0$ if $j \neq k$ and $t_{j\,-j}^{(\ell)}(\pi) = i^{2\ell}$.

If $\theta = 0$, then $r_z(0)$ is the identity matrix I_2, and $t^{(\ell)}(0)$ is the identity matrix $I_{2\ell+1}$. If $0 \leq \theta < \pi$, then we can find the matrix $t^{(\ell)}(\theta)$ using equation $(*_{25})$ in we which we set $\alpha = \delta = \cos\frac{\theta}{2} \neq 0$ (since $0 \leq \theta < \pi$), and $\beta = \gamma = i\sin\frac{\theta}{2}$. We obtain the following formula.

Proposition 5.24. *The elements of the matrix $t^{(\ell)}(\theta) = t^{(\ell)}(r_z(\theta/2))$ ($0 \leq \theta < \pi$) are given by the formula*

$$t_{jk}^{(\ell)}(\theta) = i^{-(j+k)} \sqrt{\frac{(\ell-j)!(\ell-k)!}{(\ell+j)!(\ell+k)!}} \left(\cos\frac{\theta}{2}\right)^{j+k} \times \sum_{h=\max(j,k)}^{\ell} \frac{(\ell+h)!\, i^{2h}}{(\ell-h)!(h-j)!(h-k)!} \left(\sin\frac{\theta}{2}\right)^{2h-(j+k)}. \tag{$*_{32}$}$$

If ℓ is a half integer, then h is also a half integer. For $\theta = 0$, we must have $h = j = k$, and $t^{(\ell)}(0)$ is the identity matrix $I_{2\ell+1}$, as we already know.

If we assume that $0 < \theta < \pi$, then we obtain the following formula given in Vilenkin:

$$t_{jk}^{(\ell)}(\theta) = i^{-(j+k)} \sqrt{\frac{(\ell-j)!(\ell-k)!}{(\ell+j)!(\ell+k)!}} \left(\cot\frac{\theta}{2}\right)^{j+k} \times \sum_{h=\max(j,k)}^{\ell} \frac{(\ell+h)!\, i^{2h}}{(\ell-h)!(h-j)!(h-k)!} \left(\sin\frac{\theta}{2}\right)^{2h}. \tag{$*_{33}$}$$

If we recall from (†) that if $j = -k$ then

$$\frac{(\ell+h)!}{(\ell-h)!(h+k)!(h-k)!} = \binom{\ell+h}{2h}\binom{2h}{h-k},$$

we obtain

$$t^{(\ell)}_{k\,-k}(\theta) = t^{(\ell)}_{-k\,k}(\theta) = \sum_{h=\max(-k,k)}^{\ell} \binom{\ell+h}{2h}\binom{2h}{h-k} i^{2h}\left(\sin\frac{\theta}{2}\right)^{2h}. \qquad (*_{34})$$

Even though this equation was derived assuming that $\theta < \pi$, it is still correct for $\theta = \pi$, namely the following equation holds

$$\sum_{h=\max(-k,k)}^{\ell} \binom{\ell+h}{2h}\binom{2h}{h-k} i^{2h} = i^{2\ell},$$

or equivalently, since we may assume that $k \geq 0$,

$$\sum_{h=k}^{\ell}(-1)^{\ell-h}\binom{\ell+h}{2h}\binom{2h}{h-k} = 1. \qquad (\dagger\dagger)$$

This can be proven using an identity due to Euler. As a first step, we can prove that

$$\sum_{h=k}^{\ell}(-1)^{\ell-h}\binom{\ell+h}{2h}\binom{2h}{h-k} = \sum_{h=k}^{\ell}(-1)^{\ell-h}\binom{\ell+h}{\ell-k}\binom{\ell-k}{\ell-h}.$$

Next, there are two cases depending on ℓ being an integer or a half integer. The second case reduces to the first by writing $\ell = ll+1/2, k = kk+1/2, h = hh+1/2$, where ll, kk, hh are integers. The details are left as an exercise. If ℓ is an integer, then by changing the index of summation, we have

$$\begin{aligned}\sum_{h=k}^{\ell}(-1)^{\ell-h}\binom{\ell+h}{\ell-k}\binom{\ell-k}{\ell-h} &= \sum_{h=0}^{\ell-k}(-1)^{\ell-h-k}\binom{\ell-k}{h}\binom{\ell+h+k}{\ell-k}\\ &= (-1)^N\sum_{h=0}^{N}(-1)^h\binom{N}{h}\binom{N+2k+h}{N},\end{aligned}$$

with $N = \ell - k$. At this stage, we use an identity known as *Euler's finite difference formula*, namely

$$\sum_{h=0}^{n}(-1)^h\binom{n}{h}\binom{x+hy}{n} = (-1)^n y^n.$$

Remarkably, the result is independent of x. Finally, we let $n = N$, $x = N + 2k$ and $y = 1$ to match Euler's formula. We leave the details as an exercise.

Because there is a surjective homomorphism $\rho\colon \mathbf{SU}(2) \to \mathbf{SO}(3)$ whose kernel is $\{I, -I\}$ (see Theorem 3.9), Proposition 3.10, Proposition 5.1, and the fact that the representation $U_{2\ell}\colon \mathbf{SU}(2) \to \mathbf{GL}(\mathcal{P}^{\mathbb{C}}_{2\ell}(2))$ is equivalent to the representation $T_\ell\colon \mathbf{SU}(2) \to \mathbf{GL}(\mathcal{P}^{\mathbb{C}}_{\ell})$ (see the end of Section 5.5), imply that the irreducible unitary representations of $\mathbf{SO}(3)$ are of the form $W_\ell\colon \mathbf{SO}(3) \to \mathbf{U}(\mathcal{P}^{\mathbb{C}}_{\ell})$, with

$$W_\ell(\rho_q) = T_\ell(q) \quad q \in \mathbf{SU}(2), \quad \ell \in \mathbb{N},$$

and where $T_{\ell'}\colon \mathbf{SU}(2) \to \mathbf{U}(\mathcal{P}^{\mathbb{C}}_{\ell'})$ are the irreducible unitary representations of $\mathbf{SU}(2)$ (with ℓ' a half integer or an integer). So the irreducible representations of $\mathbf{SO}(3)$ constitute only half of the representations of $\mathbf{SU}(2)$, those that correspond to nonnegative *integer values* of ℓ. Therefore, all the formulae obtained for the matrices $t^{(\ell)}_{jk}(q)$ apply and *the matrix* $w^{(\ell)}_{jk}(\rho_q)$ *associated with the unitary map* $W_\ell(\rho_q)$ *is* $t^{(\ell)}_{jk}(q)$, *with* $\ell \in \mathbb{N}$.

Remarkably, if $q \in \mathbf{SU}(2)$ is expressed in terms of the Euler angles as $q = u(\varphi, \theta, \psi) = r_x(\varphi/2) r_z(\theta/2) r_x(\psi/2)$, then the corresponding rotation matrix $R = \rho_q$ is given by $R = R_x(\varphi) R_z(\theta) R_x(\psi)$, where we may assume that $0 \le \varphi < 2\pi$, $0 \le \theta \le \pi$, $0 \le \psi < 2\pi$ (see Section 5.3). Consequently, if we express a rotation matrix $R \in \mathbf{SO}(3)$ in terms of Euler angles as $R = R_x(\varphi) R_z(\theta) R_x(\psi)$, we find that the matrix $w^{(\ell)}(R)$ associated with the unitary map $W_\ell(R)$ is $t^{(\ell)}(u(\varphi, \theta, \psi))$, with $\ell \in \mathbb{N}$. Using Proposition 5.23 and since by Definition 5.11, $t^{(\ell)}(\theta) = t^{(\ell)}(r_z(\theta/2))$, we obtain the following result.

Proposition 5.25. *For any matrix* $R \in \mathbf{SO}(3)$ *expressed in terms of the Euler angles as* $R = R_x(\varphi) R_z(\theta) R_x(\psi)$, *with respect to the orthonormal basis* (ψ_k) *of* $\mathcal{P}^{\mathbb{C}}_{\ell}$, *the matrix* $w^{(\ell)}(R)$ *of the unitary map* $W_\ell(R)$ *associated with the irreducible representation* $W_\ell\colon \mathbf{SO}(3) \to \mathbf{U}(\mathcal{P}^{\mathbb{C}}_{\ell})$ *is given by*

$$w^{(\ell)}_{jk}(R) = e^{-i(j\varphi + k\psi)}\, t^{(\ell)}_{jk}(\theta), \quad \ell \in \mathbb{N}. \qquad (*_{31'})$$

Formula $(*_{31'})$ still gives the matrix elements $T_\ell(q)$ (with $q \in \mathbf{SU}(2)$) of the irreducible representation T_ℓ of $\mathbf{SU}(2)$ when ℓ is a positive half integer, but this is *not* a representation of $\mathbf{SO}(3)$. This point is a notorious source of confusion.

The functions $e^{-i(j\varphi + k\psi)}\, t^{(\ell)}_{jk}(\theta)$ arise in quantum mechanics, but physicists prefer the functions $t^{(\ell)}_{jk}(\theta)$ to be real. In his famous book first published in German in 1931 and then in English in 1959 (translated by

J.J. Griffin), E. Wigner [Wigner (1959)] introduced the matrices $d^{\ell}(\theta)$ given by

$$d^{\ell}_{jk}(\theta) = (-1)^{j-k} i^{j-k} t^{(\ell)}_{jk}(\theta).$$

The reason for the factor $(-1)^{j-k} i^{j-k}$ is that by using Formula $(*_{24})$ with $\alpha = \delta = \cos\frac{\theta}{2}$ and $\beta = \gamma = i\sin\frac{\theta}{2}$, we obtain

$$\begin{aligned} t^{(\ell)}_{jk}(\theta) &= i^{j-k}\sqrt{(\ell-j)!(\ell+j)!(\ell-k)!(\ell+k)!} \\ &\times \left[\sum_{h=M}^{N} (-1)^h (h!(\ell-j-h)!(\ell+k-h)!(j-k+h)!)^{-1} \right. \\ &\left. \left(\cos\frac{\theta}{2}\right)^{2\ell+k-j-2h} \left(\sin\frac{\theta}{2}\right)^{2h+j-k}\right], \end{aligned}$$

with $M = \max(0, k-j)$, $N = \min(\ell-j, \ell+k)$ and $0 \le \theta \le \pi$. When we multiply the above expression by $(-1)^{j-k} i^{j-k}$, we obtain the term

$$(-1)^{j-k} i^{j-k} i^{j-k} = (-1)^{j-k} i^{2(j-k)} = (-1)^{j-k}(-1)^{j-k} = +1.$$

The above amounts to performing the following operations on the matrix $t^{(\ell)}(\theta)$: multiply the jth row by $(-1)^j i^j$ and multiply the kth column by $(-1)^{-k} i^{-k}$. The resulting matrix $d^{(\ell)}(\theta)$ remains unitary. In fact, it becomes a real orthogonal matrix.

Definition 5.12. The *Wigner's d-matrices* $d^{(\ell)}(\theta)$ are given by

$$\begin{aligned} d^{(\ell)}_{jk}(\theta) &= \sqrt{(\ell-j)!(\ell+j)!(\ell-k)!(\ell+k)!} \\ &\times \left[\sum_{h=M}^{N} (-1)^h (h!(\ell-j-h)!(\ell+k-h)!(j-k+h)!)^{-1} \right. \\ &\left. \left(\cos\frac{\theta}{2}\right)^{2\ell+k-j-2h} \left(\sin\frac{\theta}{2}\right)^{2h+j-k}\right], \qquad (*_{35}) \end{aligned}$$

with $M = \max(0, k-j)$, $N = \min(\ell-j, \ell+k)$; see Wigner [Wigner (1959)], Formula (15.27).

The d-matrices $d^{(\ell)}(\theta)$ are real orthogonal matrices. However, beware that besides the fact that the indices ℓ, j, k, h are denoted j, μ', μ, κ and the angles φ, θ, ψ are denoted α, β, γ, the angles α, β, γ *have a different meaning*. Indeed, Wigner factors a unit quaternion as $q = r_x(-\alpha/2) r_y(\beta/2) r_x(-\gamma/2)$ (where r_x and r_y are defined in Section 5.3), and the x-axis and the z-axis are swapped, which means that in our notation, the rotation matrix R associated with q is

$$R = R_z(-\alpha) R_y(\beta) R_z(-\gamma).$$

Wigner uses $r_y(\beta/2)$ instead of $r_z(\beta/2)$ because it is a real matrix. As a consequence, *Wigner's $\mathcal{D}$-matrices* (see Wigner [Wigner (1959)], Formula (15.8) and Formula (15.27)) are the matrices $\mathcal{D}^{(\ell)}$ given by

$$\mathcal{D}^{(\ell)}_{jk}(\alpha,\beta,\gamma) = e^{i(j\alpha+k\gamma)}d^{(\ell)}_{jk}(\beta).$$

As earlier, the matrices $\mathcal{D}^{(\ell)}$ correspond to the irreducible unitary representations U_ℓ of $\mathbf{SU}(2)$ when ℓ assumes all nonnegative integer and half integer values, and when ℓ is restricted to be a nonnegative integer, they correspond to the irreducible unitary representations W_ℓ of $\mathbf{SO}(3)$.

According to Wigner, the method for determining the irreducible representations of $\mathbf{SO}(3)$ as the irreducible representations of $\mathbf{SU}(2)$ corresponding to nonnegative *integer values* of ℓ is due to H. Weyl, who also discovered the irreducible representations of $\mathbf{SU}(2)$. The irreducible representations of $\mathbf{SU}(2)$ corresponding to half integer values of ℓ are often called *double-valued representations* of $\mathbf{SO}(3)$, an unfortunate terminology since they are *not* representations of $\mathbf{SO}(3)$, but instead representations of $\mathbf{SU}(2)$.

Wigner's sign conventions is not always the sign convention used in the physics literature. For example, using our notation, Sakurai and Napolitano [Sakurai and Napolitano (2021)] factor a rotation matrix in terms of the Euler angles as

$$R = R_z(\alpha)R_y(-\beta)R_z(\gamma),$$

where $R_z(\alpha)$, $R_y(-\beta)$, $R_z(\gamma)$ are expressed as in Definition 5.5 by

$$R_z(\alpha) = e^{-i\alpha J_z}, \quad R_y(-\beta) = e^{-i\beta J_y}, \quad R_z(\gamma) = e^{-i\gamma J_z}.$$

They also add the sign factor $(-1)^{j-k}$ to the Wigner d-matrix entry $d^{(\ell)}_{jk}(\theta)$ (see Formula (3.426)) and they define the $\mathcal{D}$-matrix as

$$\mathcal{D}^{(\ell)}_{jk}(\alpha,\beta,\gamma) = e^{-i(j\alpha+k\gamma)}(-1)^{j-k}d^{(\ell)}_{jk}(\beta),$$

where $d^{(\ell)}_{jk}(\beta)$ is given by $(*_{35})$; see Formula (3.202) in Sakurai and Napolitano [Sakurai and Napolitano (2021)].

5.11 Representations of $\mathbf{SL}(2,\mathbb{C})$ and $\mathbf{SU}(2)$ Using Finite Fourier Series

There is one more method for computing the matrix elements $t^{(\ell)}_{jk}(A)$ (where $A \in \mathbf{SL}(2,\mathbb{C})$) based on integration. The idea is to use another representing space for the representation T_ℓ, namely the vector space (of dimension $2\ell+1$) of finite Fourier series

$$\Phi(e^{i\varphi}) = \sum_{k=-\ell}^{\ell} c_k e^{-ik\varphi},$$

with $c_k \in \mathbb{C}$. Observe that if $Q(z)$ is the polynomial of degree 2ℓ given by

$$Q(z) = \sum_{k=-\ell}^{\ell} c_k z^{\ell-k},$$

so that the powers appears in the order $z^{2\ell}, z^{2\ell-1}, \ldots, z, 1$, the Fourier series $\Phi(e^{i\varphi})$ with the same coefficients is given by

$$\Phi(e^{i\varphi}) = e^{-i\ell\varphi} Q(e^{i\varphi}).$$

Denote the space of Fourier series of dimension $2\ell+1$ as $\mathfrak{F}_\ell$. We would like to define a representation of $\mathbf{SL}(2,\mathbb{C})$ in $\mathfrak{F}_\ell$. By analogy with what we did when we defined the representation T_ℓ in the space $\mathcal{P}_\ell^{\mathbb{C}}$ from the representation U_ℓ in the space $\mathcal{P}_{2\ell}^{\mathbb{C}}(2)$, observe that

$$\begin{aligned}
&e^{-i\ell\varphi}(ae^{i\varphi}+c)^\ell(be^{i\varphi}+d)^\ell \Phi\left(\frac{ae^{i\varphi}+c}{be^{i\varphi}+d}\right)\\
&\quad = e^{-i\ell\varphi}(ae^{i\varphi}+c)^\ell(be^{i\varphi}+d)^\ell \sum_{k=-\ell}^{\ell} c_k \left(\frac{ae^{i\varphi}+c}{be^{i\varphi}+d}\right)^{-k}\\
&\quad = e^{-i\ell\varphi} \sum_{k=-\ell}^{\ell} c_k (ae^{i\varphi}+c)^{\ell-k}(be^{i\varphi}+d)^{\ell+k}\\
&\quad = e^{-i\ell\varphi} S(e^{i\varphi}),
\end{aligned}$$

where $S(z)$ is the polynomial of degree 2ℓ given by

$$S(z) = \sum_{k=-\ell}^{\ell} c_k (az+c)^{\ell-k}(bz+d)^{\ell+k},$$

and so $e^{-i\ell\varphi}S(e^{i\varphi})$ is indeed a Fourier series in the space $\mathfrak{F}_\ell$. Consequently, we make the following definition.

Definition 5.13. The map $\mathcal{T}_\ell \colon \mathbf{SL}(2,\mathbb{C}) \to \mathbf{GL}(\mathfrak{F}_\ell)$ is defined by

$$\mathcal{T}_\ell(A)(\Phi(e^{i\varphi})) = e^{-i\ell\varphi}(ae^{i\varphi}+c)^\ell(be^{i\varphi}+d)^\ell \Phi\left(\frac{ae^{i\varphi}+c}{be^{i\varphi}+d}\right) \qquad (\mathcal{T}_\ell)$$

for every matrix $A = \begin{pmatrix} a & b \\ c & d \end{pmatrix} \in \mathbf{SL}(2,\mathbb{C})$ and every Fourier series $\Phi(e^{i\varphi}) \in \mathfrak{F}_\ell$.

It is easily verified that $\mathcal{T}_\ell\colon \mathbf{SL}(2,\mathbb{C}) \to \mathbf{GL}(\mathfrak{F}_\ell)$ is a representation. We also check immediately that the linear map defined on basis vectors by $z^{\ell-k} \mapsto e^{-ik\varphi}$ is an isomorphism between the vector spaces $\mathcal{P}_\ell^{\mathbb{C}}$ and $\mathfrak{F}_\ell$, and we make it an isometry by declaring that the inner product on $\mathfrak{F}_\ell$ is defined such that

$$\langle e^{-im\varphi}, e^{-in\varphi}\rangle = \begin{cases} 0, & \text{if } m \neq n \\ (\ell - m)!(\ell + m)!, & \text{if } m = n. \end{cases}$$

Observe the very useful facts that

$$\langle e^{-im\varphi}, e^{-in\varphi}\rangle = \frac{1}{2\pi}\int_0^{2\pi} e^{-im\varphi}e^{in\varphi}d\varphi = 0, \quad m \neq n,$$

and

$$\langle e^{-im\varphi}, e^{-im\varphi}\rangle = (\ell - m)!\ell + m)! = \frac{(\ell - m)!\ell + m)!}{2\pi}\int_0^{2\pi} e^{-im\varphi}e^{im\varphi}\, d\varphi.$$

Therefore, for any Fourier series $\Phi(e^{i\varphi}) \in \mathfrak{F}_\ell$, we have

$$\langle \Phi(e^{i\varphi}), e^{-im\varphi}\rangle = \frac{(\ell - m)!(\ell + m)!}{2\pi}\int_0^{2\pi} \Phi(e^{i\varphi})e^{im\varphi}\, d\varphi. \tag{$*36$}$$

To show that the representation $\mathcal{T}_\ell\colon \mathbf{SL}(2,\mathbb{C}) \to \mathbf{GL}(\mathfrak{F}_\ell)$ is equivalent to the representation $T_\ell\colon \mathbf{SL}(2,\mathbb{C}) \to \mathbf{GL}(\mathcal{P}_\ell^{\mathbb{C}})$ we proceed as follows. Let $F\colon \mathcal{P}_\ell^{\mathbb{C}} \to \mathfrak{F}_\ell$ be the linear map given by

$$F((Q(z)) = e^{-i\ell\varphi}Q(e^{i\varphi}) = \Phi(e^{i\varphi}),$$

which on the basis $(z^{\ell-k})$ is given by $F(z^{\ell-k}) = e^{-ik\varphi}$. These equations show that F is an isomorphism. Moreover, it is a unitary map because it preserves the hermitian inner product (we defined the hermitian product on $\mathfrak{F}_\ell$ to achieve this).

We need to prove that

$$F \circ T_\ell(A) = \mathcal{T}_\ell(A) \circ F, \quad \text{for all } A \in \mathbf{SL}(2,\mathbb{C}).$$

For any $Q \in \mathcal{P}_\ell^{\mathbb{C}}$, if we write

$$Q(z) = \sum_{k=-\ell}^{\ell} c_k z^{\ell-k},$$

then we have

$$\begin{aligned} T_\ell(A)(Q(z)) &= (bz+d)^{2\ell} Q\left(\frac{az+c}{bz+d}\right) \\ &= \sum_{k=-\ell}^{\ell} c_k (bz+d)^{2\ell} \left(\frac{az+c}{bz+d}\right)^{\ell-k} \\ &= \sum_{k=-\ell}^{\ell} c_k (az+c)^{\ell-k}(bz+d)^{\ell+k} = S(z). \end{aligned}$$

Using the above equation, we have

$$F(T_\ell(A)(Q(z))) = e^{-i\ell\varphi} S(e^{i\varphi}) = e^{-i\ell\varphi} \sum_{k=-\ell}^{\ell} c_k (ae^{i\varphi}+c)^{\ell-k}(be^{i\varphi}+d)^{\ell+k}.$$

We also proved earlier that

$$\begin{aligned} \mathcal{T}_\ell(A)(\Phi(e^{i\varphi})) &= e^{-i\ell\varphi}(ae^{i\varphi}+c)^\ell (be^{i\varphi}+d)^\ell \Phi\left(\frac{ae^{i\varphi}+c}{be^{i\varphi}+d}\right) \\ &= e^{-i\ell\varphi} S(e^{i\varphi}), \end{aligned}$$

with

$$S(z) = \sum_{k=-\ell}^{\ell} c_k (az+c)^{\ell-k}(bz+d)^{\ell+k}.$$

But by definition $\Phi(e^{i\varphi}) = F(Q(z))$ for Q as above, so we proved that

$$F \circ T_\ell(A) = \mathcal{T}_\ell(A) \circ F \quad \text{for all } A \in \mathbf{SL}(2,\mathbb{C}),$$

which shows that $\mathcal{T}_\ell \colon \mathbf{SL}(2,\mathbb{C}) \to \mathbf{GL}(\mathfrak{F}_\ell)$ is a representation equivalent to the representation $T_\ell \colon \mathbf{SL}(2,\mathbb{C}) \to \mathbf{GL}(\mathcal{P}_\ell^{\mathbb{C}})$. Because the linear map $F \colon \mathcal{P}_\ell^{\mathbb{C}} \to \mathfrak{F}_\ell$ is unitary, we claim that the matrix of $T_\ell(A)$ in the basis $(\psi_k(z))$ is identical to the matrix of $\mathcal{T}_\ell(A)$ in the basis $(\psi'_k(\varphi))$ with $\psi'_k(\varphi) = \dfrac{e^{-ik\varphi}}{\sqrt{(\ell-k)!(\ell+k)!}}$. Recall that $\psi_k(z) = \dfrac{z^{\ell-k}}{\sqrt{(\ell-k)!(\ell+k)!}}$. This is because $\mathcal{T}_\ell(A) = F \circ T_\ell(A) \circ F^{-1}$, $F(\psi_k(z)) = \psi'_k(\varphi)$, and the (j,k)-entry $t_{jk}^{(\ell)'}(A)$ of the matrix of $\mathcal{T}_\ell(A)$ in the basis $(\psi'_k(\varphi))$ is given by

$$t_{jk}^{(\ell)'}(A) = \langle \mathcal{T}_\ell(A)(\psi'_k(\varphi)), \psi'_j(\varphi)\rangle,$$

which is rewritten as

$$t_{jk}^{(\ell)'}(A) = \langle (F \circ T_\ell(A) \circ F^{-1})(\psi'_k(\varphi)), F(\psi_j(z))\rangle,$$

and then as

$$t_{jk}^{(\ell)'}(A) = \langle F(T_\ell(A)(\psi_k(z))),\, F(\psi_j(z))\rangle.$$

Since F is unitary, we obtain

$$t_{jk}^{(\ell)'}(A) = \langle F(T_\ell(A)(\psi_k(z))),\, F(\psi_j(z))\rangle = \langle T_\ell(A)(\psi_k(z)),\, \psi_j(z)\rangle = t_{jk}^{(\ell)}(A),$$

establishing our claim.

We now compute $\mathcal{T}_\ell(A)(\Phi(e^{i\varphi}))$ with $\Phi(e^{i\varphi}) = (e^{i\varphi})^{-k}$ using Formula $(\mathcal{T}_\ell)$. We get

$$\begin{aligned}\mathcal{T}_\ell(A)(e^{-ik\varphi}) &= e^{-i\ell\varphi}(ae^{i\varphi}+c)^\ell(be^{i\varphi}+d)^\ell\left(\frac{ae^{i\varphi}+c}{be^{i\varphi}+d}\right)^{-k}\\ &= (ae^{i\varphi}+c)^{\ell-k}(be^{i\varphi}+d)^{\ell+k}\,e^{-i\ell\varphi}.\end{aligned}$$

As a consequence the matrix elements are given by

$$\begin{aligned}t_{jk}^{(\ell)}(A) &= \langle \mathcal{T}_\ell(A)(\psi_k'(\varphi)),\, \psi_j'(\varphi)\rangle = \frac{\langle \mathcal{T}_\ell(A)(e^{-ik\varphi}),\, e^{-ij\varphi}\rangle}{\sqrt{(\ell-j)!(\ell+j)!(\ell-k)!(\ell+k)!}}\\ &= \frac{\langle (ae^{i\varphi}+c)^{\ell-k}(be^{i\varphi}+d)^{\ell+k}\,e^{-i\ell\varphi},\, e^{-ij\varphi}\rangle}{\sqrt{(\ell-j)!(\ell+j)!(\ell-k)!(\ell+k)!}}.\end{aligned}$$

Using $(*_{36})$, we obtain the following result.

Proposition 5.26. *The matrix elements* $t_{jk}^{(\ell)}(A)$ *are given by the following formula:*

$$\begin{aligned}t_{jk}^{(\ell)}(A) &= \frac{1}{2\pi}\sqrt{\frac{(\ell-j)!(\ell+j)!}{(\ell-k)!(\ell+k)!}}\\ &\quad \int_0^{2\pi}(ae^{i\varphi}+c)^{\ell-k}(be^{i\varphi}+d)^{\ell+k}\,e^{i(j-\ell)\varphi}\,d\varphi. \qquad (*_{37})\end{aligned}$$

We obtain another useful formula for computing $t_{jk}^{(\ell)}(\theta)$ by applying the above formula to the matrix

$$r_z(\theta/2) = \begin{pmatrix}\cos\frac{\theta}{2} & i\sin\frac{\theta}{2}\\ i\sin\frac{\theta}{2} & \cos\frac{\theta}{2}\end{pmatrix} \in \mathbf{SU}(2).$$

We get

$$\begin{aligned}t_{jk}^{(\ell)}(\theta) &= \frac{1}{2\pi}\sqrt{\frac{(\ell-j)!(\ell+j)!}{(\ell-k)!(\ell+k)!}}\\ &\quad \int_0^{2\pi}\left(\cos\frac{\theta}{2}e^{i\varphi}+i\sin\frac{\theta}{2}\right)^{\ell-k}\left(i\sin\frac{\theta}{2}e^{i\varphi}+\cos\frac{\theta}{2}\right)^{\ell+k}e^{i(j-\ell)\varphi}\,d\varphi,\end{aligned}$$

and since $e^{-i\ell\varphi} = e^{-\frac{i(\ell+k)\varphi}{2}}e^{-\frac{i(\ell-k)\varphi}{2}}$, the above formula is also written as stated below.

Proposition 5.27. *The matrix elements* $t^{(\ell)}_{jk}(\theta)$ $(0 \leq \theta \leq \pi)$ *are given by the following formula:*

$$t^{(\ell)}_{jk}(\theta) = \frac{1}{2\pi}\sqrt{\frac{(\ell-j)!(\ell+j)!}{(\ell-k)!(\ell+k)!}}$$
$$\int_0^{2\pi}\left(\cos\frac{\theta}{2}e^{\frac{i\varphi}{2}} + i\sin\frac{\theta}{2}e^{-\frac{i\varphi}{2}}\right)^{\ell-k}\left(i\sin\frac{\theta}{2}e^{\frac{i\varphi}{2}} + \cos\frac{\theta}{2}e^{-\frac{i\varphi}{2}}\right)^{\ell+k} e^{ij\varphi}\,d\varphi. \tag{$*38$}$$

For small values of ℓ, this equation is quite practical. For example, here is a list of the matrices $t^{\ell}(\theta)$ for $\ell = 0, 1/2, 1, 3/2$ as in Vilenkin [Vilenkin (1968)] (Chapter III, Section 3.7).

$$t^{(0)}(\theta) = (1), \quad t^{(1/2)}(\theta) = r_z(\theta/2) = \begin{pmatrix} \cos\frac{\theta}{2} & i\sin\frac{\theta}{2} \\ i\sin\frac{\theta}{2} & \cos\frac{\theta}{2} \end{pmatrix},$$

$$t^{(1)}(\theta) = \begin{pmatrix} \cos^2\frac{\theta}{2} & \frac{i}{\sqrt{2}}\sin\frac{\theta}{2} & -\sin^2\frac{\theta}{2} \\ \frac{i}{\sqrt{2}}\sin\frac{\theta}{2} & \cos\theta & \frac{i}{\sqrt{2}}\sin\frac{\theta}{2} \\ -\sin^2\frac{\theta}{2} & \frac{i}{\sqrt{2}}\sin\frac{\theta}{2} & \cos^2\frac{\theta}{2} \end{pmatrix},$$

and

$$t^{(3/2)}(\theta) =$$
$$\begin{pmatrix} \cos^3\frac{\theta}{2} & i\sqrt{3}\sin\frac{\theta}{2}\cos^2\frac{\theta}{2} & -\sqrt{3}\sin^2\frac{\theta}{2}\cos\frac{\theta}{2} & -i\sin^3\frac{\theta}{2} \\ i\sqrt{3}\sin\frac{\theta}{2}\cos^2\frac{\theta}{2} & \cos^3\frac{\theta}{2} - 2\cos\frac{\theta}{2}\sin^2\frac{\theta}{2} & 2i\cos^2\frac{\theta}{2}\sin\frac{\theta}{2} - i\sin^3\frac{\theta}{2} & -\sqrt{3}\sin^2\frac{\theta}{2}\cos\frac{\theta}{2} \\ -\sqrt{3}\sin^2\frac{\theta}{2}\cos\frac{\theta}{2} & 2i\cos^2\frac{\theta}{2}\sin\frac{\theta}{2} - i\sin^3\frac{\theta}{2} & \cos^3\frac{\theta}{2} - 2\cos\frac{\theta}{2}\sin^2\frac{\theta}{2} & i\sqrt{3}\sin\frac{\theta}{2}\cos^2\frac{\theta}{2} \\ -i\sin^3\frac{\theta}{2} & -\sqrt{3}\sin^2\frac{\theta}{2}\cos\frac{\theta}{2} & i\sqrt{3}\sin\frac{\theta}{2}\cos^2\frac{\theta}{2} & \cos^3\frac{\theta}{2} \end{pmatrix}.$$

5.12 Matrix Elements of $T_\ell(q)$ and Jacobi Polynomials

In this section, we assume again that $q \in \mathbf{SU}(2)$ is given in terms of the Euler angles as $q = u(\varphi, \theta, \psi) = r_x(\varphi/2)r_z(\theta/2)r_x(\psi/2)$. Since $\cos\theta = 2\cos^2\frac{\theta}{2} - 1$ and $\sin^2\frac{\theta}{2} + \cos^2\frac{\theta}{2} = 1$, for $0 \leq \theta \leq \pi$, we have $0 \leq \cos\frac{\theta}{2} \leq 1$ and $0 \leq \sin\frac{\theta}{2} \leq 1$, so

$$\cos\frac{\theta}{2} = \sqrt{\frac{1+\cos\theta}{2}}, \quad \sin\frac{\theta}{2} = \sqrt{\frac{1-\cos\theta}{2}}, \quad \cot\frac{\theta}{2} = \sqrt{\frac{1+\cos\theta}{1-\cos\theta}}, \tag{$*39$}$$

with $\theta > 0$ for the third formula. Thus, we see that $t^{(\ell)}_{jk}(\theta)$ is a function of $\cos\theta$ for $0 \leq \theta < \pi$. Therefore there is a function $P^{\ell}_{jk}(z)$, such that

$$t^{(\ell)}_{jk}(\theta) = P^{\ell}_{jk}(\cos\theta), \quad 0 \leq \theta < \pi,$$

and $(*_{31})$ is also written as

$$t_{jk}^{(\ell)}(q) = e^{-i(j\varphi+k\psi)}\, P_{jk}^{\ell}(\cos\theta).$$

By equation $(*_{32})$ and the above trigonometric identities, we obtain the following result.

Proposition 5.28. *The polynomial $P_{jk}^{\ell}(z)$ $(-1 < z \leq 1)$ given by*

$$P_{jk}^{\ell}(z) = i^{-(j+k)} \sqrt{\frac{(\ell-j)!(\ell-k)!}{(\ell+j)!(\ell+k)!}} \left(\frac{1+z}{2}\right)^{\frac{j+k}{2}} \times \sum_{h=\max(j,k)}^{\ell} \frac{(\ell+h)!\, i^{2h}}{(\ell-h)!(h-j)!(h-k)!} \left(\frac{1-z}{2}\right)^{\frac{2h-(j+k)}{2}} \qquad (*_{40})$$

has the property that

$$t_{jk}^{(\ell)}(\theta) = P_{jk}^{\ell}(\cos\theta), \quad 0 \leq \theta < \pi, \qquad (*_{41})$$

and

$$t_{jk}^{(\ell)}(q) = e^{-i(j\varphi+k\psi)}\, P_{jk}^{\ell}(\cos\theta). \qquad (*_{42})$$

If ℓ is a half integer, then h is also a half integer. It is understood that if $z = 1$, then $P_{jk}^{\ell}(1) = 1$ iff $j = k$, and $P_{jk}^{\ell}(1) = 0$ otherwise.

If $0 < \theta < \pi$, since $\alpha = \delta = \cos\frac{\theta}{2}$ and $\beta = \gamma = i\sin\frac{\theta}{2}$ are all nonzero, we obtain another formula from equation $(*_{30})$ recalled below:

$$t_{jk}^{(\ell)}(q) = \sqrt{\frac{(\ell+j)!}{(\ell-k)!(\ell+k)!(\ell-j)!}}\, \frac{\beta^{k-j}}{\alpha^{k+j}} \frac{d^{\ell-j}}{dz^{\ell-j}} [z^{\ell-k}(z+1)^{\ell+k}]_{z=\beta\gamma}.$$

We perform the change of variable $z = (y-1)/2$, so $y = 2z+1$ and since $\beta\gamma = -\sin^2\frac{\theta}{2}$, the condition $z = \beta\gamma$ becomes $y = -2\sin^2\frac{\theta}{2} + 1 = \cos\theta$, and

$$\frac{d^{\ell-j}}{dy^{\ell-j}} = 2^{\ell-j} \frac{d^{\ell-j}}{dz^{\ell-j}}.$$

We also have $z = -\frac{1-y}{2}$, $z+1 = \frac{1+y}{2}$,

$$z^{\ell-k}(z+1)^{\ell+k} = (-1)^{\ell-k} 2^{-2\ell} (1-y)^{\ell-k}(1+y)^{\ell+k}.$$

Using the trigonometric identities in equations $(*_{39})$, we obtain

$$\begin{aligned}\frac{\beta^{k-j}}{\alpha^{k+j}} &= \frac{\left(i\sin\frac{\theta}{2}\right)^{k-j}}{\left(\cos\frac{\theta}{2}\right)^{k+j}} = i^{k-j}\left(\frac{1-\cos\theta}{2}\right)^{\frac{k-j}{2}} \left(\frac{1+\cos\theta}{2}\right)^{-\frac{(k+j)}{2}} \\ &= i^{k-j}\, 2^{j} (1+\cos\theta)^{-\frac{(k+j)}{2}} (1-\cos\theta)^{\frac{k-j}{2}},\end{aligned}$$

and with $z = \cos\theta$ we obtain the following result.

Proposition 5.29. *If $0 < \theta < \pi$, so that $-1 < z < 1$, then we have*

$$P^{\ell}_{jk}(z) = \frac{(-1)^{\ell-k} i^{k-j}}{2^{\ell}} \sqrt{\frac{(\ell+j)!}{(\ell-k)!(\ell+k)!(\ell-j)!}} \times (1+z)^{-\frac{(j+k)}{2}} (1-z)^{\frac{k-j}{2}} \frac{d^{\ell-j}}{dy^{\ell-j}} [(1-y)^{\ell-k}(1+y)^{\ell+k}]_{y=z}. \quad (*_{43})$$

The polynomials $P^{\ell}_{jk}(z)$ enjoy some symmetry relations. For example, Formula $(*_{40})$ shows that

$$P^{\ell}_{jk}(z) = P^{\ell}_{kj}(z), \quad -1 < z \leq 1.$$

Since $r_z(\theta/2)$ and $r_z(\pi/2) = \begin{pmatrix} 0 & i \\ i & 0 \end{pmatrix}$ commute, it can be shown that

$$P^{\ell}_{jk}(z) = P^{\ell}_{-j\,-k}(z), \quad -1 < z \leq 1.$$

We leave the proof as an exercise. By Formula $(*_{43})$, we see immediately that

$$P^{\ell}_{jk}(-z) = i^{2(\ell-j-k)} P^{\ell}_{j\,-k}(z), \quad -1 < z < 1.$$

It is also immediately verified that

$$r_z(\theta/2)^{-1} = r_z(-\theta/2) = r_x(\pi/2) r_z(\theta/2) r_x(-\pi/2),$$

where

$$r_x(\pi/2) = \begin{pmatrix} i & 0 \\ 0 & -i \end{pmatrix}, \quad r_x(-\pi/2) = \begin{pmatrix} -i & 0 \\ 0 & i \end{pmatrix},$$

so by $(*_{42})$ (with $\varphi = \pi, \psi = -\pi$) we obtain

$$t^{(\ell)}_{jk}(-\theta) = (-1)^{k-j} P^{\ell}_{jk}(\cos\theta).$$

Formula $(*_{43})$ also reveals a relationship with the Jacobi polynomials.

Definition 5.14. The *Jacobi polynomials* $P^{\lambda,\mu}_{h}(z)$, with $\lambda, \mu \in \mathbb{R}$, $h \in \mathbb{N}$, are defined by the formula

$$P^{\lambda,\nu}_{h}(z) = \frac{(-1)^h}{2^h h!} (1-z)^{-\lambda} (1+z)^{-\mu} \frac{d^h}{dz^h} [(1-z)^{\lambda+h}(1+z)^{\mu+h}]. \quad \text{(Ja)}$$

To show that the $P^{\ell}_{jk}(z)$ are related to the Jacobi polynomials, taking a cue from Vilenkin we compute

$$D = 2^j i^{k-j} \sqrt{\frac{(\ell-k)!(\ell+k)!}{(\ell-j)!(\ell+j)!}} (1-z)^{\frac{k-j}{2}} (1+z)^{-\frac{(k+j)}{2}} P^{\ell}_{jk}(z). \quad (*_{44})$$

We get

$$\begin{aligned}
D &= 2^j i^{k-j}\sqrt{\frac{(\ell-k)!(\ell+k)!}{(\ell-j)!(\ell+j)!}}\,(1-z)^{\frac{k-j}{2}}(1+z)^{-\frac{(k+j)}{2}} \\
&\quad\times\left[\frac{(-1)^{\ell-k}i^{k-j}}{2^\ell}\sqrt{\frac{(\ell+j)!}{(\ell-k)!(\ell+k)!(\ell-j)!}}\right. \\
&\quad\left.\times(1+z)^{-\frac{(k+j)}{2}}(1-z)^{\frac{k-j}{2}}\frac{d^{\ell-j}}{dy^{\ell-j}}[(1-y)^{\ell-k}(1+y)^{\ell+k}]_{y=z}\right] \\
&= \frac{(-1)^{\ell-j}}{2^{\ell-j}(\ell-j)!}(1-z)^{k-j}(1+z)^{-(k+j)}\frac{d^{\ell-j}}{dz^{\ell-j}}[(1-z)^{\ell-k}(1+z)^{\ell+k}].
\end{aligned}$$

To match D with a Jacobi polynomial, we need to find h, λ, μ, such that

$$h = \ell - j, \quad \lambda = -(k-j), \quad \mu = k+j, \quad \lambda + h = \ell - k, \quad \mu + h = \ell + k.$$

We see that h, λ, μ are uniquely determined by

$$h = \ell - j, \quad \lambda = j - k, \quad \mu = k + j,$$

and that the last two equations are also satisfied. Observe that λ and μ are integers. Thus, we proved the following result.

Proposition 5.30. *The polynomials $P^\ell_{jk}(z)$ and the Jacobi polynomials are related by the equation*

$$P^{j-k,k+j}_{\ell-j}(z) = 2^j i^{k-j}\sqrt{\frac{(\ell-k)!(\ell+k)!}{(\ell-j)!(\ell+j)!}}\,(1-z)^{\frac{k-j}{2}}(1+z)^{-\frac{(k+j)}{2}}P^\ell_{jk}(z). \tag{$*_{45}$}$$

As we noted earlier, if ℓ is a half integer then j and k cannot be zero. If ℓ is an integer, then $j = 0$ or $k = 0$ is allowed, and so $\lambda = 0$ and $\mu = 0$ are also allowed. In this case, the Jacobi polynomial $P^{0,0}_\ell(z)$, simply denoted as $P_\ell(z)$, is given by

$$P_\ell(z) = \frac{(-1)^\ell}{2^\ell \ell!}\frac{d^\ell}{dz^\ell}(1-z^2)^\ell,$$

or equivalently

$$P_\ell(z) = \frac{1}{2^\ell \ell!}\frac{d^\ell}{dz^\ell}(z^2-1)^\ell.$$

This is a *Legendre* polynomial.

Similarly, if ℓ is an integer, then for $k = 0$ the polynomials $P^\ell_{m0}(z)$ are related to polynomials $P^m_\ell(z)$ known as the associated Legendre polynomials.

Definition 5.15. The *Legendre polynomial* $P_\ell(z)$ are defined by

$$P_\ell(z) = \frac{(-1)^\ell}{2^\ell \ell!}\frac{d^\ell}{dz^\ell}(1-z^2)^\ell,$$

and the *associated Legendre polynomials* are defined by

$$P_\ell^m(z) = \frac{(-1)^{m+\ell}}{2^\ell \ell!}(1-z^2)^{\frac{m}{2}}\frac{d^{m+\ell}}{dz^{m+\ell}}(1-z^2)^\ell = (-1)^m(1-z^2)^{\frac{m}{2}}\frac{d^m}{dz^m}P_\ell(z),$$

with $\ell, m \in \mathbb{N}$.

Some authors omit the sign $(-1)^m$ in the definition of the associated Legendre polynomials. We see immediately that

$$P_{00}^\ell(z) = P_\ell(z). \tag{$*_{46}$}$$

It is not hard to show that

$$P_\ell^j(z) = i^j\sqrt{\frac{(\ell+j)!}{(\ell-j)!}}\,P_{j0}^\ell(z). \tag{$*_{47}$}$$

See Vilenkin [Vilenkin (1968)] (Chapter III, Section 3.9). Since by ($*_{42}$), we have

$$t_{j0}^{(\ell)}(q) = e^{-ij\varphi}\,P_{j0}^\ell(\cos\theta),$$

we obtain

$$t_{j0}^{(\ell)}(q) = i^{-j}\sqrt{\frac{(\ell-j)!}{(\ell+j)!}}\,e^{-ij\varphi}\,P_\ell^j(\cos\theta), \quad -\ell \le j \le \ell. \tag{$*_{48}$}$$

Recall that ℓ is an integer.

Following Vilenkin [Vilenkin (1968)] (Chapter III, Section 2.7) we show how the function $t_{j0}^{(\ell)}(q)$ (with $q = r_x(\varphi/2)r_z(\theta/2)$), which does not depend on ψ, can be viewed as a function on the sphere S^2.

5.13 Harmonic Functions on the Sphere S^2

First, recall that the group **SO**(3) acts transitively in the sphere S^2 and that the stabilizer of the point $e_1 = (1,0,0)$ is the subgroup H_x of rotations

$$R_x(\varphi) = \begin{pmatrix} 1 & 0 & 0 \\ 0 & \cos\varphi & -\sin\varphi \\ 0 & \sin\varphi & \cos\varphi \end{pmatrix}$$

around the x-axis, so the sphere S^2 is homeomorphic to the quotient space $\mathbf{SO}(3)/H_x$. It follows that the functions $f \in \mathrm{L}^2(\mathbf{SO}(3))$, such that

$f(RQ) = f(R)$ for all $R \in \mathbf{SO}(3)$ and all $Q \in H_x$ correspond bijectively to the functions in $\mathrm{L}^2(S^2)$. From Section 5.3, since every rotation R can be factored as

$$R = R_x(\varphi)R_z(\theta)R_x(\psi),$$

with $R_x(\varphi), R_x(\psi) \in H_x$, we see that a representative of the left coset RH_x is given by

$$R_x(\varphi)R_z(\theta).$$

Therefore the points of S^2 are the orbit of $e_1 = (1,0,0)$ under all rotations $R_x(\varphi)R_z(\theta)$.

But the group H_x corresponds to the subgroup Ω_x defined below.

Definition 5.16. The subgroup Ω_x of $\mathbf{SU}(2)$ is given by

$$\Omega_x = \left\{ H(t) = r_x(t/2) = \begin{pmatrix} e^{\frac{it}{2}} & 0 \\ 0 & e^{-\frac{it}{2}} \end{pmatrix} \;\middle|\; 0 \leq t \leq 2\pi \right\}. \tag{Ω_x}$$

In fact, we claim that $\mathbf{SU}(2)/\Omega_x$ is a homogeneous space homeomorphic to S^2 so that the functions $f \in \mathrm{L}^2(\mathbf{SU}(2))$ such that $f(qH) = f(q)$ for all $q \in \mathbf{SU}(2)$ and all $H \in \Omega_x$ also correspond bijectively to the functions in $\mathrm{L}^2(S^2)$.

The group $\mathbf{SU}(2)$ acts on the sphere S^2 by rotations, which means that for any skew-hermitian matrix

$$X = \begin{pmatrix} ix & y+iz \\ -y+iz & -ix \end{pmatrix}, \quad (x,y,z) \in S^2,$$

and any $q \in \mathbf{SU}(2)$, we have the action

$$q \cdot X = qXq^*.$$

Since this action is a rotation of S^2, it is transitive. The stabilizer of $e_1 = (1,0,0)$ is the subgroup consisting of all unit quaternions

$$q = \begin{pmatrix} \alpha & \beta \\ -\overline{\beta} & \overline{\alpha} \end{pmatrix}$$

such that

$$\begin{pmatrix} \alpha & \beta \\ -\overline{\beta} & \overline{\alpha} \end{pmatrix} \begin{pmatrix} i & 0 \\ 0 & -i \end{pmatrix} \begin{pmatrix} \overline{\alpha} & -\beta \\ \overline{\beta} & \alpha \end{pmatrix} = \begin{pmatrix} i & 0 \\ 0 & -i \end{pmatrix},$$

or equivalently

$$\begin{pmatrix} \alpha & \beta \\ -\overline{\beta} & \overline{\alpha} \end{pmatrix} \begin{pmatrix} i & 0 \\ 0 & -i \end{pmatrix} = \begin{pmatrix} i & 0 \\ 0 & -i \end{pmatrix} \begin{pmatrix} \alpha & \beta \\ -\overline{\beta} & \overline{\alpha} \end{pmatrix},$$

which means that

$$\begin{pmatrix} i\alpha & -i\beta \\ -i\overline{\beta} & -i\overline{\alpha} \end{pmatrix} = \begin{pmatrix} i\alpha & i\beta \\ i\overline{\beta} & -i\overline{\alpha} \end{pmatrix},$$

and so $\beta = 0$. Therefore, the stabilizer of $e_1 = (1, 0, 0)$ is indeed the subgroup Ω_x. From Section 5.3, since every unit quaternion q can be factored as

$$q = r_x(\varphi/2)r_z(\theta/2)r_x(\psi/2),$$

with $r_x(\varphi/2), r_x(\psi/2) \in \Omega_x$, we see that a representative of the left coset $q\Omega_x$ is given by

$$r_x(\varphi/2)r_z(\theta/2).$$

Therefore, the points of S^2 are the orbit of $e_1 = (1, 0, 0)$ under all rotations $r_x(\varphi/2)r_z(\theta/2)$, and from Section 5.3, since the corresponding rotation matrices are

$$Q = \begin{pmatrix} 1 & 0 & 0 \\ 0 & \cos\varphi & -\sin\varphi \\ 0 & \sin\varphi & \cos\varphi \end{pmatrix} \begin{pmatrix} \cos\theta & -\sin\theta & 0 \\ \sin\theta & \cos\theta & 0 \\ 0 & 0 & 1 \end{pmatrix},$$

by reading of the first column of the matrix Q, we see that the corresponding orbit points on the sphere S^2 have coordinates

$$(\cos\theta,\ \sin\theta\cos\varphi,\ \sin\theta\sin\varphi).$$

According to the physical convention, the spherical coordinates of a point p with respect to the (azimuthal) angle φ measured from the x-axis in the xy-plane and (polar) angle θ measured from the z-axis in the plane containing the z-axis an passing through the point p are given by

$$(\sin\theta\cos\varphi,\ \sin\theta\sin\varphi,\ \cos\theta).$$

Thus, we see that the coordinates

$$(\cos\theta,\ \sin\theta\cos\varphi,\ \sin\theta\sin\varphi)$$

are "funny" spherical coordinates for which the x-axis and the z-axis are swapped and φ is changed to $\pi/2 - \varphi$.

Following Vilenkin (Chapter III, Section 3.10), we make the following definition.

Definition 5.17. For any j, such that $-\ell \leq j \leq \ell$, the function $t^{(\ell)}_{j0}(q)$, which does not depend on ψ (with $q = r_x(\varphi/2)r_z(\theta/2)r_x(\psi/2)$), can be viewed as a function on the sphere S^2, and is denoted $Y_{\ell j}(\varphi, \theta)$, with $0 \leq \varphi < 2\pi$ and $0 \leq \theta < \pi$. The function $Y_{\ell j}(\varphi, \theta)$ is called a *spherical function*.

Observe that the $2\ell+1$ functions $Y_{\ell j}(\varphi,\theta) = t^{(\ell)}_{j0}(q)$ $(-\ell \leq j \leq \ell)$ constitute the *middle column* of the matrix $t^{(\ell)}(q)$.

In view of Proposition 5.25 and $(*_{41})$, for any matrix $R \in \mathbf{SO}(3)$ expressed in terms of the Euler angles as $R = R_x(\varphi)R_z(\theta)R_x(\psi)$, with respect to the orthonormal basis (ψ_k) of $\mathcal{P}^{\mathbb{C}}_\ell$, the matrix $w^{(\ell)}(R)$ of the unitary map $W_\ell(R)$ associated with the irreducible representation $W_\ell\colon \mathbf{SO}(3) \to \mathbf{U}(\mathcal{P}^{\mathbb{C}}_\ell)$ is given by

$$w^{(\ell)}_{jk}(R) = e^{-i(j\varphi+k\psi)}\, t^{(\ell)}_{jk}(\theta) = e^{-i(j\varphi+k\psi)}\, P^\ell_{jk}(\theta) = t^{(\ell)}_{jk}(q), \quad \ell \in \mathbb{N},$$

where $q = r_x(\varphi/2)r_z(\theta/2)r_x(\psi/2)$. In particular, for $k=0$, we see that

$$w^{(\ell)}_{j0}(R) = t^{(\ell)}_{j0}(q) = Y_{\ell j}(\varphi,\theta).$$

Thus, we have shown the following result.

Proposition 5.31. *The following facts hold.*

(1) *For any matrix $R \in \mathbf{SO}(3)$ expressed as $R = R_x(\varphi)R_z(\theta)R_x(\psi)$ in terms of the Euler angles, with respect to the orthonormal basis (ψ_k) of $\mathcal{P}^{\mathbb{C}}_\ell$, the matrix $w^{(\ell)}(R)$ of the unitary map $W_\ell(R)$ associated with the irreducible representation $W_\ell\colon \mathbf{SO}(3) \to \mathbf{U}(\mathcal{P}^{\mathbb{C}}_\ell)$ is equal to the matrix $t^{(\ell)}(q)$ of the unitary map $T_\ell(q)$ associated with the irreducible representation $T_\ell\colon \mathbf{SU}(2) \to \mathbf{U}(\mathcal{P}^{\mathbb{C}}_\ell)$, where $q = r_x(\varphi/2)r_z(\theta/2)r_x(\psi/2)$ $(\ell \in \mathbb{N})$.*

(2) *Viewed as functions on S^2, the $2\ell+1$ functions $t^{(\ell)}_{j0}(q)$ (with $q = r_x(\varphi/2)r_z(\theta/2)r_x(\psi/2)$) constitute the middle column of the matrix $t^{(\ell)}(q)$ and the $2\ell+1$ functions $w^{(\ell)}_{j0}(R)$ (with $R = R_x(\varphi)R_z(\theta)R_x(\psi)$) constitute the middle column of the matrix $w^{(\ell)}(R)$.*

(3) *Viewed as a function on S^2 in spherical coordinates*

$$(x,y,z) = (\cos\theta,\ \sin\theta\cos\varphi,\ \sin\theta\sin\varphi),$$

we have

$$Y_{\ell j}(x,y,z) = Y_{\ell j}(\varphi,\theta) = t^{(\ell)}_{j0}(q) = w^{(\ell)}_{j0}(R),$$

with $q = r_x(\varphi/2)r_z(\theta/2)$ and $R = R_x(\varphi)R_z(\theta)$.

As we observed earlier, the matrices $t^{(\ell)}(\theta)$, and so the polynomials $P^\ell_{jk}(z)$, are not all real. And indeed equation $(*_{48})$ shows that the functions $Y_{\ell j}(\varphi,\theta) = t^{(\ell)}_{j0}(q)$ are not all real. A way to fix this is to multiply $Y_{\ell j}(\varphi,\theta)$ by i^j. It turns out that $i^j\sqrt{2\ell+1}\,Y_{\ell j}(\varphi,\theta)$ is a function known as the classical spherical harmonic, (unfortunately) denoted $Y^j_\ell(\theta,\varphi)$.

Definition 5.18. The function $Y_\ell^j(\theta,\varphi)$ called *Laplace spherical harmonic* by Dieudonné is given by

$$Y_\ell^j(\theta,\varphi) = \sqrt{\frac{(2\ell+1)(\ell-j)!}{(\ell+j)!}}\,e^{-ij\varphi}P_\ell^j(\cos\theta).$$

If we recall that the motivation for introducing the Wigner d-matrices was to deal with real orthogonal matrices instead of complex unitary matrices, we can use the Wigner d-matrices instead of the matrices $t^{(\ell)}(\theta)$, but there is an annoying sign issue. Wigner defines his d-matrices as

$$d_{jk}^{\ell}(\theta) = (-1)^{j-k}i^{j-k}t_{jk}^{(\ell)}(\theta),$$

so for $k=0$, the factor i^j makes the term real, but now we have an extra factor $(-1)^j$, so the middle column of the d-matrix consists of the entries $(-1)^jP_\ell^j(\cos\theta)$ instead of $P_\ell^j(\cos\theta)$. The remedy is to redefine the Wigner d-matrices by omitting the factor $(-1)^{j-k}$ in the above formula, or equivalently to define the Wigner $\mathcal{D}$-matrix $\mathcal{D}^{(\ell)}(R) = \mathcal{D}^{(\ell)}(\varphi,\theta,\psi)$ as follows.

Definition 5.19. The Wigner $\mathcal{D}$-matrix $\mathcal{D}^{(\ell)}(R)$ is defined as

$$\begin{aligned}\mathcal{D}_{jk}^{(\ell)}(R) = \mathcal{D}_{jk}^{(\ell)}(\varphi,\theta,\psi) &= e^{-i(j\varphi+k\psi)}(-1)^{j-k}d_{jk}^{(\ell)}(\theta)\\ &= e^{-i(j\varphi+k\psi)}i^{j-k}t_{jk}^{(\ell)}(\theta),\end{aligned}$$

where $R = R_x(\varphi)R_z(\theta)R_x(\psi)$.

Of course the Wigner $\mathcal{D}$-matrix $\mathcal{D}^{(\ell)}$ defines an irreducible representation $\mathcal{D}^{(\ell)}\colon \mathbf{SO}(3)\to\mathbf{U}(\mathcal{P}_\ell^{\mathbb{C}})$ equivalent to the irreducible representation $W_\ell\colon \mathbf{SO}(3)\to\mathbf{U}(\mathcal{P}_\ell^{\mathbb{C}})$. Also now the middle column of $\mathcal{D}^{(\ell)}(\varphi,\theta,\psi)$ consists of the rescaled functions $1/\sqrt{2\ell+1}\,Y_\ell^j(\theta,\varphi)$, as desired. Note that Sakurai and Napolitano [Sakurai and Napolitano (2021)] also add the factor $(-1)^{j-k}$ in their definition of the $\mathcal{D}$-matrix. We will prove in Section 5.15 that the family of functions $(Y_\ell^j(\theta,\varphi))_{\ell\in\mathbb{N},\,-\ell\leq j\leq\ell}$ forms a Hilbert basis for the functions in $\mathrm{L}^2(S^2)$.

There is another property of the functions $Y_\ell^j(\theta,\varphi)$ worth stating because it plays a role in equivariant deep learning in cnns. Here we assume that $Y_\ell^j(\theta,\varphi)$ is viewed as a function on $\mathbf{SO}(3)/H_x$. Since the group $\mathbf{SO}(3)$ acts on S^2, it is natural to wonder how the function $\lambda_R Y_\ell^j$ is related to Y_ℓ^j, for $R\in\mathbf{SO}(3)$. Here is more natural to write $Y_\ell^j(x,y,z)$, where $(x,y,z)\in S^2$ are expressed in spherical coordinates in terms of the Euler angles φ and θ as in Proposition 5.31.

Proposition 5.32. *Denote the column vector consisting of the $2\ell+1$ functions Y_ℓ^j by Y_ℓ ($\ell \in \mathbb{N}$). For every rotation $R \in \mathbf{SO}(3)$ expressed as $R = R_x(\varphi)R_z(\theta)R_x(\psi)$, we have*

$$Y_\ell(R \cdot (x, y, z)) = \mathcal{D}^{(\ell)}(R)Y_\ell(x, y, z) = \mathcal{D}^{(\ell)}(\varphi, \theta, \psi)Y_\ell(x, y, z),$$

for all $(x, y, z) \in S^2$. As a corollary, we also have

$$\overline{Y_\ell}(R^{-1} \cdot (x, y, z)) = (\mathcal{D}^{(\ell)}(R))^\top\, \overline{Y_\ell}(x, y, z) = (\mathcal{D}^{(\ell)}(\varphi, \theta, \psi))^\top\, \overline{Y_\ell}(x, y, z),$$

for all $(x, y, z) \in S^2$.

Proof. Since $\mathcal{D}^{(\ell)} \colon \mathbf{SO}(3) \to \mathbf{U}(\mathcal{P}_\ell^{\mathbb{C}})$ is a representation, we have $\mathcal{D}^{(\ell)}(RS) = \mathcal{D}^{(\ell)}(R)\mathcal{D}^{(\ell)}(S)$ for all $R, S \in \mathbf{SO}(3)$. Since (x, y, z) is expressed in terms of the rotation matrix $S = R_x(\varphi_1)R_z(\theta_1)$ for some Euler angles φ_1, θ_1 and since the middle column of the matrix $\mathcal{D}^{(\ell)}(RS)$ consists of the column vector $(1/\sqrt{2\ell+1})Y_\ell(R \cdot (x, y, z))$ and the middle column of the matrix $\mathcal{D}^{(\ell)}(S)$ consists of the column vector $(1/\sqrt{2\ell+1})Y_\ell(x, y, z)$, our result follows immediately by multiplying both sides of the equation $\mathcal{D}^{(\ell)}(RS) = \mathcal{D}^{(\ell)}(R)\mathcal{D}^{(\ell)}(S)$ by $\sqrt{2\ell+1}$. There is actually a subtle point, which is that $R \cdot (x, y, z) \in S^2$ is generally not represented by RS, but by some rotation of the form $R_x(\varphi_2)R_z(\theta_2)$ in the coset RSH_x. The formulae of Section 5.4 can be used to factor $RS = R_x(\varphi)R_z(\theta)R_x(\psi)R_x(\varphi_1)R_z(\theta_1)$ as $R_x(\varphi_2)R_z(\theta_2)R_x(\psi_2)$ and then $R_x(\varphi_2)R_z(\theta_2)$ is a representative in $\mathbf{SO}(3)/H_x$ of the coset RSH_x. However, as a function on $\mathbf{SO}(3)$, the functions in the middle column of $\mathcal{D}^{(\ell)}(RS)$ and in the middle column of $\mathcal{D}^{(\ell)}(R_x(\varphi_2)R_z(\theta_2))$ are identical!

If we replace R by R^{-1}, we get

$$Y_\ell(R^{-1} \cdot (x, y, z)) = \mathcal{D}^{(\ell)}(R^{-1})\, Y_\ell(x, y, z).$$

But $\mathcal{D}^{(\ell)}(R^{-1}) = \overline{(\mathcal{D}^{(\ell)}(R))^\top}$, so by conjugating on both sides of the above equation, we get

$$\overline{Y_\ell}(R^{-1} \cdot (x, y, z)) = (\mathcal{D}^{(\ell)}(R))^\top\, \overline{Y_\ell}(x, y, z),$$

as claimed. ∎

In special case where $j = 0$ the function $t_{00}^{(\ell)}(q) = P_\ell(\cos\theta)$ depends only on θ and is called a *zonal spherical function*.

More properties of the Legendre and Jacobi polynomials and functional relations and generating functions for the functions $P_{jk}^\ell(z)$, can be found in Vilenkin [Vilenkin (1968)], Chapter III, Sections 3–5.

We will now derive an explicit formula for an invariant Haar measure on $\mathbf{SU}(2)$ in terms of the Euler angles φ, θ, ψ. Then since the representations $T_\ell \colon \mathbf{SU}(2) \to \mathbf{GL}(\mathcal{P}_\ell^{\mathbb{C}})$ form a complete set of irreducible representations of $\mathbf{SU}(2)$, they are equivalent to the representation M_ρ of Peter–Weyl I (with $\rho = \ell$), so we will able to obtain a Hilbert sum decomposition of $\mathrm{L}^2(\mathbf{SU}(2))$ in terms of the functions $t_{jk}^{(\ell)}(q)$, with $q \in \mathbf{SU}(2)$. We will also obtain a Hilbert sum decomposition of $\mathrm{L}^2(S^2)$.

5.14 Integration on SU(2) and SO(3)

In this section, we derive explicit formulae for the normalized Haar measures on $\mathbf{SU}(2)$ and $\mathbf{SO}(3)$ when these groups are parametrized by the Euler angles. Technically, these parametrizations are injective only on open subsets of $\mathbf{SU}(2)$ and $\mathbf{SO}(3)$, but the complements of these open sets have measure zero so from the point of view integration we obtain formulae for integrating all functions in $\mathrm{L}^2(\mathbf{SU}(2))$ and all functions in $\mathrm{L}^2(\mathbf{SO}(3))$ (respectively equipped with these left and right invariant Haar measures).

As a first step, we will need to derive a formula for an $\mathbf{SU}(2)$-invariant volume form on $\mathbf{SU}(2)$ as a pull-back of the $\mathbf{SO}(4)$-invariant volume form ω_{S^3} on S^3. The reader may want to review volume forms and integration on manifolds before reading this section. These topics are covered in Gallier and Quaintance [Gallier and Quaintance (2020b)] (Chapters 4 and 6).

Definition 5.20. The bijection $\Sigma \colon \mathbb{H} \to \mathbb{R}^4$ from the space $\mathbb{H}$ of quaternions to $\mathbb{R}^4$ is defined as follows: for every quaternion $A \in \mathbb{H}$, with

$$A = \begin{pmatrix} a+ib & c+id \\ -c+id & a-ib \end{pmatrix}, \quad a,b,c,d \in \mathbb{R},$$

we have

$$\Sigma(A) = (a,b,c,d),$$

where, as usual, we view (a,b,c,d) as a column vector. The bijection Σ restricts to a bijection $\Sigma \colon \mathbf{SU}(2) \to S^3$ from $\mathbf{SU}(2)$ to the sphere S^3 (in $\mathbb{R}^4$).

It is clear that the map $\Sigma \colon \mathbf{SU}(2) \to S^3$ is a homeomorphism. In fact, it is a smooth diffeomorphism. We will compute the tangent map $d\Sigma_A \colon T_A\mathbf{SU}(2) \to T_{\Sigma(A)}S^3$, with $A \in \mathbf{SU}(2)$. Then we will use Σ to define a volume form ω on $\mathbf{SU}(2)$ by pulling back a volume form ω_{S^3} on S^3.

We warn our readers that in this section we do not follow our usual notational convention that a unit quaternion, an element of $\mathbf{SU}(2)$, is denoted by a lower-case letter, typically q. Since we also need to denote points on the sphere S^3, to avoid potential confusion we denote unit quaternions using capital letters, A, A', *etc.*

The volume form ω_{S^3} on S^3 is $\mathbf{SO}(4)$-invariant but, to prove that $\omega = \Sigma^*\omega_{S^3}$ is $\mathbf{SU}(2)$-invariant we need to understand how the left (or right) action of $\mathbf{SU}(2)$ on itself translates into an action on S^3. Here we use the "ancient" fact that left and right translation in $\mathbf{SU}(2)$ translate into a rotation in $\mathbb{R}^4$ restricted to S^3 *via* Σ. This fact can be found as far back as Veblen and Young [Veblen and Young (1946)] and also in Gallier [Gallier (2011)] (Chapter 9).

Given two matrices $A, A' \in \mathbf{SU}(2)$, if $\Sigma(A) = (a,b,c,d)$ and $\Sigma(A') = (a',b',c,d')$, by multiplying the matrices A and A', we obtain the following identities:

$$\Sigma(AA') = \Sigma(L_A A') = \begin{pmatrix} a & -b & -c & -d \\ b & a & -d & c \\ c & d & a & -b \\ d & -c & b & a \end{pmatrix} \begin{pmatrix} a' \\ b' \\ c' \\ d' \end{pmatrix},$$

and

$$\Sigma(AA') = \Sigma(R_{A'} A) = \begin{pmatrix} a' & -b' & -c' & -d' \\ b' & a' & d' & -c' \\ c' & -d' & a' & b' \\ d' & c' & -b' & a' \end{pmatrix} \begin{pmatrix} a \\ b \\ c \\ d \end{pmatrix}.$$

Definition 5.21. Let $M(L_A)$ and $M(R_{A'})$ be the matrices

$$M(L_A) = \begin{pmatrix} a & -b & -c & -d \\ b & a & -d & c \\ c & d & a & -b \\ d & -c & b & a \end{pmatrix}, \quad M(R_{A'}) = \begin{pmatrix} a' & -b' & -c' & -d' \\ b' & a' & d' & -c' \\ c' & -d' & a' & b' \\ d' & c' & -b' & a' \end{pmatrix}. \tag{M1}$$

In summary, we proved that

$$\Sigma(AA') = \Sigma(L_A A') = M(L_A)\Sigma(A') = \Sigma(R_{A'} A) = M(R_{A'})\Sigma(A). \tag{M2}$$

Proposition 5.33. *If A and A' are unit quaternions, then $M(L_A)$ and $M(R_{A'})$ belong to $\mathbf{SO}(4)$; that is, they are rotation matrices.*

Proof. Observe that the columns (and the rows) of the matrices $M(L_A)$ and $M(R_{A'})$ are orthogonal. Thus, when A and A' are unit quaternions, both $M(L_A)$ and $M(R_{A'})$ are orthogonal matrices. Furthermore, it is obvious that $M(L_{A^*}) = M(L_A)^\top$, the transpose of $M(L_A)$, and similarly, $M(R_{(A')^*}) = M(R_{A'})^\top$. Since $AA^* = (a^2+b^2+c^2+d^2)I_2 = N(A)I_2$, the matrix $M(L_A)M(L_A)^\top$ is the diagonal matrix $N(A)I$ (where I is the identity 4×4 matrix), and similarly the matrix $M(R_{A'})M(R_{A'})^\top$ is the diagonal matrix $N(A')I$. Since $M(L_A)$ and $M(L_A)^\top$ have the same determinant, we deduce that $\det(M(L_A))^2 = N(A)^4$, and thus $\det(M(L_A)) = \pm N(A)^2$. However, it is obvious that one of the terms in $\det(M(L_A))$ is a^4, and thus

$$\det(M(L_A)) = (a^2+b^2+c^2+d^2)^2.$$

This shows that when A is a unit quaternion, $M(L_A) \in \mathbf{SO}(4)$, that is, $M(L_A)$ a rotation matrix, and similarly when A' is a unit quaternion, $M(R_{A'}) \in \mathbf{SO}(4)$ (see Veblen and Young [Veblen and Young (1946)]). ☐

We also need an explicit formula for the derivative $d\Sigma_A\colon T_A\mathbf{SU}(2) \to T_{\Sigma(A)}S^3$.

Proposition 5.34. *For all $A \in \mathbf{SU}(2)$ and all $Y \in T_A\mathbf{SU}(2)$, we have*

$$d\Sigma_A(Y) = \Sigma(Y). \qquad (d\Sigma)$$

Proof. Since the tangent space $T_A\mathbf{SU}(2)$ is equal to $A\mathfrak{su}(2)$, every $Y \in T_A\mathbf{SU}(2)$ is of the form $Y = A\theta X$ for some $X \in \mathfrak{su}(2)$ given by

$$X = \begin{pmatrix} ix_1 & x_2 + ix_3 \\ -x_2 + ix_3 & -ix_1 \end{pmatrix}$$

with $x_1^2 + x_2^2 + x_3^2 = 1$ and some $\theta \in \mathbb{R}$, and we have a curve

$$c(t) = Ae^{t\theta X}$$

such that $c(0) = A$ and $c'(0) = A\theta X = Y$. But

$$e^{t\theta X} = \cos(t\theta)I_2 + \sin(t\theta)X,$$

so by the chain rule

$$\begin{aligned}
d\Sigma_A(Y) &= d\Sigma_{c(0)}(c'(0)) = (\Sigma(c(t)))'_{t=0} \\
&= (\Sigma(A(\cos(t\theta)I_2 + \sin(t\theta)X))'_{t=0} \\
&= (M(L_A)(\Sigma(\cos(t\theta)I_2 + \sin(t\theta)X))'_{t=0} \\
&= (M(L_A)(\cos(t\theta),\, \sin(t\theta)x_1,\, \sin(t\theta)x_2,\, \sin(t\theta)x_3))'_{t=0} \\
&= (M(L_A)(-\theta\sin(t\theta),\, \theta\cos(t\theta)x_1,\, \theta\cos(t\theta)x_2,\, \theta\cos(t\theta)x_3))_{t=0} \\
&= M(L_A)(0,\, \theta x_1,\, \theta x_2,\, \theta x_3) \\
&= M(L_A)\Sigma(\theta X) = \Sigma(A\theta X) = \Sigma(AA^{-1}Y) = \Sigma(Y).
\end{aligned}$$

In summary, for all $A \in \mathbf{SU}(2)$ and all $Y \in T_A\mathbf{SU}(2)$, we have

$$d\Sigma_A(Y) = \Sigma(Y),$$

as claimed. $\square$

Since Σ is linear, the pull-back $\Sigma^*\omega_{S^3}$ is given by

$$\Sigma^*(\omega_{S^3})_A(Y) = (\omega_{S^3})_{\Sigma(A)}(d\Sigma_A(Y)) = (\omega_{S^3})_{\Sigma(A)}(\Sigma(Y)).$$

Definition 5.22. The *volume form* ω *on* $\mathbf{SU}(2)$ is defined as $\omega = \Sigma^*(\omega_{S^3})$; that is, for all $A \in \mathbf{SU}(2)$ and all $Y \in T_A\mathbf{SU}(2)$, we have

$$\omega_A(Y) = (\omega_{S^3})_{\Sigma(A)}(\Sigma(Y)). \tag{ω}$$

Since $\Sigma(A) = \Sigma(AI_2) = M(L_A)\Sigma(I_2) = M(L_A)e_1$, since $Y = A\theta X$ with $X \in \mathfrak{su}(2)$ and $M(L_A)$ is a rotation matrix, $\Sigma(A) = M(L_A)e_1$ and $\Sigma(Y) = M(L_A)\Sigma(\theta X)$ are indeed orthogonal because e_1 and $\Sigma(\theta X)$ are orthogonal since θX has no real part.

Since the volume form ω_{S^3} is invariant under $\mathbf{SO}(4)$, we can prove that the 3-form ω is $\mathbf{SU}(2)$-invariant.

Proposition 5.35. *The volume form* ω *is invariant under* $\mathbf{SU}(2)$.

Proof. First, we verify left-invariance. We need to prove that $L_A^*\omega = \omega$ for all $A \in \mathbf{SU}(2)$. Since L_A given by $L_A(A') = AA'$ is linear, we have $(dL_A)_{A'} = L_A$ for all $A' \in \mathbf{SU}(2)$ and similarly since $L_{M(L_A)}$ is linear, $d(L_{M(L_A)})_Q = L_{M(L_A)}$ for all $Q \in \mathbf{SO}(3)$, and so

$$\begin{aligned}
(L_A^*\omega)_{A'}(Y) &= \omega_{L_A(A')}((dL_A)_{A'}(Y)) = \omega_{L_A(A')}(L_A(Y)) \\
&= (\omega_{S^3})_{\Sigma(AA')}(\Sigma(L_A(Y))) \\
&= (\omega_{S^3})_{M(L_A)\Sigma(A')}(M(L_A)\Sigma(Y)) \\
&= (\omega_{S^3})_{M(L_A)\Sigma(A')}(d(L_{M(L_A)})_{\Sigma(A')}(\Sigma(Y))),
\end{aligned}$$

and since $M(L_A) \in \mathbf{SO}(4)$ and ω_{S^3} is invariant under $\mathbf{SO}(4)$, we conclude that

$$\begin{aligned}
(L_A^*\omega)_{A'}(Y) &= (\omega_{S^3})_{M(L_A)\Sigma(A')}(d(L_{M(L_A)})_{\Sigma(A')}(\Sigma(Y))) \\
&= (\omega_{S^3})_{\Sigma(A')}(\Sigma(Y)) = \omega_{A'}(Y),
\end{aligned}$$

as claimed. Next, we verify right-invariance, which means that we need to check that $R_A^*\omega = \omega$. Since R_A given by $R_A(A') = A'A$ is linear, we have

$(dR_A)_{A'} = R_A$ for all $A' \in \mathbf{SU}(2)$ and similarly since $R_{M(R_A)}$ is linear, $d(R_{M(R_A)})_Q = R_{M(R_A)}$ for all $Q \in \mathbf{SO}(3)$, and so

$$\begin{aligned}(R_A^*\omega)_{A'}(Y) &= \omega_{R_A(A')}((dR_A)_{A'}(Y)) = \omega_{R_A(A')}(R_A(Y))\\ &= (\omega_{S^3})_{\Sigma(A'A)}(\Sigma(R_A(Y)))\\ &= (\omega_{S^3})_{M(R_A)\Sigma(A')}(M(R_A)\Sigma(Y))\\ &= (\omega_{S^3})_{M(R_A)\Sigma(A')}(d(R_{M(R_A)})_{\Sigma(A')}(\Sigma(Y))),\end{aligned}$$

and since $M(R_A) \in \mathbf{SO}(4)$ and ω_{S^3} is invariant under $\mathbf{SO}(4)$, we conclude that

$$\begin{aligned}(R_A^*\omega)_{A'}(Y) &= (\omega_{S^3})_{M(R_A)\Sigma(A')}(d(R_{M(R_A)})_{\Sigma(A')}(\Sigma(Y)))\\ &= (\omega_{S^3})_{\Sigma(A')}(\Sigma(Y)) = \omega_{A'}(Y),\end{aligned}$$

establishing right-invariance. ☐

It is a standard result of differential geometry that the restriction ω_{S^3} of the differential 3-form $\widetilde{\omega}$ on $\mathbb{R}^4$ to S^3 given by

$$\widetilde{\omega}_p = a dx_2\wedge dx_3\wedge dx_4 - b dx_1\wedge dx_3\wedge dx_4 + c dx_1\wedge dx_2\wedge dx_4 - dx_1\wedge dx_2\wedge dx_3 \quad (\widetilde{\omega})$$

is a volume form on S^3, with

$$(\omega_{S^3})_p(u_1, u_2, u_3) = \det(p, u_1, u_2, u_3), \qquad (\widetilde{\omega}_{S^3})$$

where $p = (a, b, c, d) \in S^3$ and $u_1, u_2, u_3 \in T_pS^3$. See Gallier and Quaintance [Gallier and Quaintance (2020b)] (Chapter 6). Invariance under $\mathbf{SO}(4)$ follows from the fact that the determinant is preserved under $\mathbf{SO}(4)$. Consequently, using the diffeomorphism $\Sigma\colon \mathbf{SU}(2) \to S^3$, we obtain the volume form $\omega_{\mathbf{SU}(2)}$ on $\mathbf{SU}(2)$, for short ω, given by

$$\begin{aligned}(\omega_{\mathbf{SU}(2)})_A = \omega_A &= a dx_2 \wedge dx_3 \wedge dx_4 - b dx_1 \wedge dx_3 \wedge dx_4\\ &\quad + c dx_1 \wedge dx_2 \wedge dx_4 - dx_1 \wedge dx_2 \wedge dx_3, \qquad (\omega_{\mathbf{SU}(2)})\end{aligned}$$

where

$$A = \begin{pmatrix} a+ib & c+id \\ -c+id & a-ib \end{pmatrix}.$$

In the above formula we abused notation because we identified $T_A\mathbf{SU}(2)$ with $\mathbb{R}^4$ using Σ. As we showed earlier, the 3-form ω is $\mathbf{SU}(2)$-invariant. After all this work, it is nice to see that "things" are basically the same as if we were dealing with S^3, but some justifications are required, in particular invariance under $\mathbf{SU}(2)$. After all, $\mathbf{SU}(2)$ consists of *complex* matrices, but $\mathbf{SO}(4)$ consists of *real* matrices.

Definition 5.23. Let $\Omega \subseteq \mathbb{R}^3$ be the open subset

$$\Omega = (0, 2\pi) \times (0, \pi) \times (-2\pi, 2\pi). \tag{Ω}$$

By Proposition 5.4, the map $u \colon \Omega \to \mathbf{SU}(2)$ given by

$$\begin{aligned} u(\varphi, \theta, \psi) &= r_x(\varphi/2) r_z(\theta/2) r_x(\psi/2) \\ &= \begin{pmatrix} e^{\frac{i\varphi}{2}} & 0 \\ 0 & e^{-\frac{i\varphi}{2}} \end{pmatrix} \begin{pmatrix} \cos\frac{\theta}{2} & i\sin\frac{\theta}{2} \\ i\sin\frac{\theta}{2} & \cos\frac{\theta}{2} \end{pmatrix} \begin{pmatrix} e^{\frac{i\psi}{2}} & 0 \\ 0 & e^{-\frac{i\psi}{2}} \end{pmatrix} \\ &= \begin{pmatrix} \cos\frac{\theta}{2}\, e^{\frac{i(\varphi+\psi)}{2}} & i\sin\frac{\theta}{2}\, e^{\frac{i(\varphi-\psi)}{2}} \\ i\sin\frac{\theta}{2}\, e^{-\frac{i(\varphi-\psi)}{2}} & \cos\frac{\theta}{2}\, e^{-\frac{i(\varphi+\psi)}{2}} \end{pmatrix} \end{aligned}$$

is a diffeomorphism onto an open subset of $\mathbf{SU}(2)$ that omits a subset of measure zero in its range. We need to find the pull-back $u^*\omega$ of the volume form ω on $\mathbf{SU}(2)$, and since $\omega = \Sigma^*(\omega_{S^3})$, we need to find

$$\omega_\Omega = u^*\omega = u^*(\Sigma^*(\omega_{S^3})) = (\Sigma \circ u)^*(\omega_{S^3}).$$

Definition 5.24. Let $\Phi \colon \Omega \to S^3$ be the composed map $\Phi = \Sigma \circ u$ from Ω onto an open subset of S^3, given by

$$\Phi(\varphi, \theta, \psi) = (\Phi_1(\varphi, \theta, \psi),\ \Phi_2(\varphi, \theta, \psi),\ \Phi_3(\varphi, \theta, \psi),\ \Phi_4(\varphi, \theta, \psi)),$$

with

$$\begin{aligned} \Phi_1(\varphi, \theta, \psi) &= \cos\frac{\theta}{2}\cos\frac{(\varphi+\psi)}{2} & \Phi_2(\varphi, \theta, \psi) &= \cos\frac{\theta}{2}\sin\frac{(\varphi+\psi)}{2} \\ \Phi_3(\varphi, \theta, \psi) &= -\sin\frac{\theta}{2}\sin\frac{(\varphi-\psi)}{2} & \Phi_4(\varphi, \theta, \psi) &= \sin\frac{\theta}{2}\cos\frac{(\varphi-\psi)}{2}. \end{aligned}$$

The map Φ is a diffeomorphism.

Definition 5.25. Let ω_Ω be the pull-back form $\omega_\Omega = \Phi^*\omega_{S^3}$.

The pull-back form $\omega_\Omega = \Phi^*\omega_{S^3}$ is a volume form on Ω, and from the point of integration, since a only subset of measure zero is omitted we can use it to define integration on $\mathbf{SU}(2)$.

Proposition 5.36. *The volume form ω_Ω is given by*

$$\omega_\Omega = \frac{1}{8}\sin\theta\, d\theta \wedge d\varphi \wedge d\psi.$$

Proof. We need to compute

$$\begin{aligned} \Phi^*\omega_{S^3} &= \Phi_1(\varphi, \theta, \psi) d\Phi_2 \wedge d\Phi_3 \wedge d\Phi_4 - \Phi_2(\varphi, \theta, \psi) d\Phi_1 \wedge d\Phi_3 \wedge d\Phi_4 \\ &\quad + \Phi_3(\varphi, \theta, \psi) d\Phi_1 \wedge d\Phi_2 \wedge d\Phi_4 - \Phi_4(\varphi, \theta, \psi) d\Phi_1 \wedge d\Phi_2 \wedge d\Phi_3. \end{aligned}$$

It turns out that the computation is simpler if we let $\sigma = \dfrac{\varphi+\psi}{2}$ and $\tau = \dfrac{\varphi-\psi}{2}$. Then we have

$$d\sigma \wedge d\tau = \frac{1}{4}(d\varphi + d\psi) \wedge (d\varphi - d\psi) = -\frac{1}{2} d\varphi \wedge d\psi.$$

Since

$$d\Phi_1 = \frac{\partial \Phi_1}{\partial \sigma} d\sigma + \frac{\partial \Phi_1}{\partial \theta} d\theta + \frac{\partial \Phi_1}{\partial \tau} d\tau \quad d\Phi_2 = \frac{\partial \Phi_2}{\partial \sigma} d\sigma + \frac{\partial \Phi_2}{\partial \theta} d\theta + \frac{\partial \Phi_2}{\partial \tau} d\tau$$
$$d\Phi_4 = \frac{\partial \Phi_3}{\partial \sigma} d\sigma + \frac{\partial \Phi_3}{\partial \theta} d\theta + \frac{\partial \Phi_3}{\partial \tau} d\tau \quad d\Phi_4 = \frac{\partial \Phi_4}{\partial \sigma} d\sigma + \frac{\partial \Phi_4}{\partial \theta} d\theta + \frac{\partial \Phi_4}{\partial \tau} d\tau,$$

after a moment of reflexion, we see that

$$\Phi^* \omega_{S^3} = \begin{vmatrix} \Phi_1 & \dfrac{\partial \Phi_1}{\partial \sigma} & \dfrac{\partial \Phi_1}{\partial \theta} & \dfrac{\partial \Phi_1}{\partial \tau} \\ \Phi_2 & \dfrac{\partial \Phi_2}{\partial \sigma} & \dfrac{\partial \Phi_2}{\partial \theta} & \dfrac{\partial \Phi_2}{\partial \tau} \\ \Phi_3 & \dfrac{\partial \Phi_3}{\partial \sigma} & \dfrac{\partial \Phi_3}{\partial \theta} & \dfrac{\partial \Phi_3}{\partial \tau} \\ \Phi_4 & \dfrac{\partial \Phi_4}{\partial \sigma} & \dfrac{\partial \Phi_4}{\partial \theta} & \dfrac{\partial \Phi_4}{\partial \tau} \end{vmatrix} d\sigma\, d\theta\, d\tau.$$

We find that

$$\begin{aligned}
\frac{\partial \Phi_1}{\partial \theta} &= -\frac{1}{2}\sin\frac{\theta}{2}\cos\sigma & \frac{\partial \Phi_1}{\partial \sigma} &= -\cos\frac{\theta}{2}\sin\sigma & \frac{\partial \Phi_1}{\partial \tau} &= 0 \\
\frac{\partial \Phi_2}{\partial \theta} &= -\frac{1}{2}\sin\frac{\theta}{2}\sin\sigma & \frac{\partial \Phi_2}{\partial \sigma} &= \cos\frac{\theta}{2}\cos\sigma & \frac{\partial \Phi_2}{\partial \tau} &= 0 \\
\frac{\partial \Phi_3}{\partial \theta} &= -\frac{1}{2}\cos\frac{\theta}{2}\sin\tau & \frac{\partial \Phi_3}{\partial \sigma} &= 0 & \frac{\partial \Phi_3}{\partial \tau} &= -\sin\frac{\theta}{2}\cos\tau \\
\frac{\partial \Phi_4}{\partial \theta} &= \frac{1}{2}\cos\frac{\theta}{2}\cos\tau & \frac{\partial \Phi_4}{\partial \sigma} &= 0 & \frac{\partial \Phi_4}{\partial \tau} &= -\sin\frac{\theta}{2}\sin\tau.
\end{aligned}$$

Since

$$d\sigma \wedge d\theta \wedge d\tau = -d\theta \wedge d\sigma \wedge d\tau = \frac{1}{2} d\theta \wedge d\varphi \wedge d\psi,$$

we find that $\Phi^* \omega_{S^3}$ is equal to

$$\frac{1}{4}\cos\frac{\theta}{2}\sin\frac{\theta}{2} \begin{vmatrix} \cos\frac{\theta}{2}\cos\sigma & -\sin\frac{\theta}{2}\cos\sigma & -\sin\sigma & 0 \\ \cos\frac{\theta}{2}\sin\sigma & -\sin\frac{\theta}{2}\sin\sigma & \cos\theta & 0 \\ -\sin\frac{\theta}{2}\sin\tau & -\cos\frac{\theta}{2}\sin\tau & 0 & -\cos\tau \\ \sin\frac{\theta}{2}\cos\tau & \cos\frac{\theta}{2}\cos\tau & 0 & -\sin\tau \end{vmatrix} d\theta \wedge d\varphi \wedge d\psi.$$

Observe that the matrix corresponding to the determinant is orthogonal, so

$$\Phi^*\omega_{S^3} = \pm\frac{1}{4}\cos\frac{\theta}{2}\sin\frac{\theta}{2}\,d\theta\wedge d\varphi\wedge d\psi = \pm\frac{1}{8}\sin\theta\,d\theta\wedge d\varphi\wedge d\psi.$$

In fact, it can be verified that the determinant has the value $+1$, so we get

$$\omega_\Omega = \Phi^*\omega_{S^3} = \frac{1}{8}\sin\theta\,d\theta\wedge d\varphi\wedge d\psi,$$

as claimed. □

Finally, given a continuous function $f\colon \Omega\to\mathbb{C}$ the integral $\int_\Omega f\omega_\Omega$ is defined by

$$\int_\Omega f\omega_\Omega = \frac{1}{8}\int_{-2\pi}^{2\pi}\int_0^{2\pi}\int_0^{\pi} f(\varphi,\theta,\psi)\sin\theta\,d\theta\,d\varphi\,d\psi.$$

Since this integral yields the volume $2\pi^2$ (for $f\equiv 1$), we normalize the measure ν associated with ω_Ω so that $\int_\Omega 1\omega_\Omega = 1$. Since $\Sigma\colon \mathbf{SU}(2)\to S^3$ is a diffeomorphism we also have $\omega_{S^3} = (\Sigma^{-1})^*\omega$, and so for any continuous function $f\colon \mathbf{SU}(2)\to\mathbb{C}$,

$$\int_{\mathbf{SU}(2)} f\omega = \int_{S^3}(f\circ\Sigma^{-1})\omega_{S^3},$$

and since for any continuous function $f\colon S^3\to\mathbb{C}$, we have

$$\int_{S^3} f\omega_{S^3} = \int_\Omega (f\circ\Phi)\omega_\Omega = \int_\Omega (f\circ\Sigma\circ u)\omega_\Omega,$$

we obtain

$$\int_{\mathbf{SU}(2)} f\omega = \int_{S^3}(f\circ\Sigma^{-1})\omega_{S^3} = \int_\Omega (f\circ\Sigma^{-1}\circ\Sigma\circ u)\omega_\Omega = \int_\Omega (f\circ u)\omega_\Omega.$$

In summary, we obtained the following result.

Proposition 5.37. *The pull-back volume form $\omega_\Omega = \Phi^*\omega_{S^3}$ on Ω is given by*

$$\omega_\Omega = \frac{1}{8}\sin\theta\,d\theta\wedge d\varphi\wedge d\psi. \qquad (\omega_\Omega)$$

For a continuous function $f\colon \Omega\to\mathbb{C}$, the integral $\int_\Omega f\omega_\Omega$ is given by

$$\int_\Omega f\omega_\Omega = \frac{1}{16\pi^2}\int_{-2\pi}^{2\pi}\int_0^{2\pi}\int_0^{\pi} f(\varphi,\theta,\psi)\sin\theta\,d\theta\,d\varphi\,d\psi. \qquad \text{(INT-}\Omega)$$

For any continuous function $f\colon \mathbf{SU}(2) \to \mathbb{C}$, *the integral* $\int_{\mathbf{SU}(2)} f\omega$ *(which is equal to* $\int_\Omega (f \circ u)\omega_\Omega$*) is given by*

$$\int_{\mathbf{SU}(2)} f\omega = \frac{1}{16\pi^2} \int_{-2\pi}^{2\pi} \int_0^{2\pi} \int_0^{\pi} f(u(\varphi, \theta, \psi)) \sin\theta \, d\theta d\varphi d\psi. \quad \text{(INT-}\mathbf{SU}(2))$$

For any continuous function $f\colon S^3 \to \mathbb{C}$, *the integral* $\int_{S^3} f\omega = \int_\Omega (f \circ \Phi)\omega_\Omega$ *is given by*

$$\int_{S^3} f\omega_{S^3} = \frac{1}{16\pi^2} \int_{-2\pi}^{2\pi} \int_0^{2\pi} \int_0^{\pi} f(\Phi(\varphi, \theta, \psi)) \sin\theta \, d\theta \, d\varphi \, d\psi. \qquad \text{(INT-}S^3)$$

We also write $\int_{\mathbf{SU}(2)} f \, d\nu$ for $\int_{\mathbf{SU}(2)} f\omega$.

Remark: Formula (INT$-\Omega$) is stated in Vilenkin [Vilenkin (1968)] (Chapter III, Section 6) and in Kosmann-Schwarzbach [Kosmann-Schwarzbach (2010)], see Exercise 5.6.

It is remarkable that we can also obtain the normalized Haar measure on **SO**(3) in terms of the Euler angles from the normalized Haar measure on **SU**(2) *without any additional computation.* Recall that as a corollary of Proposition 5.4, the map $\rho \circ u_0$ from $[0, 2\pi) \times [0, \pi] \times [0, 2\pi)$ to **SO**(3) is surjective, where $\rho\colon \mathbf{SU}(2) \to \mathbf{SO}(3)$ is a surjective homomorphism whose kernel is $\{-I, I\}$ and u_0 is the restriction of $u\colon [0, 2\pi) \times [0, \pi] \times [-2\pi, 2\pi) \to \mathbf{SU}(2)$ to $[0, 2\pi) \times [0, \pi] \times [0, 2\pi)$.

Definition 5.26. Let $\Omega_0 \subseteq \mathbb{R}^3$ be the open subset

$$\Omega_0 = (0, 2\pi) \times (0, \pi) \times (0, 2\pi), \qquad (\Omega_0)$$

a proper open subset of $\Omega = (0, 2\pi) \times (0, \pi) \times (-2\pi, 2\pi)$, and let $u_0\colon \Omega_0 \to \mathbf{SU}(2)$ be the restriction of $u\colon \Omega \to \mathbf{SU}(2)$ to Ω_0.

Definition 5.27. The map $R_0\colon \Omega_0 \to \mathbf{SO}(3)$ is given by

$$R_0(\varphi, \theta, \psi) = R_x(\varphi) R_z(\theta) R_x(\psi),$$

or more explicitly,

$$R_0(\varphi, \theta, \psi) = \begin{pmatrix} 1 & 0 & 0 \\ 0 & \cos\varphi & -\sin\varphi \\ 0 & \sin\varphi & \cos\varphi \end{pmatrix} \begin{pmatrix} \cos\theta & -\sin\theta & 0 \\ \sin\theta & \cos\theta & 0 \\ 0 & 0 & 1 \end{pmatrix} \begin{pmatrix} 1 & 0 & 0 \\ 0 & \cos\psi & -\sin\psi \\ 0 & \sin\psi & \cos\psi \end{pmatrix}.$$

Proposition 5.4 implies that u is injective on $\Omega = (0, 2\pi) \times (0, \pi) \times (-2\pi, 2\pi)$, and thus u_0 is injective on Ω_0.

Definition 5.28. Let $U_0 = u_0(\Omega_0)$ be the image of Ω_0 by u_0, an open subset of $\mathbf{SU}(2)$.

Proposition 5.38. *The restriction ρ_0 of ρ to U_0 is injective. As a consequence, the map $R_0 \colon \Omega_0 \to \mathbf{SO}(3)$ is also injective.*

Proof. This is because if $\rho(q_1) = \rho(q_2)$ for some $q_1, q_2 \in \Omega_0$, such that $q_1 \neq q_2$, then $q_2 = -q_1$, but if $q_1 = r_x(\varphi/2)r_z(\theta/2)r_x(\psi/2)$, then

$$\begin{aligned} r_x(\varphi/2)r_z(\theta/2)r_x((\psi - 2\pi)/2) &= r_x(\varphi/2)r_z(\theta/2)r_x(\psi/2 - \pi) \\ &= -r_x(\varphi/2)r_z(\theta/2)r_x(\psi/2) = -q_1, \end{aligned}$$

so $q_2 = r_x(\varphi/2)r_z(\theta/2)r_x((\psi - 2\pi)/2)$ with $-2\pi < \psi < 0$ contradicting the fact that $q_2 \in \Omega_0$. $\square$

As a consequence, if we let $V_0 = \rho_0(U_0) \subseteq \mathbf{SO}(3)$, the bijective map $\rho_0 \colon U_0 \to V_0$ has an inverse $s_0 \colon V_0 \to U_0$.

Definition 5.29. Let $V_0 = \rho_0(U_0) \subseteq \mathbf{SO}(3)$ and $s_0 \colon V_0 \to U_0$ be the inverse of $\rho_0 \colon U_0 \to V_0$. Also let Σ_0 be the composition $\Sigma_0 = \Sigma \circ s_0$.

Consider the commutative diagram

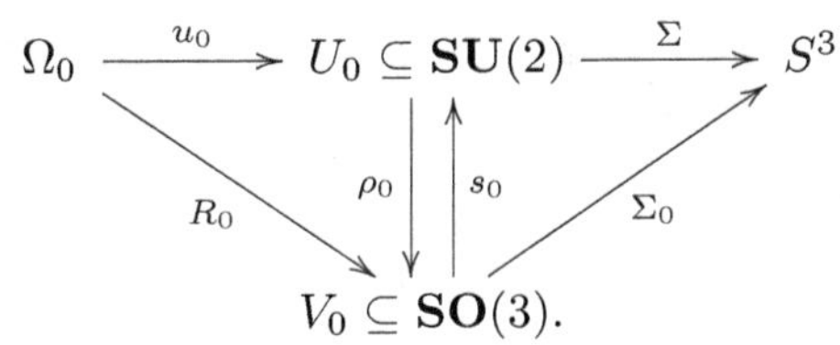

Since s_0 is a diffeomorphism, we can pull back the volume form ω on $\mathbf{SU}(2)$ (really on U_0) to $\mathbf{SO}(3)$ (really V_0).

Proposition 5.39. *The 3-form $s_0^*\omega$ is $\mathbf{SO}(3)$-invariant.*

Proof. We check left-invariance, leaving right-invariance as an exercise. For this, for any $Q, R \in V_0 \subseteq \mathbf{SO}(3)$ and any $Y \in T_R\mathbf{SO}(3)$, we compute

$$\begin{aligned} (L_Q^*(s_0^*\omega))_R(Y) &= ((s_0 \circ L_Q)^*\omega)_R(Y) = \omega_{(s_0 \circ L_Q)(R)}(d(s_0 \circ L_Q)_R(Y)) \\ &= \omega_{s_0(QR)}(d(s_0 \circ L_Q)_R(Y)) \\ &= \omega_{s_0(Q)s_0(R)}(d(s_0 \circ L_Q)_R(Y)). \end{aligned}$$

But since $(s_0 \circ L_Q)(R) = s_0(QR) = s_0(Q)s_0(R) = L_{s_0(Q)}(s_0(R))$, we see that $d(s_0 \circ L_Q)_R = d(L_{s_0(Q)})_{s_0(R)} \circ (ds_0)_R$, so we have

$$\begin{aligned} (L_Q^*(s_0^*\omega))_R(Y) &= \omega_{s_0(Q)s_0(R)}(d(s_0 \circ L_Q)_R(Y)) \\ &= \omega_{L_{s_0(Q)}(s_0(R))}(d(L_{s_0(Q)})_{s_0(R)}((ds_0)_R(Y))). \end{aligned}$$

Since ω is left-invariant, we obtain

$$\begin{aligned}(L_Q^*(s_0^*\omega))_R(Y) &= \omega_{L_{s_0(Q)}(s_0(R))}(d(L_{s_0(Q)})_{s_0(R)}((ds_0)_R(Y))) \\ &= \omega_{(s_0(R)}((ds_0)_R(Y)) = (s_0^*\omega)_R(Y),\end{aligned}$$

establishing left-invariance. □

It follows that $s_0^*\omega$ is an invariant volume form on V_0, and thus a volume form $\omega_{\mathbf{SO}(3)}$ on $\mathbf{SO}(3)$ up to a set of measure zero.

Remark: It can be shown that there is an invariant volume form $\omega_{\mathbf{SO}(3)}$ on $\mathbf{SO}(3)$ such that $\omega_{\mathbf{SO}(3)}$ and the volume form $\omega_{\mathbf{SU}(2)}$ on $\mathbf{SU}(2)$ are related by $\omega_{\mathbf{SU}(2)} = \rho^*\omega_{\mathbf{SO}(3)}$, where ρ is the covering homomorphism $\rho\colon \mathbf{SU}(2) \to \mathbf{SO}(3)$. This is because $\mathbf{SO}(3)$ is orientable. The proof is essentially the same as the proof that $\mathbb{RP}^n$ is orientable iff n is odd; see Madsen and Tornehave [Madsen and Tornehave (1998)], Example 9.19. Thus the volume form $s_0^*\omega$ on $V_0 \subseteq \mathbf{SO}(3)$ is a piece of the volume form $\omega_{\mathbf{SO}(3)}$ corresponding to a section s_0 of ρ. The complement of the domain of s_0 is a subset of measure zero in $\mathbf{SO}(3)$. The volume form $\omega_{\mathbf{SO}(3)}$ defines the Haar measure on $\mathbf{SO}(3)$, and up to a subset of measure zero, so does $s_0^*\omega$.

The pull-back $\omega_{\Omega_0} = R_0^*(s_0^*\omega) = (s_0 \circ R_0)^*\omega$ of the volume form $s_0^*\omega$ by R_0 is the volume form on Ω_0 that we seek. However, the commutativity of the above diagram and the fact that by definition $\omega = \Sigma^*\omega_{S^3}$ show that

$$\omega_{\Omega_0} = (s_0 \circ R_0)^*\omega = u_0^*\omega = u_0^*(\Sigma^*\omega_{S^3}) = (\Sigma \circ u_0)^*\omega_{S^3}.$$

Definition 5.30. Let $\Phi_0\colon \Omega_0 \to S^3$ be the composed map $\Phi_0 = \Sigma \circ u_0$ and let ω_{Ω_0} be the volume form on Ω_0 defined as the pull-back $\Phi_0^*\omega_{S^3}$ of ω_{S^3}.

Since u_0 is the restriction of u to Ω_0, we conclude that $\omega_{\Omega_0} = \Phi_0^*\omega_{S^3}$ is given as in the $\mathbf{SU}(2)$ case, by

$$\omega_{\Omega_0} = \frac{1}{8}\sin\theta\, d\theta \wedge d\varphi \wedge d\psi.$$

Proposition 5.40. *The pull-back volume form* $\omega_{\Omega_0} = \Phi_0^*\omega_{S^3}$ *on* Ω_0 *is given by*

$$\omega_{\Omega_0} = \frac{1}{8}\sin\theta\, d\theta \wedge d\varphi \wedge d\psi. \tag{ω_{Ω_0}}$$

For a continuous function $f\colon \Omega_0 \to \mathbb{C}$, *the integral* $\int_{\Omega_0} f\omega_{\Omega_0}$ *is given by*

$$\int_{\Omega_0} f\omega_{\Omega_0} = \frac{1}{8\pi^2}\int_0^{2\pi}\int_0^{2\pi}\int_0^{\pi} f(\varphi,\theta,\psi)\sin\theta\, d\theta d\varphi d\psi. \tag{INT-Ω_0}$$

For a continuous function $f\colon \mathbf{SO}(3) \to \mathbb{C}$, *the integral* $\int_{\mathbf{SO}(3)} f\omega_{\mathbf{SO}(3)}$ *is given by*

$$\int_{\mathbf{SO}(3)} f\omega_{\mathbf{SO}(3)} = \frac{1}{8\pi^2}\int_0^{2\pi}\int_0^{2\pi}\int_0^{\pi} f(R_0(\varphi,\theta,\psi))\sin\theta\, d\theta\, d\varphi\, d\psi. \tag{INT-$\mathbf{SO}(3)$}$$

We also write $\int_{\mathbf{SO}(3)} f\, d\nu_0$ for $\int_{\mathbf{SO}(3)} f\omega_{\mathbf{SO}(3)}$. Observe that the measure ν_0 associated with ω_{Ω_0} is already normalized.

Remark: Formula (INT$-\Omega_0$) is also stated in Kosmann-Schwarzbach [Kosmann-Schwarzbach (2010)], see Exercise 5.5.

5.15 Fourier Series of Functions in $L^2(\mathbf{SU}(2))$, $L^2(\mathbf{SO}(3))$ and $L^2(S^2)$

In the preceding sections we computed explicitly several matrix representations $t^{(\ell)}(q)$ for the irreducible representations $T_\ell\colon \mathbf{SU}(2) \to \mathbf{U}(\mathcal{P}_\ell^{\mathbb{C}})$ with respect to an invariant hermitian inner product on $\mathcal{P}_\ell^{\mathbb{C}}$. In terms of the general results presented in Sections 4.1–4.3, especially Theorem 4.6, $\rho = \ell$, $n_\rho = 2\ell+1$, $M_\ell(q) = t^{(\ell)}(q)$, and since

$$M_\ell(q) = \left(\frac{1}{n_\ell} m_{ij}^{(\ell)}(q)\right),$$

the functions $m_{ij}^{(\ell)}(q)$ are given by $m_{ij}^{(\ell)}(q) = (2\ell+1)t_{ij}^{(\ell)}(q)$, where ℓ ranges through the set $R = \{0, 1/2, 1, 3/2, 2, 5/2, 3, \ldots\}$ of all nonnegative integer and half integer values. By Peter–Weyl I (Theorem 4.2), the $n_\ell^2 = (2\ell+1)^2$ functions $\frac{1}{\sqrt{n_\ell}} m_{ij}^{(\ell)} = \sqrt{2\ell+1}\, t_{ij}^{(\ell)}$ in the matrix $\sqrt{2\ell+1}\, t^{(\ell)}$ form an orthonormal basis of the minimal two-sided ideal $\mathfrak{a}_\ell$ arising in the Hilbert sum

$$L^2(\mathbf{SU}(2)) = \bigoplus_\ell \mathfrak{a}_\ell,$$

and thus the family of functions

$$\left(\sqrt{2\ell+1}\, t_{ij}^{(\ell)}\right)_{-\ell\le i,j\le\ell,\, \ell\in R},$$

with $R = \{0, 1/2, 1, 3/2, 2, \ldots\}$, is a Hilbert basis of $L^2(\mathbf{SU}(2))$. By the results of Section 4.6 on the Fourier transform and the Fourier cotransform, by Definition 4.17 of the Fourier transform $\mathcal{F}(f)$ and equation (FI) (see also Theorem 4.35),

$$f(s) = \sum_{\rho\in R} n_\rho \operatorname{tr}\left(\mathcal{F}(f)(\rho) M_\rho(s)\right), \qquad f \in L^2(G),\ s \in G,$$

since $M_\ell(q) = t^{(\ell)}(q)$, for every $\ell \in R$, the $(2\ell+1)\times(2\ell+1)$ matrix $\alpha^{(\ell)} = \mathcal{F}(f)(\ell)$ of Fourier coefficients of $f \in \mathrm{L}^2(\mathbf{SU}(2))$ is given by

$$\alpha^{(\ell)} = \int_{\mathbf{SU}(2)} f(q)(t^{(\ell)}(q))^*\, d\nu(q),$$

where ν is the normalized Haar measure on $\mathbf{SU}(2)$, and by the Fourier inversion formula (FI), we have

$$f(q) = \sum_{\ell\in R}(2\ell+1)\operatorname{tr}\big(\alpha^{(\ell)}t^{(\ell)}(q)\big), \quad q \in \mathbf{SU}(2).$$

Written in terms of matrix elements, we obtain the equations

$$\alpha^{(\ell)}_{jk} = \int_{\mathbf{SU}(2)} f(q)\overline{t^{(\ell)}_{kj}(q)}\, d\nu(q), \tag{FC1}$$

and

$$f(q) = \sum_{\ell\in R}(2\ell+1)\sum_{j=-\ell}^{\ell}\sum_{k=-\ell}^{\ell} \alpha^{(\ell)}_{kj}\, t^{(\ell)}_{jk}(q), \quad q \in \mathbf{SU}(2). \tag{FS1}$$

Using the Euler angles, Proposition 5.28 (in particular, $(*_{41}), (*_{42})$), namely

$$\begin{aligned} t^{(\ell)}_{jk}(q) &= t^{(\ell)}_{jk}(u(\varphi,\theta,\psi)) = e^{-i(j\varphi+k\psi)}\, t^{(\ell)}_{jk}(\theta) \\ &= e^{-i(j\varphi+k\psi)}\, P^{\ell}_{jk}(\cos\theta), \quad \ell \in R. \end{aligned}$$

Proposition 5.37, and the fact that $\overline{P^{\ell}_{jk}(\cos\theta)} = (-1)^{j-k}P^{\ell}_{jk}(\cos\theta)$ (left as an exercise), by swapping j and k in (FC1), we obtain the following series expansion for the functions in $\mathrm{L}^2(\mathbf{SU}(2))$.

Proposition 5.41. *Every function $f \in \mathrm{L}^2(\mathbf{SU}(2))$ expressed in terms of the Euler angles $(0 \le \varphi < 2\pi, 0 \le \theta < \pi, -2\pi \le \psi < 2\pi)$ can be written as the Fourier series*

$$f(u(\varphi,\theta,\psi)) = \sum_{\ell\in R}(2\ell+1)\sum_{j=-\ell}^{\ell}\sum_{k=-\ell}^{\ell} \alpha^{(\ell)}_{kj}\, e^{-i(j\varphi+k\psi)}P^{\ell}_{jk}(\cos\theta), \tag{FS2}$$

where the Fourier coefficients are given by

$$\alpha^{(\ell)}_{kj} = \frac{(-1)^{j-k}}{16\pi^2}\int_{-2\pi}^{2\pi}\int_0^{2\pi}\int_0^{\pi} f(u(\varphi,\theta,\psi))e^{i(j\varphi+k\psi)}P^{\ell}_{jk}(\cos\theta)\sin\theta\, d\theta\, d\varphi\, d\psi. \tag{FC2}$$

Remark: Vilenkin uses a different definition for the Fourier coefficients, namely he uses the matrix $(2\ell+1)(\alpha^{(\ell)})^\top$; see Vilenkin [Vilenkin (1968)], Chapter III, Section 6.3. As a consequence, his formulae are slightly different.

Remark: The Parseval identity is given by

$$\sum_{\ell}(2\ell+1)\sum_{j=-\ell}^{\ell}\sum_{k=-\ell}^{\ell}|\alpha_{jk}^{(\ell)}|^2 = \frac{1}{16\pi^2}\int_{-2\pi}^{2\pi}\int_0^{2\pi}\int_0^{\pi}|f(u(\varphi,\theta,\psi))|^2\sin\theta\, d\theta\, d\varphi\, d\psi. \tag{PS1}$$

The above discussion applies to $\mathbf{SO}(3)$ and its irreducible representations $W_\ell\colon \mathbf{SO}(3)\to\mathbf{U}(\mathcal{P}_\ell^{\mathbb{C}})$, which are now indexed by the set $\mathbb{N}$ of natural numbers. By Peter–Weyl I (Theorem 4.2), the $n_\ell^2=(2\ell+1)^2$ functions $\frac{1}{\sqrt{n_\ell}}m_{ij}^{(\ell)}=\sqrt{2\ell+1}\,w_{ij}^{(\ell)}$ in the matrix $\sqrt{2\ell+1}\,w^{(\ell)}$, where $w^{(\ell)}(R)$ is the matrix associated with $W^\ell(R)$ for $R\in\mathbf{SO}(3)$, form an orthonormal basis of the minimal two-sided ideal $\mathfrak{a}_\ell$ arising in the Hilbert sum

$$\mathrm{L}^2(\mathbf{SO}(3))=\bigoplus_{\ell}\mathfrak{a}_\ell,$$

and thus the family of functions

$$\left(\sqrt{2\ell+1}\,w_{ij}^{(\ell)}\right)_{-\ell\le i,j\le\ell,\,\ell\in\mathbb{N}}$$

is a Hilbert basis of $\mathrm{L}^2(\mathbf{SO}(3))$. It follows that for every $\ell\in\mathbb{N}$, the $(2\ell+1)\times(2\ell+1)$ matrix $\alpha^{(\ell)}=\mathcal{F}(f)(\ell)$ of Fourier coefficients of $f\in\mathrm{L}^2(\mathbf{SO}(3))$ is given by

$$\alpha^{(\ell)}=\int_{\mathbf{SO}(3)}f(R)(w^{(\ell)}(R))^*\,d\nu_0(R),$$

where ν_0 is the normalized Haar measure on $\mathbf{SO}(3)$, and by the Fourier inversion formula (FI), we have

$$f(R)=\sum_{\ell\in\mathbb{N}}(2\ell+1)\,\mathrm{tr}\big(\alpha^{(\ell)}w^{(\ell)}(R)\big),\quad R\in\mathbf{SO}(3).$$

Written in terms of matrix elements, we obtain the equations

$$\alpha_{jk}^{(\ell)}=\int_{\mathbf{SO}(3)}f(R)\overline{w_{kj}^{(\ell)}(R)}\,d\nu_0(R), \tag{FC1'}$$

and

$$f(R)=\sum_{\ell\in\mathbb{N}}(2\ell+1)\sum_{j=-\ell}^{\ell}\sum_{k=-\ell}^{\ell}\alpha_{kj}^{(\ell)}\,w_{jk}^{(\ell)}(q),\quad R\in\mathbf{SO}(3). \tag{FS1'}$$

Using the Euler angles, Proposition 5.28 (in particular, equations $(*_{41})$ and $(*_{42})$), Proposition 5.40, that by Proposition 5.25, we have

$$w_{jk}^{(\ell)}(R_0(\varphi,\theta,\psi)) = e^{-i(j\varphi+k\psi)}\, t_{jk}^{(\ell)}(\theta) = e^{-i(j\varphi+k\psi)}\, P_{jk}^{\ell}(\cos\theta), \quad \ell \in \mathbb{N},$$

and using the fact that $\overline{P_{jk}^{\ell}(\cos\theta)} = (-1)^{j-k} P_{jk}^{\ell}(\cos\theta)$ (left as an exercise), we obtain the following series expansion for the functions in $\mathrm{L}^2(\mathbf{SO}(3))$.

Proposition 5.42. *Every function* $f \in \mathrm{L}^2(\mathbf{SO}(3))$ *expressed in terms of the Euler angles* $(0 \le \varphi < 2\pi, 0 \le \theta < \pi, 0 \le \psi < 2\pi)$ *can be written as the Fourier series*

$$f(R_0(\varphi,\theta,\psi)) = \sum_{\ell\in\mathbb{N}} (2\ell+1) \sum_{j=-\ell}^{\ell} \sum_{k=-\ell}^{\ell} \alpha_{kj}^{(\ell)}\, e^{-i(j\varphi+k\psi)} P_{jk}^{\ell}(\cos\theta), \qquad \text{(FS2')}$$

where the Fourier coefficients are given by

$$\alpha_{kj}^{(\ell)} = \frac{(-1)^{j-k}}{8\pi^2} \int_0^{2\pi}\int_0^{2\pi}\int_0^{\pi} f(R_0(\varphi,\theta,\psi)) e^{i(j\varphi+k\psi)} P_{jk}^{\ell}(\cos\theta)\sin\theta\, d\theta\, d\varphi\, d\psi. \qquad \text{(FC2')}$$

Remarks:

(1) If the functions f are real-valued, it may be preferable to use the Wigner d-matrices $d^{(\ell)}(\theta)$ of Definition 5.12, which are real orthogonal, instead of the complex matrices $t^{(\ell)}(\theta)$, which amounts to using $(-1)^{j-k} i^{j-k} t_{jk}^{(\ell)}(\theta)$ instead of $t_{jk}^{(\ell)}(\theta)$, that is, the real polynomials $(-1)^{j-k} i^{j-k} P_{jk}^{\ell}$ instead of P_{jk}^{ℓ} in (FS2') and (FC2'). This is common practice in computer vision.

(2) A variant of the definition of the Fourier transform and of the Fourier cotransform occurs in the computer vision community. In these formula, $w^{(\ell)}(R)$ is replaced by $(w^{(\ell)}(R))^*$, namely

$$\alpha^{(\ell)} = \int_{\mathbf{SO}(3)} f(R) w^{(\ell)}(R)\, d\nu_0(R),$$

and

$$f(R) = \sum_{\ell\in\mathbb{N}} (2\ell+1)\operatorname{tr}\big(\alpha^{(\ell)} (w^{(\ell)}(R))^*\big), \quad R \in \mathbf{SO}(3).$$

Our version is consistent with the definition of the Fourier transform in the case where G is abelian.

Vilenkin investigates the expansion in Fourier series for two subspaces $\mathfrak{L}_k^2$ and ${}_j\mathfrak{L}^2$ of $\mathrm{L}^2(\mathbf{SU}(2))$.

Definition 5.31. The subspace $\mathfrak{L}^2_k$ consists of the functions $f \in \mathrm{L}^2(\mathbf{SU}(2))$, such that

$$f(qH(\psi)) = e^{-ik\psi} f(q), \quad q \in \mathbf{SU}(2),$$

for all $H(\psi)$ in the subgroup Ω_x of $\mathbf{SU}(2)$ given by

$$\Omega_x = \left\{ H(t) = r_x(t/2) = \begin{pmatrix} e^{\frac{it}{2}} & 0 \\ 0 & e^{-\frac{it}{2}} \end{pmatrix} \;\middle|\; 0 \leq t \leq 2\pi \right\}. \qquad (\Omega_x)$$

It is easy to see that $\mathfrak{L}^2_k$ consists of the functions $f \in \mathrm{L}^2(\mathbf{SU}(2))$, such that

$$f(u(\varphi, \theta, \psi)) = e^{-ik\psi} f(u(\varphi, \theta, 0)),$$

for all φ, θ, ψ.

Definition 5.32. The subspace ${}_j\mathfrak{L}^2$ consists of the set of functions $f \in \mathrm{L}^2(\mathbf{SU}(2))$, such that

$$f(H(\varphi)q) = e^{-ij\varphi} f(q), \quad q \in \mathbf{SU}(2),$$

for all $H(\varphi)$ in the subgroup Ω_x of $\mathbf{SU}(2)$.

It is easy to see that ${}_j\mathfrak{L}^2$ consists of the functions $f \in \mathrm{L}^2(\mathbf{SU}(2))$, such that

$$f(u(\varphi, \theta, \psi)) = e^{-ij\varphi} f(u(0, \theta, \psi)),$$

for all φ, θ, ψ.

Observe that the functions $t^{(\ell)}_{jk}(q)$ (with q expressed in terms of the Euler angles) for k fixed,

$$t^{(\ell)}_{jk}(q) = e^{-ik\psi} e^{-ij\varphi} P^{\ell}_{jk}(\cos\theta),$$

for $\ell = |k|, |k|+1, \ldots, |k|+m, \ldots,\ -\ell \leq j \leq \ell$, belong to $\mathfrak{L}^2_k$. In fact, it is shown in Vilenkin [Vilenkin (1968)] (Chapter III, Section 6.4) that these functions form an orthogonal basis of $\mathfrak{L}^2_k$, more precisely, every function $f \in \mathfrak{L}^2_k$ can written as the Fourier series

$$f(u(\varphi, \theta, \psi)) = e^{-ik\psi} \sum_{\ell=|k|}^{\infty} (2\ell+1) \sum_{j=-\ell}^{\ell} \alpha^{(\ell)}_j \, e^{-ij\varphi} P^{(\ell)}_{jk}(\cos\theta), \qquad \text{(FS3)}$$

where the Fourier coefficients are given by

$$\alpha^{(\ell)}_j = \frac{(-1)^{j-k}}{4\pi} \int_0^{2\pi} \int_0^{\pi} f(u(\varphi, \theta, 0)) e^{ij\varphi} P^{\ell}_{jk}(\cos\theta) \sin\theta \, d\theta \, d\varphi. \qquad \text{(FC3)}$$

In particular, for functions $f \in \mathrm{L}^2(\mathbf{SU}(2))$, such that

$$f(qH) = f(q), \quad \text{for all } q \in \mathbf{SU}(2) \text{ and all } H \in \Omega_x,$$

namely, functions which do not depend on the Euler angle ψ, we have

$$f(u(\varphi,\theta,0)) = \sum_{\ell=0}^{\infty}(2\ell+1)\sum_{j=-\ell}^{\ell} \alpha_j^{(\ell)}\, e^{-ij\varphi} P_{j0}^{(\ell)}(\cos\theta), \tag{FS4}$$

with

$$\alpha_j^{(\ell)} = \frac{(-1)^j}{4\pi}\int_0^{2\pi}\int_0^{\pi} f(u(\varphi,\theta,0))e^{ij\varphi}P_{j0}^{\ell}(\cos\theta)\sin\theta\, d\theta\, d\varphi. \tag{FC4}$$

It is shown in Vilenkin [Vilenkin (1968)] (Chapter III, Section 3.9) that $(*_{47})$, namely

$$P_\ell^k(z) = i^k\sqrt{\frac{(\ell+k)!}{(\ell-k)!}}\, P_{k0}^{\ell}(z),$$

implies that we have

$$P_\ell^{-j}(z) = (-1)^j\frac{(\ell-j)!}{(\ell+j)!}P_\ell^j(z),$$

so we obtain an expansion in terms of the associated Legendre functions $P_\ell^j(\cos\theta)$,

$$f(u(\varphi,\theta,0)) = \sum_{\ell=0}^{\infty}(2\ell+1)\sum_{j=-\ell}^{\ell} \beta_\ell^j\, e^{-ij\varphi} P_\ell^j(\cos\theta), \tag{FS5}$$

with

$$\beta_\ell^j = \frac{1}{4\pi}\frac{(\ell-j)!}{(\ell+j)!}\int_0^{2\pi}\int_0^{\pi} f(u(\varphi,\theta,0))e^{ij\varphi}P_\ell^j(\cos\theta)\sin\theta\, d\theta\, d\varphi. \tag{FC5}$$

Similarly, it can be shown (see Vilenkin [Vilenkin (1968)], Chapter III, Section 6.4) that every function $f \in {}_j\mathfrak{L}^2$ can written as the Fourier series

$$f(u(\varphi,\theta,\psi)) = e^{-ij\varphi}\sum_{\ell=|j|}^{\infty}(2\ell+1)\sum_{k=-\ell}^{\ell} \alpha_k^{(\ell)}\, e^{-ik\psi} P_{jk}^{(\ell)}(\cos\theta), \tag{FS6}$$

where the Fourier coefficients are given by

$$\alpha_k^{(\ell)} = \frac{(-1)^{j-k}}{8\pi}\int_{-2\pi}^{2\pi}\int_0^{\pi} f(u(0,\theta,\psi))e^{ik\psi}P_{jk}^{\ell}(\cos\theta)\sin\theta\, d\theta\, d\psi. \tag{FC6}$$

In particular, for functions $f \in \mathrm{L}^2(\mathbf{SU}(2))$, such that

$$f(Hq) = f(q), \quad \text{for all } q \in \mathbf{SU}(2) \text{ and all } H \in \Omega_x,$$

namely, functions which do not depend on the Euler angle φ, we have

$$f(u(0,\theta,\psi)) = \sum_{\ell=0}^{\infty}(2\ell+1) \sum_{k=-\ell}^{\ell} \beta_\ell^k \, e^{-ik\psi} P_\ell^k(\cos\theta), \tag{FS7}$$

with

$$\beta_\ell^k = \frac{1}{8\pi}\frac{(\ell-k)!}{(\ell+k)!}\int_{-2\pi}^{2\pi}\int_0^{\pi} f(u(0,\theta,\psi))e^{ik\psi} P_\ell^k(\cos\theta)\sin\theta \, d\theta \, d\psi. \tag{FC7}$$

Fourier expansion formulae for the functions in ${}_j\mathfrak{L}^2 \cap \mathfrak{L}_k^2$ can also be obtained, as well as formulae for functions $f \in L^2(\mathbf{SU}(2))$, such that

$$f(H_1 q H_2) = f(q), \quad \text{for all } q \in \mathbf{SU}(2) \text{ and all } H_1, H_2 \in \Omega_x,$$

namely functions that do not depend on the Euler angles φ and ψ, but we leave these as exercises (see Vilenkin [Vilenkin (1968)], Chapter III, Section 6.4).

The expansion in Fourier series of the function in $L^2(\mathbf{SU}(2))$ that are independent of ψ yields a Fourier series expansion of functions in $L^2(S^2)$.

This is because, as explained in Section 5.12, $\mathbf{SU}(2)/\Omega_x$ (see Definition 5.16 for the definition of Ω_x) is a homogeneous space homeomorphic to S^2 and the functions $f \in L^2(\mathbf{SU}(2))$ such that $f(qH) = f(q)$ for all $q \in \mathbf{SU}(2)$ and all $H \in \Omega_x$ correspond bijectively to the functions in $L^2(S^2)$.

We can use (FS5) and (FC5) to obtain the following Fourier series expansion for every function $f \in L^2(S^2)$ in terms of the associated Legendre functions,

$$f(\varphi,\theta) = \sum_{\ell=0}^{\infty}(2\ell+1) \sum_{j=-\ell}^{\ell} \beta_\ell^j \, e^{-ij\varphi} P_\ell^j(\cos\theta), \tag{FS8}$$

with

$$\beta_\ell^j = \frac{1}{4\pi}\frac{(\ell-j)!}{(\ell+j)!}\int_0^{2\pi}\int_0^{\pi} f(\varphi,\theta)e^{ij\varphi} P_\ell^j(\cos\theta)\sin\theta \, d\theta \, d\varphi. \tag{FC8}$$

We also have the Parseval identity

$$\int_0^{2\pi}\int_0^{\pi} |f(\varphi,\theta)|^2 d\nu = \sum_{\ell=0}^{\infty}(2\ell+1) \sum_{j=-\ell}^{\ell} \frac{(\ell+j)!}{(\ell-j)!}|\beta_\ell^j|^2, \tag{PS2}$$

where $d\nu = (1/4)\sin\theta \, d\theta \, d\varphi$ is the normalized measure on S^2 in spherical coordinates; among other sources, see Gallier and Quaintance [Gallier and Quaintance (2020b)] (Section 6.4).

Recall from Definition 5.17 and ($*_{48}$) that

$$Y_{\ell j}(\varphi,\theta) = t^{(\ell)}_{j0}(q) = i^{-j}\sqrt{\frac{(\ell-j)!}{(\ell+j)!}}\, e^{-ij\varphi}\, P^j_\ell(\cos\theta), \quad -\ell \le j \le \ell,$$

with $\ell \in \mathbb{N}$, so we have

$$\sqrt{\frac{(2\ell+1)(\ell-j)!}{(\ell+j)!}} e^{-ij\varphi} P^j_\ell(\cos\theta) = i^j\sqrt{2\ell+1}\, Y_{\ell j}(\varphi,\theta),$$

for $\ell \in \mathbb{N}$ and $-\ell \le j \le \ell$, and in view of (FS8) and (FS2), *the above functions form a Hilbert basis for the functions in* $\mathrm{L}^2(S^2)$. As we explained just after Proposition 5.31, the functions $i^j\sqrt{2\ell+1}\; Y_{\ell j}(\varphi,\theta)$ are (a version of) the Laplace spherical harmonics $Y^j_\ell(\theta,\varphi)$, namely

$$Y^j_\ell(\theta,\varphi) = \sqrt{\frac{(2\ell+1)(\ell-j)!}{(\ell+j)!}} e^{-ij\varphi} P^j_\ell(\cos\theta).$$

Remark: Some authors include $1/\sqrt{4\pi}$ in the leading constant.

The associated Legendre functions can be computed starting with the Legendre polynomials using some recurrence equations; see Gallier and Quaintance [Gallier and Quaintance (2020b)] (Section 7.3).

5.16 Decomposition of Fields on the Sphere S^2

In various applications it is necessary to decompose not only scalar-valued but also vector-valued functions on the sphere S^2 into Fourier series that behave nicely under rotations of the sphere. Vilenkin suggests a way to do this that we now discuss (see Vilenkin [Vilenkin (1968)], Chapter III, Section 6.6). To simplify notation, we will write $\mathcal{P}_\ell$ instead of $\mathcal{P}^{\mathbb{C}}_\ell$.

Let $T_\ell \colon \mathbf{SU}(2) \to \mathbf{U}(\mathcal{P}_\ell)$ be the irreducible representation of $\mathbf{SU}(2)$ associated with $\ell \in R = \{0, 1/2, 1, 3/2, 2, 5/2, 3, \ldots\}$.

Definition 5.33. Let $\mathfrak{F}^S_\ell$ be the Hilbert space of functions $f\colon S^2 \to \mathcal{P}_\ell$ defined by the isomorphism

$$\mathfrak{F}^S_\ell \simeq \bigoplus_{j=-\ell}^{\ell} \mathrm{L}^2(S^2)\psi_j,$$

where the ψ_j constitute an orthonormal basis of $\mathcal{P}_\ell$ for an $\mathbf{SU}(2)$-invariant hermitian inner product defined in Section 5.8 ($\mathcal{P}_\ell$ is a complex vector space of dimension $2\ell + 1$). More precisely, the inner product of two functions $f, g \in \mathfrak{F}_\ell^S$ is given by

$$\langle f, g \rangle = \int_{S^2} \langle f(\xi), g(\xi) \rangle \, d\sigma(\xi),$$

where $\langle -, - \rangle$ is the $\mathbf{SU}(2)$-invariant hermitian inner product on $\mathcal{P}_\ell$ defined earlier and σ is the normalized $\mathbf{SO}(3)$-invariant measure on S^2.

Remark: In computer vision, the vectors in the vector space $\mathcal{P}_\ell$ are called *steerable vectors*. The group $\mathbf{SU}(2)$ acts on $\mathcal{P}_\ell$ *via* the action

$$q \cdot v = T_\ell(q)(v), \quad v \in \mathcal{P}_\ell, \; q \in \mathbf{SU}(2).$$

As in the previous section, $\mathbf{SU}(2)$ acts on S^2 by rotations, but for simplicity of notation, we write qX instead of $q \cdot X = qXq^*$, where $q \in \mathbf{SU}(2)$, X is the skew-hermitian matrix corresponding to the point (x, y, z) on the sphere S^2, and we write $X \in S^2$. We define the following representation.

Definition 5.34. For every $\ell \in R$, let $V_\ell \colon \mathbf{SU}(2) \to \mathbf{GL}(\mathfrak{F}_\ell^S)$ be the representation given by

$$[V_\ell(q)(f)](X) = [T_\ell(q)](f(q^{-1}X)), \qquad q \in \mathbf{SU}(2), \; (f \colon S^2 \to \mathcal{P}_\ell) \in \mathfrak{F}_\ell^S, \; X \in S^2.$$

Vilenkin calls the functions in $\mathfrak{F}_\ell^S$ *fields of quantities on the sphere transforming according to the irreducible representation T_ℓ*. For example, for $\ell = 1$, since $2\ell + 1 = 3$, we get a vector field on the sphere.

Observe that for any two functions $(f \colon S^2 \to \mathcal{P}_\ell), (g \colon S^2 \to \mathcal{P}_\ell) \in \mathfrak{F}_\ell^S$, since T_ℓ is unitary and σ is rotation invariant, we have

$$\begin{aligned}
\langle V_\ell(q)(f), V_\ell(q)(g) \rangle &= \int_{S^2} \langle [V_\ell(q)(f)](X), [V_\ell(q)(g)](X) \rangle \, d\sigma(X) \\
&= \int_{S^2} \langle [T_\ell(q)](f(q^{-1}X), [T_\ell(q)](g(q^{-1}X)) \rangle \, d\sigma(X) \\
&= \int_{S^2} \langle f(q^{-1}X), g(q^{-1}X) \rangle \, d\sigma(X) \\
&= \int_{S^2} \langle f(X), g(X) \rangle \, d\sigma(X) = \langle f, g \rangle.
\end{aligned}$$

Therefore the representation V_ℓ is unitary, that is, we have $V_\ell \colon \mathbf{SU}(2) \to \mathbf{U}(\mathfrak{F}_\ell^S)$.

For technical reasons, we need to convert the functions in $\mathfrak{F}_\ell^S$, which are functions on the sphere, to functions on $\mathbf{SU}(2)$. Let X_0 be the skew-hermitian matrix

$$X_0 = \begin{pmatrix} i & 0 \\ 0 & -i \end{pmatrix}$$

corresponding to $e_1 = (1, 0, 0) \in S^2$.

Definition 5.35. For every function $(f\colon S^2 \to \mathcal{P}_\ell) \in \mathfrak{F}_\ell^S$, let $\widehat{f}\colon \mathbf{SU}(2) \to \mathcal{P}_\ell$ be the function defined by

$$\widehat{f}(q) = [V_\ell(q^{-1})(f)](X_0) = [T_\ell(q^{-1})](f(qX_0)), \quad q \in \mathbf{SU}(2).$$

The functions $\widehat{f}\colon \mathbf{SU}(2) \to \mathcal{P}_\ell$ belong to the Hilbert space defined below.

Definition 5.36. Let $\mathfrak{F}_\ell^{\mathbf{SU}}$ be the Hilbert space of functions $f\colon \mathbf{SU}(2) \to \mathcal{P}_\ell$ defined by the isomorphism

$$\mathfrak{F}_\ell^{\mathbf{SU}} \simeq \bigoplus_{j=-\ell}^{\ell} \mathrm{L}^2(\mathbf{SU}(2))\psi_j.$$

More precisely, the inner product of two functions $f, g \in \mathfrak{F}_\ell^{\mathbf{SU}}$ is given by

$$\langle f, g \rangle = \int_{\mathbf{SU}(2)} \langle f(q), g(q) \rangle \, d\nu(q),$$

where $\langle -, - \rangle$ is the $\mathbf{SU}(2)$-invariant hermitian inner product on $\mathcal{P}_\ell$ defined earlier and ν is the normalized Haar measure on $\mathbf{SU}(2)$.

Proposition 5.43. *The map* $f \mapsto \widehat{f}$ *is an injection from* $\mathfrak{F}_\ell^S$ *to* $\mathfrak{F}_\ell^{\mathbf{SU}}$.

Proof. Indeed, if $\widehat{f}(q) = \widehat{g}(q)$ for all $q \in \mathbf{SU}(2)$, then $[T_\ell(q^{-1})](f(qX_0)) = [T_\ell(q^{-1})](g(qX_0))$ for all $q \in \mathbf{SU}(2)$, and since $T_\ell(q^{-1})$ is a bijection, $f(qX_0) = g(qX_0)$ for all $q \in \mathbf{SU}(2)$, and since the action of $\mathbf{SU}(2)$ on S^2 is transitive, we must have $f = g$. $\square$

Definition 5.37. The image of $\mathfrak{F}_\ell^S$ in $\mathfrak{F}_\ell^{\mathbf{SU}}$ by the map $\widehat{\ }$ is denoted by $\widehat{\mathfrak{F}_\ell^S}$.

Observe that f can be recovered from $\widehat{f}$ as follows:

$$f(qX_0) = [T_\ell(q)](\widehat{f}(q)), \quad q \in \mathbf{SU}(2).$$

Since Ω_x is the stabilizer of X_0, for every $h \in \Omega_x$ we have $hX_0 = X_0$, and so

$$\begin{aligned} \widehat{f}(qh) &= [T_\ell((qh)^{-1})](f(qhX_0)) \\ &= T_\ell(h^{-1})([T_\ell(q^{-1})](f(qX_0))) \\ &= [T_\ell(h)^{-1}](\widehat{f}(q)), \end{aligned}$$

which we record as the equation

$$\widehat{f}(qh) = [T_\ell(h)^{-1}](\widehat{f}(q)), \quad (f\colon S^2 \to \mathcal{P}_\ell) \in \mathfrak{F}_\ell^S,\ h \in \Omega_x. \tag{fhat}$$

Let us figure out what is the function in $\widehat{\mathfrak{F}_\ell^S} \subseteq \mathfrak{F}_\ell^{\mathbf{SU}}$ corresponding to the function $[V_\ell(q_0)](f) \in \mathfrak{F}_\ell^S$, with $(f\colon S^2 \to \mathcal{P}_\ell) \in \mathfrak{F}_\ell^S$. We have

$$\begin{aligned}([V_\ell(q_0)](f))\widehat{\ }(q) &= [V_\ell(q^{-1})(V_\ell(q_0)(f))](X_0)\\ &= [V_\ell(q^{-1}q_0)(f)](X_0) = \widehat{f}(q_0^{-1}q),\end{aligned}$$

for all $q_0, q \in \mathbf{SU}(2)$. This suggests the following definition.

Definition 5.38. The representation $\widehat{V}_\ell\colon \mathbf{SU}(2) \to \mathbf{U}(\widehat{\mathfrak{F}_\ell^S})$ is given by

$$[\widehat{V}_\ell(q_0)(\widehat{f}\,)](q) = \widehat{f}(q_0^{-1}q) = [T_\ell(q^{-1}q_0)](f(q_0^{-1}qX_0)),$$

for all $(f\colon S^2 \to \mathcal{P}_\ell) \in \mathfrak{F}_\ell^S$ and all $q_0, q \in \mathbf{SU}(2)$.

The definition of $\widehat{V}_\ell$ implies that

$$\widehat{V}_\ell(q_0)(\widehat{f}\,) = ([V_\ell(q_0)](f))\widehat{\ },$$

so that the following diagram commutes,

$$\begin{array}{ccc} \mathfrak{F}_\ell^S & \xrightarrow{\ V_\ell(q_0)\ } & \mathfrak{F}_\ell^S \\ \Big\downarrow \widehat{\ } & & \Big\downarrow \widehat{\ } \\ \widehat{\mathfrak{F}_\ell^S} & \xrightarrow[\ \widehat{V}_\ell(q_0)\ ]{} & \widehat{\mathfrak{F}_\ell^S}, \end{array}$$

and since $\widehat{\ }$ is an isomorphism between $\mathfrak{F}_\ell^S$ and $\widehat{\mathfrak{F}_\ell^S}$, the representations $V_\ell\colon \mathbf{SU}(2) \to \mathbf{U}(\mathfrak{F}_\ell^S)$ and $\widehat{V}_\ell\colon \mathbf{SU}(2) \to \mathbf{U}(\widehat{\mathfrak{F}_\ell^S})$ are equivalent.

The trick is now to decompose the space $\widehat{\mathfrak{F}_\ell^S}$ into a direct sum of $2\ell + 1$ subspaces $(\widehat{\mathfrak{F}_\ell^S})_k$, which are related to the spaces $\mathfrak{L}^2_{-k}$ introduced in Section 5.15.

Definition 5.39. For k with $-\ell \leq k \leq \ell$, for every function $(f\colon S^2 \to \mathcal{P}_\ell) \in \mathfrak{F}_\ell^S$, define the function $\widehat{f}_k \in \widehat{\mathfrak{F}_\ell^S}$ as

$$\widehat{f}_k(q) = \frac{1}{2\pi}\int_0^{2\pi} \widehat{f}(qh(t))e^{-ikt}\, dt, \quad q \in \mathbf{SU}(2),$$

with

$$h(t) = \begin{pmatrix} e^{\frac{it}{2}} & 0 \\ 0 & e^{-\frac{it}{2}} \end{pmatrix} \in \Omega_x.$$

Proposition 5.44. *For k with $-\ell \leq k \leq \ell$, for every function $(f\colon S^2 \to \mathcal{P}_\ell) \in \mathfrak{F}_\ell^S$, the following properties hold:*

(1)

$$\widehat{f}_k(qh(s)) = e^{iks}\widehat{f}_k(q), \quad q \in \mathbf{SU}(2),\, h(s) \in \Omega_x.$$

Consequently, $\widehat{f}_k \in \bigoplus_{j=-\ell}^{\ell} \mathfrak{L}^2_{-k}\psi_j$.

(2)

$$\widehat{f}(q) = \sum_{k=-\ell}^{\ell} \widehat{f}_k(q), \quad q \in \mathbf{SU}(2).$$

(3)

$$[\widehat{V}_\ell(q_0)(\widehat{f}\,)]_k(q) = \widehat{f}_k(q_0^{-1}q), \quad q_0, q \in \mathbf{SU}(2).$$

Proof. (1) We have

$$\begin{aligned}
\widehat{f}_k(qh(s)) &= \frac{1}{2\pi}\int_0^{2\pi} \widehat{f}(qh(s)h(t))e^{-ikt}\, dt \\
&= e^{iks}\frac{1}{2\pi}\int_0^{2\pi} \widehat{f}(qh(s+t))e^{-ik(s+t)}\, dt \\
&= e^{iks}\frac{1}{2\pi}\int_0^{2\pi} \widehat{f}(qh(t_1))e^{-ikt_1}\, dt_1 \\
&= e^{iks}\widehat{f}_k(q),
\end{aligned}$$

as claimed.

(2) Using (fhat), we have

$$\begin{aligned}
\sum_{k=-\ell}^{\ell} \widehat{f}_k(q) &= \frac{1}{2\pi}\sum_{k=-\ell}^{\ell}\int_0^{2\pi} \widehat{f}(qh(t))e^{-ikt}\, dt \\
&= \frac{1}{2\pi}\int_0^{2\pi}\left(\sum_{k=-\ell}^{\ell} e^{-ikt}T_\ell(h(t)^{-1})\right)(\widehat{f}(q))\, dt.
\end{aligned}$$

Here we need to recall from Proposition 5.15 that in the basis $(\psi_m)_{-\ell\leq m\leq\ell}$, since $h(t) = r_x(t/2)$, the matrix of $T_\ell(h(t)^{-1})$ is the diagonal matrix

$$\begin{pmatrix}
e^{-i\ell t} & & & & \\
 & e^{-i(\ell-1)t} & & & \\
 & & \ddots & & \\
 & & & e^{i(\ell-1)t} & \\
 & & & & e^{i\ell t}
\end{pmatrix}.$$

Consequently, the entries of $e^{-ikt}T_\ell(h(t)^{-1})$ are of the form $e^{-i(\ell-j+k)t}$ with $j = 0, 1, \ldots, 2\ell$ and $\ell + k - j$ an integer, but

$$\frac{1}{2\pi}\int_0^{2\pi} e^{-i(\ell+k-j)t}\,dt = \delta_{\ell+k,j}$$

so the only entry that survives corresponds to $j = \ell + k$ and its contribution is 1, so in fact

$$\sum_{k=-\ell}^{\ell} e^{-ikt}T_\ell(h(t)^{-1}) = I_{2\ell+1},$$

and thus

$$\frac{1}{2\pi}\int_0^{2\pi}\left(\sum_{k=-\ell}^{\ell} e^{-ikt}T_\ell(h(t)^{-1})\right)(\widehat{f}(q))\,dt = \widehat{f}(q),$$

as claimed.

(3) We have

$$\begin{aligned}
[\widehat{V}(q_0)(\widehat{f}\,)]_k(q) &= \frac{1}{2\pi}\int_0^{2\pi} [\widehat{V}(q_0)(\widehat{f}\,)](qh(t))e^{-ikt}\,dt \\
&= \frac{1}{2\pi}\int_0^{2\pi} \widehat{f}(q_0^{-1}qh(t))e^{-ikt}\,dt \\
&= \widehat{f}_k(q_0^{-1}q),
\end{aligned}$$

which concludes the proof of the proposition. ☐

As a corollary, we have the following result.

Proposition 5.45. *Denote by $(\widehat{\mathfrak{F}_\ell^S})_k$ the image of $\widehat{\mathfrak{F}_\ell^S}$ by the linear map $\widehat{f} \mapsto \widehat{f}_k$, with $(f\colon S^2 \to \mathcal{P}_\ell) \in \mathfrak{F}_\ell^S$.*

(1) We have a direct sum

$$\widehat{\mathfrak{F}_\ell^S} = \bigoplus_{k=-\ell}^{\ell} (\widehat{\mathfrak{F}_\ell^S})_k,$$

where every function $\widehat{f}_k \in (\widehat{\mathfrak{F}_\ell^S})_k$ satisfies the equation

$$\widehat{f}_k(qh(s)) = e^{iks}\widehat{f}_k(q), \quad q \in \mathbf{SU}(2),\ h(s) \in \Omega_x.$$

(2) The map $(\widehat{V}_\ell)_k$ defined by

$$[\widehat{V}_\ell(q_0)(\widehat{f}\,)]_k(q) = \widehat{f}_k(q_0^{-1}q), \quad q_0, q \in \mathbf{SU}(2)$$

is a representation $(\widehat{V}_\ell)_k\colon \mathbf{SU}(2) \to \mathbf{U}((\widehat{\mathfrak{F}_\ell^S})_k)$, and we have

$$\widehat{V}_\ell(q_0) = \bigoplus_{k=-\ell}^{\ell} (\widehat{V}_\ell)_k(q_0).$$

Proof. Since by (2), $\widehat{\mathfrak{F}_\ell^S}$ is the sum of the subspaces $(\widehat{\mathfrak{F}_\ell^S})_k$, and by (1), if there was a nonzero function such that $\widehat{f}_{k_1} = \widehat{f}_{k_2}$ for some $k_1 \neq k_2$, then we would have

$$\widehat{f}_{k_1}(qh(s)) = e^{ik_1 s}\widehat{f}_{k_1}(q) = \widehat{f}_{k_2}(qh(s)) = e^{ik_2 s}\widehat{f}_{k_2}(q),$$

and so we would have $e^{ik_1 s} = e^{ik_2 s}$ for all s, which implies $k_1 = k_2$, a contradiction.

The fact that

$$\widehat{V}_\ell(q_0) = \bigoplus_{k=-\ell}^{\ell} (\widehat{V}_\ell)_k(q_0),$$

follows from parts (2) and (3). □

For every function $\sum_{j=-\ell}^{\ell} \widehat{(f_j)}\psi_j$ in $(\widehat{\mathfrak{F}_\ell^S})_k$, we have $\widehat{(f_j)} \in \mathfrak{L}^2_{-k}$ (with $f_j \in \mathrm{L}^2(S^2)$), so as shown in Section 5.15, the function $\widehat{(f_j)}$ can be expanded in Fourier series according to Formulae (FS3) and (FC3) (with k changed to $-k$).

We leave it as an exercise to the reader to use the isomorphism $\widehat{\ }\colon \mathfrak{F}_\ell^S \to \widehat{\mathfrak{F}_\ell^S}$ to define a direct sum decomposition of $\mathfrak{F}_\ell^S$ of the form

$$\mathfrak{F}_\ell^S = \bigoplus_{k=-\ell}^{\ell} (\mathfrak{F}_\ell^S)_k$$

and to translate the results obtained for functions in $\widehat{\mathfrak{F}_\ell^S}$ and the representations $\widehat{V}_\ell$ to the functions in $\mathfrak{F}_\ell^S$ and to the representations V_ℓ.

5.17 The Clebsch–Gordan Coefficients

The Clebsch–Gordan coefficients have to do with tensor products of the irreducible representations T_ℓ of **SU**(2) (see Definition 4.11 for the definition of the tensor product of representations). In general, the tensor product $T_{\ell_1} \otimes T_{\ell_2}$ of two irreducible representations T_{ℓ_1} and T_{ℓ_2} of **SU**(2) is not irreducible, so according to the Peter–Weyl theorem (Theorem 4.16) it splits as a direct sum of irreducible representations. Since the character associated with the representation $T_{\ell_1} \otimes T_{\ell_2}$ is equal to the product $\chi_{T_{\ell_1}}\chi_{T_{\ell_2}}$ of the characters $\chi_{T_{\ell_1}}$ and $\chi_{T_{\ell_2}}$ associated with T_{ℓ_1} and T_{ℓ_2}, by Proposition 4.18, this splitting as a direct sum decomposition translates into a decomposition

$$\chi_{T_{\ell_1}}\chi_{T_{\ell_2}} = \sum_\ell c^\ell_{\ell_1,\ell_2}\chi_{T_\ell},$$

where $c^{\ell}_{\ell_1,\ell_2}$ is the number of times that the irreducible representation T_ℓ occurs in the representation $T_{\ell_1} \otimes T_{\ell_2}$ (see Section 4.3).

The natural numbers $c^{\ell}_{\ell_1,\ell_2}$ can be determined from the expression of the characters that was obtained in Section 5.1. However this expression was obtained for the representations U_m in the space $\mathcal{P}^{\mathbb{C}}_m(2)$ of homogeneous polynomials of degree m in two variables, so we work out the expression of the characters in terms of the representations T_ℓ in the spaces $\mathcal{P}^{\mathbb{C}}_\ell$ of polynomials of degree 2ℓ in one variable (in particular, ℓ is now an integer or a half integer). To simplify notation, we will write $\mathcal{P}_\ell$ instead of $\mathcal{P}^{\mathbb{C}}_\ell$.

Recall that we showed in the proof of Proposition 5.1 that every unitary matrix $q \in \mathbf{SU}(2)$ is diagonalizable as

$$q = Rr_x(t/2)R^*$$

for some unitary matrix $R \in \mathbf{SU}(2)$, where

$$r_x(t/2) = \begin{pmatrix} e^{\frac{it}{2}} & 0 \\ 0 & e^{-\frac{it}{2}} \end{pmatrix}$$

is uniquely determined if $0 \leq t \leq 2\pi$. If the matrix q is given by

$$q = \begin{pmatrix} \alpha & \beta \\ -\overline{\beta} & \overline{\alpha} \end{pmatrix}, \quad |\alpha|^2 + |\beta|^2 = 1,$$

then its eigenvalues are the zeros of the equation

$$\begin{vmatrix} \lambda - \alpha & -\beta \\ \overline{\beta} & \lambda - \overline{\alpha} \end{vmatrix} = 0,$$

that is,

$$\lambda^2 - 2\Re(\alpha)\lambda + 1 = 0,$$

(since $\alpha + \overline{\alpha} = 2\Re(\alpha)$), whose zeros are

$$\lambda = \Re(\alpha) \pm i\sqrt{1 - (\Re(\alpha))^2}.$$

Since we assumed that the eigenvalues of q are $e^{\pm \frac{it}{2}}$, we have

$$\Re(\alpha) = \cos\frac{t}{2}.$$

If q is expressed in terms of the Euler angles as $q = u(\varphi, \theta, \psi)$, then from the formulae just before Proposition 5.4, we have

$$\alpha = \cos\frac{\theta}{2}\, e^{\frac{i(\varphi+\psi)}{2}},$$

and so

$$\Re(\alpha) = \cos\frac{t}{2} = \cos\frac{\theta}{2}\,\cos\frac{(\varphi+\psi)}{2}.$$

Since the characters are central functions, that is, constant on conjugacy classes, we have

$$\chi_{T_\ell}(q) = \chi_{T_\ell}(r_x(t/2)) = \mathrm{tr}(T_\ell(r_x(t/2))).$$

Since we showed in Proposition 5.15 that in the basis $(z^{\ell-k})_{-\ell\leq k\leq \ell}$, the matrix of $T_\ell(r_x(t/2))$ is the diagonal matrix

$$\begin{pmatrix} e^{i\ell t} & & & & \\ & e^{i(\ell-1)t} & & & \\ & & \ddots & & \\ & & & e^{-i(\ell-1)t} & \\ & & & & e^{-i\ell t} \end{pmatrix},$$

we obtain

$$\chi_{T_\ell}(q) = \chi_{T_\ell}(r_x(t/2)) = \sum_{k=-\ell}^{\ell} e^{-ikt} = \frac{\epsilon^{\ell+1}-\epsilon^{-\ell}}{\epsilon-1},$$

with $\epsilon = e^{-it}$, and we showed in the proof of Proposition 5.1 that we obtain the following expression (in that formula we make $m = 2\ell$ and $\varphi = t/2$):

$$\chi_{T_\ell}(q) = \frac{\epsilon^{\ell+1}-\epsilon^{-\ell}}{\epsilon-1} = \frac{\sin\left(\ell+\frac{1}{2}\right)t}{\sin\frac{t}{2}},$$

with $\epsilon = e^{-it}$. Compare Vilenkin [Vilenkin (1968)], Chapter III, Section 7.1. Using the above formula, we obtain the following result.

Proposition 5.46. *For any two irreducible representations T_{ℓ_1} and T_{ℓ_2} of $\mathbf{SU}(2)$, we have*

$$\chi_{T_{\ell_1}}(q)\chi_{T_{\ell_2}}(q) = \sum_{\ell=|\ell_1-\ell_2|}^{\ell_1+\ell_2} \chi_{T_\ell}(q), \quad q \in \mathbf{SU}(2). \tag{CG1}$$

Proof. We follow Vilenkin [Vilenkin (1968)], Chapter III, Section 8.1. First assume that $\ell_1 \geq \ell_2$. With $\epsilon = e^{-it}$ as above, we have

$$\begin{aligned}\chi_{T_{\ell_1}}(q)\chi_{T_{\ell_2}}(q) &= \sum_{k=-\ell_2}^{\ell_2} \epsilon^k \frac{(\epsilon^{\ell_1+1}-\epsilon^{-\ell_1})}{\epsilon-1} \\ &= \sum_{k=-\ell_2}^{\ell_2} \frac{\epsilon^{\ell_1+k+1}-\epsilon^{k-\ell_1}}{\epsilon-1} \\ &= \frac{1}{\epsilon-1}(\epsilon^{\ell_1+\ell_2+1}+\cdots+\epsilon^{\ell_1-\ell_2+1}-\epsilon^{\ell_2-\ell_1}-\cdots-\epsilon^{-\ell_1-\ell_2}) \\ &= \frac{1}{\epsilon-1}(\epsilon^{\ell_1+\ell_2+1}-\epsilon^{\ell_2-\ell_1}+\cdots+\epsilon^{\ell_1-\ell_2+1}-\epsilon^{\ell_2-\ell_1}),\end{aligned}$$

where the last line is obtained by combining pairwise positive and negative terms, the sum of whose indices is equal to 1, we obtain

$$\chi_{T_{\ell_1}}(q)\chi_{T_{\ell_2}}(q) = \sum_{\ell=\ell_1-\ell_2}^{\ell_1+\ell_2} \frac{\epsilon^{\ell+1} - \epsilon^{-\ell}}{\epsilon - 1} = \sum_{\ell=\ell_1-\ell_2}^{\ell_1+\ell_2} \chi_{T_\ell}(q).$$

If $\ell_2 \geq \ell_1$, the proof is similar but the sum starts with $\ell_2 - \ell_1$. ☐

Remark: The above computation essentially appears in Wigner [Wigner (1959)], Chapter 17, pages 186–187.

The above proposition shows the somewhat unexpected fact that in the decomposition of the tensor product representation $T_{\ell_1} \otimes T_{\ell_2}$, those representations T_ℓ that occur correspond to values of ℓ such that $|\ell_1 - \ell_2| \leq \ell \leq \ell_1 + \ell_2$ where ℓ is an integer or a half integer as $\ell_1 + \ell_2$ is, and each such representation occurs exactly once. Thus

$$T_{\ell_1} \otimes T_{\ell_2} = \bigoplus_{\ell=|\ell_1-\ell_2|}^{\ell_1+\ell_2} T_\ell. \tag{CG2}$$

We also have an isomorphism

$$\mathcal{P}_{\ell_1} \otimes \mathcal{P}_{\ell_2} \simeq \bigoplus_{\ell=|\ell_1-\ell_2|}^{\ell_1+\ell_2} \mathcal{P}_\ell. \tag{CG3}$$

The space $\mathcal{P}_{\ell_1} \otimes \mathcal{P}_{\ell_2}$ has dimension $(2\ell_1 + 1)(2\ell_2 + 1)$ and each summand $\mathcal{P}_\ell$ has dimension $2\ell + 1$. The reader should check that

$$(2\ell_1 + 1)(2\ell_2 + 1) = \sum_{\ell=|\ell_1-\ell_2|}^{\ell_1+\ell_2} (2\ell + 1).$$

Recall from Proposition 5.16 that each vector space $\mathcal{P}_\ell$ has an orthonormal basis (ψ_k) $(-\ell \leq k \leq \ell)$ invariant under the action of $\mathbf{SU}(2)$. Following Vilenkin [Vilenkin (1968)] (Chapter III, Section 8.2), we denote the basis of $\mathcal{P}_{\ell_1}$ as $(\mathbf{f}_j)$ $(-\ell_1 \leq j \leq \ell_1)$ and the basis of $\mathcal{P}_{\ell_2}$ as $(\mathbf{h}_k)$ $(-\ell_2 \leq k \leq \ell_2)$. Then the family of tensor products

$$\mathbf{f}_j \otimes \mathbf{h}_k, \quad -\ell_1 \leq j \leq \ell_1, \ -\ell_2 \leq k \leq \ell_2$$

is a basis of $\mathcal{P}_{\ell_1} \otimes \mathcal{P}_{\ell_2}$. If we give $\mathcal{P}_{\ell_1} \otimes \mathcal{P}_{\ell_2}$ the inner product defined in Definition 4.10 induced by the inner products associated with the bases $(\mathbf{f}_j)$ and $(\mathbf{h}_k)$, then the vectors $(\mathbf{f}_j \otimes \mathbf{h}_k)$ form an orthonormal basis of $\mathcal{P}_{\ell_1} \otimes \mathcal{P}_{\ell_2}$.

Since we have the direct sum

$$\mathcal{P}_{\ell_1} \otimes \mathcal{P}_{\ell_2} \simeq \bigoplus_{\ell=|\ell_1-\ell_2|}^{\ell_1+\ell_2} \mathcal{P}_\ell,$$

we also have a basis of $\mathcal{P}_{\ell_1} \otimes \mathcal{P}_{\ell_2}$ consisting of the union of the bases associated with each of the summand $\mathcal{P}_l$, which Vilenkin denotes by

$$\mathbf{a}_m^\ell, \quad |\ell_1 - \ell_2| \leq \ell \leq \ell_1 + \ell_2, \ -\ell \leq m \leq \ell,$$

where for ℓ fixed, $(\mathbf{a}_m^\ell)$ $(-\ell \leq m \leq \ell)$ is the basis of $\mathcal{P}_\ell$. Since both bases are orthonormal bases of $\mathcal{P}_{\ell_1} \otimes \mathcal{P}_{\ell_2}$, there is a unitary matrix C expressing the basis $(\mathbf{f}_j \otimes \mathbf{h}_k)$ in terms of the basis $(\mathbf{a}_m^\ell)$, and the entries of the matrix C are called the *Clebsch–Gordan coefficients*.

Amazingly, these coefficients can be computed explicitly, but the formulae are very complicated and the technical details of the computations are quite involved. Complete details can be found in Vilenkin [Vilenkin (1968)] (Chapter III, Section 8). We will content ourselves by providing an outline of these computations.

The first observation is that the matrix of $(T_{\ell_1} \otimes T_{\ell_2})(q) = T_{\ell_1}(q) \otimes T_{\ell_2}(q)$ with respect to the basis $(\mathbf{f}_j \otimes \mathbf{h}_k)$ is the Kronecker product of the matrices $t^{(\ell_1)}(q)$ and $t^{(\ell_2)}(q)$. Following Vilenkin we denote this matrix as $\alpha(q) = (\alpha_{(jk),(j'k')}(q))$, and we have

$$\alpha_{(jk),(j'k')}(q) = t_{jj'}^{(\ell_1)}(q) t_{kk'}^{(\ell_2)}(q), \tag{CG4}$$

with $-\ell_1 \leq j,\, j' \leq \ell_1$, $-\ell_2 \leq k,\, k' \leq \ell_2$.

On the other hand, in the basis $(\mathbf{a}_m^\ell)$, the matrix representing the map $T_{\ell_1}(q) \otimes T_{\ell_2}(q)$ is a block-diagonal matrix whose blocks are the matrices $t^{(\ell)}(q)$. Again, following Vilenkin, we denote this matrix as $\beta(q) = (\beta_{(\ell m),(\ell' m')}(q))$, with $|\ell_1 - \ell_2| \leq \ell, \ell' \leq \ell_1 + \ell_2$, $-\ell \leq m \leq \ell$ and $-\ell' \leq m' \leq \ell'$. Since this matrix is block-diagonal we must have

$$\beta_{(\ell m),(\ell' m')}(q) = 0 \quad \text{if}\, \ell \neq \ell',$$

and if $\ell = \ell'$, then $\beta_{(\ell m),(\ell m')}(q) = t_{mm'}^{(\ell)}(q)$, so we have

$$\beta_{(\ell m),(\ell' m')}(q) = \delta_{\ell\ell'} t_{mm'}^{(\ell)}(q), \tag{CG5}$$

with $-\ell \leq m \leq \ell$, $-\ell' \leq m' \leq \ell'$.

The change of basis matrix $C = (C_{(\ell m),(jk)})$ is the unitary matrix defined such that the (jk)th column of C consists of the coefficients of $\mathbf{f}_j \otimes \mathbf{h}_k$ over the basis $(\mathbf{a}_m^\ell)$, namely

$$\mathbf{f}_j \otimes \mathbf{h}_k = \sum_{\ell=|\ell_1-\ell_2|}^{\ell_1+\ell_2} \sum_{m=-\ell}^{\ell} C_{(\ell m),(jk)} \mathbf{a}_m^\ell, \tag{CG6}$$

with $-\ell_1 \leq j \leq \ell_1$, $-\ell_2 \leq k \leq \ell_2$. Since $\beta(q)$ is the matrix of $T_{\ell_1}(q)\otimes T_{\ell_2}(q)$ in the "old" basis $(\mathbf{a}_m^\ell)$ and $\alpha(q)$ is the matrix of $T_{\ell_1}(q)\otimes T_{\ell_2}(q)$ in the "new" basis $(\mathbf{f}_j \otimes \mathbf{h}_k)$, we have

$$\alpha(q) = C^* \beta(q) C. \tag{CG7}$$

It turns out that it is often desirable to indicate explicitly the dependence of C on the indices ℓ_1 and ℓ_2, so we also write $C(\ell_1, \ell_2, \ell; j, k, m)$ instead of $C_{(\ell m),(jk)}$. To be more concise, we introduce the following notation.

Definition 5.40. The coefficients $C(\ell_1, \ell_2, \ell; j, k, m)$ are also written as $C(\mathbf{l}, \mathbf{j})$, with $\mathbf{l} = (\ell_1, \ell_2, \ell)$ and $\mathbf{j} = (j, k, m)$ and are called the *Clebsch–Gordan coefficients*.

Remark: In Wigner [Wigner (1959)] (Chapter 17, Section Vector Addition Model), the matrix C is introduced but it is denoted S; see Formula (17.16), which is the analog of equation (CG7). The coefficients $C_{(\ell m),(jk)}$ are called *vector coupling coefficients* instead of Clebsch–Gordan coefficients, although in the book index, there is an entry for the Clebsch–Gordan coefficients.

In terms of matrix elements, (CG7) yields

$$\alpha_{(jk),(j'k')}(q) = \sum_{\ell,\ell'=|\ell_1-\ell_2|}^{\ell_1+\ell_2} \sum_{m=-\ell}^{\ell} \sum_{m'=-\ell'}^{\ell'} \overline{C_{(\ell m),(jk)}} \beta_{(\ell m),(\ell' m')} C_{(\ell' m'),(j'k')}. \tag{CG8}$$

Using (CG4) and (CG5), we obtain

$$t_{jj'}^{(\ell_1)}(q) t_{kk'}^{(\ell_2)}(q) = \sum_{\ell=|\ell_1-\ell_2|}^{\ell_1+\ell_2} \sum_{m,m'=-\ell}^{\ell} C(\mathbf{l}, \mathbf{j}') \overline{C(\mathbf{l}, \mathbf{j})} t_{mm'}^{(\ell)}(q), \tag{CG9}$$

with $\mathbf{l} = (\ell_1, \ell_2, \ell)$, $\mathbf{j} = (j, k, m)$, $\mathbf{j}' = (j', k', m')$.

Equation (CG9) is the key to the computation of the coefficients $C(\mathbf{l}, \mathbf{j})$. By a clever use of the fact that the functions $\sqrt{2\ell+1}\, t_{mm'}^{(\ell)}(q)$ $(-\ell \leq m, m' \leq \ell)$ form a Hilbert basis of $\mathrm{L}^2(\mathbf{SU}(2))$ (see the beginning of Section 5.15) and the expression of $t_{mm'}^{(\ell)}(q)$ in terms of the Euler angles given by Proposition 5.28 as

$$t_{mm'}^{(\ell)}(q) = e^{-i(m\varphi + m'\psi)}\, P_{mm'}^{\ell}(\cos\theta) \tag{$*$}$$

(with $q = u(\varphi, \theta, \psi)$), it is possible to find (more or less explicit) formulae for $C(\mathbf{l}, \mathbf{j})$.

The first step is to multiply both sides of (CG9) by $\overline{t^{(\ell)}_{mm'}(q)}$ and integrate over $\mathbf{SU}(2)$. Since the $\sqrt{2\ell+1}\, t^{(\ell)}_{mm'}(q)$ $(-\ell \leq m, m' \leq \ell)$ form a Hilbert basis, we obtain

$$C(\mathbf{l}, \mathbf{j}')\overline{C(\mathbf{l}, \mathbf{j})} = (2\ell+1)\int_{\mathbf{SU}(2)} t^{(\ell_1)}_{jj'}(q)t^{(\ell_2)}_{kk'}(q)\overline{t^{(\ell)}_{mm'}(q)}\, d\mu(q). \qquad \text{(CG10)}$$

Using $(*)$ and the volume form

$$\frac{1}{16\pi^2}\sin\theta\, d\theta\, d\varphi\, d\psi$$

(see Proposition 5.37), we find that *the integral in (CG10) is nonzero if and only if* $j+k=m$ *and* $j'+k'=m'$. Therefore we only need to compute the Clebsch–Gordan coefficients $C(\ell_1, \ell_2, \ell; j, k, j+k)$. Among those, it turns out that the coefficients $C(\ell_1, \ell_2, \ell; \ell_1, -\ell_2, \ell_1-\ell_2)$ play a special role. They can be computed before the arbitrary coefficients $C(\ell_1, \ell_2, \ell; j, k, j+k)$ and it can be arranged that $C(\ell_1, \ell_2, \ell; \ell_1, -\ell_2, \ell_1-\ell_2) \geq 0$, which implies that all $C(\ell_1, \ell_2, \ell; j, k, j+k)$ are real, even though a priori they are complex numbers.

The points is that for each ℓ, we can multiply all the basis vectors $\mathbf{a}^{\ell}_m$ by a complex number of unit length and still obtain a Hilbert basis, and ensure that $C(\ell_1, \ell_2, \ell; \ell_1, -\ell_2, \ell_1-\ell_2) \geq 0$. From now on, we assume that this normalization has been made.

We now go back to (CG10) in which we set $m = j+k$ and $m' = j'+k'$ and use $(*)$ to integrate (making the substitution $x = \cos\theta$) to obtain

$$C(\mathbf{l}, \mathbf{j}')\overline{C(\mathbf{l}, \mathbf{j})} = \frac{(2\ell+1)}{2}\int_{-1}^{1} P^{\ell_1}_{jj'}(x)P^{\ell_2}_{kk'}(x)\overline{P^{\ell}_{j+k\, j'+k'}(x)}\, dx, \qquad \text{(CG11)}$$

with $\mathbf{l} = (\ell_1, \ell_2, \ell)$, $\mathbf{j} = (j, k, j+k)$, $\mathbf{j}' = (j', k', j'+k')$.

In order to compute $C(\ell_1, \ell_2, \ell; \ell_1, -\ell_2, \ell_1-\ell_2)$ we set $j' = \ell_1$ and $k' = -\ell_2$ in (CG11). Then we can use the special case of $(*_{40})$ in which $j = \ell$, the symmetry equations $P^{\ell}_{mn}(z) = P^{\ell}_{-m-n}(z)$ and $P^{\ell}_{mn}(z) = P^{\ell}_{nm}(z)$ (just after Proposition 5.29), and $(*_{43})$, and after a bit of work on (CG11) (see Vilenkin [Vilenkin (1968)], Chapter III, Section 8.3, page 179), we obtain the following formidable equation:

$$C(\ell_1,\ell_2,\ell;\ell_1,-\ell_2,\ell_1-\ell_2)\overline{C(\ell_1,\ell_2,\ell;j,k,j+k)}$$
$$= \left[\frac{(-1)^{-\ell+\ell_1+k}(2\ell+1)}{2^{\ell+\ell_1+\ell_2+1}}\sqrt{\frac{(2\ell_1)!(2\ell_2)!(\ell+j+k)!}{(\ell_1-j)!(\ell_1+j)!(\ell_2-k)!(\ell_2+k)!}}\right.$$
$$\sqrt{\frac{1}{(\ell-j-k)!(\ell+\ell_1-\ell_2)!(\ell-\ell_1+\ell_2)!}}$$
$$\left.\int_{-1}^{1}(1-x)^{\ell_1-j}(1+x)^{\ell_2-k}\frac{d^{\ell-j-k}}{dx^{\ell-j-k}}[(1-x)^{\ell-\ell_1+\ell_2}(1+x)^{\ell+\ell_1-\ell_2}]\,dx\right]. \tag{CG12}$$

To find $C(\ell_1,\ell_2,\ell;\ell_1,-\ell_2,\ell_1-\ell_2)$ we set $j=\ell_1$ and $k=-\ell_2$. Sparing the reader some details found in Vilenkin and using integration by parts $\ell-\ell_1+\ell_2$ times, we find that

$$|C(\ell_1,\ell_2,\ell;\ell_1,-\ell_2,\ell_1-\ell_2)|^2 = \frac{(2\ell+1)(2\ell_1)!(2\ell_2)!}{(\ell_1+\ell_2-\ell)!(\ell_1+\ell_2+\ell+1)!}.$$

Since we normalized our bases so that $C(\ell_1,\ell_2,\ell;\ell_1,-\ell_2,\ell_1-\ell_2)\geq 0$, we obtain

$$C(\ell_1,\ell_2,\ell;\ell_1,-\ell_2,\ell_1-\ell_2) = \sqrt{\frac{(2\ell+1)(2\ell_1)!(2\ell_2)!}{(\ell_1+\ell_2-\ell)!(\ell_1+\ell_2+\ell+1)!}}. \tag{CG13}$$

Observe that (CG13) implies that $C(\ell_1,\ell_2,\ell;j,k,j+k)$ is real, so plugging (CG13) into (CG12), we finally obtain the "master equation"

$$C(\ell_1,\ell_2,\ell;j,k,j+k) = \frac{(-1)^{-\ell+\ell_1+k}}{2^{\ell+\ell_1+\ell_2+1}}$$
$$\sqrt{\frac{(2\ell+1)(\ell+j+k)!(\ell_1+\ell_2-\ell)!(\ell_1+\ell_2+\ell+1)!}{(\ell_1-j)!(\ell_1+j)!(\ell_2-k)!(\ell_2+k)!(\ell-j-k)!(\ell+\ell_1-\ell_2)!(\ell-\ell_1+\ell_2)!}}$$
$$\int_{-1}^{1}(1-x)^{\ell_1-j}(1+x)^{\ell_2-k}\frac{d^{\ell-j-k}}{dx^{\ell-j-k}}[(1-x)^{\ell-\ell_1+\ell_2}(1+x)^{\ell+\ell_1-\ell_2}]\,dx. \tag{CG14}$$

Remark: Another master equation is obtained from (CG11) as follows. When we set $j'=\ell_1$ and $k'=-\ell_2$, the polynomial $P^{\ell}_{j+k,\ell_1-\ell_2}(x)$ appears, but $P^{\ell}_{j+k,\ell_1-\ell_2}(x) = P^{\ell}_{\ell_1-\ell_2,j+k}(x)$, so we can use $P^{\ell}_{\ell_1-\ell_2,j+k}(x)$ and obtain another version of (CG1); see Vilenkin [Vilenkin (1968)], Chapter III, Section 8.3, page 181.

Equation (CG14) still does not give an explicit formula but such formulae can be obtained. By using the product rule in (CG14) and integrating term by term, it is shown in Vilenkin [Vilenkin (1968)] (also page 181) that we have

$$C(\ell_1,\ell_2,\ell;j,k,j+k) = (-1)^{-\ell+\ell_1+k}$$
$$\sqrt{\frac{(2\ell+1)(\ell+j+k)!(\ell_1+\ell_2-\ell)!(\ell-j-k)!(\ell+\ell_1-\ell_2)!(\ell-\ell_1+\ell_2)!}{(\ell_1+\ell_2+\ell+1)!(\ell_1-j)!(\ell_1+j)!(\ell_2-k)!(\ell_2+k)!}}$$
$$\sum_{s=M}^{N} \frac{(-1)^s(\ell+\ell_2-j-s)!(\ell_1+j+s)!}{s!(\ell-j-k-s)!(\ell-\ell_1+\ell_2-s)!(\ell_1-\ell_2+j+k+s)!} \tag{CG15}$$

with $M = \max(0, \ell_2-\ell_1-j-k)$, $N = \min(\ell-j-k, \ell-\ell_1+\ell_2)$.

Two more explicit formulae for $C(\ell_1,\ell_2,\ell;j,k,j+k)$ in terms of sums are given in Vilenkin [Vilenkin (1968)], Chapter III, Section 8.3, pages 181–182.

The Clebsch–Gordan coefficients enjoy several symmetry relations. These relations are discussed in Vilenkin [Vilenkin (1968)], Chapter III, Section 8.4. For example, it can be shown that

$$\begin{aligned} C(\ell_1,\ell_2,\ell;j,k,j+k) &= (-1)^{\ell-\ell_1-\ell_2}C(\ell_1,\ell_2,\ell;-j,-k,-j-k)\\ C(\ell_1,\ell_2,\ell;j,k,j+k) &= (-1)^{\ell-\ell_1-\ell_2}C(\ell_1,\ell_2,\ell;k,j,j+k). \end{aligned}$$

Wigner came up with an ingenious device to formulate these symmetry relations. The *Wigner symbol* (also known as *3j-symbol*)

$$\begin{pmatrix} \ell_1 & \ell_2 & \ell_3 \\ m_1 & m_2 & m_3 \end{pmatrix},$$

which is zero unless $m_3 = -m_1 - m_2$, is defined in terms of the Clebsch–Gordan coefficients by the equation

$$C(\ell_1,\ell_2,\ell;j,k,j+k) = (-1)^{\ell_1-\ell_2+j+k}\sqrt{2\ell+1}\begin{pmatrix} \ell_1 & \ell_2 & \ell \\ j & k & -j-k \end{pmatrix}. \tag{CG16}$$

In Wigner [Wigner (1959)] (Chapter 24), the 3j-symbol is defined on page 290 in Equation (24.9a) (see also Equation (24.9)). The Wigner symbol enjoys a total of 72 symmetries that can be formulated as follows. If we associate to the Wigner symbol the 3×3 matrix shown below,

$$\begin{pmatrix} j_1 & j_2 & j_3 \\ m_1 & m_2 & m_3 \end{pmatrix} \mapsto \begin{pmatrix} -j_1+j_2+j_3 & j_1-j_2+j_3 & j_1+j_2-j_3 \\ j_1-m_1 & j_2-m_2 & j_3-m_3 \\ j_1+m-1 & j_2+m_2 & j_3+m_3 \end{pmatrix},$$

then for an even permutation of the rows or columns of the 3×3 matrix or under transposition, the Wigner symbol is unchanged, and for an odd permutation of the rows or columns it is multiplied by $(-1)^{j_1+j_2+j_3}$; see Vilenkin [Vilenkin (1968)], Chapter III, Section 8.4.

More properties of the Clebsch–Gordan coefficients, including special values, expansions of products of the functions $P^\ell_{mn}(z)$, connections with Jacobi polynomials, recurrence formulae, and generating functions, can be found in Vilenkin [Vilenkin (1968)], Chapter III, Sections 8.5–8.9.

5.18 Problems

Problem 5.1. Check that the polynomials listed at the end of Section 5.2 are indeed harmonic.

Problem 5.2. In Section 5.4, we would like to determine some angles α, β, γ, such that

$$r_x(\alpha/2)r_z(\beta/2)r_x(\gamma/2) = q = r_z(\theta_1/2)r_x(\varphi_2/2)r_z(\theta_2/2).$$

(1) Prove that α, β, γ must satisfy the equations

$$\begin{aligned} \cos\beta &= \cos\theta_1\cos\theta_2 - \sin\theta_1\sin\theta_2\cos\varphi_2 \\ e^{i\alpha} &= \frac{\sin\theta_1\cos\theta_2 + \cos\theta_1\sin\theta_2\cos\varphi_2 + i\sin\theta_2\sin\varphi_2}{\sin\beta} \\ e^{\frac{i(\alpha+\gamma)}{2}} &= \frac{\cos\frac{\theta_1}{2}\cos\frac{\theta_2}{2}e^{\frac{i\varphi_2}{2}} - \sin\frac{\theta_1}{2}\sin\frac{\theta_2}{2}e^{\frac{-i\varphi_2}{2}}}{\cos\frac{\beta}{2}}. \end{aligned}$$

(2) (Advanced) Prove that the following equations can be deduced from the equations of part (1):

$$\begin{aligned} \tan\alpha &= \frac{\sin\theta_2\sin\varphi_2}{\cos\theta_1\sin\theta_2\cos\varphi_2 + \sin\theta_1\cos\theta_2} \\ \tan\gamma &= \frac{\sin\theta_1\sin\varphi_2}{\sin\theta_1\cos\theta_2\cos\varphi_2 + \cos\theta_1\sin\theta_2}, \end{aligned}$$

and β is determined by the equation

$$\cos\beta = \cos\theta_1\cos\theta_2 - \sin\theta_1\sin\theta_2\cos\varphi_2,$$

with $0 \leq \theta \leq \pi$.

Problem 5.3. Prove that T_ℓ as given by Definition 5.3 is a representation $T_\ell\colon \mathbf{SL}(2,\mathbb{C}) \to \mathbf{GL}(\mathcal{P}^{\mathbb{C}}_\ell)$, which yields a representation $T_\ell\colon \mathbf{SU}(2) \to \mathbf{GL}(\mathcal{P}^{\mathbb{C}}_\ell)$ when restricted to the subgroup $\mathbf{SU}(2)$ of $\mathbf{SL}(2,\mathbb{C})$.

Problem 5.4. Let Y be the matrix

$$Y = \begin{pmatrix} b & d \\ -a & -c \end{pmatrix}.$$

(1) For any matrix

$$A = \begin{pmatrix} a & b \\ c & d \end{pmatrix}, \quad a, b, c, d \in \mathbb{C}, \ ad - bc = 1,$$

in $\mathbf{SL}(2,\mathbb{C})$, verify that

$$YA^\top = A^{-1}Y,$$

and $\det(Y) = ad - bc = 1$. Then prove that Y defines a linear isomorphism of $\mathcal{P}^{\mathbb{C}}_{2\ell}(2)$ given by $Q(z_1, z_2) \mapsto Q(bz_1 + dz_2, -az_1 - cz_2)$ and that this map is an equivalence between the representations U_ℓ and U^v_ℓ (see just after Definition 5.3).

(2) Check that the representation $U_{2\ell}\colon \mathbf{SL}(2,\mathbb{C}) \to \mathbf{GL}(\mathcal{P}^{\mathbb{C}}_{2\ell}(2))$ is equivalent to the representation $T_\ell\colon \mathbf{SL}(2,\mathbb{C}) \to \mathbf{GL}(\mathcal{P}^{\mathbb{C}}_\ell)$, and similarly the representation $U_{2\ell}\colon \mathbf{SU}(2) \to \mathbf{GL}(\mathcal{P}^{\mathbb{C}}_{2\ell}(2))$ is equivalent to the representation $T_\ell\colon \mathbf{SU}(2) \to \mathbf{GL}(\mathcal{P}^{\mathbb{C}}_\ell)$.

Hint. Use the dehomogenization and the homogenization maps, which are linear isomorphisms.

Problem 5.5. Consider the basis (ξ_1, ξ_2, ξ_3) of $\mathfrak{sl}(2,\mathbb{C})$ and $\mathfrak{su}(2)$ given in Definition 5.4.

(1) Prove the following cyclic equations regarding Lie brackets:

$$[\xi_1,\, \xi_2] = \xi_3, \quad [\xi_2,\, \xi_3] = \xi_1, \quad [\xi_3,\, \xi_1] = \xi_2.$$

(2) Prove the following equation:

$$e^{\varphi\xi_3} = r_x(\varphi/2), \quad e^{\theta\xi_2} = r_y(\theta/2), \quad e^{\psi\xi_1} = r_z(\psi/2).$$

Problem 5.6. Let (E_1, E_2, E_3) be the basis for $\mathfrak{so}(3)$ given by

$$E_1 = \begin{pmatrix} 0 & 0 & 0 \\ 0 & 0 & -1 \\ 0 & 1 & 0 \end{pmatrix}, \quad E_2 = \begin{pmatrix} 0 & 0 & 1 \\ 0 & 0 & 0 \\ -1 & 0 & 0 \end{pmatrix}, \quad E_3 = \begin{pmatrix} 0 & -1 & 0 \\ 1 & 0 & 0 \\ 0 & 0 & 0 \end{pmatrix}.$$

(1) Prove the equations

$$[E_1,\, E_2] = E_3, \quad [E_2,\, E_3] = E_1, \quad [E_3,\, E_1] = E_2.$$

(2) Prove that

$$e^{\varphi E_1} = R_x(\varphi), \quad e^{-\theta E_2} = R_y(\theta), \quad e^{\psi E_3} = R_z(\psi).$$

Problem 5.7. Consider the basis (X, Y, H) of $\mathfrak{sl}(2, \mathbb{C})$ (see Section 5.6) given by

$$X = \begin{pmatrix} 0 & 1 \\ 0 & 0 \end{pmatrix}, \quad Y = \begin{pmatrix} 0 & 0 \\ 1 & 0 \end{pmatrix}, \quad H = \begin{pmatrix} 1 & 0 \\ 0 & -1 \end{pmatrix}.$$

Prove that

$$[X, Y] = H, \quad [H, X] = 2X, \quad [H, Y] = -2Y.$$

(The Lie bracket $[A, B]$ of two square matrices A and B is defined as $[A, B] = AB - BA$.)

Problem 5.8. Fill in the details of the computations needed to prove equations $(*_{23})$ and $(*_{24})$ in Section 5.9.

Problem 5.9. Consider the map $\mathcal{T}_\ell \colon \mathbf{SL}(2, \mathbb{C}) \to \mathbf{GL}(\mathfrak{F}_\ell)$ from Definition 5.13.

(1) Prove that $\mathcal{T}_\ell$ is a representation.
(2) Check that the linear map defined on basis vectors by $z^{\ell-k} \mapsto e^{-ik\varphi}$ is an isomorphism between the vector spaces $\mathcal{P}_\ell^{\mathbb{C}}$ and $\mathfrak{F}_\ell$.

Problem 5.10. Use Proposition 5.27 to derive explicit formulae $t^\ell(\theta)$ for $\ell = 0, 1/2, 1, 3/2$. Derive a formula for $\ell = 2$.

Problem 5.11. Prove that the polynomials $P^\ell_{jk}(z)$ defined in Proposition 5.28 satisfies the symmetry relation

$$P^\ell_{jk}(z) = P^\ell_{-j\,-k}(z), \quad -1 < z \leq 1.$$

Problem 5.12. Prove the equation

$$P^j_\ell(z) = i^j \sqrt{\frac{(\ell + j)!}{(\ell - j)!}}\, P^\ell_{j0}(z). \tag{$*_{47}$}$$

Problem 5.13. Supply the missing computations in the proof of Proposition 5.36.

Problem 5.14. Prove that

$$\overline{P^\ell_{jk}(\cos\theta)} = (-1)^{j-k} P^\ell_{jk}(\cos\theta).$$

Problem 5.15. Check that

$$(2\ell_1 + 1)(2\ell_2 + 1) = \sum_{\ell = |\ell_1 - \ell_2|}^{\ell_1 + \ell_2} (2\ell + 1).$$

Problem 5.16. Provide the details of the proofs of equations (CG12), (CG13) and (CG15).

Hint. See Vilenkin [Vilenkin (1968)], Chapter III, Section 8.3.

Problem 5.17. Prove the following equations:

$$C(\ell_1, \ell_2, \ell; j, k, j+k) = (-1)^{\ell-\ell_1-\ell_2} C(\ell_1, \ell_2, \ell; -j, -k, -j-k)$$
$$C(\ell_1, \ell_2, \ell; j, k, j+k) = (-1)^{\ell-\ell_1-\ell_2} C(\ell_1, \ell_2, \ell; k, j, j+k).$$

Chapter 6

Induced Representations

If G is a locally compact group and if H is a closed subgroup of G, under certain conditions, it is possible to construct a Hilbert space $\mathcal{H}$ and a unitary representation $\Pi\colon G \to \mathbf{U}(\mathcal{H})$ of G in $\mathcal{H}$ from a unitary representation $U\colon H \to \mathbf{U}(E)$ of H in a (separable) Hilbert space E. The representation Π is called an *induced representation*. In particular, this construction can be used to define unitary representations of the group $\mathbf{SL}(2, \mathbb{R})$, which would be hard to find if we did not have this method.

There are two approaches for the construction of the Hilbert space $\mathcal{H}$:

(1) The Hilbert space $\mathcal{H}$ is a set of functions from $X = G/H$ to E.
(2) The Hilbert space $\mathcal{H}$ is a set of functions from G to E.

In the first approach, we will construct unitary representations of G in $\mathcal{H}$ using certain functions $\alpha\colon G \times (G/H) \to \mathbf{GL}(E)$ called *cocycles*. In the second approach, the construction of the Hilbert space $\mathcal{H}$ is more complicated, but the definition of the operator Π_s is simpler.

The general construction (in the first approach) consists of seven steps, where the first four are purely algebraic and do not deal with continuous unitary representations, but instead linear representations (group homomorphisms $U\colon G \to \mathbf{GL}(E)$, where G is a group not equipped with any topology and E is just a vector space with no additional structure):

(1) Let G be a group acting (on the left) on a set X, and let E be a vector space. In Section 6.1, we define the notion of *equilinear action* of G on $X \times E$, which is an action of the form

$$s \cdot (x, z) = (s \cdot x, \alpha(s, x)(z)), \qquad s \in G, x \in X, z \in E,$$

where $\alpha(s, x)$ is a linear automorphism of E satisfying the conditions

(a) For all $x \in X$

$$\alpha(e, x) = \mathrm{id}_E.$$

(b) For all $x \in X$ and all $s, t \in G$,

$$\alpha(st, x) = \alpha(s, t \cdot x) \circ \alpha(t, x).$$

A map $\alpha\colon G \times X \to \mathbf{GL}(E)$ satisfying conditions (a) and (b) is called a *cocycle of* G *with values in* $\mathbf{GL}(E)$. Conversely, an action of G on X and a cocycle $\alpha\colon G \times X \to \mathbf{GL}(E)$ determine an equilinear action of G on $X \times E$. Then, we show that an equilinear action of G on $X \times E$ induces a homomorphism $\Pi\colon G \to \mathbf{GL}(E^X)$, where E^X is the vector space of all functions from X to E. More precisely, for every function $f\colon X \to E$, for every $s \in G$, $\Pi_s(f)\colon X \to E$ is function given by

$$(\Pi_s(f))(x) = \alpha(s, s^{-1} \cdot x)(f(s^{-1} \cdot x)), \qquad \text{for every } x \in X.$$

(2) In Section 6.2, we specialize the construction to the homogeneous space $X = G/H$ of left cosets. Then G acts on G/H on the left by

$$s \cdot (gH) = sgH.$$

By choosing a set of representatives $(r_x)_{x \in G/H}$ in the cosets of $X = G/H$ (with $x_0 = H$ and $r_{x_0} = e$), a cocycle $\alpha\colon G \times X \to \mathbf{GL}(E)$ determines a homomorphism $\sigma\colon H \to \mathbf{GL}(E)$ given by $\sigma(h) = \alpha(h, x_0)$ and a map $\beta\colon X \to \mathbf{GL}(E)$ given by $\beta(x) = \alpha(r_x, x_0)$. Conversely, a homomorphism $\sigma\colon H \to \mathbf{GL}(E)$ and a map $\beta\colon X \to \mathbf{GL}(E)$ determine a cocycle $\alpha\colon G \times X \to \mathbf{GL}(E)$. In fact, we may restrict ourselves to the map β given by $\beta(x) = \mathrm{id}_E$, and if we define $u\colon G \times X \to H$ by $u(s, x) = r_{s \cdot x}^{-1} s r_x$, the map $\alpha\colon G \times X \to \mathbf{GL}(E)$ given by $\alpha(s, x) = \sigma(u(s, x))$ is a cocycle. The induced representation is given by

$$(\Pi_s(f))(x) = \sigma(u(s, s^{-1} \cdot x))(f(s^{-1} \cdot x)), \;\; f \in E^X, \; x \in X.$$

This step is the most important application of step 1, and E is an arbitrary vector space.

(3) For a given homomorphism $\sigma\colon H \to \mathbf{GL}(E)$, the homomorphisms $\Pi\colon G \to \mathbf{GL}(E^X)$ corresponding to cocycles associated with different maps β are equivalent.

(4) In Section 6.3, we show that a cocycle $\alpha\colon G \times X \to \mathbf{GL}(E)$ determines a bijection between E^X and a subspace L^α of the set E^G of maps from G to E defined by

$$L^\alpha = \{f \in E^G \mid f(sh) = \sigma(h^{-1})(f(s)), \;\; s \in G, h \in H\}.$$

As a consequence, the representation $\Pi\colon G \to \mathbf{GL}(E^X)$ corresponding to a cocycle α is equivalent to the representation $\Pi_{L^\alpha}\colon G \to \mathbf{GL}(L^\alpha)$ given by

$$((\Pi_{L^\alpha})_s(g))(t) = g(s^{-1}t), \qquad \text{for all } g \in L^\alpha \text{ and all } s, t \in G.$$

Observe that this is simply the left regular representation of L^α. The issue of choosing between representations in the space E^X or representations in the space L^α comes up in Chapter 8.

This completes the purely algebraic construction. The next steps use topology and analysis to construct *unitary* representations.

(5) In Section 6.4, we assume that G is a locally compact group and H is a closed subgroup of G, in which case G/H is also locally compact. Let μ be a positive measure on $X = G/H$, and assume that E is a separable Hilbert space. We then define a Hilbert space $\mathcal{L}^2_\mu(X;E)$ consisting of measurable functions from X to E.

(6) In Section 6.5, given a unitary representation U of H in E, we assume that the measure μ on $X = G/H$ is G-invariant and that the cocycle α satisfies the conditions:

   (i) The linear automorphisms $\alpha(s,x)$ of E are unitary operators of E for all $s \in G$ and all $x \in G/H$, and $\alpha(h, x_0) = U(h)$ for all $h \in H$ (where x_0 denotes the coset H).

   (ii) For every $s \in G$, for every $f \in \mathcal{L}^2_\mu(X;E)$, the map $x \mapsto \alpha(s,x)(f(x))$ from X to E is μ-measurable.

   (iii) For every $f \in \mathcal{L}^2_\mu(X;E)$, the map $s \mapsto [\Pi_s(f)]$ from G to $\mathrm{L}^2_\mu(X;E)$ is continuous.

   Then the homomorphism $s \mapsto \Pi_s([f]) = [\Pi_s(f)]$ is a unitary representation of G in $\mathrm{L}^2_\mu(X;E) = \mathcal{H}$.

(7) In Sections 6.6 and 6.7, we generalize the previous construction to certain measure called *quasi-invariant*. If the measure μ on G/H is quasi-invariant and another technical condition is satisfied, then the homomorphism $s \mapsto \Pi_s([f]) = [\Pi_s(f)]$ is a unitary representation of G in $\mathrm{L}^2_\mu(X;E)$. Quasi-invariant measures on G/H always exist and can be constructed using rho-functions.

In Section 6.8, we illustrate the method of Section 6.7 by showing how to construct unitary representations of $\mathbf{SL}(2,\mathbb{R})$ using induced representations. One example involves the action of $\mathbf{SL}(2,\mathbb{R})$ on the projective line $\mathbb{RP}^1$, and the other example involves the action of $\mathbf{SL}(2,\mathbb{R})$ on the upper half plane.

In Section 6.9, we consider a compact (metrizable) group G and a closed subgroup H of G, and our goal is to determine the canonical (unitary) representation of G in $\mathrm{L}^2_\mu(G/H;\mathbb{C})$ induced by the trivial representation of H in $E=\mathbb{C}$ (see Definition 6.13), where μ is the G-invariant measure on G/H induced by a Haar measure λ on G. For simplicity of notation, we write $\mathrm{L}^2_\mu(G/H)$ instead of $\mathrm{L}^2_\mu(G/H;\mathbb{C})$. To do this, it is necessary to understand what is the restriction of the representation $M_\rho\colon G\to\mathbf{U}(\mathbb{C}^{n_\rho})$ to H, with $\rho\in R(G)$.

In Proposition 6.18, we show that the space $\mathrm{L}^2_\mu(G/H)$ is the Hilbert sum of subspaces $L_\rho\subseteq\mathfrak{a}_\rho$. If the trivial representation σ_0 of H is contained $d=(\rho:\sigma_0)\geq 1$ times in the restriction of M_ρ to H, then L_ρ is the direct sum of the first d columns of the matrix $M_\rho^{(H)}=P^*M_\rho P$, where P is a suitable change of basis matrix, namely,

$$L_\rho=\bigoplus_{j=1}^{d}\mathfrak{l}_j^{(\rho,H)}\quad\text{and}\quad \mathfrak{l}_j^{(\rho,H)}=\bigoplus_{k=1}^{n_\rho}\mathbb{C}m_{kj}^{(\rho,H)}.$$

If $d=0$, then $L_\rho=(0)$.

Then, we consider the space $H\backslash G$ of right cosets Hs of G ($s\in G$). If $\pi\colon G\to H\backslash G$ is the quotient map $\pi(s)=Hs$, the fact that the Haar measure λ on a compact group is left and right invariant implies immediately that there is a G-invariant measure μ' on $H\backslash G$. We show in Proposition 6.19 that the space $\mathrm{L}^2_{\mu'}(H\backslash G)$ is the Hilbert sum of subspaces $\check{\overline{L}}_\rho\subseteq\mathfrak{a}_\rho$. If the trivial representation σ_0 of H is contained $d=(\rho:\sigma_0)\geq 1$ times in the restriction of M_ρ to H, then $\check{\overline{L}}_\rho$ is the direct sum of the first d rows of $M_\rho^{(H)}$; that is

$$\check{\overline{L}}_\rho=\bigoplus_{i=1}^{d}\bigoplus_{j=1}^{n_\rho}\mathbb{C}m_{ij}^{(\rho,H)}.$$

In preparation for Chapter 9, we consider the intersection $\mathrm{L}^2_\mu(G/H)\cap\mathrm{L}^2_{\mu'}(H\backslash G)$. This is a closed involutive subalgebra of $\mathrm{L}^2(G)$, thus a complete Hilbert algebra. We can view a function $g\in\mathrm{L}^2_\mu(G/H)\cap\mathrm{L}^2_{\mu'}(H\backslash G)$ as a function $g\in\mathrm{L}^2(G)$, such that

$$g(tst')=g(s),\qquad\text{for all } t,t'\in H \text{ and all } s\in G. \tag{$*_{H\backslash G/H}$}$$

We can also think of the functions $g\in\mathrm{L}^2_\mu(G/H)\cap\mathrm{L}^2_{\mu'}(H\backslash G)$ as functions defined on the *double classes (or double cosets)* HsH of G with respect to H.

We denote the algebra of functions in $\mathrm{L}^2(G)$ satisfying $(*_{H\backslash G/H})$ as $\mathrm{L}^2(H\backslash G/H)$. Then we show in Proposition 6.20 that the algebra $\mathrm{L}^2(H\backslash G/H)$ is the Hilbert sum of the minimal two-sided ideals

$$\mathfrak{a}_{\rho,\sigma_0} = L_\rho \cap \check{L}_\rho = \bigoplus_{i=1}^{d}\bigoplus_{j=1}^{d} \mathbb{C} m_{ij}^{(\rho,H)}.$$

Each $\mathfrak{a}_{\rho,\sigma_0}$ is a matrix algebra of dimension d^2 having the family $(m_{ij}^{(\rho,H)})_{1\leq i,j\leq d}$ as a basis.

Again, in preparation for Chapter 9 on Gelfand pairs, we show in Proposition 6.21 that the algebra $\mathrm{L}^2(H\backslash G/H)$ is commutative if and only if $(\rho : \sigma_0) \leq 1$ for all $\rho \in R(G)$. If so, then for every $\rho \in R(G)$, such that $(\rho : \sigma_0) = 1$, the ideal $\mathfrak{a}_{\rho,\sigma_0}$ is one-dimensional and is spanned by the function

$$\omega_\rho(s) = \frac{1}{n_\rho} m_{11}^{(\rho,H)}(s),$$

which is continuous and of positive type. Thus

$$\mathrm{L}^2(H\backslash G/H) = \bigoplus_{\rho \mid (\rho:\sigma_0)=1} \mathbb{C}\omega_\rho.$$

The function ω_ρ also satisfies the following equations:

$$\begin{aligned} \omega_\rho(tst') &= \omega_\rho(s), \qquad \text{for all } s \in G \text{ and all } t, t' \in H, \\ \omega_\rho(e) &= 1. \end{aligned}$$

The function ω_ρ is called a *(zonal) spherical function*. Such functions are crucial in generalizing the notion of Fourier transform to a homogeneous space G/H. In Chapter 9, we will see how to achieve this when G is not compact (but H is compact). The key point is to consider pairs (G, H) for which $\mathrm{L}^2(H\backslash G/H)$ is commutative. Actually, we can't quite work with $\mathrm{L}^2(H\backslash G/H)$ because this space is not closed under convolution, but we will be able to work with another commutative algebra $\mathrm{L}^1(H\backslash G/H)$.

In Section 6.10, we present a nice example of the above situation for $G = \mathbf{SO}(n+1)$ and $H = \mathbf{SO}(n)$. In this case, $G/H = \mathbf{SO}(n+1)/\mathbf{SO}(n) \simeq S^n$, the sphere in $\mathbb{R}^{n+1}$. As a consequence, we obtain a decomposition of $\mathrm{L}^2(S^n)$ as a Hilbert sum of the classical spaces $\mathcal{H}_k^{\mathbb{C}}(S^n)$ of spherical harmonics on S^n.

In Section 6.11, we present a method due to Blattner to deal with the situation where G/H has no G-invariant measure. This is a modification of the construction of the Hilbert space $\mathcal{H}$ and of the inner product described at the end of Section 6.5. This can be done in two ways. These constructions yield induced unitary representations of G from a unitary representation $U\colon H \to \mathbf{U}(E)$ of H and do not involve cocycles.

In Section 6.12, we explain how the spaces of functions L^α (from Definition 6.8), and the spaces $\mathcal{H}_0$ and $\mathcal{H}^0$ from Section 6.11 can be viewed as sections of spaces that are similar to vector bundles but have less structure. More precisely, such structures have no trivialization maps.

We begin with the simplest situation where we have a group G without any topology on it, a subgroup H of G, a vector space $\mathcal{H}_\sigma$, and a linear representation $\sigma\colon H \to \mathbf{GL}(\mathcal{H}_\sigma)$. As usual, write $X = G/H$ and $\pi\colon G \to G/H$ for the quotient map. Let L^σ be the subspace of $(\mathcal{H}_\sigma)^G$ consisting of all functions $f\colon G \to \mathcal{H}_\sigma$, such that

$$f(gh) = \sigma(h^{-1})(f(g)), \qquad \text{for all } g \in G \text{ and all } h \in H.$$

The key point is to construct a space $E = G \times_H \mathcal{H}_\sigma$, together with a surjective map $p\colon E \to X$, such that for every $x \in X = G/H$, the fibre $E_x = p^{-1}(x)$ is isomorphic to the vector space $\mathcal{H}_\sigma$, and the space of sections from X to E is in bijection with L^σ. This is a special case of the so-called Borel construction used to construct a vector bundle from a principal bundle; see Gallier and Quaintance [Gallier and Quaintance (2020b)] (Chapter 9, Section 9.9). Then the main point of this section is to define two maps $\mathcal{S}\colon L^\sigma \to \Gamma(E)$ and $\mathcal{L}\colon \Gamma(E) \to L^\sigma$, which are mutual inverses, where $\Gamma(E)$ is the space of sections of E, namely the set of functions $s\colon X \to E$, such that $p \circ s = \mathrm{id}_X$, where p is the projection $p\colon E \to X$.

The last important ingredient is that G acts (on the left) on $E = G \times_H \mathcal{H}_\sigma$ in an *equilinear* fashion; this is explained in Proposition 6.23.

In Section 6.13, we show how induced representations can be recovered from certain kinds of vector bundles E over the base space $X = G/H$ (actually a more basic notion of vector bundle) equipped with an equilinear action of a group G on E. Such bundles, called *G-bundles*, are equipped with an equilinear action of the group G and generalize the notion of bundle introduced in the previous section. If x_0 denotes the coset $H = eH$ in G, then the action of G on the fibre E_0 above x_0 defines a representation $\sigma\colon H \to \mathbf{GL}(E_0)$. Again, the main point is to define a space of functions L^σ and two maps $\mathcal{S}\colon L^\sigma \to \Gamma(E)$ and $\mathcal{L}\colon \Gamma(E) \to L^\sigma$, which are mutual inverses, where $\Gamma(E)$ is the space of sections of E. The induced representation of G induced by the representation σ of H can then be recovered from the action of G on sections of E in terms of $\mathcal{L}$ and $\mathcal{S}$.

The sections in $\Gamma(E)$, called *feature fields* in group equivariant deep learning in computer vision, are functions whose domain transforms under the action of G and whose codomain transforms by representations of H equivalent to $\sigma\colon H \to \mathbf{GL}(E_0)$.

The above definitions and constructions are adapted to deal with unitary representations in Section 6.14. In this case, G is a locally compact group, H is a closed subgroup of G, and $\sigma\colon H \to \mathbf{U}(\mathcal{H}_\sigma)$ is a unitary representation, where $\mathcal{H}_\sigma$ is a separable Hilbert space. These bundles are called *hermitian G-bundles*. We treat the special case where $\mathcal{H}_\sigma$ is finite-dimensional in detail.

Unfortunately, in general the maps $\mathcal{L}$ and $\mathcal{S}$ are no longer well-defined. To remedy this problem, we assume that our G-bundles are locally trivializable, that is, they are (smooth) vector bundles.

Consequently in Section 6.15, we review principal H-bundles and hermitian vector bundles. We then define *hermitian G-vector bundles*, which are simultaneously hermitian vector bundles and hermitian G-bundles. We discuss the construction of a hermitian vector bundle from a principal H-bundle obtained by replacing the fibre H by a vector space $\mathcal{H}_\sigma$, which is the space of a unitary representation $\sigma\colon H \to \mathbf{U}(\mathcal{H}_\sigma)$; see Theorem 6.27.

The generalization to hermitian G-vector bundles of infinite rank is sketched in Section 6.16, but we do not know how to proceed when G/H does not have a G-invariant measure.

6.1 Cocycles and Induced Representations

As a warm up and as an example of the second approach, we consider the case where G is compact, H is a closed subgroup of G, and U is a linear representation of H in a *finite-dimensional* vector space E. This means that U is a homomorphism $U\colon H \to \mathbf{GL}(E)$ and that condition (C) of Definition 3.1 is dropped.

Consider the Hilbert space $\mathrm{L}^2(G; E)$ consisting of all functions $f\colon G \to E$, such that for any orthonormal basis $(e_1, \ldots, e_n)$ of E, $f = f_1 e_1 + \cdots + f_n e_n$, where the f_i are functions in $\mathrm{L}^2(G)$; equivalently, $\mathrm{L}^2(G; E)$ is the finite Hilbert sum $\mathrm{L}^2(G; E) = \bigoplus_{i=1}^n \mathrm{L}^2(G) e_i$. The inner product of two functions $f = \sum_{i=1} f_i e_i$ and $g = \sum_{i=1} g_i e_i$ is

$$\langle f, g\rangle = \sum_{i=1}^{n} \int_G f_i(s)\overline{g_i(s)}\, d\lambda(s),$$

where λ is a Haar measure on G. This construction will be generalized in Section 6.4 to an infinite-dimensional Hilbert space. Consider the subspace $\mathcal{H}$ of $L^2(G; E)$ consisting of all functions f, such that

$$f(sh) = U(h^{-1})(f(s)), \qquad \text{for all } s \in G \text{ and all } h \in H. \qquad (*)$$

It is easy to check that $\mathcal{H}$ is closed in $L^2(G; E)$, so it is a Hilbert space. For any $f \in \mathcal{H}$, as before, let $\lambda_s f$ be the function given by

$$(\lambda_s f)(t) = f(s^{-1}t), \qquad s, t \in G.$$

For $s \in G$ fixed, the map $f \mapsto \lambda_s f$ is obviously linear. Observe that by $(*)$, for all $s, t \in G$, all $h \in H$, and all $f \in \mathcal{H}$, we have

$$(\lambda_t f)(sh) = f(t^{-1}sh) = U(h^{-1})(f(t^{-1}s)) = U(h^{-1})((\lambda_t f)(s)),$$

so $\lambda_t f \in \mathcal{H}$. For all $s, t, t' \in G$, we also have

$$(\lambda_{tt'} f)(s) = f((tt')^{-1}s) = f(t'^{-1}t^{-1}s) = (\lambda_{t'} f)(t^{-1}s) = \lambda_t((\lambda_{t'} f))(s).$$

If we define the map $\Pi \colon G \to \mathbf{GL}(\mathcal{H})$ by

$$\Pi_s(f) = \lambda_s f, \qquad s \in G,\ f \in \mathcal{H},$$

equivalently

$$(\Pi_s(f))(t) = f(s^{-1}t), \qquad s, t \in G,\ f \in \mathcal{H},$$

then we see that Π is a linear representation of G in $\mathcal{H}$ (condition (C) of Definition 3.1 may fail, but here we are not considering continuous representations). Since the Haar measure is left and right invariant, the maps $\lambda_t f$ are unitary ($f \in \mathcal{H}$), so $\Pi \colon G \to \mathbf{GL}(\mathcal{H})$ is a unitary representation of G in $\mathcal{H}$, called the representation *induced* by $U \colon H \to \mathbf{GL}(E)$.

It is easy to see that if we replace U by an equivalent representation $h \mapsto PU(h)P^{-1}$, where P is a unitary transformation $P \colon E \to E'$, then the corresponding induced representation is $s \mapsto f_P \Pi_s f_P^{-1}$, a unitary representation equivalent to Π, where f_P is the linear map from $\mathcal{H}$ to $\mathcal{H}'$ given by $f_P(f) = P \circ f$. Therefore, the above construction defines a class of unitary representations of G induced by a class of linear representations of H.

Let us now consider a more general situation. Our first construction is purely algebraic and does not assume that the group G or the vector space E have any topology. As a consequence, until Section 6.4 we consider linear representations of G in E; these are simply homomorphisms $U \colon G \to \mathbf{GL}(E)$, with no continuity requirement.

Definition 6.1. If we have a left group action $\cdot\colon G\times X\to X$ of a group G on a set X, for any vector space E, a left action $\cdot\colon G\times(X\times E)\to X\times E$ is *equilinear* if there is some function $\alpha\colon G\times X\to \mathbf{GL}(E)$, such that

$$s\cdot(x,z)=(s\cdot x,\alpha(s,x)(z)),\qquad \text{for all } s\in G, \text{ all } x\in X, \text{ and all } z\in E.$$

The crucial property of an equilinear action is that the second component $pr_2(s\cdot(x,z))$ of the action of $s\in G$ on $(x,z)\in X\times E$ given by

$$s\cdot(x,z)=(s\cdot x,pr_2(s\cdot(x,z)))$$

is *linear* in z. This is the reason for introducing the linear isomorphism $\alpha(s,x)$ given by $\alpha(s,x)(z)=pr_2(s\cdot(x,z))$.

If we have an equilinear action $\cdot\colon G\times(X\times E)\to X\times E$, then the conditions for being a left action are

$$\begin{aligned}e\cdot(x,z)&=(x,z)\\(st)\cdot(x,z)&=s\cdot(t\cdot(x,z)),\end{aligned}$$

which translate to

$$\begin{aligned}(e\cdot x,\alpha(e,x)(z))&=(x,z)\\((st)\cdot x,\alpha(st,x)(z))&=s\cdot(t\cdot x,\alpha(t,x)(z))\\&=(s\cdot(t\cdot x),\alpha(s,t\cdot x)(\alpha(t,x)(z))),\end{aligned}$$

so we must have

$$\begin{aligned}\alpha(e,x)(z)&=z\\\alpha(st,x)(z)&=(\alpha(s,t\cdot x)\circ\alpha(t,x))(z),\end{aligned}$$

for all $s,t\in G$, all $x\in X$, and all $z\in E$. By reversing the above computations, we see that if a function $\alpha\colon G\times X\to\mathbf{GL}(E)$ satisfies the above two conditions, then the map given by

$$s\cdot(x,z)=(s\cdot x,\alpha(s,x)(z)),\qquad \text{for all } s\in G, \text{ all } x\in X, \text{ and all } z\in E,$$

is an equilinear action. In summary, we proved the following proposition.

Proposition 6.1. *Given a left group action $\cdot\colon G\times X\to X$ and a vector space E, for any function $\alpha\colon G\times X\to\mathbf{GL}(E)$, the map $\cdot\colon G\times(X\times E)\to X\times E$ given by*

$$s\cdot(x,z)=(s\cdot x,\alpha(s,x)(z)),\qquad \textit{for all } s\in G, \textit{ all } x\in X, \textit{ and all } z\in E,$$

is an equilinear action if and only if the following two conditions hold:

(a) For all $x \in X$,

$$\alpha(e, x) = \mathrm{id}_E.$$

(b) For all $x \in X$ and all $s, t \in G$,

$$\alpha(st, x) = \alpha(s, t \cdot x) \circ \alpha(t, x).$$

In view or Proposition 6.1, we make the following definition.

Definition 6.2. Let G be a left action of a group G on a set X, and let E be a vector space. Let $\alpha \colon G \times X \to \mathbf{GL}(E)$ be a function and assume that the following conditions hold:

(a) For all $x \in X$,

$$\alpha(e, x) = \mathrm{id}_E.$$

(b) For all $x \in X$ and all $s, t \in G$,

$$\alpha(st, x) = \alpha(s, t \cdot x) \circ \alpha(t, x).$$

A map $\alpha \colon G \times X \to \mathbf{GL}(E)$ satisfying conditions (a) and (b) is called a *cocycle of G with values in $\mathbf{GL}(E)$*.

The point of equilinear actions is that they yield homomorphisms $\Pi \colon G \to \mathbf{GL}(E^X)$, that is, linear representations of G in the vector space $[X \to E] = E^X$. We just explained before Definition 6.2 how a cocycle defines an equilinear action. The reader may wonder where cocycles come from. The answer will be given in the next section; they are induced by linear representations of subgroups of G.

Given an equilinear action $\cdot \colon G \times (X \times E) \to X \times E$, we obtain an action Π of G on E^X as follows: for every $s \in G$, for every $f \in E^X$, the function $\Pi_s(f) \in E^X$ is given by the equation

$$s \cdot (x, f(x)) = (s \cdot x, (\Pi_s(f))(s \cdot x)), \quad \text{for all } x \in X.$$

Using Definition 6.1, the above equation is equivalent to

$$(\Pi_s(f))(s \cdot x) = \alpha(s, x)(f(x)), \quad \text{for all } x \in X,$$

which is equivalent to

$$(\Pi_s(f))(x) = \alpha(s, s^{-1} \cdot x)(f(s^{-1} \cdot x)), \quad \text{for all } x \in X.$$

We are led to the following definition.

Definition 6.3. Let G be a left action of a group G on a set X, and let E be a vector space. For every equilinear action $\cdot\colon G \times (X \times E) \to X \times E$ defined by a cocycle $\alpha\colon G \times X \to \mathbf{GL}(E)$, for every function $f\colon X \to E$, for every $s \in G$, let $\Pi^\alpha_s(f)\colon X \to E$ be the function given by

$$(\Pi^\alpha_s(f))(x) = \alpha(s, s^{-1} \cdot x)(f(s^{-1} \cdot x)), \qquad \text{for every } x \in X. \qquad (\Pi^\alpha_s)$$

The above equation defines a map $\Pi^\alpha_s\colon E^X \to E^X$. The map $\Pi^\alpha\colon G \to \mathbf{GL}(E^X)$ given by $s \mapsto \Pi^\alpha_s$ is the *(linear) representation of G in E^X induced by the cocycle α*. For simplicity of notation, we write Π instead of Π^α.

The following proposition confirms that the map Π is a linear representation of G in the vector space E^X.

Proposition 6.2. *Let G be a left action of a group G on a set X, and let E be a vector space. For every equilinear action $\cdot\colon G \times (X \times E) \to X \times E$ defined by a cocycle $\alpha\colon G \times X \to \mathbf{GL}(E)$, for every $s \in G$, the map $\Pi_s\colon E^X \to E^X$ is a linear isomorphism, and the map $\Pi\colon G \to \mathbf{GL}(E^X)$ given by $s \mapsto \Pi_s$ is a homomorphism, that is, a linear representation of G in the vector space E^X.*

Proof. Since $\alpha(s, s^{-1} \cdot x)$ is a linear automorphism of E, we have

$$\begin{aligned}
(\Pi_s(f_1 + f_2))(x) &= \alpha(s, s^{-1} \cdot x)((f_1 + f_2)(s^{-1} \cdot x)) \\
&= \alpha(s, s^{-1} \cdot x)(f_1(s^{-1} \cdot x) + f_2(s^{-1} \cdot x)) \\
&= \alpha(s, s^{-1} \cdot x)(f_1(s^{-1} \cdot x)) + \alpha(s, s^{-1} \cdot x)(f_2(s^{-1} \cdot x)) \\
&= (\Pi_s(f_1))(x) + (\Pi_s(f_2))(x),
\end{aligned}$$

and for every $\lambda \in \mathbb{C}$,

$$\begin{aligned}
(\Pi_s(\lambda f))(x) &= \alpha(s, s^{-1} \cdot x)((\lambda f)(s^{-1} \cdot x)) \\
&= \alpha(s, s^{-1} \cdot x)(\lambda f(s^{-1} \cdot x)) \\
&= \lambda \alpha(s, s^{-1} \cdot x)(f(s^{-1} \cdot x)) \\
&= \lambda(\Pi_s(f))(x),
\end{aligned}$$

so the map $f \mapsto \Pi_s(f)$ from E^X to itself is linear. Given any fixed $s \in G$, for every function $g\colon X \to E$, we have $\Pi_s(f) = g$ iff $(\Pi_s(f))(x) = g(x)$ for all $x \in X$ iff

$$\alpha(s, s^{-1} \cdot x)(f(s^{-1} \cdot x)) = g(x), \quad \text{for all } x \in X,$$

and since $\alpha(s, s^{-1} \cdot x)$ is an invertible linear map, we must have

$$f(s^{-1} \cdot x) = (\alpha(s, s^{-1} \cdot x))^{-1}(g(x)), \quad \text{for all } x \in X,$$

so if we write $y = s^{-1} \cdot x$, then $x = s \cdot y$ and since the map $y \mapsto s \cdot y$ is a bijection (because $\cdot$ is a group action of G on X), we have

$$f(y) = (\alpha(s,y))^{-1}(g(s \cdot y)), \quad \text{for all } y \in X,$$

which shows that f is uniquely determined and thus that Π_s is a bijection.

For all $s, t \in G$, we have

$$g(y) = (\Pi_t(f))(y) = \alpha(t, t^{-1} \cdot y)(f(t^{-1} \cdot y)),$$

so

$$\begin{aligned}
\big(\Pi_s(\Pi_t(f))\big)(x) &= (\Pi_s(g))(x) \\
&= \alpha(s, s^{-1} \cdot x)(g(s^{-1} \cdot x)) \\
&= \big(\alpha(s, s^{-1} \cdot x) \circ \alpha(t, t^{-1} \cdot (s^{-1} \cdot x))\big)(f(t^{-1} \cdot (s^{-1} \cdot x))),
\end{aligned}$$

and we also have

$$\begin{aligned}
(\Pi_{st}(f))(x) &= \alpha(st, (st)^{-1} \cdot x)(f((st)^{-1} \cdot x)) \\
&= \big(\alpha(s, t \cdot ((t^{-1}s^{-1}) \cdot x)) \circ \alpha(t, (t^{-1}s^{-1}) \cdot x)\big)(f(t^{-1}s^{-1} \cdot x)) \\
&= \big(\alpha(s, s^{-1} \cdot x) \circ \alpha(t, t^{-1} \cdot (s^{-1} \cdot x))\big)(f(t^{-1} \cdot (s^{-1} \cdot x))) \\
&= \big(\Pi_s(\Pi_t(f))\big)(x),
\end{aligned}$$

which proves that $\Pi_{st}(f) = (\Pi_s \circ \Pi_t)(f)$, that is, Π is a homomorphism. □

If we let $t = s^{-1}$ in (b) of Definition 6.2, we obtain

$$\alpha(s^{-1}, x) = (\alpha(s, s^{-1} \cdot x))^{-1},$$

so $\Pi_s(f)$ can also be written as

$$(\Pi_s(f))(x) = (\alpha(s^{-1}, x))^{-1}(f(s^{-1} \cdot x)). \tag{Π_s}$$

6.2 Cocycles on a Homogeneous Space $X = G/H$

We now consider the special case where $X = G/H$ is the homogeneous space of left cosets for some subgroup H of G, and the left action of G acts on G/H given by

$$s \cdot (gH) = sgH.$$

Definition 6.4. Given a group G and a subgroup H of G, a *set of representatives* $(r_x)_{x \in G/H}$ for the cosets of G/H is the choice for every coset $x \in G/H$ of some element $r_x \in G$, so that $x = r_x H$. Then every element g of $x = r_x H$ is written uniquely as $g = r_x h$, with $h \in H$. We denote the coset H by x_0 and pick $r_{x_0} = e$. For any $s \in G$, the representative of $s \cdot x = s \cdot r_x H = s r_x H$ is denoted by $r_{s \cdot x}$.

If we denote the quotient map by $\pi\colon G \to G/H$, then picking a set of representatives $(r_x)_{x\in G/H}$ in the cosets of G/H is equivalent to picking a *section of* π, that is, a map $r\colon G/H \to G$, such that $\pi \circ r = \mathrm{id}_{G/H}$.

Since for every coset $x \in G/H$ we have $x = r_x \cdot x_0$ (the class $r_x H$, which is x), condition (b) of Definition 6.2 yields

$$\alpha(sr_x, x_0) = \alpha(s, r_x \cdot x_0) \circ \alpha(r_x, x_0) = \alpha(s, x) \circ \alpha(r_x, x_0),$$

and so

$$\alpha(s, x) = \alpha(sr_x, x_0) \circ (\alpha(r_x, x_0))^{-1}. \tag{$*_1$}$$

Equation $(*_1)$ shows that the automorphisms $\alpha(s, x_0)$ of E determine the $\alpha(s, x)$ for all $x \in X$.

Denote $\alpha(s, x_0)$ by $\alpha_0(s)$. Conditions (a) and (b) of Definition 6.2 imply that

$$\begin{aligned} \alpha_0(e) &= \mathrm{id}_E \\ \alpha(sh, x_0) &= \alpha(s, h \cdot x_0) \circ \alpha(h, x_0), \end{aligned}$$

for all $s \in G$ and all $h \in H$, and since $h \cdot x_0 = x_0$, we get

$$\alpha_0(sh) = \alpha_0(s) \circ \alpha_0(h), \qquad \text{for all } s \in G \text{ and all } h \in H. \tag{$*_2$}$$

Now sr_x belongs to the coset $sr_x H = s \cdot r_x H = s \cdot x = r_{s\cdot x} H$, so there is a unique element of H, denoted $u(s, x)$, such that

$$sr_x = r_{s\cdot x} u(s, x), \tag{$\dagger$}$$

and by $(*_2)$,

$$\alpha_0(sr_x) = \alpha_0(r_{s\cdot x} u(s, x)) = \alpha_0(r_{s\cdot x}) \circ \alpha_0(u(s, x)),$$

so $(*_1)$ can be written as

$$\alpha(s, x) = \alpha_0(r_{s\cdot x}) \circ \alpha_0(u(s, x)) \circ (\alpha_0(r_x))^{-1}. \tag{$*_3$}$$

Definition 6.5. Given $\alpha\colon G \times X \to \mathbf{GL}(E)$ as in Definition 6.2, for all $s \in G$, all $h \in H$, and all $x \in X$, define $\alpha_0(s)$, $\sigma(h)$, $\beta(x)$ and $u(s, x)$ by

$$\begin{aligned} \alpha_0(s) &= \alpha(s, x_0) \\ \sigma(h) &= \alpha(h, x_0) = \alpha_0(h) \\ \beta(x) &= \alpha(r_x, x_0) = \alpha_0(r_x) \\ u(s, x) &= r_{s\cdot x}^{-1} s r_x \in H. \end{aligned} \tag{u}$$

Then $(*_3)$ becomes

$$\alpha(s,x) = \beta(s\cdot x)\circ\sigma(u(s,x))\circ(\beta(x))^{-1}, \qquad (*_4)$$

and $(*_2)$ implies that

$$\sigma(h_1h_2) = \sigma(h_1)\circ\sigma(h_2), \qquad \text{for all } h_1,h_2\in H, \qquad (*_5)$$

which shows that $\sigma\colon H\to \mathbf{GL}(E)$ is a homomorphism. Thus, the restriction of the cocycle α to $H\times\{x_0\}$ is a representation of H in E.

Note that for $x = x_0 = H$, since $r_{x_0} = e$, Equation (u) yields

$$u(s,x_0) = r_{s\cdot x_0}^{-1} s r_{x_0} = r_x^{-1}s,$$

so we get

$$s = r_x u(s,x_0), \quad s\in G, \quad x = sH. \qquad \text{(s)}$$

In other words, $u(s,x_0)$ is the unique element $h\in H$, such that $s\in G$ is expressed as $s = r_x h$ in terms of the coset representative r_x in sH.

Conversely, let $\sigma\colon H\to\mathbf{GL}(E)$ be any homomorphism, and let $\beta\colon X\to \mathbf{GL}(E)$ be any function. Then we define the function $u\colon G\times G/H\to H$ using the equation

$$u(s,x) = r_{s\cdot x}^{-1} s r_x$$

of Definition 6.5, and the function $\alpha\colon G\times X\to\mathbf{GL}(E)$ given by $(*_4)$, namely

$$\alpha(s,x) = \beta(s\cdot x)\circ\sigma(u(s,x))\circ(\beta(x))^{-1}.$$

Observe that for all $s,t\in G$ and all $x\in X$, we have

$$u(st,x) = u(s,t\cdot x)u(t,x), \qquad (*_h)$$

because by (†)

$$\begin{aligned} str_x &= r_{(st)\cdot x}u(st,x)\\ sr_{t\cdot x} &= r_{s\cdot(t\cdot x)}u(s,t\cdot x)\\ &= r_{(st)\cdot x}u(s,t\cdot x)\\ tr_x &= r_{t\cdot x}u(t,x), \end{aligned}$$

so we have

$$r_{(st)\cdot x}u(st,x) = str_x = sr_{t\cdot x}u(t,x) = r_{(st)\cdot x}u(s,t\cdot x)u(t,x),$$

and since $r_{(st)\cdot x}\in G$, it is invertible, which proves $(*_h)$. The verification that $\alpha(e,x) = \mathrm{id}_E$ is immediate, because we have $e\cdot x = x$, so $u(e,x) = e$,

$\beta(e \cdot x) = \beta(x)$, and $\sigma(e) = \mathrm{id}_E$. Using $(*_h)$ and the fact that σ is a homomorphism, we have

$$\begin{aligned}\alpha(st, x) &= \beta(st \cdot x) \circ \sigma(u(st, x)) \circ (\beta(x))^{-1} \\ &= \beta(st \cdot x) \circ \sigma(u(s, t \cdot x)u(t, x)) \circ (\beta(x))^{-1} \\ &= \beta(s \cdot (t \cdot x)) \circ \sigma(u(s, t \cdot x)) \circ \sigma(u(t, x)) \circ (\beta(x))^{-1} \\ &= \beta(s \cdot (t \cdot x)) \circ \sigma(u(s, t \cdot x)) \circ \beta(t \cdot x)^{-1} \circ \beta(t \cdot x) \circ \sigma(u(t, x)) \circ (\beta(x))^{-1} \\ &= \alpha(s, t \cdot x) \circ \alpha(t, x),\end{aligned}$$

which shows that α is a cocycle. In summary, we obtained the following result.

Proposition 6.3. *Let G be a group, H be a subgroup of G, and E be a vector space. Choose a set $(r_x)_{x \in G/H}$ of representatives for the cosets of $X = G/H$ as explained above, with $x_0 = H$ and $r_{x_0} = e$. Every cocycle $\alpha\colon G \times X \to \mathbf{GL}(E)$ determines a homomorphism $\sigma\colon H \to \mathbf{GL}(E)$ with $\sigma(h) = \alpha(h, x_0)$ for all $h \in H$, a map $\beta\colon X \to \mathbf{GL}(E)$ given by $\beta(x) = \alpha(r_x, x_0)$ for all $x \in X$, and a map $u\colon G \times G/H \to H$ given by $u(s, x) = r_{s \cdot x}^{-1} s r_x \in H$, such that*

$$\alpha(s, x) = \beta(s \cdot x) \circ \sigma(u(s, x)) \circ (\beta(x))^{-1}.$$

Conversely, given a homomorphism $\sigma\colon H \to \mathbf{GL}(E)$ and a map $\beta\colon X \to \mathbf{GL}(E)$, if we set

$$u(s, x) = r_{s \cdot x}^{-1} s r_x \tag{u}$$

and

$$\alpha(s, x) = \beta(s \cdot x) \circ \sigma(u(s, x)) \circ (\beta(x))^{-1}, \tag{α}$$

then $\alpha\colon G \times X \to \mathbf{GL}(E)$ is a cocycle.

Remarks: Kirillov [Kirillov (2004)] (Appendix V, Section 2.1) calls (u) the *Master equation*. See also Proposition 5, Lemma 2, and Lemma 3. This material is also discussed in Kririllov [Kirillov (1976)] (Sections 13.1 and 13.2).

In view of Proposition 6.3, we make the following definition.

Definition 6.6. Given a homomorphism $\sigma\colon H \to \mathbf{GL}(E)$ and a map $\beta\colon X \to \mathbf{GL}(E)$, if α is the cocycle associated with σ and β, we say that the representation Π^α of G in E^X defined by α is the *representation induced by σ and β*.

Remarkably, for a given homomorphism $\sigma\colon H \to \mathbf{GL}(E)$, the representations $\Pi_1\colon G \to \mathbf{GL}(E^X)$ and $\Pi_2\colon G \to \mathbf{GL}(E^X)$ corresponding to the cocycles α_1 and α_2 associated with two maps β_1 and β_2 are equivalent, in the sense that there is an automorphism γ of E^X, such that

$$\Pi_2 = \gamma \circ \Pi_1 \circ \gamma^{-1}.$$

This is proven as follows.

Proposition 6.4. *Let G be a group, H be a subgroup of G, and E be a vector space. Choose a set $(r_x)_{x \in G/H}$ of representatives for the cosets of $X = G/H$ as explained above, with $x_0 = H$ and $r_{x_0} = e$. Let $\sigma\colon H \to \mathbf{GL}(E)$ be a homomorphism, $\beta\colon X \to \mathbf{GL}(E)$ be a map, and α be the cocycle determined by σ and β as in Proposition 6.3, and $\Pi\colon G \to \mathbf{GL}(E^X)$ be the corresponding representation. If $c(x) = \beta(x)^{-1}$ for all $x \in X$, then define the automorphism γ of E^X by*

$$(\gamma(f))(x) = c(x)(f(x)), \qquad f \in E^X,\ x \in X.$$

Then the representation

$$\Pi' = \gamma \circ \Pi \circ \gamma^{-1}$$

is associated with the cocycle α' given by

$$\alpha'(s, x) = \sigma(u(s, x)), \tag{α'}$$

with

$$u(s, x) = r_{s\cdot x}^{-1} s r_x. \tag{u}$$

Thus, the representation Π induced by σ and β is equivalent to the representation induced by σ and β', with $\beta'(x) = \mathrm{id}_E$ for all $x \in X$. The induced representation Π' associated with α' is given by

$$\begin{aligned}(\Pi'_s(f))(x) &= \sigma(u(s^{-1}, x)^{-1})(f(s^{-1}\cdot x)) \\ &= \sigma(u(s, s^{-1}\cdot x))(f(s^{-1}\cdot x)),\ f \in E^X,\ x \in X.\end{aligned} \tag{Π'}$$

Proof. Consider any map $c\colon X \to \mathbf{GL}(E)$, such that $c(x_0) = \mathrm{id}_E$. Define the automorphism γ of E^X by

$$(\gamma(f))(x) = c(x)(f(x)), \qquad f \in E^X,\ x \in X,$$

and let

$$\Pi' = \gamma \circ \Pi \circ \gamma^{-1}.$$

Since γ is an automorphism of E^X, the map Π' is a linear representation of G in E^X. Clearly, the inverse of γ is given by

$$(\gamma^{-1}(f))(x) = c(x)^{-1}(f(x)), \qquad f \in E^X,\ x \in X,$$

and since for any $g \in E^X$, we have

$$(\Pi_s(g))(x) = \alpha(s, s^{-1} \cdot x)(g(s^{-1} \cdot x)),$$

with $g = \gamma^{-1}(f)$, we obtain

$$\begin{aligned}(\Pi_s(\gamma^{-1})(f))(x) &= \alpha(s, s^{-1} \cdot x)(\gamma^{-1}(f)(s^{-1} \cdot x)) \\ &= \alpha(s, s^{-1} \cdot x)\big(c(s^{-1} \cdot x)^{-1}(f(s^{-1} \cdot x))\big),\end{aligned}$$

and so

$$c(x)\big((\Pi_s(\gamma^{-1})(f))(x)\big) = c(x)\big(\alpha(s, s^{-1} \cdot x)\big(c(s^{-1} \cdot x)^{-1}(f(s^{-1} \cdot x))\big)\big);$$

that is,

$$(\Pi'_s(f))(x) = \big(c(x) \circ \alpha(s, s^{-1} \cdot x) \circ c(s^{-1} \cdot x)^{-1}\big)(f(s^{-1} \cdot x)), \tag{$*_6$}$$

which shows that Π' is obtained from α' as Π is obtained from α, with

$$\alpha'(s, x) = c(s \cdot x) \circ \alpha(s, x) \circ c(x)^{-1}. \tag{$*_7$}$$

If we write $\alpha'_0(s) = \alpha'(s, x_0)$ and $\beta'(x) = \alpha'_0(r_x)$ as before, then the hypothesis $c(x_0) = \mathrm{id}_E$ implies that

$$\begin{aligned}\alpha'_0(h) = \alpha'(h, x_0) &= c(h \cdot x_0) \circ \alpha(h, x_0) \circ c(x_0)^{-1} \\ &= c(x_0) \circ \sigma(h) \circ c(x_0)^{-1} = \sigma(h)\end{aligned}$$

for all $h \in H$, and

$$\begin{aligned}\beta'(x) &= \alpha'(r_x, x_0) \\ &= c(r_x \cdot x_0) \circ \alpha(r_x, x_0) \circ c(x_0)^{-1} \\ &= c(x) \circ \alpha(r_x, x_0) \\ &= c(x) \circ \beta(x);\end{aligned}$$

that is,

$$\beta'(x) = c(x) \circ \beta(x). \tag{$*_8$}$$

Since $\beta(x_0) = \mathrm{id}_E$, we can pick $c(x) = \beta(x)^{-1}$, and then

$$\beta'(x) = \mathrm{id}_E(x) \qquad \text{for all } x \in X, \tag{$*_9$}$$

and since

$$\alpha(s, x) = \beta(s \cdot x) \circ \sigma(u(s, x)) \circ (\beta(x))^{-1},$$

from $(*_7)$ we obtain

$$\alpha'(s, x) = \sigma(u(s, x)). \tag{$*_{10}$}$$

Therefore, Π is equivalent to Π' with $\beta'(x) = \mathrm{id}_E$ for all $x \in X$. ☐

It is also easy to check that if σ is replaced by an equivalent representation σ' of H in E^X, then the corresponding representations Π and Π' of G in E^X are equivalent.

Therefore, the process for making a representation Π of G in E^X from a representation σ of H in E and a function $\beta\colon X \to \mathbf{GL}(E)$ defines a class of representations of G in E^X. Furthermore, there is a special representation associated with σ and the constant function β given by $\beta(x) = \mathrm{id}_E$, for all $x \in X$.

In summary, the method is find a set $(r_x)_{x\in G/H}$ of representatives for the cosets of G/H, then to construct u given by $u(s,x) = r_{s\cdot x}^{-1} s r_x$ as in Equation (u), and then to define α by $\alpha(s,x) = \sigma(u(s,x))$. The induced representation is given by

$$(\Pi_s(f))(x) = \sigma(u(s, s^{-1}\cdot x))(f(s^{-1}\cdot x)), \;\; f \in E^X,\, x \in X. \qquad (*)$$

Vilenkin [Vilenkin (1968)] (Chapter 1, Section 7) calls such a representation a *representation with operator factor*.

From a theoretical point of view, a cocycle α is equivalent to a pair (σ,β) as in Proposition 6.3, but from a practical point of view, it may be very hard (if not impossible) to find constructively a set $(r_x)_{x\in G/H}$ of representatives for the cosets of G/H. Thus, we use cocycles α that agree with a given representation $\sigma\colon H \to \mathbf{GL}(E)$, in the sense that $\alpha(h, x_0) = \sigma(h)$ for all $h \in H$.

A case of practical interest in equivariant machine learning is the case where $G = \mathbf{SE}(3)$ and $H = \mathbf{SO}(3)$.

Example 6.1. Let $G = \mathbf{SE}(3)$ and $H = \mathbf{SO}(3)$. The group $\mathbf{SE}(3)$ is the group of affine rigid motions of $\mathbb{R}^3$ consisting of rotations and translations. Here we view $\mathbf{SE}(3)$ as the group of matrices

$$s = \begin{pmatrix} Q & a \\ 0 & 1 \end{pmatrix}, \qquad Q \in \mathbf{SO}(3),\; a \in \mathbb{R}^3$$

under multiplication. For short, we denote the above matrix by (a, Q). The group $\mathbf{SE}(3)$ acts on $\mathbb{R}^3$ by

$$(a, Q)\cdot x = Qx + a, \quad x \in \mathbb{R}^3.$$

Multiplication in $\mathbf{SE}(n)$ is given by

$$(a, Q)(b, R) = (a + Qb, QR),$$

and the inverse of (a, Q) is

$$(a, Q)^{-1} = (-Q^\top a, Q^\top).$$

For details on $\mathbf{SE}(3)$ and the fact that it is a semi-direct product of $\mathbb{R}^3$ and $\mathbf{SO}(3)$, see Example 7.1. It is easy to see that the homogeneous space $\mathbf{SE}(3)/\mathbf{SO}(3)$ is $\mathbb{R}^3$. Indeed, $\mathbf{SE}(3)$ acts on $\mathbb{R}^3$, and the stabilizer of the origin 0_3 is $\mathbf{SO}(3)$ viewed as the set of matrices

$$\begin{pmatrix} Q & 0 \\ 0 & 1 \end{pmatrix}, \quad Q \in \mathbf{SO}(3).$$

We now use the method based on Proposition 6.3 and Proposition 6.4 to construct an induced representation of $\mathbf{SE}(3)$ from a representation $\sigma\colon \mathbf{SO}(3) \to \mathbf{GL}(E)$ of $\mathbf{SO}(3)$. For this, we need to find a set of representative for the cosets of $\mathbb{R}^3 = \mathbf{SE}(3)/\mathbf{SO}(3)$ in order to define u, and then $\alpha(s, x)$ is given by $\alpha(s, x) = \sigma(u(s, x))$ and the induced representation Π is given by $(*)$. This is a case where it is easy to pick a set of coset representatives, namely for each $x \in \mathbb{R}^3$, $r_x \in \mathbf{SE}(3)$ is the matrix

$$\begin{pmatrix} I_3 & x \\ 0 & 1 \end{pmatrix},$$

the translation by x. The coset $x\mathbf{SO}(3)$ consists of the matrices

$$\begin{pmatrix} Q & x \\ 0 & 1 \end{pmatrix} = \begin{pmatrix} I_3 & x \\ 0 & 1 \end{pmatrix} \begin{pmatrix} Q & 0 \\ 0 & 1 \end{pmatrix}$$

with x fixed. Let us compute $u(s, x) = r_{s\cdot x}^{-1} s r_x$. First, $s \cdot x = (a, Q) \cdot x = Qx + a$, so

$$r_{s\cdot x} = \begin{pmatrix} I_3 & Qx + a, \\ 0 & 1 \end{pmatrix}, \quad r_{s\cdot x}^{-1} = \begin{pmatrix} I_3 & -Qx - a, \\ 0 & 1 \end{pmatrix},$$

and finally,

$$\begin{aligned} u(s, x) = r_{s\cdot x}^{-1} s r_x &= \begin{pmatrix} I_3 & -Qx - a, \\ 0 & 1 \end{pmatrix} \begin{pmatrix} Q & a \\ 0 & 1 \end{pmatrix} \begin{pmatrix} I_3 & x \\ 0 & 1 \end{pmatrix} \\ &= \begin{pmatrix} Q & -Qx \\ 0 & 1 \end{pmatrix} \begin{pmatrix} I_3 & x \\ 0 & 1 \end{pmatrix} = \begin{pmatrix} Q & 0 \\ 0 & 1 \end{pmatrix}. \end{aligned}$$

Consequently, if $\sigma\colon \mathbf{SO}(3) \to \mathbf{GL}(E)$ is any representation of $\mathbf{SO}(3)$ is a finite-dimensional (nontrivial) vector space E, the above shows that $u(s, x)$ is independent of x and given by

$$u(s, x) = u((a, Q), x) = Q,$$

and so $\alpha((a, Q), x)$ is given by

$$\alpha((a, Q), x) = \sigma(u((a, Q), x)) = \sigma(Q).$$

Then by $(*)$ we obtain the representation $\Pi\colon \mathbf{SE}(3) \to \mathbf{GL}(E^{\mathbb{R}^3})$ of $\mathbf{SE}(3)$ in $E^{\mathbb{R}^3}$ given by

$$\begin{aligned}(\Pi_{(a,Q)}(f))(x) &= \sigma(u(s, s^{-1} \cdot x))(f(s^{-1} \cdot x)) \\ &= \sigma(Q)(f((a, Q)^{-1} \cdot x)) = \sigma(Q)(f(Q^\top (x - a))),\end{aligned}$$

that is,

$$(\Pi_{(a,Q)}(f))(x) = \sigma(Q)(f(Q^\top (x - a))), \quad f \in E^{\mathbb{R}^3},\ x \in \mathbb{R}^3.$$

This representation is reducible because the subspace of constant functions from $\mathbb{R}^3$ to E is invariant.

6.3 Converting Induced Representations of G From E^X to E^G

We can also show that a cocycle $\alpha\colon G \times X \to \mathbf{GL}(E)$ defines an isomorphism τ between the space E^X and a subspace L^α of the space E^G.

Definition 6.7. Let G be a group, H be a subgroup of G, E be a vector space, and write $X = G/H$. Given any cocycle $\alpha\colon G \times X \to \mathbf{GL}(E)$, for any function $f\colon X \to E$, the function $f^\alpha\colon G \to E$ is given by

$$f^\alpha(s) = \alpha(s^{-1}, s \cdot x_0)(f(s \cdot x_0)) = (\alpha(s, x_0))^{-1}(f(s \cdot x_0)), \qquad \text{for all } s \in G, \tag{$*_{\alpha_1}$}$$

with $x_0 = H$.

Recall from Definition 6.5 that $\sigma(h) = \alpha(h, x_0)$ for all $h \in H$.

Proposition 6.5. *With the hypotheses of Definition 6.7, the function f^α satisfies the equation*

$$f^\alpha(sh) = \sigma(h^{-1})(f^\alpha(s)), \qquad \textit{for all } h \in H \textit{ and all } s \in G. \tag{$*_{\alpha_2}$}$$

Proof. By (b) of Definition 6.2 and since $h \cdot x_0 = x_0$, we have

$$\begin{aligned}f^\alpha(sh) &= \alpha((sh)^{-1}, (sh) \cdot x_0)(f((sh) \cdot x_0)) \\ &= \alpha(h^{-1}s^{-1}, (sh) \cdot x_0)(f((sh) \cdot x_0)) \\ &= \big(\alpha(h^{-1}, s^{-1} \cdot (s \cdot (h \cdot x_0))) \circ \alpha(s^{-1}, s \cdot (h \cdot x_0))\big)(f(s \cdot (h \cdot x_0))) \\ &= \big(\alpha(h^{-1}, x_0) \circ \alpha(s^{-1}, s \cdot x_0)\big)(f(s \cdot x_0)) \\ &= \sigma(h^{-1})(f^\alpha(s)),\end{aligned}$$

establishing the proposition. ☐

Definition 6.8. Let G be a group, H be a subgroup of G, E be a vector space, and write $X = G/H$. Given any cocycle $\alpha\colon G \times X \to \mathbf{GL}(E)$, let L^α be the subspace of E^G consisting of all functions $g\colon G \to E$, such that

$$g(sh) = \sigma(h^{-1})(g(s)), \qquad \text{for all } s \in G \text{ and all } h \in H, \qquad (*_{\alpha 3})$$

where $\sigma(h) = \alpha(h, x_0)$, for all $h \in H$ (with $x_0 = H$).

Proposition 6.6. *With the hypotheses of Definition 6.7, for every $g \in L^\alpha$, there is a unique function $f\colon E \to X$, such that $g = f^\alpha$. Therefore, the map $\tau\colon E^X \to L^\alpha$ given by $\tau(f) = f^\alpha$ is an isomorphism.*

Proof. Note that the function $s \mapsto \alpha(s, x_0)(g(s))$ has the same value if s is replaced by sh for every $h \in H$, since by (b) of Definition 6.2, $(*_{\alpha 3})$, and the facts that $\sigma(h) = \alpha(h, x_0)$ and $h \cdot x_0 = x_0$ for $h \in H$,

$$\begin{aligned}
\alpha(sh, x_0)(g(sh)) &= \big(\alpha(s, h\cdot x_0) \circ \alpha(h, x_0)\big)(g(sh))\\
&= \big(\alpha(s, x_0) \circ \alpha(h, x_0)\big)\big(\sigma(h^{-1})(g(s))\big)\\
&= \big(\alpha(s, x_0) \circ \sigma(h) \circ \sigma(h^{-1})\big)(g(s))\\
&= \alpha(s, x_0)(g(s)).
\end{aligned}$$

Therefore, we have a well-defined function $f\colon X \to E$ given by

$$f(x) = f(s\cdot x_0) = \alpha(s, x_0)(g(s)), \qquad (*_f)$$

and by definition of f^α, we have

$$\begin{aligned}
f^\alpha(s) &= (\alpha(s, x_0))^{-1}(f(s\cdot x_0))\\
&= (\alpha(s, x_0))^{-1}\big(\alpha(s, x_0)(g(s))\big)\\
&= g(s),
\end{aligned}$$

that is, $f^\alpha = g$, which shows that τ is surjective.

Since $\alpha(s, x)$ is an automorphism and since the map $s \mapsto s\cdot x_0$ from G to G/H is surjective, for any two functions $f_1, f_2 \in E^X$, if $f_1^\alpha = f_2^\alpha$, then

$$\alpha(s^{-1}, s\cdot x_0)(f_1(s\cdot x_0)) = \alpha(s^{-1}, s\cdot x_0)(f_2(s\cdot x_0))$$

for all $s \in G$, so $f_1 = f_2$, which shows that τ is injective. ∎

Observe that in the proof of Proposition 6.6, Equation $(*_f)$ and the fact that $\tau(f) = f^\alpha = g$ show that if $g \in L^\alpha$, then

$$(\tau^{-1}(g))(s\cdot x_0) = \alpha(s, x_0)(g(s)). \qquad (*_{\tau^{-1}(g)})$$

For any cocycle $\alpha\colon G \times X \to \mathbf{GL}(E)$, we can use the isomorphism $\tau\colon E^X \to L^\alpha$ to convert the representation $\Pi\colon G \to \mathbf{GL}(E^X)$ defined by α into the equivalent representation Π_{L^α} given by $\Pi_{L^\alpha}(s) = \tau \circ \Pi(s) \circ \tau^{-1}$.

Proposition 6.7. *For every cocycle $\alpha\colon G \times X \to \mathbf{GL}(E)$, if $\Pi\colon G \to \mathbf{GL}(E^X)$ is the representation defined by α, then the equivalent representation $\Pi_{L^\alpha}\colon G \to \mathbf{GL}(L^\alpha)$ defined by $\Pi_{L^\alpha}(s) = \tau \circ \Pi(s) \circ \tau^{-1}$ is given by*

$$((\Pi_{L^\alpha})_s(g))(t) = g(s^{-1}t), \qquad \text{for all } g \in L^\alpha \text{ and all } s, t \in G. \qquad (\Pi_{L^\alpha})$$

Proof. For any $g \in L^\alpha$, since by $(*_{\tau^{-1}(g)})$

$$(\tau^{-1}(g))(u \cdot x_0) = \alpha(u, x_0)(g(u)),$$

and

$$(\Pi_s(f))(x) = \alpha(s, s^{-1} \cdot x_0)(f(s^{-1} \cdot x)),$$

with $x = t \cdot x_0$, we have

$$(\Pi_s(f))(t \cdot x_0) = \alpha(s, s^{-1} \cdot x_0)(f(s^{-1} \cdot (t \cdot x_0))),$$

and by setting $f = \tau^{-1}(g)$, we get

$$\begin{aligned}\big(\Pi_s(\tau^{-1}(g))\big)(t \cdot x_0) &= \alpha(s, s^{-1} \cdot (t \cdot x_0))\big((\tau^{-1}(g))(s^{-1} \cdot (t \cdot x_0))\big)\\ &= \alpha(s, (s^{-1}t) \cdot x_0)\big((\tau^{-1}(g))((s^{-1}t) \cdot x_0)\big)\\ &= \alpha(s, (s^{-1}t) \cdot x_0)\big(\alpha(s^{-1}t, x_0)(g(s^{-1}t))\big)\\ &= \alpha(ss^{-1}t, x_0)(g(s^{-1}t))\\ &= \alpha(t, x_0)(g(s^{-1}t)).\end{aligned}$$

Since

$$(\tau(h))(t) = (\alpha(t, x_0))^{-1}(h(t \cdot x_0)),$$

for any $h \in E^X$, with $h = \Pi_s(\tau^{-1}(g))$, we obtain

$$\tau\big(\Pi_s(\tau^{-1}(g))\big)(t) = (\alpha(t, x_0))^{-1}\big(\alpha(t, x_0)(g(s^{-1}t))\big) = g(s^{-1}t),$$

as claimed. ∎

Remark: Observe that L^α only depends on σ, so we may write L^σ instead of L^α, and Π_{L^α} depends only on σ, so we may also write Π_{L^σ} instead of Π_{L^α}.

The representation Π_{L^σ}, which is simply the left regular representation of G on L^α, is more intrinsic than the representations Π^α acting on the space of functions in E^X. The representations Π^α acting on the space of functions in E^X require for their construction the choice of a set of coset representatives $(r_x)_{x \in G/H}$ in addition to the representation $\sigma\colon H \to \mathbf{GL}(E)$ in order to define a cocycle α. However, if $X = G/H$ is a lot "smaller" than G, then the space of functions in L^α (a space of functions from G to E) is very redundant and from a practical point of view, it might be better to use the representations defined on the smaller space of functions from X to E. This issue will come up in Chapter 8.

We have concluded our discussion of algebraic methods for constructing representations of G from representations of a subgroup H of G.

6.4 Construction of the Hilbert Space $\mathbf{L}^2_\mu(X; E)$

We now assume that G is a locally compact group and that H is a closed subgroup of G. By Vol. I, Proposition 8.6(1), the space $X = G/H$ is also locally compact. If G is separable, then so is G/H, and if G is metrizable, then so G/H; see Dieudonné [Dieudonné (2005)] (Chapter XII, Sections 10 and 11).

Given a unitary representation $U\colon H \to \mathbf{U}(E)$ of H, we would like to construct a unitary representation $\Pi\colon G \to \mathbf{U}(\mathcal{H})$ of G. This is possible under certain conditions on H and G and on measures on $X = G/H$. Note that unlike in the previous sections, we are now considering *continuous* unitary representations.

The first step is to construct a Hilbert space $\mathcal{H}$ that will be the representation space of a unitary representation of G. There are two approaches:

(1) The Hilbert space $\mathcal{H}$ is a set of functions from $X = G/H$ to E.
(2) The Hilbert space $\mathcal{H}$ is a set of functions from G to E, analogous to the space L^α of Section 6.3.

The second step is to define the operators Π_s (for $s \in G$), so that they are unitary operators of $\mathcal{H}$. This involves defining an inner product in $\mathcal{H}$ that makes the operators Π_s unitary. In the first approach that makes use of cocycles, the definition of the inner product on $\mathcal{H}$ is straightforward. To ensure that the operators Π_s are unitary, a Borel measure μ on $X = G/H$ is needed, and the cocycles must satisfy some additional conditions with respect to the measure μ. The case where the measure μ is G-invariant is simpler than the case where μ is only quasi-invariant.

In the second approach, the definition of the Hilbert space $\mathcal{H}$ is more complicated and requires a completion. We will sketch two variants of this method at the end of Section 6.7.

A good candidate for the first approach is a subspace $\mathrm{L}^2_\mu(X;E)$ of the vector space E^X, where μ is positive Borel measure on G/H. In the special case where H is compact, given a cocycle α on $G \times X$ satisfying some suitable conditions, the space L^α will be a subspace of $\mathrm{L}^2_\lambda(G;E) \subseteq E^G$, where λ is a left-invariant Haar measure on G.

Whether μ is G-invariant is an issue that will come up later, but for the time being we can ignore it.

Let E be a separable Hilbert space, and let (a_n) be a Hilbert basis of E. Every function $f\colon X \to E$ can be written uniquely as $f = \sum_n f_n a_n$, where $f_n\colon X \to \mathbb{C}$, and such that the series $\sum_n |f_n(x)|^2$ converges for all $x \in X$. By definition, we let

$$\|f(x)\|_E^2 = \sum_n |f_n(x)|^2.$$

We claim that a function $f\colon X \to E$ is μ-measurable iff all the f_n are μ-measurable.

If f is μ-measurable, since $f_n(x) = \langle f(x), a_n\rangle$, the f_n are μ-measurable. Conversely, if the f_n are μ-measurable, then Egoroff's theorem implies that f is μ-measurable; see Dieudonné [Dieudonné (2005)] (Chapter XIII, Theorem 13.9.10).

Definition 6.9. Let G be a locally compact group, H be a closed subgroup of G, μ be a positive Borel measure on $X = G/H$, and E be a separable Hilbert space. For any Hilbert basis (a_n) of E, let $\mathcal{L}^2_\mu(X;E)$ be the space of all μ-measurable functions $f\colon X \to E$ with $f = \sum_n f_n a_n$, such that the function $x \mapsto \sum_n |f_n(x)|^2 = \|f(x)\|_E^2$ is μ-integrable.

It is easy to see that if $f = \sum_n f_n a_n$, then $f_n \in \mathcal{L}^2_\mu(X;\mathbb{C})$, and

$$\int_{G/H} \|f\|_E^2 \, d\mu = \sum_n \int_{G/H} |f_n|^2 \, d\mu = \sum_n \|f_n\|_2^2;$$

see Dieudonné [Dieudonné (2005)] (Chapter XIII, Sections 8 and 9). As a consequence, given two functions $f = \sum_n f_n a_n$ and $g = \sum_n g_n a_n$ in $\mathcal{L}^2_\mu(X;E)$, by Vol. I, Proposition 5.41, the function $x \mapsto \langle f(x), g(x)\rangle$ is integrable and

$$\int_{G/H} \langle f(x), g(x)\rangle \, d\mu(x) = \sum_n \int_{G/H} f_n(x)\overline{g_n(x)} \, d\mu(x).$$

Definition 6.10. We say that a function $f \in \mathcal{L}^2_\mu(X;E)$ is *negligible* if the function $x \mapsto \|f(x)\|_E^2$ is zero almost everywhere.

The quotient of the space $\mathcal{L}^2_\mu(X;E)$ by the subspace of negligible functions is denoted by $\mathrm{L}^2_\mu(X;E)$. It is a hermitian space under the inner product

$$\langle [f],[g]\rangle = \int_{G/H} \langle f(x), g(x)\rangle \, d\mu(x),$$

and we have the norm N_2[1] given by

$$N_2([f]) = \sqrt{\langle [f],[f]\rangle}.$$

If $[f]$ is represented by $f = \sum_n f_n a_n$, then

$$N_2([f])^2 = \int_{G/H} \|f\|_E^2 \, d\mu = \sum_n \|f_n\|_2^2 .$$

Actually, it turns out that the hermitian space $\mathrm{L}^2_\mu(X;E)$ is complete, that is, it is a Hilbert space. In fact, it is a separable Hilbert space.

Proposition 6.8. *Let G be a locally compact group, H be a closed subgroup of G, μ be a positive Borel measure on $X = G/H$, and E be a separable Hilbert space. The space $\mathrm{L}^2_\mu(X;E)$ is a separable Hilbert space.*

Proof. Let $(f^{(m)})$ be a Cauchy sequence in $\mathrm{L}^2_\mu(X;E)$, with $f^{(m)} = \sum_n f_n^{(m)} a_n$. For every $\epsilon > 0$, there is some m_0, such that for all $p, q \geq m_0$, we have

$$N_2(f^{(p)} - f^{(q)})^2 = \sum_n \int_{G/H} |f_n^{(p)} - f_n^{(q)}|^2 \, d\mu \leq \epsilon, \tag{$*_1$}$$

and this implies that for every n, the sequence $(f_n^{(m)})_{m\geq 1}$ is a Cauchy sequence in $\mathrm{L}^2_\mu(X;\mathbb{C})$. Therefore, each sequence $(f_n^{(m)})_{m\geq 1}$ has a limit $g_n \in \mathrm{L}^2_\mu(X;\mathbb{C})$, since $\mathrm{L}^2_\mu(X;\mathbb{C})$ is complete by Fischer–Riesz. For every integer $N > 0$, if we let q tend to $+\infty$ in $(*_1)$, we see that

$$\sum_{n=1}^N \left\| g_n - f_n^{(p)} \right\|_2^2 \leq \epsilon, \tag{$*_2$}$$

so

$$\sum_{n=1}^N \|g_n\|_2^2 \leq \sum_{n=1}^N \left\| g_n - f_n^{(p)} \right\|_2^2 + \sum_{n=1}^N \left\| f_n^{(p)} \right\|_2^2 \leq \epsilon + \left\| f^{(p)} \right\|_2^2,$$

[1]We are using the notation N_2 for the norm on $\mathrm{L}^2_\mu(X;E)$ to avoid a confusion with the norm $\| \, \|_2$ on $\mathrm{L}^2_\mu(X;\mathbb{C})$.

which proves that the series $\sum_{n=1}^{\infty} \|g_n\|_2^2$ converges. Since (by definition)

$$\sum_{n=1}^{\infty} \|g_n\|_2^2 = N_2(g)^2,$$

it follows that $g = \sum_n g_n a_n \in \mathrm{L}^2_\mu(X; E)$, and by $(*_2)$

$$N_2(g - f^{(p)})^2 = \sum_n \left\| g_n - f_n^{(p)} \right\|_2^2 \leq \epsilon$$

for all $p \geq m_0$, and so g is the limit of the sequence $(f^{(m)})$ in $\mathrm{L}^2_\mu(X; E)$.

If D is a countable dense subset of $\mathrm{L}^2_\mu(X; \mathbb{C})$, then we can check that the set of functions $f = \sum f_n a_n$, such that $f_n \in D$ for all n and $f_n = 0$ but all for finitely many values of n is dense in $\mathrm{L}^2_\mu(X; E)$. $\square$

6.5 Induced Representations, I; G/H has a G-Invariant Measure

In the rest of this chapter, by unitary representation, we mean *continuous* unitary representation.

We will now assume that the positive Borel measure μ on $X = G/H$ is G-invariant. Recall from Vol. I, Section 8.10 (Definition 8.18) that

$$(\lambda_s(\mu))(A) = \mu(s^{-1} \cdot A),$$

for every Borel subset A of X, so μ is G-invariant if for every Borel subset A of X,

$$\mu(s^{-1} \cdot A) = \mu(A), \qquad \text{for all } s \in G.$$

In this case,

$$\int_{G/H} f(s \cdot x)\, d\mu(x) = \int_{G/H} f(x)\, d\mu(x), \qquad \text{for all } s \in G.$$

Let E be a separable Hilbert space, and $U\colon H \to \mathbf{U}(E)$ be a unitary representation of H.

Theorem 6.9. *Let G be a locally compact group, H be a closed subgroup of G, E be a separable Hilbert space, and $U\colon H \to \mathbf{U}(E)$ be a unitary representation of H. If $X = G/H$ admits a G-invariant σ-Radon measure μ, and for any cocycle $\alpha\colon G \times X \to \mathbf{U}(E)$, if the following conditions hold*

(1) We have $\alpha(h, x_0) = U(h)$ for all $h \in H$;

(2) For every $s \in G$, for every $f \in \mathrm{L}^2_\mu(X;E)$, the map $x \mapsto \alpha(s,x)(f(x))$ from X to E is μ-measurable;

(3) For every $f \in \mathrm{L}^2_\mu(X;E)$, the map $s \mapsto \Pi_s(f)$ is a continuous map from G to $\mathrm{L}^2_\mu(X;E)$, where Π is the homomorphism $\Pi\colon G \to \mathbf{GL}(E^X)$ induced by the cocycle α;

then the homomorphism $\Pi\colon G \to \mathbf{U}(\mathrm{L}^2_\mu(X;E))$ induced by the cocycle α given by

$$(\Pi_s(f))(x) = (\alpha(s^{-1},x))^{-1}(f(s^{-1}\cdot x)), \qquad f \in \mathrm{L}^2_\mu(X;E),\ x \in X,$$

(see Definition 6.3) is a unitary representation of G.

Proof. We simply have to prove that

$$N_2(\Pi_s(f)) = N_2(f), \qquad \text{for all } f \in \mathcal{L}^2_\mu(X;E) \text{ and all } s \in G,$$

which implies that $\Pi_s(f) \in \mathcal{L}^2_\mu(X;E)$, and the other conditions imply that the homomorphism $\Pi\colon G \to \mathbf{GL}(\mathrm{L}^2_\mu(X;E))$ induced by α is a unitary representation of G. Since by hypothesis $\alpha(s,s^{-1}\cdot x)$ is a unitary operator, we have

$$\|(\Pi_s(f))(x)\|_E = \left\|\alpha(s,s^{-1}\cdot x)(f(s^{-1}\cdot x))\right\|_E = \left\|f(s^{-1}\cdot x)\right\|_E,$$

and since μ is G-invariant,

$$\int_{G/H} \left\|f(s^{-1}\cdot x)\right\|_E^2 d\mu(x) = \int_{G/H} \|f(x)\|_E^2\, d\mu(x),$$

and so $N_2(\Pi_s(f)) = N_2(f)$. $\square$

Definition 6.11. The unitary representation $\Pi\colon G \to \mathbf{U}(\mathrm{L}^2_\mu(X;E))$ induced by the cocycle α (and the unitary representation $U\colon H \to \mathbf{U}(E)$) is denoted $\mathrm{Ind}_H^G\alpha$, or by abuse of notation even $\mathrm{Ind}_H^G U$.

Remark: To be very precise, the representing space $\mathrm{L}^2_\mu(X;E)$ of this representation should be specified, for example as in $\mathrm{Ind}^G_{H,\mathrm{L}^2_\mu(X;E)}\alpha$, because there are variants of this construction that use a different representation space.

If U is the trivial representation of H in E, and if we choose $\alpha(s,x) = \mathrm{id}_E$ for all $(s,x) \in G \times (G/H)$, then it can be verified that the hypotheses of Theorem 6.9 are satisfied. To verify condition (3), we use the fact that the family of maps $f \mapsto \Pi_s(f)$ $(s \in G)$ is equicontinuous; see Vol. I, Proposition 2.13. Then we use Vol. I, Proposition 2.12; for details, see Dieudonné [Dieudonné (2003b)] (Chapter XXII, Section 3). In this case, the subspace L^α corresponding to $\mathcal{L}^2_\mu(X;E)$ consists of all functions of the form $f \circ \pi$ with $f \in \mathcal{L}^2_\mu(X;E)$, where $\pi\colon G \to G/H$ is the projection map.

If H is a (closed) compact subgroup of G, then by Vol. I, Proposition 8.43, the space G/H has G-invariant measures (unique up to a scalar). This is a special case of particular interest. A good illustration of this situation is provided by Example 6.1 that we now revisit.

Example 6.2. As in Example 6.1, consider the groups $G = \mathbf{SE}(3)$ and $H \approx \mathbf{SO}(3)$, where G is locally compact and H is compact and closed in G. Consequently, $X = G/H \approx \mathbb{R}^3$ has an $\mathbf{SE}(3)$-invariant Radon measure μ. Consider any unitary representation $\sigma\colon \mathbf{SO}(3) \to \mathbf{U}(E)$ of $\mathbf{SO}(3)$ in a separable Hilbert space E. We showed in Example 6.1 that we have a cocycle $\alpha\colon \mathbf{SE}(3) \times \mathbb{R}^3 \to \mathbf{U}(E)$ given by

$$\alpha((a, Q), x) = \sigma(Q), \quad a, x \in \mathbb{R}^3,\ Q \in \mathbf{SO}(3),$$

and the homomorphism $\Pi\colon \mathbf{SE}(3) \to \mathbf{GL}(E^{\mathbb{R}^3})$ induced by α is given by

$$(\Pi_{(a,Q)}(f))(x) = \sigma(Q) f(Q^\top (x - a)), \quad f \in E^{\mathbb{R}^3},\ x \in \mathbb{R}^3.$$

We leave it as an exercise to check that conditions (1)–(3) of Theorem 6.9 are satisfied, and so Π is a unitary representation $\Pi\colon \mathbf{SE}(3) \to \mathbf{U}(\mathrm{L}^2_\mu(\mathbb{R}^3;E))$ of $\mathbf{SE}(3)$ in the Hilbert space $\mathrm{L}^2_\mu(\mathbb{R}^3;E)$. If E is finite-dimensional, say of dimension $n \geq 1$, then the Hilbert space $\mathrm{L}^2_\mu(\mathbb{R}^3;E)$ is isomorphic to the direct sum of n copies of $\mathrm{L}^2_\mu(\mathbb{R}^3;\mathbb{C})$. Then every function $f \in \mathrm{L}^2_\mu(\mathbb{R}^3;E)$ is identified with the n-tuple $f = (f_1, \ldots, f_n)$ where $f_i \in \mathrm{L}^2_\mu(\mathbb{R}^3;\mathbb{C})$, with the inner product of $f = (f_1, \ldots, f_n)$ and $g = (g_1, \ldots, g_n)$ given by

$$\langle f, g\rangle = \sum_{i=1}^n \int_{\mathbb{R}^3} f_n(x)\overline{g(x)}\, d\mu(x).$$

Another example of induced representations of $G = \mathbf{SE}(n)$ arises from the normal abelian subgroup $H = \mathbb{R}^n$.

Example 6.3. Consider the groups $G = \mathbf{SE}(n)$ and $H \approx \mathbb{R}^n$, where G is locally compact and H is a closed normal abelian group in G. Here, $G = \mathbf{SE}(n)$ consists of all matrices

$$s = \begin{pmatrix} Q & a \\ 0 & 1 \end{pmatrix}, \quad \text{with } Q \in \mathbf{SO}(n) \text{ and } a \in \mathbb{R}^n,$$

$H \approx \mathbb{R}^n$ is the normal subgroup of $\mathbf{SE}(n)$ consisting of all matrices

$$h = \begin{pmatrix} I_n & b \\ 0 & 1 \end{pmatrix}, \quad \text{with } b \in \mathbb{R}^n,$$

and $X = G/H \approx \mathbf{SO}(n)$ is the compact subgroup of $\mathbf{SE}(n)$ consisting of all matrices

$$\begin{pmatrix} R & 0 \\ 0 & 1 \end{pmatrix}, \quad \text{with } R \in \mathbf{SO}(n).$$

Recall that since $\mathbb{R}^n$ is abelian, its irreducible representations are one-dimensional. Therefore, the irreducible representations of $\mathbb{R}^n$ are determined by the characters of $\mathbb{R}^n$, which by Vol. I, Corollary 10.93 are of the form $\chi_y : \mathbb{R}^n \to \mathbb{T}$ for any $y \in \mathbb{R}^n$, with

$$\chi_y(x) = e^{iy \cdot x}, \quad x \in \mathbb{R}^n.$$

Consequently, the irreducible representations $\rho : \mathbb{R}^n \to \mathbf{U}(1)$ of $\mathbb{R}^n$ are of the form

$$(\rho(x))(z) = \chi_y(x)z, \quad x \in \mathbb{R}^n,\ z \in \mathbb{C}$$

for any fixed $y \in \mathbb{R}^n$, namely, multiplication by $\chi_y(x)$. Since for

$$s = (a, Q) = \begin{pmatrix} Q & a \\ 0 & 1 \end{pmatrix} \in \mathbf{SE}(n) \quad \text{and} \quad h = (b, I) = \begin{pmatrix} I & b \\ 0 & 1 \end{pmatrix} \in H \approx \mathbb{R}^n,$$

we have

$$\begin{aligned} sH = (a, Q)H = \{(a, Q)h \mid h \in H\} &= \left\{ \begin{pmatrix} Q & a + Qb \\ 0 & 1 \end{pmatrix} \mid b \in \mathbb{R}^n \right\} \\ &= \left\{ \begin{pmatrix} Q & c \\ 0 & 1 \end{pmatrix} \mid c \in \mathbb{R}^n \right\}, \end{aligned}$$

and we have an isomorphism between $\mathbf{SO}(n)$ and $X = \mathbf{SE}(n)/H$ given by

$$Q \mapsto (a, Q)H = \left\{ \begin{pmatrix} Q & c \\ 0 & 1 \end{pmatrix} \mid c \in \mathbb{R}^n \right\}.$$

Since each matrix in the coset $(a, Q)H$ can be written uniquely as

$$\begin{pmatrix} Q & c \\ 0 & 1 \end{pmatrix} = \begin{pmatrix} Q & 0 \\ 0 & 1 \end{pmatrix} \begin{pmatrix} I_n & Q^\top c \\ 0 & 1 \end{pmatrix},$$

it is very easy to pick a coset representative in $\mathbf{SE}(n)$, namely

$$r_Q = \begin{pmatrix} Q & 0 \\ 0 & 1 \end{pmatrix}, \quad Q \in \mathbf{SO}(n).$$

The coset H as point in $X = \mathbf{SE}(n)/H \approx \mathbf{SO}(n)$ is $x_0 = I_n$. Since the action of $\mathbf{SE}(n)$ on $X = \mathbf{SE}(n)/H \approx \mathbf{SO}(n)$ is given by

$$s_1(sH) = (s_1 s)H, \quad s_1, s \in \mathbf{SE}(n),$$

we have

$$s_1(sH) = (s_1s)H = (a_1, Q_1)(a, Q)H = (a_1 + Q_1a, Q_1Q)H$$
$$= \left\{ \begin{pmatrix} Q_1Q & c \\ 0 & 1 \end{pmatrix} \mid c \in \mathbb{R}^n \right\},$$

and using our isomorphism between $\mathbf{SO}(n)$ and $X = \mathbf{SE}(n)/H$, the above equation becomes

$$s_1 \cdot Q = (a_1, Q_1) \cdot Q = Q_1Q, \quad Q, Q_1 \in \mathbf{SO}(n),\ a_1 \in \mathbb{R}^n.$$

Then, since

$$s \cdot R = (a, Q) \cdot R = QR,$$

$u(s, R) = (r_{s\cdot R})^{-1} s r_R$ is given by

$$u(s, R) = (r_{s\cdot R})^{-1} s r_R = \begin{pmatrix} R^\top Q^\top & 0 \\ 0 & 1 \end{pmatrix} \begin{pmatrix} Q & a \\ 0 & 1 \end{pmatrix} \begin{pmatrix} R & 0 \\ 0 & 1 \end{pmatrix}$$
$$= \begin{pmatrix} R^\top & R^\top Q^\top a \\ 0 & 1 \end{pmatrix} \begin{pmatrix} R & 0 \\ 0 & 1 \end{pmatrix} = \begin{pmatrix} I_n & R^\top Q^\top a \\ 0 & 1 \end{pmatrix}.$$

Technically, we prefer dealing with representations $\sigma\colon \mathbb{R}^n \to \mathbf{U}(1)$ rather than $\sigma\colon H \to \mathbf{U}(1)$, so using the isomorphism $\mathbb{R}^n \approx H$, we have

$$u((a, Q), R) = R^\top Q^\top a.$$

Consequently, for every irreducible representation $\sigma = \chi_y\colon \mathbb{R}^n \to \mathbf{U}(1)$, we have the cocycle $\alpha\colon \mathbf{SE}(n) \times \mathbf{SO}(n) \to \mathbf{U}(1)$ given by

$$\alpha((a, Q), R) = \sigma(u((a, Q), R)) = \sigma(R^\top Q^\top a) = \chi_y(R^\top Q^\top a).$$

Observe that if $(a, Q) \in H \approx \mathbb{R}^n$, that is, $Q = I$, and $R = x_0 = I$, we have $\alpha((a, I), I) = \sigma(a) = \chi_y(a)$. Since

$$s^{-1} \cdot R = (a, Q)^{-1} \cdot R = Q^\top R,$$

the representation $\Pi\colon \mathbf{SE}(n) \to \mathbf{GL}(\mathbb{C}^{\mathbf{SO}(n)})$ of $\mathbf{SE}(n)$ in $\mathbb{C}^{\mathbf{SO}(n)}$ induced by the representation $\sigma = \chi_y\colon \mathbb{R}^n \to \mathbf{U}(1)$ is defined such that for all $s = (a, Q) \in \mathbf{SE}(n)$, $R \in \mathbf{SO}(n)$ and all functions $f\colon \mathbf{SO}(n) \to \mathbb{C}$,

$$\begin{aligned}
(\Pi_{(a,Q)}(f))(R) &= \alpha(s, s^{-1} \cdot R)(f(s^{-1} \cdot R)) \\
&= \alpha((a, Q), Q^\top R)(f((a, Q)^{-1} \cdot R)) \\
&= \sigma((Q^\top R)^\top Q^\top a)(f((a, Q)^{-1} \cdot R)) \\
&= \sigma(R^\top a) f((a, Q)^{-1} \cdot R) \\
&= \chi_y(R^\top a) f((a, Q)^{-1} \cdot R) \\
&= e^{i(y\cdot(R^\top a))} f((a, Q)^{-1} \cdot R) \\
&= e^{i((Ry)\cdot a)} f(Q^\top R).
\end{aligned}$$

Since the action of $\mathbf{SE}(n)$ on $\mathbf{SO}(n)$ is identical to the action of $\mathbf{SO}(n)$ on $\mathbf{SO}(n)$, the homogeneous space $X = \mathbf{SO}(n)$ has an $\mathbf{SE}(n)$-invariant Radon measure, namely the Haar measure μ on $\mathbf{SO}(n)$. We already checked that the cocycle

$$\alpha((a,Q),R) = \chi_y(R^\top Q^\top a) = e^{i(y\cdot(R^\top Q^\top a))} = e^{i((Ry)\cdot(Q^\top a))}$$

satisfies condition (1) of Theorem 6.9, and we leave it as an exercise to prove that conditions (2) and (3) are also satisfied. As a consequence, we obtain a unitary representation $\Pi\colon \mathbf{SE}(n) \to \mathbf{U}(\mathrm{L}^2_\mu(\mathbf{SO}(n);\mathbb{C}))$ of $\mathbf{SE}(n)$ in the Hilbert space $\mathrm{L}^2_\mu(\mathbf{SO}(n);\mathbb{C})$ given by

$$(\Pi_{(a,Q)}(f))(R) = e^{i((Ry)\cdot a)} f(Q^\top R), \quad (a,Q) \in \mathbf{SE}(n),\ R \in \mathbf{SO}(n),$$
$$f \in \mathrm{L}^2_\mu(\mathbf{SO}(n);\mathbb{C}),\ y \in \mathbb{R}^n.$$

The above formula suggests that it might be possible to define a representation of $\mathbf{SE}(n)$ in the smaller Hilbert space $\mathrm{L}^2_\lambda(S^{n-1};\mathbb{C})$, where λ is an $\mathbf{SO}(n)$-invariant Radon measure on S^{n-1}, which exists since S^{n-1} is a homogeneous space obtained by making $\mathbf{SO}(n)$ act on S^{n-1} by the action $R\cdot x = Rx$, where $R \in \mathbf{SO}(n)$ and $x \in S^{n-1}$, so $S^{n-1} \approx \mathbf{SO}(n)/\mathbf{SO}(n-1)$ with $\mathbf{SO}(n-1)$ compact. Before proceeding any further, the reader may want to review Vol. I, Section C.2 and Section C.3. We may assume that $y \neq 0$, because when $y = 0$, we have

$$(\Pi_{(a,Q)}(f))(R) = f(Q^\top R),$$

a reducible representation called a *quasi-regular representation* of $\mathbf{SE}(n)$. Here, we pick the base point to be

$$x_0 = (1/r)y \in S^{n-1}, \quad \text{with } r = \|y\|.$$

The stabilizer $\mathbf{SO}(n)_{x_0} \approx \mathbf{SO}(n-1)$ of x_0 is given by

$$\mathbf{SO}(n)_{x_0} = \{R \in \mathbf{SO}(n) \mid Rx_0 = x_0\},$$

and so, for any $R_1, R_2 \in \mathbf{SO}(n)$, the two cosets $R_1\mathbf{SO}(n)_{x_0}$ and $R_2\mathbf{SO}(n)_{x_0}$ are identical iff $R_2^\top R_1 \in \mathbf{SO}(n)_{x_0}$ iff $R_2^\top R_1 x_0 = x_0$ iff $R_1x_0 = R_2x_0$. The isomorphism between $\mathbf{SO}(n)/\mathbf{SO}(n)_{x_0}$ and the orbit $\mathbf{SO}(n)x_0 = S^{n-1}$ is given by $R\,\mathbf{SO}(n)_{x_0} \mapsto Rx_0 = (1/r)(Ry)$, where $R \in \mathbf{SO}(n)$. Consider the map $\widetilde{\Pi}\colon \mathbf{SE}(n) \to \mathbf{U}(\mathrm{L}^2_\lambda(S^{n-1};\mathbb{C}))$ given by

$$(\widetilde{\Pi}_{(a,Q)}(f))([R]) = e^{i((Ry)\cdot a)} f(Q^\top [R]), \quad (a,Q) \in \mathbf{SE}(n),\ y \in \mathbb{R}^n, \qquad (*_1)$$

with $[R] \in \mathbf{SO}(n)/\mathbf{SO}_{x_0}$ and $f \in \mathrm{L}^2_\lambda(S^{n-1};\mathbb{C})$, and where $[R]$ denotes the coset $R\,\mathbf{SO}_{x_0}$. Since by definition of the stabilizer $\mathbf{SO}_{x_0}$, if $[R_1] = [R_2]$, then $R_1y = R_2y$, the right-hand side of $(*_1)$ does not depend on the representative chosen in the coset $[R]$, so $\widetilde{\Pi}_{(a,Q)}$ is well-defined, and if we write $x = (1/r)(Ry) \in S^{n-1}$, since $\mathbf{SO}(n)/\mathbf{SO}_{x_0} \approx S^{n-1}$ under the map $[R] \mapsto Rx_0 = (1/r)Ry = x$, we have

$$(\widetilde{\Pi}_{(a,Q)}(f))(x) = e^{ir(x\cdot a)} f(Q^\top x), \qquad (a,Q) \in \mathbf{SE}(n),\ x \in S^{n-1}, \quad f \in \mathrm{L}^2_\lambda(S^{n-1}; \mathbb{C}),\ r > 0. \qquad (*_2)$$

The above also shows that the representation $\Pi\colon \mathbf{SE}(n) \to \mathbf{U}(\mathrm{L}^2_\mu(\mathbf{SO}(n); \mathbb{C}))$ of $\mathbf{SE}(n)$ in the Hilbert space $\mathrm{L}^2_\mu(\mathbf{SO}(n); \mathbb{C})$ is *reducible* because the subspace of $\mathrm{L}^2_\mu(\mathbf{SO}(n); \mathbb{C})$ consisting of the functions $f \in \mathrm{L}^2_\mu(\mathbf{SO}(n); \mathbb{C})$, such that

$$f(RT) = f(R), \quad \text{for all } R \in \mathbf{SO}(n) \text{ and all } T \in \mathbf{SO}(n)_{x_0},$$

is invariant under $\Pi_{(a,Q)}$, because for all $Q, R \in \mathbf{SO}(n)$ and all $T \in \mathbf{SO}(n)_{x_0}$, we have

$$\begin{aligned} e^{i((RTy)\cdot a)} f(Q^\top RT) &= e^{ir((RTx_0)\cdot a)} f(Q^\top R) \\ &= e^{ir((Rx_0)\cdot a)} f(Q^\top R) = e^{i((Ry)\cdot a)} f(Q^\top R), \end{aligned}$$

since $Tx_0 = x_0$ and $f(Q^\top RT) = f(Q^\top R)$.

The representations given by $(*_2)$ are half of the representations of $\mathbf{SE}(n)$ discussed in Vilenkin [Vilenkin (1968)] (Chapter XI, Section 2), the other half corresponding to $r < 0$. However, it is easy to see that each representation given by

$$(\widetilde{\Pi}_{(a,Q)}(f))(x) = e^{-ir(x\cdot a)} f(Q^\top x), \qquad (a,Q) \in \mathbf{SE}(n),\ x \in S^{n-1}, \quad f \in \mathrm{L}^2_\lambda(S^{n-1}; \mathbb{C}),\ r > 0 \qquad (*_3)$$

is equivalent to the corresponding representation given by $(*_2)$ (with no negative sign in front of $ir > 0$) using the isometry T of $\mathrm{L}^2_\lambda(S^{n-1}; \mathbb{C})$ given by

$$T(f)(x) = f(-x), \quad x \in S^{n-1},$$

in other words, $T(f) = \check{f}$ (see Vol. I, Definition 8.11). It is proven in Vilenkin [Vilenkin (1968)] (Chapter XI, Section 2) that the representations given by $(*_2)$ (and thus by $(*_3)$) are irreducible.

Actually, if we allow ir to be *any* nonzero complex number $z = ir$, then Vilenkin proves that

$$(\widetilde{\Pi}_{(a,Q)}(f))(x) = e^{z(x\cdot a)} f(Q^\top x), \qquad (a,Q) \in \mathbf{SE}(n),\ x \in S^{n-1}, \quad f \in \mathrm{L}^2_\lambda(S^{n-1}; \mathbb{C}),\ z \in \mathbb{C}^* \qquad (*_4)$$

still defines an irreducible representation, but it is not unitary unless $z = ir$ with $r \in \mathbb{R}$ and $r \neq 0$.

The representations of $\mathbf{SE}(n)$ given by $(*_2)$ have the following interesting property. If we consider their restriction to $\mathbf{SO}(n)$, so that $s = (0, Q)$, then we see that they are given by

$$(\widetilde{\Pi}_{(0,Q)}(f))(x) = f(Q^\top x), \quad Q \in \mathbf{SO}(n), x \in S^{n-1}, f \in \mathrm{L}^2_\lambda(S^{n-1}; \mathbb{C}). \quad (*_5)$$

The constant function $f_0\colon S^{n-1} \to \mathbb{C}$ with value 1 is invariant under $\mathbf{SO}(n)$, in the sense that

$$(\widetilde{\Pi}_{(0,Q)}(f_0))(x) = f_0(Q^\top x) = 1 \quad \text{for all } Q \in \mathbf{SO}(n) \text{ and all } x \in S^{n-1},$$

which means that

$$\widetilde{\Pi}_{(0,Q)}(f_0) = f_0, \quad \text{for all } Q \in \mathbf{SO}(n).$$

This is an instance of what is called a representation of class 1 relative to $\mathbf{SO}(n)$.

Definition 6.12. Let G be a locally compact group and H be a closed subgroup of G. A unitary representation $U\colon G \to \mathbf{U}(E)$ of G in a Hilbert space E is a *representation of class* 1 *relative to* H if there is some nonzero vector $x \in E$ invariant relative to H, which means that

$$U_h(x) = x, \quad \text{for all } h \in H.$$

Remark: Vilenkin [Vilenkin (1968)] (Chapter I, Section 2) allows U to be nonunitary, but in this case the restriction of U to H must be unitary.

The representations of Example 6.3 given by $(*_2)$ are of class 1 relative to $\mathbf{SO}(n)$.

One of the reasons why representations of class 1 are interesting is the following. Suppose $a \in E$ is a nonzero vector invariant under H as above. For every $x \in E$, we define the function $f_x\colon G \to \mathbb{C}$ given by

$$f_x(s) = \langle U_s(x), a\rangle, \quad s \in G.$$

The functions f_x are called *spherical functions of* U *relative to* H. We claim that the functions f_x are constant on right cosets Hs.

Indeed, for all $s \in G$ and all $h \in H$, we have

$$\begin{aligned} f_x(hs) &= \langle U_{hs}(x), a\rangle \\ &= \langle U_h(U_s(x)), a\rangle \\ &= \langle U_s(x), U_h^*(a)\rangle \\ &= \langle U_s(x), U_{h^{-1}}(a)\rangle \\ &= \langle U_s(x), a\rangle = f_x(s), \end{aligned}$$

so

$$f_x(hs) = f(s), \quad \text{for all } s \in G \text{ and all } h \in H.$$

In particular, for $x = a$, we claim that the function f_a, called a *zonal spherical function*, is constant on the two-sided cosets HsH $(s \in G)$.

Since we already know that $f_a(h_1 s) = f_a(s)$ for all $h_1 \in H$, it suffices to show that $f_a(sh_2) = f_a(s)$ for all $h_2 \in H$. We have

$$\begin{aligned} f_a(sh_2) &= \langle U_{sh_2}(a), a\rangle \\ &= \langle U_s(U_{h_2}(a)), a\rangle \\ &= \langle U_s(a), a\rangle = f_a(s). \end{aligned}$$

Thus, we proved that

$$f_a(h_1 s h_2) = f_a(s), \quad \text{for all } h_1, h_2 \in H \text{ and all } s \in G,$$

which means that f_a is constant on the double cosets HsH. Geometrically, this means that f_a is constant on "spheres." In particular, if $G = \mathbf{SO}(3)$ and $H = \mathbf{SO}(2)$, then the spherical functions are the well-known spherical harmonics $Y_l^m(\theta, \varphi)$ and the zonal spherical functions are the Legendre polynomials $P_l(\cos\theta)$. If $G = \mathbf{SO}(n)$ and $H = \mathbf{SO}(n-1)$, then the zonal spherical functions are given in terms of Gegenbaur polynomials; see Gallier and Quaintance [Gallier and Quaintance (2020b)] (Chapter 7, Sections 3, 5–7).

Under some mild additional conditions, induced unitary representations of G in $\mathrm{L}^2_\mu(X; E)$ can be converted to unitary representations of G in a closed subspace of $\mathrm{L}^2_\lambda(G; E)$ (where λ is a left Haar measure on G).

Suppose that the unitary cocycle α has the property that the map

$$s \mapsto f^\alpha(s) = \alpha(s^{-1}, s \cdot x_0)(f(s \cdot x_0))$$

from G to E is λ-measurable for every $f \in \mathcal{L}^2_\mu(X; E)$. If so, using Proposition 6.5, we have

$$\|f^\alpha(sh)\|_E = \|f^\alpha(s)\|_E = \|f(s \cdot x_0)\|_E$$

for all $s \in G$ and all $h \in H$, and since by Vol. I, Proposition 8.43 and Theorem 7.10, for any $g \in \mathrm{L}^2(G/H; \mathbb{C})$, we have

$$\int_{G/H} g\, d\mu = \int_G (g \circ \pi)\, d\lambda,$$

so we obtain

$$N_2(f^\alpha)^2 = \int_G \|f^\alpha(s)\|_E^2 \, d\lambda(s) = \int_G \|f(s\cdot x_0)\|_E^2 \, d\lambda(s) = N_2(f\circ\pi)^2$$

$$N_2(f\circ\pi)^2 = \int_G \|f(s\cdot x_0)\|_E^2 \, d\lambda(s) = \int_{G/H} \|f(x)\|_E \, d\mu(x) = N_2(f)^2,$$

that is, $N_2(f^\alpha) = N_2(f)$, and we conclude that $f^\alpha \in \mathcal{L}^2_\lambda(G;E)$.

Conversely, if $g \in \mathcal{L}^2_\lambda(G;E)$ satisfies the property

$$g(sh) = U(h^{-1})(g(s)), \qquad \text{for all } s\in G \text{ and all } h\in H, \qquad (*_U)$$

and if the map $s \mapsto \alpha(s,x_0)(g(s))$ from G to E is λ-measurable, then as in Proposition 6.6, we can write this map as $f\circ\pi$ for some $f\in \mathrm{L}^2_\mu(X;E)$, and we have $g = f^\alpha$.

In this case, up to equivalence, we can consider the unitary representation $\mathrm{Ind}^G_{H,F}\,\alpha$ induced by α as a unitary representation of G in the closed subspace F of $\mathrm{L}^2_\lambda(G;E)$ spanned by the functions $g\in\mathcal{L}^2_\lambda(G;E)$ satisfying property $(*_U)$. Then for all $s\in G$,

$$(\mathrm{Ind}^G_{H,F}\,\alpha)_s(g) = \lambda_s g, \qquad \text{for all } g\in F, \qquad (\mathrm{Ind}_G)$$

equivalently, for all $s,t\in G$,

$$((\mathrm{Ind}^G_{H,F}\,\alpha)_s(g))(t) = g(s^{-1}t), \qquad \text{for all } g\in F.$$

Notice the analogy with Proposition 6.7.

Note that $\mathrm{Ind}^G_{H,F}\,\alpha$ depends only on U, so we usually write $\mathrm{Ind}^G_{H,F}\,U$ instead of $\mathrm{Ind}^G_{H,F}\,\alpha$.

If $E=\mathbb{C}$, then $\mathrm{Ind}^G_{H,F}\,U$ is a subrepresentation of the regular representation of G in $\mathrm{L}^2(G)$.

Definition 6.13. If we choose U to be the trivial representation of H in E, then the functions $g\in\mathrm{L}^2_\lambda(G;E)$ satisfying condition $(*_U)$ are constant on the classes sH, so we can identify F with $\mathrm{L}^2_\mu(X;E)$. In this case, we say that the induced representation $\mathrm{Ind}^G_H\,U$ of G in $\mathrm{L}^2_\mu(X;E)$ is the *canonical representation* of G corresponding to the compact subgroup H and to its trivial representation in E.

If $H=(e)$ and $E=\mathbb{C}$, then the induced representation is the regular representation of G in $\mathrm{L}^2(G)$.

Going back to the case where H is an arbitrary closed subgroup of G, and where there is a G-invariant measure on G/H, there is another method, not using cocyles, for defining a unitary induced representation of G from a unitary representation $U\colon H \to \mathbf{U}(E)$. We can define a Hilbert space $\mathcal{H}$ such that formula (Ind_G) defines a unitary induced representation $\mathrm{Ind}_{H,\mathcal{H}}^G U$ of G in $\mathcal{H}$. This method is described in Folland [Folland (1995)] (Chapter 6, Section 1), and we briefly describe it.

Given a unitary representation $U\colon H \to \mathbf{U}(E)$, let $\mathcal{H}_0$ be the following set of functions:

$$\begin{aligned}\mathcal{H}_0 = \{f \in \mathcal{C}(G,E) \mid\ & \pi(\mathrm{supp}(f)) \text{ is compact and} \\ & f(sh) = U(h^{-1})(f(s)) \quad \text{for all } s \in G \text{ and all } h \in H\}.\end{aligned}$$

The problem is that it is not obvious that $\mathcal{H}_0$ is nonempty! However, the following result proven in Folland [Folland (1995)] (Chapter 6, Proposition 6.1) shows that this is not the case.

Proposition 6.10. *If $\varphi\colon G \to E$ is a continuous function with compact support, then the function f_φ from G to E given by*

$$f_\varphi(s) = \int_H U(h)(\varphi(hs))\, d\lambda_H(h)$$

belongs to $\mathcal{H}_0$ and is uniformly continuous on G. Moreover, every element of $\mathcal{H}_0$ is of the form f_φ for some $\varphi \in \mathcal{K}(G,E)$.

The group G acts on the left on $\mathcal{H}_0$ by $f \mapsto \lambda_s f$. In order to act by unitary maps, we need to define an inner product on $\mathcal{H}_0$ with respect to which these left translations are isometries. Since G/H has G-invariant measures, this is easy to achieve. If $f, g \in \mathcal{H}_0$, then the map $s \mapsto \langle f(s), g(s)\rangle_E$ depends only on the coset sH, so we can define the inner product $\langle f, g\rangle$ by

$$\langle f, g\rangle = \int_{G/H} \langle f(s), g(s)\rangle_E\, d\mu(sH).$$

This is clearly a positive hermitian form, and it is positive definite because $\mu(A) > 0$ for every nonempty open set A. This inner product is invariant under the left translations λ_s because μ is G-invariant. Therefore, with respect to this inner product, the maps $f \mapsto \lambda_s f$ are unitary. If $\mathcal{H}$ is the Hilbert space, which is the completion of $\mathcal{H}_0$, then the maps $f \mapsto \lambda_s f$ extend to unitary operators on $\mathcal{H}$. It follows from Proposition 6.10 that the map $s \mapsto \lambda_s f$ from G to $\mathcal{H}$ are continuous for every $f \in \mathcal{H}_0$. Therefore, they define a unitary representation of G in $\mathcal{H}$ given by

$$(\mathrm{Ind}_{H,\mathcal{H}}^G U)_s(f) = \lambda_s(f), \qquad f \in \mathcal{H}.$$

This unitary representation has the advantage that it depends only on U, but one should not neglect the fact that the construction involving cocycles allows more flexibility. The Hilbert space $\mathcal{H}$ is also more complicated than the Hilbert space $\mathrm{L}^2_\mu(X; E)$.

When G/H admits no G-invariant measure, then we need to use a weaker notion of invariance. It turns out that the notion of (strong) quasi-invariance does the job.

6.6 Quasi-Invariant Measures on G/H

The notion of quasi-invariance was first introduced by Mackey and Bruhat in the early 1950's. It also occurs in Bourbaki [Bourbaki (1965)] (Chapter VII, §2, No. 5). We follow the exposition in Folland [Folland (1995)] (Chapter 2, Section 2.6, and Chapter 6, Section 1).

As we said in Section 6.5, given any measure μ on $X = G/H$, for any $s \in G$, the measure $\lambda_s(\mu)$ is given by

$$(\lambda_s(\mu))(A) = \mu(s^{-1} \cdot A),$$

for every Borel subset A of $X = G/H$. We say μ is G-invariant if for every Borel subset A of X,

$$\mu(s^{-1} \cdot A) = \mu(A), \qquad \text{for all } s \in G.$$

In this case,

$$\int_{G/H} g(s \cdot x)\, d\mu(x) = \int_{G/H} g(x)\, d\mu(x), \qquad \text{for all } s \in G$$

and for all $g \in \mathrm{L}^1_\mu(G/H)$. It is not hard to prove an analog of Vol. I, Proposition 8.16(3), namely

$$\int_{G/H} g(s \cdot x)\, d\mu(x) = \int_{G/H} g(x)\, d\lambda_s(\mu)(x)$$

for all $g \in \mathrm{L}^1_\mu(G/H)$ and all $s \in G$. A weaker requirement than G-invariance is that

$$\int_{G/H} g(s \cdot x)\, d\mu(x) = \int_{G/H} g(x)\, d\lambda_s(\mu)(x) = \int_{G/H} \varrho(s, x) g(x)\, d\mu(x)$$

for some continuous function $\varrho\colon G \times (G/H) \to (0, \infty)$, for all $g \in \mathcal{K}_{\mathbb{C}}(G/H)$ and all $s \in G$. The above discussion suggests the following definition.

Definition 6.14. A measure μ on G/H is *(strongly) quasi-invariant* if there is a continuous function $\varrho\colon G \times (G/H) \to (0, \infty)$, such that

$$\int_{G/H} g(s \cdot x)\, d\mu(x) = \int_{G/H} \varrho(s, x) g(x)\, d\mu(x), \quad \text{for all } g \in \mathcal{K}_{\mathbb{C}}(G/H) \text{ and all } s \in G. \qquad (\text{qi}_\varrho)$$

The key to quasi-invariance is the existence of certain functions from G to $(0, \infty)$ called *rho-functions*.

Definition 6.15. A function $\rho\colon G \to (0, \infty)$ is a *rho-function* for the pair (G, H) if it is a continuous function, such that

$$\rho(sh) = \frac{\Delta_H(h)}{\Delta_G(h)}\, \rho(s), \qquad s \in G, h \in H, \qquad (*_\rho)$$

where Δ_G is the modular function on G and Δ_H is the modular function on H.

Proposition 6.11. *If G is any locally compact group and H is any closed subgroup of G, then (G, H) admits rho-functions.*

Proposition 6.11 is proven in Folland [Folland (1995)] (Chapter 2, Proposition 2.54). One first proves [Folland (1995)] (Chapter 2, Lemma 2.53) that there is a continuous function $\varphi\colon G \to (0, \infty)$, such that the following properties hold:

(i) $\{y \in G \mid \varphi(y) > 0\} \cap sH \neq \emptyset$ for all $s \in G$.
(ii) $\mathrm{supp}(\varphi) \cap KH$ is compact for every compact subset K of G.

Then define ρ by

$$\rho(s) = \int_H \frac{\Delta_G(h)}{\Delta_H(h)}\, \varphi(sh)\, d\lambda_H(h).$$

It is not hard to check that the above function is a rho-function.

Recall from Vol. I, Definition 8.20 that the definition of the projection map $P\colon \mathcal{K}_{\mathbb{C}}(G) \to \mathcal{K}_{\mathbb{C}}(G/H)$ defined as follow: for every $f \in \mathcal{K}_{\mathbb{C}}(G)$, for every $s \in G$, let

$$(P(f))(sH) = \int_H f(sh)\, d\lambda_H(h).$$

By Vol. I, Proposition 8.40, the map P is surjective.

The next proposition is proven in Folland [Folland (1995)] (Chapter 2, Lemma 2.55).

Proposition 6.12. *For any function $f \in \mathcal{K}_{\mathbb{C}}(G)$, if $P(f) = 0$, then $\int f\rho\, d\lambda = 0$, for any rho-function ρ.*

The proof of Proposition 6.12 is very similar to the argument given in the proof of Vol. I, Theorem 8.42. Then we have our first main theorem. Recall that $\pi\colon G \to G/H$ denotes the quotient map.

Theorem 6.13. *Let G be any locally compact group and H be any closed subgroup of G. For every rho-function ρ for (G,H), there is a unique σ-Radon measure μ on G/H, such that*

$$\int_{G/H} P(f)(x)\, d\mu(x) = \int_G f(s)\rho(s)\, d\lambda(s), \qquad \textit{for all } f \in \mathcal{K}_{\mathbb{C}}(G). \qquad \text{(qi)}$$

Furthermore, if we let $\varrho\colon G \times (G/H) \to (0,\infty)$ be the continuous function given by

$$\varrho(s, \pi(t)) = \frac{\rho(s^{-1}t)}{\rho(t)} \qquad s,t \in G,$$

then for every $g \in \mathcal{K}_{\mathbb{C}}(G/H)$, we have

$$\int_{G/H} g(s\cdot x)\, d\mu(x) = \int_{G/H} \varrho(s,x) g(x)\, d\mu(x), \qquad \textit{for all } s \in G, \qquad (\mathrm{qi}_{\varrho})$$

which means that μ is strongly quasi-invariant.

Proof. Theorem 6.13 is proven in Folland [Folland (1995)] (Chapter 2, Theorem 2.56). For any $f \in \mathcal{K}_{\mathbb{C}}(G)$, since P is surjective and since by Proposition 6.12, if $P(f) = P(g)$, then $\int_G f\rho\, d\lambda = \int_G g\rho\, d\lambda$, the map Φ given by $\Phi(P(f)) = \int_G f\rho\, d\lambda$ is a well-defined positive linear functional on $\mathcal{K}_{\mathbb{C}}(G/H)$. By Radon–Riesz I, it defines a unique σ-Radon measure μ on G/H satisfying (qi).

The equation

$$\rho(sh) = \frac{\Delta_H(h)}{\Delta_G(h)}\rho(s), \qquad s \in G, h \in H,$$

satisfied by a rho-function shows that the ratio $\rho(st)/\rho(t)$ depends only on the coset $\pi(t) = tH$, because

$$\frac{\rho(sth)}{\rho(th)} = \frac{\Delta_H(h)}{\Delta_G(h)}\frac{\rho(st)}{\rho(th)} = \frac{\Delta_H(h)}{\Delta_G(h)}\frac{\Delta_G(h)}{\Delta_H(h)}\frac{\rho(st)}{\rho(t)} = \frac{\rho(st)}{\rho(t)},$$

so we obtain a continuous function $\varrho\colon G \times (G/H) \to (0,\infty)$ given by

$$\varrho(s, \pi(t)) = \frac{\rho(s^{-1}t)}{\rho(t)}, \qquad s,t \in G.$$

First, by expanding both integrals as double integrals, it is easy to show that

$$\int_{G/H} P(f)(s\cdot x)\, d\mu(x) = \int_{G/H} P(\lambda_{s^{-1}} f)(x)\, d\mu(x).$$

Then, we have

$$\begin{aligned}
\int_{G/H} P(f)(s\cdot x)\, d\mu(x) &= \int_{G/H} P(\lambda_{s^{-1}} f)(x)\, d\mu(x) \\
&= \int_G f(st)\rho(t)\, d\lambda(t) \\
&= \int_G f(t)\rho(s^{-1}t)\, d\lambda(t) \\
&= \int_G \varrho(s,\pi(t)) f(t)\rho(t)\, d\lambda(t) \\
&= \int_{G/H} P(\varrho(s,\pi(-))f)(x)\, d\mu(x) \\
&= \int_{G/H} \varrho(s,x) P(f)(x)\, d\mu(x),
\end{aligned}$$

where we used Vol. I, Proposition 8.38(3) to prove the last step, which concludes the proof. □

Remark: The map $x \mapsto \varrho(s,x)$ is the Radon–Nikodym derivative of $\lambda_s(\mu)$ with respect to μ.

The following converse of Theorem 6.13 is proven in Folland [Folland (1995)] (Chapter 2, Theorem 2.59).

Theorem 6.14. *Let G be any locally compact group and H be any closed subgroup of G. Every quasi-invariant measure μ on G/H arises from a rho-function as in* (qi) *and* (qi$_\varrho$). *Furthermore, any two such measures μ and μ' are strongly equivalent, which means that there is a continuous function $\varphi\colon G/H \to (0,\infty)$, such that $\int_{G/H} g(x)\, d\mu'(x) = \int_{G/H} \varphi(x) g(x)\, d\mu(x)$ for all $g \in \mathcal{K}_{\mathbb{C}}(G/H)$.*

The following proposition shows that ϱ behaves like a cocycle.

Proposition 6.15. *Let G be any locally compact group and H be any closed subgroup of G. For any quasi-invariant measure μ on G/H associated with the continuous function $\varrho\colon G\times (G/H) \to (0,\infty)$, we have*

$$\varrho(st,x) = \varrho(s,t\cdot x)\varrho(t,x), \qquad \text{for all } s,t\in G \text{ and all } x \in G/H. \qquad (*_\varrho)$$

Proof. Using (qi_ϱ), for every function $g \in \mathcal{K}_{\mathbb{C}}(G/H)$, we have

$$\begin{aligned}\int_{G/H} g(x)\varrho(st,x)\,d\mu(x) &= \int_{G/H} g((st)\cdot x)\,d\mu(x) = \int_{G/H} g(s\cdot(t\cdot x))\,d\mu(x)\\ &= \int_{G/H} \varrho(s,t\cdot x)g(t\cdot x)\,d\mu(x)\\ &= \int_{G/H} g(x)\varrho(s,t\cdot x)\varrho(t,x)\,d\mu(x),\end{aligned}$$

which proves that

$$\varrho(st,x) = \varrho(s,t\cdot x)\varrho(t,x),$$

as claimed. ☐

Remark: Dieudonné denotes $\varrho(s,x)$ by $J_s(x)$; see Dieudonné [Dieudonné (2003b)] (Chapter XXII, Section 3, No. 22.3.8.1–22.3.8.2).

We now use quasi-invariant measures to generalize the construction of Section 6.5.

6.7 Induced Representations, II; G/H has a Quasi-Invariant Measure

If μ is a quasi-invariant measure on G/H, then by making a simple modification to condition (1) of Theorem 6.9, we obtain the following result.

Theorem 6.16. *Let G be a locally compact group, H be a closed subgroup of G, E be a separable Hilbert space, and $U\colon H \to \mathbf{U}(E)$ be a unitary representation of H. For any quasi-invariant measure μ on $X = G/H$ associated with the continuous function $\varrho\colon G\times(G/H)\to(0,\infty)$, for any cocycle $\alpha\colon G\times X\to\mathbf{U}(E)$, if the following conditions hold*

(1) The map $\varrho(s^{-1},x)^{1/2}\alpha(s,x)$ is a unitary map of E for all $s\in G$ and all $x\in X$, such that $\varrho(h^{-1},x_0)^{1/2}\alpha(h,x_0) = U(h)$ for all $h\in H$;
(2) For every $s\in G$, for every $f\in \mathrm{L}^2_\mu(X;E)$, the map $x\mapsto \alpha(x,s)(f(x))$ from X to E is μ-measurable;
(3) For every $f\in \mathrm{L}^2_\mu(X;E)$, the map $s\mapsto \Pi_s(f)$ is a continuous map from G to $\mathrm{L}^2_\mu(X;E)$, where Π is the homomorphism $\Pi\colon G\to\mathbf{GL}(E^X)$ induced by the cocycle α;

then the homomorphism $\Pi\colon G\to\mathbf{U}(\mathrm{L}^2_\mu(X;E))$ induced by the cocycle α given by

$$(\Pi_s(f))(x) = (\alpha(s^{-1},x))^{-1}(f(s^{-1}\cdot x)), \qquad f\in\mathrm{L}^2_\mu(X;E),\ x\in X,$$

(see Definition 6.3) is a unitary representation of G.

Proof. We simply have to prove that

$$N_2(\Pi_s(f)) = N_2(f), \qquad \text{for all } f \in \mathcal{L}^2_\mu(X;E) \text{ and all } s \in G,$$

which implies that $\Pi_s(f) \in \mathcal{L}^2_\mu(X;E)$, and the other conditions imply that the homomorphism $\Pi\colon G \to \mathbf{GL}(\mathrm{L}^2_\mu(X;E))$ induced by α is a unitary representation of G. Since by hypothesis $\varrho(s^{-1},x)^{1/2}\alpha(s,x)$ is a unitary operator, using (qi$_\varrho$), we have

$$\begin{aligned}
N_2(\Pi_s(f))^2 &= \int_{G/H} \left\|\alpha(s,s^{-1}\cdot x)(f(s^{-1}\cdot x))\right\|^2 d\mu(x)\\
&= \int_{G/H} \varrho(s^{-1},x)\left\|\alpha(s,x)(f(x))\right\|^2 d\mu(x)\\
&= \int_{G/H} \left\|\varrho(s^{-1},x)^{1/2}\alpha(s,x)(f(x))\right\|^2 d\mu(x)\\
&= \int_{G/H} \left\|f(x)\right\|^2 d\mu(x),
\end{aligned}$$

and so $N_2(\Pi_s(f)) = N_2(f)$. $\square$

As an application of Theorem 6.16, we can pick

$$\alpha(s,x) = (\varrho(s^{-1},x))^{-1/2+ri}\,\mathrm{id}_E,$$

with $r \in \mathbb{R}$. By Proposition 6.15, the function α is a cocycle. Condition (2) is satisfied because ϱ is measurable (in fact, continuous). The maps $\varrho(s^{-1},x)^{1/2}\alpha(s,x) = (\varrho(s^{-1},x))^{ri}\,\mathrm{id}_E$ are unitary, since they are multiplication by a complex number of modulus 1. It remains to check condition (3). This verification is performed in Dieudonné [Dieudonné (2003b)] (Chapter XXII, Section 3, No. 22.3.8.3).

6.8 Examples of Induced Representations Via Method II

We will now give several examples of the application of Theorem 6.16 to the group $\mathbf{SL}(2,\mathbb{R})$. It turns out that the group $\mathbf{SL}(2,\mathbb{R})$ has no finite-dimensional unitary representations except the trivial one, and Theorem 6.16 can be used to produce nontrivial unitary representations.

Example 6.4. Let $G = \mathbf{SL}(2,\mathbb{R})$ and $H = S_1$ be the subgroup

$$S_1 = \left\{ \begin{pmatrix} a & b \\ 0 & a^{-1} \end{pmatrix} \;\middle|\; a,b \in \mathbb{R}, a \neq 0 \right\},$$

and let $E = \mathbb{C}$. We claim that the homogeneous space $\mathbf{SL}(2,\mathbb{R})/S_1$ is homeomorphic to $\mathbb{P}^1(\mathbb{R}) = \mathbb{RP}^1$, the real projective line. Indeed, there is an action of $\mathbf{SL}(2,\mathbb{R})$ on $\mathbb{RP}^1$ viewed as $\mathbb{R} \cup \{\infty\}$ given by

$$s \cdot z = \begin{pmatrix} a & b \\ c & d \end{pmatrix} \cdot z = \frac{az+b}{cz+d}, \qquad z \in \mathbb{RP}^1,$$

with the convention that when $z = -d/c$, then the result is ∞, and when $z = \infty$, then the result if a/c. It is easy to check that this action is transitive and that the stabilizer of ∞ is the subgroup S_1. We give $\mathbb{RP}^1$ the measure μ, which is the Lebesgue measure extended so that $\{\infty\}$ has measure zero. Since

$$\begin{pmatrix} a & b \\ c & d \end{pmatrix}^{-1} = \begin{pmatrix} d & -b \\ -c & a \end{pmatrix},$$

(recall that $ad - bc = 1$), and since the derivative of the function

$$x \mapsto \frac{dx - b}{-cx + a}$$

is

$$\frac{d(-cx+a) - (dx-b)(-c)}{(-cx+a)^2} = \frac{1}{(cx-a)^2},$$

we see that for any function $f \in \mathrm{L}^2_\mu(\mathbb{RP}^1;\mathbb{C})$, using the change of variable $x = \frac{az+b}{cz+d}$,

$$\int_{-\infty}^{+\infty} f(s\cdot z)\, d\mu(z) = \int_{-\infty}^{+\infty} f\left(\frac{az+b}{cz+d}\right) d\mu(z) = \int_{-\infty}^{+\infty} \frac{1}{(cx-a)^2} f(x)\, d\mu(x).$$

It follows that μ is quasi-invariant with

$$\varrho(s,x) = \frac{1}{(cx-a)^2}, \quad \text{where} \quad s = \begin{pmatrix} a & b \\ c & d \end{pmatrix}.$$

The method of Theorem 6.16 with

$$\alpha(s,x) = (\varrho(s^{-1},x))^{-1/2+(r/2)i}\, \mathrm{id}_{\mathbb{C}},$$

where $r \in \mathbb{R}$, and with

$$(\Pi_s(f))(x) = (\alpha(s^{-1},x))^{-1}(f(s^{-1}\cdot x)),$$

yields the unitary representations of $\mathbf{SL}(2,\mathbb{R})$ in $\mathrm{L}^2_\mu(\mathbb{RP}^1;\mathbb{C})$ given by

$$\Pi_s(f)(x) = |cx-a|^{-1+ri} f\left(\frac{b-dx}{cx-a}\right), \qquad f \in \mathrm{L}^2_\mu(\mathbb{RP}^1;\mathbb{C}),$$

where

$$s = \begin{pmatrix} a & b \\ c & d \end{pmatrix}.$$

It is also easy to check that the cocycles

$$\alpha(s, x) = \left(\frac{1}{(cx - a)^2}\right)^{-1/2+(r/2)i} \operatorname{sign}(cx - a)\,\mathrm{id}_{\mathbb{C}}$$

with $r \in \mathbb{R}$ also work, and we get the representations of $\mathbf{SL}(2, \mathbb{R})$ in $\mathrm{L}^2_\mu(\mathbb{RP}^1; \mathbb{C})$ given by

$$\Pi_s(f)(x) = |cx - a|^{-1+ri} \operatorname{sign}(cx - a) f\left(\frac{b - dx}{cx - a}\right), \quad f \in \mathrm{L}^2_\mu(\mathbb{RP}^1; \mathbb{C}),$$

where

$$s = \begin{pmatrix} a & b \\ c & d \end{pmatrix}.$$

It can be shown that all these representations are irreducible and pairwise inequivalent for $r > 0$. These representations constitute the *principal series* of irreducible unitary representations of $\mathbf{SL}(2, \mathbb{R})$.

Example 6.5. Let $G = \mathbf{SL}(2, \mathbb{R})$ and $H = \mathbf{SO}(2)$ be the subgroup

$$\mathbf{SO}(2) = \left\{ \begin{pmatrix} \cos\theta & -\sin\theta \\ \sin\theta & \cos\theta \end{pmatrix} \,\middle|\, 0 \le \theta \le 2\pi, \right\},$$

and let $E = \mathbb{C}$. We claim that the homogeneous space $\mathbf{SL}(2, \mathbb{R})/\mathbf{SO}(2)$ is homeomorphic to the upper half plane $P = \{z = x + iy \in \mathbb{C} \mid y > 0\}$. Indeed, there is an action of $\mathbf{SL}(2, \mathbb{R})$ on P given by

$$s \cdot z = \begin{pmatrix} a & b \\ c & d \end{pmatrix} \cdot z = \frac{az + b}{cz + d}, \qquad z = x + iy \in P.$$

It is easy to check that this action is transitive and that the stabilizer of $z = i$ is $\mathbf{SO}(2)$. Since the group $\mathbf{SO}(2)$ is compact, the space $P = \mathbf{SL}(2, \mathbb{R})/\mathbf{SO}(2)$ admits $\mathbf{SL}(2, \mathbb{R})$-invariant measures. In fact, the measure μ corresponding to the positive Radon functional

$$h \mapsto \int_P h(x + iy) \frac{dx\,dy}{y^2} = \int_{y>0} \int_{x=-\infty}^{+\infty} h(x + iy) \frac{dx\,dy}{y^2}, \qquad h \in \mathcal{K}_{\mathbb{C}}(P),$$

is such a measure.

We will need a method for picking a representative in every coset of $\mathbf{SL}(2, \mathbb{R})/\mathbf{SO}(2)$ that corresponds in a one-to-one fashion to an element

$z = x + iy \in P$. For this, we use the fact that every matrix $s \in \mathbf{SL}(2, \mathbb{R})$ can be uniquely factored as

$$\begin{aligned}\begin{pmatrix} a & b \\ c & d \end{pmatrix} &= \begin{pmatrix} \lambda & \mu \\ 0 & \lambda^{-1} \end{pmatrix} \begin{pmatrix} \cos\theta & -\sin\theta \\ \sin\theta & \cos\theta \end{pmatrix} \\ &= \begin{pmatrix} \lambda\cos\theta + \mu\sin\theta & -\lambda\sin\theta + \mu\cos\theta \\ \lambda^{-1}\sin\theta & \lambda^{-1}\cos\theta \end{pmatrix},\end{aligned}$$

with $\lambda, \mu \in \mathbb{R}, \lambda > 0$, and $0 \leq \theta < 2\pi$.

Indeed, if there is such a decomposition, then

$$c = \lambda^{-1}\sin\theta, \qquad d = \lambda^{-1}\cos\theta,$$

so

$$\sin\theta = \lambda c, \qquad \cos\theta = \lambda d,$$

and since

$$\begin{pmatrix} \lambda & \mu \\ 0 & \lambda^{-1} \end{pmatrix} = \begin{pmatrix} a & b \\ c & d \end{pmatrix} \begin{pmatrix} \cos\theta & \sin\theta \\ -\sin\theta & \cos\theta \end{pmatrix} = \begin{pmatrix} a\cos\theta - b\sin\theta & a\sin\theta + b\cos\theta \\ c\cos\theta - d\sin\theta & c\sin\theta + d\cos\theta \end{pmatrix},$$

we see that

$$\lambda^{-1} = c\sin\theta + d\cos\theta = \lambda(c^2 + d^2),$$

and since $ad - bc = 1$, we have $c^2 + d^2 \neq 0$, so we can pick

$$\lambda = \frac{1}{\sqrt{c^2 + d^2}},$$

and then $\theta \in [0, 2\pi)$ is uniquely determined by

$$\cos\theta = \frac{d}{\sqrt{c^2 + d^2}}, \qquad \sin\theta = \frac{c}{\sqrt{c^2 + d^2}},$$

and μ is determined by

$$\mu = a\sin\theta + b\cos\theta = \lambda(ac + bd) = \frac{ac + db}{\sqrt{c^2 + d^2}}.$$

Observe that the group

$$S_0 = \left\{ \begin{pmatrix} a & b \\ 0 & a^{-1} \end{pmatrix} \,\middle|\, a, b \in \mathbb{R}, a > 0 \right\}$$

is a subgroup of the group S_1 of Example 6.4.

Given any $z = x + iy \in P$, there is a unique coset $s\mathbf{SO}(2) \subseteq \mathbf{SL}(2, \mathbb{R})$ (where $s \in \mathbf{SL}(2, \mathbb{R})$) that maps i to z, and in view of the above factorization of matrices in $\mathbf{SL}(2,\mathbb{R})$, we can pick as a representative of this coset $s\mathbf{SO}(2)$ the matrix $r_z \in S_0$, such that

$$r_z \cdot i = z = x + iy,$$

namely,

$$r_z = \begin{pmatrix} \sqrt{y} & x/\sqrt{y} \\ 0 & 1/\sqrt{y} \end{pmatrix}.$$

We now determine $u(s,z)$, such that $sr_z = r_{s\cdot z}u(s,z)$ (see Definition 6.5), with $u(s,z) \in \mathbf{SO}(2)$ and

$$s = \begin{pmatrix} a & b \\ c & d \end{pmatrix},$$

as in Section 6.2 (see Propositions 6.3 and 6.4). Since the imaginary part of $s \cdot z = (az+b)/(cz+d)$ is $y/|cz+d|^2$, we have

$$r_{s\cdot z} = \begin{pmatrix} \sqrt{y}/|cz+d| & * \\ 0 & |cz+d|/\sqrt{y} \end{pmatrix},$$

so the equation $sr_z = r_{s\cdot z}u(s,z)$ with

$$u(s,z) = \begin{pmatrix} \cos\theta & -\sin\theta \\ \sin\theta & \cos\theta \end{pmatrix},$$

namely,

$$\begin{pmatrix} a & b \\ c & d \end{pmatrix} \begin{pmatrix} \sqrt{y} & x/\sqrt{y} \\ 0 & 1/\sqrt{y} \end{pmatrix} = \begin{pmatrix} \sqrt{y}/|cz+d| & * \\ 0 & |cz+d|/\sqrt{y} \end{pmatrix} \begin{pmatrix} \cos\theta & -\sin\theta \\ \sin\theta & \cos\theta \end{pmatrix},$$

yields in particular

$$c\sqrt{y} = \frac{|cz+d|}{\sqrt{y}} \sin\theta, \qquad \frac{cx+d}{\sqrt{y}} = \frac{|cz+d|}{\sqrt{y}} \cos\theta.$$

Therefore,

$$\cos\theta + i\sin\theta = \frac{cx+d+ciy}{|cz+d|} = \frac{cz+d}{|cz+d|},$$

equivalently,

$$e^{i\theta} = \frac{cz+d}{|cz+d|}.$$

The group $\mathbf{SO}(2)$ is abelian, and since its unitary representations in $\mathbb{C}$ are characters, by Vol. I, Proposition 10.9, they are of the form

$$h \mapsto \sigma_n(h) = e^{ni\theta}, \qquad n \in \mathbb{Z},$$

with

$$h = \begin{pmatrix} \cos\theta & -\sin\theta \\ \sin\theta & \cos\theta \end{pmatrix}.$$

By the method of Section 6.2, since $e^{i\theta} = \frac{cz+d}{|cz+d|}$, we can pick the cocycle α to be

$$\alpha(s,z) = \sigma_n(u(s,z)) = \frac{(cz+d)^n}{|cz+d|^n}\,\mathrm{id}_{\mathbb{C}},$$

and then

$$(\Pi_s(f))(z) = (\alpha(s^{-1},x))^{-1}(f(s^{-1}\cdot z)), \qquad f \in \mathrm{L}^2_\mu(P;\mathbb{C}),\ z \in P.$$

Since

$$s^{-1} = \begin{pmatrix} d & -b \\ -c & a \end{pmatrix},$$

we get

$$\alpha(s^{-1},z) = \frac{(cz-a)^n}{|cz-a|^n}\,\mathrm{id}_{\mathbb{C}},$$

and thus

$$(\Pi_s(f))(z) = \frac{(cz-a)^{-n}}{|cz-a|^{-n}} f\left(\frac{b-dz}{cz-a}\right),\ \ f \in \mathrm{L}^2_\mu(P;\mathbb{C}),\ z \in P,\ s = \begin{pmatrix} a & b \\ c & d \end{pmatrix},$$

and for $n \in \mathbb{Z}$.

These representations are not always irreducible. A way to see this is to construct an equivalent representation by using the cocycles $\alpha'(s,z) = c(s\cdot z)\circ\alpha(s,z)\circ c(z)^{-1}$, as in Section 6.2, with

$$c(z) = c(x+iy) = y^{-n/2}.$$

The image of $\mathrm{L}^2_\mu(P;\mathbb{C})$ under the map $f \mapsto cf$ is the space E_n of functions $g\colon P \to \mathbb{C}$, such that the map $z \mapsto y^n g(z)^2$ is μ-integrable. One can then show that the equivalent unitary representation is given by

$$(\Pi_s(f))(z) = (cz-a)^{-n} g\left(\frac{b-dz}{cz-a}\right), \qquad g \in E_n,\ z \in P,\ s = \begin{pmatrix} a & b \\ c & d \end{pmatrix}.$$

It can be shown that for $n > 1$, the space H^2_n of holomorphic functions on P is nonempty and invariant under Π. Furthermore, these representations in H^2_n are irreducible. The complex conjugates of these representations are also irreducible and not equivalent to the previous ones; see Dieudonné [Dieudonné (2003b)] (Chapter XXII, Section 3).

These irreducible unitary representations of $\mathbf{SL}(2,\mathbb{R})$ constitute the *discrete series*. There are other irreducible unitary representations of $\mathbf{SL}(2,\mathbb{R})$ called the *complementary series*.

For comprehensive treatments of the irreducible representations of $\mathbf{SL}(2,\mathbb{R})$ and other semisimple Lie groups, see Knapp [Knapp (2001)], Vilenkin [Vilenkin (1968)], and Taylor [Taylor (1986)].

6.9 Partial Traces, Induced Representations of Compact Groups

In this section, we consider a compact (metrizable) group G and a closed subgroup H of G, and our goal is to determine the canonical (unitary) representation of G in $\mathrm{L}^2_\mu(G/H;\mathbb{C})$ induced by the trivial representation of H in $E = \mathbb{C}$ (see Definition 6.13), where μ is the G-invariant measure on G/H induced by a Haar measure λ on G. For simplicity of notation, we write $\mathrm{L}^2_\mu(G/H)$ instead of $\mathrm{L}^2_\mu(G/H;\mathbb{C})$. To do this, it is necessary to understand what is the restriction of the representation $M_\rho\colon G \to \mathbf{U}(\mathbb{C}^{n_\rho})$ to H, with $\rho \in R(G)$.

We will denote the complete set of the irreducible representations of G given by the Peter–Weyl theorem I (Theorem 4.2) by $\rho \in R(G)$, the corresponding representations by $M_\rho\colon G \to \mathbf{U}(\mathbb{C}^{n_\rho})$, and the identity element of $\mathfrak{a}_\rho$ by $u_\rho = \frac{1}{n_\rho}\chi_\rho$, where χ_ρ is the character associated with ρ. Similarly, we will denote the complete set of irreducible representations of H given by the Peter–Weyl theorem I by $\sigma \in R(H)$, the corresponding representations by $M_\sigma\colon H \to \mathbf{U}(\mathbb{C}^{n_\sigma})$, and the identity element of $\mathfrak{a}_\sigma$ by $u_\sigma = \frac{1}{n_\sigma}\chi_\sigma$, where χ_σ is the character associated with σ. The Haar measure on G is denoted by λ_G, and the Haar measure on H is denoted by λ_H.

Consider the restriction $V\colon H \to \mathbf{U}(\mathbb{C}^{n_\rho})$ of the representation $M_\rho\colon G \to \mathbf{U}(\mathbb{C}^{n_\rho})$ to H. Recall that for any function $f \in \mathrm{L}^2(H)$ and any $x \in \mathbb{C}^{n_\rho}$, $V_{\mathrm{ext}}(f)(x)$ is the weak integral of the function $t \mapsto V(t)(x)$ with respect to $fd\lambda_H$ ($t \in H$). We will write $M_\rho(f)$ for $V_{\mathrm{ext}}(f)$. By the Peter–Weyl theorem II (Theorem 4.16), for every $\sigma \in R(H)$, the map $\pi_\sigma^{M_\rho} = M_\rho(u_{\overline{\sigma}})$ given by

$$\pi_\sigma^{M_\rho}(x) = \frac{1}{n_\sigma}\int_H \overline{\chi_\sigma(t)} M_\rho(t)(x)\, d\lambda_H(t), \quad x \in \mathbb{C}^{n_\rho},$$

is the orthogonal projection of $\mathbb{C}^{n_\rho}$ onto a closed subspace E_σ of $\mathbb{C}^{n_\rho}$, and we have a Hilbert sum

$$\mathbb{C}^{n_\rho} = \bigoplus_{\sigma \in R(H)} E_\sigma.$$

Recall from Section 4.6 that the integral defining $\pi_\sigma^{M_\rho}(x)$ can be computed by integrating the matrix $\overline{\chi_\sigma(t)} M_\rho(t)(x)$, term by term. Furthermore, for each subspace $E_\sigma \neq (0)$, each irreducible representation M_σ of H is contained a certain number of times in the restriction of M_ρ to H, which we denote $d_\sigma = (\rho : \sigma)$, so E_σ is a finite Hilbert sum

$$E_\sigma = \bigoplus_{k=1}^{d_\sigma} F_k^\sigma,$$

of subspaces $F_1^\sigma, F_2^\sigma, \ldots, F_{d_\sigma}^\sigma$ of dimension n_σ, invariant under $M_\rho(t)$ for every $t \in H$, and such that the restriction of M_ρ to H and to each F_k^σ is equivalent to the irreducible representation M_σ. Thus, E_σ has dimension $p_\sigma = d_\sigma n_\sigma$.

We can pick an orthonormal basis of $\mathbb{C}^{n_\rho}$ consisting of the union of orthonormal bases of each of the F_j^σ and of a basis of the orthogonal complement F' of E_σ in $\mathbb{C}^{n_\rho}$. Let P be the change of basis matrix, which is unitary. For the basis of E_σ consisting of the first $p_\sigma = d_\sigma n_\sigma$ vectors of this basis, the matrix $M_{\rho,\sigma}(t)$ of the restriction of $P^* M_\rho(t) P$ to E_σ is a block diagonal matrix (consisting of d_σ blocks) of the form

$$M_{\rho,\sigma}(t) = \begin{pmatrix} M_\sigma(t) & 0 & \cdots & 0 \\ 0 & M_\sigma(t) & \cdots & 0 \\ \vdots & \vdots & \ddots & \vdots \\ 0 & 0 & \cdots & M_\sigma(t) \end{pmatrix},$$

for every $t \in H$.

The automorphism $M_\rho(s)$ of $\mathbb{C}^{n_\rho} = E_\sigma \oplus F'$ is defined by four linear maps $P_{\rho,\sigma}(s)\colon E_\sigma \to E_\sigma$, $M_2(s)\colon F' \to E_\sigma$, $M_3(s)\colon E_\sigma \to F'$, and $M_4(s)\colon F' \to F'$, such that for any $(u, v) \in E_\sigma \times F'$, we have

$$M_\rho(s)\begin{pmatrix} u \\ v \end{pmatrix} = \begin{pmatrix} P_{\rho,\sigma}(s) & M_2(s) \\ M_3(s) & M_4(s) \end{pmatrix}\begin{pmatrix} u \\ v \end{pmatrix} = \begin{pmatrix} P_{\rho,\sigma}(s)u + M_2(s)v \\ M_3(s)u + M_4(s)v \end{pmatrix}. \tag{M1}$$

Since $M_\rho(u_{\overline{\sigma}})$ is the orthogonal projection of $\mathbb{C}^{n_\rho} = E_\sigma \oplus F'$ onto E_σ, the endomorphism $M_\rho(u_{\overline{\sigma}}) M_\rho(s) M_\rho(u_{\overline{\sigma}})$ of $\mathbb{C}^{n_\rho}$ $(s \in G)$ is defined by

$$M_\rho(u_{\overline{\sigma}}) M_\rho(s) M_\rho(u_{\overline{\sigma}}) \begin{pmatrix} u \\ v \end{pmatrix} = \begin{pmatrix} P_{\rho,\sigma}(s)u \\ 0 \end{pmatrix},$$

and so we can write

$$M_\rho(u_{\overline{\sigma}}) M_\rho(s) M_\rho(u_{\overline{\sigma}}) = \begin{pmatrix} P_{\rho,\sigma}(s) & 0 \\ 0 & 0 \end{pmatrix}. \tag{M2}$$

In the new orthonormal basis of $E_\sigma \oplus F' = \mathbb{C}^{n_\rho}$, the matrix of the endomorphism $M_\rho(u_{\overline{\sigma}}) M_\rho(s) M_\rho(u_{\overline{\sigma}})$ is the block matrix

$$P^* M_\rho(u_{\overline{\sigma}}) M_\rho(s) M_\rho(u_{\overline{\sigma}}) P = \begin{pmatrix} Q(s) & 0 \\ 0 & 0 \end{pmatrix},$$

with $Q(s)$ the $p_\sigma \times p_\sigma$ matrix

$$Q(s) = \frac{1}{n_\rho} \begin{pmatrix} m_{11}^{(\rho,\sigma)}(s) & m_{12}^{(\rho,\sigma)}(s) & \dots & m_{1p_\sigma}^{(\rho,\sigma)}(s) \\ m_{21}^{(\rho,\sigma)}(s) & m_{22}^{(\rho,\sigma)}(s) & \dots & m_{2p_\sigma}^{(\rho,\sigma)}(s) \\ \vdots & \vdots & \ddots & \vdots \\ m_{p_\sigma 1}^{(\rho,\sigma)}(s) & m_{p_\sigma 2}^{(\rho,\sigma)}(s) & \dots & m_{p_\sigma p_\sigma}^{(\rho,\sigma)}(s) \end{pmatrix}.$$

Note that since we have made a change of basis, the entries $m_{ij}^{(\rho,\sigma)}(s)$ are *not equal* to the original entries $m_{ij}^{(\rho)}(s)$ occurring in $M_\rho(s)$.

For any $t \in H$, the matrix $Q(t)$ of $P_{\rho,\sigma}(t)$ is the block diagonal matrix $M_{\rho,\sigma}(t)$, because the subspaces F_k^σ are invariant under $M_\rho(t)$ for $t \in H$. Since E_σ is invariant under $M_\rho(t)$ for $t \in H$, by (M1), for any $(u, v) \in E_\sigma \times F'$,

$$M_\rho(t)M_\rho(u_{\overline{\sigma}}) \begin{pmatrix} u \\ v \end{pmatrix} = M_\rho(t) \begin{pmatrix} u \\ 0 \end{pmatrix} = \begin{pmatrix} P_{\rho,\sigma}(t)u \\ 0 \end{pmatrix},$$

and

$$M_\rho(u_{\overline{\sigma}})M_\rho(t) \begin{pmatrix} u \\ v \end{pmatrix} = M_\rho(u_{\overline{\sigma}}) \begin{pmatrix} P_{\rho,\sigma}(t)u \\ M_3(t)u + M_4(t)v \end{pmatrix} = \begin{pmatrix} P_{\rho,\sigma}(t)u \\ 0 \end{pmatrix},$$

so

$$M_\rho(t)M_\rho(u_{\overline{\sigma}}) = M_\rho(u_{\overline{\sigma}})M_\rho(t) = \begin{pmatrix} P_{\rho,\sigma}(t) & 0 \\ 0 & 0 \end{pmatrix}, \tag{M3}$$

and since

$$M_\rho(tst') = M_\rho(t)M_\rho(s)M_\rho(t'),$$

by (M3) and (M2), we obtain the equation

$$\begin{aligned} M_\rho(u_{\overline{\sigma}})M_\rho(tst')M_\rho(u_{\overline{\sigma}}) &= M_\rho(u_{\overline{\sigma}})M_\rho(t)M_\rho(s)M_\rho(t')M_\rho(u_{\overline{\sigma}}) \\ &= M_\rho(t)M_\rho(u_{\overline{\sigma}})M_\rho(s)M_\rho(u_{\overline{\sigma}})M_\rho(t') \\ &= M_\rho(t)M_\rho(u_{\overline{\sigma}})M_\rho(u_{\overline{\sigma}})M_\rho(s) \\ &\quad M_\rho(u_{\overline{\sigma}})M_\rho(u_{\overline{\sigma}})M_\rho(t') \\ &= \begin{pmatrix} P_{\rho,\sigma}(t) & 0 \\ 0 & 0 \end{pmatrix} \begin{pmatrix} P_{\rho,\sigma}(s) & 0 \\ 0 & 0 \end{pmatrix} \begin{pmatrix} P_{\rho,\sigma}(t') & 0 \\ 0 & 0 \end{pmatrix} \\ &= \begin{pmatrix} P_{\rho,\sigma}(t)P_{\rho,\sigma}(s)P_{\rho,\sigma}(t') & 0 \\ 0 & 0 \end{pmatrix}, \end{aligned}$$

and since by (M2), we have

$$M_\rho(u_{\overline{\sigma}})M_\rho(tst')M_\rho(u_{\overline{\sigma}}) = \begin{pmatrix} P_{\rho,\sigma}(tst') & 0 \\ 0 & 0 \end{pmatrix},$$

we obtain the equation

$$P_{\rho,\sigma}(tst') = P_{\rho,\sigma}(t)P_{\rho,\sigma}(s)P_{\rho,\sigma}(t'), \qquad \text{for all } s \in G \text{ and all } t, t' \in H. \quad (*)$$

Since $\mathrm{tr}(AB) = \mathrm{tr}(BA)$ and $M_\rho(u_{\overline{\sigma}})M_\rho(u_{\overline{\sigma}}) = M_\rho(u_{\overline{\sigma}})$ since $M_\rho(u_{\overline{\sigma}})$ is a projection, we have

$$\mathrm{tr}(M_\rho(u_{\overline{\sigma}})M_\rho(s)M_\rho(u_{\overline{\sigma}})) = \mathrm{tr}(M_\rho(u_{\overline{\sigma}})M_\rho(u_{\overline{\sigma}})M_\rho(s)) = \mathrm{tr}(M_\rho(u_{\overline{\sigma}})M_\rho(s)).$$

Definition 6.16. The *partial trace* of ρ with respect to σ is the function

$$s \mapsto \theta_{\rho,\sigma}(s) = \mathrm{tr}(M_\rho(u_{\overline{\sigma}})M_\rho(s)M_\rho(u_{\overline{\sigma}})) = \mathrm{tr}(M_\rho(u_{\overline{\sigma}})M_\rho(s)),$$

which can also be expressed as

$$\theta_{\rho,\sigma}(s) = \frac{1}{n_\rho}(m_{11}^{(\rho,\sigma)}(s) + \cdots + m_{p_\sigma p_\sigma}^{(\rho,\sigma)}(s)).$$

The function $\theta_{\rho,\sigma}$ is continuous, no longer central, and not identically zero if σ is contained in the restriction of M_ρ to H. This function depends on ρ and σ, and we have

$$\begin{aligned} \theta_{\rho,\sigma}(t) &= (\rho : \sigma)\chi_\sigma(t), && \text{for all } t \in H, \\ \chi_\rho(s) &= \sum_\sigma \theta_{\rho,\sigma}(s), && \text{for all } s \in G. \end{aligned}$$

It will be shown in Section 9.1 (see Example 9.6) that the partial traces for which $p = 1$ are the spherical functions when (G, H) is a Gelfand pair.

We have the following proposition.

Proposition 6.17. *The following properties hold.*

(1) We have

$$\theta_{\rho,\sigma}(tst^{-1}) = \theta_{\rho,\sigma}(s), \qquad \textit{for all } s \in G \textit{ and all } t \in H.$$

(2) When ρ ranges over $R(G)$ and σ ranges over $R(H)$, the partial traces $\theta_{\rho,\sigma}$ are pairwise orthogonal. In particular, $\theta_{\rho,\sigma}$ and $\theta_{\rho',\sigma'}$ can only be proportional if $\rho' = \rho$ and $\sigma' = \sigma$.

(3) The partial traces $\theta_{\rho,\sigma}$ are continuous functions of positive type.

Proof. (1) This equation follows immediately from $(*)$ and the commutativity of the trace.

(2) This follows from the equation

$$\theta_{\rho,\sigma}(s) = \frac{1}{n_\rho}(m_{11}^{(\rho,\sigma)}(s) + \cdots + m_{p_\sigma p_\sigma}^{(\rho,\sigma)}(s))$$

and the orthogonality properties of the $m_{ij}^{(\rho,\sigma)}$; see Proposition 4.9.

(3) This follows from the properties $m_{ii}^{(\rho,\sigma)} = \check{\overline{m}}_{ii}^{(\rho,\sigma)} = m_{ii}^{(\rho,\sigma)} * m_{ii}^{(\rho,\sigma)}$ of Proposition 4.9, the fact that $f * \check{\overline{f}}$ is of positive type for every $f \in \mathcal{L}^2(G)$, and the equation

$$\theta_{\rho,\sigma}(s) = \frac{1}{n_\rho}(m_{11}^{(\rho,\sigma)}(s) + \cdots + m_{p_\sigma p_\sigma}^{(\rho,\sigma)}(s)). \qquad \square$$

Since G and H are compact, G/H has a G-invariant measure μ induced by a Haar measure on G. We now try to understand what the canonical unitary representation of G in $\mathrm{L}^2_\mu(G/H)$ induced by the trivial representation of H in $E = \mathbb{C}$ looks like. With the notations as above, we have $n_{\sigma_0} = 1$, and $p_{\sigma_0} = d$.

First, let us observe that a function $g \in \mathcal{L}^2_\mu(G/H)$ can be viewed as a function $g \in \mathcal{L}^2(G)$, such that

$$g(st) = g(s), \qquad \text{for all } t \in H \text{ and all } s \in G. \qquad (*_{G/H})$$

Since $(g * \delta_t)(s) = g(st)$, the above condition is equivalent to

$$g * \delta_t = g, \qquad \text{for all } t \in H, \qquad (*'_{G/H})$$

and thus for any measure $\nu \in \mathcal{M}^1(G)$, the function $\nu * g \in \mathcal{L}^2_\mu(G/H)$ also satisfies the equation

$$(\nu * g) * \delta_t = \nu * g,$$

so we deduce that $\mathrm{L}^2_\mu(G/H)$ is a closed left ideal in $\mathcal{M}^1(G)$, which implies that $\mathrm{L}^2_\mu(G/H)$ is a closed left ideal in $\mathrm{L}^2(G)$. In particular, for every $\rho \in R(G)$, the projection $g \mapsto u_\rho * g$ of $\mathrm{L}^2(G)$ onto the ideal $\mathfrak{a}_\rho$ maps $\mathrm{L}^2_\mu(G/H)$ onto itself, so $\mathrm{L}^2_\mu(G/H)$ is the Hilbert sum of the subspaces

$$L_\rho = \mathrm{L}^2_\mu(G/H) \cap \mathfrak{a}_\rho.$$

It remains to determine what the L_ρ are. We explained that by applying Peter–Weyl II (Theorem 4.16) to the restriction of the representation

$M_\rho\colon G \to \mathbf{U}(C^{n_\rho})$ to H, we obtain a decomposition of $\mathbb{C}^{n_\rho}$ as a finite Hilbert sum

$$\mathbb{C}^{n_\rho} = E_{\sigma_1} \oplus \cdots \oplus E_{\sigma_q},$$

with each E_{σ_i} a direct sum

$$E_{\sigma_i} = \bigoplus_{k=1}^{d_{\sigma_i}} F_k^{\sigma_i}$$

of subspaces $F_1^{\sigma_i}, F_2^{\sigma_i}, \ldots, F_{d_{\sigma_i}}^{\sigma_i}$ of dimension n_{σ_i}, invariant under $M_\rho(t)$ for every $t \in H$, and such that the restriction of M_ρ to each $F_k^{\sigma_i}$ is equivalent to the irreducible representation M_{σ_i}. Let us pick for an orthonormal basis of $\mathbb{C}^{n_\rho}$ the union of orthonormal bases of the $F_k^{\sigma_i}$, and let P be the change of basis matrix, which is unitary. Then for any $t \in H$, we have

$$P^* M_\rho(t) P = \begin{pmatrix} M_{\rho,\sigma_1}(t) & 0 & \cdots & 0 \\ 0 & M_{\rho,\sigma_2}(t) & \cdots & 0 \\ \vdots & \vdots & \ddots & \vdots \\ 0 & 0 & \cdots & M_{\rho,\sigma_q}(t) \end{pmatrix},$$

where $M_{\rho,\sigma_i}(t)$ is the block matrix

$$M_{\rho,\sigma_i}(t) = \begin{pmatrix} M_{\sigma_i}(t) & 0 & \cdots & 0 \\ 0 & M_{\sigma_i}(t) & \cdots & 0 \\ \vdots & \vdots & \ddots & \vdots \\ 0 & 0 & \cdots & M_{\sigma_i}(t) \end{pmatrix}$$

(consisting of d_{σ_i} blocks) defined earlier. Thus, the matrix $M_\rho^{(H)}(t) = P^* M_\rho(t) P$ (with $t \in H$) is the block matrix consisting of the blocks $M_{\sigma_i}(t)$, each one repeated d_{σ_i} times. We also define the matrices $M_\rho^{(H)}(s) = (m_{ij}^{(\rho,H)}(s))$ for *all* $s \in G$ by

$$M_\rho^{(H)}(s) = P^* M_\rho(s) P, \quad s \in G.$$

Beware that if $s \in G$ but $s \notin H$, then the matrix $M_\rho^{(H)}(s)$ does *not* have the nice block structure enjoyed by the matrices $M_\rho^{(H)}(t)$ when $t \in H$. The representations of G in $\mathbb{C}^{n_\rho}$ defined by the matrices $M_\rho(s)$ and $M_\rho^{(H)}(s)$ ($s \in G$) are equivalent. The matrix $M_\rho^{(H)}$ denotes the matrix of n_ρ^2 functions $m_{ij}^{(\rho,H)}$ given by $s \mapsto m_{ij}^{(\rho,H)}(s)$ and we also write $M_\rho^{(H)} = P^* M_\rho P$. By Proposition 4.9, the matrix $M_\rho^{(H)}$ defines n_ρ^2 functions $m_{ij}^{(\rho,H)}$ that form an orthonormal basis of $\mathfrak{a}_\rho$ and satisfy the same properties as the functions $m_{i,j}^{(\rho)}$ defined by the matrix M_ρ.

Proposition 6.18. *The space $\mathrm{L}^2_\mu(G/H)$ is the Hilbert sum of subspaces $L_\rho \subseteq \mathfrak{a}_\rho$. If the trivial representation σ_0 of H is contained $d = (\rho : \sigma_0) \geq 1$ times in the restriction of M_ρ to H, then L_ρ is the direct sum of the first d columns of $M_\rho^{(H)} = P^* M_\rho P$,*

$$L_\rho = \bigoplus_{j=1}^{d} \mathfrak{l}_j^{(\rho,H)} \quad \textit{and} \quad \mathfrak{l}_j^{(\rho,H)} = \bigoplus_{k=1}^{n_\rho} \mathbb{C} m_{kj}^{(\rho,H)}.$$

If $d = 0$, then $L_\rho = (0)$. The subrepresentation $\Pi\colon G \to \mathbf{U}(L_\rho)$ in L_ρ of the canonical representation $\Pi\colon G \to \mathbf{U}(\mathrm{L}^2_\mu(G/H))$ of G in $\mathrm{L}^2_\mu(G/H)$ induced by the trivial representation of H in $\mathbb{C}$ is the Hilbert sum of $d = (\rho : \sigma_0)$ irreducible representations equivalent to $M_{\overline{\rho}}$.

Proof. Since any function $g \in \mathrm{L}^2(G/H) \cap \mathfrak{a}_\rho$ can be written as $g = \sum_{ij} c_{ij} m_{ij}^{(\rho,H)}$ (for some $c_{ij} \in \mathbb{C}$), the equation $(*_{G/H})$ and Proposition 4.9(4) yield

$$\sum_{i,j,k} c_{ij} m_{ik}^{(\rho,H)}(s) m_{kj}^{(\rho,H)}(t) = \sum_{i,k} c_{ik} m_{ik}^{(\rho,H)}(s),$$

with $1 \leq i,j,k \leq n_\rho$, all $s \in G$ and all $t \in H$, and since the n_ρ^2 functions $m_{ij}^{(\rho,H)}$ are linearly independent, we get

$$\sum_{j=1}^{n_\rho} c_{ij} m_{kj}^{(\rho,H)}(t) = c_{ik}, \tag{$*_1$}$$

for all i,k with $1 \leq i,k \leq n_\rho$ and for all $t \in H$.

Suppose that the trivial representation σ_0 of H is contained $d \geq 1$ times in the restriction of M_ρ to H, which means that the first d matrices $M_{\sigma_0}(t)$ in $M_{\rho,\sigma_0}(t)$ are just one-dimensional matrices equal to 1, the other matrices being at least two-dimensional. We then have $m_{kj}^{(\rho,H)}(t) = \delta_{kj}$ for $k \leq d$ or $j \leq d$, hence $(*_1)$ is trivially verified for $k \leq d$, and we are left with the equations

$$\sum_{j=d+1}^{n_\rho} c_{ij} m_{kj}^{(\rho,H)}(t) = c_{ik}, \qquad k > d \text{ and } t \in H. \tag{$*_2$}$$

Consider one of the matrices $M_{\sigma_h}(t)$ and assume it corresponds to the lines of index k, such that $k'_h \leq k \leq k''_h$. Then we have $m_{kj}^{(\rho,H)}(t) = 0$ for $k'_h \leq k \leq k''_h$ and all j except those for which $k'_h \leq j \leq k''_h$; in addition, since $\sigma \neq \sigma_0$, by the fact stated just after Definition 4.3, we have

$$\int_H m_{kj}^{(\rho,H)}(t)\, d\lambda_H(t) = 0$$

for all these indices. Integrating both sides of $(*_2)$, we see that $c_{ik} = 0$ for all indices i and all $k > d$.

Therefore L_ρ is the subspace of $\mathfrak{a}_\rho$, of dimension dn_ρ, spanned by the $m_{ij}^{(\rho,H)}$ such that $j \leq d$, equivalently, the *direct sum of the first d columns of $M_\rho^{(H)}$*,

$$L_\rho = \bigoplus_{j=1}^{d} \mathfrak{l}_j^{(\rho,H)} \quad \text{and} \quad \mathfrak{l}_j^{(\rho,H)} = \bigoplus_{k=1}^{n_\rho} \mathbb{C} m_{kj}^{(\rho,H)}.$$

If $d = 0$, then the above reasoning shows that $L_\rho = (0)$.

The canonical representation $\Pi\colon G \to \mathbf{U}(\mathrm{L}^2_\mu(G/H))$ of G in $\mathrm{L}^2_\mu(G/H)$ induced by the trivial representation of H in $\mathbb{C}$ is a subrepresentation of the regular representation of G. We know from the discussion just after Definition 4.7 that on $\mathfrak{a}_\rho$, the regular representation $\mathbf{R}$ splits into n_ρ irreducible representations all equivalent to $M_{\overline{\rho}}$, and we can view these representations as acting on the columns of M_ρ, the left ideals $\mathfrak{l}_j^{(\rho)}$. Therefore, the subrepresentation in L_ρ of the canonical representation Π of G induced by the trivial representation of H in $\mathbb{C}$ is the Hilbert sum of $(\rho : \sigma_0)$ irreducible representations equivalent to $M_{\overline{\rho}}$. ☐

Remark: It is possible to describe the unitary representations of G induced by the nontrivial irreducible representations M_{σ_i} of H; see Dieudonné's [Dieudonné (2003b)] Chapter XXII, Section 5, Problem 1.

We can also consider the space $H\backslash G$ of right cosets Hs of G ($s \in G$). If $\pi\colon G \to H\backslash G$ is the quotient map $\pi(s) = Hs$, the fact that the Haar measure λ on a compact group is left and right invariant implies immediately that there is a G-invariant measure μ' on $H\backslash G$, such that

$$\int_{G/H} g(x)\, d\mu'(x) = \int_G (g \circ \pi)\, d\lambda,$$

and

$$\int_{G/H} g(x \cdot s)\, d\mu'(x) = \int_{G/H} g(x)\, d\mu'(x), \qquad \text{for all } s \in G,$$

with

$$(Ht) \cdot s = Hts, \qquad s, t \in G.$$

Every function $g \in \mathcal{L}^2_{\mu'}(H\backslash G)$ can be viewed as a function $g \in \mathcal{L}^2(G)$, such that

$$g(ts) = g(s), \qquad \text{for all } t \in H \text{ and all } s \in G. \qquad (*_{H\backslash G})$$

Since $(\delta_t * g)(s) = g(t^{-1}s)$, the above condition is equivalent to

$$\delta_t * g = g, \qquad \text{for all } t \in H. \qquad (*'_{H\backslash G})$$

The space $\mathrm{L}^2_{\mu'}(H\backslash G)$ is the image of the space $\mathrm{L}^2_{\mu}(G/H)$ under the isomorphism $g \mapsto \check{g}$ (here we use the fact that G is unimodular). Therefore $\mathrm{L}^2_{\mu'}(H\backslash G)$ is a closed right ideal in $\mathrm{L}^2(G)$, and it is the Hilbert sum of the images $\check{L}_\rho$ of the L_ρ; since by Theorem 4.6(2) we have $m_{ji} = \check{\overline{m}}_{ij}$, we deduce that $\check{L}_\rho$ is the direct sum of the first d rows of $M_\rho^{(H)}$ (with $d = (\rho : \sigma_0)$).

Let us record this fact.

Proposition 6.19. *The space* $\mathrm{L}^2_{\mu'}(H\backslash G)$ *is the Hilbert sum of subspaces* $\check{L}_\rho \subseteq \mathfrak{a}_\rho$. *If the trivial representation* σ_0 *of* H *is contained* $d = (\rho : \sigma_0) \geq 1$ *times in the restriction of* M_ρ *to* H, *then* $\check{L}_\rho$ *is the direct sum of the first* d *rows of* $M_\rho^{(H)}$; *that is,*

$$\check{L}_\rho = \bigoplus_{i=1}^{d} \bigoplus_{j=1}^{n_\rho} \mathbb{C} m_{ij}^{(\rho,H)}.$$

Let us now consider the intersection $\mathrm{L}^2_{\mu}(G/H) \cap \mathrm{L}^2_{\mu'}(H\backslash G)$. This is a closed involutive subalgebra of $\mathrm{L}^2(G)$, thus a complete Hilbert algebra. We can view a function $g \in \mathrm{L}^2_{\mu}(G/H) \cap \mathrm{L}^2_{\mu'}(H\backslash G)$ as a function $g \in \mathrm{L}^2(G)$, such that

$$g(tst') = g(s), \qquad \text{for all } t, t' \in H \text{ and all } s \in G, \qquad (*_{H\backslash G/H})$$

or equivalently,

$$\delta_t * g * \delta_{t'} = g, \qquad \text{for all } t, t' \in H. \qquad (*'_{H\backslash G/H})$$

We can also think of the functions $g \in \mathrm{L}^2_{\mu}(G/H) \cap \mathrm{L}^2_{\mu'}(H\backslash G)$ as functions defined on the *double classes (or double cosets)* HsH of G with respect to H. In this case, if $\pi\colon G \to H\backslash G/H$ is the quotient map $\pi(s) = HsH$, the fact that the Haar measure λ on a compact group is left and right invariant implies that there is a G-invariant measure μ on $H\backslash G/H$, such that

$$\int_{H\backslash G/H} g(x)\, d\mu(x) = \int_G (g \circ \pi)\, d\lambda.$$

We denote the algebra of functions in $\mathrm{L}^2(G)$ satisfying $(*_{H\backslash G/H})$ as $\mathrm{L}^2_{\mu}(H\backslash G/H)$, or simply as $\mathrm{L}^2(H\backslash G/H)$. The following proposition follows immediately from the previous two propositions.

Proposition 6.20. *The algebra* $\mathrm{L}^2(H\backslash G/H)$ *is the Hilbert sum of the minimal two-sided ideals*

$$\mathfrak{a}_{\rho,\sigma_0} = L_\rho \cap \check{\bar{L}}_\rho = \bigoplus_{i=1}^{d}\bigoplus_{j=1}^{d} \mathbb{C} m_{ij}^{(\rho,H)}.$$

Each $\mathfrak{a}_{\rho,\sigma_0}$ *is a matrix algebra of dimension* d^2 *having the family* $(m_{ij}^{(\rho,H)})_{1\leq i,j\leq d}$ *as a basis. The center of* $\mathfrak{a}_{\rho,\sigma_0}$ *is the one-dimensional subspace*

$$\mathbb{C}(m_{11}^{(\rho,H)} + \cdots + m_{dd}^{(\rho,H)}) = \mathbb{C} n_\rho \theta_{\rho,\sigma_0},$$

and $u_{\rho,\sigma_0} = m_{11}^{(\rho,H)} + \cdots + m_{dd}^{(\rho,H)}$ *is the unit of* $\mathfrak{a}_{\rho,\sigma_0}$. *The map* $g \mapsto u_{\rho,\sigma_0} * g = g * u_{\rho,\sigma_0}$ *is the orthogonal projection of* $\mathrm{L}^2(H\backslash G/H)$ *onto* $\mathfrak{a}_{\rho,\sigma_0}$.

The subspace

$$\mathfrak{l}_{\sigma_0,1}^{(\rho,H)} = \mathfrak{l}_1^{(\rho,H)} \cap \mathfrak{a}_{\rho,\sigma_0} = \mathbb{C} m_{11}^{(\rho,H)} \oplus \cdots \oplus \mathbb{C} m_{d1}^{(\rho,H)}$$

is a minimal left ideal of $\mathrm{L}^2(H\backslash G/H)$. By Theorem 2.36, this ideal defines the irreducible representation $W_\rho\colon \mathrm{L}^2(H\backslash G/H) \to \mathcal{L}(\mathfrak{l}_{\sigma_0,1}^{(\rho,H)})$ of the algebra $\mathrm{L}^2(H\backslash G/H)$ in $\mathfrak{l}_{\sigma_0,1}^{(\rho,H)}$ (of dimension d), given by

$$(W_\rho(g))(f) = g * f, \qquad g \in \mathrm{L}^2(H\backslash G/H),\ f \in \mathfrak{l}_{\sigma_0,1}^{(\rho,H)}.$$

From Theorem 2.36(2), up to equivalence, we obtain all irreducible representations of the space $\mathrm{L}^2(H\backslash G/H)$ in $\mathfrak{l}_{\sigma_0,1}^{(\rho,H)}$ in this fashion. We can describe the representation W_ρ explicitly as follows. For every $g \in \mathrm{L}^2(H\backslash G/H)$, by Proposition 4.9, we can write

$$g * u_{\rho,\sigma_0} = \sum_{1\leq i,k\leq d} c_{ik}(g) m_{ik}^{(\rho,H)} \in \mathfrak{a}_{\rho,\sigma_0},$$

and the jth column of the matrix $W_\rho(g)$ consists of the coordinates of $W_\rho(g)(m_{j1}^{(\rho,H)}) = g * m_{j1}^{(\rho,H)}$ over the basis $(m_{11}^{(\rho,H)}, \ldots, m_{d1}^{(\rho,H)})$, and since u_{ρ,σ_0} is the unit of $\mathfrak{a}_{\rho,\sigma_0}$, by Proposition 4.9,

$$\begin{aligned} g * m_{1j}^{(\rho,H)} &= g * u_{\rho,\sigma_0} * m_{1j}^{(\rho,H)} = \left(\sum_{1\leq i,k\leq d} c_{ik}(g) m_{ik}^{(\rho,H)}\right) * m_{j1}^{(\rho,H)} \\ &= \sum_{1\leq i\leq d} c_{ij}(g) m_{i1}^{(\rho,H)}, \end{aligned}$$

so $W_\rho(g) = (c_{ij}(g))$, a $d\times d$ matrix.

The above facts imply the following proposition.

Proposition 6.21. *The algebra* $L^2(H\backslash G/H)$ *is commutative if and only if* $(\rho : \sigma_0) \leq 1$ *for all* $\rho \in R(G)$. *If so, then for every* $\rho \in R(G)$, *such that* $(\rho : \sigma_0) = 1$, *the ideal* $\mathfrak{a}_{\rho,\sigma_0}$ *is one-dimensional and is spanned by the function*

$$\omega_\rho(s) = \theta_{\rho,\sigma_0} = \frac{1}{n_\rho} m_{11}^{(\rho,H)}(s),$$

which is continuous and of positive type. Thus,

$$L^2(H\backslash G/H) = \bigoplus_{\rho \mid (\rho:\sigma_0)=1} \mathbb{C}\omega_\rho.$$

The orthogonal projection of $L^2(H\backslash G/H)$ *onto* $\mathbb{C}\omega_\rho$ *is given by*

$$g \mapsto \omega_\rho * g = g * \omega_\rho.$$

The function ω_ρ *also satisfies the following equations:*

$$\begin{aligned} \omega_\rho(tst') &= \omega_\rho(s), \qquad \text{for all } s \in G \text{ and all } t, t' \in H \\ \omega_\rho(e) &= 1. \end{aligned}$$

The function ω_ρ is called a *(zonal) spherical function.*

The irreducible unitary representation W_ρ is one-dimensional, which implies that for every $g \in \mathcal{L}^2(H\backslash G/H)$ (since $g * \omega_\rho$ is continuous), we have

$$g * \omega_\rho = \zeta(g)\omega_\rho,$$

where ζ must be a character of $L^2(H\backslash G/H)$ with values in $\mathbb{T}$ (because $\omega_\rho * \omega_\rho = \omega_\rho$ and $g * \omega_\rho = \omega_\rho * g$). Since ζ is an algebra homomorphism and $\zeta(g) \in \mathbb{T}$, we conclude that ζ is a hermitian character of $L^2(H\backslash G/H)$.

Since $(\rho : \sigma_0) = 1$, the left ideal L_ρ is equal to the ideal $\mathfrak{l}_1^{(\rho,H)}$, which by Proposition 4.9(5) is a minimal ideal in $\mathfrak{a}_\rho$, and by Proposition 4.4, it is spanned by the elements of the form $\lambda_s\omega_\rho = \delta_s * \omega_\rho$, for all $s \in G$.

6.10 Spherical Harmonics on S^n and $L^2(S^n)$

A nice example of the above situation arises if $G = \mathbf{SO}(n+1)$ and $H = \mathbf{SO}(n)$. In this case, $G/H = \mathbf{SO}(n+1)/\mathbf{SO}(n) \simeq S^n$. By Proposition 6.18, the space $L^2(\mathbf{SO}(n+1)/\mathbf{SO}(n)) \simeq L^2(S^n)$ is the Hilbert sum of the subspaces $L_\rho \subseteq \mathfrak{a}_\rho$ for which $(\rho : \sigma_0) \geq 1$, where

$$L^2(\mathbf{SO}(n+1)) = \bigoplus_\rho \mathfrak{a}_\rho$$

is the Hilbert sum given by Peter–Weyl I and where $d = (\rho : \sigma_0) \geq 1$ is the number of times that the trivial representation σ_0 of $\mathbf{SO}(n)$ is contained in the restriction of M_ρ to $\mathbf{SO}(n)$. Then L_ρ is the direct sum of the first d columns of $M_\rho^{(H)}$,

$$L_\rho = \bigoplus_{j=1}^{d} \mathfrak{l}_j^{(\rho,H)} \quad \text{and} \quad \mathfrak{l}_j^{(\rho,H)} = \bigoplus_{k=1}^{n_\rho} \mathbb{C} m_{kj}^{(\rho,H)}.$$

The subrepresentation $\Pi\colon \mathbf{SO}(n+1) \to \mathbf{U}(L_\rho)$ of the canonical representation (see Definition 6.13) $\Pi\colon \mathbf{SO}(n+1) \to \mathbf{U}(\mathrm{L}^2(S^n))$ of $\mathbf{SO}(n+1)$ in $\mathrm{L}^2(S^n)$ induced by the trivial representation of $\mathbf{SO}(n)$ in $\mathbb{C}$ is the Hilbert sum of $d = (\rho : \sigma_0)$ irreducible representations equivalent to $M_{\overline{\rho}}$. Recall (see (Ind_G) before Definition 6.13) that

$$(\Pi_Q(f))(x) = f(Q^{-1}x) = f(Q^\top x), \;\; Q \in \mathbf{SO}(n+1), \; f \in \mathrm{L}^2(S^n), \; x \in S^n.$$

However, $(\mathbf{SO}(n+1), \mathbf{SO}(n))$ is one of examples of a Gelfand pair given in Section 9.7, Case 1, so $\mathrm{L}^2(H\backslash G/H)$ is commutative. We need to exhibit $\mathbf{SO}(n)$ as a subgroup of the fixed points of an involution σ of $\mathbf{SO}(n+1)$. To do this, let $s\colon \mathbb{R}^{n+1} \to \mathbb{R}^{n+1}$ be the reflection about the hyperplane $x_1 = 0$, which is given by

$$s(x_1, x_2, \ldots, x_{n+1}) = (-x_1, x_2, \ldots, x_{n+1}).$$

Obviously $s^{-1} = s$. Then let $\sigma\colon \mathbf{SO}(n+1) \to \mathbf{SO}(n+1)$ be the automorphism given by

$$\sigma(Q) = sQs, \quad Q \in \mathbf{SO}(n+1).$$

Since $s^2 = I$, we also have $\sigma^2 = \mathrm{id}$. In matrix form

$$\sigma(Q) = \begin{pmatrix} -1 & 0 \\ 0 & I_n \end{pmatrix} Q \begin{pmatrix} -1 & 0 \\ 0 & I_n \end{pmatrix}.$$

The groups $\mathbf{SO}(n+1)^\sigma$ of fixed points of σ are the rotations $Q \in \mathbf{SO}(n+1)$, such that

$$Q = \begin{pmatrix} -1 & 0 \\ 0 & I_n \end{pmatrix} Q \begin{pmatrix} -1 & 0 \\ 0 & I_n \end{pmatrix},$$

and if we write

$$Q = \begin{pmatrix} q_{11} & u \\ v & Q_1 \end{pmatrix},$$

we must have

$$\begin{pmatrix} q_{11} & u \\ v & Q_1 \end{pmatrix} = \begin{pmatrix} -1 & 0 \\ 0 & I_n \end{pmatrix} \begin{pmatrix} q_{11} & u \\ v & Q_1 \end{pmatrix} \begin{pmatrix} -1 & 0 \\ 0 & I_n \end{pmatrix} = \begin{pmatrix} q_{11} & -u \\ -v & Q_1 \end{pmatrix},$$

and so $u = v = 0$. Consequently, $\mathbf{SO}(n+1)^\sigma = S(\mathbf{O}(1) \times \mathbf{O}(n))$, with

$$S(\mathbf{O}(1) \times \mathbf{O}(n)) = \left\{ \begin{pmatrix} \lambda & 0 \\ 0 & Q_1 \end{pmatrix} \;\middle|\; \lambda = \pm 1,\ Q_1 \in \mathbf{O}(n),\ \lambda \det(Q_1) = 1 \right\}.$$

The stabilizer of $e_1 = (1, 0, \ldots, 0)$ corresponds to $\lambda = +1$, and it is indeed isomorphic to $\mathbf{SO}(n)$.

Since $(\mathbf{SO}(n+1), \mathbf{SO}(n))$ is a Gelfand pair, $\mathrm{L}^2(H\backslash G/H)$ is commutative (with $G = \mathbf{SO}(n+1)$, $H = \mathbf{SO}(n)$), so by Proposition 6.21, we have $d = (\rho : \sigma_0) \leq 1$ for all ρ.

It can be shown that the L_ρ for which $(\rho : \sigma_0) = 1$ are exactly the spaces $\mathcal{H}_k^{\mathbb{C}}(S^n)$ of spherical harmonics on S^n; see Definition 5.1. Thus, we have a Hilbert sum

$$\mathrm{L}^2(S^n) = \bigoplus_{k \geq 0} \mathcal{H}_k^{\mathbb{C}}(S^n).$$

We also obtain a decomposition of the representation $\mathbf{R}\colon \mathbf{SO}(n+1) \to \mathbf{U}(\mathrm{L}^2(S^n))$ into irreducible representations $\mathbf{R}_k\colon \mathbf{SO}(n+1) \to \mathbf{U}(\mathcal{H}_k^{\mathbb{C}}(S^n))$ of $\mathbf{SO}(n+1)$ in the spaces $\mathcal{H}_k^{\mathbb{C}}(S^n)$ of spherical harmonics on S^n. The above facts are proven in Dieudonné [Dieudonné (2003c)] (Chapter XXIII, Section 38). A different proof is given in Gallier and Quaintance [Gallier and Quaintance (2020b)] (Chapter 7). One of the technical results used in these proofs is that

$$\begin{aligned} \mathcal{P}_k^{\mathbb{C}}(n) = \mathcal{H}_k^{\mathbb{C}}(n) \oplus \|x\|^2 \mathcal{H}_{k-2}^{\mathbb{C}}(n) \oplus \cdots \oplus \|x\|^{2j} \mathcal{H}_{k-2j}^{\mathbb{C}}(n) \\ \oplus \cdots \oplus \|x\|^{2[k/2]} \mathcal{H}_{[k/2]}^{\mathbb{C}}(n), \end{aligned}$$

with the understanding that only the first term occurs on the right-hand side when $k < 2$ (the spaces $\mathcal{P}_k^{\mathbb{C}}(n)$ and $\mathcal{H}_k^{\mathbb{C}}(n)$ are described in Definition 5.1).

It is shown in Vilenkin [Vilenkin (1968)] (Chapter IX, Sections 2.10, 2.11) that the irreducible representations $\mathbf{R}_k\colon \mathbf{SO}(n+1) \to \mathbf{U}(\mathcal{H}_k^{\mathbb{C}}(S^n))$ are irreducible representations of class 1 relative to $\mathbf{SO}(n)$ (see Definition 6.12) and that they form a complete set of representations of class 1 of $\mathbf{SO}(n+1)$ relative to the subgroup $\mathbf{SO}(n)$. For $n = 2$, these are actually all the irreducible representations of $\mathbf{SO}(3)$ (see Proposition 5.3).

The space $\mathcal{H}_k^{\mathbb{C}}(S^n)$ is also the eigenspace associated to the eigenvalue $-k(n+k-1)$ of the Laplacian Δ_{S^n} on S^n. The unique zonal spherical function $\omega_\rho = \frac{1}{n_\rho} m_{11}^{(\rho,H)}$ in $\mathcal{H}_k^{\mathbb{C}}(S^n)$ is given in terms of Gegenbaur polynomials; see Gallier and Quaintance [Gallier and Quaintance (2020b)] (Chapter 7, Sections 3, 5–7).

6.11 Induced Representations, III; Blattner's Method

It is possible to modify the construction of the Hilbert space $\mathcal{H}$ and the inner product described at the end of Section 6.5 to deal with the situation where G/H has no G-invariant measure. This can be done in two ways as explained in Folland [Folland (1995)] (Chapter 6, Section 6.1). These constructions yield induced unitary representations of G from a unitary representation $U\colon H \to \mathbf{U}(E)$ of H and do not involve cocycles.

First method. In the first construction, the space $\mathcal{H}_0$ and the inner product are defined as before, namely

$$\begin{aligned}\mathcal{H}_0 = \{f \in \mathcal{C}(G,E) \mid\ & \pi(\mathrm{supp}(f)) \text{ is compact and} \\ & f(sh) = U(h^{-1})(f(s)) \quad \text{for all } s \in G \text{ and all } h \in H\},\end{aligned}$$

and

$$\langle f, g\rangle = \int_{G/H} \langle f(s), g(s)\rangle_E \, d\mu(sH).$$

Recall that $\pi\colon G \to G/H$ denotes the quotient map. The Hilbert space $\mathcal{H}$ is the completion of $\mathcal{H}_0$. The new ingredient is that to make the operators Π_s unitary, we use a quasi-invariant measure μ associated with a continuous function $\varrho\colon G \times (G/H) \to (0,\infty)$. We define Π_s^μ by

$$(\Pi_s^\mu(f))(t) = \varrho(s, tH)^{1/2} f(s^{-1}t), \qquad f \in \mathcal{H},\ s,t \in G. \tag{indv1}$$

Then, as in the proof of Theorem 6.16, we check that the operators Π_s^μ are unitary with respect to the inner product on $\mathcal{H}$ defined above, and we obtain a unitary representation $\Pi^\mu\colon G \to \mathbf{U}(\mathcal{H})$, also denoted $\mathrm{Ind}_{H,\mathcal{H}}^{G,\mu} U$ (for short $\mathrm{Ind}_H^G U$). This representation depends on μ, but it can be shown that if μ' is another quasi-invariant measure on G/H associated with $\varrho'\colon G \times (G/H) \to (0,\infty)$, then the automorphism $f \mapsto (\rho'/\rho)^{1/2} f$ is a unitary equivalence of ρ^μ and $\rho^{\mu'}$, where ρ and ρ' are the rho-functions associated with μ and μ' (see Theorem 6.14).

Blattner's method. The second construction, due to Blattner, does not make use of quasi-invariant measures, but instead modifies the definition of the space $\mathcal{H}_0$. In this sense, it is more intrinsic. Define the space $\mathcal{H}^0$ as

$$\mathcal{H}^0 = \Bigg\{f \in \mathcal{C}(G,E) \mid \pi(\mathrm{supp}(f)) \text{ is compact and} \\ f(sh) = \left(\frac{\Delta_H(h)}{\Delta_G(h)}\right)^{1/2} U(h^{-1})(f(s)) \text{ for all } s \in G \text{ and all } h \in H\Bigg\}.$$

The map $\pi\colon G \to G/H$ is the quotient map. Again, it is not obvious that $\mathcal{H}^0$ is not empty, but Proposition 6.10 can be modified (by adding the factor $\left(\frac{\Delta_G(h)}{\Delta_H(h)}\right)^{1/2}$ under the integral) to show that $\mathcal{H}^0$ is nonempty.

Next, we need to define an inner product on $\mathcal{H}^0$, so that the operators Π_s become unitary. The construction of such an inner product is a bit eccentric. For every $f \in \mathcal{H}^0$, the map $s \mapsto \|f(s)\|_E^2$ is almost a rho-function, except that it is not strictly positive. However, it is still possible to prove that the map

$$P\colon \varphi \mapsto \int_G \varphi(s)\, \|f(s)\|_E^2\, d\lambda(s), \qquad \varphi \in \mathcal{K}_{\mathbb{C}}(G)$$

is a positive Radon functional on $\mathcal{K}_{\mathbb{C}}(G/H)$, so by Radon–Riesz I, there is σ-Radon measure μ_f on G/H, such that

$$\int_{G/H} P(\varphi)\, d\mu_f = \int_G \varphi(s)\, \|f(s)\|_E^2\, d\lambda(s),$$

for all $\varphi \in \mathcal{K}_{\mathbb{C}}(G)$. Furthermore, the support of μ_f is contained in $\pi(\mathrm{supp}(f))$, hence compact. Therefore, $\mu_f(G/H)$ is finite. Then, given $f, g \in \mathcal{H}^0$, define the complex measure $\mu_{f,g}$ by polarization as

$$\mu_{f,g} = \frac{1}{4}(\mu_{f+g} - \mu_{f-g} + i\mu_{f+ig} - i\mu_{f-ig}),$$

so that

$$\int_{G/H} P(\varphi)\, d\mu_{f,g} = \int_G \varphi(s)\langle f(s), g(s)\rangle_E\, d\lambda(s), \qquad \varphi \in \mathcal{K}_{\mathbb{C}}(G).$$

The inner product on $\mathcal{H}^0$ is defined as

$$\langle f, g\rangle = \mu_{f,g}(G/H).$$

It can be verified that we obtain a hermitian inner product, and we let $\mathcal{H}'$ be the Hilbert space completion of $\mathcal{H}^0$. Finally, it can be verified that the operators Π_s are unitary with respect to the inner product on $\mathcal{H}'$ defined above, where

$$(\Pi_s(f))(t) = f(s^{-1}t), \qquad f \in \mathcal{H}',\ s, t \in G. \tag{indv2}$$

Therefore we obtain a unitary representation $\Pi'\colon G \to \mathbf{U}(\mathcal{H}')$, also denoted $\mathrm{Ind}_{H,\mathcal{H}'}^G U$ (for short $\mathrm{Ind}_H^G U$).

It can also be shown that for any quasi-invariant measure μ on G/H, the representations $\Pi^\mu\colon G \to \mathbf{U}(\mathcal{H})$ and $\Pi'\colon G \to \mathbf{U}(\mathcal{H}')$ are equivalent (the isomorphism between $\mathcal{H}$ and $\mathcal{H}'$ is given by $f \mapsto \rho^{1/2} f$).

There is also an interpretation of the representations Π^{μ} and Π' as representations on sections of homogeneous hermitian vector bundles over G/H. Such description is discussed in Folland [Folland (1995)] (Chapter 6) and Kirillov [Kirillov (1976)] (Section 13) and we present this approach in the following sections.

6.12 The Borel Construction from a Representation

In this section, we explain how the spaces of functions L^{α} (from Definition 6.8), and the spaces $\mathcal{H}_0$ and $\mathcal{H}^0$ from Section 6.11 can be viewed as sections of spaces that are similar to vector bundles but have less structure. More precisely, such structures have no trivialization maps.

We begin with the simplest situation where we have a group G without any topology on it, a subgroup H of G, a vector space $\mathcal{H}_{\sigma}$, and a linear representation $\sigma\colon H \to \mathbf{GL}(\mathcal{H}_{\sigma})$. As usual, we write $X = G/H$ and $\pi\colon G \to G/H$ for the quotient map. Let L^{σ} be the subspace of $(\mathcal{H}_{\sigma})^G$ consisting of all functions $f\colon G \to \mathcal{H}_{\sigma}$, such that

$$f(gh) = \sigma(h^{-1})(f(g)), \qquad \text{for all } g \in G \text{ and all } h \in H.$$

The key point is to construct a space $E = G \times_H \mathcal{H}_{\sigma}$ together with a surjective map $p\colon E \to X$, such that for every $x \in X = G/H$, the fibre $E_x = p^{-1}(x)$ is isomorphic to the vector space $\mathcal{H}_{\sigma}$, and the space of sections from X to E is in bijection with L^{σ}. This is a special case of the so-called Borel construction used to construct a vector bundle from a principal bundle; see Gallier and Quaintance [Gallier and Quaintance (2020b)] (Chapter 9, Section 9.9).

Definition 6.17. Consider a group G, a subgroup H of G, a vector space $\mathcal{H}_{\sigma}$, and a linear representation $\sigma\colon H \to \mathbf{GL}(\mathcal{H}_{\sigma})$. As usual, write $X = G/H$, $\pi\colon G \to G/H$ for the quotient map, and denote the coset $H = eH$ by x_0. The group H acts on $G \times \mathcal{H}_{\sigma}$ *on the right* by the action

$$(g, u) \cdot h = (gh, \sigma(h^{-1})(u)), \quad g \in G,\ u \in \mathcal{H}_{\sigma},\ h \in H. \tag{act1}$$

The space $E = G \times_H \mathcal{H}_{\sigma}$ is the orbit space of $G \times \mathcal{H}_{\sigma}$ under the above action, namely the set of equivalence classes

$$[(g, u)] = \{(gh, \sigma(h^{-1})(u)) \mid h \in H\} \ (g \in G, u \in \mathcal{H}_{\sigma})$$

of $G \times \mathcal{H}_{\sigma}$ under the equivalence relation $\sim$ defined such that for all $g_1, g_2 \in G$ and $u_1, u_2 \in \mathcal{H}_{\sigma}$,

$$(g_1, u_1) \sim (g_2, u_2) \quad \text{iff} \quad (\exists h \in H)(g_2 = g_1 h,\ u_2 = \sigma(h^{-1})(u_1)). \tag{$\sim_1$}$$

The projection $p\colon E \to X$ is defined as $\pi \circ pr_1$, namely for every equivalence class $z = [(g, u)] = \{(gh, \sigma(h^{-1})(u)) \mid h \in H\}$,

$$p(z) = gH = \pi(pr_1(z)).$$

It is immediately verified that the above definition does not depend on the choice of g in the coset gH.

For every $x = gH \in G/H = X$, the fibre $E_x = p^{-1}(x)$ can be given the structure of a vector space isomorphic to $\mathcal{H}_\sigma$. If we pick a section $r\colon X \to G$, namely a set of representatives $(r_x)_{x \in X}$ (with $r_x \in G$) for the cosets $x \in X = G/H$, with $r_{x_0} = e$,[2] then the map $\theta_{r_x}\colon \mathcal{H}_\sigma \to E_x$ given by

$$\theta_{r_x}(u) = [(r_x, u)] \tag{θ_{r_x}}$$

is injective, because if $\theta_{r_x}(u_1) = \theta_{r_x}(u_2)$, then

$$\{(r_x h, \sigma(h^{-1})(u_1)) \mid h \in H\} = \{(r_x h, \sigma(h^{-1})(u_2)) \mid h \in H\},$$

which implies that $\sigma(h^{-1})(u_1) = \sigma(h^{-1})(u_2)$ for all $h \in H$. In particular, for $h = e$, since $\sigma(e) = \mathrm{id}$, we get $u_1 = u_2$. The map θ_{r_x} is also surjective since for any equivalence class $[(r_x, u)] \in E_x$, by construction, $\theta_{r_x}(u) = [(r_x, u)]$.

Note that the above shows that the equivalence classes in the fibre E_x are the subsets $[(r_x, u)] = \{(r_x h, \sigma(h^{-1})(u)) \mid h \in H\}$ and that any two such classes are disjoint for distinct vectors u_1, u_2 in $\mathcal{H}_\sigma$.

We can transfer the vector space structure on $\mathcal{H}_\sigma$ to E_x using the bijection θ_{r_x}, namely

$$\begin{aligned} [(r_x, u_1)] + [(r_x, u_2)] &= [(r_x, u_1 + u_2)] \\ \lambda[(r_x, u)] &= [(r_x, \lambda u)], \end{aligned}$$

for all $u, u_1, u_2 \in \mathcal{H}_\sigma$, and $\lambda \in \mathbb{C}$. If a different section $r_2\colon G/H \to G$ is used, then $(r_2)_x = r_x h_x$ for some $h_x \in H$. Then we have (with $h = h_x^{-1}$)

$$\theta_{(r_2)_x}(u) = [(r_x h_x, u)] = [(r_x h_x h_x^{-1}, \sigma(h_x)(u))] = [(r_x, \sigma(h_x)(u))].$$

The vector space structure on E_x defined by the section r_2 is now given by

$$\begin{aligned} [(r_x, u_1)] + [(r_x, u_2)] &= [(r_x, \sigma(h_x)(u_1) + \sigma(h_x)(u_2))] \\ \lambda[(r_x, u)] &= [(r_x, \lambda\sigma(h_x)(u))], \end{aligned}$$

[2]We always assume that for every set of coset representatives $(r_x)_{x \in X}$ we chose $r_{x_0} = e$.

and since $\sigma(h_x)$ is linear, we have

$$\begin{aligned}[(r_x, u_1)] + [(r_x, u_2)] &= [(r_x, \sigma(h_x)(u_1 + u_2))] \\ \lambda[(r_x, u)] &= [(r_x, \sigma(h_x)(\lambda u))].\end{aligned}$$

Since $\sigma(h_x)$ is a linear isomorphism, we see that $\sigma(h_x)$ is a linear isomorphism between E_x as a vector space, whose structure is induced by r and as a vector space, whose structure is induced by r_2. In any case, there is a linear isomorphism between $\mathcal{H}_\sigma$ and E_x (although noncanonical).

Looking ahead, if $\mathcal{H}_\sigma$ is a (separable) Hilbert space, G is locally compact, H is a closed subgroup of G, and if $\sigma\colon H \to \mathbf{U}(\mathcal{H}_\sigma)$ is a unitary representation, then the map θ_{r_x} can be used to transfer the Hilbert space structure of $\mathcal{H}_\sigma$ to the fibre E_x by setting

$$\langle [(r_x, u_1)], [(r_x, u_2)] \rangle = \langle u_2, u_2 \rangle, \quad u_1, u_2 \in \mathcal{H}_\sigma.$$

If a different section r_2 is used, since the maps $\sigma(h_x)$ are unitary, $\langle \sigma(h_x)(u_2), \sigma(h_x)(u_2) \rangle = \langle u_2, u_2 \rangle$, and so the hermitian inner products on E_x induced by r and r_2 are identical. There are unitary isomorphisms between the Hilbert spaces $\mathcal{H}_\sigma$ and E_x.

The fact that the space L^σ is realized by the space of sections of E is shown in the next proposition.

Definition 6.18. A *section of* E is any function $s\colon X \to E$, such that $p \circ s = \mathrm{id}_X$, where p is the projection $p\colon E \to X$, or equivalently a function $s\colon X \to E$, such that $s(x) \in E_x$ for every $x \in X = G/H$. The set of sections $s\colon X \to E$ is denoted $\Gamma(E)$.

Remark: At this stage, $X = G/H$ is just a set without any topology, so a section is just a function. Later, when G and H are locally compact groups, it will make sense to consider continuous sections.

Given a set of coset representatives $(r_x)_{x \in X}$, recall from Definition 6.5 (Equation (u)) that we define $u(g, x)$ as

$$u(g, x) = r_{g \cdot x}^{-1} g r_x,$$

and that by Equation (s), we have

$$g = r_x u(g, x_0), \quad s \in G, \quad x = gH.$$

Proposition 6.22. *Let* $E = G \times_H \mathcal{H}_\sigma$, $X = G/H$, $p\colon E \to X$, *and* $\sigma\colon H \to \mathbf{GL}(\mathcal{H}_\sigma)$, *as in Definition 6.17. Also let* L^σ *be the set consisting of all functions* $f\colon G \to \mathcal{H}_\sigma$, *such that*

$$f(gh) = \sigma(h^{-1})(f(g)), \qquad \textit{for all } g \in G \textit{ and all } h \in H.$$

The maps $\mathcal{S}\colon L^\sigma \to \Gamma(E)$ *and* $\mathcal{L}\colon \Gamma(E) \to L^\sigma$ *are defined as follows. Pick any set of representatives* $(r_x)_{x\in X}$ *(with* $r_x \in G$*) for the cosets* $x \in X = G/H$*. For every function* $f\colon G \to \mathcal{H}_\sigma$*, for any coset* $x = r_x H$*, define the section* $\mathcal{S}(f)$ *by*

$$\mathcal{S}(f)(x) = [(r_x, f(r_x))], \tag{$\mathcal{S}$}$$

and for every section $s\colon X \to E$*, for every coset* $x = r_x H$*, if* $s(x) = [(r_x, u)]$ *for some* $u \in \mathcal{H}_\sigma$*, define the function* $\mathcal{L}(s)$ *on* G *by*

$$\mathcal{L}(s)(r_x h) = \sigma(h^{-1})(u), \quad h \in H, \tag{$\mathcal{L}$}$$

or equivalently

$$\mathcal{L}(s)(g) = \sigma(u(g, x_0)^{-1})(u), \quad g \in G. \tag{$\mathcal{L}'$}$$

Then, $\mathcal{S}(f)$ *does not depend on the set of coset representatives* $(r_x)_{x\in X}$*,* $\mathcal{S}(f) \in \Gamma(E)$*,* $\mathcal{L}(s) \in L^\sigma$*, and* $\mathcal{S}$ *and* $\mathcal{L}$ *are mutual inverses. Therefore,* $\mathcal{S}$ *is a bijection between* L^σ *and* $\Gamma(E)$*.*

Proof. If another set $(r_x h_x)_{x\in X}$ of coset representatives is used (with $h_x \in H$ for all $x \in X$), since $f \in L^\sigma$, we have $f(r_x h_x) = \sigma(h_x^{-1})(f(r_x))$, so by the definition of the equivalence relation on $G \times \mathcal{H}_\sigma$, we have

$$\mathcal{S}(f)(x) = [(r_x h_x, f(r_x h_x))] = [(r_x h_x, \sigma(h_x^{-1})(f(r_x)))] = [(r_x, f(r_x))],$$

which shows that the definition of $\mathcal{S}(f)$ does not depend on set of coset representatives $(r_x)_{x\in X}$. By definition, $\mathcal{S}(f)(x) \in E_x$, so $\mathcal{S}(f)$ is a section of E.

If we write $g = r_x h$, for every $h_2 \in H$, since σ is a representation, we have

$$\begin{aligned}\mathcal{L}(s)(gh_2) &= \mathcal{L}(s)(r_x h h_2) = \sigma((hh_2)^{-1})(u) = \sigma(h_2^{-1}h^{-1})(u)\\ &= \sigma(h_2^{-1})(\sigma(h^{-1})(u)) = \sigma(h_2^{-1})(\mathcal{L}(s)(g)),\end{aligned}$$

which shows that $\mathcal{L}(s) \in L^\sigma$.

Given any $f \in L^\sigma$, for every $x \in X$, we have

$$\mathcal{S}(f)(x) = [(r_x, f(r_x))],$$

and then since $f \in L^\sigma$, for all $h \in H$, we have

$$\mathcal{L}(\mathcal{S}(f))(r_x h) = \sigma(h^{-1})(f(r_x)) = f(r_x h),$$

namely, $\mathcal{L}(\mathcal{S}(f)) = f$.

Given any $s \in \Gamma(E)$, for every coset $x = r_x H$, if $s(x) = [(r_x, u)]$, for every $h \in H$, we have

$$\mathcal{L}(s)(r_x h) = \sigma(h^{-1})(u),$$

so in particular $\mathcal{L}(s)(r_x) = u$, and then

$$\mathcal{S}(\mathcal{L}(s))(x) = [(r_x, \mathcal{L}(s)(r_x))] = [(r_x, u)] = s(x),$$

that is, $\mathcal{S}(\mathcal{L}(s)) = s$. Since $\mathcal{L}(\mathcal{S}(f)) = f$ and $\mathcal{S}(\mathcal{L}(s)) = s$, the maps $\mathcal{S}$ and $\mathcal{L}$ are mutual inverses. □

Remark: If we use the isomorphisms $\theta_{r_x} : \mathcal{H}_\sigma \to E_x$ given by

$$\theta_{r_x}(u) = [(r_x, u)], \quad u \in \mathcal{H}_\sigma,$$

then the maps $\mathcal{S} \colon L^\sigma \to \Gamma(E)$ and $\mathcal{L} \colon \Gamma(E) \to L^\sigma$ are defined as follows. For every function $f \colon G \to \mathcal{H}_\sigma$ in L^σ and for any coset $x = r_x H$,

$$\mathcal{S}(f)(x) = \theta_{r_x}(f(r_x)), \tag{$\mathcal{S}_2$}$$

and for every section $s \colon X \to E$ and any $g \in G$,

$$\mathcal{L}(s)(g) = \sigma(u(g, x_0)^{-1})(\theta_{r_x}^{-1}(s(gH))). \tag{$\mathcal{L}_2$}$$

The isomorphisms θ_{r_x} are omitted by some authors but this is not quite right.

The last important ingredient is that G acts (on the left) on the space $E = G \times_H \mathcal{H}_\sigma$ in an *equilinear* fashion.

Definition 6.19. Under the same conditions as in Definition 6.17, we define a left action of G on $E = G \times_H \mathcal{H}_\sigma$ by

$$g_1 \cdot [(g, u)] = [(g_1 g, u)], \quad g_1, g \in G,\ u \in \mathcal{H}_\sigma.$$

That this action is equilinear means the following.

Proposition 6.23. *Under the same conditions as in Definition 6.19, the following facts hold:*

(1) The action of G on $E = G \times_H \mathcal{H}_\sigma$ is equivariant, which means that

$$p(g \cdot [(g_1, u)]) = g \cdot p([(g_1, u)]), \quad g, g_1 \in G,\ u \in \mathcal{H}_\sigma.$$

The action on the right-hand side is the action on cosets in G/H given by $g \cdot (g_2 H) = (g g_2) H$.

(2) The restriction of the action of G to the fibre E_x is a linear isomorphism between E_x and $E_{g \cdot x}$ ($x \in G/H$). In particular, every fibre E_x is isomorphic to E_{x_0}.

Proof. We have

$$\begin{aligned} p((g\cdot[(g_1,u)]) &= p([(gg_1,u)]) \\ &= \pi(gg_1) = (gg_1)H = g\cdot(g_1H) = g\cdot p([(g_1,u)]). \end{aligned}$$

We will show in the next proposition that an equivariant action as above is a bijection between each fibre E_x and the fibre $E_{g\cdot x}$. Since the fibre E_x consists of the equivalence classes $[(r_x, u)]$ ($u \in \mathcal{H}_\sigma$) for some coset representative r_x of $x \in G/H$, the action of G on E_x is given by

$$g\cdot[(r_x,u)] = [(gr_x,u)], \quad g\in G,$$

which is obviously linear in u. Thus, the map induced by the action of G on the fibre E_x is a linear isomorphism. □

Observe that the action of H on the fibre E_{x_0} is a representation $\sigma^0\colon H \to \mathbf{GL}(E_{x_0})$ equivalent to the representation $\sigma\colon H \to \mathbf{GL}(\mathcal{H}_\sigma)$, since E_{x_0} consists of the equivalence classes of the form $[(e,u)]$ (recall that $r_{x_0} = e$), with $u \in \mathcal{H}_\sigma$, so for every $h \in H$,

$$h\cdot[(e,u)] = [(h,u)] = [(hh^{-1},\sigma(h)(u))] = [(e,\sigma(h)(u))].$$

The linear isomorphisms between the fibres E_{x_0} and E_x induce representations $\sigma^x\colon H \to \mathbf{GL}(E_x)$ equivalent to the representation $\sigma\colon H \to \mathbf{GL}(\mathcal{H}_\sigma)$.

6.13 Induced Representations and G-Bundles

Next, what we would like to do is to show how induced representations can be recovered from certain kinds of vector bundles (actually a more basic notion of vector bundle) equipped with an equilinear action. Such construction is given as an exercise in Dieudonné [Dieudonné (2003b)] (Chapter XXII, Section 3, Problem 16). It is also discussed in Kirillov [Kirillov (1976)] (Section 13) and sketched without details in Folland [Folland (1995)] (Chapter 6). Following Kirillov, we adopt the terminology of G-bundle. First, we introduce a weaker notion that we call pre-G-bundle (for the lack of a better name).

Definition 6.20. Let G be a group, H be a subgroup of G, E be some set, and $p\colon E \to X$ be a surjective map, where as usual, we write $X = G/H$. We say that E (really $p\colon E \to X$) is a *pre-G-bundle* if there is an *equivariant* left action $\cdot$ of G on E, which means that

$$p(g\cdot z) = g\cdot p(z), \quad g\in G,\ z\in E.$$

Proposition 6.24. *If $p\colon E \to X$ is a pre-G-bundle, then for every $x \in X = G/H$, for every $g \in G$, the map $z \mapsto g \cdot z$ ($z \in E_x$) is a bijection from E_x to $E_{g\cdot x}$.*

Proof. Equivariance means that if $z \in E_x$, that is, $p(z) = x$, then

$$p(g \cdot z) = g \cdot p(z) = g \cdot x,$$

so $g \cdot z \in E_{g\cdot x}$. Thus

$$g \cdot E_x \subseteq E_{g\cdot x}. \tag{1}$$

The above inclusion holds for all $g \in G$ and all $x \in X = G/H$, and in particular for g^{-1} and $g \cdot x$, which yields

$$g^{-1} \cdot E_{g\cdot x} \subseteq E_{g^{-1}\cdot(g\cdot x)} = E_x.$$

This last equation is equivalent to

$$E_{g\cdot x} \subseteq g \cdot E_x. \tag{2}$$

By (1) and (2), we obtain

$$g \cdot E_x = E_{g\cdot x}. \tag{3}$$

The maps induced by group actions are bijective, so our result is proven. □

We finally come to the desired concept by requiring that the fibres are vector spaces and that the bijections between fibres are linear isomorphisms. The key concept is the notion of *equilinear action* which occurs in Dieudonné [Dieudonné (2007)], Chapter XIX, Section 1.

Definition 6.21. Let G be a group, H be a subgroup of G, E be some set, and $p\colon E \to X$ be a surjective map, where as usual, we write $X = G/H$. We say that E (really $p\colon E \to X$) is a *G-bundle* if each fibre E_x ($x \in X = G/H$) is a vector space and if there is an *equilinear* left action $\cdot$ of G on E, which means that:

(1) The action is equivariant, that is,
$$p(g \cdot z) = g \cdot p(z), \quad g \in G,\ z \in E.$$
(2) For every $x \in X = G/H$, for every $g \in G$, the map $z \mapsto g \cdot z$ ($z \in E_x$) is a linear isomorphism between E_x and $E_{g\cdot x}$.

Let x_0 denote the coset $H = eH$. Proposition 6.24 implies that every fibre is isomorphic to E_{x_0}. Then the restriction of the action of G to H on the fibre E_{x_0}, for simplicity also denoted as E_0, maps E_0 to E_0 (since $h\cdot x_0 = x_0$ for all $h \in H$). Since the maps $z \mapsto h \cdot z$ ($z \in E_0$) are linear isomorphisms, we have a representation $\sigma\colon H \to \mathbf{GL}(E_0)$ given by

$$\sigma(h)(z) = h \cdot z. \tag{σ}$$

Observe that $E = G \times_H \mathcal{H}_\sigma$ with the projection $p\colon E \to X$ is a G-bundle. If E is an abstract G-bundle as in Definition 6.21, then *the fibre E_0 plays the role of the vector space $\mathcal{H}_\sigma$*, which occurs in the representation $\sigma\colon H \to \mathbf{GL}(\mathcal{H}_\sigma)$ and is involved in the construction of the G-bundle $G \times_H \mathcal{H}_\sigma$. So it is natural to also refer to E_0 as $\mathcal{H}_\sigma$, which we will do except when confusion arises.

Let $(r_x)_{x \in X}$ be any set of coset representatives of $X = G/H$. The map σ_x from E_x to itself given by

$$\sigma_x(h)(z) = (r_x h r_x^{-1}) \cdot z = r_x \cdot \sigma(h)(r_x^{-1} \cdot z), \quad h \in H,\ z \in E_x,$$

is a linear isomorphism of E_x, in other words, a representation $\sigma_x\colon H \to \mathbf{GL}(E_x)$. The representation $\sigma_x\colon H \to \mathbf{GL}(E_x)$ is equivalent to the representation $\sigma\colon H \to \mathbf{GL}(E_0)$ *via* the linear isomorphism from E_0 to E_x given by $z \mapsto r_x \cdot z$.

It is easy to see that if another set of coset representatives $(r_x h_x)_{x \in X}$ is used, then

$$\sigma_x(h)(z) = (r_x h_x h h_x^{-1} r_x^{-1}) \cdot z;$$

in other words, we obtain a representation equivalent to $\sigma\colon H \to \mathbf{GL}(E_0)$, where the linear isomorphism from E_0 to E_x is given by $z \mapsto r_x h_x \cdot z$.

Consequently, we see that the sections in $\Gamma(E)$, called *feature fields* in group equivariant deep learning in computer vision, are functions whose domain transforms under the action of G and whose codomain transforms by representations of H equivalent to $\sigma\colon H \to \mathbf{GL}(E_0)$; more precisely, each fibre E_x transforms under the representation σ_x.

The space L^σ and the representation of G in L^σ induced by $\sigma\colon H \to \mathbf{GL}(E_0)$ can be recovered from the G-vector bundle as we now explain. We use inspiration from Proposition 6.22.

Definition 6.22. Let $p\colon E \to X$ be a G-bundle, with $X = G/H$. As before, let $x_0 = H$, E_0 be the fibre $E_0 = p^{-1}(x_0)$, and $\sigma\colon H \to \mathbf{GL}(E_0)$ be the representation given by $\sigma(h)(z) = h \cdot z$ for all $z \in E_0$ and all $h \in H$. Also, let L^σ be the set consisting of all functions $f\colon G \to E_0$, such that

$$f(gh) = \sigma(h^{-1})(f(g)) = h^{-1} \cdot f(g), \qquad \text{for all } g \in G \text{ and all } h \in H. \quad (*_{\dagger_1})$$

There is an action of G on the set $\Gamma(E)$ of section $s\colon X \to E$ given by

$$(g \cdot s)(x) = g \cdot (s(g^{-1} \cdot x)), \qquad g \in G,\ x \in X. \qquad (\dagger_\Gamma)$$

In the above equation, G acts on $X = G/H$ in $g^{-1} \cdot x$, and G acts on E in $g \cdot (s(g^{-1} \cdot x))$.

Define the maps $\mathcal{S}$ and $\mathcal{L}$ as follows. For every function $f\colon G \to E_0 \in L^\sigma$, for every coset $x \in X = G/H$ and any coset representative $r_x \in G$ of x, let

$$\mathcal{S}(f)(x) = r_x \cdot f(r_x), \qquad (\mathcal{S}_3)$$

where the action is the action of G on E. For every section $s\colon X \to E$, for every $g \in G$, let

$$\mathcal{L}(s)(g) = g^{-1} \cdot s(gH) = g^{-1} \cdot s(g \cdot x_0), \qquad (\mathcal{L}_3)$$

where the action is the action of G on E.

It is instructive to see how the formulae for the maps $\mathcal{L}$ and $\mathcal{S}$ given in Definition 6.22 are equivalent to the formulae $(\mathcal{S}_2)$ and $(\mathcal{L}_2)$ when specialized to the G-bundles of the form $p\colon G \times_H \mathcal{H}_\sigma \to G/H$.

First, note that the fibre E_0 consists of the equivalence classes $[(e, u)]$ with $u \in \mathcal{H}_\sigma$, and since the action of G on $E = G \times_H \mathcal{H}_\sigma$ is given by $g_1 \cdot [(g, u)] = [(g_1 g, u)]$, we get

$$r_x \cdot [(e, u)] = [r_x, u] = \theta_{r_x}(u).$$

Because $E = G \times_H \mathcal{H}_\sigma$ is constructed from the space $\mathcal{H}_\sigma$, in Proposition 6.22, the functions in L^σ are functions from G to $\mathcal{H}_\sigma$, but in the more general setting there is no such "privileged" vector space, so in Definition 6.22, the functions in L^σ are functions from G to E_0. If we denote the set of functions $f\colon X \to \mathcal{H}_\sigma$ by $L^{\sigma,\mathcal{H}}$, then the map $f \mapsto \theta_e \circ f$ is an isomorphism between $L^{\sigma,\mathcal{H}}$ and L^σ (recall that θ_e is the isomorphism $\theta_e\colon \mathcal{H}_\sigma \to E_0$). We also denote the map $\mathcal{S}$ from $L^{\sigma,\mathcal{H}}$ to $\Gamma(E)$ defined by Equation $(\mathcal{S}_2)$ as $\mathcal{S}^{\mathcal{H}}$. It follows that for every function $f\colon G \to \mathcal{H}_\sigma$ in $L^{\sigma,\mathcal{H}}$, since $\theta_e \circ f$ is a function from G to E_0, we have

$$\mathcal{S}(\theta_e \circ f)(x) = r_x \cdot (\theta_e \circ f)(r_x) = r_x \cdot [(e, f(r_x))] = [(r_x, f(r_x))] = \theta_{r_x}(f(r_x)).$$

The above formula shows that

$$\mathcal{S}^{\mathcal{H}}(f) = \mathcal{S}(\theta_e \circ f).$$

It is often convenient to identify $\mathcal{H}_\sigma$ and E_0 using the isomorphism θ_e.

To go from sections to functions in L^σ, we need to remember that when we use the G-bundle $E = G \times_H \mathcal{H}_\sigma$, σ is a representation $\sigma\colon H \to \mathbf{GL}(\mathcal{H}_\sigma)$ and $L^{\sigma,\mathcal{H}}$ consists of functions $f\colon G \to \mathcal{H}_\sigma$, but in the abstract version of G-bundles, σ is a representation $\sigma\colon H \to \mathbf{GL}(E_0)$ and L^σ consists of functions $f\colon G \to E_0$. Let us denote the representation from H to $\mathcal{H}_\sigma$ by $\sigma^{\mathcal{H}}\colon H \to \mathbf{GL}(\mathcal{H}_\sigma)$. Then if E is the G-bundle $E = G \times_H \mathcal{H}_\sigma$, we have

$$\sigma(h) = \theta_e \circ \sigma^{\mathcal{H}}(h) \circ \theta_e^{-1}, \quad h \in H.$$

Let us denote the map $\mathcal{L}$ from $\Gamma(E)$ to $L^{\sigma,\mathcal{H}}$ defined by Equation ($\mathcal{L}_2$) as $\mathcal{L}^{\mathcal{H}}$. For any section $s\colon X \to E$, since $g = r_x u(g, x_0)$ with $x = gH$, we have

$$\begin{aligned}\mathcal{L}(s)(g) = g^{-1} \cdot s(gH) &= (r_x u(g,x_0))^{-1} \cdot s(gH) = u(g,x_0)^{-1} \cdot (r_x^{-1} \cdot s(gH)) \\ &= \sigma(u(g,x_0)^{-1})(r_x^{-1} \cdot s(gH)) \\ &= \theta_e \circ \sigma^{\mathcal{H}}(u(g,x_0)^{-1})(\theta_e^{-1}(r_x^{-1} \cdot s(gH))).\end{aligned}$$

We now have to work on the term $\theta_e^{-1}(r_x^{-1} \cdot s(gH))$. But we know that the section $s\colon X \to E$ is of the form $s(x) = [(r_x, u)]$ for some $u \in E_0$ (with $x = gH$), so

$$r_x^{-1} \cdot s(gH) = r_x^{-1} \cdot s(x) = r_x^{-1} \cdot [(r_x, u)] = [(e, u)] = \theta_e(u),$$

and thus

$$\theta_e^{-1}(r_x^{-1} \cdot s(gH)) = u = \theta_{r_x}^{-1}([(r_x, u)]) = \theta_{r_x}^{-1}(s(x)).$$

In summary, we proved that

$$\mathcal{L}(s)(g) = \theta_e \circ \sigma^{\mathcal{H}}(u(g,x_0)^{-1})(\theta_{r_x}^{-1}(s(gH))) = \theta_e \circ \mathcal{L}^{\mathcal{H}}(s)(g),$$

or equivalently

$$\mathcal{L}^{\mathcal{H}}(s)(g) = \theta_e^{-1} \circ \mathcal{L}(s)(g).$$

Note that we have also proved that

$$\begin{aligned}\mathcal{L}(s)(g) &= \sigma(u(g,x_0)^{-1})(r_x^{-1} \cdot s(x)) \\ &= (\sigma(u(g,x_0)))^{-1}(r_x^{-1} \cdot s(g \cdot x_0)), \quad x = gH = g \cdot x_0. \end{aligned} \qquad (\mathcal{L}_3')$$

In Definition 6.7 of Section 6.3, given a cocycle α, given a section $s\colon X \to E$, we defined the function $s^\alpha\colon G \to E$ by

$$s^\alpha(g) = (\alpha(g, x_0))^{-1}(s(g \cdot x_0)), \qquad \text{for all } g \in G. \qquad (*_{\alpha_1})$$

There, E plays the role of the vector space E_0 involved in the representation $\sigma\colon H \to \mathbf{GL}(E_0)$. For any set $(r_x)_{x \in X}$ of coset representatives of $X = G/H$, we know that $\alpha(g,x) = \sigma(u(g,x))$ ($g \in G, x \in G/H$) is a cocycle and Equation ($\mathcal{L}_3'$) is basically Equation ($*_{\alpha_1}$), except that $r_x^{-1} \cdot s(g \cdot x_0)$ occurs instead of $s(g \cdot x_0)$. But as we explained earlier, in the case of the G-bundle $G \times_H \mathcal{H}_\sigma$, since $s(g \cdot x_0) \in E_{g \cdot x_0}$, we have $r_x^{-1} \cdot s(g \cdot x_0) \in E_0 \approx \mathcal{H}_\sigma$, so the two formulae are equivalent.

Furthermore, since $g = r_x u(g, x_0)$ with $x = gH = g \cdot x_0$, we have

$$\begin{aligned}\mathcal{S}(f)(g \cdot x_0) &= g \cdot f(g) = (r_x u(g, x_0)) \cdot f(g) = r_x \cdot (u(g, x_0) \cdot f(g)) \\ &= r_x \cdot \sigma(u(g, x_0))(f(g)),\end{aligned}$$

namely

$$\mathcal{S}(f)(g \cdot x_0) = r_x \cdot \sigma(u(g, x_0))(f(g)). \qquad (\mathcal{S}_3')$$

In Proposition 6.6, we proved that for any function $f\colon G \to E \in L^\alpha$, the function $F\colon X \to E$ given by

$$F(x) = f(g \cdot x_0) = \alpha(g, x_0)(f(g)) \qquad (*_f)$$

has the property that $F^\alpha = f$. Again, if α is the cocycle given by $\alpha(g, x) = \sigma(u(g, x))$ $(g \in G, x \in G/H)$, then Equations $(*_f)$ and $(\mathcal{S}_3')$ only differ by the presence of the term r_x, but in the case of the G-bundle $G \times_H \mathcal{H}_\sigma$, we have $\sigma(u(g, x_0))(f(g)) \in E_0 \approx \mathcal{H}_\sigma$ and the purpose of r_x is move $\sigma(u(g, x_0))(f(g))$ to the fibre $E_{g \cdot x_0}$, so the two formulae are equivalent.

Proposition 6.25. *The following facts hold.*

(1) The map $\mathcal{S}(f)$ is independent of the choice of the representative g chosen in the coset $x = gH = g \cdot x_0$ and $\mathcal{S}(f) \in \Gamma(E)$; that is,

$$\mathcal{S}(f)(gH) = \mathcal{S}(f)(g \cdot x_0) = g \cdot f(g), \quad g \in G. \qquad (\mathcal{S}_3'')$$

(2) We have $\mathcal{L}(s) \in L^\sigma$.

(3) The maps $\mathcal{S}\colon L^\sigma \to \Gamma(E)$ and $\mathcal{L}\colon \Gamma(E) \to L^\sigma$ are mutual inverses. Thus, $\mathcal{S}$ is an isomorphism between L^σ and $\Gamma(E)$.

Proof. (1) If another representative $g_2 = gh$ in the coset $x = gH \in G/H$ is used (with $h \in H$), since $f \in L^\sigma$ and by definition of σ, $\sigma(h^{-1})(f(g)) = h^{-1} \cdot f(g)$, we have

$$\begin{aligned}\mathcal{S}(f)(g_2 H) &= (gh) \cdot f(gh) = (gh) \cdot \sigma(h^{-1})(f(g)) = (gh) \cdot (h^{-1} \cdot f(g)) \\ &= ((gh) \cdot h^{-1}) \cdot f(g) = (g \cdot (h \cdot h^{-1})) \cdot f(g) = g \cdot f(g).\end{aligned}$$

Therefore,

$$\mathcal{S}(f)(g_2 H) = g_2 \cdot f(g_2) = g \cdot f(g) = \mathcal{S}(f)(gH),$$

namely, the definition of $\mathcal{S}(f)$ does not depend on the choice of representatives in the coset $x = gH$. For any $x = gH \in G/H$, since $f(g) \in E_0 = E_{x_0}$, we have $g \cdot f(g) \in E_{g \cdot x_0} = E_x$, since g is a representative in the coset $x = gH \in X = G/H$. Thus, $\mathcal{S}(f) \in \Gamma(E)$.

(2) If $x = gH$, then $s(x) \in E_x$, and so $g^{-1} \cdot s(x) \in E_{g^{-1}\cdot x} = E_{g^{-1}\cdot gH} = E_H = E_0$. For any $h \in H$, by definition of σ in terms of the action of H on E_0, we have

$$\begin{aligned}\mathcal{L}(s)(gh) &= (gh)^{-1} \cdot s((gh)H) = (h^{-1}g^{-1}) \cdot s(g(hH)) = (h^{-1}g^{-1}) \cdot s(gH)\\ &= h^{-1} \cdot (g^{-1} \cdot s(gH)) = h^{-1} \cdot \mathcal{L}(s)(g) = \sigma(h^{-1})(\mathcal{L}(s)(g)),\end{aligned}$$

which shows that $\mathcal{L}(s) \in L^\sigma$.

(3) For any $f \in L^\sigma$, if $x = gH$, we have

$$\mathcal{S}(f)(gH) = g \cdot f(g),$$

and then

$$\mathcal{L}(\mathcal{S}(f))(g) = g^{-1} \cdot \mathcal{S}(f)(gH) = g^{-1} \cdot (g \cdot f(g)) = (g^{-1} \cdot g) \cdot f(g) = f(g),$$

that is, $\mathcal{L}(\mathcal{S}(f)) = f$.

For any $s\colon X \to E_0$, for every $g \in G$, we have

$$\mathcal{L}(s)(g) = g^{-1} \cdot s(gH),$$

and then we have

$$\begin{aligned}\mathcal{S}(\mathcal{L}(s))(gH) &= g \cdot \mathcal{L}(s)(g) = g \cdot (g^{-1} \cdot s(gH))\\ &= (g \cdot g^{-1}) \cdot s(gH)) = s(gH),\end{aligned}$$

that is, $\mathcal{S}(\mathcal{L}(s)) = s$. Consequently, the maps $\mathcal{S}\colon L^\sigma \to \Gamma(E)$ and $\mathcal{L}\colon \Gamma(E) \to L^\sigma$ are mutual inverses and $\mathcal{S}$ is an isomorphism between L^σ and $\Gamma(E)$. ☐

We can also recover the representation $\mathrm{Ind}_H^G \sigma\colon G \to \mathbf{GL}(L^\sigma)$ induced by the representation $\sigma\colon H \to \mathbf{GL}(E_0)$.

Proposition 6.26. *Define the map* $\rho\colon G \to \mathbf{GL}(L^\sigma)$ *by*

$$\rho(g)(f) = \mathcal{S}^{-1}(g \cdot \mathcal{S}(f)) = \mathcal{L}(g \cdot \mathcal{S}(f)), \quad g \in G,\ f \in L^\sigma. \tag{$\dagger_2$}$$

In the above equation, $\mathcal{S}(f) \in \Gamma(E)$ *and the action of* G *is the action of* G *on* $\Gamma(E)$ *from Definition 6.22. For all* $g, g_1 \in G$ *and all* $f \in L^\sigma$*, we have*

$$[\rho(g)(f)](g_1) = f(g^{-1}g_1), \tag{$\dagger_3$}$$

that is, $\rho\colon G \to \mathbf{GL}(L^\sigma)$ *is the representation* $\mathrm{Ind}_H^G \sigma$ *induced from the representation* $\sigma\colon H \to \mathbf{GL}(E_0)$*.*

Proof. By the definitions of $\mathcal{S}$ and $\mathcal{L}$ and of the action of G on sections in $\Gamma(E)$ (see ($\dagger_\Gamma$)) in Definition 6.22, and using Proposition 6.25(1), we have

$$\begin{aligned}\mathcal{L}(g \cdot \mathcal{S}(f))(g_1) &= g_1^{-1} \cdot (g \cdot \mathcal{S}(f))(g_1H) = g_1^{-1} \cdot (g \cdot \mathcal{S}(f)(g^{-1} \cdot (g_1H)))\\ &= (g_1^{-1}g) \cdot \mathcal{S}(f)(g^{-1}g_1H) = (g_1^{-1}g) \cdot ((g^{-1}g_1) \cdot f(g^{-1}g_1))\\ &= f(g^{-1}g_1),\end{aligned}$$

as claimed. ☐

6.14 Hermitian G-Bundles

The above definitions and constructions can be adapted to deal with unitary representations. In this case, G is a locally compact group, H is a closed subgroup of G, and $\sigma\colon H \to \mathbf{U}(\mathcal{H}_\sigma)$ is a unitary representation, where $\mathcal{H}_\sigma$ is a separable Hilbert space. As we explained earlier, up to linear isomorphisms, we can endow the fibres E_x of the G-bundle $E = G \times_H \mathcal{H}_\sigma$ with a Hilbert space structure, so that each fibre E_x is isometric to $\mathcal{H}_\sigma$ *via* a unitary isomorphism. The action of G on E has the property that each map $z \mapsto g \cdot z$ from the fibre E_x to the fibre $E_{g\cdot x}$ is *unitary*.

Definition 6.23. Let G be a locally compact group, H a closed subgroup of G, E be some topological Hausdorff space, and $p\colon E \to X$ be a surjective *continuous* map, where as usual, we write $X = G/H$. We say that E (really $p\colon E \to X$) is a *hermitian G-bundle* if each fibre E_x ($x \in X = G/H$) is a separable Hilbert space and if there is an *equilinear continuous* left action $\cdot\colon G \times E \to E$ of G on E, which means that:

(1) The action is equivariant, that is,

$$p(g \cdot z) = g \cdot p(z), \quad g \in G,\ z \in E.$$

(2) For every $x \in X = G/H$, for every $g \in G$, the map $z \mapsto g \cdot z$ ($z \in E_x$) is a unitary isomorphism between E_x and $E_{g\cdot x}$.

Let x_0 denote the coset $H = eH$. Every fibre is isomorphic to E_{x_0} and the restriction of the action of G to H on the fibre E_{x_0}, for simplicity also denoted as E_0, maps E_0 to E_0 (since $h \cdot x_0 = x_0$ for all $h \in H$). Since the action of G on E is continuous, for every $z \in E_0$ the map $h \mapsto h \cdot z$ is a continuous map from H to E_0, and since the maps $z \mapsto h \cdot z$ ($z \in E_0$) are unitary, we have a unitary representation $\sigma\colon H \to \mathbf{U}(E_0)$ given by

$$\sigma(h)(z) = h \cdot z. \tag{σ}$$

If the fibres are finite-dimensional vector spaces equipped with hermitian inner products, we say that E has *finite rank*, and the common dimension of these vector spaces is called the *rank* of E.

Assume that E is a hermitian G-bundle of rank n, and pick some orthonormal basis $(e_1, \ldots, e_n)$ of E_0. Since the map $z \mapsto g \cdot z$ ($z \in E_0, g \in G$) is a unitary map from E_0 to $E_{g\cdot x_0}$, the n-tuple $(g \cdot e_1, \ldots, g \cdot e_n)$ is an orthonormal basis of $E_{g\cdot x_0}$. Inspired by Section 6.1, we make the following definition.

Definition 6.24. Let E be a hermitian G-bundle of rank n and pick some orthonormal basis $(e_1, \ldots, e_n)$ of E_0. The Hilbert space $\mathrm{L}^2(G; E_0)$ consists of all functions $f\colon G \to E_0$, such that $f = f_1 e_1 + \cdots + f_n e_n$, where the f_i are functions in $\mathrm{L}^2(G)$; equivalently, $\mathrm{L}^2(G; E_0)$ is the finite Hilbert sum $\mathrm{L}^2(G; E_0) = \bigoplus_{i=1}^n \mathrm{L}^2(G) e_i$. The inner product of two functions $f = \sum_{i=1} f_i e_i$ and $g = \sum_{i=1} g_i e_i$ is

$$\langle f, g \rangle = \sum_{i=1}^{n} \int_G f_i(s) \overline{g_i(s)}\, d\lambda_G(s),$$

where λ_G is a left Haar measure on G. Let L^σ be the subspace of $\mathrm{L}^2(G; E_0)$ given by

$$L^\sigma = \Big\{ f \in \mathrm{L}^2(G; E_0) \mid f(gh) = \sigma(h^{-1})(f(g)), \quad \text{for all } g \in G \quad \text{and all } h \in H \Big\}. \qquad (\dagger_4)$$

It is easy to check that L^σ is closed in $\mathrm{L}^2(G; E_0)$, so it is a Hilbert space. If $A(h)$ is the unitary matrix representing $\sigma(h)$ with respect to the basis $(e_1, \ldots, e_n)$, with

$$\sigma(h)(e_j) = \sum_{i=1}^{n} a_{ij} e_i,$$

we leave it as an exercise to prove that the condition $f(gh) = \sigma(h^{-1})(f(g))$ translates into

$$\begin{pmatrix} f_1(gh) \\ \vdots \\ f_n(gh) \end{pmatrix} = A(h)^* \begin{pmatrix} f_1(g) \\ \vdots \\ f_n(g) \end{pmatrix}.$$

Since $(g \cdot e_1, \ldots, g \cdot e_n)$ is an orthonormal basis of $E_{g \cdot x_0}$ for every g, every section $s\colon X \to E$ is uniquely determined by n functions $s_i\colon X \to \mathbb{C}$ defined by

$$s(g \cdot x_0) = s_1(g \cdot x_0)(g \cdot e_1) + \cdots + s_n(g \cdot x_0)(g \cdot e_n), \quad g \in G.$$

By analogy with the definition of $\mathrm{L}^2(G; E_0)$, we have the following definition.

Definition 6.25. Let E be a hermitian G-bundle of rank n and pick some orthonormal basis $(e_1, \ldots, e_n)$ of E_0. The subspace $\mathrm{L}^2(X; E)$ of $\Gamma(E)$ is defined as the Hilbert space of sections $s\colon X \to E$ that can be expressed as

$$s(g \cdot x_0) = s_1(g \cdot x_0)(g \cdot e_1) + \cdots + s_n(g \cdot x_0)(g \cdot e_n), \quad g \in G,$$

for some functions $s_i \in \mathrm{L}^2_\mu(X)$, where μ is the G-invariant measure (unique up to a scalar) on $X = G/H$ induced by λ_G. Then for two sections $s, t \in \mathrm{L}^2(X; E)$ determined by the n-tuples $(s_1, \ldots, s_n)$ and $(t_1, \ldots, t_n)$ of functions in $\mathrm{L}^2_\mu(X)$, the inner product is given by

$$\langle s, t\rangle = \sum_{i=1}^{n} \int_X s_i(x)\overline{t_i(x)}\, d\mu(x).$$

Note that the induced representation $\rho\colon G \to \mathbf{GL}(L^\sigma)$ of Proposition 6.26 is now a unitary representation $\rho\colon G \to \mathbf{U}(L^\sigma)$.

A difficulty that arises because sections now belong to $\mathrm{L}^2(X; E)$ and functions in L^σ now belong to $\mathrm{L}^2(G; E_0)$ is that, in general, if $r\colon X \to G$ is a section specifying a set of coset representatives of $X = G/H$, the maps $\mathcal{L}$ and $\mathcal{S}$ as defined by

$$\mathcal{L}(s)(g) = (\sigma(u(g, x_0)))^{-1}(r_x^{-1} \cdot s(g \cdot x_0)), \quad x = gH = g \cdot x_0, \tag{$\mathcal{L}_3'$}$$

and

$$\mathcal{S}(f)(g \cdot x_0) = r_x \cdot \sigma(u(g, x_0))(f(g)), \tag{$\mathcal{S}_3'$}$$

may yield a function $\mathcal{L}(s)$ *not* in L^σ for some section $s \in \mathrm{L}^2(X; E)$, or a section $\mathcal{S}(f)$ *not* in $\mathrm{L}^2(X; E)$ for some function $f \in L^\sigma$. If they do for all $s \in \mathrm{L}^2(X; E)$ and all $f \in L^\sigma$, they are mutual inverse maps from L^σ to $\mathrm{L}^2(X; E)$, so we can figure out what is the induced representation $\Pi\colon G \to \mathbf{U}(\mathrm{L}^2(X; E))$ from the definition of the representation $\rho\colon G \to \mathbf{U}(L^\sigma)$ using the fact that the following diagram commutes

$$\begin{array}{ccc} L^\sigma & \xrightarrow{\ \rho(g)\ } & L^\sigma \\ {\scriptstyle \mathcal{L}}\big\uparrow & & \big\downarrow{\scriptstyle \mathcal{S}} \\ \mathrm{L}^2(X; E) & \xrightarrow[\ \Pi_g\ ]{} & \mathrm{L}^2(X; E) \end{array}$$

for every $g \in G$. For any $g \in G$, any $x \in X = G/H$, and any $s \in \mathrm{L}^2(X; E)$, since $\rho(g)(f) = \mathcal{L}(g \cdot \mathcal{S}(f))$ for any $f \in L^\sigma$, we have

$$\begin{aligned} (\Pi_g(s))(x) &= [\mathcal{S}(\rho(g)(\mathcal{L}(s)))](x) = [\mathcal{S}(\mathcal{L}(g \cdot \mathcal{S}(\mathcal{L}(s))))](x) \\ &= (g \cdot s)(x) = g \cdot s(g^{-1} \cdot x), \end{aligned}$$

which we record as

$$(\Pi_g(s))(x) = g \cdot s(g^{-1} \cdot x). \tag{$\dagger_5$}$$

Now, since by definition of $u(g,y)$,

$$r_{g\cdot y}u(g,y) = gr_y,$$

with $y = g^{-1}\cdot x$, we get

$$r_x u(g, g^{-1}\cdot x) = gr_{g^{-1}\cdot x},$$

which implies that

$$g = r_x u(g, g^{-1}\cdot x)(r_{g^{-1}\cdot x})^{-1}, \tag{$\dagger_6$}$$

and substituting the right-hand side expression for the leftmost occurrence of g in $g\cdot s(g^{-1}\cdot x)$, we deduce that

$$\begin{aligned}(\Pi_g(s))(x) &= g\cdot s(g^{-1}\cdot x) = [r_x u(g,g^{-1}\cdot x)(r_{g^{-1}\cdot x})^{-1}]\cdot s(g^{-1}\cdot x)\\ &= r_x\cdot [u(g,g^{-1}\cdot x)\cdot((r_{g^{-1}\cdot x})^{-1}\cdot s(g^{-1}\cdot x))]\\ &= r_x\cdot \sigma(u(g,g^{-1}\cdot x))((r_{g^{-1}\cdot x})^{-1}\cdot s(g^{-1}\cdot x)),\end{aligned}$$

that is,

$$(\Pi_g(s))(x) = r_x\cdot \sigma(u(g,g^{-1}\cdot x))((r_{g^{-1}\cdot x})^{-1}\cdot s(g^{-1}\cdot x)). \tag{$\dagger_7$}$$

If we compare with Formula (Π^α_s) in Definition 6.3, namely,

$$(\Pi^\alpha_g(s))(x) = \alpha(g,g^{-1}\cdot x)(s(g^{-1}\cdot x)), \tag{Π^α_s}$$

since $\alpha(g,y) = \sigma(u(g,y))$ is a cocycle, we have

$$(\Pi^\sigma_g(s))(x) = \sigma(u(g,g^{-1}\cdot x))(s(g^{-1}\cdot x)), \tag{Π^σ_s}$$

and as we explained earlier, in the case of the G-bundle $G\times_H \mathcal{H}_\sigma$, the map $z\mapsto r_x\cdot z$ sends the fibre E_0 to the fibre E_x and the map $x\mapsto g^{-1}\cdot x$ sends the fibre $E_{g\cdot x}$ to the fibre E_0, and since σ is a representation in E_0, we see that $(\dagger_7)$ and (Π^σ_s) are indeed equivalent.

We still have the issue that, in general, the representation Π may not be continuous. This depends on the existence of suitable sections $r\colon X\to G$. A case where continuous sections exist is when $G = N\rtimes H$ is a semi-direct product with N abelian; see Section 8.13.

6.15 Hermitian G-Vector Bundles

A way to deal with the problem that continuous sections $r\colon X\to G$ may not exist is to assume that $p\colon E\to G/H$ is locally trivial, namely to assume the existence of local trivializations. In other words, we assume that E is a vector bundle. We will recall the definition of vector bundles and principal bundles below. Vector bundles and principal bundles are discussed in Gallier and Quaintance [Gallier and Quaintance (2020b)], Bott and Tu [Bott and Tu (1986)], Morita [Morita (2001)], Bröcker and tom Dieck [Bröcker and tom Dieck (1985)], Duistermaat and Kolk [Duistermaat and Kolk (2000)] and Dieudonné [Dieudonné (2006)].

To avoid technical complications, we assume that G is a Lie group and that H is a closed Lie subgroup of G. Now, because G is a Lie group and H is a closed Lie subgroup of G, the quotient space $X = G/H$ is a smooth manifold and $\pi\colon G \to G/H$ is a principal H-bundle, whose definition is recalled below; see Gallier and Quaintance [Gallier and Quaintance (2020b)] (Section 9.9, Proposition 9.2) and Duistermaat and Kolk [Duistermaat and Kolk (2000)] (Appendix A).

Definition 6.26. A *principal H-bundle* is a quadruple $\xi = (\mathcal{E}, \pi, \mathcal{E}/H, H)$, where $\mathcal{E}$ be a smooth manifold, H is Lie group, and $\cdot\colon \mathcal{E} \times H \to \mathcal{E}$ is a smooth right action of H on $\mathcal{E}$ satisfying the following properties:

(1) The right action of H on $\mathcal{E}$ is free;
(2) The orbit space $X = \mathcal{E}/H$ is a smooth manifold under the quotient topology, and the projection $\pi\colon \mathcal{E} \to \mathcal{E}/H$ is smooth;
(3) There is some open cover $\mathcal{U} = (U_\alpha)_{\alpha\in I}$ of $X = \mathcal{E}/H$ and a family $\psi = (\psi_\alpha)_{\alpha\in I}$ of diffeomorphisms called *(local) trivializations*

$$\psi_\alpha\colon \pi^{-1}(U_\alpha) \to U_\alpha \times H,$$

such that

(a) (local triviality) the diagram

$$\begin{array}{ccc} \pi^{-1}(U_\alpha) & \xrightarrow{\psi_\alpha} & U_\alpha \times H \\ & {\scriptstyle\pi}\searrow \quad \swarrow{\scriptstyle pr_1} & \\ & U_\alpha & \end{array}$$

commutes.

(b) Every map $\psi_\alpha\colon \pi^{-1}(U_\alpha) \to U_\alpha \times H$ is an equivariant diffeomorphism, which means that

$$\psi_\alpha(z \cdot h) = \psi_\alpha(z) \cdot h$$

for all $z \in \pi^{-1}(U_\alpha)$ and all $h \in H$, where the right action of H on $U_\alpha \times H$ is $(x, h_1)\cdot h = (x, h_1 h)$. Observe that if $\psi_\alpha(z) = (x, h_1)$, then since $\psi_\alpha(z) \cdot h = (x, h_1 h)$, we have $pr_1(\psi_\alpha(z) \cdot h) = pr_1(\psi_\alpha(z)) = x = \pi(z)$.

Recall that the action $\cdot\colon \mathcal{E}\times H \to \mathcal{E}$ is *free* if it acts without fixed points, that is, for every $h \in H$, if $h \neq 1$, then $x \cdot h \neq x$ for all $x \in \mathcal{E}$.

By conditions (a) and (b), and the definition of the right action of H on $U_\alpha \times H$, for all $z \in \pi^{-1}(U_\alpha)$ and all $h \in H$, we have

$$\pi(z \cdot h) = pr_1(\psi_\alpha(z \cdot h)) = pr_1(\psi_\alpha(z) \cdot h) = pr_1(\psi_\alpha(z)) = \pi(z),$$

so for any $x \in X = \mathcal{E}/H$ and any $z \in \mathcal{E}_x = \pi^{-1}(x)$, we have $z \cdot h \in \mathcal{E}_x$. In fact, for any $z \in \mathcal{E}_x$, we claim that

$$\mathcal{E}_x = \{z \cdot h \mid h \in H\},$$

namely *the orbits of the right action of H on $\mathcal{E}$ are the fibres $\mathcal{E}_x$, with $x \in X$.* Since the action of H on $\mathcal{E}$ is free, the action of H on $\mathcal{E}_x$ is also free.

To prove this, first observe that the restriction of ψ_α to $\mathcal{E}_x$ for any $x \in U_\alpha$ is a diffeomorphism from $\mathcal{E}_x$ onto $\{x\} \times H$ given by

$$\psi_\alpha(z) = (x, \psi_{\alpha,x}(z)),$$

where $\psi_{\alpha,x}\colon \mathcal{E}_x \to H$ is a diffeomorphism between $\mathcal{E}_x$ and H.

Secondly, by definition of the right action of H on $U_\alpha \times H$, for any $z \in \mathcal{E}_x$, if $\psi_\alpha(z) = (x, h_1)$, we have

$$\begin{aligned}\{\psi_\alpha(z \cdot h) \mid h \in H\} &= \{\psi_\alpha(z) \cdot h \mid h \in H\} = \{(x, h_1) \cdot h \mid h \in H\} \\ &= \{(x, h_1 h) \mid h \in H\} = \{x\} \times H,\end{aligned}$$

and so

$$\{z \cdot h \mid h \in H\} = \psi_\alpha^{-1}(\{x\} \times H) = \mathcal{E}_x.$$

For all α, β such that $U_\alpha \cap U_\beta \neq \emptyset$, for every $x \in U_\alpha \cap U_\beta$, we have a diffeomorphism

$$\psi_{\alpha,x} \circ \psi_{\beta,x}^{-1}\colon H \longrightarrow H,$$

which yields the map $g_{\alpha\beta}\colon U_\alpha \cap U_\beta \to \mathrm{Diff}(H)$ called a *transition map* given by

$$g_{\alpha\beta}(x) = \psi_{\alpha,x} \circ \psi_{\beta,x}^{-1}, \quad x \in U_\alpha \cap U_\beta.$$

Intuitively, the transition functions express how the fibre $\mathcal{E}_x$ twists as x moves in $U_\alpha \cap U_\beta$. From the definition above, the isomorphism $\psi_\alpha \circ \psi_\beta^{-1}\colon (U_\alpha \cap U_\beta) \times H \to (U_\alpha \cap U_\beta) \times H$ is given by

$$(\psi_\alpha \circ \psi_\beta^{-1})(x, h) = (x, g_{\alpha\beta}(x)(h)), \quad x \in U_\alpha \cap U_\beta,\ h \in H.$$

A priori, the map $g_{\alpha\beta}(x)$ is a diffeomorphism of the Lie group H, but because the transition maps ψ_α are equivariant, it is shown in Gallier and Quaintance [Gallier and Quaintance (2020b)] (Chapter 9, Proposition 9.21) that $g_{\alpha\beta}(x)$ is the left translation by $g_{\alpha\beta}(x)(1) \in H$, that is,

$$g_{\alpha\beta}(x)(h) = g_{\alpha\beta}(x)(1)h, \quad x \in U_\alpha \cap U_\beta,\ h \in H.$$

Since the group of left translations of H (the maps $L_h\colon H \to H$ given by $L_h(h_1) = hh_1$ $(h, h_1 \in H)$) is isomorphic to H, we usually view the map $g_{\alpha\beta}(x)$ as the element $g_{\alpha\beta}(x)(1)$ of H, and thus we view the transition function $g_{\alpha\beta}$ as a map $g_{\alpha\beta}\colon U_\alpha \cap U_\beta \to H$, such that

$$(\psi_\alpha \circ \psi_\beta^{-1})(x, h) = (x, g_{\alpha\beta}(x)h), \quad x \in U_\alpha \cap U_\beta,\ h \in H.$$

Another technical issue is that Definition 6.26 is too restrictive because it does not allow for the addition of compatible local trivializations. We can fix this problem as follows.

Definition 6.27. Let $\xi = (\mathcal{E}, \pi, X, H)$ be principal bundle, with $X = \mathcal{E}/H$. Given a trivializing cover $\{(U_\alpha, \psi_\alpha)\}$ for ξ, for any open U of X and any diffeomorphism

$$\varphi\colon \pi^{-1}(U) \to U \times H,$$

we say that (U, φ) is *compatible with the trivializing cover* $\{(U_\alpha, \psi_\alpha)\}$ iff whenever $U \cap U_\alpha \neq \emptyset$, there is some smooth map $g_\alpha\colon U \cap U_\alpha \to H$, so that

$$\varphi \circ \psi_\alpha^{-1}(x, h) = (x, g_\alpha(x)(h)),$$

for all $x \in U \cap U_\alpha$ and all $h \in H$. Two trivializing covers are *equivalent* iff every local trivialization of one cover is compatible with the other cover. This is equivalent to saying that the union of two trivializing covers is a trivializing cover.

Definition 6.27 yields the official definition of a principal bundle $\xi = (\mathcal{E}, \pi, X, H)$ in which $\{(U_\alpha, \psi_\alpha)\}$ is an equivalence class of trivializing covers. As for manifolds, given a trivializing cover $\{(U_\alpha, \psi_\alpha)\}$, the set of all bundle charts compatible with $\{(U_\alpha, \psi_\alpha)\}$ is a maximal trivializing cover equivalent to $\{(U_\alpha, \psi_\alpha)\}$.

In the special case where $\mathcal{E}$ is equal to a Lie group G and H is a closed Lie subgroup of G, as we said above, $(G, \pi\colon G \to X, X, H)$ is principal H-bundle (with $X = G/H$).

Definition 6.28. A *hermitian vector bundle with fibre* $\mathcal{H}$ is a quadruple $\xi = (\mathcal{E}, p, X, \mathcal{H})$, where $\mathcal{E}$ and X are smooth manifold, $p\colon \mathcal{E} \to X$ is a surjective smooth map, and $\mathcal{H}$ is a finite-dimensional complex vector space with a hermitian inner product, such that the following conditions hold:

(1) Each fibre $\mathcal{E}_x$ ($x \in X$) is a finite-dimensional space equipped with a hermitian inner product $\langle -, - \rangle_x$.

(2) There is some open cover $\mathcal{U} = (U_\alpha)_{\alpha \in I}$ of X and a family $\varphi = (\varphi_\alpha)_{\alpha \in I}$ of diffeomorphisms called *(local) trivializations*

$$\varphi_\alpha\colon p^{-1}(U_\alpha) \to U_\alpha \times \mathcal{H},$$

such that:

(a) (local triviality) the diagram

$$\begin{array}{ccc} p^{-1}(U_\alpha) & \xrightarrow{\quad\varphi_\alpha\quad} & U_\alpha \times \mathcal{H} \\ & {\scriptstyle p}\searrow \quad \swarrow{\scriptstyle pr_1} & \\ & U_\alpha & \end{array}$$

commutes.

(b) For every $x \in U_\alpha$, the map $\varphi_{\alpha,x}\colon \mathcal{E}_x \to \mathcal{H}$ is a unitary isomorphism.

Since the maps $\varphi_{\alpha,x}$ are unitary, the maps

$$\varphi_{\alpha,x} \circ \varphi_{\beta,x}^{-1}\colon \mathcal{H} \longrightarrow \mathcal{H}$$

are also unitary, so the transition maps are of the form $g_{\alpha\beta}\colon U_\alpha \cap U_\beta \to \mathbf{U}(\mathcal{H})$.

We also need to be able to add compatible trivializations.

Definition 6.29. Let $\xi = (\mathcal{E}, p, X, \mathcal{H})$ be a hermitian vector bundle. Given a trivializing cover $\{(U_\alpha, \varphi_\alpha)\}$ for ξ, for any open U of X and any diffeomorphism

$$\varphi\colon p^{-1}(U) \to U \times \mathcal{H},$$

we say that (U, φ) is *compatible with the trivializing cover* $\{(U_\alpha, \varphi_\alpha)\}$ iff whenever $U \cap U_\alpha \neq \emptyset$, there is some smooth map $g_\alpha\colon U \cap U_\alpha \to \mathbf{U}(\mathcal{H})$, so that

$$\varphi \circ \varphi_\alpha^{-1}(x, u) = (x, g_\alpha(x)(u))$$

for all $x \in U \cap U_\alpha$ and all $u \in \mathcal{H}$. Two trivializing covers are *equivalent* iff every local trivialization of one cover is compatible with the other cover. This is equivalent to saying that the union of two trivializing covers is a trivializing cover.

The official definition of a hermitian vector bundle $\xi = (\mathcal{E}, p, X, \mathcal{H})$ requires $\{(U_\alpha, \varphi_\alpha)\}$ to be an equivalence class of trivializing covers. As earlier, given a trivializing cover $\{(U_\alpha, \varphi_\alpha)\}$, the set of all bundle charts compatible with $\{(U_\alpha, \varphi_\alpha)\}$ is a maximal trivializing cover equivalent to $\{(U_\alpha, \varphi_\alpha)\}$.

Technically, a hermitian vector bundle has the property that the hermitian inner product $\langle -, - \rangle_x$ on the fibre $\mathcal{E}_x$ varies smoothly with $x \in X$. This is formalized as follows. For any open subset U of X, a *frame over* U is an n-tuple $(s_1, \ldots, s_n)$ of smooth sections $s_i \colon U \to \mathcal{E}$, such that $(s_1(x), \ldots, s_n(x))$ is a basis of the fibre $\mathcal{E}_x$ for all $x \in U$ (where n is the dimension of $\mathcal{H}$ and all the $\mathcal{E}_x$). The hermitian inner products $\langle -, - \rangle_x$ have the property that for every U_α, for every frame $(s_1, \ldots, s_n)$ over U_α, the maps

$$x \mapsto \langle s_i(x), s_j(x) \rangle_x, \quad 1 \leq i, j \leq n,\ x \in U_\alpha,$$

are smooth. For details, see Gallier and Quaintance [Gallier and Quaintance (2020b)] (Section 9.8) and Morita [Morita (2001)] (Section 5.1).

Remark: Since X is a manifold, for any local trivialization $\varphi_\alpha \colon p^{-1}(U_\alpha) \to U_\alpha \times \mathcal{H}$, of ξ, since U_α is an open subset of X, there is some open set V in the maximal atlas defining X, such that $V \subseteq U_\alpha$ and a chart $\theta \colon V \to \mathbb{R}^m$ (where m is the dimension of the manifold X), so if $\rho \colon \mathcal{H} \to \mathbb{R}^n$ is an isomorphism (obtained by picking a basis on $\mathcal{H}$), the map

$$(\theta \times \rho) \circ \varphi_\alpha \colon p^{-1}(V) \to \mathbb{R}^{m+n}$$

is a chart of $\mathcal{E}$ viewed as a manifold.

We can now define a hermitian G-vector bundle as a hermitian G-bundle, which is also a hermitian vector bundle in the special case where $X = G/H$.

Definition 6.30. Let G be a Lie group, H a closed subgroup of G, E a smooth manifold, $\mathcal{H}$ a finite-dimensional complex vector space equipped with a hermitian inner product, and $p \colon E \to X$ be a surjective *smooth* map, where as usual, we write $X = G/H$. We say that E, more precisely $(E, p, X, \mathcal{H}, G)$, is a *hermitian G-vector bundle with fibre $\mathcal{H}$* if

(1) Each fibre E_x ($x \in X = G/H$) is a finite-dimensional vector space equipped with a hermitian inner product $\langle -, - \rangle_x$ and there is an *equilinear smooth* left action $\cdot$ of G on E.

(2) There is some open cover $\mathcal{U} = (U_\alpha)_{\alpha \in I}$ of $X = G/H$ and a family $\varphi = (\varphi_\alpha)_{\alpha \in I}$ of diffeomorphisms called *(local) trivializations*

$$\varphi_\alpha \colon p^{-1}(U_\alpha) \to U_\alpha \times \mathcal{H},$$

such that:

(a) (local triviality) the diagram

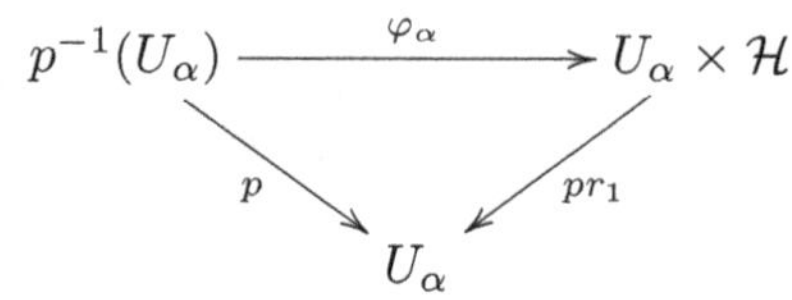

commutes.

(b) For every $x \in U_\alpha$, the map $\varphi_{\alpha,x} \colon \mathcal{E}_x \to \mathcal{H}$ is a unitary isomorphism.

As in the case of a hermitian vector bundle, we require that the hermitian inner product $\langle -, - \rangle_x$ varies smoothly with x.

If $\xi = (\mathcal{E}, \pi, X, H)$ is a principal H-bundle (with $X = \mathcal{E}/H$) and $\sigma \colon H \to \mathbf{U}(\mathcal{H}_\sigma)$ is a unitary representation of H in a finite-dimensional hermitian vector space $\mathcal{H}_\sigma$, then the Borel construction of Section 6.12 can be adapted to produce a hermitian vector bundle $(E, p \colon E \to X, X, \mathcal{H}_\sigma)$, with $E = \mathcal{E} \times_H \mathcal{H}_\sigma$ and $X = \mathcal{E}/H$. The following theorem is a special case of a construction in which $\mathcal{H}_\sigma$ is replaced by any smooth manifold F equipped with a smooth left action of H on F (technically, an effective action);[3] see Dieudonné [Dieudonné (2006)] (Theorem 16.14.7).

Theorem 6.27. *Let $\xi = (\mathcal{E}, \pi, X, H)$ be a principal H-bundle (with $X = \mathcal{E}/H$) and $\sigma \colon H \to \mathbf{U}(\mathcal{H}_\sigma)$ be a unitary representation of H in a finite-dimensional hermitian vector space $\mathcal{H}_\sigma$. Consider the right action of H on $\mathcal{E} \times \mathcal{H}_\sigma$ given by*

$$(z, u) \cdot h = (z \cdot h, \sigma(h^{-1})(u)), \quad z \in \mathcal{E},\ u \in \mathcal{H}_\sigma,\ h \in H. \tag{act2}$$

Here, the right action of H on $\mathcal{E}$ is the action arising from the fact that $\mathcal{E}$ is a principal H-bundle, so it is free, and thus the action (act2) is also free. Then, the orbit space $E = \mathcal{E} \times_H \mathcal{H}_\sigma$ is a smooth manifold. Furthermore, the following properties hold.

(1) The quadruple $(E, p, \mathcal{E}/H, \mathcal{H}_\sigma)$ is a hermitian vector bundle with fibre $\mathcal{H}_\sigma$, where the projection $p \colon \mathcal{E} \times_H \mathcal{H}_\sigma \to \mathcal{E}/H$ is given by $p([(z, u)]) = \pi(z)$, $z \in \mathcal{E}, u \in \mathcal{H}_\sigma$, and with $\pi \colon \mathcal{E} \to \mathcal{E}/H$.

---

[3]Recall that an action $\cdot \colon H \times F \to F$ is effective if for any $h \in H$, if $h \cdot x = x$ for all $x \in F$, then $h = 1$.

(2) *If* $\mathcal{U} = (U_\alpha)_{\alpha\in I}$ *is an open cover of* $X = \mathcal{E}/H$ *and* $\psi = (\psi_\alpha)_{\alpha\in I}$ *is a family of local trivializations* $\psi_\alpha\colon \pi^{-1}(U_\alpha) \to U_\alpha \times H$ *for* $\mathcal{E}$, *then for any smooth section* $s\colon U_\alpha \to \pi^{-1}(U_\alpha)$, *the inverse* φ_α^{-1} *of a local trivialization* $\varphi_\alpha\colon p^{-1}(U_\alpha) \to U_\alpha \times \mathcal{H}_\sigma$ *of* E *is given by*

$$\varphi_\alpha^{-1}(x, u) = [(s(x), u)], \quad x \in U_\alpha,\ u \in \mathcal{H}_\sigma.$$

(3) *For every* $z \in \mathcal{E}$, *the map* $u \mapsto [(z, u)]$ *is a unitary map from* $\mathcal{H}_\sigma$ *to the fibre* $E_{\pi(z)} = p^{-1}(\pi(z))$.

(4) *For any* $x \in X$, *for any fixed* $x_0 \in \mathcal{E}_x = \pi^{-1}(x)$, *the map given by*

$$h \cdot [(x_0, u)] = [(x_0, h \cdot u)] = [(x_0, \sigma_h(u))], \quad h \in H,\ u \in \mathcal{H}_\sigma,$$

is a unitary representation of H *on the fibre* $E_x = p^{-1}(x)$.

Proof. (1) Some details of the proof are given in Duistermaat and Kolk [Duistermaat and Kolk (2000)] (Section 2.40), and a very detailed proof is given in Dieudonné [Dieudonné (2006)] (Theorem 16.10.3, Theorem 16.14.1 and Theorem 16.14.7). The representation $\sigma\colon H \to \mathbf{U}(\mathcal{H}_\sigma)$ defines a left action of H on $\mathcal{H}_\sigma$ but it is convenient to denote a right action of H on $\mathcal{H}_\sigma$ as

$$h^{-1} \cdot u = \sigma(h^{-1})(u), \quad h \in H,\ u \in \mathcal{H}_\sigma,$$

so that the right action of H on $\mathcal{E} \times \mathcal{H}_\sigma$ is written as

$$(z, u) \cdot h = (z \cdot h, h^{-1} \cdot u), \quad z \in \mathcal{E},\ u \in \mathcal{H}_\sigma,\ h \in H. \tag{act3}$$

The above action immediately generalizes to the case where H acts (on the left) smoothly (and effectively) on a manifold F, and then the orbit space $E = \mathcal{E} \times_H F$ is a fibre bundle with fibre F.

(2) Given that it can be proven that $E = \mathcal{E} \times_H \mathcal{H}_\sigma$ is a smooth manifold (see Dieudonné [Dieudonné (2006)], Theorem 16.10.3, Theorem 16.14.1), it is interesting to see how trivializing maps are defined.

Since $\xi = (\mathcal{E}, \pi, X, H)$ is a principal H-bundle, trivializing maps exist, so there is an open cover $\mathcal{U} = (U_\alpha)_{\alpha\in I}$ of $X = \mathcal{E}/H$ and a family $\psi = (\psi_\alpha)_{\alpha\in I}$ of trivializing diffeomorphisms

$$\psi_\alpha\colon \pi^{-1}(U_\alpha) \to U_\alpha \times H. \tag{ψ_α}$$

Since $\pi = pr_1 \circ \psi_\alpha$ and ψ_α is a diffeomorphism, a smooth function $s\colon U_\alpha \to \pi^{-1}(U_\alpha)$ is a section, which means that $\pi \circ s = \mathrm{id}$, iff $\pi \circ \psi_\alpha^{-1} \circ \psi_\alpha \circ s = \mathrm{id}$ iff $pr_1 \circ \psi_\alpha \circ s = \mathrm{id}$, but since $\psi_\alpha \circ s\colon U_\alpha \to U_\alpha \times H$ and smooth sections $s_1\colon U_\alpha \to U_\alpha \times H$ are of the form $s_1(x) = (x, f_1(x))$ for some smooth function $f_1\colon U_\alpha \to H$, smooth sections $s\colon U_\alpha \to \pi^{-1}(U_\alpha)$ exist and are of the form $\psi_\alpha^{-1} \circ s_1$, where $s_1\colon U_\alpha \to U_\alpha \times H$ is a smooth section of the trivial H-bundle $U_\alpha \times H$.

For any smooth section $s\colon U_\alpha \to \pi_\alpha^{-1}(U_\alpha)$, for any $z \in \pi_\alpha^{-1}(U_\alpha)$, since $s(\pi(z)) \in \mathcal{E}_{\pi(z)}$ and H acts freely on $\mathcal{E}_{\pi(z)}$, there is a unique $h \in H$, denote it h_z, such that

$$s(\pi(z)) = z \cdot h_z. \tag{h_z}$$

Define the map $F_\alpha\colon \pi^{-1}(U_\alpha) \times \mathcal{H}_\sigma \to U_\alpha \times \mathcal{H}_\sigma$ by

$$F_\alpha(z, u) = (\pi(z), h_z^{-1} \cdot u), \quad z \in \pi^{-1}(U_\alpha),\ u \in \mathcal{H}_\sigma. \tag{F_α}$$

We claim that

$$F_\alpha(z \cdot h, h^{-1} \cdot u) = F_\alpha(z, u). \tag{$\dagger_8$}$$

Since z and $z \cdot h$ are in the same orbit, $\pi(z \cdot h) = \pi(z)$, and the unique $h_1 \in H$, such that

$$s(\pi(z)) = s(\pi(z \cdot h)) = (z \cdot h) \cdot h_1 = z \cdot (hh_1),$$

satisfies the equation $hh_1 = h_z$ because h_z is the unique element of H, such that

$$s(\pi(z)) = z \cdot h_z,$$

so $h_{z \cdot h} = h^{-1} h_z$, and then

$$\begin{aligned} F_\alpha(z \cdot h, h^{-1} \cdot u) &= (\pi(z \cdot h), h_{z \cdot h}^{-1} \cdot (h^{-1} \cdot u)) = (\pi(z), (h^{-1}h_z)^{-1} \cdot (h^{-1} \cdot u)) \\ &= (\pi(z), (h_z^{-1} h h^{-1}) \cdot u) = (\pi(z), h_z^{-1} \cdot u) = F_\alpha(z, u). \end{aligned}$$

Consequently, F_α induces a well-defined map $\varphi_\alpha\colon \pi^{-1}(U_\alpha) \times_H \mathcal{H}_\sigma \to U_\alpha \times \mathcal{H}_\sigma$ given by

$$\varphi_\alpha([(z, u)]) = F_\alpha(z, u) = (\pi(z), h_z^{-1} \cdot u), \quad z \in \pi^{-1}(U_\alpha),\ u \in \mathcal{H}_\sigma. \tag{φ_α}$$

It is easy to see that this map is smooth. The inverse $\eta_\alpha\colon \pi^{-1}(U_\alpha) \times_H \mathcal{H}_\sigma \to U_\alpha \times \mathcal{H}_\sigma$ of φ_α is defined as follows:

$$\eta_\alpha(x, u) = [(s(x), u)], \quad x \in U_\alpha,\ u \in \mathcal{H}_\sigma. \tag{η_α}$$

Since by (h_z),

$$s(\pi(z)) = z \cdot h_z,$$

and since $\pi \circ s = \mathrm{id}$ because $s\colon U_\alpha \to \pi^{-1}(U_\alpha)$ is a section, using the equivalence defined by (act3), we get

$$\begin{aligned} \eta_\alpha(\varphi_\alpha[(z, u)]) &= \eta_\alpha(\pi(z), h_z^{-1} \cdot u) = [(s(\pi(z)), h_z^{-1} \cdot u)] \\ &= [(z \cdot h_z, h_z^{-1} \cdot u)] = [(z, u)]. \end{aligned}$$

For the reverse composition, we need to observe that by definition of h_z, with $z = s(x)$, we have

$$s(x) = s(\pi(s(x))) = s(x) \cdot h_{s(x)},$$

so $h_{s(x)} = 1$, and then

$$\begin{aligned}\varphi_\alpha(\eta_\alpha(x,u)) &= \varphi_\alpha([(s(x),u)]) = (\pi(s(x)), h_{s(x)}^{-1} \cdot u)\\ &= (x, 1 \cdot u) = (x,u).\end{aligned}$$

Therefore, φ_α and η_α are mutual inverses, as claimed. It follows that $\varphi_\alpha\colon \pi^{-1}(U_\alpha) \times_H \mathcal{H}_\sigma \to U_\alpha \times \mathcal{H}_\sigma$ is a diffeomorphism. Since by definition of $p\colon \mathcal{E} \times_H \mathcal{H}_\sigma \to \mathcal{E}/H$, $p([(z,u)]) = \pi(z)$ with $\pi\colon \mathcal{E} \to \mathcal{E}/H$, we see that

$$p^{-1}(U_\alpha) = \pi^{-1}(U_\alpha) \times_H \mathcal{H}_\sigma,$$

so the diffeomorphism $\varphi_\alpha\colon p^{-1}(U_\alpha) \to U_\alpha \times \mathcal{H}_\sigma$ is a trivialization of the space $\mathcal{E} \times_H \mathcal{H}_\sigma$.

In summary, for every smooth section $s\colon U_\alpha \to \pi^{-1}(U_\alpha)$, the inverse φ_α^{-1} of a local trivialization $\varphi_\alpha\colon p^{-1}(U_\alpha) \to U_\alpha \times \mathcal{H}_\sigma$ is given by

$$\varphi_\alpha^{-1}(x,u) = [(s(x),u)], \quad x \in U_\alpha,\ u \in \mathcal{H}_\sigma. \tag{$\dagger_9$}$$

(3) As we explained in Section 6.12, for any $x \in X = \mathcal{E}/H$, for any $z \in \mathcal{E}_x = \pi^{-1}(x)$, we have $p^{-1}(x) = \{[(z,u)] \mid u \in \mathcal{H}_\sigma\}$. If $[(z,u)] = [(z,v)]$ for some $z \in \mathcal{E}$ and some $u, v \in \mathcal{H}_\sigma$, then there is some $h \in H$, such that $z = z \cdot h$ and $u = h^{-1} \cdot v$, and since H acts freely on $\mathcal{E}$, we must have $h = e$ and $v = u$. This shows that the map $u \mapsto [(z,u)]$ is injective, and since it is also surjective, it is bijective. □

If we pick the section $s_\alpha\colon U_\alpha \to \pi^{-1}(U_\alpha)$ to be the special section given by

$$s_\alpha(x) = \psi_\alpha^{-1}(x,1), \quad x \in U_\alpha, \tag{$\dagger_{10}$}$$

then we can figure out what are the corresponding transition functions. First, from ($\dagger_9$), we have

$$\varphi_\alpha^{-1}(x,u) = [(\psi_\alpha^{-1}(x,1),u)]. \tag{$\dagger_{11}$}$$

Next, for any $z \in \mathcal{E}_x$, we have

$$\varphi_\alpha([(z,u)]) = (\pi(z), h_z^{-1} \cdot u) = (x, h_z^{-1} \cdot u),$$

where h_z is the unique element of H, such that

$$s_\alpha(x) = \psi_\alpha^{-1}(x,1) = z \cdot h_z.$$

If

$$\psi_\alpha(z) = (x, h_1),$$

then since ψ_α is equivariant, we have

$$\psi_\alpha(z \cdot h_1^{-1}) = \psi_\alpha(z) \cdot h_1^{-1} = (x, h_1) \cdot h_1^{-1} = (x, 1),$$

which shows that

$$\psi_\alpha^{-1}(x, 1) = z \cdot h_1^{-1},$$

and so

$$h_z = h_1^{-1}.$$

In summary, if $\psi_\alpha(z) = (x, h_1)$, then

$$\varphi_\alpha([(z, u)]) = (x, h_1 \cdot u). \tag{$\dagger_{12}$}$$

For any α, β such that $U_\alpha \cap U_\beta \neq \emptyset$, we need to compute $(\varphi_\alpha \circ \varphi_\beta^{-1})(x, u)$. By ($\dagger_{11}$), we have

$$\varphi_\beta^{-1}(x, u) = [(\psi_\beta^{-1}(x, 1), u)].$$

If we let $z = \psi_\beta^{-1}(x, 1)$, we need to compute

$$\psi_\alpha(z) = \psi_\alpha(\psi_\beta^{-1}(x, 1)) = (x, g_{\alpha\beta}(x)(1)),$$

where the $g_{\alpha\beta}$ are the transition functions of the principal bundle $\mathcal{E}$, and so by ($\dagger_{12}$),

$$(\varphi_\alpha \circ \varphi_\beta^{-1})(x, u) = (x, g_{\alpha\beta}(x)(1) \cdot u),$$

which shows that the transition functions of the vector bundle E are also given by the $g_{\alpha\beta}(x)(1) \in H$, except that this time H acts on $\mathcal{H}$ (on the left). Going back to the original definition of the action of H on $\mathcal{H}$ given by the unitary representation σ, we have

$$(\varphi_\alpha \circ \varphi_\beta^{-1})(x, u) = (x, \sigma(g_{\alpha\beta}(x)(1))(u)).$$

Combining what we did in Section 6.12 with Theorem 6.27, we obtain the following result.

Theorem 6.28. *Let G be a Lie group, H a closed Lie subgroup of G, and $\sigma\colon H \to \mathbf{U}(\mathcal{H}_\sigma)$ a unitary representation in a finite-dimensional hermitian vector space. Then $(E, p, X, \mathcal{H}_\sigma, G)$ is a hermitian G-vector bundle, with $E = G \times_H \mathcal{H}_\sigma$ and $X = G/H$.*

6.16 The Case of G-Bundles of Infinite Rank

If the hermitian G-bundle E has infinite rank, namely the fibres are separable Hilbert spaces, a more sophisticated method is required. The solution is to restrict our attention to sections with *compact support*. There is also the problem that G/H may not have a G-invariant measure. This causes a difficulty for defining an inner product on the Hilbert space L^σ, which is the completion of a subspace of $\mathrm{L}^2(G; E_0)$. This problem can also be overcome.

Let us first assume that G/H has a G-invariant measure. In this case, L^α is replaced by the set of continuous functions $\mathcal{H}_0$ defined in Section 6.5 and given by

$$\begin{aligned}\mathcal{H}_0 = \{f \in \mathcal{C}(G, E_0) \mid \pi(\mathrm{supp}(f)) \text{ is compact and}\\ f(gh) = \sigma(h^{-1})(f(g)) \quad \text{for all } g \in G \text{ and all } h \in H\},\end{aligned}$$

where $\pi\colon G \to G/H$.

As in Section 6.5, we define a hermitian inner product on $\mathcal{H}_0$ and form the Hilbert space L^σ, which is the completion of $\mathcal{H}_0$. The induced representation in L^σ is then the usual one, namely

$$(\mathrm{Ind}_H^G\, \sigma)_g(f)(g_1) = f(g^{-1}g_1).$$

If G/H does not have a G-invariant measure, then as in Section 6.11 there are two approaches. In the first approach, we use the same space of functions E_0 as before but we use an inner product involving a quasi-invariant measure and the definition of the representation has an extra term $\varrho(g, g_1H)^{1/2}$ as in (indv1), namely

$$(\mathrm{Ind}_H^G\, \sigma)_g(f)(g_1) = \varrho(g, g_1H)^{1/2} f(g^{-1}g_1), \qquad f \in L^\sigma,\ g, g_1 \in G.$$

In the second approach, we restrict ourselves to G-bundles obtained from G and $\mathcal{H}_\sigma$ by a quotient process. In the more general case of abstract G-bundles, it should be possible to use a tensor product construction but we haven't worked out the details.

The method is to modify the (right) action of H on $G \times \mathcal{H}_\sigma$, so that it is given by

$$(g, u) \cdot h = \left(gh, \left(\frac{\Delta_H(h)}{\Delta_G(h)}\right)^{1/2} \sigma(h^{-1})(u)\right), \quad g \in G,\ u \in E_0,\ h \in H.$$

Since the modular function is a homomorphism on H, the above is indeed an action. We obtain a new orbit space that we still denote (with a slight abuse of notation) $G \times_H \mathcal{H}_\sigma$. Let $\mathcal{H}^0$ be the subspace of $\mathcal{C}(G, \mathcal{H}_\sigma)$ given by

$$\mathcal{H}^0 = \Bigg\{ f \in \mathcal{C}(G, \mathcal{H}_\sigma) \mid \pi(\mathrm{supp}(f)) \text{ is compact and}$$
$$f(gh) = \left(\frac{\Delta_H(h)}{\Delta_G(h)}\right)^{1/2} \sigma(h^{-1})(f(g)) \text{ for all } g \in G \text{ and all } h \in H \Bigg\}.$$

As in Section 6.11, a hermitian inner product on $\mathcal{H}^0$ can be defined and if L^σ is the Hilbert space, which is the completion of $\mathcal{H}^0$, then as usual the induced representation in L^σ is given by

$$(\mathrm{Ind}_H^G \sigma)_g(f)(g_1) = f(g^{-1}g_1).$$

Remark: In general, $E = G \times_H \mathcal{H}_\sigma$ does not have trivialization maps and so it is *not* a vector bundle. A counter-example is given in Folland [Folland (1995)] (Chapter 6, Section 6.8). However, if G is a Lie group and if H is a compact subgroup of G, then $G \times_H \mathcal{H}_\sigma$ is a vector bundle (in fact, a homogeneous hermitian vector bundle).

6.17 Problems

Problem 6.1. Let G be a compact group, H be a closed subgroup of G, and U be a linear representation of H in a finite-dimensional vector space E. Consider the Hilbert space $\mathrm{L}^2(G; E)$ consisting of all functions $f\colon G \to E$, such that for any orthonormal basis $(e_1, \ldots, e_n)$ of E, $f = f_1 e_1 + \cdots + f_n e_n$, where the f_i are functions in $\mathrm{L}^2(G)$. The inner product of two functions $f = \sum_{i=1} f_i e_i$ and $g = \sum_{i=1} g_i e_i$ is

$$\langle f, g \rangle = \sum_{i=1}^{n} \int_G f_i(s) \overline{g_i(s)}\, d\lambda(s),$$

where λ is a Haar measure on G. Consider the subspace $\mathcal{H}$ of $\mathrm{L}^2(G; E)$ consisting of all functions f, such that

$$f(sh) = U(h^{-1})(f(s)), \qquad \text{for all } s \in G \text{ and all } h \in H.$$

Show that $\mathcal{H}$ is (topologically) closed in $\mathrm{L}^2(G; E)$, where $\mathrm{L}^2(G; E)$ has the norm induced by the above defined inner product.

Problem 6.2. (A continuation of Problem 6.1) Define the map $\Pi\colon G \to \mathbf{GL}(\mathcal{H})$ by

$$\Pi_s(f) = \lambda_s f, \qquad s \in G,\ f \in \mathcal{H},$$

or equivalently, by

$$(\Pi_s(f))(t) = f(s^{-1}t), \qquad s, t \in G,\ f \in \mathcal{H}.$$

(1) Show that Π is a linear representation of G in $\mathcal{H}$.
(2) Show that $\Pi\colon G \to \mathbf{GL}(\mathcal{H})$ is a unitary representation of G in $\mathcal{H}$.

Now, replace U by an equivalent representation $h \mapsto PU(h)P^{-1}$, where P is a unitary transformation $P\colon E \to E'$. Show that the corresponding induced representation is $s \mapsto f_P \Pi_s f_P^{-1}$, a unitary representation equivalent to Π, where f_P is the linear map from $\mathcal{H}$ to $\mathcal{H}'$ given by $f_P(f) = P \circ f$.

Problem 6.3. Let G be a group, H be a subgroup of G, and E be a vector space. Let $X = G/H$ be the homogenous space of left coset, and $\sigma\colon H \to \mathbf{GL}(E)$ be a given homomorphism. Check that if σ is replaced by an equivalent representation σ' of H in E^X, then the corresponding representations Π and Π' of G in E^X are equivalent.

Problem 6.4. Let G be a locally compact group, H be a closed subgroup of G, E be a separable Hilbert space, and $U\colon H \to \mathbf{U}(E)$ be the trivial representation of H in E. For the choice of $\alpha(s, x) = \mathrm{id}_E$ for all $(s, x) \in G \times (G/H)$, show that the hypotheses of Theorem 6.9 are satisfied.

Hint. To verify condition (3), use the fact that the family of maps $f \mapsto \Pi_s(f)$ $(s \in G)$ is equicontinuous; see Dieudonné [Dieudonné (2003b)], Chapter XXII, Section 3.

Problem 6.5. Consider the groups $G = \mathbf{SE}(3)$ and $H \approx \mathbf{SO}(3)$ and any unitary representation $\sigma\colon \mathbf{SO}(3) \to \mathbf{U}(E)$ of $\mathbf{SO}(3)$ in a separable Hilbert space E. We showed in Example 6.1 that we have a cocycle $\alpha\colon \mathbf{SE}(3) \times \mathbb{R}^3 \to \mathbf{U}(E)$ given by

$$\alpha((a, Q), x) = \sigma(Q), \quad a, x \in \mathbb{R}^3,\ Q \in \mathbf{SO}(3),$$

and the homomorphism $\Pi\colon \mathbf{SE}(3) \to \mathbf{GL}(E^{\mathbb{R}^3})$ induced by α is given by

$$(\Pi_{(a,Q)}(f))(x) = \sigma(Q) f(Q^\top (x - a)), \quad f \in E^{\mathbb{R}^3},\ x \in \mathbb{R}^3.$$

Check that conditions (1)–(3) of Theorem 6.9 are satisfied.

Problem 6.6. In Example 6.3, we obtain a unitary representation $\Pi\colon \mathbf{SE}(n) \to \mathbf{U}(\mathrm{L}^2_\mu(\mathbf{SO}(n);\mathbb{C}))$ of $\mathbf{SE}(n)$ in the Hilbert space $\mathrm{L}^2_\mu(\mathbf{SO}(n);\mathbb{C})$ given by

$$(\Pi_{(a,Q)}(f))(R) = e^{i((Ry)\cdot a)} f(Q^\top R), \quad (a,Q) \in \mathbf{SE}(n),\ R \in \mathbf{SO}(n),\\ f \in \mathrm{L}^2_\mu(\mathbf{SO}(n);\mathbb{C}),\ y \in \mathbb{R}^n.$$

Check that the above representation satisfies conditions (2) and (3) of Theorem 6.9.

Problem 6.7. In Example 6.3, we also defined a representation $\widetilde{\Pi}\colon \mathbf{SE}(n) \to \mathbf{U}(\mathrm{L}^2_\lambda(S^{n-1};\mathbb{C}))$ of $\mathbf{SE}(n)$ in the Hilbert space $\mathrm{L}^2_\lambda(S^{n-1};\mathbb{C})$ *via*

$$(\widetilde{\Pi}_{(a,Q)}(f))(x) = e^{ir(x\cdot a)} f(Q^\top x), \qquad (a,Q) \in \mathbf{SE}(n),\ x \in S^{n-1},\\ f \in \mathrm{L}^2_\lambda(S^{n-1};\mathbb{C}),\ r > 0. \qquad (\dagger)$$

(1) Show that for each fixed r, the representation given by

$$(\widetilde{\Pi}_{(a,Q)}(f))(x) = e^{-ir(x\cdot a)} f(Q^\top x), \quad (a,Q) \in \mathbf{SE}(n),\ x \in S^{n-1},\\ f \in \mathrm{L}^2_\lambda(S^{n-1};\mathbb{C}),\ r > 0$$

is equivalent to that of (†).

(2) Prove that the representation given by (†) is irreducible. Hint: See Vilenkin [Vilenkin (1968)], Chapter XI, Section 2.

(3) (Advanced) Prove that

$$(\widetilde{\Pi}_{(a,Q)}(f))(x) = e^{z(x\cdot a)} f(Q^\top x), \quad (a,Q) \in \mathbf{SE}(n),\ x \in S^{n-1},\\ f \in \mathrm{L}^2_\lambda(S^{n-1};\mathbb{C}),\ z \in \mathbb{C}^* \quad (*_4)$$

is irreducible representation, but it is not unitary unless $z = ir$ with $r \in \mathbb{R}$ and $r \neq 0$. Hint: See Vilenkin [Vilenkin (1968)].

Problem 6.8. Prove Proposition 6.10.

Hint. See Folland [Folland (1995)], Chapter 6, Proposition 6.1.

Problem 6.9. Prove Proposition 6.11.

Hint. See Folland [Folland (1995)], Chapter 2, Proposition 2.54.

Problem 6.10. Prove Proposition 6.12.

Hint. See Folland [Folland (1995)], Chapter 2, Lemma 2.55.

Problem 6.11. (Advanced Exercise) Prove Theorem 6.14.

Hint. See Folland [Folland (1995)], Chapter 2, Theorem 2.59.

Problem 6.12. Let G be a locally compact group, H be a closed subgroup of G, E be a separable Hilbert space, and $U\colon H \to \mathbf{U}(E)$ be a unitary representation of H. For any quasi-invariant measure μ on $X = G/H$ associated with the continuous function $\varrho\colon G \times (G/H) \to (0, \infty)$, define the cocycle $\alpha\colon G \times X \to \mathbf{U}(E)$ via

$$\alpha(s, x) = (\varrho(s^{-1}, x))^{-1/2+ri} \operatorname{id}_E.$$

Verify that $\alpha(s, x)$ satisfies condition (3) of Theorem 6.16.

Hint. See Dieudonné [Dieudonné (2003b)], Chapter XXII, Section 3, No. 22.3.8.3.

Problem 6.13. Let $G = \mathbf{SL}(2, \mathbb{R})$ and $H = S_1$ be the subgroup

$$S_1 = \left\{ \begin{pmatrix} a & b \\ 0 & a^{-1} \end{pmatrix} \,\middle|\, a, b \in \mathbb{R}, a \neq 0 \right\},$$

and let $E = \mathbb{C}$.

(1) Show that the homogeneous space $\mathbf{SL}(2, \mathbb{R})/S_1$ is homeomorphic to $\mathbb{P}^1(\mathbb{R}) = \mathbb{RP}^1$, the real projective line. Recall from Example 6.4 that the action of $\mathbf{SL}(2, \mathbb{R})$ on $\mathbb{RP}^1$ viewed as $\mathbb{R} \cup \{\infty\}$ is given by

$$s \cdot z = \begin{pmatrix} a & b \\ c & d \end{pmatrix} \cdot z = \frac{az + b}{cz + d}, \qquad z \in \mathbb{RP}^1,$$

with the convention that when $z = -d/c$, then the result is ∞, and when $z = \infty$, then the result if a/c. In particular, check that this action is transitive and that the stabilizer of ∞ is the subgroup S_1.

(2) If we give $\mathbb{RP}^1$ the measure μ, which is the Lebesgue measure extended so that $\{\infty\}$ has measure zero, check that the cocycles

$$\alpha(s, x) = \left(\frac{1}{(cx - a)^2} \right)^{-1/2+(r/2)i} \operatorname{sign}(cx - a) \operatorname{id}_{\mathbb{C}},$$

with $r \in \mathbb{R}$, satisfy the conditions of Theorem 6.16.

Problem 6.14. (A continuation of Problem 6.13) Let

$$s = \begin{pmatrix} a & b \\ c & d \end{pmatrix}.$$

(1) Show that the induced representation of $\mathbf{SL}(2, \mathbb{R})$ in $\mathrm{L}^2_\mu(\mathbb{RP}^1; \mathbb{C})$ given by

$$\Pi_s(f)(x) = |cx - a|^{-1+ri} f\left(\frac{b - dx}{cx - a} \right), \qquad f \in \mathrm{L}^2_\mu(\mathbb{RP}^1; \mathbb{C}),$$

is irreducible.

(2) Show that the induced representation of $\mathbf{SL}(2,\mathbb{R})$ in $\mathrm{L}^2_\mu(\mathbb{RP}^1;\mathbb{C})$ given by

$$\Pi_s(f)(x) = |cx-a|^{-1+ri}\,\mathrm{sign}(cx-a) f\left(\frac{b-dx}{cx-a}\right), \quad f \in \mathrm{L}^2_\mu(\mathbb{RP}^1;\mathbb{C}),$$

is irreducible.

(3) Show that the above representations of $\mathbf{SL}(2,\mathbb{R})$ in $\mathrm{L}^2_\mu(\mathbb{RP}^1;\mathbb{C})$ are pairwise inequivalent for $r > 0$.

Problem 6.15. Let $G = \mathbf{SL}(2,\mathbb{R})$ and $H = \mathbf{SO}(2)$ be the subgroup

$$\mathbf{SO}(2) = \left\{ \begin{pmatrix} \cos\theta & -\sin\theta \\ \sin\theta & \cos\theta \end{pmatrix} \,\middle|\, 0 \leq \theta \leq 2\pi, \right\},$$

and let $E = \mathbb{C}$.

(1) Show that the homogeneous space $\mathbf{SL}(2,\mathbb{R})/\mathbf{SO}(2)$ is homeomorphic to the upper half plane $P = \{z = x + iy \in \mathbb{C} \mid y > 0\}$, where the action of $\mathbf{SL}(2,\mathbb{R})$ on P given by

$$s \cdot z = \begin{pmatrix} a & b \\ c & d \end{pmatrix} \cdot z = \frac{az+b}{cz+d}, \qquad z = x + iy \in P.$$

In particular, check that this action is transitive and that the stabilizer of $z = i$ is $\mathbf{SO}(2)$.

(2) In Example 6.5, we showed that

$$(\Pi_s(f))(z) = (\alpha(s^{-1},x))^{-1}(f(s^{-1}\cdot z)), \qquad f \in \mathrm{L}^2_\mu(P;\mathbb{C}),\ z \in P,$$

is an induced representation of $\mathbf{SL}(2,\mathbb{R})$ into $\mathrm{L}^2_\mu(P;\mathbb{C})$, where

$$\alpha(s,z) = \frac{(cz+d)^n}{|cz+d|^n}\,\mathrm{id}_{\mathbb{C}}.$$

Define $\alpha'(s,z) = c(s\cdot z)\circ\alpha(s,z)\circ c(z)^{-1}$, with

$$c(z) = c(x+iy) = y^{-n/2},$$

and show that image of $\mathrm{L}^2_\mu(P;\mathbb{C})$ under the map $f \mapsto cf$ is the space E_n of functions $g\colon P \to \mathbb{C}$, such that the map $z \mapsto y^n g(z)^2$ is μ-integrable.

(3) Then show that the equivalent unitary representation is given by

$$(\Pi_s(f))(z) = (cz-a)^{-n} g\left(\frac{b-dz}{cz-a}\right), \qquad g \in E_n,\ z \in P,\ s = \begin{pmatrix} a & b \\ c & d \end{pmatrix}.$$

Problem 6.16. Let G be a compact (metrizable) group and H be a closed subgroup of G. Show that $\mathrm{L}^2_{\mu'}(H\backslash G)$ is the image of $\mathrm{L}^2_\mu(G/H)$ under the isomorphism $g \mapsto \breve{g}$.

Problem 6.17. Let $G = \mathbf{SO}(n+1)$, $H = \mathbf{SO}(n)$, and $G/H = \mathbf{SO}(n+1)/\mathbf{SO}(n) \simeq S^n$. By Proposition 6.18, the space $\mathrm{L}^2(\mathbf{SO}(n+1)/\mathbf{SO}(n)) \simeq \mathrm{L}^2(S^n)$ is the Hilbert sum of the subspaces $L_\rho \subseteq \mathfrak{a}_\rho$ for which $(\rho : \sigma_0) \geq 1$, where

$$\mathrm{L}^2(\mathbf{SO}(n+1)) = \bigoplus_{\rho} \mathfrak{a}_\rho$$

is the Hilbert sum given by Peter–Weyl I and where $d = (\rho : \sigma_0) \geq 1$ is the number of times that the trivial representation σ_0 of $\mathbf{SO}(n)$ is contained in the restriction of M_ρ to $\mathbf{SO}(n)$.

(1) Show that the L_ρ for which $(\rho : \sigma_0) = 1$ are exactly the spaces $\mathcal{H}_k^{\mathbb{C}}(S^n)$ of spherical harmonics on S^n; see Definition 5.1.
(2) (Advanced) Obtain a decomposition of the regular representation $\mathbf{R}\colon \mathbf{SO}(n+1) \to \mathbf{U}(\mathrm{L}^2(S^n))$ into irreducible representations $\mathbf{R}_k\colon \mathbf{SO}(n+1) \to \mathbf{U}(\mathcal{H}_k^{\mathbb{C}}(S^n))$ of $\mathbf{SO}(n+1)$ in the spaces $\mathcal{H}_k^{\mathbb{C}}(S^n)$ of spherical harmonics on S^n. Hint: See Dieudonné [Dieudonné (2003c)], Chapter XXIII, Section 38, or Gallier and Quaintance [Gallier and Quaintance (2020b)], Chapter 7.
(3) (Advanced) Show that the irreducible representations $\mathbf{R}_k\colon \mathbf{SO}(n+1) \to \mathbf{U}(\mathcal{H}_k^{\mathbb{C}}(S^n))$ are irreducible representations of class 1 relative to $\mathbf{SO}(n)$.

Hint. See Vilenkin [Vilenkin (1968)], Chapter IX, Sections 2–3.

Problem 6.18. (Advanced Exercise) Let G be a locally compact group, H be a closed subgroup of G, E be a separable Hilbert space, and $U\colon H \to \mathbf{U}(E)$ be a unitary representation of H. Define

$$\begin{aligned}\mathcal{H}_0 = \{f \in \mathcal{C}(G, E) \mid\ & \pi(\mathrm{supp}(f)) \text{ is compact and} \\ & f(sh) = U(h^{-1})(f(s)) \quad \text{for all } s \in G \text{ and all } h \in H\},\end{aligned}$$

and define an inner product on $\mathcal{H}_0$ via

$$\langle f, g\rangle = \int_{G/H} \langle f(s), g(s)\rangle_E \, d\mu(sH).$$

Let $\mathcal{H}$ be the Hilbert space completion of $\mathcal{H}_0$. Given any quasi-invariant measure μ associated with a continuous function $\varrho\colon G \times (G/H) \to (0, \infty)$, define Π_s^μ by

$$(\Pi_s^\mu(f))(t) = \varrho(s, tH)^{1/2} f(s^{-1}t), \qquad f \in \mathcal{H},\ s, t \in G.$$

Show that the operators Π_s^μ are unitary with respect to the inner product on $\mathcal{H}$ defined above.

Hint. See Folland [Folland (1995)], Chapter 6, Section 6.1.

Problem 6.19. (Advanced Exercise) Let G be a locally compact group, H be a closed subgroup of G, E be a separable Hilbert space, and $U\colon H \to \mathbf{U}(E)$ be a unitary representation of H. Define

$$\mathcal{H}^0 = \Bigg\{ f \in \mathcal{C}(G, E) \mid \pi(\text{supp}(f)) \text{ is compact and}$$

$$f(sh) = \left(\frac{\Delta_H(h)}{\Delta_G(h)}\right)^{1/2} U(h^{-1})(f(s)) \text{ for all } s \in G \text{ and all } h \in H \Bigg\}.$$

Show that $\mathcal{H}^0$ is not empty.

Hint. Modify Proposition 6.10 by adding the factor $\left(\frac{\Delta_G(h)}{\Delta_H(h)}\right)^{1/2}$ under the integral.

Problem 6.20. Show that $p\colon E \to X$ as defined in Definition 6.17 does not depend on the choice of g in the coset gH.

Problem 6.21. Verify that $E = G \times_H \mathcal{H}_\sigma$ with the projection $p\colon E \to X$ as defined in Definition 6.17 is a G-bundle (see Definition 6.21).

Problem 6.22. Let $\mathrm{L}^2(G; E_0)$ be the Hilbert space defined in Definition 6.24. Let L^σ be the subspace of $\mathrm{L}^2(G; E_0)$ given by

$$L^\sigma = \{f \in \mathrm{L}^2(G; E_0) \mid f(gh) = \sigma(h^{-1})(f(g)), \text{for all } g \in G \text{ and all } h \in H\}.$$

Show that L^σ is closed in $\mathrm{L}^2(G; E_0)$.

Problem 6.23. Let G be a locally compact group, H be a closed subgroup of G, E be some topological Hausdorff space, and $p\colon E \to X$ be a surjective continuous map, where as usual, we write $X = G/H$. Furthermore, assume that E (really $p\colon E \to X$) is a hermitian G-bundle of finite rank as defined in Definition 6.23. This means we have a unitary representation $\sigma\colon H \to \mathbf{U}(E_0)$ given by

$$\sigma(h)(z) = h \cdot z.$$

If $A(h)$ is the unitary matrix representing $\sigma(h)$ with respect to the basis $(e_1, \ldots, e_n)$, with

$$\sigma(h)(e_j) = \sum_{i=1}^{n} a_{ij} e_i,$$

prove that the condition $f(gh) = \sigma(h^{-1})(f(g))$ translates into

$$\begin{pmatrix} f_1(gh) \\ \vdots \\ f_n(gh) \end{pmatrix} = A(h)^* \begin{pmatrix} f_1(g) \\ \vdots \\ f_n(g) \end{pmatrix}.$$

Chapter 7

Constructing Induced Representations a la Mackey

One of the most important contributions to the theory of unitary representations is a method due to Mackey for constructing all irreducible representations of a locally compact group as induced irreducible representations from "small" subgroups H_ν of G. This method is often referred to as the "Mackey machine." In its most general form, the method is very complicated but in the case where G has an abelian normal subgroup N, it is tractable. The basic reason is that because N is abelian its irreducible representations are given by the characters of N. There is also a natural action $\cdot\colon G \times \widehat{N} \to \widehat{N}$ of G on the dual group $\widehat{N}$ (the group of characters of N). The key to the construction is that because N is an abelian locally compact group, by Theorem 3.20, for any unitary representation $U\colon G \to \mathbf{U}(\mathcal{H}_U)$ of G, since the restriction of U to N is a unitary representation, there is a unique regular projection-valued measure P on the dual group $\widehat{N}$, such that

$$U(n) = \int_{\widehat{N}} \chi(n)\, dP(\chi), \quad n \in N.$$

Moreover, the projection-valued measure P on $\widehat{N}$ satisfies two properties (see Proposition 7.1):

(1) We have

$$U(s)P(E)U(s)^{-1} = P(s \cdot E), \qquad \text{for all Borel subsets } E \subseteq \widehat{N} \text{ and all } s \in G. \tag{imp}$$

(2) If U is irreducible, then for every G-invariant Borel set $E \subseteq \widehat{N}$ (which means that $\{s\cdot\chi \mid \chi \in E\} = s\cdot E = E$ for every $s \in G$), either $P(E) = I$ or $P(E) = 0$.

If the action of G on $\widehat{N}$ is nice enough (the space of orbits of this action is countably separated, see Definition 7.2), then P is identically zero except on a single orbit $\mathcal{O}_\nu$, so we can consider P as living on G/G_ν (where G_ν is the stabilizer of ν), and G acts transitively on this space. Then the data (G, U, X, P) consisting of the unitary representation $U\colon G \to \mathbf{U}(\mathcal{H}_U)$, of a transitive action of G on the homogeneous space $X = G/G_\nu$ (for some fixed $\nu \in \widehat{N}$), and of a regular projection-valued measure P on G/G_ν, such that

$$U(s)P(E)U(s)^{-1} = P(s \cdot E), \quad \text{for all Borel sets } E \subseteq G/G_\nu \text{ and all } s \in G,$$

constitute a *transitive system of imprimitivity* (see Definition 7.3). Technically, there are two equivalent ways of defining systems of imprimitivity but in this introduction we can ignore the second definition. The relevance of systems of imprimitivity is Mackey's *imprimitivity theorem* (Theorem 7.3), which implies that U is equivalent to a representation obtained by the induction method from some irreducible representation of G_ν. Technically, Mackey's *imprimitivity theorem* says more, namely that any transitive system of imprimitivity is equivalent to a system of imprimitivity arising by induction from the subgroup defining the homogeneous space X. If the action of G on $\widehat{N}$ is regular (see Definition 7.6), then Mackey's imprimitivity theorem implies Theorem 7.4, which is the result that shows that for every irreducible representation $U\colon G \to \mathbf{U}(\mathcal{H}_U)$ of G, there is a unique orbit $\mathcal{O}$, such that for any $\nu \in \mathcal{O}$ (so that $\mathcal{O} = \mathcal{O}_\nu$), there is an irreducible unitary representation $\sigma\colon G_\nu \to \mathbf{U}(\mathcal{H}_\sigma)$, such that U is equivalent to $\mathrm{Ind}_{G_\nu}^G \sigma$, the induced representation obtained from σ.

Unfortunately, the subgroups G_ν may still not be small enough. However, if for some $\nu \in \widehat{N}$ there is a continuous homomorphism $\widetilde{\nu}\colon G_\nu \to \mathbf{U}(1)$ extending ν, then for every irreducible representation $\rho\colon G_\nu/N \to \mathbf{U}(\mathcal{H}_\rho)$ of G_ν/N, the map $\sigma\colon G_\nu \to \mathbf{U}(\mathcal{H}_\rho)$ given by

$$\sigma(s) = \widetilde{\nu}(s)\rho(sN), \quad s \in G_\nu,$$

defines an irreducible representation of G_ν in $\mathcal{H}_\rho$ (see Proposition 7.6). In this case, we can use irreducible representations of the "little groups" $H_\nu = G_\nu/N$ in the inducing process of Theorem 7.4.

The above extension condition is satisfied by semi-direct products $G = N \rtimes H$, where N is a normal abelian subgroup of G. Then every irreducible representation of G is obtained in terms of the characters of N and of the irreducible representations of the little groups H_ν associated with the characters $\nu \in \widehat{N}$; see Theorem 7.7. Using this method, we describe all irreducible representations of $\mathbf{SE}(n)$; see Example 7.1. We also determine all irreducible representations of $\mathbf{O}(2)$ (see Example 7.2) and indicate how all irreducible representations of $\mathbf{E}(2)$ and $\mathbf{E}(3)$ can be obtained.

Historically, the little group method was first used by Wigner in a famous paper (1939) on the representations of the *Poincaré group* $\mathbb{R}^4 \rtimes \mathbf{SO}_0(3,1)$, where $\mathbf{SO}_0(3,1)$ is the so-called *restricted Lorentz group.*

A thorough exposition of Mackey's method is given in Folland [Folland (1995)] (Chapter 6). A concise but very clear description of Mackey's method is also provided in Warner [Warner (1972a)] (Chapter 5, Section 5.4). The reader interested in the history and the applications to physics (in particular, quantum mechanics) of harmonic analysis should consult Mackey [Mackey (1992)].

7.1 Introduction to the Mackey Machine

The reader may want to review the notion of projection-valued measure discussed in Section 2.12. Let G be a locally compact group and N be a nontrivial closed *abelian normal* subgroup of G. The group G acts by conjugation on the normal subgroup N, namely for every $s \in G$, we let C_s be the automorphism of N given by $C_s(t) = sts^{-1}$ for all $t \in N$. Then the map $s \mapsto C_s$ is a homomorphism from G to $\mathrm{Aut}(N)$. Since G is a locally compact group and N is a closed abelian subgroup of G, the dual group $\widehat{N}$, namely the group of characters $\chi\colon N \to \mathbb{C}$ of N, is well-defined. But then we can define an action of G on $\widehat{N}$ as follows.

Definition 7.1. With G, N and $\widehat{N}$ as above, we define an action $\cdot\colon G\times\widehat{N} \to \widehat{N}$, such that for all $s \in G$, $n \in N$, and $\chi \in \widehat{N}$,

$$(s \cdot \chi)(n) = \chi(s^{-1}ns). \tag{act}$$

To simplify notation, we often denote $s \cdot \chi$ as $s\chi$.

Note that

$$\begin{aligned}((st) \cdot \chi)(n) &= \chi((st)^{-1}nst) = \chi(t^{-1}s^{-1}nst) \\ &= (t \cdot \chi)(s^{-1}ns) = (s \cdot (t \cdot \chi))(n),\end{aligned}$$

and obviously

$$(e \cdot \chi)(n) = \chi(e^{-1}ne) = \chi(n),$$

so $\cdot \colon G \times \widehat{N} \to \widehat{N}$ is indeed an action of G on $\widehat{N}$. Then, as usual, for every $\chi \in \widehat{N}$, we define the stabilizer G_χ of χ and the orbit $\mathcal{O}_\chi \subseteq \widehat{N}$ of χ as

$$\begin{aligned} G_\chi &= \{s \in G \mid s \cdot \chi = \chi\} \\ \mathcal{O}_\chi &= \{s \cdot \chi \mid s \in G\}. \end{aligned}$$

The subgroup G_χ is closed in G. Since N is abelian, we have $N \subseteq G_\chi$, and recall that there is a bijection between $\mathcal{O}_\chi$ and G/G_χ. The action $\cdot \colon G \times \widehat{N} \to \widehat{N}$ is never transitive (for instance, $\mathcal{O}_1 = \{1\}$) and the orbits can be complicated. What is remarkable is the fact that under certain conditions on the action of G on $\widehat{N}$, an irreducible unitary representation U of G arises from some irreducible representation ρ of G_ν for some $\nu \in \widehat{N}$ as an induced representation from G_ν to G.

The key to the construction is that because N is an abelian locally compact group, by Theorem 3.20, for any unitary representation $U \colon G \to \mathbf{U}(\mathcal{H}_U)$ of G, since the restriction of U to N is a unitary representation, there is a unique regular projection-valued measure P on the dual group $\widehat{N}$, such that

$$U(n) = \int_{\widehat{N}} \chi(n)\, dP(\chi), \quad n \in N.$$

The crucial step is to figure out for every fixed $s \in G$ what are the projection-valued measures associated with the representations

$$n \mapsto U(s)U(n)U(s^{-1})$$

and

$$n \mapsto U(sns^{-1}).$$

Since U is a representation, $U(s)U(n)U(s^{-1}) = U(sns^{-1})$, so we obtain a condition on P.

Proposition 7.1. *Let G be a locally compact group and N be a closed abelian normal subgroup of G. For any unitary representation $U \colon G \to \mathbf{U}(\mathcal{H}_U)$ of G, let P be the unique regular projection-valued measure on $\widehat{N}$, such that for the restriction $U \colon N \to \mathbf{U}(\mathcal{H}_U)$ of U to N, we have*

$$U(n) = \int_{\widehat{N}} \chi(n)\, dP(\chi), \quad n \in N.$$

The following properties hold.

(1) The projection-valued measure P on $\widehat{N}$ satisfies the equation

$$U(s)P(E)U(s)^{-1} = P(s \cdot E), \quad \text{for all Borel subsets } E \subseteq \widehat{N} \text{ and all } s \in G. \qquad \text{(imp)}$$

(2) *If U is irreducible, then for every G-invariant Borel set $E \subseteq \widehat{N}$ (which means that $\{s \cdot \chi \mid \chi \in E\} = s \cdot E = E$ for every $s \in G$), either $P(E) = I$ or $P(E) = 0$. We say that P is ergodic.*

Proof. Since $U(s)$ is a unitary map for every $s \in G$ and since each $P(E)$ is a self-adjoint idempotent linear map, it is immediately verified that $U(s)P(E)U(s)^{-1} = U(s)P(E)U(s)^*$ is also a self-adjoint idempotent linear map. It is not hard to check that for any fixed $s \in G$, the map Q defined on the Borel subsets of $\widehat{N}$ by

$$Q(E) = U(s)P(E)U(s)^{-1}$$

is a regular projection-valued measure (see Definition 2.22). For all $u, v \in \mathcal{H}_U$, for all $n \in N$, since $U(s)$ is unitary, we have

$$\langle U(s)U(n)U(s)^{-1}(u), v\rangle = \langle U(n)U(s)^{-1}(u), U(s)^{-1}(v)\rangle,$$

so by definition,

$$\langle U(n)U(s)^{-1}(u), U(s)^{-1}(v)\rangle = \int_{\widehat{N}} \chi(n)\, dP_{U(s)^{-1}(u),U(s)^{-1}(v)}(\chi).$$

But by definition, for any Borel set E in $\widehat{N}$,

$$\begin{aligned} P_{U(s)^{-1}(u),U(s)^{-1}(v)}(E) &= \langle P(E)U(s)^{-1}(u), U(s)^{-1}(v)\rangle \\ &= \langle U(s)P(E)U(s)^{-1}(u), v\rangle \\ &= \langle Q(E)(u), v\rangle = Q_{u,v}(E), \end{aligned}$$

so we deduce that

$$\langle U(s)U(n)U(s)^{-1}(u), v\rangle = \int_{\widehat{N}} \chi(n)\, dQ_{u,v}(\chi),$$

and thus

$$U(s)U(n)U(s)^{-1} = \int_{\widehat{N}} \chi(n)\, dQ(\chi). \tag{$*_1$}$$

We also have

$$\langle U(sns^{-1})(u), v\rangle = \int_{\widehat{N}} \chi(sns^{-1})\, dP_{u,v}(\chi).$$

Since by (act),

$$(s \cdot \chi)(n) = \chi(s^{-1}ns),$$

we obtain

$$\langle U(sns^{-1})(u), v\rangle = \int_{\widehat{N}} (s^{-1} \cdot \chi)(n)\, dP_{u,v}(\chi).$$

We now need to go back to Vol. I, Section 8.10. We have an action $\cdot\colon G \times \widehat{N} \to \widehat{N}$ and a σ-Radon measure μ on $\widehat{N}$ (which is locally compact). Recall from Vol. I, Definition 8.18 that for any $s \in G$ and any Borel subset E of $\widehat{N}$, we define $s \cdot E$ as

$$s \cdot E = \{s \cdot \chi \mid \chi \in E\},$$

for any function $f\colon \widehat{N} \to \mathbb{C}$, the function $\lambda_s(f)$ by

$$(\lambda_s(f))(\chi) = f(s^{-1} \cdot \chi),$$

and the measure $\lambda_s(\mu)$ by

$$(\lambda_s(\mu))(E) = \mu(s^{-1} \cdot E).$$

The proof of Vol. I, Proposition 8.16 is immediately adapted to show that for any $f \in \mathrm{L}^1(\widehat{N})$, we have

$$\int_{\widehat{N}} \lambda_s(f)\, d\mu = \int_{\widehat{N}} f\, d\lambda_{s^{-1}}(\mu),$$

which can also be written as

$$\int_{\widehat{N}} f(s^{-1} \cdot \chi)\, d\mu(\chi) = \int_{\widehat{N}} f(\chi)\, d(\lambda_{s^{-1}}(\mu))(\chi).$$

If we apply the above equation to the function f given by

$$f(\chi) = \chi(n)$$

for some fixed $n \in N$ and to the positive measure $P_{u,u}$, we obtain

$$\int_{\widehat{N}} (s^{-1} \cdot \chi)(n)\, dP_{u,u}(\chi) = \int_{\widehat{N}} \chi(n)\, d(\lambda_{s^{-1}}(P_{u,u}))(\chi).$$

Using the polarization method of Section 2.12, since

$$\begin{aligned}\int f\, dP_{u,v} = \frac{1}{4}\biggl(\int f\, dP_{u+v,u+v} &- \int f\, dP_{u-v,u-v}\\ &+ i\left(\int f\, dP_{u+iv,u+iv} - \int f\, dP_{u-iv,u-iv}\right)\biggr),\end{aligned}$$

we obtain

$$\begin{aligned}\langle U(sns^{-1})(u), v\rangle &= \int_{\widehat{N}} (s^{-1} \cdot \chi)(n)\, dP_{u,v}(\chi)\\ &= \int_{\widehat{N}} \chi(n)\, d(\lambda_{s^{-1}}(P_{u,v}))(\chi),\end{aligned}$$

where $(\lambda_{s^{-1}}(P_{u,v}))(E) = P_{u,v}(s \cdot E)$, so

$$\langle U(sns^{-1})(u), v\rangle = \int_{\widehat{N}} \chi(n)\, d(\lambda_{s^{-1}}(P_{u,v}))(\chi). \tag{$*_2$}$$

Now by definition, for any Borel subset E of $\widehat{N}$,

$$P_{u,v}(s \cdot E) = \langle P(s \cdot E)(u), v\rangle,$$

so

$$(\lambda_{s^{-1}}(P_{u,v}))(E) = P_{u,v}(s \cdot E) = \langle P(s \cdot E)(u), v\rangle.$$

We can check quickly that the map $E \mapsto P(s \cdot E)$ is a regular projection-valued measure, since the map $E \mapsto s \cdot E$ is a bijection on Borel sets, such that $s \cdot \widehat{N} = \widehat{N}$ and $s \cdot \emptyset = \emptyset$. Consequently, the map $\lambda_{s^{-1}}(P)$ given by $(\lambda_{s^{-1}}(P))(E) = P(s \cdot E)$ is a projection-valued measure, and by $(*_2)$, we have

$$U(sns^{-1}) = \int_{\widehat{N}} \chi(n)\, d(\lambda_{s^{-1}}(P))(\chi). \tag{$*_3$}$$

Since U is a representation, $U(s)U(n)U(s)^{-1} = U(sns^{-1})$, so by $(*_1)$ and $(*_3)$, we obtain

$$U(sns^{-1}) = \int_{\widehat{N}} \chi(n)\, dQ(\chi) = \int_{\widehat{N}} \chi(n)\, d(\lambda_{s^{-1}}(P))(\chi). \tag{$*_4$}$$

By uniqueness of the projection-valued measure defining a unitary representation, we conclude that

$$Q = \lambda_{s^{-1}}(P),$$

which more explicitly means that

$$U(s)P(E)U(s)^{-1} = P(s \cdot E), \quad \text{for all Borel sets } E \subseteq \widehat{N}.$$

If U is irreducible and if the Borel set E is G-invariant, that is, $s \cdot E = E$, then

$$U(s)P(E)U(s)^{-1} = P(E), \quad \text{for all } s \in G,$$

so $U(s)P(E) = P(E)U(s)$ for all $s \in G$, which means that $P(E) \in \mathcal{C}(U)$, where $\mathcal{C}(U)$ is the commutant of U (see Definition 3.9). By Schur's Lemma, $P(E)$ is a scalar multiple of the identity, and since it is a projection, either $P(E) = I$ or $P(E) = 0$. $\square$

In summary, for any unitary representation $U \colon G \to \mathbf{U}(\mathcal{H}_U)$ of G, there is some regular projection-valued measure P on $\widehat{N}$, such that

$$U(s)P(E)U(s)^{-1} = P(s \cdot E), \quad \text{for all Borel sets } E \subseteq \widehat{N} \text{ and all } s \in G,$$

and there is an action of the group G on $\widehat{N}$. These are just the ingredients that constitute Mackey's systems of imprimitivity! However, before defining systems of imprimitivity, we note that if the representation U is irreducible, it would be nice if P was identically zero except on a single orbit $\mathcal{O}_\nu$ (for some $\nu \in \widehat{N}$) because then we could consider P as living on G/G_ν, and G acts transitively on this space. Furthermore, in this case, Mackey's *imprimitivity theorem* applies, which implies that U is equivalent to a representation obtained by the induction method from some irreducible representation of G_ν. This is the essence of the Mackey machine for constructing induced representations.

Definition 7.2. Let G be a locally compact group and N be a closed normal abelian subgroup of G. Consider the action of G on $\widehat{N}$ as in Definition 7.1. The space of orbits of this action is *countably separated* if there is a countable family (E_j) of G-invariant Borel subsets of $\widehat{N}$, such that for each orbit $\mathcal{O}$, we have

$$\mathcal{O} = \bigcap \{E_j \mid \mathcal{O} \subseteq E_j\};$$

in other words, each orbit is the intersection of the E_j that contain it.

Proposition 7.2. *If U is an irreducible unitary representation $U\colon G \to \mathbf{U}(\mathcal{H}_U)$ and if the space of orbits of the action of Definition 7.1 is countably separated, then there is a single orbit $\mathcal{O} = \mathcal{O}_\nu$ in $\widehat{N}$, such that $P(\mathcal{O}_\nu) = I$.*

Proof. Let (E_j) be a countable family of G-invariant Borel subsets of $\widehat{N}$ with the property of Definition 7.2, so that for every orbit $\mathcal{O}$, there is some countable index set J, such that

$$\mathcal{O} = \bigcap_{j \in J} E_j.$$

It follows that $P(\mathcal{O})$ is the projection onto the intersection of the ranges of the $P(E_j)$, with $j \in J$. Since U is irreducible, by Proposition 7.1, either $P(E_j) = I$ or $P(E_j) = 0$. Consequently, if $P(E_j) = I$ for all $j \in J$, then $P(\mathcal{O}) = I$, or else $P(\mathcal{O}) = 0$ if $P(E_j) = 0$ for some $j \in J$. We claim that there is some orbit $\mathcal{O}$, such that $P(\mathcal{O}) = I$. Otherwise, for every orbit $\mathcal{O}$ there is some index $j_{\mathcal{O}}$, such that $\mathcal{O} \subseteq E_{j_{\mathcal{O}}}$ and $P(E_{j_{\mathcal{O}}}) = 0$. Since $\widehat{N}$ is the union of the orbits, by property (4) of Definition 2.22, we obtain $P(\widehat{N}) = 0$, which is absurd since $P(\widehat{N}) = 1$. Finally, suppose that there are two disjoint orbits $\mathcal{O}_1$ and $\mathcal{O}_2$, such that $P(\mathcal{O}_1) = P(\mathcal{O}_2) = I$. But then by property (3) of Definition 2.22,

$$I = P(\mathcal{O}_1) \circ P(\mathcal{O}_2) = P(\mathcal{O}_1 \cap \mathcal{O}_2) = P(\emptyset) = 0,$$

a contradiction. $\square$

If the action of G on $\widehat{N}$ is nice enough so that the space of orbits of this action is countably separated and if G/G_χ is homeomorphic to $\mathcal{O}_\chi$ for all $\chi \in \widehat{N}$, then the data consisting of the unitary representation $U: G \to \mathbf{U}(H)$, of a transitive action of G on the homogeneous space $X = G/G_\nu$ (for some fixed $\nu \in \widehat{N}$), and of a regular projection-valued measure P on G/G_ν, such that

$$U(s)P(E)U(s)^{-1} = P(s \cdot E), \text{ for all Borel sets } E \subseteq G/G_\nu \text{ and all } s \in G,$$

constitute a *transitive system of imprimitivity*. Mackey's *imprimitivity theorem* applies to such a system, and this theorem is the key to defining irreducible representations obtained by the induced representation method. We now define (transitive) systems of imprimitivity and state Mackey's famous imprimitivity theorem.

7.2 Systems of Imprimitivity and the Imprimitivity Theorem

There are two equivalent ways of defining systems of imprimitivity. The first definition makes explicit use of a projection-valued measure and the second one uses a representation of the algebra $\mathcal{C}_0(S;\mathbb{C})$. The second definition is often technically easier to work with.

Definition 7.3. A *system of imprimitivity, version 1,* is a quadruple $\Sigma = (G, U, X, P)$, where

(1) G is a locally compact group.
(2) $U: G \to \mathbf{U}(\mathcal{H}_U)$ is a unitary representation of G in a Hilbert space $\mathcal{H}_U$.
(3) X is a G-space, which means that X is a locally compact Hausdorff space and there is a continuous action $\cdot: G \times X \to X$.
(4) P is projection-valued measure on X with values in $\mathcal{L}(\mathcal{H}_U)$ satisfying the equation

$$U(s)P(E)U(s)^{-1} = P(s \cdot E), \qquad \text{for all Borel subsets } E \subseteq X \text{ and all } s \in G. \tag{impl}$$

The system of imprimitivity $\Sigma = (G, U, X, P)$ is *transitive* if X is a homogeneous G-space. This means that $X = G/H$ for some closed subgroup H of G (with the action $g \cdot (sH) = (gs)H$, for all $g, s \in G$).

The projection-valued measure P on X determines a non-degenerate representation

$V\colon \mathcal{C}_0(X;\mathbb{C}) \to \mathcal{L}(\mathcal{H}_U)$ of the algebra $\mathcal{C}_0(X;\mathbb{C})$ defined by

$$V(f) = \int_X f\, dP, \quad f \in \mathcal{C}_0(X;\mathbb{C}).$$

As in the proof of Proposition 7.1, for all $u, v \in \mathcal{H}_U$ and all $f \in \mathcal{C}_0(X;\mathbb{C})$, since $U(s)$ is unitary, we have

$$\langle U(s)V(f)U(s)^{-1}(u), v\rangle = \langle V(f)U(s)^{-1}(u), U(s)^{-1}(v)\rangle,$$

so by definition

$$\langle V(f)U(s)^{-1}(u), U(s)^{-1}(v)\rangle = \int_X f\, dP_{U(s)^{-1}(u), U(s)^{-1}(v)}.$$

But by definition, for any Borel set E in X,

$$\begin{aligned} P_{U(s)^{-1}(u),U(s)^{-1}(v)}(E) &= \langle P(E)U(s)^{-1}(u), U(s)^{-1}(v)\rangle \\ &= \langle U(s)P(E)U(s)^{-1}(u), v\rangle \\ &= \langle P(s\cdot E)(u), v\rangle \qquad \text{by (imp1)} \\ &= P_{u,v}(s\cdot E) = \lambda_{s^{-1}}(P_{u,v})(E). \end{aligned}$$

Consequently,

$$\langle U(s)V(f)U(s)^{-1}(u), v\rangle = \int_X f\, d\lambda_{s^{-1}}(P_{u,v}) = \int_X \lambda_s(f)\, dP_{u,v}.$$

The above equation says that

$$U(s)V(f)U(s)^{-1} = \int_X \lambda_s(f)\, dP,$$

which, by definition of V, means that

$$U(s)V(f)U(s)^{-1} = V(\lambda_s(f)), \quad f \in \mathcal{C}_0(X;\mathbb{C}),\ s \in G.$$

Conversely, if we have a nondegenerate representation $V\colon \mathcal{C}_0(X;\mathbb{C}) \to \mathcal{L}(\mathcal{H}_U)$ satisfying the above equation, then by Theorem 2.61, there is a projection-valued measure P on X, such that

$$V(f) = \int_X f\, dP, \quad f \in \mathcal{C}_0(X;\mathbb{C}).$$

Since the equation

$$U(s)V(f)U(s)^{-1} = V(\lambda_s(f)), \quad f \in \mathcal{C}_0(X;\mathbb{C}),\ s \in G$$

holds, a reasoning similar to the one used in the proof of Proposition 7.1 shows that Equation (imp1) holds. We are led to the following definition, which, by the above reasoning, is equivalent to Definition 7.3.

Definition 7.4. A *system of imprimitivity, version 2,* is a quadruple $\Sigma = (G, U, X, V)$, where

(1) G is a locally compact group.
(2) $U\colon G \to \mathbf{U}(\mathcal{H}_U)$ is a unitary representation of G in a Hilbert space $\mathcal{H}_U$.
(3) X is a G-space, which means that X is a locally compact Hausdorff space and there is a continuous action $\cdot\colon G \times X \to X$.
(4) V is a nondegenerate representation $V\colon \mathcal{C}_0(X;\mathbb{C}) \to \mathcal{L}(\mathcal{H}_U)$ satisfying the equation

$$U(s)V(f)U(s)^{-1} = V(\lambda_s(f)), \quad f \in \mathcal{C}_0(X;\mathbb{C}),\ s \in G. \qquad \text{(imp2)}$$

As before, the system of imprimitivity $\Sigma = (G, U, X, V)$ is *transitive* if X is a homogeneous G-space, $X = G/H$, for some closed subgroup H of G.

One of the main sources of systems of imprimitivity is from induced representations. In fact, we obtain transitive systems of imprimitivity.

Technically, it is better to use Blattner's method for constructing an induced unitary representation $\Pi'\colon G \to \mathbf{U}(\mathcal{H}')$ of G from a unitary representation U of H, where H is a closed subgroup of G, as described in Section 6.11, because the definition of Π' in Formula (indv2) is simpler than Formula (indv1).

Since we denote the subgroup of G by H and since we use E to denote Borel sets, to avoid a notational clash we denote the Hilbert space involved in the unitary representation U of H by $\mathcal{H}_U$, so that our representation of H is written $U\colon H \to \mathbf{U}(\mathcal{H}_U)$. Also, since we are using Blattner's construction instead of the first method from Section 6.11, we will drop the prime superscript and write Π instead of Π' and $\mathcal{H}$ instead of $\mathcal{H}'$. The Hilbert space $\mathcal{H}_\Pi$ associated with the induced unitary representation $\Pi\colon G \to \mathbf{U}(\mathcal{H}_\Pi)$ of the representation $U\colon H \to \mathbf{U}(\mathcal{H}_U)$, usually denoted $\mathcal{H}$ unless confusion arises, is the completion of a space $\mathcal{H}^0$ defined as

$$\mathcal{H}^0 = \Bigg\{ f \in \mathcal{C}(G, \mathcal{H}_U) \mid \pi(\mathrm{supp}(f)) \text{ is compact and}$$
$$f(sh) = \left(\frac{\Delta_H(h)}{\Delta_G(h)}\right)^{1/2} U(h^{-1})(f(s)) \text{ for all } s \in G \text{ and all } h \in H \Bigg\}.$$

Here, $\pi\colon G \to G/H$ denotes the quotient map.

Given a unitary representation $U\colon H \to \mathbf{U}(\mathcal{H}_U)$, the induced unitary representation $\Pi\colon G \to \mathbf{U}(\mathcal{H})$, also denoted $\mathrm{Ind}_{H,\mathcal{H}}^G U$ or even $\mathrm{Ind}_H^G U$, is given by

$$(\Pi_s(f))(t) = f(s^{-1}t), \qquad f \in \mathcal{H},\ s, t \in G.$$

A natural candidate for a projection-valued measure P^U on $X = G/H$ is to set

$$P^U(E)(f) = (\chi_E \circ \pi)(f), \quad E \subseteq G/H,\ f \in \mathcal{H}.$$

Here, $P^U(E) \in \mathcal{L}(\mathcal{H})$ and $(\chi_E \circ \pi)(f)$ is the pointwise-multiplication of the functions $\chi_E \circ \pi$ and f, both defined on G. However, it is not obvious that this definition makes sense and that condition (imp1) is satisfied, so we circumvent these difficulties by using the definition of a system of imprimitivity given by Definition 7.4. We need to define a representation $V\colon \mathcal{C}_0(G/H,\mathbb{C}) \to \mathcal{L}(\mathcal{H})$ satisfying condition (imp2).

If we take a close look at the definition of $\mathcal{H}^0$ in Section 6.11, we can check that for any $\varphi \in \mathcal{C}_0(G/H;\mathbb{C})$ and any $f \in \mathcal{H}^0$, since $f\colon G \to \mathcal{H}_U$ and $\varphi\circ\pi\colon G \to \mathbb{C}$, the function $(\varphi\circ\pi)f$ from G to $\mathcal{H}_U$ given by

$$((\varphi\circ\pi)f)(s) = (\varphi\circ\pi)(s)f(s)$$

belongs to $\mathcal{H}^0$ and that

$$\|(\varphi\circ\pi)f\|_{\mathcal{H}} \leq \|\varphi\|_\infty \|f\|_{\mathcal{H}}.$$

As a consequence, since $\mathcal{H}^0$ is dense in $\mathcal{H}$, if we set

$$V(\varphi)(f) = (\varphi\circ\pi)f, \quad f \in \mathcal{H},$$

we obtain a representation $V\colon \mathcal{C}_0(G/H,\mathbb{C}) \to \mathcal{L}(\mathcal{H})$, and it is easy to see that V is nondegenerate. It remains to that prove that (indv2) hold (with respect to Π). For all $f \in \mathcal{H}^0$ and all $s,t \in G$ (recall that functions in $\mathcal{H}^0$ have domain G), we have

$$\begin{aligned}
((\Pi_s V(\varphi)\Pi_s^{-1})(f))(t) &= \Pi_s(V(\varphi)(\Pi_{s^{-1}}(f(t)))) \\
&= V(\varphi)(\Pi_{s^{-1}}(f(s^{-1}t))) && \text{by definition of } \Pi \\
&= \varphi(\pi(s^{-1}t))\Pi_{s^{-1}}(f(s^{-1}t)) && \text{by definition of } V \\
&= \varphi(\pi(s^{-1}t))f(t) && \text{by definition of } \Pi \\
&= \varphi(s^{-1}\cdot\pi(t))f(t) && \text{by the action on } G/H \\
&= \lambda_s(\varphi)(\pi(t))f(t) \\
&= (V(\lambda_s(\varphi))(f))(t), && \text{by definition of } V
\end{aligned}$$

proving that

$$(\Pi_s V(\varphi)\Pi_s^{-1})(f) = V(\lambda_s(\varphi))(f),$$

for all $f \in \mathcal{H}^0$. Since $\mathcal{H}^0$ is dense in $\mathcal{H}$, we deduce that (ind2) holds, and so $(G,\Pi,G/H,V)$ is a transitive system of imprimitivity, version 2.

By Theorem 2.61, there is a unique projection-valued measure P^U on X, such that

$$V(f) = \int_X f\, dP^U, \quad f \in \mathcal{C}_0(X; \mathbb{C}),$$

and $(G, \mathrm{Ind}_H^G U, G/H, P^U)$ is called the *canonical system of imprimitivity* associated to $\Pi = \mathrm{Ind}_H^G U$. It can be verified that

$$P^U(E)(f) = (\chi_E \circ \pi)(f), \quad E \subseteq G/H,\, f \in \mathcal{H},$$

as we said earlier, but using the representation V, we verified that such a definition is legitimate.

Remark: The projection-valued measure arising from V is denoted by P^U. We prefer the notation P^U to the notation P^V even though P^U arises from V, because V is a representation in a Hilbert space $\mathcal{H}$ obtained as the completion of a space $\mathcal{H}^0$, which consists of certain functions from G to $\mathcal{H}_U$, where $\mathcal{H}_U$ is the Hilbert space of the representation U.

The *raison d'être* for all this is that every transitive system of imprimitivity is equivalent to the canonical system of imprimitivity arising from some induced representation. This theorem originally due to Mackey is one of the greatest results in the theory of unitary representations. First, we define the notion of equivalence of systems of imprimivity.

Definition 7.5. Two systems of imprimitivity $\Sigma = (G, U, X, P)$ and $\Sigma' = (G, U', X, P')$ (with the same group G and the same space X), where $U\colon G \to \mathbf{U}(\mathcal{H}_U)$ and $U'\colon G \to \mathbf{U}(\mathcal{H}_{U'})$ are two unitary representations, are *equivalent* if there is a unitary map $T\colon \mathcal{H}_U \to \mathcal{H}_{U'}$, such that

$$\begin{aligned} TU(s)T^{-1} &= U'(s), \quad \text{for all} \in G \\ TP(E)T^{-1} &= P'(E), \quad \text{for all Borel sets } E \subseteq X. \end{aligned}$$

We now state the celebrated imprimitivity theorem.

Theorem 7.3. *(Mackey's Imprimitivity Theorem, 1949–1953) Let G be a locally compact group and H be a closed subgroup of G. Every transitive system of imprimitivity $\Sigma = (G, U, G/H, P)$, where $U\colon G \to \mathbf{U}(\mathcal{H}_U)$ is a unitary representation of G, is equivalent to a transitive system of imprimitivity of the form $(G, \Pi, G/H, P^\sigma)$, where $\Pi = \mathrm{Ind}_H^G \sigma\colon G \to \mathbf{U}(\mathcal{H})$ is the representation induced by some unitary representation $\sigma\colon H \to \mathbf{U}(\mathcal{H}_\sigma)$ of H. Thus, there is a unitary map $T\colon \mathcal{H}_U \to \mathcal{H}$, such that*

$$\begin{aligned} TU(s)T^{-1} &= (\mathrm{Ind}_H^G \sigma)(s), \quad \textit{for all } s \in G \\ TP(E)T^{-1} &= P^\sigma(E), \quad \textit{for all Borel sets } E \subseteq G/X. \end{aligned}$$

Moreover, the unitary representation $\sigma\colon H \to \mathbf{U}(\mathcal{H}_\sigma)$ *is determined by* Σ *up to equivalence.*

The proof of Theorem 7.3 is long and very technical. The version of the proof given in Folland [Folland (1995)] requires two sections (Sections 6.4 and 6.5) and stretches from page 167 to page 182. A key idea due to Blatter is to use an algebra $L(X \times G)$ and to extend a unitary representation of G to this algebra, by analogy with the method of extending a representation of G to a representation of $\mathrm{L}^1(G)$. If G is a Lie group, the proof is significantly simpler. A version of the imprimitivity theorem for Lie groups called the *Mackey Inducibility Criterion* by Kirillov is proven in Kirillov [Kirillov (2004)]; see Appendix V, Section 2.4. A sketch of proof for Lie groups is also given by Taylor [Taylor (1986)]; see Chapter V, Section 1. The good news is that we now have all the machinery needed to tackle the problem introduced in Section 7.1.

7.3 The Mackey Machine

Let G be a locally compact group and N be a nontrivial closed *abelian normal* subgroup of G. As introduced in Definition 7.1, there is an action $\cdot\colon G \times \widehat{N} \to \widehat{N}$, such that for all $s \in G$, $n \in N$, and $\chi \in \widehat{N}$,

$$(s \cdot \chi)(n) = \chi(s^{-1}ns). \tag{act}$$

Recall that for every $\chi \in \widehat{N}$, we define the stabilizer G_χ of χ and the orbit $\mathcal{O}_\chi$ of χ as

$$\begin{aligned} G_\chi &= \{s \in G \mid s \cdot \chi = \chi\} \\ \mathcal{O}_\chi &= \{s \cdot \chi \mid s \in G\}. \end{aligned}$$

Our goal is to show that if the action of G on $\widehat{N}$ is nice enough, then every irreducible representation $U\colon G \to \mathbf{U}(\mathcal{H}_U)$ of G arises as an induced representation of some irreducible representation σ of G_ν for some $\nu \in \widehat{N}$. The notion of nice action is formalized as follows.

Definition 7.6. We say the action $\cdot\colon G \times \widehat{N} \to \widehat{N}$ is *regular*, or that G *acts regularly on* $\widehat{N}$, if the following two conditions hold:

(1) The orbit space of our action is countably separated, as in Definition 7.2. Recall that this means that there is a countable family (E_j) of G-invariant Borel subsets of $\widehat{N}$, such that for each orbit $\mathcal{O}$, we have

$$\mathcal{O} = \bigcap\{E_j \mid \mathcal{O} \subseteq E_j\},$$

(2) For every $\nu \in \widehat{N}$, the map from G/G_ν to $\mathcal{O}_\nu$ given by $gG_\nu \mapsto g \cdot \nu$ is a homeomorphism.

Remark: When condition (1) of Definition 7.6 holds, Kirillov called the orbit space *tame*; see Kirillov [Kirillov (2004)], Appendix V, Section 2.4.

Now, given an irreducible representation $U\colon G \to \mathbf{U}(\mathcal{H}_U)$ of G, the "miracle" is that $(G, U, G/G_\nu, P)$ is a transitive system of imprimitivity for some $\nu \in \widehat{N}$, where P is the projection-valued measure arising from Proposition 7.1 and where the unique orbit $\mathcal{O}_\nu$ exists by Proposition 7.2. Then the imprimitivity theorem applies and yields a unitary representation $\sigma\colon G_\nu \to \mathcal{H}_\sigma$, such that U is equivalent to $\mathrm{Ind}_{G_\nu}^G \sigma\colon G \to \mathbf{U}(\mathcal{H})$. This yields most of the first main theorem of this section.

Theorem 7.4. *Let G be a locally compact group and N be a nontrivial closed abelian normal subgroup of G. Suppose the action $\cdot\colon G \times \widehat{N} \to \widehat{N}$ is regular. For every irreducible representation $U\colon G \to \mathbf{U}(\mathcal{H}_U)$ of G, there is a unique orbit $\mathcal{O}$, such that for any $\nu \in \mathcal{O}$ (so that $\mathcal{O} = \mathcal{O}_\nu$), there is an irreducible unitary representation $\sigma\colon G_\nu \to \mathbf{U}(\mathcal{H}_\sigma)$, such that U is equivalent to $\mathrm{Ind}_{G_\nu}^G \sigma$. Moreover, we have*

$$\sigma(n) = \nu(n)\mathrm{id}_{\mathcal{H}_\sigma}$$

for all $n \in N$.

Proof sketch. Since the unitary representation $U\colon G \to \mathbf{U}(\mathcal{H}_U)$ is irreducible, by Proposition 7.1, the projection-valued measure P induced by the restriction of U to N has property (indv1). Since the action of G on $\widehat{N}$ is regular, the orbit space is countably separated, so by Proposition 7.2, there is a single orbit $\mathcal{O}$, such that $P(\mathcal{O}) = I$, and so P is identically zero on $\widehat{N} - \mathcal{O}$. Pick any $\nu \in \widehat{N}$, such that $\mathcal{O} = \mathcal{O}_\nu$. The fact that the action of G on $\widehat{N}$ is regular implies that G/G_ν is homeomorphic to $\mathcal{O}_\nu$ and we may pull P back to G/G_ν. Now $(G, U, G/G_\nu, P)$ is a transitive system of imprimitivity so we can apply the imprimitivity theorem (Theorem 7.3), which tells us that $(G, U, G/G_\nu, P)$ is equivalent to a transitive system of imprimitivity of the form $(G, \mathrm{Ind}_{G_\nu}^G \sigma, G/G_\nu, P^\sigma)$, where $\mathrm{Ind}_{G_\nu}^G \sigma\colon G \to \mathbf{U}(\mathcal{H})$ is the representation induced by some unitary representation $\sigma\colon G_\nu \to \mathbf{U}(\mathcal{H}_\sigma)$ of G_ν. In particular, the unitary representations $U\colon G \to \mathbf{U}(\mathcal{H}_U)$ and $\mathrm{Ind}_{G_\nu}^G \sigma\colon G \to \mathbf{U}(\mathcal{H})$ are equivalent, and since U is irreducible, so is $\mathrm{Ind}_{G_\nu}^G \sigma$. This also implies that σ is irreducible. The last part of the theorem is proven in Folland [Folland (1995)]; see Proposition 6.37. $\square$

The orbit $\mathcal{O}$ in Theorem 7.4 is unique, but ν may be chosen arbitrarily in $\mathcal{O}$. If ν' is another element of $\mathcal{O} = \mathcal{O}_\nu$, then $\nu' = s \cdot \nu$ for some $s \in G$ and the stabilizers G_ν and $G_{\nu'}$ are isomorphic; in fact, $G_{\nu'} = s \cdot G_\nu \cdot s^{-1}$; see Vol. I, Section C.3. But then, given any unitary representation $\sigma\colon G_\nu \to \mathbf{U}(\mathcal{H}_\sigma)$ of G_ν, we obtain the representation $\sigma'\colon G_{\nu'} \to \mathbf{U}(\mathcal{H}_\sigma)$ of $G_{\nu'}$ given by $\sigma'(t) = \sigma(s^{-1}ts)$, for all $t \in G_\nu$, and this map is obviously bijective. We can also check that the unitary transformation $T\colon \mathcal{H} \to \mathcal{H}'$ given by $(Tf)(t) = f(s^{-1}ts)$ (where $f \in \mathcal{H}$) is an equivalence of the unitary representations $\mathrm{Ind}_{G_\nu}^G \sigma\colon G \to \mathbf{U}(\mathcal{H})$ and $\mathrm{Ind}_{G_{\nu'}}^G \sigma'\colon G \to \mathbf{U}(\mathcal{H}')$. Hence, the choice of ν in $\mathcal{O}$ is not essential.

Theorem 7.4 has the following converse.

Theorem 7.5. *Let G be a locally compact group and N be a nontrivial closed abelian normal subgroup of G. Suppose the action $\cdot\colon G \times \widehat{N} \to \widehat{N}$ is regular. For any $\nu \in \widehat{N}$ and for any irreducible unitary representation $\sigma\colon G_\nu \to \mathbf{U}(\mathcal{H}_\sigma)$ of G_ν, such that $\sigma(n) = \nu(n)\mathrm{id}_{\mathcal{H}_\sigma}$ for all $n \in N$, the unitary representation $\mathrm{Ind}_{G_\nu}^G \sigma\colon G \to \mathbf{U}(\mathcal{H})$ is irreducible. If $\sigma'\colon G_\nu \to \mathbf{U}(\mathcal{H}_{\sigma'})$ is another unitary representation of G_ν, such that $\mathrm{Ind}_{G_\nu}^G \sigma\colon G \to \mathbf{U}(\mathcal{H})$ and $\mathrm{Ind}_{G_\nu}^G \sigma'\colon G \to \mathbf{U}(\mathcal{H}')$ are equivalent, then σ and σ' are equivalent.*

Theorem 7.5 is proven in Folland [Folland (1995)]; see Theorem 6.39.

Theoretically, Theorem 7.4 and Theorem 7.5 settle our problem, but in many cases these results are not useful because the groups G_ν may be rather large and their representations may not be much easier to analyze than the representations of G itself. For example, if $\nu = 1$ (the constant character with value 1), then $\mathcal{O}_\nu = \{1\}$ and $G_\nu = G$. In this case, Theorem 7.4 yields an irreducible representation $\sigma\colon G \to \mathbf{U}(\mathcal{H}_\sigma)$, such that $\sigma(n) = \mathrm{id}_{\mathcal{H}_\sigma}$ for all $n \in N$ equivalent to the original representation $U\colon G \to \mathbf{U}(\mathcal{H}_U)$. Since σ is trivial on N, it follows that σ yields an irreducible representation $\rho\colon G/N \to \mathbf{U}(\mathcal{H}_\sigma)$ of G/N, and σ is a lift of the representation ρ of the smaller group G/N to G, in the sense that $\sigma = \rho \circ q$, where $q\colon G \to G/N$ is the quotient map as illustrated below.

$$\begin{array}{ccc} G & \xrightarrow{\ q\ } & G/N \\ & {\scriptstyle\sigma}\searrow & \downarrow{\scriptstyle\rho} \\ & & \mathbf{U}(\mathcal{H}_\sigma). \end{array}$$

Recall that $N \subseteq G_\nu$. There are a number of examples where the character $\nu \in \widehat{N}$ can be extended "nicely" to a representation of G_ν. In this case, we can lift an irreducible representation of the smaller group G_ν/N to G_ν.

Proposition 7.6. *Let G be a locally compact group and N be a nontrivial closed abelian normal subgroup of G. Suppose that for some $\nu \in \widehat{N}$ there is a continuous homomorphism $\widetilde{\nu}\colon G_\nu \to \mathbb{T}$ extending ν. For every irreducible representation $\rho\colon G_\nu/N \to \mathbf{U}(\mathcal{H}_\rho)$ of G_ν/N, the map $\sigma\colon G_\nu \to \mathbf{U}(\mathcal{H}_\rho)$ given by*

$$\sigma(s) = \widetilde{\nu}(s)\rho(sN), \quad s \in G_\nu$$

defines an irreducible representation of G_ν in $\mathcal{H}_\rho$, such that $\sigma(n) = \nu(n)\mathrm{id}_{\mathcal{H}_\rho}$ for all $n \in N$. Furthermore, every irreducible unitary representation σ of G_ν as above arises in this way.

Proposition 7.6 is proven in Folland [Folland (1995)]; see Proposition 6.40. The proof is simple but not illuminating. An orbit $\mathcal{O}$ such that some character $\nu \in \mathcal{O}$ can be extended to a continuous homomorphism $\widetilde{\nu}\colon G_\nu \to \mathbb{T}$ is called *accommodating*; see Warner [Warner (1972a)] (Section 5.4).

It turns out that an interesting class of groups to which Proposition 7.6 applies is the class of semi-direct products $N \rtimes H$, in which N is an abelian group. In this case, the groups G_ν/N are isomorphic to the groups $H_\nu = G_\nu \cap H$, called *little groups*. The Mackey machine yields *all* irreducible representations of $G = N \rtimes H$ as induced representations obtained by combining characters of N and irreducible representations of the little groups H_ν. The little group method was first used by Wigner in a famous paper (1939) on the representations of the *Poincaré group* $\mathbb{R}^4 \rtimes \mathbf{SO}_0(3,1)$, where $\mathbf{SO}_0(3,1)$ is the so-called *restricted Lorentz group*.

7.4 Irreducible Representations of Semi-Direct Products

For our purposes, it is more convenient to adopt the "internal" view of a semi-direct product where a group G is already given as well as two subgroups N and H, such that

(1) N is a normal subgroup of G.
(2) $G = NH$.
(3) $N \cap H = \{e\}$.

Then (2) and (3) imply that the map $N \times H \mapsto G$ given by $(n, h) \mapsto nh$ is a bijection. The multiplication operation in G is given by

$$(n_1h_1)(n_2h_2) = (n_1[h_1n_2h_1^{-1}])(h_1h_2), \quad n_1, n_2 \in N,\ h_1, h_2 \in H.$$

So H acts on N by conjugation on the left. It is immediately verified that the inverse of $nh \in G$ is given by

$$(nh)^{-1} = (h^{-1}n^{-1}h)h^{-1}.$$

Since we also assume that G is locally compact, we require N and H to be closed, and that the map $N \times H \mapsto G$ given by $(n, h) \mapsto nh$ is a homeomorphism. The standard notation for a semi-direct product is $G = N \rtimes H$.[1] The multiplication operation in $G = N \rtimes H$ makes it clear that the map $q\colon N \rtimes H \to H$ given by $q(nh) = h$ $(n \in N, h \in H)$ is a surjective homomorphism with kernel N.

Now if N is also abelian, as before the group of characters $\widehat{N}$ (the dual group) makes sense. For any $\nu \in \widehat{N}$, since $N \subseteq G_\nu$, the quotient group G_ν/N is well-defined. Since $G = NH$, if we let $H_\nu = G_\nu \cap H$, we check immediately that

$$G_\nu = N \rtimes H_\nu.$$

Since G_ν/N is the group of cosets $sN = Ns$ (since N is normal) with $s \in G_\nu$, the map $Ns \mapsto s$ (with $s \in H_\nu$) is an isomorphism from G_ν/N to H_ν.

Definition 7.7. Given any semi-direct product $G = N \rtimes H$ (with N normal and abelian) as above, for any $\nu \in \widehat{N}$, the group H_ν given by

$$H_\nu = G_\nu \cap H$$

is called the *little group* associated with ν. As observed above,

$$G_\nu = N \rtimes H_\nu, \quad H_\nu \approx G_\nu/N.$$

The reason why little groups are interesting is that Proposition 7.6 applies. Indeed, given any character $\nu\colon N \to \mathbb{T}$, we can extend ν to a homomorphism $\widetilde{\nu}\colon G_\nu \to \mathbb{T}$ as follows:

$$\widetilde{\nu}(nh) = \nu(n), \quad h \in H_\nu,\ n \in N. \tag{$\widetilde{\nu}$}$$

We need to check that $\widetilde{v}$ is a homomorphism. First, we have

$$\begin{aligned}\widetilde{\nu}((n_1h_1)(n_2h_2)) &= \widetilde{\nu}((n_1[h_1n_2h_1^{-1}])(h_1h_2)) \\ &= \nu(n_1[h_1n_2h_1^{-1}]) \\ &= \nu(n_1)\nu(h_1n_2h_1^{-1}),\end{aligned}$$

[1]Curiously, Folland uses the notation $N \ltimes H$; see Folland [Folland (1995)].

that is

$$\widetilde{\nu}((n_1h_1)(n_2h_2)) = \nu(n_1)\nu(h_1n_2h_1^{-1}). \tag{1}$$

However, by definition of the action of G on $\widehat{N}$ (see Definition 7.1), for any $\chi \in \widehat{N}$ and any $s \in G$, we have

$$(s \cdot \chi)(n) = \chi(s^{-1}ns),$$

so

$$\nu(h_1n_2h_1^{-1}) = (h_1^{-1} \cdot \nu)(n_2). \tag{2}$$

But $h_1 \in H_\nu = G_\nu \cap H$ and since H_ν is a group, $h_1^{-1} \in H_\nu$, so as $h_1^{-1} \in H_\nu$ is a stabilizer of ν, we have

$$h_1^{-1} \cdot \nu = \nu, \tag{3}$$

and thus by (2) and (3),

$$\nu(h_1n_2h_1^{-1}) = \nu(n_2). \tag{4}$$

Finally, by (1) and (4), and by definition of $\widetilde{\nu}$, we have

$$\widetilde{\nu}((n_1h_1)(n_2h_2)) = \nu(n_1)\nu(n_2) = \widetilde{\nu}(n_1h_1)\widetilde{\nu}(n_2h_2),$$

that is,

$$\widetilde{\nu}((n_1h_1)(n_2h_2)) = \widetilde{\nu}(n_1h_1)\widetilde{\nu}(n_2h_2), \tag{5}$$

which shows that $\widetilde{\nu}\colon G_\nu \to \mathbb{T}$ is a homomorphism extending ν.

We can now apply Proposition 7.6. Since $H_\nu \approx G_\nu/N$, for every irreducible representation $\rho\colon H_\nu \to \mathbf{U}(\mathcal{H}_\rho)$ of H_ν, the map $\sigma\colon G_\nu \to \mathbf{U}(\mathcal{H}_\rho)$ given by

$$\sigma(nh) = \widetilde{\nu}(nh)\rho(h) = \nu(n)\rho(h), \quad n \in N,\, h \in H_\nu,$$

defines an irreducible representation of G_ν in $\mathcal{H}_\rho$, such that $\sigma(n) = \nu(n)\mathrm{id}_{\mathcal{H}_\rho}$ for all $n \in N$.

Definition 7.8. For any $\nu \in \widehat{N}$ and any irreducible representation $\rho\colon H_\nu \to \mathbf{U}(\mathcal{H}_\rho)$ of H_ν, the irreducible representation $\sigma\colon G_\nu \to \mathbf{U}(\mathcal{H}_\rho)$ given by

$$\sigma(nh) = \nu(n)\rho(h), \quad n \in N,\, h \in H_\nu$$

is denoted by $\nu\rho$.

Since the restriction of $\nu\rho$ to H_ν is equal to σ, it is easy to see that $\nu\rho$ is equivalent to $\nu\rho'$ iff ρ is equivalent to ρ'.

Remark: (Serre) Since $H_\nu \approx G_\nu/N$, there is a surjective quotient map $q_\nu\colon G_\nu \to H_\nu$, so any representation $\rho\colon H_\nu \to \mathbf{U}(\mathcal{H}_\rho)$ of H_ν lifts to the representation $q_\nu \circ \rho\colon G_\nu \to \mathbf{U}(\mathcal{H}_\rho)$ of G_ν. It is also clear that if ρ is irreducible, then so is $q_\nu \circ \rho$. Write $\widetilde{\rho} = q_\nu \circ \rho$. Since $\widetilde{\nu}\colon G_\nu \to \mathbf{U}(1)$ is also an irreducible representation of G_ν, we deduce that $\nu\rho$ is equivalent to the tensor product representation $\widetilde{\nu} \otimes \widetilde{\rho}$ (recall that $\mathbb{C} \otimes_{\mathbb{C}} W \approx W$ for any complex vector space W).

We can now apply Theorem 7.4 and Theorem 7.5 to the above situation to obtain a complete characterization of the irreducible representations of a semi-direct product $G = N \rtimes H$ (with N normal an abelian) in terms of the characters of N and of the irreducible representations of the little groups H_ν associated with the characters $\nu \in \widehat{N}$.

Theorem 7.7. *Let G be locally compact group which is a semi-direct product $G = N \rtimes H$ with N normal and abelian. Suppose that G acts regularly on $\widehat{N}$.*

(1) For any $\nu \in \widehat{N}$, if $\rho\colon H_\nu \to \mathbf{U}(\mathcal{H}_\rho)$ is any irreducible representation of the little group H_ν, then the induced representation $\mathrm{Ind}_{G_\nu}^G \nu\rho$ of G (with $\nu\rho$ as in Definition 7.8) is irreducible.
(2) Every irreducible representation $U\colon G \to \mathbf{U}(\mathcal{H}_U)$ of G is equivalent to some irreducible induced representation $\mathrm{Ind}_{G_\nu}^G \nu\rho$ as in (1).
(3) Two induced representations $\mathrm{Ind}_{G_\nu}^G \nu\rho$ and $\mathrm{Ind}_{G_{\nu'}}^G \nu'\rho'$ are equivalent iff $\nu' = s\cdot\nu$ for some $s \in G$ (ν and ν' belong to the same orbit), and the representation ρ and $h \mapsto \rho'(s^{-1}hs)$ are equivalent.

We are now ready for some examples.

Example 7.1. Consider the group $\mathbf{SE}(n)$ of rigid motions of $\mathbb{R}^n$ defined as the group of $(n+1)\times(n+1)$ matrices

$$\mathbf{SE}(n) = \left\{ \begin{pmatrix} Q & u \\ 0 & 1 \end{pmatrix} \mid Q \in \mathbf{SO}(n),\ u \in \mathbb{R}^n \right\}.$$

We assume that $n \geq 2$, since $\mathbf{SE}(1) \approx \mathbb{R}$ is abelian so its irreducible unitary representations are one-dimensional, and thus are of the form $z \mapsto \chi(x)z$ for all $x \in \mathbb{R}$ and all $z \in \mathbb{C}$, where χ is any character of $\mathbb{R}$. We know from Vol. I, Proposition 10.9 that the characters of $\mathbb{R}$ are of the form $x \mapsto e^{iyx}$, for any fixed $y \in \mathbb{R}$. The subgroups N and H are defined as follows:

$$N = \left\{ \begin{pmatrix} I_n & u \\ 0 & 1 \end{pmatrix} \mid u \in \mathbb{R}^n \right\}, \quad H = \left\{ \begin{pmatrix} Q & 0 \\ 0 & 1 \end{pmatrix} \mid Q \in \mathbf{SO}(n) \right\}.$$

We have

$$\begin{pmatrix} Q & u \\ 0 & 1 \end{pmatrix} = \begin{pmatrix} I_n & u \\ 0 & 1 \end{pmatrix} \begin{pmatrix} Q & 0 \\ 0 & 1 \end{pmatrix},$$

which shows that $\mathbf{SE}(n) = NH$, and

$$\begin{pmatrix} Q & u \\ 0 & 1 \end{pmatrix} \begin{pmatrix} R & v \\ 0 & 1 \end{pmatrix} = \begin{pmatrix} QR & u + Qv \\ 0 & 1 \end{pmatrix},$$

and so

$$\begin{pmatrix} Q & u \\ 0 & 1 \end{pmatrix}^{-1} = \begin{pmatrix} Q^\top & -Q^\top u \\ 0 & 1 \end{pmatrix}.$$

Clearly, $N \cap H = \{I_{n+1}\}$ and N is abelian. We also have

$$\begin{aligned}
\begin{pmatrix} Q & u \\ 0 & 1 \end{pmatrix} \begin{pmatrix} I_n & v \\ 0 & 1 \end{pmatrix} \begin{pmatrix} Q & u \\ 0 & 1 \end{pmatrix}^{-1} &= \begin{pmatrix} Q & u \\ 0 & 1 \end{pmatrix} \begin{pmatrix} I_n & v \\ 0 & 1 \end{pmatrix} \begin{pmatrix} Q^\top & -Q^\top u \\ 0 & 1 \end{pmatrix} \\
&= \begin{pmatrix} Q & u + Qv \\ 0 & 1 \end{pmatrix} \begin{pmatrix} Q^\top & -Q^\top u \\ 0 & 1 \end{pmatrix} \\
&= \begin{pmatrix} I_n & Qv \\ 0 & 1 \end{pmatrix},
\end{aligned}$$

so N is a normal subgroup (which is also abelian). Consequently, $\mathbf{SE}(n)$ is the semidirect product $\mathbf{SE}(n) = N \rtimes H$. It is also clear that H is isomorphic to $\mathbf{SO}(n)$ and that N is isomorphic to $\mathbb{R}^n$, so we may write $\mathbf{SE}(n) = \mathbb{R}^n \rtimes \mathbf{SO}(n)$. It is often convenient to use a more concise notation for the element of $\mathbf{SE}(n) = \mathbb{R}^n \rtimes \mathbf{SO}(n)$, namely we denote the matrix

$$\begin{pmatrix} Q & u \\ 0 & 1 \end{pmatrix}$$

by (u, Q). Multiplication in $\mathbf{SE}(n)$ is then given by

$$(u, Q)(v, R) = (u + Qv, QR).$$

The action of $\mathbf{SE}(n)$ on $\mathbb{R}^n$ is given by

$$(u, Q)x = Qx + u, \quad x \in \mathbb{R}^n,$$

namely rotate x by Q and then translate by u. This is equivalent to the usual trick of embedding $\mathbb{R}^n$ in $\mathbb{R}^{n+1}$ by mapping x to $\binom{x}{1}$, and then

$$\begin{pmatrix} Q & u \\ 0 & 1 \end{pmatrix} \begin{pmatrix} x \\ 1 \end{pmatrix} = \begin{pmatrix} Qx + u \\ 1 \end{pmatrix}.$$

We have to figure out how $\mathbf{SE}(n)$ acts on $\widehat{N}$ to determine its orbits. Observe that

$$\begin{aligned}\begin{pmatrix} Q & u \\ 0 & 1 \end{pmatrix}^{-1} \begin{pmatrix} I_n & v \\ 0 & 1 \end{pmatrix} \begin{pmatrix} Q & u \\ 0 & 1 \end{pmatrix} &= \begin{pmatrix} Q^\top & -Q^\top u \\ 0 & 1 \end{pmatrix} \begin{pmatrix} I_n & v \\ 0 & 1 \end{pmatrix} \begin{pmatrix} Q & u \\ 0 & 1 \end{pmatrix} \\ &= \begin{pmatrix} Q^\top & -Q^\top u + Q^\top v \\ 0 & 1 \end{pmatrix} \begin{pmatrix} Q & u \\ 0 & 1 \end{pmatrix} \\ &= \begin{pmatrix} I_n & Q^\top v \\ 0 & 1 \end{pmatrix},\end{aligned}$$

that is

$$(u, Q)^{-1}(v, I_n)(u, Q) = (Q^\top v, I_n). \tag{†}$$

To describe the action of $G = \mathbf{SE}(n)$ on $\widehat{N}$, we introduce the isomorphism $t\colon N \to \mathbb{R}^n$ given by

$$t(x, I_n) = x, \quad x \in \mathbb{R}^n.$$

Then we have an isomorphism between $\widehat{\mathbb{R}^n}$ and $\widehat{N}$ given by $\chi \mapsto \chi \circ t$, with $\chi \in \widehat{\mathbb{R}^n}$. By Vol. I, Corollary 10.11, the characters in $\widehat{\mathbb{R}^n}$ are the homomorphims χ_y from $\mathbb{R}^n$ to $\mathbb{T}$ (with $y \in \mathbb{R}^n$) given by

$$\chi_y(x) = e^{iy\cdot x}, \qquad x \in \mathbb{R}^n,$$

where $y \cdot x$ is the Euclidean product in $\mathbb{R}^n$ ($y \cdot x = \sum_{k=1}^n y_k x_k$). By composing the isomorphism from $\mathbb{R}^n$ to $\widehat{\mathbb{R}^n}$ given by $y \mapsto \chi_y$ and the isomorphism between $\widehat{\mathbb{R}^n}$ and $\widehat{N}$ given by $\chi \mapsto \chi \circ t$, we obtain the isomorphism between $\mathbb{R}^n$ and $\widehat{N}$ given by $y \mapsto \chi_y \circ t$ (with $y \in \mathbb{R}^n$).

Thus the characters in $\widehat{N}$ are of the form $(x, I_n) \mapsto \chi_y(t(x, I_n)) = \chi_y(x)$. By Definition 7.1, the action of $G = \mathbf{SE}(n)$ on $\widehat{N}$ is given by

$$((u, Q) \cdot \chi_y)(x, I_n) = \chi_y(t((u, Q)^{-1}(x, I_n)(u, Q))), \quad x \in \mathbb{R}^n,$$

which by (†) yields

$$\begin{aligned}((u, Q) \cdot \chi_y)(x, I_n) &= \chi_y(t(Q^\top x, I_n)) = \chi_y(Q^\top x) = e^{iy\cdot(Q^\top x)} \\ &= e^{i(Qy)\cdot x} = \chi_{Qy}(t(x, I_n)).\end{aligned}$$

Therefore, under the isomorphism between $\mathbb{R}^n$ and $\widehat{N}$ given by $y \mapsto \chi_y \circ t$ (with $y \in \mathbb{R}^n$), we see that the action of $G = \mathbf{SE}(n)$ on $\widehat{N}$ is the action of $G = \mathbf{SE}(n)$ on $\mathbb{R}^n$ given by

$$(u, Q)(y) = Qy, \quad y \in \mathbb{R}^n; \tag{††}$$

in other words, only the rotation Q is applied. This is the usual action of $\mathbf{SO}(n)$ on $\mathbb{R}^n$.

Remark: Note the subtle point that the action of $G = \mathbf{SE}(n)$ on $\widehat{N}$ uses a *right conjugation* of (x, I_n) by (u, Q), namely $(u, Q)^{-1}(x, I_n)(u, Q)$, and this yields $(Q^\top x, I_n)$. The appearance of $Q^\top$ seems wrong, but it is compensated by the fact that in the argument of χ_y, we now have the inner product $y \cdot (Q^\top x)$, and in order to make the input x appear, we transpose again to obtain $Qy \cdot x = y \cdot (Q^\top x)$.

Remember that we have an isomorphism between $\mathbb{R}^n$ and $\widehat{N}$ given by $y \mapsto \chi_y \circ t$ (with $y \in \mathbb{R}^n$), so a character $\nu \in \widehat{N}$ may be denoted by ν_y. Using this isomorphism, it is easy to determine the orbits and the little groups. By (††), *the orbits of the action of* $G = \mathbf{SE}(n)$ *on* $\widehat{N}$ *can be viewed as the orbits of the action of* $\mathbf{SO}(n)$ *on* $\mathbb{R}^n$, namely for every $r \in [0, +\infty)$, the sphere $S_r(0)$ of radius r centered at the origin,

$$\mathcal{O}_r = S_r(0) = \{x \in \mathbb{R}^n \mid \|x\|_2 = r\}.$$

For $r = 0$, we have $\mathcal{O}_0 = \{0_n\}$. For the countable separation property, we use the G-invariant annuli

$$\{x \in \mathbb{R}^n \mid \alpha < \|x\|_2 < \beta\}$$

with $\alpha < \beta$ rational. For $r > 0$, we pick the special representative re_1 on the sphere $\mathcal{O}_r$ with $e_1 = (1, 0, \ldots, 0) \in \mathbb{R}^n$. Then we see immediately that the little group H_{re_1} is isomorphic to $\mathbf{SO}(n-1)$. For $r = 0$, the little group H_0 is $\mathbf{SO}(n)$ and $G_0 = G = \mathbf{SE}(n)$. For $r > 0$, the characters $\nu \in \widehat{N}$ corresponding to points in $\mathcal{O}_r$ are of the form $\chi_y \circ t$, with $y \in \mathbb{R}^n$ and $\|y\|_2 = r$. Earlier, we denoted them by ν_y.

Theorem 7.7 yields all irreducible representations of $\mathbf{SE}(n)$.

(1) For $r = 0$, we have $G_0 = G$ and $H_0 = \mathbf{SO}(n)$. We obtain the finite-dimensional irreducible representations $q \circ \sigma$ by lifting the irreducible representations σ of $\mathbf{SO}(n)$ to $\mathbf{SE}(n)$ by composition with the quotient map $q \colon \mathbf{SE}(n) \to \mathbf{SO}(n)$.

(2) For every $r > 0$, for every $y \in \mathbb{R}^n$ with $\|y\|_2 = r$, we have the character $\nu_{r,y}$ given by $\nu_{r,y}(x) = e^{i(y \cdot x)}$. We also have $H_{r,y} = \mathbf{SO}(n-1)$ and $G_{r,y} = \mathbb{R}^n \rtimes \mathbf{SO}(n-1)$. Then for every irreducible representation $\rho \colon \mathbf{SO}(n-1) \to \mathbf{U}(\mathcal{H}_\rho)$ of $\mathbf{SO}(n-1)$, we have the irreducible representation $\nu_{r,y}\rho \colon \mathbb{R}^n \rtimes \mathbf{SO}(n-1) \to \mathbf{U}(\mathcal{H}_\rho)$ given by

$$(\nu_{r,y}\rho)(xQ) = \nu_{r,y}(x)\rho(Q) = e^{iy \cdot x}\rho(Q), \quad x \in \mathbb{R}^n,\ Q \in \mathbf{SO}(n-1),\ \|y\|_2 = r.$$

The induced representation $\mathrm{Ind}_{\mathbb{R}^n \rtimes \mathbf{SO}(n-1)}^{\mathbf{SE}(n)} \nu_{r,y}\rho$ of $\mathbf{SE}(n)$ is irreducible.

In the special case of (2), where $\rho\colon \mathbf{SO}(n-1) \to \mathbf{U}(1)$ is the trivial representation ($\rho(Q) = 1$ for all $Q \in \mathbf{SO}(n-1)$), it can be shown that the induced representation $\mathrm{Ind}_{\mathbb{R}^n \rtimes \mathbf{SO}(n-1)}^{\mathbf{SE}(n)} \nu_{r,y}\rho$ is equivalent to the induced representation $\mathrm{Ind}_{\mathbb{R}^n}^{\mathbf{SE}(n)} \nu_{r,y}$ (see Folland [Folland (1995)], Section 6.3). But we have determined such induced representations in Example 6.3. We found that these are the irreducible representations $\widetilde{\Pi}\colon \mathbf{SE}(n) \to \mathbf{U}(\mathrm{L}^2_\lambda(S^{n-1};\mathbb{C}))$ of class 1 described in Vilenkin [Vilenkin (1968)] (Chapter XI, Section 2) given by

$$(\widetilde{\Pi}_{(a,Q)}(f))(x) = e^{ir(x\cdot a)} f(Q^\top x), \qquad (a,Q) \in \mathbf{SE}(n),\ x \in S^{n-1},\ f \in \mathrm{L}^2_\lambda(S^{n-1};\mathbb{C}),\ r > 0.$$

For $n = 2, 3$, we can be more precise.

(1) For $n = 2$, we have $\mathbf{SO}(1) = \{1\}$. Thus, for $r > 0$ the irreducible representations of $\mathbf{SE}(2)$ are of the form

$$\mathrm{Ind}_{\mathbb{R}^2}^{\mathbf{SE}(2)} \nu_{r,y}, \quad \text{with } \nu_{r,y}(x) = e^{iy\cdot x},\ x, y \in \mathbb{R}^2,\ \|y\|_2 = r.$$

According to the above discussion, they are equivalent to the irreducible representations $\widetilde{\Pi}\colon \mathbf{SE}(2) \to \mathbf{U}(\mathrm{L}^2_\lambda(S^1;\mathbb{C}))$ of class 1 described in Vilenkin [Vilenkin (1968)] (Chapter IV, Section 2) given by

$$(\widetilde{\Pi}_{(a,Q)}(f))(x) = e^{ir(x\cdot a)} f(Q^\top x), \qquad (a,Q) \in \mathbf{SE}(2),\ x \in S^1,\ f \in \mathrm{L}^2_\lambda(S^1;\mathbb{C}),\ r > 0.$$

For $r = 0$, $G_0 = \mathbf{SO}(2)$. The group $\mathbf{SO}(2)$ is abelian and $\mathbf{SO}(2) \approx \mathbf{U}(1) \approx \mathbb{T}$, so we know from Vol. I, Proposition 10.9 that the irreducible representations of $\mathbf{SO}(2)$ are the homomorphisms $\rho_k\colon \mathbf{SO}(2) \to \mathbf{U}(1)$ given by

$$\rho_k(e^{i\theta})(z) = e^{ik\theta} z, \quad k \in \mathbb{Z},\ 0 \le \theta < 2\pi,\ z \in \mathbb{C}.$$

We obtain irreducible representations of $\mathbf{SE}(2)$ obtained by lifting the irreducible representations ρ_k of $\mathbf{SO}(2)$ to $\mathbf{SE}(2)$ by composing with the projection map $q\colon \mathbf{SE}(2) \to \mathbf{SO}(2)$.

(2) For $n = 3$,if $r > 0$ then $H_{re_1} = \mathbf{SO}(2)$. As in (1), the irreducible representations of $\mathbf{SO}(2) \approx \mathbf{U}(1)$ are the homomorphisms $\rho_k : \mathbf{SO}(2) \to \mathbf{U}(1)$ given by

$$\rho_k(e^{i\theta})(z) = e^{ik\theta} z, \quad k \in \mathbb{Z},\ 0 \le \theta < 2\pi,\ z \in \mathbb{C}.$$

We obtain the irreducible representations $\nu_{r,y}\rho_k \colon \mathbb{R}^3 \rtimes \mathbf{SO}(2) \to \mathbf{U}(1)$ given by

$$(\nu_{r,y}\rho_k)(xe^{i\theta})(z) = e^{i(y\cdot x+k\theta)}z, \quad x, y \in \mathbb{R}^3, \|y\|_2 = r,\ 0 \le \theta < 2\pi,\ k \in \mathbb{Z},\ z \in \mathbb{C},$$

which yield the irreducible representations $\mathrm{Ind}_{\mathbb{R}^3 \rtimes \mathbf{SO}(2)}^{\mathbf{SE}(3)} \nu_{r,y}\rho_k$ of $\mathbf{SE}(3)$. In the special case $k = 0$, these are equivalent to the irreducible representations $\widetilde{\Pi} \colon \mathbf{SE}(3) \to \mathbf{U}(\mathrm{L}^2_\lambda(S^2; \mathbb{C}))$ of class 1 given by

$$(\widetilde{\Pi}_{(a,Q)}(f))(x) = e^{ir(x\cdot a)} f(Q^\top x), \qquad (a, Q) \in \mathbf{SE}(3),\ x \in S^2,\ f \in \mathrm{L}^2_\lambda(S^2; \mathbb{C}),\ r > 0.$$

If $r = 0$, we obtain the irreducible representations of $\mathbf{SE}(3)$ by lifting the irreducible representations of $\mathbf{SO}(3)$ to $\mathbf{SE}(3)$ by composing with the projection map $q \colon \mathbf{SE}(3) \to \mathbf{SO}(3)$.

In the next example, we find all irreducible representations of $\mathbf{O}(2)$.

Example 7.2. In Section 4.4, we claimed that $\mathbf{O}(2)$ is isomorphic to the semi-direct product $\mathbf{SO}(2) \rtimes \{I_2, J\}$, where

$$J = \begin{pmatrix} -1 & 0 \\ 0 & 1 \end{pmatrix}.$$

Here, $N = \mathbf{SO}(2)$ and $H = \{I_2, J\} \simeq \mathbb{Z}/2\mathbb{Z}$. Clearly, $\mathbf{SO}(2) \cap H = \{I_2\}$ and if $Q \in \mathbf{O}(2)$ with $\det(Q) = -1$, then $Q = (QJ)J$ with $QJ \in \mathbf{SO}(2)$, a unique factorization of Q in NH. The subgroup $\mathbf{SO}(2)$ is normal in $\mathbf{O}(2)$ since for any $Q \in \mathbf{O}(2)$ and any $R \in \mathbf{SO}(2)$, we have

$$\det(Q^\top RQ) = \det(Q^\top)\det(R)\det(Q) = \det(Q)^2 \det(R) = 1 \times 1 = 1,$$

so $Q^\top RQ \in \mathbf{SO}(2)$. The above argument can be immediately adapted to prove that $\mathbf{O}(2m)$ is isomorphic to the semi-direct product $\mathbf{SO}(2m) \rtimes \{I_{2m}, J\}$ for any $m \ge 1$, where J is any reflection in $\mathbf{O}(2m)$, for instance $J = \mathrm{diag}(-1, 1, \ldots, 1)$. Unfortunately, H is *not* normal in $\mathbf{O}(2m)$, even for $m = 1$ (we leave this fact as an exercise).

Using the equation $JR_\theta J = R_{-\theta}$ proven in $(*_J)$ below, multiplication in $\mathbf{SO}(2) \rtimes \{I_2, J\}$ is given by

$$R_\theta R_\varphi = R_{\theta+\varphi},$$

$$R_\theta(R_\varphi J) = R_{\theta+\varphi} J,$$

and the two nontrivial cases

$$(R_\theta J)R_\varphi = R_\theta(JR_\varphi J)J = R_\theta(R_{-\varphi})J = R_{\theta-\varphi}J,$$

and

$$(R_\theta J)(R_\varphi J) = R_\theta(JR_\varphi J) = R_\theta R_{-\varphi} = R_{\theta-\varphi}.$$

We will use the isomorphism from $\mathbf{SO}(2)$ to $\mathbf{U}(1)$ given by

$$R_\theta = \begin{pmatrix} \cos\theta & -\sin\theta \\ \sin\theta & \cos\theta \end{pmatrix} \mapsto e^{i\theta}.$$

Since $\mathbf{SO}(2) \approx \mathbf{U}(1)$ is abelian, its irreducible representations are given by its characters, namely the homomorphisms $\chi_m \colon \mathbf{SO}(2) \to \mathbf{U}(1)$ given by

$$\chi_m(R_\theta) = e^{im\theta}, \quad m \in \mathbb{Z},\ 0 \leq \theta < 2\pi.$$

Let us figure out the action of $\mathbf{O}(2)$ on the group of characters $\widehat{\mathbf{SO}(2)} \simeq \mathbb{Z}$. For any character χ_m and any $Q \in \mathbf{O}(2)$, by (act), we have

$$Q \cdot \chi_m(R_\theta) = \chi_m(Q^\top R_\theta Q).$$

There are two cases.

(1) If $Q \in \mathbf{O}(2)$ and $\det(Q) = +1$, then $Q = R_\varphi \in \mathbf{SO}(2)$, so

$$Q^\top R_\theta Q = R_{-\varphi} R_\theta R_\varphi = R_{-\varphi} R_\varphi R_\theta = R_\theta,$$

since $\mathbf{SO}(2)$ is abelian.

(2) If $Q \in \mathbf{O}(2)$ and $\det(Q) = -1$, then $Q = R_\varphi J$ with $R_\varphi \in \mathbf{SO}(2)$, so

$$Q^\top R_\theta Q = JR_{-\varphi} R_\theta R_\varphi J = JR_\theta J.$$

But

$$\begin{aligned} JR_\theta J &= \begin{pmatrix} -1 & 0 \\ 0 & 1 \end{pmatrix} \begin{pmatrix} \cos\theta & -\sin\theta \\ \sin\theta & \cos\theta \end{pmatrix} \begin{pmatrix} -1 & 0 \\ 0 & 1 \end{pmatrix} \\ &= \begin{pmatrix} -\cos\theta & \sin\theta \\ \sin\theta & \cos\theta \end{pmatrix} \begin{pmatrix} -1 & 0 \\ 0 & 1 \end{pmatrix} \\ &= \begin{pmatrix} \cos\theta & \sin\theta \\ -\sin\theta & \cos\theta \end{pmatrix} = R_{-\theta}, \end{aligned}$$

which we record as

$$JR_\theta J = R_{-\theta}. \tag{$*_J$}$$

In summary, since $\chi_m(R_{-\theta}) = \chi_{-m}(R_\theta)$, we have

$$Q \cdot \chi_m(R_\theta) = \begin{cases} \chi_m(R_\theta), & \text{if } \det(Q) = +1 \\ \chi_{-m}(R_\theta), & \text{if } \det(Q) = -1. \end{cases}$$

We can now determine the stabilizers of the characters. If $\nu = \chi_0$, then

$$Q \cdot \chi_0 = \chi_0, \quad \text{for all } Q \in \mathbf{O}(2),$$

so $G_{\chi_0} = \mathbf{O}(2)$ and $H_{\chi_0} = H$.

If $\nu = \chi_m$ with $m \neq 0$, then

$$Q \cdot \chi_m = \chi_m \quad \text{iff} \quad \det(Q) = +1,$$

so $G_\nu = \mathbf{SO}(2)$, and in this case, $H_\nu = \{I_2\}$.

The orbits are given by

$$\mathcal{O}_{\chi_0} = \{\chi_0\},$$

and for $m \neq 0$,

$$\mathcal{O}_{\chi_m} = \{Q \cdot \chi_m \mid Q \in \mathbf{O}(2)\} = \{\chi_m, \chi_{-m}\}.$$

Since $\widehat{\mathbf{SO}(2)} \simeq \mathbb{Z}$ is discrete, $\mathbf{O}(2)$ acts regularly on $\widehat{\mathbf{SO}(2)}$.

According to Theorem 7.7, we obtain the following irreducible representations.

(1) If $\nu = \chi_0$, since $G_{\chi_0} = \mathbf{O}(2)$ and $H = \{I_2, J\}$, for any irreducible representation ρ of H, $\chi_0\rho = \rho$ is an irreducible representation of $\mathbf{O}(2)$. Since $H = \{I_2, J\}$ is abelian and finite it has two irreducible representations $\rho_0 \colon H \to \mathbf{U}(1)$ and $\rho_1 \colon H \to \mathbf{U}(1)$ given by $\rho_0(I_2) = \rho_0(J) = 1$ and $\rho_1(I_2) = 1$, $\rho_1(J) = -1$, so we obtain the irreducible unitary representations of $\mathbf{O}(2)$ in $\mathbf{U}(1)$ given by

$$\rho_k(QX) = \rho_k(X), \quad Q \in \mathbf{SO}(2),\ X \in \{I_2, J\},\ k \in \{0, 1\}.$$

Since $\rho_1(X) = \det(X) = \det(QX)$, ρ_1 is the determinant representation of $\mathbf{O}(2)$ in $\mathbf{U}(1)$, and ρ_0 is the trivial representation in $\mathbf{U}(1)$.

(2) If $\nu = \chi_m$ with $m \neq 0$, since $G_\nu = \mathbf{SO}(2)$ and $H_\nu = \{I_2\}$, the only irreducible representation of H_ν is the trivial representation $\rho_0 \colon H_\nu \to \mathbf{U}(1)$ given by $\rho_0(I_2) = 1$, so we have the irreducible unitary representation $\chi_m\rho_0 = \chi_m$ of $\mathbf{SO}(2)$ in $\mathbf{U}(1)$, and the induced representation $\mathrm{Ind}_{\mathbf{SO}(2)}^{\mathbf{O}(2)} \chi_m$ is an irreducible representation of $\mathbf{O}(2)$.

The space $\mathcal{H}_m$ of the representation $\mathrm{Ind}_{\mathbf{SO}(2)}^{\mathbf{O}(2)} \chi_m$ can be determined. Recall that $\mathcal{H}_m$ is the Hilbert space, which is the completion of the space $\mathcal{H}_m^0$ defined as

$$\mathcal{H}_m^0 = \Big\{ f \in \mathcal{C}(\mathbf{O}(2), \mathbb{C}) \mid \pi(\mathrm{supp}(f)) \text{ is compact and}$$
$$f(sh) = \left(\frac{\Delta_{\mathbf{SO}(2)}(h)}{\Delta_{\mathbf{O}(2)}(h)}\right)^{1/2} \chi_m(h^{-1})(f(s))$$
$$\text{for all } s \in \mathbf{O}(2) \text{ and all } h \in \mathbf{SO}(2) \Big\},$$

where $\pi\colon \mathbf{O}(2) \to \mathbf{O}(2)/\mathbf{SO}(2)$ is the quotient map. Here, $U = \chi_m$ is an irreducible representation of $\mathbf{SO}(2)$ with $m \neq 0$, namely a character of $\mathbf{SO}(2)$, so $\mathcal{H}_U = \mathbb{C}$, and since both $\mathbf{SO}(2)$ and $\mathbf{O}(2)$ are compact, the terms $\Delta_{\mathbf{SO}(2)}(h)$ and $\Delta_{\mathbf{O}(2)}(h)$ are both equal to 1, and so

$$\mathcal{H}_m^0 = \Big\{ f \in \mathcal{C}(\mathbf{O}(2), \mathbb{C}) \mid f(sh) = \chi_m(h^{-1})(f(s))$$
$$\text{for all } s \in \mathbf{O}(2) \text{ and all } h \in \mathbf{SO}(2) \Big\}.$$

If $\det(s) = 1$, so that $s = R_\theta \in \mathbf{SO}(2)$, then we must have

$$f(sh) = \chi_m(h^{-1})(f(s)), \quad \text{for all } h \in \mathbf{SO}(2),$$

which means that if we write $h = R_\varphi$, then

$$f(R_\theta R_\varphi) = e^{-im\varphi} f(R_\theta) \quad \text{for all } \varphi,$$

that is (for $\varphi = -\theta$),

$$f(R_\theta) = e^{-im\theta} f(I_2).$$

If $\det(s) = -1$, so that $s = R_\theta J$ with $R_\theta \in \mathbf{SO}(2)$, then we must have

$$f(sh) = \chi_m(h^{-1})(f(s)), \quad \text{for all } h \in \mathbf{SO}(2),$$

which means that if we write $h = R_\varphi$, then

$$f(R_\theta J R_\varphi) = e^{-im\varphi} f(R_\theta J) \quad \text{for all } \varphi.$$

But

$$\begin{aligned}
R_\theta J R_\varphi &= \begin{pmatrix} \cos\theta & -\sin\theta \\ \sin\theta & \cos\theta \end{pmatrix} \begin{pmatrix} -1 & 0 \\ 0 & 1 \end{pmatrix} \begin{pmatrix} \cos\varphi & -\sin\varphi \\ \sin\varphi & \cos\varphi \end{pmatrix} \\
&= \begin{pmatrix} -\cos\theta & -\sin\theta \\ -\sin\theta & \cos\theta \end{pmatrix} \begin{pmatrix} \cos\varphi & -\sin\varphi \\ \sin\varphi & \cos\varphi \end{pmatrix} \\
&= \begin{pmatrix} -\cos\theta\cos\varphi - \sin\theta\sin\varphi & \cos\theta\sin\varphi - \sin\theta\cos\varphi \\ -\sin\theta\cos\varphi + \cos\theta\sin\varphi & \sin\theta\sin\varphi + \cos\theta\cos\varphi \end{pmatrix} \\
&= \begin{pmatrix} -\cos(\theta - \varphi) & -\sin(\theta - \varphi) \\ -\sin(\theta - \varphi) & \cos(\theta - \varphi) \end{pmatrix}.
\end{aligned}$$

It follows that for $\varphi = \theta$, we have

$$f(R_\theta J) = e^{im\theta} f(J).$$

In summary, $f \in \mathcal{C}(\mathbf{O}(2), \mathbb{C})$ must satisfy the conditions

$$f(s) = \begin{cases} e^{-im\theta} f(I_2), & \text{if } s = R_\theta \\ e^{im\theta} f(J), & \text{if } s = R_\theta J. \end{cases}$$

Then $f(I_2), f(J)$ are two arbitrary scalars z_1 and z_2 in $\mathbb{C}$, so $\mathcal{H}_m = \mathcal{H}_m^0$ is a two-dimensional space isomorphic to $\mathbb{C}^2$, where the isomorphism is given by $f \mapsto (f(I_2), f(J)) \in \mathbb{C}^2$. We can also describe the space $\mathcal{H}_m$ as the space of functions $f\colon \mathbf{O}(2) \to \mathbb{C}$ given by

$$f(s) = \begin{cases} e^{-im\theta} z_1, & \text{if } s = R_\theta \\ e^{im\theta} z_2, & \text{if } s = R_\theta J, \end{cases}$$

for some $(z_1, z_2) \in \mathbb{C}^2$.

The induced representation $\Pi_m\colon \mathbf{O}(2) \to \mathbf{U}(\mathcal{H}_m)$ is the left-regular representation,

$$(\Pi_m(s)(f))(t) = f(s^{-1}t), \quad f \in \mathcal{H}_m,\ s, t \in \mathbf{O}(2).$$

Given f defined by $(z_1, z_2) \in \mathbb{C}^2$, we can find out which vector in $\mathbb{C}^2$ corresponds to the function $\Pi_m(s)(f)$ given by $t \mapsto f(s^{-1}t)$ (for s fixed).

(1) If $s = R_\theta$, then $s^{-1}t = R_{\varphi-\theta}$ if $t = R_\varphi$, and $s^{-1}t = R_{\varphi-\theta}J$ if $t = R_\varphi J$, so

$$(\Pi_m(R_\theta)(f))(t) = \begin{cases} e^{-im\varphi} e^{im\theta} z_1, & \text{if } t = R_\varphi \\ e^{im\varphi} e^{-im\theta} z_2, & \text{if } t = R_\varphi J, \end{cases}$$

so the function $\Pi_m(s)(f)$ is determined by $(e^{im\theta} z_1, e^{-im\theta} z_2)$.

(2) If $s = R_\theta J$ and $t = R_\varphi$, then $s^{-1}t = JR_{-\theta}R_\varphi = JR_{\varphi-\theta}$. Since by $(*_J)$ we have $JR_{\varphi-\theta}J = R_{\theta-\varphi}$ and $J^2 = I_2$, we obtain $JR_{\varphi-\theta} = R_{\theta-\varphi}J$, and so

$$s^{-1}t = R_{\theta-\varphi}J.$$

If $s = R_\theta J$ and $t = R_\varphi J$, then $s^{-1}t = JR_{-\theta}R_\varphi J = JR_{\varphi-\theta}J = R_{\theta-\varphi}$. It follows that

$$(\Pi_m(R_\theta J)(f))(t) = \begin{cases} e^{-im\varphi} e^{im\theta} z_2, & \text{if } t = R_\varphi \\ e^{im\varphi} e^{-im\theta} z_1, & \text{if } t = R_\varphi J, \end{cases}$$

so the function $\Pi_m(R_\theta J)(f)$ is determined by $(e^{im\theta} z_2, e^{-im\theta} z_1)$.

In summary, if f is given by $(z_1, z_2) \in \mathbb{C}^2$, for any nonzero $m \in \mathbb{Z}$, we have

$$\Pi_m(s)\begin{pmatrix} z_1 \\ z_2 \end{pmatrix} = \begin{pmatrix} e^{im\theta} & 0 \\ 0 & e^{-im\theta} \end{pmatrix}\begin{pmatrix} z_1 \\ z_2 \end{pmatrix}, \quad \text{if } s = R_\theta$$

$$\Pi_m(s)\begin{pmatrix} z_1 \\ z_2 \end{pmatrix} = \begin{pmatrix} 0 & e^{im\theta} \\ e^{-im\theta} & 0 \end{pmatrix}\begin{pmatrix} z_1 \\ z_2 \end{pmatrix}$$
$$= \begin{pmatrix} 0 & 1 \\ 1 & 0 \end{pmatrix}\begin{pmatrix} e^{im\theta} & 0 \\ 0 & e^{-im\theta} \end{pmatrix}\begin{pmatrix} z_1 \\ z_2 \end{pmatrix}, \quad \text{if } s = R_\theta J.$$

This expresses the representation Π_m as an irreducible unitary representation of $\mathbf{O}(2)$ in $\mathbb{C}^2$. We can also express the above as

$$\Pi_m(R_\theta X) = \begin{pmatrix} 0 & 1 \\ 1 & 0 \end{pmatrix}^{\frac{1-\det(X)}{2}} \begin{pmatrix} e^{im\theta} & 0 \\ 0 & e^{-im\theta} \end{pmatrix}, \quad X \in \{I_2, J\},$$

or

$$\Pi_m(R_\theta X) = \begin{pmatrix} 0 & 1 \\ 1 & 0 \end{pmatrix}^{\frac{1-\det(X)}{2}} e^{im\theta\begin{pmatrix} 1 & 0 \\ 0 & -1 \end{pmatrix}}, \quad X \in \{I_2, J\},$$

for any nonzero $m \in \mathbb{Z}$.

For $m = 1$, it is an easy exercise to prove that the set of matrices

$$\left\{ \begin{pmatrix} e^{i\theta} & 0 \\ 0 & e^{-i\theta} \end{pmatrix}, \begin{pmatrix} 0 & e^{i\theta} \\ e^{-i\theta} & 0 \end{pmatrix} \,\middle|\, 0 \le \theta < 2\pi \right\}$$

is a subgroup of $\mathbf{SU}(2)$ isomorphic to $\mathbf{O}(2)$ under the map

$$R_\theta \mapsto \begin{pmatrix} e^{i\theta} & 0 \\ 0 & e^{-i\theta} \end{pmatrix}, \ R_\theta J \mapsto \begin{pmatrix} 0 & e^{i\theta} \\ e^{-i\theta} & 0 \end{pmatrix},$$

so the representation Π_1 is equivalent to the standard action of $\mathbf{O}(2)$ on $\mathbb{C}^2$ by multiplication.

Using the results of Example 7.2, we can determine all the irreducible representations of $\mathbf{E}(2) = \mathbb{R}^2 \rtimes \mathbf{O}(2)$, by a method analogous to the method used for $\mathbf{SE}(2) = \mathbb{R}^2 \rtimes \mathbf{SO}(2)$ in Example 7.1. Similarly, using the results of Example 4.6 to determine the irreducible representations of $\mathbf{O}(3)$, we can determine all the irreducible representations of $\mathbf{E}(3) = \mathbb{R}^3 \rtimes \mathbf{O}(3)$, also by the method used for $\mathbf{SE}(3) = \mathbb{R}^3 \rtimes \mathbf{SO}(3)$ in Example 7.1. We leave the details as exercises.

7.5 Problems

Problem 7.1. Let G be a locally compact group, N be a closed abelian subgroup of G, and $\widehat{N}$ be the group of characters $\chi\colon N \to \mathbb{C}$ of N. Recall from Definition 7.1 that the action of G on $\widehat{N}$ is given by

$$(s \cdot \chi)(n) = \chi(s^{-1}ns).$$

The stabilizer G_χ of χ and the orbit $\mathcal{O}_\chi \subseteq \widehat{N}$ of χ are defined as

$$\begin{aligned} G_\chi &= \{s \in G \mid s \cdot \chi = \chi\} \\ \mathcal{O}_\chi &= \{s \cdot \chi \mid s \in G\}. \end{aligned}$$

(1) Show that subgroup G_χ is closed in G. Also show that $N \subseteq G_\chi$.
(2) Construct an explicit bijection between $\mathcal{O}_\chi$ and G/G_χ.

Problem 7.2. Let G be a locally compact group and N be a closed abelian normal subgroup of G. For any unitary representation $U\colon G \to \mathbf{U}(\mathcal{H}_U)$ of G, let P be the unique regular projection-valued measure on $\widehat{N}$, such that for the restriction $U\colon N \to \mathbf{U}(\mathcal{H}_U)$ of U to N, we have

$$U(n) = \int_{\widehat{N}} \chi(n)\, dP(\chi), \quad n \in N.$$

(1) Show that for any fixed $s \in G$, the map Q defined on the Borel subsets of $\widehat{N}$ by

$$Q(E) = U(s)P(E)U(s)^{-1}$$

is a regular projection-valued measure.
(2) For any $f \in \mathrm{L}^1(\widehat{N})$, show that

$$\int_{\widehat{N}} f(s^{-1} \cdot \chi)\, d\mu(\chi) = \int_{\widehat{N}} f(\chi)\, d(\lambda_{s^{-1}}(\mu))(\chi).$$

Hint. Adapt the proof of Vol. I, Proposition 8.16.
(3) For any Borel subset E of $\widehat{N}$, show that map $E \mapsto P(s \cdot E)$ is a regular projection-valued measure.

Problem 7.3. Let $(G, \mathrm{Ind}_H^G U, G/H, P^U)$ be the canonical system of imprimitivity associated to $\Pi = \mathrm{Ind}_H^G U$; see Definition 7.4. Show that

$$P^U(E)(f) = (\chi_E \circ \pi)(f), \quad E \subseteq G/H,\ f \in \mathcal{H}.$$

Problem 7.4. (Advanced Exercise) Study Sections 6.4 and 6.5 of Folland [Folland (1995)] and write up a detailed proof sketch of Theorem 7.3, Mackey's imprimitivity theorem.

Problem 7.5. Complete the proof sketch of Theorem 7.4.

Hint. See Proposition 6.37 of Folland [Folland (1995)].

Problem 7.6. (Advanced Exercise) Prove Theorem 7.5.

Hint. See Theorem 6.39 of Folland [Folland (1995)].

Problem 7.7. Let G be a locally compact group, H be a closed subgroup of G, and N be closed normal subgroup of G. Let $G = N \rtimes H$ be the semi-direct product as defined at the beginning of Section 7.4.

(1) Show that the map $q\colon N \rtimes H \to H$ given by $q(nh) = h$ $(n \in N, h \in H)$ is a surjective homomorphism with kernel N.
(2) Let $H_\nu = G_\nu \cap H$ and check that $G_\nu = N \rtimes H_\nu$.
(3) Prove that the map $Ns \mapsto s$ (with $s \in H_\nu$) is an isomorphism from G_ν/N to H_ν.

Problem 7.8. Review Definition 7.8 and show that $\nu\rho$ is equivalent to $\nu\rho'$ iff ρ is equivalent to ρ'.

Problem 7.9. Recall that $\mathbf{SE}(n)$ is the group of rigid motions of $\mathbb{R}^n$ defined as the group of $(n+1) \times (n+1)$ matrices

$$\mathbf{SE}(n) = \left\{ \begin{pmatrix} Q & u \\ 0 & 1 \end{pmatrix} \mid Q \in \mathbf{SO}(n),\ u \in \mathbb{R}^n \right\};$$

see Example 7.1. The subgroups N and H are defined as follows:

$$N = \left\{ \begin{pmatrix} I_n & u \\ 0 & 1 \end{pmatrix} \mid u \in \mathbb{R}^n \right\}, \quad H = \left\{ \begin{pmatrix} Q & 0 \\ 0 & 1 \end{pmatrix} \mid Q \in \mathbf{SO}(n) \right\},$$

in which case $\mathbf{SE}(n) = N \rtimes H$.

The orbits of the action of $G = \mathbf{SE}(n)$ on $\widehat{N}$ can be viewed as the orbits of the action of $\mathbf{SO}(n)$ on $\mathbb{R}^n$, namely for every $r \in [0, +\infty)$, the sphere $S_r(0)$ of radius r centered at the origin,

$$\mathcal{O}_r = S_r(0) = \{x \in \mathbb{R}^n \mid \|x\|_2 = r\}.$$

For every $r > 0$, for every $y \in \mathbb{R}^n$ with $\|y\|_2 = r$, we have the character $\nu_{r,y}$ given by $\nu_{r,y}(x) = e^{i(y \cdot x)}$. Then for every irreducible representation $\rho\colon \mathbf{SO}(n-1) \to \mathbf{U}(\mathcal{H}_\rho)$ of $\mathbf{SO}(n-1)$, we have the irreducible representation $\nu_{r,y}\rho\colon \mathbb{R}^n \rtimes \mathbf{SO}(n-1) \to \mathbf{U}(\mathcal{H}_\rho)$ given by

$$(\nu_{r,y}\rho)(xQ) = \nu_{r,y}(x)\rho(Q) = e^{iy\cdot x}\rho(Q), \quad x \in \mathbb{R}^n,\ Q \in \mathbf{SO}(n-1),\ \|y\|_2 = r;$$

see (2) of Example 7.1.

In the special case where $\rho\colon \mathbf{SO}(n-1) \to \mathbf{U}(1)$ is the trivial representation ($\rho(Q) = 1$ for all $Q \in \mathbf{SO}(n-1)$), show that the induced representation $\mathrm{Ind}_{\mathbb{R}^n \rtimes \mathbf{SO}(n-1)}^{\mathbf{SE}(n)} \nu_{r,y}\rho$ is equivalent to the induced representation $\mathrm{Ind}_{\mathbb{R}^n}^{\mathbf{SE}(n)} \nu_{r,y}$.

Hint. See Section 6.3 in Folland [Folland (1995)].

Problem 7.10.

(1) Verify that $\mathbf{O}(2)$ is isomorphic to the semi-direct product $\mathbf{SO}(2) \rtimes \{I_2, J\}$, where

$$J = \begin{pmatrix} -1 & 0 \\ 0 & 1 \end{pmatrix}.$$

(2) Prove that $\mathbf{O}(2m)$ is isomorphic to the semi-direct product $\mathbf{SO}(2m) \rtimes \{I_{2m}, J\}$ for any $m \geq 1$, where J is any reflection in $\mathbf{O}(2m)$, for instance $J = \mathrm{diag}(-1, 1, \ldots, 1)$. Also show that H is not normal in $\mathbf{O}(2m)$.

(3) Verify that $\mathbf{O}(2)$ acts regularly on $\widehat{\mathbf{SO}(2)}$. See Example 7.2 and Definition 7.6.

(4) Prove that the set of matrices

$$\left\{ \begin{pmatrix} e^{i\theta} & 0 \\ 0 & e^{-i\theta} \end{pmatrix}, \begin{pmatrix} 0 & e^{i\theta} \\ e^{-i\theta} & 0 \end{pmatrix} \,\middle|\, 0 \leq \theta < 2\pi \right\}$$

is a subgroup of $\mathbf{SU}(2)$ isomorphic to $\mathbf{O}(2)$ under the map

$$R_\theta \mapsto \begin{pmatrix} e^{i\theta} & 0 \\ 0 & e^{-i\theta} \end{pmatrix}, \; R_\theta J \mapsto \begin{pmatrix} 0 & e^{i\theta} \\ e^{-i\theta} & 0 \end{pmatrix}.$$

Problem 7.11. Use the results of Example 7.2 to determine all the irreducible representations of $\mathbf{E}(2) = \mathbb{R}^2 \rtimes \mathbf{O}(2)$.

Hint. Use a method analogous to the method used for $\mathbf{SE}(2) = \mathbb{R}^2 \rtimes \mathbf{SO}(2)$ in Example 7.1.

Problem 7.12. Use the results of Example 4.6 to determine all the irreducible representations of $\mathbf{E}(3) = \mathbb{R}^3 \rtimes \mathbf{O}(3)$ *via* the techniques of Example 7.1.

Chapter 8

Equivariant Convolutional Neural Networks

8.1 Introduction

Most of the material in this chapter is heavily inspired by the work of Bekkers, Boomsma, Cesa, Cohen, Forré, Geiger, Lang, Verlinder, Weiler, and Welling. The general theme is to develop a theory of equivariant convolutional neural networks (CNNs). Such neural networks process spatially structured data like images, audio, or videos. The purpose and the need for such neural networks is very clearly articulated in the preface of the recent book by Weiler, Forré, Verlinde, and Welling [Weiler *et al.* (2023)] that we highly recommend. Eric Bekkers' Lectures available on YouTube also provide an excellent coverage of this topic (Group Equivariant Deep Learning, UvA-2022). Our goal in this chapter is to show how many of the fairly abstract concepts discussed earlier (representations, analysis on compact groups, Peter–Weyl theorems, Fourier transform, induced representations) are used to tackle very practical problems. Sections 8.7, 8.9 and 8.10 do contain original results.

In Section 8.2, motivated by the problem of matching a pattern k (also called a correlation kernel or template kernel) in an image f, we define the notion of *cross-correlation*, for short *correlation*, given by

$$(k \star f)(x) = \int_{\mathbb{R}^2} f(t)k(t-x)\,dt.$$

Technically, a correlation is a convolution with the reflected kernel $\check{k}$, which is the function defined by $\check{k}(s) = k(-s)$. However, for our purpose, the notion of correlation is more natural.

Since images and correlation kernels are viewed as functions from $\mathbb{R}^2$ to $\mathbb{R}$, it is natural to view the action of a group G on images as given by the regular representation $\mathbf{R}$ of G on $\mathrm{L}^2(\mathbb{R}^2)$ induced by an action of G in $\mathbb{R}^2$, namely

$$[\mathbf{R}_g(f)](x) = \lambda_g(f)(x) = f(g^{-1}\cdot x), \quad g \in G,\, x \in \mathbb{R}^2,\, f \in \mathrm{L}^2(\mathbb{R}^2).$$

If we want to be more precise, we denote this representation by $\mathbf{R}^{G\to \mathrm{L}^{\mathbb{R}^2}}$. Observe that

$$[\mathbf{R}_g(f)](g\cdot x) = f(g^{-1}\cdot (g\cdot x)) = f(x),$$

so the color $f(x)$ of the pixel originally at location x is now the color at location $g\cdot x$ in the image $\mathbf{R}_g(f)$, which means the image defined by $\mathbf{R}_g(f)$ is obtained by applying the transformation g to the image defined by f. In computer vision, this is called *image warping*. For example, if $G = \mathbf{SE}(2)$, the image f is translated and rotated by $g = (x, R) \in \mathbf{SE}(2)$.

In the special case where $G = \mathbb{R}^2$, the group of translations of $\mathbb{R}^2$ itself, because the Lebesgue measure on $\mathbb{R}^2$ is translation-invariant, we have the following commutative diagram expressing that the linear map $\Phi\colon \mathrm{L}^2(\mathbb{R}^2) \to \mathrm{L}^2(\mathbb{R}^2)$ given by $\Phi(f) = k \star f$ is *translation-invariant*:

$$\begin{array}{ccc} \mathrm{L}^2(\mathbb{R}^2) & \xrightarrow{\Phi} & \mathrm{L}^2(\mathbb{R}^2) \\ {\scriptstyle \mathbf{R}_x^{\mathbb{R}^2\to \mathrm{L}^2(\mathbb{R}^2)}}\big\downarrow & & \big\downarrow{\scriptstyle \mathbf{R}_x^{\mathbb{R}^2\to \mathrm{L}^2(\mathbb{R}^2)}} \\ \mathrm{L}^2(\mathbb{R}^2) & \xrightarrow[\Phi]{} & \mathrm{L}^2(\mathbb{R}^2) \end{array}$$

commutes for all $x \in \mathbb{R}^2$. See Figure 8.3.

However, if G is the group $\mathbf{SO}(2)$, the rotations in the plane $\mathbb{R}^2$, if the image f is rotated by an angle θ, we have the new image given by

$$(\mathbf{R}_{R_\theta}^{\mathbf{SO}(2)\to \mathrm{L}^2(\mathbb{R}^2)} f)(t) = f(R_{-\theta}(t)), \quad t \in \mathbb{R}^2,\ R_\theta \in \mathbf{SO}(2),$$

but the diagram

$$\begin{array}{ccc} \mathrm{L}^2(\mathbb{R}^2) & \xrightarrow{\Phi} & \mathrm{L}^2(\mathbb{R}^2) \\ {\scriptstyle \mathbf{R}_{R_\theta}^{\mathbf{SO}(2)\to \mathrm{L}^2(\mathbb{R}^2)}}\big\downarrow & & \big\downarrow{\scriptstyle \mathbf{R}_{R_\theta}^{\mathbf{SO}(2)\to \mathrm{L}^2(\mathbb{R}^2)}} \\ \mathrm{L}^2(\mathbb{R}^2) & \xrightarrow[\Phi]{} & \mathrm{L}^2(\mathbb{R}^2) \end{array}$$

does *not* commute. The linear map Φ is *not* rotation-equivariant. See Figure 8.4.

This is unfortunate, because in general, we would like to know whether the pattern k occurs in f, translated or rotated. More generally, if G is a group of transformations of $\mathbb{R}^2$, we would like our transform Φ to be *G-equivariant*, which means that the diagram

$$\begin{array}{ccc} \mathrm{L}^2(\mathbb{R}^2) & \xrightarrow{\Phi} & \mathrm{L}^2(\mathbb{R}^2) \\ {\scriptstyle \mathbf{R}_g^{G\to \mathrm{L}^2(\mathbb{R}^2)}}\downarrow & & \downarrow{\scriptstyle \mathbf{R}_g^{G\to \mathrm{L}^2(\mathbb{R}^2)}} \\ \mathrm{L}^2(\mathbb{R}^2) & \xrightarrow[\Phi]{} & \mathrm{L}^2(\mathbb{R}^2) \end{array}$$

commutes for all $g \in G$.

As we just explained, equivariance fails beyond translation-equivariance, so what can we do to remedy this problem?

A solution is to define a *lifted correlation*. The basic idea presented in Section 8.3 is that instead of rotating the input image, we apply a *rotated kernel* to the image. We first illustrate this process in the case of $\mathbf{SE}(2) = \mathbb{R}^2 \rtimes \mathbf{SO}(2)$, but this method works for semi-direct products of the form $G = \mathbb{R}^d \rtimes H$. We will denote an element of $\mathbf{SE}(2)$ as $g = (x, \theta)$, where $x \in \mathbb{R}^2$ and $\theta \in \mathbb{R} \pmod{2\pi}$. Then we define the lifted correlation $k \widetilde{\star} f$ by

$$(k \widetilde{\star} f)(x, \theta) = \int_{\mathbb{R}^2} f(t)(\lambda_{(x,\theta)} k)(t)\, dt = \int_{\mathbb{R}^2} f(t) k(R_{-\theta}(t - x))\, dt.$$

We are now using the lifted (rotated) kernel $\lambda_{R_\theta} k$, but observe that our transform now takes an input function f (image, signal) in $\mathrm{L}^2(\mathbb{R}^2)$, but yields an output function $\Phi(f) = k \widetilde{\star} f$ in the larger function space $\mathrm{L}^2(\mathbf{SE}(2))$ of functions defined on the *group* $\mathbf{SE}(2)$. Such functions are called *feature maps*. In this situation, a feature map in $\mathrm{L}^2(\mathbf{SE}(2))$ can be viewed as a stack of rectangular grids, one for each $\theta \in \mathbb{R} \pmod{2\pi}$. We can obtain a final score of the occurrence of the template k moved over the image f in all positions determined by the rotation $R_\theta \in \mathbf{SO}(2)$ *via* some projection process over the θ-axis; this is often called *pooling* in machine learning (max-pooling being a common instance of pooling).

The major benefit of lifted kernels is that we recover equivariance under the group $\mathbf{SO}(2)$.

All this is generalized to semi-direct products of the form $G = \mathbb{R}^d \rtimes H$, where H is a compact group. Correlation on feature maps (functions in $\mathrm{L}^2(G)$), called group correlation, is discussed in Section 8.4. However, for $d > 2$, it is usually not practically possible to discretize the group H, so a different approach is needed. A solution is to use *steerable families*, which are discussed in Section 8.6. The notion of steerability occurred first in the seminal paper of Freeman and Adelson [Freeman and Adelson (1991)].

The idea behind steerability is that if a function f is defined on some measure space X and if there is an action of a group H on X, then it would be nice if $f(h^{-1} \cdot x)$ could be expressed in a simple way in terms of $f(x)$. In general, this is asking for too much, but if we consider a *family* of linearly independent functions $(Y_1, \ldots, Y_L)$ in $\mathrm{L}^2(X)$, then we say that they form an *H-steerable family* if there is representation $\Sigma\colon H \to \mathbf{U}(L)$, such that

$$Y(h^{-1} \cdot x) = \Sigma(h)^{\top} Y(x), \quad h \in H,\, x \in X,$$

where $Y(x)$ denotes the column vector

$$Y(x) = \begin{pmatrix} Y_1(x) \\ \vdots \\ Y_L(x) \end{pmatrix} \in \mathbb{C}^L;$$

see Definition 8.5. A typical case is $X = S^1$ (the circle) and $H = \mathrm{SO}(2)$ (the group of rotation in the plane), in which case, for any integers $n_1, \ldots, n_L$, the circular harmonics $(Y_1(\alpha) = e^{-in_1\alpha}, \ldots, Y_L(\alpha) = e^{-in_L\alpha})$ form a steerable family. If a correlation kernel k can be expressed as a linear combination of a steerable family Y, then the lifted convolution $k \widetilde{\star} f$ can be computed in a cheap way in terms of the vector-valued function

$$f^Y(x) = \int_{\mathbb{R}^d} f(t) Y(t - x)\, dt.$$

We can think of $f^Y(x)$ as some kinds of Fourier coefficients.

In Section 8.7, we present a method for finding steerable families on a suitable space X equipped with a continuous action of a compact group H. The trick is to consider the unitary representation $V\colon H \to \mathbf{U}(\mathrm{L}^2(X))$ given by

$$(V(h)f)(x) = f(h^{-1} \cdot x), \quad h \in H,\, f \in \mathrm{L}^2(X),\, x \in X. \qquad (V)$$

According to the Peter–Weyl theorem, Version II, the space $\mathrm{L}^2(X)$ is the Hilbert sum of closed subspaces E_ρ with $\rho \in R(H)$. Furthermore, each subspace E_ρ is a finite or countably infinite Hilbert sum of d_ρ (where $d_\rho = \infty$ is possible) closed finite-dimensional subspaces $E_\rho^{k_\rho}$ $(1 \leq k_\rho \leq d_\rho)$, such that for every ρ and every k_ρ, each subrepresentation $V_\rho^{k_\rho}\colon H \to \mathbf{U}(E_\rho^{k_\rho})$ is equivalent to the irreducible representation $M_\rho\colon H \to \mathbf{U}(\mathbb{C}^{n_\rho})$. We can find steerable families for each of the spaces $E_\rho^{k_\rho}$ (the families Y_{ρ,k_ρ}; see Theorem 8.7). We give many examples of steerable families, including $\mathbf{SE}(d)$ and homogeneous spaces $X = H/H_0$ with H compact.

In Section 8.8, we introduce the notion of *feature field*, which as Cesa, Lang and Weiler [Cesa *et al.* (2022)] say, "is the fundamental design choice underlying steerable CNNs." Such functions already arise when steerable kernels are used. Feature fields are vector-valued functions $f\colon \mathbb{R}^d \to \mathcal{H}$ whose domain transforms under the action of a group $G = \mathbb{R}^d \rtimes H$ and whose codomain transforms under a representation $\sigma\colon H \to \mathbf{GL}(\mathcal{H})$, in most cases actually a unitary representation. Thus, the space of feature fields transforms under the induced representation $\mathrm{Ind}_H^G\,\sigma$, namely for any feature field f,

$$[(\mathrm{Ind}_H^G\,\sigma)_{(x,h)} f](t) = \sigma(h)(f(h^{-1}\cdot(t-x))), \quad (x,h) \in \mathbb{R}^d \rtimes H,\ t \in \mathbb{R}^d.$$

We know how to transform G-feature maps using group correlation defined in Definition 8.4. This defines a transform Φ on $\mathrm{L}^2(G)$ (where $G = \mathbb{R}^d \rtimes H$) given by $f_{\mathrm{out}} = \Phi(f_{\mathrm{in}}) = k \star f_{\mathrm{in}}$. Since it is too expensive to compute $\Phi(f_{\mathrm{in}}) = k \star f_{\mathrm{in}}$, it would be nice if we could define a vector space of Fourier coefficients $\mathrm{L}^2(\mathbb{R}^d, \widehat{H})$ consisting of matrix-valued functions on $\mathbb{R}^d$, and a new Fourier transform $\mathcal{F}^\tau\colon \mathrm{L}^2(G) \to \mathrm{L}^2(\mathbb{R}^d, \widehat{H})$ and cotransform $\overline{\mathcal{F}^\tau}\colon \mathrm{L}^2(\mathbb{R}^d, \widehat{H}) \to \mathrm{L}^2(G)$ that promote the Fourier transform $\mathcal{F}$ on H (and are cheap to compute), so that we have the following diagram

$$\begin{array}{ccc} \mathrm{L}^2(G) & \xrightarrow{\quad\Phi\quad} & \mathrm{L}^2(G) \\ \mathcal{F}^\tau \Big\downarrow \Big\uparrow \overline{\mathcal{F}^\tau} & & \mathcal{F}^\tau \Big\downarrow \Big\uparrow \overline{\mathcal{F}^\tau} \\ \mathrm{L}^2(\mathbb{R}^d, \widehat{H}) & \xrightarrow[?]{\quad\cdots\quad} & \mathrm{L}^2(\mathbb{R}^d, \widehat{H}). \end{array}$$

The missing map $\widehat{\Phi}$, a notion of correlation on feature fields, would allow us to recover $k \star f_{\mathrm{in}}$ by Fourier inversion. We simply define $\widehat{\Phi}$ as

$$\widehat{\Phi} = \mathcal{F}^\tau \circ \Phi \circ \overline{\mathcal{F}^\tau},$$

using $\mathcal{F}^\tau$ and $\overline{\mathcal{F}^\tau}$. The problem is then to define the space $\mathrm{L}^2(\mathbb{R}^d, \widehat{H})$ and the Fourier transform and Fourier cotransform on it. To do this rigorously is nontrivial.

A function $f \in \mathrm{L}^2(\mathbb{R}^d \rtimes H)$ can be viewed as a function $f^H\colon \mathbb{R}^d \to \mathrm{L}^2(H)$, and when H is a compact group, f^H corresponds to a family $(\widehat{f}_\rho)$ of functions defined by the Fourier transforms of the functions $f^H(x)$. Furthermore, the functions $\widehat{f}_\rho$ are feature fields $\widehat{f}_\rho\colon \mathbb{R}^d \to \mathrm{M}_{n_\rho}(\mathbb{C})$. The original function $f \in \mathrm{L}^2(\mathbb{R}^d \rtimes H)$ can be recovered pointwise by Fourier inversion from the family of functions $\widehat{f}_\rho$. However, the new twist is that the Fourier coefficients of f are now tuples $(\widehat{f}_\rho)_{\rho\in R(H)}$ of *functions* $\widehat{f}_\rho\colon \mathbb{R}^d \to \mathrm{M}_{n_\rho}(\mathbb{C})$. This causes new problems to reconstruct a function from its Fourier coefficients because even if the functions $\widehat{f}_\rho$ belong to $\mathrm{L}^2(\mathbb{R}^d, \mathrm{M}_{n_\rho}(\mathbb{C}))$, there is no guarantee that the function obtained from the inverse Fourier transform belongs to $\mathrm{L}^2(G)$. Some additional condition is required on the functions $\widehat{f}_\rho$.

We provide a solution to this problem in Section 8.9 by constructing a Hilbert space $\mathrm{L}^2(\mathbb{R}^d, \widehat{H})$, such that the new Fourier transform $\mathcal{F}^\tau\colon \mathrm{L}^2(G) \to \mathrm{L}^2(\mathbb{R}^d, \widehat{H})$ and the Fourier cotransform $\overline{\mathcal{F}^\tau}\colon \mathrm{L}^2(\mathbb{R}^d, \widehat{H}) \to \mathrm{L}^2(G)$ are mutual inverses; see Theorem 8.8.

We denote the projection of $\mathrm{L}^2(\mathbb{R}^d, \widehat{H})$ on the ρth factor, by $\mathrm{L}^2(\mathbb{R}^d, \widehat{H})_\rho$. Then the maps $\mathcal{F}^\tau\colon \mathrm{L}^2(G) \to \mathrm{L}^2(\mathbb{R}^d, \widehat{H})$ and $\overline{\mathcal{F}^\tau}\colon \mathrm{L}^2(\mathbb{R}^d, \widehat{H}) \to \mathrm{L}^2(G)$ define the family of maps $\mathcal{F}^\tau_\rho\colon \mathrm{L}^2(G) \to \mathrm{L}^2(\mathbb{R}^d, \widehat{H})_\rho$ and $\overline{\mathcal{F}^\tau}_\rho\colon \mathrm{L}^2(\mathbb{R}^d, \widehat{H})_\rho \to \mathrm{L}^2(G)$. For every $\rho \in R(H)$, let $\sigma_\rho\colon H \to \mathbf{U}(\mathrm{M}_{n_\rho}(\mathbb{C}))$ be the representation

$$\sigma_\rho = \mathrm{Hom}(M_\rho, \mathrm{id})$$

associated with the representation $M_\rho\colon H \to \mathbf{U}(\mathbb{C}^{n_\rho})$ as in Definition 8.8. Then the diagrams

$$\begin{array}{ccc}
\mathrm{L}^2(G) & \xrightarrow{\ \mathcal{F}^\tau_\rho\ } & \mathrm{L}^2(\mathbb{R}^d, \widehat{H})_\rho \\
{\scriptstyle \mathbf{R}^{G\to\mathrm{L}^2(G)}_{(x,h)}}\big\downarrow & & \big\downarrow{\scriptstyle (\mathrm{Ind}_H^G\,\sigma_\rho)_{(x,h)}} \\
\mathrm{L}^2(G) & \xrightarrow[\ \mathcal{F}^\tau_\rho\ ]{} & \mathrm{L}^2(\mathbb{R}^d, \widehat{H})_\rho
\end{array}$$

and

$$\begin{array}{ccc}
\mathrm{L}^2(\mathbb{R}^d, \widehat{H})_\rho & \xrightarrow{\ \overline{\mathcal{F}^\tau}_\rho\ } & \mathrm{L}^2(G) \\
{\scriptstyle (\mathrm{Ind}_H^G\,\sigma_\rho)_{(x,h)}}\big\downarrow & & \big\downarrow{\scriptstyle \mathbf{R}^{G\to\mathrm{L}^2(G)}_{(x,h)}} \\
\mathrm{L}^2(\mathbb{R}^d, \widehat{H})_\rho & \xrightarrow[\ \overline{\mathcal{F}^\tau}_\rho\ ]{} & \mathrm{L}^2(G)
\end{array}$$

commute, with

$$[(\mathrm{Ind}_H^G\,(\sigma_\rho)_{(x,h)}\,\widehat{f}_\rho](x_1) = \widehat{f}_\rho(h^{-1}\cdot(x_1 - x))\,M_\rho(h)^*.$$

Since a group correlation Φ on $\mathrm{L}^2(G)$ is equivariant with respect to the regular representation $\mathbf{R}^{G\to\mathrm{L}^2(G)}$, we see that if we define $\widehat{\Phi}_{\rho_2,\rho_1}$ as

$$\widehat{\Phi}_{\rho_2,\rho_1}(\widehat{f}_{\rho_1}) = \mathcal{F}^\tau_{\rho_2}(\Phi(\overline{\mathcal{F}^\tau}_{\rho_1}(\widehat{f}_{\rho_1}))),$$

using the two commutative diagrams above, we have the following commutative diagram

$$\begin{array}{ccc} \mathrm{L}^2(\mathbb{R}^d, \widehat{H})_{\rho_1} & \xrightarrow{\widehat{\Phi}_{\rho_2,\rho_1}} & \mathrm{L}^2(\mathbb{R}^d, \widehat{H})_{\rho_2} \\ {\scriptstyle (\mathrm{Ind}_H^G \sigma_{\rho_1})_{(x,h)}} \Big\downarrow & & \Big\downarrow {\scriptstyle (\mathrm{Ind}_H^G \sigma_{\rho_2})_{(x,h)}} \\ \mathrm{L}^2(\mathbb{R}^d, \widehat{H})_{\rho_1} & \xrightarrow[\widehat{\Phi}_{\rho_2,\rho_1}]{} & \mathrm{L}^2(\mathbb{R}^d, \widehat{H})_{\rho_2}, \end{array}$$

which shows that $\widehat{\Phi}_{\rho_2,\rho_1}$ is equivariant with respect to the representations $\mathrm{Ind}_H^G \sigma_{\rho_1}$ and $\mathrm{Ind}_H^G \sigma_{\rho_2}$. If we define $\widehat{\Phi}_{\rho_1}$ as

$$\widehat{\Phi}_{\rho_1}(\widehat{f}_{\rho_1}) = \sum_{\rho_1 \in R(H)} \widehat{\Phi}_{\rho_2,\rho_1}(\widehat{f}_{\rho_1}),$$

then we obtain our desired correlation on $\mathrm{L}^2(\mathbb{R}^d, \widehat{H})$ given by

$$\widehat{\Phi}((\widehat{f}_{\rho_1})_{\rho_1 \in R(H)}) = \big(\widehat{\Phi}_{\rho_1}(\widehat{f}_{\rho_1})\big)_{\rho_2 \in R(H)},$$

and in view of the previous discussion, it is equivariant with respect to the induced representations $\mathrm{Ind}_H^G \sigma_{\rho_1}$ and $\mathrm{Ind}_H^G \sigma_{\rho_2}$, more precisely the representation defined by the family of these representations. Of course, this is the main point of group correlation!

In Section 8.10, we show that if the group correlation $\Phi\colon \mathrm{L}^2(G) \to \mathrm{L}^2(G)$ (with $G = \mathbb{R}^d \rtimes H$) is given by a kernel k as

$$\Phi(f)(x,h) = \int_{\mathbb{R}^d \rtimes H} k(h^{-1} \cdot (x_1 - x), h^{-1}h_1) f(x_1, h_1)\, d\lambda_H(h_1)\, dx_1,$$

then

$$[\widehat{\Phi}_{\rho_2,\rho_1}(\widehat{f}_{\rho_1})](x) = \mathcal{F}^{\tau}_{\rho_2}[\Phi(\overline{\mathcal{F}^{\tau}}_{\rho_1}(\widehat{f}_{\rho_1}))](x) = \int_{\mathbb{R}^d} \Phi_{\rho_2,\rho_1}(x_1 - x, \widehat{f}_{\rho_1}(x_1))\, dx_1, \tag{$\widehat{\Phi}^{\text{bis}}_{\rho_2,\rho_1}$}$$

where $\Phi_{\rho_2,\rho_1}\colon \mathbb{R}^d \times \mathrm{M}_{n_{\rho_1}}(\mathbb{C}) \to \mathrm{M}_{n_{\rho_2}}(\mathbb{C})$ is given by

$$\Phi_{\rho_2,\rho_1}(x_1 - x, A) = \int_H \int_H n_{\rho_1} \mathrm{tr}\Big(A M_{\rho_1}(h_1)\Big) k(h^{-1} \cdot (x_1 - x), h^{-1}h_1) M_{\rho_2}(h)^*\, d\lambda_H(h)\, d\lambda_H(h_1), \tag{Φ_{ρ_2,ρ_1}}$$

with $A \in \mathrm{M}_{n_{\rho_1}}(\mathbb{C})$.

In order to go further, we need to expres1s the kernel $\Phi_{\rho_2,\rho_1}(x, A)$ in terms of H-steerable functions on $\mathrm{L}^2(\mathbb{R}^d \rtimes H)$. We show how to proceed with

$H = \mathbf{SO}(d)$ using the steerable basis of $\mathrm{L}^2(\mathbf{SE}(d))$ given in Example 8.9. We obtain

$$\Phi_{\rho_2,\rho_1}(x_1 - x, A) = \int_H n_{\rho_1} \operatorname{tr}\Big(AM_{\rho_1}(h_1)\Big) W_{\rho_2}(h_1^{-1}(x_1 - x)) M^*_{\rho_2}(h_1)\, d\lambda_H(h_1), \qquad (*_{\Phi_{\rho_2,\rho_1}})$$

and

$$\begin{aligned}[\widehat{\Phi}_{\rho_2,\rho_1}(\widehat{f}_{\rho_1})](x) &= \mathcal{F}^{\tau}_{\rho_2}[\Phi(\overline{\mathcal{F}^{\tau}}_{\rho_1}(\widehat{f}_{\rho_1}))](x) \\ &= \int_{\mathbb{R}^d} \Phi_{\rho_2,\rho_1}(x_1 - x, \widehat{f}_{\rho_1}(x_1))\, dx_1, \qquad (*_{\widehat{\Phi}_{\rho_2,\rho_1}})\end{aligned}$$

where $W_{\rho_2}(x_1)$ is an $n_{\rho_2} \times n_{\rho_2}$ matrix, whose $(k_{\rho_2}, \ell_{\rho_2})$ entry is a function in $\mathrm{L}^2(\mathbb{R}^d)$.

The above construction is performed entirely in Section 8.11 for the group $\mathbf{SE}(2) = \mathbb{R}^2 \rtimes \mathbf{SO}(2)$. The corresponding CNNs are known as *harmonic nets*.

Because the group correlation Φ is equivariant with respect to the left regular representation $\mathbf{R}$ (on $\mathrm{L}^2(G)$), the components $\widehat{\Phi}_{\rho_2,\rho_1}$ of $\widehat{\Phi}$ are equivariant with respect to the representations $\mathrm{Ind}_H^G\, \sigma_{\rho_1}$ and $\mathrm{Ind}_H^G\, \sigma_{\rho_2}$, namely the following diagram commutes.

$$\begin{array}{ccc} \mathrm{L}^2(\mathbb{R}^d, \widehat{H})_{\rho_1} & \xrightarrow{\widehat{\Phi}_{\rho_2,\rho_1}} & \mathrm{L}^2(\mathbb{R}^d, \widehat{H})_{\rho_2} \\ {\scriptstyle (\mathrm{Ind}_H^G \sigma_{\rho_1})_{(x,h)}} \downarrow & & \downarrow {\scriptstyle (\mathrm{Ind}_H^G \sigma_{\rho_2})_{(x,h)}} \\ \mathrm{L}^2(\mathbb{R}^d, \widehat{H})_{\rho_1} & \xrightarrow[\widehat{\Phi}_{\rho_2,\rho_1}]{} & \mathrm{L}^2(\mathbb{R}^d, \widehat{H})_{\rho_2}. \end{array}$$

Practice shows that it is desirable to design more general group correlations that are equivariant with respect to other representations besides the left regular representation and to consider feature fields that transform under representations other than the representations $\mathrm{Hom}(M_\rho, \mathrm{id})$.

A first generalization is to have two feature fields spaces $\mathbf{FF}(\mathbb{R}^d, H, \sigma_{\mathrm{in}} \colon H \to \mathbf{U}(\mathcal{H}_{\mathrm{in}}))$ and $\mathbf{FF}(\mathbb{R}^d, H, \sigma_{\mathrm{out}} \colon H \to \mathbf{U}(\mathcal{H}_{\mathrm{out}}))$ associated with an input representation σ_{in} and an output representation σ_{out}, where $\mathcal{H}_{\mathrm{in}}$ and $\mathcal{H}_{\mathrm{out}}$ are two finite-dimensional vector spaces equipped with a hermitian inner product, and what we are seeking is a linear G-equivariant map $\widehat{\Phi}$ between these spaces. We assume that feature fields $f \colon \mathbb{R}^d \to \mathcal{H}_{\mathrm{in}}$ are functions in $\mathrm{L}^2(\mathbb{R}^d, \mathcal{H}_{\mathrm{in}})$, and similarly for feature fields $f \colon \mathbb{R}^d \to \mathcal{H}_{\mathrm{out}}$

(see Definition 6.25). To say that $\widehat{\Phi}$ is G-equivariant means that the following diagram commutes

$$\begin{array}{ccc} \mathbf{FF}(\mathbb{R}^d, H, \sigma_{\mathrm{in}}) & \xrightarrow{\widehat{\Phi}} & \mathbf{FF}(\mathbb{R}^d, H, \sigma_{\mathrm{out}}) \\ {\scriptstyle (\mathrm{Ind}_H^G\, \sigma_{\mathrm{in}})_{(x,h)}} \downarrow & & \downarrow {\scriptstyle (\mathrm{Ind}_H^G\, \sigma_{\mathrm{out}})_{(x,h)}} \\ \mathbf{FF}(\mathbb{R}^d, H, \sigma_{\mathrm{in}}) & \xrightarrow[\widehat{\Phi}]{} & \mathbf{FF}(\mathbb{R}^d, H, \sigma_{\mathrm{out}}) \end{array}$$

for all $(x, h) \in G = \mathbb{R}^d \rtimes H$.

A complete solution to this problem was given in a sequence of remarkable papers by Weiler, Geiger, Weilling, Boomsma and Cohen [Weiler *et al.* (2018)] (for $\mathbf{SE}(3)$), Weiler and Cesa [Weiler and Cesa (2019)] (for $\mathbf{E}(2)$), Lang and Weiler [Lang and Weiler (2019)] (for a homogeneous space X induced by a transitive action of a compact group H), Cesa, Lang and Weiler [Cesa *et al.* (2022)] (for $\mathbf{E}(3)$), and Cohen, Geiger and Weiler [Cohen *et al.* (2019)] (feature fields on homogeneous spaces).

In the case where $H = \mathbf{SO}(d)$, it is shown in Section 8.12 that such a map is given by a kernel $K\colon \mathbb{R}^d \to \mathrm{Hom}(\mathcal{H}_{\mathrm{in}}, \mathcal{H}_{\mathrm{out}})$ *via*

$$\widehat{\Phi}(f)(t) = \int_{\mathbb{R}^d} K(y - t)(f(y))\, dy, \quad f\colon \mathbb{R}^d \to \mathcal{H}_{\mathrm{in}}, \quad t \in \mathbb{R}^d, \tag{K1}$$

and the kernel K satisfies the *equivariance constraint*

$$K(h \cdot t) = \sigma_{\mathrm{out}}(h) \circ K(t) \circ \sigma_{\mathrm{in}}(h)^{-1}, \quad h \in \mathbf{SO}(d),\ t \in \mathbb{R}^d. \tag{EC$_1$}$$

Functions $K\colon \mathbb{R}^d \to \mathrm{Hom}(\mathcal{H}_{\mathrm{in}}, \mathcal{H}_{\mathrm{out}})$ satisfying the equivariance constraint (EC$_1$) are called *equivariant convolution kernels* or *G-steerable kernels*. The above result is often referred to by the slogan "correlation is all you need."

Until now, we have been assuming that we are dealing with feature fields defined on $X = \mathbb{R}^d$ and that the group G is a semi-direct product $G = \mathbb{R}^d \rtimes H$ with $H = \mathbf{SO}(d)$, and more generally a compact group. It is possible to deal with the more general situation, where X is a homogeneous space of the form $X = G/H$ with G locally compact and unimodular and H compact equipped with a unitary representation $\sigma\colon H \to \mathbf{U}(\mathcal{H}_\sigma)$. The main problem is to define the "right" notion of feature field. This issue is addressed in Section 8.13.

Cohen, Geiger and Weiler [Cohen *et al.* (2019)] propose to use the G-bundle $E = G \times_H \mathcal{H}_\sigma$ introduced in Section 6.13; see Definition 6.17. But then we might as well use the hermitian G-bundles of finite rank of Definition 6.23 (see Section 6.13) and *the natural choice for the space of feature fields is the subspace* $\mathrm{L}^2(X; E)$ *of the space of sections of the hermitian* G*-bundle* $p\colon E \to X$, with $X = G/H$ (see Definition 6.25). Recall that the restriction of the action of G to H on the fibre E_0 is a unitary representation $\sigma\colon H \to \mathbf{U}(E_0)$. For the time being, we will assume that there exists a section $r\colon X \to G$, such that the maps $\mathcal{L}\colon \mathrm{L}^2(X;E) \to L^\sigma$ and $\mathcal{S}\colon L^\sigma \to \mathrm{L}^2(X;E)$ define isomorphisms between $\mathrm{L}^2(X;E)$ and L^σ. Recall from Equation ($\dagger_4$) of Definition 6.24 that L^σ is the set consisting of all functions $f \in \mathrm{L}^2(G; E_0)$, such that

$$f(gh) = \sigma(h^{-1})(f(g)) = h^{-1} \cdot f(g), \qquad \text{for all } g \in G \text{ and all } h \in H.$$

In view of the isomorphism between $\mathrm{L}^2(X;E)$ and L^σ, the induced representation $\mathrm{Ind}_H^G \sigma$ is equivalent to the left regular representation of G in L^σ. We also assume that the section $r\colon X \to G$ makes the representation Π continuous.

Inspired by Cohen, Geiger and Weiler [Cohen *et al.* (2019)] we consider the more general situation in which we have two hermitian G-bundles of finite rank $p_{\mathrm{in}}\colon E_{\mathrm{in}} \to X_{\mathrm{in}}$ and $p_{\mathrm{out}}\colon E_{\mathrm{out}} \to X_{\mathrm{out}}$, where $X_{\mathrm{in}} = G/H_{\mathrm{in}}$ and $X_{\mathrm{out}} = G/H_{\mathrm{out}}$ for the *same* group G, input and output representations σ_{in} and σ_{out}, and determine what are the linear maps $\Phi\colon L^{\sigma_{\mathrm{in}}} \to L^{\sigma_{\mathrm{out}}}$ that are equivariant with respect to the representations $\mathrm{Ind}_{H_{\mathrm{in}}}^G \sigma_{\mathrm{in}}$ and $\mathrm{Ind}_{H_{\mathrm{out}}}^G \sigma_{\mathrm{out}}$, which means that the following diagram commutes

$$\begin{array}{ccc}
L^{\sigma_{\mathrm{in}}} & \xrightarrow{\ \Phi\ } & L^{\sigma_{\mathrm{out}}} \\
{\scriptstyle (\mathrm{Ind}_{H_{\mathrm{in}}}^G \sigma_{\mathrm{in}})(g)}\Big\downarrow & & \Big\downarrow{\scriptstyle (\mathrm{Ind}_{H_{\mathrm{out}}}^G \sigma_{\mathrm{out}})(g)} \\
L^{\sigma_{\mathrm{in}}} & \xrightarrow[\ \Phi\ ]{} & L^{\sigma_{\mathrm{out}}}
\end{array}$$

for all $g \in G$ (for simplicity of notation, we use Φ instead of $\widehat{\Phi}$).

To reduce the amount of subscripts, we will denote the fibre $(E_{\mathrm{in}})_0$ above $x_0^{\mathrm{in}} = H_{\mathrm{in}}$ by E_0^{in} and the fibre $(E_{\mathrm{out}})_0$ above $x_0^{\mathrm{out}} = H_{\mathrm{out}}$ by E_0^{out}. Note that E_0^{in} plays the role of $\mathcal{H}_{\mathrm{in}}$ and E_0^{out} plays the role of $\mathcal{H}_{\mathrm{out}}$. Then our representations σ_{in} and σ_{out} are $\sigma_{\mathrm{in}}\colon H_{\mathrm{in}} \to \mathbf{U}(E_0^{\mathrm{in}})$ and $\sigma_{\mathrm{out}}\colon H_{\mathrm{out}} \to \mathbf{U}(E_0^{\mathrm{out}})$. Technically, the equivariant linear maps Φ from $L^{\sigma_{\mathrm{in}}}$ to $L^{\sigma_{\mathrm{out}}}$ are the maps in the space $\mathrm{Hom}_{H_{\mathrm{in}},H_{\mathrm{out}}}(\mathrm{Ind}_{H_{\mathrm{in}}}^G \sigma_{\mathrm{in}}, \mathrm{Ind}_{H_{\mathrm{out}}}^G \sigma_{\mathrm{out}})$ of maps between the representations $\mathrm{Ind}_{H_{\mathrm{in}}}^G \sigma_{\mathrm{in}}$ and $\mathrm{Ind}_{H_{\mathrm{out}}}^G \sigma_{\mathrm{out}}$.

Proposition 8.12 generalizes results proven in Cohen, Geiger and Weiler [Cohen *et al.* (2019)] (see Theorem 3.1 and Theorem 3.2) and shows that the equivariant maps Φ as above are determined by the space of equivariant G-kernels given by

$$\begin{aligned}\mathrm{Hom}_{H_{\mathrm{in}},H_{\mathrm{out}}}(G,\mathrm{Hom}(E_0^{\mathrm{in}},E_0^{\mathrm{out}})) = \{K\colon G \to \mathrm{Hom}(E_0^{\mathrm{in}},E_0^{\mathrm{out}}) \mid \\ K(h_2 g h_1) = \sigma_{\mathrm{out}}(h_2)\circ K(g)\circ\sigma_{\mathrm{in}}(h_1), \\ g\in G,\ h_1\in H_{\mathrm{in}},\ h_2\in H_{\mathrm{out}}\}. \qquad (\mathrm{EC}_2)\end{aligned}$$

The above condition is more complicated than (EC_1), and these kernels are defined on G, which makes them rather impractical.

In Section 8.14, we give another characterizations originally due to Cohen, Geiger and Weiler [Cohen *et al.* (2019)] of the space $\mathrm{Hom}_{H_{\mathrm{in}},H_{\mathrm{out}}}(\mathrm{Ind}_{H_{\mathrm{in}}}^G \sigma_{\mathrm{in}}, \mathrm{Ind}_{H_{\mathrm{out}}}^G \sigma_{\mathrm{out}})$ in terms of kernels defined on $X_{\mathrm{in}} = G/H_{\mathrm{in}}$. More precisely, we prove that there is a bijection between the space of equivariant G-kernels $\mathrm{Hom}_{H_{\mathrm{in}},H_{\mathrm{out}}}(G,\mathrm{Hom}(E_0^{\mathrm{in}},E_0^{\mathrm{out}}))$ and the space $\mathrm{Hom}_{H_{\mathrm{out}}}(X_{\mathrm{in}},\mathrm{Hom}(E_0^{\mathrm{in}},E_0^{\mathrm{out}}))$ of equivariant X_{in}-kernels, which are maps $\kappa\colon X_{\mathrm{in}} \to \mathrm{Hom}(E_0^{\mathrm{in}},E_0^{\mathrm{out}})$ satisfying a certain condition; see Proposition 8.13.

The G-equivariant maps in $\mathrm{Hom}_{H_{\mathrm{in}},H_{\mathrm{out}}}(\mathrm{Ind}_{H_{\mathrm{in}}}^G \sigma_{\mathrm{in}},\ \mathrm{Ind}_{H_{\mathrm{out}}}^G \sigma_{\mathrm{out}})$ are functions from $L^{\sigma_{\mathrm{in}}}$ to $L^{\sigma_{\mathrm{out}}}$ and still require integration over G to be computed using equivariant kernels in the space $\mathrm{Hom}_{H_{\mathrm{in}},H_{\mathrm{out}}}(G,\mathrm{Hom}(E_0^{\mathrm{in}},E_0^{\mathrm{out}}))$. It would be nice if we could transform the integration over G to a more practically computable integration over X_{in}. This can be achieved by using the maps $\mathcal{S}_{\mathrm{out}}\colon L^{\sigma_{\mathrm{out}}} \to \mathrm{L}^2(X_{\mathrm{out}},E_{\mathrm{out}})$ and $\mathcal{L}_{\mathrm{in}}\colon \mathrm{L}^2(X_{\mathrm{in}},E_{\mathrm{in}}) \to L^{\sigma_{\mathrm{in}}}$ given by $(\mathcal{S}_3'')$ and $(\mathcal{L}_3')$ of Section 6.13. When these maps are well-defined, which is our assumption, they can be used to define maps from $\mathrm{L}^2(X,E_{\mathrm{in}})$ to $\mathrm{L}^2(X,E_{\mathrm{out}})$ from functions from $L^{\sigma_{\mathrm{in}}}$ to $L^{\sigma_{\mathrm{out}}}$. This process is explained in Section 8.15.

Pick a set of coset representatives $(r_x^{\mathrm{in}})_{x\in G/H_{\mathrm{in}}}$ for $X_{\mathrm{in}} = G/H_{\mathrm{in}}$ and a set of coset representatives $(r_x^{\mathrm{out}})_{x\in G/H_{\mathrm{out}}}$ for $X_{\mathrm{out}} = G/H_{\mathrm{out}}$. Then for every section $s\in \mathrm{L}^2(X_{\mathrm{in}},E_{\mathrm{in}})$, for every $x\in X_{\mathrm{out}}$, observe that for every equivariant kernel $K\in \mathrm{Hom}_{H_{\mathrm{in}},H_{\mathrm{out}}}(G,\mathrm{Hom}(E_0^{\mathrm{in}},E_0^{\mathrm{out}}))$, the function $\widetilde{\Phi}$ given by

$$\widetilde{\Phi}(s) = \mathcal{S}_{\mathrm{out}}(K \star (\mathcal{L}_{\mathrm{in}}(s)))$$

maps $L^2(X_{in}, E_{in})$ to $L^2(X_{out}, E_{out})$, as illustrated in the following diagram.

$$\begin{array}{ccc} L^{\sigma_{in}} & \xrightarrow{\Phi = K\star -} & L^{\sigma_{out}} \\ {\scriptstyle \mathcal{L}_{in}} \Big\uparrow & & \Big\downarrow {\scriptstyle \mathcal{S}_{out}} \\ L^2(X_{in}, E_{in}) & \xrightarrow[\widetilde{\Phi}]{} & L^2(X_{out}, E_{out}). \end{array}$$

We obtain formulae expressing $\mathcal{S}_{out}(K \star (\mathcal{L}_{in}(s)))$ in terms of an integral over X_{in}; see Formulae ($\dagger_8$) and ($\dagger_9$). Special cases of this formula are also discussed. The issue of finding G-equivariant kernels still remains and is addressed in Section 8.16.

As in Lang and Weiler [Lang and Weiler (2019)] and Cesa, Lang and Weiler [Cesa *et al.* (2022)] we now assume that $H_{in} = H_{out} = H$, so $X_{in} = X_{out} = X = G/H$, and we have two Hermitian G-bundles E_{in} and E_{out}. The Hermitian G-bundles define two representations $\sigma_{in} \colon H \to \mathbf{U}(\mathcal{H}_{in})$ and $\sigma_{out} \colon H \to \mathbf{U}(\mathcal{H}_{out})$, where we denote the fibres E_0^{in} and E_0^{out} as $\mathcal{H}_{in}$ and $\mathcal{H}_{out}$, which is closer to the notation used by the above authors. We consider the space of *equivariant X-kernels* defined as

$$\begin{aligned} \mathrm{Hom}_H(X, \mathrm{Hom}(\mathcal{H}_{in}, \mathcal{H}_{out})) = \{\kappa \colon X \to \mathrm{Hom}(\mathcal{H}_{in}, \mathcal{H}_{out}) \mid \\ \kappa(h \cdot x) = \sigma_{out}(h) \circ \kappa(x) \circ \sigma_{in}(h)^{-1}, \\ x \in X,\ h \in H\}. \end{aligned} \tag{EC$_6$}$$

Remarkably, Lang and Weiler [Lang and Weiler (2019)] and Cesa, Lang and Weiler [Cesa *et al.* (2022)] completely characterized the kernels in $\kappa \in \mathrm{Hom}_H(X, \mathrm{Hom}(\mathcal{H}_{in}, \mathcal{H}_{out}))$, when H is a compact group acting on a topological Hausdorff space X equipped with the σ-algebra of Borel sets and an H-invariant measure μ.

A key ingredient is the analog of the left regular representation $V \colon H \to \mathbf{U}(L^2(X))$ of $L^2(X)$ induced by the action of H on X already introduced in Section 8.7 and given by

$$(V(h)f)(x) = f(h^{-1} \cdot x), \quad h \in H,\ f \in L^2(X),\ x \in X.$$

For the sake of consistency of notation, we will also denote the representation V as $\mathbf{R}^{H \to L^2(X)}$.

The other key ingredient is the set of H-maps

$$\mathrm{Hom}_H(\mathbf{R}^{H \to L^2(X)}, \mathrm{Hom}(\sigma_{in}, \sigma_{out})),$$

which is the space of linear maps $\mathcal{K}\colon \mathrm{L}^2(X) \to \mathrm{Hom}(\mathcal{H}_{\mathrm{in}}, \mathcal{H}_{\mathrm{out}})$, such that the following diagram commutes

$$\begin{array}{ccc} \mathrm{L}^2(X) & \xrightarrow{\mathcal{K}} & \mathrm{Hom}(\mathcal{H}_{\mathrm{in}}, \mathcal{H}_{\mathrm{out}}) \\ {\scriptstyle \mathbf{R}^{H \to \mathrm{L}^2(X)}(h)}\big\downarrow & & \big\downarrow{\scriptstyle \mathrm{Hom}(\sigma_{\mathrm{in}}, \sigma_{\mathrm{out}})(h)} \\ \mathrm{L}^2(X) & \xrightarrow[\mathcal{K}]{} & \mathrm{Hom}(\mathcal{H}_{\mathrm{in}}, \mathcal{H}_{\mathrm{out}}) \end{array}$$

for every $h \in H$; see Definition 3.9. Cesa, Lang and Weiler [Cesa *et al.* (2022)] call the maps $\mathcal{K}$ *kernel operators*.

The main result is that there is a bijection between the space $\mathrm{Hom}_H(X, \mathrm{Hom}(\mathcal{H}_{\mathrm{in}}, \mathcal{H}_{\mathrm{out}}))$ of equivariant X-kernels and the space $\mathrm{Hom}_H(\mathbf{R}^{H \to \mathrm{L}^2(X)}, \mathrm{Hom}(\sigma_{\mathrm{in}}, \sigma_{\mathrm{out}}))$ of kernel operators; see Theorem 8.14. This isomorphism is a kind of linearization of the first space.

But now Proposition 4.23 tells us that the representations $\mathrm{Hom}(\sigma_{\mathrm{in}}, \sigma_{\mathrm{out}})$ and $\overline{\sigma_{\mathrm{in}}} \otimes \sigma_{\mathrm{out}}$ are equivalent, so we obtain an isomorphism

$$\mathrm{Hom}_H(X, \mathrm{Hom}(\mathcal{H}_{\mathrm{in}}, \mathcal{H}_{\mathrm{out}})) \approx \mathrm{Hom}_H(\mathbf{R}^{H \to \mathrm{L}^2(X)}, \overline{\sigma_{\mathrm{in}}} \otimes \sigma_{\mathrm{out}}). \qquad (\dagger_{15})$$

Since H is a compact group, we can now use Theorem 8.7 (a direct consequence of Peter–Weyl II) to express $\mathrm{L}^2(X)$ as a Hilbert sum of spaces corresponding to irreducible representations of H and the decomposition of the tensor product representation $\overline{\sigma_{\mathrm{in}}} \otimes \sigma_{\mathrm{out}}$ as a Hilbert sum of irreducible representations of H (see Proposition 4.18 and Equation ($\otimes$) in Section 4.4). Such a decomposition is achieved in a theorem referred to as *Wigner–Eckart theorem for steerable kernels* by Cesa, Lang and Weiler [Cesa *et al.* (2022)] (Theorem B.5); see Theorem 8.15. The more general theorem that also applies to real representations is proven in Cesa, Lang and Weiler [Cesa *et al.* (2022)].

Cesa, Lang and Weiler [Cesa *et al.* (2022)] also prove a version of the above result in which a basis of $\mathrm{Hom}_H(X, \mathrm{Hom}(\mathcal{H}_{\mathrm{in}}, \mathcal{H}_{\mathrm{out}}))$ is exhibited. The formulae are a bit messy, so we will not give details here; see Theorem B.6 and Theorem B.7 in Cesa, Lang and Weiler [Cesa *et al.* (2022)]. The idea is clear though. A steerable basis for $\mathrm{L}^2(X)$ is provided by Theorem 8.7; these are the functions Y_{ρ, k_ρ}. Matrices $CG_j^{c^{\rho_1}_{\mathrm{in,out}}}$ of Clebsch–Gordan coefficients expressing the change of basis required when decomposing the representation $\overline{\sigma_{\mathrm{in}}} \otimes \sigma_{\mathrm{out}}$ into irreducibles are needed. The commutants $\mathcal{C}(M_{\rho_1})$ of the representations M_{ρ_1} also play a role in the real case; see Section 8.16.

8.2 Cross-Correlation as Template Matching

Template matching is one of the central problems arising in computer vision. Instances of this problem are (1) recognizing a face or shape in a scene, say a group of people; (2) detecting the presence of an abnormal cell in a tissue.

A way to describe the problem in mathematical terms is to represent images, more generally "signals," as functions $f\colon \mathbb{R}^2 \to \mathbb{R}$. Actually, such functions have compact support, which is typically a rectangular grid, and they are only defined on the finite number of grid points. Mathematically, it is convenient to assume that these functions belong to the Hilbert space of real valued functions in $\mathrm{L}^2(\mathbb{R}^2)$. The pattern to be detected is also given by a function $k \in \mathrm{L}^1(\mathbb{R}^2)$ with compact support usually known as a *kernel*.[1] To detect whether the pattern k occurs in the image f, we slide k over f using all possible translations $x \in \mathbb{R}^2$, namely create the translate functions $\lambda_x k$ given by $(\lambda_x k)(t) = k(t-x)$ $(t \in \mathbb{R}^2)$, and test whether we find a match at any location $t \in \mathbb{R}^2$ between $f(t)$ and $(\lambda_x k)(t)$ by forming the product $f(t)k(t-x)$ and then by averaging these scores by computing the integral

$$\int_{\mathbb{R}^2} f(t)k(t-x)\,dt, \tag{$*_1$}$$

where dt is the Lebesgue measure on $\mathbb{R}^2$. See Figures 8.1 and 8.2. If k and f are discrete functions defined on the same grid, then the above integral is a finite sum.

Now the expression in $(*_1)$ is almost the convolution of f and k, except for a wrong sign. The convolution $f * k$ is given by

$$(f * k)(x) = \int_{\mathbb{R}^2} f(t)k(x-t)\,dt. \tag{$*_2$}$$

As we said earlier in Vol. I, Section 8.12, the expression in $(*_1)$ is the convolution $f * \check{k}$, where $\check{k}$ is the function defined by $\check{k}(s) = k(-s)$. We can think of $\check{k}$ as a reflected kernel. Since for our purpose expression $(*_1)$ is more natural than a convolution with a reflected kernel, we define the notion of cross-correlation as follows.

Definition 8.1. For any function $f \in \mathrm{L}^2(\mathbb{R}^2)$ and for any correlation kernel $k \in \mathrm{L}^1(\mathbb{R}^2)$ the *cross-correlation* of k and f, denoted by $k \star f$, is defined so that for all $x \in \mathbb{R}^2$,

$$(k \star f)(x) = \int_{\mathbb{R}^2} f(t)k(t-x)\,dt. \tag{$*_3$}$$

If we need to be more precise, we write $k \star_{\mathbb{R}^2} f$.

[1]Unfortunately, the term kernel has multiple meanings, but it seems universally adopted in signal processing.

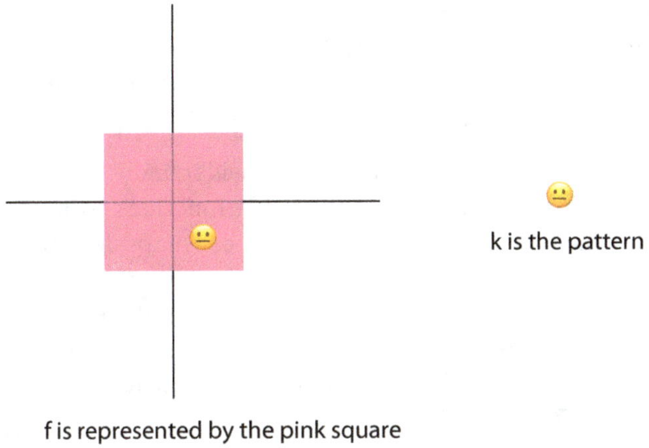

Fig. 8.1 Let $f(x)$ represent the color of the pixel at location x. The graph of $f\colon \mathbb{R}^2 \to \mathbb{R}$ is represented by the pink square with a single smiley face. We want to locate the smiley face $k\colon \mathbb{R}^2 \to \mathbb{R}$ within this square.

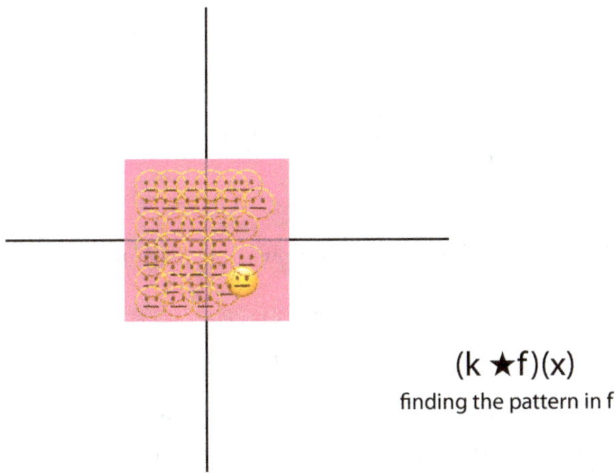

Fig. 8.2 To detect whether the smiley face occurs in the square, we form the product $f(t)k(t-x)$ and compute $(k \star f)(x) = \int_{\mathbb{R}^2} f(t)k(t-x)\,dt$.

The function k is often called the convolution kernel, but it is really the *correlation kernel* (or *template kernel*).

Note that we did not specify what kind of condition needs to be imposed on the kernel k. We certainly would like $k \star f$ to belong to $\mathrm{L}^2(\mathbb{R}^2)$. By a result of Folland [Folland (1995)] (Chapter 2, Proposition 2.39), if $k \in \mathrm{L}^1(\mathbb{R}^2)$ and if k has compact support, then $k \star f \in \mathrm{L}^2(\mathbb{R}^2)$.

However, later we will also need to consider more general notions of correlation from some space of L^2-integrable functions to itself and it will be crucial that such maps Φ be continuous linear maps definable in terms of kernels. This implies that these correlations are *Hilbert–Schmidt operators*, as discussed in Section 2.5. Recall the fundamental result that if $\mathrm{L}^2_\mu(X)$ is separable, then a linear map $T\colon \mathrm{L}^2_\mu(X) \to \mathrm{L}^2_\mu(X)$ is a Hilbert–Schmidt operator iff there is a kernel $\mathcal{K}\colon X \times X \to \mathbb{C}$, which is a function in the space $\mathrm{L}^2_{\mu\otimes\mu}(X \times X)$, such that

$$(T(f))(x) = \int_X \mathcal{K}(x, y) f(y)\, d\mu(y), \quad f \in \mathrm{L}^2_\mu(X),\ x \in X. \tag{HS1}$$

To assert that $\mathcal{K} \in \mathrm{L}^2_{\mu\otimes\mu}(X \times X)$ means that $\mathcal{K}$ is a $\mu \otimes \mu$-measurable map and that

$$\|\mathcal{K}\|_2^2 = \int_X \int_X |\mathcal{K}(x, y)|^2\, d\mu(x)\, d\mu(y) < \infty. \tag{HS2}$$

If we define $\mathcal{K}$ as $\mathcal{K}(x, y) = k(y - x)$, we see that the map $f \mapsto k \star f$ is Hilbert-Schmidt if $k \in \mathrm{L}^2(\mathbb{R}^2)$ and if

$$\int_{\mathbb{R}^2} \int_{\mathbb{R}^2} |k(y - x)|^2\, d_{\lambda_{\mathbb{R}^2}}(x)\, d_{\lambda_{\mathbb{R}^2}}(y) < \infty,$$

where $\lambda_{\mathbb{R}^2}$ is the Lebesgue measure on $\mathbb{R}^2$.

The right-hand side of $(*_3)$ can be viewed as an inner product in $\mathrm{L}^2(\mathbb{R}^2)$, namely

$$\langle f, \lambda_x(k)\rangle = \int_{\mathbb{R}^2} f(t) k(t - x)\, dt. \tag{$*_4$}$$

The above observation is the key to generalizing cross-correlation to groups more general than $\mathbb{R}^2$. The key observation is that the function $\lambda_x k$ is the result of *applying the regular left representation* $\mathbf{R}$ *of* $\mathbb{R}^2$ *in* $\mathrm{L}^2(\mathbb{R}^2)$, since

$$(\mathbf{R}_x(k))(t) = (\lambda_x k)(t) = k(-x + t) = k(t - x),$$

as the group $\mathbb{R}^2$ is abelian. In order to define the cross-correlation for more general groups G, we use the regular left representation $\mathbf{R}$ of G in $\mathrm{L}^2(G)$ or possibly $\mathrm{L}^2(X)$ for some space X on which G acts.

If we wish to be more precise, we denote this representation by

$$\mathbf{R}^{G \to \mathrm{L}^2(X)}.$$

The machine learning community tends to use the notation $\mathcal{L}^{G \to \mathrm{L}^2(X)}$.

Now because the Lebesgue measure is translation invariant, cross-correlation is *equivariant under translation*. What this means is that if we define the (linear) map $\Phi \colon \mathrm{L}^2(\mathbb{R}^2) \to \mathrm{L}^2(\mathbb{R}^2)$ by

$$\Phi(f) = k \star f,$$

since $\mathbb{R}^2$ acts on $\mathrm{L}^2(\mathbb{R}^2)$ by the regular left representation $\mathbf{R}^{\mathbb{R}^2 \to \mathrm{L}^2(\mathbb{R}^2)}$, we have the commutative diagram

$$\begin{array}{ccc} \mathrm{L}^2(\mathbb{R}^2) & \xrightarrow{\ \Phi\ } & \mathrm{L}^2(\mathbb{R}^2) \\ {\scriptstyle \mathbf{R}_x^{\mathbb{R}^2 \to \mathrm{L}^2(\mathbb{R}^2)}}\big\downarrow & & \big\downarrow{\scriptstyle \mathbf{R}_x^{\mathbb{R}^2 \to \mathrm{L}^2(\mathbb{R}^2)}} \\ \mathrm{L}^2(\mathbb{R}^2) & \xrightarrow[\ \Phi\ ]{} & \mathrm{L}^2(\mathbb{R}^2) \end{array}$$

for all $x \in \mathbb{R}^2$, whose proof is left as an exercise. The above diagram expresses the fact that the transform Φ is $\mathbb{R}^2$-equivariant. See Figure 8.3.

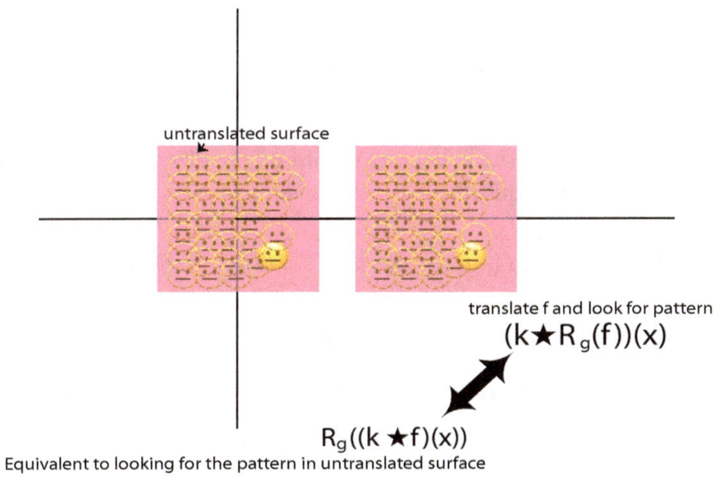

Fig. 8.3 If we translate the surface and look for the pattern, we must calculate $(k \star R_g(f))(x) = \int_{\mathbb{R}^2} f(t-g)k(t-x)\,dt$. This is equivalent to looking for the pattern in the original surface via $R_g((k \star f)(x)) = \int_{\mathbb{R}^2} f(t)k(t-x+g)\,dt$.

Now what happens if the image is rotated by an angle θ, so that we have the new image given by

$$(\mathbf{R}_{R_\theta}^{\mathbf{SO}(2)\to\mathrm{L}^2(\mathbb{R}^2)} f)(t) = f(R_{-\theta}(t)), \quad t \in \mathbb{R}^2,\ R_\theta \in \mathbf{SO}(2)?$$

The problem is that the new diagram

$$\begin{array}{ccc} \mathrm{L}^2(\mathbb{R}^2) & \xrightarrow{\ \Phi\ } & \mathrm{L}^2(\mathbb{R}^2) \\ {\scriptstyle \mathbf{R}_{R_\theta}^{\mathbf{SO}(2)\to\mathrm{L}^2(\mathbb{R}^2)}}\big\downarrow & & \big\downarrow{\scriptstyle \mathbf{R}_{R_\theta}^{\mathbf{SO}(2)\to\mathrm{L}^2(\mathbb{R}^2)}} \\ \mathrm{L}^2(\mathbb{R}^2) & \xrightarrow[\ \Phi\ ]{} & \mathrm{L}^2(\mathbb{R}^2) \end{array}$$

does *not* commute; the transform

$$[\Phi(\mathbf{R}_{R_\theta}^{\mathbf{SO}(2)\to\mathrm{L}^2(\mathbb{R}^2)} f)](x) = \int_{\mathbb{R}^2} f(R_{-\theta}(t))k(t-x)\,dt$$

of the rotated image f is *not* equal to the rotated transform

$$[\mathbf{R}_{R_\theta}^{\mathbf{SO}(2)\to\mathrm{L}^2(\mathbb{R}^2)}(\Phi(f))](x) = \int_{\mathbb{R}^2} f(t)k(t - R_{-\theta}(x))\,dt.$$

This is because we can make the change of variable $t_1 = R_{-\theta}(t)$ in the integral

$$[\Phi(\mathbf{R}_{R_\theta}^{\mathbf{SO}(2)\to\mathrm{L}^2(\mathbb{R}^2)} f)](x) = \int_{\mathbb{R}^2} f(R_{-\theta}(t))k(t-x)\,dt,$$

so $t = R_\theta(t_1)$ and since the diffeomorphism of $\mathbb{R}^2$ given by $t \mapsto R_{-\theta}(t)$ has Jacobian determinant equal to $+1$, we obtain

$$\int_{\mathbb{R}^2} f(t_1)k(R_\theta(t_1) - x)\,dt_1 \neq \int_{\mathbb{R}^2} f(t)k(t - R_{-\theta}(x))\,dt.$$

Equivariance under rotation fails; see Figure 8.4.

This is a very unfortunate problem, because in general, we would like to know whether the pattern k occurs in f, translated or rotated.

One possible remedy is *data augmentation*, which means creating rotated templates, but this is an expensive technique which often fails to identify rotated patterns.

A better solution is to define a *lifted correlation*.

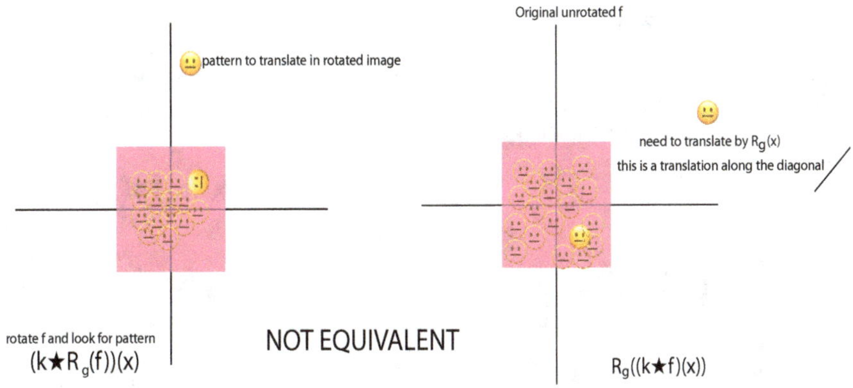

Fig. 8.4 If we rotate the surface and look for the pattern, we must calculate $(k \star R_g(f))(x) = \int_{\mathbb{R}^2} f(R_{-\theta}(t))k(t-x)\,dt$. This is not the same as the rotated transformation $R_g((k \star f)(x)) = \int_{\mathbb{R}^2} f(t)k(t - R_{-\theta}(x))\,dt$.

8.3 Lifted Correlation

The basic idea is that instead of rotating the input image, we apply a rotated kernel to the image. We first illustrate this process in the case of $\mathbf{SE}(2) = \mathbb{R}^2 \rtimes \mathbf{SO}(2)$, but this method works for semi-direct products of the form $G = \mathbb{R}^d \rtimes H$. We will denote an element of $\mathbf{SE}(2)$ as $g = (x, \theta)$, where $x \in \mathbb{R}^2$ and $\theta \in \mathbb{R} \pmod{2\pi}$. Then we define the lifted correlation $k \mathbin{\widetilde{\star}} f$ by

$$
\begin{aligned}
(k \mathbin{\widetilde{\star}} f)(x, \theta) = \langle f, \lambda_{(x,\theta)} k \rangle &= \int_{\mathbb{R}^2} f(t)(\lambda_{(x,\theta)} k)(t)\,dt \\
&= \int_{\mathbb{R}^2} f(t)k(R_{-\theta}(t - x))\,dt; \qquad (*_5)
\end{aligned}
$$

see Figure 8.5.

Observe that we are still *integrating over* $\mathbb{R}^2$, but we are using the *rotated* kernel $\lambda_{R_\theta} k$, and we can still use efficient methods available to compute the last integral.

But observe that our transform now takes an input function f (image, signal) in $L^2(\mathbb{R}^2)$, but yields an output function $\Phi(f) = k \mathbin{\widetilde{\star}} f$ in the larger function space $L^2(\mathbf{SE}(2))$ of functions defined on the *group* $\mathbf{SE}(2)$.

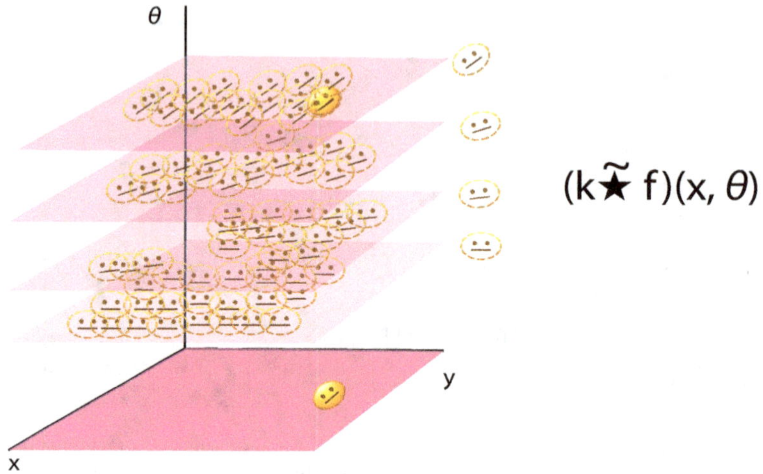

Fig. 8.5 An illustration of the lifted correlation $(k \widetilde{\star} f)(x, \theta)$. As before, f is represented by the pink square in the horizontal xy-plane. Each θ layer contains a rotated copy of the smiley face pattern which is then translated over the image of f in the xy-plane.

Equivariance is verified as follows. We have

$$\begin{aligned}(k \widetilde{\star} \lambda_{R_\varphi} f)(x, \theta) &= \int_{\mathbb{R}^2} (\lambda_{R_\varphi} f)(t) k(R_{-\theta}(t - x))\, dt \\ &= \int_{\mathbb{R}^2} f(R_{-\varphi} t) k(R_{-\theta}(t - x))\, dt.\end{aligned}$$

We can make the change of variable $t_1 = R_{-\varphi} t$, so $t = R_\varphi t_1$ and since the diffeomorphism of $\mathbb{R}^2$ given by $t \mapsto R_{-\varphi} t$ has Jacobian determinant equal to $+1$, we obtain

$$\begin{aligned}(k \widetilde{\star} \lambda_{R_\varphi} f)(x, \theta) &= \int_{\mathbb{R}^2} f(R_{-\varphi} t) k(R_{-\theta}(t - x))\, dt \\ &= \int_{\mathbb{R}^2} f(t_1) k(R_{-\theta}(R_\varphi t_1 - x))\, dt_1.\end{aligned}$$

On the other hand, the left regular action of $\mathbf{SO}(2)$ in $\mathrm{L}^2(\mathbf{SE}(2))$ is given by

$$\lambda_{R_\varphi}(g)(x, \theta) = g(R_{-\varphi} x, \theta - \varphi), \quad g \in \mathrm{L}^2(\mathbf{SE}(2)),$$

so we have

$$\begin{aligned}\lambda_{R_\varphi}(k \,\widetilde{\star}\, f)(x,\theta) &= (k \,\widetilde{\star}\, f)(R_{-\varphi}x, \theta - \varphi)\\ &= \int_{\mathbb{R}^2} f(t)k(R_{-(\theta-\varphi)}(t - R_{-\varphi}x))\, dt\\ &= \int_{\mathbb{R}^2} f(t)k(R_{-\theta}(R_\varphi t - x))\, dt.\end{aligned}$$

Therefore

$$(k \,\widetilde{\star}\, \lambda_{R_\varphi} f)(x,\theta) = \lambda_{R_\varphi}(k \,\widetilde{\star}\, f)(x,\theta),$$

or equivalently, the diagram

$$\begin{array}{ccc} \mathrm{L}^2(\mathbb{R}^2) & \xrightarrow{\ \Phi\ } & \mathrm{L}^2(\mathbf{SE}(2)) \\ {\scriptstyle \mathbf{R}_{R_\varphi}^{\mathbf{SO}(2)\to \mathrm{L}^2(\mathbb{R}^2)}}\Big\downarrow & & \Big\downarrow {\scriptstyle \mathbf{R}_{R_\varphi}^{\mathbf{SO}(2)\to \mathrm{L}^2(\mathbf{SE}(2))}} \\ \mathrm{L}^2(\mathbb{R}^2) & \xrightarrow[\ \Phi\ ]{} & \mathrm{L}^2(\mathbf{SE}(2)) \end{array}$$

commutes for all $R_\varphi \in \mathbf{SO}(2)$, which shows that equivariance holds.

The above definition can be generalized to a group G, which is a semidirect product $G = \mathbb{R}^d \rtimes H$. At first, it is more convenient to assume that $G = NH$, where N is normal in G and isomorphic to $\mathbb{R}^d$ (thus abelian) and that $N \cap H = \{e\}$; see Section 7.4 for details. Given that multiplication in G is given by

$$(n_1h_1)(n_2h_2) = (n_1[h_1n_2h_1^{-1}])(h_1h_2), \tag{mult1}$$

and that the inverse of n_1h_1 is $(h_1^{-1}n_1^{-1}h_1)h_1^{-1}$, since we are assuming that N is abelian, it is convenient to use the additive notation $+$ for the group operation on N, so that multiplication in G is given by

$$(n_1h_1)(n_2h_2) = (n_1 + h_1n_2h_1^{-1})(h_1h_2),$$

and the inverse of (n_1h_1) is $(h_1^{-1}(-n_1)h_1)h_1^{-1}$. But because $N \cap H = \{e\}$, the additive identity of $N \simeq \mathbb{R}^d$ is denoted e instead of 0. The group $G = N \rtimes H \simeq \mathbb{R}^d \rtimes H$ acts on $\mathbb{R}^d$ as follows:

$$(nh) \cdot t = n + hth^{-1},\ nh \in \mathbb{R}^d \rtimes H,\ t \in \mathbb{R}^d. \tag{$*_6$}$$

Since $(nh)^{-1} = (h^{-1}(-n)h)h^{-1}$, we obtain

$$(nh)^{-1} \cdot t = h^{-1}(-n)h + h^{-1}th = h^{-1}(t-n)h,\ nh \in \mathbb{R}^d \rtimes H,\ t \in \mathbb{R}^d. \tag{$*_7$}$$

In the special case where $n = e$, we obtain the action of H on $\mathbb{R}^d$ given by

$$h \cdot t = hth^{-1},\ h \in H,\ t \in \mathbb{R}^d, \qquad (*_8)$$

and in the special case when $h = e$, we obtain the action of $\mathbb{R}^d$ on $\mathbb{R}^d$ given by

$$n \cdot t = n + t,\ n, t \in \mathbb{R}^d. \qquad (*_9)$$

We have the corresponding actions of H and $\mathbb{R}^d$ on $\mathrm{L}^2(\mathbb{R}^d)$ given by

$$\begin{aligned}(\lambda_h k)(t) &= k(h^{-1} \cdot t) = k(h^{-1}th)\\ (\lambda_x k)(t) &= k((-x) \cdot t) = k(t - x),\end{aligned}$$

which induce the left regular representations $\mathbf{R}^{H \to \mathrm{L}^2(\mathbb{R}^d)}$ of H in $\mathbb{R}^d$ and $\mathbf{R}^{\mathbb{R}^d \to \mathrm{L}^2(\mathbb{R}^d)}$ of $\mathbb{R}^d$ in $\mathrm{L}^2(\mathbb{R}^d)$ given by

$$(\mathbf{R}_h^{H \to \mathrm{L}^2(\mathbb{R}^d)} k)(t) = k(h^{-1} \cdot t) = k(h^{-1}th) \qquad (\mathbf{R}^{H \to \mathrm{L}^2(\mathbb{R}^d)})$$

$$(\mathbf{R}_x^{\mathbb{R}^d \to \mathrm{L}^2(\mathbb{R}^d)} k)(t) = k((-x) \cdot t) = k(t - x). \qquad (\mathbf{R}^{\mathbb{R}^d \to \mathrm{L}^2(\mathbb{R}^d)})$$

It follows that for any $t \in \mathbb{R}^d$ and any $xh \in \mathbb{R}^d \rtimes H$, we have

$$(\lambda_{(xh)} k)(t) = k((xh)^{-1} \cdot t) = k(h^{-1}(t - x)h) = k(h^{-1} \cdot (t - x)) = (\lambda_x(\lambda_h k))(t).$$

Definition 8.2. For any function $f \in \mathrm{L}^2(\mathbb{R}^d)$, for any correlation kernel $k \in \mathrm{L}^2(\mathbb{R}^d)$ with

$$\int_{\mathbb{R}^d}\int_{\mathbb{R}^d} |k(y - x)|^2 \, d_{\lambda_{\mathbb{R}^d}}(x) \, d_{\lambda_{\mathbb{R}^d}}(y) < \infty,$$

where $\lambda_{\mathbb{R}^d}$ is the Lebesgue measure on $\mathbb{R}^d$, and for any semi-direct product $\mathbb{R}^d \rtimes H$ (H is usually a compact group), the *lifted correlation* $k \,\widetilde{\star}\, f$ is defined by

$$(k \,\widetilde{\star}\, f)(xh) = \langle f, \lambda_{(xh)} k \rangle = \int_{\mathbb{R}^d} f(t) k(h^{-1}(t - x)h) \, dt, \quad xh \in \mathbb{R}^d \rtimes H. \ (*_{10})$$

Since $k(h^{-1}(t - x)h) = (\lambda_x(\lambda_h k))(t)$, we see that this formula is similar to $(*_3)$ except that we use the correlation kernel $\lambda_h k$ instead of k.

Given an input function $f \in \mathrm{L}^2(\mathbb{R}^d)$, we obtain an output function $k \,\widetilde{\star}\, f \in \mathrm{L}^2(\mathbb{R}^d \rtimes H)$. In the special case where $H = \mathbf{SO}(d)$, the formula in $(*_{10})$ simplifies as shown in the next example.

Example 8.1. In the special case $H = \mathbf{SO}(d)$, we have $G = \mathbf{SE}(d) \simeq \mathbb{R}^d \rtimes \mathbf{SO}(d)$, where $\mathbf{SE}(d)$ consists of the $(d + 1) \times (d + 1)$ matrices

$$\begin{pmatrix} Q & x \\ 0 & 1 \end{pmatrix}, \quad Q \in \mathbf{SO}(d),\ x \in \mathbb{R}^d,$$

and since

$$\begin{pmatrix} Q & x \\ 0 & 1 \end{pmatrix} = \begin{pmatrix} I_d & x \\ 0 & 1 \end{pmatrix} \begin{pmatrix} Q & 0 \\ 0 & 1 \end{pmatrix},$$

we have $\mathbf{SE}(d) = N \rtimes H$, where $N \simeq \mathbb{R}^d$ consists of all matrices n of the form

$$n = \begin{pmatrix} I_d & x \\ 0 & 1 \end{pmatrix}, \quad x \in \mathbb{R}^d,$$

and $H \simeq \mathbf{SO}(d)$ consists of all matrices h of the form

$$h = \begin{pmatrix} Q & 0 \\ 0 & 1 \end{pmatrix}, \quad Q \in \mathbf{SO}(d).$$

If we denote an element of $\mathbf{SE}(d)$ as (x, Q), since

$$h \begin{pmatrix} I_d & t \\ 0 & 1 \end{pmatrix} h^{-1} = \begin{pmatrix} I_d & Qt \\ 0 & 1 \end{pmatrix},$$

using the isomorphism

$$\begin{pmatrix} I_d & y \\ 0 & 1 \end{pmatrix} \mapsto y$$

between N and $\mathbb{R}^d$,

$$hth^{-1} = Qt,$$

so the action in $(*_6)$ of $\mathbf{SE}(d)$ on $\mathbb{R}^d$ (with $n = x$) is given by

$$(x, Q) \cdot t = x + Qt, \quad t, x \in \mathbb{R}^d,\ Q \in \mathbf{SO}(d). \tag{$*_{11}$}$$

With this notation, when $G = \mathbf{SE}(d) = \mathbb{R}^d \rtimes \mathbf{SO}(d)$, Equation $(*_{10})$ becomes

$$\begin{aligned} (k \mathbin{\widetilde{\star}} f)(x, Q) &= \langle f, \lambda_{(x,Q)} k \rangle \\ &= \int_{\mathbb{R}^d} f(t) k(Q^{-1}(t - x))\, dt,\ (x, Q) \in \mathbb{R}^d \rtimes \mathbf{SO}(d). \end{aligned} \tag{$*_{12}$}$$

Note that viewing the elements of $\mathbf{SE}(d)$ are pairs (x, Q) with $x \in \mathbb{R}^d$ and $Q \in \mathbf{SO}(d)$ corresponds to defining the semi-direct product $G = N \rtimes H$ as the Cartesian product $N \times H$ with the multiplication operation

$$(n_1, h_1)(n_2, h_2) = (n_1 + (h_1 \cdot n_2), h_1 h_2), \tag{mult2}$$

where $\cdot\colon H \times N \to N$ is an action of H on N, such that for each $h \in H$, the map $n \mapsto h \cdot n$ is an automorphism of N; equivalently, the action $\cdot$ is defined by a homomorphism $\tau\colon H \to \mathrm{Aut}(N)$, with $h \cdot n = \tau(h)(n)$. Since N and

H are no longer given as subgroups of a common group, we denote the identity element of N as 0 and the identity element of H as e. The inverse of an element $(n, h) \in G = N \rtimes H$ is equal to $(-h^{-1} \cdot n, h^{-1})$. For details, see Gallier and Quaintance [Gallier and Quaintance (2020a)] (Chapter 19, Section 19.5). With this point of view, the action of $\mathbf{SO}(d)$ on $\mathbb{R}^d$ is

$$Q \cdot x = Qx,$$

and multiplication is given by

$$(x_1, Q_1)(x_2, Q_2) = (x_1 + Q_1 x_2, Q_1 Q_2).$$

Furthermore, Equation $(*_{10})$ becomes

$$(k \widetilde{\star} f)(x, h) = \int_{\mathbb{R}^d} f(t) k(h^{-1} \cdot (t - x))\, dt, \quad (x, h) \in \mathbb{R}^d \times H. \qquad (*_{10'})$$

Actually, the above equation is also valid if $G = N \rtimes H$ is viewed as NH, as in our first interpretation, since the action of H on $N \simeq \mathbb{R}^d$ is given by $h \cdot n = hnh^{-1}$.

From now on, unless specified otherwise, we will use the version of a semi-direct product $N \rtimes H$ as the Cartesian product $N \times H$ with the multiplication operation given by (mult2). Elements of $N \rtimes H$ are denoted by pairs $(x, h) \in N \times H$.

The functions arising from lifted convolutions are given a name as below.

Definition 8.3. Functions $f\colon \mathbb{R}^d \times H \to \mathbb{C}$ in $\mathrm{L}^2(\mathbb{R}^d \rtimes H) = \mathrm{L}^2(G)$ are called *G-feature maps*.

If the group H can be easily discretized, then we can compute a discretization of the function $k \widetilde{\star} f$. This is the case when $d = 2$ and $H = \mathbf{SO}(2)$. In this situation, a feature map in $\mathrm{L}^2(\mathbf{SE}(2))$ can be viewed as a stack of rectangular grids, one for each $\theta \in \mathbb{R} \pmod{2\pi}$. We can obtain a final score of the occurrence of the template k moved over the image f in all positions determined by the group elements $h \in H$, in this case, rotations $R_\theta \in \mathbf{SO}(2)$, by some projection process over the "H-axis," in this case, the θ-axis; this is often called *pooling* in machine learning (max-pooling being a common instance of pooling).

The major benefit of lifted kernels is that we recover equivariance under the group H. Since H is a subgroup of $G = \mathbb{R}^d \rtimes H$, we have the left regular representation of H in $\mathrm{L}^2(\mathbb{R}^d \rtimes H)$, namely for every $f \in \mathrm{L}^2(\mathbb{R}^d \rtimes H)$, we have

$$(\mathbf{R}_h^{H \to \mathrm{L}^2(\mathbb{R}^d \rtimes H)}(f))(x, h_1) = f(h^{-1} \cdot (x, h_1)) = f(h^{-1} \cdot x, h^{-1} h_1),\ h \in H \qquad (\mathbf{R}^{H \to \mathrm{L}^2(\mathbb{R}^d \rtimes H)})$$

for all $(x, h_1) \in \mathbb{R}^d \rtimes H$.

The following result reveals a sufficient condition on the action of H on $\mathbb{R}^d$ to ensure equivariance.

Proposition 8.1. *Let* $\Phi\colon \mathrm{L}^2(\mathbb{R}^d) \to \mathrm{L}^2(\mathbb{R}^d \rtimes H)$ *be given by*

$$\Phi(f) = k \,\widetilde{\star}\, f$$

as in $(*_{10'})$. *We have*

$$\begin{aligned}(\mathbf{R}_h^{H\to \mathrm{L}^2(\mathbb{R}^d\rtimes H)}(k \,\widetilde{\star}\, f))(x,h_1) &= \int_{\mathbb{R}^d} f(t)k(h_1^{-1}\cdot(h\cdot t - x))\,dt\\ (k \,\widetilde{\star}\, (\mathbf{R}_h^{H\to \mathrm{L}^2(\mathbb{R}^d)} f))(x,h_1) &= \int_{\mathbb{R}^d} f(h^{-1}\cdot t)k(h_1^{-1}\cdot(t - x))\,dt,\end{aligned}$$

for all $h, h_1 \in H$ *and all* $x \in \mathbb{R}^d$. *If we denote by* $J_h = \det d(L_h)_0$ *the Jacobian determinant of the linear map* $L_h\colon \mathbb{R}^d \to \mathbb{R}^d$ *defined by* $L_h(t) = h{\cdot}t$ $(h \in H, t \in \mathbb{R}^d)$,[2] *then*

$$(k \,\widetilde{\star}\, (\mathbf{R}_h^{H\to \mathrm{L}^2(\mathbb{R}^d)} f))(x,h_1) = |J_h|(\mathbf{R}_h^{H\to \mathrm{L}^2(\mathbb{R}^d\rtimes H)}(k \,\widetilde{\star}\, f))(x,h_1) \qquad (\dagger_1)$$

for all $h, h_1 \in H$ *and all* $x \in \mathbb{R}^d$. *Consequently, if* $|J_h| = 1$ *for all* $h \in H$, *then the following diagram*

$$\begin{array}{ccc}\mathrm{L}^2(\mathbb{R}^d) & \xrightarrow{\ \Phi\ } & \mathrm{L}^2(\mathbb{R}^d \rtimes H)\\ {\scriptstyle \mathbf{R}_h^{H\to \mathrm{L}^2(\mathbb{R}^d)}}\big\downarrow & & \big\downarrow{\scriptstyle \mathbf{R}_h^{H\to \mathrm{L}^2(\mathbb{R}^d\rtimes H)}}\\ \mathrm{L}^2(\mathbb{R}^d) & \xrightarrow[\ \Phi\ ]{} & \mathrm{L}^2(\mathbb{R}^d \rtimes H)\end{array}$$

commutes for all $h \in H$, *which shows that* Φ *is* H*-equivariant.*

Proof. We have

$$\begin{aligned}(\mathbf{R}_h^{H\to \mathrm{L}^2(\mathbb{R}^d\rtimes H)}(k \,\widetilde{\star}\, f))(x,h_1) &= (k \,\widetilde{\star}\, f)(h^{-1}\cdot x, h^{-1}h_1)\\ &= \int_{\mathbb{R}^d} f(t)k((h^{-1}h_1)^{-1}\cdot(t - h^{-1}\cdot x))\,dt\\ &= \int_{\mathbb{R}^d} f(t)k(h_1^{-1}\cdot(h\cdot t - x))\,dt,\end{aligned}$$

and

$$(k \,\widetilde{\star}\, (\mathbf{R}_h^{H\to \mathrm{L}^2(\mathbb{R}^d)} f))(x,h_1) = \int_{\mathbb{R}^d} f(h^{-1}\cdot t)k(h_1^{-1}\cdot(t - x))\,dt.$$

[2] Recall that the derivative $d(L_h)_x$ of the linear map L_h at $x \in \mathbb{R}^d$ is independent of $x \in \mathbb{R}^d$.

Since the maps L_h are injective and C^1, by making the change of variable $t = h \cdot t_1$ (so $h^{-1} \cdot t = t_1$) in the integral

$$\int_{\mathbb{R}^d} f(h^{-1} \cdot t)k(h_1^{-1} \cdot (t - x))\, dt,$$

by the change of variable equation in an integral, we get

$$\int_{\mathbb{R}^d} f(h^{-1} \cdot t)k(h_1^{-1} \cdot (t - x))\, dt = |J_h| \int_{\mathbb{R}^d} f(t_1)k(h_1^{-1} \cdot (h \cdot t_1 - x))\, dt_1,$$

which proves Equation ($\dagger_1$). $\square$

Under the conditions of Proposition 8.1, by construction Φ is also $\mathbb{R}^d$-equivariant, so Φ is actually $\mathbb{R}^d \rtimes H$-equivariant.

8.4 Group Correlation on Locally Compact Groups

Now that we are dealing with functions defined on the group $G = \mathbb{R}^d \rtimes H$, in order to proceed to different layers of a convolutional neural network, we take the last step in generalizing the notion of cross-correlation, which is to define the notion of group correlation. This notion actually makes sense for any locally compact group.

Definition 8.4. Let G be a locally compact group with left Haar measure λ_G. For any $f \in \mathrm{L}^2(G)$ and for any correlation kernel $k \in \mathrm{L}^2(G)$ with

$$\int_G \int_G |k(g_1^{-1} g_2)|^2\, d\lambda_G(g_1)\, d\lambda_G(g_2) < \infty,$$

where λ_G is the left Haar measure on G, the *cross-correlation operator*, or simply *correlation operator* of k and f, denoted $k \star f$, is defined by

$$\begin{aligned}(k \star f)(s) = \langle f, \lambda_s(k)\rangle &= \int_G f(t)\lambda_s(k)(t)\, d\lambda_G(t) \\ &= \int_G f(t)k(s^{-1}t)\, d\lambda_G(t), \quad s \in G.\end{aligned}$$

The function k is called the *correlation kernel*.

Recall the left regular representation of G in $\mathrm{L}^2(G)$ given by

$$(\mathbf{R}_g^{G \to \mathrm{L}^2(G)} f)(t) = (\lambda_g f)(t) = f(g^{-1}t), \quad g, t \in G,\ f \in \mathrm{L}^2(G).$$

We have

$$(k \star \lambda_g(f))(s) = \int_G (\lambda_g f)(t)k(s^{-1}t)\, d\lambda_G(t) = \int_G f(g^{-1}t)k(s^{-1}t)\, d\lambda_G(t),$$

and

$$\begin{aligned}(\lambda_g(k \star f))(s) = (k \star f)(g^{-1}s) &= \int_G f(t)k((g^{-1}s)^{-1}t)\, d\lambda_G(t) \\ &= \int_G f(t)k(s^{-1}gt)\, d\lambda_G(t).\end{aligned}$$

Since the Haar measure λ_G is left-invariant, by substituting $t = g^{-1}t$ for t, we get

$$\begin{aligned}(\lambda_g(k \star f))(s) &= \int_G f(t)k(s^{-1}gt)\, d\lambda_G(t) \\ &= \int_G f(g^{-1}t)k(s^{-1}t)\, d\lambda_G(t) = (k \star \lambda_g(f))(s).\end{aligned}$$

This shows that the map $\Phi\colon \mathrm{L}^2(G) \to \mathrm{L}^2(G)$ given by

$$\Phi(f) = k \star f$$

is G-equivariant, which means that the following diagram

$$\begin{array}{ccc} \mathrm{L}^2(G) & \xrightarrow{\ \Phi\ } & \mathrm{L}^2(G) \\ {\scriptstyle \mathbf{R}_s^{G\to \mathrm{L}^2(G)}}\Big\downarrow & & \Big\downarrow{\scriptstyle \mathbf{R}_s^{G\to \mathrm{L}^2(G)}} \\ \mathrm{L}^2(G) & \xrightarrow[\ \Phi\ ]{} & \mathrm{L}^2(G) \end{array}$$

commutes for all $s \in G$. In group correlation (Definition 8.4), the template k is moved around by all group elements $s \in G$, and the moved template $\lambda_s k$ is matched against the image f.

Example 8.2. In the special case where $G = \mathbb{R}^d \rtimes \mathbf{SO}(d)$ $(d = 2, 3)$, Definition 8.4 becomes

$$\begin{aligned}(k \star f)(x, Q) &= \langle f, \lambda_{(x,Q)}(k)\rangle \\ &= \int_{\mathbb{R}^d}\int_{\mathbf{SO}(d)} f(y, R)\lambda_{(x,Q)}(k)(y, R) d\lambda_{\mathbb{R}^d}(y) d\lambda_{\mathbf{SO}(d)}(R) \\ &= \int_{\mathbb{R}^d}\int_{\mathbf{SO}(d)} f(y, R)k(Q^{-1}(y - x), Q^{-1}R) d\lambda_{\mathbb{R}^d}(y) d\lambda_{\mathbf{SO}(d)}(R),\end{aligned}$$

with $x \in \mathbb{R}^d, Q \in \mathbf{SO}(d)$. In the above example, since f and k are functions from $G = \mathbb{R}^d \rtimes \mathbf{SO}(d)$ to $\mathbb{R}$ and since we denote the elements of G as pairs (x, Q), we write f and k as functions of two arguments.

For $d = 2$ and $H = \mathbf{SO}(2)$, Definition 8.2 and the above formula can be effectively computed.

A typical equivariant convolutional neural network (CNN) consists of several layers, starting with an input image $f_0 \in \mathrm{L}^2(\mathbb{R}^2)$. The first layer is a lifting layer, which produces the first $\mathbf{SE}(2)$-feature map

$$f_1 = k_1 \,\widetilde{\star}\, f_0$$

using a correlation kernel $k_1 \in \mathrm{L}^2(\mathbb{R}^2)$. The subsequent layers are group correlation layers, such that

$$f_{l+1} = k_{l+1} \star f_l,$$

with $f_l, f_{l+1} \in \mathrm{L}^2(\mathbf{SE}(2))$ and $k_{l+1} \in \mathrm{L}^2(\mathbf{SE}(2))$. After each correlation transform, some activation function is applied, possibly a pointwise RELU. The last layer is a projection layer which produces a function in $\mathrm{L}^2(\mathbb{R}^2)$ using some pooling process over the θ-axis. The use of correlation kernels in $\mathrm{L}^2(\mathbf{SE}(2))$ allows the detection of patterns using features at relative poses.

For $d = 3$, there are serious computational issues. A way around this difficulty is to use *steerable kernels* discussed in Section 8.6. But first, we need to discuss how to equip a semi-direct product with a Haar measure, because there are some subtle issues.

8.5 Haar Measures on Semi-Direct Products

Again, it is more convenient to assume that we consider semi-direct products $G = N \rtimes H = NH$, where N and H are subgroups of G with N normal, and $N \cap H = \{e\}$; see Section 7.4. We also assume that N and H are locally compact and that the map $\varphi\colon N \times H \to G$ given by $\varphi(n, h) = nh$ is a homeomorphism. Then by Definition 3.16, we obtain the measure $\mu_G = \varphi_*(\mu_N \otimes \mu_H)$ on G as the direct image of the product measure $\mu_N \otimes \mu_H$ on $N \times H$, where μ_N is a Haar measure on N and μ_H is a Haar measure on H. By Proposition 3.19, for any $f \in \mathrm{L}^1_{\mu_G}(G)$, we have $f \circ \varphi \in \mathrm{L}^1_{\mu_N \otimes \mu_H}(G)$, and

$$\begin{aligned}\int_G f(g) d\mu_G(g) &= \int_{N\times H} f(nh)\, d(\mu_N \otimes \mu_H)(n, h)\\ &= \int_H \left(\int_N f(nh)\, d\mu_N(n)\right) d\mu_H(h). \qquad (\mu_1)\end{aligned}$$

Curiously, if μ_N and μ_H are right invariant, then μ_G is also right-invariant, but if μ_N and μ_H are left invariant, then μ_G is not necessarily left-invariant. We will construct both a left Haar measure and a right Haar measure for G from a left Haar measure λ_N on N and a left Haar measure λ_H on H and obtain a formula for the modular function on $G = N \rtimes H$. In general, even if N and H are unimodular, G is not. The problem is that H acts on N by conjugation, and these automorphisms may not have a modulus equal to 1.

Recall from Vol. I, Section 8.8 that if G is a locally compact group with a left Haar measure λ_G, for any automorphism $u\colon G \to G$ of G, there is a unique positive number $\mathrm{mod}(u)$ called the *modulus* of u, such that

$$\int_G f(u(s))\, d\lambda_G(s) = (\mathrm{mod}(u))^{-1} \int_G f(s)\, d\lambda_G(s) \qquad (\mu_2)$$

for all $f \in \mathcal{L}^1_{\lambda_G}(G)$; see Vol. I, Definition 8.17 and Vol. I, Proposition 8.28. By Vol. I, Proposition 8.29, the above property also holds for a right Haar measure since $\mathrm{mod}(u)$ is the same as in the case of a left Haar measure. The map $u \mapsto \mathrm{mod}(u)$ is continuous. This is shown in Hewitt and Ross [Hewitt and Ross (1979)], Chapter IV, Section 15, Section 15.29, and in Bourbaki [Bourbaki (1965)], Chapter VII, Section 1, no. 10, Proposition 12. It is also easy to see that if G is discrete or compact, then $\mathrm{mod}(u) = 1$ for any automorphism u of G.

Also recall from Vol. I, Section 8.6 that if λ_G is a left Haar measure, then there is a homomorphism $\Delta\colon G \to \mathbb{R}^*_+$ (where $\mathbb{R}^*_+$ denotes the set of positive reals) called the *modular function of* G, such that

$$\int_G f(xs)\, d\lambda_G(x) = \Delta(s)^{-1} \int_G f(x)\, d\lambda_G(x), \quad s \in G, \qquad (\mu_3)$$

for all $f \in \mathcal{L}^1_{\lambda_G}(G)$; see Vol. I, Definition 8.12 and Vol. I, Proposition 8.22. It is shown in Bourbaki [Bourbaki (1965)], Chapter VII, Section 1, no. 1, Proposition 1, that Δ is continuous. This fact is also proven in Hewitt and Ross [Hewitt and Ross (1979)], Chapter IV, Section 15, Theorem 15.11. By Vol. I, Proposition 8.27, if λ_G is a left Haar measure, then $\Delta_G^{-1} \cdot \lambda_G$ is a right Haar measure defined, such that

$$\int_G f(x)\, d(\Delta_G^{-1} \cdot \lambda_G)(x) = \int_G \Delta(x)^{-1} f(x)\, d\lambda_G(x). \qquad (\mu_4)$$

This can also be verified directly as follows.

$$\begin{aligned}
\int_G f(xs)\, d(\Delta_G^{-1} \cdot \lambda_G)(x) &= \int_G \Delta(x)^{-1} f(xs)\, d\lambda_G(x) && (1)\\
&= \int_G \Delta(s)\Delta(s)^{-1}\Delta(x)^{-1} f(xs)\, d\lambda_G(x) && (2)\\
&= \Delta(s) \int_G \Delta(xs)^{-1} f(xs)\, d\lambda_G(x) && (3)\\
&= \Delta(s)\Delta(s)^{-1} \int_G \Delta(x)^{-1} f(x)\, d\lambda_G(x) && (4)\\
&= \int_G f(x)\, d(\Delta_G^{-1} \cdot \lambda_G)(x), && (5)
\end{aligned}$$

where (1) follows by definition, (2) by inserting $\Delta(s)\Delta(s)^{-1}$, (3) since Δ is a group homomorphism, (4) by (μ_3) applied to the function $x \mapsto \Delta(x)^{-1} f(x)$, and (5) by definition.

Now since Haar measures (left or right) are unique up to a constant, we deduce that every right Haar measure ν on G if of the form $\nu = \Delta^{-1} \cdot \lambda$ for some left Haar measure λ on G, and thus $\Delta \cdot \nu = \Delta \cdot \Delta^{-1} \cdot \lambda = \lambda$ is a left Haar measure. Furthermore, if λ is a left Haar measure on G, we have

$$\begin{aligned}
\int_G f(sx)\, d(\Delta_G^{-1} \cdot \lambda)(x) &= \int_G \Delta(x)^{-1} f(sx)\, d\lambda(x) && (1)\\
&= \int_G \Delta(s)\Delta(sx)^{-1} f(sx)\, d\lambda(x) && (2)\\
&= \Delta(s) \int_G \Delta(x)^{-1} f(x)\, d\lambda(x) && (3)\\
&= \Delta(s) \int_G f(x)\, d(\Delta_G^{-1} \cdot \lambda)(x), && (4)
\end{aligned}$$

where (1) follows by definition, (2) by replacing $\Delta(x)^{-1}$ by $\Delta(s)\Delta(sx)^{-1}$, (3) by left-invariance of λ, and (4) by definition. This proves that the modular function of the right Haar measure $\Delta_G^{-1} \cdot \lambda$ is also Δ_G, and thus the modular function of all right Haar measures on G is Δ_G. If ν_G is a right Haar measure on G, then we have the following analog of (μ_3):

$$\int_G f(sx)\, d\nu_G(x) = \Delta(s) \int_G f(x)\, d\nu_G(x), \quad s \in G, \qquad (\mu_3')$$

for all $f \in \mathcal{L}^1_{\nu_G}(G)$.

The image of the product measure $\lambda_N \otimes \lambda_H$ by φ given by (μ_1) is a natural candidate for a left Haar measure on G, but the following result shows that left-invariance fails. However, this result also suggests how to repair the lack of left-invariance. The repair involves the term $\mathrm{mod}(i_h)$, where $i_h\colon N \to N$ is the automorphism given by

$$i_h(n) = hnh^{-1}, \quad n \in N,\, h \in H. \tag{μ_6}$$

Proposition 8.2. *Let $G = N \rtimes H = NH$ be the semi-direct product of two locally compact subgroups N and H, and μ_G be the image of the product measure $\lambda_N \otimes \lambda_H$ by φ, where λ_N and λ_H are left Haar measures on N and H, respectively. Then, we have*

$$\int_G f((n_1h_1)(nh))\, d\mu_G(nh) = \mathrm{mod}(i_{h_1})^{-1} \int_G f(nh)\, d\mu_G(nh) \tag{μ_5}$$

for all $n_1h_1 \in G$ and all $f \in \mathcal{L}^1_{\mu_G}(G)$, where $i_{h_1}\colon N \to N$ is the automorphism given by

$$i_{h_1}(n) = h_1nh_1^{-1}, \quad n \in N.$$

Proof. Since $(n_1h_1)(nh) = (n_1[h_1nh_1^{-1}])(h_1h)$, using Fubini's theorem, we have

$$\begin{aligned}
&\int_G f((n_1h_1)(nh))\, d\mu_G(nh)\\
&\quad= \int_G f((n_1[h_1nh_1^{-1}])(h_1h))\, d\mu_G(nh) &(1)\\
&\quad= \int_H \left(\int_N f((n_1[h_1nh_1^{-1}])(h_1h))\, d\lambda_N(n)\right) d\lambda_H(h) &(2)\\
&\quad= \mathrm{mod}(i_{h_1})^{-1} \int_H \left(\int_N f((n_1n)(h_1h))\, d\lambda_N(n)\right) d\lambda_H(h) &(3)\\
&\quad= \mathrm{mod}(i_{h_1})^{-1} \int_N \left(\int_H f(n(h_1h))\, d\lambda_H(h)\right) d\lambda_N(n) &(4)\\
&\quad= \mathrm{mod}(i_{h_1})^{-1} \int_N \left(\int_H f(nh)\, d\lambda_H(h)\right) d\lambda_N(n) &(5)\\
&\quad= \mathrm{mod}(i_{h_1})^{-1} \int_G f(nh)\, d\mu_G(nh), &(6)
\end{aligned}$$

where (3) follows by (μ_2), (4) follows by Fubini and left-invariance of λ_N, (5) follows by left invariance of λ_H, and (2) and (6) by Fubini and by definition of the measure $d\mu_G$. ∎

Proposition 8.2 shows that the measure μ_G is not left-invariant in general. The culprit is the term $\mathrm{mod}(i_{h_1})^{-1}$, which reflects the action of H on

N. Following Bourbaki [Bourbaki (1965)], Chapter VII, Section 2, Number 9, we can fix the problem by considering the image of the measure $\lambda_N \otimes \mathrm{mod}(i_h)^{-1}\lambda_H$ by φ, which means that we define the measure λ_G, such that for all $f \in \mathcal{L}^1_{\lambda_G}(G)$,

$$\int_G f(g) d\lambda_G(g) = \int_H \mathrm{mod}(i_h)^{-1} \left(\int_N f(nh)\, d\lambda_N(n) \right) d\lambda_H(h). \tag{μ_7}$$

Technically, we need to view (μ_7) as defining a functional on $\mathcal{K}(G)$ and use a version of the Radon–Riesz theorem to construct the measure λ_G but we will not delve into this matter here and leave the details as an exercise. The following result shows that λ_G is left-invariant.

Proposition 8.3. *Let $G = N \rtimes H = NH$ be the semi-direct product of two locally compact subgroups N and H with λ_N and λ_H left Haar measures on N and H, respectively, and λ_G be image of the measure $\lambda_N \otimes \mathrm{mod}(i_h)^{-1}\lambda_H$ by φ, which is defined such that*

$$\int_G f(nh) d\lambda_G(nh) = \int_H \mathrm{mod}(i_h)^{-1} \left(\int_N f(nh)\, d\lambda_N(n) \right) d\lambda_H(h)$$

for all $f \in \mathcal{L}^1_{\mu_G}(G)$. The measure λ_G is a left-invariant Haar measure.

Proof. Using Fubini and the equation

$$\mathrm{mod}(i_{h_1 h})^{-1} = \mathrm{mod}(i_h)^{-1}\mathrm{mod}(i_{h_1})^{-1},$$

we have

$$\begin{aligned}
\int_G f((n_1h_1)(nh))\, d\lambda_G(nh) &= \int_G f((n_1[h_1nh_1^{-1}])(h_1h))\, d\lambda_G(nh) && (1)\\
&= \int_H \mathrm{mod}(i_h)^{-1} \left(\int_N f((n_1[h_1nh_1^{-1}])(h_1h))\, d\lambda_N(n) \right) d\lambda_H(h) && (2)\\
&= \int_H \mathrm{mod}(i_h)^{-1}\mathrm{mod}(i_{h_1})^{-1} \left(\int_N f(n(h_1h))\, d\lambda_N(n) \right) d\mu_H(h) && (3)\\
&= \int_N \left(\int_H \mathrm{mod}(i_{h_1h})^{-1} f(n(h_1h)) d\lambda_H(h) \right) d\mu_N(n) && (4)\\
&= \int_N \left(\int_H \mathrm{mod}(i_h)^{-1} f(nh)\, d\lambda_H(h) \right) d\mu_N(n) && (5)\\
&= \int_H \mathrm{mod}(i_h)^{-1} \left(\int_N f(nh)\, d\lambda_N(n) \right) d\lambda_H(h) && (6)\\
&= \int_G f(nh)\, d\lambda_G(nh), && (7)
\end{aligned}$$

where (1), (2) and (7) follow by definition, (3) by (μ_2) and by left-invariance of λ_N, (4) by Fubini and the fact that mod is a group homomorphism, (5) by left invariance of λ_H, and (6) by Fubini. □

The following result shows how right-invariance fails but it also reveals what is the modular function of $G = N \rtimes H$. Recall that

$$(nh)(n_1h_1) = (n[hn_1h^{-1}])(hh_1).$$

Proposition 8.4. *Let $G = N \rtimes H = NH$ be the semi-direct product of two locally compact subgroups N and H, and λ_G be the left Haar measure defined by (μ_7), with λ_N and λ_H left Haar measures on N and H, respectively. We have*

$$\begin{aligned}&\int_G f((nh)(n_1h_1))\, d\lambda_G(nh)\\&= \operatorname{mod}(i_{h_1})\Delta_N(n_1)^{-1}\Delta_H(h_1)^{-1}\int_G f(nh) d\lambda_N(n)\, d\lambda_H(h).\end{aligned} \tag{μ_8}$$

Consequently, the modular function of the group G is given by

$$\Delta_G(nh) = \operatorname{mod}(i_h)^{-1}\Delta_N(n)\Delta_H(h). \tag{μ_9}$$

Proof. We need the fact that since N is a closed normal subgroup of G, then

$$\Delta_N(n) = \Delta_G(n), \quad n \in N,$$

where Δ_G is the modular function of the left Haar measure λ_G given by (μ_7). This is Proposition 10 in Bourbaki [Bourbaki (1965)], Chapter VII, Section 2, No. 7. We have

$$\begin{aligned}
&\int_G f((nh)(n_1h_1))\, d\lambda_G(nh) = \int_G f((n[hn_1h^{-1}])(hh_1))\, d\lambda_G(nh) && (1)\\
&= \int_H \left(\operatorname{mod}(i_h)^{-1}\int_N f((n[hn_1h^{-1}])(hh_1)) d\lambda_N(n)\right) d\lambda_H(h) && (2)\\
&= \int_H \left(\operatorname{mod}(i_h)^{-1}\int_N \Delta_N(hn_1h^{-1})^{-1} f(n(hh_1)) d\lambda_N(n)\right) d\lambda_H(h) && (3)\\
&= \int_H \left(\operatorname{mod}(i_h)^{-1}\int_N \Delta_N(hh_1h_1^{-1}n_1h_1(hh_1)^{-1})^{-1} f(n(hh_1))\right.\\
&\qquad \left. d\lambda_N(n)\right) d\lambda_H(h) && (4)\\
&= \operatorname{mod}(i_{h_1})\int_N \left(\int_H \operatorname{mod}(i_{hh_1})^{-1}\Delta_N(hh_1h_1^{-1}n_1h_1(hh_1)^{-1})^{-1}\right.\\
&\qquad \left. f(n(hh_1)) d\lambda_H(h)\right) d\lambda_N(n) && (5)
\end{aligned}$$

$$= \operatorname{mod}(i_{h_1})\Delta_H(h_1)^{-1} \int_N \left(\int_H \operatorname{mod}(i_h)^{-1} \Delta_N(hh_1^{-1}n_1h_1h^{-1})^{-1} \right.$$
$$\left. f(nh) d\lambda_H(h) \right) d\lambda_N(n) \tag{6}$$
$$= \operatorname{mod}(i_{h_1})\Delta_H(h_1)^{-1} \int_N \left(\int_H \operatorname{mod}(i_h)^{-1} \Delta_G(hh_1^{-1}n_1h_1h^{-1})^{-1} \right.$$
$$\left. f(nh) d\lambda_H(h) \right) d\lambda_N(n) \tag{7}$$
$$= \operatorname{mod}(i_{h_1})\Delta_H(h_1)^{-1} \int_N \left(\int_H \operatorname{mod}(i_h)^{-1} \Delta_G(n_1)^{-1} \right.$$
$$\left. f(nh) d\lambda_H(h) \right) d\lambda_N(n) \tag{8}$$
$$= \operatorname{mod}(i_{h_1})\Delta_N(n_1)^{-1}\Delta_H(h_1)^{-1} \int_N \left(\int_H \operatorname{mod}(i_h)^{-1} \right.$$
$$\left. f(nh) d\lambda_H(h) \right) d\lambda_N(n) \tag{9}$$
$$= \operatorname{mod}(i_{h_1})\Delta_N(n_1)^{-1}\Delta_H(h_1)^{-1} \int_H \operatorname{mod}(i_h)^{-1}$$
$$\left(\int_N f(nh)\, d\lambda_N(n) \right) d\lambda_H(h) \tag{10}$$
$$= \operatorname{mod}(i_{h_1})\Delta_N(n_1)^{-1}\Delta_H(h_1)^{-1} \int_G f(nh) d\lambda_G(nh), \tag{11}$$

where (1), (2) and (11) follow by definition, (3) by (μ_3) for λ_N, (4) by inserting $h_1h_1^{-1}$ twice in $\Delta_N(hn_1h^{-1})^{-1}$, (5) by replacing $\operatorname{mod}(i_h)^{-1}$ by $\operatorname{mod}(i_{h_1})\operatorname{mod}(i_{hh_1})^{-1}$ and Fubini, (6) by (μ_3) for λ_H applied to the entire integrand, (7) by replacing Δ_N by Δ_G, (8) since Δ_G is a group homomorphism, (9) by moving the constant term $\Delta_N(n_1)^{-1} = \Delta_G(n_1)^{-1}$ outside of the integrals, and (10) by Fubini. □

A right Haar measure can be obtained by using the right Haar measures $\Delta_N^{-1} \cdot \lambda_N$ and $\Delta_H^{-1} \cdot \lambda_H$.

Proposition 8.5. *Let $G = N \rtimes H = NH$ be the semi-direct product of two locally compact subgroups N and H, and ν_G be the image of the measure $(\Delta_N^{-1} \cdot \lambda_N) \otimes (\Delta_H^{-1} \cdot \lambda_H)$ by φ, where λ_N is a left Haar measure on N and λ_H is a left Haar measure on H, defined such that*

$$\int_G f(nh)\,d\nu_G(nh) = \int_H \Delta_H(h)^{-1}\left(\int_N \Delta_N(n)^{-1} f(nh)\,d\lambda_N(n)\right) d\lambda_H(h) \tag{μ_{10}}$$

for all $f \in \mathcal{L}^1_{\mu_G}(G)$. The measure ν_G is a right-invariant Haar measure.

Proof. Using Fubini and left-invariance, we have

$$\begin{aligned}
&\int_G f((nh)(n_1h_1))\,d\nu_G(nh)\\
&= \int_H \Delta_H(h)^{-1}\Big(\int_N \Delta_N(n)^{-1} f((n(hn_1h^{-1}))(hh_1))\,d\lambda_N(n)\Big)d\lambda_H(h) && (1)\\
&= \int_H \Delta_H(h)^{-1}\Big(\int_N \Delta_N(hn_1h^{-1})\Delta_N(n(hn_1h^{-1}))^{-1}\\
&\qquad f((n(hn_1h^{-1}))(hh_1))\,d\lambda_N(n)\Big)d\lambda_H(h) && (2)\\
&= \int_H \Delta_H(h)^{-1}\Big(\int_N \Delta_N(hn_1h^{-1})\Delta_N(hn_1h^{-1})^{-1}\Delta_N(n)^{-1}\\
&\qquad f(n(hh_1))\,d\lambda_N(n)\Big)d\lambda_H(h) && (3)\\
&= \int_N \Delta_N(n)^{-1}\Big(\int_H \Delta_H(h)^{-1} f(n(hh_1))\,d\lambda_H(h)\Big)d\lambda_N(n) && (4)\\
&= \int_N \Delta_N(n)^{-1}\Big(\int_H \Delta_H(h_1)\Delta_H(hh_1)^{-1} f(n(hh_1))\,d\lambda_H(h)\Big)d\lambda_N(n) && (5)\\
&= \int_N \Delta_N(n)^{-1}\Big(\int_H \Delta_H(h_1)\Delta_H(h_1)^{-1}\Delta_H(h)^{-1} f(nh)\,d\lambda_H(h)\Big)\\
&\qquad d\lambda_N(n) && (6)\\
&= \int_N \Delta_N(n)^{-1}\Big(\int_H \Delta_H(h)^{-1} f(nh)\,d\lambda_H(h)\Big)d\lambda_N(n)\\
&= \int_G f(nh)\,d\nu_G(nh), && (7)
\end{aligned}$$

where (1) and (7) follow by definition, (2) by replacing $\Delta_N(n)^{-1}$ by the term $\Delta_N(hnh^{-1})\,\Delta_N(n(hnh^{-1}))^{-1}$, (3) by ($\mu_3$) for λ_N, (4) by simplification and Fubini, (5) by replacing $\Delta_H(h)^{-1}$ by $\Delta_H(h_1)\Delta_H(hh_1)^{-1}$, (6) by ($\mu_3$) for λ_H, and (7) by simplification and Fubini. □

Remark: If ν_N and ν_H are right Haar measures, a similar but simpler proof shows that the image of the measure $\nu_N \otimes \nu_H$ by φ is right-invariant. This fact is also stated in Hewitt and Ross [Hewitt and Ross (1979)], Chapter IV, Section 15.

The right Haar measure ν_G is not left-invariant in general.

Proposition 8.6. *Let $G = N \rtimes H = NH$ be the semi-direct product of two locally compact subgroups N and H, and ν_G be the image of the measure $(\Delta_N^{-1} \cdot \lambda_N) \otimes (\Delta_H^{-1} \cdot \lambda_H)$ by φ, where λ_N is a left Haar measure on N and λ_H is a left Haar measure on H, defined such that*

$$\int_G f(nh)\, d\nu_G(nh) = \int_H \Delta_H(h)^{-1} \left(\int_N \Delta_N(n)^{-1} f(nh)\, d\lambda_N(n) \right) d\lambda_H(h)$$

for all $f \in \mathcal{L}^1_{\mu_G}(G)$. We have

$$\int_G f((n_1h_1)(nh))\, d\nu_G(nh) = \mathrm{mod}(i_{h_1})^{-1} \Delta_N(n_1) \Delta_H(h_1) \int_G f(nh)\, d\nu_G(nh). \qquad (\mu_{11})$$

Proof. Using Fubini and left-invariance, we have

$$\begin{aligned}
&\int_G f((n_1h_1)(nh))\, d\nu_G(nh) \\
&= \int_H \Delta_H(h)^{-1} \left(\int_N \Delta_N(n)^{-1} f((n_1(h_1nh_1^{-1}))(h_1h))\, d\lambda_N(n) \right) d\lambda_H(h) && (1) \\
&= \int_H \Delta_H(h)^{-1} \left(\int_N \Delta_N(h_1nh_1^{-1})^{-1} f((n_1(h_1nh_1^{-1}))(h_1h))\, d\lambda_N(n) \right) \\
&\quad d\lambda_H(h) && (2) \\
&= \mathrm{mod}(i_{h_1})^{-1} \int_H \Delta_H(h)^{-1} \left(\int_N \Delta_N(n)^{-1} f((n_1n)(h_1h))\, d\lambda_N(n) \right) \\
&\quad d\lambda_H(h) && (3) \\
&= \mathrm{mod}(i_{h_1})^{-1} \int_H \Delta_H(h)^{-1} \left(\int_N \Delta_N(n_1) \Delta_N(n_1n)^{-1} \right. \\
&\quad \left. f((n_1n)(h_1h))\, d\lambda_N(n) \right) d\lambda_H(h) && (4) \\
&= \mathrm{mod}(i_{h_1})^{-1} \Delta_N(n_1) \int_N \Delta_N(n)^{-1} \left(\int_H \Delta_H(h)^{-1} \right. \\
&\quad \left. f(n(h_1h))\, d\lambda_H(h) \right) d\lambda_N(n) && (5) \\
&= \mathrm{mod}(i_{h_1})^{-1} \Delta_N(n_1) \int_N \Delta_N(n)^{-1} \left(\int_H \Delta_H(h_1) \Delta_H(h_1h)^{-1} \right. \\
&\quad \left. f(n(h_1h))\, d\lambda_H(h) \right) d\lambda_N(n) && (6)
\end{aligned}$$

$$= \mathrm{mod}(i_{h_1})^{-1}\Delta_N(n_1)\Delta_H(h_1)\int_N \Delta_N(n)^{-1}\Big(\int_H \Delta_H(h)^{-1}$$
$$f(nh)\, d\lambda_H(h)\Big) d\lambda_N(n) \tag{7}$$
$$= \mathrm{mod}(i_{h_1})^{-1}\Delta_N(n_1)\Delta_H(h_1)\int_G f(nh)\, d\nu_G(nh), \tag{8}$$

where (1) and (8) follow by definition, (2) by replacing $\Delta_N(n)^{-1}$ by $\Delta_N(h_1nh_1^{-1})^{-1}$, (3) by (μ_2), (4) by replacing $\Delta_N(n)^{-1}$ by $\Delta_N(n_1)\Delta_N(n_1n)^{-1}$, (5) by left invariance for λ_N and Fubini, (6) by replacing $\Delta_H(h)^{-1}$ by $\Delta_H(h_1)\Delta_H(h_1h)^{-1}$, and (7) by left invariance for λ_H. ☐

In general, the left-invariant measure λ_G, image of the measure $\lambda_N \otimes \mathrm{mod}(i_h)^{-1}\lambda_H$ by φ is not right-invariant, and the right Haar measure ν_G, image of the measure $(\Delta_N^{-1}\cdot\lambda_N)\otimes(\Delta_H^{-1}\cdot\lambda_H)$ by φ, is not left-invariant. The modular function of the right measure ν_G is

$$\Delta_{\nu_G}(nh) = \mathrm{mod}(i_h)^{-1}\Delta_N(n)\Delta_H(h), \tag{μ_{12}}$$

where $i_h\colon N \to N$ is the automorphism given by

$$i_h(n) = hnh^{-1}, \quad n \in N, h \in H.$$

This confirms the result of Proposition 8.4.

As a consequence, even if N and G are unimodular, which means that $\Delta_N(n) = 1$ for all $n \in N$ and $\Delta_H(h) = 1$ for all $h \in H$, neither λ_G nor ν_G is both left and right-invariant. However, if $\mathrm{mod}(i_h) = 1$ for all $h \in H$, then the Haar measures λ_G and ν_G are both left and right-invariant and $G = N \rtimes H$ is unimodular. This is the case when N is compact. This is also the case where $N = \mathbb{R}^d$, H is a closed subgroup of $\mathbf{GL}(d)$, and the automorphisms i_h are linear maps of determinant ± 1. In particular, $\mathbf{SE}(d) = \mathbb{R}^d \rtimes \mathbf{SO}(d)$ and $\mathbf{E}(d) = \mathbb{R}^d \rtimes \mathbf{O}(d)$ satisfy this condition, so they are unimodular. Indeed, by Vol. I, Proposition 8.32, if $u \in \mathbf{GL}(d)$, then

$$\mathrm{mod}(u) = |\det(u)|. \tag{μ_{13}}$$

As shown in Example 8.1, the action of $H \simeq \mathbf{SO}(d)$ on $N \simeq \mathbb{R}^d$ given by

$$n \mapsto hnh^{-1}, \quad n = \begin{pmatrix} I_d & x \\ 0 & 1 \end{pmatrix}, \; h = \begin{pmatrix} Q & 0 \\ 0 & 1 \end{pmatrix}, \quad Q \in \mathbf{SO}(d),\; x \in \mathbb{R}^d,$$

yields

$$hnh^{-1} = \begin{pmatrix} I_d & Qx \\ 0 & 1 \end{pmatrix},$$

i_h is indeed a linear map of determinant $+1$, and thus so by (μ_{13}), we have $\mathrm{mod}(i_h) = 1$. More generally, if $\det(Q) = \pm 1$, so that H is isomorphic to a closed subgroup of $\mathbf{GL}(d)$ consisting of matrices of determinant ± 1, then $\det(i_h) = \pm 1$, and again by (μ_{13}), we have $\mathrm{mod}(i_h) = 1$.

All of the results of this section also hold for L^2 instead of L^1. The proof is left as an exercise.

8.6 Steerable Families

Since it is not practical to use the definition of cross-correlation involving integration over the group $G = \mathbb{R}^d \rtimes H$, we go back to the notion of lifted correlation. As we said earlier, it is more convenient to assume that the semi-direct product $G = \mathbb{R}^d \rtimes H$ is defined by an action of H on $\mathbb{R}^d$ by automorphisms, so that elements of G are denoted as pairs $(x, h) \in \mathbb{R}^d \times H$. Then as in Definition 8.2, for any function $f \in \mathrm{L}^2(\mathbb{R}^d)$ and any correlation kernel k, the *lifted correlation* $k \widetilde{\star} f$ is defined by ($*_{10'}$), namely

$$(k \widetilde{\star} f)(x, h) = \int_{\mathbb{R}^d} f(t) k(h^{-1} \cdot (t - x))\, dt, \quad (x, h) \in \mathbb{R}^d \times H.$$

Observe that $k \widetilde{\star} f$ is a function with domain $\mathbb{R}^d \times H$. Computing $(k \widetilde{\star} f)(x, h)$ requires sampling the group H, which is too expensive if $d \geq 3$. A way around this problem is to express the kernels k in terms of a basis of "steerable functions." Intuitively, this means using some kind of generalized harmonic functions. In our case, we need to find bases of functions in $\mathrm{L}^2(\mathbb{R}^d)$ that are H-steerable. In applications, H is a compact group so as we will see shortly, we can use the Peter–Weyl theorem, actually Version II, namely Theorem 4.3, to find steerable bases.

The problem is the presence of the term $k(h^{-1} \cdot (t - x))$ in the integral defining $k \widetilde{\star} f$. The key point is that if we can express the kernel k as a linear combination of linearly independent functions $Y_1, \ldots, Y_L$ in $\mathrm{L}^2(\mathbb{R}^d)$ that are "nice," which means that for every $h \in H$ and every $x \in \mathbb{R}^d$, each $Y_j(h^{-1} \cdot x)$ can be expressed as a linear combination of $Y_1(x), \ldots, Y_L(x)$, then it is possible to express $(k \widetilde{\star} f)(x, h)$ in a linear fashion in terms of the vector

$$f^Y(x) = \int_{\mathbb{R}^d} f(t) Y(t - x)\, dt,$$

where $Y(x)$ denotes the column vector

$$Y(x) = \begin{pmatrix} Y_1(x) \\ \vdots \\ Y_L(x) \end{pmatrix} \in \mathbb{C}^L.$$

So let us assume that for every $h \in H$, there is an invertible matrix $\Sigma(h) \in \mathbf{GL}(L, \mathbb{C})$, such that

$$\begin{pmatrix} Y_1(h^{-1} \cdot x) \\ Y_2(h^{-1} \cdot x) \\ \vdots \\ Y_L(h^{-1} \cdot x) \end{pmatrix} = \begin{pmatrix} \Sigma(h)_{11} & \Sigma(h)_{2,1} & \cdots & \Sigma(h)_{L1} \\ \Sigma(h)_{12} & \Sigma(h)_{2,2} & \cdots & \Sigma(h)_{L2} \\ \vdots & \vdots & \ddots & \vdots \\ \Sigma(h)_{1L} & \Sigma(h)_{2,L} & \cdots & \Sigma(h)_{LL} \end{pmatrix} \begin{pmatrix} Y_1(x) \\ Y_2(x) \\ \vdots \\ Y_L(x) \end{pmatrix},$$

or more concisely,

$$Y(h^{-1} \cdot x) = \Sigma(h)^\top Y(x), \quad x \in \mathbb{R}^d. \tag{steer1}$$

If Equation (steer1) holds, we say that $(Y_1, \ldots, Y_L)$ is an *H-steerable family* (or *H-steerable basis*). For short, we often drop H. In fact, we will see later that the map $\Sigma \colon H \to \mathbf{U}(L)$ is a representation of H. The reason for using $\Sigma(h)^\top$ instead of $\Sigma(h)$ is technical and will become clear later when we explain how to create steerable families. The notion of steerability occurred first in the seminal paper of Freeman and Adelson [Freeman and Adelson (1991)].

Next, assume that the kernel k can be expressed as a linear combination of the Y_i using some coefficients $w_i \in \mathbb{C}$ that we call *weights*. Let us write

$$k(x; w) = \sum_{i=1}^{L} \overline{w_i}\, Y_i(x) = w^* Y(x),\ x \in X, \tag{kw1}$$

where $w \in \mathbb{C}^L$ is the column vector consisting of the w_i. The reason for using conjugate weights will become apparent in the computation below.

Let us compute $k(h^{-1} \cdot x; w)$. Since Y is a steerable family, we have

$$\begin{aligned} k(h^{-1} \cdot x; w) &= w^* Y(h^{-1} \cdot x) = w^* \Sigma(h)^\top Y(x) \\ &= (\overline{\Sigma(h)} w)^* Y(x) = k(x; \overline{\Sigma(h)} w), \end{aligned}$$

namely

$$k(h^{-1} \cdot x; w) = k(x; \overline{\Sigma(h)} w). \tag{kw2}$$

So the new kernel is obtained by simply modifying the weights using the matrix $\overline{\Sigma(h)}$.

And now let us compute the *lifted correlation* $k \widetilde{\star} f$ given by

$$(k \widetilde{\star} f)(x, h) = \int_{\mathbb{R}^d} f(t)k(h^{-1} \cdot (t - x); w)\, dt, \quad (x, h) \in \mathbb{R}^d \rtimes H.$$

Using the fact that k is a steerable family, we have

$$\begin{aligned}(k \widetilde{\star} f)(x, h) &= \int_{\mathbb{R}^d} f(t)k(h^{-1} \cdot (t - x); w)\, dt \\ &= \int_{\mathbb{R}^d} f(t)w^* \Sigma(h)^\top Y(t - x)\, dt \\ &= w^* \Sigma(h)^\top \int_{\mathbb{R}^d} f(t)Y(t - x)\, dt.\end{aligned}$$

Let f^Y be the function $f^Y : \mathbb{R}^d \to \mathbb{C}^L$ given by

$$f^Y(x) = \begin{pmatrix} \int_{\mathbb{R}^d} f(t)Y_1(t - x)\, dt \\ \vdots \\ \int_{\mathbb{R}^d} f(t)Y_L(t - x)\, dt \end{pmatrix} = \int_{\mathbb{R}^d} f(t)Y(t - x)\, dt. \qquad (f^Y)$$

Then using the trick that for any two column vectors $u, v \in \mathbb{C}^n$, we have

$$u^\top v = \mathrm{tr}(vu^\top),$$

we obtain

$$\begin{aligned}(k \widetilde{\star} f)(x, h) &= \int_{\mathbb{R}^d} f(t)k(h^{-1} \cdot (t - x); w)\, dt \\ &= w^* \Sigma(h)^\top \int_{\mathbb{R}^d} f(t)Y(t - x)\, dt \\ &= w^* \Sigma(h)^\top f^Y(x) = \mathrm{tr}(f^Y(x)w^* \, \Sigma(h)^\top),\end{aligned}$$

where we use the identity $u^\top v = \mathrm{tr}(vu^\top)$ with $u = (w^* \Sigma(h)^\top)^\top$ and $v = f^Y(x)$. Observe that

$$\widehat{f}(x) = f^Y(x)w^*$$

is an $L \times L$ matrix that can be thought of as some kind of Fourier coefficients of f. The formula

$$(k \widetilde{\star} f)(x, h) = \mathrm{tr}\left(f^Y(x)w^* \, \Sigma(h)^\top\right) = \mathrm{tr}\left(\widehat{f}(x)\, \Sigma(h)^\top\right)$$

shows that $(k \widetilde{\star} f)(x, h)$ is similar to a Fourier cotransform with respect to the representation $\Sigma^\top$, where $\widehat{f}(x)/L$ plays the role of $\mathcal{F}(f)(\rho)$ and $\Sigma^\top$ plays the role of M_ρ (see Formula (FI) in Section 4.6 and Theorem 4.35), except that $\Sigma^\top$ is not necessarily irreducible. The function $\widehat{f} \colon \mathbb{R}^d \to \mathrm{M}_L(\mathbb{C})$ given by $\widehat{f}(x) = f^Y(x)w^*$ is a matrix-valued function.

What we have gained is that when we compute the integral

$$f^Y(x) = \int_{\mathbb{R}^d} f(t)Y(t-x)\,dt,$$

we incorporate all the information about the action of H on $\mathbb{R}^d$ into f^Y *without having to sample* the group H. The functions $(Y_1, \ldots, Y_L)$ package all the information about the group H needed to compute the essential part of $(k \widetilde{\star} f)(x,h)$. The outer product $\widehat{f}(x) = f^Y(x)w^*$ incorporates all the information in the kernel k using the weight vector w. Computing

$$(k \widetilde{\star} f)(x,h) = \operatorname{tr}\left(\widehat{f}(x)\,\Sigma(h)^\top\right) = \operatorname{tr}\left(\Sigma(h)\widehat{f}(x)^\top\right)$$

is then very cheap, since it is a linear operation only involving the matrix $\Sigma(h)$.

Another important observation is that starting with an input function $f \in \mathrm{L}^2(\mathbb{R}^d)$, the lifted correlation $k \widetilde{\star} f$ is a *scalar-valued* function (with codomain $\mathbb{C}$) defined on the *augmented domain* $\mathbb{R}^d \times H$, but $\widehat{f}$ is a *vector-valued* function from $\mathbb{R}^d$ to the *augmented codomain* $\mathrm{M}_L(\mathbb{C})$. The group $\mathbb{R}^d \times H$ acts on the domain $\mathbb{R}^d$ of $\widehat{f}$, and the group H acts on its codomain $\mathrm{M}_L(\mathbb{C})$ in terms of the representation Σ by multiplication on the left by $\Sigma(h)$. This is one of the motivations for introducing certain vector-valued functions called *feature fields*, discussed in Section 8.8.

The notion of steerability is easily generalized to any measure space X, such that $\mathrm{L}^2(X)$ is separable and H acts continuously on X. For example, any locally compact, metrizable, separable space X equipped with a σ-regular, locally finite, Borel measure μ will do; see Vol. I, Theorem 7.11.

Definition 8.5. Let X be any measure space, such that $\mathrm{L}^2(X)$ is separable and H acts continuously on X. Some linearly independent functions $(Y_1, \ldots, Y_L)$ in $\mathrm{L}^2(X)$ form an *H-steerable family* (or *H-steerable basis*) if there is a representation $\Sigma\colon H \to \mathbf{U}(L)$, such that

$$Y(h^{-1}\cdot x) = \Sigma(h)^\top Y(x), \quad h \in H,\ x \in X, \tag{steer2}$$

where $Y(x)$ denotes the column vector

$$Y(x) = \begin{pmatrix} Y_1(x) \\ \vdots \\ Y_L(x) \end{pmatrix} \in \mathbb{C}^L.$$

This more general notion will be needed in Section 8.16 to construct equivariant kernels.

The simplest example (simpler that $X = \mathbb{R}^2$) is the circle, $X = S^1$, with $H = \mathbf{SO}(2)$, the group of rotations in the plane.

Example 8.3. Let $H = \mathbf{SO}(2)$ and $X = S^1 \approx \mathbf{SO}(2)$. For any L-tuple of integers $(n_1, \ldots, n_L)$, we claim that

$$Y(\alpha) = (e^{-in_1\alpha}, \ldots, e^{-in_L\alpha})$$

is a steerable family (where the expression on the right-hand side denotes a column vector).

As we saw in Proposition 3.14, every unitary representation $\Sigma\colon \mathbf{SO}(2) \to \mathbf{U}(L)$ is given by a matrix of the form

$$\Sigma(\alpha) = \begin{pmatrix} e^{ik_1\alpha} & 0 & \ldots & 0 \\ 0 & e^{ik_2\alpha} & \ldots & 0 \\ \vdots & \vdots & \ddots & \vdots \\ 0 & \ldots & \ldots & e^{ik_L\alpha} \end{pmatrix}$$

with $k_1, \ldots, k_L \in \mathbb{Z}$, so if we pick $k_j = n_j$, for $j = 1, \ldots, L$ and

$$Y_j(\alpha) = e^{-in_j\alpha},$$

since

$$Y_j(\alpha - \theta) = e^{-in_j(\alpha-\theta)} = e^{in_j\theta}e^{-in_j\alpha} = e^{in_j\theta}Y_j(\alpha),$$

we see that

$$\begin{pmatrix} Y_1(\alpha-\theta) \\ \vdots \\ Y_L(\alpha-\theta) \end{pmatrix} = \begin{pmatrix} e^{in_1\theta} & 0 & \ldots & 0 \\ 0 & e^{in_2\theta} & \ldots & 0 \\ \vdots & \vdots & \ddots & \vdots \\ 0 & \ldots & \ldots & e^{in_L\theta} \end{pmatrix} \begin{pmatrix} Y_1(\alpha) \\ \vdots \\ Y_L(\alpha) \end{pmatrix},$$

which confirms that $(Y_1(\alpha) = e^{-in_1\alpha}, \ldots, Y_L(\alpha) = e^{-in_L\alpha})$ is a steerable family (again, the expression on the right-hand side denotes a column vector).

8.7 Construction of H-Steerable Families

We now present a method for finding steerable families on a space X as above equipped with a continuous action of a compact group H. The trick is to consider the unitary representation $V\colon H \to \mathbf{U}(\mathrm{L}^2(X))$ given by

$$(V(h)f)(x) = f(h^{-1} \cdot x), \quad h \in H,\ f \in \mathrm{L}^2(X),\ x \in X. \qquad (V)$$

According to the Peter–Weyl theorem, Version II (Theorem 4.16), the space $\mathrm{L}^2(X)$ is the Hilbert sum of closed subspaces E_ρ with $\rho \in R(H)$

(which may be reduced to zero), where E_ρ is the projection of $\mathrm{L}^2(X)$ under the projection π_ρ^V given by

$$\pi_\rho^V(f) = n_\rho \int_H \overline{\chi_\rho(h)}(V(h)f)\, d\lambda(h),$$

where $f \in \mathrm{L}^2(X)$ and λ is a left Haar measure on H.

First, we need to take care of a technicality. As stated the theorem involves the projection $\pi_\rho^V(f)$ of a function $f \in \mathrm{L}^2(X)$ and it is defined as a weak integral. For our purposes, we need a formula defining $(\pi_\rho^V(f))(x)$ for every $x \in H$. This can be achieved as follows. By definition of $\pi_\rho^V(f)$ as a weak integral, it is the unique function (given by the Riesz representation theorem, Theorem 3.16(2)), such that

$$\langle \pi_\rho^V(f), g\rangle = n_\rho \int_H \overline{\chi_\rho(h)}\langle V(h)f, g\rangle\, d\lambda(h)$$

for all $g \in \mathrm{L}^2(X)$, and using the definition of the inner product on $\mathrm{L}^2(X)$ and Fubini, the above is expressed as

$$\begin{aligned}
&\int_X (\pi_\rho^V(f))(x)\overline{g(x)}\, d\mu_X(x) \\
&\quad = n_\rho \int_H \overline{\chi_\rho(h)} \int_X (V(h)f)(x)\overline{g(x)}\, d\mu_X(x)\, d\lambda(h) \\
&\quad = n_\rho \int_X \left(\int_H \overline{\chi_\rho(h)}(V(h)f)(x)\, d\lambda(h) \right) \overline{g(x)}\, d\mu_X(x),
\end{aligned}$$

and since it holds for all $g \in \mathrm{L}^2(X)$, we must have

$$\begin{aligned}
(\pi_\rho^V(f))(x) &= n_\rho \int_H \overline{\chi_\rho(h)}(V(h)f)(x)\, d\lambda(h) \\
&= n_\rho \int_H \overline{\chi_\rho(h)} f(h^{-1}\cdot x)\, d\lambda(h). \qquad (\pi_\rho^V)
\end{aligned}$$

So the projection $\pi_\rho^V(f)$ of the function $f \in \mathrm{L}^2(X)$ can be defined pointwise by (π_ρ^V), but it is not obvious *a priori* that this yields a function in $\mathrm{L}^2(X)$, which is guaranteed by the weak integral argument.

Going back to Peter–Weyl II, each subspace E_ρ is a finite or countably infinite Hilbert sum of d_ρ (where $d_\rho = \infty$ is possible) closed finite-dimensional subspaces $E_\rho^{k_\rho}$ $(1 \le k_\rho \le d_\rho)$, such that for every ρ and every

k_ρ, each subrepresentation $V_\rho^{k_\rho}\colon H \to \mathbf{U}(E_\rho^{k_\rho})$ is equivalent to the irreducible representation $M_\rho\colon H \to \mathbf{U}(\mathbb{C}^{n_\rho})$. Thus there are linear isomorphisms $\theta_\rho^{k_\rho}\colon E_\rho^{k_\rho} \to \mathbb{C}^{n_\rho}$, such that the following diagram commutes

$$\begin{array}{ccc} E_\rho^{k_\rho} & \xrightarrow{\theta_\rho^{k_\rho}} & \mathbb{C}^{n_\rho} \\ {\scriptstyle V_\rho^{k_\rho}(h)}\big\downarrow & & \big\downarrow{\scriptstyle M_\rho(h)} \\ E_\rho^{k_\rho} & \xrightarrow[\theta_\rho^{k_\rho}]{} & \mathbb{C}^{n_\rho} \end{array}$$

for all $h \in H$. Since $(V(h)f)(x) = f(h^{-1}\cdot x)$, we have

$$f(h^{-1}\cdot x) = (V_\rho^{k_\rho}(h)f)(x), \quad h \in H,\ f \in E_\rho^{k_\rho},\ x \in X, \qquad \text{(steer3)}$$

for all $\rho \in R(H)$ and all k_ρ. If we pick an orthonormal basis (orthogonal works too) $(Y_{\rho,k_\rho}^1, \ldots, Y_{\rho,k_\rho}^{n_\rho})$ in each $E_\rho^{k_\rho}$ so that the family $(Y_{\rho,k_\rho}^j)_{\rho\in R(H), 1\leq k_\rho\leq d_\rho, 1\leq j\leq n_\rho}$ is a Hilbert basis of $\mathrm{L}^2(X)$, then there is an $n_\rho \times n_\rho$ unitary matrix $M^{\rho,k_\rho}(h)$ representing the linear map $V_\rho^{k_\rho}(h)$ with respect to the basis $(Y_{\rho,k_\rho}^1, \ldots, Y_{\rho,k_\rho}^{n_\rho})$ defined by

$$Y_{\rho,k_\rho}^j(h^{-1}\cdot x) = \sum_{i=1}^{n_\rho} M_{ij}^{\rho,k_\rho}(h) Y_{\rho,k_\rho}^i(x).$$

If we stack the $Y_{\rho,k_\rho}^i(x)$ into a column vector $Y_{\rho,k_\rho}(x)$ and the $Y_{\rho,k_\rho}^j(h^{-1}\cdot x)$ into a column vector $Y_{\rho,k_\rho}(h^{-1}\cdot x)$, we can write

$$Y_{\rho,k_\rho}(h^{-1}\cdot x) = (M^{\rho,k_\rho}(h))^\top Y_{\rho,k_\rho}(x). \qquad \text{(steer4)}$$

Remark: The presence of the transposition is the familiar artifact of linear algebra caused by the fact that $Y_{\rho,k_\rho}(x)$ is a column vector.

Replacing h by h^{-1} in (steer4), we get

$$\begin{aligned} Y_{\rho,k_\rho}(h\cdot x) &= (M^{\rho,k_\rho}(h^{-1}))^\top Y_{\rho,k_\rho}(x) = ((M^{\rho,k_\rho}(h))^*)^\top Y_{\rho,k_\rho}(x) \\ &= \overline{M^{\rho,k_\rho}(h)} Y_{\rho,k_\rho}(x), \end{aligned}$$

so conjugating on both sides, we get

$$\overline{Y_{\rho,k_\rho}}(h\cdot x) = M^{\rho,k_\rho}(h)\overline{Y_{\rho,k_\rho}}(x). \qquad \text{(steer5)}$$

Observe that the unitary representation $V_\rho^{k_\rho}\colon H \to \mathbf{U}(E_\rho^{k_\rho})$ define a representation in matrix form $M^{\rho,k_\rho}\colon H \to \mathbf{U}(\mathbb{C}^{n_\rho})$ equivalent to the irreducible representation M_ρ. The equations (steer4) express the fact that the basis functions $(Y_{\rho,k_\rho}^1, \ldots, Y_{\rho,k_\rho}^{n_\rho})$ are steerable.

Definition 8.6. If H is a compact group and X is a locally compact, metrizable, separable space equipped with a σ-regular, locally finite, Borel measure μ, given any continuous action of H on X, some linearly independent functions $(Y_1, \ldots, Y_L)$ in $\mathrm{L}^2(X)$ form an *H-steerable family* (or *H-steerable basis*) if there is a representation $\Sigma\colon H \to \mathbf{U}(L)$, such that

$$Y(h^{-1} \cdot x) = \Sigma(h)^\top Y(x), \quad h \in H,\, x \in X, \tag{steer6}$$

or equivalently,

$$\overline{Y}(h \cdot x) = \Sigma(h)\, \overline{Y}(x), \quad h \in H,\, x \in X, \tag{steer7}$$

where $Y(x)$ denotes the column vector

$$Y(x) = \begin{pmatrix} Y_1(x) \\ \vdots \\ Y_L(x) \end{pmatrix} \in \mathbb{C}^L.$$

Remark: Steerability as defined above is equivalent to the notion of steerability as defined in Lang and Weiler [Lang and Weiler (2019)]. In Cesa, Lang and Weiler [Cesa *et al.* (2022)] as well as Bekkers [Bekkers (2020)], the notion of steerability is defined using $Y(h \cdot x)$ instead of $Y(h^{-1} \cdot x)$. We pass from one version to the other by conjugation of the functions. In the papers mentioned above, steerable families are also called *harmonic basis functions*.

Due to its importance, the preceding discussion is summarized in the following theorem.

Theorem 8.7. *Let X be a locally compact, metrizable, separable space equipped with a σ-regular, locally finite, Borel measure μ. If H is a compact group acting continuously on X (not necessarily in a transitive fashion), consider the unitary representation $V\colon H \to \mathbf{U}(\mathrm{L}^2(X))$ given by*

$$(V(h)f)(x) = f(h^{-1} \cdot x), \quad h \in H,\, f \in \mathrm{L}^2(X),\, x \in X.$$

According to the Peter–Weyl theorem, Version II, the space $\mathrm{L}^2(X)$ is the Hilbert sum of closed subspaces E_ρ with $\rho \in R(H)$ (which may be reduced to zero), where E_ρ is the projection of $\mathrm{L}^2(X)$ under the projection π^V_ρ given by

$$(\pi^V_\rho(f))(x) = n_\rho \int_H \overline{\chi_\rho(h)} f(h^{-1} \cdot x)\, d\lambda(h),$$

where $f \in \mathrm{L}^2(X)$, $x \in X$, and λ is a left Haar measure on H. Each subspace E_ρ is a finite or countably infinite Hilbert sum of d_ρ (where $d_\rho = \infty$

is possible) closed finite-dimensional subspaces $E_\rho^{k_\rho}$ $(1 \leq k_\rho \leq d_\rho)$, *such that for every* ρ *and every* k_ρ, *each subrepresentation* $V_\rho^{k_\rho}\colon H \to \mathbf{U}(E_\rho^{k_\rho})$ *is equivalent to the irreducible representation* $M_\rho\colon H \to \mathbf{U}(\mathbb{C}^{n_\rho})$. *Furthermore, each space* $E_\rho^{k_\rho}$ *has an* H*-steerable orthonormal basis with respect to an irreducible representation equivalent to* M_ρ *(the functions specified by the column vectors* Y_{ρ,k_ρ}*). The union of these* H*-steerable families for all* $\rho \in R(h)$ *and all* k_ρ *is an* H*-steerable Hilbert basis of* $\mathrm{L}^2(X)$.

A similar result is proven in Lang and Weiler [Lang and Weiler (2019)] and in Cesa Lang and Weiler [Cesa *et al.* (2022)]. In fact, a version of this result applying to real representations is also shown.

We now consider several examples.

Example 8.4. Let $H = \mathbf{SO}(2)$ and $X = S^1 \approx \mathbf{SO}(2)$. In this case, $R = \mathbb{Z}$, all irreducible representations are one-dimensional and of the form $z \mapsto e^{in\theta}z$, and the characters are given by $\chi_n(e^{i\theta}) = e^{in\theta}$. Given a function $f \in \mathrm{L}^2(S^1)$, we have

$$\begin{aligned}
\pi_n(f)(e^{i\alpha}) &= \int_H \overline{\chi_n(e^{i\theta})} f((e^{i\theta})^{-1}e^{i\alpha})\, d\lambda(h) = \int_{-\pi}^{\pi} e^{-in\theta} f(e^{-i\theta}e^{i\alpha})\, \frac{d\theta}{2\pi} \\
&= \frac{1}{2\pi}\int_{-\pi}^{\pi} e^{-in\theta} f(e^{i(\alpha-\theta)})\, d\theta = \frac{1}{2\pi}\int_{\pi}^{-\pi} e^{-in(\alpha-\varphi)} f(e^{i\varphi})\,(-d\varphi) \\
&= \frac{1}{2\pi} e^{-in\alpha}\int_{-\pi}^{\pi} e^{in\varphi} f(e^{i\varphi})\, d\varphi \\
&= \frac{1}{2\pi} e^{i(-n)\alpha}\int_{-\pi}^{\pi} e^{-i(-n)\varphi} f(e^{i\varphi})\, d\varphi = e^{-in\alpha}c_{-n},
\end{aligned}$$

where

$$c_n = \frac{1}{2\pi}\int_{-\pi}^{\pi} e^{-in\varphi} f(e^{i\varphi})\, d\varphi$$

is the nth Fourier coefficient of f. Thus

$$\pi_n(f)(e^{i\alpha}) = e^{-in\alpha}c_{-n}. \tag{str1}$$

The index n is flipped to $-n$ due to the fact that the projection operator uses $\overline{\chi_\rho(h)}$. The space E_n is one-dimensional and has the function

$$Y_n(e^{i\alpha}) = e^{-in\alpha} \tag{str2}$$

as a basis. It is steerable since

$$Y_n(e^{i(\alpha-\theta)}) = e^{-in(\alpha-\theta)} = e^{in\theta}e^{-in\alpha} = e^{in\theta}Y_n(e^{i\alpha}),$$

and $\chi_n(e^{i\theta}) = e^{in\theta}$ is a character.

Example 8.5. Let $H = \mathbf{SO}(2)$ and $X = \mathbb{R}^2$. In this case, again $R = \mathbb{Z}$, all irreducible representations are one-dimensional and the characters are of the form $\chi_n(e^{i\theta}) = e^{in\theta}$. Given any function $f \in \mathrm{L}^2(\mathbb{R}^2)$, we have

$$\pi_n(f)(x) = \int_H \overline{\chi_n(e^{i\theta})} f(R_\theta^{-1} x)\, d\lambda(h) = \int_{-\pi}^{\pi} e^{-in\theta} f((R_{-\theta})x)\, \frac{d\theta}{2\pi},$$

where R_θ is the rotation matrix

$$R_\theta = \begin{pmatrix} \cos\theta & -\sin\theta \\ \sin\theta & \cos\theta \end{pmatrix}.$$

This time E_n is the Hilbert sum of countably many subspaces of dimension 1. Let us compute $f_n(R_\varphi x)$ where $f_n(x) = \pi_n(f)(x)$. We have

$$\begin{aligned} f_n(R_\varphi x) &= \int_{-\pi}^{\pi} e^{-in\theta} f(R_{-\theta} R_\varphi x)\, \frac{d\theta}{2\pi} \\ &= \int_{-\pi}^{\pi} e^{-in\theta} f(R_{-(\theta-\varphi)} x)\, \frac{d\theta}{2\pi} = \int_{-\pi}^{\pi} e^{-in(\psi+\varphi)} f((R_{-\psi})x)\, \frac{d\psi}{2\pi} \\ &= e^{-in\varphi} \int_{-\pi}^{\pi} e^{-in\psi} f((R_{-\psi})x)\, \frac{d\psi}{2\pi} = e^{-in\varphi} f_n(x). \end{aligned}$$

In summary,

$$f_n(R_\varphi x) = e^{-in\varphi} f_n(x). \tag{str3}$$

For $x = re_1$, $r \in \mathbb{R}_+$, where $e_1 = (1, 0)$, we get

$$f_n(rR_\varphi e_1) = e^{-in\varphi} f_n^{\mathrm{rad}}(r),$$

with

$$f_n^{\mathrm{rad}}(r) = f_n(re_1) = \int_{-\pi}^{\pi} e^{-in\theta} f(r(R_{-\theta})e_1)\, \frac{d\theta}{2\pi}. \tag{str4}$$

The function f_n^{rad} is called a *radial function*. It is a function defined on $\mathbb{R}_+$. We see that in polar coordinates (r, φ),

$$f_n(r, \varphi) = e^{-in\varphi} f_n^{\mathrm{rad}}(r). \tag{str5}$$

Thus, we are reduced to finding a Hilbert basis of $\mathrm{L}^2(\mathbb{R}_+)$. There are many candidates but the Hilbert basis involving the Hermite functions is particularly elegant. These are the functions

$$\psi_m(x) = e^{-\frac{x^2}{2}} H_m(x), \tag{str6}$$

where the $H_m(x)$ are Hermite polynomials. The functions ψ_m are also a Hilbert basis of $\mathrm{L}^2(\mathbb{R})$; see Sansone [Sansone (1991)], Chapter IV, Section 7 and Folland [Folland (1992)], Chapter 6, Section 6.4.

The Hermite polynomials are real polynomials given by the equations

$$H_m(x) = (-1)^m e^{x^2} \frac{d^m}{dx^m} e^{-x^2}. \tag{str7}$$

They are also defined by the recurrence relations

$$\begin{aligned} H_{n+2}(x) &= 2xH_{n+1}(x) - 2(n+1)H_n(x) \\ H_1(x) &= 2x \\ H_0(x) &= 1. \end{aligned}$$

From these equations, the following explicit formula can be derived:

$$H_m(x) = \sum_{k=0}^{\lfloor \frac{m}{2} \rfloor} (-1)^k \frac{m(m-1)(m-2)\cdots(m-2k+1)}{k!} (2x)^{m-2k};$$

see Sansone [Sansone (1991)], Chapter IV, Section 2 and Folland [Folland (1992)], Chapter 6, Section 6.4. The first six Hermite polynomials are

$$\begin{aligned} H_0(x) &= 1 & H_1(x) &= 2x \\ H_2(x) &= 4x^2 - 2 & H_3(x) &= 8x^2 - 12x \\ H_4(x) &= 16x^4 - 48x^2 + 12 & H_5(x) &= 32x^5 - 160x^3 + 120x. \end{aligned}$$

The Hermite polynomials are orthogonal with respect to the inner product

$$\langle f, g \rangle = \int_{\mathbb{R}} e^{-x^2} f(x)g(x)\, dx,$$

and so the functions ψ_m are orthogonal with respect to the usual inner product on $\mathrm{L}^2(\mathbb{R})$. They are not orthonormal because

$$\int_{-\infty}^{\infty} H_m^2(x)e^{-x^2}\, dx = \sqrt{\pi}\, 2^m m!.$$

See Sansone [Sansone (1991)], Chapter IV, Section 2. The purpose of the term $e^{-\frac{x^2}{2}}$ is to insure that the functions ψ_m are square integrable over $\mathbb{R}$.

The Hermite polynomials are discussed quite extensively in Sansone [Sansone (1991)], Chapter IV, Sections 2–5 and 7, and in Folland [Folland (1992)], Chapter 6, Section 6.4. Note that Sansone omits the factor $(-1)^m$. As a consequence, for m odd, Sansone's version of the Hermite polynomials is $-H_m(x)$, and so the term of degree m has a negative coefficient. The fact that the Hermite functions $e^{-\frac{x^2}{2}} H_m(x)$ form a Hilbert basis of $\mathrm{L}^2(\mathbb{R})$ is proven in Sansone [Sansone (1991)], Chapter IV, Section 7 and in Folland [Folland (1992)], Chapter 6, Section 6.4.

Then the functions

$$Y_{m,n}(r,\varphi) = e^{-in\varphi}e^{-\frac{r^2}{2}}H_m(r), \ \ m \geq 0, \tag{str8}$$

form a steerable Hilbert basis of E_n ($n \in \mathbb{Z}$). Indeed, we see immediately that

$$Y_{m,n}(r,\varphi-\theta) = e^{in\theta}Y_{m,n}(r,\varphi).$$

This case was also investigated by Weiler and Cesa [Lang and Weiler (2019)] in a more informal fashion.

Example 8.6. Let $H = \mathbf{SO}(2)$ and $X = \mathrm{L}^2(\mathbf{SE}(2))$. The action of $\mathbf{SO}(2)$ on $\mathrm{L}^2(\mathbf{SE}(2))$ is the left regular action $\mathbf{R}^{\mathbf{SO}(2)\to\mathrm{L}^2(\mathbf{SE}(2))}$ given by

$$\mathbf{R}^{\mathbf{SO}(2)\to\mathrm{L}^2(\mathbf{SE}(2))}_{R_\varphi}(f)(x,\psi) = f(R_{-\varphi}x, \psi-\varphi), \quad f \in \mathrm{L}^2(\mathbf{SE}(2)),$$
$$x \in \mathbb{R}^2, \ R_\varphi \in \mathbf{SO}(2).$$

In this case, again $R = \mathbb{Z}$, all irreducible representations are one-dimensional and the characters are of the form $\chi_n(e^{i\varphi}) = e^{in\varphi}$. Given any function $f \in \mathrm{L}^2(\mathbf{SE}(2))$, we have

$$\pi_n(f)(x,\psi) = \int_{-\pi}^{\pi} e^{-in\varphi} f(R_{-\varphi}x, \psi-\varphi)\,\frac{d\varphi}{2\pi},$$

where R_φ is the rotation matrix

$$R_\varphi = \begin{pmatrix} \cos\varphi & -\sin\varphi \\ \sin\varphi & \cos\varphi \end{pmatrix}.$$

The space E_n is the Hilbert sum of countably many subspaces of dimension 1. Write $f_n(x,\psi) = \pi_n(f)(x,\psi)$. If we let $\varphi = \psi+\varphi_1$, so that $\psi-\varphi = -\varphi_1$, we obtain

$$\begin{aligned} f_n(x,\psi) &= \int_{-\pi}^{\pi} e^{-in\varphi} f(R_{-\varphi}x, \psi-\varphi)\,\frac{d\varphi}{2\pi} \\ &= \int_{-\pi}^{\pi} e^{-in(\psi+\varphi_1)} f(R_{-(\psi+\varphi_1)}x, -\varphi_1)\,\frac{d\varphi_1}{2\pi} \\ &= e^{-in\psi}\int_{-\pi}^{\pi} e^{-in\varphi_1} f(R_{-\varphi_1}R_{-\psi}x, -\varphi_1)\,\frac{d\varphi_1}{2\pi} \\ &= e^{-in\psi}\int_{-\pi}^{\pi} e^{in\varphi_1} f(R_{\varphi_1}R_{-\psi}x, \varphi_1)\,\frac{d\varphi_1}{2\pi}. \end{aligned}$$

In summary, we proved that

$$f_n(x,\psi) = e^{-in\psi}\int_{-\pi}^{\pi} e^{in\varphi_1} f(R_{\varphi_1}R_{-\psi}x, \varphi_1)\,\frac{d\varphi_1}{2\pi}. \tag{str9}$$

As a consequence, we have

$$\begin{aligned}f_n(R_\alpha x, \psi + \alpha) &= e^{-in(\psi+\alpha)} \int_{-\pi}^{\pi} e^{in\varphi_1} f(R_{\varphi_1} R_{-(\psi+\alpha)} R_\alpha x, \varphi_1)\, \frac{d\varphi_1}{2\pi} \\ &= e^{-in\alpha} e^{-in\psi} \int_{-\pi}^{\pi} e^{in\varphi_1} f(R_{\varphi_1} R_{-\psi} x, \varphi_1)\, \frac{d\varphi_1}{2\pi} \\ &= e^{-in\alpha} f_n(x, \psi).\end{aligned}$$

Thus, we proved that

$$f_n(R_\alpha x, \psi + \alpha) = e^{-in\alpha} f_n(x, \psi). \tag{str10}$$

For $x = re_1$, $r \in \mathbb{R}_+$, with $e_1 = (1, 0)$, from (str9), we get

$$f_n(rR_\alpha e_1, \theta) = e^{-in\theta} f_n^{\mathrm{rad}}(r, \theta - \alpha), \tag{str11}$$

with

$$f_n^{\mathrm{rad}}(r, \psi) = f_n(re_1, \psi) = \int_{-\pi}^{\pi} e^{in\varphi} f(rR_\varphi R_{-\psi} e_1, \varphi)\, \frac{d\varphi}{2\pi}. \tag{str12}$$

In polar coordinates (r, α),

$$f_n((r, \alpha), \theta) = e^{-in\theta} f_n^{\mathrm{rad}}(r, \theta - \alpha). \tag{str13}$$

Observe that since in polar coordinates the effect of a rotation $R_{-\varphi}$ is to transform (r, α) to $(r, \alpha - \varphi)$, we have

$$\begin{aligned}f_n((r, \alpha - \varphi), \theta - \varphi) &= e^{-in(\theta-\varphi)} f_n^{\mathrm{rad}}(r, \theta - \varphi - (\alpha - \varphi)) \\ &= e^{in\varphi} e^{-in\theta} f_n^{\mathrm{rad}}(r, \theta - \alpha) = e^{in\varphi} f_n((r, \alpha), \theta),\end{aligned}$$

confirming that the functions f_n are steerable.

The functions f_n^{rad} belong to $\mathrm{L}^2(\mathbb{R}_+ \times \mathbf{SO}(2))$. For every fixed ψ they are radial function of $x = re_1$, and for fixed r they are functions of ψ. Since the functions $e^{-\frac{r^2}{2}} H_m(r)$ form a Hilbert basis of $\mathrm{L}^2(\mathbb{R}_+)$ and the functions $e^{-ik\psi}$ form a Hilbert basis of $\mathrm{L}^2(\mathbf{SO}(2))$, it can be shown that the family of functions $e^{-\frac{r^2}{2}} H_m(r) e^{-ik\psi}$ form a Hilbert basis of $\mathrm{L}^2(\mathbb{R}_+ \times \mathbf{SO}(2))$; see Lang [Lang (1996)], Chapter XVII, Problem 9.

At first glance, it is not obvious that the functions $f_n^{\mathrm{rad}}(r, \psi)$ yield all the functions in the Hilbert basis of $\mathrm{L}^2(\mathbb{R}_+ \times \mathbf{SO}(2))$. In fact they do and this is shown as follows. If we define the function f by

$$f(rR_\alpha e_1, \psi) = e^{-\frac{r^2}{2}} H_m(r) e^{-ik\alpha} e^{-in\psi} e^{ik\psi},$$

then we have

$$f(rR_{\varphi-\psi} e_1, \varphi) = e^{-\frac{r^2}{2}} H_m(r) e^{-ik(\varphi-\psi)} e^{-in\varphi} e^{ik\varphi} = e^{-\frac{r^2}{2}} H_m(r) e^{-in\varphi} e^{ik\psi},$$

and then by (str12) and the previous equation

$$\begin{aligned} f_n^{\mathrm{rad}}(r,\psi) &= \int_{-\pi}^{\pi} e^{in\varphi} f(rR_{\varphi-\psi}e_1,\varphi)\,\frac{d\varphi}{2\pi} \\ &= \int_{-\pi}^{\pi} e^{-\frac{r^2}{2}} H_m(r) e^{in\varphi} e^{-in\varphi} e^{ik\psi}\,\frac{d\varphi}{2\pi} = e^{-\frac{r^2}{2}} H_m(r) e^{ik\psi}. \end{aligned}$$

By (str13) and the above reasoning, for n fixed, the functions $e^{-in\theta}e^{ik(\theta-\alpha)}e^{-\frac{r^2}{2}}H_m(r)$ form a Hilbert basis of E_n, and thus the functions

$$Y_{k,m,n}((r,\alpha),\theta) = e^{-in\theta}e^{ik(\theta-\alpha)}e^{-\frac{r^2}{2}}H_m(r) = e^{-i(n-k)\theta}e^{-ik\alpha}e^{-\frac{r^2}{2}}H_m(r) \tag{str14}$$

form a steerable basis of $\mathrm{L}^2(\mathbf{SE}(2))$, with $n,k\in\mathbb{Z}$ and $m\geq 0$. In Section 8.11, it will be more convenient to change the index k to $n-k$, in which case the term $e^{-i(n-k)\theta}e^{-ik\alpha}$ becomes

$$e^{-ik\theta}e^{-i(n-k)\alpha} = e^{-in\alpha}e^{-ik(\theta-\alpha)},$$

and so we also have the steerable basis of functions

$$e^{-in\alpha}e^{-ik(\theta-\alpha)}e^{-\frac{r^2}{2}}H_m(r), \quad n,k\in\mathbb{Z},\ m\geq 0. \tag{str15}$$

Example 8.7. Let H be any compact group and let $X = G$ with G acting on itself by left multiplication. Since the M_ρ are (irreducible) representations of H, we have $M_\rho(s^{-1}t) = M_\rho(s^{-1})M_\rho(t) = M_\rho(s)^*M_\rho(t)$, so the jth column $(1/n_\rho)m_{*j}^{(\rho)}(s^{-1}t)$ of the matrix $M_\rho(s^{-1}t)$ can be expressed as

$$(1/n_\rho)m_{*j}^{(\rho)}(s^{-1}t) = M_\rho(s)^*(1/n_\rho)m_{*j}^{(\rho)}(t) = (\overline{M_\rho(s)})^\top(1/n_\rho)m_{*j}^{(\rho)}(t),$$

and so

$$\overline{m_{*j}^{(\rho)}}(s^{-1}t) = (M_\rho(s))^\top\overline{m_{*j}^{(\rho)}}(t). \tag{str16}$$

Since the family of functions

$$\left(\frac{1}{\sqrt{n_\rho}}m_{ij}^{(\rho)}\right)_{1\leq i,j\leq n_\rho,\ \rho\in R(H)}$$

is a Hilbert basis of $\mathrm{L}^2(G)$, it follows that according to Definition 8.6, $(\overline{m_{1j}},\ldots,\overline{m_{n_\rho,j}})$ forms a steerable basis of $\overline{\mathfrak{l}_j^{(\rho)}}$ for $j=1,\ldots,n_\rho$, using the notation of Section 4.1. Note that in terms of the notation used in Theorem 8.7, $d_\rho = n_\rho$. Recall that by Peter–Weyl I, $\mathrm{L}^2(H)$ is the Hilbert sum of minimal two-sided ideals $\mathfrak{a}_\rho$ isomorphic to the matrix algebra $\mathrm{M}_{n_\rho}(\mathbb{C})$, and $\mathfrak{a}_\rho$ is expressed as the finite Hilbert sum of n_ρ minimal left ideals $\mathfrak{l}_j^{(\rho)}$.

Observe that we can also obtain the above result by considering the left regular representation $V = \mathbf{R}$ of G. As noted just after Definition 4.7, the projection π_ρ^V maps $\mathrm{L}^2(G)$ onto $\overline{\mathfrak{a}_\rho}$, so the functions $(\overline{m_{1j}}, \ldots, \overline{m_{n_\rho,j}})$ are indeed in $\overline{\mathfrak{a}_\rho}$ and form a basis of $\overline{\mathfrak{l}_j^{(\rho)}}$, the jth column of $\overline{M}_\rho$. Thus Equation (steer6) holds.

Example 8.8. Let $H = \mathbf{SO}(n+1)$ and $X = S^n$ $(n \geq 2)$. It is shown in Section 6.10 that we have a Hilbert sum

$$\mathrm{L}^2(S^n) = \bigoplus_{\ell \geq 0} \mathcal{H}_\ell^{\mathbb{C}}(S^n)$$

and that we obtain a decomposition of the left regular representation $\mathbf{R}\colon \mathbf{SO}(n+1) \to \mathbf{U}(\mathrm{L}^2(S^n))$ into a family of irreducible representations $\mathbf{R}_\ell\colon \mathbf{SO}(n+1) \to \mathbf{U}(\mathcal{H}_\ell^{\mathbb{C}}(S^n))$ of $\mathbf{SO}(n+1)$ in the spaces $\mathcal{H}_\ell^{\mathbb{C}}(S^n)$ of spherical harmonics on S^n. The space $\mathcal{H}_k^{\mathbb{C}}(S^n)$ has dimension

$$a_{k,n+1} = \binom{n+k}{k} - \binom{n+k-2}{k-2}$$

if $n \geq 1$, $k \geq 2$, with $a_{0,n+1} = 1$ and $a_{1,n+1} = n$. For $n = 2$, we get $a_{k,3} = 2k+1$. Since $\mathbf{R}_\ell$ is the left regular representation,

$$f(Q^{-1}x) = ((\mathbf{R}_\ell)_Q(f))(x), \quad Q \in \mathbf{SO}(n+1),\ f \in \mathcal{H}_\ell^{\mathbb{C}}(S^n),\ x \in S^n.$$

This shows that by Definition 8.6, the space $\mathcal{H}_\ell^{\mathbb{C}}(S^n)$ is steerable. In theory, by picking an orthogonal basis $(Y_1^\ell, \ldots, Y_{a_{\ell,n+1}}^\ell)$ of size $a_{\ell,n+1}$ in $\mathcal{H}_\ell^{\mathbb{C}}(S^n)$ and expressing the representation $\mathbf{R}_\ell$ in matrix form over this basis, we obtain a steerable basis. The basis functions Y_j^ℓ can be expressed in terms of the Gegenbauer polynomials and some suitable points on the sphere S^n; see Theorem 7.29 in Gallier and Quaintance [Gallier and Quaintance (2020b)].

When $n = 2$, the spherical harmonics in $\mathcal{H}_\ell^{\mathbb{C}}(S^2)$ can be expressed in spherical coordinates in terms of the associated Legendre functions P_ℓ^j. These are the Laplace spherical harmonics Y_ℓ^j, which are functions of the form $e^{-ij\varphi} P_\ell^j(\cos\theta)$, with $\ell \in \mathbb{N}$ and $-\ell \leq j \leq \ell$, up to a constant; see Definition 5.17 in Section 5.12 and Section 5.15. We know from Section 5.12 that the representations $\mathcal{D}^{(\ell)}\colon \mathbf{SO}(3) \to \mathbf{U}(\mathcal{P}_\ell^{\mathbb{C}})$, where $\mathcal{D}^{(\ell)}$ is the Wigner $\mathcal{D}$-matrix defined in Definition 5.19, are irreducible, and by Proposition 5.32, the column vector $\overline{Y_\ell}$ consisting of the $2\ell+1$ functions $\overline{Y_\ell^j}$ is a steerable basis. The case $n = 2$ is also treated in Lang and Weiler [Lang and Weiler (2019)], but they do not state explicitly what is their definition of the Wigner $\mathcal{D}$-matrices, and it appears that they are missing a conjugation.

Example 8.9. In this example, we describe a method generalizing the method of Example 8.6 to decompose $\mathrm{L}^2(\mathbf{SE}(n))$ using the representation V and the projections

$$\begin{aligned}(\pi_\rho^V(f))(x,h_1) &= n_\rho \int_{\mathbf{SO}(n)} \overline{\chi_\rho(h)} f(h^{-1}x, h^{-1}h_1)\, d\lambda(h)\\ &= \int_{\mathbf{SO}(n)} \overline{u_\rho(h)} f(h^{-1}x, h^{-1}h_1)\, d\lambda(h),\end{aligned}$$

with $(x,h_1) \in \mathbf{SE}(n)$, for all $f \in \mathrm{L}^2(\mathbf{SE}(n))$ and all $\rho \in R(\mathbf{SO}(n))$.

Write $f^\rho(x,h_1) = (\pi_\rho^V(f))(x,h_1)$. Since $\mathbf{SO}(n)$ is unimodular (because it is compact), we have

$$\begin{aligned}f^\rho(x,h_1) &= \int_{\mathbf{SO}(n)} \overline{u_\rho(h)} f(h^{-1}x, h^{-1}h_1)\, d\lambda(h),\\ &= \int_{\mathbf{SO}(n)} \overline{u_\rho(h_1h_2)} f(h_2^{-1}h_1^{-1}x, h_2^{-1})\, d\lambda(h_2), \qquad h = h_1h_2\\ &= \int_{\mathbf{SO}(n)} \overline{u_\rho(h_1h_2^{-1})} f(h_2h_1^{-1}x, h_2)\, d\lambda(h_2). \qquad (*_{13})\end{aligned}$$

Recall that $u_\rho = m_{11}^{(\rho)} + \cdots + m_{n_\rho n_\rho}^{(\rho)}$, which is n_ρ times the trace of the matrix M_ρ corresponding to the irreducible representation of $\mathbf{SO}(n)$ indexed by ρ. Since $M_\rho = (1/n_\rho)\left(m_{ij}^{(\rho)}\right)$ and it is a representation, because $(1/n_\rho)m_{ii}^{(\rho)}(h_1h_3)$ is the (i,i)-entry in the matrix $M_\rho(h_1h_3)$, it is equal to the inner product of the ith row of $M_\rho(h_1)$ by the ith column of $M_\rho(h_3)$, so

$$(1/n_\rho)m_{ii}^{(\rho)}(h_1h_3) = \sum_{j=1}^{n_\rho}(1/n_\rho)m_{ij}^{(\rho)}(h_1)(1/n_\rho)m_{ji}^{(\rho)}(h_3),$$

and by multiplying both sides by n_ρ, we get

$$u_\rho(h_1h_3) = \sum_{i=1}^{n_\rho} m_{ii}^{(\rho)}(h_1h_3) = (1/n_\rho)\sum_{i,j=1}^{n_\rho} m_{ij}^{(\rho)}(h_1)m_{ji}^{(\rho)}(h_3). \qquad (*_{14})$$

The calculations in $(*_{14})$ and $(*_{13})$ imply that

$$\begin{aligned}f^\rho(x,h_1) &= \int_{\mathbf{SO}(n)} \overline{u_\rho(h_1h_2^{-1})} f(h_2h_1^{-1}x, h_2)\, d\lambda(h_2)\\ &= n_\rho \sum_{i,j=1}^{n_\rho} \overline{\frac{1}{n_\rho}m_{ij}^{(\rho)}(h_1)} \int_{\mathbf{SO}(n)} \overline{\frac{1}{n_\rho}m_{ji}^{(\rho)}(h_2^{-1})} f(h_2h_1^{-1}x, h_2)\, d\lambda(h_2)\\ &= n_\rho \sum_{i,j=1}^{n_\rho} \overline{\frac{1}{n_\rho}m_{ij}^{(\rho)}(h_1)} \int_{\mathbf{SO}(n)} \frac{1}{n_\rho}m_{ij}^{(\rho)}(h_2) f(h_2h_1^{-1}x, h_2)\, d\lambda(h_2).\end{aligned}$$

$$(*_{15})$$

Using the fact that if A and B are any two $n \times n$ matrices, then

$$\sum_{i,j=1}^{n} a_{ij} b_{ij} = \mathrm{tr}(AB^{\top}),$$

and observing that the matrix whose entries are the terms

$$\int_{\mathbf{SO}(n)} \frac{1}{n_\rho} m_{ij}^{(\rho)}(h_2) f(h_2 h_1^{-1} x, h_2)\, d\lambda(h_2)$$

is the matrix

$$\begin{aligned} &\int_{\mathbf{SO}(n)} M_\rho(h_2) f(h_2 h_1^{-1} x, h_2)\, d\lambda(h_2) \\ &\qquad = \int_{\mathbf{SO}(n)} \left(\left(\overline{M_\rho(h_2)}\right)^*\right)^{\top} f(h_2 h_1^{-1} x, h_2)\, d\lambda(h_2), \end{aligned}$$

we obtain

$$f^\rho(x, h_1) = n_\rho \,\mathrm{tr}\left(\overline{M_\rho(h_1)} \int_{\mathbf{SO}(n)} \left(\overline{M_\rho(h_2)}\right)^* f(h_2 h_1^{-1} x, h_2)\, d\lambda(h_2)\right). \tag{$*16$}$$

Observe that this is the generalization of (str9). Also, $\left(\overline{M_\rho(h_2)}\right)^* = \left(M_\rho(h_2)\right)^{\top}$. If for any fixed (x, h_1) we define the function $f_{(x,h_1)}$ given by

$$f_{(x,h_1)}(h_2) = f(h_2 h_1^{-1} x, h_2), \tag{$f_{(x,h_1)}$}$$

then the value of the Fourier transform of $f_{(x,h_1)}$ at $\overline{\rho}$ is

$$\mathcal{F}(f_{(x,h_1)})_{\overline{\rho}} = \int_{\mathbf{SO}(n)} \left(\overline{M_\rho(h_2)}\right)^* f(h_2 h_1^{-1} x, h_2)\, d\lambda(h_2), \tag{$\mathcal{F}(f_{(x,h_1)})_{\overline{\rho}}$}$$

and thus

$$f^\rho(x, h_1) = n_\rho \,\mathrm{tr}\left(\overline{M_\rho(h_1)}\, \mathcal{F}(f_{(x,h_1)})_{\overline{\rho}}\right) = n_\rho \mathrm{tr}\left(M_{\overline{\rho}}(h_1)\, \mathcal{F}(f_{(x,h_1)})_{\overline{\rho}}\right). \tag{$*17$}$$

We also define $f_\rho^{\mathrm{rad}} \colon \mathbb{R}^d \to \mathrm{M}_{n_\rho}(\mathbb{C})$ by

$$\begin{aligned} f_\rho^{\mathrm{rad}}(x) &= \int_{\mathbf{SO}(n)} \left(\overline{M_\rho(h_2)}\right)^* f(h_2 x, h_2)\, d\lambda(h_2) \\ &= \int_{\mathbf{SO}(n)} M_\rho(h_2)^{\top} f(h_2 x, h_2)\, d\lambda(h_2), \end{aligned} \tag{f_ρ^{rad}}$$

and so we have

$$f^\rho(x, h_1) = n_\rho \,\mathrm{tr}\left(\overline{M_\rho(h_1)}\, f_\rho^{\mathrm{rad}}(h_1^{-1} x)\right). \tag{f^ρ}$$

Observe that

$$f^\rho(hx, hh_1) = n_\rho \operatorname{tr}\Big(\overline{M_\rho(hh_1)}\, f_\rho^{\mathrm{rad}}((hh_1)^{-1}hx)\Big)$$
$$= n_\rho \operatorname{tr}\Big(\overline{M_\rho(h)}\,\overline{M_\rho(h_1)}\, f_\rho^{\mathrm{rad}}(h_1^{-1}x)\Big), \qquad \text{(str17)}$$

which expresses steerability with respect to $\mathbf{SO}(n)$.

Using Lang [Lang (1996)], Chapter XVII, Problem 9, since the family of functions $e^{-\frac{x^2}{2}}H_m(x)$ is a Hilbert basis of $\mathrm{L}^2(\mathbb{R})$, the family of functions

$$e^{-\frac{x_1^2}{2}}H_{k_1}(x_1)\cdots e^{-\frac{x_n^2}{2}}H_{k_n}(x_n) = e^{-\frac{\|x\|_2^2}{2}}H_{k_1}(x_1)\cdots H_{k_n}(x_n),$$
$$k_1, \ldots, k_n \geq 0,$$

with $x = (x_1, \ldots, x_n) \in \mathbb{R}^n$, is a Hilbert basis of $\mathrm{L}^2(\mathbb{R}^n)$.

For $f \in \mathrm{L}^2(\mathbf{SE}(n))$ given by

$$f(x, h_2) = e^{-\|x\|^2/2}H_{k_1}(x_1)\cdots H_{k_n}(x_n)\,\overline{m_{k\ell}^{(\rho)}(h_2)},$$

we find that $f_\rho^{\mathrm{rad}}(x)$ is the $n_\rho \times n_\rho$-matrix whose (ℓ, k) entry is $e^{-\|x\|^2/2}H_{k_1}(x_1)\cdots H_{k_n}(x_n)$, and all other entries are 0, which implies that

$$f^\rho(x, h_1) = \overline{m_{k\ell}^{(\rho)}(h_1)}\, e^{-\|x\|^2/2}H_{k_1}((h_1^{-1}x)_1)\cdots H_{k_n}((h_1^{-1}x)_n)$$

belongs to the subspace E_ρ, the projection of $\mathrm{L}^2(\mathbf{SE}(n))$ by π_ρ^V.

The Hilbert space $\mathrm{L}^2(\mathbf{SE}(n))$ is isomorphic to $\mathrm{L}^2(\mathbf{SO}(n) \times \mathbb{R}^n)$, and since by Peter–Weyl I, the Hilbert space $\mathrm{L}^2(\mathbf{SO}(n))$ is the Hilbert sum of the minimal two-sided ideals $\mathfrak{a}_\rho$, which have the n_ρ^2 functions $\overline{m_{k\ell}^{(\rho)}}$ as an orthogonal basis, we conclude that the family of functions

$$\Big(\overline{m_{k\ell}^{(\rho)}(h_1)}\, e^{-\frac{\|x\|^2}{2}}H_{k_1}((h_1^{-1}x)_1)\cdots H_{k_n}((h_1^{-1}x)_n)\Big) \;\Big|$$
$$\rho \in R(\mathbf{SO}(n)),\ 1 \leq k, \ell \leq n_\rho,\ k_1, \ldots, k_n \geq 0, \qquad \text{(str18)}$$

with $h_1 \in \mathbf{SO}(n)$ and $x \in \mathbb{R}^n$, is an $\mathbf{SO}(n)$-steerable Hilbert basis of $\mathrm{L}^2(\mathbf{SE}(n))$. More precisely, for any fixed $\rho \in R(\mathbf{SO}(n))$, $1 \leq \ell \leq n_\rho$, $k_1, \ldots, k_n \geq 0$, if we write $\mathbf{k} = (k_1, \ldots, k_n)$, by (str16), the column vector $Y_{\ell,\mathbf{k}}^\rho(h_1, x)$ of dimension n_ρ with

$$Y_{k,\ell,\mathbf{k}}^\rho(h_1, x) = \overline{m_{k\ell}^{(\rho)}(h_1)}\, e^{-\frac{\|x\|^2}{2}}H_{k_1}((h_1^{-1}x)_1)\cdots H_{k_n}((h_1^{-1}x)_n), 1 \leq k \leq n_\rho,$$

satisfies the steerability equation

$$Y_{\ell,\mathbf{k}}^\rho(h^{-1}h_1, h^{-1}x) = (M_\rho(h))^\top Y_{\ell,\mathbf{k}}^\rho(h_1, x). \qquad \text{(str19)}$$

If $n = 2$, then $R(\mathbf{SO}(2)) = \mathbb{Z}$, $m_\ell(\theta) = e^{i\ell\theta}$, so we find that the family

$$\left(e^{-i\ell\theta}\, e^{-\frac{x^2+y^2}{2}} H_{k_1}(x\cos\theta + y\sin\theta) H_{k_2}(-x\sin\theta + y\cos\theta)\right)_{\ell\in\mathbb{Z},\, k_1,k_2\geq 0} \tag{str20}$$

is steerable basis of $\mathrm{L}^2(\mathbf{SE}(2))$.

If $n = 3$, then $R(\mathbf{SO}(3)) = \mathbb{N}$, $\rho = \ell$, $n_\rho = 2\ell+1$, the functions $\sqrt{2\ell+1}\,\overline{w^{(\ell)}_{jk}(R)}$ ($R \in \mathbf{SO}(3)$) of Section 5.15 (see also Section 5.10) from a Hilbert basis of $\mathrm{L}^2(\mathbf{SO}(3))$, so we find that the family

$$\left(\sqrt{2\ell+1}\,\overline{w^{(\ell)}_{jk}(R)}\, e^{-\frac{x_1^2+x_2^2+x_3^2}{2}} H_{k_1}((R^{-1}x)_1) H_{k_2}((R^{-1}x)_2) H_{k_3}((R^{-1}x)_3)\right) \Bigm| \ell\in\mathbb{N},\ -\ell\leq j,k\leq\ell,\ k_1,k_2,k_3\geq 0, \tag{str21}$$

with $R \in \mathbf{SO}(3)$ and $x \in \mathbb{R}^3$, is steerable basis of $\mathrm{L}^2(\mathbf{SE}(3))$. For fixed $\ell \in \mathbb{N}, -\ell \leq k \leq \ell, k_1.k_2, k_3 \geq 0$, if we write $\mathbf{k} = (k_1, k_2, k_3)$ and if $Y^\ell_{k,\mathbf{k}}(R, x)$ is the column vector given by

$$\begin{aligned} &Y^\ell_{j,k,\mathbf{k}}(R,x) \\ &= \sqrt{2\ell+1}\,\overline{w^{(\ell)}_{jk}(R)}\, e^{-\frac{x_1^2+x_2^2+x_3^2}{2}} H_{k_1}((R^{-1}x)_1) H_{k_2}((R^{-1}x)_2) H_{k_3}((R^{-1}x)_3), \end{aligned}$$

with $-\ell \leq j \leq \ell$, then we have

$$Y^\ell_{k,\mathbf{k}}(Q^{-1}R, Q^{-1}x) = (w^{(\ell)}(Q))^\top Y^\ell_{k,\mathbf{k}}(R,x). \tag{str22}$$

We can also express the matrices $w^{(\ell)}(R)$ in terms of the Euler angles and the Wigner d-matrices as in Section 5.15; see the remark just after Proposition 5.42.

Example 8.10. In Example 8.8, we considered the compact group $H = \mathbf{SO}(n+1)$, the subgroup $H_0 = \mathbf{SO}(n)$, and the homogeneous space $X = S^n = H/H_0$. We now consider the more general situation, where H is a compact (metrizable and separable) group, H_0 is any closed subgroup of H, and X is the homogeneous space $X = H/H_0$. This is a situation where we can use Proposition 6.18.

If the trivial representation σ_0 of H_0 is contained d_ρ times in the restriction of M_ρ to H_0, then $L_\rho = \mathrm{L}^2(H/H_0) \cap \mathfrak{a}_\rho$ is the direct sum of the first d_ρ columns of $M^{(H_0)}_\rho$, where $M^{(H_0)}_\rho$ is a matrix similar to M_ρ defined in Section 6.9. Proposition 6.18 also implies that the functions $m^{(\rho,H_0)}_{ij}$ satisfy the equations

$$m^{(\rho,H_0)}_{ij}(hh_0) = m^{(\rho,H_0)}_{ij}(h), \quad h\in H,\ h_0\in H_0,\ 1\leq i\leq n_\rho,\ 1\leq j\leq d_\rho,$$

which means that they can be viewed as functions on $\mathrm{L}^2(H/H_0) = \mathrm{L}^2(X)$. From now on, we will use the representations defined by the matrices $M_\rho^{(H_0)}$, which are equivalent to the representations M_ρ. For simplicity of notation (and with a slight abuse of notation), we will denote the matrix $M_\rho^{(H_0)}$ as M_ρ.

Since the functions $m_{k\ell}^{(\rho)}$ for $1 \leq k \leq n_\rho$ and $1 \leq \ell \leq d_\rho$ are functions defined on H/H_0, equivalently functions defined on H constant on cosets in H/H_0, for any $x = h_1 H_0 \in H/H_0$, we have

$$m_{k\ell}^{(\rho)}(hx) = m_{k\ell}^{(\rho)}((hh_1)H_0) = m_{k\ell}^{(\rho)}(hh_1). \tag{str23}$$

By (str16), we have the following equation between column vectors

$$\overline{m_{*\ell}^{(\rho)}}(h^{-1}h_1) = (M_\rho(h))^\top \overline{m_{*\ell}^{(\rho)}}(h_1),$$

and so by (str23) and the above equation, we also have the equation

$$\overline{m_{*\ell}^{(\rho)}}(h^{-1}x) = (M_\rho(h))^\top \overline{m_{*\ell}^{(\rho)}}(x), \quad 1 \leq \ell \leq d_\rho, \ x \in H/H_0. \tag{str24}$$

Then for fixed $\rho \in R(H), 1 \leq \ell \leq d_\rho$, if $Y_\ell^\rho(x)$ is the column vector given by

$$Y_{k,\ell}^\rho(x) = \overline{m_{k\ell}^{(\rho)}(x)}, \ 1 \leq k \leq n_\rho,$$

by (str24), we have the steerability equations

$$Y_\ell^\rho(h^{-1}x) = (M_\rho(h))^\top Y_\ell^\rho(x), \quad x \in H/H_0 \approx X. \tag{str25}$$

In the special case where $H = \mathbf{SO}(n+1)$ and $H_0 = \mathbf{SO}(n)$, the homogeneous space is $X = S^n$, and we know from Section 6.10 that $d_\rho = 1$ when the trivial representation of $\mathbf{SO}(n)$ occurs in the restriction of M_ρ (the irreducible representation of $\mathbf{SO}(n+1)$ corresponding to ρ) to $\mathbf{SO}(n)$. In this case, the first column of $\overline{M_\rho}$ is a steerable family for S^n.

Also note that Example 8.7 corresponds to the special case where H_0 is the trivial group $\{e\}$.

Example 8.11. In this example, we describe a general method to decompose $\mathrm{L}^2(X)$ using the representation V and the projections

$$\begin{aligned}(\pi_\rho^V(f))(x) &= n_\rho \int_H \overline{\chi_\rho(h)} f(h^{-1} \cdot x)\, d\lambda(h)\\ &= \int_H \overline{u_\rho(h)} f(h^{-1} \cdot x)\, d\lambda(h), \ x \in X,\end{aligned}$$

for all $f \in L^2(X)$ and all $\rho \in R(H)$. As before, we assume that X is a locally compact, metrizable, separable space equipped with a σ-regular, locally finite, Borel measure μ.

Recall that $u_\rho = m_{11}^{(\rho)} + \cdots + m_{n_\rho n_\rho}^{(\rho)}$, n_ρ times the trace of the matrix M_ρ. Our goal is to compute $(\pi_\rho^V(f))(h_1 \cdot x)$ with $h_1 \in H$. Since H is compact, it is unimodular, so we have

$$\begin{aligned}
(\pi_\rho^V(f))(h_1 \cdot x) &= \int_H \overline{u_\rho(h)} f(h^{-1} \cdot (h_1 \cdot x))\, d\lambda(h) \\
&= \int_H \overline{u_\rho(h)} f((h^{-1}h_1) \cdot x)\, d\lambda(h), \qquad\qquad h = h_1 h_2 \\
&= \int_H \overline{u_\rho(h_1 h_2)} f(h_2^{-1} \cdot x)\, d\lambda(h_2) \\
&= \int_H \overline{u_\rho(h_1 h_2^{-1})} f(h_2 \cdot x)\, d\lambda(h_2).
\end{aligned}$$

Since M_ρ is a representation

$$u_\rho(h_1 h_3) = \sum_{i=1}^{n_\rho} m_{ii}^{(\rho)}(h_1 h_3) = (1/n_\rho) \sum_{i,j=1}^{n_\rho} m_{ij}^{(\rho)}(h_1) m_{ji}^{(\rho)}(h_3),$$

so we get

$$\begin{aligned}
(\pi_\rho^V(f))(h_1 \cdot x) &= \sum_{i,j=1}^{n_\rho} \overline{m_{ij}^{(\rho)}(h_1)} \int_H (1/n_\rho) \overline{m_{ji}^{(\rho)}(h_2^{-1})} f(h_2 \cdot x)\, d\lambda(h_2) \\
&= \sum_{i,j=1}^{n_\rho} \overline{m_{ij}^{(\rho)}(h_1)} \int_H (1/n_\rho) m_{ij}^{(\rho)}(h_2) f(h_2 \cdot x)\, d\lambda(h_2). \quad (*_{18})
\end{aligned}$$

The action of H on X defines an orbit space Ω also denoted X/G, where Ω is the topological space, which is the quotient of X by the equivalence relation defined such that $x \sim y$ iff $y = h \cdot x$ for some $h \in H$ $(x, y \in X)$. For any $x \in X$, the orbit $H \cdot x$ (also denoted by $O(x)$) of x is defined as $H \cdot x = \{h \cdot x \mid h \in H\}$. Each orbit $\omega \in \Omega$ is an equivalence class of the equivalence relation $\sim$, and because the action of H is transitive on each orbit, the orbit ω is equal to $H \cdot x$ for all x in ω. In general, Ω is not a manifold; in fact, it may not even be Hausdorff, although it is since H is compact, because in this case the action is proper; see Gallier and Quaintance [Gallier and Quaintance (2020a)] (Chapter 23, see Proposition 23.5 and Corollary 23.8). However, the projection map $p\colon X \to X/G$ is open and continuous, so if X is locally compact, then so is $\Omega = X/G$; see Duistermaat and Kolk [Duistermaat and Kolk (2000)], Section 2.5. In fact, Duistermaat and Kolk [Duistermaat and Kolk (2000)] contains a thorough treatment of the properties of the orbit space of a proper but not necessarily free action of a Lie group on a topological space.

If we pick some base-point x_0^ω in each orbit ω and if $H_{x_0^\omega}$ is the stabilizer of the transitive action of H on ω (namely, $H_{x_0^\omega} = \{h \in H \mid h \cdot x_0^\omega = x_0^\omega\}$), the projection map $\pi_{x_0^\omega} \colon H \to \omega$ given by $\pi_{x_0^\omega}(h) = h \cdot x_0^\omega$ induces the bijection $\overline{\pi_{x_0^\omega}} \colon H/H_{x_0^\omega} \to \omega$ defined by $\overline{\pi_{x_0^\omega}}(hH_{x_0^\omega}) = h \cdot x_0^\omega$ for all $h \in H$. It follows that for any *fixed orbit* ω, for any $h \cdot x_0^\omega \in \omega$, we have

$$(\pi_\rho^V(f))(h \cdot x_0^\omega) = \sum_{i,j=1}^{n_\rho} \overline{m_{ij}^{(\rho)}(h)} \int_H (1/n_\rho) m_{ij}^{(\rho)}(h_2) f(h_2 \cdot x_0^\omega)\, d\lambda(h_2), h \in H,$$

and in view of the bijection $\overline{\pi_{x_0^\omega}} \colon H/H_{x_0^\omega} \to \omega$, *we can view the restriction of* $\pi_\rho^V(f) \colon X \to \mathbb{C}$ *to* ω *as a function on* $H/H_{x_0^\omega}$ defined by

$$(\pi_\rho^V(f))(hH_{x_0^\omega}) = \sum_{i,j=1}^{n_\rho} \overline{m_{ij}^{(\rho)}(h)} \int_H (1/n_\rho) m_{ij}^{(\rho)}(h_2) f(h_2 \cdot x_0^\omega)\, d\lambda(h_2), h \in H. \tag{$*_{19}$}$$

We can also view a function $g \colon H/H_{x_0^\omega} \to \mathbb{C}$ as a function on H constant on cosets, which means that

$$g(hh_\omega) = g(h), \quad \text{for all } h_\omega \in H_{x_0^\omega} \text{ and all } h \in H,$$

and since we view $\pi_\rho^V(f)$ as a function on $H/H_{x_0^\omega}$, we can view $\pi_\rho^V(f)$ as a function on H satisfying the equation

$$\pi_\rho^V(f)(hh_\omega) = \pi_\rho^V(f)(h), \quad h \in H,\ h_\omega \in H_{x_0^\omega},$$

and for any $h_1 \in hH_{x_0^\omega}$, we can write

$$(\pi_\rho^V(f))(h_1) = \sum_{i,j=1}^{n_\rho} \overline{m_{ij}^{(\rho)}(h_1)} \int_H (1/n_\rho) m_{ij}^{(\rho)}(h_2) f(h_2 \cdot x_0^\omega)\, d\lambda(h_2). \tag{$*_{20}$}$$

Equation ($*_{20}$) implies that $\pi_\rho^V(f) \in \overline{L_\rho} = \mathrm{L}^2(H/H_{x_0^\omega}) \cap \overline{\mathfrak{a}_\rho}$ so as in Example 8.10 we can use Proposition 6.18. If the trivial representation σ_0 of $H_{x_0^\omega}$ is contained d_ρ times in the restriction of $\overline{M_\rho}$ to $H_{x_0^\omega}$, then $\overline{L_\rho} = \mathrm{L}^2(H/H_{x_0^\omega}) \cap \overline{\mathfrak{a}_\rho}$ is the direct sum of the first d_ρ columns of $\overline{M_\rho^{(H_{x_0^\omega})}}$, where $M_\rho^{(H_{x_0^\omega})}$, the matrix similar to M_ρ defined in Section 6.9. Proposition 6.18 also implies that the functions $\overline{m_{ij}^{(\rho,H_{x_0^\omega})}}$ satisfy the equations

$$\overline{m_{ij}^{(\rho,H_{x_0^\omega})}}(hh_\omega) = \overline{m_{ij}^{(\rho,H_{x_0^\omega})}}(h), \quad h \in H,\ h_\omega \in H_{x_0^\omega},\ 1 \le i \le n_\rho,\ 1 \le j \le d_\rho,$$

which means that they can be viewed as functions on $\mathrm{L}^2(H/H_{x_0^\omega}) = \mathrm{L}^2(\omega)$. From now on, we will use the representations defined by the matrices $M_\rho^{(H_{x_0^\omega})}$, which are equivalent to the representations M_ρ, so $(*_{20})$ becomes

$$(\pi_\rho^V(f))(h_1) = \sum_{i,j=1}^{n_\rho} \overline{m_{ij}^{(\rho,H_{x_0^\omega})}(h_1)} \int_H (1/n_\rho) m_{ij}^{(\rho,H_{x_0^\omega})}(h_2) f(h_2 \cdot x_0^\omega)\, d\lambda(h_2),$$
$$h_1 \in hH_{x_0^\omega}, \qquad (*_{20'})$$

which implies that

$$\int_H (1/n_\rho) m_{ij}^{(\rho,H_{x_0^\omega})}(h_2) f(h_2 \cdot x_0^\omega)\, d\lambda(h_2) = 0,\ 1 \le i \le n_\rho,\ d_\rho + 1 \le j \le n_\rho.$$

For simplicity of notation (and with a slight abuse of notation), we will denote the matrix $M_\rho^{(H_{x_0^\omega})}$ as M_ρ. By $(*_{20'})$, for any $x_\omega = hH_{x_0^\omega} \in H/H_{x_0^\omega} \approx \omega$, we have

$$(\pi_\rho^V(f))(x_\omega) = \sum_{i=1}^{n_\rho} \sum_{j=1}^{d_\rho} \overline{m_{ij}^{(\rho)}(x_\omega)} \int_H (1/n_\rho) m_{ij}^{(\rho)}(h_2) f(h_2 \cdot x_0^\omega)\, d\lambda(h_2).$$

If $\overline{M_\rho}[n_\rho; d_\rho](h)$ is the matrix whose first d_ρ columns are the entries $\overline{(1/n_\rho) m_{ij}^{(\rho)}(h)}$ for $i = 1, \ldots, n_\rho$ and $j = 1, \ldots, d_\rho$ and whose last $n_\rho - d_\rho$ columns are zero columns, and if $\left(\overline{M_\rho}[n_\rho; d_\rho]\right)^*(h_2)$, a matrix whose last $n_\rho - d_\rho$ rows are zero rows, is the conjugate transpose of the matrix $\overline{M_\rho}[n_\rho; d_\rho](h_2)$, then

$$\begin{aligned}
&(\pi_\rho^V(f))(x_\omega)\\
&\quad = n_\rho \operatorname{tr}\left(\overline{M_\rho}[n_\rho; d_\rho](x_\omega) \int_H \left(\overline{M_\rho}[n_\rho; d_\rho]\right)^*(h_2) f(h_2 \cdot x_0^\omega)\, d\lambda(h_2)\right)\\
&\quad = n_\rho \operatorname{tr}\left(\overline{M_\rho}[n_\rho; d_\rho](x_\omega) \int_H \left(M_\rho[n_\rho; d_\rho](h_2)\right)^\top f(h_2 \cdot x_0^\omega)\, d\lambda(h_2)\right)
\end{aligned}$$

for all $x_\omega = hH_{x_0^\omega} \in H/H_{x_0^\omega} \approx \omega$.

The next step is to get a better understanding of the function $f_\rho^{\mathrm{rad}} \colon \Omega \to \mathrm{M}_{n_\rho}(\mathbb{C})$ given by

$$\begin{aligned}
f_\rho^{\mathrm{rad}}(\omega) &= \int_H \left(\overline{M_\rho}[n_\rho; d_\rho]\right)^*(h_2) f(h_2 \cdot x_0^\omega)\, d\lambda(h_2)\\
&= \int_H \left(M_\rho[n_\rho; d_\rho](h_2)\right)^\top f(h_2 \cdot x_0^\omega)\, d\lambda(h_2),
\end{aligned}$$

which actually yields matrices whose last $n_\rho - d_\rho$ rows are zero rows.

At this point, we can't say anything without making assumptions on the orbit space Ω, which could be very complicated. For example, some orbits could be discrete, as in the case where $X = \mathbb{R}^d$ and $H = \mathbf{SO}(d)$, where the orbit space can be identified with $\mathbb{R}_+ = \{\omega \in \mathbb{R} \mid \omega \geq 0\}$, and the orbits are spheres of radius $\omega > 0$ and the set $\{0\}$ (where $0 \in \mathbb{R}^d$) for $\omega = 0$.

If we assume that X is a manifold and that the action of H on X is free, since H is compact this action is also proper, and thus the quotient space $\Omega = X/H$ is a manifold. In fact, the canonical projection is a submersion and X/H is a principal (right) H-bundle; see Theorem 9.31 in Gallier and Quaintance [Gallier and Quaintance (2020b)] and Theorem 23.11 in Gallier and Quaintance [Gallier and Quaintance (2020a)]. But as we already said earlier, since we are assuming that G is compact and metrizable and that X is locally compact, metrizable, and separable, the orbit space $\Omega = X/G$ is also Hausdorff, locally compact, and separable. Then $\mathrm{L}^2(\Omega)$ is also separable, so we can pick a Hilbert basis of $\mathrm{L}^2(\Omega)$, say $(\varphi_i)_{i\in I}$ for some countable index set I.

Since the orbits $\omega \in \Omega$ form a partition of X and since we have bijections $\overline{\pi_{x_0^\omega}}\colon H/H_{x_0^\omega} \to \omega$, *we can view X as the disjoint union of the homogeneous spaces* $H/H_{x_0^\omega}$, and so every function $f\colon X \to \mathbb{C}$ can be expressed *locally in a unique way* in terms of a family of functions $f_\omega\colon H/H_{x_0^\omega} \to \mathbb{C}$, with $H/H_{x_0^\omega} \approx \omega \in \Omega$. For every $\omega \in \Omega$, we have

$$f(x_\omega) = f_\omega(x_\omega), \quad \text{for all } x_\omega \in H/H_{x_0^\omega} \approx \omega.$$

We can view x_ω as a generalization of the polar coordinates, which in $\mathbb{R}^d$ are specified by the vectors rx, with $x \in S^{d-1}(1)$ and $r \in \mathbb{R}_+ - \{0\}$, where $S^{d-1}(1)$ is the unit sphere in $\mathbb{R}^d$. These are commonly used in differential geometry. In $\mathbb{R}^2$, the polar coordinates are given by $(r\cos\theta, r\sin\theta)$.

Consider the *particular function* f defined such that for every $\omega \in \Omega$,

$$f(x_\omega) = \overline{m_{k\ell}^{(\rho)}(x_\omega)}\, \varphi_i(\omega), \quad \text{for all } x_\omega \in H/H_{x_0^\omega} \approx \omega,$$

where φ_i is some function in a Hilbert basis of $\mathrm{L}^2(\Omega)$. Using $(*_{20'})$ and the orthogonality relations of the functions $m_{ij}^{(\rho)}$, we find that for all $\omega \in \Omega$, we have

$$(\pi_\rho^V(f))(x_\omega) = \overline{m_{k\ell}^{(\rho)}(x_\omega)}\, \varphi_i(\omega), \quad \text{for all } x_\omega \in H/H_{x_0^\omega} \approx \omega.$$

Remark: We can also show that $f_\rho^{\mathrm{rad}}(\omega)$ is the $n_\rho \times n_\rho$ matrix whose (ℓ, k)-entry is $\varphi_i(\omega)$ and all other entries are 0. This fact is left as an exercise.

Therefore, the family of functions

$$\left(\overline{m^{(\rho)}_{k\ell}(x_\omega)\,\varphi_i(\omega)}\right), \quad x_\omega \in H/H_{x_0^\omega} \approx \omega,\ \omega \in \Omega,$$
$$\rho \in R(H), 1 \leq k \leq n_\rho,\ 1 \leq \ell \leq d_\rho,\ i \in I,$$

is a steerable Hilbert basis of $\mathrm{L}^2(X)$. More precisely, recall that the action of the group H on the homogeneous space $H/H_{x_0^\omega}$ is given by $h\cdot(h_1 H_{x_0^\omega}) = (hh_1)H_{x_0^\omega}$. Since the functions $\overline{m^{(\rho)}_{k\ell}}$ for $1 \leq k \leq n_\rho$ and $1 \leq \ell \leq d_\rho$ are functions defined on $H/H_{x_0^\omega}$, equivalently functions defined on H constant on cosets in $H/H_{x_0^\omega}$, for any $x_\omega = h_1 H_{x_0^\omega} \in H/H_{x_0^\omega}$, we have

$$\overline{m^{(\rho)}_{k\ell}}(hx_\omega) = \overline{m^{(\rho)}_{k\ell}}((hh_1)H_{x_0^\omega}) = \overline{m^{(\rho)}_{k\ell}}(hh_1).$$

By (str16), we have the following equation between column vectors

$$\overline{m^{(\rho)}_{*\ell}}(h^{-1}h_1) = (M_\rho(h))^\top \overline{m^{(\rho)}_{*\ell}}(h_1),$$

and so we also have the equation

$$\overline{m^{(\rho)}_{*\ell}}(h^{-1}x_\omega) = (M_\rho(h))^\top \overline{m^{(\rho)}_{*\ell}}(x_\omega), \quad 1 \leq \ell \leq d_\rho. \tag{str26}$$

Then for fixed $\rho \in R(H), 1 \leq \ell \leq d_\rho, i \in I$, if $Y^\rho_{\ell,i}(x_\omega,\omega)$ is the column vector given by

$$Y^\rho_{k,\ell,i}(x_\omega) = \overline{m^{(\rho)}_{k\ell}(x_\omega)\,\varphi_i(\omega)},\ 1 \leq k \leq n_\rho,$$

by (str26), we have the steerability equations

$$Y^\rho_{\ell,i}(h^{-1}x_\omega) = (M_\rho(h))^\top Y^\rho_{\ell,i}(x_\omega,\omega), \quad x_\omega \in H/H_{x_0^\omega} \approx \omega,\ \omega \in \Omega. \tag{str27}$$

In the special case where $X = \mathbb{R}^d$ and $H = \mathbf{SO}(d)$, as we said earlier, the orbit space is $\Omega = \mathbb{R}_+$ and the orbits are spheres $S^{d-1}(\omega)$ of radius $\omega > 0$ together with $\{0\}$; see Figure 8.6. Note that the action of $\mathbf{SO}(d)$ on $\mathbb{R}^d$ is *not* free. The origin is a fixed point of every rotation, and for d odd, the eigenspace associated with the eigenvalue 1 has dimension at least 1. And indeed, $\mathbb{R}_+$ is not a manifold (but it is a manifold with boundary $\{0\}$).

We can pick $x_0^\omega = \omega e_0$ (with $e_0 = (1, 0, \ldots, 0)$) and all subgroups $H_{x_0^\omega}$ are all isomorphic to $\mathbf{SO}(d-1)$; see Figure 8.7. In this case, the spherical harmonics constitute a Hilbert basis for S^{d-1} and the Hermite functions give us a Hilbert basis for $\mathrm{L}^2(\mathbb{R}_+)$. Using the notation of Example 8.8, if $(Y_1^\ell, \ldots, Y_{a_{\ell,d}}^\ell)$ is an orthogonal basis of size $a_{\ell,d}$ in $\mathcal{H}_\ell^{\mathbb{C}}(S^{d-1})$, the family of functions

$$\left(Y_j^\ell\left(\frac{x}{\|x\|}\right) e^{-\|x\|^2/2} H_m(\|x\|)\right) \;\Big|\; \ell, m \in \mathbb{N},\ 1 \leq j \leq a_{\ell,d}$$

for all $x \in \mathbb{R}^d - \{0\}$, is a steerable basis of $L^2(\mathbb{R}^d)$. Here we are using the polar coordinates $x = ry$, with $y = \frac{x}{\|x\|} \in S^{d-1}$ and $r = \|x\| \in \mathbb{R}_+$. For $d = 3$, due to the change of indexing the spherical harmonics associated with the index ℓ constitute the middle column of the matrix $w^{(\ell)}(R)$ ($R \in \mathbf{SO}(d)$). This case is also dealt with in Weiler, Geiger, Welling, Boomsma, and Cohen [Weiler *et al.* (2018)].

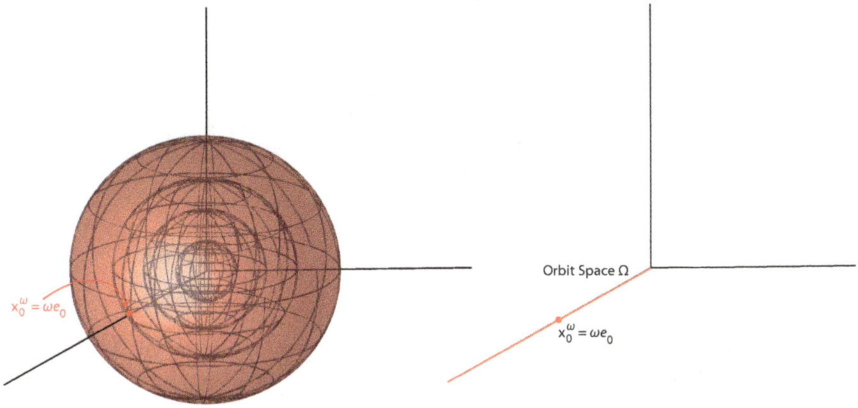

Fig. 8.6 For $X = \mathbb{R}^3$ and $H = \mathbf{SO}(3)$, the orbits are spheres $S^2(\omega)$ and the orbit space is the x-axis.

Example 8.11 suggests an alternative to the steerable family found in Example 8.9.

Example 8.12. Since the Hilbert space $L^2(\mathbf{SE}(n))$ is isomorphic to $L^2(\mathbf{SO}(n) \times \mathbb{R}^n)$, we can use the steerable basis of $L^2(\mathbb{R}^n)$ found in Example 8.11, and so we obtain the steerable family

$$\left(\overline{m^{(\rho)}_{j_1k_1}(h_1)}\, Y^{\ell_2}_{j_2}\Big(h_1^{-1}\frac{x}{\|x\|}\Big)e^{-\frac{\|x\|^2}{2}}H_m(\|x\|)\right) \Big| \quad \rho \in R(\mathbf{SO}(n)),\ 1 \leq j_1, k_1 \leq n_\rho,\ \ell_2, m \in \mathbb{N},\ 1 \leq j_2 \leq a_{\ell_2,n}, \tag{str28}$$

with $h_1 \in \mathbf{SO}(n)$ and $x \in \mathbb{R}^n$, is an $\mathbf{SO}(n)$-steerable Hilbert basis of $L^2(\mathbf{SE}(n))$.

For $n = 3$, the family

$$\left(\sqrt{2\ell_1+1}\,\overline{w^{(\ell_1)}_{j_1k_1}(R)}\, Y^{j_2}_{\ell_2}\Big(R^{-1}\frac{x}{\|x\|}\Big)e^{-\frac{\|x\|^2}{2}}H_m(\|x\|)\right) \Big| \quad \ell_1, \ell_2, m \in \mathbb{N},\ -\ell_1 \leq j_1, k_1 \leq \ell_1,\ -\ell_2 \leq j_2 \leq \ell_2, \tag{str29}$$

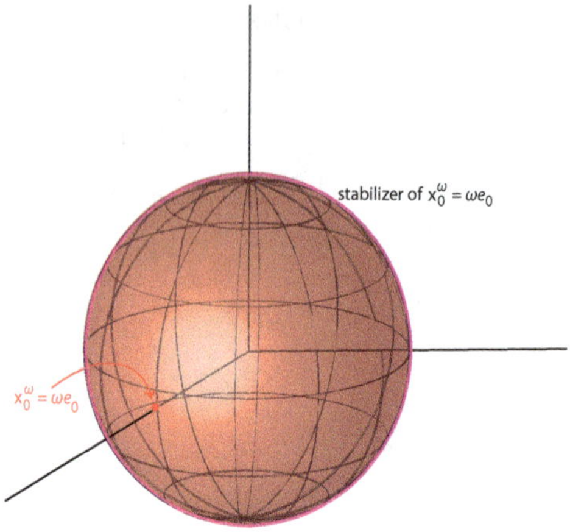

Fig. 8.7 The stabilizer of $\omega(1,0,0) \in S^2(\omega)$ is the circle in the yz-plane with equation $y^2 + z^2 = \omega$. Such a circle is isomorphic to $\mathbf{SO}(2)$.

with $R \in \mathbf{SO}(3)$ and $x \in \mathbb{R}^3$, is an $\mathbf{SO}(3)$-steerable basis of $\mathrm{L}^2(\mathbf{SE}(3))$. In the above formula, the $Y_{\ell_2}^{j_2}$ are the Laplace spherical harmonics. We can also express the matrices $w^{(\ell)}(R)$ in terms of the Euler angles and the Wigner d-matrices as in Section 5.15; see the remark just after Proposition 5.42. Details are left as an exercise.

In Section 8.6, we noticed that the functions $\widehat{f}$ are *vector-valued* functions from $\mathbb{R}^d$ to the codomain $\mathrm{M}_L(\mathbb{C})$ and that the group $G = \mathbb{R}^d \rtimes H$ acts on their domain $\mathbb{R}^d$, whereas the group H acts on their codomain $\mathrm{M}_L(\mathbb{C})$ in terms of the representation Σ. Experience has shown that the design of efficient convolution neural networks (CNN) is greatly facilitated if they operate on functions having the properties of the $\widehat{f}$ listed above. Such functions are known as *feature fields*.

8.8 Feature Fields

We begin with the definition of feature fields involving a semi-direct product group $G = \mathbb{R}^d \rtimes H$. This definition will be generalized later to a G-bundle on a homogenous space X (see Section 6.13).

To help intuition, suppose that $G = \mathbb{R}^2 \rtimes \mathbf{SO}(2)$. A scalar-valued function $f \colon \mathbb{R}^2 \to \mathbb{R}$ (more generally $f \colon \mathbb{R}^2 \to \mathbb{C}$) can be viewed as a gray-scale image, or temperature field, or pressure field. The group $G = \mathbb{R}^2 \rtimes \mathbf{SO}(2)$

acts on such an image by moving each pixel at t to the new position $Rt+x$, since $f \mapsto \mathbf{R}_{(x,R)}f$, with $(\mathbf{R}_{(x,R)}f)(t) = f((x,R)^{-1}\cdot t) = f(R^{-1}(t-x))$, where $g=(x,R)\in \mathbb{R}^2 \rtimes \mathbf{SO}(2)$, so $(\mathbf{R}_{(x,R)}f)(Rt+x) = f(R^{-1}(Rt+x-x)) = f(t)$; see Figure 8.8.

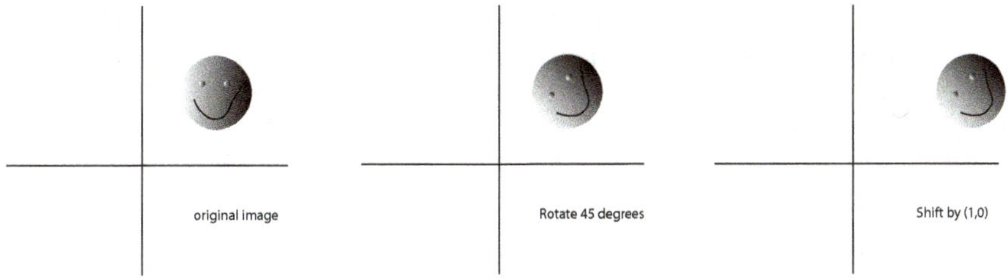

Fig. 8.8 The image of $f(t)$ is the gray-scaled smiley face. The action of $G=\mathbb{R}^2\rtimes\mathbf{SO}(2)$ on this image moves each pixel to $Rt+x$, where R is a rotation by 45 degrees counter-clockwise and x is a translation by $[1\ 0]^\top$.

On the other hand, a function $f\colon \mathbb{R}^2 \to \mathbb{R}^2$ defines a vector field, such as a velocity field, an optical flow, or a gradient image. This time, such a vector field transforms under the action of $G=\mathbb{R}^2\rtimes\mathbf{SO}(2)$ as follows: the vector $v=f(t)$ originally located at t is moved to the location $Rt+x$, and then *rotated* by R, so that the overall action results in the vector Rv in location $Rt+x$. See Figure 8.9.

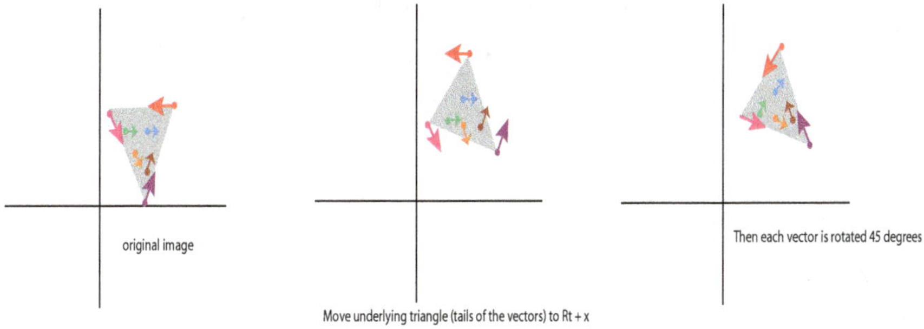

Fig. 8.9 The image of $f(t)$ is the vectorized triangular smiley face. The action of $G=\mathbb{R}^2\rtimes\mathbf{SO}(2)$ on this image moves each pixel to $Rt+x$, (where R is a rotation by 45 degrees counter-clockwise and x is a translation by $[1\ 0]^\top$), and then rotates the vector by 45 degrees counter-clockwise.

Given a more general vector field $f\colon \mathbb{R}^2 \to E$, where E is some hermitian vector space of finite-dimension. it is useful to generalize the action on a vector $v = f(t)$ so that it is specified by a representation $\sigma\colon \mathbf{SO}(2) \to \mathbf{U}(E)$ as

$$\sigma(R)(v) \text{ in location } Rt + x.$$

Observe that the expression $\sigma(R)(f(R^{-1}(t - x)))$ is exactly the formula obtained in Example 6.1 for the induced representation $[(\mathrm{Ind}_{\mathbf{SO}(2)}^{G} \sigma)_{(x,R)}](f) = \Pi_{(x,R)}(f)$, except that here $\mathbf{SO}(3)$ has been replaced by $\mathbf{SO}(2)$. The preceding discussion suggests the following definition.

Definition 8.7. Let $G = \mathbb{R}^d \rtimes H$ be a semi-direct product with H a compact group and $\sigma\colon H \to \mathbf{GL}(\mathcal{H})$ be a representation, where $\mathcal{H}$ is any complex vector space (possibly infinite dimensional). If $\mathcal{H}$ is finite-dimensional or a separable Hilbert space, we assume that $\sigma\colon H \to \mathbf{U}(\mathcal{H})$ is a unitary representation. A *feature field* is any function $f\colon \mathbb{R}^d \to \mathcal{H}$. The space of such feature fields is denoted by $\mathbf{FF}(\mathbb{R}^d, H, \sigma\colon H \to \mathbf{GL}(\mathcal{H}))$. The representation σ is called the *type* of the feature field. The group G acts on feature fields *via* the induced representation $\mathrm{Ind}_H^G \sigma$, namely

$$[(\mathrm{Ind}_H^G \sigma)_{(x,h)} f](t) = \sigma(h)(f(h^{-1}\cdot(t-x))), \quad (x,h) \in \mathbb{R}^d \rtimes H,\, t \in \mathbb{R}^d. \quad (\dagger_2)$$

Note that $(\dagger_2)$ is the immediate generalization of the formula obtained in Example 6.1, for the induced representation $[(\mathrm{Ind}_H^G \sigma)_{(x,Q)}](f) = \Pi_{(x,Q)}(f)$. Most authors use ρ instead of σ. This clashes with our notation used for indexing the irreducible representations of the group H, so we use σ instead.

A scalar field, namely a function $f\colon \mathbb{R}^d \to \mathbb{C}$ in $\mathrm{L}^2(\mathbb{R}^d)$, is the special case corresponding to $\mathcal{H} = \mathbb{C}$ and representation $\sigma\colon H \to \mathbf{U}(1)$ given by $\sigma(h) = \mathrm{id}_{\mathbb{C}}$ for all $h \in H$. In this case, $\mathrm{Ind}_H^G \sigma = \mathbf{R}^{G \to \mathrm{L}^2(\mathbb{R}^d)}$, the left regular representation of G.

A vector field $f\colon \mathbb{R}^d \to \mathbb{C}^d$ corresponds to the case where H is a closed subgroup of $\mathbf{GL}(d, \mathbb{C})$ and the representation $\sigma\colon H \to \mathbf{GL}(d, \mathbb{C})$ is the standard representation given by $\sigma(h) = h$, namely $\sigma(h)(x) = hx$ for any $x \in \mathbb{C}^d$, where h is a matrix in H.

Example 8.13. Let us show how G-feature maps $f\colon \mathbb{R}^d \times H \to \mathbb{C}$ in $\mathrm{L}^2(\mathbb{R}^d \rtimes H)$ can be viewed as feature fields $f^H\colon \mathbb{R}^d \to \mathrm{L}^2(H)$ (with $G = \mathbb{R}^d \rtimes H$). The left regular representation $\mathbf{R}^{G \to \mathrm{L}^2(\mathbb{R}^d \rtimes H)}$ acts on G-feature maps *via*

$$(\mathbf{R}_{(x,h)}^{G \to \mathrm{L}^2(\mathbb{R}^d \rtimes H)} f)(x_1, h_1) = f(h^{-1}\cdot(x_1 - x), h^{-1}h_1), \quad x, x_1 \in \mathbb{R}^d,\, h, h_1 \in H.$$

A G-feature map can be converted into a feature field as follows. Given $f\colon \mathbb{R}^d \times H \to \mathbb{C}$ in $\mathrm{L}^2(\mathbb{R}^d \rtimes H)$, let $f^H\colon \mathbb{R}^d \to \mathrm{L}^2(H)$, where

$$(f^H(x))(h) = f(x,h), \quad x \in \mathbb{R}^d,\ h \in H.$$

From an intuitive point of view, for $h \in H$ fixed, the map $x \mapsto f(x,h)$ can be viewed as a sort of image based on $\mathbb{R}^d$, where the value $f(x,h)$ is the color at the location $x \in \mathbb{R}^d$; see Figure 8.10. These images can be thought of as parallel layers, and for x fixed, as h varies the color $f(x,h)$ moves along a sort of fibre that passes through each of the layers "above x." For $d = 2$ and $H = \mathbf{SO}(2)$, it is possible to visualize these fibres. They are circles, but it is simpler to view them as line segments of height 2π with both endpoints identified. See Figure 8.11.

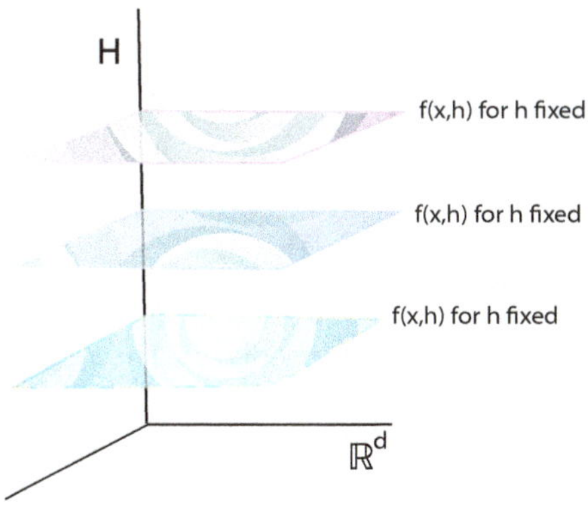

Fig. 8.10 A schematic illustration of $f^H(x) = f(x,h)$, where $H = \mathbf{SO}(2)$. For each fixed $h \in H$, the image of $f(x,h)$ is the horizontal colored layer.

The left regular representation $\mathbf{R}^{H\to\mathrm{L}^2(H)}$ acts on $\mathrm{L}^2(H)$ in the usual way, namely

$$(\mathbf{R}_h^{H\to\mathrm{L}^2(H)} g)(h_1) = g(h^{-1}h_1), \quad g \in \mathbb{C}^H,\ h, h_1 \in H.$$

Then the induced representation $\mathrm{Ind}_H^G \mathbf{R}^{H\to\mathrm{L}^2(H)}$ (here $\sigma = \mathbf{R}^{H\to\mathrm{L}^2(H)}$) acts on the feature fields $f^H\colon \mathbb{R}^d \to \mathrm{L}^2(H)$ by

$$[(\mathrm{Ind}_H^G \mathbf{R}^{H\to\mathrm{L}^2(H)})_{(x,h)} f^H](x_1) = \mathbf{R}_h^{H\to\mathrm{L}^2(H)}(f^H(h^{-1}\cdot(x_1 - x))).$$

By definition of $\mathbf{R}^{H\to\mathrm{L}^2(H)}$, we get

$$\begin{aligned}(\mathbf{R}_h^{H\to\mathrm{L}^2(H)}(f^H(h^{-1}\cdot(x_1 - x))))(h_1) &= (f^H(h^{-1}\cdot(x_1 - x)))(h^{-1}h_1)\\ &= f(h^{-1}\cdot(x_1 - x), h^{-1}h_1)\\ &= (\mathbf{R}_{(x,h)}^{G\to\mathrm{L}^2(\mathbb{R}^d\rtimes H)} f)(x_1, h_1).\end{aligned}$$

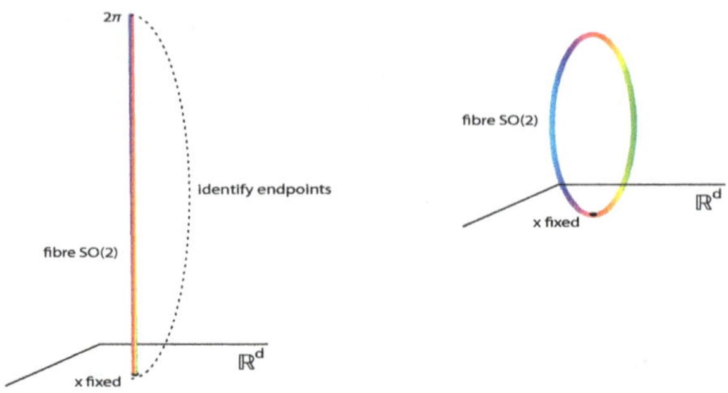

Fig. 8.11 Two illustrations of the fibre $\mathbf{SO}(2)$ above a fixed $x \in \mathbb{R}^2$.

Therefore,

$$(\mathrm{Ind}_H^G \mathbf{R}^{H \to \mathrm{L}^2(H)})_{(x,h)} f^H = \mathbf{R}_{(x,h)}^{G \to \mathrm{L}^2(\mathbb{R}^d \rtimes H)} f,$$

which shows that G-feature maps $f\colon \mathbb{R}^d \times H \to \mathbb{C}$ can be viewed as feature fields $f^H\colon \mathbb{R}^d \to \mathrm{L}^2(H)$, using the left regular representations $\mathbf{R}^{H \to \mathrm{L}^2(H)}$. In this case, $\mathcal{H} = \mathrm{L}^2(H)$ and $\sigma = \mathbf{R}^{H \to \mathrm{L}^2(H)}$.

The following definition will be needed in the next section.

Definition 8.8. Let $\sigma\colon H \to \mathbf{GL}(F)$ be a representation with F finite-dimensional. Define the function $\mathrm{Hom}(\sigma, \mathrm{id})$ by

$$\mathrm{Hom}(\sigma, \mathrm{id})_h f = f \circ \sigma_{h^{-1}}, \quad f \in \mathrm{Hom}(F, F),\ h \in H.$$

Since both $\sigma_{h^{-1}}\colon F \to F$ and $f\colon F \to F$ are linear, $\mathrm{Hom}(\sigma, \mathrm{id})_h f\colon F \to F$ is also a linear map for all $h \in H$. Furthermore, we have

$$\begin{aligned}\mathrm{Hom}(\sigma, \mathrm{id})_{h_1 h_2} f &= f \circ \sigma_{(h_1 h_2)^{-1}} = f \circ \sigma_{h_2^{-1}} \circ \sigma_{h_1^{-1}} \\ &= (\mathrm{Hom}(\sigma, \mathrm{id})_{h_2} f) \circ \sigma_{h^{-1}} = \mathrm{Hom}(\sigma, \mathrm{id})_{h_1}(\mathrm{Hom}(\sigma, \mathrm{id})_{h_2} f) \\ &= (\mathrm{Hom}(\sigma, \mathrm{id})_{h_1} \circ \mathrm{Hom}(\sigma, \mathrm{id})_{h_2}) f,\end{aligned}$$

which proves that $\mathrm{Hom}(\sigma, \mathrm{id})$ is a representation $\mathrm{Hom}(\sigma, \mathrm{id})\colon H \to \mathbf{GL}(\mathrm{Hom}(F, F))$.

Actually, the representation $\mathrm{Hom}(\sigma, \mathrm{id})$ is a special case of the Hom representation in Definition 4.16 with $\sigma_1\colon H \to \mathbf{GL}(F)$ the representation $\sigma_1 = \sigma$ and σ_2 the trivial representation given by $\sigma_2(h) = \mathrm{id}_F$ for all $h \in H$.

If $F = \mathbb{C}^n$, then $\mathrm{Hom}(\mathbb{C}^n, \mathbb{C}^n)$ is isomorphic to the space $\mathrm{M}_n(\mathbb{C})$ of $n \times n$ matrices, and if H is a closed subgroup of $\mathbf{GL}(n, \mathbb{C})$, then $\mathrm{Hom}(\sigma, \mathrm{id})$ acts on $\mathrm{M}_n(\mathbb{C})$ by multiplication on the right by the matrix σ_h^{-1}, namely

$$\mathrm{Hom}(\sigma, \mathrm{id})_h(A) = A\sigma_h^{-1}, \quad A \in \mathrm{M}_n(\mathbb{C}). \tag{$*_{22}$}$$

This is the situation that occurs in practice. If $\mathbb{C}^n$ is equipped with its standard hermitian inner product and if $\sigma\colon H \to \mathbf{U}(n)$ is a unitary representation, so that σ_h is a unitary matrix, if we give $\mathrm{M}_n(\mathbb{C})$ the hermitian inner product $\langle A, B\rangle = \mathrm{tr}(B^*A)$, then the representation $\mathrm{Hom}(\sigma, \mathrm{id})$ is unitary because using the fact that $\mathrm{tr}(XY) = \mathrm{tr}(YX)$, we have

$$\begin{aligned}\langle A\sigma_h^{-1}, B\sigma_h^{-1}\rangle &= \langle A\sigma_h^*, B\sigma_h^*\rangle = \mathrm{tr}((B\sigma_h^*)^*(A\sigma_h^*)) = \mathrm{tr}(\sigma_h B^* A\sigma_h^*)\\ &= \mathrm{tr}(\sigma_h^*\sigma_h B^*A) = \mathrm{tr}(B^*A) = \langle A, B\rangle.\end{aligned}$$

In the next section, we show how to construct a Fourier transform on a semi-direct product $G = \mathbb{R}^d \rtimes H$, where H is compact in terms of the Fourier transform $\mathcal{F}$ on H.

8.9 Promoting the Fourier Transform from H to $\mathbb{R}^d \rtimes H$

If we view a function defined on $G = \mathbb{R}^d \rtimes H$ as a function $f\colon \mathbb{R}^d \rtimes H \to \mathbb{C}$, the new twist is that the Fourier coefficients of f are now tuples $(\widehat{f}_\rho)_{\rho\in R(H)}$ of *functions* $\widehat{f}_\rho\colon \mathbb{R}^d \to \mathrm{M}_{n_\rho}(\mathbb{C})$. This causes new problems to reconstruct a function from its Fourier coefficients because even if the functions $\widehat{f}_\rho$ belong to $\mathrm{L}^2(\mathbb{R}^d, \mathrm{M}_{n_\rho}(\mathbb{C}))$, there is no guarantee that the function obtained from the inverse Fourier transform belongs to $\mathrm{L}^2(G)$. Some additional condition is required on the functions $\widehat{f}_\rho$. We provide a solution to this problem below by constructing a Hilbert space $\mathrm{L}^2(\mathbb{R}^d, \widehat{H})$, such that the new Fourier transform $\mathcal{F}^\tau\colon \mathrm{L}^2(G) \to \mathrm{L}^2(\mathbb{R}^d, \widehat{H})$ and the Fourier cotransform $\overline{\mathcal{F}^\tau}\colon \mathrm{L}^2(\mathbb{R}^d, \widehat{H}) \to \mathrm{L}^2(G)$ are mutual inverses. We found the key idea in a paper by Mensah and Awussi [Mensah and Awussi (2013)], who investigate the situation of a semi-direct product $H \rtimes \mathbb{R}^d$, where $\mathbb{R}^d$ acts on H by automorphisms.

The first crucial observation is that for any function $f \in \mathrm{L}^2(\mathbb{R}^d \rtimes H)$, by Fubini, for any fixed $x \in \mathbb{R}^d$, we have $f^H(x) \in \mathrm{L}^2(H)$, where f^H is the function defined in Example 8.13. Since H is a compact group, the Fourier transform $\mathcal{F}(f^H(x))$ is well-defined. For every $\rho \in R(H)$ and every *fixed* $x \in \mathbb{R}^d$, recall that $\mathcal{F}(f^H(x))(\rho)$ is the $n_\rho \times n_\rho$ matrix given by

$$\mathcal{F}(f^H(x))(\rho) = \int_H (f^H(x))(h)M_\rho(h)^*\, d\lambda(h) = \int_H f(x,h)M_\rho(h)^*\, d\lambda(h),$$

where M_ρ is an irreducible representation of H in $\mathbb{C}^{n_\rho}$. To reduce the amount of superscripts, we also denote $f^H(x)$ as $f(x, -)$.

Technically, $\mathcal{F}\colon \mathrm{L}^2(H) \to \mathrm{L}^2(\widehat{H})$ is defined for functions with domain H, with

$$\mathrm{L}^2(\widehat{H}) = \left\{ F \in \prod_{\rho \in R(H)} \mathrm{M}_{n_\rho}(\mathbb{C}) \mid \|F\|_{\mathrm{L}^2(\widehat{H})} < \infty \right\},$$

and

$$\|F\|_{\mathrm{L}^2(\widehat{H})} = \Bigg(\sum_{\rho \in R(H)} n_\rho \, \|F(\rho)\|_{\mathrm{HS}}^2 \Bigg)^{1/2} = \Bigg(\sum_{\rho \in R(H)} n_\rho \operatorname{tr}\Big(F(\rho)^* F(\rho) \Big) \Bigg)^{1/2};$$

see Definition 4.20 and Definition 4.21. The vector space $\mathrm{L}^2(\widehat{H})$ is a Hilbert space under the inner product

$$\langle F_1, F_2 \rangle_{\mathrm{L}^2(\widehat{H})} = \sum_{\rho \in R(H)} n_\rho \langle F_1(\rho), F_2(\rho) \rangle_{\mathrm{HS}} = \sum_{\rho \in R(H)} n_\rho \operatorname{tr}\Big(F_2(\rho)^* F_1(\rho) \Big);$$

see Theorem 4.27.

We would like to define a notion of Fourier transform on functions in $\mathrm{L}^2(\mathbb{R}^d \rtimes H)$ that makes use of the Fourier transform $\mathcal{F}$ defined on H, so to avoid confusion, we will denote this new Fourier transform by $\mathcal{F}^\tau$. The motivation is that $\tau\colon H \to \mathbf{GL}(n)$ is the action of H on $\mathbb{R}^d$, with $\tau(h)(x) = hxh^{-1}$.

Definition 8.9. For any *fixed* $x \in \mathbb{R}^d$ and any G-feature map $f \in \mathrm{L}^2(\mathbb{R}^d \rtimes H)$, we define $\mathcal{F}(f(x,-)) = (\mathcal{F}(f(x,-))_\rho)_{\rho \in R(H)} \in \mathrm{L}^2(\widehat{H})$, also denoted $\widehat{f}(x)$, by

$$\mathcal{F}(f(x,-))_\rho = \widehat{f}(x)_\rho = \int_H f(x,h) M_\rho(h)^* \, d\lambda(h), \quad \rho \in R(H). \qquad (\widehat{f}(x))$$

Then if we let x vary in $\mathbb{R}^d$, for any *fixed* ρ, we obtain a function $\widehat{f}_\rho\colon \mathbb{R}^d \to \mathrm{M}_{n_\rho}(\mathbb{C})$ given by

$$\widehat{f}_\rho(x) = \widehat{f}(x)_\rho = \int_H f(x,h) M_\rho(h)^* \, d\lambda(h), \quad x \in \mathbb{R}^d. \qquad (\widehat{f}_\rho)$$

By Fubini, since $f \in \mathrm{L}^2(\mathbb{R}^d \rtimes H)$, we have $\widehat{f}_\rho \in \mathrm{L}^2(\mathbb{R}^d, \mathrm{M}_{n_\rho}(\mathbb{C}))$. This step requires a justification that we leave as an exercise.

The function $\widehat{f}_\rho$ is called a *Fourier coefficients feature field of type* ρ or *steerable feature field of type* ρ. The $R(H)$-indexed family $(\widehat{f}_\rho)_{\rho \in R(H)}$ is denoted by $\widehat{f}$ and is called the *family of Fourier coefficients feature fields of* f or *family of steerable feature fields of* f.

Observe that

$$\widehat{f}(x) = (\widehat{f}_\rho(x))_{\rho \in R(H)} \in \mathrm{L}^2(\widehat{H}), \quad \text{for every } x \in \mathbb{R}^d,$$

and consequently, $(\widehat{f}_\rho)_{\rho \in R(H)}$ belongs to the space $\mathfrak{C}^\tau(\widehat{H})$ defined next.

Definition 8.10. The vector space $\mathfrak{C}^\tau(\widehat{H})$ is defined by

$$\mathfrak{C}^\tau(\widehat{H}) = \left\{ F \in \prod_{\rho \in R(H)} \mathrm{L}^2(\mathbb{R}^d, \mathrm{M}_{n_\rho}(\mathbb{C})) \mid (F_\rho(x))_{\rho \in R(H)} \in \mathrm{L}^2(\widehat{H}),\, x \in \mathbb{R}^d \right\}. \tag{$\mathfrak{C}^\tau(\widehat{H})$}$$

Note the analogy with the space $\mathfrak{C}(\widehat{H})$ of Definition 4.21.

Definition 8.11. We define the map $\mathcal{F}^\tau$ from $\mathrm{L}^2(\mathbb{R}^d \rtimes H)$ to $\mathfrak{C}^\tau(\widehat{H})$ by setting

$$\mathcal{F}^\tau(f) = (\mathcal{F}^\tau_\rho(f))_{\rho \in R(H)}, \quad f \in \mathrm{L}^2(G), \text{ with}$$

$$\mathcal{F}^\tau_\rho(f)(x) = \widehat{f}_\rho(x) = \mathcal{F}(f(x, -))_\rho = \int_H f(x,h) M_\rho(h)^* \, d\lambda(h) \tag{$\mathcal{F}^\tau$}$$

for all $x \in \mathbb{R}^d$, $\rho \in R(H)$. Observe that by line $(\widehat{f}(x))$, for every fixed $x \in \mathbb{R}^d$, we have

$$\mathcal{F}^\tau(f)(x) = (\mathcal{F}^\tau_\rho(f)(x))_{\rho \in R(H)} = \mathcal{F}(f(x,-)). \tag{$\mathcal{F}^\tau(f)(x)$}$$

We will see shortly that steerable feature fields of type ρ transform under the representation $\mathrm{Hom}(M_\rho, \mathrm{id})$. For this reason, the space of steerable feature fields of type ρ is denoted by $\mathbf{FF}(\mathbb{R}^d, H, \mathrm{Hom}(M_\rho, \mathrm{id}))$. These are matrix-valued functions $\widehat{f}_\rho \colon \mathbb{R}^d \to \mathrm{M}_{n_\rho}(\mathbb{C})$ that belong to $\mathrm{L}^2(\mathbb{R}^d, \mathrm{M}_{n_\rho}(\mathbb{C}))$. Actually, we will see below (see Definition 8.12) that there is some extra condition on the family $(\widehat{f}_\rho)_{\rho \in R(H)}$ that ensures that Fourier inversion yields a function in $\mathrm{L}^2(G)$.

For every *fixed* $x \in \mathbb{R}^d$, the function $f^H(x) \in \mathrm{L}^2(H)$ can be recovered by Fourier inversion using the Fourier cotransform $\overline{\mathcal{F}}$ from $\mathrm{L}^2(\widehat{H})$ to $\mathrm{L}^2(H)$ from the family of Fourier coefficients feature fields $\widehat{f} = (\widehat{f}_\rho)_{\rho \in R(H)} \in \mathfrak{C}^\tau(\widehat{H})$ evaluated at x, namely the $R(H)$-indexed family $\widehat{f}(x) = (\widehat{f}_\rho(x))_{\rho \in R(H)} \in \mathrm{L}^2(\widehat{H})$, using the formula

$$(f^H(x))(h) = [\overline{\mathcal{F}}(\widehat{f}(x))](h) = \sum_{\rho \in R(H)} n_\rho \operatorname{tr}\left(\widehat{f}_\rho(x) M_\rho(h)\right), \quad h \in H.$$

Thus the G-feature map $f \colon \mathbb{R}^d \times H \to \mathbb{C}$ can also be recovered *pointwise, via*

$$f(x,h) = [\overline{\mathcal{F}}(\widehat{f}(x))](h) = \sum_{\rho \in R(H)} n_\rho \operatorname{tr}\left(\widehat{f}_\rho(x) M_\rho(h)\right). \tag{$\overline{\mathcal{F}}(\widehat{f}(x))$}$$

The definition of a map $\overline{\mathcal{F}^\tau}$ from $\mathfrak{E}^\tau(\widehat{H})$ to $\mathrm{L}^2(\mathbb{R}^d \rtimes H)$ is more delicate. The space $\mathfrak{E}^\tau(\widehat{H})$ is actually too big to ensure that the resulting functions belong to $\mathrm{L}^2(\mathbb{R}^d \rtimes H)$. Inspired by Mensah and Awussi [Mensah and Awussi (2013)], we define the following space.

Definition 8.12. Define the vector space $\mathrm{L}^2(\mathbb{R}^d, \widehat{H})$ by

$$\mathrm{L}^2(\mathbb{R}^d, \widehat{H}) = \left\{ F \in \mathfrak{E}^\tau(\widehat{H}) \;\middle|\; \|F(-)\|_{\mathrm{L}^2(\widehat{H})} \in \mathrm{L}^2(\mathbb{R}^d) \right\}, \qquad (\mathrm{L}^2(\mathbb{R}^d, \widehat{H}))$$

where $\|F(-)\|_{\mathrm{L}^2(\widehat{H})}$ is the function defined, such that if $F = (F_\rho)_{\rho \in R(H)}$, then

$$\|F(x)\|_{\mathrm{L}^2(\widehat{H})} = \left(\sum_{\rho \in R(H)} n_\rho \, \|F_\rho(x)\|_{\mathrm{HS}}^2 \right)^{1/2}. \qquad (\|F(-)\|_{\mathrm{L}^2(\widehat{H})})$$

Note that $\|F(-)\|_{\mathrm{L}^2(\widehat{H})} \in \mathrm{L}^2(\mathbb{R}^d)$ implies that

$$\int_{\mathbb{R}^d} \|F(x)\|_{\mathrm{L}^2(\widehat{H})}^2 \, dx < \infty.$$

The vector space $\mathrm{L}^2(\mathbb{R}^d, \widehat{H})$ is equipped with the norm $\| \;\|_{\mathrm{L}^2(\mathbb{R}^d, \widehat{H})}$ given by

$$\|F\|_{\mathrm{L}^2(\mathbb{R}^d, \widehat{H})}^2 = \int_{\mathbb{R}^d} \|F(x)\|_{\mathrm{L}^2(\widehat{H})}^2 \, dx = (\|F(-)\|_{\mathrm{L}^2(\widehat{H})}^2)_{\mathrm{L}^2(\mathbb{R}^d)}^2. \qquad (\|F\|_{\mathrm{L}^2(\mathbb{R}^d, \widehat{H})})$$

Note the analogy with the definition of the space $\mathrm{L}^2(\widehat{H})$ in Definition 4.21. We also define an inner product on $\mathrm{L}^2(\mathbb{R}^d, \widehat{H})$ as follows.

Definition 8.13. For any two sequences of functions $F_1, F_2 \in \mathrm{L}^2(\mathbb{R}^d, \widehat{H})$, let $\langle F_1, F_2 \rangle_{\mathrm{L}^2(\mathbb{R}^d, \widehat{H})}$ be given by

$$\langle F_1, F_2 \rangle_{\mathrm{L}^2(\mathbb{R}^d, \widehat{H})} = \int_{\mathbb{R}^d} \sum_{\rho \in R(H)} n_\rho \operatorname{tr}\Big(((F_2)_\rho(x))^* (F_1)_\rho(x) \Big) \, dx. \qquad (\langle -, - \rangle)$$

Observe that

$$\|F\|_{\mathrm{L}^2(\mathbb{R}^d, \widehat{H})}^2 = \langle F, F \rangle_{\mathrm{L}^2(\mathbb{R}^d, \widehat{H})},$$

but we still need to prove that the integral in $(\langle -, - \rangle)$ is well defined. Using the Cauchy–Schwarz inequality both in $\mathrm{L}^2(\widehat{H})$ and $\mathrm{L}^2(\mathbb{R}^d)$, we have

$$
\begin{aligned}
&|\langle F_1, F_2\rangle_{\mathrm{L}^2(\mathbb{R}^d,\widehat{H})}| \\
&\quad = \left| \int_{\mathbb{R}^d} \sum_{\rho \in R(H)} n_\rho \, \langle (F_1)_\rho(x), (F_2)_\rho(x)\rangle_{\mathrm{HS}} \, dx \right| && (1)\\
&\quad \le \int_{\mathbb{R}^d} \left| \sum_{\rho \in R(H)} n_\rho \, \langle (F_1)_\rho(x), (F_2)_\rho(x)\rangle_{\mathrm{HS}} \right| dx && (2)\\
&\quad = \int_{\mathbb{R}^d} \left| \langle F_1(x), F_2(x)\rangle_{\mathrm{L}^2(\widehat{H})} \right| dx && (3)\\
&\quad \le \int_{\mathbb{R}^d} \|F_1(x)\|_{\mathrm{L}^2(\widehat{H})} \, \|F_2(x)\|_{\mathrm{L}^2(\widehat{H})} \, dx && (4)\\
&\quad \le \left(\int_{\mathbb{R}^d} \|F_1(x)\|^2_{\mathrm{L}^2(\widehat{H})} \, dx \right)^{1/2} \left(\int_{\mathbb{R}^d} \|F_2(x)\|^2_{\mathrm{L}^2(\widehat{H})} \, dx \right)^{1/2} && (5)\\
&\quad = \|F_1\|_{\mathrm{L}^2(\mathbb{R}^d,\widehat{H})} \, \|F_2\|_{\mathrm{L}^2(\mathbb{R}^d,\widehat{H})}, && (6)
\end{aligned}
$$

where (1) holds by definition, (2) by a standard property of the integral, (3) by definition of the inner product in $\mathrm{L}^2(\widehat{H})$, (4) by the Cauchy–Schwarz inequality in $\mathrm{L}^2(\widehat{H})$, (5) by the Cauchy–Schwarz inequality in $\mathrm{L}^2(\mathbb{R}^d)$, and (6) by definition (see ($\|F\|_{\mathrm{L}^2(\mathbb{R}^d,\widehat{H})}$)).

We will also need the projection $\mathrm{L}^2(\mathbb{R}^d,\widehat{H})_\rho$ of $\mathrm{L}^2(\mathbb{R}^d,\widehat{H})$ on the ρth factor, that is,

$$
\mathrm{L}^2(\mathbb{R}^d,\widehat{H})_\rho = \{F_\rho \mid (F_\rho)_{\rho \in R(H)} \in \mathrm{L}^2(\mathbb{R}^d,\widehat{H})\}. \qquad (\mathrm{L}^2(\mathbb{R}^d,\widehat{H})_\rho)
$$

We have the following important version of Plancherel theorem for our Fourier transform $\mathcal{F}^\tau \colon \mathrm{L}^2(G) \to \mathrm{L}^2(\mathbb{R}^d,\widehat{H})$.

Theorem 8.8. *The map $\mathcal{F}^\tau \colon \mathrm{L}^2(G) \to \mathrm{L}^2(\mathbb{R}^d,\widehat{H})$ (with $G = \mathbb{R}^d \rtimes H$) is an isometric isomorphism of Hilbert spaces. That is, it is bijective and*

$$
\langle \mathcal{F}^\tau(f), \mathcal{F}^\tau(g)\rangle_{\mathrm{L}^2(\mathbb{R}^d,\widehat{H})} = \langle f, g\rangle_{\mathrm{L}^2(G)}, \quad f, g \in \mathrm{L}^2(\mathbb{R}^d \rtimes H).
$$

In particular, it is continuous.

Proof. First, we prove that the map $\mathcal{F}^\tau$ is an isometry. Since $\mathrm{L}^2(G)$ is a Hilbert space, this proves that $\mathrm{L}^2(\mathbb{R}^d,\widehat{H})$ is also a Hilbert space. Since the norm on $\mathrm{L}^2(\mathbb{R}^d,\widehat{H})$ is induced by the inner product on $\mathrm{L}^2(\mathbb{R}^d,\widehat{H})$, it suffices to prove that the norm is preserved. This is a standard result of linear algebra; for example, see Gallier and Quaintance [Gallier and Quaintance (2020c)] (Chapter 13, Proposition 13.1). For any $f \in \mathrm{L}^2(\mathbb{R}^d \rtimes H)$, we have

$$\begin{aligned} \|\mathcal{F}^\tau(f)\|^2_{\mathrm{L}^2(\mathbb{R}^d,\widehat{H})} &= \int_{\mathbb{R}^d} \|\mathcal{F}^\tau(f)(x)\|^2_{\mathrm{L}^2(\widehat{H})}\, dx && \text{by definition} \\ &= \int_{\mathbb{R}^d} \|\mathcal{F}(f(x,-))\|^2_{\mathrm{L}^2(\widehat{H})}\, dx && \text{by } (\mathcal{F}^\tau(f)(x)). \end{aligned}$$

However, for fixed x, $\mathcal{F}(f(x,-))$ is the Fourier transform of the function $f(x,-) \in \mathrm{L}^2(H)$. By Plancherel Theorem (Theorem 4.32), we have

$$\|\mathcal{F}(f(x,-))\|^2_{\mathrm{L}^2(\widehat{H})} = \|f(x,-)\|^2_{\mathrm{L}^2(H)}.$$

Since $f \in \mathrm{L}^2(\mathbb{R}^d \rtimes H)$, by Fubini

$$\|f\|^2_{\mathrm{L}^2(G)} = \int_G |f(x,h)|^2\, d\lambda_G(x,h) = \int_{\mathbb{R}^d}\int_H |f(x,h)|^2\, d\lambda_H(h)\, dx < \infty,$$

but

$$\int_{\mathbb{R}^d}\int_H |f(x,h)|^2\, d\lambda_H(h)\, dx = \int_{\mathbb{R}^d} \|f(x,-)\|^2_{\mathrm{L}^2(H)}\, dx,$$

which shows that the function $x \mapsto \|\mathcal{F}(f(x,-))\|_{\mathrm{L}^2(\widehat{H})}$ is in $\mathrm{L}^2(\mathbb{R}^d)$. Consequently, we have

$$\begin{aligned} \|\mathcal{F}^\tau(f)\|^2_{\mathrm{L}^2(\mathbb{R}^d,\widehat{H})} &= \int_{\mathbb{R}^d} \|\mathcal{F}(f(x,-))\|^2_{\mathrm{L}^2(\widehat{H})}\, dx \\ &= \int_{\mathbb{R}^d} \|f(x,-)\|^2_{\mathrm{L}^2(H)}\, dx && \text{by Plancherel} \\ &= \int_{\mathbb{R}^d}\int_H |f(x,h)|^2\, d\lambda_H(h)\, dx && \mathrm{L}^2(H)\text{-norm} \\ &= \|f\|^2_{\mathrm{L}^2(G)}. && \text{by Fubini} \end{aligned}$$

Since $\mathcal{F}^\tau$ is an isometry, it is injective. It remains to prove that it is surjective.

For any $F = (F_\rho)_{\rho \in R(H)} \in \mathrm{L}^2(\mathbb{R}^d, \widehat{H})$ and for every fixed $x \in \mathbb{R}^d$, we have $F(x) = (F_\rho(x))_{\rho \in R(H)} \in \mathrm{L}^2(\widehat{H})$. By Plancherel applied to the Fourier transform $\mathcal{F}$ between $\mathrm{L}^2(H)$ and $\mathrm{L}^2(\widehat{H})$, there is a unique function $f_x \in \mathrm{L}^2(H)$, such that

$$\mathcal{F}(f_x) = F(x) \quad \text{and} \quad \|f_x\|_{\mathrm{L}^2(H)} = \|F(x)\|_{\mathrm{L}^2(\widehat{H})}. \tag{$*_{23}$}$$

Define the function $f\colon \mathbb{R}^d \rtimes H \to \mathbb{C}$ by

$$f(x,h) = f_x(h), \quad x \in \mathbb{R}^d,\ h \in H. \tag{$*_{24}$}$$

Observe that

$$f(x, -) = f_x, \tag{$*_{25}$}$$

so we get

$$\begin{aligned}
\|f\|_{\mathrm{L}^2(G)} &= \int_{\mathbb{R}^d}\int_H |f(x,h)|^2\, d\lambda_H(h)\, dx && \text{by definition of } \|f\|^2_{\mathrm{L}^2(\mathrm{G})}\\
&= \int_{\mathbb{R}^d}\int_H |f_x(h)|^2\, d\lambda_H(h)\, dx && \text{by } (*_{24})\\
&= \int_{\mathbb{R}^d} \|f_x\|^2_{\mathrm{L}^2(\mathrm{H})}\, dx && \text{by definition of } \|f_x\|^2_{\mathrm{L}^2(\mathrm{H})}\\
&= \int_{\mathbb{R}^d} \|F(x)\|^2_{\mathrm{L}^2(\widehat{\mathrm{H}})}\, dx && \text{by } (*_{23})\\
&= \|F\|^2_{\mathrm{L}^2(\mathbb{R}^d,\widehat{H})} < \infty, && \text{by definition of } \|F\|^2_{\mathrm{L}^2(\mathbb{R}^d,\widehat{H})}
\end{aligned}$$

and the last step because $F \in \mathrm{L}^2(\mathbb{R}^d, \widehat{H})$. Therefore, $f \in \mathrm{L}^2(G)$. Then by $(\mathcal{F}^\tau(f)(x))$, $(*_{25})$ and $(*_{23})$, we have

$$\mathcal{F}^\tau(f)(x) = \mathcal{F}(f(x,-)) = \mathcal{F}(f_x) = F(x), \quad x \in \mathbb{R}^d,$$

which means that $\mathcal{F}^\tau(f) = F$, and thus $\mathcal{F}^\tau$ is surjective. □

Since we already know that functions in $\mathrm{L}^2(G)$ can be recovered pointwise using the Fourier transform on H, we can exhibit the inverse $\overline{\mathcal{F}^\tau}$ of the Fourier transform $\mathcal{F}^\tau$.

Definition 8.14. Define the map $\overline{\mathcal{F}^\tau}_\rho\colon \mathrm{L}^2(\mathbb{R}^d, \widehat{H})_\rho \to \mathrm{L}^2(G)$ for every $\rho \in R(H)$ by

$$\overline{\mathcal{F}^\tau}_\rho(\widehat{f}_\rho)(x,h) = n_\rho \operatorname{tr}\left(\widehat{f}_\rho(x) M_\rho(h)\right), \qquad \widehat{f}_\rho \in \mathrm{L}^2(\mathbb{R}^d, \widehat{H})_\rho, \quad x \in \mathbb{R}^d,\ h \in H, \tag{$\overline{\mathcal{F}^\tau}_\rho$}$$

and the map $\overline{\mathcal{F}^\tau}\colon \mathrm{L}^2(\mathbb{R}^d, \widehat{H}) \to \mathrm{L}^2(G)$ by

$$\overline{\mathcal{F}^\tau}((\widehat{f}_\rho)_{\rho\in R(H)})(x,h) = \sum_{\rho\in R(H)} \overline{\mathcal{F}^\tau}_\rho(\widehat{f}_\rho)(x,h), \quad (\widehat{f}_\rho)_{\rho\in R(H)} \in \mathrm{L}^2(\mathbb{R}^2, \widehat{H}), \quad x \in \mathbb{R}^d,\ h \in H. \tag{$\overline{\mathcal{F}^\tau}$}$$

Then $\mathcal{F}^\tau\colon \mathrm{L}^2(G) \to \mathrm{L}^2(\mathbb{R}^d, \widehat{H})$ and $\overline{\mathcal{F}^\tau}\colon \mathrm{L}^2(\mathbb{R}^d, \widehat{H}) \to \mathrm{L}^2(G)$ are mutual inverses.

We claim that the map $\widehat{f}_\rho \in \mathrm{L}^2(\mathbb{R}^d, \widehat{H})_\rho$ is indeed a feature field, with $\mathcal{H} = \mathrm{M}_{n_\rho}(\mathbb{C})$ and $\sigma = \mathrm{Hom}(M_\rho, \mathrm{id})$. For this, we need to see how the function $\widehat{f}_\rho$ changes when $G = \mathbb{R}^d \rtimes H$ acts on f *via* the left regular action $\mathbf{R}^{G\to \mathrm{L}^2(G)}$ given by

$$\mathbf{R}^{G\to\mathrm{L}^2(G)}_{(x,h)}(f)(x_1, h_1) = f(h^{-1}\cdot(x_1 - x), h^{-1}h_1).$$

Proposition 8.9. *For every $\rho \in R(H)$, let $\sigma_\rho \colon H \to \mathbf{U}(\mathrm{M}_{n_\rho}(\mathbb{C}))$ be the representation $\sigma_\rho = \mathrm{Hom}(M_\rho, \mathrm{id})$ associated with the representation $M_\rho \colon H \to \mathbf{U}(\mathbb{C}^{n_\rho})$ as in Definition 8.8. For every function $\widehat{f}_\rho \in \mathrm{L}^2(\mathbb{R}^d, \widehat{H})_\rho$, we have*

$$\begin{aligned}\mathcal{F}^\tau_\rho[\mathbf{R}^{G\to\mathrm{L}^2(G)}_{(x,h)}(f)](x_1) &= [(\mathrm{Ind}^G_H(\sigma_\rho)_{(x,h)}\,\widehat{f}_\rho](x_1)\\ &= \widehat{f}_\rho(h^{-1}\cdot(x_1 - x))\,M_\rho(h)^*. \qquad (*_{26})\end{aligned}$$

Proof. Using the fact that the Haar measure λ is left (and right) invariant and the fact that M_ρ is a representation, we have

$$\begin{aligned}&\mathcal{F}^\tau_\rho[\mathbf{R}^{G\to\mathrm{L}^2(G)}_{(x,h)}(f)](x_1)\\ &= \int_H \mathbf{R}^{G\to\mathrm{L}^2(G)}_{(x,h)}(f)(x_1, h_1)M_\rho(h_1)^*\,d\lambda(h_1)\\ &= \int_H f(h^{-1}\cdot(x_1 - x), h^{-1}h_1)M_\rho(h_1)^*\,d\lambda(h_1)\\ &= \int_H f(h^{-1}\cdot(x_1 - x), h_2)M_\rho(hh_2)^*\,d\lambda(h_2), \qquad h_1 = hh_2\\ &= \left(\int_H f(h^{-1}\cdot(x_1 - x), h_2)M_\rho(h_2)^*\,d\lambda(h_2)\right)M_\rho(h)^*\\ &= \widehat{f}_\rho(h^{-1}\cdot(x_1 - x))\,M_\rho(h)^*.\end{aligned}$$

The above computation shows that

$$\mathcal{F}^\tau_\rho[\mathbf{R}^{G\to\mathrm{L}^2(G)}_{(x,h)}(f)](x_1) = \widehat{f}_\rho(h^{-1}\cdot(x_1 - x))\,M_\rho(h)^*. \qquad \square$$

Equation $(*_{26})$ shows that the group $G = \mathbb{R}^d \rtimes H$ acts on the feature fields of type ρ *via*

$$[(\mathrm{Ind}^G_H(\sigma_\rho)_{(x,h)}\,\widehat{f}_\rho](x_1) = \widehat{f}_\rho(h^{-1}\cdot(x_1 - x))\,M_\rho(h)^*, \qquad (\sigma_\rho)$$

for all $(x, h) \in \mathbb{R}^d \rtimes H$ and all $x_1 \in \mathbb{R}^d$, and $(*_{26})$ is equivalent to the commutativity of the following diagram

$$\begin{array}{ccc}\mathrm{L}^2(G) & \xrightarrow{\mathcal{F}^\tau_\rho} & \mathrm{L}^2(\mathbb{R}^d, \widehat{H})_\rho\\ {\scriptstyle \mathbf{R}^{G\to\mathrm{L}^2(G)}_{(x,h)}}\downarrow & & \downarrow{\scriptstyle (\mathrm{Ind}^G_H\,\sigma_\rho)_{(x,h)}}\\ \mathrm{L}^2(G) & \xrightarrow[\mathcal{F}^\tau_\rho]{} & \mathrm{L}^2(\mathbb{R}^d, \widehat{H})_\rho\end{array}$$

for all $(x, h) \in G = \mathbb{R}^d \rtimes H$.

We also package the representations $\mathrm{Ind}_H^G \sigma_\rho \colon G \to \mathbf{U}(\mathrm{L}^2(\mathbb{R}^d, \widehat{H})_\rho)$ in the map

$$\mathrm{Ind}_H^G \sigma \colon G \times \mathrm{L}^2(\mathbb{R}^d, \widehat{H}) \to \mathrm{L}^2(\mathbb{R}^d, \widehat{H})$$

defined such that for any $\widehat{f} = (\widehat{f}_\rho)_{\rho \in R(H)}$,

$$[(\mathrm{Ind}_H^G \sigma)_{(x,h)} \widehat{f}]_\rho(x_1) = [(\mathrm{Ind}_H^G \sigma_\rho)_{(x,h)} \widehat{f}_\rho](x_1), \; x_1 \in \mathbb{R}^d, \; \rho \in R(H). \quad (\sigma)$$

The following result should not be too surprising.

Proposition 8.10. *The following diagram commutes*

$$\begin{array}{ccc} \mathrm{L}^2(\mathbb{R}^d, \widehat{H})_\rho & \xrightarrow{\overline{\mathcal{F}^\tau}_\rho} & \mathrm{L}^2(G) \\ {\scriptstyle (\mathrm{Ind}_H^G \sigma_\rho)_{(x,h)}} \big\downarrow & & \big\downarrow {\scriptstyle \mathbf{R}^{G \to \mathrm{L}^2(G)}_{(x,h)}} \\ \mathrm{L}^2(\mathbb{R}^d, \widehat{H})_\rho & \xrightarrow[\overline{\mathcal{F}^\tau}_\rho]{} & \mathrm{L}^2(G) \end{array}$$

for all $(x,h) \in G = \mathbb{R}^d \rtimes H$.

Proof. For any $\widehat{f}_\rho \in \mathrm{L}^2(\mathbb{R}^d, \widehat{H})_\rho$ we have

$$\begin{aligned} &\overline{\mathcal{F}^\tau}_\rho((\mathrm{Ind}_H^G \sigma_\rho)_{(x,h)} \widehat{f}_\rho)(x_1, h_1) \\ &\qquad = n_\rho \operatorname{tr}\Big(((\mathrm{Ind}_H^G \sigma_\rho)_{(x,h)} \widehat{f}_\rho)(x_1) M_\rho(h_1)\Big) && \text{by } (\overline{\mathcal{F}^\tau}_\rho) \\ &\qquad = n_\rho \operatorname{tr}\Big(\widehat{f}_\rho(h^{-1} \cdot (x_1 - x)) M_\rho(h)^* M_\rho(h_1)\Big) && \text{by } (*_{26}) \\ &\qquad = n_\rho \operatorname{tr}\Big(\widehat{f}_\rho(h^{-1} \cdot (x_1 - x)) M_\rho(h^{-1} h_1)\Big). \end{aligned}$$

We also have

$$\begin{aligned} &\mathbf{R}_{(x,h)}(\overline{\mathcal{F}^\tau}_\rho(\widehat{f}_\rho))(x_1, h_1) \\ &\quad = \overline{\mathcal{F}^\tau}_\rho(\widehat{f}_\rho)(h^{-1} \cdot (x_1 - x)), h^{-1} h_1) && \text{by definition of } \mathbf{R}_{(x,h)} \\ &\quad = n_\rho \operatorname{tr}\Big(\widehat{f}_\rho(h^{-1} \cdot (x_1 - x)) M_\rho(h^{-1} h_1)\Big). && \text{by } (\overline{\mathcal{F}^\tau}_\rho) \end{aligned}$$

Consequently

$$\overline{\mathcal{F}^\tau}_\rho((\mathrm{Ind}_H^G \sigma_\rho)_{(x,h)} \widehat{f}_\rho)(x_1, h_1) = \mathbf{R}_{(x,h)}(\overline{\mathcal{F}^\tau}_\rho(\widehat{f}_\rho))(x_1, h_1),$$

as claimed. □

Remark: We also have the representation $\mathrm{Hom}(\mathrm{id}, M_\rho)$, which acts on $\mathrm{M}_{n_\rho}(\mathbb{C})$ by multiplication on the left by $M_\rho(h)$ for every $h \in H$. The induced representation $\mathrm{Hom}(\mathrm{id}, M_\rho)$ of $\mathbb{R}^d \rtimes H$ on the feature fields of type ρ is then given by

$$\begin{aligned}[(\mathrm{Ind}_H^G \mathrm{Hom}(\mathrm{id}, M_\rho))_{(x,h)} \widehat{f}_\rho](x_1) &= \mathrm{Hom}(\mathrm{id}, M_\rho)(h)(\widehat{f}_\rho(h^{-1} \cdot (x_1 - x))) \\ &= M_\rho(h)\widehat{f}_\rho(h^{-1} \cdot (x_1 - x))\end{aligned}$$

for all $(x, h) \in \mathbb{R}^d \rtimes H$ and all $x_1 \in \mathbb{R}^d$. It is a bit more natural than the representation induced by $\mathrm{Hom}(M_\rho, \mathrm{id})$.[3]

Example 8.14. Let $H = \mathbf{SO}(2)$ so that $G = \mathbb{R}^2 \rtimes \mathbf{SO}(2) = \mathbf{SE}(2)$. In this case, $R(\mathbf{SO}(2)) = \mathbb{Z}$ and $n_\rho = 1$. We will denote ρ as ℓ. For any $f \in \mathrm{L}^2(\mathbf{SE}(2))$, for every $x \in \mathbb{R}^2$, the Fourier transform $\mathcal{F}^\tau(f)$ of f is the $\mathbb{Z}$-indexed sequence $(\widehat{f}_\ell)_{\ell \in \mathbb{Z}}$ of functions given by

$$\widehat{f}_\ell(x) = \mathcal{F}^\tau(f(x, -))_\ell = \int_{S^1} e^{-i\ell\theta} f(x, \theta)\, d\theta, \quad x \in \mathbb{R}^2,\ \ell \in \mathbb{Z}.$$

The functions $\widehat{f}_\ell$ are the feature fields associated with ℓ. Observe that this is an example of $(\widehat{f}_\rho)$.

Given a family $\widehat{f} = (\widehat{f}_m)_{m\in\mathbb{Z}}$ of function $\widehat{f}_m \in \mathrm{L}^2(\mathbb{R}^2, \mathbb{Z})_m$, such that $\widehat{f}(x) = (\widehat{f}_m(x))_{m\in\mathbb{Z}} \in \ell^2(\mathbb{Z})$ for all $x \in \mathbb{R}^2$ and

$$\left(\sum_{m=-\infty}^{\infty} |\widehat{f}_m(-)|^2\right)^{1/2} \in \mathrm{L}^2(\mathbb{R}^2),$$

the Fourier cotransform $\overline{\mathcal{F}^\tau}(\widehat{f})(x, \theta)$ is given by

$$\overline{\mathcal{F}^\tau}(\widehat{f})(x, \theta) = \sum_{m=-\infty}^{\infty} \widehat{f}_m(x) e^{im\theta}.$$

It is instructive to see in this more concrete case how the function $\widehat{f}_\ell$ changes when $\mathbf{SE}(2) = \mathbb{R}^2 \rtimes \mathbf{SO}(2)$ acts on f *via* the left regular action $\mathbf{R}^{\mathbf{SE}(2)\to \mathrm{L}^2(\mathbf{SE}(2))}$ given by

$$\mathbf{R}^{\mathbf{SE}(2)\to \mathrm{L}^2(\mathbf{SE}(2))}_{(x,\theta)}(f)(x_1, \theta_1) = f(R_{-\theta}(x_1 - x), \theta_1 - \theta).$$

[3]Which representation arises naturally depends on the definition of the Fourier transform. The literature is not consistent on this matter. For example, Bekkers uses M_ρ instead of M_ρ^*.

Using the fact that the Haar measure on $\mathbf{SO}(2)$ is left (and right) invariant, we have

$$\begin{aligned}
&\mathcal{F}^{\tau}[\mathbf{R}_{(x,\theta)}^{\mathbf{SE}(2)\to \mathrm{L}^2(\mathbf{SE}(2))}(f)(x_1, -)]_\ell \\
&\quad = \int_{\mathbf{SO}(2)} \mathbf{R}_{(x,\theta)}^{\mathbf{SE}(2)\to \mathrm{L}^2(\mathbf{SE}(2))}(f)(x_1, \theta_1) e^{-i\ell\theta_1}\, d\theta_1 \\
&\quad = \int_{\mathbf{SO}(2)} f(R_{-\theta}(x_1 - x), \theta_1 - \theta) e^{-i\ell\theta_1}\, d\theta_1 \\
&\quad = \int_{\mathbf{SO}(2)} f(R_{-\theta}(x_1 - x), \theta_2) e^{-i\ell(\theta+\theta_2)}\, d\theta_2, \qquad \theta_1 = \theta + \theta_2 \\
&\quad = e^{-i\ell\theta} \int_{\mathbf{SO}(2)} f(R_{-\theta}(x_1 - x), \theta_2) e^{-i\ell\theta_2}\, d\theta_2 \\
&\quad = e^{-i\ell\theta} \widehat{f}_\ell(R_{-\theta}(x_1 - x)).
\end{aligned}$$

Thus, we have

$$\widehat{\mathbf{R}_{(x,\theta)}(f)}_\ell(x_1) = e^{-i\ell\theta} \widehat{f}_\ell(R_{-\theta}(x_1 - x)),$$

so the representation that needs to be associated with the feature fields corresponding to ℓ is $e^{-i\ell\theta}$, and not $e^{i\ell\theta}$. Since multiplication in $\mathbb{C}$ is commutative, given a character $\chi_\ell(\theta) = e^{i\ell\theta}$, the representation $\mathrm{Hom}(\chi_\ell, \mathrm{id})$ is just multiplication by $e^{-i\ell\theta}$ and the representation $\mathrm{Hom}(\mathrm{id}, \chi_\ell)$ is just multiplication by $e^{i\ell\theta}$.

Example 8.15. Given a function $f\colon \mathbb{R}^d \to \mathbb{C}$ in $\mathrm{L}^2(\mathbb{R}^d)$, in the case of a steerable kernel $k(x; w)$ expressed in terms of a steerable family $Y = (Y_1, \ldots, Y_L)$ as in Section 8.6, by lifted correlation we obtain the G-feature map $k \mathbin{\widetilde{\star}} f\colon \mathbb{R}^d \times H \to \mathbb{C}$ given by

$$(k \mathbin{\widetilde{\star}} f)(x, h) = \mathrm{tr}\left(\widehat{f}(x)\, \Sigma(h)^\top\right),$$

using the Fourier coefficients

$$\widehat{f}(x) = f^Y(x) w^* \in \mathrm{M}_L(\mathbb{C}),$$

with

$$f^Y(x) = \int_{\mathbb{R}^d} f(t) Y(t - x)\, dt, \;\; x \in \mathbb{R}^d.$$

The map $\widehat{f}\colon \mathbb{R}^d \to \mathrm{M}_L(\mathbb{C})$ is a feature field that transforms under $\mathrm{Ind}_H^G \mathrm{Hom}(\mathrm{id}, \Sigma)$. In this case $\mathcal{H} = \mathrm{M}_L(\mathbb{C})$, $\sigma = \mathrm{Hom}(\mathrm{id}, \Sigma)$, and

$$[(\mathrm{Ind}_H^G \mathrm{Hom}(\mathrm{id}, \Sigma))_{(x,h)}\, \widehat{f}_\rho](t) = \Sigma(h) \widehat{f}_\rho(h^{-1} \cdot (t - x)).$$

8.10 Correlation on the Space of Feature Fields $\mathbf{L^2(\mathbb{R}^d, \widehat{H})}$

As we mentioned at the end of Section 8.4, a typical CNN consists of layers, starting with a lifting layer followed by group correlation layers (often called group convolution layers). The last layer is typically a projection layer involving some pooling process. This is a simpler process that we will not discuss here.

The lifting layer takes as input a function $f_{\text{in}} \in L^2(\mathbb{R}^d)$ and produces an output function $f_{\text{out}} \in L^2(\mathbb{R}^d \rtimes H)$ given by a lifted correlation with

$$f_{\text{out}}(x, h) = (k \mathbin{\widetilde{\star}} f_{\text{in}})(x, h),$$

where

$$(k \mathbin{\widetilde{\star}} f_{\text{in}})(x, h) = \int_{\mathbb{R}^d} f_{\text{in}}(t) k(h^{-1} \cdot (t - x))\, dt, \quad (x, h) \in \mathbb{R}^d \times H.$$

Computing $(k \mathbin{\widetilde{\star}} f_{\text{in}})(x, h)$ requires discretizing the group H, which is not possible in practice if $d > 2$. If the kernel k can be expressed in terms of an H-steerable family Y of L functions in $L^2(\mathbb{R}^d)$ and a representation $\Sigma \colon H \to \mathbf{U}(L)$, then $f_{\text{out}}(x, h)$ can be computed a lot cheaply in terms of a feature field $\widehat{f}_{\text{out}} \colon \mathbb{R}^d \to \mathrm{M}_L(\mathbb{C})$ defined from f_{in} and Y as

$$f_{\text{out}}(x, h) = (k \mathbin{\widetilde{\star}} f_{\text{in}})(x, h) = \operatorname{tr}\left(\widehat{f}_{\text{out}}(x)\, \Sigma(h)^{\top}\right),$$

where $\widehat{f}_{\text{out}}(x)$ is a matrix of Fourier coefficients.

A group correlation layer takes as input a function $f_{\text{in}} \in L^2(\mathbb{R}^d \rtimes H)$ and produces as output a function $f_{\text{out}} \in L^2(\mathbb{R}^d \rtimes H)$ using a group correlation

$$f_{\text{out}}(s) = (k \star f_{\text{in}})(s) = \int_G f_{\text{in}}(t) k(s^{-1}t)\, d\lambda_G(t), \quad s \in G = \mathbb{R}^d \rtimes H.$$

We saw in the previous section that a G-feature map $f \in L^2(\mathbb{R}^d \rtimes H)$ yields a family $\widehat{f} = (\widehat{f}_\rho)_{\rho \in R(H)}$ of feature fields $\widehat{f}_\rho \in L^2(\mathbb{R}^d, \widehat{H})_\rho$ and that f can be recovered pointwise by Fourier inversion, namely

$$f(x, h) = \sum_{\rho \in R(H)} n_\rho \operatorname{tr}\left(\widehat{f}_\rho(x) M_\rho(h)\right).$$

We know how to transform G-feature maps using group correlation defined in Definition 8.4. This defines a transform Φ on $L^2(G)$ (where

$G = \mathbb{R}^d \rtimes H$) given by $f_{\text{out}} = \Phi(f_{\text{in}}) = k \star f_{\text{in}}$. We can summarize the situation by the following diagram:

$$\begin{array}{ccc}
\mathrm{L}^2(G) & \xrightarrow{\Phi} & \mathrm{L}^2(G) \\
{\scriptstyle \mathcal{F}^{\mathcal{T}}} \downarrow \uparrow {\scriptstyle \overline{\mathcal{F}^{\mathcal{T}}}} & & {\scriptstyle \mathcal{F}^{\mathcal{T}}} \downarrow \uparrow {\scriptstyle \overline{\mathcal{F}^{\mathcal{T}}}} \\
\mathrm{L}^2(\mathbb{R}^d, \widehat{H}) & \xrightarrow[?]{\cdots\cdots} & \mathrm{L}^2(\mathbb{R}^d, \widehat{H}).
\end{array}$$

Since it is too expensive to compute $\Phi(f_{\text{in}}) = k \star f_{\text{in}}$, it would be nice if we could define the missing map $\widehat{\Phi}$, a notion of correlation

$$\widehat{\Phi}\colon \mathrm{L}^2(\mathbb{R}^d, \widehat{H}) \to \mathrm{L}^2(\mathbb{R}^d, \widehat{H})$$

on feature fields, and then we would recover $k \star f_{\text{in}}$ by Fourier inversion. In theory, this is possible; we simply define $\widehat{\Phi}$ as

$$\widehat{\Phi} = \mathcal{F}^{\mathcal{T}} \circ \Phi \circ \overline{\mathcal{F}^{\mathcal{T}}},$$

using $\mathcal{F}^{\mathcal{T}}$ and $\overline{\mathcal{F}^{\mathcal{T}}}$. We can push this approach further, using the fact that the Fourier transform $\mathcal{F}^{\mathcal{T}}_{\rho}$ is continuous and that Φ is a continuous linear map. The following proposition is needed.

Proposition 8.11. *If E and F are two normed vector spaces and if $\Phi\colon E \to F$ is a continuous linear map, then the following properties hold:*

(1) For any convergent series $\sum_{n=1}^{\infty} u_n$ (with $u_n \in E$), the series $\sum_{n=1}^{\infty} \Phi(u_n)$ converges in F and

$$\Phi\left(\sum_{n=1}^{\infty} u_n\right) = \sum_{n=1}^{\infty} \Phi(u_n).$$

(2) For any countable index set Λ, for any summable series $\sum_{\ell \in \Lambda} u_\ell$ (with $u_\ell \in E$), the series $\sum_{\ell \in \Lambda} \Phi(u_\ell)$ is summable in F and

$$\Phi\left(\sum_{\ell \in \Lambda} u_\ell\right) = \sum_{\ell \in \Lambda} \Phi(u_\ell).$$

See Vol. I, Definition D.6 for the definition of a summable series.

Proof. We prove (1) leaving the proof of (2) as an exercise. If $S_n = \sum_{k=1}^{n} u_k$ is a partial sum, since Φ is linear,

$$\Phi(S_n) = \Phi\left(\sum_{k=1}^{\infty} u_k\right) = \sum_{k=1}^{n} \Phi(u_k).$$

Since the sequence (S_n) converges to $\sum_{n=1}^{\infty} u_n$ and since Φ is continuous, the sequence $(\Phi(S_n))$ converges to $\Phi\Big(\sum_{n=1}^{\infty} u_n\Big)$, and thus the sequence of partial sums $\sum_{k=1}^{n} \Phi(u_k)$ also converges to the same limit $\Phi\Big(\sum_{n=1}^{\infty} u_n\Big)$ as claimed. $\square$

Proposition 8.11(2) applies to series indexed by $\mathbb{Z}$. In particular, a summable series $\sum_{n\in\mathbb{Z}} u_n$ corresponds to the case below.

Then for any family $(\widehat{f}_{\rho_1})_{\rho_1\in R(H)}$ of feature fields in $\mathrm{L}^2(\mathbb{R}^d, \widehat{H})$, we have

$$
\begin{aligned}
&\widehat{\Phi}((\widehat{f}_{\rho_1})_{\rho_1\in R(H)}) \\
&= \mathcal{F}^\tau\bigg(\Phi\bigg(\sum_{\rho_1\in R(H)} \overline{\mathcal{F}^\tau}_{\rho_1}(\widehat{f}_{\rho_1})\bigg)\bigg) && \text{by } (\overline{\mathcal{F}^\tau}) \\
&= \mathcal{F}^\tau\bigg(\sum_{\rho_1\in R(H)} \Phi(\overline{\mathcal{F}^\tau}_{\rho_1}(\widehat{f}_{\rho_1}))\bigg) && \text{by Proposition 8.11 for } \Phi \\
&= \bigg(\mathcal{F}^\tau_{\rho_2}\bigg(\sum_{\rho_1\in R(H)} \Phi(\overline{\mathcal{F}^\tau}_{\rho_1}(\widehat{f}_{\rho_1}))\bigg)\bigg)_{\rho_2\in R(H)} && \text{by definition of } \mathcal{F}^\tau \\
&= \bigg(\sum_{\rho_1\in R(H)} \mathcal{F}^\tau_{\rho_2}(\Phi(\overline{\mathcal{F}^\tau}_{\rho_1}(\widehat{f}_{\rho_1})))\bigg)_{\rho_2\in R(H)} && \text{by Proposition 8.11 for } \mathcal{F}^\tau_{\rho_2}.
\end{aligned}
$$

Define $\widehat{\Phi}_{\rho_1}$ and $\widehat{\Phi}_{\rho_2,\rho_1}$ as

$$
\widehat{\Phi}_{\rho_2,\rho_1}(\widehat{f}_{\rho_1}) = \mathcal{F}^\tau_{\rho_2}(\Phi(\overline{\mathcal{F}^\tau}_{\rho_1}(\widehat{f}_{\rho_1}))) \tag{$\widehat{\Phi}_{\rho_2,\rho_1}$}
$$

$$
\widehat{\Phi}_{\rho_1}(\widehat{f}_{\rho_1}) = \sum_{\rho_1\in R(H)} \widehat{\Phi}_{\rho_2,\rho_1}(\widehat{f}_{\rho_1}), \tag{$\widehat{\Phi}_{\rho_1}$}
$$

so that

$$
\widehat{\Phi}((\widehat{f}_{\rho_1})_{\rho_1\in R(H)}) = \big(\widehat{\Phi}_{\rho_1}(\widehat{f}_{\rho_1})\big)_{\rho_2\in R(H)}. \tag{$\widehat{\Phi}$}
$$

It is an interesting and useful fact that the transforms $\widehat{\Phi}_{\rho_2,\rho_1}$ are equivariant with respect to the representations $\mathrm{Ind}_H^G\,\sigma_{\rho_1}$ and $\mathrm{Ind}_H^G\,\sigma_{\rho_2}$. Consider the diagram

$$
\begin{array}{ccccccc}
\mathrm{L}^2(\mathbb{R}^d, \widehat{H})_{\rho_1} & \xrightarrow{\overline{\mathcal{F}^\tau}_{\rho_1}} & \mathrm{L}^2(G) & \xrightarrow{\Phi} & \mathrm{L}^2(G) & \xrightarrow{\mathcal{F}^\tau_{\rho_2}} & \mathrm{L}^2(\mathbb{R}^d, \widehat{H})_{\rho_2} \\
\Big\downarrow{\scriptstyle (\mathrm{Ind}_H^G\,\sigma_{\rho_1})_{(x,h)}} & & \Big\downarrow{\scriptstyle \mathbf{R}_{(x,h)}} & & \Big\downarrow{\scriptstyle \mathbf{R}_{(x,h)}} & & \Big\downarrow{\scriptstyle (\mathrm{Ind}_H^G\,\sigma_{\rho_2})_{(x,h)}} \\
\mathrm{L}^2(\mathbb{R}^d, \widehat{H})_{\rho_1} & \xrightarrow[\overline{\mathcal{F}^\tau}_{\rho_1}]{} & \mathrm{L}^2(G) & \xrightarrow[\Phi]{} & \mathrm{L}^2(G) & \xrightarrow[\mathcal{F}^\tau_{\rho_2}]{} & \mathrm{L}^2(\mathbb{R}^d, \widehat{H})_{\rho_2}.
\end{array}
$$

Since the three squares commute, the outer square also commutes, so we have the following commutative diagram

$$\begin{array}{ccc} L^2(\mathbb{R}^d, \widehat{H})_{\rho_1} & \xrightarrow{\widehat{\Phi}_{\rho_2,\rho_1}} & L^2(\mathbb{R}^d, \widehat{H})_{\rho_2} \\ {\scriptstyle (\mathrm{Ind}_H^G \sigma_{\rho_1})_{(x,h)}}\downarrow & & \downarrow{\scriptstyle (\mathrm{Ind}_H^G \sigma_{\rho_2})_{(x,h)}} \\ L^2(\mathbb{R}^d, \widehat{H})_{\rho_1} & \xrightarrow[\widehat{\Phi}_{\rho_2,\rho_1}]{} & L^2(\mathbb{R}^d, \widehat{H})_{\rho_2}, \end{array}$$

which shows that $\widehat{\Phi}_{\rho_2,\rho_1}$ is equivariant with respect to the representations $\mathrm{Ind}_H^G \sigma_{\rho_1}$ and $\mathrm{Ind}_H^G \sigma_{\rho_2}$.

Suppose the group correlation $\Phi\colon L^2(G) \to L^2(G)$ is given by a kernel k as

$$\Phi(f)(x,h) = \int_{\mathbb{R}^d \rtimes H} k(h^{-1}\cdot(x_1 - x), h^{-1}h_1) f(x_1,h_1)\, d\lambda_H(h_1)\, dx_1.$$

Since

$$[\overline{\mathcal{F}^\tau}_{\rho_1}(\widehat{f}_{\rho_1})](x_1,h_1) = n_{\rho_1}\mathrm{tr}\Big(\widehat{f}_{\rho_1}(x_1)M_{\rho_1}(h_1)\Big),$$

we have

$$\begin{aligned}&\Phi(\overline{\mathcal{F}^\tau}_{\rho_1}(\widehat{f}_{\rho_1}))(x,h)\\ &= \int_{\mathbb{R}^d \rtimes H} k(h^{-1}\cdot(x_1 - x), h^{-1}h_1)\, n_{\rho_1}\mathrm{tr}\Big(\widehat{f}_{\rho_1}(x_1)M_{\rho_1}(h_1)\Big)\, d\lambda_H(h_1)\, dx_1,\end{aligned}$$

and then using Fubini, we have

$$\begin{aligned}&\mathcal{F}^\tau_{\rho_2}[\Phi(\overline{\mathcal{F}^\tau}_{\rho_1}(\widehat{f}_{\rho_1}))](x)\\ &= \int_H\int_{\mathbb{R}^d \rtimes H} k(h^{-1}\cdot(x_1 - x), h^{-1}h_1)\, n_{\rho_1}\mathrm{tr}\Big(\widehat{f}_{\rho_1}(x_1)M_{\rho_1}(h_1)\Big)\, d\lambda_H(h_1)\, dx_1\\ &\quad M_{\rho_2}(h)^*\, d\lambda_H(h)\\ &= \int_{\mathbb{R}^d}\int_H\int_H n_{\rho_1}\mathrm{tr}\Big(\widehat{f}_{\rho_1}(x_1)M_{\rho_1}(h_1)\Big)\, k(h^{-1}\cdot(x_1 - x), h^{-1}h_1)\\ &\quad M_{\rho_2}(h)^*\, d\lambda_H(h)\, d\lambda_H(h_1)\, dx_1.\end{aligned}$$

This suggests defining $\Phi_{\rho_2,\rho_1}\colon \mathbb{R}^d \times \mathrm{M}_{n_{\rho_1}}(\mathbb{C}) \to \mathrm{M}_{n_{\rho_2}}(\mathbb{C})$ by

$$\begin{aligned}&\Phi_{\rho_2,\rho_1}(x_1 - x, A)\\ &\quad = \int_H\int_H n_{\rho_1}\mathrm{tr}\Big(AM_{\rho_1}(h_1)\Big)k(h^{-1}\cdot(x_1 - x), h^{-1}h_1)M_{\rho_2}(h)^*\\ &\qquad d\lambda_H(h)\, d\lambda_H(h_1), \qquad (\Phi_{\rho_2,\rho_1})\end{aligned}$$

where $A \in \mathrm{M}_{n_{\rho_1}}(\mathbb{C})$, so that

$$[\widehat{\Phi}_{\rho_2,\rho_1}(\widehat{f}_{\rho_1})](x) = \mathcal{F}^{\tau}_{\rho_2}[\Phi(\overline{\mathcal{F}^{\tau}}_{\rho_1}(\widehat{f}_{\rho_1}))](x)$$
$$= \int_{\mathbb{R}^d} \Phi_{\rho_2,\rho_1}(x_1 - x, \widehat{f}_{\rho_1}(x_1))\, dx_1. \qquad (\widehat{\Phi}^{\mathrm{bis}}_{\rho_2,\rho_1})$$

In order to go further, we need to express the kernel $\Phi_{\rho_2,\rho_1}(x, A)$ in terms of H-steerable functions on $\mathrm{L}^2(\mathbb{R}^d \rtimes H)$. We will do this explicitly for $\mathbf{SE}(2)$ in Section 8.11. Next, we show how to proceed with $H = \mathbf{SO}(d)$.

By (f^{ρ}) and (str18) in Example 8.9, the Hilbert space $\mathrm{L}^2(\mathbf{SE}(d))$ has a Hilbert basis consisting of functions of the form

$$\left(\overline{m^{(\rho)}_{k_\rho \ell_\rho}(h_1)}\, w_{\rho,k_\rho,\ell_\rho}(h_1^{-1}x)\right)_{1 \le k_\rho, \ell_\rho \le n_\rho,\ \rho \in R(\mathbf{SO}(d))}, \qquad \text{(str30)}$$

with $h_1 \in \mathbf{SO}(d)$ and $x \in \mathbb{R}^d$, where $w_{\rho,k_\rho,\ell_\rho}$ is the sum of a series in the functions

$$e^{-\frac{\|x\|^2}{2}} H_{k_1}(x_1) \cdots H_{k_n}(x_d). \qquad \text{(str31)}$$

Thus, the kernel $k(x_1, h_1)$ can be expressed as the sum of a series

$$k(x_1, h_1) = \sum_{1 \le k_\rho, \ell_\rho \le n_\rho,\ \rho \in R(\mathbf{SO}(d))} \overline{m^{(\rho)}_{k_\rho \ell_\rho}(h_1)}\, w_{\rho,k_\rho,\ell_\rho}(h_1^{-1}x_1). \qquad (k(x_1,h_1))$$

The result to be presented next makes use of the $n_{\rho_2} \times n_{\rho_2}$ matrix $W_{\rho_2}(x_1)$ whose $(k_{\rho_2}, \ell_{\rho_2})$ entry is $w_{\rho_2,k_{\rho_2},\ell_{\rho_2}}(x_1)$. We need to find an expression for $k(h^{-1}(x_1 - x), h^{-1}h_1)$. We have

$$\begin{aligned} &k(h^{-1}(x_1 - x), h^{-1}h_1) \\ &= \sum_{1 \le k_\rho, \ell_\rho \le n_\rho,\ \rho \in R(\mathbf{SO}(d))} \overline{m^{(\rho)}_{k_\rho \ell_\rho}(h^{-1}h_1)}\, w_{\rho,k_\rho,\ell_\rho}((h^{-1}h_1)^{-1}h^{-1}(x_1 - x)) \\ &= \sum_{1 \le k_\rho, \ell_\rho \le n_\rho,\ \rho \in R(\mathbf{SO}(d))} \overline{m^{(\rho)}_{k_\rho \ell_\rho}(h^{-1}h_1)}\, w_{\rho,k_\rho,\ell_\rho}(h_1^{-1}(x_1 - x)). \end{aligned}$$

As explained earlier and using Theorem 4.6(2), we have

$$\begin{aligned} \overline{m^{(\rho)}_{k_\rho \ell_\rho}(h^{-1}h_1)} &= (1/n_\rho) \sum_{j_\rho = 1}^{n_\rho} \overline{m^{(\rho)}_{k_\rho j_\rho}(h^{-1})}\, \overline{m^{(\rho)}_{j_\rho \ell_\rho}(h_1)} \\ &= (1/n_\rho) \sum_{j_\rho = 1}^{n_\rho} m^{(\rho)}_{j_\rho k_\rho}(h)\, \overline{m^{(\rho)}_{j_\rho \ell_\rho}(h_1)}, \end{aligned}$$

which yields

$$
\begin{aligned}
&k(h^{-1}(x_1 - x), h^{-1}h_1) \\
&= \sum_{1 \le k_\rho, \ell_\rho \le n_\rho,\, \rho \in R(\mathbf{SO}(d))} \overline{m^{(\rho)}_{k_\rho \ell_\rho}(h^{-1}h_1)}\, w_{\rho,k_\rho,\ell_\rho}(h_1^{-1}(x_1 - x)) \\
&= \sum_{\substack{1 \le k_\rho, \ell_\rho, j_\rho \le n_\rho \\ \rho \in R(\mathbf{SO}(d))}} (1/n_\rho)\, m^{(\rho)}_{j_\rho k_\rho}(h)\, \overline{m^{(\rho)}_{j_\rho \ell_\rho}(h_1)}\, w_{\rho,k_\rho,\ell_\rho}(h_1^{-1}(x_1 - x)).
\end{aligned}
$$

Plugging the above expression in

$$
\begin{aligned}
&\Phi_{\rho_2,\rho_1}(x_1 - x, A) \\
&= \int_H \int_H n_{\rho_1} \mathrm{tr}\Big(AM_{\rho_1}(h_1)\Big) k(h^{-1}(x_1 - x), h^{-1}h_1) M_{\rho_2}(h)^* \\
&\quad d\lambda_H(h)\, d\lambda_H(h_1),
\end{aligned}
$$

we get

$$
\begin{aligned}
&\Phi_{\rho_2,\rho_1}(x_1 - x, A) \\
&= \sum_{\substack{1 \le k_\rho, \ell_\rho, j_\rho \le n_\rho \\ \rho \in R(\mathbf{SO}(d))}} \int_H \int_H n_{\rho_1}\, \mathrm{tr}\Big(AM_{\rho_1}(h_1)\Big) (1/n_\rho)\, \overline{m^{(\rho)}_{j_\rho \ell_\rho}(h_1)} \\
&\qquad w_{\rho,k_\rho,\ell_\rho}(h_1^{-1}(x_1 - x))\, m^{(\rho)}_{j_\rho k_\rho}(h)\, M_{\rho_2}(h)^*\, d\lambda_H(h)\, d\lambda_H(h_1).
\end{aligned}
$$

Since the functions $m^{(\rho)}_{j_\rho k_\rho}$ and $m^{(\rho_2)}_{j_{\rho_2} k_{\rho_2}}$ are orthogonal for $\rho \neq \rho_2$ by Theorem 4.6(1), only the terms for which $\rho = \rho_2$ survive, so we get

$$
\begin{aligned}
&\Phi_{\rho_2,\rho_1}(x_1 - x, A) \\
&= \sum_{1 \le k_{\rho_2}, \ell_{\rho_2}, j_{\rho_2} \le n_{\rho_2}} \int_H n_{\rho_1}\, \mathrm{tr}\Big(AM_{\rho_1}(h_1)\Big)\, \overline{m^{(\rho_2)}_{j_{\rho_2} \ell_{\rho_2}}(h_1)} \\
&\quad w_{\rho_2,k_{\rho_2},\ell_{\rho_2}}(h_1^{-1}(x_1 - x)) \int_H (1/n_{\rho_2})\, m^{(\rho_2)}_{j_{\rho_2} k_{\rho_2}}(h)\, M_{\rho_2}(h)^*\, d\lambda_H(h)\, d\lambda_H(h_1).
\end{aligned}
$$

Now the $(k'_{\rho_2}, j'_{\rho_2})$-entry in the matrix $M_{\rho_2}(h)^*$ is $\overline{m^{(\rho_2)}_{j'_{\rho_2} k'_{\rho_2}}(h)}$, and since by Theorem 4.6(1,3) the functions $m^{(\rho_2)}_{j'_{\rho_2} k'_{\rho_2}}$ and $m^{(\rho_2)}_{j_{\rho_2} k_{\rho_2}}$ are orthogonal unless $k'_{\rho_2} = k_{\rho_2}$ and $j'_{\rho_2} = j_{\rho_2}$, in which case $\langle m^{(\rho_2)}_{j_{\rho_2} k_{\rho_2}}, m^{(\rho_2)}_{j_{\rho_2} k_{\rho_2}} \rangle = n_{\rho_2}$, the inner integral evaluates to

$$
\int_H (1/n_{\rho_2})\, m^{(\rho_2)}_{j_{\rho_2} k_{\rho_2}}(h)\, M_{\rho_2}(h)^*\, d\lambda_H(h) = E_{k_{\rho_2} j_{\rho_2}},
$$

the matrix with 1 in the (k_{ρ_2}, j_{ρ_2}) entry and 0 otherwise, so

$$\begin{aligned}
&\Phi_{\rho_2,\rho_1}(x_1 - x, A)\\
&= \int_H n_{\rho_1} \operatorname{tr}\Big(AM_{\rho_1}(h_1)\Big) \sum_{\ell_{\rho_2}=1}^{n_{\rho_2}} \sum_{1\leq k_{\rho_2}, j_{\rho_2}\leq n_{\rho_2}} w_{\rho_2,k_{\rho_2},\ell_{\rho_2}}(h_1^{-1}(x_1 - x))\\
&\quad \overline{m^{(\rho_2)}_{j_{\rho_2}\ell_{\rho_2}}(h_1)}\, E_{k_{\rho_2}j_{\rho_2}}\, d\lambda_H(h_1)\\
&= \int_H n_{\rho_1} \operatorname{tr}\Big(AM_{\rho_1}(h_1)\Big) \sum_{1\leq k_{\rho_2}, j_{\rho_2}\leq n_{\rho_2}} \sum_{\ell_{\rho_2}=1}^{n_{\rho_2}} w_{\rho_2,k_{\rho_2},\ell_{\rho_2}}(h_1^{-1}(x_1 - x))\\
&\quad m^{(\rho_2)}_{\ell_{\rho_2}j_{\rho_2}}(h_1^{-1})\, E_{k_{\rho_2}j_{\rho_2}}\, d\lambda_H(h_1)\\
&= \int_H n_{\rho_1} \operatorname{tr}\Big(AM_{\rho_1}(h_1)\Big)\\
&\quad \left(\sum_{\ell_{\rho_2}=1}^{n_{\rho_2}} w_{\rho_2,k_{\rho_2},\ell_{\rho_2}}(h_1^{-1}(x_1 - x))\, m^{(\rho_2)}_{\ell_{\rho_2}j_{\rho_2}}(h_1^{-1})\right)_{1\leq k_{\rho_2}, j_{\rho_2}\leq n_{\rho_2}} d\lambda_H(h_1)\\
&= \int_H n_{\rho_1} \operatorname{tr}\Big(AM_{\rho_1}(h_1)\Big) W_{\rho_2}(h_1^{-1}(x_1 - x)) M^*_{\rho_2}(h_1)\, d\lambda_H(h_1).
\end{aligned}$$

In summary,

$$\begin{aligned}
&\Phi_{\rho_2,\rho_1}(x_1 - x, A)\\
&\quad = \int_H n_{\rho_1} \operatorname{tr}\Big(AM_{\rho_1}(h_1)\Big) W_{\rho_2}(h_1^{-1}(x_1 - x)) M^*_{\rho_2}(h_1)\, d\lambda_H(h_1), \qquad (*_{\Phi_{\rho_2,\rho_1}})
\end{aligned}$$

and

$$\begin{aligned}
[\widehat{\Phi}_{\rho_2,\rho_1}(\widehat{f}_{\rho_1})](x) &= \mathcal{F}^{\tau}_{\rho_2}[\Phi(\overline{\mathcal{F}^{\tau}}_{\rho_1}(\widehat{f}_{\rho_1}))](x)\\
&= \int_{\mathbb{R}^d} \Phi_{\rho_2,\rho_1}(x_1 - x, \widehat{f}_{\rho_1}(x_1))\, dx_1, \qquad (*_{\widehat{\Phi}_{\rho_2,\rho_1}})
\end{aligned}$$

where $W_{\rho_2}(x_1)$ is the $n_{\rho_2} \times n_{\rho_2}$ matrix whose $(k_{\rho_2}, \ell_{\rho_2})$ entry is $w_{\rho_2,k_{\rho_2},\ell_{\rho_2}}(x_1)$ introduced just after $(k(x_1, h_1))$.

It is not hard to show that the above results can be generalized to the situation where H is a compact matrix group acting on $\mathbb{R}^d$ by multiplication.

In the special case where $d = 2$ and $H = \mathbf{SO}(2)$, we can use polar coordinates and view the functions in $\mathrm{L}^2(\mathbf{SE}(2))$ as functions $f((\|x\|, \alpha), \theta)$. In this case, by (str14) from Example 8.6, a Hilbert basis consists of the functions of the form

$$e^{-im\theta} e^{ik(\theta-\alpha_x)} w_{m,k}(\|x\|), \quad m, k \in \mathbb{Z}.$$

In this special case $\ell_\rho = \rho \in \mathbb{Z}$, there is no index k_ρ since $n_\rho = 1$, $h_1 = e^{i\theta'}$, $m^{(\rho)}_{k_\rho \ell_\rho}(h_1) = e^{-i\rho\theta'}$, $M_\rho(\theta') = e^{i\rho\theta'}$, $m = \rho_2$, and by (str30) and (str31) the matrix $W_{\rho_2}(\|x_1 - x\|, \alpha_{x_1-x} - \theta')$ consists of the series

$$\sum_{k=\infty}^{\infty} e^{-ik(\theta' - \alpha_{x_1-x})} w_{\rho_2,k}(\|x_1 - x\|).$$

It follows that we need to evaluate the integral $(*_{\Phi_{\rho_2,\rho_1}})$;

$$\begin{aligned}\sum_{k=\infty}^{\infty} &\int_{\mathbf{SO}(2)} e^{i\rho_1\theta'} e^{-ik(\theta' - \alpha_{x_1-x})} w_{\rho_2,k}(\|x_1 - x\|) e^{-i\rho_2\theta'}\, d\theta' \\ &= \sum_{k=\infty}^{\infty} e^{ik\alpha_{x_1-x}} w_{\rho_2,k}(\|x_1 - x\|) \int_{\mathbf{SO}(2)} e^{i(\rho_1 - \rho_2 - k)\theta'}\, d\theta' \\ &= e^{-i(\rho_2 - \rho_1)\alpha_{x_1-x}} w_{\rho_2,\rho_1-\rho_2}(\|x_1 - x\|).\end{aligned}$$

In conclusion, we obtain the kernel

$$\Phi_{\rho_2,\rho_1}(x_1 - x, A) = Ae^{-i(\rho_2 - \rho_1)\alpha_{x_1-x}} w_{\rho_2,\rho_1-\rho_2}(\|x_1 - x\|).$$

Since this is a scalar kernel that simply multiplies by A, we can express it as

$$\Phi_{\rho_2,\rho_1}(x_1 - x) = e^{-i(\rho_2 - \rho_1)\alpha_{x_1-x}} w_{\rho_2,\rho_1-\rho_2}(\|x_1 - x\|).$$

We derive this formula in full detail in the next section. The second index $\rho_1 - \rho_2$ is different from what we get in the next section because the computation makes use of polar coordinates early on. If we index $w_{m,k}$ as $w_{m,k+m}$, we find the same term $w_{\rho_2,\rho_1}(\|x_1 - x\|)$.

8.11 Harmonic Nets

In the special case where $X = \mathbb{R}^2$, $H = \mathbf{SO}(2)$ and $G = \mathbf{SE}(2) = \mathbb{R}^2 \rtimes \mathbf{SO}(2)$, it is possible to construct the transform $\widehat{\Phi}$ explicitly. This case is known in the literature as *harmonic nets*. We follow Eric Bekkers' YouTube video Lecture 2.7 with a few corrections.

Recall that group correlation on $\mathbf{SE}(2)$ is given by

$$\Phi(f)(x,\theta) = \int_{\mathbf{SE}(2)} k(R_{-\theta}(x' - x), \theta' - \theta) f(x', \theta')\, dx'\, d\theta'. \qquad (*_{27})$$

Given a sequence $\widehat{f} = (\widehat{f}_\ell)_{\ell\in\mathbb{Z}}$ of functions $\widehat{f}_\ell \colon \mathbb{R}^2 \to \mathbb{C}$, such that $(\widehat{f}_\ell(x))_{\ell\in\mathbb{Z}} \in \ell^2(\mathbb{C})$ for all $x \in \mathbb{R}^2$ and

$$\left(\sum_{\ell=-\infty}^{\infty} |\widehat{f}_\ell(-)|^2\right)^{1/2} \in \mathrm{L}^2(\mathbb{R}^2),$$

we define the Fourier cotransform $\overline{\mathcal{F}^\tau}(\widehat{f})$ of $\widehat{f}$ by

$$\overline{\mathcal{F}^\tau}(\widehat{f})(x', \theta') = \sum_{\ell=-\infty}^{\infty} \widehat{f}_\ell(x') e^{i\ell\theta'}. \qquad (*_{28})$$

Now $\overline{\mathcal{F}^\tau}(\widehat{f}) \in \mathrm{L}^2(\mathbf{SE}(2))$ so $\Phi(\overline{\mathcal{F}^\tau}(\widehat{f}))$ makes sense, and then for every $x \in \mathbb{R}^2$, we can compute the Fourier transform $\mathcal{F}^\tau(\Phi(\overline{\mathcal{F}^\tau}(\widehat{f}))(x, -))$ of $\Phi(\overline{\mathcal{F}^\tau}(\widehat{f}))(x, -)$ by

$$\mathcal{F}^\tau(\Phi(\overline{\mathcal{F}^\tau}(\widehat{f}))(x, -))_m = \int_{\mathbf{SO}(2)} \Phi(\overline{\mathcal{F}^\tau}(\widehat{f}))(x, \theta) e^{-im\theta}\, d\theta, \quad m \in \mathbb{Z}. \qquad (*_{29})$$

Then the correlation transform $\widehat{\Phi}$ defined on sequences of functions $\widehat{f} = (\widehat{f}_\ell)_{\ell\in\mathbb{Z}}$ (with $\widehat{f}_\ell \colon \mathbb{R}^2 \to \mathbb{C}$) that we are seeking is the family of functions $(\widehat{\Phi}(\widehat{f})_m)_{m\in\mathbb{Z}}$ given by

$$\widehat{\Phi}(\widehat{f})_m(x) = \mathcal{F}^\tau(\Phi(\overline{\mathcal{F}^\tau}(\widehat{f}))(x, -))_m.$$

Substituting the expression $(*_{28})$ in $(*_{29})$, we obtain

$$\begin{aligned}\mathcal{F}^\tau(\Phi(\overline{\mathcal{F}^\tau}(\widehat{f}))(x, -))_m &= \int_{\mathbf{SO}(2)} \Phi\Big(\sum_{\ell=-\infty}^{\infty} \widehat{f}_\ell(x') e^{i\ell\theta'}\Big)(x, \theta) e^{-im\theta}\, d\theta, \ m \in \mathbb{Z} \\ &= \sum_{\ell=-\infty}^{\infty} \int_{\mathbf{SO}(2)} \Phi(\widehat{f}_\ell(x') e^{i\ell\theta'})(x, \theta) e^{-im\theta}\, d\theta, \ \ m \in \mathbb{Z},\end{aligned}$$

where we used Proposition 8.11 to swap the infinite sum ($\mathcal{F}^\tau$ and Φ are linear and continuous). Then, using $(*_{27})$ and Fubini, we obtain

$$\begin{aligned}&\sum_{\ell=-\infty}^{\infty} \int_{\mathbf{SO}(2)} \Phi(\widehat{f}_\ell(x') e^{i\ell\theta'})(x, \theta) e^{-im\theta}\, d\theta \\ &= \sum_{\ell=-\infty}^{\infty} \int_{\mathbb{R}^2}\int_{\mathbf{SO}(2)}\int_{\mathbf{SO}(2)} k(R_{-\theta}(x' - x), \theta' - \theta) \widehat{f}_\ell(x') e^{i\ell\theta'} e^{-im\theta}\, d\theta\, d\theta'\, dx'.\end{aligned} \qquad (*_{30})$$

This shows that we need to figure out what is the term

$$\Phi_{m,\ell}(x' - x) = \int_{\mathbf{SO}(2)}\int_{\mathbf{SO}(2)} k(R_{-\theta}(x' - x), \theta' - \theta) e^{-im\theta}\, d\theta\, e^{i\ell\theta'}\, d\theta', \quad (*_{31})$$

since we obtain

$$\widehat{\Phi}(\widehat{f})_m(x) = \sum_{\ell=-\infty}^{\infty} \int_{\mathbb{R}^2} \Phi_{m,\ell}(x' - x) \widehat{f}_\ell(x')\, dx'. \qquad (*_{32})$$

At this stage, we use the result of Example 8.6 to express the kernel $k \in \mathrm{L}^2(\mathbf{SE}(2))$ as a sum of $\mathbf{SO}(2)$-steerable functions using polar coordinates $x = (\|x\|_2, \alpha_x)$, in the form

$$k(x,\psi) = \sum_{n=-\infty}^{\infty} k_n(x,\psi) \tag{$*_{33}$}$$

$$k_n(x,\psi) = e^{-in\alpha_x} w_n(\|x\|, \psi - \alpha_x) \tag{$*_{34}$}$$

as in (str15), where w_n is a function on $\mathbb{R}^+ \times \mathbf{SO}(2)$ that can also be expanded in terms of the basis functions $e^{-ik\theta}e^{-\frac{r^2}{2}}H_m(r)$, but right now we do need to do this. By (str10), steerability means that[4]

$$k_n(R_{-\theta}x, \psi - \theta) = e^{in\theta}k_n(x,\psi), \tag{$*_{35}$}$$

which yields[5]

$$\begin{aligned} k(R_{-\theta}(x'-x), \theta'-\theta) &= \sum_{n=-\infty}^{\infty} k_n(R_{-\theta}(x'-x), \theta'-\theta) \\ &= \sum_{n=-\infty}^{\infty} e^{in\theta}k_n(x'-x, \theta') \\ &= \sum_{n=-\infty}^{\infty} e^{in\theta}e^{-in\alpha_{x'-x}} w_n(\|x'-x\|, \theta' - \alpha_{x'-x}). \end{aligned} \tag{$*_{36}$}$$

Going back to the expression for $\Phi_{m,\ell}(x'-x)$ in $(*_{28})$, we obtain

$$\begin{aligned} &\Phi_{m,\ell}(x'-x) \\ &= \int_{\mathbf{SO}(2)}\int_{\mathbf{SO}(2)}\left(\sum_{n=-\infty}^{\infty} e^{in\theta}e^{-in\alpha_{x'-x}} w_n(\|x'-x\|, \theta'-\alpha_{x'-x})\right) \\ &\quad e^{-im\theta}\, d\theta\, e^{i\ell\theta'}\, d\theta' \\ &= \int_{\mathbf{SO}(2)}\left(\sum_{n=-\infty}^{\infty}\int_{\mathbf{SO}(2)} w_n(\|x'-x\|, \theta'-\alpha_{x'-x})e^{-in\alpha_{x'-x}}e^{-i(m-n)\theta}\, d\theta\right) \\ &\quad e^{i\ell\theta'}\, d\theta' \end{aligned} \tag{$*_{37}$}$$

using Proposition 8.11 to swap the infinite sum ($\mathcal{F}^\tau$ is linear and continuous). Since

$$\int_{\mathbf{SO}(2)} e^{-i(m-n)\theta}\, d\theta = \begin{cases} 0, & \text{if } m \neq n \\ 1, & \text{if } m = n, \end{cases}$$

[4]Bekkers has $\psi - \theta$ instead of ψ in the term $k_n(x,\psi)$, so his k_n is not steerable.
[5]Bekkers has an extra θ and a missing $\alpha_{x'-x}$ in the term $w_n(\|x'-x\|, \theta' - \alpha_{x'-x})$.

there is only one term in the sum for $n = m$, so we get

$$\Phi_{m,\ell}(x' - x) = \left(\int_{\mathbf{SO}(2)} w_m(\|x' - x\|, \theta' - \alpha_{x'-x}) e^{i\ell\theta'} \, d\theta' \right) e^{-im\alpha_{x'-x}}. \tag{$*_{38}$}$$

Observe that the term

$$\int_{\mathbf{SO}(2)} w_m(\|x' - x\|, \theta' - \alpha_{x'-x}) e^{i\ell\theta'} \, d\theta'$$

is the Fourier cotransform of the function $\lambda_{\alpha_{x'-x}} w_m(\|x' - x\|, -)$, so the shift property (see Vol. I, Proposition 10.19(3)), implies that

$$\begin{aligned} &\int_{\mathbf{SO}(2)} w_m(\|x' - x\|, \theta' - \alpha_{x'-x}) e^{i\ell\theta'} \, d\theta' \\ &\qquad = e^{i\ell\alpha_{x'-x}} \int_{\mathbf{SO}(2)} w_m(\|x' - x\|, \theta') e^{i\ell\theta'} \, d\theta', \end{aligned} \tag{$*_{39}$}$$

which is also easy to verify by making the change of variable $\psi = \theta' - \alpha_{x'-x}$. But the Fourier cotransform and the Fourier transform of a function g are related by the equation

$$\overline{\mathcal{F}^{\tau}}(g)_\ell = \overline{\mathcal{F}^{\tau}(\overline{g})_\ell},$$

(see Vol. I, Proposition 10.19(1)), so we have

$$\int_{\mathbf{SO}(2)} w_m(\|x' - x\|, \theta') e^{i\ell\theta'} \, d\theta' = \overline{\mathcal{F}^{\tau}(\overline{w_m(\|x' - x\|, -)})_\ell},$$

It is simpler to stick to the Fourier cotransform since it will be evaluated anyway when we express $w_n(\|x' - x\|, \theta')$ in terms of the basis functions $e^{-ik\theta'}$ and $e^{-\frac{r^2}{2}} H_q(r)$ as in Example 8.6. From $(*_{38})$ and $(*_{39})$, we obtain

$$\Phi_{m,\ell}(x' - x) = \overline{\mathcal{F}^{\tau}}(w_m(\|x' - x\|, -))_\ell \, e^{-i(m-\ell)\alpha_{x'-x}}. \tag{$*_{40}$}$$

In summary, we proved that

$$\widehat{\Phi}(\widehat{f})_m(x) = \sum_{\ell=-\infty}^{\infty} \int_{\mathbb{R}^2} \Phi_{m,\ell}(x' - x) \widehat{f}_\ell(x') \, dx' \tag{$*_{41}$}$$

$$\Phi_{m,\ell}(x' - x) = \overline{\mathcal{F}^{\tau}}(w_m(\|x' - x\|, -))_\ell \, e^{-i(m-\ell)\alpha_{x'-x}}. \tag{$*_{42}$}$$

The above formula can be further simplified using the fact shown in Example 8.6, namely that the function $(r, \theta') \mapsto w_m(r, \theta')$ can be expressed as a series of the form

$$w_m(r, \theta') = \sum_{k=-\infty}^{\infty} e^{-ik\theta'} w_{m,k}^{\text{rad}}(r), \tag{$*_{43}$}$$

where $w^{\mathrm{rad}}_{m,k}(r)$ is a series whose terms are the functions $e^{-\frac{r^2}{2}}H_q(r)$; see (str15). Then we can compute the Fourier cotransform $\overline{\mathcal{F}^{\tau}}(w_m(\|x'-x\|,-))_\ell$, which is given by

$$\begin{aligned}
&\overline{\mathcal{F}^{\tau}}(w_m(\|x'-x\|,-))_\ell \\
&\quad= \int_{\mathbf{SO}(2)} w_m(\|x'-x\|,\theta')e^{i\ell\theta'}\,d\theta' \\
&\quad= \int_{\mathbf{SO}(2)} \left(\sum_{k=-\infty}^{\infty} e^{-ik\theta'} w^{\mathrm{rad}}_{m,k}(\|x'-x\|)\right) e^{i\ell\theta'}\,d\theta' \\
&\quad= \sum_{k=-\infty}^{\infty} w^{\mathrm{rad}}_{m,k}(\|x'-x\|)\int_{\mathbf{SO}(2)} e^{i(\ell-k)\theta'}\,d\theta' = w^{\mathrm{rad}}_{m,\ell}(\|x'-x\|),
\end{aligned}$$

where we used Proposition 8.11 to swap the infinite sum (the Fourier cotransform is linear and continuous). Finally, we obtain the formulae

$$\widehat{\Phi}(\widehat{f})_m(x) = \sum_{\ell=-\infty}^{\infty}\int_{\mathbb{R}^2}\Phi_{m,\ell}(x'-x)\widehat{f}_\ell(x')\,dx' \tag{$*_{44}$}$$

$$\Phi_{m,\ell}(x'-x) = w^{\mathrm{rad}}_{m,\ell}(\|x'-x\|)\,e^{-i(m-\ell)\alpha_{x'-x}}. \tag{$*_{45}$}$$

It is instructive to see how the original group correlation $\Phi(\widehat{f})(x,\theta)$ is expressed for an input function given as a Fourier series

$$\widehat{f}(x,\theta) = \sum_{\ell=-\infty}^{\infty}\widehat{f}_\ell(x)e^{i\ell\theta}$$

when we express the kernel k in terms of steerable functions as we did before. Since we are not applying a Fourier transform the blue terms involving $e^{-im\theta}\,d\theta$ disappear, and using Proposition 8.11, we get

$$\begin{aligned}
\Phi(\widehat{f})(x,\theta) &= \int_{\mathbf{SE}(2)} k(R_{-\theta}(x'-x),\theta'-\theta)\left(\sum_{\ell=-\infty}^{\infty}\widehat{f}_\ell(x')e^{i\ell\theta'}\right)dx'\,d\theta' \\
&= \sum_{\ell=-\infty}^{\infty}\int_{\mathbb{R}^2}\int_{\mathbf{SO}(2)} k(R_{-\theta}(x'-x),\theta'-\theta)\widehat{f}_\ell(x')e^{i\ell\theta'}\,d\theta'\,dx'.
\end{aligned} \tag{$*_{46}$}$$

This shows that we need to figure out what is the term

$$\Phi_\ell(x'-x,\theta) = \int_{\mathbf{SO}(2)} k(R_{-\theta}(x'-x),\theta'-\theta)e^{i\ell\theta'}\,d\theta', \tag{$*_{47}$}$$

since we have

$$\Phi(\widehat{f})(x,\theta) = \sum_{\ell=-\infty}^{\infty} \int_{\mathbb{R}^2} \Phi_\ell(x'-x,\theta)\widehat{f}_\ell(x')\,dx'. \qquad (*_{48})$$

Retracing our steps, we leave it as an exercise to show that

$$\Phi_\ell(x'-x,\theta) = \sum_{n=-\infty}^{\infty} w_{n,\ell}^{\mathrm{rad}}(\|x'-x\|)\, e^{-i(n-\ell)\alpha_{x'-x}} e^{in\theta}. \qquad (*_{49})$$

When we take the Fourier transform of $\Phi_\ell(x'-x,-)$ (with x, x' fixed), we have

$$\begin{aligned}\Phi_\ell(x'-x,-)_m &= \int_{\mathbf{SO}(2)} \sum_{n=-\infty}^{\infty} w_{n,\ell}^{\mathrm{rad}}(\|x'-x\|)\, e^{-i(n-\ell)\alpha_{x'-x}} e^{in\theta} e^{-im\theta}\, d\theta \\ &= w_{m,\ell}^{\mathrm{rad}}(\|x'-x\|)\, e^{-i(m-\ell)\alpha_{x'-x}} = \Phi_{m,\ell}(x'-x), \qquad (*_{50})\end{aligned}$$

confirming that we go back and forth from the group correlation Φ to the steerable group correlation $\widehat{\Phi}$ *via* the Fourier transform and the Fourier cotransform. Also note that steerable correlation is much cheaper than group correlation since to compute the group correlation term $\Phi_\ell(x'-x,\theta)$ requires integration involving the radial functions $w_{n,\ell}^{\mathrm{rad}}(\|x'-x\|)$ for all n.

8.12 Equivariant Correlation G-Kernels When $G = \mathbb{R}^d \rtimes H$

In Section 8.10, we solved the problem of finding a notion of equivariant group correlation for feature fields $\widehat{f}_\rho \in \mathrm{L}^2(\mathbb{R}^d, \widehat{H})$, which are functions $\widehat{f}_\rho \colon \mathbb{R}^d \to \mathrm{M}_{n_\rho}(\mathbb{C})$ that transform under the representation $\sigma_\rho \colon H \to \mathbf{U}(\mathrm{M}_{n_\rho}(\mathbb{C}))$, with $\sigma_\rho = \mathrm{Hom}(M_\rho, \mathrm{id})$ (see Proposition 8.9). For this, we used the Fourier transform $\mathcal{F}^\tau$ and the Fourier cotransform $\overline{\mathcal{F}^\tau}$ defined in Section 8.9. Recall that given a correlation kernel k on $\mathrm{L}^2(G)$, we have the group correlation Φ on $\mathrm{L}^2(G)$ (where $G = \mathbb{R}^d \rtimes H$) given by $f_{\mathrm{out}} = \Phi(f_{\mathrm{in}}) = k \star f_{\mathrm{in}}$. The correlation $\widehat{\Phi}$ on feature fields in $\mathrm{L}^2(\mathbb{R}^d, \widehat{H})$ is the map that makes the following diagram commute:

$$\begin{array}{ccc} \mathrm{L}^2(G) & \xrightarrow{\;\Phi\;} & \mathrm{L}^2(G) \\ \mathcal{F}^\tau \Big\downarrow \Big\uparrow \overline{\mathcal{F}^\tau} & & \mathcal{F}^\tau \Big\downarrow \Big\uparrow \overline{\mathcal{F}^\tau} \\ \mathrm{L}^2(\mathbb{R}^d, \widehat{H}) & \xrightarrow[\;\widehat{\Phi}\;]{} & \mathrm{L}^2(\mathbb{R}^d, \widehat{H}). \end{array}$$

In Section 8.10, we showed how to construct $\widehat{\Phi}$ by expressing the kernel k in terms of a basis of steerable functions in $\mathrm{L}^2(G)$.

Because the group correlation Φ is equivariant with respect to the left regular representation $\mathbf{R}$ (on $\mathrm{L}^2(G)$), the components $\widehat{\Phi}_{\rho_2,\rho_1}$ of $\widehat{\Phi}$ are equivariant with respect to the representations $\mathrm{Ind}_H^G\,\sigma_{\rho_1}$ and $\mathrm{Ind}_H^G\,\sigma_{\rho_2}$, namely the following diagram commutes.

$$\begin{array}{ccc} \mathrm{L}^2(\mathbb{R}^d, \widehat{H})_{\rho_1} & \xrightarrow{\widehat{\Phi}_{\rho_2,\rho_1}} & \mathrm{L}^2(\mathbb{R}^d, \widehat{H})_{\rho_2} \\ {\scriptstyle (\mathrm{Ind}_H^G\,\sigma_{\rho_1})_{(x,h)}}\Big\downarrow & & \Big\downarrow{\scriptstyle (\mathrm{Ind}_H^G\,\sigma_{\rho_2})_{(x,h)}} \\ \mathrm{L}^2(\mathbb{R}^d, \widehat{H})_{\rho_1} & \xrightarrow[\widehat{\Phi}_{\rho_2,\rho_1}]{} & \mathrm{L}^2(\mathbb{R}^d, \widehat{H})_{\rho_2}. \end{array}$$

Practice shows that it is desirable to design more general group correlations that are equivariant with respect to other representations besides the left regular representation and to consider feature fields that transform under representations other than the representations $\mathrm{Hom}(M_\rho, \mathrm{id})$.

A first generalization is to have two feature fields spaces $\mathbf{FF}(\mathbb{R}^d, H, \sigma_{\mathrm{in}} \colon H \to \mathbf{U}(\mathcal{H}_{\mathrm{in}}))$ and $\mathbf{FF}(\mathbb{R}^d, H, \sigma_{\mathrm{out}} \colon H \to \mathbf{U}(\mathcal{H}_{\mathrm{out}}))$ associated with an input representation σ_{in} and an output representation σ_{out}, where $\mathcal{H}_{\mathrm{in}}$ and $\mathcal{H}_{\mathrm{out}}$ are two finite-dimensional vector spaces equipped with a hermitian inner product, and what we are seeking is a linear G-equivariant map $\widehat{\Phi}$ between these spaces. We assume that feature fields $f \colon \mathbb{R}^d \to \mathcal{H}_{\mathrm{in}}$ are functions in $\mathrm{L}^2(\mathbb{R}^d, \mathcal{H}_{\mathrm{in}})$, and similarly, for feature fields $f \colon \mathbb{R}^d \to \mathcal{H}_{\mathrm{out}}$ (see Definition 6.25).

It turns out that in order to guarantee the existence of G-equivariant continuous linear maps $\widehat{\Phi} \colon \mathbf{FF}(\mathbb{R}^d, H, \sigma_{\mathrm{in}}) \to \mathbf{FF}(\mathbb{R}^d, H, \sigma_{\mathrm{out}})$ specified by kernels, we need to consider the special continuous linear maps consisting of the *Hilbert–Schmidt operators* discussed in Section 2.5.

To say that $\widehat{\Phi}$ is G-equivariant means that the following diagram

$$\begin{array}{ccc} \mathbf{FF}(\mathbb{R}^d, H, \sigma_{\mathrm{in}}) & \xrightarrow{\widehat{\Phi}} & \mathbf{FF}(\mathbb{R}^d, H, \sigma_{\mathrm{out}}) \\ {\scriptstyle (\mathrm{Ind}_H^G\,\sigma_{\mathrm{in}})_{(x,h)}}\Big\downarrow & & \Big\downarrow{\scriptstyle (\mathrm{Ind}_H^G\,\sigma_{\mathrm{out}})_{(x,h)}} \\ \mathbf{FF}(\mathbb{R}^d, H, \sigma_{\mathrm{in}}) & \xrightarrow[\widehat{\Phi}]{} & \mathbf{FF}(\mathbb{R}^d, H, \sigma_{\mathrm{out}}) \end{array}$$

commutes for all $(x, h) \in G = \mathbb{R}^d \rtimes H$, with

$$[(\mathrm{Ind}_H^G\,\sigma_{\mathrm{in}})_{(x,h)} f_{\mathrm{in}}](t) = \sigma_{\mathrm{in}}(h)(f_{\mathrm{in}}(h^{-1} \cdot (t - x))), \quad f_{\mathrm{in}} \colon \mathbb{R}^d \to \mathcal{H}_{\mathrm{in}}$$
$$[(\mathrm{Ind}_H^G\,\sigma_{\mathrm{out}})_{(x,h)} f_{\mathrm{out}}](t) = \sigma_{\mathrm{out}}(h)(f_{\mathrm{out}}(h^{-1} \cdot (t - x))), \; f_{\mathrm{out}} \colon \mathbb{R}^d \to \mathcal{H}_{\mathrm{out}},$$

for all $t \in \mathbb{R}^d$, as in $(\dagger_2)$.

A complete solution to this problem was given in a sequence of remarkable papers by Weiler, Geiger, Weilling, Boomsma and Cohen [Weiler *et al.* (2018)] (for $\mathbf{SE}(3)$), Weiler and Cesa [Weiler and Cesa (2019)] (for $\mathbf{E}(2)$), Lang and Weiler [Lang and Weiler (2019)] (for a homogeneous space X induced by a transitive action of a compact group H), Cesa, Lang and Weiler [Cesa *et al.* (2022)] (for $\mathbf{E}(3)$), and Cohen, Geiger and Weiler [Cohen *et al.* (2019)] (feature fields on homogeneous spaces).[6]

It is shown by Weiler, Geiger, Weilling, Boomsma and Cohen [Weiler *et al.* (2018)] that in the case where $H = \mathbf{SO}(d)$, such a map is given by a kernel $K\colon \mathbb{R}^d \to \mathrm{Hom}(\mathcal{H}_{\mathrm{in}}, \mathcal{H}_{\mathrm{out}})$ *via*

$$\widehat{\Phi}(f)(t) = \int_{\mathbb{R}^d} K(y-t)(f(y))\, dy, \quad f\colon \mathbb{R}^d \to \mathcal{H}_{\mathrm{in}}, \quad t \in \mathbb{R}^d, \tag{K1}$$

and the kernel K satisfies the *equivariance constraint*

$$K(h\cdot t) = \sigma_{\mathrm{out}}(h) \circ K(t) \circ \sigma_{\mathrm{in}}(h)^{-1}, \quad h \in \mathbf{SO}(d),\ t \in \mathbb{R}^d. \tag{EC$_1$}$$

Functions $K\colon \mathbb{R}^d \to \mathrm{Hom}(\mathcal{H}_{\mathrm{in}}, \mathcal{H}_{\mathrm{out}})$ satisfying the equivariance constraint (EC$_1$) are called *equivariant convolution kernels* or *G-steerable kernels*. The above result is often referred to by the slogan "correlation is all you need."

It is instructive to give the proof since it is prototypical of this kind of argument.

Proof. The first step is to make use of a result of functional analysis that says that any Hilbert–Schmidt operator $\widehat{\Phi}\colon \mathrm{L}^2(\mathbb{R}^d, \mathcal{H}_{\mathrm{in}}) \to \mathrm{L}^2(\mathbb{R}^d, \mathcal{H}_{\mathrm{out}})$ can be expressed in terms of a so-called kernel $\mathcal{K}\colon \mathbb{R}^d \times \mathbb{R}^d \to \mathrm{Hom}(\mathcal{H}_{\mathrm{in}}, \mathcal{H}_{\mathrm{out}})$, as

$$\widehat{\Phi}(f)(t) = \int_{\mathbb{R}^d} \mathcal{K}(t,y)(f(y))\, dy, \quad f \in \mathrm{L}^2(\mathbb{R}^d, \mathcal{H}_{\mathrm{in}}),\ t, y \in \mathbb{R}^d, \tag{$*_{\mathcal{K}_1}$}$$

with $\mathcal{K} \in \mathrm{L}^2(\mathbb{R}^d \times \mathbb{R}^d, \mathrm{Hom}(\mathcal{H}_{\mathrm{in}}, \mathcal{H}_{\mathrm{out}}))$. This result is shown in Section 2.5.

The next step is to find the conditions for a linear continuous map $\widehat{\Phi}$ as above to be equivariant, which means that

$$(\mathrm{Ind}_H^G \sigma_{\mathrm{out}})_{(x,h)} \circ \widehat{\Phi} = \widehat{\Phi} \circ (\mathrm{Ind}_H^G \sigma_{\mathrm{in}})_{(x,h)}$$

for all $g = (x,h) \in \mathbb{R}^d \rtimes H$ (with $H = \mathbf{SO}(d)$).

Since $H = \mathbf{SO}(d)$ and $\mathbf{SO}(d)$ acts on $\mathbb{R}^d$ by multiplication, we simply write hy for $h\cdot y$, where $h \in \mathbf{SO}(d)$ and $y \in \mathbb{R}^d$. The action of $G =$

[6]The fact that we must consider correlations that are Hilbert–Schmidt is not mentioned in the above papers.

$\mathbb{R}^d \rtimes \mathbf{SO}(d)$ on $\mathbb{R}^d$ is given by $g \cdot y = hy + x$, where $g = (x, h) \in \mathbb{R}^d \rtimes \mathbf{SO}(d)$ and $y \in \mathbb{R}^d$. Using $(*_{\mathcal{K}_1})$, we have

$$\widehat{\Phi}[(\mathrm{Ind}_H^G \sigma_{\mathrm{in}})_{(x,h)} f](t) = \int_{\mathbb{R}^d} \mathcal{K}(t, y)(\sigma_{\mathrm{in}}(h)(f(h^{-1}(y - x)))) \, dy,$$

and since $g^{-1} = (x, h)^{-1} = (-h^{-1}x, h^{-1})$, if we make the change of variable $y \mapsto hy + x = g \cdot y$, since the determinant of the Jacobian matrix of this affine map is $+1$, by the change of variable formula, we get

$$\int_{\mathbb{R}^d} \mathcal{K}(t, y)(\sigma_{\mathrm{in}}(h)(f(h^{-1}(y - x)))) \, dy = \int_{\mathbb{R}^d} \mathcal{K}(t, g \cdot y)(\sigma_{\mathrm{in}}(h)(f(y))) \, dy.$$

Using $(*_{\mathcal{K}_1})$ and since $\sigma_{\mathrm{out}}(h)$ is linear, by Vol. I, Proposition 5.24(7), we also have

$$\begin{aligned}
[(\mathrm{Ind}_H^G \sigma_{\mathrm{out}})_{(x,h)} \widehat{\Phi}](t) &= \sigma_{\mathrm{out}}(h)(\widehat{\Phi}(h^{-1}(t - x))) \\
&= \sigma_{\mathrm{out}}(h) \left(\int_{\mathbb{R}^d} \mathcal{K}(h^{-1}(t - x), y)(f(y)) \, dy \right) \\
&= \int_{\mathbb{R}^d} \sigma_{\mathrm{out}}(h) \left(\mathcal{K}(g^{-1} \cdot t, y)(f(y)) \right) dy.
\end{aligned}$$

Consequently, we must have

$$\mathcal{K}(t, g \cdot y) \circ \sigma_{\mathrm{in}}(h) = \sigma_{\mathrm{out}}(h) \circ \mathcal{K}(g^{-1} \cdot t, y)$$

for all $g \in G = \mathbb{R}^d \rtimes H$ and all $t, y \in \mathbb{R}^d$, which by replacing t by $g \cdot t$ is equivalent to

$$\mathcal{K}(g \cdot t, g \cdot y) = \sigma_{\mathrm{out}}(h) \circ \mathcal{K}(t, y) \circ \sigma_{\mathrm{in}}(h)^{-1}, \quad g \in G,\ h \in H,\ t, y \in \mathbb{R}^d. \quad (\mathcal{K}_1)$$

In particular, for $g = -t$ and $h = e$, we get

$$\mathcal{K}(0, y - t) = \mathcal{K}(t, y), \qquad (\mathcal{K}_1')$$

so we define K, such that

$$K(y) = \mathcal{K}(0, y),$$

and since $\mathcal{K}(t, y) = \mathcal{K}(0, y - t) = K(y - t)$, $(*_{\mathcal{K}_1})$ becomes

$$\widehat{\Phi}(f)(t) = \int_{\mathbb{R}^d} K(y - t)(f(y)) \, dy,$$

as claimed.

By setting $t = 0$ in $(\mathcal{K}_1)$, we see that K satisfies the condition

$$K(g \cdot y) = \sigma_{\mathrm{out}}(h) \circ K(y) \circ \sigma_{\mathrm{in}}(h)^{-1}, \quad g \in G,\ h \in H,\ y \in \mathbb{R}^d.$$

Since the expression given by $(K1)$ is already translation invariant, it suffices to require the above condition for $g \in H = \mathbf{SO}(d)$, which is (EC_1). $\square$

Note that a crucial point of the proof is that we are using the Lebesgue measure on $\mathbb{R}^d$ and that the determinant of the Jacobian of the change of variable is $+1$, because we are considering transformations in the affine group of rigid motions $\mathbf{SE}(d)$.

Earlier, Bekkers [Bekkers (2020)] considered a situation which is less general in a way, because no representations are involved, but more general in another way, because he is dealing with two homogeneous spaces $X_{\text{in}} = G/H_{\text{in}}$ and $X_{\text{out}} = G/H_{\text{out}}$, where G is a locally compact group which is not necessarily a semi-direct product. In this case, we would like to know when a Hilbert–Schmidt operator Φ from $\mathrm{L}^2(X_{\text{in}})$ to $\mathrm{L}^2(X_{\text{out}})$ is equivariant with respect to the regular representations $\mathbf{R}^{G\to \mathrm{L}^2(X_{\text{in}})}$ and $\mathbf{R}^{G\to \mathrm{L}^2(X_{\text{out}})}$ induced by G on $\mathrm{L}^2(X_{\text{in}})$ and $\mathrm{L}^2(X_{\text{out}})$. A new difficulty that now comes up is that X_{in} may not have a G-invariant measure. Although Bekkers [Bekkers (2020)] does not make use of quasi-invariant measures, he proves a result in terms of Radon–Nikodym derivatives of measures, which can be translated as follows using ϱ-functions. Let x_0^{out} be a chosen point in $X_{\text{out}} = G/H_{\text{out}}$, so that H_{out} is the stabilizer of x_0^{out}.

Suppose that ϱ defines a quasi-invariant measure μ on $X_{\text{in}} = G/H_{\text{in}}$. First, we have the fact that every equivariant Hilbert–Schmidt operator Φ from $\mathrm{L}^2(X_{\text{in}})$ to $\mathrm{L}^2(X_{\text{out}})$ is given by

$$\Phi(f)(y) = \int_{X_{\text{in}}} \mathcal{K}(x,y) f(x)\, d\mu(x),\ y \in X_{\text{out}},\ f \in \mathrm{L}^2(X_{\text{in}}), \qquad (*_{\mathcal{K}_2})$$

for some kernel $\mathcal{K} \in \mathrm{L}^2_{\mu_{\text{in}}\otimes\mu_{\text{out}}}(X_{\text{in}} \times X_{\text{out}})$. This result is shown in Section 2.5. Observe that the roles of x and y have been switched in $\mathcal{K}(x,y)$ since we now have $f(x)$, so this is correct. To say that Φ is G-equivariant means that the following diagram commutes

$$\begin{array}{ccc} \mathrm{L}^2(X_{\text{in}}) & \xrightarrow{\ \Phi\ } & \mathrm{L}^2(X_{\text{out}}) \\ {\scriptstyle \mathbf{R}_g^{G\to \mathrm{L}^2(X_{\text{in}})}}\big\downarrow & & \big\downarrow{\scriptstyle \mathbf{R}_g^{G\to \mathrm{L}^2(X_{\text{out}})}} \\ \mathrm{L}^2(X_{\text{in}}) & \xrightarrow[\ \Phi\ ]{} & \mathrm{L}^2(X_{\text{out}}) \end{array}$$

for all $g \in G$. For any $f \in \mathrm{L}^2(X_{\text{in}})$ and any $y \in X_{\text{out}}$, we have

$$[(\Phi \circ \mathbf{R}_g^{G\to \mathrm{L}^2(X_{\text{in}})})(f)](y) = \int_{X_{\text{in}}} \mathcal{K}(x,y) f(g^{-1}\cdot x)\, d\mu(x),$$

and

$$
\begin{aligned}
&[(\mathbf{R}_g^{G\to \mathrm{L}^2(X_{\mathrm{out}})} \circ \Phi)(f)](y) \\
&\quad = \int_{X_{\mathrm{in}}} \mathcal{K}(x, g^{-1}\cdot y) f(x)\, d\mu(x) \\
&\quad = \int_{X_{\mathrm{in}}} \varrho(g^{-1}, x)\mathcal{K}(g^{-1}\cdot x, g^{-1}\cdot y) f(g^{-1}\cdot x)\, d\mu(x).
\end{aligned}
$$

The equation

$$
[(\Phi \circ \mathbf{R}_g^{G\to \mathrm{L}^2(X_{\mathrm{in}})})(f)](y) = [(\mathbf{R}_g^{G\to \mathrm{L}^2(X_{\mathrm{out}})} \circ \Phi)(f)](y)
$$

asserting the commutativity of the above diagram implies that $\mathcal{K}$ satisfies the equation

$$
\mathcal{K}(x, y) = \varrho(g^{-1}, x)\mathcal{K}(g^{-1}\cdot x, g^{-1}\cdot y),\ g\in G,\ x\in X_{\mathrm{in}},\ y \in X_{\mathrm{out}}. \quad (\mathcal{K}_2)
$$

If we define $K\colon X_{\mathrm{in}} \to \mathbb{C}$ by

$$
K(x) = \mathcal{K}(x, x_0^{\mathrm{out}}),
$$

then for any $g_y \in G$ such that $y = g_y \cdot x_0^{\mathrm{out}}$,

$$
\begin{aligned}
\mathcal{K}(x, y) &= \mathcal{K}(x, g_y\cdot x_0^{\mathrm{out}}) = \varrho(g_y^{-1}, x)\mathcal{K}(g_y^{-1}\cdot x, g_y^{-1}\cdot y) \\
&= \varrho(g_y^{-1}, x)\mathcal{K}(g_y^{-1}\cdot x, x_0^{\mathrm{out}}) = \varrho(g_y^{-1}, x)K(g_y^{-1}\cdot x).
\end{aligned}
$$

Consequently, every equivariant continuous linear map Φ from $\mathrm{L}^2(X_{\mathrm{in}})$ to $\mathrm{L}^2(X_{\mathrm{out}})$ is given by

$$
\Phi(f)(y) = \int_{X_{\mathrm{in}}} \varrho(g_y^{-1}, x)K(g_y^{-1}\cdot x) f(x)\, d\mu(x),\ y\in X_{\mathrm{out}},\ f\in \mathrm{L}^2(X_{\mathrm{in}}), \quad (\mathrm{K2})
$$

where $g_y \in G$ is any element such that $y = g_y \cdot x_0^{\mathrm{out}}$. Since $h \cdot x_0^{\mathrm{out}} = x_0^{\mathrm{out}}$ for all $h \in H_{\mathrm{out}}$, by setting $g = h \in H_{\mathrm{out}}$ and $y = x_0^{\mathrm{out}}$ in $(\mathcal{K}_2)$, we deduce that the map $K\colon X_{\mathrm{in}} \to \mathbb{C}$ satisfies the condition

$$
K(x) = \varrho(h^{-1}, x)K(h^{-1}\cdot x), \quad h\in H_{\mathrm{out}},\ x\in X_{\mathrm{in}}. \quad (\mathcal{K}_3)
$$

The factor involving ϱ disappears or is replaced by a more tractable term in many practical cases. This is the case when G is unimodular. If $X_{\mathrm{in}} = \mathbb{R}^d$ and $G = \mathbb{R}^d \rtimes H_{\mathrm{out}}$ with H_{out} a closed subgroup of $\mathbf{GL}(d)$, then if $g = (x, h) \in G$, the condition on K becomes

$$
K(x) = \frac{1}{|\det(h)|} K(h^{-1}\cdot x), \quad h\in H_{\mathrm{out}},\ x\in X_{\mathrm{in}}, \quad (\mathcal{K}_4)
$$

where $\det(h)$ is the determinant of the matrix representing h. For more details, see Bekkers [Bekkers (2020)].

8.13 Equivariant Correlation G-Kernels; General Case

Until now, we have been assuming that we are dealing with feature fields defined on $X = \mathbb{R}^d$ and that the group G is a semi-direct product $G = \mathbb{R}^d \rtimes H$ with $H = \mathbf{SO}(d)$, and more generally a compact group. It is possible to deal with the more general situation, where X is a homogeneous space of the form $X = G/H$ with G locally compact and unimodular and H compact equipped with a unitary representation $\sigma\colon H \to \mathbf{U}(\mathcal{H}_\sigma)$. The main problem is to define the "right" notion of feature field.

Cohen, Geiger and Weiler [Cohen *et al.* (2019)] propose to use the G-bundle $E = G \times_H \mathcal{H}_\sigma$ introduced in Section 6.13; see Definition 6.17. But then, we might as well use the hermitian G-bundles of finite rank of Definition 6.23 (see Section 6.13) and *the natural choice for the space of feature fields is the subspace* $\mathrm{L}^2(X; E)$ *of the space of sections of the hermitian G-bundle* $p\colon E \to X$, with $X = G/H$ (see Definition 6.25). Recall that the restriction of the action of G to H on the fibre E_0 is a unitary representation $\sigma\colon H \to \mathbf{U}(E_0)$, and that for every fibre E_x, there is a representation $\sigma_x\colon H \to \mathbf{U}(E_x)$ equivalent to the representation $\sigma\colon H \to \mathbf{U}(E_0)$. For the time being, we will assume that there exists a section $r\colon X \to G$, such that the maps $\mathcal{L}\colon \mathrm{L}^2(X; E) \to L^\sigma$ and $\mathcal{S}\colon L^\sigma \to \mathrm{L}^2(X; E)$ define isomorphisms between $\mathrm{L}^2(X; E)$ and L^σ. Recall from Equation ($\dagger_4$) of Definition 6.24 that L^σ is the set consisting of all functions $f \in \mathrm{L}^2(G; E_0)$, such that

$$f(gh) = \sigma(h^{-1})(f(g)) = h^{-1} \cdot f(g), \qquad \text{for all } g \in G \text{ and all } h \in H.$$

We will assume that the representations $\sigma\colon H \to \mathbf{U}(E_0)$ are irreducible. Then the feature fields with values in the fibre E_x transform according to the induced representation $\mathrm{Ind}_H^G \sigma_x = \Pi$; see Equation ($\dagger_7$) in Section 6.13. In view of the isomorphism between $\mathrm{L}^2(X; E)$ and L^σ given by the map $\mathcal{L}\colon \mathrm{L}^2(X; E) \to L^\sigma$ (see Definition 6.22, Equation ($\mathcal{L}_3$)), with

$$\mathcal{L}(s)(g) = g^{-1} \cdot s(g \cdot x_0), \quad s \in \mathrm{L}^2(X; E),\ g \in G,$$

the induced representation $\mathrm{Ind}_H^G \sigma_x = \Pi$ is equivalent to the left regular representation of G in L^σ. We also assume that the section $r\colon X \to G$ makes the representation Π continuous.

Inspired by Cohen, Geiger and Weiler [Cohen *et al.* (2019)] we consider the more general situation in which we have two hermitian G-bundles of finite rank $p_{\mathrm{in}}\colon E_{\mathrm{in}} \to X_{\mathrm{in}}$ and $p_{\mathrm{out}}\colon E_{\mathrm{out}} \to X_{\mathrm{out}}$, where $X_{\mathrm{in}} = G/H_{\mathrm{in}}$ and $X_{\mathrm{out}} = G/H_{\mathrm{out}}$ for the *same* group G, input and output representations σ_{in} and σ_{out}, and determine what are the linear maps $\Phi\colon L^{\sigma_{\mathrm{in}}} \to L^{\sigma_{\mathrm{out}}}$ that are equivariant with respect to the representations $\mathrm{Ind}_{H_{\mathrm{in}}}^G \sigma_{\mathrm{in}}$ and $\mathrm{Ind}_{H_{\mathrm{out}}}^G \sigma_{\mathrm{out}}$, which means that the following diagram commutes

$$\begin{array}{ccc} L^{\sigma_{\mathrm{in}}} & \xrightarrow{\Phi} & L^{\sigma_{\mathrm{out}}} \\ {\scriptstyle (\mathrm{Ind}_{H_{\mathrm{in}}}^G \sigma_{\mathrm{in}})(g)} \downarrow & & \downarrow {\scriptstyle (\mathrm{Ind}_{H_{\mathrm{out}}}^G \sigma_{\mathrm{out}})(g)} \\ L^{\sigma_{\mathrm{in}}} & \xrightarrow[\Phi]{} & L^{\sigma_{\mathrm{out}}} \end{array}$$

for all $g \in G$, where

$$[(\mathrm{Ind}_{H_{\mathrm{in}}}^G \sigma_{\mathrm{in}})(g)](f_{\mathrm{in}})(g_1) = f_{\mathrm{in}}(g^{-1}g_1), \quad g, g_1 \in G,\ f_{\mathrm{in}} \in L^{\sigma_{\mathrm{in}}}$$
$$[(\mathrm{Ind}_{H_{\mathrm{out}}}^G \sigma_{\mathrm{out}})(g)](f_{\mathrm{out}})(g_1) = f_{\mathrm{out}}(g^{-1}g_1), \quad g, g_1 \in G,\ f_{\mathrm{out}} \in L^{\sigma_{\mathrm{out}}}.$$

To reduce the amount of subscripts, we will denote the fibre $(E_{\mathrm{in}})_0$ above $x_0^{\mathrm{in}} = H_{\mathrm{in}}$ by E_0^{in} and the fibre $(E_{\mathrm{out}})_0$ above $x_0^{\mathrm{out}} = H_{\mathrm{out}}$ by E_0^{out}. Then our representations σ_{in} and σ_{out} are $\sigma_{\mathrm{in}} \colon H_{\mathrm{in}} \to \mathbf{U}(E_0^{\mathrm{in}})$ and $\sigma_{\mathrm{out}} \colon H_{\mathrm{out}} \to \mathbf{U}(E_0^{\mathrm{out}})$.

The following proposition generalizes results proven in Cohen, Geiger and Weiler [Cohen *et al.* (2019)] (see Theorem 3.1 and Theorem 3.2). In the sequel, we assume that *all hermitian G-bundles have finite rank.*

Proposition 8.12. *Let* $p_{\mathrm{in}} \colon E_{\mathrm{in}} \to X_{\mathrm{in}}$ *and* $p_{\mathrm{out}} \colon E_{\mathrm{out}} \to X_{\mathrm{out}}$ *be two hermitian G-bundles, where* $X_{\mathrm{in}} = G/H_{\mathrm{in}}$ *and* $X_{\mathrm{out}} = G/H_{\mathrm{out}}$ *for the same locally compact and unimodular group G. If the space of equivariant G-kernels is defined as*

$$\begin{aligned} &\mathrm{Hom}_{H_{\mathrm{in}},H_{\mathrm{out}}}(G, \mathrm{Hom}(E_0^{\mathrm{in}}, E_0^{\mathrm{out}})) \\ &\quad = \{K \colon G \to \mathrm{Hom}(E_0^{\mathrm{in}}, E_0^{\mathrm{out}}) \mid \\ &\qquad K(h_2 g h_1) = \sigma_{\mathrm{out}}(h_2) \circ K(g) \circ \sigma_{\mathrm{in}}(h_1), \\ &\qquad \int_G \int_G \left\| K(g_1^{-1} g_2) \right\|^2 d\lambda_G(g_1)\, d\lambda_G(g_2) < \infty, \\ &\qquad g \in G,\ h_1 \in H_{\mathrm{in}},\ h_2 \in H_{\mathrm{out}}\}, \end{aligned} \tag{EC$_2$}$$

then an equivariant Hilbert–Schmidt operator $\Phi \in \mathrm{Hom}_{H_{\mathrm{in}},H_{\mathrm{out}}}(\mathrm{Ind}_{H_{\mathrm{in}}}^G \sigma_{\mathrm{in}}, \mathrm{Ind}_{H_{\mathrm{out}}}^G \sigma_{\mathrm{out}})$ *is of the form*

$$(\Phi(f_{\mathrm{in}}))(g) = \int_G K(g^{-1}t)(f_{\mathrm{in}}(t))\, d\lambda_G(t) = (K \star f_{\mathrm{in}})(g),\ f_{\mathrm{in}} \in L^{\sigma_{\mathrm{in}}},\ g \in G, \tag{Φ}$$

for a unique $K \in \mathrm{Hom}_{H_{\mathrm{in}},H_{\mathrm{out}}}(G, \mathrm{Hom}(E_0^{\mathrm{in}}, E_0^{\mathrm{out}}))$.

Proof. As usual, the first step is to use the fact from functional analysis that any Hilbert–Schmidt operator Φ from $L^{\sigma_{\text{in}}}$ to $L^{\sigma_{\text{out}}}$ is of the form

$$\Phi(f_{\text{in}})(g_1) = \int_G \mathcal{K}(g_1, g_2)(f_{\text{in}}(g_2))\, d\lambda_G(g_2), \quad f_{\text{in}} \in L^{\sigma_{\text{in}}},\ g_1 \in G, \quad (*_{\mathcal{K}_3})$$

for some kernel $\mathcal{K} \in \mathrm{L}^2_{\lambda_G \otimes \lambda_G}(G \times G, \mathrm{Hom}(E^{\text{in}}, E^{\text{out}}))$. See Section 2.5. The second step is to assert that Φ is equivariant, which is expressed by the equation

$$(\mathrm{Ind}_{H_{\text{out}}}^G \sigma_{\text{out}})(g) \circ \Phi = \Phi \circ (\mathrm{Ind}_{H_{\text{in}}}^G \sigma_{\text{in}})(g),$$

for all $g \in G$. Since

$$[(\mathrm{Ind}_{H_{\text{in}}}^G \sigma_{\text{in}})(g)](f_{\text{in}})(g_2) = f_{\text{in}}(g^{-1}g_2), \quad g, g_2 \in G,\ f_{\text{in}} \in L^{\sigma_{\text{in}}}$$
$$[(\mathrm{Ind}_{H_{\text{out}}}^G \sigma_{\text{out}})(g)](f_{\text{out}})(g_1) = f_{\text{out}}(g^{-1}g_1), \quad g, g_1 \in G,\ f_{\text{out}} \in L^{\sigma_{\text{out}}},$$

using the fact that λ_G is left invariant, we obtain

$$\begin{aligned}
&\Phi[[(\mathrm{Ind}_{H_{\text{in}}}^G \sigma_{\text{in}})(g)](f_{\text{in}})](g_1)\\
&\quad = \int_G \mathcal{K}(g_1, g_2)([(\mathrm{Ind}_{H_{\text{in}}}^G \sigma_{\text{in}})(g)](f_{\text{in}})(g_2))\, d\lambda_G(g_2)\\
&\quad = \int_G \mathcal{K}(g_1, g_2)(f_{\text{in}}(g^{-1}g_2))\, d\lambda_G(g_2)\\
&\quad = \int_G \mathcal{K}(g_1, gg_2)(f_{\text{in}}(g_2))\, d\lambda_G(g_2),
\end{aligned}$$

and

$$[[(\mathrm{Ind}_{H_{\text{out}}}^G \sigma_{\text{out}})(g)](\Phi(f_{\text{in}}))](g_1) = \int_G \mathcal{K}(g^{-1}g_1, g_2)(f_{\text{in}}(g_2))\, d\lambda_G(g_2).$$

Consequently, equivariance is equivalent to

$$\mathcal{K}(g_1, gg_2) = \mathcal{K}(g^{-1}g_1, g_2), \quad g, g_1, g_2 \in G,$$

which is equivalent to

$$\mathcal{K}(gg_1, gg_2) = \mathcal{K}(g_1, g_2), \quad g, g_1, g_2 \in G. \qquad (\mathcal{K}_5)$$

If we define $K\colon G \to \mathrm{Hom}(E^{\text{in}}, E^{\text{out}})$ by

$$K(g) = \mathcal{K}(e, g),$$

then we have

$$\mathcal{K}(g_1, g_2) = \mathcal{K}(e, g_1^{-1}g_2) = K(g_1^{-1}g_2),$$

so $(*_{\mathcal{K}_3})$ becomes

$$\Phi(f_{\text{in}})(g_1) = \int_G K(g_1^{-1}g_2)(f_{\text{in}}(g_2))\, d\lambda_G(g_2), \quad f_{\text{in}} \in L^{\sigma_{\text{in}}},\ g_1 \in G. \quad (*_{\mathcal{K}_4})$$

Now, since Φ maps $L^{\sigma_{\mathrm{in}}}$ to $L^{\sigma_{\mathrm{out}}}$ and the functions in these spaces satisfy the conditions

$$\begin{aligned} f_{\mathrm{in}}(gh_1) &= \sigma_{\mathrm{in}}(h_1^{-1})(f_{\mathrm{in}}(g)), \quad h_1 \in H_{\mathrm{in}},\ g \in G, \\ f_{\mathrm{out}}(gh_2) &= \sigma_{\mathrm{out}}(h_2^{-1})(f_{\mathrm{out}}(g)), \quad h_2 \in H_{\mathrm{out}},\ g \in G, \end{aligned}$$

this imposes certain conditions on the kernel K. Since $f_{\mathrm{in}} \in L^{\sigma_{\mathrm{in}}}$ and λ_G is right-invariant because G is assumed to be unimodular, for any $h_1 \in H_{\mathrm{in}}$, we have

$$\begin{aligned} \int_G K(g_1^{-1}g_2h_1)(f_{\mathrm{in}}(g_2))\, d\lambda_G(g_2) &= \int_G K(g_1^{-1}g_2)(f_{\mathrm{in}}(g_2h_1^{-1}))\, d\lambda_G(g_2) \\ &= \int_G K(g_1^{-1}g_2)(\sigma_{\mathrm{in}}(h_1)(f_{\mathrm{in}}(g_2)))\, d\lambda_G(g_2), \end{aligned}$$

which implies that

$$K(g_1^{-1}g_2h_1) = K(g_1^{-1}g_2) \circ \sigma_{\mathrm{in}}(h_1), \quad g_1, g_2 \in G,\ h_1 \in H_{\mathrm{in}},$$

and thus,

$$K(gh_1) = K(g) \circ \sigma_{\mathrm{in}}(h_1), \quad g \in G,\ h_1 \in H_{\mathrm{in}}. \tag{$\mathcal{K}_6$}$$

Observe that if G is not unimodular, in which case the Haar measure λ_G is only left-invariant, the modular term $\Delta(h_1)$ needs to be added, namely we have the equation

$$K(gh_1) = \Delta(h_1)K(g) \circ \sigma_{\mathrm{in}}(h_1), \quad g \in G,\ h_1 \in H_{\mathrm{in}}. \tag{$\mathcal{K}_6'$}$$

We also need to assert that $\Phi(f_{\mathrm{in}}) \in L^{\sigma_{\mathrm{out}}}$, namely $\Phi(f_{\mathrm{in}})(gh_2) = \sigma_{\mathrm{out}}(h_2^{-1})(\Phi(f_{\mathrm{in}})(g))$, for all $h_2 \in H_{\mathrm{out}}$. We have

$$\begin{aligned} \Phi(f_{\mathrm{in}})(gh_2) &= \int_G K((gh_2)^{-1}g_2)(f_{\mathrm{in}}(g_2))\, d\lambda_G(g_2) \\ &= \int_G K(h_2^{-1}g^{-1}g_2)(f_{\mathrm{in}}(g_2))\, d\lambda_G(g_2), \end{aligned}$$

and then (by Vol. I, Proposition 5.24(7), since $\sigma(h_2^{-1})$ is linear)

$$\begin{aligned} \sigma_{\mathrm{out}}(h_2^{-1})(\Phi(f_{\mathrm{in}})(g)) &= \sigma_{\mathrm{out}}(h_2^{-1}) \left(\int_G K(g^{-1}g_2)(f_{\mathrm{in}}(g_2))\, d\lambda_G(g_2) \right) \\ &= \int_G \sigma_{\mathrm{out}}(h_2^{-1})(K(g^{-1}g_2)(f_{\mathrm{in}}(g_2)))\, d\lambda_G(g_2), \end{aligned}$$

so we deduce that

$$K(h_2^{-1}g^{-1}g_2) = \sigma_{\mathrm{out}}(h_2^{-1}) \circ K(g^{-1}g_2), \quad g, g_2 \in G,\ h_2 \in H_{\mathrm{out}},$$

which is equivalent to

$$K(h_2g) = \sigma_{\mathrm{out}}(h_2) \circ K(g), \quad g, \in G,\ h_2 \in H_{\mathrm{out}}. \tag{$\mathcal{K}_7$}$$

But $(\mathcal{K}_6)$ and $(\mathcal{K}_7)$ together are equivalent to (EC_2), which concludes the proof. $\square$

Observe that $\Phi(f_{\text{in}})$ is a generalization of group correlation as defined in Definition 8.4 to vector valued-functions. Since we are dealing with finite-dimensional vector spaces, we don't need the notion of weak integral and in (Φ) we use component-wise integration. Recall that $\text{Ind}_{H_{\text{in}}}^{G} \sigma_{\text{in}} \colon G \to \mathbf{U}(L^{\sigma_{\text{in}}})$ and $\text{Ind}_{H_{\text{out}}}^{G} \sigma_{\text{out}} \colon G \to \mathbf{U}(L^{\sigma_{\text{out}}})$ and that $L^{\sigma_{\text{in}}}$ is a space of functions from G to E_0^{in} and that $L^{\sigma_{\text{out}}}$ is a space of functions from G to E_0^{out}.

The equivariance condition (EC_2) is a bit awkward since it involves the two-sided term $h_2 g h_1$, with $h_1 \in H_{\text{in}}$ and $h_2 \in H_{\text{out}}$. Lang and Weiler [Lang and Weiler (2019)] showed that by considering the group $H = H_{\text{out}} \times H_{\text{in}}$, condition ($\text{EC}_2$) can be reduced to the familiar condition

$$K(h \cdot g) = \sigma_{\text{out}}(h) \circ K(g) \circ \sigma_{\text{in}}(h)^{-1}, \quad h \in H,\ g \in G. \tag{EC}$$

The key observation is that as subgroups of G, H_{in} and H_{out} act on G, but we can consider the more general situation where a compact group H acts on G and seek kernels $K \colon G \to \text{Hom}(E_0^{\text{in}}, E_0^{\text{out}})$ satisfying the condition (EC). Indeed, if we let $H = H_{\text{out}} \times H_{\text{in}}$ and define the left action of $H = H_{\text{out}} \times H_{\text{in}}$ on G by

$$(h_2, h_1) \cdot g = h_2 g h_1^{-1}, \quad h_1 \in H_{\text{in}},\ h_2 \in H_{\text{out}},$$

and the representations $\sigma_{\text{in}}^H \colon H \to \mathbf{U}(E_0^{\text{in}})$ and $\sigma_{\text{out}}^H \colon H \to \mathbf{U}(E_0^{\text{out}})$ by

$$\begin{aligned} \sigma_{\text{in}}^H(h_2, h_1) &= \sigma_{\text{in}}(h_1) \\ \sigma_{\text{out}}^H(h_2, h_1) &= \sigma_{\text{out}}(h_2), \end{aligned}$$

then the condition (EC), namely

$$K(h \cdot g) = \sigma_{\text{out}}^H(h) \circ K(g) \circ (\sigma_{\text{in}}^H(h))^{-1}, \quad h = (h_2, h_1) \in H,\ g \in G,$$

is equivalent to

$$K(h_2 g h_1^{-1}) = \sigma_{\text{out}}(h_2) \circ K(g) \circ \sigma_{\text{in}}(h_1)^{-1}, \quad h_1 \in H_{\text{in}},\ h_2 \in H_{\text{out}},\ g \in H,$$

which is equivalent to

$$K(h_2 g h_1) = \sigma_{\text{out}}(h_2) \circ K(g) \circ \sigma_{\text{in}}(h_1),\ h_1 \in H_{\text{in}},\ h_2 \in H_{\text{out}},\ g \in H, \tag{EC$_2$}$$

since in the quantification over $h_1 \in H_{\text{in}}$, we can replace h_1^{-1} by h_1.

Unlike the previous cases, the kernels K are defined on the group G and the formula (Φ) expressing $K \star f_{\text{in}}$ as an integral requires integration over G. This is more expensive than the previous cases that only required integration over $\mathbb{R}^d$ or more generally, over X_{in}. The technical reason is that the definition of the induced representations $\text{Ind}_{H_{\text{in}}}^{G} \sigma_{\text{in}}$ and $\text{Ind}_{H_{\text{out}}}^{G} \sigma_{\text{out}}$ is a lot simpler when they are acting on the spaces $L^{\sigma_{in}}$ and $L^{\sigma_{out}}$, since they are simply the regular representations. The representations $\sigma_{\rm in}$ and $\sigma_{\rm out}$ are hidden in the definition of the spaces $L^{\sigma_{in}}$ and $L^{\sigma_{out}}$.

To define these representations on functions defined on $X_{\rm in}$ or $X_{\rm out}$ is more complicated because this requires picking some sets of coset representatives $(r_x^{\rm in})_{x\in G/H_{\rm in}}$ and $(r_x^{\rm out})_{x\in G/H_{\rm out}}$ but then, there is no guarantee that the corresponding sections are continuous. We will assume in the sequel that the maps $\mathcal{L}_{\rm in}$ and $\mathcal{S}_{\rm in}$ are continuous, and similarly for the maps $\mathcal{L}_{\rm out}$ and $\mathcal{S}_{\rm out}$.

8.14 Equivariant Correlation $X_{\rm in}$-Kernels

Cohen, Geiger and Weiler [Cohen *et al.* (2019)] give other characterizations of the space $\mathrm{Hom}_{H_{\rm in},H_{\rm out}}(\mathrm{Ind}_{H_{\rm in}}^{G}\sigma_{\rm in}, \mathrm{Ind}_{H_{\rm out}}^{G}\sigma_{\rm out})$; one in terms of kernels defined on $X_{\rm in} = G/H_{\rm in}$, and the other in terms of kernels on the space $H_{\rm out}\backslash G/H_{\rm in}$ of double cosets. We discuss the solution in terms of kernels on $X_{\rm in} = G/H_{\rm in}$ and refer the reader to Cohen, Geiger and Weiler [Cohen *et al.* (2019)] for the third solution (see Theorem 3.4).

The key is to pick a set of coset representatives $(r_x^{\rm in})_{x\in G/H_{\rm in}}$. Here $x_0^{\rm in} = H_{\rm in}$, and as usual $r_{x_0^{\rm in}}^{\rm in} = e$. As we said earlier, we assume that maps $\mathcal{L}_{\rm in}$ and $\mathcal{S}_{\rm in}$ are continuous. Then, recall from Definition 6.5 that for every coset $x \in X_{\rm in} = G/H_{\rm in}$ and every $g \in G$, we set

$$u^{\rm in}(g,x) = (r_{g\cdot x}^{\rm in})^{-1} g r_x^{\rm in} \in H_{\rm in}, \tag{u}$$

and that by Equation (s), if $x = gH_{\rm in} = g\cdot x_0^{\rm in}$, we have

$$g = r_x^{\rm in} u^{\rm in}(g, x_0^{\rm in}).$$

Then for any $g \in G$, if $x = gH_{\rm in} = g\cdot x_0^{\rm in}$, by setting $h_2 = e$ in (EC_2), we have $K(g_1h_1) = K(g_1)\circ\sigma_{\rm in}(h_1)$ for all $g_1 \in G$ and all $h_1 \in H_{\rm in}$, so we can write

$$K(g) = K(r_x^{\rm in} u^{\rm in}(g, x_0^{\rm in})) = K(r_x^{\rm in})\circ\sigma_{\rm in}(u^{\rm in}(g, x_0^{\rm in})).$$

This suggests defining $\kappa\colon X_{\rm in} \to \mathrm{Hom}(E_0^{\rm in}, E_0^{\rm out})$ by

$$\kappa(x) = K(r_x^{\rm in}), \quad x \in X_{\rm in} = G/H_{\rm in}. \tag{κ}$$

By setting $h_1 = e$, the left $H_{\rm out}$-equivariance of K says that $K(h_2g_1) = \sigma_{\rm out}(h_2)\circ K(g_1)$ for all $g_1 \in G$ and all $h_2 \in H_{\rm out}$. Then if $x = gH_{\rm in}$, using $(*_h)$ just after Definition 6.5 (namely $u^{\mathrm{in}}(h_2g, x_0^{\mathrm{in}}) = u^{\mathrm{in}}(h_2, g \cdot x_0^{\mathrm{in}})u^{\mathrm{in}}(g, x_0^{\mathrm{in}})$) and since $h_2gH_{\mathrm{in}} = h_2 \cdot (gH_{\mathrm{in}}) = h_2 \cdot x$, we have

$$\begin{aligned} K(h_2g) &= K(r^{\mathrm{in}}_{h_2\cdot x}u^{\mathrm{in}}(h_2g, x_0^{\mathrm{in}})) = K(r^{\mathrm{in}}_{h_2\cdot x}u^{\mathrm{in}}(h_2, g \cdot x_0^{\mathrm{in}})u^{\mathrm{in}}(g, x_0^{\mathrm{in}})) \\ &= K(r^{\mathrm{in}}_{h_2\cdot x}u^{\mathrm{in}}(h_2, x)u^{\mathrm{in}}(g, x_0^{\mathrm{in}})) \\ &= K(r^{\mathrm{in}}_{h_2\cdot x}) \circ \sigma_{\mathrm{in}}(u^{\mathrm{in}}(h_2, x)u^{\mathrm{in}}(g, x_0^{\mathrm{in}})) \\ &= K(r^{\mathrm{in}}_{h_2\cdot x}) \circ \sigma_{\mathrm{in}}(u^{\mathrm{in}}(h_2, x)) \circ \sigma_{\mathrm{in}}(u^{\mathrm{in}}(g, x_0^{\mathrm{in}})), \end{aligned}$$

and since

$$K(g) = K(r^{\mathrm{in}}_x) \circ \sigma_{\mathrm{in}}(u^{\mathrm{in}}(g, x_0^{\mathrm{in}})),$$

the equation $K(h_2g) = \sigma_{\mathrm{out}}(h_2) \circ K(g)$ yields

$$\begin{aligned} &K(r^{\mathrm{in}}_{h_2\cdot x}) \circ \sigma_{\mathrm{in}}(u^{\mathrm{in}}(h_2, x)) \circ \sigma_{\mathrm{in}}(u^{\mathrm{in}}(g, x_0^{\mathrm{in}})) \\ &\qquad = \sigma_{\mathrm{out}}(h_2) \circ K(r^{\mathrm{in}}_x) \circ \sigma_{\mathrm{in}}(u^{\mathrm{in}}(g, x_0^{\mathrm{in}})), \end{aligned}$$

that is

$$K(r^{\mathrm{in}}_{h_2\cdot x}) = \sigma_{\mathrm{out}}(h_2) \circ K(r^{\mathrm{in}}_x) \circ \sigma_{\mathrm{in}}(u^{\mathrm{in}}(h_2, x)^{-1}).$$

Since by definition $\kappa(x) = K(r^{\mathrm{in}}_x)$ and $\kappa(h_2 \cdot x) = K(r^{\mathrm{in}}_{h_2\cdot x})$, we obtain

$$\kappa(h_2 \cdot x) = \sigma_{\mathrm{out}}(h_2) \circ \kappa(x) \circ \sigma_{\mathrm{in}}(u^{\mathrm{in}}(h_2, x)^{-1}). \qquad (\kappa_{H_{\mathrm{out}}})$$

Now given a function $\kappa\colon X_{\mathrm{in}} \to \mathrm{Hom}(E_0^{\mathrm{in}}, E_0^{\mathrm{out}})$, define $K\colon G \to \mathrm{Hom}(E_0^{\mathrm{in}}, E_0^{\mathrm{out}})$, such that for every $g \in G$, if $x = gH_{\mathrm{in}}$, then

$$K(g) = \kappa(x) \circ \sigma_{\mathrm{in}}(u^{\mathrm{in}}(g, x_0^{\mathrm{in}})). \qquad (*_K)$$

For any $h_1 \in H_{\mathrm{in}}$, since $h_1 \cdot x_0^{\mathrm{in}} = x_0^{\mathrm{in}}$ and $r^{\mathrm{in}}_{x_0^{\mathrm{in}}} = e$, we have

$$u^{\mathrm{in}}(h_1, x_0^{\mathrm{in}}) = (r^{\mathrm{in}}_{h_1\cdot x_0^{\mathrm{in}}})^{-1}h_1r^{\mathrm{in}}_{x_0^{\mathrm{in}}} = (r^{\mathrm{in}}_{x_0^{\mathrm{in}}})^{-1}h_1 = h_1.$$

Then by $(*_h)$,

$$u^{\mathrm{in}}(gh_1, x_0^{\mathrm{in}}) = u^{\mathrm{in}}(g, h_1 \cdot x_0^{\mathrm{in}})u^{\mathrm{in}}(h_1, x_0^{\mathrm{in}}) = u^{\mathrm{in}}(g, x_0^{\mathrm{in}})h_1,$$

and since $x = gH_{\mathrm{in}} = gh_1H_{\mathrm{in}}$, we have

$$\begin{aligned} K(gh_1) &= \kappa(x) \circ \sigma_{\mathrm{in}}(u^{\mathrm{in}}(gh_1, x_0^{\mathrm{in}})) = \kappa(x) \circ \sigma_{\mathrm{in}}(u^{\mathrm{in}}(g, x_0^{\mathrm{in}})h_1) \\ &= \kappa(x) \circ \sigma_{\mathrm{in}}(u^{\mathrm{in}}(g, x_0^{\mathrm{in}})) \circ \sigma_{\mathrm{in}}(h_1) = K(g) \circ \sigma_{\mathrm{in}}(h_1). \end{aligned}$$

This shows that K is right H_{in}-equivariant. Now assume that κ satisfies $(\kappa_{H_{\mathrm{out}}})$. If $x = g \cdot x_0^{\mathrm{in}}$, then $h_2g \cdot x_0^{\mathrm{in}} = h_2 \cdot (g \cdot x_0^{\mathrm{in}}) = h_2 \cdot x$, so

$$K(h_2g) = \kappa(h_2 \cdot x) \circ \sigma_{\mathrm{in}}(u^{\mathrm{in}}(h_2g, x_0^{\mathrm{in}})),$$

which by ($\kappa_{H_{\text{out}}}$) yields

$$K(h_2 g) = \sigma_{\text{out}}(h_2) \circ \kappa(x) \circ \sigma_{\text{in}}(u^{\text{in}}(h_2, x)^{-1}) \circ \sigma_{\text{in}}(u^{\text{in}}(h_2 g, x_0^{\text{in}})).$$

Using the fact that

$$u^{\text{in}}(h_2 g, x_0^{\text{in}}) = u^{\text{in}}(h_2, g \cdot x_0^{\text{in}}) u^{\text{in}}(g, x_0^{\text{in}}),$$

we have

$$\begin{aligned}\sigma_{\text{in}}(u^{\text{in}}(h_2 g, x_0^{\text{in}})) &= \sigma_{\text{in}}(u^{\text{in}}(h_2, g \cdot x_0^{\text{in}})) \circ \sigma_{\text{in}}(u^{\text{in}}(g, x_0^{\text{in}})) \\ &= \sigma_{\text{in}}(u^{\text{in}}(h_2, x)) \circ \sigma_{\text{in}}(u^{\text{in}}(g, x_0^{\text{in}})),\end{aligned}$$

and we obtain

$$\begin{aligned}&K(h_2 g) \\ &\quad = \sigma_{\text{out}}(h_2) \circ \kappa(x) \circ \sigma_{\text{in}}(u^{\text{in}}(h_2, x)^{-1}) \circ \sigma_{\text{in}}(u^{\text{in}}(h_2, x)) \circ \sigma_{\text{in}}(u^{\text{in}}(g, x_0^{\text{in}})) \\ &\quad = \sigma_{\text{out}}(h_2) \circ \kappa(x) \circ \sigma_{\text{in}}(u^{\text{in}}(g, x_0^{\text{in}})) = \sigma_{\text{out}}(h_2) \circ K(g),\end{aligned}$$

which shows that K is left H_{out}-equivariant. As a consequence of the above argument, we obtain the following proposition which generalizes a result originally proven in Cohen, Geiger and Weiler [Cohen *et al.* (2019)] (Theorem 3.3).

Proposition 8.13. *Let $p_{\text{in}} \colon E_0^{\text{in}} \to X_{\text{in}}$ and $p_{\text{out}} \colon E_0^{\text{out}} \to X_{\text{out}}$ be two hermitian G-bundles where $X_{\text{in}} = G/H_{\text{in}}$ and $X_{\text{out}} = G/H_{\text{out}}$ for the same locally compact and unimodular group G. If the space of equivariant G-kernels is defined as*

$$\begin{aligned}&\mathrm{Hom}_{H_{\text{in}}, H_{\text{out}}}(G, \mathrm{Hom}(E_0^{\text{in}}, E_0^{\text{out}})) \\ &\qquad = \{K \colon G \to \mathrm{Hom}(E_0^{\text{in}}, E_0^{\text{out}}) \mid \\ &\qquad\qquad K(h_2 g h_1) = \sigma_{\text{out}}(h_2) \circ K(g) \circ \sigma_{\text{in}}(h_1), \\ &\qquad\qquad \int_G \int_G \left\| K(g_1^{-1} g_2) \right\|^2 d\lambda_G(g_1)\, d\lambda_G(g_2) < \infty, \\ &\qquad\qquad g \in G,\ h_1 \in H_{\text{in}},\ h_2 \in H_{\text{out}}\},\end{aligned}$$

and the space of equivariant X_{in}-kernels is defined as

$$\begin{aligned}&\mathrm{Hom}_{H_{\text{out}}}(X_{\text{in}}, \mathrm{Hom}(E_0^{\text{in}}, E_0^{\text{out}})) \\ &\qquad = \{\kappa \colon X_{\text{in}} \to \mathrm{Hom}(E_0^{\text{in}}, E_0^{\text{out}}) \mid \\ &\qquad\qquad \kappa(h_2 \cdot x) = \sigma_{\text{out}}(h_2) \circ \kappa(x) \circ \sigma_{\text{in}}(u^{\text{in}}(h_2, x)^{-1}), \\ &\qquad\qquad \int_G \int_G \left\| \kappa((g_1^{-1} g_2) \cdot x_0^{\text{in}}) \right\|^2 d\lambda_G(g_1)\, d\lambda_G(g_2) < \infty, \\ &\qquad\qquad x \in X_{\text{in}},\ h_2 \in H_{\text{out}}\},\end{aligned} \tag{EC$_3$}$$

then the map that assigns to every X_{in}-kernel κ the G-kernel K defined such that for every $g \in G$, if $x = gH_{\text{in}} = g \cdot x_0^{\text{in}}$, then

$$K(g) = \kappa(x) \circ \sigma_{\text{in}}(u^{\text{in}}(g, x_0^{\text{in}}))$$

is a bijection.

The dependency on x of the term $\sigma_{\text{in}}(u^{\text{in}}(h_2, x)^{-1})$ is a problem. It would be nice if H_{in} had the property that we could find a section (a set of coset representatives) $r^{\text{in}} \colon G/H_{\text{in}} \to G$ satisfying the property

$$r^{\text{in}}_{h_2 \cdot x} = h_2 r^{\text{in}}_x h_2^{-1}, \quad x \in X_{\text{in}} = G/H_{\text{in}},\ h_2 \in H_{\text{out}}. \tag{$\dagger_3$}$$

Indeed, in this case, from (u) rewritten as

$$r^{\text{in}}_{h_2 \cdot x} u^{\text{in}}(h_2, x) = h_2 r^{\text{in}}_x,$$

we get

$$h_2 r^{\text{in}}_x h_2^{-1} u^{\text{in}}(h_2, x) = h_2 r^{\text{in}}_x,$$

that is,

$$u^{\text{in}}(h_2, x) = h_2. \tag{$\dagger_4$}$$

It follows that

$$\sigma_{\text{in}}(u^{\text{in}}(h_2, x)^{-1}) = \sigma_{\text{in}}(h_2)^{-1},$$

and (EC_3) is then the more friendly condition

$$\kappa(h_2 \cdot x) = \sigma_{\text{out}}(h_2) \circ \kappa(x) \circ \sigma_{\text{in}}(h_2)^{-1}, \quad h_2 \in H_{\text{out}},\ x \in X_{\text{in}}. \tag{EC_4}$$

Equation ($\dagger_3$) holds in the case where $H = H_{\text{in}} = H_{\text{out}}$ and G is a semi-direct product. Technically, it is preferable to view $G = N \rtimes H$ in the flavor where $G = NH$ with N and H subgroups of G. Then every $g \in G$ can be written uniquely as $g = nh$ with $n \in N$ and $h \in H$, so $X = G/H$ is isomorphic to N and we identify them. We can pick the set of coset representatives

$$r_n = n, \quad n \in N \approx G/H. \tag{$\dagger_5$}$$

The action of N on $G/H = N$ is just multiplication, so for any $n, n_1 \in N$, we have

$$r_{n_1 \cdot n} = n_1 n = n_1 r_n. \tag{$\dagger_6$}$$

If $x = nH$, since $h_2 \cdot x = h_2 nH$ and

$$h_2 n = h_2 n h_2^{-1} h_2,$$

as N is normal in G, $h_2nh_2^{-1} \in N$, and since $h_2 \cdot x = h_2nH = h_2nh_2^{-1}h_2H = h_2nh_2^{-1}H$ with $h_2nh_2^{-1} \in N$, we have $r_{h_2 \cdot x} = h_2nh_2^{-1} = h_2r_xh_2^{-1}$, namely ($\dagger_3$) holds. Consequently, by ($\dagger_4$), the equation satisfied by $X_{\rm in}$-kernels is

$$\kappa(h_2 \cdot x) = \sigma_{\rm out}(h_2) \circ \kappa(x) \circ \sigma_{\rm in}(h_2)^{-1}, \quad h_2 \in H, \, x \in X, \qquad (\mathrm{EC}_5)$$

which gives us back the condition satisfied by kernels in the case where G is a semi-direct product $G = \mathbb{R}^d \rtimes H$.

In general, there does not appear to be a simple way to find conditions for which the term $\sigma^{\rm in}(u^{\rm in}(h_2, x)^{-1})$ goes away. Cohen, Geiger and Weiler [Cohen *et al.* (2019)] (Theorem 3.4) show that by considering kernels defined on the double coset space $H_{\rm out}\backslash G/H_{\rm in}$, condition ($\mathrm{EC}_3$) almost becomes condition (EC_5), but the analog of the representation $\sigma_{\rm in}$ *depends on* x, so this is not a reduction to (EC_5).

8.15 Passing from $L^{\sigma_{\rm in}}$ and $L^{\sigma_{\rm out}}$ to $\mathrm{L}^2(X_{\rm in}, E_{\rm in})$ and $\mathrm{L}^2(X_{\rm out}, E_{\rm out})$

The G-equivariant maps in $\mathrm{Hom}_{H_{\rm in}, H_{\rm out}}(\mathrm{Ind}_{H_{\rm in}}^G \sigma_{\rm in}, \mathrm{Ind}_{H_{\rm out}}^G \sigma_{\rm out})$ are functions from $L^{\sigma_{\rm in}}$ to $L^{\sigma_{\rm out}}$ and still require integration over G to be computed using equivariant kernels in the space $\mathrm{Hom}_{H_{\rm in}, H_{\rm out}}(G, \mathrm{Hom}(E_0^{\rm in}, E_0^{\rm out}))$. It would be nice if we could transform the integration over G to a more practically computable integration over $X_{\rm in}$. This can be achieved by using the maps $\mathcal{S}_{\rm out}\colon L^{\sigma_{\rm out}} \to \mathrm{L}^2(X_{\rm out}, E_{\rm out})$ and $\mathcal{L}_{\rm in}\colon \mathrm{L}^2(X_{\rm in}, E_{\rm in}) \to L^{\sigma_{\rm in}}$ given by ($\mathcal{S}_3''$) and ($\mathcal{L}_3'$) of Section 6.13. When these maps are well-defined, which is our assumption, they can be used to define maps from $\mathrm{L}^2(X, E_{\rm in})$ to $\mathrm{L}^2(X, E_{\rm out})$ from functions from $L^{\sigma_{\rm in}}$ to $L^{\sigma_{\rm out}}$. Recall that ($\mathcal{L}_3'$) is given by

$$\mathcal{L}(s)(g) = \sigma(u(g, x_0)^{-1})(r_x^{-1} \cdot s(x)), x = gH = g \cdot x_0, \, g \in G, \, s \in \mathrm{L}^2(X, E),$$

and ($\mathcal{S}_3''$) is given by

$$\mathcal{S}(f)(gH) = \mathcal{S}(f)(g \cdot x_0) = g \cdot f(g), \quad g \in G, \, f \in L^\sigma.$$

Pick a set of coset representatives $(r_x^{\rm in})_{x \in G/H_{\rm in}}$ for $X_{\rm in} = G/H_{\rm in}$ and a set of coset representatives $(r_x^{\rm out})_{x \in G/H_{\rm out}}$ for $X_{\rm out} = G/H_{\rm out}$. Then for every section $s \in \mathrm{L}^2(X_{\rm in}, E_{\rm in})$, for every $x \in X_{\rm out}$, observe that for every equivariant kernel $K \in \mathrm{Hom}_{H_{\rm in}, H_{\rm out}}(G, \mathrm{Hom}(E_0^{\rm in}, E_0^{\rm out}))$, the function $\widetilde{\Phi}$ given by

$$\widetilde{\Phi}(s) = \mathcal{S}_{\rm out}(K \star (\mathcal{L}_{\rm in}(s)))$$

maps $\mathrm{L}^2(X_{\mathrm{in}}, E_{\mathrm{in}})$ to $\mathrm{L}^2(X_{\mathrm{out}}, E_{\mathrm{out}})$, because $\mathcal{L}_{\mathrm{in}}(s) \in L^{\sigma_{\mathrm{in}}}$, $K \star (\mathcal{L}_{\mathrm{in}}(s)) \in L^{\sigma_{\mathrm{out}}}$, and $\mathcal{S}_{\mathrm{out}}(K\star(\mathcal{L}_{\mathrm{in}}(s))) \in \mathrm{L}^2(X_{\mathrm{out}}, E_{\mathrm{out}})$, as illustrated in the following diagram:

$$\begin{array}{ccc} L^{\sigma_{\mathrm{in}}} & \xrightarrow{\Phi = K\star -} & L^{\sigma_{\mathrm{out}}} \\ \mathcal{L}_{\mathrm{in}} \Big\uparrow & & \Big\downarrow \mathcal{S}_{\mathrm{out}} \\ \mathrm{L}^2(X_{\mathrm{in}}, E_{\mathrm{in}}) & \xrightarrow[\widetilde{\Phi}]{} & \mathrm{L}^2(X_{\mathrm{out}}, E_{\mathrm{out}}). \end{array}$$

We now work out several explicit formulae for $\mathcal{S}_{\mathrm{out}}(K \star (\mathcal{L}_{\mathrm{in}}(s)))$, the most general ones being ($\dagger_8$) and ($\dagger_9$). Since for any $s \in \mathrm{L}^2(X_{\mathrm{in}}, E_{\mathrm{in}})$,

$$\mathcal{L}_{\mathrm{in}}(s)(t) = \sigma_{\mathrm{in}}(u^{\mathrm{in}}(t, x_0^{\mathrm{in}})^{-1})((r_y^{\mathrm{in}})^{-1} \cdot s(y)),$$
$$y = t \cdot x_0^{\mathrm{in}} \in X_{\mathrm{in}},\ x_0^{\mathrm{in}} = H_{\mathrm{in}},\ t \in G,$$

by (Φ) of Proposition 8.12, for any $g \in G$, we have

$$\begin{aligned}(K \star (\mathcal{L}_{\mathrm{in}}(s)))(g) &= \int_G K(g^{-1}t)[\mathcal{L}_{\mathrm{in}}(s)(t)]\, d\lambda_G(t) \\ &= \int_G K(g^{-1}t)[\sigma_{\mathrm{in}}(u^{\mathrm{in}}(t, x_0^{\mathrm{in}})^{-1})((r_y^{\mathrm{in}})^{-1} \cdot s(y))]\, d\lambda_G(t),\end{aligned}$$

and since

$$\mathcal{S}_{\mathrm{out}}(f)(x) = r_x^{\mathrm{out}} \cdot f(r_x^{\mathrm{out}}), \quad x = r_x^{\mathrm{out}} \cdot x_0^{\mathrm{out}} \in X_{\mathrm{out}},\ x_0^{\mathrm{out}} = H_{\mathrm{out}},$$

with $f = K \star (\mathcal{L}_{\mathrm{in}}(s))$, we get

$$\begin{aligned}&[\mathcal{S}_{\mathrm{out}}(K \star (\mathcal{L}_{\mathrm{in}}(s)))](x) \\ &\quad = r_x^{\mathrm{out}} \cdot \int_G K((r_x^{\mathrm{out}})^{-1}t)[\sigma_{\mathrm{in}}(u^{\mathrm{in}}(t, x_0^{\mathrm{in}})^{-1})((r_y^{\mathrm{in}})^{-1} \cdot s(y))]\, d\lambda_G(t).\end{aligned}$$

Since by Proposition 8.13,

$$K(g_1) = \kappa(g_1 \cdot x_0^{\mathrm{in}}) \circ \sigma_{\mathrm{in}}(u^{\mathrm{in}}(g_1, x_0^{\mathrm{in}})),$$

and $y = t \cdot x_0^{\mathrm{in}}$, we get

$$\begin{aligned}&[\mathcal{S}_{\mathrm{out}}(K \star (\mathcal{L}_{\mathrm{in}}(s)))](x) \\ &= r_x^{\mathrm{out}} \cdot \int_G \kappa((r_x^{\mathrm{out}})^{-1}y)[[\sigma_{\mathrm{in}}(u^{\mathrm{in}}((r_x^{\mathrm{out}})^{-1}t, x_0^{\mathrm{in}})) \circ \sigma_{\mathrm{in}}(u^{\mathrm{in}}(t, x_0^{\mathrm{in}})^{-1})] \\ &\qquad ((r_y^{\mathrm{in}})^{-1} \cdot s(y))]\, d\lambda_G(t) \\ &= r_x^{\mathrm{out}} \cdot \int_G \kappa((r_x^{\mathrm{out}})^{-1}y)[\sigma_{\mathrm{in}}(u^{\mathrm{in}}((r_x^{\mathrm{out}})^{-1}t, x_0^{\mathrm{in}})u^{\mathrm{in}}(t, x_0^{\mathrm{in}})^{-1}) \\ &\qquad ((r_y^{\mathrm{in}})^{-1} \cdot s(y))]\, d\lambda_G(t).\end{aligned}$$

We now proceed by simplifying the expression in the square bracket. Using the equation

$$u(st, x) = u(s, t \cdot x)u(t, x), \tag{$*_h$}$$

we have

$$u^{\rm in}((r_x^{\rm out})^{-1}t, x_0^{\rm in}) = u^{\rm in}((r_x^{\rm out})^{-1}, t \cdot x_0^{\rm in})u^{\rm in}(t, x_0^{\rm in}),$$

so

$$u^{\rm in}((r_x^{\rm out})^{-1}t, x_0^{\rm in})u^{\rm in}(t, x_0^{\rm in})^{-1} = u^{\rm in}((r_x^{\rm out})^{-1}, t \cdot x_0^{\rm in}) = u^{\rm in}((r_x^{\rm out})^{-1}, y).$$

Consequently, for any $s \in \mathrm{L}^2(X_{\rm in}, E_{\rm in})$, we obtain

$$\begin{aligned}&[\mathcal{S}_{\rm out}(K \star (\mathcal{L}_{\rm in}(s)))](x)\\ &= r_x^{\rm out} \cdot \int_G \kappa((r_x^{\rm out})^{-1}y)[\sigma_{\rm in}(u^{\rm in}((r_x^{\rm out})^{-1}, y))((r_y^{\rm in})^{-1} \cdot s(y))]\, d\lambda_G(t),\end{aligned} \tag{$\dagger_7$}$$

with $y = t \cdot x_0^{\rm in}$, $t \in G$, and $x = r_x^{\rm out} \cdot x_0^{\rm out} \in X_{\rm out}$.

By Vol. I, Proposition 8.43, since G is a locally compact group and $H_{\rm in}$ is a compact subgroup of G, the space $X_{\rm in} = G/H_{\rm in}$ admits a G-invariant σ-Radon measure γ so that for any $s \in \mathrm{L}^2(X_{\rm in}, E_{\rm in})$ and any $x = r_x^{\rm out} \cdot x_0^{\rm out} \in X_{\rm out}$,

$$\begin{aligned}&[\mathcal{S}_{\rm out}(K \star (\mathcal{L}_{\rm in}(s)))](x)\\ &= r_x^{\rm out} \cdot \int_{X_{\rm in}} \kappa((r_x^{\rm out})^{-1}y)[\sigma_{\rm in}(u^{\rm in}((r_x^{\rm out})^{-1}, y))((r_y^{\rm in})^{-1} \cdot s(y))]\, d\gamma(y).\end{aligned} \tag{$\dagger_8$}$$

The condition on κ in (EC_3) translates to

$$\int_{X_{\rm in}} \int_{X_{\rm in}} \left\|\kappa((r_x^{\rm out})^{-1}y)\right\|^2 d\lambda_{X_{\rm in}}(x)\, d\lambda_{X_{\rm in}}(y) < \infty.$$

Formula ($\dagger_8$) is the main formula of this section. It uses a cheaper integration over $X_{\rm in}$ and the simpler kernel κ. This formula holds in the general framework of hermitian G-bundles of finite rank. A similar formula is given in Cohen, Geiger and Weiler [Cohen *et al.* (2019)] (Formula (14)), but with the term $u^{\rm in}((r_x^{\rm out})^{-1}r_y^{\rm in}, x_0^{\rm in})$ instead of the term $u^{\rm in}((r_x^{\rm out})^{-1}, y)$. In fact, these terms are equal. This is because by ($*_h$),

$$\begin{aligned}u^{\rm in}((r_x^{\rm out})^{-1}r_y^{\rm in}, x_0^{\rm in}) &= u^{\rm in}((r_x^{\rm out})^{-1}, r_y^{\rm in} \cdot x_0^{\rm in})u^{\rm in}(r_y^{\rm in}, x_0^{\rm in})\\ &= u^{\rm in}((r_x^{\rm out})^{-1}, y),\end{aligned}$$

since $u^{\rm in}(r_y^{\rm in}, x_0^{\rm in}) = e$, which follows from Equation (u), since

$$u^{\rm in}(r_y^{\rm in}, x_0^{\rm in}) = (r^{\rm in}_{r_y^{\rm in} \cdot x_0^{\rm in}})^{-1} r_y^{\rm in} r_{x_0^{\rm in}}^{\rm in} = (r_y^{\rm in})^{-1} r_y^{\rm in} e = e.$$

We finish this section by considering two special cases of the main formula.

Example 8.16. If the hermitian G-bundles E_{in} and E_{out} arise from the Borel construction (see Section 6.12) from the representations $\sigma_{\mathrm{in}}\colon H_{\mathrm{in}} \to \mathbf{U}(\mathcal{H}_{\mathrm{in}})$ and $\sigma_{\mathrm{out}}\colon H_{\mathrm{out}} \to \mathbf{U}(\mathcal{H}_{\mathrm{out}})$, then the fibres $E_{x_{\mathrm{in}}}$ (with $x_{\mathrm{in}} \in X_{\mathrm{in}}$) consists of equivalence classes $\{[(r^{\mathrm{in}}_{x_{\mathrm{in}}}, u_{\mathrm{in}})] \mid u_{\mathrm{in}} \in \mathcal{H}_{\mathrm{in}}\}$, and the fibres $E_{x_{\mathrm{out}}}$ (with $x_{\mathrm{out}} \in X_{\mathrm{out}}$) consists of equivalence classes $\{[(r^{\mathrm{out}}_{x_{\mathrm{out}}}, u_{\mathrm{out}})] \mid u_{\mathrm{out}} \in \mathcal{H}_{\mathrm{out}}\}$. The fibre E^{in}_0 above $x^{\mathrm{in}}_0 = H_{\mathrm{in}}$ consists of equivalence classes of the form $[(e, u_{\mathrm{in}})]$, and the fibre E^{out}_0 above $x^{\mathrm{out}}_0 = H_{\mathrm{out}}$ consists of equivalence classes of the form $[(e, u_{\mathrm{out}})]$. The fibre E^{in}_0 is isomorphic to $\mathcal{H}_{\mathrm{in}}$, and the fibre E^{out}_0 is isomorphic to $\mathcal{H}_{\mathrm{out}}$; see the discussion just after Definition 6.21. We also explained in Section 6.13 that the definition of the action of G on these hermitian G-bundles implies that

$$(r^{\mathrm{in}}_{x_{\mathrm{in}}})^{-1} \cdot [(r^{\mathrm{in}}_{x_{\mathrm{in}}}, u_{\mathrm{in}})] = [(e, u_{\mathrm{in}})],$$

and

$$r^{\mathrm{out}}_{x_{\mathrm{out}}} \cdot [(e, u_{\mathrm{out}})] = [(r^{\mathrm{out}}_{x_{\mathrm{out}}}, u_{\mathrm{out}})],$$

so the above maps provide isomorphisms from $E_{x_{\mathrm{in}}}$ to E^{in}_0 and from E^{out}_0 to $E_{x_{\mathrm{out}}}$. Since the sections in $\Gamma(E_{\mathrm{in}})$ are of the form

$$s_{\mathrm{in}}(x_{\mathrm{in}}) = [(r^{\mathrm{in}}_{x_{\mathrm{in}}}, u_{\mathrm{in}})],$$

and the sections in $\Gamma(E_{\mathrm{out}})$ are of the form

$$s_{\mathrm{out}}(x_{\mathrm{out}}) = [(r^{\mathrm{out}}_{x_{\mathrm{out}}}, u_{\mathrm{out}})],$$

and since $\kappa(x_{\mathrm{in}})$ maps the fibre E^{in}_0 to the fibre E^{out}_0, we see that if we identify *all* the fibres $E_{x_{\mathrm{in}}}$ with E^{in}_0 and *all* the fibres $E_{x_{\mathrm{out}}}$ with E^{out}_0, then we can view sections in $\Gamma(E_{\mathrm{in}})$ as functions from X_{in} to $E^{\mathrm{in}}_0 \approx \mathcal{H}_{\mathrm{in}}$ and sections in $\Gamma(E_{\mathrm{out}})$ as functions from X_{out} to $E^{\mathrm{out}}_0 \approx \mathcal{H}_{\mathrm{out}}$, so we can drop the terms r^{out}_x and $(r^{\mathrm{in}}_y)^{-1}$ and we get the formula

$$[\mathcal{S}_{\mathrm{out}}(K\star(\mathcal{L}_{\mathrm{in}}(s)))](x) = \int_{X_{\mathrm{in}}} \kappa((r^{\mathrm{out}}_x)^{-1}y)[\sigma_{\mathrm{in}}(u^{\mathrm{in}}((r^{\mathrm{out}}_x)^{-1}, y))(s(y))]\, d\gamma(y), \tag{$\dagger_9$}$$

for all $s \in \mathrm{L}^2(X_{\mathrm{in}}, E_{\mathrm{in}})$, with $y \in X_{\mathrm{in}}$ and $x \in X_{\mathrm{out}}$.

The second special case deals with semi-direct products.

Example 8.17. If $H = H_{\mathrm{in}} = H_{\mathrm{out}}$ and G is a semi-direct product $G = N \rtimes H$, then $X = G/H \approx N$. By ($\dagger_6$), $r_{n\cdot y} = nr_y$ when $n \in N$, and from

$$r_{n\cdot y}u(n, y) = nr_y,$$

we get $nr_yu(n, y) = nr_y$, that is

$$u(n, y) = e. \tag{$\dagger_{10}$}$$

Consequently, by setting $n = (r_x)^{-1} \in N$, we have $u(r_x^{-1}, y) = e$, and since $r_x = x$ and $r_y = y$, by ($\dagger_{10}$) and ($\dagger_8$), we obtain

$$[\mathcal{S}(K \star (\mathcal{L}(s)))](x) = x \cdot \int_N \kappa(x^{-1}y)(y^{-1} \cdot s(y))\, d\gamma(y), \quad x, y \in N, \quad (\dagger_{11})$$

for all $s \in \mathrm{L}^2(X, E)$.

If the hermitian G-bundles are constructed from representations $\sigma_{\text{in}} \colon H \to \mathbf{U}(\mathcal{H}_{\text{in}})$ and $\sigma_{\text{out}} \colon H \to \mathbf{U}(\mathcal{H}_{\text{out}})$, the above formula becomes

$$[\mathcal{S}(K \star (\mathcal{L}(s)))](x) = \int_N \kappa(x^{-1}y)(s(y))\, d\gamma(y), \quad x, y \in N, \qquad (\dagger_{12})$$

for all $s \in \mathrm{L}^2(X, E)$. Note the analogy of ($\dagger_{12}$) and (Φ) from Proposition 8.12.

The issue of finding G-equivariant kernels still remains.

8.16 Equivariant Kernels and Kernel Operators

As in Lang and Weiler [Lang and Weiler (2019)] and Cesa, Lang and Weiler [Cesa *et al.* (2022)], we now assume that $H_{\text{in}} = H_{\text{out}} = H$, so $X_{\text{in}} = X_{\text{out}} = X = G/H$, and we have two Hermitian G-bundles E_{in} and E_{out}. The Hermitian G-bundles define two representations $\sigma_{\text{in}} \colon H \to \mathbf{U}(\mathcal{H}_{\text{in}})$ and $\sigma_{\text{out}} \colon H \to \mathbf{U}(\mathcal{H}_{\text{out}})$, where we denote the fibres E_0^{in} and E_0^{out} as $\mathcal{H}_{\text{in}}$ and $\mathcal{H}_{\text{out}}$, which is closer to the notation used by the above authors. We consider the space of *equivariant X-kernels* defined below.

Definition 8.15. The space of *equivariant X-kernels* $\mathrm{Hom}_H(X, \mathrm{Hom}(\mathcal{H}_{\text{in}}, \mathcal{H}_{\text{out}}))$ is given by

$$\begin{aligned} \mathrm{Hom}_H(X, \mathrm{Hom}(\mathcal{H}_{\text{in}}, \mathcal{H}_{\text{out}})) = \{\kappa \colon X \to \mathrm{Hom}(\mathcal{H}_{\text{in}}, \mathcal{H}_{\text{out}}) \mid \\ \kappa(h \cdot x) = \sigma_{\text{out}}(h) \circ \kappa(x) \circ \sigma_{\text{in}}(h)^{-1}, \\ \int_X \int_X \left\|\kappa((r_x)^{-1}y)\right\|^2 d\lambda_X(x)\, d\lambda_X(y) < \infty, \\ x \in X,\ h \in H\}. \qquad (\mathrm{EC}_6) \end{aligned}$$

Remarkably, Lang and Weiler [Lang and Weiler (2019)] and Cesa, Lang and Weiler [Cesa *et al.* (2022)] completely characterized the kernels in $\kappa \in \mathrm{Hom}_H(X, \mathrm{Hom}(\mathcal{H}_{\text{in}}, \mathcal{H}_{\text{out}}))$ when H is a compact group acting on a topological Hausdorff space X equipped with the σ-algebra of Borel sets and an H-invariant measure μ. This does not cover all the cases of condition (EC_3) but it does cover the case where $X = G$ and $H = H_{\text{out}} \times H_{\text{in}}$, since condition ($EC_2$) can be reduced to condition (EC), and the case where $X = X_{\text{in}} = G/H_{\text{in}}$ if condition (EC_4) holds (with $H = H_{\text{out}}$), which happens if $H_{\text{in}} = H_{\text{out}}$ and G is a semi-direct product $G = N \rtimes H$. Note that typically H does not act transitively on X. In all of these cases, X arises from a transitive action of the group G, but the results described below hold for any space X. We assume that $\sigma_{\text{in}} \colon H \to \mathbf{U}(\mathcal{H}_{\text{in}})$ and $\sigma_{\text{out}} \colon H \to \mathbf{U}(\mathcal{H}_{\text{out}})$ are irreducible. This is not a restriction as explained in Lang and Weiler [Lang and Weiler (2019)].

If we pick bases for the (finite-dimensional) spaces $\mathcal{H}_{\text{in}}$ and $\mathcal{H}_{\text{out}}$, then the representations σ_{in} and σ_{out} are represented by matrices Σ_{in} and Σ_{out}, and $\kappa(x)$ is also represented by a matrix $\widetilde{\kappa}(x)$. If we vectorize the matrix $\widetilde{\kappa}(x)$ by making it into a vector $\text{vec}(\widetilde{\kappa}(x))$ obtained by concatenating its rows, then we can use the fact from linear algebra that for any $m \times m$ matrix A, any $n \times n$ matrix B, and $m \times n$ matrix Z, we have the identity

$$\text{vec}(AZB) = (B^\top \otimes A)\text{vec}(Z),$$

where $\otimes$ denotes the Kronecker product of matrices. Then (since $(\Sigma_{\text{in}}^{-1})^\top = \overline{\Sigma_{\text{in}}}$), the equation

$$\kappa(h \cdot x) = \sigma_{\text{out}}(h) \circ \kappa(x) \circ \sigma_{\text{in}}(h)^{-1}$$

becomes the following equation in matrix form

$$\text{vec}(\widetilde{\kappa})(h \cdot x) = [(\overline{\Sigma_{\text{in}}} \otimes \Sigma_{\text{out}})(h)]\text{vec}(\widetilde{\kappa}(x)). \tag{$*_{51}$}$$

Observe that ($*_{51}$) is the analog of (steer7) with $\text{vec}(\widetilde{\kappa}(x))$ instead of $\overline{Y_{\rho,k_\rho}}$ and $\overline{\Sigma_{\text{in}}} \otimes \Sigma_{\text{out}}$ instead of $M^{\rho,k_\rho}(h)$. If the representation $\overline{\sigma_{\text{in}}} \otimes \sigma_{\text{out}}$ is reducible, we need to decompose it as a direct sum of irreducibles. Otherwise, some steerable family $\overline{Y_{\rho,k_\rho}}$ is a solution of ($*_{51}$). We now explain how a basis of solutions of ($*_{51}$) can be found.

A key ingredient is the analog of the left regular representation $V \colon H \to \mathbf{U}(\mathrm{L}^2(X))$ of $\mathrm{L}^2(X)$ induced by the action of H on X already introduced in Section 8.7 and given by

$$(V(h)f)(x) = f(h^{-1} \cdot x), \quad h \in H,\ f \in \mathrm{L}^2(X),\ x \in X.$$

For the sake of consistency of notation, we will also denote the representation V as $\mathbf{R}^{H \to \mathrm{L}^2(X)}$.

The other key ingredient is the set of H-maps

$$\mathrm{Hom}_H(\mathbf{R}^{H \to \mathrm{L}^2(X)}, \mathrm{Hom}(\sigma_{\text{in}}, \sigma_{\text{out}})),$$

which is the space of continuous linear maps $\mathcal{K}\colon \mathrm{L}^2(X) \to \mathrm{Hom}(\mathcal{H}_{\mathrm{in}}, \mathcal{H}_{\mathrm{out}})$, such that the following diagram commutes

$$\begin{array}{ccc}
\mathrm{L}^2(X) & \xrightarrow{\ \mathcal{K}\ } & \mathrm{Hom}(\mathcal{H}_{\mathrm{in}}, \mathcal{H}_{\mathrm{out}}) \\
{\scriptstyle \mathbf{R}^{H \to \mathrm{L}^2(X)}(h)} \Big\downarrow & & \Big\downarrow {\scriptstyle \mathrm{Hom}(\sigma_{\mathrm{in}}, \sigma_{\mathrm{out}})(h)} \\
\mathrm{L}^2(X) & \xrightarrow[\ \mathcal{K}\ ]{} & \mathrm{Hom}(\mathcal{H}_{\mathrm{in}}, \mathcal{H}_{\mathrm{out}})
\end{array}$$

for every $h \in H$; see Definition 3.9. Cesa, Lang and Weiler [Cesa *et al.* (2022)] call the maps $\mathcal{K}$ *kernel operators*. Recall from Definition 4.16 that the representation $\mathrm{Hom}(\sigma_{\mathrm{in}}, \sigma_{\mathrm{out}})$ is defined such that

$$[\mathrm{Hom}(\sigma_{\mathrm{in}}, \sigma_{\mathrm{out}})(h)](f) = \sigma_{\mathrm{out}}(h) \circ f \circ \sigma_{\mathrm{in}}(h^{-1}), \qquad f \in \mathrm{Hom}(\mathcal{H}_{\mathrm{in}}, \mathcal{H}_{\mathrm{out}}),\ h \in H.$$

The main result is that there is a bijection between the space $\mathrm{Hom}_H(X, \mathrm{Hom}(\mathcal{H}_{\mathrm{in}}, \mathcal{H}_{\mathrm{out}}))$ of equivariant X-kernels and the space $\mathrm{Hom}_H(\mathbf{R}^{H \to \mathrm{L}^2(X)}, \mathrm{Hom}(\sigma_{\mathrm{in}}, \sigma_{\mathrm{out}}))$ of kernel operators. This isomorphism is a kind of linearization of the first space. The following result is shown in Cesa, Lang and Weiler [Cesa *et al.* (2022)] (Theorem B2). See also Lang and Weiler [Lang and Weiler (2019)] (Theorem C7).

Theorem 8.14. *There is an isomorphism between the space* $\mathrm{Hom}_H(X, \mathrm{Hom}(\mathcal{H}_{\mathrm{in}}, \mathcal{H}_{\mathrm{out}}))$ *of equivariant X-kernels and the space* $\mathrm{Hom}_H(\mathbf{R}^{H \to \mathrm{L}^2(X)}, \mathrm{Hom}(\sigma_{\mathrm{in}}, \sigma_{\mathrm{out}}))$ *of kernel operators.*

The isomorphism is defined as follows. The map from $\mathrm{Hom}_H(X, \mathrm{Hom}(\mathcal{H}_{\mathrm{in}}, \mathcal{H}_{\mathrm{out}}))$ to $\mathrm{Hom}_H(\mathbf{R}^{H \to \mathrm{L}^2(X)}, \mathrm{Hom}(\sigma_{\mathrm{in}}, \sigma_{\mathrm{out}}))$ is defined such that for any kernel $\kappa\colon X \to \mathrm{Hom}(\mathcal{H}_{\mathrm{in}}, \mathcal{H}_{\mathrm{out}})$, we define the extension $\widehat{\kappa}$ of κ to $\mathrm{L}^2(X)$ as

$$\widehat{\kappa}(f) = \int_X f(x)\kappa(x)\, d\mu(x), \quad f \in \mathrm{L}^2(X), \tag{$\dagger_{13}$}$$

and the map from $\mathrm{Hom}_H(\mathbf{R}^{H \to \mathrm{L}^2(X)}, \mathrm{Hom}(\sigma_{\mathrm{in}}, \sigma_{\mathrm{out}}))$ to $\mathrm{Hom}_H(X, \mathrm{Hom}(\mathcal{H}_{\mathrm{in}}, \mathcal{H}_{\mathrm{out}}))$ is defined such that for every kernel operator $\mathcal{K}\colon \mathrm{L}^2(X) \to \mathrm{Hom}(\mathcal{H}_{\mathrm{in}}, \mathcal{H}_{\mathrm{out}})$, we define the restriction $\mathcal{K}|_X$ of $\mathcal{K}$ to X as

$$(\mathcal{K}|_X)(x) = \mathcal{K}(\delta_x), \quad x \in X, \tag{$\dagger_{14}$}$$

where δ_x is the Dirac δ-function. Technically, this does not make sense and to make this correct it is necessary to use a form of regularization similar to the method explained in Vol. I, Section 8.16; see Corollary 8.52. This what is done in Lang and Weiler [Lang and Weiler (2019)] (Appendix C, Theorem C7). It is shown in Cesa, Lang and Weiler [Cesa *et al.* (2022)] that the above maps are mutual inverses (see also Lang and Weiler [Lang and Weiler (2019)] for some of the details, Appendix C, Part C2).

But now Proposition 4.23 tells us that the representations $\mathrm{Hom}(\sigma_{\mathrm{in}}, \sigma_{\mathrm{out}})$ and $\overline{\sigma_{\mathrm{in}}} \otimes \sigma_{\mathrm{out}}$ are equivalent, so we obtain an isomorphism

$$\mathrm{Hom}_H(X, \mathrm{Hom}(\mathcal{H}_{\mathrm{in}}, \mathcal{H}_{\mathrm{out}})) \approx \mathrm{Hom}_H(\mathbf{R}^{H \to \mathrm{L}^2(X)}, \overline{\sigma_{\mathrm{in}}} \otimes \sigma_{\mathrm{out}}). \qquad (\dagger_{15})$$

Since H is a compact group, we can now use Theorem 8.7 (a direct consequence of Peter–Weyl II) to express $\mathrm{L}^2(X)$ as a Hilbert sum of spaces corresponding to irreducible representations of H and the decomposition of the tensor product representation $\overline{\sigma_{\mathrm{in}}} \otimes \sigma_{\mathrm{out}}$ as a Hilbert sum of irreducible representations of H (see Proposition 4.18 and Equation ($\otimes$) in Section 4.4) to obtain the following decomposition of $\mathrm{Hom}_H(\mathbf{R}^{H \to \mathrm{L}^2(X)}, \overline{\sigma_{\mathrm{in}}} \otimes \sigma_{\mathrm{out}})$, and thus of $\mathrm{Hom}_H(X, \mathrm{Hom}(\mathcal{H}_{\mathrm{in}}, \mathcal{H}_{\mathrm{out}}))$. We obtain

$$\begin{aligned}\mathrm{Hom}_H(\mathbf{R}^{H \to \mathrm{L}^2(X)}, \overline{\sigma_{\mathrm{in}}} \otimes \sigma_{\mathrm{out}}) &\approx \bigoplus_{\rho_1 \in R(H)} \bigoplus_{k_{\rho_1}=1}^{d_{\rho_1}} \mathrm{Hom}_H(M_{\rho_1}, \overline{\sigma_{\mathrm{in}}} \otimes \sigma_{\mathrm{out}}) \\ &\approx \bigoplus_{\rho_1 \in R(H)} \bigoplus_{k_{\rho_1}=1}^{d_{\rho_1}} \bigoplus_{\rho_2 \in R(H)} \bigoplus_{j=1}^{c^{\rho_2}_{\mathrm{in,out}}} \mathrm{Hom}_H(M_{\rho_1}, M_{\rho_2}),\end{aligned}$$

where $c^{\rho_2}_{\mathrm{in,out}}$ is the number of times that the irreducible representation M_{ρ_2} occurs in the representation $\overline{\sigma_{\mathrm{in}}} \otimes \sigma_{\mathrm{out}}$ (which is equal to 0 if it does not occur). Recall that the coefficients $c^{\rho_2}_{\mathrm{in,out}}$ are the Clebsch–Gordan coefficients (see Definition 4.11). Now, since we have been dealing with complex representations all along, by Schur's Lemma (see Lemma 3.2(2) or Theorem 3.11(2)), since M_{ρ_1} and M_{ρ_2} are irreducible,

$$\mathrm{Hom}_H(M_{\rho_1}, M_{\rho_2}) = \begin{cases} \{0\}, & \text{if } \rho_1 \neq \rho_2 \\ \mathbb{C}, & \text{if } \rho_1 = \rho_2. \end{cases}$$

Therefore, we obtain

$$\begin{aligned}\mathrm{Hom}_H(\mathbf{R}^{H \to \mathrm{L}^2(X)}, \overline{\sigma_{\mathrm{in}}} \otimes \sigma_{\mathrm{out}}) &\approx \bigoplus_{\rho_1 \in R(H)} \bigoplus_{k_{\rho_1}=1}^{d_{\rho_1}} \bigoplus_{j=1}^{c^{\rho_1}_{\mathrm{in,out}}} \mathrm{Hom}_H(M_{\rho_1}, M_{\rho_1}) \\ &\approx \bigoplus_{\rho_1 \in R(H)} \bigoplus_{k_{\rho_1}=1}^{d_{\rho_1}} \bigoplus_{j=1}^{c^{\rho_1}_{\mathrm{in,out}}} \mathcal{C}(M_{\rho_1}) = \bigoplus_{\rho_1 \in R(H)} \bigoplus_{k_{\rho_1}=1}^{d_{\rho_1}} \bigoplus_{j=1}^{c^{\rho_1}_{\mathrm{in,out}}} \mathbb{C},\end{aligned}$$

where $\mathcal{C}(M_{\rho_1})$ is commutant (or centralizer) of M_{ρ_1}, see Definition 3.9.

Cesa, Lang and Weiler [Cesa *et al.* (2022)] also consider real representations. In this case, Schur's Lemma (Lemma 3.2(2)) is not as strong and the centralizers may have dimension greater than 1. The situation where real representations are considered is actually quite subtle and fairly involved. Among other things, a version of the Peter–Weyl theorems in the real case is required. Cesa, Lang and Weiler [Cesa *et al.* (2022)] and Lang and Weiler [Lang and Weiler (2019)] address these issues in great depth.

In summary, we have shown the following basis independent version in the case of complex representations of a theorem referred to as *Wigner–Eckart theorem for steerable kernels* by Cesa, Lang and Weiler [Cesa *et al.* (2022)] (Theorem B.5). The more general theorem that also applies to real representations is proven in Cesa, Lang and Weiler [Cesa *et al.* (2022)].

Theorem 8.15. *There is an isomorphism of vector spaces*

$$\mathrm{Hom}_H(X, \mathrm{Hom}(\mathcal{H}_{\mathrm{in}}, \mathcal{H}_{\mathrm{out}})) \approx \bigoplus_{\rho_1 \in R(H)} \bigoplus_{k_{\rho_1}=1}^{d_{\rho_1}} \bigoplus_{j=1}^{c^{\rho_1}_{\mathrm{in,out}}} \mathcal{C}(M_{\rho_1}). \tag{$\dagger_{16}$}$$

If the representations are complex, then $\mathcal{C}(M_{\rho_1}) = \mathbb{C}$ *for all* $\rho_1 \in R(H)$.

Cesa, Lang and Weiler [Cesa *et al.* (2022)] also prove a version of the above result in which a basis of $\mathrm{Hom}_H(X, \mathrm{Hom}(\mathcal{H}_{\mathrm{in}}, \mathcal{H}_{\mathrm{out}}))$ is exhibited. The formulae are a bit messy so we will not give details here; see Theorem B.6 and Theorem B.7 in Cesa, Lang and Weiler [Cesa *et al.* (2022)]. The idea is clear though. A steerable basis for $\mathrm{L}^2(X)$ is provided by Theorem 8.7; these are the functions Y_{ρ,k_ρ}. Matrices $CG_j^{c^{\rho_1}_{\mathrm{in,out}}}$ of Clebsch–Gordan coefficients expressing the change of basis required when decomposing the representation $\overline{\sigma_{\mathrm{in}}} \otimes \sigma_{\mathrm{out}}$ into irreducibles are needed. If $\mathcal{C}(M_{\rho_1})$ is not one-dimensional, then a basis $C^{j\rho_1}$ (in matrix form) of each copy of $\mathcal{C}(M_{\rho_1})$ is needed. In matrix form, a basis of $\mathrm{Hom}_H(X, \mathrm{Hom}(\mathcal{H}_{\mathrm{in}}, \mathcal{H}_{\mathrm{out}}))$ is given by

$$K_{\rho k_\rho j \rho_1} = (CG_j^{c^{\rho_1}_{\mathrm{in,out}}})^* C^{j\rho_1} \overline{Y_{\rho,k_\rho}}; \tag{$\dagger_{17}$}$$

see Theorem B.7 in Cesa, Lang and Weiler [Cesa *et al.* (2022)] and Theorem D13 (Formula (20)) in Lang and Weiler [Lang and Weiler (2019)]. These two papers use different definitions of steerability and since we use (steer7), as in Lang and Weiler [Lang and Weiler (2019)], we need a conjugation over Y (see the remark just before Theorem 8.7).

With Formula ($\dagger_{17}$), we finally achieved the goal of Section 8.12, which was to define a notion of G-equivariant correlation directly on feature fields.

For this, we introduced G-equivariant kernels (for $G = \mathbb{R}^d \rtimes H$) and we showed that they must satisfy the equivariance constraints (EC_1). In Section 8.13, we considered the more general situation of two hermitian G-bundles. In Proposition 8.12, we characterized equivariant correlation G-kernels (denoted K), whose domain in G. These are not practical so we showed in Section 8.14 that the previous G-kernels can be expressed in terms of X_{in}-kernels (denoted κ) which are defined on X_{in}; see Proposition 8.13. They satisfy condition (EC_3), which is simpler than (EC_1) but still contains the term $u^{\mathrm{in}}(h_2, x)^{-1}$. This term disappears if G is a semi-direct product. In Section 8.15, we show how to replace integration over G by integration over X_{in}, using the X_{in}-kernels instead of G-kernels; see Formula ($\dagger_8$). Finally, in this section, we obtained a complete characterization as well as steerable bases of X_{in}-kernels in the simper case where $H_{\mathrm{in}} = H_{\mathrm{out}} = H$ (but we still have input and output representations σ_{in} and σ_{out}). This result (Theorem 8.15) is a generalization of the results shown in Sections 8.6 and 8.7, where the steerable families Y_{ρ,k_ρ} are replaced by a basis of steerable kernels given by ($\dagger_{17}$).

As explained in Section 6 and Section B4 of Lang and Weiler [Lang and Weiler (2019)] and in Cesa, Lang and Weiler [Cesa *et al.* (2022)] (Algorithm 1 in Section 3), in order to actually compute a basis of a steerable kernel, the following steps need to be carried out.

(1) Compute the irreducible representations M_ρ of H (or at least those are needed).
(2) Compute a steerable basis of functions Y_{ρ,k_ρ} for $\mathrm{L}^2(X)$ (provided by Theorem 8.7). See Appendix B of Cesa, Lang and Weiler [Cesa *et al.* (2022)], in particular, Section B.4.
(3) Find the Clebsch–Gordan decomposition (in matrix form) into irreducibles for the tensor product representation $\overline{\sigma_{\mathrm{in}}} \otimes \sigma_{\mathrm{out}}$. Appendix E of Cesa, Lang and Weiler [Cesa *et al.* (2022)] presents numerical methods to do this. If the group H is finite then this can be done by solving for the null space of some suitable linear system. If the group H is infinite, then this can often be done by random sampling and then solving for the null space of the linear system obtained from the sampling process.
(4) Find a basis for the commutant $\mathcal{C}(M_\rho)$ of M_ρ for all $\rho \in R(H)$ (or at least those that are needed). This problem is addressed in Appendix C of Cesa, Lang and Weiler [Cesa *et al.* (2022)].

Explicit examples are given in Cesa, Lang and Weiler [Cesa *et al.* (2022)].

8.17 Problems

Problem 8.1. Let $\Phi\colon \mathrm{L}^2(\mathbb{R}^2) \to \mathrm{L}^2(\mathbb{R}^2)$ be the linear operator given by

$$\Phi(f) = k \star f,$$

as in Section 8.2. Recall that $\mathbb{R}^2$ acts on $\mathrm{L}^2(\mathbb{R}^2)$ by the regular left representation $\mathbf{R}^{\mathbb{R}^2 \to \mathrm{L}^2(\mathbb{R}^2)}$. Prove that we have the commutative diagram

$$\begin{array}{ccc} \mathrm{L}^2(\mathbb{R}^2) & \xrightarrow{\Phi} & \mathrm{L}^2(\mathbb{R}^2) \\ {\scriptstyle \mathbf{R}_x^{\mathbb{R}^2 \to \mathrm{L}^2(\mathbb{R}^2)}}\big\downarrow & & \big\downarrow{\scriptstyle \mathbf{R}_x^{\mathbb{R}^2 \to \mathrm{L}^2(\mathbb{R}^2)}} \\ \mathrm{L}^2(\mathbb{R}^2) & \xrightarrow[\Phi]{} & \mathrm{L}^2(\mathbb{R}^2) \end{array}$$

for all $x \in \mathbb{R}^2$, which shows that the transform Φ is translation-equivariant.

Problem 8.2. Consider the function φ on $\mathcal{K}(G)$ given by

$$\varphi(f) = \int_H \mathrm{mod}(i_h)^{-1} \left(\int_N f(nh)\, d\lambda_N(n) \right) d\lambda_H(h). \tag{μ_7}$$

Using the Radon–Riesz theorem II (see Vol. I, Theorem 7.15) to show that φ defines a Radon measure λ_G on G, such that $\int_G f(g) d\lambda_G(g) = \varphi(f)$ for all $f \in \mathcal{K}(G)$.

Problem 8.3. Prove that if ν_N and ν_H are right Haar measures, then the image of the measure $\nu_N \otimes \nu_H$ by φ is also a right Haar measure.

Problem 8.4. Find the Hermite polynomials $H_0, \ldots, H_5$ using the recurrence relations given in Example 8.5 just after (srt7).

Problem 8.5. Prove that if $\widehat{f}_\rho$ is the function defined in Definition 8.9, then $\widehat{f}_\rho \in \mathrm{L}^2(\mathbb{R}^d, \mathrm{M}_{n_\rho}(\mathbb{C}))$.

Hint. Use Fubini.

Problem 8.6. Prove part (2) of Proposition 8.11.

Problem 8.7. Study the proof of Theorem 8.14 given in Cesa, Lang and Weiler [Cesa *et al.* (2022)] (Theorem B2). See also Lang and Weiler [Lang and Weiler (2019)] (Theorem C7), in particular, the argument involving approximating the Dirac function.

Problem 8.8. Study the examples given in Cesa, Lang and Weiler [Cesa *et al.* (2022)], as well as Appendix C.

Chapter 9

Harmonic Analysis on Gelfand Pairs

This chapter is the culmination of all of the theories discussed in this book. We are able to present a very general version of the Fourier transform on a homogeneous space G/K, where (G, K) is a Gelfand pair. This chapter presents material discussed in Dieudonné [Dieudonné (2003b)] (Chapter XXII, Sections 6–10).

We saw in Section 6.9 that if G is a compact group and if H is a closed subgroup of G, then the algebra $\mathrm{L}^2(H\backslash G/H)$ is commutative if and only if $(\rho : \sigma_0) \leq 1$ for all $\rho \in R(G)$ (where σ_0 is the class of the trivial representation of H). If so, then for every $\rho \in R(G)$, such that $(\rho : \sigma_0) = 1$, the ideal $\mathfrak{a}_{\rho,\sigma_0}$ is one-dimensional and is spanned by the function

$$\omega_\rho(s) = \theta_{\rho,\sigma_0} = \frac{1}{n_\rho} m_{11}^{(\rho,H)}(s),$$

which is continuous and of positive type. The function ω_ρ is called a (zonal) spherical function.

The goal of this chapter is to generalize the above results for a compact group to a locally compact (metrizable and separable) unimodular group G and to a compact subgroup K of G.

The first difficulty is that if G is not compact, then $\mathrm{L}^2(G)$ is not closed under convolution (in general, $\mathrm{L}^2(G)$ is not contained in $\mathrm{L}^1(G)$). So we have to work with $\mathcal{K}(G)$ instead (recall that $\mathcal{K}(G)$ is the subset of $\mathcal{C}(G)$ consisting of the continuous functions with compact support $f\colon G \to \mathbb{C}$).

There is a bijection between the space $\mathcal{C}(G/K)$ of continuous functions $f\colon G/K \to \mathbb{C}$ and the subspace of continuous functions $g\colon G \to \mathbb{C}$, such that

$$g(st) = g(s), \qquad \text{for all } s \in G \text{ and all } t \in K.$$

We also have a bijection between the space $\mathcal{C}(K\backslash G)$ of continuous functions $f\colon K\backslash G \to \mathbb{C}$ and the subspace of continuous functions $g\colon G \to \mathbb{C}$, such that

$$g(ts) = g(s), \qquad \text{for all } s \in G \text{ and all } t \in K.$$

Let $\mathcal{C}(K\backslash G/K) = \mathcal{C}(G/K) \cap \mathcal{C}(K\backslash G)$, which consists of the continuous functions $f\colon G \to \mathbb{C}$ which are constant on double cosets KsK ($s \in G$), and let $\mathcal{K}(K\backslash G/K)$ be the subspace of $\mathcal{C}(K\backslash G/K)$ consisting of the continuous functions with compact support. The space $\mathcal{K}(K\backslash G/K)$ is an involutive subalgebra of $\mathcal{K}(G)$, and thus of $\mathrm{L}^1(G)$.

The key ingredient is the Banach algebra $\mathrm{L}^1(K\backslash G/K)$, the closure of $\mathcal{K}(K\backslash G/K)$ in $\mathrm{L}^1(G)$. Gelfand's remarkable discovery is that much of the harmonic analysis on abelian locally compact groups and compact groups can be generalized to a pair (G, K), where G is a noncommutative locally compact unimodular (metrizable and separable) group, and K is a compact subgroup of G, *if the algebra* $\mathrm{L}^1(K\backslash G/K)$ *is commutative*. In this case, (G, K) is called a *Gelfand pair*.

Fortunately, there is a sufficient criterion for a pair (G, K) to be a Gelfand pair involving an involutive isomorphism $\sigma\colon G \to G$, such that K is a closed subgroup of the group G^σ of fixed points of σ (see Theorem 9.2). This criterion is reminiscent of Élie Cartan's notion of symmetric space (see Helgason [Helgason (2001)] or Gallier and Quaintance [Gallier and Quaintance (2020a)]), and indeed, many kinds of symmetric spaces are Gelfand pairs. The proof that a pair satisfying this criterion is a Gelfand pair is given in Section 9.1. The conditions of this criterion are flexible enough to apply to three broad classes of pairs (G, K); see Section 9.7.

The purpose of Section 9.2 is to characterize the characters of the algebra $\mathrm{L}^1(K\backslash G/K)$ in terms of certain functions in $\mathcal{C}(K\backslash G/K)$, called *spherical functions*. Every character ζ of the commutative Banach algebra $\mathrm{L}^1(K\backslash G/K)$ is given by a unique function $\omega \in \mathcal{C}(K\backslash G/K)$, which is bounded and continuous on G, with

$$\zeta(f) = (f, \omega) = \int_G f(x)\omega(x)\, d\lambda_G(x), \qquad f \in \mathcal{L}^1(K\backslash G/K);$$

see Proposition 9.4. A function $\omega \in \mathcal{C}(K\backslash G/K)$ as above is called a *spherical function*.

Two criteria for a bounded function $\omega \in \mathcal{C}(K\backslash G/K)$ (different from the zero function) to be a spherical function are given in Theorem 9.6. In particular, the function ω is a spherical function on G relative to K iff

$$\int_K \omega(xty)\, d\lambda_K(t) = \omega(x)\omega(y), \qquad \text{for all } x, y \in G. \tag{s_1}$$

The space of spherical functions on the Gelfand pair (G, K) is denoted $\mathbf{S}(G/K)$. The subspace of characters of the commutative involutive Banach algebra $A = \mathrm{L}^1(K\backslash G/K) \oplus \mathbb{C}\delta_e$ whose restriction to $\mathrm{L}^1(K\backslash G/K)$ is not the zero function is denoted by $\mathsf{X}_0(A)$. This subspace is locally compact in the weak*-topology (metrizable and separable).

The map $\omega \mapsto \zeta_\omega = (f, \omega)$ is a homeomorphism of $\mathbf{S}(G/K)$ equipped with the induced topology of Fréchet space of $\mathcal{C}(G)$ onto $\mathsf{X}_0(A)$ equipped with the topology induced by the weak*-topology of the dual A' of A. Consequently, $\mathbf{S}(G/K)$ is locally compact.

An important class of Lie groups that yield Gelfand pairs are the real forms of a complex semi-simple Lie group, and Sections 9.3–9.6 are devoted to a discussion of these Lie groups. One needs to understand how to find the *real forms* $\mathfrak{g}_0$ of a complex Lie algebra $\mathfrak{g}$, that is, the real Lie algebras $\mathfrak{g}_0$, such that

$$\mathfrak{g} = \mathfrak{g}_0 \oplus i\mathfrak{g}_0,$$

namely, $\mathfrak{g}$ is the complexification of $\mathfrak{g}_0$. Finding such real algebras $\mathfrak{g}_0$ is equivalent to finding certain semilinear idempotent maps on $\mathfrak{g}$ called *conjugations* (see Proposition 9.11). Now, if we happen to have some semi-simple real form $\mathfrak{g}_u$, such that

$$\mathfrak{g} = \mathfrak{g}_u \oplus i\mathfrak{g}_u,$$

and if c_u is the corresponding conjugation, then it can be shown that *all* the other real forms $\mathfrak{g}_0$ of $\mathfrak{g}$ are given by conjugations c_0 that commute with c_u. Then all this has to be promoted to Lie groups (essentially by using the exponential map).

Three examples of Gelfand pairs are discussed in Section 9.7. In the first example, G is a compact Lie group that has an involutive automorphism σ; this corresponds to symmetric spaces of compact type. In the second example, G_1 arises as a real form of a complex, semi-simple, simply-connected Lie group G, and G_1 has finite center. This corresponds to a symmetric space of noncompact type. The third example is a certain kind of semi-direct product; a typical illustration of this case is the group of rigid motions $\mathbf{SE}(n, \mathbb{R})$.

The Fourier transform and the Fourier cotransform for a Gelfand pair are introduced in Section 9.8. For every function $f \in \mathcal{L}^1(K\backslash G/K)$, the *Fourier cotransform* $\overline{\mathcal{F}}f$ of f is the function $\overline{\mathcal{F}}f\colon \mathbf{S}(G/K) \to \mathbb{C}$ given by

$$(\overline{\mathcal{F}}f)(\omega) = (f, \omega) = \int_G f(x)\omega(x)\, d\lambda_G(x), \qquad \omega \in \mathbf{S}(G/K),$$

and the *Fourier transform* $\mathcal{F}f$ of f is the function $\mathcal{F}f\colon \mathbf{S}(G/K) \to \mathbb{C}$ given by

$$(\mathcal{F}f)(\omega) = (\check{f}, \omega) = \int_G f(x^{-1})\omega(x)\, d\lambda_G(x) = \int_G f(x)\omega(x^{-1})\, d\lambda_G(x)$$

for all $\omega \in \mathbf{S}(G/K)$. Using the space $\mathbf{S}(G/K)$ of spherical functions as the domain of $\mathcal{F}$ and $\overline{\mathcal{F}}$ instead of characters yields a simultaneous generalization of the case where G is commutative and the case where G is compact. On $\mathcal{L}^1(K\backslash G/K)$, we have the familiar equations

$$\mathcal{F}(f * g) = (\mathcal{F}f)(\mathcal{F}g), \qquad \overline{\mathcal{F}}(f * g) = (\overline{\mathcal{F}}f)(\overline{\mathcal{F}}g). \tag{$*$}$$

In Section 9.9, we generalize Fourier inversion. For this, we use the construction of certain positive σ-Radon measures from measures of positive type (recall Definition 3.21) using the *Plancherel transform*.

When G is not compact, the spherical functions in $\mathbf{S}(G/K)$ are not necessarily of positive type. The subset of $\mathbf{S}(G/K)$ consisting of the *spherical functions of positive type* is denoted by $\mathbf{Z}(G/K)$.

Theorem 9.21 states that given a measure μ of positive type on G, there is a unique (positive) Radon measure $\mu^{\triangle}$ on $\mathbf{Z}(G/K)$, such that for every function $f \in \mathcal{K}(K\backslash G/K)$, the Fourier cotransform $\overline{\mathcal{F}}f$ belong to $\mathcal{L}^2_{\mu^{\triangle}}(\mathbf{Z}(G/K);\mathbb{C})$, and for any two functions $f, g \in \mathcal{K}(K\backslash G/K)$, we have

$$\int_G (g^* * f) d\mu = \int_{\mathbf{Z}(G/K)} (\overline{\mathcal{F}}f)(\omega)\overline{(\overline{\mathcal{F}}g)(\omega)}\, d\mu^{\triangle}(\omega).$$

The measure $\mu^{\triangle}$ is called the *Plancherel transform* of μ.

In particular, if $\mu = \delta_e$, the Dirac measure, we obtain a measure $m_{\mathbf{Z}} = \delta_e^{\triangle}$, called the *canonical measure* on $\mathbf{Z}(G/K)$, and then the linear maps $f \mapsto \mathcal{F}f$ and $f \mapsto \overline{\mathcal{F}}f$, with $f, g \in \mathcal{K}(K\backslash G/K)$, are isometries, and by Theorem 9.21, these maps extend to isomorphisms from the Hilbert space $\mathrm{L}^2(K\backslash G/K)$ onto the Hilbert space $\mathrm{L}^2_{m_{\mathbf{Z}}}(\mathbf{Z}(G/K))$. This is a generalization of the Plancherel theorem (Vol. I, Theorem 10.27).

Another type of Fourier inversion formula is given by Proposition 9.28. The map $p \mapsto (p\lambda_G)^{\triangle}$ is a bijection between the space $\mathcal{P}_+(K\backslash G/K)$

of functions in $\mathcal{C}(K\backslash G/K)$, which are of positive type onto the space $\mathcal{M}^1_+(\mathbf{Z}(G/K))$ of bounded positive measures on $\mathbf{Z}(G/K)$.

Section 9.10 discusses an extension of the Plancherel transform to the space $\mathbf{P}(G)$ which is the complex vector space spanned by the union of the complex measures and the Radon measures. In Proposition 9.34, we obtain a Fourier inversion formula which yields the inversion formula of the Pontrjagin duality theorem, Vol. I, Theorem 10.30, as a special case.

Finally, in Section 9.11, we show that functions of positive type induce irreducible representations; see Theorem 9.35. We also state a theorem of Stone characterizing the unitary representations of $\mathbb{R}$ in a separable Hilbert space (Theorem 9.40).

9.1 Gelfand Pairs

In the rest of this chapter, we assume that G is locally compact, metrizable, separable and unimodular group, and that K is a compact subgroup of G. Recall that there is a bijection between the space $\mathcal{C}(G/K)$ of continuous functions $f\colon G/K \to \mathbb{C}$ and the subspace of continuous functions $g\colon G \to \mathbb{C}$, such that

$$g(st) = g(s), \qquad \text{for all } s \in G \text{ and all } t \in K,$$

equivalently,

$$g * \delta_t = g, \qquad \text{for all } t \in K.$$

This bijection is given by the map $f \mapsto f \circ \pi$, where $\pi\colon G \to G/K$ is the projection map. Observe that $(g * \delta_t)(s) = g(st^{-1})$ (see $(*_{\rho_{s^{-1}}})$ after Vol. I, Definition 8.26; since G is unimodular, the term $\Delta(t^{-1})$ is equal to 1), so the condition $g * \delta_t = g$ is equivalent to $g(st^{-1}) = g(s)$ for all $t \in K$, but this is equivalent to $g(st) = g(s)$ for all $t \in K$ since K is a group. From now on, we identify $\mathcal{C}(G/K)$ with the subspace of $\mathcal{C}(G)$ satisfying the above equivalent properties.

We also have a bijection between the space $\mathcal{C}(K\backslash G)$ of continuous functions $f\colon K\backslash G \to \mathbb{C}$ and the subspace of continuous functions $g\colon G \to \mathbb{C}$, such that

$$g(ts) = g(s), \qquad \text{for all } s \in G \text{ and all } t \in K,$$

equivalently,

$$\delta_t * g = g, \qquad \text{for all } t \in K,$$

and we identify $\mathcal{C}(K\backslash G)$ with the subspace of $\mathcal{C}(G)$ satisfying the above equivalent properties. Observe that $(\delta_t * g)(s) = g(t^{-1}s)$, so $\delta_t * g = g$ is equivalent to $g(t^{-1}s) = g(s)$ for all $t \in K$, which is equivalent to $g(ts) = g(s)$ for all $t \in K$.

Let $\mathcal{C}(K\backslash G/K) = \mathcal{C}(G/K) \cap \mathcal{C}(K\backslash G)$, which consists of the continuous functions $f: G \to \mathbb{C}$ which are constant on double cosets KsK ($s \in G$).

Since K is compact, for every compact subset A of G/K, the subset $\pi^{-1}(A)$ is compact in G. The map $f \mapsto f \circ \pi$ is thus a bijection between the subspace $\mathcal{K}(G/K)$ of $\mathcal{C}(G/K)$ onto a subspace of $\mathcal{K}(G)$. This subspace is denoted by $\mathcal{K}(G) \cap \mathcal{C}(G/K)$. Similarly, there is a bijection between the space $\mathcal{K}(K\backslash G)$ and the space $\mathcal{K}(G) \cap \mathcal{C}(K\backslash G)$, and a bijection between the space $\mathcal{K}(K\backslash G/K)$ and the space $\mathcal{K}(G) \cap \mathcal{C}(K\backslash G/K)$.

We saw earlier that $\mathcal{K}(G)$ is an involutive subalgebra under convolution of the algebra $\mathrm{L}^1(G)$ (see Vol. I, Example 9.6(4)). It follows that $\mathcal{K}(G/K)$ is a left ideal in $\mathcal{K}(G)$ and that $\mathcal{K}(K\backslash G)$ is a right ideal in $\mathcal{K}(G)$, and the involution $f \mapsto \check{\overline{f}}$ maps $\mathcal{K}(G/K)$ onto $\mathcal{K}(K\backslash G)$. As a consequence, $\mathcal{K}(K\backslash G/K)$ is an involutive subalgebra of $\mathcal{K}(G)$ (and so of $\mathrm{L}^1(G)$).

Let λ_K be the Haar measure on K normalized so that $\lambda_K(K) = 1$, and since G is assumed to be unimodular, let λ_G be a left and right-invariant Haar measure on G (since K is compact, it is unimodular so λ_K is also left and right-invariant). In order to study the characters of the Banach algebra $\mathrm{L}^1(K\backslash G/K)$, which is the closure of $\mathcal{K}(K\backslash G/K)$ in $\mathrm{L}^1(G)$, we need to project $\mathcal{C}(G)$ onto $\mathcal{C}(K\backslash G/K)$.

Definition 9.1. We define a projection map from $\mathcal{C}(G)$ onto $\mathcal{C}(K\backslash G/K)$ by

$$f^\sharp(s) = \int_K \int_K f(tst')\, d\lambda_K(t)\, d\lambda_K(t'), \qquad s \in G,$$

for any $f \in \mathcal{C}(G)$.

It is easily checked that if $f \in \mathcal{C}(G)$, then $f^\sharp(t_1 s t_1') = f^\sharp(s)$ for all $t_1, t_1' \in K$, so $f^\sharp \in \mathcal{C}(K\backslash G/K)$. As a consequence, since $\lambda(K) = 1$,

$$\begin{aligned} f^{\sharp\sharp}(s) &= \int_K \int_K f^\sharp(tst')\, d\lambda_K(t)\, d\lambda_K(t') \\ &= \int_K \int_K f^\sharp(s)\, d\lambda_K(t)\, d\lambda_K(t') = f^\sharp(s), \end{aligned}$$

and the map $f \mapsto f^\sharp$ is indeed a projection. It is also easy to check that for all $f \in \mathcal{C}(K\backslash G/K)$ and for all $g \in \mathcal{C}(G)$, we have

$$(fg)^\sharp = fg^\sharp.$$

Proposition 9.1. *The restriction of the projection $f \mapsto f^\sharp$ to $\mathcal{K}(G)$ maps $\mathcal{K}(G)$ onto $\mathcal{K}(K\backslash G/K)$, and we have*

$$(f * g)^\sharp = f * g^\sharp, \qquad (g * f)^\sharp = g^\sharp * f$$

for all $f \in \mathcal{K}(K\backslash G/K)$, and all $g \in \mathcal{C}(G)$.

Proof. We leave the first statement as an exercise and prove the first of the two equations. We have

$$(f * g)(x) = \int_G f(s)g(s^{-1}x)\, d\lambda_G(s),$$

and using the left invariance of λ_G and the fact that $f(ts) = f(s)$ and $\lambda_K(K) = 1$, we have

$$\begin{aligned}
(f * g)^\sharp(x) &= \int_K \int_K \int_G f(s)g(s^{-1}txt')\, d\lambda_G(s)\, d\lambda_K(t)\, d\lambda_K(t') \\
&= \int_K \int_K \int_G f(ts)g(s^{-1}xt')\, d\lambda_G(s)\, d\lambda_K(t)\, d\lambda_K(t') \\
&= \int_K \int_K \int_G f(s)g(s^{-1}xt')\, d\lambda_G(s)\, d\lambda_K(t)\, d\lambda_K(t') \\
&= \int_K \int_G f(s)g(s^{-1}xt')\, d\lambda_G(s)\, d\lambda_K(t').
\end{aligned}$$

We also have

$$g^\sharp(x) = \int_K \int_K g(txt')\, d\lambda_K(t)\, d\lambda_K(t'),$$

and using the right invariance of λ_G and the fact that $f(st) = f(s)$ and $\lambda_K(K) = 1$, we have

$$\begin{aligned}
(f * g^\sharp)(x) &= \int_G \int_K \int_K f(s)g(ts^{-1}xt')\, d\lambda_K(t)\, d\lambda_K(t')\, d\lambda_G(s) \\
&= \int_G \int_K \int_K f(st)g(s^{-1}xt')\, d\lambda_K(t)\, d\lambda_K(t')\, d\lambda_G(s) \\
&= \int_G \int_K \int_K f(s)g(s^{-1}xt')\, d\lambda_K(t)\, d\lambda_K(t')\, d\lambda_G(s) \\
&= \int_G \int_K f(s)g(s^{-1}xt')\, d\lambda_K(t')\, d\lambda_G(s),
\end{aligned}$$

and by Fubini, we can interchange the integrations, which shows that we have $(f * g)^\sharp = f * g^\sharp$. The proof that $(g * f)^\sharp = g^\sharp * f$ is analogous. ☐

A Gelfand pair (G, K) is defined as follows.

Definition 9.2. Let G be a locally compact (metrizable and separable) unimodular group, and let K be a compact subgroup. We say that the pair (G, K) is a *Gelfand pair* if the algebra $\mathcal{K}(K\backslash G/K)$ is commutative (under convolution).

Obviously, if G is abelian, then (G, K) is a Gelfand pair. The following theorem, which is one of the key theorems of the theory of Gelfand pairs, gives a sufficient criterion for a pair (G, K) to be a Gelfand pair. Using this criterion, we will show later that there are lots of Gelfand pairs.

Theorem 9.2. *(Gelfand) Let G be locally compact (metrizable and separable) unimodular group, $\sigma\colon G \to G$ be an involutive isomorphism of G ($\sigma \circ \sigma = \mathrm{id}_G$), and $G^\sigma = \{s \in G \mid \sigma(s) = s\}$ be the subgroup of elements of G left fixed by σ. Let K be any closed subgroup of G^σ and assume the following properties:*

(1) The subgroup G^σ is compact.
(2) Every $x \in G$ can be written (possibly not uniquely) as $x = yz$, with $y \in K$, $z \in G$, and $\sigma(z) = z^{-1}$.

Then (G, K) is Gelfand pair.

Proof. Since $\sigma^2 = \mathrm{id}_G$, we have $\mathrm{mod}(\sigma)^2 = \mathrm{mod}(\sigma^2) = \mathrm{mod}(\mathrm{id}_G) = 1$, and since $\mathrm{mod}(\sigma) > 0$, we get $\mathrm{mod}(\sigma) = 1$ (see Vol. I, Definition 8.17 and Proposition 8.31). It follows that σ leaves any Haar measure λ_G of G invariant. For every $f \in \mathcal{K}(G)$, let f^σ be the function given by

$$f^\sigma(s) = f(\sigma(s)) = f(\sigma^{-1}(s)), \qquad s \in G.$$

We check immediately that the map $\widehat{\sigma}\colon \mathcal{K}(G) \to \mathcal{K}(G)$ given by $\widehat{\sigma}(f) = f^\sigma$ is an involutive automorphism of the vector space $\mathcal{K}(G)$, and since σ leaves any Haar measure on G invariant, it is an automorphism of the *algebra* $\mathcal{K}(G)$ (under convolution). Since $\sigma(t) = t$ for all $t \in K$, if $f(x) = f(txt')$ for all $t, t \in K$, then $f(\sigma(txt')) = f(t\sigma(x)t') = f(\sigma(x))$, so if $f \in \mathcal{K}(K\backslash G/K)$ then $f^\sigma \in \mathcal{K}(K\backslash G/K)$, which means that the automorphism $\widehat{\sigma}$ leaves the algebra $\mathcal{K}(K\backslash G/K)$ invariant, and its restriction to $\mathcal{K}(K\backslash G/K)$ is an automorphism of this subalgebra. If we can prove that

$$f^\sigma * g^\sigma = g^\sigma * f^\sigma, \qquad \text{for all } f, g \in \mathcal{K}(K\backslash G/K),$$

then we will have proven that $\mathcal{K}(K\backslash G/K)$ is commutative.

The trick is that for any function $f \in \mathcal{K}(K\backslash G/K)$, we have $f^\sigma = \check{f}$. Every $x \in G$ can be written as $x = yz$ with $y \in K$ and $\sigma(z) = z^{-1}$, which yields

$$\sigma(x) = \sigma(yz) = \sigma(y)\sigma(z) = yz^{-1} = y(z^{-1}y^{-1})y,$$

and for every $f \in \mathcal{K}(K\backslash G/K)$, we have

$$f(\sigma(x)) = f(y(z^{-1}y^{-1})y) = f(z^{-1}y^{-1}) = f(x^{-1}) = \check{f}(x);$$

that is, $f^\sigma = \check{f}$. Since G is unimodular, the Haar measure is right-invariant, so for any two function $f, g \in \mathcal{K}(G)$, we easily verify that

$$(\check{f} * \check{g})\check{} = g * f.$$

Then for $f, g \in \mathcal{K}(K\backslash G/K)$, we have

$$f^\sigma * g^\sigma = \check{f} * \check{g} = (g * f)\check{} = (g * f)^\sigma = g^\sigma * f^\sigma,$$

where we used Vol. I, Proposition 8.28 and the fact that $\mathrm{mod}(\sigma) = 1$ to prove the equality $(g * f)^\sigma = g^\sigma * f^\sigma$. The details are left as an exercise. $\square$

From now on, *we assume that (G, K) is a Gelfand pair.*

Definition 9.3. The closure of $\mathcal{K}(K\backslash G/K)$ in $\mathrm{L}^1(G)$ is denoted $\mathrm{L}^1(K\backslash G/K)$. The space $\mathrm{L}^1(K\backslash G/K)$ is an involutive, commutative, Banach subalgebra of $\mathrm{L}^1(G)$. Similarly, denote by $\mathrm{L}^2(K\backslash G/K)$ the closure of $\mathcal{K}(K\backslash G/K)$ in $\mathrm{L}^2(G)$.

The following results are proven in Dieudonné [Dieudonné (2003b)] (Chapter XXII, Section 6).

Proposition 9.3. *The projection $f \mapsto f^\sharp$ of $\mathcal{K}(G)$ onto $\mathcal{K}(K\backslash G/K)$ extends to a continuous projection of $\mathrm{L}^1(G)$ onto $\mathrm{L}^1(K\backslash G/K)$. For any $f \in \mathcal{L}^1(G)$, we have $\|f^\sharp\|_1 \leq \|f\|_1$, and the class $[f^\sharp]$ is the class of the function $f^\sharp$ equal almost everywhere to the function*

$$s \mapsto \int_K \int_K f(tst')\, d\lambda_K(t)\, d\lambda_K(t').$$

Similarly, the projection $f \mapsto f^\sharp$ of $\mathcal{K}(G)$ onto $\mathcal{K}(K\backslash G/K)$ extends to a continuous projection of $\mathrm{L}^2(G)$ onto $\mathrm{L}^2(K\backslash G/K)$. For any $f \in \mathcal{L}^2(G)$, we have $\|f^\sharp\|_2 \leq \|f\|_2$, and the class $[f^\sharp]$ is the class of the function $f^\sharp$ equal almost everywhere to the function

$$s \mapsto \int_K \int_K f(tst')\, d\lambda_K(t)\, d\lambda_K(t').$$

Definition 9.4. We denote by $\mathcal{L}^1(G/K)$ and $\mathcal{L}^2(G/K)$ the subspaces of $\mathcal{L}^1(G)$ and $\mathcal{L}^2(G)$ consisting of the functions f, such that for almost all $s \in G$, we have

$$f(st) = f(s), \qquad \text{for all } t \in K.$$

Similarly, we denote by $\mathcal{L}^1(K\backslash G)$ and $\mathcal{L}^2(K\backslash G)$ the subspaces of $\mathcal{L}^1(G)$ and $\mathcal{L}^2(G)$ consisting of the functions f, such that for almost all $s \in G$, we have

$$f(ts) = f(s), \qquad \text{for all } t \in K.$$

Let $\mathcal{L}^1(K\backslash G/K) = \mathcal{L}^1(G/K) \cap \mathcal{L}^1(K\backslash G)$ and $\mathcal{L}^2(K\backslash G/K) = \mathcal{L}^2(G/K) \cap \mathcal{L}^2(K\backslash G)$.

If $f \in \mathcal{L}^1(G)$ (resp. $f \in \mathcal{L}^2(G)$), then $f^\sharp \in \mathcal{L}^1(K\backslash G/K)$ (resp. $f^\sharp \in \mathcal{L}^2(K\backslash G/K)$), and $\mathrm{L}^1(K\backslash G/K)$ (resp. $\mathrm{L}^2(K\backslash G/K)$) is the canonical image in $\mathrm{L}^1(G)$ (resp. $\mathrm{L}^2(G)$) of $\mathcal{L}^1(K\backslash G/K)$ (resp. $\mathcal{L}^2(K\backslash G/K)$).

We also obtain an alternative description of $\mathrm{L}^1(K\backslash G/K)$ (resp. $\mathrm{L}^2(K\backslash G/K)$) in terms of $\mathcal{L}^1(G/K)$ and $\mathcal{L}^1(K\backslash G)$ (resp. $\mathcal{L}^2(G/K)$ and $\mathcal{L}^2(K\backslash G)$).

If we denote by $\mathrm{L}^1(G/K)$, $\mathrm{L}^1(K\backslash G)$ (resp. $\mathrm{L}^2(G/K)$, $\mathrm{L}^2(K\backslash G)$) the canonical images in $\mathrm{L}^1(G)$ (resp. $\mathrm{L}^2(G)$) of $\mathcal{L}^1(G/K), \mathcal{L}^1(K\backslash G)$ (resp. $\mathcal{L}^2(G/K), \mathcal{L}^2(K\backslash G)$), then we have $\mathrm{L}^1(K\backslash G/K) = \mathrm{L}^1(G/K) \cap \mathrm{L}^1(K\backslash G)$ (resp. $\mathrm{L}^2(K\backslash G/K) = \mathrm{L}^2(G/K) \cap \mathrm{L}^2(K\backslash G)$).

9.2 Spherical Functions

Our next goal is to characterize the characters of the algebra $\mathrm{L}^1(K\backslash G/K)$ in terms of certain functions in $\mathcal{C}(K\backslash G/K)$ called spherical functions. In this chapter, we will use the notation (f, g) as an abbreviation for

$$\int_G f(x)g(x)\, d\lambda_G(x),$$

whenever such an integral makes sense for some functions $f, g\colon G \to \mathbb{C}$.

Theorem 9.4. *Every nonzero character ζ of the commutative Banach algebra $\mathrm{L}^1(K\backslash G/K)$ is given by a unique function $\omega \in \mathcal{C}(K\backslash G/K)$, which is bounded and continuous on G, with*

$$\zeta(f) = (f, \omega) = \int_G f(x)\omega(x)\, d\lambda_G(x), \qquad f \in \mathcal{L}^1(K\backslash G/K).$$

Furthermore, ω is uniformly continuous for every left-invariant metric on G, and $|\omega(s)| \leq \omega(e) = 1$ for all $s \in G$.

Proof. Every character ζ of the commutative subalgebra $\mathrm{L}^1(K\backslash G/K)$ of $\mathcal{M}^1(G)$ can be extended to a character ζ' of the unital commutative Banach algebra $\mathrm{L}^1(K\backslash G/K) \oplus \mathbb{C}\delta_e$ by setting $\zeta'(f + \lambda\delta_e) = \zeta(f) + \lambda$. By Vol. I, Theorem 9.19, we have $|\zeta(f)| \le \|f\|_1$ for all $f \in \mathrm{L}^1(K\backslash G/K)$. As a consequence, the map $\Phi\colon \mathrm{L}^1(G) \to \mathbb{C}$ given by

$$\Phi(f) = \zeta(f^\sharp), \qquad f \in \mathrm{L}^1(G)$$

is a linear form of norm ≤ 1 because by Proposition 9.3, $\|f^\sharp\|_1 \le \|f\|_1$, and by Vol. I, Theorem 5.51, there is a unique function $\omega_0 \in \mathcal{L}^\infty(G)$ with $\|\omega_0\|_\infty \le 1$, such that

$$\Phi(f) = \zeta(f^\sharp) = \int_G f(x)\omega_0(x)\, d\lambda_G(x). \tag{Φ1}$$

The problem is that $\omega_0 \in \mathcal{L}^\infty(G)$ is in the wrong space because we need it to be in $\mathcal{C}(K\backslash G/K)$ (and to be bounded by 1). To remedy this problem, we define another function ω in terms of ω_0, which we regularize by integrating against some function $f_0 \in \mathcal{K}(K\backslash G/K)$. In the end, we will see that ω_0 and ω are equal almost everywhere, but this will take some work.

First, observe that for all $t, t' \in K$ and all $f \in \mathcal{K}(G)$, we have

$$\int f(txt')\omega_0(x)\, d\lambda_G(x) = \int f(x)\omega_0(x)\, d\lambda_G(x). \tag{$*$}$$

To prove this, if we let $h_{t,t'}$ be the function given by $h_{t,t'}(x) = f(txt')$, then will prove that $h^\sharp_{t,t'} = f^\sharp$, and so $\zeta(f^\sharp) = \zeta(h^\sharp_{t,t'})$. Indeed, using the left and right invariance of the Haar measure λ_K, we have

$$\begin{aligned} h^\sharp_{t,t'}(x) &= \int_K \int_K h_{t,t'}(t_1 x t_2)\, d\lambda_K(t_1)\, d\lambda_K(t_2) \\ &= \int_K \int_K f(tt_1 x t_2 t')\, d\lambda_K(t_1)\, d\lambda_K(t_2) \\ &= \int_K \int_K f(t_1 x t_2)\, d\lambda_K(t_1)\, d\lambda_K(t_2) = f^\sharp(x). \end{aligned}$$

By hypothesis, the character ζ is not identically zero, so there is some function $f_0 \in \mathcal{K}(K\backslash G/K)$, such that $\zeta(f_0) \neq 0$. Since ζ is a character of $\mathcal{L}^1(K\backslash G/K)$, for every function $g \in \mathcal{K}(K\backslash G/K)$, since $f_0 \in \mathcal{K}(K\backslash G/K)$ we also have $g * f_0 \in \mathcal{K}(K\backslash G/K)$, so $(g * f_0)^\sharp = g * f_0$, and using Fubini's theorem and the fact that ζ is a character of $\mathcal{K}(K\backslash G/K)$, we have

$$\begin{aligned}\zeta(g) &= \zeta(f_0)^{-1}\zeta(f_0)\zeta(g) = \zeta(f_0)^{-1}\zeta(g)\zeta(f_0)\\ &= \zeta(f_0)^{-1}\zeta(g * f_0) = \zeta(f_0)^{-1}\zeta((g * f_0)^{\sharp})\\ &= \zeta(f_0)^{-1}\iint_{G\times G} f_0(s^{-1}x)g(s)\omega_0(x)\,d\lambda_G(s)\,d\lambda_G(x)\\ &= \int_G g(s)\omega(s)\,d\lambda_G(s),\end{aligned}$$

with

$$\begin{aligned}\omega(s) &= \zeta(f_0)^{-1}\int_G f_0(s^{-1}x)\omega_0(x)\,d\lambda_G(x)\\ &= \zeta(f_0)^{-1}\int_G f_0(x)\omega_0(sx)\,d\lambda_G(x). \qquad (\omega 1)\end{aligned}$$

It follows by Theorem 14.10.6(ii) of Dieudonné [Dieudonné (2005)] (Chapter XIV, Section 10) that ω is bounded in G and uniformly continuous for every left-invariant distance on G. Observe that for the integral on the right-hand side of

$$\zeta(g) = \int_G g(s)\omega(s)\,d\lambda_G(s)$$

to make sense, we need $g \in \mathcal{K}(K\backslash G/K)$, since we only know that $\omega_0 \in \mathcal{L}^{\infty}(G)$. We will extend the above equation to functions in $\mathrm{L}^1(K\backslash G/K)$ by density.

Next, we need to prove that $\omega \in \mathcal{C}(K\backslash G/K)$. For all $t, t' \in K$, since $f_0 \in \mathcal{K}(K\backslash G/K)$, by $(\omega 1)$, we have

$$\begin{aligned}\omega(tst') &= \zeta(f_0)^{-1}\int_G f_0(t'^{-1}s^{-1}t^{-1}x)\omega_0(x)\,d\lambda_G(x)\\ &= \zeta(f_0)^{-1}\int_G f_0(s^{-1}t^{-1}x)\omega_0(x)\,d\lambda_G(x)\\ &= \zeta(f_0)^{-1}\int_G f_0(s^{-1}x)\omega_0(x)\,d\lambda_G(x) = \omega(s),\end{aligned}$$

where the first equation holds because $f_0 \in \mathcal{K}(K\backslash G/K)$ and we used $(*)$ with $f(x) = f_0(s^{-1}x)$ in the last step. Since $\mathcal{K}(K\backslash G/K)$ is dense in $\mathrm{L}^1(K\backslash G/K)$, we proved that

$$\zeta(f) = (f, \omega) = \int_G f(x)\omega(x)\,d\lambda_G(x), \qquad f \in \mathcal{L}^1(K\backslash G/K). \qquad (\dagger_1)$$

We still need to prove that $|\omega(s)| \leq 1$ for all $s \in G$. Since we know that this is true of ω_0, we prove that $[\omega] = [\omega_0]$. By Vol. I, Theorem 7.10 and Theorem 5.51, it suffices to prove that

$$\int_G f(x)\omega(x)\, d\lambda(x) = \int_G f(x)\omega_0(x)\, d\lambda(x) = \zeta(f^\sharp), \qquad \text{for all } f \in \mathcal{K}(G).$$

Observe that since $\omega^\sharp = \omega$ (because $\omega \in \mathcal{C}(K\backslash G/K)$), and

$$\zeta(g) = \int_G g(s)\omega(s)\, d\lambda_G(s), \qquad g \in \mathrm{L}^1(K\backslash G/K), \tag{$\zeta 1$}$$

if we can prove that

$$(f^\sharp, \psi) = \int_G f^\sharp(x)\psi(x)\, d\lambda_G(x) = \int_G f(x)\psi^\sharp(x)\, d\lambda_G(x) = (f, \psi^\sharp), \tag{$**$}$$

for all $f \in \mathcal{K}(G)$ and all $\psi \in \mathcal{C}(G)$, we will be done, because by ($\Phi 1$), ($\zeta 1$) and ($**$),

$$\begin{aligned}\int_G f(x)\omega_0(x)\, d\lambda(x) = \zeta(f^\sharp) &= \int_G f^\sharp(s)\omega(s)\, d\lambda_G(s) \\ &= \int_G f(s)\omega^\sharp(s)\, d\lambda_G(s) \\ &= \int_G f(s)\omega(s)\, d\lambda_G(s),\end{aligned}$$

as claimed. To prove ($**$), using Fubini, and the fact that G is unimodular, we have

$$\begin{aligned}\int_G f^\sharp(x)\psi(x)\, d\lambda_G(x) &= \int_G\int_K\int_K f(tst')\psi(s)\, d\lambda_K(t)\, d\lambda_K(t')\, d\lambda_G(s) \\ &= \int_G\int_K\int_K f(s)\psi(t^{-1}st'^{-1})\, d\lambda_K(t)\, d\lambda_K(t')\, d\lambda_G(s) \\ &= \int_G\int_K\int_K f(s)\psi(tst')\, d\lambda_K(t)\, d\lambda_K(t')\, d\lambda_G(s) \\ &= \int_G f(x)\psi^\sharp(x)\, d\lambda_G(x).\end{aligned}$$

Now, since $\omega = \omega_0$ almost everywhere and since ω is continuous, we have proven that there is unique function $\omega \in \mathcal{C}(K\backslash G/K)$ satisfying the condition of the proposition and that $|\omega(s)| \leq 1$, since $\|\omega_0\|_\infty \leq 1$. If we let $s = e$ in

$$\omega(s) = \zeta(f_0)^{-1}\int_G f_0(s^{-1}x)\omega_0(x)\, d\lambda_G(x),$$

since

$$\zeta(f_0^\sharp) = \int_G f_0(x)\omega_0(x)\, d\lambda_G(x),$$

and $f_0^\sharp = f_0$ because $f_0 \in \mathcal{K}(K\backslash G/K)$, we get

$$\begin{aligned}\omega(e) &= \zeta(f_0)^{-1} \int_G f_0(x)\omega_0(x)\, d\lambda_G(x) \\ &= \zeta(f_0)^{-1}\zeta(f_0^\sharp) = \zeta(f_0)^{-1}\zeta(f_0) = 1,\end{aligned}$$

as claimed. □

Remark: The above proof shows that the function $\omega \in \mathcal{C}(K\backslash G/K)$ of Theorem 9.4 is given by

$$\omega(s) = (f_0, \omega)^{-1} \int_G f_0(s^{-1}x)\omega(x)\, d\lambda_G(x),$$

with $(f_0, \omega) = \int_G f_0(s)\omega(s)\, d\lambda_G(s)$, for any function $f_0 \in \mathcal{K}(K\backslash G/K)$, such that $(f_0, \omega) \neq 0$.

Definition 9.5. A bounded function $\omega \in \mathcal{C}(K\backslash G/K)$ is a *spherical (or zonal spherical) function* on G relative to K, if the function

$$f \mapsto (f, \omega) = \int_G f(s)\omega(s)\, d\lambda_G(s), \qquad f \in \mathrm{L}^1(K\backslash G/K)$$

is a nonzero character of $\mathrm{L}^1(K\backslash G/K)$, which means that the map $f \mapsto (f, \omega)$ is linear in $f \in \mathrm{L}^1(K\backslash G/K)$ and that

$$(f * g, \omega) = (f, \omega)(g, \omega), \qquad \text{for all } f, g \in \mathrm{L}^1(K\backslash G/K). \tag{$\dagger_2$}$$

Theorem 9.4 shows that if ω is a spherical function, then $\omega(e) = 1$ and $|\omega(s)| \leq 1$ for all $s \in G$.

Proposition 9.5. *If ω is a spherical function, then $\overline{\omega}$ and $\check{\omega}$ are also spherical functions.*

Proof. For every $f \in \mathcal{K}(K\backslash G/K)$, by Vol. I, Proposition 7.24, we have

$$\int f(s)\overline{\omega}(s)\, d\lambda_G(s) = \overline{\int \overline{f}(s)\omega(s)\, d\lambda_G(s)},$$

and for all $f, g \in \mathcal{K}(K\backslash G/K)$, we have

$$\overline{f * g} = \overline{f} * \overline{g},$$

which proves that $\overline{\omega}$ induces a character, because

$$\begin{aligned}\int (f*g)(s)\overline{\omega}(s)\,d\lambda_G(s) &= \overline{\int \overline{(f*g)}(s)\omega(s)\,d\lambda_G(s)}\\ &= \overline{\int \overline{f}*\overline{g}(s)\omega(s)\,d\lambda_G(s)}\\ &= \overline{\int \overline{f}(s)\omega(s)\,d\lambda_G(s)}\;\overline{\int \overline{g}(s)\omega(s)\,d\lambda_G(s)} \text{ by } (\dagger_2)\\ &= \int f(s)\overline{\omega}(s)\,d\lambda_G(s)\int g(s)\overline{\omega}(s)\,d\lambda_G(s).\end{aligned}$$

Since G is unimodular, we also have

$$\begin{aligned}\int_G f(s)\check{\omega}(s)\,d\lambda_G(s) &= \int_G f(s)\omega(s^{-1})\,d\lambda_G(s)\\ &= \int_G f(s^{-1})\omega(s)\,d\lambda_G(s)\\ &= \int_G \check{f}(s)\omega(s)\,d\lambda_G(s),\end{aligned}$$

and if $f, g \in \mathcal{K}(K\backslash G/K)$, we have

$$(f*g)\check{} = \check{g}*\check{f} = \check{f}*\check{g},$$

so $\check{\omega}$ induces a character. ☐

Observe that Theorem 9.4 shows that a spherical function is uniformly continuous for every left-invariant as well as every right-invariant distance on G. In general, a spherical function *does not have compact support*.

The next theorem gives criteria for a bounded function in $\mathcal{C}(K\backslash G/K)$ (different from the zero function) to be a spherical function.

Theorem 9.6. *Let ω be a bounded function in $\mathcal{C}(K\backslash G/K)$ not equal to the zero function. The following properties are equivalent:*

(1) The function ω is a spherical function on G relative to K.

(2) We have

$$\int_K \omega(xty)\,d\lambda_K(t) = \omega(x)\omega(y), \qquad \text{for all } x, y \in G. \qquad (s_1)$$

(3) We have $\omega(e) = 1$, and for every function $f \in \mathcal{K}(K\backslash G/K)$, there is some $\lambda_f \in \mathbb{C}$, such that

$$f*\omega = \lambda_f\omega. \qquad (s_2)$$

In fact

$$\lambda_f = (\check{f}, \omega) = \int_G \check{f}(x)\omega(x)\,d\lambda_G(x).$$

Similarly, $\omega * f = \lambda_f\omega$, for the same λ_f.

Proof. First we prove that (3) implies (1). Assume that $f * \omega = \lambda_f \omega$ for every $f \in \mathcal{K}(K\backslash G/K)$. Since $\omega(e) = 1$, by (s_2), we have

$$\lambda_{\check{f}} = (\check{f} * \omega)(e) = \int_G f(s)\omega(s)\, d\lambda_G(s) = (f, \omega).$$

Therefore,

$$\lambda_f = (\check{f}, \omega).$$

For all $f, g \in \mathcal{K}(K\backslash G/K)$, we have $(f * g)\check{} * \omega = (\check{g} * \check{f}) * \omega = \check{g} * (\check{f} * \omega)$, which by (s_2) implies that

$$\lambda_{\check{g} * \check{f}} = \lambda_{\check{g}} \lambda_{\check{f}},$$

and so

$$(f * g, \omega) = (f, \omega)(g, \omega),$$

which shows that the map $f \mapsto (f, \omega)$ is a character. The proof is similar if $\omega * f = \lambda_f \omega$.

We now prove that (1) implies (3). We claim that for all $f \in \mathcal{K}(K\backslash G/K)$, we have

$$f * \omega = (\check{f}, \omega)\omega = \left(\int_G f(s^{-1})\omega(s)\, d\lambda_G(s) \right) \omega. \tag{$\dagger_3$}$$

Since $f * \omega \in \mathcal{C}(K\backslash G/K)$, in view of $(**)$, namely

$$(h^\sharp, \psi) = (h, \psi^\sharp), \qquad \text{for all } h \in \mathcal{K}(G) \text{ and all } \psi \in \mathcal{K}(G),$$

and since $\omega^\sharp = \omega$, it suffices to prove that

$$(g, f * \omega) = (\check{f}, \omega)(g, \omega), \qquad \text{for all } g \in \mathcal{K}(K\backslash G/K), \tag{$\dagger_4$}$$

because by Proposition 9.1, $(f * \omega)^\sharp = f * \omega^\sharp = f * \omega$, so by $(**)$ and $(\dagger_4)$,

$$\begin{aligned}
(h, f * \omega) &= (h, (f * \omega)^\sharp) \\
&= (h^\sharp, f * \omega) \\
&= (\check{f}, \omega)(h^\sharp, \omega) \\
&= (\check{f}, \omega)(h, \omega^\sharp) \\
&= (\check{f}, \omega)(h, \omega) = (h, (\check{f}, \omega)\omega)
\end{aligned}$$

for all $h \in \mathcal{K}(G)$, and thus $f * \omega = (\check{f}, \omega)\omega$. In that last step, we used the fact that the map $f_1, f_2 \mapsto (f_1, f_2)$ is bilinear, so for every $\lambda \in \mathbb{C}$, we have $\lambda(f_1, f_2) = (\lambda f_1, f_2) = (f_1, \lambda f_2)$.

Using Fubini and the fact that G is unimodular, we have

$$(g, f * \omega) = (\check{f} * g, \omega),$$

since

$$\begin{aligned}\int_G \int_G & g(s) f(t) \omega(t^{-1}s) d\lambda_G(t)\, d\lambda_G(s) \\ &= \int_G \int_G g(ts) f(t) \omega(s) d\lambda_G(t)\, d\lambda_G(s) \\ &= \int_G \int_G f(t^{-1}) g(t^{-1}s) \omega(s) d\lambda_G(t)\, d\lambda_G(s).\end{aligned}$$

Since ω is a spherical function, by ($\dagger_2$),

$$(g, f * \omega) = (\check{f} * g, \omega) = (\check{f}, \omega)(g, \omega).$$

Since $\check{\overline{\omega}}$ is also spherical (by Proposition 9.5), by ($\dagger_3$) and since G is unimodular, using Vol. I, Proposition 8.27, we have

$$\omega * f = \overline{(\check{\overline{f}} * \check{\overline{\omega}})}^{\vee} = \overline{(\overline{f}, \check{\overline{\omega}})\check{\overline{\omega}}}^{\vee} = \overline{(\overline{f}, \check{\overline{\omega}})}\omega = (\check{f}, \omega)\omega.$$

Let us now prove that (2) and (3) are equivalent. For any $\omega \in \mathcal{C}(K\backslash G/K)$, define the function h by

$$h(x, y) = \int_K \omega(xty)\, d\lambda_K(t), \qquad x, y \in G.$$

A simple adaptation of the proof of Vol. I, Proposition 8.20 shows the map $x \mapsto h(x, y)$ is continuous. For all $t' \in K$, since $\omega(t'xty) = \omega(xty)$, we have $h(t'x, y) = h(x, y)$, and due to the invariance of λ_K, we have $h(xt', y) = h(x, y)$. It follows that the function $x \mapsto h(x, y)$ is in $\mathcal{C}(K\backslash G/K)$. Let us show that for every function $f \in \mathcal{K}(K\backslash G/K)$, we have

$$\int_G \check{f}(x) h(x, y)\, d\lambda_G(x) = (f * \omega)(y). \tag{$\dagger_5$}$$

Since $f * \omega \in \mathcal{C}(K\backslash G/K)$ and G is unimodular, by Fubini, we have

$$\begin{aligned}\int_G & \check{f}(x) h(x, y)\, d\lambda_G(x) \\ &= \int_G \int_K \check{f}(x) \omega(xty)\, d\lambda_K(t)\, d\lambda_G(x) \\ &= \int_G \int_K f(x) \omega(x^{-1}ty)\, d\lambda_G(x)\, d\lambda_K(t) \\ &= \int_K (f * \omega)(ty)\, d\lambda_K(t) = \int_K (f * \omega)(y)\, d\lambda_K(t) = (f * \omega)(y).\end{aligned}$$

If (2) holds, Equation (s_1), namely

$$\int_K \omega(xty)\,d\lambda_K(t) = \omega(x)\omega(y),$$

implies that $h(x,y) = \omega(x)\omega(y)$, so by $(\dagger_5)$, we have

$$\begin{aligned}(f*\omega)(y) &= \int_G \check{f}(x)h(x,y)\,d\lambda_G(x) = \int_G \check{f}(x)\omega(x)\omega(y)d\lambda_G(x)\\ &= \omega(y)\int_G \check{f}(x)\omega(x)d\lambda_G(x) = (\check{f},\omega)\omega(y),\end{aligned}$$

which proves (3).

Conversely, if (3) holds, we saw in proving that (3) implies (1) that $f*\omega = (\check{f},\omega)\omega$. We prove that this implies that $h(x,y) = \omega(x)\omega(y)$, which is (s_1). For any function $g \in \mathcal{K}(G)$, by $(**)$, using the fact that $h,\omega \in \mathcal{C}(K\backslash G/K)$ and $(\dagger_3)$, we have

$$\begin{aligned}(g,h(-,y)) &= (g,h(-,y)^\sharp) = (g^\sharp,h(-,y)) = (\check{g}^\sharp*\omega)(y)\\ &= (g^\sharp,\omega)\omega(y) = (g,\omega^\sharp)\omega(y) = (g,\omega)\omega(y) = (g,\omega(y)\omega),\end{aligned}$$

and so $h(x,y) = \omega(x)\omega(y)$, as claimed. ☐

Remark: If a bounded continuous function ω on G not equal to the zero function satisfies the equation

$$\int_K \omega(xty)\,d\lambda_K(t) = \omega(x)\omega(y)$$

for all $x,y \in G$, then it must belong to $\mathcal{C}(K\backslash G/K)$, and thus is a spherical function. Indeed, since λ_K is left and right invariant, for any $t' \in K$, we have

$$\omega(xt')\omega(y) = \omega(x)\omega(y) = \omega(x)\omega(t'y)$$

for all $x,y \in G$ and all $t' \in K$, which shows that $\omega(xt') = \omega(x)$ and $\omega(t'y) = \omega(y)$.

Definition 9.6. Let $\mathbf{S}(G/K)$, or simply $\mathbf{S}$, denote the space of spherical functions on G relative to K. This is a subspace of $\mathcal{C}(K\backslash G/K) \cap \mathrm{L}^\infty(G)$.

Let $A = \mathrm{L}^1(K\backslash G/K) \oplus \mathbb{C}\delta_e$, a commutative, involutive, unital Banach algebra. In the degenerate case where $\delta_e \in \mathrm{L}^1(K\backslash G/K)$, the group G is a discrete group and δ_e is invariant by translation by the elements of K. This implies that $K = \{e\}$, and that G is commutative and discrete. Otherwise, there is exactly one character of the algebra $A = \mathrm{L}^1(K\backslash G/K) \oplus \mathbb{C}\delta_e$, whose restriction to $\mathrm{L}^1(K\backslash G/K)$ is the zero function.

Definition 9.7. The subspace of characters of the commutative involutive Banach algebra $A = \mathrm{L}^1(K\backslash G/K) \oplus \mathbb{C}\delta_e$, whose restriction to $\mathrm{L}^1(K\backslash G/K)$ is not the zero function is denoted by $\mathsf{X}_0(A)$. This subspace is locally compact (metrizable and separable).

For every spherical function $\omega \in \mathbf{S}(G/K)$, let $\zeta_\omega \in \mathsf{X}_0(A)$ be the character given by

$$\zeta_\omega(f) = (f, \omega) = \int_G f(x)\omega(x)\, d\lambda_G(x), \qquad f \in \mathrm{L}^1(K\backslash G/K).$$

The map $\omega \mapsto \zeta_\omega$ is a bijection between $\mathbf{S}(G/K)$ and $\mathsf{X}_0(A)$. In fact, we have the following stronger result.

Theorem 9.7. *The following properties hold:*

(1) The map $\omega \mapsto \zeta_\omega$ is a homeomorphism of $\mathbf{S}(G/K)$ equipped with the induced topology of Fréchet space of $\mathcal{C}(G)$ (see Vol. I, Definition 2.18 and Proposition 2.23) onto $\mathsf{X}_0(A)$ equipped with the topology induced by the weak-topology of the dual A' of A (see Vol. I, Definition 9.13). Consequently, $\mathbf{S}(G/K)$ is locally compact.*
(2) Every compact subset L of $\mathbf{S}(G/K)$ is equicontinuous.
(3) The map $(x, \omega) \mapsto \omega(x)$ from $G \times \mathbf{S}(G/K)$ to $\mathbb{C}$ is continuous.

Proof sketch. The proof of (1) is given in Dieudonné [Dieudonné (2003b)] (Chapter XXII, Section 6, no. 22.6.9). This proof is very technical and makes use of the following results proven in Dieudonné [Dieudonné (2003b)] (Chapter XXII, Section 1, no. 22.1.11.1 and 22.1.11.2).

Proposition 9.8. *For every subset B of $\mathcal{C}(G)$ consisting of uniformly bounded functions, the topology induced by the weak*-topology on $\mathrm{L}^\infty(G)$ (see Definition 3.19) is coarser than the topology induced by the topology of $\mathcal{C}(G)$ as a Fréchet space.*

Using Proposition 9.8, the following result can be shown.

Proposition 9.9. *Let B be a subset of $\mathcal{C}(G)$ consisting of uniformly bounded functions and having the following property: for every $p_0 \in B$, for every compact subset K of G, for every $\epsilon > 0$, there is some neighborhood U of p_0 in B for the weak*-topology on $\mathrm{L}^\infty(G)$ and some compact neighborhood W of e in G, such that for every function $p \in U$, we have*

$$|(a^{-1}\chi_W * p)(s) - p(s)| \leq \epsilon \qquad \textit{for all } s \in K,$$

where $a = \lambda_G(W)$. Then the topology on B induced by the weak-topology on $\mathrm{L}^\infty(G)$ is identical to the topology induced by the topology of $\mathcal{C}(G)$ as a Fréchet space.*

Proposition 9.9 follows from the following result also proven in Dieudonné [Dieudonné (2003b)] (Chapter XXII, Section 1, no. 22.1.11.5).

Proposition 9.10. *For every function $f \in \mathcal{L}^1(G)$ and for every bounded subset B of the Banach space $\mathrm{L}^\infty(G)$, the map $g \mapsto f * g$ is a continuous map from B equipped with the weak*-topology on $\mathrm{L}^\infty(G)$ to the Fréchet space $\mathcal{C}(G)$.*

The proof of Proposition 9.10 uses the trick that

$$\begin{aligned}(f * g)(s) &= \int_G f(t)g(t^{-1}s)\, d\lambda_G(t) = \int_G f(st)g(t^{-1})\, d\lambda_G(t)\\ &= \int_G (\lambda_{s^{-1}} f)(t)\check{g}(t)\, d\lambda_G(t) = (\check{g}, \lambda_{s^{-1}} f).\end{aligned}$$

(2) Pick any $x_0 \in G$. For every compact neighborhood V_0 of x_0, by definition of the Fréchet topology on $\mathcal{C}(G)$, the restriction map $f \mapsto f|V_0$ from $\mathcal{C}(G)$ to $\mathcal{C}(V; \mathbb{C})$ is continuous, thus the image L_0 of L under this map is compact. By Ascoli III (Vol. I, Theorem 2.14, Dieudonné's version), L_0 is equicontinuous. Consequently, for every $\epsilon > 0$ there is a neighborhood $V \subseteq V_0$ of x_0, such that $|\omega(x) - \omega(x_0)| \leq \epsilon$ for all $x \in V$ and all $\omega \in L$.

(3) Let (x_0, ω_0) be an element of $G \times \mathbf{S}(G/K)$. By (2), for every $\epsilon > 0$, there is a compact neighborhood V of x_0 in G and a compact neighborhood W of ω_0 in $\mathbf{S}(G/K)$, such that $|\omega(x) - \omega(x_0)| \leq \epsilon$ for all $x \in V$ and all $\omega \in W$. By definition of the Fréchet topology on $\mathcal{C}(G)$, there is a neighborhood $U \subseteq W$ of ω_0 in $\mathbf{S}(G/K)$, such that $|\omega(x) - \omega_0(x)| \leq \epsilon$ for all $x \in V$ and all $\omega \in U$. We deduce that

$$|\omega(x) - \omega_0(x_0)| \leq |\omega(x) - \omega(x_0)| + |\omega(x_0) - \omega_0(x_0)| \leq 2\epsilon$$

for all $x \in V$ and all $\omega \in U$. $\square$

A theory of spherical functions for Lie groups, in particular symmetric spaces, not based on Gelfand pairs but instead on certain invariant differential operators is discussed in Helgason [Helgason (2000)] (Chapter 4).

In order to present some of the examples of Gelfand pairs involving Lie groups, we need to discuss some material about semi-simple Lie groups.

9.3 Real Forms of a Complex Semi-Simple Lie Algebra

This section assumes some background of Lie algebras and Lie groups. Such material is discussed extensively in Carter, Segal and Macdonald [Carter *et al.* (1995)], Dieudonné [Dieudonné (2003a)], Duistermaat and Kolk [Duistermaat and Kolk (2000)], Fulton and Harris [Fulton and Harris (1991)], Gallier and Quaintance [Gallier and Quaintance (2020a,b)], Hall [Hall (2003)], Helgason [Helgason (2001)], Humphreys [Humphreys (1972)], Knapp [Knapp (2002, 2001)], Samelson [Samelson (1990)], Serre [Serre (2000, 1992)], and Varadarajan [Varadarajan (1984)]. The most elementary presentations occur in Carter, Segal and Macdonald [Carter *et al.* (1995)], Hall [Hall (2003)], and Gallier and Quaintance [Gallier and Quaintance (2020a)]. We need to review the process of "complexifying" a real vector space V. But first, we recall how to view a complex vector space as a real vector space.

Definition 9.8. If V is a complex vector space, then we denote by $V|_{\mathbb{R}}$ the vector space whose underlying abelian group is V, but with the scalar multiplication *restricted* to $\mathbb{R}$.

We can define the complexification $V_{\mathbb{C}}$ of the real vector space V as the complex vector space whose carrier is the tensor product $\mathbb{C} \otimes_{\mathbb{R}} V$, but more directly as $V \times V$, with the addition operation

$$(u_1, v_1) + (u_2, v_2) = (u_1 + u_2, v_1, +v_2), \qquad u_1, u_2, v_1, v_2 \in V,$$

and the scalar product given by

$$(a + ib)(u, v) = (au - bv, av + bu), \qquad a, b \in \mathbb{R}, u, v \in V.$$

Observe that

$$(0, v) = i(v, 0),$$

so we can write

$$(u, v) = (u, 0) + i(v, 0),$$

and if $j\colon V \to V_{\mathbb{C}}$ is the injection given by $j(u) = (u, 0)$, for all $u \in V$, then we have an isomorphism

$$V_{\mathbb{C}} \cong V \oplus iV,$$

as a direct sum of real subspaces. More precisely, the injection j induces an isomorphism

$$(V_{\mathbb{C}})|_{\mathbb{R}} \cong V \oplus iV,$$

but by abuse of notation, we usually write $V_{\mathbb{C}} = V \oplus iV$. Using the above isomorphism, the scalar multiplication of a vector $u+iv \in V_{\mathbb{C}}$ by a complex number $a+ib$ is given by

$$(a+ib)(u+iv) = au - bv + i(av + bu).$$

The map $c_V : V_{\mathbb{C}} \to V_{\mathbb{C}}$ given by

$$c_V(u+iv) = u - iv, \qquad u, v \in V,$$

is semi-linear and an involution; that is, $c_V^2 = \mathrm{id}_{V_{\mathbb{C}}}$. The map c_V is called the *conjugation* of $V_{\mathbb{C}}$ associated with V. Observe that

$$V = \{w \in V_{\mathbb{C}} \mid c_V(w) = w\}, \quad iV = \{w \in V_{\mathbb{C}} \mid c_V(w) = -w\}.$$

To simplify notation, we usually write c instead of c_V.

If $\mathfrak{g}$ is a real Lie algebra, then its complexification $\mathfrak{g}_{\mathbb{C}}$ is the complex Lie algebra whose carrier is the complex vector space, such that $(\mathfrak{g}_{\mathbb{C}})|_{\mathbb{R}} = \mathfrak{g} \oplus i\mathfrak{g}$ as a direct sum of real subspaces, with the Lie bracket given by

$$[u+iv, x+iy]_{\mathbb{C}} = [u,x] - [v,y] + i([u,y] + [v,x]).$$

If c is the conjugation of $\mathfrak{g}_{\mathbb{C}}$ associated with $\mathfrak{g}$, then

$$\begin{aligned} c([u+iv, x+iy]_{\mathbb{C}}) &= c([u,x] - [v,y] + i([u,y] + [v,x])) \\ &= c([u,x]) - c([v,y]) - i(c([u,y]) + c([v,x])) \\ &= [u,x] - [v,y] - i([u,y] + [v,x]) \\ &= [u-iv, x-iy]_{\mathbb{C}} = [c(u+iv), c(x+iy)]_{\mathbb{C}}. \end{aligned}$$

This shows that c is an automorphism of the real Lie algebra $(\mathfrak{g}_{\mathbb{C}})|_{\mathbb{R}}$.

Definition 9.9. Given a complex Lie algebra $\mathfrak{g}$, a real Lie algebra $\mathfrak{g}_0$, such that

$$\mathfrak{g}|_{\mathbb{R}} \cong \mathfrak{g}_0 \oplus i\mathfrak{g}_0$$

as a direct sum of real subspaces is called a *real form* of $\mathfrak{g}$. The complex Lie algebra $\mathfrak{g}$ is the *complexification* of $\mathfrak{g}_0$.

The following proposition shows that finding a real form of a complex Lie algebra $\mathfrak{g}$ is equivalent to finding an automorphism c of the real Lie algebra $\mathfrak{g}|_{\mathbb{R}}$ satisfying the properties stated in the following proposition.

Proposition 9.11. *Let $\mathfrak{g}$ be a complex Lie algebra, $\mathfrak{g}_0$ be a real Lie algebra, and assume that $\mathfrak{g}|_{\mathbb{R}} = \mathfrak{g}_0 \oplus i\mathfrak{g}_0$; that is, $\mathfrak{g}$ is the complexification of $\mathfrak{g}_0$. Then the conjugation $c_0 : \mathfrak{g} \to \mathfrak{g}$ given by $c_0(u+iv) = u - iv$ $(u, v \in \mathfrak{g}_0)$ has the following properties:*

(a) *The map* c_0 *is semi-linear, which means that* $c_0(x+y) = c_0(x)+c_0(y)$, *and* $c_0(ix) = -ic_0(x)$, *for all* $x, y \in \mathfrak{g}$.
(b) *The map* c_0 *is idempotent; that is,* $c_0^2 = \mathrm{id}_{\mathfrak{g}}$.
(c) *We have*

$$c_0([x,y]) = [c_0(x), c_0(y)], \qquad \textit{for all } x, y \in \mathfrak{g}.$$

Conversely, if a map $c\colon \mathfrak{g} \to \mathfrak{g}$ *has properties (a), (b), (c), then if we consider the linear automorphism* $c\colon \mathfrak{g}|_{\mathbb{R}} \to \mathfrak{g}|_{\mathbb{R}}$ *of* $\mathfrak{g}$ *viewed as a real vector space, and if we let*

$$\mathfrak{g}_1 = \{x \in \mathfrak{g} \mid c(x) = x\},$$

the subspace $\mathfrak{g}_1$ *is a real Lie subalgebra of* $\mathfrak{g}|_{\mathbb{R}}$, *and we have*

$$\mathfrak{g}|_{\mathbb{R}} = \mathfrak{g}_1 \oplus i\mathfrak{g}_1;$$

that is, $\mathfrak{g}$ *is the complexification of* $\mathfrak{g}_1$.

Proof. We already proved the first part of the proposition. Conversely, since c is an involutive automorphism of $\mathfrak{g}|_{\mathbb{R}}$, we know by linear algebra that

$$\mathfrak{g}|_{\mathbb{R}} = \mathfrak{g}_1 \oplus \mathfrak{g}_{-1}$$

where $\mathfrak{g}_1$ and $\mathfrak{g}_{-1}$ are the real eigenspaces of c given by

$$\mathfrak{g}_1 = \{x \in \mathfrak{g} \mid c(x) = x\}, \qquad \mathfrak{g}_{-1} = \{x \in \mathfrak{g} \mid c(x) = -x\}.$$

Since c is semi-linear, for every $x \in \mathfrak{g}_1$, we have

$$c(ix) = -ic(x) = -ix,$$

so $ix \in \mathfrak{g}_{-1}$, which shows that

$$i\mathfrak{g}_1 \subseteq \mathfrak{g}_{-1}. \tag{$*$}$$

For every $x \in \mathfrak{g}_{-1}$, we have

$$c(ix) = -ic(x) = ix,$$

so $ix \in \mathfrak{g}_1$, which shows that

$$i\mathfrak{g}_{-1} \subseteq \mathfrak{g}_1.$$

But the above inclusion implies that

$$\mathfrak{g}_{-1} \subseteq i\mathfrak{g}_1. \tag{$**$}$$

By $(*)$ and $(**)$, we get

$$\mathfrak{g}_{-1} = i\mathfrak{g}_1,$$

and so

$$\mathfrak{g}|_{\mathbb{R}} = \mathfrak{g}_1 \oplus i\mathfrak{g}_1,$$

as claimed. Since c is the identity on $\mathfrak{g}_1$ and satisfies (c), we conclude that $\mathfrak{g}_1$ is a (real) subalgebra of $\mathfrak{g}$. ☐

Definition 9.10. Given a complex Lie algebra $\mathfrak{g}$, a map $c_0\colon \mathfrak{g} \to \mathfrak{g}$ satisfying conditions (a), (b), (c) of Proposition 9.11 is called a *conjugation*.

Recall that a *semi-simple* Lie algebra has no commutative ideals other than (0). A Lie group is *semi-simple* if its Lie algebra is semi-simple; see Gallier and Quaintance [Gallier and Quaintance (2020a)], Section 21.5, Definition 21.8. If the reader is familiar with the notion of Killing form, Cartan's criterion says that a Lie algebra is semi-simple iff its Killing form is nondegenerate; see Gallier and Quaintance [Gallier and Quaintance (2020a)], Section 21.6, Theorem 21.26, Knapp [Knapp (2002)], and Serre [Serre (2000)].

Let G_u be a real compact semi-simple simply-connected Lie group, $\mathfrak{g}_u$ be its real semi-simple Lie algebra, and $\mathfrak{g} = \mathfrak{g}_u \oplus i\mathfrak{g}_u$ be the complex Lie algebra which is its complexification. Then it is possible to determine all the real forms $\mathfrak{g}_0$ of $\mathfrak{g}$ (up to isomorphism).

Recall that the conjugation $c_u\colon \mathfrak{g} \to \mathfrak{g}$ associated with $\mathfrak{g}_u$ is given by

$$c_u(x + iy) = x - iy, \qquad x, y \in \mathfrak{g}_u.$$

It can be proven that in order to find all conjugations $c_0\colon \mathfrak{g} \to \mathfrak{g}$ of $\mathfrak{g}$, it suffices to consider conjugations that *commute* with c_u; see Dieudonné [Dieudonné (2003a)] (Chapter XXI, no. 21.18.3). The key to the proof is that the Killing form associated with the Lie algebra of a compact semi-simple connected Lie group is negative definite; see Gallier and Quaintance [Gallier and Quaintance (2020a)], Section 21.6, Theorem 21.27. Technically, we have the following result.

Theorem 9.12. *Let G_u be a real compact semi-simple simply-connected Lie group, $\mathfrak{g}_u$ be its real semi-simple Lie algebra, and $\mathfrak{g} = \mathfrak{g}_u \oplus i\mathfrak{g}_u$ be its complexification. For any conjugation c of $\mathfrak{g}$, there is an automorphism $\varphi\colon \mathfrak{g} \to \mathfrak{g}$, such that c_u and $\varphi \circ c \circ \varphi^{-1}$ commute.*

Proof. The proof assumes some familiarity with the properties of the Killing form and may be safely omitted. The Killing form is the bilinear form on $\mathfrak{g}$ defined by

$$B_{\mathfrak{g}}(u, v) = \mathrm{tr}(\mathrm{ad}_u \circ \mathrm{ad}_v), \quad u, v \in \mathfrak{g},$$

where $\mathrm{ad}_u(w) = [u, w]$. The first fact is that the mapping

$$(u, v) \mapsto \langle u, v\rangle = -B_{\mathfrak{g}}(u, c_u(v))$$

is a Hermitian inner product on $\mathfrak{g}$, where $B_{\mathfrak{g}}$ is the Killing form of $\mathfrak{g}$; see Dieudonné [Dieudonné (2003a)] (Chapter XXI, no. 21.17.2.1). Consider the map $h = c \circ c_u$. For simplicity of notation, we suppress the composition symbol. Since c and c_u are semi-linear maps that preserve the Lie bracket, $h = cc_u$ is actually linear, and since $h^{-1} = (cc_u)^{-1} = c_u c$, we see that cc_u is an automorphism of the Lie algebra $\mathfrak{g}$. We will prove that h is self-adjoint with respect to the inner product $\langle -, - \rangle$. As a consequence, h is diagonalizable and has real eigenvalues, so $S = h^2$ has strictly positive eigenvalues. Then we will see that $\varphi = S^{-1/4}$ does the job.

First, observe that

$$h^{-1} c_u = c_u c c_u = c_u h.$$

Since h and h^{-1} preserve the Lie bracket, they also preserve the Killing form, and since the Killing form is invariant under automorphisms (see Proposition 21.25 in Gallier and Quaintance [Gallier and Quaintance (2020a)]), so we have

$$\begin{aligned} \langle h(u), v \rangle &= -B_{\mathfrak{g}}(h(u), c_u(v)) \\ &= -B_{\mathfrak{g}}(u, h^{-1}(c_u(v))) \\ &= -B_{\mathfrak{g}}(u, c_u(h(v))) \\ &= \langle u, h(v) \rangle, \end{aligned}$$

which shows that h is self-adjoint. Thus, h is diagonalizable with respect to an orthonormal basis of eigenvectors and its eigenvalues are real. But then $S = h^2$ is a self-adjoint linear map with strictly positive eigenvalues, so with respect to an orthonormal basis $(e_1, \ldots, e_n)$ of eigenvectors, S is represented by a diagonal matrix

$$\Lambda = \mathrm{diag}(\lambda_1, \ldots, \lambda_n),$$

with $\lambda_i > 0$ for $i = 1, \ldots, n$. For any real number $t > 0$, define the matrix Λ^t as

$$\Lambda^t = \mathrm{diag}(\lambda_1^t, \ldots, \lambda_n^t).$$

Obviously, the linear isomorphisms S^t represented by the matrices Λ^t commute with h. We claim that they are also Lie algebra isomorphisms of $\mathfrak{g}$. The Lie bracket on $\mathfrak{g}$ is determined by its values on the basis $(e_1, \ldots, e_n)$, that is, by equations

$$[e_j, e_k] = \sum_{l=1}^{n} a_{jkl} e_l, \quad 1 \leq j, k \leq n,$$

for some $a_{jkl} \in \mathbb{C}$. To express that S is an automorphism of $\mathfrak{g}$ is equivalent to stating the equations

$$[S(e_j), S(e_k)] = S([e_j, e_k]), \quad 1 \leq j, k \leq n,$$

and since

$$[S(e_j), S(e_k)] = \lambda_j\lambda_k[e_j, e_k], \qquad S(e_l) = \lambda_l e_l,$$

to stating the equations

$$\lambda_j\lambda_k \sum_{l'=1}^{n} a_{jkl'}e_{l'} = \sum_{l=1}^{n} \lambda_l a_{jkl}e_l,$$

that is, to stating the equations

$$\lambda_j\lambda_k a_{jkl} = a_{jkl}\lambda_l, \quad 1 \leq j, k, l \leq n. \tag{$*_1$}$$

These equations are nontrivial only when $a_{jkl} \neq 0$, in which case

$$\lambda_j\lambda_k = \lambda_l.$$

These equations imply that

$$\lambda_j^t\lambda_k^t = \lambda_l^t$$

for all $t \geq 0$, so

$$\lambda_j^t\lambda_k^t a_{jkl} = a_{jkl}\lambda_l^t, \quad 1 \leq j, k, l \leq n,$$

which shows that the S^t are Lie algebra isomorphisms of $\mathfrak{g}$. If we let $S^{-t} = (S^t)^{-1}$ for $t > 0$, consider the conjugation of $\mathfrak{g}$ given by

$$c^{(t)} = S^t c_u S^{-t}. \tag{$c^{(t)}$}$$

Beware that in general $c^{(1)} \neq c$, this is why we use the notation $c^{(t)}$ instead of c^t. Observe that since $h = cc_u$,

$$c_u h c_u^{-1} = c_u c c_u c_u^{-1} = c_u c = h^{-1},$$

so

$$c_u h c_u^{-1} c_u h c_u^{-1} = h^{-1}h^{-1},$$

namely

$$c_u h^2 c_u^{-1} = h^{-2},$$

and since $S = h^2$, we get $c_u S c_u^{-1} = S^{-1}$. Since $c_u^2 = \mathrm{id}$, the equation $c_u S c_u^{-1} = S^{-1}$ is equivalent to

$$S c_u = c_u S^{-1}. \tag{$*_2$}$$

We prove that the above equation implies that

$$S^t c_u = c_u S^{-t} \quad \text{for all } t > 0. \tag{$*_3$}$$

The map c_u is linear over $\mathbb{R}$, so we can express $(*_2)$ in terms of matrices over the basis $(e_1, \ldots, e_n)$. If (c_{ij}) is the matrix representing c_u, we know that S is represented by the diagonal matrix Λ, so $(*_2)$ is equivalent to the equations

$$\lambda_i c_{ij} = c_{ij}\lambda_j^{-1}, \qquad 1 \leq i, j \leq n. \tag{$*_4$}$$

These equations are trivially satisfied if $c_{ij} = 0$ and otherwise they imply

$$\lambda_i = \lambda_j^{-1},$$

which in turn imply

$$\lambda_i^t = \lambda_j^{-t}, \qquad \text{for all } t > 0,$$

and thus

$$\lambda_i^t c_{ij} = c_{ij}\lambda_j^{-t}, \qquad 1 \leq i, j \leq n, \tag{$*_5$}$$

which means that

$$S^t c_u = c_u S^{-t},$$

as claimed. Using the equation $S^t c_u = c_u S^{-t}$ from $(*_3)$ and Equation $(c^{(t)})$, we have

$$\begin{aligned} cc^{(t)} &= cS^t c_u S^{-t} = cc_u S^{-2t} = hS^{-2t} \\ c^{(t)}c &= (cc^{(t)})^{-1} = S^{2t}h^{-1} = h^{-1}S^{2t}. \end{aligned}$$

If we let $t = 1/4$, then

$$h^{-1}S^{1/2} = cc^{(t)} = c^{(t)}c,$$

since $S = h^2$ and so

$$hS^{-1/2} = hS^{-1}S^{1/2} = hh^{-2}S^{1/2} = h^{-1}S^{1/2}.$$

Thus, with $\varphi = S^{-1/4}$, we see that $c^{(1/4)} = S^{1/4}c_u S^{-1/4} = \varphi^{-1} \circ c_u \circ \varphi$ commutes with c, namely

$$c \circ \varphi^{-1} \circ c_u \circ \varphi = \varphi^{-1} \circ c_u \circ \varphi \circ c,$$

which implies

$$\varphi \circ c \circ \varphi^{-1} \circ c_u = c_u \circ \varphi \circ c \circ \varphi^{-1},$$

that is, c_u and $\varphi \circ c \circ \varphi^{-1}$ commute.

Beware that even though $S = h^2$, in general, $S^{1/2} \neq h$ because h may have some negative eigenvalues, but S is positive definite and so are all of its powers S^t. $\square$

Let $c_0 \colon \mathfrak{g} \to \mathfrak{g}$ be a conjugation of $\mathfrak{g}$ that commutes with c_u. In this case, for any $x \in \mathfrak{g}_u$, since

$$c_0(x) = c_0(c_u(x)) = c_u(c_0(x)),$$

we see that $c_0(x) \in \mathfrak{g}_u$, so $\mathfrak{g}_u$ is invariant under c_0, and similarly, for any $x \in \mathfrak{g}_u$, since

$$c_0(ix) = -c_0(c_u(ix)) = -c_u(c_0(ix)),$$

so $i\mathfrak{g}_u$ is also invariant under c_0.

Since the restriction of c_0 to $\mathfrak{g}_u$ is an involutive automorphism of $\mathfrak{g}_u$, we know by linear algebra that

$$\mathfrak{g}_u = E_1 \oplus E_{-1},$$

where E_1 and E_{-1} are the real eigenspaces of c_0 given by

$$E_1 = \{x \in \mathfrak{g}_u \mid c_0(x) = x\}, \qquad E_{-1} = \{x \in \mathfrak{g}_u \mid c_0(x) = -x\}.$$

It is customary to denote E_1 by $\mathfrak{k}_0$ and E_{-1} by $i\mathfrak{p}_0$, where both are real vector spaces, so that

$$\mathfrak{g}_u = \mathfrak{k}_0 \oplus i\mathfrak{p}_0,$$

and $i\mathfrak{g}_u = i\mathfrak{k}_0 \oplus \mathfrak{p}_0$. Since c_0 is semi-linear, for any $ix \in i\mathfrak{p}_0$, we have

$$-ix = c_0(ix) = -ic_0(x),$$

so $c_0(x) = x$ if $x \in \mathfrak{p}_0$, and for $ix \in i\mathfrak{k}_0$, we have

$$c_0(ix) = -ic_0(x) = -ix,$$

since $c_0(x) = x$ for $x \in \mathfrak{k}_0$. Consequently, the (real) Lie algebra $\mathfrak{g}_0$ of fixed points of c_0 is

$$\mathfrak{g}_0 = \mathfrak{k}_0 \oplus \mathfrak{p}_0,$$

and

$$\mathfrak{g} = \mathfrak{g}_0 + i\mathfrak{g}_0 = \mathfrak{k}_0 \oplus \mathfrak{p}_0 \oplus i\mathfrak{k}_0 \oplus i\mathfrak{p}_0 = \mathfrak{k}_0 \oplus i\mathfrak{p}_0 \oplus i\mathfrak{k}_0 \oplus \mathfrak{p}_0 = \mathfrak{g}_u \oplus i\mathfrak{g}_u,$$

so $\mathfrak{g}_0$ is a semi-simple Lie algebra.

Definition 9.11. The decomposition

$$\mathfrak{g}_0 = \mathfrak{k}_0 \oplus \mathfrak{p}_0$$

is called a *Cartan decomposition* of the real Lie algebra $\mathfrak{g}_0$ (with respect to the conjugation c_0).

Next, we give several examples of Cartan decompositions. The group

$$G = \mathbf{SL}(n, \mathbb{C}) = \{X \in \mathrm{M}_n(\mathbb{C}) \mid \det(X) = 1\}$$

is one of the simplest and most important examples of a complex semi-simple Lie group, and the Lie group

$$G_u = \mathbf{SU}(n) = \{X \in \mathrm{M}_n(\mathbb{C}) \mid XX^* = X^*X = I_n,\ \det(X) = 1\}$$

is one of the simplest and most important example of a real semi-simple Lie group, so we use these groups in our examples. The Lie group $\mathbf{SL}(n, \mathbb{C})$ has Lie algebra $\mathfrak{sl}(n, \mathbb{C})$ and the Lie group $\mathbf{SU}(n)$ has Lie algebra $\mathfrak{su}(n)$, both defined in the next section.

9.4 Examples of Cartan Decompositions

Example 9.1. Consider the real Lie algebra $\mathfrak{g}_u = \mathfrak{su}(n)$ of $n \times n$ complex skew-hermitian matrices with zero trace,

$$\mathfrak{su}(n) = \{X \in \mathrm{M}_n(\mathbb{C}) \mid X^* = -X,\ \mathrm{tr}(X) = 0\}.$$

We claim that the complexification $\mathfrak{g} = \mathfrak{su}(n)_{\mathbb{C}}$ of $\mathfrak{su}(n)$ is the complex Lie algebra $\mathfrak{sl}(n, \mathbb{C})$ of $n \times n$ complex matrices with zero trace,

$$\mathfrak{sl}(n, \mathbb{C}) = \{X \in \mathrm{M}_n(\mathbb{C}) \mid \mathrm{tr}(X) = 0\}.$$

First, observe that $i\mathfrak{su}(n)$ is the (real) vector space (not a Lie algebra) of hermitian matrices with zero trace,

$$i\mathfrak{su}(n) = \{X \in \mathrm{M}_n(\mathbb{C}) \mid X^* = X,\ \mathrm{tr}(X) = 0\}.$$

Indeed, if $X^* = -X$, then $(iX)^* = -iX^* = iX$, and if $\mathrm{tr}(X) = 0$, then $\mathrm{tr}(iX) = i\mathrm{tr}(X) = 0$, so

$$i\mathfrak{su}(n) \subseteq \{X \in \mathrm{M}_n(\mathbb{C}) \mid X^* = X,\ \mathrm{tr}(X) = 0\}.$$

Conversely, if $X^* = X$, then $(iX)^* = -iX^* = -iX$, and if $\mathrm{tr}(X) = 0$, then $\mathrm{tr}(iX) = i\mathrm{tr}(X) = 0$, so

$$i\{X \in \mathrm{M}_n(\mathbb{C}) \mid X^* = X,\ \mathrm{tr}(X) = 0\} \subseteq \mathfrak{su}(n).$$

But the above equation implies that

$$\{X \in \mathrm{M}_n(\mathbb{C}) \mid X^* = X,\ \mathrm{tr}(X) = 0\} \subseteq i\mathfrak{su}(n),$$

so

$$i\mathfrak{su}(n) = \{X \in \mathrm{M}_n(\mathbb{C}) \mid X^* = X,\ \mathrm{tr}(X) = 0\}.$$

Every complex matrix $X \in \mathrm{M}_n(\mathbb{C})$ can be written as

$$X = \frac{1}{2}(X + X^*) + \frac{1}{2}(X - X^*),$$

and we have

$$\frac{1}{2}(X + X^*)^* = \frac{1}{2}(X^* + X^{**}) = \frac{1}{2}(X^* + X) = \frac{1}{2}(X + X^*),$$

and

$$\frac{1}{2}(X - X^*)^* = \frac{1}{2}(X^* - X^{**}) = \frac{1}{2}(X^* - X) = -\frac{1}{2}(X - X^*).$$

Also, if $\mathrm{tr}(X) = 0$, then

$$\mathrm{tr}\left(\frac{1}{2}(X + X^*)\right) = \frac{1}{2}\left(\mathrm{tr}(X) + \mathrm{tr}(X^*)\right) = \frac{1}{2}\left(\mathrm{tr}(X) + \mathrm{tr}(X)\right) = 0,$$

and

$$\mathrm{tr}\left(\frac{1}{2}(X - X^*)\right) = \frac{1}{2}\left(\mathrm{tr}(X) - \mathrm{tr}(X^*)\right) = \frac{1}{2}\left(\mathrm{tr}(X) - \mathrm{tr}(X)\right) = 0.$$

Thus, if $X \in \mathfrak{sl}(n, \mathbb{C})$, then $\frac{1}{2}(X + X^*) \in \mathfrak{su}(n)$ and $\frac{1}{2}(X - X^*) \in i\mathfrak{su}(n)$, which proves that

$$\mathfrak{sl}(n, \mathbb{C}) = \mathfrak{su}(n) \oplus i\mathfrak{su}(n). \tag{$*$}$$

The sum is a direct sum because the only matrix such that $X^* = -X$ and $X^* = X$ is the matrix $X = 0$.

Since $\mathfrak{su}(n) = \{X \in \mathfrak{sl}(n, \mathbb{C}) \mid X^* = -X\}$, the conjugation c_u of $\mathfrak{sl}(n, \mathbb{C})$ associated with $\mathfrak{su}(n)$ is the map given by $c_u(X) = -X^*$.

We now consider three types of conjugations on $\mathfrak{g} = \mathfrak{sl}(n, \mathbb{C})$ which lead to interesting real forms of $\mathfrak{sl}(n, \mathbb{C})$.

Example 9.2. Consider the conjugation c_0 of $\mathfrak{sl}(n, \mathbb{C})$ given by $c_0(X) = \overline{X}$. Obviously, c_0 commutes with c_u. The restriction of c_0 to $\mathfrak{g}_u = \mathfrak{su}(n)$ is also c_0, and we obtain

$$\begin{aligned} \mathfrak{k}_0 &= \mathfrak{so}(n) = \{X \in \mathfrak{su}(n) \mid \overline{X} = X\} \\ i\mathfrak{p}_0 &= \{X \in \mathfrak{su}(n) \mid \overline{X} = -X\}, \end{aligned}$$

where $\mathfrak{so}(n)$ is the Lie algebra of $n \times n$ real skew-symmetric matrices.

But for any $X = (x_{jk}) \in \mathfrak{su}(n)$, if $x_{jk} = a_{jk} + ib_{jk}$, with $a_{jk}, b_{jk} \in \mathbb{R}$, since $X^* = -X$, we have $x_{kj} = -a_{jk} + ib_{jk}$, so $a_{kk} = 0$. If we also have

$\overline{X} = -X$, then $a_{jk} + ib_{jk} = -a_{jk} + ib_{jk}$, so $a_{jk} = 0$ for all j, k, which means that $X = (ib_{jk})$, with (b_{jk}) a real symmetric matrix. Thus

$$\mathfrak{p}_0 = \mathfrak{s}(n) = \{X \in \mathrm{M}_n(\mathbb{R}) \mid X^\top = X,\ \mathrm{tr}(X) = 0\},$$

the vector space of $n \times n$ real symmetric matrices with zero trace (not a Lie algebra). We obtain

$$\mathfrak{g}_0 = \mathfrak{k}_0 \oplus \mathfrak{p}_0 = \mathfrak{so}(n) \oplus \mathfrak{s}(n) = \mathfrak{sl}(n, \mathbb{R}),$$

with

$$\mathfrak{sl}(n, \mathbb{R}) = \{X \in \mathrm{M}_n(\mathbb{R}) \mid \mathrm{tr}(X) = 0\}.$$

It turns out that a Gelfand pair arises from two Lie groups G_0 and K_0, whose Lie algebras are $\mathfrak{sl}(n, \mathbb{R})$ and $\mathfrak{sl}(n, \mathbb{R}) \cap \mathfrak{su}(n) = \mathfrak{so}(n)$. To describe these groups, first we need to consider the complex simply-connected Lie group $G = \mathbf{SL}(n, \mathbb{C})$, whose complex Lie algebra is $\mathfrak{g} = \mathfrak{sl}(n, \mathbb{C})$.

Definition 9.12. A complex Lie group G of (complex) dimension n can also be viewed as a real Lie group of (real) dimension $2n$ denoted $G|_\mathbb{R}$, by viewing a holomorphic chart $\varphi\colon U \to \mathbb{C}^n$ of G as a real smooth function $\varphi\colon U \to \mathbb{R}^{2n}$. More generally, a complex manifold M of (complex) dimension n can be viewed as a real manifold of dimension $2n$ denoted $M|_\mathbb{R}$.

Using the correspondence between simply-connected real Lie groups and real Lie algebras (see Gallier and Quaintance [Gallier and Quaintance (2020a)], Section 19.4), there is a unique automorphism $\sigma_0\colon \mathbf{SL}(n, \mathbb{C})|_\mathbb{R} \to \mathbf{SL}(n, \mathbb{C})|_\mathbb{R}$ such that $d(\sigma_0)_I = c_0$, where $c_0\colon \mathfrak{sl}(n, \mathbb{C})_\mathbb{R} \to \mathfrak{sl}(n, \mathbb{C})|_\mathbb{R}$ is the map $c_0(X) = \overline{X}$, and σ_0 is also given by

$$\sigma_0(X) = \overline{X}.$$

The real Lie group G_0 is the set of fixed points of $\mathbf{SL}(n, \mathbb{C})|_\mathbb{R}$ under the automorphism σ_0, given by

$$G_0 = \{X \in \mathbf{SL}(n, \mathbb{C})_\mathbb{R} \mid \overline{X} = X\} = \mathbf{SL}(n, \mathbb{R}).$$

Note that the Lie algebra of $G_0 = \mathbf{SL}(n, \mathbb{R})$ is $\mathfrak{g}_0 = \mathfrak{sl}(n, \mathbb{R})$. The simply-connected real Lie group G_u whose Lie algebra is $\mathfrak{su}(n)$ is $G_u = \mathbf{SU}(n)$. We define K_0 by

$$K_0 = G_0 \cap G_u = \mathbf{SL}(n, \mathbb{R}) \cap \mathbf{SU}(n) = \mathbf{SO}(n).$$

Note that

$$\begin{aligned} \mathbf{SL}(n, \mathbb{R}) &= \{X \in \mathrm{M}_n(\mathbb{R}) \mid \det(X) = 1\} \\ \mathbf{SO}(n) &= \{X \in \mathrm{M}_n(\mathbb{R}) \mid XX^\top = X^\top X = I_n,\ \det(X) = 1\}, \end{aligned}$$

and that the Lie algebra of the compact Lie group $K_0 = \mathbf{SO}(n)$ is

$$\mathfrak{k}_0 = \mathfrak{so}(n) = \{X \in \mathrm{M}_n(\mathbb{R}) \mid X^\top = -X\}.$$

The following paragraph is meant for readers well acquainted with Lie groups and Lie algebras and can be safely omitted. The real Lie group $G_0 = \mathbf{SL}(n, \mathbb{R})$ is semi-simple and the real Lie group $K_0 = \mathbf{SO}(n)$ is semi-simple for $n \geq 3$. These groups are connected but not simply-connected. For $n = 2$, the universal cover of $\mathbf{SL}(2, \mathbb{R})$ is $\mathbb{R}^3$ and the universal cover of $\mathbf{SO}(2)$ is $\mathbb{R}$. For $n \geq 3$, the universal cover $\widetilde{G}_0 = \widetilde{\mathbf{SL}}(n, \mathbb{R})$ of $G_0 = \mathbf{SL}(n, \mathbb{R})$ is not a matrix group and the universal cover of $K_0 = \mathbf{SO}(n)$ is $\widetilde{K}_0 = \mathbf{Spin}(n)$. The real semi-simple (connected) Lie group $G_0 = \mathbf{SL}(n, \mathbb{R})$ is called a *real form* of the complex semi-simple (simply-connected) Lie group $G = \mathbf{SL}(n, \mathbb{C})$. We will show later that the pair $(G_0, K_0) = (\mathbf{SL}(n, \mathbb{R}), \mathbf{SO}(n))$ is a Gelfand pair.

Example 9.3. Again, consider the real Lie algebra $\mathfrak{g}_u = \mathfrak{su}(n)$ of $n \times n$ complex skew-hermitian matrices with zero trace. In this example, we also need the Lie algebra $\mathfrak{u}(n)$ of the real Lie group

$$\mathbf{U}(n) = \{X \in \mathrm{M}_n(\mathbb{C}) \mid XX^* = X^*X = I_n\},$$

given by

$$\mathfrak{u}(n) = \{X \in \mathrm{M}_n(\mathbb{C}) \mid X^* = -X\}.$$

In other words, $\mathfrak{u}(n)$ consists of all skew-Hermitian complex matrices. Observe that

$$\mathfrak{su}(n) = \{X \in \mathfrak{u}(n) \mid \mathrm{tr}(X) = 0\} = \mathfrak{u}(n) \cap \mathfrak{sl}(n, \mathbb{C}).$$

We showed in Example 9.2 that the complexification $\mathfrak{su}(n)_{\mathbb{C}}$ of $\mathfrak{su}(n)$ is the complex Lie algebra $\mathfrak{g} = \mathfrak{sl}(n, \mathbb{C})$ of $n \times n$ complex matrices with zero trace. This time, let $c_0 \colon \mathfrak{sl}(n, \mathbb{C}) \to \mathfrak{sl}(n, \mathbb{C})$ be the conjugation given by

$$c_0(X) = -I_{p,n-p} X^* I_{p,n-p},$$

where

$$I_{p,n-p} = \begin{pmatrix} I_p & 0_{p,n-p} \\ 0_{n-p,p} & -I_{p,n-p} \end{pmatrix},$$

with $1 \leq p \leq n-1$. Obviously, $I_{p,n-p}^* = I_{p,n-p}$ and $I_{p,n-p}^2 = I_n$, and c_0 commutes with c_u (given by $c_u(X) = -X^*$). Since matrices in $\mathfrak{su}(n)$ satisfy the property $X^* = -X$, the restriction of c_0 to $\mathfrak{su}(n)$ is given by $c_0(X) = I_{p,n-p} X I_{p,n-p}$. If we write

$$X = \begin{pmatrix} U & B \\ A & V \end{pmatrix},$$

where $U \in \mathrm{M}_p(\mathbb{C}), V \in \mathrm{M}_{n-p}(\mathbb{C}), A \in \mathrm{M}_{n-p,p}(\mathbb{C}), B \in \mathrm{M}_{p,n-p}(\mathbb{C})$, then

$$\begin{aligned} I_{p,n-p}XI_{p,n-p} &= \begin{pmatrix} I_p & 0_{p,n-p} \\ 0_{n-p,p} & -I_{n-p} \end{pmatrix} \begin{pmatrix} U & B \\ A & V \end{pmatrix} \begin{pmatrix} I_p & 0_{p,n-p} \\ 0_{n-p,p} & -I_{n-p} \end{pmatrix} \\ &= \begin{pmatrix} U & -B \\ -A & V \end{pmatrix}. \end{aligned}$$

Therefore

$$\begin{aligned} \mathfrak{k}_0 &= \{X \in \mathfrak{su}(n) \mid I_{p,n-p}XI_{p,n-p} = X\} \\ &= \left\{ \begin{pmatrix} U & 0 \\ 0 & V \end{pmatrix} \,\middle|\, U^* = -U,\ V^* = -V,\ \mathrm{tr}(U) + \mathrm{tr}(V) = 0 \right\} \\ &= \left\{ \begin{pmatrix} U & 0 \\ 0 & V \end{pmatrix} \,\middle|\, U \in \mathfrak{u}(p),\ V \in \mathfrak{u}(n-p),\ \mathrm{tr}(U) + \mathrm{tr}(V) = 0 \right\}, \end{aligned}$$

and

$$\begin{aligned} i\mathfrak{p}_0 &= \{X \in \mathfrak{su}(n) \mid -I_{p,n-p}XI_{p,n-p} = X\} \\ &= \left\{ \begin{pmatrix} 0 & -A^* \\ A & 0 \end{pmatrix} \,\middle|\, A \in \mathrm{M}_{n-p,p}(\mathbb{C}) \right\}, \end{aligned}$$

so

$$\begin{aligned} \mathfrak{p}_0 &= \left\{ \begin{pmatrix} 0 & -iA^* \\ iA & 0 \end{pmatrix} \,\middle|\, A \in \mathrm{M}_{n-p,p}(\mathbb{C}) \right\} \\ &= \left\{ \begin{pmatrix} 0 & (iA)^* \\ iA & 0 \end{pmatrix} \,\middle|\, A \in \mathrm{M}_{n-p,p}(\mathbb{C}) \right\} \\ &= \left\{ \begin{pmatrix} 0 & A^* \\ A & 0 \end{pmatrix} \,\middle|\, A \in \mathrm{M}_{n-p,p}(\mathbb{C}) \right\}. \end{aligned}$$

Consequently, the real Lie algebra $\mathfrak{g}_0$ corresponding to the conjugation c_0 is given by

$$\begin{aligned} \mathfrak{g}_0 = \mathfrak{k}_0 \oplus \mathfrak{p}_0 = \Bigg\{ \begin{pmatrix} U & A^* \\ A & V \end{pmatrix} \,\Bigg|\, & U \in \mathfrak{u}(p),\ V \in \mathfrak{u}(n-p),\ A \in \mathrm{M}_{n-p,p}(\mathbb{C}), \\ & \mathrm{tr}(U) + \mathrm{tr}(V) = 0 \Bigg\}. \end{aligned}$$

Thus, $\mathfrak{g}_0 = \mathfrak{su}(p, n-p)$, the Lie algebra of the real Lie group $\mathbf{SU}(p, n-p)$ defined by

$$\mathbf{SU}(p, n-p) = \{X \in \mathrm{M}_n(\mathbb{C}) \mid X^* I_{p,n-p} X = I_{p,n-p}, \quad \det(X) = 1\}.$$

Given any $p \times p$ matrix U, if $\alpha = \mathrm{tr}(U)$, we let

$$U_1 = U - \frac{\alpha}{p} I_p,$$

and then we have

$$\mathrm{tr}(U_1) = \mathrm{tr}(U) - p\frac{\alpha}{p} = \alpha - \alpha = 0,$$

so we can write

$$U = U_1 + \frac{\alpha}{p} I_p$$

with $\mathrm{tr}(U_1) = 0$, and since $\mathrm{tr}(U) + \mathrm{tr}(V) = 0$, we also let

$$V_1 = V + \frac{\alpha}{n-p} I_{n-p},$$

so that $\mathrm{tr}(V_1) = 0$ and

$$V = V_1 - \frac{\alpha}{n-p} I_{n-p},$$

which shows that we can also write every matrix $X \in \mathfrak{k}_0$ as

$$X = \begin{pmatrix} U_1 & 0 \\ 0 & 0 \end{pmatrix} + i \begin{pmatrix} \frac{\alpha}{p} I_p & 0 \\ 0 & -\frac{\alpha}{n-p} I_{n-p} \end{pmatrix} + \begin{pmatrix} 0 & 0 \\ 0 & V_1 \end{pmatrix},$$
$$U_1 \in \mathfrak{su}(p),\ V_1 \in \mathfrak{su}(n-p),\ \alpha \in \mathbb{R},$$

and

$$\mathfrak{k}_0 \cong \mathfrak{su}(p) \oplus i\mathbb{R} \oplus \mathfrak{su}(n-p),$$

where all the summands are ideals.

The derivative $d(\sigma_0)_I$ of the automorphism $\sigma_0 \colon \mathbf{SL}(n, \mathbb{C})|_{\mathbb{R}} \to \mathbf{SL}(n, \mathbb{C})|_{\mathbb{R}}$ given by

$$\sigma_0(X) = I_{p,n-p}(X^*)^{-1} I_{p,n-p}$$

is the map $c_0 \colon \mathfrak{sl}(n, \mathbb{C})_{\mathbb{R}} \to \mathfrak{sl}(n, \mathbb{C})_{\mathbb{R}}$ also given by $c_0(X) = -I_{p,n-p} X^* I_{p,n-p}$. The real Lie group G_0 is the set of fixed points of $\mathbf{SL}(n, \mathbb{C})_{\mathbb{R}}$ under the automorphism σ_0, given by

$$\begin{aligned} G_0 &= \{X \in \mathbf{SL}(n, \mathbb{C})|_{\mathbb{R}} \mid X = I_{p,n-p}(X^*)^{-1} I_{p,n-p}\} \\ &= \{X \in \mathbf{SL}(n, \mathbb{C})|_{\mathbb{R}} \mid X^* I_{p,n-p} X = I_{p,n-p}\} = \mathbf{SU}(p, n-p). \end{aligned}$$

If we write

$$X = \begin{pmatrix} U & B \\ A & V \end{pmatrix}$$

for any $X \in \mathbf{SU}(p, n-p)$, it is easy to check that if $X \in \mathbf{SU}(p, n-p) \cap \mathbf{SU}(n)$, then

$$\begin{aligned} U^*U - A^*A &= I_p \\ V^*V - B^*B &= I_{n-p} \\ U^*U + A^*A &= I_p \\ V^*V + B^*B &= I_{n-p} \\ U^*B - A^*V &= 0 \\ U^*B + A^*V &= 0, \end{aligned}$$

which implies that $A = 0$ and $B = 0$. Therefore,

$$\begin{aligned} K_0 &= G_0 \cap G_u = \mathbf{SU}(p, n-p) \cap \mathbf{SU}(n) \\ &= \left\{ \begin{pmatrix} U & 0 \\ 0 & V \end{pmatrix} \;\middle|\; U \in \mathbf{U}(p),\ V \in \mathbf{U}(n-p),\ \det(U)\det(V) = 1 \right\}. \end{aligned}$$

This group is usually denoted $S(\mathbf{U}(p) \times \mathbf{U}(n-p))$. For any $X \in S(\mathbf{U}(p) \times \mathbf{U}(n-p))$, we can write

$$X = \begin{pmatrix} U_1 & 0 \\ 0 & I_{n-p} \end{pmatrix} \begin{pmatrix} \lambda & 0 & 0 & 0 \\ 0 & I_{p-1} & 0 & 0 \\ 0 & 0 & \lambda^{-1} & 0 \\ 0 & 0 & 0 & I_{n-p-1} \end{pmatrix} \begin{pmatrix} I_p & 0 \\ 0 & V_1 \end{pmatrix},$$

with $U_1 \in \mathbf{SU}(p), V_1 \in \mathbf{SU}(n-p)$ and $\lambda \in \mathbb{C}$ with $|\lambda| = 1$. Therefore

$$K_0 = S(\mathbf{U}(p) \times \mathbf{U}(n-p)) \cong \mathbf{SU}(p) \times \mathbf{U}(1) \times \mathbf{SU}(n-p).$$

Again, the following paragraph is meant for readers well acquainted with Lie groups and Lie algebras and can be safely omitted. The Lie algebra of the real compact Lie group $K_0 = \mathbf{SU}(p) \times \mathbf{U}(1) \times \mathbf{SU}(n-p)$ is $\mathfrak{k}_0 = \mathfrak{su}(p) \oplus i\mathbb{R} \oplus \mathfrak{su}(n-p)$. The real Lie groups $G_0 = \mathbf{SU}(p, n-p)$ and $K_0 = \mathbf{SU}(p) \times \mathbf{U}(1) \times \mathbf{SU}(n-p)$ are semi-simple and connected but not simply-connected. The universal cover of K_0 is $\widetilde{K}_0 = \mathbf{SU}(p) \times \mathbb{R} \times \mathbf{SU}(n-p)$, and the universal cover of $G_0 = \mathbf{SU}(p, n-p)$ is $G_0 = \mathbf{Spin}(p, n-p)$. The real semi-simple (connected) Lie group $G_0 = \mathbf{SU}(p, n-q)$ is called a real form of the complex semi-simple (simply-connected) Lie group $G = \mathbf{SL}(n, \mathbb{C})$. We will show later that the pair $(G_0, K_0) = (\mathbf{SU}(p, n-p), \mathbf{SU}(p) \times \mathbf{U}(1) \times \mathbf{SU}(n-p))$ is a Gelfand pair.

9.5 Quaternionic and Complex Symplectic Lie Groups

Again, consider the real Lie algebra $\mathfrak{g}_u = \mathfrak{su}(n)$ of $n \times n$ complex skew-hermitian matrices with zero trace and its complexification $\mathfrak{g} = \mathfrak{sl}(n, \mathbb{C})$. This time assume $n = 2m$, and let $c_0 : \mathfrak{sl}(n, \mathbb{C}) \to \mathfrak{sl}(n, \mathbb{C})$ be the conjugation given by

$$c_0(X) = -J_m \overline{X} J_m,$$

with

$$J_m = \begin{pmatrix} 0_m & I_m \\ -I_m & 0_m \end{pmatrix}.$$

Since $J_m^2 = -I_{2m}$, $\overline{J_m} = J_m$, and $J_m^\top = -J_m$, the conjugation c_0 commutes with c_u. The automorphism $\sigma_0 : \mathbf{SL}(2m, \mathbb{C})|_\mathbb{R} \to \mathbf{SL}(2m, \mathbb{C})_\mathbb{R}$ such that $(d\sigma_0)_I = c_0$ is also $\sigma_0(X) = -J_m \overline{X} J_m$. In this example, since we also consider matrices whose entries are quaternions, we denote the group $\mathbf{SU}(n)$ as $\mathbf{SU}(n, \mathbb{C})$ and the Lie algebra $\mathfrak{su}(n)$ as $\mathfrak{su}(n, \mathbb{C})$ to avoid confusion. We will determine the Lie algebras $\mathfrak{k}_0 = \{X \in \mathfrak{su}(2m, \mathbb{C}) \mid c_0(X) = X\}$, the group $G_0 = \{X \in \mathbf{SL}(2m, \mathbb{C})|_\mathbb{R} \mid \sigma_0(X) = X\}$, and the group $K_0 = G_0 \cap \mathbf{SU}(2m, \mathbb{C})$. The group G_0 is a Lie group known as $\mathbf{SU}^*(2m)$. We give another description of the group $\mathbf{SU}^*(2m)$ as a group of matrices with quaternion entries ($\mathbf{SL}(m, \mathbb{H})$). We also give two descriptions of the group K_0; one as a group of matrices with quaternion entries ($\mathbf{SU}(m, \mathbb{H})$), and the other as a symplectic group ($\mathbf{Sp}(m)$).

If we write

$$X = \begin{pmatrix} U & V \\ A & B \end{pmatrix} \in \mathbf{SL}(2m, \mathbb{C}),$$

where $U, V, A, B \in \mathrm{M}_m(\mathbb{C})$, then we have

$$\begin{aligned} -J_m \overline{X} J_m &= \begin{pmatrix} 0 & -I_m \\ I_m & 0 \end{pmatrix} \begin{pmatrix} \overline{U} & \overline{V} \\ \overline{A} & \overline{B} \end{pmatrix} \begin{pmatrix} 0 & I_m \\ -I_m & 0 \end{pmatrix} \\ &= \begin{pmatrix} 0 & -I_m \\ I_m & 0 \end{pmatrix} \begin{pmatrix} -\overline{V} & \overline{U} \\ -\overline{B} & \overline{A} \end{pmatrix} = \begin{pmatrix} \overline{B} & -\overline{A} \\ -\overline{V} & \overline{U} \end{pmatrix}, \end{aligned}$$

so $X = -J_m \overline{X} J_m$ iff

$$U = \overline{B}, \quad V = -\overline{A}, \quad A = -\overline{V}, \quad B = \overline{U},$$

which simplifies to

$$B = \overline{U}, \qquad A = -\overline{V}.$$

Therefore, X is of the form

$$X = \begin{pmatrix} U & V \\ -\overline{V} & \overline{U} \end{pmatrix},$$

and the real Lie group $G_0 = \{X \in \mathbf{SL}(2m, \mathbb{C})|_{\mathbb{R}} \mid X = -J_m \overline{X} J_m\}$ is given by

$$G_0 = \left\{ \begin{pmatrix} U & V \\ -\overline{V} & \overline{U} \end{pmatrix} \in \mathbf{SL}(2m, \mathbb{C}) \;\middle|\; U, V \in \mathrm{M}_m(\mathbb{C}) \right\}.$$

Definition 9.13. The real Lie group $\mathbf{SU}^*(2m)$ is defined by

$$\mathbf{SU}^*(2m) = \left\{ \begin{pmatrix} U & V \\ -\overline{V} & \overline{U} \end{pmatrix} \in \mathbf{SL}(2m, \mathbb{C}) \;\middle|\; U, V \in \mathrm{M}_m(\mathbb{C}) \right\}.$$

The notation $\mathbf{SU}^*(2m)$ for this real Lie group is introduced in Helgason [Helgason (2001)], Chapter X, Section 2. We will show later that the (real) Lie group $\mathbf{SU}^*(2m)$ is isomorphic to the quaternionic (real) Lie group $\mathbf{SL}(m, \mathbb{H})$.

Since $X \in \mathbf{SU}^*(2m)$ is invertible, for any vector $z = \binom{x}{y} \in \mathbb{C}^{2m}$, if $Xz = 0$ then $z = 0$, so in particular, for $y = 0$, since

$$Xz = \begin{pmatrix} U & V \\ -\overline{V} & \overline{U} \end{pmatrix} \begin{pmatrix} x \\ y \end{pmatrix} = \begin{pmatrix} Ux + Vy \\ -\overline{V}x + \overline{U}y \end{pmatrix},$$

we have $Xz = 0$ iff $Ux + Vy = 0$ and $\overline{V}x + \overline{U}y = 0$. So with $y = 0$, if $Ux = 0$, then $x = 0$, and with $x = 0$, if $Vy = 0$, then $y = 0$ which implies that U and V are invertible. Therefore, the group G_0, the set of fixed points of σ_0, is also given by

$$G_0 = \left\{ X = \begin{pmatrix} U & V \\ -\overline{V} & \overline{U} \end{pmatrix} \;\middle|\; U, V \in \mathbf{GL}(m, \mathbb{C})|_{\mathbb{R}},\ \det(X) = 1 \right\}.$$

Since the conjugation c_0 on $\mathfrak{sl}(2m, \mathbb{C})$ has the same expression as the conjugation σ_0 on $\mathbf{SL}(2m, \mathbb{C})|_{\mathbb{R}}$, the same computation as above shows that

$$\mathfrak{su}^*(2m) = \left\{ \begin{pmatrix} U & V \\ -\overline{V} & \overline{U} \end{pmatrix} \;\middle|\; U, V \in \mathrm{M}_m(\mathbb{C}),\ \mathrm{tr}(U) + \mathrm{tr}(\overline{U}) = 0 \right\}.$$

We also have

$$\begin{aligned} \mathfrak{k}_0 &= \{X \in \mathfrak{su}(2m, \mathbb{C}) \mid X = -J_m \overline{X} J_m\} \\ &= \left\{ \begin{pmatrix} U & V \\ -\overline{V} & \overline{U} \end{pmatrix} \;\middle|\; U \in \mathfrak{u}(m, \mathbb{C}),\ V \in \mathrm{M}_m(\mathbb{C}),\ V^\top = V \right\} \\ &= \left\{ \begin{pmatrix} U & V \\ -\overline{V} & -U^\top \end{pmatrix} \;\middle|\; U \in \mathfrak{u}(m, \mathbb{C}),\ V \in \mathrm{M}_m(\mathbb{C}),\ V^\top = V \right\}, \end{aligned}$$

and

$$\begin{aligned} i\mathfrak{p}_0 &= \{X \in \mathfrak{su}(2m,\mathbb{C}) \mid X = J_m\overline{X}J_m\} \\ &= \left\{ \begin{pmatrix} U & V \\ \overline{V} & -\overline{U} \end{pmatrix} \;\middle|\; U \in \mathfrak{u}(m,\mathbb{C}),\ V \in \mathrm{M}_m(\mathbb{C}),\ V^\top = -V,\ \mathrm{tr}(U) = 0 \right\}. \end{aligned}$$

Consequently, we have

$$\begin{aligned} \mathfrak{p}_0 &= \left\{ \begin{pmatrix} iU & iV \\ i\overline{V} & -i\overline{U} \end{pmatrix} \;\middle|\; U \in \mathfrak{u}(m,\mathbb{C}),\ V \in \mathrm{M}_m(\mathbb{C}),\ V^\top = -V,\ \mathrm{tr}(U) = 0 \right\} \\ &= \left\{ \begin{pmatrix} A & B \\ -\overline{B} & \overline{A} \end{pmatrix} \;\middle|\; A \in i\mathfrak{u}(m,\mathbb{C}),\ B \in \mathrm{M}_m(\mathbb{C}),\ B^\top = -B,\ \mathrm{tr}(A) = 0 \right\} \\ &= \left\{ \begin{pmatrix} A & B \\ -\overline{B} & \overline{A} \end{pmatrix} \;\middle|\; A, B \in \mathrm{M}_m(\mathbb{C}),\ A^* = A,\ B^\top = -B,\ \mathrm{tr}(A) = 0 \right\} \\ &= \left\{ \begin{pmatrix} A & B \\ -\overline{B} & A^\top \end{pmatrix} \;\middle|\; A, B \in \mathrm{M}_m(\mathbb{C}),\ A^* = A,\ B^\top = -B,\ \mathrm{tr}(A) = 0 \right\}. \end{aligned}$$

Observe that if $X \in \mathfrak{k}_0$, then automatically $\mathrm{tr}(X) = 0$. We immediately check that

$$\mathfrak{su}^*(2m) = \mathfrak{k}_0 \oplus \mathfrak{p}_0.$$

The (real) Lie group $\mathbf{SU}^*(2m)$ can also be viewed as the (real) Lie group $\mathbf{SL}(m,\mathbb{H})$, where $\mathbb{H}$ is the skew-field of quaternions. To see this, it is convenient to view the real algebra $\mathbb{H}$ as

$$\mathbb{H} = \mathbb{C} \oplus \mathbb{C}j.$$

using the fact that every quaternion can be written uniquely as

$$q = a + xi + yj + zk = a + xi + (y + zi)j, \quad a, x, y, z \in \mathbb{R},$$

since $ij = k$. Since the conjugate $\overline{q}$ of q is $\overline{q} = a - xi - yj - zk = a - xi - (y + zi)j$, if we write $q = \alpha + \beta j$ with $\alpha, \beta \in \mathbb{C}$, then

$$\overline{q} = \overline{\alpha} - \beta j.$$

Also, since $ij = -ji$, for any $\alpha = a + bi \in \mathbb{C}$, we have

$$j\alpha = j(a + bi) = ja + bji = aj - bij = (a - bi)j = \overline{\alpha}j.$$

In summary, we have

$$\overline{\alpha + j\beta} = \overline{\alpha} - \beta j, \qquad j\alpha = \overline{\alpha}j. \tag{conj}$$

Every $m \times m$ matrix $X = (\alpha_{k\ell} + \beta_{k\ell}j) \in \mathrm{M}_m(\mathbb{H})$ can be written uniquely as

$$X = U + Vj,$$

with $U = (\alpha_{k\ell}) \in \mathrm{M}_m(\mathbb{C})$ and $V = (\beta_{k\ell}) \in \mathrm{M}_m(\mathbb{C})$. Then in view of the equations (conj), if $X = (\alpha_{k\ell} + \beta_{k\ell} j) \in \mathrm{M}_m(\mathbb{H})$ and if we define

$$\overline{X} = (\overline{\alpha_{\ell k} + \beta_{\ell k} j}) \quad \text{and} \quad X^* = \overline{X}^\top,$$

then we have

$$X^* = (U + Vj)^* = U^* - V^\top j.$$

We also have

$$jV = \overline{V} j.$$

In summary, we have

$$(U + Vj)^* = U^* - V^\top j, \qquad jV = \overline{V} j. \tag{conj'}$$

The (real) Lie group $\mathbf{GL}(m, \mathbb{H})$ is the group of matrices $X \in \mathrm{M}_m(\mathbb{H})$ such that there is some $Y \in \mathrm{M}_m(\mathbb{H})$ with

$$XY = YX = I_m.$$

Observe that since $j^2 = -1$, we have

$$\begin{aligned}
(U_1 + V_1 j)(U_2 + V_2 j) &= U_1U_2 + U_1V_2 j + V_1 j U_2 + V_1 j V_2 j \\
&= U_1U_2 + U_1V_2 j + V_1\overline{U_2} j + V_1\overline{V_2} j^2 \\
&= U_1U_2 - V_1\overline{V_2} + (U_1V_2 + V_1\overline{U_2}) j.
\end{aligned}$$

Since

$$\begin{pmatrix} U_1 & V_1 \\ -\overline{V_1} & \overline{U_1} \end{pmatrix} \begin{pmatrix} U_2 & V_2 \\ -\overline{V_2} & \overline{U_2} \end{pmatrix} = \begin{pmatrix} U_1U_2 - V_1\overline{V_2} & U_1V_2 + V_1\overline{U_2} \\ -\overline{U_1}\,\overline{V_2} - \overline{V_1}U_2 & \overline{U_1}\,\overline{U_2} - \overline{V_1}V_2 \end{pmatrix},$$

the map $\varphi\colon \mathrm{M}_m(\mathbb{H}) \to \mathrm{M}_{2m}(\mathbb{C})$ given by

$$\varphi(U + Vj) = \begin{pmatrix} U & V \\ -\overline{V} & \overline{U} \end{pmatrix}$$

is an injective $\mathbb{R}$-algebra homomorphism. Observe that this is the matrix in the definition of G_0.

What we did previously shows that this homomorphism restricts to an injective homomorphism $\varphi\colon \mathbf{GL}(m, \mathbb{H}) \to \mathbf{GL}(2m, \mathbb{C})$, which allows us to view the group $\mathbf{GL}(m, \mathbb{H})$ as

$$\varphi(\mathbf{GL}(m, \mathbb{H})) = \left\{ X = \begin{pmatrix} U & V \\ -\overline{V} & \overline{U} \end{pmatrix} \;\middle|\; X \in \mathbf{GL}(2m, \mathbb{C}),\, U, V \in \mathbf{GL}(m, \mathbb{C}) \right\}.$$

Definition 9.14. We define the (real) Lie group $\mathbf{SL}(m, \mathbb{H})$ as

$$\begin{aligned}
\mathbf{SL}(m, \mathbb{H}) &= \{X \in \varphi(\mathbf{GL}(m, \mathbb{H})) \mid \det(X) = 1\} \\
&= \varphi(\mathbf{GL}(m, \mathbb{H})) \cap \mathbf{SL}(2m, \mathbb{C}).
\end{aligned}$$

Therefore, we conclude that

$$G_0 = \mathbf{SU}^*(2m) = \mathbf{SL}(m, \mathbb{H}).$$

This is a real, semi-simple, simply-connected Lie group.

Technically, $\mathbf{SL}(m, \mathbb{H})$ should be defined as the subgroup of $\mathbf{GL}(m, \mathbb{H})$ given by

$$\{X \in \mathbf{GL}(m, \mathbb{H}) \mid \det(\varphi(X)) = 1\},$$

but for our purpose, it is more convenient to view $\mathbf{GL}(m, \mathbb{H})$ and its various subgroups as subgroups of $\mathbf{GL}(2m, \mathbb{C})$.

With this identification in mind, the Lie algebras $\mathfrak{gl}(m, \mathbb{H})$ of $\mathbf{GL}(m, \mathbb{H})$ and $\mathfrak{sl}(m, \mathbb{H})$ of $\mathbf{SL}(m, \mathbb{H})$ are given by

$$\mathfrak{gl}(m, \mathbb{H}) = \left\{ \begin{pmatrix} U & V \\ -\overline{V} & \overline{U} \end{pmatrix} \,\middle|\, U, V \in \mathbf{M}_m(\mathbb{C}) \right\},$$

and

$$\mathfrak{sl}(m, \mathbb{H}) = \left\{ \begin{pmatrix} U & V \\ -\overline{V} & \overline{U} \end{pmatrix} \,\middle|\, U, V \in \mathbf{M}_m(\mathbb{C}),\ \mathrm{tr}(U) + \mathrm{tr}(\overline{U}) = 0 \right\}.$$

Observe that

$$\mathfrak{sl}(m, \mathbb{H}) = \mathfrak{su}^*(2m),$$

as it should be.

We can also identify

$$K_0 = G_0 \cap \mathbf{SU}(2m, \mathbb{C}) = \mathbf{SU}^*(2m) \cap \mathbf{SU}(2m, \mathbb{C}) = \mathbf{SL}(m, \mathbb{H}) \cap \mathbf{SU}(2m, \mathbb{C}).$$

There are two descriptions for the Lie group K_0; one as a quaternionic group; another in terms of the complex symplectic group $\mathbf{Sp}(m, \mathbb{C})$.

For the quaternionic description, define $\mathbf{U}(m, \mathbb{H})$ as the (real) Lie group

$$\mathbf{U}(m, \mathbb{H}) = \{X = U + Vj \in \mathbf{GL}(m, \mathbb{H}) \mid X^*X = XX^* = I_m\}.$$

It is easy to see that only the first equation is needed. Using (conj'), this is equivalent to

$$\begin{aligned}
(U + Vj)^*(U + Vj) &= (U^* - V^\top j)(U + Vj) \\
&= U^*U + U^*Vj - V^\top jU - V^\top jVj \\
&= U^*U + U^*Vj - V^\top \overline{U}j - V^\top \overline{V}j^2 \\
&= U^*U + V^\top \overline{V} + (U^*V - V^\top \overline{U})j = I_m,
\end{aligned}$$

so we get

$$U^*U + V^\top \overline{V} = I_m, \qquad U^*V - V^\top \overline{U} = 0.$$

On the other hand, for

$$X = \begin{pmatrix} U & V \\ -\overline{V} & \overline{U} \end{pmatrix},$$

we have $X^*X = I_{2m}$ iff

$$\begin{pmatrix} U^* & -V^\top \\ V^* & U^\top \end{pmatrix} \begin{pmatrix} U & V \\ -\overline{V} & \overline{U} \end{pmatrix} = \begin{pmatrix} U^*U + V^\top \overline{V} & U^*V - V^\top \overline{U} \\ V^*U - U^\top \overline{V} & V^*V + U^\top \overline{U} \end{pmatrix} = \begin{pmatrix} I_m & 0 \\ 0 & I_m \end{pmatrix},$$

and these are equivalent to the same conditions as above,

$$U^*U + V^\top \overline{V} = I_m, \qquad U^*V - V^\top \overline{U} = 0.$$

We conclude that

$$\varphi(\mathbf{U}(m, \mathbb{H})) = \varphi(\mathbf{GL}(m, \mathbb{H})) \cap \mathbf{U}(2m, \mathbb{C}).$$

Definition 9.15. The (real) Lie group $\mathbf{SU}(m, \mathbb{H})$ is defined as

$$\mathbf{SU}(m, \mathbb{H}) = \varphi(\mathbf{GL}(m, \mathbb{H})) \cap \mathbf{SU}(2m, \mathbb{C}).$$

Since

$$\mathbf{SL}(m, \mathbb{H}) = \varphi(\mathbf{GL}(m, \mathbb{H})) \cap \mathbf{SL}(2m, \mathbb{C}),$$

we have

$$\mathbf{SU}(m, \mathbb{H}) = \mathbf{SL}(m, \mathbb{H}) \cap \mathbf{SU}(2m, \mathbb{C}).$$

Again, for our purpose, it is more convenient to view $\mathbf{SL}(m, \mathbb{H})$ and $\mathbf{SU}(m, \mathbb{H})$ as subgroups of $\mathbf{GL}(2m, \mathbb{C})$.

In summary, we see that

$$\begin{aligned} K_0 &= \mathbf{SU}^*(2m) \cap \mathbf{SU}(2m, \mathbb{C}) = \mathbf{SL}(m, \mathbb{H}) \cap \mathbf{SU}(2m, \mathbb{C}) \\ &= \mathbf{SU}(m, \mathbb{H}). \end{aligned}$$

This is a real, semi-simple, simply-connected Lie group.

The Lie algebra $\mathfrak{u}(m, \mathbb{H})$ of $\mathbf{U}(m, \mathbb{H})$ consists of the space of matrices

$$\mathfrak{u}(m, \mathbb{H}) = \{X \in \mathrm{M}_m(\mathbb{H}) \mid X^* = -X\}.$$

If we write $X = U + Vj$ (with $U, V \in \mathrm{M}_n(\mathbb{C})$), then $X^* = -X$ is equivalent to

$$(U + Vj)^* = U^* - V^\top j = -U - Vj,$$

that is,

$$U^* = -U, \qquad V^\top = V.$$

Thus, we can also write

$$\begin{aligned}\mathfrak{u}(m, \mathbb{H}) &= \left\{ \begin{pmatrix} U & V \\ -\overline{V} & \overline{U} \end{pmatrix} \,\middle|\, U, V \in \mathbf{M}_m(\mathbb{C}),\, U^* = -U,\, V^\top = V \right\} \\ &= \left\{ \begin{pmatrix} U & V \\ -\overline{V} & -U^\top \end{pmatrix} \,\middle|\, U \in \mathfrak{u}(m, \mathbb{C}),\, V \in \mathbf{M}_m(\mathbb{C}),\, V^\top = V \right\},\end{aligned}$$

and so

$$\mathfrak{u}(m, \mathbb{H}) = \mathfrak{k}_0.$$

Since

$$\mathfrak{su}(m, \mathbb{H}) = \{X \in \mathfrak{u}(m, \mathbb{H}) \mid \mathrm{tr}(X) = 0\},$$

we find that

$$\mathfrak{su}(m, \mathbb{H}) = \mathfrak{u}(m, \mathbb{H}) = \mathfrak{k}_0,$$

as it should be.

The real Lie group $\mathbf{SU}(m, \mathbb{H})$ has another description in terms of the complex symplectic group $\mathbf{Sp}(m, \mathbb{C})$. Since

$$\mathbf{SU}(m, \mathbb{H}) = \{X \in \mathrm{M}_{2m}(\mathbb{C}) \mid X^*X = I_{2m},\ X = -J_m\overline{X}J_m,\ \det(X) = 1\},$$

first from $X^*X = I_{2m}$, we get

$$I_{2m} = X^*X = -X^*J_m\overline{X}J_m;$$

since $J_m^2 = -I_{2m}$, we get

$$X^*J_m\overline{X} = J_m,$$

and since $\overline{J_m} = J_m$, by conjugating both sides, we get

$$X^\top J_m X = J_m.$$

Definition 9.16. The *complex symplectic group* $\mathbf{Sp}(m, \mathbb{C})$ is defined as

$$\mathbf{Sp}(m, \mathbb{C}) = \{X \in \mathrm{M}_{2m}(\mathbb{C}) \mid X^\top J_m X = J_m\}.$$

It can be shown that the complex Lie algebra $\mathfrak{sp}(m, \mathbb{C})$ of the complex Lie group $\mathbf{Sp}(m, \mathbb{C})$ is given by

$$\mathfrak{sp}(m, \mathbb{C}) = \left\{ \begin{pmatrix} U & V_1 \\ V_2 & -U^\top \end{pmatrix} \,\middle|\, U, V_1, V_2 \in \mathbf{M}_m(\mathbb{C}),\ V_1^\top = V_1,\ V_2^\top = V_2 \right\};$$

see Helgason [Helgason (2001)] (Chapter X, Section 2, page 446).

The preceding argument showed that

$$\mathbf{SU}(m, \mathbb{H}) = \mathbf{SL}(m, \mathbb{H}) \cap \mathbf{SU}(2m, \mathbb{C}) \subseteq \mathbf{Sp}(m, \mathbb{C}) \cap \mathbf{U}(2m, \mathbb{C}).$$

It can also be shown that the group $\mathbf{Sp}(m, \mathbb{C})$ is connected; see Helgason [Helgason (2001)] (Chapter X, Section 2, Lemma 2.4) or Knapp [Knapp (2002)] (Chapter I, Proposition 1.145). Since J_m is invertible ($J_m^{-1} = -J_m$), the equation $X^\top J_m X = J_m$ shows that $\det(X) = \pm 1$. Since $\det(I) = 1$ and $\mathbf{Sp}(m, \mathbb{C})$ is connected, we deduce that for every $X \in \mathbf{Sp}(m, \mathbb{C})$, we have $\det(X) = 1$. (There are also purely algebraic proofs of this property using the fact that the symplectic transvections generate $\mathbf{Sp}(m, \mathbb{C})$ and that they have determinant 1.) Thus, if $X \in \mathbf{Sp}(m, \mathbb{C}) \cap \mathbf{U}(2m, \mathbb{C})$, then in fact $X \in \mathbf{Sp}(m, \mathbb{C}) \cap \mathbf{SU}(2m, \mathbb{C})$, and from $X^\top J_m X = J_m$ and $X^* X = I_{2m}$, by reversing the above argument, we deduce that $X = -J_m \overline{X} J_m$, and so

$$\mathbf{Sp}(m, \mathbb{C}) \cap \mathbf{U}(2m, \mathbb{C}) \subseteq \mathbf{SL}(m, \mathbb{H}) \cap \mathbf{SU}(2m, \mathbb{C}) = \mathbf{SU}(m, \mathbb{H}).$$

We just showed that

$$\mathbf{SU}(m, \mathbb{H}) = \mathbf{SL}(m, \mathbb{H}) \cap \mathbf{SU}(2m, \mathbb{C}) = \mathbf{Sp}(m, \mathbb{C}) \cap \mathbf{U}(2m, \mathbb{C}).$$

Definition 9.17. We define the real Lie group $\mathbf{Sp}(m)$ as

$$\mathbf{Sp}(m) = \mathbf{Sp}(m, \mathbb{C}) \cap \mathbf{U}(2m, \mathbb{C});$$

see Helgason [Helgason (2001)] (Chapter X, Section 2).

Since we showed that $\det(X) = 1$ for all $X \in \mathbf{Sp}(m, \mathbb{C})$, we also have

$$\mathbf{Sp}(m) = \mathbf{Sp}(m, \mathbb{C}) \cap \mathbf{SU}(2m, \mathbb{C}).$$

In view of Definition 9.17, we just proved that

$$\mathbf{Sp}(m) = \mathbf{SU}(m, \mathbb{H}),$$

and the Lie algebra $\mathfrak{sp}(m)$ of $\mathbf{Sp}(m)$ is equal to $\mathfrak{sp}(m, \mathbb{C}) \cap \mathfrak{u}(2m, \mathbb{C})$, so it is given by

$$\begin{aligned} \mathfrak{sp}(m) &= \left\{ \begin{pmatrix} U & V \\ -V^* & -U^\top \end{pmatrix} \,\middle|\, U \in \mathfrak{u}(m, \mathbb{C}),\ V \in \mathrm{M}_m(\mathbb{C}),\ V^\top = V \right\} \\ &= \left\{ \begin{pmatrix} U & V \\ -\overline{V} & -U^\top \end{pmatrix} \,\middle|\, U \in \mathfrak{u}(m, \mathbb{C}),\ V \in \mathrm{M}_m(\mathbb{C}),\ V^\top = V \right\}. \end{aligned}$$

Again,

$$\mathfrak{sp}(m) = \mathfrak{k}_0,$$

as it should be.

The real semi-simple simply-connected Lie group $G_0 = \mathbf{SL}(m, \mathbb{H}) = \mathbf{SU}^*(2m)$ is another real form of the complex semi-simple (simply-connected) Lie group $G = \mathbf{SL}(2m, \mathbb{C})$. We will show later that the pair $(G_0, K_0) = (\mathbf{SL}(m, \mathbb{H}), \mathbf{SU}(m, \mathbb{H})) = (\mathbf{SU}^*(2m), \mathbf{Sp}(m))$ is a Gelfand pair.

It can be shown that up to isomorphism, $\mathbf{SL}(n, \mathbb{R})$, $\mathbf{SU}(p, n-p)$, and $\mathbf{SL}(m, \mathbb{H}) = \mathbf{SU}^*(2m)$ (when $n = 2m$), are the only real forms of $\mathbf{SL}(n, \mathbb{C})$; see Helgason [Helgason (2001)] and Dieudonné [Dieudonné (2003a)] (Chapter XXI, Section 21.18.11).

9.6 Real Forms of Complex Semi-Simple Simply-Connected Lie Groups

A general method to find real forms of a complex, semi-simple, connected, simply-connected Lie group G with complex semi-simple Lie algebra $\mathfrak{g}$ goes as follows. Suppose we have real, compact, semi-simple, connected, simply-connected Lie group G_u with Lie algebra $\mathfrak{g}_u$, such that $\mathfrak{g} = \mathfrak{g}_u \oplus i\mathfrak{g}_u$ is the complexification of the real Lie algebra $\mathfrak{g}_u$. In our previous examples, $\mathfrak{g} = \mathfrak{sl}(n, \mathbb{C})$, $G = \mathbf{SL}(n, \mathbb{C})$, $\mathfrak{g}_u = \mathfrak{su}(n, \mathbb{C})$, and $G_u = \mathbf{SU}(n, \mathbb{C})$. By a famous theorem of Hermann Weyl, such a real semi-simple Lie algebra $\mathfrak{g}_u$ and such a compact, semi-simple, connected, simply-connected, Lie group G_u always exist; see Dieudonné [Dieudonné (2003a)] (Chapter XXI, no. 21.20.7).

Also assume that we have a conjugation $c_0 \colon \mathfrak{g} \to \mathfrak{g}$ that commutes with the conjugation c_u associated with $\mathfrak{g}_u$. Using the correspondence between (real) connected, simply-connected Lie groups and (real) Lie algebras, there is a unique involutive automorphism $\sigma_0 \colon G|_{\mathbb{R}} \to G|_{\mathbb{R}}$ such that $d(\sigma_0)_e = c_0$, where $c_0 \colon \mathfrak{g}|_{\mathbb{R}} \to \mathfrak{g}|_{\mathbb{R}}$. If P is the closed submanifold of $G|_{\mathbb{R}}$ which is the image of the real vector space $i\mathfrak{g}_u$ by the exponential map $\exp_G \colon \mathfrak{g} \to G$ ($ix \mapsto \exp(ix), x \in \mathfrak{g}_u$), then the following result can be shown.

Proposition 9.13. *The exponential map* $\exp_G$ *is a diffeomorphism of* $i\mathfrak{g}_u$ *onto* P*, and the map* $(x, y) \mapsto xy$ *is a diffeomorphism of* $G_u \times P$ *onto* $G|_{\mathbb{R}}$ *(as real manifolds).*

The automorphism σ_0 of $G|_{\mathbb{R}}$ leaves G_u and $P|_{\mathbb{R}}$ invariant since c_0 leaves $\mathfrak{g}_u$ and $i\mathfrak{g}_u$ invariant.

Example 9.4. If $G = \mathbf{SL}(n,\mathbb{C})$, $\mathfrak{g} = \mathfrak{sl}(n,\mathbb{C})$, $G_u = \mathbf{SU}(n,\mathbb{C})$, and $\mathfrak{g}_u = \mathfrak{su}(n,\mathbb{C})$, then

$$i\mathfrak{g}_u = i\mathfrak{su}(n,\mathbb{C}) = \{X \in \mathrm{M}_n(\mathbb{C}) \mid X^* = X,\ \mathrm{tr}(X) = 0\},$$

and $P = \exp_G(i\mathfrak{g}_u)$ is the manifold of hermitian positive definite matrices with determinant $+1$. The factorization $\mathbf{SL}(n,\mathbb{C}) = \mathbf{SU}(n,\mathbb{C})P$ is the polar form for matrices in $\mathbf{SL}(n,\mathbb{C})$.

Let G_0 be the real Lie subgroup of $G|_{\mathbb{R}}$ given by

$$G_0 = \{s \in G|_{\mathbb{R}} \mid \sigma_0(s) = s\}.$$

It is shown in Dieudonné [Dieudonné (2007)] (Proposition 20.4.3) that the Lie algebra $\mathfrak{g}_0$ of G_0 is the subalgebra of $\mathfrak{g}$ consisting of the set of fixed points of $d(\sigma_0)_e$. Since G is semi-simple, it is not hard to show (using the Killing form of $\mathfrak{g}$ and the fact that the Killing form is invariant under automorphisms) that $\mathfrak{g}_0$ is semi-simple. Thus, the group G_0 is a real semi-simple Lie group. Since σ_0 leaves G_u invariant, the group

$$K_0 = G_0 \cap G_u = \{s \in G_u \mid \sigma_0(s) = s\}$$

is a real compact subgroup of G_u consisting of the fixed points of G_u under σ_0. The group G_0 also contains the image P_0 of $\mathfrak{p}_0 \subseteq i\mathfrak{g}_u$ under the exponential map $v \mapsto \exp_{G_0}(v)$, and since $\exp_{G_0}(c_0(v)) = \sigma_0(\exp_{G_0}(v))$ (see Proposition 19.7 in Gallier and Quaintance [Gallier and Quaintance (2020a)]), the manifold P_0 is the set of fixed points of $P|_{\mathbb{R}}$ under σ_0. We have the following result whose proof is given in Dieudonné [Dieudonné (2007)] (Proposition 21.18.5.1).

Proposition 9.14. *The map $v \mapsto \exp_{G_0}(v)$ is a diffeomorphism of $\mathfrak{p}_0$ onto the closed manifold P_0, and the map $(y,z) \mapsto yz$ from $K_0 \times P_0$ to G_0 is a diffeomorphism.*

Example 9.5. If $G = \mathbf{SL}(n,\mathbb{C})$, $\mathfrak{g} = \mathfrak{sl}(n,\mathbb{C})$, $G_u = \mathbf{SU}(n,\mathbb{C})$, and $\mathfrak{g}_u = \mathfrak{su}(n,\mathbb{C})$, and $c_0(X) = \overline{X}$, as in Example 9.2, we have

$$\begin{aligned} &G_0 = \mathbf{SL}(n,\mathbb{R}), \quad K_0 = \mathbf{SO}(n), \quad \mathfrak{k}_0 = \mathfrak{so}(n),\\ &\mathfrak{p}_0 = \mathfrak{s}(n) = \{X \in \mathrm{M}_n(\mathbb{R}) \mid X^\top = X\}, \end{aligned}$$

and $P_0 = \exp_{G_0}(\mathfrak{p}_0)$ is the manifold of symmetric positive definite matrices with determinant $+1$. The factorization $\mathbf{SL}(n,\mathbb{R}) = \mathbf{SO}(n)P_0$ is the polar form for matrices in $\mathbf{SL}(n,\mathbb{R})$.

Thus G_0 is a real, noncompact, semi-simple, connected Lie group, diffeomorphic to the product of a compact semi-simple Lie group K_0 and some $\mathbb{R}^N$. It can be shown that the group K_0 is a maximal compact subgroup of G_0, is connected, but in general not semi-simple nor simply-connected.

Definition 9.18. Given a complex, semi-simple, connected, simply-connected Lie group G, a real, semi-simple, connected, simply-connected, compact Lie group G_u such that $\mathfrak{g} = \mathfrak{g}_u \oplus i\mathfrak{g}_u$ ($\mathfrak{g}$ is the complexification of the $\mathfrak{g}_u$) is called a *compact form* of G. The real, semi-simple, connected Lie group G_0 corresponding to σ_0 as above is a *real form* of G.

The pair (G_0, K_0) is a Gelfand pair, but not with respect to the involution σ_0, because σ_0 is the identity on P_0. However, there is also a unique involutive automorphism $\sigma_u \colon G|_{\mathbb{R}} \to G|_{\mathbb{R}}$ such that $d(\sigma_u)_e = c_u$, and this involution makes (G_0, K_0) a Gelfand pair.

Since $\mathfrak{k}_0 \subseteq \mathfrak{g}_u$ and since $\mathfrak{g}_u$ is the set of fixed points of c_u, the Lie group K_0 is a set of fixed points of σ_u. Similarly, since $\mathfrak{p}_0 \subseteq i\mathfrak{g}_u$, and since $c_u(ix) = -ix$ for all $x \in \mathfrak{g}_u$, we see that $\sigma_u(\exp_{G_0}(ix)) = \exp_{G_0}(c_u(ix)) = \exp_{G_0}(-ix) = (\exp_{G_0}(ix))^{-1}$ for all $ix \in \mathfrak{p}_0$, so $\sigma_u(s) = s^{-1}$ for all $s \in P_0$ (since $P_0 = \exp_{G_0}(\mathfrak{p}_0)$).

Proposition 9.15. *We have $G_0 = K_0P_0$, $\sigma_u(s) = s$ for all $s \in K_0$, and $\sigma_u(s) = s^{-1}$ for all $s \in P_0$. The group K_0 is the set of fixed points of σ_u,*

$$K_0 = \{s \in G_0 \mid \sigma_u(s) = s\}.$$

Proof. Since every $s \in G_0 = K_0P_0$ can be written as $s = xy$ with $x \in K_0$ and $y \in P_0$, we have $\sigma_u(s) = s$ iff $\sigma_u(xy) = xy$, but then

$$xy = \sigma_u(xy) = \sigma_u(x)\sigma_u(y) = xy^{-1},$$

so we deduce that

$$y^2 = e.$$

However, $y \in P_0 = \exp_{G_0}(\mathfrak{p}_0)$, so for $y = \exp_{G_0}(w)$ we have $y^2 = \exp_{G_0}(2w) = e$, and since $\exp_{G_0}$ is a diffeomorphism on $\mathfrak{p}_0$, we must have $w = 0$, and thus $y = e$. □

In summary, since K_0 is compact, we proved that (G_0, K_0) *is a Gelfand pair with involution* σ_u.

The real, semi-simple, connected Lie group G_0 is called a real form of the complex, semi-simple, connected, simply-connected Lie group G because its real Lie algebra $\mathfrak{g}_0$ is a real form of the complex Lie algebra $\mathfrak{g}$ of G.

There are other real, semi-simple, connected Lie groups having $\mathfrak{g}_0$ as Lie algebra, and they can all be found (up to isomorphism) as follows; see Dieudonné [Dieudonné (2003a)] (Chapter XXI, no. 21.18.8–21.18.12).

Proposition 9.16. *Let $\widetilde{G}_0$ be the universal cover of G_0, $\pi\colon \widetilde{G}_0 \to G_0$ be the covering map, and $\widetilde{K}_0 = \pi^{-1}(K_0)$. Then $\widetilde{K}_0$ is isomorphic to the universal cover of the compact Lie group K_0, the exponential map $\exp_{\widetilde{G}_0}$ is a diffeomorphism of $\mathfrak{p}_0$ onto a closed submanifold $\widetilde{P}_0$ of $\widetilde{G}_0$, such that $\widetilde{K}_0 \cap \widetilde{P}_0 = \{e\}$, and the map $(x, y) \mapsto xy$ is a diffeomorphism of $\widetilde{K}_0 \times \widetilde{P}_0$ onto $\widetilde{G}_0$. The center Z of $\widetilde{G}_0$ is a discrete subgroup contained in the center of $\widetilde{K}_0$.*

If K_0 is not semi-simple, then Z is not equal to the center of $\widetilde{K}_0$, and $\widetilde{K}_0$ is not compact.

Theorem 9.17. *Every real, semi-simple, connected Lie group G_1 having $\mathfrak{g}_0$ as real Lie algebra is of the form $G_1 = \widetilde{G}_0/D$, where D is a (discrete) subgroup of the center Z of $\widetilde{G}_0$. The center C_1 of G_1 is given by $C_1 = Z/D$. The Lie group $K_1 = \widetilde{K}_0/D$ is a connected subgroup of G_1 which contains C_1 (in general, C_1 is not equal to the center of G_1), and whose Lie algebra is $\mathfrak{k}_0$. The Lie group K_1 is compact iff C_1 is finite. The exponential map $\exp_{G_1}$ is a diffeomorphism of $\mathfrak{p}_0$ onto a closed submanifold P_1 of G_1, such that $K_1 \cap P_1 = \{e\}$, and the map $(x, y) \mapsto xy$ is a diffeomorphism of $K_1 \times P_1$ onto G_1.*

Definition 9.19. The factorization $G_1 = K_1P_1$ is called a *Cartan decomposition* of G_1.

The Cartan decomposition is a generalization of the polar form for invertible matrices.

It can also be shown that K_1 is isomorphic to the product of a compact Lie group with some $\mathbb{R}^m$. Thus, G_1 is diffeomorphic to the product of a compact Lie group with some $\mathbb{R}^M$ (in fact, this compact group is maximal in G_1).

The reasoning in the proof of Proposition 9.15 involving the conjugation σ_u can be used to show that the conjugation σ_u on $\widetilde{G}_0$, such that $d(\sigma_u)_e = c_u$ induces a conjugation on $G_1 = \widetilde{G}_0/D$ by passing to the quotient, and because Z is contained in $\widetilde{K}_0$, that

$$K_1 = \{s \in G_1 \mid \sigma_u(s) = s\},$$

and that $\sigma_u(s) = s^{-1}$ for all $s \in P_1$. Using the proof of Proposition 21.18.8 in Dieudonné [Dieudonné (2003a)], we can show that if the center C_1 of G_1 is finite, then K_1 is compact, and so (G_1, K_1) is a Gelfand pair with involution σ_u.

9.7 Examples of Gelfand Pairs

There are three important cases for which Gelfand's theorem (Theorem 9.2) applies.

(1) Let G be a compact, connected real Lie group, and σ be an involutive automorphism of G ($\sigma \neq \mathrm{id}_G$); see Dieudonné [Dieudonné (2003a)] (Chapter XXI, no. 21.18.13). Let G^σ be the closed (thus compact) Lie subgroup of G consisting of the fixed points of σ,

$$G^\sigma = \{s \in G \mid \sigma(s) = s\}.$$

The derivative $\theta = d\sigma_e$ of σ is an involution of the Lie algebra $\mathfrak{g}$ of G. Then as in Section 9.3, we know by linear algebra that

$$\mathfrak{g} = \mathfrak{g}_1 \oplus \mathfrak{g}_{-1},$$

where $\mathfrak{g}_1$ and $\mathfrak{g}_{-1}$ are the real eigenspaces of θ given by

$$\mathfrak{g}_1 = \{x \in \mathfrak{g} \mid \theta(x) = x\}, \qquad \mathfrak{g}_{-1} = \{x \in \mathfrak{g} \mid \theta(x) = -x\}.$$

Let G_0^σ be the connected component of e in G^σ. For any closed subgroup K (thus compact) such that $G_0^\sigma \subseteq K \subseteq G^\sigma$, using some differential geometry, it can be shown that $\mathfrak{g}_1$ is the Lie algebra of K; see O'Neill [O'Neill (1983)] or Gallier and Quaintance [Gallier and Quaintance (2020a)] (Proposition 23.33). Let $P = \exp(\mathfrak{g}_{-1})$. Then $\sigma(s) = s$ for all $s \in K$ and $\sigma(s) = s^{-1}$ for all $s \in P$ (since $\theta(x) = -x$ for all $x \in \mathfrak{g}_{-1}$). It remains to prove that $G = KP$. For this, again we use some differential geometry.

Since K is compact, G/K has some G-invariant metric. In fact, G/K is a naturally reductive homogeneous space. Since G is compact, G/K is compact, and by Hopf–Rinov, it is geodesically complete. But since G/K is naturally reductive, the tangent space $T_e(G/K) \cong \mathfrak{g}_{-1}$, and every geodesic γ_x through e with initial velocity $x \in \mathfrak{g}_{-1}$ is given by

$$\gamma_x(t) = \pi(\exp_G(tx));$$

see Proposition 23.27 in Gallier and Quaintance [Gallier and Quaintance (2020a)]. Consequently, $\pi(P) = G/K$, or equivalently, $G = PK$.

But since $P = \exp(\mathfrak{g}_{-1})$, we see that P is closed under the map $s \mapsto s^{-1}$, and since $G = PK$, for every $s \in G$, we have $s = xy$ with $x \in P$ and $y \in K$, so $s^{-1} = y^{-1}x^{-1} \in KP$, and since this holds for any $s \in G$, we have $G = KP$. Therefore, (G, K) is a Gelfand pair for the involution σ.

If the compact Lie group G is also semi-simple, then its Killing form is negative definite, so G/K is a symmetric space of compact type.

(2) Let G_u be a real, compact, semi-simple, connected, simply-connected Lie group, $\mathfrak{g}_u$ be its Lie algebra, $\mathfrak{g} = \mathfrak{g}_u \oplus i\mathfrak{g}_u$ be the complexification of $\mathfrak{g}_u$, and G be the complex, semi-simple, connected, simply-connected Lie group with Lie algebra $\mathfrak{g}$. For any conjugation $c_0 \colon \mathfrak{g} \to \mathfrak{g}$ that commutes with the conjugation c_u associated with $\mathfrak{g}_u$, let $\mathfrak{g}_0$ be the real form of $\mathfrak{g}$ induced by c_0. For any real form G_1 of G with Lie algebra $\mathfrak{g}_0$, let σ_u be the involution of G_1, such that $d(\sigma_u)_e = c_u$, as explained in Section 9.6. If the center C_1 of G_1 is finite, then $K_1 = \{s \in G_1 \mid \sigma_u(s) = s\}$ is a compact subgroup of G_1, such that (G_1, K_1) is Gelfand pair. The space G_1/K_1 is a symmetric space of non-compact type.

A typical example is given by $G = \mathbf{SL}(n, \mathbb{C})$, $G_u = \mathbf{SU}(n)$, $G_1 = \mathbf{SL}(n, \mathbb{R})$, and $K_1 = \mathbf{SO}(n, \mathbb{R})$; the maps c_0, c_u, and σ_0, are given by $c_0(X) = \overline{X}$, $c_u(X) = -X^*$, $\sigma_0(X) = \overline{X}$, $\sigma_u(X) = -X^*$.

(3) The group G is unimodular and contains

(a) A closed *commutative, normal* subgroup A such that $s^2 = e$ implies $s = e$ for all $s \in A$, and

(b) A compact subgroup K such that the mapping $(t, s) \mapsto ts$ from $K \times A$ to G is a homeomorphism. This implies that G is a semi-direct product of K and A, with A the normal factor. But beware that due to the order of the factors, since every element $g \in G = KA$ is written uniquely as $g = ka$ with $k \in K$ and $a \in A$, the multiplication in $G = KA$ is given by

$$(k_1a_1)(k_2a_2) = (k_1k_2)([k_2^{-1}a_1k_2]a_2),$$

where $k_1, k_2 \in K$ and $a_1, a_2 \in A$. So K acts on A by conjugation *on the right*. See Section 7.4 and Gallier and Quaintance [Gallier and Quaintance (2020a)], Section 19.5, Definition 19.20 and the remarks that follow. A typical example is $G = \mathbf{SE}(n, \mathbb{R})$.

Let σ be given by $\sigma(ts) = ts^{-1}$, for all $t \in K$ and all $s \in A$. Obviously σ is continuous, and we have

$$\sigma^2(ts) = \sigma(ts^{-1}) = ts,$$

so $\sigma^2 = \mathrm{id}_G$. For $t, t \in K$ and $s, s' \in A$, since A is a normal subgroup of G and $s \in A$, we have $t'^{-1}s^{-1}t' \in A$, and since $s' \in A$, so we also have $t'^{-1}st's' \in A$, and we have

$$\sigma(tst's') = \sigma((tt')(t'^{-1}st's')) = tt'(s'^{-1}t'^{-1}s^{-1}t'),$$

and

$$\sigma(ts)\sigma(t's') = ts^{-1}t's'^{-1} = tt'(t'^{-1}s^{-1}t's'^{-1}).$$

Since from above $t'^{-1}s^{-1}t' \in A$, and since A is abelian, $t'^{-1}s^{-1}t's'^{-1} = s'^{-1}t'^{-1}s^{-1}t'$, and so $\sigma(tst's') = \sigma(ts)\sigma(t's')$. Thus, σ is an involutive automorphism of G. By definition, $K = \{t \in G \mid \sigma(t) = t\}$, and $\sigma(s) = s^{-1}$ for $s \in A$. Therefore, (G, K) is a Gelfand pair.

Example 9.6. Assume the group G is compact. If (G, K) is a Gelfand pair, then the closure $\mathrm{L}^2(K\backslash G/K)$ of $\mathcal{K}(K\backslash G/K)$ in $\mathrm{L}^2(G)$ is commutative, which corresponds to the situation considered in Proposition 6.21. Furthermore, the restriction of every character of $\mathrm{L}^1(K\backslash G/K)$ to $\mathrm{L}^2(K\backslash G/K)$ is a (continuous) character of the algebra $\mathrm{L}^2(K\backslash G/K)$. We know that the direct sum

$$B = \bigoplus_{(\rho:\sigma_0)=1} \mathbb{C}\omega_\rho$$

is a dense algebra in $\mathrm{L}^2(K\backslash G/K)$, and it is easy to see that the only homomorphisms from B to $\mathbb{C}$ different from the zero function are the maps of the form $f \mapsto (f, \overline{\omega}_\rho)$, which shows that the spherical functions of G relative to K are the partial traces ω_ρ for all $\rho \in R(G)$, such that $(\rho : \sigma_0) = 1$ (see Definition 6.16 and Proposition 6.21). Since $\omega_\rho \in \mathrm{L}^1(G)$, the set of functions $f \in \mathrm{L}^\infty(G)$ such that $|(f - \omega_\rho, \overline{\omega}_\rho)| \leq \frac{1}{2}$ is a neighborhood of ω_ρ in the weak*-topology of $\mathrm{L}^\infty(G)$. Since $(\omega_\rho, \overline{\omega}_\rho) = 1$ and $(\omega_{\rho'}, \overline{\omega}_\rho) = 0$ when $\rho \neq \rho'$, we conclude that the space $\mathbf{S}(G/K)$ is *discrete*.

The Grassmannians constitute a very good example. Let $G = \mathbf{SO}(n)$ (with $n \geq 2$), let

$$I_{k,n-k} = \begin{pmatrix} I_k & 0 \\ 0 & -I_{n-k} \end{pmatrix},$$

where I_k is the $k \times k$-identity matrix, and let σ be given by

$$\sigma(P) = I_{k,n-k} P I_{k,n-k}, \quad P \in \mathbf{SO}(n).$$

It is clear that σ is an involutive automorphism of G. Let us find the set G^σ of fixed points of σ. If we write

$$P = \begin{pmatrix} Q & U \\ V & R \end{pmatrix}, \qquad Q \in M_{k,k}(\mathbb{R}),\ U \in M_{k,n-k}(\mathbb{R}),$$
$$V \in M_{n-k,k}(\mathbb{R}),\ R \in M_{n-k,n-k}(\mathbb{R}),$$

then $P = I_{k,n-k} P I_{k,n-k}$ iff

$$\begin{pmatrix} Q & U \\ V & R \end{pmatrix} = \begin{pmatrix} I_k & 0 \\ 0 & -I_{n-k} \end{pmatrix} \begin{pmatrix} Q & U \\ V & R \end{pmatrix} \begin{pmatrix} I_k & 0 \\ 0 & -I_{n-k} \end{pmatrix},$$

iff

$$\begin{pmatrix} Q & U \\ V & R \end{pmatrix} = \begin{pmatrix} Q & -U \\ -V & R \end{pmatrix},$$

so $U = 0$, $V = 0$, $Q \in \mathbf{O}(k)$ and $R \in \mathbf{O}(n-k)$. Since $P \in \mathbf{SO}(n)$, we conclude that $\det(Q)\det(R) = 1$, so

$$G^\sigma = \left\{ \begin{pmatrix} Q & 0 \\ 0 & R \end{pmatrix} \;\middle|\; Q \in \mathbf{O}(k),\ R \in \mathbf{O}(n-k),\ \det(Q)\det(R) = 1 \right\};$$

that is,

$$G^\sigma = S(\mathbf{O}(k) \times \mathbf{O}(n-k)).$$

We also have

$$G_0^\sigma = \mathbf{SO}(k) \times \mathbf{SO}(n-k).$$

For $K = G^\sigma$, the homogeneous space

$$G/K = \mathbf{SO}(n)/(S(\mathbf{O}(k) \times \mathbf{O}(n-k)))$$

is the Grassmannian $G(k,n)$ of k-subspaces in $\mathbb{R}^n$. For $K = \mathbf{SO}(k) \times \mathbf{SO}(n-k)$, the homogeneous space

$$G/K = \mathbf{SO}(n)/(\mathbf{SO}(k) \times \mathbf{SO}(n-k))$$

is the Grassmannian $G^0(k,n)$ of oriented k-subspaces in $\mathbb{R}^n$. In particular, for $k = 1$, $G^0(1, n-1) = S^{n-1}$ and $G(1, n-1) = \mathbb{RP}^{n-1}$.

Example 9.7. Let G be a real, semi-simple, connected, noncompact Lie group with finite center and K be a maximal compact subgroup of K, so that (G,K) is a Gelfand pair, as in (2) above. Note that we are now denoting G_1 as G and K_1 as K, we hope that this does not cause any confusion.

We showed that $G = KP$ where P is closed manifold in G, and P is closed under the map $s \mapsto s^{-1}$, but in general is not a group. However, it is known that there is closed solvable subgroup S of G such that $G = KS$, and that the map $(x, y) \mapsto xy$ from $K \times S$ to G is a diffeomorphism; for the definition of a solvable lie algebra and a solvable Lie group, see Gallier and Quaintance [Gallier and Quaintance (2020a)], Section 21.5, Definition 21.12. This is a corollary of the Iwasawa decomposition, which is a generalization of the QR-decomposition for invertible matrices; see Dieudonné [Dieudonné (2003a)] (Chapter XXI, no. 21.21.10). Since $(yx)^{-1} = x^{-1}y^{-1}$, the map $(y, x) \mapsto yx$ from $S \times K$ to G is also a diffeomorphism since it is the composition of the diffeomorphisms $(y, x) \mapsto (x^{-1}, y^{-1})$ from $S \times K$ to $K \times S$, the map $(x, y) \mapsto xy$ from $K \times S$ to G, and the map $s \mapsto s^{-1}$ from G to G. A way to construct spherical functions goes as follows.

Suppose we have a continuous homomorphism $\alpha\colon S \to \mathbb{C}^*$ (called an *exponential* of S). Then we can extend α to G as follows:

$$\alpha(st) = \alpha(s), \qquad \text{for all } s \in S \text{ and all } t \in K. \tag{$*$}$$

We claim that the following properties hold

$$\begin{aligned} \alpha(xt) &= \alpha(x), && \text{for all } x \in G \text{ and all } t \in K \\ \alpha(sx) &= \alpha(s)\alpha(x), && \text{for all } s \in S \text{ and all } x \in G. \end{aligned}$$

Since for $x \in G$, we can write $x = st'$ with $s \in S$ and $t' \in K$, by $(*)$, we have

$$\alpha(xt) = \alpha(st't) = \alpha(s) = \alpha(st') = \alpha(x).$$

Similarly, we can write $x = s't$ with $s' \in S$ and $t \in K$, so by $(*)$, we have

$$\alpha(sx) = \alpha(ss't) = \alpha(ss') = \alpha(s)\alpha(s') = \alpha(s)\alpha(s't) = \alpha(s)\alpha(x).$$

Define the function $\omega\colon G \to \mathbb{C}$ by

$$\omega(x) = \int_K \alpha(tx)\, d\lambda_K(t). \tag{$\dagger_4$}$$

The function ω is continuous, and we claim that if it is bounded, then it is a spherical function. By the remark just after Theorem 9.6, it suffices to prove that the equation

$$\int_K \omega(xt'y)\, d\lambda_K(t') = \omega(x)\omega(y), \qquad \text{for all } x, y \in G \tag{s_1}$$

holds. The left-hand side of this equation is

$$\int_K \omega(xt'y)\,d\lambda_K(t') = \int_K \int_K \alpha(txt'y)\,d\lambda_K(t)\,d\lambda_K(t').$$

We can also write for every fixed $x \in G$ and all $t \in K$, $tx = s(t)u(t)$, with $s(t) \in S$ and $u(t) \in K$, where s and u are continuous in $t \in K$, and using Fubini, the invariance of λ_K, and the equations just after $(*)$,

$$\begin{aligned}\int_K \int_K & \alpha(txt'y)\,d\lambda_K(t)\,d\lambda_K(t') \\ &= \int_K \int_K \alpha(s(t)u(t)t'y)\,d\lambda_K(t)\,d\lambda_K(t') \\ &= \int_K \int_K \alpha(s(t)t'y)\,d\lambda_K(t)\,d\lambda_K(t') \\ &= \int_K \int_K \alpha(s(t))\alpha(t'y)\,d\lambda_K(t)\,d\lambda_K(t') \\ &= \int_K \int_K \alpha(s(t)u(t))\alpha(t'y)\,d\lambda_K(t)\,d\lambda_K(t') \\ &= \int_K \int_K \alpha(tx)\alpha(t'y)\,d\lambda_K(t)\,d\lambda_K(t') = \omega(x)\omega(y).\end{aligned}$$

Harish–Chandra has shown that *all* continuous solutions of the functional equation (s_1) are given by $(\dagger_4)$. Such functions may be called generalized spherical functions. He also determined explicitly the exponentials $\alpha\colon S \to \mathbb{C}^*$ of S by a very deep study of the Lie algebra of G. One also knows exactly when the generalized spherical functions are bounded, and thus the spherical functions are completely known.

Example 9.8. Consider the groups $G = \mathbf{SL}(2,\mathbb{R})$, $K = \mathbf{SO}(2)$, and

$$S = S_0 = \left\{ \begin{pmatrix} a & b \\ 0 & a^{-1} \end{pmatrix} \;\middle|\; a, b \in \mathbb{R},\ a > 0 \right\},$$

from Example 6.5. We know that $G = \mathbf{SL}(2,\mathbb{R})$ acts transitively on the upper half plane $P = \{z = x + iy \in \mathbb{C} \mid y > 0\}$ by the action given by

$$X \cdot z = \begin{pmatrix} a & b \\ c & d \end{pmatrix} \cdot z = \frac{az+b}{cz+d}, \qquad z = x + iy \in P,$$

and that the stabilizer of i is $\mathbf{SO}(2)$. Given any $z = x + iy \in P$, there is a unique coset $X\mathbf{SO}(2) \subseteq \mathbf{SL}(2,\mathbb{R})$ (where $X \in \mathbf{SL}(2,\mathbb{R})$) that maps i to z, and in view of the unique factorization of a matrix X in $\mathbf{SL}(2,\mathbb{R})$ as $X = st$ with $s \in S_0$ and $t \in K$, we can pick as a representative of this coset $X\mathbf{SO}(2)$ the matrix $s_z \in S_0$, such that

$$s_z \cdot i = z = x + iy,$$

namely

$$s_z = \begin{pmatrix} \sqrt{y} & x/\sqrt{y} \\ 0 & 1/\sqrt{y} \end{pmatrix}.$$

We will show that the functions f in $\mathcal{C}(K\backslash G/K)$ are those which may be written as $f(1/2(a^2+b^2+c^2+d^2))$, where f is a continuous function defined on the interval $[1, +\infty)$. The proof is fairly tedious and involves a geometric argument which identifies a double coset KXK (with $X \in \mathbf{SL}(2, \mathbb{R})$) as a circle, the orbit of a point ir on the imaginary axis $(0 < r \leq 1)$ under the action of K. We will also determine the exponentials α in terms of the continuous homomorphisms from $\mathbb{R}_+^*$ to $\mathbb{C}^*$.

Since every coset XK (with $X \in \mathbf{SL}(2, \mathbb{R})$) corresponds to a unique point $z \in P$ in the upper half plane, and since there is a unique matrix $s_z \in S_0$, such that $s_z \cdot i = z = x + iy$, the coset XK is uniquely represented by the matrix $s_z \in S_0$. It follows that the double coset KXK is uniquely determined by the set of matrices Ks_z, and geometrically this set of matrices corresponds to the orbit in P of the (left) action of $K = \mathbf{SO}(2)$ on the point z. Although this is not obvious, such an orbit is a circle centered on the y-axis. To show this, we will prove that it suffices to prove that the orbit of a point ir on the imaginary axis $(0 < r \leq 1)$ under the action of K is a circle of center iv and radius R, with

$$v = \frac{1}{2}\left(r + \frac{1}{r}\right), \qquad R = \frac{1}{2}\left|\frac{1}{r} - r\right|;$$

see Figure 9.1.

If so, the equation of this circle is $x^2 + (y - v)^2 = R^2$, that is,

$$x^2 + y^2 - 2yv = R^2 - v^2.$$

But

$$v^2 - 1 = \frac{1}{4}r^2 + \frac{1}{2} + \frac{1}{4r^2} - 1 = \frac{1}{4}r^2 - \frac{1}{2} + \frac{1}{4r^2} = \left(\frac{1}{2}\left(\frac{1}{r} - r\right)\right)^2 = R^2,$$

so $R^2 - v^2 = -1$, and the equation of our circle is

$$x^2 + y^2 + 1 = 2yv. \tag{$*_1$}$$

We also showed that $R = \sqrt{v^2 - 1}$.

We can now prove that the orbit of any point $z = x + iy$ in the upper half plane under the action of K is a circle centered on the y-axis, which is also the orbit of a point ir with $0 < r \leq 1$ under the action of K. See Figure 9.2.

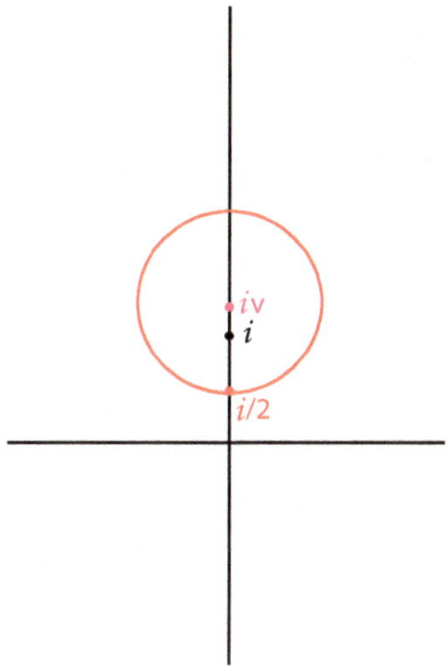

Fig. 9.1 The orbit of $i/2$ is the red circle with center $iv = 5i/4$ and radius $R = 3/4$.

Observe that since $R = \sqrt{v^2 - 1}$, $R > 0$ and $v > 0$, we have $v \geq 1$. We have to find $r > 0$, such that

$$r + \frac{1}{r} = 2v,$$

that is

$$r^2 - 2vr + 1 = 0,$$

and the zeros of this equation are

$$r = v \pm \sqrt{v^2 - 1}.$$

If $r = v - \sqrt{v^2 - 1}$, then $0 < r \leq 1$. Therefore, we found that $z \in P$ is on the circle of center iv and radius R, the orbit of ir by the action of K, with

$$v = \frac{x^2 + y^2 + 1}{2y}, \qquad R = \sqrt{v^2 - 1}, \qquad r = v - \sqrt{v^2 - 1} = v - R.$$

It remains to prove that the orbit of the point ir with $0 < r \leq 1$ under the action of K is the circle of center iv and radius R, with

$$v = \frac{1}{2}\left(r + \frac{1}{r}\right), \qquad R = \sqrt{v^2 - 1}.$$

Since a rotation in $K = \mathbf{SO}(2)$ is of the form

$$t_\theta = \begin{pmatrix} \cos\theta & -\sin\theta \\ \sin\theta & \cos\theta \end{pmatrix},$$

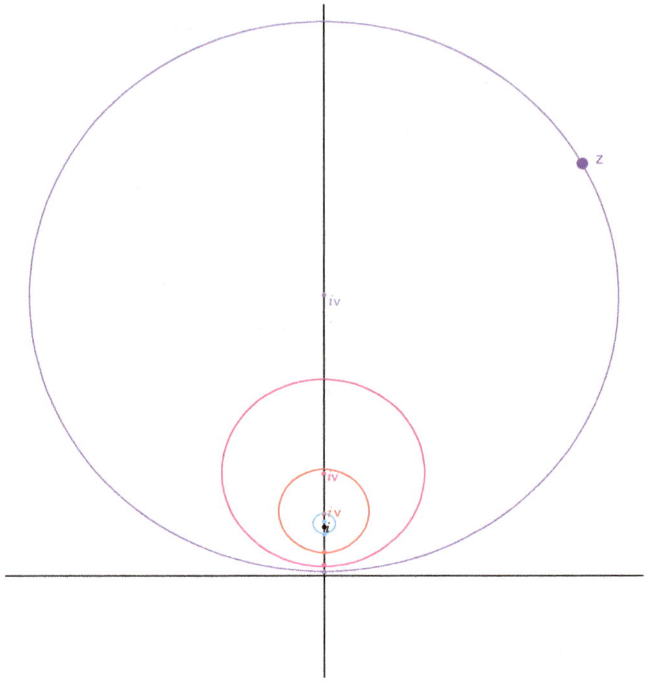

Fig. 9.2 The upper half plane partitioned into the circular orbits of ir, for $0 < r \leq 1$. If $r = 1$, the orbit is the single point i. When r is close to zero, the radius of the orbit is very large.

the orbit of ir consists of the points $x_\theta + iy_\theta = t_\theta \cdot ir$, with

$$x_\theta + iy_\theta = \frac{ir\cos\theta - \sin\theta}{ir\sin\theta + \cos\theta}. \tag{$*_2$}$$

From $(*_2)$, we have

$$\begin{aligned} x_\theta + iy_\theta &= \frac{ir\cos\theta - \sin\theta}{ir\sin\theta + \cos\theta} = \frac{ir\cos\theta - \sin\theta}{ir\sin\theta + \cos\theta} \times \frac{-ir\sin\theta + \cos\theta}{-ir\sin\theta + \cos\theta} \\ &= \frac{(r^2 - 1)\sin\theta\cos\theta + ir}{r^2\sin^2\theta + \cos^2\theta} = \frac{\frac{(r^2-1)}{2}(2\sin\theta\cos\theta) + ir}{(1 - r^2)\cos^2\theta + r^2} \\ &= \frac{\frac{(r^2-1)}{2}\sin 2\theta + ir}{\frac{(1-r^2)}{2}(2\cos^2\theta - 1) + r^2 + \frac{1-r^2}{2}} \\ &= \frac{-(1 - r^2)\sin 2\theta + i2r}{(1 - r^2)\cos 2\theta + r^2 + 1}. \end{aligned}$$

Let us compute $x_\theta + iy_\theta - iv$. We have

$$\begin{aligned}
&x_\theta + iy_\theta - iv \\
&\quad = \frac{-(1-r^2)\sin 2\theta + i2r}{(1-r^2)\cos 2\theta + r^2 + 1} - \frac{i(r^2+1)}{2r} \\
&\quad = \frac{-2r(1-r^2)\sin 2\theta + i4r^2 - i(r^2+1)((1-r^2)\cos 2\theta + r^2 + 1)}{2r((1-r^2)\cos 2\theta + r^2 + 1)} \\
&\quad = \frac{-2r(1-r^2)\sin 2\theta + i4r^2 - i(r^2+1)^2 - i(r^2+1)(1-r^2)\cos 2\theta}{2r((1-r^2)\cos 2\theta + r^2 + 1)} \\
&\quad = \frac{(1-r^2)(-2r\sin 2\theta - i(1-r^2+(r^2+1)\cos 2\theta))}{2r((1-r^2)\cos 2\theta + r^2 + 1)}.
\end{aligned}$$

The numerator of $|x_\theta + i(y_\theta - v)|^2$ is

$$\begin{aligned}
N &= (1-r^2)^2(4r^2\sin^2 2\theta + (1-r^2)^2 + 2(1-r^2)(r^2+1)\cos 2\theta \\
&\quad + (r^2+1)^2\cos^2 2\theta) \\
&= (1-r^2)^2(4r^2(1-\cos^2 2\theta) + (1-r^2)^2 + 2(1-r^2)(r^2+1)\cos 2\theta \\
&\quad + (r^2+1)^2\cos^2 2\theta) \\
&= (1-r^2)^2((1-r^2)^2\cos^2 2\theta + 2(1-r^2)(r^2+1)\cos 2\theta \\
&\quad + (1+r^2)^2) \\
&= (1-r^2)^2((1-r^2)\cos 2\theta + (1+r^2))^2.
\end{aligned}$$

The denominator of $|x_\theta + i(y_\theta - v)|^2$ is

$$4r^2((1-r^2)\cos 2\theta + (1+r^2))^2.$$

Therefore

$$|x_\theta + i(y_\theta - v)|^2 = \frac{(1-r^2)^2}{4r^2} = \left(\frac{1}{2}\left(\frac{1}{r} - r\right)\right)^2 = R^2 = v^2 - 1,$$

or equivalently

$$x_\theta^2 + (y_\theta - v)^2 = v^2 - 1,$$

which confirms our claim that the point $x_\theta + iy_\theta$ is on the circle of center v and radius $R = \sqrt{v^2-1}$.

Therefore the matrices of the double class KXK are exactly the matrices ts_θ, where $t \in K = \mathbf{SO}(2)$ and $s_\theta \in S_0$ is the matrix

$$s_\theta = \begin{pmatrix} \sqrt{y_\theta} & x_\theta/\sqrt{y_\theta} \\ 0 & 1/\sqrt{y_\theta} \end{pmatrix}$$

corresponding to the point $x_\theta + iy_\theta = t_\theta \cdot ir$ in $(*_2)$. The matrices s_θ satisfy the property

$$\mathrm{tr}(s_\theta^\top s_\theta) = \frac{x_\theta^2 + y_\theta^2 + 1}{y_\theta} = 2v,$$

which characterizes them. Now one has $(ts_\theta)^\top(ts_\theta) = s_\theta^\top t^\top t s_\theta = s_\theta^\top s_\theta$, so we get

$$\mathrm{tr}((ts_\theta)^\top(ts_\theta)) = \mathrm{tr}(s_\theta^\top s_\theta).$$

We deduce that the matrices

$$X = \begin{pmatrix} a & b \\ c & d \end{pmatrix} \in G = \mathbf{SL}(2, \mathbb{R}), \quad \text{with} \quad X = ts_\theta,$$

which form the double class KXK, are exactly those for which

$$\mathrm{tr}(X^\top X) = a^2 + b^2 + c^2 + d^2 = 2v,$$

with $v \geq 1$. Hence the functions $f \in \mathcal{C}(K \backslash G/K)$ are those which may be written as $f(1/2(a^2+b^2+c^2+d^2))$, where f is a continuous function defined on the interval $[1, +\infty)$.

We can also determine the exponentials $\alpha\colon S \to \mathbb{C}^*$ of S. Given any group G, recall that for any two elements $a, b \in G$, the element $a^{-1}b^{-1}ab$ is the *commutator* of a and b, and that the subgroup of G generated by the commutators is called the *commutator subgroup* of G and is denoted by DG. If $\alpha\colon S \to \mathbb{C}^*$ is a (continuous) homomorphism, then obviously α has the value 1 on the commutator subgroup DS of S (since $\alpha(a^{-1}b^{-1}ab) = \alpha(a^{-1})\alpha(b^{-1})\alpha(a)\alpha(b) = \alpha(a)^{-1}\alpha(b)^{-1}\alpha(a)\alpha(b) = 1$). But since

$$\begin{pmatrix} a_1 & b_1 \\ 0 & a_1^{-1} \end{pmatrix} \begin{pmatrix} a_2 & b_2 \\ 0 & a_2^{-1} \end{pmatrix} = \begin{pmatrix} a_1 a_2 & a_1 b_2 + b_1 a_2^{-1} \\ 0 & a_1^{-1} a_2^{-1} \end{pmatrix}$$

for any two matrices in S, and

$$\begin{pmatrix} a_1 & b_1 \\ 0 & a_1^{-1} \end{pmatrix}^{-1} = \begin{pmatrix} a_1^{-1} & -b_1 \\ 0 & a_1 \end{pmatrix},$$

we see that

$$DS = \left\{ \begin{pmatrix} 1 & b \\ 0 & 1 \end{pmatrix} \;\middle|\; b \in \mathbb{R} \right\}.$$

Furthermore, every matrix $X \in S$ can be factored as

$$X = \begin{pmatrix} a & b \\ 0 & a^{-1} \end{pmatrix} = \begin{pmatrix} a & 0 \\ 0 & a^{-1} \end{pmatrix} \begin{pmatrix} 1 & a^{-1}b \\ 0 & 1 \end{pmatrix},$$

with $a > 0$, so the homomorphism α is determined by its restriction to the subgroup

$$\left\{ \begin{pmatrix} a & 0 \\ 0 & a^{-1} \end{pmatrix} \;\middle|\; a > 0 \right\} \cong \mathbb{R}_+^*,$$

and thus it corresponds to a continuous homomorphism from $\mathbb{R}_+^*$ to $\mathbb{C}^*$, which is well-known to be of the form $t \mapsto t^\lambda = e^{\lambda \log t}$, for some $\lambda \in \mathbb{C}$. This is because the map $x \mapsto e^x$ is a continuous homomorphism from $(\mathbb{R}, +)$ to $\mathbb{R}_+^*$, and every continuous homomorphism from $(\mathbb{R}, +)$ to $\mathbb{C}^*$ is of the form $x \mapsto e^{\lambda x}$ for some $\lambda \in \mathbb{C}$, as the proof of Vol. I, Proposition 10.9(4) shows. We showed earlier (see $(*_2)$ and the calculations that follow) that

$$\begin{aligned} x + iy &= \frac{ir\cos\theta - \sin\theta}{ir\sin\theta + \cos\theta} \\ &= \frac{(r^2 - 1)\sin\theta\cos\theta + ir}{r^2\sin^2\theta + \cos^2\theta}, \end{aligned}$$

so we get

$$\frac{1}{y} = r\sin^2\theta + \frac{1}{r}\cos^2\theta.$$

Using the fact that

$$v = \frac{1}{2}\left(r + \frac{1}{r}\right),$$

we get

$$\begin{aligned} \frac{1}{y} &= r\sin^2\theta + \frac{1}{r}\cos^2\theta \\ &= v + r\sin^2\theta - \frac{1}{2}r + \frac{1}{r}\cos^2\theta - \frac{1}{2r} \\ &= v + \frac{r}{2}(2\sin^2\theta - 1) + \frac{1}{2r}(2\cos^2\theta - 1) \\ &= v + \frac{r}{2}(1 - 2\cos^2\theta) + \frac{1}{2r}(2\cos^2\theta - 1) = v + \frac{1}{2}\left(\frac{1}{r} - r\right)\cos 2\theta \\ &= v + \sqrt{v^2 - 1}\cos 2\theta. \end{aligned}$$

Since

$$X = \begin{pmatrix} \sqrt{y} & x/\sqrt{y} \\ 0 & 1/\sqrt{y} \end{pmatrix},$$

the above reasoning shows that

$$\alpha(XU) = \alpha(X) = e^{\lambda \log(\sqrt{y})} = e^{\frac{1}{2}\lambda \log y},$$

for some complex number $\lambda \in \mathbb{C}$, where $U \in \mathbf{SO}(2)$. However, it is more convenient to express $\alpha(X)$ in terms of $1/y$, so we write

$$\alpha(X) = e^{-\rho \log y} = y^{-\rho},$$

with $\rho = -\frac{1}{2}\lambda$, so finally, the generalized spherical function given by ($\dagger_4$) is

$$P_\rho(v) = \frac{1}{2\pi}\int_0^{2\pi} (v + \sqrt{v^2 - 1}\cos\varphi)^\rho \, d\varphi, \qquad v \geq 1,\ \rho \in \mathbb{C}.$$

The function P_ρ a *Legendre function of (complex) index* ρ. When ρ is a positive integer n, it can be shown that up to a constant, P_n is a Legendre polynomial.

In order for the function P_ρ to be bounded when $v \geq 1$, some conditions must be imposed on ρ. If ρ is purely imaginary, then P_ρ is bounded, but if ρ is a positive real, then it is not bounded.

One can check that the functional equation (s_1) becomes

$$\frac{1}{2\pi}\int_0^{2\pi} P_\rho(\cosh t \cosh u + \sinh t \sinh u \cos\varphi)\, d\varphi = P_\rho(\cosh t) P_\rho(\cosh u)$$

for all $t, u \in \mathbb{R}$.

Example 9.9. Let us now consider case (3) above, where G is a unimodular group containing an abelian normal subgroup A and a compact subgroup K, such that the map $(t, s) \mapsto ts$ is a homeomorphism from $K \times A$ to G. Let $\alpha\colon A \to \mathbb{C}^*$ be a continuous homomorphism (an exponential of A). By analogy with Example 9.7, define the function $\omega\colon G \to \mathbb{C}$ by

$$\omega(x) = \int_K \alpha(usu^{-1})\, d\lambda_K(u), \qquad x = ts,\ t \in K,\ s \in A. \tag{$\dagger_5$}$$

The function ω is continuous, and we claim that if ω is bounded, then it is a spherical function for (G, K). For this, we verify that the functional equation (s_1) holds.

Let $x = t_1 s_1, y = t_2 s_2$, with $t_1, t_2 \in K$, $s_1, s_2 \in A$. For $v \in K$, we may write

$$xvy = t_1 s_1 v t_2 s_2 = (t_1 v t_2)(((v t_2)^{-1} s_1 (v t_2)) s_2),$$

with $t_1 v t_2 \in K$ and $((vt_2)^{-1} s_1 (vt_2)) s_2 \in A$, because A is normal, so $(vt_2)^{-1} s_1 (vt_2) \in A$, and $((vt_2)^{-1} s_1 (vt_2)) s_2 \in A$. Consequently, since α

is a homomorphism of A, the subgroup A is a normal subgroup, and by Fubini, we have

$$\begin{aligned}
&\int_K \omega(xvy)\, d\lambda_K(v) \\
&\quad = \int_K \int_K \alpha(u((vt_2)^{-1}s_1(vt_2))s_2u^{-1})\, d\lambda_K(u)\, d\lambda_K(v) \\
&\quad = \int_K \int_K \alpha(((vt_2u^{-1})^{-1}s_1(vt_2u^{-1}))(us_2u^{-1}))\, d\lambda_K(u)\, d\lambda_K(v) \\
&\quad = \int_K \int_K \alpha((vt_2u^{-1})^{-1}s_1(vt_2u^{-1}))\alpha(us_2u^{-1})\, d\lambda_K(u)\, d\lambda_K(v) \\
&\quad = \int_K \alpha(us_2u^{-1}) \int_K \alpha((vt_2u^{-1})^{-1}s_1(vt_2u^{-1}))\, d\lambda_K(v)\, d\lambda_K(u).
\end{aligned}$$

But since K is unimodular, we have

$$\begin{aligned}
\int_K \alpha((vt_2u^{-1})^{-1}s_1(vt_2u^{-1}))\, d\lambda_K(v) &= \int_K \alpha(v^{-1}s_1v)\, d\lambda_K(v) \\
&= \int_K \alpha(vs_1v^{-1})\, d\lambda_K(v) = \omega(x),
\end{aligned}$$

and thus

$$\begin{aligned}
&\int_K \omega(xvy)\, d\lambda_K(v) \\
&\quad = \int_K \alpha(us_2u^{-1}) \int_K \alpha((vt_2u^{-1})^{-1}s_1(vt_2u^{-1}))\, d\lambda_K(v)\, d\lambda_K(u) \\
&\quad = \int_K \alpha(us_2u^{-1})\omega(x)\, d\lambda_K(u) = \omega(x)\omega(y),
\end{aligned}$$

as claimed.

Conversely, it can be shown that all spherical functions are given by ($\dagger_5$). A proof is sketched in Dieudonné [Dieudonné (1980)] (Chapter 16).

Example 9.10. Consider the example $G = \mathbf{SE}(2, \mathbb{R})$ of the group of rigid motions of $\mathbb{R}^2$. Since we view this group as the semi-direct product of $\mathbf{SO}(2)$ and $\mathbb{R}^2$ (instead of $\mathbb{R}^2$ and $\mathbf{SO}(2)$), we want a matrix representation in which every rigid motion is written as the product of a rotation and a translation, so we view $\mathbf{SE}(2, \mathbb{R})$ as

$$\mathbf{SE}(2, \mathbb{R}) = \left\{ \begin{pmatrix} \cos\theta & \sin\theta & 0 \\ -\sin\theta & \cos\theta & 0 \\ a & b & 1 \end{pmatrix} \;\middle|\; a, b \in \mathbb{R},\ 0 \leq \theta < 2\pi \right\},$$

instead of

$$\begin{pmatrix} \cos\theta & -\sin\theta & a \\ \sin\theta & \cos\theta & b \\ 0 & 0 & 1 \end{pmatrix}.$$

Clearly,

$$\begin{pmatrix} \cos\theta & \sin\theta & 0 \\ -\sin\theta & \cos\theta & 0 \\ a & b & 1 \end{pmatrix} = \begin{pmatrix} \cos\theta & \sin\theta & 0 \\ -\sin\theta & \cos\theta & 0 \\ 0 & 0 & 1 \end{pmatrix} \begin{pmatrix} 1 & 0 & 0 \\ 0 & 1 & 0 \\ a & b & 1 \end{pmatrix},$$

so $\mathbf{SE}(2,\mathbb{R}) = KA$, where K (the rotations) is isomorphic to $\mathbf{SO}(2)$ and A (the translations) is isomorphic to $\mathbb{R}^2$. Note that in this representation of $\mathbf{SE}(2,\mathbb{R}^2)$, we use matrices

$$s = \begin{pmatrix} Q^\top & 0 \\ u^\top & 1 \end{pmatrix}$$

with $u \in \mathbb{R}^2$ and $Q \in \mathbf{SO}(2)$, and we have the right action given by

$$x \cdot s = x^\top s = x^\top Q^\top + u^\top = (Qx + u)^\top, \quad x \in \mathbb{R}^2,$$

which corresponds to the matrix equation

$$(y^\top \;\; 1) = (x^\top \;\; 1) \begin{pmatrix} Q^\top & 0 \\ u^\top & 1 \end{pmatrix}.$$

The choice of this representation forces everything to be transposed. In particular, if we denote the matrix

$$\begin{pmatrix} Q^\top & 0 \\ u^\top & 1 \end{pmatrix}$$

by $(Q^\top, u^\top)$, since the product of the matrices

$$s = \begin{pmatrix} Q^\top & 0 \\ u^\top & 1 \end{pmatrix}, \quad t = \begin{pmatrix} R^\top & 0 \\ v^\top & 1 \end{pmatrix}$$

is

$$st = \begin{pmatrix} Q^\top R^\top & 0 \\ u^\top R^\top + v^\top & 1 \end{pmatrix},$$

the multiplication operation is given by

$$(Q^\top, u^\top)(R^\top, v^\top) = (Q^\top R^\top, v^\top + u^\top R^\top) = ((RQ)^\top, (v + Ru)^\top).$$

When $\mathbf{SE}(2,\mathbb{R})$ is viewed as the semi-direct product of $\mathbb{R}^2$ and $\mathbf{SO}(2)$, we use the representation (u, Q), and multiplication is given by

$$(v, R)(u, Q) = (v + Ru, RQ),$$

which corresponds to $(Q^\top, u^\top)(R^\top, v^\top)$ by transposition, but note the *reversal* of the arguments in the multiplication (which must take place since transposition of a product of matrices switches the order of the arguments).

For any matrix $s \in \mathbf{SE}(2, \mathbb{R})$ and any matrices $t_1, t_2 \in \mathbf{SO}(2)$, if we write

$$s = \begin{pmatrix} R^\top & 0 \\ u^\top & 1 \end{pmatrix}, \; t_1 = \begin{pmatrix} Q_1^\top & 0 \\ 0 & 1 \end{pmatrix}, \; t_2 = \begin{pmatrix} Q_2^\top & 0 \\ 0 & 1 \end{pmatrix}, \; Q_1, Q_2, R \in \mathbf{SO}(2), \; u \in \mathbb{R}^2,$$

then we have

$$\begin{aligned} t_1 s t_2 &= \begin{pmatrix} Q_1^\top & 0 \\ 0 & 1 \end{pmatrix} \begin{pmatrix} R^\top & 0 \\ u^\top & 1 \end{pmatrix} \begin{pmatrix} Q_2^\top & 0 \\ 0 & 1 \end{pmatrix} = \begin{pmatrix} Q_1^\top R^\top & 0 \\ u^\top & 1 \end{pmatrix} \begin{pmatrix} Q_2^\top & 0 \\ 0 & 1 \end{pmatrix} \\ &= \begin{pmatrix} Q_1^\top R^\top Q_2^\top & 0 \\ u^\top Q_2^\top & 1 \end{pmatrix}. \end{aligned}$$

If we pick $Q_1 = (Q_2 R)^\top$, then we see that the matrix

$$\begin{pmatrix} I & 0 \\ u^\top Q_2^\top & 1 \end{pmatrix}$$

belongs to the class KsK. Since $\mathbf{SO}(2)$ acts transitively on $\mathbb{R}^2$, it follows that the matrix

$$\begin{pmatrix} I & 0 \\ v^\top & 1 \end{pmatrix}$$

also belongs to the double class KsK for any vector v, such that $\|v\| = \|u\| = r^2$ $(r \geq 0)$. Therefore, every double class KsK corresponds bijectively to some $r \in \mathbb{R}$ with $r \geq 0$. We can also view such a double class KsK as any vector (a, b) for which $a^2 + b^2 = r^2$ for a fixed $r \geq 0$ in $\mathbb{R}$; see Figure 9.3. The functions in $\mathcal{C}(K \backslash G / K)$ are those of the form $\psi(r)$, where $\psi \colon [0, +\infty) \to \mathbb{C}$ is any continuous function. We showed in Vol. I, Corollary 10.11 that the continuous homomorphisms $\alpha \colon \mathbb{R}^2 \to \mathbb{C}^*$ are of the form

$$\alpha(a, b) = e^{\lambda a + \mu b}, \qquad \lambda, \mu \in \mathbb{C}.$$

For any $u \in K$ and any $s \in A$, we have

$$\begin{aligned} usu^{-1} &= \begin{pmatrix} \cos\theta & \sin\theta & 0 \\ -\sin\theta & \cos\theta & 0 \\ 0 & 0 & 1 \end{pmatrix} \begin{pmatrix} 1 & 0 & 0 \\ 0 & 1 & 0 \\ a & b & 1 \end{pmatrix} \begin{pmatrix} \cos\theta & -\sin\theta & 0 \\ \sin\theta & \cos\theta & 0 \\ 0 & 0 & 1 \end{pmatrix} \\ &= \begin{pmatrix} \cos\theta & \sin\theta & 0 \\ -\sin\theta & \cos\theta & 0 \\ a & b & 1 \end{pmatrix} \begin{pmatrix} \cos\theta & -\sin\theta & 0 \\ \sin\theta & \cos\theta & 0 \\ 0 & 0 & 1 \end{pmatrix} \\ &= \begin{pmatrix} 1 & 0 & 0 \\ 0 & 1 & 0 \\ a\cos\theta + b\sin\theta & -a\sin\theta + b\cos\theta & 1 \end{pmatrix}. \end{aligned}$$

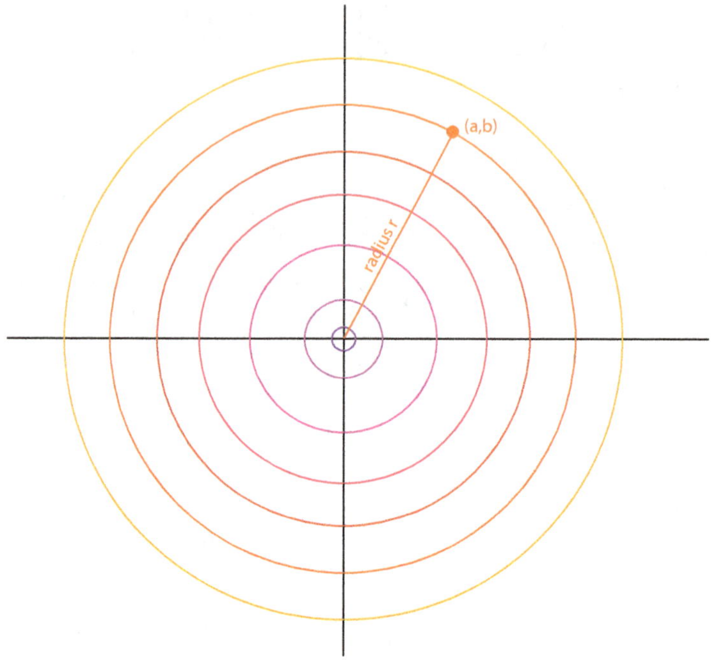

Fig. 9.3 The partition of $\mathbb{R}$ into circular orbits, each of which corresponds to a double coset.

For $a = r\cos\varphi$ and $b = r\sin\varphi$, so that $a^2 + b^2 = r^2$, we have

$$usu^{-1} = \begin{pmatrix} 1 & 0 & 0 \\ 0 & 1 & 0 \\ r\cos(\varphi-\theta) & r\sin(\varphi-\theta) & 1 \end{pmatrix}.$$

Consequently, according to ($\dagger_5$), for

$$x = ts = t\begin{pmatrix} 1 & 0 & 0 \\ 0 & 1 & 0 \\ r\cos\varphi & r\sin\varphi & 1 \end{pmatrix}, \qquad t \in K,$$

we have

$$\begin{aligned} \omega(x) &= \int_K \alpha(usu^{-1})\, d\lambda_K(u) \\ &= \frac{1}{2\pi}\int_0^{2\pi} e^{r(\lambda\cos(\varphi-\theta)+\mu\sin(\varphi-\theta))}\, d\theta = \frac{1}{2\pi}\int_0^{2\pi} e^{r(\lambda\cos\theta+\mu\sin\theta)}\, d\theta. \end{aligned}$$

It follows that the generalized spherical functions are the continuous functions on $[0, +\infty)$ given by

$$\psi(r) = \frac{1}{2\pi} \int_0^{2\pi} e^{r(\lambda \cos\theta + \mu \sin\theta)}\, d\theta$$

for any $\lambda, \mu \in \mathbb{C}$. For λ and μ imaginary, these functions are bounded, hence they really are spherical functions. In the special case where $\lambda = 0$ and $\mu = i$, the function ψ is the *Bessel function* J_0. By a change of variable, if both λ and μ are imaginary, the function ψ becomes J_0.

The irreducible unitary representations of $\mathbf{SE}(2, \mathbb{R})$ are completely determined; see Section 7.4 and Vilenkin [Vilenkin (1968)] (Chapter IV, Section 2). The matrix elements of these representations can be expressed by means of Bessel functions; see Vilenkin [Vilenkin (1968)] (Chapter IV, Section 3).

In the general case where $G = \mathbf{SE}(n, \mathbb{R})$, with $K = \mathbf{SO}(n)$ and $A = \mathbb{R}^n$, with

$$\mathbf{SE}(n, \mathbb{R}) = \left\{ \begin{pmatrix} Q^\top & 0 \\ w^\top & 1 \end{pmatrix} \,\middle|\, Q \in \mathbf{SO}(n),\ w \in \mathbb{R}^n \right\},$$

the irreducible unitary representations of $\mathbf{SE}(n, \mathbb{R})$ are also completely determined; see Section 7.4. One can also determine the generalized spherical functions, and they are now expressed in terms of the Bessel functions $J_{(n-2)/2}$; see Vilenkin [Vilenkin (1968)] (Chapter XI).

In general, if G is a connected unimodular Lie group and K is a compact subgroup of G such that (G, K) is a Gelfand pair, it can be shown that the spherical functions are not only continuous but also *smooth*. The proof is not difficult but not that informative, so we omit it. This proof can be found in Dieudonné [Dieudonné (1980)] (Chapter 12).

It is also possible to figure out how a differential operator on G that is invariant by left translations by G and invariant by right translations by K operates on spherical functions. In this case, the spherical functions are *eigenfunctions* of all such differential operators. If G is also semisimple, then more can be said (there are *elliptic* operators, in particular, the *Casimir operator*), but will not go into this right now. These topics are discussed in Dieudonné [Dieudonné (2003c)] (Chapter XXIII, Sections 36 and 37).

9.8 The Fourier Transform

Again, let (G, K) be a Gelfand pair. Recall from Definition 9.7 that every spherical function $\omega \in \mathbf{S}(G/K)$ defines the character $\zeta_\omega \in \mathsf{X}_0(A)$ given by

$$\zeta_\omega(f) = (f, \omega) = \int_G f(x)\omega(x)\, d\lambda_G(x), \qquad f \in \mathrm{L}^1(K\backslash G/K),$$

where $A = \mathrm{L}^1(K\backslash G/K) \oplus \mathbb{C}\delta_e$, a commutative, involutive, unital Banach algebra. By Theorem 9.7, the map $\omega \mapsto \zeta_\omega$ is a homeomorphism of $\mathbf{S}(G/K)$ equipped with the induced topology of Fréchet space of $\mathcal{C}(G)$ onto $\mathsf{X}_0(A)$ equipped with the topology induced by the weak *-topology of the dual A' of A.

It follows that the restriction of the Gelfand transform to $\mathsf{X}_0(A)$ of an element $f \in \mathrm{L}^1(K\backslash G/K)$ can be identified with the function $\overline{\mathcal{F}}f \colon \mathbf{S}(G/K) \to \mathbb{C}$ given by

$$(\overline{\mathcal{F}}f)(\omega) = (f, \omega) = \int_G f(x)\omega(x)\, d\lambda_G(x).$$

The above can be viewed as the Fourier cotransform of f. Thus, we are led to the following definition.

Definition 9.20. Let (G, K) be a Gelfand pair (recall that G is unimodular). For every function $f \in \mathcal{L}^1(K\backslash G/K)$, the *Fourier cotransform* $\overline{\mathcal{F}}f$ of f is the function $\overline{\mathcal{F}}f \colon \mathbf{S}(G/K) \to \mathbb{C}$ given by

$$(\overline{\mathcal{F}}f)(\omega) = (f, \omega) = \int_G f(x)\omega(x)\, d\lambda_G(x), \qquad \omega \in \mathbf{S}(G/K),$$

and the *Fourier transform* $\mathcal{F}f$ of f is the function $\mathcal{F}f \colon \mathbf{S}(G/K) \to \mathbb{C}$ given by

$$\begin{aligned}(\mathcal{F}f)(\omega) = (\check{f}, \omega) &= \int_G f(x^{-1})\omega(x)\, d\lambda_G(x) \\ &= (f, \check{\omega}) = \int_G f(x)\omega(x^{-1})\, d\lambda_G(x), \qquad \omega \in \mathbf{S}(G/K).\end{aligned}$$

Observe that

$$\mathcal{F}f = \overline{\mathcal{F}}\check{f}.$$

It is also clear that $\mathcal{F}f$ and $\overline{\mathcal{F}}f$ depend only on the equivalence class $[f] \in \mathrm{L}^1(K\backslash G/K)$, so the Fourier transform and the Fourier cotransform are also defined on $\mathrm{L}^1(K\backslash G/K)$.

Definition 9.20 also applies to arbitrary functions $f \in \mathcal{L}^1(G)$. By (**) of Section 9.2, namely

$$(f^\sharp, \psi) = \int_G f^\sharp(x)\psi(x)\, d\lambda_G(x) = \int_G f(x)\psi^\sharp(x)\, d\lambda_G(x) = (f, \psi^\sharp), \qquad (**)$$

for all $f \in \mathcal{K}(G)$ and all $\psi \in \mathcal{C}(G)$, since $\omega^\sharp = \omega$, we obtain

$$\mathcal{F}f = \mathcal{F}(f^\sharp), \qquad \overline{\mathcal{F}}f = \overline{\mathcal{F}}(f^\sharp), \qquad \text{for all } f \in \mathcal{L}^1(G).$$

The Fourier transform and the Fourier cotransform are also defined on $\mathrm{L}^1(G)$.

As a consequence of the properties of the Gelfand transform, we have the following results.

Proposition 9.18. *Let (G, K) be a Gelfand pair. For every function $f \in \mathcal{L}^1(G)$, the Fourier transform $\mathcal{F}f$ and the Fourier cotransform $\overline{\mathcal{F}}f$ are continuous functions that tend to zero at infinity. For all $f, g \in \mathcal{L}^1(K\backslash G/K)$, we have*

$$\mathcal{F}(f * g) = (\mathcal{F}f)(\mathcal{F}g),^1 \qquad \overline{\mathcal{F}}(f * g) = (\overline{\mathcal{F}}f)(\overline{\mathcal{F}}g).^2 \qquad (*)$$

For all $f \in \mathcal{L}^1(G)$, we have

$$\|\mathcal{F}f\| \le \|f\|_1, \qquad \|\overline{\mathcal{F}}f\| \le \|f\|_1.$$

Therefore, $\mathcal{F}$ and $\overline{\mathcal{F}}$ are continuous linear maps from $\mathrm{L}^1(G)$ to $\mathcal{C}_0(\mathbf{S}(G/K); \mathbb{C})$.

Beware that the equations in $(*)$ *generally fail* if $f, g \in \mathcal{L}^1(G)$. Also, in general, even if $f \in \mathcal{K}_{\mathbb{C}}(G)$, the functions $\mathcal{F}f$ and $\overline{\mathcal{F}}f$ *do not* have compact support. However, we have the following properties (see Dieudonné [Dieudonné (2003b)] (Chapter XXII, Proposition 22.6.4.7).

Proposition 9.19. *Let f and g be two functions in $\mathcal{L}^1(G)$. If either $f \in \mathcal{L}^1(G/K)$ or $g \in \mathcal{L}^1(K\backslash G)$, then*

$$\mathcal{F}(f * g) = (\mathcal{F}f)(\mathcal{F}g) \qquad \text{and} \qquad \overline{\mathcal{F}}(f * g) = (\overline{\mathcal{F}}f)(\overline{\mathcal{F}}g).$$

Proof. We prove the first equation assuming that $f \in \mathcal{L}^1(G/K)$, the proof of the other equations being similar. By left-invariance, we have

$$\begin{aligned}\mathcal{F}(f * g)(\omega) &= \int_G\int_G \omega(x^{-1}) f(s) g(s^{-1}x)\, d\lambda_G(s)\, d\lambda_G(x)\\ &= \int_G\int_G \omega(x^{-1}s^{-1}) f(s) g(x)\, d\lambda_G(s)\, d\lambda_G(x),\end{aligned}$$

[1]Here, $(\mathcal{F}f)(\mathcal{F}g)$ is the pointwise multiplication of the functions $\mathcal{F}f$ and $\mathcal{F}g$.
[2]Similarly, $(\overline{\mathcal{F}}f)(\overline{\mathcal{F}}g)$ is the pointwise multiplication of the functions $\overline{\mathcal{F}}f$ and $\overline{\mathcal{F}}g$.

and by right-invariance, since $f(st^{-1}) = f(s)$ for all $t \in K$, $\lambda_K(K) = 1$, the fact that $\mathcal{F}(f * g)(\omega)$ is independent of t, and by (s_1) of Theorem 9.6, we have

$$\begin{aligned}
&\mathcal{F}(f * g)(\omega)\\
&\quad = \int_G \int_G \omega(x^{-1}s^{-1}) f(s) g(x)\, d\lambda_G(s)\, d\lambda_G(x)\\
&\quad = \int_G \int_G \omega(x^{-1}ts^{-1}) f(st^{-1}) g(x)\, d\lambda_G(s)\, d\lambda_G(x)\\
&\quad = \int_G \int_G \omega(x^{-1}ts^{-1}) f(s) g(x)\, d\lambda_G(s)\, d\lambda_G(x)\\
&\quad = \int_G \int_G \int_K \omega(x^{-1}ts^{-1})\, d\lambda_K(t) f(s) g(x)\, d\lambda_G(s)\, d\lambda_G(x)\\
&\quad = \int_G \int_G \omega(x^{-1})\omega(s^{-1}) f(s) g(x)\, d\lambda_G(s)\, d\lambda_G(x)\\
&\quad = \int_G \omega(s^{-1}) f(s)\, d\lambda_G(s) \int_G \omega(x^{-1}) g(x)\, d\lambda_G(x) = (\mathcal{F}f)(\omega)(\mathcal{F}g)(\omega),
\end{aligned}$$

as claimed. $\square$

In general, $\mathcal{F}(f * g) \neq (\mathcal{F}f)(\mathcal{F}g)$ if $f \notin \mathcal{L}^1(G/K)$ and $g \notin \mathcal{L}^1(K\backslash G)$.

The following properties also hold.

Proposition 9.20. *Let (G, K) be a Gelfand pair. For every function $f \in \mathcal{L}^1(K\backslash G)$, every $s \in G$, and every $\omega \in \mathbf{S}(G/K)$, we have*

$$\mathcal{F}(\lambda_s f)(\omega) = \omega(s^{-1})(\mathcal{F}f)(\omega), \qquad \overline{\mathcal{F}}(\lambda_s f)(\omega) = \omega(s)(\overline{\mathcal{F}}f)(\omega),$$

and for every function $f \in \mathcal{L}^1(G/K)$, every $s \in G$, and every $\omega \in \mathbf{S}(G/K)$, we have

$$\mathcal{F}(\rho_s f)(\omega) = \omega(s)(\mathcal{F}f)(\omega), \qquad \overline{\mathcal{F}}(\rho_s f)(\omega) = \omega(s^{-1})(\overline{\mathcal{F}}f)(\omega).$$

Proof. We prove that $\overline{\mathcal{F}}(\lambda_s f)(\omega) = \omega(s)(\overline{\mathcal{F}}f)(\omega)$, the proof for the other formulae being similar. Since $f(tx) = f(x)$ for all $t \in K$ and almost all $x \in G$, and since λ_G is left-invariant, we have

$$\begin{aligned}
\overline{\mathcal{F}}(\lambda_s f)(\omega) &= \int_G f(s^{-1}x)\omega(x)\, d\lambda_G(x) = \int_G f(x)\omega(sx)\, d\lambda_G(x)\\
&= \int_G f(tx)\omega(stx)\, d\lambda_G(x) = \int_G f(x)\omega(stx)\, d\lambda_G(x).
\end{aligned}$$

Then, since the rightmost integral above is independent of t because $\overline{\mathcal{F}}(\lambda_s f)(\omega)$ is independent of t, $\overline{\mathcal{F}}(\lambda_s f)(\omega)$ is independent of t, by (s_1) (from Theorem 9.6) and Fubini, we have

$$\begin{aligned}\int_G f(x)\omega(stx)\,d\lambda_G(x) &= \int_G\int_K f(x)\omega(stx)\,d\lambda_G(x)\,d\lambda_K(t)\\ &= \int_G f(x)\int_K \omega(stx)\,d\lambda_K(t)\,d\lambda_G(x)\\ &= \omega(s)\int_G f(x)\omega(x)\,d\lambda_G(x) = \omega(s)(\overline{\mathcal{F}}f)(\omega),\end{aligned}$$

as claimed. $\square$

In the next section, we try to generalize Fourier inversion.

9.9 The Plancherel Transform

As in the previous section, let (G, K) be a Gelfand pair. If G is compact, then by Example 9.6 and Proposition 6.21, the spherical functions in $\mathbf{S}(G/K)$ are of positive type (recall Definition 3.17). However, when G is not compact, the spherical functions in $\mathbf{S}(G/K)$ are not necessarily of positive type. The subspace of spherical functions of positive type is deeply related to the measures of positive type (recall Definition 3.21) and is the domain of certain positive measures that yield a kind of Fourier inversion. The reader may want to review Section 2.9 before proceeding with this section.

Definition 9.21. The subset of $\mathbf{S}(G/K)$ consisting of the *spherical functions of positive type* is denoted by $\mathbf{Z}(G/K)$. This space is equipped with the induced topology of Fréchet space of $\mathcal{C}(G)$.

In view of Theorem 3.22(b), the space $\mathbf{Z}(G/K)$ is closed in $\mathbf{S}(G/K)$, and thus it is locally compact.

Given a measure of positive type μ on G (see Definition 3.21), recall from Section 3.7 that the linear map $\varphi_\mu\colon \mathcal{K}_{\mathbb{C}}(G) \to \mathbb{C}$ given by

$$\varphi_\mu(f) = \int f(s)\,d\mu(s)$$

is a positive linear form in the sense of Definition 2.10. As in Section 3.5, the set

$$\mathfrak{n} = \{f \in \mathcal{K}_{\mathbb{C}}(G) \mid \varphi_\mu(f^* * f) = 0\}$$

is a left ideal in $\mathcal{K}_{\mathbb{C}}(G)$, and $H_0 = \mathcal{K}_{\mathbb{C}}(G)/\mathfrak{n}$ is a hermitian space with the hermitian inner product

$$\langle \pi(f), \pi(g) \rangle_\mu = \varphi_\mu(g^* * f) = \int (g^* * f)(s)\, d\mu(s), \qquad (\dagger_6)$$

where $\pi\colon \mathcal{K}_{\mathbb{C}}(G) \to \mathcal{K}_{\mathbb{C}}(G)/\mathfrak{n}$ is the quotient map. Since

$$\begin{aligned}\int (g^* * f)(s)\, d\mu(s) &= \int\int \overline{g(t^{-1})} f(t^{-1}s)\, d\lambda(t)\, d\mu(s)\\ &= \int\int \overline{g(t)} f(ts)\, d\lambda(t)\, d\mu(s),\end{aligned}$$

we have

$$\langle \pi(f), \pi(g) \rangle_\mu = \varphi_\mu(g^* * f) = \int\int \overline{g(t)} f(ts)\, d\lambda(t)\, d\mu(s). \qquad (\dagger_7)$$

The hermitian space $H_0 = \mathcal{K}_{\mathbb{C}}(G)/\mathfrak{n}$ is separable, and we let H_μ be the Hilbert space which is the completion of H_0. By Theorem 3.30, the measure of positive type μ defines a unitary representation $U_\mu\colon G \to \mathbf{U}(H_\mu)$, where $U_\mu(s) \in \mathbf{U}(H_\mu)$ is the extension of the map $U_\mu(s) \in \mathbf{U}(H_0)$ defined by

$$U_\mu(s)(\pi(f)) = \pi(\delta_s * f), \qquad \text{for all } s \in G \text{ and all } f \in \mathcal{K}_{\mathbb{C}}(G).$$

By Theorem 3.17, the unitary representation $U_\mu\colon G \to \mathbf{U}(H_\mu)$ extends to a non-degenerate algebra representation $(U_\mu)_{\mathrm{ext}}\colon \mathrm{L}^1(G) \to \mathcal{L}(H_\mu)$.

The map γ defined on $\mathcal{K}(G) \times \mathcal{K}(G)$ *via* $(\dagger_6)$ by

$$\gamma(g, h) = \varphi_\mu(h^* * g)$$

satisfies the conditions (U) and (N) of Section 2.9, and thus the restriction of γ to the involutive and commutative subalgebra $\mathcal{K}(K\backslash G/K)$ is a bitrace (see Definition 2.13), which also satisfies condition (U). Actually, this bitrace also satisfies condition (N). This can be shown using a regularization argument that we omit. For details, see Dieudonné [Dieudonné (2003b)] (Chapter XXII, Section 7). Then $\mathcal{K}(K\backslash G/K)$ and $\pi(\mathcal{K}(K\backslash G/K))$ are commutative Hilbert algebras.

Let $\mathcal{H}_\mu$ be the closure of $\pi(\mathcal{K}(K\backslash G/K))$ in H_μ. Then the map $f \mapsto (U_\mu)_{\mathrm{ext}}(f)|\mathcal{H}_\mu$ is a representation of $\mathcal{K}(K\backslash G/K)$ in the separable Hilbert space $\mathcal{H}_\mu$ that we denote V_μ. Thus, we have

$$V_\mu(f)(\pi(g)) = \pi(f * g), \qquad f, g \in \mathcal{K}(K\backslash G/K)$$

and by Proposition 3.15, we have

$$\|V_\mu(f)\| \le \|f\|_1 .$$

This means that V_μ is a continuous algebra homomorphism from $\mathcal{K}(K\backslash G/K)$ (with the topology induced by the topology of $\mathrm{L}^1(K\backslash G/K)$) to the algebra $\mathcal{L}(\mathcal{H}_\mu)$ of continuous linear operators of $\mathcal{H}_\mu$. Since $\mathrm{L}^1(G)$ is a separable Banach algebra, we deduce that the closure $\mathcal{A}_\mu$ of $V_\mu(K\backslash G/K)$ is a C^* commutative separable subalgebra of $\mathcal{L}(\mathcal{H}_\mu)$.

Thus we see that the bitrace obtained by restricting the bitrace γ to $\mathcal{K}(K\backslash G/K)$ satisfies all the hypotheses of the Plancherel–Godement theorem (Theorem 2.43). To be more specific, in terms of the notations of Theorem 2.43, we have $A = \mathcal{K}(K\backslash G/K)$, $g = \gamma$, $U_g = V_\mu$, $H_g = \mathcal{H}_\mu$, and $\mathcal{A}_g = \mathcal{A}_\mu$. The Plancherel–Godement theorem yields the following result.

Theorem 9.21. *(Plancherel Transform Theorem) Let (G, K) be a Gelfand pair. For every measure of positive type μ on G, there is a unique (positive) Radon measure $\mu^\triangle$ defined on the locally compact space $\mathbf{Z}(G/K)$ of spherical functions of positive type, such that for every function $f \in \mathcal{K}(K\backslash G/K)$, the Fourier cotransform $\overline{\mathcal{F}}f$ belongs to $\mathcal{L}^2_{\mu^\triangle}(\mathbf{Z}(G/K); \mathbb{C})$, and for any two functions $f, g \in \mathcal{K}(K\backslash G/K)$, we have*

$$\int_G (g^* * f)d\mu = \int_{\mathbf{Z}(G/K)} (\overline{\mathcal{F}}f)(\omega)\overline{(\overline{\mathcal{F}}g)(\omega)}\, d\mu^\triangle(\omega).$$

The map $\Phi\colon f \mapsto [\overline{\mathcal{F}}f]$ from $\mathcal{K}(K\backslash G/K)$ to $\mathrm{L}^2_{\mu^\triangle}(\mathbf{Z}(G/K); \mathbb{C})$ factors as

$$\Phi = T_0 \circ \pi,$$

with $T_0\colon \pi(\mathcal{K}(K\backslash G/K)) \to \mathrm{L}^2_{\mu^\triangle}(\mathbf{Z}(G/K); \mathbb{C}) \cap \mathcal{C}_0(\mathbf{Z}(G/K); \mathbb{C})$, and T_0 extends to an isomorphism T between the Hilbert space $\mathcal{H}_\mu$ and the Hilbert space $\mathrm{L}^2_{\mu^\triangle}(\mathbf{Z}(G/K); \mathbb{C})$, as illustrated below:

$$\begin{array}{ccccc}
\mathcal{K}(K\backslash G/K) & \xrightarrow{\ \pi\ } & \pi(\mathcal{K}(K\backslash G/K)) & \longrightarrow & \mathcal{H}_\mu \\
& \searrow^{\Phi} & \downarrow{\scriptstyle T_0} & & \downarrow{\scriptstyle T} \\
& & \mathrm{L}^2_{\mu^\triangle}(\mathbf{Z}(G/K); \mathbb{C}) \cap \mathcal{C}_0(\mathbf{Z}(G/K); \mathbb{C}) & \longrightarrow & \mathrm{L}^2_{\mu^\triangle}(\mathbf{Z}(G/K); \mathbb{C}).
\end{array}$$

The only points which need clarification are the facts that the space $S_{\underset{\wedge}{g}}$ of Theorem 2.43 is homeomorphic to $\mathbf{Z}(G/K)$ and that the map $f \mapsto \widehat{f}$, with $f \in A = \mathcal{K}(K\backslash G/K)$, is simply the Fourier cotransform $\overline{\mathcal{F}}(f)$. The details require some knowledge of the proof of the Plancherel–Godement theorem and are omitted. The reader is referred to Dieudonné [Dieudonné (2003b)] (Chapter XXII, Section 7, Theorem 22.7.4).

Definition 9.22. Let (G, K) be a Gelfand pair. For every complex measure μ of positive type on G, the (positive) Radon measure $\mu^\triangle$ on $\mathbf{Z}(G/K)$ given by Theorem 9.21 is called the *Plancherel transform* of μ.

It is also useful to define the projection of a complex measure $\mu \in \mathbb{C}\mathcal{M}^1(G)$ onto the subspace $\mathbb{C}\mathcal{M}^1(K\backslash G/K)$ of complex measures invariant by left and right translations by elements of K.

First, assume that μ is a positive finite measure. We define the linear functional $\Phi^\sharp_\mu$ by

$$\Phi^\sharp_\mu(f) = \int_G f^\sharp\, d\mu, \qquad f \in \mathcal{K}(G).$$

Since μ is a positive measure, the functional $\Phi^\sharp_\mu$ is positive, so by Radon–Riesz I (Vol. I, Theorem 7.8), there is a unique σ-Radon measure $\mu^\sharp$, such that

$$\int f\, d\mu^\sharp = \int f^\sharp\, d\mu, \qquad f \in \mathcal{K}(G).$$

Going back to an arbitrary complex measure μ and expressing it as $\mu = \mu_1^+ - \mu_1^- + i(\mu_2^+ - \mu_2^-)$, where the four measures on the right-hand side are positive, we obtain a complex measure $\mu^\sharp$, such that

$$\int f\, d\mu^\sharp = \int f^\sharp\, d\mu, \qquad f \in \mathcal{K}(G). \tag{$*_\sharp$}$$

Definition 9.23. Given any complex measure $\mu \in \mathbb{C}\mathcal{M}^1(G)$, the complex measure $\mu^\sharp \in \mathbb{C}\mathcal{M}^1(K\backslash G/K)$ defined by $(*_\sharp)$ is called the *projection* of μ.

Consequently, we see that

$$\lambda_t\mu^\sharp = \rho_t\mu^\sharp = \mu^\sharp, \qquad \text{for all } t \in K.$$

Conversely, the above equations imply that $\mu = \mu^\sharp$. *Thus, the map* $\mu \mapsto \mu^\sharp$ *is a projection of* $\mathbb{C}\mathcal{M}^1(G)$ *onto the subspace* $\mathbb{C}\mathcal{M}^1(K\backslash G/K)$.

The following result is not hard to prove.

Proposition 9.22. *If* μ *is a measure of positive type on* G*, then for every* $f \in \mathcal{K}(G)$*, we have*

$$\int_G (f^* * f)^\sharp\, d\mu \geq 0.$$

Proposition 9.22 is proven in Dieudonné [Dieudonné (2003b)] (Chapter XXII, Section 7, Lemma 22.7.4.3). Using Proposition 9.22, we see that if μ is of positive type, then so is $\mu^\sharp$. Also, $\mu^\sharp = 0$ means that μ vanishes on the subspace $\mathcal{K}(K\backslash G/K)$ of $\mathcal{K}(G)$. Thus by the uniqueness clause in Theorem 9.21, we have the following result.

Proposition 9.23. *If μ is a measure of positive type on G, then $(\mu^\sharp)^\triangle = \mu^\triangle$. For any two measures of positive type μ and ν, we have $\mu^\triangle = \nu^\triangle$ iff $\mu^\sharp = \nu^\sharp$.*

Proposition 9.24. *For all $\omega \in \mathbf{Z}(G/K)$ and for every $f \in \mathcal{K}(K\backslash G/K)$, we have*

$$(\overline{\mathcal{F}}f)(\omega) = \overline{(\mathcal{F}\overline{f})(\omega)}.$$

Proof. By Theorem 3.22(3), if p is a function of positive type, then $\overline{\check{p}} = p$, so $\overline{\check{\omega}} = \omega$ for all $\omega \in \mathbf{Z}(G/K)$, and for every $f \in \mathcal{K}(K\backslash G/K)$, we have

$$\begin{aligned}(\overline{\mathcal{F}}f)(\omega) &= \int f(x)\omega(x)\, d\lambda_G(x) = \int f(x)\overline{\omega(x^{-1})}\, d\lambda_G(x)\\ &= \overline{\int \overline{f(x)}\omega(x^{-1})\, d\lambda_G(x)} = \overline{(\mathcal{F}\overline{f})(\omega)}.\end{aligned}$$

Therefore,

$$(\overline{\mathcal{F}}f)(\omega) = \overline{(\mathcal{F}\overline{f})(\omega)},$$

as claimed. ☐

In general, given a function $f \in \mathcal{K}(K\backslash G/K)$, by Theorem 9.21, $\overline{\mathcal{F}}f \in \mathrm{L}^2_{\mu^\triangle}(\mathbf{Z}(G/K))$, but $\overline{\mathcal{F}}f \notin \mathrm{L}^1_{\mu^\triangle}(\mathbf{Z}(G/K))$. However, if g is another function in $\mathcal{K}(K\backslash G/K)$, then Theorem 9.21 also shows that $\overline{\mathcal{F}}(f * g) = (\overline{\mathcal{F}}f)(\overline{\mathcal{F}}g) \in \mathrm{L}^1_{\mu^\triangle}(\mathbf{Z}(G/K))$.

Proposition 9.25. *For all $f, g \in \mathcal{K}(K\backslash G/K)$, for any measure μ of positive type, we have*

$$\int_G (f * g)\, d\mu = \int_{\mathbf{Z}(G/K)} (\overline{\mathcal{F}}f)(\omega)(\overline{\mathcal{F}}g)(\omega)\, d\mu^\triangle(\omega).$$

Consequently, $\overline{\mathcal{F}}(f * g) = (\overline{\mathcal{F}}f)(\overline{\mathcal{F}}g) \in \mathrm{L}^1_{\mu^\triangle}(\mathbf{Z}(G/K))$.

Proof. Since $\overline{\mathcal{F}}g = \overline{\mathcal{F}\overline{g}}$ and $\mathcal{F}g = \overline{\mathcal{F}}\check{g}$, we have $\overline{\overline{\mathcal{F}}\,\overline{g}} = \mathcal{F}g$ and $\mathcal{F}\check{g} = \overline{\mathcal{F}}g$, and so

$$\overline{\overline{\mathcal{F}}g^*} = \overline{\overline{\mathcal{F}}\,\overline{\check{g}}} = \mathcal{F}\check{g} = \overline{\mathcal{F}}g.$$

If we recall that $\mathcal{K}(K\backslash G/K)$ is a commutative algebra (under convolution), from Theorem 9.21 with g replaced by g^*, we deduce that

$$\int_G (f * g)\, d\mu = \int_G (g * f)\, d\mu = \int_{\mathbf{Z}(G/K)} (\overline{\mathcal{F}}f)(\omega)(\overline{\mathcal{F}}g)(\omega)\, d\mu^\triangle(\omega),$$

as claimed. ☐

Example 9.11. One of the main examples of Plancherel transform is the Dirac measure $\mu = \delta_e$. As a corollary of Proposition 3.28, the Dirac measure δ_e is of positive type.

Definition 9.24. The Plancherel transform $\delta_e^{\triangle}$ of the Dirac measure δ_e is called the *canonical measure* on $\mathbf{Z}(G/K)$ and is denoted $m_{\mathbf{Z}}$.

Because G is unimodular, we have

$$\int_G (g^* * f)\, d\delta_e = (g^* * f)(e) = \int_G f(s^{-1})\overline{g(s^{-1})}\, d\lambda_G(s) = \int_G f(s)\overline{g(s)}\, d\lambda_G(s),$$

and for any two functions $f, g \in \mathcal{K}(K\backslash G/K)$, by Theorem 9.21, we have

$$\int_G f(s)\overline{g(s)}\, d\lambda_G(s) = \int_G g^* * f\, d\delta_e = \int_{\mathbf{Z}(G/K)} (\overline{\mathcal{F}}f)(\omega)\overline{(\overline{\mathcal{F}}g)(\omega)}\, dm_{\mathbf{Z}}(\omega). \tag{$\dagger_8$}$$

Write $f = f_1 + if_2$ and $g = g_1 + ig_2$, where f_1, f_2, g_1, g_2 are all real-valued. Then (since conjugation has no effect on real-valued functions), we have

$$\begin{aligned}\int_{\mathbf{Z}(G/K)} (\overline{\mathcal{F}}f)(\omega)\overline{(\overline{\mathcal{F}}g)(\omega)}\, dm_{\mathbf{Z}}(\omega) &= \int_{\mathbf{Z}(G/K)} (\overline{\mathcal{F}}f_1)(\omega)\overline{(\overline{\mathcal{F}}g_1)(\omega)}\, dm_{\mathbf{Z}}(\omega)\\ &\quad + \int_{\mathbf{Z}(G/K)} (\overline{\mathcal{F}}f_2)(\omega)\overline{(\overline{\mathcal{F}}g_2)(\omega)}\, dm_{\mathbf{Z}}(\omega)\\ &\quad - i\int_{\mathbf{Z}(G/K)} (\overline{\mathcal{F}}f_1)(\omega)\overline{(\overline{\mathcal{F}}g_2)(\omega)}\, dm_{\mathbf{Z}}(\omega)\\ &\quad + i\int_{\mathbf{Z}(G/K)} (\overline{\mathcal{F}}g_1)(\omega)\overline{(\overline{\mathcal{F}}f_2)(\omega)}\, dm_{\mathbf{Z}}(\omega).\end{aligned}$$

Since $(\overline{\mathcal{F}}f)(\omega) = \overline{(\mathcal{F}\overline{f})(\omega)}$, and since f_1, f_2, g_1, g_2 are real-valued, we have

$$\begin{aligned}\int_G f_1 g_1\, d\lambda_G &= \int_{\mathbf{Z}(G/K)} (\overline{\mathcal{F}}f_1)(\omega)\overline{(\overline{\mathcal{F}}g_1)(\omega)}\, dm_{\mathbf{Z}}(\omega)\\ &= \int_{\mathbf{Z}(G/K)} (\mathcal{F}f_1)(\omega)\overline{(\mathcal{F}g_1)(\omega)}\, dm_{\mathbf{Z}}(\omega)\\ \int_G f_2 g_2\, d\lambda_G &= \int_{\mathbf{Z}(G/K)} (\overline{\mathcal{F}}f_2)(\omega)\overline{(\overline{\mathcal{F}}g_2)(\omega)}\, dm_{\mathbf{Z}}(\omega)\\ &= \int_{\mathbf{Z}(G/K)} (\mathcal{F}f_2)(\omega)\overline{(\mathcal{F}g_2)(\omega)}\, dm_{\mathbf{Z}}(\omega)\\ \int_G f_1 g_2\, d\lambda_G &= \int_{\mathbf{Z}(G/K)} (\overline{\mathcal{F}}f_1)(\omega)\overline{(\overline{\mathcal{F}}g_2)(\omega)}\, dm_{\mathbf{Z}}(\omega)\\ &= \int_{\mathbf{Z}(G/K)} (\mathcal{F}f_1)(\omega)\overline{(\mathcal{F}g_2)(\omega)} dm_{\mathbf{Z}}(\omega)\end{aligned}$$

$$\begin{aligned}\int_G g_1 f_2 \, d\lambda_G &= \int_{\mathbf{Z}(G/K)} (\overline{\mathcal{F}} g_1)(\omega) \overline{(\overline{\mathcal{F}} f_2)(\omega)} \, dm_{\mathbf{Z}}(\omega) \\ &= \int_{\mathbf{Z}(G/K)} (\mathcal{F} g_1)(\omega) \overline{(\mathcal{F} f_2)(\omega)} \, dm_{\mathbf{Z}}(\omega),\end{aligned}$$

because the left integrals are real, and thus

$$\int_G f(s) \overline{g(s)} \, d\lambda_G(s) = \int_{\mathbf{Z}(G/K)} (\overline{\mathcal{F}} f)(\omega) \overline{(\overline{\mathcal{F}} g)(\omega)} \, dm_{\mathbf{Z}}(\omega) \qquad (*_1)$$

$$= \int_{\mathbf{Z}(G/K)} (\mathcal{F} f)(\omega) \overline{(\mathcal{F} g)(\omega)} \, dm_{\mathbf{Z}}(\omega). \qquad (*_2)$$

Observe that the above integrals are inner products. As a consequence, we have the following result.

Proposition 9.26. *For any two functions $f, g \in \mathcal{K}(K\backslash G/K)$, we have*

$$\begin{aligned}\int_G f(s) \overline{g(s)} \, d\lambda_G(s) &= \int_{\mathbf{Z}(G/K)} (\mathcal{F} f)(\omega) \overline{(\mathcal{F} g)(\omega)} \, dm_{\mathbf{Z}}(\omega) \\ &= \int_{\mathbf{Z}(G/K)} (\overline{\mathcal{F}} f)(\omega) \overline{(\overline{\mathcal{F}} g)(\omega)} \, dm_{\mathbf{Z}}(\omega).\end{aligned}$$

The linear maps $f \mapsto \mathcal{F} f$ and $f \mapsto \overline{\mathcal{F}} f$, with $f, g \in \mathcal{K}(K\backslash G/K)$, are isometries, and by Theorem 9.21, these maps extend to isomorphisms from the Hilbert space $\mathrm{L}^2(K\backslash G/K)$ onto the Hilbert space $\mathrm{L}^2_{m_{\mathbf{Z}}}(\mathbf{Z}(G/K))$.

We can further extend these maps to linear maps of $\mathrm{L}^2(G)$ onto $\mathrm{L}^2_{m_{\mathbf{Z}}}(\mathbf{Z}(G/K))$, by setting $\mathcal{F}([f]) = \mathcal{F}([f^\sharp])$ and $\overline{\mathcal{F}}([f]) = \overline{\mathcal{F}}([f^\sharp])$. The equation $\overline{\mathcal{F}}([f]) = \overline{\mathcal{F}([\overline{f}])}$ holds. By abuse of notations, we write $\mathcal{F} f$ (resp. $\overline{\mathcal{F}} f$) for any function in the class $\mathcal{F}([f])$ (resp. $\overline{\mathcal{F}}([f])$). With these notation, $(*_1)$ and $(*_2)$ hold for $f, g \in \mathcal{L}^2(K\backslash G/K)$.

Proposition 9.26 is a generalization of the Plancherel theorem (Vol. I, Theorem 10.27), as we will see in Example 9.12.

Example 9.12. Another important example is the case where G is commutative and $K = \{e\}$. In this case, the functional equation (s_1) characterizing spherical functions reduces to

$$\omega(xy) = \omega(x)\omega(y),$$

so $\omega \colon G \to \mathbb{C}^*$ is a continuous homomorphism such that (see Theorem 9.4) $\omega(e) = 1$ and $|\omega(x)| \leq 1$ for all $x \in G$. Since $\omega(x^{-1}) = \omega(x)^{-1}$, we conclude that $|\omega(x)| = 1$ for all $x \in G$, which means that ω is a *group character*.

The function ω is of positive type. This is because $\omega(y^{-1}x) = \overline{\omega(y)}\omega(x)$, so for every function $f \in \mathcal{K}(G)$, we have

$$\begin{aligned}\int_G (f^* * f)\omega\, d\lambda_G &= \int_G \int_G f^*(y) f(y^{-1}x)\omega(x)\, d\lambda_G(y)\, d\lambda_G(x)\\ &= \int_G \int_G \overline{f(y^{-1})} f(y^{-1}x)\omega(x)\, d\lambda_G(y)\, d\lambda_G(x)\\ &= \int_G \int_G \overline{f(y)} f(x)\omega(y^{-1}x)\, d\lambda_G(y)\, d\lambda_G(x)\\ &= \int_G \int_G \overline{f(y)\omega(y)} f(x)\omega(x)\, d\lambda_G(y)\, d\lambda_G(x)\\ &= \left|\int_G f(x)\omega(x)\, d\lambda_G(x)\right|^2,\end{aligned}$$

so

$$\int_G (f^* * f)\omega\, d\lambda_G = \left|\int_G f(x)\omega(x)\, d\lambda_G(x)\right|^2 \geq 0.$$

Consequently, $\mathbf{S}(G/\{e\}) = \mathbf{Z}(G/\{e\})$ is the space of group characters of G, and this space is homeomorphic to $\mathsf{X}_0(A)$, where $A = \mathrm{L}^1(G) \oplus \mathbb{C}\delta_e$.

The topological space $\widehat{G} = \mathbf{S}(G/\{e\})$ of group characters is a group, and it can be shown that the topology of $\mathbf{S}(G/\{e\})$ is compatible with the group structure; see Dieudonné [Dieudonné (2003b)] (Chapter XXII, Section 10, Lemma 22.10.2). Therefore, $\widehat{G}$ is a commutative topological group which is locally compact, metrizable and separable. Given a (left) Haar measure λ_G on G, it can be shown that the Plancherel transform $\lambda_{\widehat{G}} = m_{\mathbf{Z}} = \delta_e^{\triangle}$ is a (left) Haar measure on $\widehat{G}$; see Dieudonné [Dieudonné (2003b)] (Chapter XXII, Section 10, Lemma 22.10.5).

Definition 9.25. The Haar measure $\lambda_{\widehat{G}}$ on $\widehat{G}$ and the Haar measure λ_G on G are said to be *associated*.

If λ_G is replaced by $a\lambda_G$ with $a > 0$, then $\lambda_{\widehat{G}}$ is replaced by $a^{-1}\lambda_{\widehat{G}}$.

Proposition 9.26 shows that the equation

$$\int_G f(s)\overline{g(s)}\, d\lambda_G(s) = \int_{\widehat{G}} (\mathcal{F}f)(\omega)\overline{(\mathcal{F}g)(\omega)}\, d\lambda_{\widehat{G}}(\omega)$$

holds, and that $\mathcal{F}$ has an extension which is an isometry from $\mathrm{L}^2(G)$ to $\mathrm{L}^2(\widehat{G})$, providing another proof of the Plancherel theorem, Vol. I, Theorem 10.27.

We now return to an arbitrary locally compact group G (metrizable, separable, and unimodular). For any function $\omega \in \mathbf{Z}(G/K)$ of positive type, the measure $\omega\, d\lambda_G$ is a measure of positive type. It can be shown that

$$(\omega\, d\lambda_G)^{\triangle} = \delta_\omega,$$

the Dirac measure at ω; see Dieudonné [Dieudonné (2003b)] (Chapter XXII, Section 7, Lemma 22.7.6.1). The result is a sort of Fourier inversion formula.

Proposition 9.27. *If μ is a measure of positive type on G, then for every function $f \in \mathcal{K}(K\backslash G/K)$, if the Fourier cotransform $\overline{\mathcal{F}}f$ belongs to $\mathrm{L}^1_{\mu^\triangle}(\mathbf{Z}(G/K))$, then*

$$\int_G f\, d\mu = \int_{\mathbf{Z}(G/K)} (\overline{\mathcal{F}}f)(\omega)\, d\mu^{\triangle}(\omega).$$

Proposition 9.27 is proven in Dieudonné [Dieudonné (2003b)] (Chapter XXII, Section 7, Lemma 22.7.8).

In general, given a function $f \in \mathcal{K}(K\backslash G/K)$, by Theorem 9.21, $\overline{\mathcal{F}}f \in \mathrm{L}^2_{\mu^\triangle}(\mathbf{Z}(G/K))$, but $\overline{\mathcal{F}}f \notin \mathrm{L}^1_{\mu^\triangle}(\mathbf{Z}(G/K))$. Using Proposition 9.25, the proof of Proposition 9.27 can be adapted to use the technique of regularization. If (g_n) is a sequence of positive functions in $\mathcal{K}(K\backslash G/K)$ having a compact support that tends to $\{e\}$, and such that $\int g_n\, d\mu = 1$, then $\int (f * g_n) d\mu$ tends to $\int f d\mu$, and we have

$$\int_G f d\mu = \lim_{n\mapsto\infty} \int_G (f * g_n) d\mu = \lim_{n\mapsto\infty} \int_{\mathbf{Z}(G/K)} (\overline{\mathcal{F}}f)(\omega)(\overline{\mathcal{F}}g_n)(\omega) d\mu^{\triangle}(\omega).$$

In particular, if we apply the above formula to $\mu = \delta_e$, if we compute $\int_G (\lambda_s f)\, d\delta_e = f(s^{-1})$, using Proposition 9.20, we find that for every $f \in \mathcal{K}(K\backslash G/K)$, we have

$$f(s) = \lim_{n\mapsto\infty} \int_{\mathbf{Z}(G/K)} \omega(s^{-1})(\overline{\mathcal{F}}f)(\omega)(\overline{\mathcal{F}}g_n)(\omega)\, dm_{\mathbf{Z}}(\omega).$$

The above process for the inversion of the Fourier cotransform is usually used when G is abelian and $K = \{e\}$.

We now take a closer look at the space $\mathcal{P}_+(K\backslash G/K)$ of functions in $\mathcal{C}(K\backslash G/K)$ which are of positive type.

Proposition 9.28. *The map $p \mapsto (p\, \lambda_G)^{\triangle}$ is a bijection between the space $\mathcal{P}_+(K\backslash G/K)$ of functions in $\mathcal{C}(K\backslash G/K)$ which are of positive type onto* *the space $\mathcal{M}^1_+(\mathbf{Z}(G/K))$ of bounded positive measures on $\mathbf{Z}(G/K)$. The inverse $\overline{\mathcal{F}}'$ of the above map ($p \mapsto (p\,\lambda_G)^{\triangle}$) is given by*

$$(\overline{\mathcal{F}}'\mu)(x) = \int_{\mathbf{Z}(G/K)} \omega(x)\, d\mu(\omega), \qquad \mu \in \mathcal{M}^1_+(\mathbf{Z}(G/K)).$$

For every $f \in \mathcal{L}^1(G)$, we have

$$\int_G f(x)(\overline{\mathcal{F}}'\mu)(x)\, d\lambda_G(x) = \int_{\mathbf{Z}(G/K)} (\overline{\mathcal{F}}f)(\omega)\, d\mu(\omega).$$

Proposition 9.28 is proven in Dieudonné [Dieudonné (2003b)] (Chapter XXII, Section 7, Lemma 22.7.10). The proof uses the Bochner–Godement theorem (Theorem 2.44).

In the special case where G is commutative and $K = \{e\}$, the space $\mathcal{P}_+(G)$ is the set of all continuous functions of positive type on G, and Proposition 9.28 implies that every function p of positive type can be written uniquely as

$$p(x) = \int_{\widehat{G}} \omega(x)\, d\mu(\omega)$$

for some positive measure μ on $\widehat{G}$, a result known as *Bochner's theorem*; see also Folland [Folland (1995)], Chapter 4, Theorem 4.18.

We conclude with three remarks.

(1) If the Haar measure λ_G on G is replaced by $a\lambda_G$ with $a > 0$, then the space $\mathbf{S}(G/K)$ of spherical functions is unchanged. The Fourier transform and the Fourier cotransform are multiplied by a, and the Plancherel transform (see Definition 9.22) is multiplied by a^{-1}.
(2) If (G, K) is a Gelfand pair and if G is compact, then by Example 9.6, Proposition 6.17, and Proposition 6.21, *all spherical functions are of positive type*. However, if G is not compact, this is generally false. For instance, the functions P_ρ of Example 9.7 do not satisfy the property $\overline{\check{\omega}} = \omega$ unless $\Re\rho = -1/2$. It can be shown that these functions are of positive type if $\Re\rho = -1/2$.
(3) If G is compact, then we saw that the space $\mathbf{S}(G/K)$ is discrete and in bijection with the subset of $R(G)$ (of irreducible representations of G) consisting of those $\rho \in R(G)$ such that $(\rho : \sigma_0) = 1$. We can view the Fourier transform $\mathcal{F}f$ of a function $f \in \mathcal{L}^1(G)$ as the family

$$\left(c_\rho = \frac{1}{n_\rho}\langle f, m^{(\rho)}_{11}\rangle\right)_{\rho \in \mathbf{S}(G/K)}.$$

The Fourier transform and the Plancherel measures are discussed from a different point of view for symmetric spaces in Helgason [Helgason (2000)] (Chapter 4) and Helgason [Helgason (1994)] (Chapter 3, especially Section 12).

9.10 Extension of the Plancherel Transform; $\mathbf{P}(G)$ and $\mathbf{P}'(\mathbf{Z})$ ⊛

The purpose of this section is to extend the Plancherel transform to a bigger set of measures and to define the notion of Fourier transform of a measure. The reader may want to refer to Vol. I, Chapter 7 for some of the measure-theoretic definitions. Let X be a locally compact space. Recall that the space of σ-Radon measures on X is denoted by $\mathcal{M}_\sigma^+(X)$ (see Vol. I, Definition 7.5) and the space of Radon measures on X is denoted $\mathcal{M}^+(X)$ (see Vol. I, Definition 7.7). The space of complex measures on X is denoted by $\mathbb{C}\mathcal{M}^1(X)$ and is called the space of *bounded measures*. The space of regular complex Borel measures is denoted by $\mathcal{M}^1(X)$ (which is an abbreviation for $\mathcal{M}^1_{\mathrm{reg},\mathbb{C}}(X)$) (see Vol. I, Definition 7.22). The space $\mathbb{C}\mathcal{M}^1(X)$ contains the space of positive bounded Borel measures, and of course, $\mathcal{M}^1(X) \subseteq \mathbb{C}\mathcal{M}^1(X)$.

Let G be a locally compact, metrizable, separable, and unimodular group.

Definition 9.26. The complex vector space spanned by the union of the complex measures and the σ-Radon measures is denoted by $\mathcal{M}_\mathbb{C}(G)$. Let $\mathbf{P}_+(G)$ be the set of measures of positive type, and $\mathbf{P}(G)$ be the complex subspace of $\mathcal{M}_\mathbb{C}(G)$ spanned by $\mathbf{P}_+(G)$, which consists of all combinations $\mu_1 - \mu_2 + i\mu_3 - i\mu_4$, where the μ_i belong to $\mathbf{P}_+(G)$.

As a general rule, the subscript $+$ indicates that we are dealing with functions or measures of positive type, and the suppression of the subscript $+$ that we are considering the vector space spanned by that set.

It is easy to check that if $\mu \in \mathbf{P}(G)$, then $\mu^\sharp \in \mathbf{P}(G)$. The image of $\mathbf{P}(G)$ by the map $\mu \mapsto \mu^\sharp$ is $\mathbf{P}(G) \cap \mathcal{M}_\mathbb{C}(K\backslash G/K)$.

Let (G, K) be a Gelfand pair. The Plancherel transform $\mu \mapsto \mu^\triangle$ is a map from $\mathbf{P}_+(G)$ to $\mathcal{M}^+(\mathbf{Z}(G/K))$, the space of positive measures on $\mathbf{Z}(G/K)$. We have $(\mu+\nu)^\triangle = \mu^\triangle + \nu^\triangle$ and $(c\mu)^\triangle = c\mu^\triangle$, for any $c > 0$. From this, it is easy to show that for any combination $\mu = \mu_1 - \mu_2 + i\mu_3 - i\mu_4$

of measures $\mu_i \in \mathbf{P}_+(G)$, the sum $\mu_1^{\triangle} - \mu_2^{\triangle} + i\mu_3^{\triangle} - i\mu_4^{\triangle}$ is a measure on $\mathbf{Z}(G/K)$ that depends only on μ and not on its decomposition.

Definition 9.27. The $\mathbb{C}$-linear map $\mu \mapsto \mu^{\triangle}$ from $\mathbf{P}(G)$ to $\mathcal{M}_{\mathbb{C}}(\mathbf{Z}(G/K))$ is also called the *Plancherel transform*.

It is clear that Proposition 9.25 also applies to measures in $\mathbf{P}(G)$; that is, for any $\mu \in \mathbf{P}(G)$, we have

$$\int_G (f * g)\, d\mu = \int_{\mathbf{Z}(G/K)} (\overline{\mathcal{F}}f)(\omega)(\overline{\mathcal{F}}g)(\omega)\, d\mu^{\triangle}(\omega), \text{for all } f, g \in \mathcal{K}(K\backslash G/K).$$

By regularization, it can be shown that for any two measures $\mu, \nu \in \mathbf{P}(G) \cap \mathcal{M}_{\mathbb{C}}(K\backslash G/K)$, if $\mu^{\triangle} = \nu^{\triangle}$, then $\mu = \nu$. But $\mu = \mu^{\sharp}$ and $\nu = \nu^{\sharp}$, so it follows that the kernel of the Plancherel transform is the subspace of measures $\mu \in \mathbf{P}(G)$ such that $\mu^{\sharp} = 0$.

Definition 9.28. The image of $\mathbf{P}(G)$ (or $\mathbf{P}(G) \cap \mathcal{M}_{\mathbb{C}}(K\backslash G/K)$) by the Plancherel transform is denoted by $\mathbf{P}'(\mathbf{Z})$. Let $\mathcal{P}(G)$ be the complex vector space spanned by the functions of positive type on G.

The space $\mathcal{P}(G)$ is a subspace of both $\mathrm{L}^{\infty}(G)$ and $\mathbf{P}(G)$, by viewing $f \in \mathcal{P}(G)$ as $f\, d\lambda_G$. By Proposition 9.28, the image of $\mathcal{P}(G)$ by the Plancherel transform is $\mathbb{C}\mathcal{M}^1(\mathbf{Z}(G/K))$, the space of bounded measures on $\mathbf{Z}(G/K)$. Consequently, we obtain the following result.

Proposition 9.29. *Let $\mathcal{P}(K\backslash G/K)$ be the subspace of $\mathcal{P}(G)$ consisting of the functions invariant by left and right translations by elements of K. Then the map $f \mapsto (f\lambda_G)^{\triangle}$ is a linear bijection between $\mathcal{P}(K\backslash G/K)$ and $\mathbb{C}\mathcal{M}^1(\mathbf{Z}(G/K))$. The inverse map is denoted by $\overline{\mathcal{F}}'$ and is given by the formula*

$$(\overline{\mathcal{F}}'\mu')(x) = \int_{\mathbf{Z}(G/K)} \omega(x)\, d\mu'(\omega), \qquad \mu' \in \mathbb{C}\mathcal{M}^1(\mathbf{Z}(G/K)).$$

We also have

$$\left\| \overline{\mathcal{F}}'\mu' \right\| \leq \|\mu'\|,$$

since $|\omega(x)| \leq 1$ for all $x \in G$ and all $\omega \in \mathbf{Z}(G/K)$.

Thus, the linear map $\overline{\mathcal{F}}'$ from $\mathbb{C}\mathcal{M}^1(\mathbf{Z}(G/K))$ to $\mathcal{P}(K\backslash G/K)$ is continuous for the topology of the Banach space $\mathcal{C}_b(G)$. However, in general, $\mathcal{P}(K\backslash G/K)$ is *not* closed in $\mathcal{C}_b(G)$.

Unfortunately, no convenient characterizations of the spaces $\mathbf{P}(G)$ and $\mathbf{P}'(\mathbf{Z})$ are known, besides their definition. There are necessary *or* sufficient conditions, but *no necessary and sufficient conditions* known. For example, a necessary condition for a measure μ' on $\mathbf{Z}(G/K)$ to belong to $\mathbf{P}'(\mathbf{Z})$ is that the Fourier cotransforms $\overline{\mathcal{F}}f$ of functions $f \in \mathcal{K}(K\backslash G/K)$ belong to $\mathrm{L}^2_{|\mu'|}(\mathbf{Z}(G/K))$, but there are counter-examples showing that this condition is not sufficient. Also, we showed (see Proposition 9.28) that $\mathbb{C}\mathcal{M}^1(\mathbf{Z}(G/K)) \subseteq \mathbf{P}'(\mathbf{Z})$, but there are unbounded measures in $\mathbf{P}'(\mathbf{Z})$ (for example, the Haar measure on $\widehat{G}$, when G is a commutative noncompact locally compact group).

The following result holds.

Proposition 9.30. *For every measure $\mu' \in \mathbf{P}'(\mathbf{Z})$, and every function $g' \in \mathcal{L}^2_{|\mu'|}(\mathbf{Z}(G/K))$, we have $g'\,d\mu' \in \mathbf{P}'(\mathbf{Z})$. If $\nu \in \mathbf{P}(G)$ is a measure such that $\nu^{\triangle} = g'\,d\mu'$, then for every $f \in \mathcal{K}(K\backslash G/K)$, we have*

$$\int_G f\,d\nu = \int_{\mathbf{Z}(G/K)} (\overline{\mathcal{F}}f)(\omega)g'(\omega)\,d\mu'(\omega).$$

Proposition 9.30 is proven in Dieudonné [Dieudonné (2003b)] (Chapter XXII, Section 8, Lemma 22.8.3).

In particular, if we take μ' to be the canonical measure $m_{\mathbf{Z}}$, then we obtain the following corollary.

Proposition 9.31. *The space $\mathrm{L}^2_{m_{\mathbf{Z}}}(\mathbf{Z}(G/K))$ (viewed as a subspace of $\mathcal{M}_{\mathbb{C}}(\mathbf{Z}(G/K))$) using the embedding $\omega \mapsto \omega dm_{\mathbf{Z}}$ is contained in $\mathbf{P}'(\mathbf{Z})$. The restriction to $\mathrm{L}^2(K\backslash G/K) \subseteq \mathbf{P}(G)$ of the Plancherel transform is identical to the extension of the Fourier transform $\mathcal{F}$ from $\mathrm{L}^2(K\backslash G/K)$ to $\mathrm{L}^2_{m_{\mathbf{Z}}}(\mathbf{Z}(G/K))$ given by Proposition 9.26. In particular, for any $f \in \mathrm{L}^2(K\backslash G/K)$, we have $(f\,d\lambda_G)^{\triangle} = (\mathcal{F}f)\,dm_{\mathbf{Z}}$.*

Proposition 9.31 is proven in Dieudonné [Dieudonné (2003b)] (Chapter XXII, Section 8, Lemma 22.8.4).

Remark: It is possible that there is some positive measure $\mu' \in \mathbf{P}'(\mathbf{Z})$, yet there are positive measures ν' with $0 \leq \nu' \leq \mu'$, and $\nu' \notin \mathbf{P}'(\mathbf{Z})$.

Here are more results about the space $\mathbb{C}\mathcal{M}^1(G)$ of bounded measures on G.

Proposition 9.32. *The space $\mathbb{C}\mathcal{M}^1(G)$ of bounded measures on G is contained in $\mathbf{P}(G)$. For every $\mu \in \mathbb{C}\mathcal{M}^1(G)$, we have*

$$\mu^{\triangle} = (\mathcal{F}\mu)\, dm_{\mathbf{Z}},$$

where $\mathcal{F}\mu$ is a continuous bounded function on $\mathbf{S}(G/K)$ given by

$$(\mathcal{F}\mu)(\omega) = \int_G \omega(x^{-1})\, d\mu(x).$$

Furthermore, for every function $f \in \mathcal{L}^1(K\backslash G/K)$ (resp. $f \in \mathcal{L}^2(K\backslash G/K)$), we have $\mu * f, f * \mu \in \mathcal{L}^1(G)$ (resp. $\mathcal{L}^2(G)$), and

$$\mathcal{F}(\mu * f) = \mathcal{F}(f * \mu) = (\mathcal{F}f)(\mathcal{F}\mu)$$

almost everywhere w.r.t. $m_{\mathbf{Z}}$.

Proposition 9.32 is proven in Dieudonné [Dieudonné (2003b)] (Chapter XXII, Section 8, Lemma 22.8.5). The following definition generalizes Vol. I, Definition 10.4.

Definition 9.29. If $\mu \in \mathbb{C}\mathcal{M}^1(G)$ is a bounded measure, then the function $\mathcal{F}\mu$ (defined on $\mathbf{S}(G/K)$) given by

$$(\mathcal{F}\mu)(\omega) = \int_G \omega(x^{-1})\, d\mu(x)$$

is called the *Fourier transform* of μ. We define the *Fourier cotransform* $\overline{\mathcal{F}}\mu$ of μ as $\mathcal{F}\check{\mu}$; that is,

$$(\overline{\mathcal{F}}\mu)(\omega) = \int_G \omega(x^{-1})\, d\check{\mu}(x) = \int_G \omega(x)\, d\mu(x).$$

For every function $f \in \mathcal{L}^1(G)$, we have $\mathcal{F}f = \mathcal{F}(f\, d\lambda_G)$, which justifies the terminology. We have

$$(\mathcal{F}\delta_x)(\omega) = \omega(x^{-1}), \qquad (\overline{\mathcal{F}}\delta_x)(\omega) = \omega(x).$$

It is clear that $\|\mathcal{F}\mu\| \leq \|\mu\|$, so $\mathcal{F}$ is a continuous linear map from the Banach space $\mathbb{C}\mathcal{M}^1(G)$ to the Banach space $\mathcal{C}_b(\mathbf{S}(G/K))$ of continuous bounded functions on $\mathbf{S}(G/K)$. However, in general, the bounded function $\mathcal{F}\mu$ does not tend to zero at infinity, as shown by $\mu = \delta_e$, for which $\mathcal{F}\delta_e$ is the constant 1.

As a corollary of Proposition 9.32, since $\mathbb{C}\mathcal{M}^1(G)$ is contained in $\mathbf{P}(G)$, we have

$$\mathrm{L}^1(G) \cap \mathrm{L}^2(G) = \mathbb{C}\mathcal{M}^1(G) \cap \mathrm{L}^2(G),$$

and by Proposition 9.31, the class of the Fourier transform $\mathcal{F}f$ of a function $f \in \mathcal{L}^1(G) \cap \mathcal{L}^2(G)$ is identical to the class $\mathcal{F}[f]$ as in Definition 9.20.

Proposition 9.33. *For any two functions $f, g \in \mathcal{L}^2(K\backslash G/K)$, the bounded continuous function $f * g$ belongs to $\mathcal{P}(K\backslash G/K)$ and we have*

$$((f * g)d\lambda_G)^{\triangle} = (\mathcal{F}f)(\mathcal{F}g)\, dm_{\mathbf{Z}}.$$

Proposition 9.33 is proven in Dieudonné [Dieudonné (2003b)] (Chapter XXII, Section 8, Lemma 22.8.8).

In general, $f * g$ is *not* integrable for the measure λ_G so its Fourier transform is not definable by the formula of Definition 9.20.

Definition 9.30. Define the spaces $\mathcal{P}^1(K\backslash G/K)$ and $\mathcal{P}^2(K\backslash G/K)$ by

$$\begin{aligned}\mathcal{P}^1(K\backslash G/K) &= \mathcal{P}(K\backslash G/K) \cap \mathrm{L}^1(K\backslash G/K)\\ \mathcal{P}^2(K\backslash G/K) &= \mathcal{P}(K\backslash G/K) \cap \mathrm{L}^2(K\backslash G/K).\end{aligned}$$

Since for any function $f \in \mathcal{L}^\infty(G) \cap \mathcal{L}^1(G)$, we have $|f(x)|^2 \leq \|f\|_\infty |f(x)|$ almost everywhere, we conclude that $f \in \mathcal{L}^2(G)$. Since the functions in $\mathcal{P}(G)$ are bounded, we have the inclusions

$$\mathcal{P}^1(K\backslash G/K) \subseteq \mathcal{P}^2(K\backslash G/K) \subseteq \mathcal{P}(K\backslash G/K).$$

Proposition 9.34. *The image of $\mathcal{P}^2(K\backslash G/K)$ (as a subspace of $\mathbb{C}\mathcal{M}^1(K\backslash G/K)$) under the Plancherel transform is the subspace $\mathrm{L}^1_{m_\mathbf{Z}}(\mathbf{Z}(G/K)) \cap \mathrm{L}^2_{m_\mathbf{Z}}(\mathbf{Z}(G/K))$ of $\mathbb{C}\mathcal{M}^1(\mathbf{Z}(G/K))$. For every function $f \in \mathcal{P}^2(K\backslash G/K)$, we have the Fourier inversion formula*

$$f = \overline{\mathcal{F}}'((\mathcal{F}f)\, dm_{\mathbf{Z}}),$$

where $\overline{\mathcal{F}}'$ is defined in Proposition 9.29. Moreover, if we also have $f \in \mathcal{P}^1(K\backslash G/K)$, then we have the Fourier inversion formula

$$f(x) = \int_{\mathbf{Z}(G/K)} \omega(x) \left(\int_G f(y)\omega(y^{-1})\, d\lambda_G(y) \right) dm_{\mathbf{Z}}(\omega).$$

Proposition 9.34 is proven in Dieudonné [Dieudonné (2003b)] (Chapter XXII, Section 8, Lemma 22.8.10).

One should be cautious that in general, f is not integrable, so we can't use the formula of Definition 9.20 to define $\mathcal{F}f$. If $f \in \mathcal{P}^1(K\backslash G/K)$, then we have the formula above, but the two integrals cannot be replaced by the double integral

$$\iint_{G \times \mathbf{Z}(G/K)} \omega(x)\omega(y^{-1}) f(y)\, d\lambda_G(y)\, dm_{\mathbf{Z}}(\omega)$$

because this integral is not defined in general.

Remark: The spaces $\mathcal{P}(K\backslash G/K), \mathcal{P}^1(K\backslash G/K), \mathcal{P}^2(K\backslash G/K)$ are generally not closed in $\mathcal{C}_b(G)$, but $\mathcal{P}^1(K\backslash G/K)$ is dense in $\mathrm{L}^1(K\backslash G/K)$, and $\mathcal{P}^2(K\backslash G/K)$ is dense in $\mathrm{L}^2(K\backslash G/K)$.

Example 9.13. In the special case where G is commutative and $K = \{e\}$, we know that $\mathbf{S}(G/\{e\}) = \widehat{G}$, and by Pontrjagin duality (Vol. I, Theorem 10.30), $\widehat{\widehat{G}}$ and G can be identified, and then the transform $\overline{\mathcal{F}}'$ from $\mathbb{C}\mathcal{M}^1(\widehat{G})$ to $\mathcal{P}(G)$ defined in Proposition 9.29 is identified with the Fourier cotransform $\overline{\mathcal{F}}$ on $\mathcal{M}^1(\widehat{G})$. The Haar measure $\lambda_{\widehat{\widehat{G}}}$ is also identified with the Haar measure λ_G. Then Proposition 9.26 and Proposition 9.34 yields the Fourier inversion formula

$$f = \overline{\mathcal{F}}(\mathcal{F}f)$$

for every function $f \in \mathcal{P}^2(G)$, and since $\mathcal{P}^2(G)$ is dense in $\mathrm{L}^2(G)$, the above formula actually holds for all $f \in \mathrm{L}^2(G)$. This gives another proof of the inversion formula of the Pontrjagin duality theorem, Vol. I, Theorem 10.30.

By Proposition 9.31 and Proposition 9.32, for any $f \in \mathrm{L}^2(G)$ and any $\mu \in \mathbb{C}\mathcal{M}^1(G)$, we have

$$(f\, d\lambda_G)^{\triangle} = (\mathcal{F}f)\, d\lambda_{\widehat{G}}, \qquad \mu^{\triangle} = (\mathcal{F}\mu)\, d\lambda_{\widehat{G}}.$$

9.11 Spherical Functions of Positive Type and Irreducible Representations

Let (G, K) be a Gelfand pair (with G a locally compact, metrizable, separable, unimodular group). From Theorem 3.22, every spherical function of positive type $\omega \in \mathbf{Z}(G/K)$ induces a cyclic unitary representation $U_\omega\colon G \to \mathbf{U}(H_\omega)$ of G in a separable Hilbert space H_ω. Recall that the map φ_ω given by

$$\varphi_\omega(\mu) = \int_G \omega(s)\, d\mu, \qquad \mu \in \mathcal{M}^1(G)$$

is a positive linear form, and so is its restriction to the unital involutive subalgebra $A = \mathrm{L}^1(G) \oplus \mathbb{C}\delta_e$. If

$$\mathfrak{n} = \{\mu \in A \mid \varphi_\omega(\mu^* * \mu) = 0\},$$

then $\mathfrak{n}$ is a left ideal in A, and $H_0 = A/\mathfrak{n}$ is a hermitian space with the inner product

$$\langle \pi(\mu), \pi(\nu)\rangle = \varphi_\omega(\nu^* * \mu) = \int_G \omega(s)\, d(\nu^* * \mu)(s), \qquad (\dagger_9)$$

where $\pi\colon A \to A/\mathfrak{n}$ is the quotient map (and $\mu^* = \overline{\tilde{\mu}}$). If H_ω is the separable Hilbert space, which is the completion of $H_0 = A/\mathfrak{n}$, then the unitary representation $U_\omega\colon G \to \mathbf{U}(H_\omega)$ is completely determined by

$$U_\omega(s)(\pi(\mu)) = \pi(\delta_s * \mu), \qquad \mu \in A, s \in G.$$

The unique unitary non-degenerate (algebra) representation $(U_\omega)_{\text{ext}}\colon A \to \mathcal{L}(H_\omega)$ extending U_ω is completely determined by

$$((U_\omega)_{\text{ext}}(\mu))(\pi(\nu)) = \pi(\mu * \nu), \qquad \mu, \nu \in A.$$

The vector $x_0 = \pi(\delta_e)$ is a cyclic vector for both representations.

Theorem 9.35. *Let (G, K) be a Gelfand pair.*

(1) For every spherical function of positive type $\omega \in \mathbf{Z}(G/K)$, the cyclic unitary representation $U_\omega\colon G \to \mathbf{U}(H_\omega)$ (with cyclic vector x_0) is irreducible, and its restriction to the compact group K contains the trivial representation of K, which means that

$$F = \{x \in H_\omega \mid U_\omega(t)(x) = x \text{ for all } t \in K\} \neq \{0\}.$$

In fact, the cyclic vector x_0 belongs to F.

(2) Conversely, every unitary representation $U\colon G \to \mathbf{U}(H)$ whose restriction to K contains the trivial representation of K is equivalent to one of the representations U_ω with $\omega \in \mathbf{Z}(G/K)$, and the multiplicity of the trivial representation of K in U is 1.

Theorem 9.35 is proven in Dieudonné [Dieudonné (2003b)] (Chapter XXII, Section 9, Lemma 22.9.2). To prove that U_ω is irreducible, it suffices to show that $P = (U_\omega)_{\text{ext}}(\chi_K \lambda_G)$ is the orthogonal projection of H_ω onto the one-dimensional subspace $\mathbb{C}x_0$. Indeed (also making use of Theorem 3.17) if F is a closed subspace of H_ω invariant under U_ω, and if F is not orthogonal to x_0, then $P(F) \subseteq F$ and $x_0 \in F$, so $F = H_\omega$ since x_0 is a cyclic vector. On the other hand, if F is orthogonal to x_0, then its orthogonal complement $F^\perp$ is also invariant under U_ω and contains x_0, and since x_0 is a cyclic vector $F^\perp = H_\omega$, and thus $F = (0)$.

The proof of Theorem 9.35 makes use of the following proposition.

Proposition 9.36. *For every irreducible representation $V\colon A \to \mathcal{L}(H)$ of an involutive commutative algebra A in a separable Hilbert space H, if $V(A)$ is separable, then V is a representation in a one-dimensional subspace (of H).*

Proposition 9.36 is proven in Dieudonné [Dieudonné (2003b)] (Chapter XXII, Section 9, Lemma 22.9.2.2).

Unlike the case where G is compact, there may not be a closed subspace F of $\mathrm{L}^2(G/K)$, invariant under the canonical representation (see Definition 6.13), and such that the sub-representation of the canonical representation to F is equivalent to some representation of the form U_ω. However, we have the following results.

Proposition 9.37. *Given a linear map $f\colon E \to E$, if f has rank 1, which means that $\dim(f(E)) = 1$, then there is a linear form $\varphi \in E^*$ and some nonzero vector $u \in E$, such that*

$$f(x) = \varphi(x)u, \qquad \text{for all } x \in E.$$

Proof. This fact is immediately obtained by picking a basis $(e_\alpha)_{\alpha\in I}$ in E. Since f has rank 1, we can pick a nonzero vector $u \in f(E)$, and then $f(e_\alpha) = \lambda_\alpha u$ for some $\lambda_\alpha \in \mathbb{C}$, so we can let φ be the linear form given by $\varphi(e_\alpha) = \lambda_\alpha$. If u is replaced by cu with $c \neq 0$, then φ is replaced by $c^{-1}\varphi$. ☐

In the situation of Proposition 9.37, we define the trace of f as

$$\mathrm{tr}(f) = \varphi(u),$$

which is independent of the choice of u. If $g\colon E \to E$ is any other linear map, then it is easy to see that $f \circ g$ and $g \circ f$ have rank 1, and that $\mathrm{tr}(f \circ g) = \mathrm{tr}(g \circ f)$.

Proposition 9.38. *The following properties hold.*

(1) For every function $f \in \mathcal{L}^1(G/K)$, the linear map $(U_\omega)_{\mathrm{ext}}(f) \in \mathcal{L}(H_\omega)$ has rank 1, and we have

$$((U_\omega)_{\mathrm{ext}}(f))(z) = \langle z, x_0\rangle ((U_\omega)_{\mathrm{ext}}(f))(x_0), \qquad z \in H_\omega,$$

where x_0 is the cyclic vector $x_0 = \pi(\delta_e)$. The trace of $(U_\omega)_{\mathrm{ext}}(f)$ is given by

$$\mathrm{tr}((U_\omega)_{\mathrm{ext}}(f)) = \overline{\mathcal{F}}f(\omega).$$

(2) For any two functions $f, g \in \mathcal{L}^1(G/K) \cap \mathcal{L}^2(G/K)$, we have

$$\mathrm{tr}((U_\omega)_{\mathrm{ext}}(f) \circ (U_\omega)_{\mathrm{ext}}(g)^*) = \overline{\mathcal{F}}(g^* * f)(\omega),$$

an integrable function for the canonical measure $m_{\mathbf{Z}}$ on $\mathbf{Z}(G/K)$, and

$$\int_G f(s)\overline{g(s)}\, d\lambda_G(s) = \int_{\mathbf{Z}(G/K)} \mathrm{tr}((U_\omega)_{\mathrm{ext}}(f) \circ (U_\omega)_{\mathrm{ext}}(g)^*)\, dm_{\mathbf{Z}}(\omega).$$

(3) *For every continuous and bounded function $f \in \mathcal{L}^1(G/K)$, such that for all $s \in G$, the function $\overline{\mathcal{F}}(\delta_s * f)$ is integrable for $m_{\mathbf{Z}}$, we have*

$$\begin{aligned} f(s) &= \int_{\mathbf{Z}(G/K)} \overline{\mathcal{F}}(\delta_s * f)(\omega)\, dm_{\mathbf{Z}}(\omega) \\ &= \int_{\mathbf{Z}(G/K)} \mathrm{tr}(U_\omega(s) \circ (U_\omega)_{\mathrm{ext}}(f))\, dm_{\mathbf{Z}}(\omega). \end{aligned}$$

Proposition 9.38 is proven in Dieudonné [Dieudonné (2003b)] (Chapter XXII, Section 9, Lemma 22.9.4).

Remark: Using Proposition 9.19, it can be shown that if $g, h \in \mathcal{K}(G/K)$, then $f = g * h$ satisfies the hypothesis of Proposition 9.38(3), namely, $\overline{\mathcal{F}}(\delta_s * f)$ is integrable for $m_{\mathbf{Z}}$. Indeed,

$$\overline{\mathcal{F}}(\delta_s * f) = \overline{\mathcal{F}}(\delta_s * (g * h)) = \overline{\mathcal{F}}((\delta_s * g) * h) = (\overline{\mathcal{F}}(\delta_s * g))(\overline{\mathcal{F}}h),$$

with both factors in $\mathcal{L}^2_{m_{\mathbf{Z}}}(\mathbf{Z}(G/K))$, so $\overline{\mathcal{F}}(\delta_s * f)$ is $m_{\mathbf{Z}}$-integrable. We also have $f = g * h \in \mathcal{K}(G/K)$.

In the special case where G is a commutative locally compact metrizable and separable group, we have the following results about unitary representations.

Proposition 9.39. *Let G be a commutative locally compact metrizable and separable group. Every unitary cyclic representation $U \colon G \to \mathbf{U}(H)$ of G in a separable Hilbert space H is equivalent to a representation $M \colon G \to \mathbf{U}(\mathrm{L}^2_\mu(\widehat{G}))$, where μ is a positive bounded measure on $\widehat{G}$ (the dual of G), and for every $s \in G$, the linear operator $M(s)$ is defined so that for every $g \in \mathcal{L}^2_\mu(\widehat{G})$, $M(s)(g)$ is the class of the function in $\mathcal{L}^2_\mu(\widehat{G})$ given by*

$$\chi \mapsto \chi(s)g(\chi), \qquad \chi \in \widehat{G}.$$

The proof of Proposition 9.39 is proven in Dieudonné [Dieudonné (2003b)] (Chapter XXII, Section 9, Lemma 22.15.1). The proof uses Bochner's theorem (see Proposition 9.28).

If $G = \mathbb{R}$, there is a more precise result due to Stone.

Theorem 9.40. *(Stone) Every unitary representation of the (additive) group $\mathbb{R}$ in a separable Hilbert space H is of the form*

$$t \mapsto e^{itA},$$

where A is a self-adjoint operator of H, not necessarily bounded. Conversely, for every self-adjoint not necessarily bounded operator A of H, the map $t \mapsto e^{itA}$ is a unitary representation of $\mathbb{R}$ in H.

Theorem 9.40 is proven in Dieudonné [Dieudonné (2003b)] (Chapter XXII, Section 9, Lemma 22.15.3).

9.12 Problems

Problem 9.1. Let G be a locally compact, metrizable, separable and unimodular group, and K be a compact subgroup of G. Let $\pi\colon G \to G/K$ be the projection map.

(i) Show that the map $f \mapsto f \circ \pi$ is a bijection between $\mathcal{K}(G/K)$ and $\mathcal{K}(G) \cap \mathcal{C}(G/K)$.
(ii) Construct a bijection between $\mathcal{K}(K\backslash G)$ and the space $\mathcal{K}(G) \cap \mathcal{C}(K\backslash G)$.
(iii) Construct a bijection between $\mathcal{K}(K\backslash G/K)$ and the space $\mathcal{K}(G) \cap \mathcal{C}(K\backslash G/K)$.
(iv) Show that $\mathcal{K}(G/K)$ is a left ideal in $\mathcal{K}(G)$, that $\mathcal{K}(K\backslash G)$ is a right ideal in $\mathcal{K}(G)$, and that the involution $f \mapsto \check{\overline{f}}$ maps $\mathcal{K}(G/K)$ onto $\mathcal{K}(K\backslash G)$.
(v) Show that $\mathcal{K}(K\backslash G/K)$ is an involutive subalgebra of $\mathcal{K}(G)$.

Problem 9.2. Recall the projection map from $\mathcal{C}(G)$ onto $\mathcal{C}(K\backslash G/K)$ given by

$$f^{\sharp}(s) = \int_K \int_K f(tst')\, d\lambda_K(t)\, d\lambda_K(t'), \qquad s \in G,$$

for any $f \in \mathcal{C}(G)$; see Definition 9.1.

(i) Check that if $f \in \mathcal{C}(G)$, then $f^{\sharp}(t_1 s t_1') = f^{\sharp}(s)$ for all $t_1, t_1' \in K$.
(ii) Check that for all $f \in \mathcal{C}(K\backslash G/K)$ and for all $g \in \mathcal{C}(G)$,

$$(fg)^{\sharp} = fg^{\sharp}.$$

Problem 9.3. Complete the proof of Proposition 9.1.

Problem 9.4. Let f^{σ} be the function given by

$$f^{\sigma}(s) = f(\sigma(s)) = f(\sigma^{-1}(s)), \qquad s \in G;$$

see the proof of Theorem 9.2. Check that the map $\widehat{\sigma}\colon \mathcal{K}(G) \to \mathcal{K}(G)$ given by $\widehat{\sigma}(f) = f^{\sigma}$ is an involutive automorphism of the vector space $\mathcal{K}(G)$. Also show that $\widehat{\sigma}\colon \mathcal{K}(G) \to \mathcal{K}(G)$ is an automorphism of the *algebra* $\mathcal{K}(G)$ (under convolution).

Problem 9.5. Prove Proposition 9.3.

Hint. See Dieudonné [Dieudonné (2003b)], Chapter XXII, Section 6.

Problem 9.6. (Advanced Exercise) Prove part (1) of Theorem 9.7.

Hint. See Dieudonné [Dieudonné (2003b)], Chapter XXII, Section 6, no. 22.6.9.

Problem 9.7. Let c and c_u be the conjugations defined in Theorem 9.12. Consider the map $h = c \circ c_u$. Show that h is a linear and an automorphism of the Lie bracket.

Problem 9.8. Recall from Example 9.1 that

$$\mathfrak{su}(n) = \{X \in \mathrm{M}_n(\mathbb{C}) \mid X^* = -X,\ \mathrm{tr}(X) = 0\},$$

and that

$$\mathfrak{sl}(n, \mathbb{C}) = \mathfrak{su}(n) \oplus i\mathfrak{su}(n).$$

Thus, the conjugation c_u of $\mathfrak{sl}(n, \mathbb{C})$ associated with $\mathfrak{su}(n)$ is given by $c_u(X) = -X^*$.

(i) Consider the conjugation c_0 of $\mathfrak{sl}(n, \mathbb{C})$ given by $c_0(X) = \overline{X}$. Show that c_0 commutes with c_u and that the restriction of c_0 to $\mathfrak{g}_u = \mathfrak{su}(n)$ is also c_0.

(ii) Now let $c_0 \colon \mathfrak{sl}(n, \mathbb{C}) \to \mathfrak{sl}(n, \mathbb{C})$ be the conjugation given by

$$c_0(X) = -I_{p,n-p} X^* I_{p,n-p},$$

where

$$I_{p,n-p} = \begin{pmatrix} I_p & 0_{p,n-p} \\ 0_{n-p,p} & -I_{p,n-p} \end{pmatrix},$$

with $1 \leq p \leq n-1$. Show that c_0 commutes with c_u. Also show that the restriction of c_0 to $\mathfrak{su}(n)$ is given by $c_0(X) = I_{p,n-p} X I_{p,n-p}$.

Problem 9.9. Consider the real Lie algebra $\mathfrak{g}_u = \mathfrak{su}(n)$ of $n \times n$ complex skew-hermitian matrices with zero trace and its complexification $\mathfrak{g} = \mathfrak{sl}(n, \mathbb{C})$. Assume that $n = 2m$, and let $c_0 \colon \mathfrak{sl}(n, \mathbb{C}) \to \mathfrak{sl}(n, \mathbb{C})$ be the conjugation given by

$$c_0(X) = -J_m \overline{X} J_m,$$

with

$$J_m = \begin{pmatrix} 0_m & I_m \\ -I_m & 0_m \end{pmatrix}.$$

(i) Show that the conjugation c_0 commutes with c_u, where $c_u(X) = -X^*$.

(ii) The real Lie group $\mathbf{SU}^*(2m)$ is defined by

$$\mathbf{SU}^*(2m) = \left\{ \begin{pmatrix} U & V \\ -\overline{V} & \overline{U} \end{pmatrix} \in \mathbf{SL}(2m, \mathbb{C}) \;\middle|\; U, V \in \mathrm{M}_m(\mathbb{C}) \right\};$$

see Definition 9.13. Show that its real lie algebra is given by

$$\mathfrak{su}^*(2m) = \left\{ \begin{pmatrix} U & V \\ -\overline{V} & \overline{U} \end{pmatrix} \;\middle|\; U, V \in \mathrm{M}_m(\mathbb{C}),\ \mathrm{tr}(U) + \mathrm{tr}(\overline{U}) = 0 \right\}.$$

(iii) Verify that $\mathfrak{su}^*(2m) = \mathfrak{k}_0 \oplus \mathfrak{p}_0$, where

$$\mathfrak{k}_0 = \{X \in \mathfrak{su}(2m, \mathbb{C}) \mid X = -J_m \overline{X} J_m\},$$

and

$$i\mathfrak{p}_0 = \{X \in \mathfrak{su}(2m, \mathbb{C}) \mid X = J_m \overline{X} J_m\}.$$

(iv) Show that the map $\varphi\colon \mathrm{M}_m(\mathbb{H}) \to \mathrm{M}_{2m}(\mathbb{C})$ given by

$$\varphi(U + Vj) = \begin{pmatrix} U & V \\ -\overline{V} & \overline{U} \end{pmatrix}$$

is an injective $\mathbb{R}$-algebra homomorphism.

Problem 9.10. (Advanced Exercise for readers familiar with Lie groups) Prove Proposition 9.14.

Hint. See Dieudonné [Dieudonné (2007)], Proposition 21.18.5.1.

Problem 9.11. (Advanced Exercise for readers familiar with Lie groups) Prove Theorem 9.17.

Hint. The reasoning in the proof of Proposition 9.15 can be used to show that the conjugation σ_u on $\widetilde{G}_0$, such that $d(\sigma_u)_e = c_u$ induces a conjugation on $G_1 = \widetilde{G}_0/D$. Then apply Proposition 21.18.8 in Dieudonné [Dieudonné (2003a)].

Problem 9.12. The Legendre function of (complex) index ρ, denoted P_ρ, is defined as

$$P_\rho(v) = \frac{1}{2\pi} \int_0^{2\pi} (v + \sqrt{v^2 - 1}\cos\varphi)^\rho \, d\varphi, \qquad v \geq 1,\ \rho \in \mathbb{C}.$$

(i) When ρ is a positive integer n, show that modulo a constant, P_n is a Legendre polynomial; see Definition 5.15.
(ii) If ρ is purely imaginary, show that P_ρ is bounded.
(iii) If ρ is a positive real number, show that P_ρ is unbounded.
(iv) Check that P_ρ satisfies the functional equation

$$\frac{1}{2\pi}\int_0^{2\pi} P_\rho(\cosh t\cosh u + \sinh t\sinh u\cos\varphi)\,d\varphi = P_\rho(\cosh t)P_\rho(\cosh u),$$

for all $t, u \in \mathbb{R}$.

Problem 9.13. Recall the generalized spherical function of Example 9.10, namely

$$\psi(r) = \frac{1}{2\pi}\int_0^{2\pi} e^{r(\lambda\cos\theta+\mu\sin\theta)}\,d\theta,$$

for any $\lambda, \mu \in \mathbb{C}$.

(i) For λ and μ imaginary, show that $\psi(r)$ is bounded.
(ii) For the special case where $\lambda = 0$ and $\mu = i$, the function ψ is the Bessel function J_0. Using a change of variable, if both λ and μ are imaginary, show that the function ψ becomes J_0.

Problem 9.14. Prove Proposition 9.18.

Problem 9.15. Complete the proof of Proposition 9.19 by showing that if either $f \in \mathcal{L}^1(G/K)$ or $g \in \mathcal{L}^1(K\backslash G)$, then

$$\overline{\mathcal{F}}(f * g) = (\overline{\mathcal{F}}f)(\overline{\mathcal{F}}g).$$

Problem 9.16. Complete the proof of Proposition 9.20 by verifying the following three identities, namely that for $f \in \mathcal{L}^1(K\backslash G)$, every $s \in G$, and every $\omega \in \mathbf{S}(G/K)$

$$\mathcal{F}(\lambda_s f)(\omega) = \omega(s^{-1})(\mathcal{F}f)(\omega),$$

and that for every function $f \in \mathcal{L}^1(G/K)$, every $s \in G$, and every $\omega \in \mathbf{S}(G/K)$,

$$\mathcal{F}(\rho_s f)(\omega) = \omega(s)(\mathcal{F}f)(\omega), \qquad \overline{\mathcal{F}}(\rho_s f)(\omega) = \omega(s^{-1})(\overline{\mathcal{F}}f)(\omega).$$

Problem 9.17. (Advanced Exercise) Read Dieudonné [Dieudonné (2003b)], Chapter XXII, Section 7, Theorem 22.7.4, and write a proof of Plancherel Transform Theorem, Theorem 9.21.

Problem 9.18. Prove Proposition 9.22.

Hint. See Dieudonné [Dieudonné (2003b)], Chapter XXII, Section 7, Lemma 22.7.4.3.

Problem 9.19. Prove Proposition 9.27.

Hint. See Dieudonné [Dieudonné (2003b)], Chapter XXII, Section 7, Lemma 22.7.8.

Problem 9.20. (Advanced Exercise) Prove Proposition 9.28.

Hint. See Dieudonné [Dieudonné (2003b)], Chapter XXII, Section 7, Lemma 22.7.10.

Problem 9.21. If the Haar measure λ_G on G is replaced by $a\lambda_G$ with $a > 0$, show that the space $\mathbf{S}(G/K)$ of spherical functions is unchanged. Show that the Fourier transform and the Fourier cotransform are multiplied by a, and that the Plancherel transform (see Definition 9.22) is multiplied by a^{-1}.

Problem 9.22. (Advanced Exercise) Prove Proposition 9.30.

Hint. See Dieudonné [Dieudonné (2003b)], Chapter XXII, Section 8, Lemma 22.8.3.

Problem 9.23. Prove Proposition 9.32.

Hint. See Dieudonné [Dieudonné (2003b)], Chapter XXII, Section 8, Lemma 22.8.5.

Problem 9.24. Prove Proposition 9.33.

Hint. See Dieudonné [Dieudonné (2003b)], Chapter XXII, Section 8, Lemma 22.8.8.

Problem 9.25. (Advanced Exercise) Prove Theorem 9.35.

Hint. See Dieudonné [Dieudonné (2003b)], Chapter XXII, Section 9, Lemma 22.9.2.

Problem 9.26. Prove Proposition 9.38.

Hint. See Dieudonné [Dieudonné (2003b)], Chapter XXII, Section 9, Lemma 22.9.4.

Problem 9.27. Prove Proposition 9.39.

Hint. See Dieudonné [Dieudonné (2003b)], Chapter XXII, Section 9, Lemma 22.15.1.

Problem 9.28. Prove Stone's theorem, Theorem 9.40.

Hint. See Dieudonné [Dieudonné (2003b)], Chapter XXII, Section 9, Lemma 22.15.3.

Bibliography

Bekkers, E. J. (2020). B-Spline CNNs on Lie Groups, in *International Conference on Learning Representations (ICLR)*, pp. 1–25.

Borel, A. (2001). *Essays in the History of Lie Groups and Algebraic Groups, History of Mathematics*, Vol. 21, 1st edn. (AMS).

Bott, R. and Tu, L. W. (1986). *Differential Forms in Algebraic Topology*, 1st edn., GTM No. 82 (Springer-Verlag).

Bourbaki, N. (1965). *Elements of Mathematics. Intégration, Chapters 7–8. Mesure de Haar, Convolution et Représentations*, 2nd edn., Eléments de Mathématiques (Hermann).

Bourbaki, N. (1968). *Algèbre, Chapitre 9*, Eléments de Mathématiques (Hermann).

Bröcker, T. and tom Dieck, T. (1985). *Representations of Compact Lie Groups*, 1st edn., GTM, Vol. 98 (Springer-Verlag).

Carter, R., Segal, G., and Macdonald, I. (1995). *Lectures on Lie Groups and Lie Algebras*, 1st edn. (Cambridge University Press).

Cesa, G., Lang, L., and Weiler, M. (2022). A program to build E(n)-equivariant steerable CNN's, in *International Conference on Learning Representations (ICLR)*, pp. 359–368, uRL:https://openreview:net /pdf?id=WE4qe9x1nQw.

Cohen, T. S., Geiger, M., and Weiler, M. (2019). A General Theory of equivariant CNN's on homogeneous spaces, in *Conference on Neural Information Processing Systems (NeurIPS)*.

Dieudonné, J. (1980). *Special Functions and Linear Representations of Lie Groups*, 1st edn., Regional Conference Series in Mathematics, No. 42 (AMS).

Dieudonné, J. (2003a). *Éléments d'Analyse, Tome V. Groupes de Lie Compacts, Groupes de Lie Semi-Simples*, 1st edn. (Edition Jacques Gabay).

Dieudonné, J. (2003b). *Éléments d'Analyse, Tome VI. Analyse Harmonique*, 1st edn. (Edition Jacques Gabay).

Dieudonné, J. (2003c). *Éléments d'Analyse, Tome VII. Équations Fonctionnelles Linéaires. Première partie, Opérateurs Pseudo-Différentiels*, 1st edn. (Edition Jacques Gabay).

Dieudonné, J. (2005). *Éléments d'Analyse, Tome II. Chapitres XII à XV*, 1st edn. (Edition Jacques Gabay).

Dieudonné, J. (2006). *Éléments d'Analyse, Tome III. Chapitres XVI et XVII*, 1st edn. (Edition Jacques Gabay).

Dieudonné, J. (2007). *Éléments d'Analyse, Tome IV. Chapitres XVIII à XX*, 1st edn. (Edition Jacques Gabay).

Dieudonné, J. (2010). *Éléments d'Analyse, Tome I. Chapitres I à XI*, 3rd edn. (Edition Jacques Gabay).

Dixmier, J. (1969). *Les C^* Algèbres et Leurs Représentations*, 2nd edn. (Edition Jacques Gabay).

Duistermaat, J. and Kolk, J. (2000). *Lie Groups*, 1st edn., Universitext (Springer-Verlag).

Dunford, N. and Schwartz, J. T. (1988). *Linear Operators, Part II: Spectral Theory*, 1st edn. (Wiley).

Folland, G. B. (1992). *Fourier Analysis and its Applications*, 1st edn. (Brooks/Cole Publishing Company).

Folland, G. B. (1995). *A Course in Abstract Harmonic Analysis*, 1st edn. (CRC Press).

Folland, G. B. (1999). *Real Analysis. Modern Techniques and Their Applications*, 2nd edn. (Wiley-Interscience).

Freeman, W. T. and Adelson, E. H. (1991). The Design and Use of Steerable Filters, *IEEE Transactions on Pattern Analysis and Machine Intelligence* **13**, pp. 891–906.

Fulton, W. and Harris, J. (1991). *Representation Theory, A first course*, 1st edn., GTM No. 129 (Springer-Verlag).

Gallier, J. H. (2011). *Geometric Methods and Applications, For Computer Science and Engineering*, 2nd edn., TAM, Vol. 38 (Springer).

Gallier, J. H. and Quaintance, J. (2020a). *Differential Geometry and Lie Groups. A Computational Perspective*, 1st edn., Geometry and Computing 12 (Springer).

Gallier, J. H. and Quaintance, J. (2020b). *Differential Geometry and Lie Groups. A Second Course*, 1st edn., Geometry and Computing 13 (Springer).

Gallier, J. H. and Quaintance, J. (2020c). *Linear Algebra and Optimization with Applications to Machine Learning, Volume I*, 1st edn. (World Scientific).

Hall, B. (2003). *Lie Groups, Lie Algebras, and Representations. An Elementary Introduction*, 1st edn., GTM No. 222 (Springer-Verlag).

Hall, M. (1976). *Theory of Groups*, 2nd edn. (AMS Chelsea).

Helgason, S. (1994). *Geometric Analysis on Symmetric Spaces*, 1st edn., SURV, Vol. 39 (AMS).

Helgason, S. (2000). *Groups and Geometric Analysis. Integral Geometry, Invariant Differential Operators and Spherical Functions*, 1st edn., MSM, Vol. 83 (AMS).

Helgason, S. (2001). *Differential Geometry, Lie Groups, and Symmetric Spaces*, 1st edn., GSM, Vol. 34 (AMS).

Hewitt, E. and Ross, K. A. (1970). *Abstract Harmonic Analysis II*, 1st edn.,

Grundlehren der Mathematischen Wissenschaften (Springer-Verlag).

Hewitt, E. and Ross, K. A. (1979). *Abstract Harmonic Analysis I*, 2nd edn., Grundlehren der Mathematischen Wissenschaften (Springer-Verlag).

Humphreys, J. E. (1972). *Introduction to Lie Algebras and Representation Theory*, 1st edn., GTM No. 9 (Springer-Verlag).

Kirillov, A. (1976). *Elements of the Theory of Representations*, 1st edn. (Springer-Verlag).

Kirillov, A. (1994). *Representation Theory and Noncommutative Harmonic Analysis*, 1st edn., Encyclopaedia of Mathematical Sciences, Vol. 22 (Springer-Verlag).

Kirillov, A. (2004). *Lectures on the Orbit Method*, 1st edn., GSM, Vol. 64 (AMS).

Knapp, A. W. (2001). *Representation Theory of Semisimple Groups*, 1st edn., Princeton Landmarks in Mathematics (Princeton University Press).

Knapp, A. W. (2002). *Lie Groups Beyond an Introduction*, 2nd edn., Progress in Mathematics, Vol. 140 (Birkhäuser).

Kosmann-Schwarzbach, Y. (2010). *Groups and Symmetries. From Finite Groups to Lie Groups*, 1st edn., Universitext (Springer-Verlag).

Lang, L. and Weiler, M. (2019). A Wigner-Eckart Theorem for Group-Equivariant Convolution Kernels, in *Conference on Neural Information Processing Systems (NeurIPS)*, uRL:https://arxiv.org/abs/2010.10952.

Lang, S. (1985). $SL_2(\mathbb{R})$, 1st edn., GTM No. 105 (Springer-Verlag).

Lang, S. (1996). *Real and Functional Analysis*, 3rd edn., GTM 142 (Springer-Verlag).

Lax, P. (2002). *Functional Analysis*, 1st edn. (Wiley).

Mackey, G. W. (1992). *The Scope and History of Commutative and Noncommutative Harmonic Analysis*, *History of Mathematics*, Vol. 5, 1st edn. (AMS).

Madsen, I. and Tornehave, J. (1998). *From Calculus to Cohomology. De Rham Cohomology and Characteristic Classes*, 1st edn. (Cambridge University Press).

Mensah, Y. and Awussi, Y. M. (2013). On the fourier transform on semi-direct products with compact factor, *International Journal of Mathematics and Computation* **21**, pp. 65–71.

Morita, S. (2001). *Geometry of Differential Forms*, 1st edn., Translations of Mathematical Monographs No. 201 (AMS).

O'Neill, B. (1983). *Semi-Riemannian Geometry With Applications to Relativity*, 1st edn., Pure and Applies Math., Vol. 103 (Academic Press).

Rudin, W. (1987). *Real and Complex Analysis*, 3rd edn. (McGraw-Hill).

Rudin, W. (1991). *Functional Analysis*, 2nd edn. (McGraw-Hill).

Sakurai, J. and Napolitano, J. (2021). *Modern Quantum Mechanics*, 3rd edn. (Cambridge University Press).

Samelson, H. (1990). *Notes on Lie Algebras*, 2nd edn., Universitext (Springer).

Sansone, G. (1991). *Orthogonal Functions*, 1st edn. (Dover).

Schwartz, L. (1980). *Topologie Générale et Analyse Fonctionnelle*, Collection Enseignement des Sciences (Hermann).

Schwartz, L. (1993). *Analyse III. Calcul Intégral*, Collection Enseignement des Sciences (Hermann).

Serre, J.-P. (1977). *Linear Representations of Finite Groups*, 1st edn., GTM No. 42 (Springer).

Serre, J.-P. (1992). *Lie Algebras and Lie Groups*, 2nd edn., Lecture Notes in Mathematics, No. 1500 (Springer).

Serre, J.-P. (2000). *Complex Semisimple Lie Algebras*, 1st edn., Springer Monographs in Mathematics (Springer).

Simon, B. (1996). *Representations of Finite and Compact Groups*, 1st edn., GSM, Vol. 10 (AMS).

Taylor, M. E. (1986). *Noncommutative Harmonic Analysis*, 1st edn., Mathematical Surveys and Monographs, No. 22 (AMS).

Varadarajan, V. (1984). *Lie Groups, Lie Algebras, and Their Representations*, 1st edn., GTM No. 102 (Springer-Verlag).

Varadarajan, V. (1999). *An Introduction to Harmonic Analysis on Semisimple Lie Groups*, 1st edn., Cambridge Studies in Advanced Mathematics (Cambridge University Press).

Veblen, O. and Young, J. W. (1946). *Projective Geometry, Vol. 2*, 1st edn. (Ginn).

Vilenkin, N. (1968). *Special Functions and the Theory of Group Representations*, 1st edn., Translations of Mathematical Monographs No. 22 (AMS).

Warner, F. (1983). *Foundations of Differentiable Manifolds and Lie Groups*, 1st edn., GTM No. 94 (Springer-Verlag).

Warner, G. (1972a). *Harmonic Analysis on Semi-Simple Lie Groups, I*, 1st edn., Grundlehren der Mathematischen Wissenschaften (Springer-Verlag).

Warner, G. (1972b). *Harmonic Analysis on Semi-Simple Lie Groups, II*, 1st edn., Grundlehren der Mathematischen Wissenschaften (Springer-Verlag).

Weiler, M. and Cesa, G. (2019). General E(2)-equivariant steerable CNNs', in *International Conference on Learning Representations (ICLR)*, uRL:https://arxiv.org/abs/1911.08251.

Weiler, M., Forré, P., Verlinde, E., and Welling, M. (2023). Equivariant and coordinate independent convolutional networks. a gauge field theory of neural networks, Tech. rep., University of Amsterdam, Amsterdam, The Netherlands.

Weiler, M., Geiger, M., Welling, M., Boomsma, W., and Cohen, T. (2018). 3D steerable CNN's: Learning rotationally equivariant features in volumetric data, in *Conference on Neural Information Processing Systems (NeurIPS)*, uRL:https://arxiv.org/abs/1807.02547.

Wigner, E. P. (1959). *Group Theory and its Applications to the Quantum Mechanics of Atomic Spectra*, 1st edn. (Academic Press).

Index

Symbol Index

www.ingramcontent.com/pod-product-compliance
Lightning Source LLC
LaVergne TN
LVHW010321220926
848484LV00004BB/822